2020 China Semiconductor Technology International Conference (CSTIC 2020)

Shanghai, China
26 June – 17 July 2020

IEEE Catalog Number:	CFP2060Y-POD
ISBN:	978-1-7281-6559-2

**Copyright © 2020 by the Institute of Electrical and Electronics Engineers, Inc.
All Rights Reserved**

Copyright and Reprint Permissions: Abstracting is permitted with credit to the source. Libraries are permitted to photocopy beyond the limit of U.S. copyright law for private use of patrons those articles in this volume that carry a code at the bottom of the first page, provided the per-copy fee indicated in the code is paid through Copyright Clearance Center, 222 Rosewood Drive, Danvers, MA 01923.

For other copying, reprint or republication permission, write to IEEE Copyrights Manager, IEEE Service Center, 445 Hoes Lane, Piscataway, NJ 08854. All rights reserved.

*** This is a print representation of what appears in the IEEE Digital Library. Some format issues inherent in the e-media version may also appear in this print version.

IEEE Catalog Number: CFP2060Y-POD
ISBN (Print-On-Demand):   978-1-7281-6559-2
ISBN (Online):   978-1-7281-6558-5

Additional Copies of This Publication Are Available From:

Curran Associates, Inc
57 Morehouse Lane
Red Hook, NY 12571 USA
Phone: (845) 758-0400
Fax: (845) 758-2633
E-mail: curran@proceedings.com
Web: www.proceedings.com

Table of Contents

Preface

Chapter I - Device Engineering and Memory Technology

Nanowire & Nanosheet FETs for Advanced Ultra-Scaled, High-Density Logic and Memory Applications**   1 (1-33)
A. Veloso, P. Matagne, G. Eneman, D. Jang, T. Huynh-Bao, A. Chasin, E. Simoen, A. De Keersgieter and N. Horiguchi
Imec, Leuven, Belgium

Design of a Novel One Transistor-DRAM Based on Bulk Silicon Substrate   5 (1-16)
K. Xiao[1], J. Liu[1], Wei Wu[2], Zheng Xu[2] and J. Wan
[1]State Key Lab of ASIC and Systems, School of Information Science and Engineering, Fudan University, Shanghai, China
[2]Fujian Electronics & Information (Group)Co., LTD, Fujian, China

Manufacturing Challenges and Cost Evaluation of New Generation 3D Memories   8 (1-34)
Belinda Langelihle Dube
System Plus Consulting, Semiconductor Devices Department, Nantes, France

A New Structure of ILD Gap Filling Improvement for Floating-Gate Memory   11 (1-27)
Zhenghong Liu, Yan Li, Ruisheng Qi, Naoki Tsuji, Guanqun Huang, Haoyu Chen and Chris Shao
Shanghai Huali Microelectronics Corporation, Shanghai, China

Neuromorphic Technology Utilizing NAND Flash Memory Cells*   14 (1-93)
Sung-Tae Lee and Jong-Ho Lee
Department of ECE and ISRC, Seoul National University, Seoul, Korea

Investigation of Multi-Level Properties of TaO$_x$-based Memristive Devices and Optimized Programming Scheme for On-Line Training   17 (1-83)
Teng Zhang[1], Yingming Lu[1], Caidie Cheng[1], Ke Yang[1], Liying Xu[1], Qingxi Duan[1], Zhaokun Jing[1], Keqin Liu[1], Rui Yuan[1], Yuchao Yang[1,2,3] and Ru Huang[1,2,3]
[1]Key Laboratory of Microelectronic Devices and Circuits (MOE), Department of Micro/nanoelectronics, Peking University, Beijing, China
[2]Academy for Artificial Intelligence, Peking University, Beijing, China
[3]Frontiers Science Center for Nano-optoelectronics, Peking University, Beijing, China

Neural Spike Detection Based on 1T1R Memristor   20 (1-79)
Zhengwu Liu, Jianshi Tang, Bin Gao, He Qian and Huaqiang Wu
Institute of Microelectronics, Beijing Innovation Center for Future Chips (ICFC), Beijing National Research Center for Information Science and Technology (BNRist), Tsinghua University, Beijing, China

Implementation of Lateral Divisive Inhibition Based on Ferroelectric FET with Ultra-Low Hardware Cost for Neuromorphic Computing
23 (1-84)

Shuhan Liu, Tianyi Liu, Zhiyuan Fu, Cheng Chen, Qianqian Huang and Ru Huang
Key Laboratory of Microelectronic Devices and Circuits (MOE), Institute of Microelectronics, Peking University, Beijing, China

High Yield, Superior Quality and Reliability of IGBT and Power Devices in the Artificial Intelligence Era*
26 (1-23)

Min-hwa Chi[1], Alicia Ding[2], Kong Tjien Lim[2] and Richard Chang[1]
[1]SiEn (QinDao) Integrated Circuits Cor., ShanDong, China
[2]ThermoFisher Scientific, Shanghai, China

True Random Number Generator (TRNG) for Secure Communications in the Era of IOT
31 (1-96)

Zhigang Ji[1], James Brown[2] and Jianfu Zhang[2]
[1]National Key Laboratory of Science and Technology on Micro/Nano Fabrication, Shanghai Jiaotong University, Shanghai, China
[2]School of Engineering, Liverpool John Moores University, Liverpool, UK

Yield Enhancement by Virtual Fabrication: Using Failure Bin Classification, Yield Prediction and Process Window Optimization to Identify and Prevent Process Failures
36 (1-68)

Qingpeng Wang, Yu De Chen, Jacky Huang, Wuping Liu and Ervin Joseph
Coventor Inc., a Lam Research Company, Pudong, Shanghai, China

A Device Design for 5 nm Logic FinFET Technology
39 (1-44)

Yu Ding, Xin Luo, Enming Shang, Shaojian Hu, Shoumian Chen and Yuhang Zhao
Shanghai Integrated Circuit Research and Development Center, Shanghai, China

A Study of FinFET Device Optimization and PPA Analysis at 5 nm Node
44 (1-50)

Xin Luo, Yu Ding, Enming Shang, Jie Sun, Shaojian Hu, Shoumian Chen and Yuhang Zhao
Shanghai IC R&D Center, Shanghai, China.

A Simple Current Test Method on Wafer Level to Pre-Verify Circuit Function
48 (1-21)

Jianrong Xu
Shanghai Huali Microelectronics Corporation, Shanghai, China

Novel Semiconductor Devices based on SOI Substrate
51 (1-29)

K. Xiao[1], J. Liu[1], JN. Deng[1], YL. Jiang[2], WZ. Bao[2], A. Zaslavsky[3], S. Cristoloveanu[4], X. Gong[5] and J. Wan[1]
[1]State key lab of ASIC and System, School of Information Science and Engineering, Fudan University, Shanghai, China
[2]State key lab ASIC and System, School of Microelectronics, Fudan University, Shanghai, China
[3]Department of Physics and School of Engineering, Brown University, Providence, Rhode Island, USA
[4]IMEP-LAHC, INP-Grenoble/Minatec, Grenoble, France
[5]Department of Electrical and Computer Engineering (ECE), National University of Singapore (NUS), Singapore

Effect of Dissolved Ozone and In-Situ Wafer Cleaning on Pre-Epitaxial Deposition for Next Generation Semiconductor Devices   53 (1-20)
Ismail Kashkoush, Darian Waugh and Gim Chen
NAURA Akrion Inc., Allentown, Pennsylvania, USA

From Microns to Nanometers: The IRDS and AMC Control   56 (1-85)
Chris Muller[1], Henry Yu[2] and David Lu[2]
[1]Muller Consulting, Lawrenceville, Georgia, USA
[2]Purafil, Inc., Doraville, Georgia, USA

Study of Shallow trench Isolation Gap Fill for 19 nm NAND Flash   60 (1-6)
Li Peng, Hongbo Li, Tiantuo Sun, Xing Gao and Qin Sun
Shanghai Huali Microelectronics Corporation, Shanghai, China

Modeling Method of Local Mismatch Model for MOS Transistors   63 (1-7)
Jinglun Gu
Shanghai Huali Microelectronics Corporation, Shanghai, China

An Application of LDMOS on ESD Protection   66 (1-8)
Li Wang, Yu Chen and Hualun Chen
Huahong Semiconductor (WUXI) Limited, Wuxi, China

HV Gate Oxide Over-Oxidation Process Optimization for SONOS 1.5T Flash Cell   68 (1-9)
Jian Zhang, Wei Xiong and Hualun Chen
Huahong Semiconductor (WUXI) Limited, Wuxi, China

Potential Applications of h-BN Crystals in Future ULSI   70 (1-10)
Guangyuan Lu, Yu Chen and Hualun Chen
Huahong Semiconductor (WUXI) Limited, Wuxi, China

Improved HCI Of Embedded High Voltage EDNMOS in Advanced CMOS Process   73 (1-11)
Junwen Liu, Hualun Chen and Yu Chen
Huahong Semiconductor (WUXI) Limited, Wuxi, China

Improved Standby Leakage of Huge Volume SRAM by Thin SiN Film of STI Liner   76 (1-13)
Xiaobing Ren, Wei Xiong and Hualun Chen
Huahong (WUXI)Semiconductor Manufacturing Corporation, Wuxi, China

Minimized Junction Leakage Current for Nanoscale MOSFET Applications   79 (1-17)
Wenqi Bai, Huishan Yang, Zhisen Huang, Kunghong Lee, Shiming Wang and Zhanyuan Hu
Huali Microelectronics Corporation, Shanghai, China

Study of Related Yield Loss and Mechanism of NOR Flash Self-Align-Source   82 (1-18)
Tian Zhi，Youhua Qin，Gu Zhen，Juanjuan Li，Qiwei Wang and Haoyu Chen
Huali Microelectronics Corporation, Shanghai, China

High Performance HVNMOS Development for Advanced Planner NAND Flash   85 (1-19)
Juanjuan Li, Zhi Tian, Xiaohua Ju, Tao Liu, Shaokang Yao, Haewan Yang and Yaoyu Chen
Huali Microelectronics Corporation, Shanghai, China

Development of Low Leakage Current in Extreme PFET Device
88 (1-22)
Wenqi Bai, Huishan Yang, Shiming Wang, Kunhong Lee, Zhisen Huang and Zhanyuan Hu
Huali Microelectronics Corporation, Shanghai, China

Fabrication and Characterization of a Novel Fully Self-Aligned Split-Gate SONOS Memory Device
91 (1-24)
Zhaozhao Xu, Jun Hu, Ning Wang, Kegang Zhang, Donghua Liu, Hualun Chen and Wensheng Qian
Huahong Grace Semiconductor Manufacturing Corporation, Shanghai, China

Investigation and Demonstration of Hot Carrier Effect in LDMOS Transistors with Ultra-Shallow Trench Isolation
95 (1-25)
Zhaozhao Xu, Jun Hu, Ziquan Fang, Wenting Duan, Donghua Liu and Wensheng Qian
Huahong Grace Semiconductor Manufacturing Corporation, Shanghai, China

A New Integration flow Study of ONO film Uniformity and Silicon Recess Improvement for 2T-SONOS Flash
99 (1-26)
Liqun Dong, Zhenghong Liu, Chris Shao, Haoyu Chen, Guanqun Huang and Ruisheng Qi
Huali Microelectronics Corporation, Shanghai, China

Study of GIDL Improvement for 2T-SONOS Flash
102 (1-28)
Zhenghong Liu, Liqun Dong, Ruisheng Qi, Shugang Dai, Guanqun Huang, Haoyu Chen and Chris Shao
Huali Microelectronics Corporation, Shanghai, China

One New Calibration Structure of MOSFET Gate Oxide Capacitor
105 (1-30)
Han Xiaojing
Huali Microelectronics Corporation, Shanghai, China

Narrow-Band Mask Synthesis with Semi-Implicit Difference
108 (1-35)
Yijiang Shen and XiaoPeng Wang
School of Automation, Guangdong University of Technology, Guangzhou, China

Wafer Edge Peeling Defect Mechanism Analysis and Reduction in IMD Process
111 (1-36)
Ya Li Feng, Qi Liang Ni, Xiao Fang Gu, Guang zhi He and Hao Guo
Huali Microelectronics Corporation, Shanghai, China

Evaluation of Pre Silicide Implant From Low Temperature to Room Temperature
114 (1-38)
Zhouchun, Caowenjie, Chengxinhua and Fangjingxun
Huali Microelectronics Corporation, Shanghai, China

Fragmentation of Square Pattern Mask with Small Corner-to-Corner Space
117 (1-40)
Yu Shirui, Chen Yanpeng, Wang Dan, Deng Guogui and Hu Yidan
Huali Integrated Circuit Corporation, Shanghai, China

Study of Low Pinch-Off Voltage JFET in 500V High Voltage Process
120 (1-41)
Wenting Duan, Ziquan Fang and Wensheng Qian
HuaHong Grace Semiconductor Manufacturing Corporation, Shanghai, China

Study of MOSFET IDVG Curve Double Hump Effect   123 (1-42)
Jun Hu, Zhaozhao Xu, Wenting Duan, Ziquan Fang, Donghua Liu and Wensheng Qian
HuaHong Grace Semiconductor Manufacturing Corporation, Shanghai, China

Effect of Implant Beam Current on Resistance of BF_2 Implanted Polysilicon   126 (1-43)
Lichao Zong, Chunling Liu, Xingjie Wang and Liming Chen
HuaHong Grace Semiconductor Manufacturing Corporation, Shanghai, China

Deep Power Down Leakage Study Caused by Poly L-shape Pattern   129 (1-45)
Chong Huang, Ming Zhang, Fangce Sun, Steam Cao, Guanghua Yang and Susanna Zheng
Huahong Grace Semiconductor Manufacturing Corporation, Shanghai, China

Virtual Source for an Odd Mathieu-Gauss Beam and Compare of the Functional Images of the Odd and Even   133 (1-46)
Xuxin Qi and Xuxin Qi
School of Science, Harbin Institute of Technology, Harbin, China

Detection of Electrical Defects by Distinguish Methodology Using an Advanced E-Beam Inspection System   136 (1-48)
Shanshan Chen, Hunglin Chen, Yin Long, Fengjia Pan and Wang Kai
Shanghai Huali Microelectronics Corporation, Shanghai, China

Study of Ultra-High Voltage BCD Process with Gate Oxide Thinning   139 (1-51)
Donghua Liu, Wenting Duan, Ziquan Fang and Wensheng Qian
HuaHong Grace Semiconductor Manufacturing Corporation, Shanghai, China

The Method of Improving ALD SiCN Film Uniformity   143 (1-52)
Yanxia Hao, Junlong Kang, Jun Yin, Yinshuai Wang, Guangyu Nie, Xinhua Cheng and Jingxun Fang
Huali Integrated Circuit Corporation, Shanghai, China

Magnetoelectric Memory Cell Based on Microsized FeGa Films on Ferroelectric 50BZT–50BCT Films   146 (1-53)
Zhi Tao[1], Yemei Han[1,2], Xianming Ren[1], Hui Li[1], Fang Wang[1] and Kailiang Zhang[1,2]
[1]*Tianjin University of Technology, Tianjin, China*
[2]*Dept.School of Electronic and Information Engineering, Tianjin Key Laboratory of Film Electronic & Communication Devices, Tianjin, China*

TaO_x Synapse Array Based on Ion Profile Engineering for High Accuracy Neuromorphic Computing   149 (1-71)
Jingjing Yang[1,2], Jiadi Zhu[2], Bingjie Dang[2], Teng Zhang[2], Qingxi Duan[2], Liying Xu[2], Keqin Liu[2], Zhiting Lin[1], Ru Huang[2] and Yuchao Yang[2]
[1]*School of Electronics and information Engineering, Anhui University, Hefei, China*
[2]*Department of Micro/nanoelectronics, Peking University, Beijing, China*

The Causation and Improvement of One Type of Particles Occurring in Batch-Clean Tool   153 (1-72)
Jing Ye, Jun Gao, Nan Lin and Jun Liu
Huahong Grace Semiconductor Manufacturing Corporation, Shanghai, China

Impact of Nanopillar-Type Electrode on HfO$_x$ -Based RRAM Performance   156 (1-73)

Baotong Zhang[1,2], Xiaokang Li[2], Yuancheng Yang[2], Haixia Li[2], Ru Huang[2], Ming Li[2] and Peimin Lu[1]

[1]*College of Physics and Information Engineering, Fuzhou University, Fuzhou, China*
[2]*Key Laboratory of Microelectronic Devices and Circuits, Institute of Microelectronics, Peking University, Beijing, China*

Improvement on Electronic Characteristics of TaO$_x$/TiO$_x$ Dual-Layer Structure Resistive Memory   159 (1-74)

Yu She, Honggang Pan, Fang Wang, Chuang Li, Zhenzhong Zhang, Yemei Han and Kailiang Zhang

Tianjin Key Laboratory of Film Electronic & Communication Devices, School of Electrical & Electronic Engineering, Tianjin University of Technology, Tianjin, China

A Novel Gate Architecture Design in STI Based LDMOS   162 (1-76)

Ziquan Fang, Zhaozhao Xu and Wensheng Qian

HuaHong Grace Semiconductor Manufacturing Corporation, Shanghai, China

Impact of Circuit Limit and Device Noise on RRAM Based Conditional Generative Adversarial Network   165 (1-77)

Shengyu Bao[1], Zongwei Wang[1,2], Tianyi Liu[1], Daqin Chen[1], Yimao Cai[1,3] and Ru Huang[1,2]

[1]*Institute of Microelectronics, Peking University, Beijing, China*
[2]*Key Laboratory of Microelectronic Devices and Circuits, Peking University, Beijing, China*
[3]*Frontiers science center for nano-optoelectronics, Peking University, Beijing, China*

Implementation of Graph Convolution Network Based on Analog RRAM   168 (1-78)

Daqin Chen[1], Zongwei Wang[1,2], Shengyu Bao[1], Yimao Cai[1,3] Ru Huang[1,2]

[1]*Institute of Microelectronics, Peking University, Beijing, China*
[2]*Key Laboratory of Microelectronic Devices and Circuits, Peking University, Beijing, China*
[3]*Frontiers science center for nano-optoelectronics, Peking University, Beijing, China*

TCAD Simulation on Random Telegraph Noise and Grain-Induced Fluctuation of 3D NAND Cell Transistors   171 (1-80)

Shijie Hu[1,2], Ming Li[2,3] and Ru Huang[1,2]

[1]*Shenzhen Graduate School, Peking University, Shenzhen, China*
[2]*Key Laboratory of Microelectronic Devices and Circuits (MOE), Institute of Microelectronics, Peking University, Beijing, China*
[3]*Frontiers Science Center for Nano-optoelectronics, Peking University, Beijing, China*

A Physical Current Model for Multi-Finger Gate Tunneling FET with Schottky Junction   174 (1-81)

Yimei Li, Jin Luo, Qianqian Huang, Xia An, Le Ye and Ru Huang

Key Laboratory of Microelectronic Devices and Circuits (MOE), Institute of Microelectronics, Peking University, Beijing, China

Origin of Steep Subthreshold Swing Within the Low Drain Current Range in Negative Capacitance Field Effect Transistor   177 (1-82)

Chang Su, Qianqian Huang, Mengxuan Yang, Liang Chen, Zhongxin Liang and Ru Huang

Key Laboratory of Microelectronic Devices and Circuits (MOE), Institute of Microelectronics, Peking University, Beijing, China

Circuit Reliability Evaluation of Approximate Computing 180 (1-86)

Yuwei Zhang[1,2], Zuodong Zhang[2], Zhe Zhang[2], Jiayang Zhang[2], Runsheng Wang[2], Zhiting Ling[1] and Ru Huang[2]

[1]*School of Electronics and Information Engineering, Anhui University, Hefei, China*
[2]*Institute of Microelectronic, Peking University, Beijing, China*

Device Modeling and Application Simulation of Ferroelectric-FETs With Dynamic Multi-Domain Behavior 183 (1-88)

Zhiyuan Fu, Cheng Chen, Jin Luo, Qianqian Huang and Ru Huang

Key Laboratory of Microelectronic Devices and Circuits (MOE), Institute of Microelectronics, Peking University, Beijing, China

A Novel Electrical Isolation Solution for Tunnel FET Integration 187 (1-89)

Ting Li[1,2], Qianqian Huang[2], Le Ye[2], Yuan Zhong[2], Mengxuan Yang[2], Yiqing Li[2], Yimei Li[2], Zhongxin Liang[2] and Ru Huang[2]

[1] *School of Electronic Information Engineering, Anhui University, Hefei, China*
[2]*Key Laboratory of Microelectronic Devices and Circuits (MOE), Institute of Microelectronics, Peking University, Beijing, China*

The Factors That Inflence The Effective Mobility in 5 nm PMOS FinFET Design 190 (1-54)

Enming Shang, Xin Luo, Yu Ding, Shaojian Hu, Shoumian Chen and Yuhang Zhao

Shanghai IC R&D Center, Zhangjiang Hi-Tech Park, Shanghai, China

Optimization of Embedded SiGe Process To Enhance PFET Performance on 28nm Low Power Platform 193 (1-56)

Wei Liu, Haibo Lei, Xuejiao Wang

Technology Development Department, Shanghai Huali Integrated Circuit Corporation, Shanghai, China

SRAM and Single Device Isolation analysis in FinFET Technology 198 (1-100)

Yijun Zhang, Li Tan, Yu Li

Technology R&D Center, SMIC, Shanghai, China

Chapter II – Lithography and Patterning

Etch Model Based on Machine Learning 202 (2-7)

Rui Chen[1], Haoru Hu[1], Xiaoting Li[2], Ying Chen[1], Xiaojing Su[1], Lisong Dong[1], Lei Qu[2], Chen Li[1], Jiang Yan[2] and Yayi Wei[1]

[1]*Key Laboratory of Microelectronics Devices and Integrated Technology, Institute of Microelectronics, Chinese Academy of Sciences, Beijing, China*
[2]*North China University of Technology, Beijing, China*

AI Computational Lithography 206 (2-34)

Xuelong Shi, Yuhang Zhao, Shoumian Chen and Chen Li

Shanghai IC Research and Development Center, Shanghai, China

Accurate Mask Model Approaches for Wafer Hot Spot Prediction and Verification 210 (2-38)

Young Ham, Colbert Lu[1], HJ Lee[1], Mohamed Ramadan, Michael Green and Chris Progler

R&D, Photronics Inc., Boise, Idaho, USA
[1]*Photronics DNP Mask Corporation (PDMC) Inc., Hsin-Chu City, Taiwan*

A Simulation Study for Typical Design Rule Patterns and Stochastic Printing Failures in a 5nm Logic Process with EUV Lithography
Yanli Li, Qiang Wu and Yuhang Zhao
Shanghai IC R&D Center, Zhangjiang Hi-Tech Park, Shanghai, China

213 (2-25)

A Study of Image Contrast, Stochastic Defectivity, and Optical Proximity Effect in EUV Photolithographic Process Under Typical 5nm Logic Design Rules
Qiang Wu, Yanli Li, Yushu Yang and Shoumian Chen
Shanghai IC R&D Center, Zhangjiang Hi-Tech Park, Shanghai, China

220 (2-20)

The Topography Effect on the Lithography Patterning Control for Implantation Layers
Dongyu Xu, Dingshuo Luo, Zhihong Wang, Wenzhan Zhou and Zhanyuan Hu
Huali Integrated Circuit Cooperation, Shanghai, China

227 (2-31)

High Speed Wafer Geometry on Silicon Wafers Using Wave Front Phase Imaging for Inline Metrology
J.M. Trujillo-Sevilla, J.M. Ramos-Rodríguez and J. Gaudestad
Wooptix, La Laguna, Tenerife, Spain

230 (2-1)

How To Improve 'Chemical Stochastic' in EUV Lithography?
Toru Fujimori
Electronic Materials Research Laboratories, R&D Management Headquarters, FUJIFILM Corporation, Yoshida-Cho, Haibara-Gun, Shizuoka, Japan

236 (2-30)

Impacts of RTP Pyrometer Offsets on Wafer Overlay Residue
Jian Lv, Qing Wang, Yetao Lu, Xing Gao and Qin Sun
Shanghai Huali Microelectronics Corporation, Shanghai, China

239 (2-3)

Applications of Sparse and Compact Resist Modeling in Advanced Node Implant Layer
Shirui Yu, Mudan Wang[*], Yiqun Tan, Juan Wei and Renyang Meng
Technology Development, Huali Integrated Circuit Corporation, Shanghai, China

242 (2-22)

Study of Alignment & Overlay Strategy in 14 nm Lithography Process
Lulu Lai, Rui Qian, Biqiu Liu, Xiaobo Guo, Cong Zhang, Jun Huang and Yu Zhang J
Huali Integrated Circuit Corporation, Shanghai, China

246 (2-26)

Litho Process Optimization to Improve Overlay Measurement in Thick PR Layer
Jiantao Wang, Jun Yu, Yuming Sun, Cong Zhang, Xiaobo Guo, Biqiu Liu, Song Gao, Shuo Liu, Jun Huang and Yu Zhang
Huali Integrated Circuit Corporation, Shanghai, China

249 (2-27)

Effects of Electron Beam on Photo Resist Shrinkage and Critical Dimension in SEM Measurement
Yuyang Bian, Hongxu Sun, Lipeng Wang, Xijun Guan, Xiaobo Guo, Biqiu Liu, Cong Zhang, Jun Huang and Yu Zhang
Huali Integrated Circuit Corporation, Shanghai, China

252 (2-28)

Enlarge Process Window of BSI in DTI Loop: A Novel OPC Approach to Add SRAF
Qiao Yanhui, Li Baoxuan, Wan Dan, Chen Yanpeng and Yu Shirui
Huali Integrated Circuit Corporation, Shanghai, China

255 (2-29)

Development of 90 nm & 5 nm High Resolution Advanced Lithographic Patterning Materials   258 (2-44)

Xuemiao Li, Zhenyu Yang, and Hai Deng
School of Microelectronics, State Key Laboratory of Molecular Engineering of Polymers，Fudan University，Shanghai，China

Chapter III – Dry & Wet Etch and Cleaning

The Law that Guides the Development of Photolithography Technology and the Methodology in the Design of Photolithographic Process   261 (3-11)

Qiang Wu, Yanli Li, Yushu Yang, Shoumian Chen and Yuhang Zhao
Shanghai IC R&D Center, Zhangjiang Hi-Tech Park, Shanghai, China

14nm Fin SADP Patterning Processes and Integration*   267 (3-10)

Chunyan Yi, Ming Li, Yongjian Luo, Weijun Wang, Zhunhua Liu，Xiaoqiang Zhou, Wen Xu and Ying Zhang
Shanghai IC R&D Center, Zhangjiang Hi-Tech Park, Shanghai, China

Towards Microstructures with Ultrahigh Aspect-Ratio and Verticality in Deep Silicon Etching   270 (3-16)

Yuanwei Lin
NAURA Technology Group Co., Ltd., Beijing, China

The Solution of AIO-ET Via Open and Process Window Improvement   273 (3-28)

Baichun Zhang, Jianguo Yang, Lei Sun，Quanbo Li，Jun Huang and Yu Zhang
Huali Integrated Circuit Corporation, Shanghai, China

The Solution of Contact Open and Short   276 (3-30)

Renhui Xu, Jie Zhang, Jianguo Yang, Lei Sun, Quanbo Li, Jun Huang and Yu Zhang
Huali Integrated Circuit Corporation, Shanghai, China

Optimized Work Function Metal Layer Damage Effect in Metal Gate BARC Etch Process by ICP Etch System   278 (3-31)

Kai Qian, Shaoxiong Liu, Lian Lu, Quanbo Li, Jun Huang and Yu Zhang
Huali Integrated Circuit Corporation, Shanghai, China

Analysis of Linewidth Uniformity and Line Edge/Width Roughness in a 5 nm Logic SAQP Process   281 (3-21)

Bowen Wang[1], Yushu Yang[1], Yibo Wang[2], Yuning Zhu[2], Yongjian Luo[2], Qiang Wu[1], Weihao Lin[1], Zhunhua Liu[1], Yanli Li[1], Qingqing Wu[1], Shoumian Chen[1] and Ying Zhang[2]
[1]Shanghai IC R&D Center, Shanghai, China
[2]Beijing NAURA Microelectronics Equipment Co., Ltd., Beijing, China

Metal Hard Mask Open Process Window Enhanced by Insertion of Polymer Deposition   285 (3-17)

Li Fei Sun[1], Qing Peng Wang[1], Ji Hong Zhang[1], Lei Sun[3], Andrew Li[2] and Yu Shan Chi[1]

[1]*Lam Research Corporation, Shanghai, China*
[2]*Lam Research Corporation, Fremont, CA, USA*
[3]*Huali Integrated Circuit Corporation, Shanghai, China*

Application of a Bevel Etch Process for Improving Particle Performance in CMOS Image Sensor Manufacture   289 (3-18)

Yiling Sun[1], Jihong Zhang[1], Yu Jiang[1], Fulong Qiao[2], Keqiang He[2], Zhigang Zhang[2], Kang Huang[2] and Yushan

[1]*Lam Research Corporation*
[2]*Huali Micro Electronic Semiconductor Corporation Ltd., Shanghai, China*

Silicon Wafer Thinning Process by Dry Etching with Low Roughness and High Uniformity   292 (3-20)

Zihan Dong, Renzhi Yuan and Yuanwei Lin

NAURA Technology Group Co., Ltd., Beijing, China

Improvement Research of Round Convex Residue in Dual Gate Layer   296 (3-25)

Mingguang Hang, Lili Jia, Fang Li, Jun Huang and Wenyan Liu

Huali Integrated Circuit Manufacturing Corporation, Shanghai, China

Advantage Timely Energized Bubble Oscillation Megasonic Nano-Spray Method to Eliminate Surface Particle Defect in Lightly Doped Drain 28nm   299 (3-26)

Hong Li [1], Fang Li[1], Wenyan Liu[1], Jun Huang[1], Yu Zhang[1], Wenjun Wang[2], Xiaoyan Zhang[2], Ting Yao[2] and David H. Wang[2]

[1]*Huali Microelectronics Corporation, Shanghai, China*
[2]*ACM Research (Shanghai), Inc., Shanghai, China*

Optimization of 28nm SiGe Sigma Shape Trench Depth Loading Effect   302 (3-27)

Lili Jia, Fang Li, Wenyan Liu and Jun Huang

Huali Integrated Circuit Manufacturing Corporation, Shanghai, China

Well CD Control and Vertical Profile BARC Etch Development and Related Theory Research   304 (3-29)

Jiang Linpeng, Zhu Yizheng, Lu Lian, Li Quanbo, Huang Jun and Zhang Yu

Huali Integrated Circuit Corporation, Shanghai, China

Impact of Rework Process to Etch Bias and the Corresponding Solution   306 (3-33)

Pengkai Xu, Penggang Han, Wenyan Sun, Sen Wu, Bin Zhao, Yi Wang and Fulong Qiao

Huali Microelectronics Corporation-Technology Development Division, Shanghai, China

Rectangular Suspended Single Crystal Si Nanowire with (001) Planes and <001> Direction Developed via TMAH Wet Chemical Etching   311 (3-40)

Shuang Sun, Baotong Zhang, Yuancheng Yang, Xia An, Xiaoyan Xu, Ru Huang and Ming Li

Key Laboratory of Microelectronic Devices and Circuits, Institute of Microelectronics, Peking University, Beijing, China

Effective Lithography Leveling Improvement Was Achieved by Retaining Wafer Back-Surface Nitride During a Novel SMT Nitride Remove Process   314 (3-46)

Weiwei Ma, Chao Sun, Xiaolin Xu and Wei Zhou

Huali Integrated Circuit Corporation, Shanghai, China

Improved Selective Silicon Nitride Etch for Advanced Logic and Memory Applications
Chien-Pin Sherman Hsu
Avantor, Chu-Bei, Hsinchu, Taiwan

317 (3-49)

Quasi-Atomic Layer Etching Technology for High Uniformity Etching Applications
Y. Zhang, J. Chong, C. Wang, Q. Xie and D. Li
NAURA Technology Group Co., Ltd.

319 (3-8)

XPS Analysis of Gallium Nitride Film after O_2/BCl_3 Digital Etch
Jiale Tang[1], Yongjie Hu[1], Yudong Zhang[1], Zhiqiang Gu[2], Dongchen Che[2], Dongdong Hu[2], Lu Chen[2], Kaidong Xu[1,2] and Shiwei Zhuang[1]
[1]School of Physics and Electronic Engineering, Jiangsu Normal University, Xuzhou, China
[2]Leuven Instruments Co. Ltd (Jiangsu), Xuzhou, China

322 (3-52)

5 nm Fin SAQP Process Development and Key Process Challenge Discussion
Yushu Yang, Bowen Wang[1], Qiang Wu[1], Yanli Li[1], Yibo Wang[2], Yuning Zhu[2], Yongjian Luo[2], Weihao Lin[1], Qingqing Wu[1], Jianjun Zhu[1], Shoumian Chen[1] and Ying Zhang[2]
[1]Shanghai IC R&D Center, Shanghai, China
[2]Beijing NAURA Microelectronics Equipment Co., Ltd., Beijing, China

326 (3-24)

Chapter IV – Thin Film, Plating and Process Integration

Titanium Silicide Anneal Process Research for 14nm FinFET Technology
Lan Jiang, Yan Gui, Yaoting Shen, Qingwei Dong, Kecheng Chen, Xinhua Cheng and Jiangxun Fang
Huali Integrated Circuit Corporation, Shanghai, China

329 (4-27)

Surface Smoothing and Roughening Effects of High-k Dielectric Materials Deposited by Atomic Layer Deposition and Their Significance for MIM Capacitors Used in DRAM Technology Part II
W.S. Lau
Zhejiang University, Department of Information Science and Electronic Engineering, Hangzhou, China

332 (4-19)

FDSOI SiGe Morphology Optimization on Boundary of AA and STI
Jiaqi Hong, Qiuming Huang, Qiang Yan, Jun Tan, Yongyue Chen, Haifeng Zhou and Jingxun Fang
Huali Integrated Circuit Corporation, Shanghai, China

335 (4-31)

Optimization of Imperfect Morphology for Selective Epitaxial SiGe Growth
Yongyue Chen, Qiang Yan, Jiaqi Hong, Qiuming Huang, Jun Tan, Haifeng Zhou and

337 (4-29)

Jingxun Fang
Huali Integrated Circuit Corporation, Shanghai, China

Mechanically Stable Ultra-Low-k Dielectric and Air-Gap Technology* 340 (4-10)
Clarissa Prawoto, Ying Xiao and Mansun Chan
Department of Electronic and Computer Engineering, the Hong Kong University of Science and Technology, Hong Kong, China

Some Key Modifications of Theory Required to Understand the Leakage Current 344 (4-18)
Mechanisms For MIM Capacitors Used in DRAM Technology
W.S. Lau
Zhejiang University, Department of Information Science and Electronic Engineering, Hangzhou, China

Thin Film Processes: Abatement of Waste Gases From Plasma Assisted Material 347 (4-22)
Processes
Christopher P. Jones
Edwards Ltd, Clevedon, North Somerset, United Kingdom

Study of Influence of STI Profile on HARP Gap-Filling Performance 350 (4-17)
Kai Wang, Zhigang Zhang, Ping Wang, Lingzhi Xu, Shenzhou Lu, Andy Tan, Zhenjie Qiao, Kang Huang, Qimeng Wang, Duo Shan, Fan Zhang, Chang Fu, Zhaoyuan Zhao and Qin Sun
Huali Microelectronics Corporation, Shanghai, China

A Novel Methodology to Monitor Wafer Placement Shift in Laser Spike Anneal 352 (4-25)
Yan Gui, Lan Jiang, Yaoting Shen, Qingwei Dong, Kecheng Chen, Xinhua Cheng and Jingxun Fang
Huali Integrated Circuit Corporation, Shanghai, China

Investigation and Characterization of Silicon Concentration in N-Free Anti-Reflective 355 (4-30)
Layer Films
Luhang Shen, Jiepeng Zhou, Chunwen Liu, Haixia Li, Yiqi Gong, Yu Bao and Jingxun Fang
Huali Integrated Circuit Corporation, Shanghai, China

The Investigation of Domestic Machines Large-Scale Production in Soak Anneal 358 (4-33)
Process
Shen Yaoting, Jiang Lan, Gui Yan, Dong Qingwei, Chen Kechen, Cheng Xinhua and Fang Jingxun
Huali Integrated Circuit Corporation, Shanghai, China

BEOL Cu Gap-fill Performance Improvement for 14nm Technology Node 361 (4-35)
Zhaoqin Zeng, Bao Yu, Yanpeng Cao, Xingkun Xue, Jianhua Xu, Yanyan Zhang, Xiaofang Wang, Jingxun Fang and Yu Zhang
Huali Integrated Circuit Corporation, Shanghai, China

Optimization on Deposition of Aluminum Nitride by Pulsed Direct Current Reactive 366 (4-39)
Magnetron Sputtering
Yu-Pu Yang, Te-Yun Lu, Song-Ho Wang, Hsueh-Er Chang, Peter J. Wang, Walter Lai, Yiin-Kuen Fuh and Tomi T. Li

Department of Mechanical Engineering, National Central University, Taoyuan City, Taiwan
Delta Electronics In., Taoyuan City, Taiwan, China

Study of PREB Process in FDSOI   370 (4-41)
Yang Song, Zhanhai Yang, Xia Tang, Yanfei Ma, Feng Niu and Changfeng Wang
Huali Microelectronics Corporation, Shanghai, China

IDT Structure Optimization Design Based on AlN/Si Substrate for SAW Devices   372 (4-43)
Kaixuan Li, Fang Wang, Shuo Yan, Meng Deng, Huanhuan Di, Wei Li and Kailiang Zhang
Tianjin Key Laboratory of Film Electronic & Communication Devices, School of Electrical
& Electronic Engineering, Tianjin University of Technology, Tianjin, China

Investigation of FDSOI Raised S/D Formation   375 (4-42)
Yanfei Ma, Yang Song and Changfeng Wang
Huali Microelectronics Corporation, Shanghai, China

Optimization of the CD Uniformity (CDU) in Silicon Oxide Spacer Process for 5 nm Fin   378 (4-45)
SAQP Process Flow
Qingqing Wu[1], Weihao Lin[1], Xiaoqiang Zhou[1], Jinhua Zhang[1], Jing Li[2], Leng Han[2], Jianjun
Zhu[1], Yushu Yang[1], Qiang Wu[1] and Shoumian Chen[1]
[1]Shanghai IC R&D Center, Zhangjiang Hi-Tech Park, Shanghai, China.
[2]Piotech Co., Ltd., Hunnan District, Shenyang, Liaoning Province, China.

Enhancing High Temperature Adhesion Performance Via a Renovated Leadframe   382 (4-36)
Surface Treatment
Din-Ghee Neoh[1], Tee Weikok[2], Liao Jinzhi Lois[3], Jia Wenping[4], Yee Boonhwa[2], Boon-
Seong Lee, Wang Bisheng[4], Zhang Xi[3], Hua Younan[3], Li Xiaomin[3], Mario Strauch[1] and
Boon-Chye Lee
[1]Atotech (Singapore) Chemicals Pte Ltd, Surface Engineering Hub, Singapore
[2]Sumitomo Bakelite Singapore Pte Ltd, Singapore
[3]WinTech Nano-Technology Services Pte. Ltd., The Alpha Science Park II, Singapore
[4]Huawei Technologies Co Ltd, Bantian Huawei Base, Shenzhen, China

Chapter V – CMP and Post-Polish Cleaning

Solving CMP Challenges for Chemically Stable Materials and 3D Shapes*   390 (5-18)
Hitoshi Morinaga
FUJIMI Incorporated, Kakamigahara, Japan

Stop on Nitride Slurry Development*   394 (5-16)
Shoutian Li, Changzhen Jia and Xiaoming Ren
Anji Microelectronics Technology (Shanghai) Co. Ltd. , Shanghai, China

Novel Abrasion-Free CMP Technology With High Performance Polishing Slurry*   398 (5-32)
Chong Luo[1] and Yuling Liu[1,2]
[1]*School of Electronic Engineering, Hebei University of Technology, Tianjin, China*
[2]*Tianjin Key Laboratory of Electronic Materials and Devices, Tianjin, China*

Study of Ceria Settling in CMP Slurry   402 (5-20)
Li Zhang[1,2], Chuangyun Wan[1], Changzhen Jia[2], Chengyao Shi[2], Xiaoming Ren[2] and Shoutian Li[2]
[1]*Shanghai Institute of Technology, Shanghai, China*
[2]*Anji Microelectronics Technology (Shanghai) Co. Ltd., Shanghai, China*

Pattern Loading Effect Optimization of BEOL Cu CMP in 14nm Technology Node   406 (5-14)
Lei Zhang, Yuanyuan Meng, Yi Xian, Wei Zhang, Haifeng Zhou and Jingxun Fang
Shanghai Huali Integrated Circuit Corporation, Shanghai, , China

The Adsorption and Removal of Corrosion Inhibitors During Metal CMP*   409 (5-47)
Jin-Goo Park[1], Heon-Yul Ryu[2], Tae-Gon Kim[3], Nagendra Prasad Yerriboina[1], Yutaka Wada[4], Satomi Hamada[4] and Hirokuni Hiyama[4]
[1]*Department of Materials Science and Chemical Engineering, Hanyang University, Ansan, Korea*
[2]*Department of Bio-Nano Technology, Hanyang University, Ansan, Korea*
[3]*Department of Smart Convergence Engineering, Hanyang University, Ansan, Korea*
[4]*EBARA Corporation, Fujisawa, Kanagawa, Japan*

Effects of Different Inhibitors on Cu-Co Galvanic Corrosion in Post CMP Cleaning   412 (5-28)
Xiaoqin Sun[1,2], Baimei Tan[1,2], Chenwei Wang[1,2], Mengrui Liu1,2, Pengcheng Gao[1,2], Qi Wang[1,2] and SiyuTian[1,2]
[1]*School of Electronic Information Engineering, Hebei University of Technology, Tianjin, China*
[2]*Tianjin Key Laboratory of Electronic Materials and Devices, Tianjin, China*

Molecular Dynamics Study on Sub-Nanoscale Removal Mechanism of 3C-SiC in a Fixed Abrasive Polishing   415 (5-29)
Piao Zhou[1], Yongwei Zhu[1] and Tao Sun[2]
[1]*College of Mechanical and Electrical Engineering, Nanjing University of Aeronautics and Astronautics, Nanjing, China*
[2]*College of Chemistry and Chemical Engineering, Shanghai University of Engineering Science, Shanghai, China*

Role of Slurry Chemistry for Defects Reduction During Barrier CMP   419 (5-31)
Chenwei Wang[1,2], Yue Li[1,2], Guoqiang Song[1,2], Zhaoqing Huo[1,2], Jia Liu[1,2] and Yuling Liu[1,2]
[1]*School of Electronic Information Engineering, Hebei University of Technology, Tianjin, China*
[2]*Tianjin Key Laboratory of Electronic Materials and Devices, Tianjin, China*

Role of Slurry Additions on Chemical Mechanical Polishing of Cu/Ru/TEOS in H_2O_2-Based Slurry   422 (5-39)
Chao Wang[1,2], Jianwei Zhou[1,2], Chenwei Wang[1,2] and Xue Zhang[1,2]

[1]School of Electronic Information Engineering, Hebei University of Technology, Tianjin, China
[2]Tianjin Key Laboratory of Electronic Materials and Devices, Tianjin, China

Effect of Various Surfactants on Surface Roughness Reduction During Cobalt "Buff Step" CMP   425 (5-34)

Yuanshen Cheng[1,2], Shengli Wang[1,2], Chenwei Wang[1,2] and Yundian Yang [1,2]
[1]School of Electronic Information Engineering, Hebei University of Technology, Tianjin, China
[2]Tianjin Key Laboratory of Electronic Materials and Devices, Tianjin, China

Mechanism Analysis of Chemical Mechanical Polishing of 4H-SiC Wafer   428 (5-43)

Gaoyang Zhao[1,2], Aoxue Xu[1,2], Fan Xu[1,2], Daohuan Feng[1,2], WeiliLiu[1,2] and Zhitang Song[1,2]
[1]Shanghai Institute of Microsystem and Information Technology, Shanghai, China
[2]University of Chinese Academy of Sciences, Beijing, China

Investigation on the Gallium Nitride Polishing Process Under Hybrid-Field Effects   431 (5-24)

Zhigang Dong[1], Liwei Ou[1], Kang Shi[2], Renke Kang[1] and Dongming Guo[1]
[1]Key Laboratory for Precision and Non-traditional Machining Technology of Ministry of Education, Dalian University of Technology, Dalian, China
[2]Department of Chemistry and State Key Laboratory of Physical Chemistry of Solid Surfaces, College of Chemistry and chemical Engineering, Xiamen University, Xiamen, China

Effect of Potassium Salts on the Chemical Mechanical Polishing Efficiency of Sapphire Substrate   434 (5-38)

Yanan Lu[1,2], Xinhuan Niu[1,2], Yaqi Cui[1,2], Xin Zhao[1,2], Zhaoqing Huo[1,2] and Chenghui Yang[1,2]
[1]School of Electronics and Information Engineering, Hebei University of Technology, Tianjin, China
[2]Tianjin Key Laboratory of Electronic Materials and Devices, Tianjin, China

A Study of Causes and Improving Methods of Chipping in BSI Process   437 (5-2)

Yurong Cao, Hu Li, Zhe Feng, Zujun Ji, Zhengyuan Zhao, Jin Chen, Youfeng Xu, Xiang Peng and Feng Ji
Huali Microelectronics Corporation, Shanghai, China

Study on the Properties of Silica Colloid Prepared by Different Processes in Silicon Wafer CMP   440 (5-3)

Weiwei Li, Zhilin Zhao, Zhen Liang and Yunqian Sun
School of Electronic Information Engineering, Hebei University of Technology, Tianjin, China

Impact of Bevel Condition on STI CMP Scratch   444 (5-15)

Yuanyuan Meng, Lei Zhang, Yibin Li, Wei Zhang, Haifeng Zhou and Jingxun Fang
Shanghai Huali Integrated Circuit Corporation, Shanghai, China

Effects of Heat Treatment in Air Environment on the Dispersivity of Nanodiamond   446 (5-25)

Menggang Lu, Xiaoguang Guo, Song Yuan, Zhuji Jin, Renke Kang and Dongming Guo
School of Mechanical Engineering, Dalian University of Technology, Dalian, China

Effect of Complexing Agent in Slurry on CMP Property for Barrier Material Cobalt   449 (5-26)

Jinsong Zuo, Fang Wang, Kai Hu, Luguang Wang, Yujie Yuan and Kailiang Zhang
Tianjin Key Laboratory of Film Electronic & Communication Devices, School of Electrical &Electronic Engineering, Tianjin University of Technology, Tianjin, China

Effect of Potassium Oleate as Inhibitor on Copper Chemical Mechanical Polishing

452 (5-36)

Chenghui Yang[1,2], Xinhuan Niu[1,2], Jiakai Zhou[1,2], Zhaoqing Huo[1,2] and Yanan Lu[1,2]
[1]School of Electronics and Information Engineering, Hebei University of Technology Tianjin, China
[2]Tianjin Key Laboratory of Electronic Materials and Devices, Tianjin, China

Effect of Chelators on the Removal of BTA in Post-CMP Cleaning

455 (5-37)

Mengrui Liu[1,2], Baimei Tan[1,2], Xiaoqin Sun[1,2], Pengcheng Gao[1,2], Qi Wang[1,2] and Siyu Tian[1,2]
[1]School of Electronic Information Engineering, Hebei University of Technology, Tianjin, China
[2]Tianjin Key Laboratory of Electronic Materials and Devices, Tianjin, China

High-K Metal Gate AL-CMP Within Die Uniformity and Selectivity Study

458 (5-12)

Ziheng_Li, Baicen_Wan, Hongdi_Wang, Andy Wang, Pujia_Shan, Zhiyang_Liang, Jian_Li and Zhijie_Zhang
Semiconductor Manufacturing North China (Beijing) Corp, Beijing, China

Impact of Pad Micro Contact Size and Distribution on the Planarization in CMP

461 (5-22)

Lin Wang, Haipeng Li, Ping Zhou and Ying Yan
Key Laboratory for Precision and Non-traditional Machining Technology of Ministry of Education, Dalian University of Technology, Dalian, China

Role of AEO-9 as Nonionic Surfactant in Barrier Slurry for Copper Interconnection CMP

465 (5-33)

Guoqiang Song[1,2], Baimei Tan[1,2], Yuling Liu[1,2] and Chenwei Wang[1,2]
[1]School of Electronic Information Engineering, Hebei University of Technology, Tianjin, China
[2]Tianjin Key Laboratory of Electronic Materials and Devices, Tianjin, China

Effect of TT-LYK on Copper CMP with Ru/Ta as Barrier/Liner

468 (5-40)

Xue Zhang[1,2], Jianwei Zhou[1,2], Chenwei Wang[1,2] and Chao Wang[1,2]
[1]School of Electronic Information Engineering, Hebei University of Technology, Tianjin, China
[2]Tianjin Key Laboratory of Electronic Materials and Devices, Tianjin, China

Effect of Various Complexing Agents for Cobalt "Bulk Step" Chemical Mechanical Planarization

471 (5-41)

Yundian Yang[1,2], Shengli Wang[1,2], Chenwei Wang[1,2], Yuanshen Cheng[1,2]
[1]School of Electronic Information Engineering, Hebei University of Technology, Tianjin, China
[2]Tianjin Key Laboratory of Electronic Materials and Devices, Tianjin, China

Lead the Future Semiconductor Evolution as Seen by CMP Manufacturers

474 (5-51)

Manabu Tsujimura
Ebara Corporation, Ohta-ku, Tokyo, Japan

Chapter VI – Metrology, Reliability and Testing

Multi-Granularity Reconfiguration Based Physical Unclonable Function Design
478 (6-30)
Jianan Mu[1,2], Jing Ye[1,2], Xiaowei Li[1,2], Huawei Li[1,2,3] and Yu Hu[1,2]
[1] *State Key Laboratory of Computer Architecture, Institute of Computing Technology, Chinese Academy of Sciences*
[2] *University of Chinese Academy of Sciences, Beijing, China*
[3] *Peng Cheng Laboratory, Shenzhen, China*

A Programming Framework of Concurrent Test on SMT7 for IPs which Share Same Access Port
481 (6-14)
Tianyu Zhang, Fang Yanfen and Kai Zhou
Advantest, Shanghai, China

Low Voltage Time-Resolved Emission (TRE) Measurements of VLSI Circuit
484 (6-17)
Shang Chih Lin and Frank Yong
Gallant Precision Machining Co., Ltd, Hsinchu, Taiwan

Unifying Yield Enhancement and Manufacturing Intelligence with Smart Sampling
487 (6-21)
Yan-Qiu Zhang and Haw-Jyue Luo
Fujian Jinhua Integrated Circuit Co., Ltd., Jinjiang, Quanzhou, China

An Adaptive Denoising System for Sub-nm Scale Failure Analysis Based on TEM Image
490 (6-22)
Chang Xu, Yi-Fu Zhang and Chi-Ren Luo
Fujian Jinhua Integrated Circuit Co., Ltd., Jinjiang, Quanzhou, China

Optical Scatterometry Modeling of 5 nm Logic Metal Gate Structures
493 (6-25)
Qi Wang, Aihua Yang, Yanli Li, Yushu Yang, Qiang Wu and Shoumian Chen
Shanghai IC R&D Center, Shanghai, China

Quality Control in Sapphire Growing: From Automated Defect Detection to Big Data Approach
497 (6-41)
Ivan Orlov and Frédéric Falise
Scientific Visual, Lausanne, Switzerland

Towards Understanding Interaction Between Hot Carrier Ageing and PBTI*
500 (6-16)
M. Duan, J. F. Zhang, Z. Ji and W. Zhang
Department of Electronics and Electrical Engineering, Liverpool John Moores University, Liverpool UK

Comprehensive Comparison of the Wire Bond Reliability Performance of Cu, PdCu and Ag Wires
503 (6-11)
Liao Jinzhi Lois[1], Yu Minglang[2], Tee Weikok[3], Wang Bisheng[2], Jia Wenping[2], Yee Boonhwa[3], Zheng Haipeng[1], Zhang Xi[1], Fu Chao[1], Li Xiaomin[1], Hua Younan[1]
[1] *WinTech Nano-Technology Services Pte. Ltd., The Alpha Science Park II, Singapore*

[2]Huawei Technologies Co Ltd, Bantian Huawei Base, Shenzhen, China
[3]Sumitomo Bakelite Singapore Pte Ltd, Singapore

Practical Carrier Lifetime Analysis in In$_{.53}$Ga$_{.47}$As Hetero-Epitaxial Layers N/A (6-7)
Eddy Simoen[1,2], Po-Chun (Brent) Hsu[1,3], Clement Merckling[1], Geert Eneman[1], Yves Mols[1], Han Han[1], AliReza Alian[1], Nadine Collaert[1] and Marc Heyns[1,3]
[1]Imec, Leuven, Belgium
[2]Department of Solid State Sciences, Ghent University, Gent, Belgium
[3]Department of Materials Engineering, KU Leuven, Leuven, Belgium

A Study of Low Temperature Al Sputter Process Electromigration Lifetime 508 (6-35)
Jun Liu, Lei Zhang, Jianmin Wang, Qinghua Liu and Di Lou
Huahong Grace Semiconductor Manufacturing Corporation, Shanghai, China

Reliability Improvement by 0.153um CMOS Using HDP-CVD at STI Edge SiN Liner 511 (6-42)
WeiYang Zhang, RenGang Qin, XiaoFeng Sun, DeJin Wang, HaoYu Wen and YaoHui Zhou
Central Semiconductor Manufacturing Corporation (CSMC), Wuxi, China

Non Linear Growth of Variance in the Process Gaps. A Cause of Adverse Cycle Times 514 (6-8)
George W Horn and William Podgorski
Middlesex Industries SA, Switzerland

Research on Improvement of Reference Voltage Shift of Wire-Bound Products 517 (6-38)
Yang Chen, Na Mei andTuobei Sun
ZTE Corporation, Zhangjiang Hi-tech Park, Shanghai, China

A Novel Vertical Closed-Loop Control Method for High Generation TFT Lithography Machine 519 (6-20)
Dan Chen
Shanghai Micro Electronics Equipment (Group) Co., Ltd., Shanghai, China

Probe Card Lifetime Control and Abrasion Coefficient Study 522 (6-9)
Lei Wang and Song Ma
Huahong Grace Semiconductor Manufacturing Corporation, Shanghai, China

The Inspection Solutions and Reduction of Extreme Tiny Poly Residue 525 (6-13)
Jianye Song, Qiliang Ni, Xiaofang Gu,Chao Han and Lijing Huang
Huali Microelectronics Corporation, Shanghai, China

An Application of Adaptive Genetic Algorithm Combining Monte Carlo Method 528 (6-15)
Wei Yu, Xu Chen,Jingjing Lu and Zhengying Wei
Huali Microelectronics Corporation, Shanghai, China

Investigation and Reduction of Systematic Defects by Wafer Backside Process in Nanometer Semiconductor Manufacturing 532 (6-18)
JianGang Zhou, Hungling Chen, Yin Long, Kai Wang and Hao Guo
Huali Integrated Circuit Manufacturing Co., Ltd, Shanghai, China

Study of High-Precision Interferometer Dynamic Switching for TFT Long-Travel- 535 (6-19)

Range Moving Stage
Dan Chen, Zhiyong Yang and Yuebin Zhu
Micro Electronics Equipment (Group) Co., Ltd., Shanghai, China

Fault Detection of Sensor Data in Semiconductor Processing with Variational Autoencoder Neural Network
Wang Yong, Chen Xu and Wei Zhengying
Huali Microelectronics Corporation, Shanghai, China

538 (6-23)

New Precision Jitter Measurement Solution on TMU - Challenge on PRBS Reconstruction
Kai Zhou, Tianyu Zhang[1], Xurong Cao and Yanyan Chang
Advantest (China) Co., Ltd, Shanghai, China

541 (6-24)

Investigation and Discovery of the Integration of FEOL Process by Electron Beam Inspections
Fengjia Pan, Hunglin Chen, Yin Long, Kai Wang and Hao Guo
Huali Integrated Circuit Corporation, China

544 (6-28)

One Comprehensive Method to Analyze Semiconductor Manufacturing Data by "Piecewise" Regression
Lin Gu and Wei Yu
Huali Microelectronics Corporation, China

547 (6-29)

The Inspection Solutions of 3bar Structure Cu Void in BEOL Advanced Semiconductor Process
Xingdi Zhang, Hunglin Chen, Yin Long and Kai Wang
Shanghai Huali Microelectronics Corporation, Shanghai, China

549 (6-32)

The Inspection and Solution of Inline CT Defect for 28nm Process Improvement
Min Wang, Hunglin Chen, Yin Long and Hao Guo
Huali Microelectronics Corporation, Shanghai, China

552 (6-33)

The Study and Investigation of Inline E-Beam Inspection for 28nm Process Window Monitor
Yin Long, Zengyi Yuan, Fengjia Pan, Kai Wang and Hunglin Chen
Huali Integrated Circuit Corporation, Shanghai, China

555 (6-37)

A Unified 4H-SiC MOSFETs TDDB Lifetime Model Based on Leakage Current Mechanism
Hua Chen, Pan Zhao, Jiahao Liu, Yusen Su, Tuo Zheng, Hao Ni and Liang He
School of Advanced Materials and Nanotechnology, Xidian University, Xi'an, China

557 (6-39)

Study on Wafer Edge Test with Optimized Test Solution
Yuxiang Zhang, Yuanyuan Zhu and Zhimin Zeng
Huahong Grace Semiconductor Manufacturing Corporation, Shanghai, China

560 (6-40)

Full Metrology Solutions for Advanced RF with Picosecond Ultrasonic Metrology

562 (6-34)

Johnny Dai[1], Priya Mukundhan[1], Johnny Mu[2], Frank Zheng[2], Cheolkyu Kim[3]
[1]*Onto Innovation, Budd Lake, NJ, USA*
[2]*Onto Innovation, Shanghai, China*
[3]*Onto Innovation, Bundang-gu, Sungnam-si,Gyunggi-do, Korea*

Monitoring Critical Process Steps in 3D NAND using Picosecond Ultrasonic Metrology with both Thickness and Sound Velocity Capabilities   565 (6-36)

Johnny Dai[1], Priya Mukundhan[1], Robin Mair[1], Manjusha Mehendale[1], Calvin Wang[2], Ewen Wang[2], Cheolkyu Kim[3]
[1]*Onto Innovation, NJ, USA*
[2]*Onto Innovation, Shanghai, China*
[3]*Onto Innovation, Bundang-gu, Sungnam-si,Gyunggi-do, Korea*

Chapter VII – Packaging and Assembly

Design and Development of 3D WLCSP for CMOS Image Sensor Using Vertical Via Technology   568 (7-8)

Tianshen Zhou[1,2], Shuying Ma[2], Fengxia Zheng[2] and Tao Hang[1]
[1]*School of Materials Science and Engineering, Shanghai Jiao Tong University, Shanghai, China*
[2] *Huatian Technology (Kunshan) Electronics Co., Ltd., Kunshan, China*

Laser-Based Full Cut Dicing Evaluations for Thin Si wafers   571 (7-2)

Peter Dijkstra, Jeroen van Borkulo and Richard van der Stam
ASM Laser Separation International B.V., Beuningen, The Netherlands

A Single-Layer Solution With Laser Debonding Technology for Temporary Bond/Debonding Applications in Wafer-Level Packaging   576 (7-6)

Xiao Liu, Lisa Kirchner, Luke Prenger, Wenkai Cheng and Rama Puligadda
Brewer Science, Inc., Rolla, MO, USA

Etching Polymer Technology for Large Number of Small Via   580 (7-10)

Kyu Jin Lee, Moon Sang You, Gun Woo Kim, Hyun Chul Han, Sang Ki Ahn, Ken Lee, Woo Jae Jeong, Jeong Hyuk Ahn, Jeong Wook Moon, Jee Hyeon Hwang, Kwang Joo Lee[1]
[1]*Simmtech Co., Ltd, Cheongju, Chungcheongbuk-do, Korea*
[2]*LG Chem Co., Ltd, Yuseong-gu, Daejeon, Korea*

A Promising Embedded Silicon Fan-Out Wafer Level Package With Laser Releasable Temporary Bonding Technology   584 (7-21)

Chengqian Wang[1,2], Aibing Zhang[1], Zhengfeng Li[1], Shouwei Li[1], Yang Li[1], Yong Ji[1], Xuefei Ming[1] and Daquan Yu[2]
[1]*The 58th Research Institute of China Electronics Technology Group Corp., Wuxi, China*
[2]*Xiamen University, Xiamen, China*

Improvement of Heat Dissipation in IPM Packaging Structure   588 (7-5)

Wenjie Xia[1], Jie Bao[1,2], Yuan Xu[2], Renxia Ning[2], Li Hou[2] and Zhenhai Chen[2]
[1]*College of Mechanical and Electrical Engineering, Huangshan University, Huangshan, China*
[2]*Engineering Technology Research Center of Intelligent Microsystem of Anhui Province, Huangshan, China*

Characterization of Multi-Scale Nanosilver Paste Reinforced With SiC Particles   591 (7-13)

Ziwei Jiang[1], Yongqian Sun[1], Qiaoran Zhang[1], Zhen Lv[1], Yanpei Wu[2], Weijuan Xi[2], Cheng Zhou[2], Shujing Chen[1], Maomao Zhang[1], Johan Liu[1,3] and Xiuzhen Lu[1]

[1]SMIT Center, School of Mechatronics Engineering and Automation Shanghai University, Shanghai, China
[2]Space Research Institute of Electronics and Information Technology, Aerospace Science and Technology Corporation, Xi'an, China
[3]Department of Microtechnology and Nanoscience Chalmers University of Technology, Gothenburg, Sweden

Investigation of Bond Pad Crystal Defect for Different Cover Transmission Rate   595 (7-9)

Chengyang Sun
Huali Microelectronics Corporation (HLMC), Shanghai, China

Effect of Bonded Ball Shape on Gold Wire Bonding Quality Based on ANSYS/LS-DYNA Simulation   599 (7-11)

Weidong Huang, Wei Wu, Jacky Wu, Grass Dong and CF Oo
Department of Research and Development, Diodes Shanghai Co., LTD., Shanghai, China

Reliability Simulation and Life Prediction of $Sn_{63}Pb_{37}$ BGa Solder Joint Under Thermal Cycling Load   602 (7-14)

Jiahao Liu[1], Liang He[1], Hua Chen[1], Pan Zhao[1], Yahui Su[2], Li chao[2] and Qin Pan[2]
[1]School of Advanced Materials and Nanotechnology, Xidian University, Xi'an, China
[2]Hua Dong Institute of Optoelectronics Devices, Bengbu, China

Yield Improvement and Cost of Test Reduction Via Automated Socket Cleaning   606 (7-3)

Jerry J. Broz and Bret A. Humphrey
International Test Solutions, Inc., Reno, Nevada, USA

A New CIS Package Process and Structure to Improve Die Clipping   609 (7-1)

Yuan Hu
Semiconductor Manufacturing International Corp, Shenzhen, Guangdong, China

A Novel Die Sorter Based on Micro Tweezer for Terahertz Schottky Barrier Diodes   612 (7-16)

Li He[1,2], Yang Kai[1,2], Zhang Jie[1,2], Zhang Hao[1,2], Zeng Jianping[1,2], An Ning[1,2], Jiang Jun[1,2] and Wang Xi[1,2]
[1]Microsystem & Terahertz Research Center, China Academy of Engineering Physics (CAEP), Chengdu, China
[2]Institute of Electronic Engineering, CAEP, Mianyang, China

A Novel Dispensing Head Fabrication Methodfor Precise Epoxy Dispensing   615 (7-17)

Zhang Jie[1,2], Li Zheng[1,2], Li ruoxue[1,2], Li He[1,2] and Wang Xi[1,2]
[1]Microsystem & Terahertz Research Center, China Academy of Engineering Physics (CAEP), Chengdu, China
[2]Institute of Electronic Engineering, CAEP, Mianyang, China

Volume Resistance of Epoxy Molding Compound   618 (7-18)

Hongjie Liu, Wei Tan, Xingming Cheng, LingLing Liu,Lanxia Li, Yangyang Duan, Dandan Fan, Xiaojuan Jiang, Liang Cui and Jianglong Han
JiangSu HuaHaiChengKe Advanced materials Co., Ltd, Lianyungang, Jiangsu, China

Warpage Control Method in Epoxy Molding Compound   621 (7-22)
Wei Tan[1], Cheng Cheng[2],Hongjie Liu[1], Yangyang Duan[1], Linlin Liu[1], Xingming Cheng[1] and Lanxia Li[1]
[1]*Jiangsu HHCK Advanced Materials Co., Ltd, China*
[2]*School of Language and Business, Jiangsu Normal University, Jiangsu, China*

Chapter VIII – MEMS, Sensors and Emerging Semiconductor Technologies

Surface Analysis and Post Thermal Treatment Process Optimization of Graphene   624 (8-8)
Oxide Thin Film for Humidity Sensor Application
Xiaoxu Kang, Ruoxi Shen and Xiaolan Zhong
Process Technology Department, Shanghai IC R&D Center, Shanghai, China

Arbitrarily Polarized CMOS Terahertz Detector With Silicon-Based Plasmonic   627 (8-10)
Antenna
Yiming Liao[1], Ke Wang[2], Jingyu Peng[2], Yaozu Guo[2], Feng Yan[2] and Xiaoli Ji[2]
[1]*School of Electronic and Optical Engineering, Nanjing University of Science and Technology, China*
[2]*D School of Electronic Science and Engineering, Nanjing University, China*

Gate Tunable Memtransistor Based on Monolayer Molybdenum Disulfide
Meng Yan, Fang Wang, Jiaqiang Shen, Xichao Di, Xin Lin, Huanhuan Di, Wei Mi and   630 (8-13)
Kailiang Zhang*
Tianjin Key Laboratory of Film Electronic & Communication Devices, School of Electrical & Electronic Engineering, Tianjin University of Technology, Tianjin, China

Ambient-Stable and High On/Off Ratio Near-Infrared Photodetector Based on   633 (8-14)
Perovskite-Treated PbS Colloidal Quantum Dots
Qingqing Wu[1], Yajie Yan[2], Ziqi Liang[2], Shaojian Hu, Jianjun Zhu[1] and Shoumian Chen[1]
[1]*Shanghai IC R&D Center, Shanghai, China.*
[2]*Department of Materials Science, Fudan University, Shanghai, China*

Synthesis of MoS_2/WS_2 Vertical Heterostructure and Its Photoelectric Properties   638 (8-12)
Xin Lin, Fang Wang, Jiaqiang Shen, Xichao Di, Huanhuan Di, Meng Yan and Kailiang Zhang
Tianjin Key Laboratory of Film Electronic & Communication Devices, School of Electrical &Electronic Engineering, Tianjin University of Technology, Tianjin, China

Effective Sparsity-Prior Image Denoising Algorithm for CMOS Image Sensor in Ultra-   641 (8-2)
Low Light Imaging Applications
Tao Zhou[1], Chen Li[1], Jiebin Duan[1], Xuan Zeng[2] and Yuhang Zhao[1]
[1]*Shanghai Integrated Circuits R&D Center Co., Ltd., Shanghai, China*
[2]*State Key Lab of ASIC & System, School of Microelectronics, Fudan University, Shanghai, China*

Chapter IX – Design Automation of Circuit and Systems

Monolithic 3D Enabled Processing-in- SRAM Memory   644 (9-4)
Vijaykrishnan Narayanan, Nagadastagiri Challapalle, Ikenna Okafor, Srivatsa Srinivasa and
Nicholas Jao
School of Electrical Engineering and Computer Science, The Pennsylvania State University,
Pennsylvania, USA,

Power Oriented CMOL Defect-Tolerant Mapping With Available Nanodevices   646 (9-30)
Shangluan Xie, Yinshui Xia, Xiaojing Zha and Xiangui Gu
EECS, Ningbo University, Ningbo, China

Signoff-Level Full-Chip ESD/Reliability Design Verification Using Logic-Driven Layout   649 (9-7)
Static Approach
Li Li, Yi-Ting Lee and Sridhar Srinivasan
Mentor, a Siemens Business, Wilsonville, Oregon USA

Analysis of ESD Effect and Ionizing Radiation Particles in Gate Oxide   653 (9-26)
C.-Z. Chen[1], David Y. Hu[2] and Hanming Wu[1]
[1]EtownIP Microelectronics, Beijing, China
[2]MetroSilicon Microsystems, Kunshan, Jiangsu, China

Statistical Wear-Leveling for Phase Change Memory   656 (9-13)
Chien Wang and Chengyu Xu
Jiangsu Advanced Memory Semiconductor Corporation, Beijing, China

The Study of Defects Auto-Classification System in Semiconductor Manufacturing   659 (9-29)
Pengfei Wang[1], Chen Li[1], Hao Fu[1], Zhengying Wei[2], Xu Chen[2], Zhounan Wang[2],
Shoumian Chen[1] and Yuhang Zhao[1]
[1]Shanghai Integrated IC R&D Center, Shanghai, China
[2]Shanghai Huali Microelectronics Corporation, Shanghai, China

A Neural-Network Approach to Better Diagnosis of Defect Pattern in Wafer Bin Map   662 (9-12)
Junjun Zhuang, Guiyun Mao, Yong Wang, Xu Chen and Zhengying Wei
Huali Microelectronics Corporation, Shanghai, China

Perceptron Algorithm and Its Verilog Design   665 (9-1)
Kainan Wang[1,2], Yingxuan Zhu[1,2] and C.-Z. Chen[2,3]
*[1]State Key Laboratory of Information Security, Institute of Information Engineering, CAS,
Beijing, China*
[2]University of Chinese Academy of Sciences, Huairou, Beijing, China
[3]EtownIP Microelectronics, Beijing, China

A clock jitter tolerant $\Sigma\Delta$ Modulator Employing a Hybrid Loop Filter in CMOS 40nm   668 (9-32)
Technology**
Negar Rashidi, Sungjun Yoon and Jose Silva-Martinez
Department of ECE, Texas A&M University, Texas, USA

Towards Optimal Logic Representations for Implication-Based Memristive Circuits   672 (9-23)
Lin Chen and Zhufei Chu
EECS, Ningbo University, Ningbo, China

Timing Violation as Dominant Reason for Failure of Clocked Digital Circuit Due to RF   675 (9-25)
Interference in Supply
Shanshan Nong and Tao Su
School of Electronics and Information Technology, Sun Yat-sen University, Guangzhou, Guangdong, China

DREAMPlace 2.0: Open-Source GPU-Accelerated Global and Detailed Placement for   678 (9-10)
Large-Scale VLSI Designs
Yibo Lin[1], David Z. Pan[2], Haoxing Ren[3] and Brucek Khailany[3]
[1]CS Department, Peking University
[2]ECE Department, The University
[3]NVIDIA, Inc.

A Real-Time Visual Tracking for Unmanned Aerial Vehicles With Dynamic Window   682 (9-24)
Jia Zhang, Tianrun Chen and Zhiguo Shi
College of Information Science & Electronic Eng., Zhejiang University, Hangzhou, China

Scalable Multi-Session TCP Offload Engine for Latency-Sensitive Applications   685 (9-14)
Jingbo Gao, Wenbo Yin, Wai-Shing Luk and Lingli Wang
School of Microelectronics, Fudan University, Shanghai, China

Advanced MOSFET Model Based on Artificial Neural Network   688 (9-11)
JH. Wei[1], W. Mao[1], H. Fang[2], Z. Zhang[2], JX. Zhang[2], BJ. La[n2] and J. Wan[1]
[1]State key lab of ASIC and System, School of Information Science and Engineering, Fudan University, Shanghai, China
[2]Suzhou Foohu Technology Co., Ltd., Suzhou, China

A Hybrid Domain Framework for Predistorter Modeling and Adaptive Digital   691 (9-15)
Predistortion Realization
Hairui Wang, Junyao Wang and Bo Wang
The Key Lab of IMS, School of ECE, Peking University Shenzhen Graduate School, Shenzhen, China

A Compiler Design for a Programmable CNN Accelerator   694 (9-17)
Jiadong Qian, Zhongcheng Huang and Lingli Wang
School of Microelectronics, Fudan University, Shanghai, China

Crystal Oscillator Frequency compensation technology of High Precision Clock   697 (9-19)
Synchronization for Time-triggered Ethernet
Haiying Yuan, Kai Zhang, Tong Zheng and Yichen Wang
Faculty of Information Technology, Beijing University of Technology, Beijing, China

A 5.5nW Voltage Reference Circuit   700 (9-27)
Kaixuan Du[1], Ziyuan Xu[3], Xiulong Wu[1], Libo Yang[2], Hao Zhang[2], Zhixuan Wang[2] and Le Ye[2]
[1]*School of Electronics and Information Engineering, Anhui University, Hefei, China*
[2]*Laboratory of Microelectronic Devices and Circuits (MOE) Institute of Microelectronic, Peking University, Beijing, China*
[3]*Jiangsu Union Technical Institute, China*

A High Linearity Readout Integrated Circuit for Uncooled IR Detector   703 (9-28)
Chang Liu, Kai Wang, Mingcheng Luo, Yaozu Guo, Feng Yan and Xiaoli Ji
Institute of the electronic Science and Engineering, Nanjing University, China

Thermal Modeling of Monolithic 3D ICS
Baoli Peng[1], Vasilis F. Pavlidis[2] and Yuanqing Cheng[1]   706 (9-16)
[1]*School of Microelectronics, Beihang University, Beijing, China*
[2]*School of Computer Science, University of Manchester, Manchester, United Kingdom*

A 2-D Capacitance Solver with Finite Difference Method   709 (9-36)
Wenjie Liang and Wenjian Yu
BNRist, Dept. Computer Science & Tech., Tsinghua University, Beijing, China

Additional Paper

Metal Trench Critical Dimension and Overlay Minor Variation Monitoring Method   712 (6-12)
with Voltage Contrast Inspection
L. Huang, Q. Ni, X. Gu, C. Han, J. Yuan
Shanghai Huali Microelectronics Corporation, Shanghai, China

China Semiconductor Technology International Conference 2020 (CSTIC 2020)

Editors:

Cor Claeys
KU Leuven
Leuven, Belgium

Steve Liang
Jiangsu Changjiang Electronics Technology Co. Ltd
Wuxi, China

Qinghuang Lin
Lam Research Corporation
Fremont, CA, USA

Ru Huang
Peking University
Beijing, China

Hanming Wu
Zhejiang University
Zhejiang, China

Peilin Song
IBM Thomas J. Watson Research Center
Yorktown Heights, NY, USA

Kafai Lai
IBM T.J. Watson Research Center
Hopewell Junction, NY, USA

Ying Zhang
Applied Materials
Santa Clara, CA, USA

Beichao Zhang
HFC Semiconductors
Shanghai, China

Xinping Qu
Fudan University
Shanghai, China

Hsiang-Lan Lung
Macronix International, Ltd.
Hsinchu, Taiwan, China

Wenjian Yu
Tsinghua University
Beijing, China

PREFACE

This issue contains a selection of the accepted papers presented at *China Semiconductor Technology International Conference 2020 (CSTIC 2020)*, March 18-19, 2020 in Shanghai, China. Due to the Covid19 pandemic the conference has been change into a virtual conference organized in the period June 29-July 17, 2020 . After reviewing a selection of the presentations have been considered for publication in IEEE Xplore.

CSTIC is the largest and the most comprehensive annual industrial semiconductor technology conference in China. It aims to provide a platform for executives, managers, engineers and researchers from around the world to exchange the latest developments in semiconductor technology and manufacturing and related fields. It also offers an opportunity for those who are interested in investing and collaboration opportunities in the semiconductor industry in Asia, particularly in China.

CSTIC covers all the aspects of semiconductor technology and manufacturing, including circuit design, system integration, devices, materials, patterning (lithography and etching), processes, integration, testing, reliability, device physics and manufacturing as well as emerging semiconductor technologies, including clean energy such as light emitting diodes (LEDs), III-V semiconductors, sensors and micro-electromechanical systems (MEMS).

CSTIC 2020, organized by Semiconductor Equipment and Material International (SEMI), imec, and The Integrated circuit Materials Industry Technology Innovation Alliance (ICMTIA) and technically sponsored by the IEEE Electron Devices Society relies on a long time tradition, which started in 2001. The original International Semiconductor Technology Conference (ISTC) merged in 2009 to become CSTIC, aiming for a broad international representation and increased paper submissions from around the world. For CSTIC 2020 the papers came from all major semiconductor manufacturing regions in the world, including Belgium, China, Denmark, Finland, France, Germany, India, Japan, Korea, Malaysia, Singapore, Spain, Sweden, Switzerland, Taiwan, The Netherlands, United Kingdom and the United States of America. About 230 papers have been selected for oral presentations and approximate 179 papers for poster presentations after careful reviews by the conference organizing committee.

In total 223 papers are included in these Proceedings after peer reviews. They represent a snapshot of the recent developments in semiconductor technology and manufacturing in the world. In particular, they offer a glimpse into the state-of-the-art of semiconductor technology and manufacturing in China. These papers are divided into nine (9) chapters according to the nine symposia of CSTIC 2020:

• Device Engineering and Memory Technology
• Lithography and Patterning
• Dry & Wet Etch and Cleaning
• Thin Film, Plating and Process Integration
• Chemical-Mechanical Polishing (CMP) and Post-Polish Cleaning
• Metrology, Reliability and Testing
• Packaging and Assembly
• MEMS, Sensors and Emerging Semiconductor Technologies
• Design and Automation of Circuits and Systems

These Proceedings are very valuable to engineers and researchers in the fast-moving and growing semiconductor industry. It will give readers a clear understanding of the status of semiconductor technology and manufacturing in China. Furthermore it will also serve as a useful reference for those who are interested in nanofabrication, micro- and nano-fluidics, micro- and nano-photonics, organic electronics, bio-chips, light emitting diodes (LEDs) and other clean energy technologies.

We thank the invited speakers and the authors, particularly the conference plenary speakers, Dr. Doug Yu, Vice-President TSMC, Taiwan, De. Anthony Yen, Vice-President ASML, The Netherlands, Dr. Ravi Mahajan, Fellow Intel, USA and Dr. Sanjay Natarajan, Vice-President Applied Materials, USA, for their valuable contributions to CSTIC 2020. We also thank the more than 120 organizing committee members, particularly the symposium chairs, for their dedication and hard work to help improve the quality and to broaden the reach of CSTIC. These committee members are experts in their respective fields of semiconductor technology and are from well-known companies or prestigious institutions. They all have demanding day jobs, yet they have volunteered to help organizing this conference and to critically review papers presented in these Proceedings. Their contributions were crucial for the success of the conference. We are also indebted to the financial support from the sponsors of CSTIC 2020. Finally, we extend our sincere thanks to SEMI for their tireless efforts and their meticulous organizational skills to help organize CSTIC 2020 and to assemble and publish these CSTIC 2020 proceedings.

Steve X. Liang, General Chair CSTIC 2020
JCET, Wuxi, China

Cor Claeys, Co-Chair, CSTIC 2020
KU Leuven, Leuven, Belgium

CSTIC 2020 Organizing Committee

July 2020, Shanghai, China

NANOWIRE & NANOSHEET FETS FOR ADVANCED ULTRA-SCALED, HIGH-DENSITY LOGIC AND MEMORY APPLICATIONS

A. Veloso[*], P. Matagne, G. Eneman, D. Jang, T. Huynh-Bao, A. Chasin, E. Simoen, A. De Keersgieter, and N. Horiguchi

Imec, Kapeldreef 75, 3001 Leuven, Belgium

*Corresponding Author's Email: Anabela.Veloso@imec.be

ABSTRACT

We report on vertically stacked lateral nanowires (NW) / nanosheets (NS) FETs with a gate-all-around (GAA) configuration as promising candidates to replace finFETs for advanced sub-5nm technology nodes, as they offer higher flexibility to obtain a better power-performance metric for logic applications. Moreover, these devices can be fabricated in a cost effective, co-integration scheme with vertical NW/NS GAA FETs for increased on-chip memory content, thanks to the latter's potential for enabling highly scaled, more energy efficient memory cells, e.g., as the SRAM cell transistors or as the selector for MRAM. Key features and technological challenges for both types of NW/NS FETs will therefore be presented and discussed in this paper.

INTRODUCTION

As conventional CMOS scaling is reaching its physical limits and facing increasingly constraining design restrictions, new, alternative solutions are required to help preserve the overall power-performance-area-cost (PPAC) logic roadmap and continue delivering profitable node-to-node scaling gains. An example is cell height reduction by decreasing the number of metal tracks to overcome the slowing down of pitch scaling. However, for ultra-scaled cells using finFET technology, this ultimately requires fin depopulation from two fins to one fin per transistor [1,2], leading to an undesirable penalty on device performance. As such, in order to keep the scaling benefits, introduction of novel transistor architectures will be needed, with NW or NS GAA FETs [2-12] being widely considered as the most promising finFET replacements for advanced nodes. These devices exhibit improved electrostatics allowing further gate length (L_{gate}) shrinkage, with NS GAA FETs being advantageous for obtaining higher drivability (I_{ON}) per layout footprint due to their larger effective widths (W_{eff}). Moreover, in the case of lateral NS FETs, they can in fact be considered a natural evolution of finFETs given the many fabrication elements shared by both transistor architectures.

Vertical NW or NS GAA FETs [4,13-18], wherein the L_{gate} is defined vertically and can thus be relaxed without impacting the layout footprint, represent a more disruptive technological change but can also open new scaling paths in the third dimension. They have been shown to have the potential to enable denser memory cells with lower power consumption [13,14,16-18]. As a result, and in view of the rising system demands for increased on-chip memory content, it is worth considering the co-fabrication of both GAA FETs (combining vertical with lateral NS FETs or finFETs; for high-performance logic or use in specialized circuits): on the same wafer [17,18] as discussed here or in two stacked wafer levels by using 3D integration [19,20].

RESULTS AND DISCUSSION

Lateral Nanowire and Nanosheet FETs

A benchmark of the power-performance values for inverter based ring oscillators (RO) is calculated in Fig. 1. A fan-out of three (FO3) with ~50× contacted-gate-pitch (CGP) length for back-end-of-line (BEOL) load, and representative design rules for various technology nodes were assumed. The data show that while finFET technology is still projected to deliver gains when scaling down from 7nm to 5nm nodes using two fins-per-device based cells, additional performance gains at iso-V_{DD} are compromised beyond that. Indeed, assuming a cell height shrank to five metal tracks at the 3nm node, this requires a change into a one fin-per-device scenario for finFET based circuits. It becomes then advantageous to introduce NS FETs as they can outperform finFETs by delivering

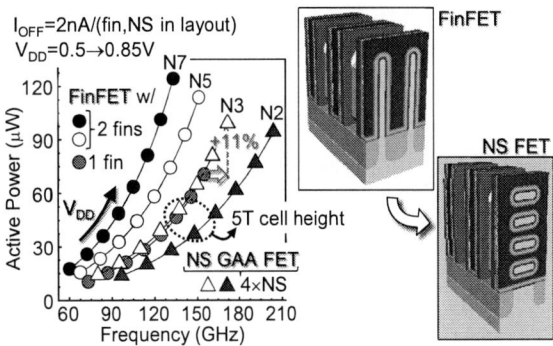

Figure 1: Simulated benchmark of ring oscillators' power-performance. FinFET based cells with two fins-per-device for 7nm (N7) and 5nm (N5) nodes are compared with cells for further scaled nodes: 3nm (N3) and 2nm (N2). The last two assume the use of buried power rail technology and: a) finFETs with one fin-per-device for N3, or b) NS GAA FETs consisting of four vertically stacked NS, each with a width of 16nm (N3) or 11nm (N2). Overall, data show that NS FETs are predicted to outperform finFETs at N3.

faster ROs at a given V_{DD} or by enabling power savings at matched performance: 15% and 19% frequency gains at V_{DD}=0.7V for the 3nm and 2nm nodes, respectively.

A key feature of NS FETs lies in the design flexibility to vary the NS width, as well as the number of vertically stacked NS (hence W_{eff}), typically targeting wider NS for high-performance computing and narrower NS for low-power applications [7,9,11]. Fig. 2a illustrates, for instance, how faster ROs, at similar V_{DD}, can be obtained with increased number of vertically stacked NS at the expense of higher power consumption. The latter is a result from the increase in capacitance occurring for a multiple-sheets structure (higher as stacking more NS). To address this issue, introduction of a new process module in the fabrication flow becomes critical and indispensable for the successful implementation of these devices: inner spacers [2,6,8]. Their integration can be done just prior to the source/drain (S/D) epitaxial growth, by performing first a controlled lateral SiGe etch (with high selectivity towards Si) in the Si/SiGe multi-layer fins, wherein the SiGe films play the role of sacrificial layers to form the Si channels. This is followed by the filling of the created cavities with the so-called inner spacers dielectric. As discussed in ref. [6], the first step may induce strain changes but epi growth in recessed S/D areas has been confirmed to still be effective for introducing a significant amount of strain in the vertically stacked Si channels for mobility boost. Another knob for RO frequency response improvement consists in reducing the vertical separation between NS to lower the device parasitics as shown in Fig. 2b. An example of hardware implementation is presented in Fig. 2c, for which the use of scaled Si/SiGe multi-layers growth prior to active level (fin) patterning is required. In addition, thinner gate stacks that can guarantee similar gate control on all NS surfaces are also needed to successfully enable a smaller NS vertical pitch.

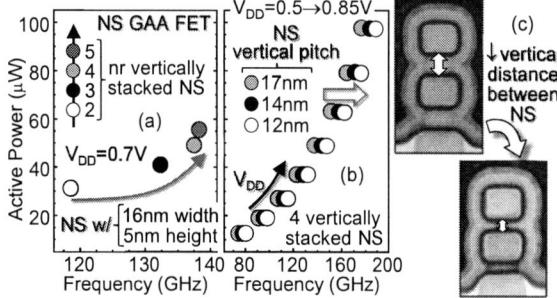

Figure 2: The speed of ring oscillators can be improved by: a) increasing the number of vertically stacked NS, though at the expense of a higher power consumption; b) reducing the vertical distance between NS. Device implementation of the latter is illustrated by the annular bright-field scanning transmission electron microscopy (ABF-STEM) images in (c).

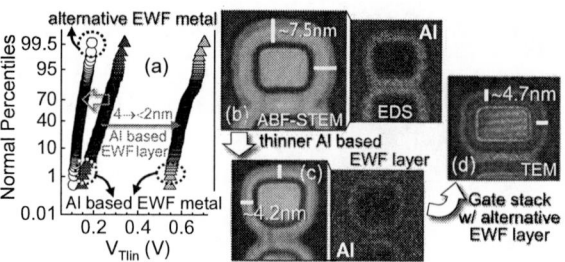

Figure 3: a) shows that increased V_{Tlin} values are obtained for n-type NS GAA FETs when thinning down the reference Al based EWF metal from 4nm (b) to <2nm (c) nominal thickness, while ~100mV lower V_{Tlin} is achieved with an alternative EWF metal, thinner than 2.5nm (d).

This poses some key technological challenges. Fig. 3 illustrates that it can be possible to thin down the reference effective work-function (EWF) metal layer (Al based for NMOS), but with a V_T penalty. Exploring alternative material options, recent promising results with a thinner EWF metal have been disclosed [12], as seen in Figs. 3a,d.

Vertical Nanowire and Nanosheet FETs

Given the wide range of functionalities expected to be required by several emerging applications, the need for hybrid scaling is increasingly gaining traction wherein devices can be customized to meet specific system requirements, rather than relying on a one-meets-all generic technology. In regard to the overall scaling roadmap, a transition from finFETs to vertically stacked lateral NS GAA FETs is projected to occur next for logic, while in SRAMs the possible change of the cell transistors to vertical NW/NS FETs has been shown to have the potential to reduce their area by more than 20-30% with improved read/write stability and reduced standby leakage [4,13,14,16]. In addition, also for MRAM, which has been emerging as a potential higher-level SRAM replacement [21], the use of vertical FETs instead of finFET selectors can further help in area shrinkage [17,18]. In this context, Fig. 4 illustrates a possible cost-effective, co-integration scheme of finFETs (or lateral NS FETs) for logic with VNS GAA FETs as the MRAM selectors. Cost adders can be minimized by sharing the maximum number of critical steps for building the two types of devices, such as: fin/VNS patterning and buried metal line (BML) formation; dummy gate patterning; epi growth (for S/D in finFETs; on top of the pillars in VNS FETs for larger contact area); dummy gate removal, gate stack deposition and metal-CMP in the replacement metal gate (RMG) module, followed by a common middle-of-line and BEOL. In contrast to finFET based MRAMs where the array of magnetic tunnel junctions (MTJ) is typically embedded in the BEOL in-between two metal levels [21], the use of VNS GAA FET selectors allows introduction of the MTJ element earlier in the flow, such that the source and word

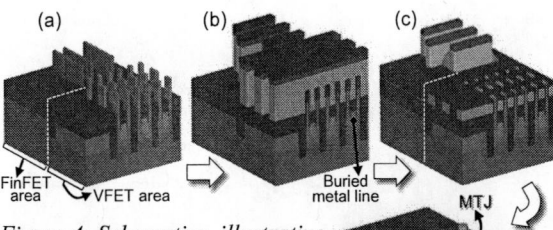

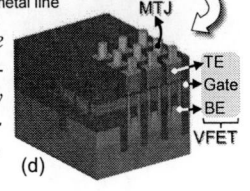

Figure 4: Schematics illustrative of process simulations of a cost-effective co-integration strategy to fabricate on the same wafer finFETs and VNS GAA FETs for use as MRAM selectors. Examples of common steps: a) fin/VNS and b) dummy gate patterning; c) epi growth (on the S/D for finFETs vs. on top of the VNS). In this scheme, the MTJ element can be introduced earlier on in the flow, just after the VNS FET's top electrode (TE) formation (d).

lines (SL, WL) can be routed with the BML defined after VNS patterning and via the gate electrode, respectively.

Doping engineering in VNW or VNS FETs can be simplified by using stacked epi layers grown and patterned into pillars such that the Si channel ends up sandwiched in-between two heavily doped layers (e.g., by boron for PMOS, phosphorus for NMOS). The junctionless concept wherein the device channel is doped with a certain doping concentration ($N_{ch} = N_{NW}$ or N_{NS}) [3,4,15,16,18] allows moving the charge distribution centroid away from the interface and achieve smaller E_{ox} values, lower for increasing N_{ch}. Fig. 5 confirms that is beneficial to get improved reliability behavior, while variability-wise, provided N_{ch} is kept low enough ($\leq 1 \times 10^{19}$ at/cm^3 for NW diameter (d_{NW}) <30nm), acceptable V_T-mismatch performance can also be extracted for these devices, with a stronger dependence of $\sigma(\Delta V_T)$ with N_{ch} seen for NMOS.

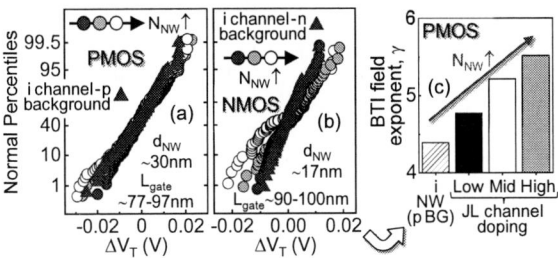

Figure 5: a) shows comparable V_T-mismatch performance for matching pairs of junctionless (JL) VNW GAA pFETs with N_{NW} up to 1×10^{19} at/cm^3, whereas a more pronounced increase of $\sigma(\Delta V_T)$ with N_{NW} is seen for n-type devices (b). In both cases, higher N_{NW} leads to substantial reliability improvement, as evidenced by the larger slope (γ) values extracted for higher N_{NW} from $|\Delta V_T|$ measured after 1s of BTI degradation vs. $V_{G,stress}$-V_T - data for PMOS in (c).

Contrary to the case for finFETs and lateral NS FETs where the S/D areas are standardly self-aligned to the gate electrode during fabrication, such feature constitutes a key integration challenge for vertical devices. Critically, its absence can have a substantial detrimental impact on the device characteristics, e.g., I_{ON} and I_{OFF}, as has been evaluated in refs. [16,18]. To avoid it, a simple scheme consisting of $Si_{0.75}Ge_{0.25}$ / Si / $Si_{0.75}Ge_{0.25}$ instead of Si-only pillars with self-aligned oxide spacers was proposed there, wherein the gate overlaps said spacers for a well-controlled L_{gate}. For such configuration, there is a small lateral consumption of the SiGe in the pillars during spacers formation - less for thinner spacers - but TCAD simulations in ref. [18] showed its impact on $R_{S/D}$ can be mitigated by SiGe and spacers thicknesses optimization.

Another important challenge for vertical FETs lies in the options regarding the fabrication of these devices with channel mobility enhancement techniques. An interesting choice can be the selection of a different substrate orientation as shown in Fig. 6. This is particularly simple when only n or p-type transistors are required as is the case for the MRAM selector. Also highlighted in Fig. 6 is the I_{ON} advantage for VNS FETs as compared to VNW FETs. This is due to the larger effective width of VNS, similarly to the case for lateral NS vs. NW GAA FETs. Overall, the ballistic current simulation results in Fig. 6 show that, for unstrained devices consisting of either VNW or VNS (with (110) for its wider sidewalls), (100) substrates are beneficial for NMOS performance while (111) substrates are more advantageous for PMOS.

To enable continuing using process-induced stress techniques for mobility boost in vertical FETs, a new integration scheme was recently proposed in ref. [18]. It allows the introduction of stress along the channel in the vertical direction, hence enhancing the device's mobility and I_{ON}. The concept is schematically illustrated in Fig. 7 for NMOS. After the pillars and spacers formation, a SiGe stressor is selectively and epitaxially grown around the vertical Si channel. The stressor is later removed at RMG module, after dummy gate removal and prior to gate stack

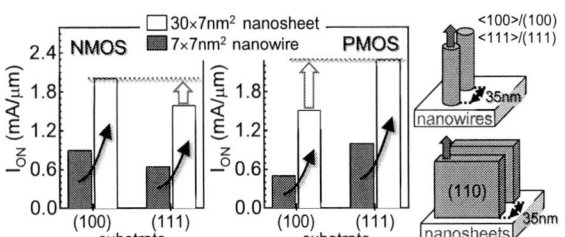

Figure 6: Similarly to the case for lateral NW/NS, ballistic current simulations yield higher I_{ON}, as normalized per NW/NS pitch, for VNS vs. VNW. The substrate orientation choice can also help to boost I_{ON} for n/p-type VFETs. (I_{ON} at $V_G = V_{G,OFF} \pm 0.6V$, $I_{OFF} = 100nA/\mu m$, 35nm NW/NS pitch.)

deposition. This leads to stress memorization occurring into its surroundings since the layer encapsulating the top of the pillars remains connected to their bottom isolation layer at certain places. Another critical layout feature is the vertical path accessing the dummy gate, needed for its and the stressor's removal, followed by gate stack filling. The amount of stress remaining in the channel after stressor removal depends on the structure dimensions, with a thicker, and with higher Ge content, stressor leading to enhanced stress and I_{ON} [18]. To illustrate some of these dependencies, Fig. 7 shows the comparison of the average stress (XX, YY, and ZZ components) induced in the channel for VNW and VNS GAA FETs with varying L_{gate} and the inferred impact on I_{ON} (shown as indicative values only) obtained by stress and ballistic current simulations [18]. For any given stressor, a higher stress remains in the ZZ direction (S_{ZZ}), along the VNS length, for VNS FETs which impacts the channel mobility and leads to a smaller I_{ON} improvement. For these devices, as shown in Fig. 7, increasing L_{gate} effectively enables a stronger relaxation of the undesirable S_{ZZ}, hence allowing higher gains in I_{ON}.

The potential benefit of vertical FETs in specialized circuits is highlighted in Fig. 8 as MRAM selectors. Assuming 3nm node design rules, considerably lower read/write energy consumption and access latency values are extracted for a 2Mbit STT-MRAM macro designed with VNS GAA FETs instead of finFETs as cell selectors. This is due to lower parasitic resistances and capacitances for the WL, SL and bit line (BL) of the cells, and thanks

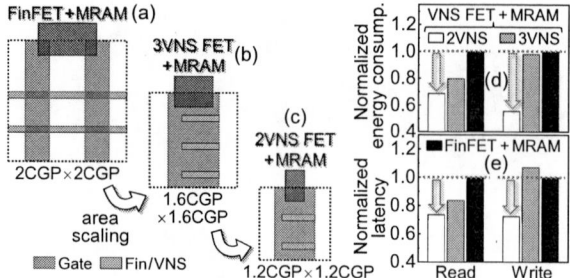

Figure 8: Assuming 3nm node design rules, a comparison of STT-MRAM cells with finFETs (4 active fins per cell, in (a)) vs. VNS FETs (3 or 2 VNS / cell in (b,c), respectively) as selectors shows that: considerable cell area reduction (a→b→c), lower read/write energy consumption (d), and smaller access latency values (e) are obtained for a 2Mbit STT-MRAM macro designed with VNS FET selectors.

also to smaller VNS FET device parasitics [14,18,22]. The substantial gains in area reduction enabled by vertical FETs are also illustrated in Fig. 8, with the MTJ pillar diameter retargeted from 34 to 24nm through a DTCO loop for obtaining further scaling from 3 to 2 VNS per cell.

CONCLUSIONS

Vertically stacked lateral NS GAA FETs are the leading candidates to succeed finFETs and maintain the PPAC logic roadmap for advanced sub-5nm technology nodes. At the same time, with hybrid scaling increasingly gaining traction to achieve higher system value/flexibility, vertical GAA FETs are also interesting to explore thanks to their potential for enabling increased and more energy efficient on-chip memory content, e.g., SRAM, MRAM.

REFERENCES

[1] M. Garcia Bardon *et al.*, *IEDM Tech. Dig.*, 2016, p. 687.
[2] H. Mertens *et al.*, *IEDM Tech. Dig.*, 2017, p. 828.
[3] A. Veloso *et al.*, *VLSI Tech. Dig.*, 2015, p. 138.
[4] A. Veloso *et al.*, *VLSI Tech. Dig.*, 2016, p. 138.
[5] H. Mertens *et al.*, *VLSI Tech. Dig.*, 2016, p. 158.
[6] S. Barraud *et al.*, *IEDM Tech. Dig.*, 2016, p. 464.
[7] D. Jang *et al.*, *IEEE Trans. Elec. Dev.*, 64(6), 2017, p. 2707.
[8] N. Loubet *et al.*, *VLSI Tech. Dig.*, 2017, p. 230.
[9] S. Barraud *et al.*, *IEDM Tech. Dig.*, 2017, p. 677.
[10] S. Barraud *et al.*, *IEDM Tech. Dig.*, 2018, p. 500.
[11] C. W. Yeung *et al.*, *IEDM Tech. Dig.*, 2018, p. 652.
[12] A. Veloso *et al.*, *SSDM Tech. Dig.*, 2019, p. 559.
[13] T. Huynh-Bao *et al.*, *SPIE Proc.*, 2016, 978102.
[14] A. Veloso *et al.*, *ECS Trans.*, 72(4), 2016, p. 31.
[15] A. Veloso *et al.*, *SSDM Tech. Dig.*, 2017, p. 221.
[16] A. Veloso *et al.*, *SSDM Tech. Dig.*, 2018, p. 159.
[17] T. Huynh-Bao *et al.*, *DAC Proc.*, 2019, 267-TY328, s13p1.
[18] A. Veloso *et al.*, *IEDM Tech. Dig.*, 2019, p. 230.
[19] P. Batude *et al.*, *IEDM Tech. Dig.*, 2017, p. 52.
[20] D. B. Ingerly *et al.*, *IEDM Tech. Dig.*, 2019, p. 466.
[21] S. Sakhare *et al.*, *IEDM Tech. Dig.*, 2018, p. 420.
[22] A. V.-Y. Thean, *VLSI Tech. Dig.*, 2015, p. 26.

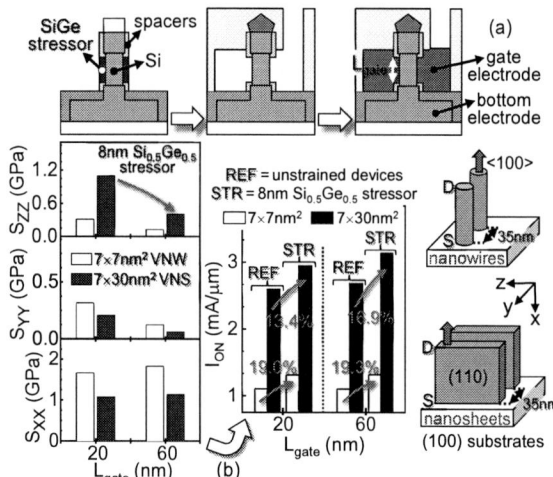

Figure 7: a) illustrates a scheme to introduce stress (S) in vertical nFETs for mobility enhancement: a SiGe stressor layer is epitaxially grown around the Si channel after spacers formation. The stressor is removed prior to gate stack deposition at RMG module leaving stress memorized into its surroundings. b) With this scheme, a longer L_{gate} can be beneficial for boosting I_{ON} in VNS FETs as it allows a stronger relaxation of the undesirable S_{zz}.

978-1-7281-6559-2/20 $31.00 © 2020 IEEE

DESIGN OF A NOVEL ONE TRANSISTOR-DRAM BASED ON BULK SILICON SUBSTRATE

K. Xiao[1], J. Liu[1], Wei Wu[2], Zheng Xu[2] and J. Wan[1]*

[1]State key lab of ASIC and System, School of Information Science and Engineering, Fudan
University, Shanghai, China
[2]Fujian Electronics & Information (Group) Co., LTD, Fujian, China
Email: jingwan@fudan.edu.cn

ABSTRACT

In this work, we propose a novel device based on bulk silicon substrate and use it as a dynamic random access memory (DRAM) with TCAD simulation. The operation principle of this device is similar to that of Z^2-FET which was demonstrated previously in SOI substrate. In our device, LDD doping and gate control are combined to build up the carrier injection barriers which enable feedback mechanism. The designed device shows similar sharp switch and gate-controlled hysteresis in its output characteristics. It is further demonstrated for DRAM application without need of extra capacitor. The DRAM operation shows high speed and reasonably long retention time.

Keywords— one-transistor dynamic random access memory; feedback mechanism; Z^2-FET; bulk substrate;

INTRODUCTION

Z^2-FET (zero impact ionization and zero subthreshold swing FET), which is essentially a gate-controlled PIN diode based on silicon-on-insulator (SOI) substrate, is a novel device operating with feedback mechanism between carrier flow and injection barriers [1]. Due to its gate-controlled hysteresis and sharp switching characteristics, it becomes very attractive in many applications, such as logic switch, high performance DRAM and ESD protection [2-4]. The conventional Z^2-FET device has the advantages of compact, single front gate footprint, undoped channel, and no impact ionization or bipolar action [5]. Besides, the conventional Z^2-FET device also finds applications in flash memory,

co-integration in flexible substrate and ion-sensitive sensor [6, 7]. Though the SOI substrate is widely used to build integrated circuits (ICs) with low power, high frequency and radiation hardness, bulk silicon substrate is still dominant in the high performance ICs due to its low cost.

Thus, in this work, we design a novel device based on bulk silicon substrate, which uses similar feedback mechanism as the Z^2-FET. The device combines lightly-doped drain (LDD) and the front gate to form the potential barriers and has symmetrical structure as the Z^2-FET. It shows similar gate-controlled hysteresis in output characteristics and used as one-transistor DRAM (1T-DRAM) showing excellent performances.

DEVICE STRUCTURE AND OPERATION PRINCIPLE

Figure 1 (a) shows the structure of the designed device. It has very similar structure as conventional p-type MOSFET, except for the opposite source and drain doping type at the same doping concentration of 1020 cm-3. The channel is n-type doped with concentration of 1018 cm-3. P-type LDD regions with doping concentration around 1019 cm-3 are formed by ion implantation self-aligned to the front gate. The distance between channel and source/drain regions is defined by the spacer in a self-aligned way. The gate length and spacer length (LG and Lspr) are both 50 nm. From source to drain, a P+-P-N-P-N+ structure is formed, which is similar to a thyristor.

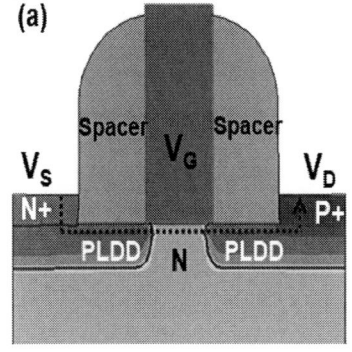

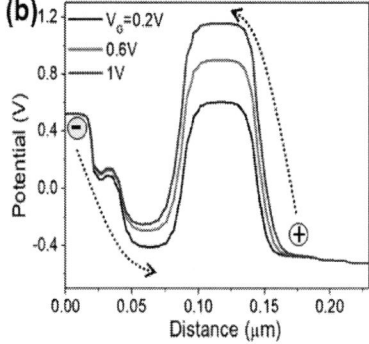

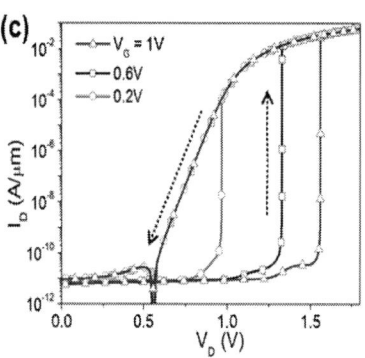

978-1-7281-6559-2/20 $31.00 © 2020 IEEE

Fig. 1. (a) Schematic view of the simulated novel device structure. (b) Potential profile from source to drain (c) I_D-V_D curves under various V_G value.

Synopsys Sentaurus is used to simulate the device. Conventional physic models are used in the simulation. The potential profile of the device from source to drain is extracted and shown in Fig. 1(b). Hole and electron injection barriers similar to that found in Z2-FET are observed. The electron injection barrier is formed by the PLDD region next to the source, and marginally affected by the gate voltage. However, the hole barrier is formed in the channel and dramatically modulated by the front gate voltage. Due to the feedback between carrier flows and their potential barriers, a sharp switch behavior is observed in the output characteristics, see Fig. 1(c). The ID-VD curves of this device also show large hysteresis window with the turn-on voltage controlled by the front gate voltage (VG). This is attributed to the hole potential barriers modulated by the VG.

APPLICATION AS 1T-DRAM

The designed device is further used as DRAM which needs no extra capacitor. It is truly a 1T-DRAM which is more compact than conventional 1T1C-DRAM. The charge is directly stored in the front gate capacitor (C_G) formed by the gate oxide to represent the logic state of the device and read out through feedback mechanism. Figure 2 shows the waveforms of the DRAM operation. In order to write logic "0", a negative V_G pulse down to 0V is used to discharge the gate capacitor. Without extra charge in the gate capacitor, the read of the device by a positive V_D pulse failed to turn on the device. Thus, the device stays in OFF state during read "0" process.

On the contrary, during write logic "1" state, the V_G reduces from 1V down to 0V, meanwhile V_D rise up from 0V to 1.2V. Since the hole injection barrier is largely reduced by $V_G = 0V$, $V_D = 1.2V$ can easily turn on the device, as the I_D curve indicates in Fig. 2. Large amount of electrons and holes are injected into the channel during ON state. Electron charges are stored in the gate capacitor as the V_G goes back to 1V. These charges can cause transient electron current during the read process when V_D pulses up to 1.2V. The transient electron current flows into the drain and is amplified by the internal feedback loop, which eventually turns on the device, as shown in Fig. 2. This process is similar to that in Z^2-FET [5]. Unlike conventional 1T1C-DRAM which needs large amount of charge to reliably drive external buffer amplifier, our novel DRAM stores only small amount of

charge to trigger the internal feedback loop and thus has very fast operation speed down to 1ns. A preliminary study is also performed on the retention time of this device. Figure 3(a) shows that after writing "0", the read of the device correctly outputs low current. However, after 1 sec, the read of the device outputs high current, which indicates the failure of stored "0". This is due to that, during holding "0", electron-hole pairs are regenerated in the depleted channel by thermal generation and tunneling. In contrast, the stored charge never loses since $V_G = 1V$ is applied during hold process.

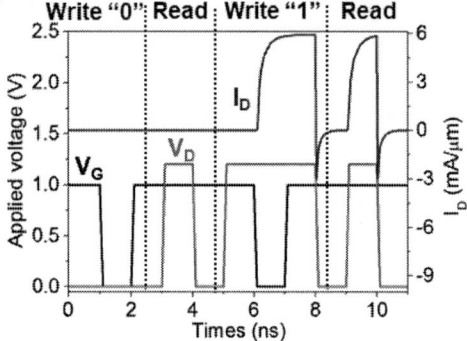

Fig. 2. The TCAD simulation showing that the operation speed of the novel DRAM can reach below 1 ns.

Thus, logic "1" is a steady state and needs no refresh. Figure 3(b) shows that reads of logic "1" correctly output high current after 1 sec and 10 secs.

CONCLUSION

A novel device is demonstrated with similar operation principle as the Z^2-FET. The device is based on bulk-Si substrate instead of SOI. By combining LDD doping and gate control, a potential profile with electron and hole injection barriers are formed in the channel, which leads to gate controlled hysteresis. This property is further used for DRAM operation. Without need of extra capacitor, the novel DRAM shows compact form and high operation speed down to 1ns. A preliminary study is also performed on the retention time of both logic "0" and "1" states. With high operation speed, high integration density, and based on widely-used bulk-Si substrate, the proposed device is expected to attract lots of interests.

978-1-7281-6559-2/20 $31.00 © 2020 IEEE

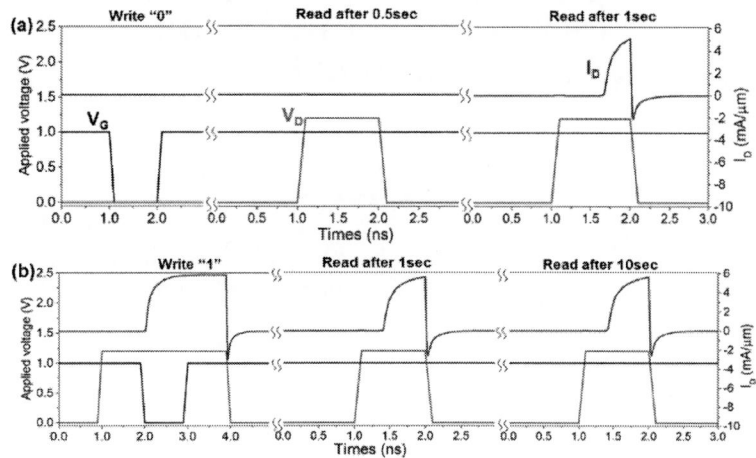

Fig. 3. Study on the retention time of the device with (a) logic "0" and (b) logic "1" states.

ACKNOWLEDGEMENTS

This work was supported by the National Key R&D Program of China (2018YFB2202800) and Natural Science Foundation of Shanghai (17ZR1146700).

REFERENCES

[1] J. Wan, S. Cristoloveanu, C. Le Royer and A. Zaslavsky, "A feedback silicon-on-insulator steep switching device with gate-controlled carrier injection," *Solid-State Electronics,* vol. 76, 2012, pp. 109-111.

[2] A. A. Salman, S. G. Beebe, M. Emam, M. M. Pelella, and D. E. Ioannou, "Field effect diode (FED): A novel device for ESD protection in deep sub-micron SOI technologies," in Tech. Dig. IEEE Int. Electron Device Meeting (IEDM), pp. 1–4, 2006.

[3] Y. Solaro, J. Wan, P. Fonteneau, C. Fenouillet-Beranger, C. Le Royer, A. Zaslavsky et al., "Z2-FET: a promising FDSOI device for ESD protection," Solid-State Electron., vol. 97, pp. 23–29, 2014.

[4] J. Wan, C. Le Royer, A. Zaslavsky and, S. Cristoloveanu, "Progress in Z2-FET 1T-DRAM: Retention time, writing modes,selective array operation, and dual bit storage," Solid State Electron., vol. 84, pp. 147–154, 2013.

[5] J. Wan, C. Le Royer, A. Zaslavsky and, S. Cristoloveanu, "A compact capacitor-less high speed DRAM using field effectcontrolled charge regeneration," IEEE Electron Device Lett., vol. 76, pp. 109–111, 2012.

[6] Y. Jeon, M. Kim, D. Lim and S. Kim, "Steep subthreshold swing n-and p-channel operation of bendable feedback field-effect transistors with p+–i–n+ nanowires by dual-top-gate voltage modulation," *Nano letters,*vol. 15, no. 8, 2015, pp. 4905-4913.

[7] S.-M. Joe, H.-J. Kang, N. Choi, M. Kang, B.-G. Park and J.-H. Lee, "Diode-type NAND flash memory cell string having super-steep switching slope based on positive feedback," *IEEE Transactions on Electron Devices,*vol. 63, no. 4, 2016, pp. 1533-1538.

978-1-7281-6559-2/20 $31.00 © 2020 IEEE

MANUFACTURING CHALLENGES AND COST EVALUATION OF NEW GENERATION 3D MEMORIES

Belinda Langelihle Dube

System Plus Consulting
22 Bd Benoni Goullin, Nantes, 44200, France
Semiconductor Devices Department
bdube@systemplus.fr

ABSTRACT

This work gives an overview of the evolution of manufacturing techniques of 3D-NAND Non-Volatile Memories (NVMs). It evaluates manufacturing challenges that directly impact manufacturing cost of memories. Included is physical analysis shown by SEM Images. The work outlines complexity of manufacturing process and consequences of adding word lines. Critical steps in the process like deposition of layers, high aspect ratio etching, and staircase formation are evaluated and their impact on wafer manufacturing cost. A proprietary software is used to simulate the fabrication steps and calculates Front-End manufacturing cost. Additional word lines increase the cost of wafer manufacture, but the cost of each Gigabit is reduced.

Keywords- 3D NAND- (three dimension), Flash Memory, Manufacturing Process, Cost, Etching, NVM-Non-Volatile-Memory, Charge-Trap Flash

INTRODUCTION

Emerging technologies have heightened the demand of dense, fast, reliable and low-cost Non-Volatile Memories (NVMs). Previous planar NAND memories had reached their scaling limit at 10nm using complex and costly multiple lithography patterning.

3D NAND overcomes the scaling limit by a method of stacking memory cells vertically with a relaxed process technology of 20nm. The stacking of worldlines allows the same process technology node to be used in different generation of 3D NAND memories. The stacking technique allows manufacturers to achieve maximum capacity in a single chip [1].

Each 3D NAND memory manufacturer strives to achieve the highest density in their NAND memory chips. Different techniques are employed to increase the memory density. New generation memories can now store more than one bit in one memory cell [2]. The examined memories carry three bits per memory cell. These are referred to as Triple Level Cells (TLC). The TLC cells account for more than 80% in the NAND Flash market and they have excellent performance and reliability. NVMs require cells that are highly reliable, pushing manufacturers to evolve their process from a floating gate process that uses polysilicon memory cells to a charge trap technique that uses high k dielectric material to store information and also prevent cell to cell interference to increase memory reliability[3]. The manufacturing process may differ, many manufacturers have opted to use high dielectric material that requires a precise deposition technique like Atomic layer Deposition as the thickness of this charge trap dielectric material is below 10nm. This step significantly increases the manufacturing cost of the 3D NAND memory because of high cost of the atomic layer deposition machine.

All manufacturers progressively add more world lines with each generation to meet the high density demands and high-performance expectations. "Fig 1" shows 3D vertical structure from Samsung's NAND Die with 92 Worldlines.

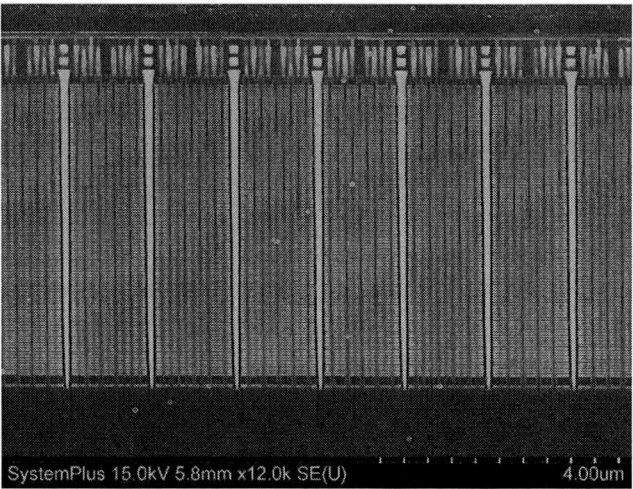

Fig 1: Samsung's 92 Layer 3D NAND Die Cross Section Overview

MANUFACTURING PROCESS AND COST EVALUATION

Four samples from different generation 3D NAND memories are analyzed. The 3D NAND studied include 64, 72, 92, and 96-layers. The Die Cross Section is analyzed. Scanning Electron Microscopy (SEM) Images are analyzed and manufacturing process steps are determined and simulated. Each process result give an estimation of the Clean Room Cost, Equipment Cost, Consumables Cost, Labor Cost and Yield Losses Cost.

PECVD technique is employed to deposit the Word lines and the insulating material between the word lines. The layers are made of two different materials consisting of 35nm SiN material that is later etched out to deposit tungsten world lines and 15nm insulating layer of SiO2. The deposition requires a precise and slow deposition rate to avoid defects. The layer deposition cost significantly contributes to the total wafer manufacturing cost. The step accounts for approximately 18% of the total wafer cost. Cost of deposition escalates due to the slow deposition rate, time spent on the deposition machine and added cost of the consumables [4]. Increase in the number of layers induces stress in the films that could result in wafer

978-1-7281-6559-2/20 $31.00 © 2020 IEEE

defects. To reduce the stress and the cost of deposition, manufacturers are using a new technique of double stacking to reduce the stress between alternating layers.

Several High Aspect Ratio (HAR) etch are carried out to form the channel hole and slit trench. The channel hole facilitates deposition of storage material. The slit trench facilitates lateral etch of the SiN and where the tungsten word line material is later deposited. From the analysis we observe that the aspect ratio increases from 42:1 for 64-layer to 45:1 for 72-layers and 52:1 for 92-layer word lines. To keep the aspect ratio minimum thinner word lines and thinner insulating material is deposited. From the physical analysis, the word lines and the insulating material from the new generation memories is measured and compared to the previous generation. The tungsten word line thickness decreases approximately by 10-15%. This reduces the height of the 3D structure and helps reduce the consumables cost. The vertical etches are a significant step in the manufacture of 3D NVM. The etch rate is reduced as the aspect ratio increases to limit defects like under etching due to weakened ion energy causing incomplete etching. Another defect like twisting of the hole could result if the etch rate is high. Slower etch rate achieves the perfect conformal vertical structure but inflates manufacturing cost [5]. It is important to avoid etching defects to keep the manufacturing yields at optimum.

Two 3D NAND manufacturers have used a double stacking technique in their 96-layer memories to reduce the 3D NAND manufacturing challenges. Half the word lines are deposited, and the memory cells are formed before depositing the rest of the Word lines. The two stacks are shown in "Fig 2". This process allows reduction of high aspect ratio etching to 20:1. The etching rate of the holes could be increased, and cost of etching reduced.

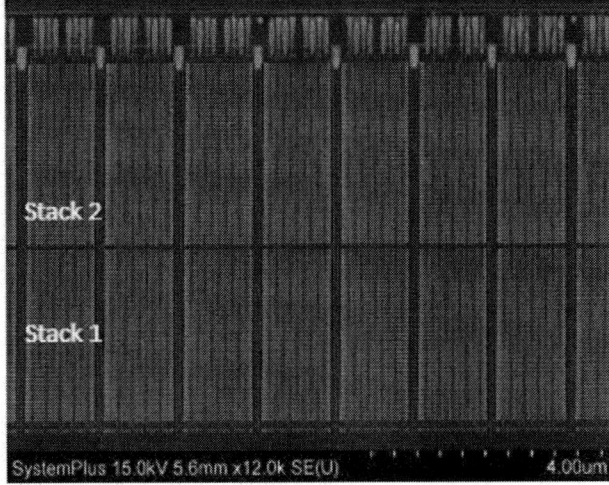

Fig 2: Toshiba's 96-layer double stacking technique

The staircase forms the contact pad to facilitate connection of the word line to the metal contacts as seen in "Fig 3". High precision and uniformity are required. Manufacturers use one lithography pattern to etch several steps at a time to reduce patterning cost. Additional word lines increase the staircase patterning. Increase of world lines has an impact on the area taken up by the staircase structure in the die and it could reduce the expected memory die density. The staircase step accounts for the highest cost in the wafer manufacturing process of 3D NAND memories. The staircase process step accounts for approximately 25% of the total wafer Cost.

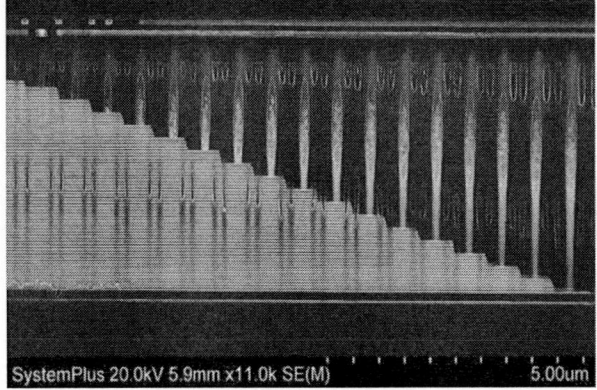

Fig 3: 64 Layer 3D NAND Die Cross Section Overview of the staircase steps and metal contacts

The manufacturing process was simulated using a software and the Front-End Cost results were studied and major steps in the process noted. The results show the impact of adding word lines on the wafer manufacturing cost, the results are shown in "Fig 4". The total Front-End cost of manufacturing new generation memories wafers increases as word lines are added due to the complexity of the manufacturing process. Lithography steps increase as manufacturers add more worldlines to the 3D structure. Increased Lithography steps elevate the final wafer Front End Cost. Added word line layers increase consumable cost, machine cost and labor cost because more time is spent on the machines therefore inflating the final wafer manufacturing cost of new generation memories. From the estimated total wafer cost each step of the manufacturing process is studied and the high cost steps are extracted.

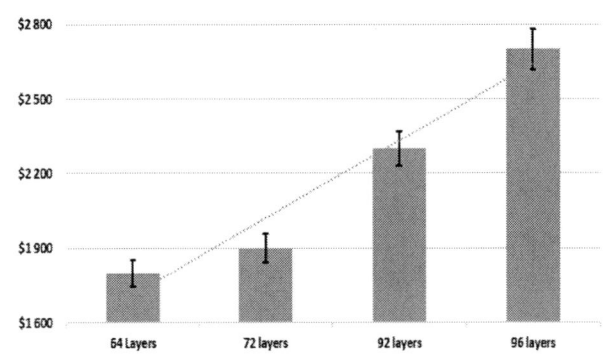

Fig 4: Estimated Total Wafer Front End Manufacturing with different Word line layers

The Wafer Cost of future generation 3D NAND Could be estimated using the above information from "Fig 4".

The final wafer Cost results show that the wafer cost is directly linked to the fabrication challenge. The manufacturing of 3D NAND continues to be a challenge as the word lines increase. The new technique of double stacking reduces the steep manufacturing cost of High aspect ratio etching. The major steps in the Front-End Process Cost are summarized in Figure "Fig 5". Etching of the staircase remains as the highest cost in the manufacture of 3D NAND.

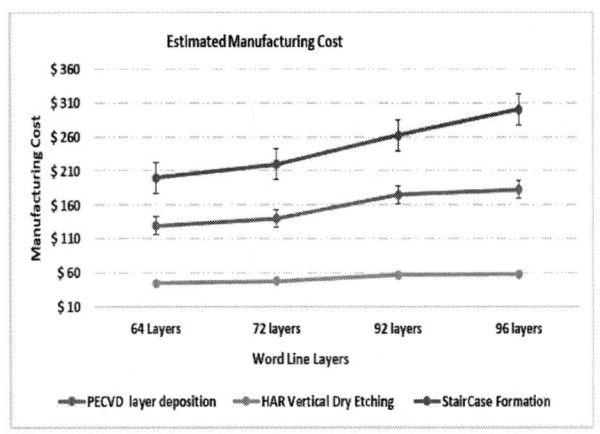

Fig 5: Wafer Front End Manufacturing Cost for major process steps

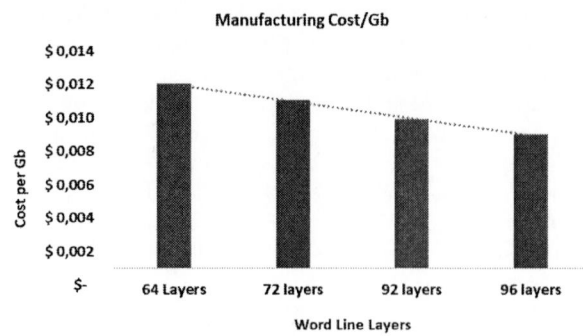

Fig 6: Evaluation of manufacturing cost of per Gb

The Gigabits produced per wafer massively increase with each generation, approximately 40% increase from 64-layer to 92-layers. The word line increase allows manufacturers to produce smaller dies and vastly improve die density. This technique further decreases the cost of the memories as more dies are produced from a single 300mm wafer.

NVM manufacturers will continue to add layers to their next generation NAND memories without altering the process technology node as this process is cost effective and highly achieves increased density [6]. The manufacturers can produce smaller dies with higher density in the new generation memories. Additional word lines result in wafers containing a higher number of memory cells. The Gigabyte (GB) produced per wafer increase as the word lines are increased. Potential GB produced per wafer for each generation are estimated "Table 1". From the results in the table it is possible to estimate the GB increase by adding word lines to the 3D NAND generation memories. [7,8]. Each word line has the potential of producing an additional 300- 400GB in a wafer.

The potential GB produced per wafer is determined by multiplying the number of dies obtained per 300mm wafer by the capacity each memory die produced.

TABLE 1. ESTIMATED GIGABYTES

Word line layers	Potential GB produced/ wafer
64 layers	~ 27 000
72 layers	~ 30 000
92 layers	~ 37 000
96 layers	~ 39 000

Table 1: Estimated GB produced per 300mm wafer. (This is an average of the top manufacturers)

The manufacturing cost per Gigabit (Gb) is evaluated. A constant decrease per Gb is realized by increasing word lines. The results of the cost estimation are shown in "Fig.6", the results are obtained by dividing the total wafer front-end cost that includes the probe test of the wafers by the number of Gb obtained per wafer. This cost does not consider the packaging cost.

RESULTS AND SIGNIFICANCE

The study of the manufacturing cost is not only advantageous in supply negotiations, but also useful in identifying main cost drivers in the manufacturing process. The results help equally in the phase of design, evaluating cost competitiveness of shifting to a new generation by predicting the cost of future generations with added Word lines. Cost limitations can be identified and resolved. The results of the study show the cost effectiveness of adding more layers. Different steps in the manufacturing process have a great impact on the final wafer cost especially the layer deposition that form the word lines, the high aspect ratio etching and staircase etching. The results show that new generation memories with added Word line layers increase the wafer manufacturing cost but have a significant increase in the number of Gb produced per wafer therefore tunneling down the cost per Gb. This entails cost reduction by producing denser and smaller dies.

REFERENCES

[1] Micheloni R., Aritome S., Crippa L. (2018) 3D NAND Flash Memories. In: Micheloni R., Marelli A., Eshghi K. (eds) Inside Solid State Drives (SSDs). [Springer Series in Advanced Microelectronics, vol 37. Springer, Singapore p.3, p.16, p.17]

[2] Dongzhe Ma, Jianhua Feng, and Guoliang Li, 2014. A survey of address translation technologies for flash memories. ACM Comput. Surv. 46, 3, Article 36 (2014),

[3] Liu Shijun, Zou Xuecheng, Analysis of 3D NAND technologies and comparison between charge-trap-based and floating-gate-based flash devices, Journal of China Universities of Posts and Telecommunications, Volume 24, Issue 3,2017,Pages 75-96,

[4] S. Wu, C. Chen, T. Luoh, L. Yang, T. Yang and K. Chen, "Virtual metrology for 3D vertical stacking processes in semiconductor manufacturing," 2016 e-Manufacturing and Design Collaboration Symposium, Hsinchu, 2016

[5] Y. Ye, Z. Xia, L. Liu and Z. Huo, "Investigation of Reducing Bow during High Aspect Ratio Trench Etching in 3D NAND Flash Memory," 2018 14th IEEE International Conference on Solid-State and Integrated Circuit Technology (ICSICT), Qingdao, 2018 p.1-3

[6] W. Feng and N. Deng, "Study on cell shape in 3D NAND flash memory," 2015 IEEE International Conference on Electron Devices and Solid-State Circuits (EDSSC), Singapore, 2015.

[7] System Plus Consulting Leading Edge 3D NAND Comparison 2018 & 2019 Report. unpublished

A NEW STRUCTURE OF ILD GAP FILLING IMPROVEMENT FOR FLOATING-GATE MEMORY

Zhenghong Liu*, Yan Li, Ruisheng Qi, Naoki Tsuji, Guanqun Huang, Haoyu Chen, Chris Shao

Shanghai Huali Microelectronics Corporation, Shanghai 201203, China
*Corresponding author. Tel.: +86-17717386771. E-mail addresses: liuzhenghong@hlmc.cn

ABSTRACT

For traditional stack gate type nonvolatile memory, high voltage apply to the control gate and well during PGM and ERS operation, this requires high break down voltage(BV) for I/O Tr . One usual method to archive high BV I/O Tr is increasing LDD energy which requests thick enough poly thickness for peripheral gate to avoid implant penetration. As the scaling of the dimension, for conventional floating gate structure poly space aspect ratio at flash array area becomes much high, in case of this, potential risk of reliability and yield loss will be induced if ILD gap fill capability is not enough among adjacent control gate. To solve this issue, in this paper, a novel dual poly structure and process flow is proposed, this dual poly structure can decrease control gate poly thickness of flash array area, while the poly thickness in the logic area keeps no change. The proposed dual poly structure decreases the memory aspect ratio effectively and independently to peripheral and improves the ILD gap filling window. Good yield and reliability results are obtained by process optimization of dual poly structure, and this novel new structure is an important candidate for further scaling of stack gate type flash.

INTRODUCTION

Floating-gate Flash cells are widely used in the analog computing [1] and modern micro and nano electronic applications because of the high performance: high speed, high density, and low power [2-4]. The classical stacked gate memory structure includes P-channel and N-channel flash memory which own same structures but different type dopant in sources/drain, as shown in Fig.1. With the development of technology, scaling the dimension of the flash cell is a very important component in the cost reduction roadmap, so the aspect ratio of the control gate spaces becomes much higher because of the stacked poly features in the array area, at this case, ILD void will be generated if the filling capacity of the tool is not further improved or other process still keeps no change, and the potential risk of reliability or yield loss will be high.

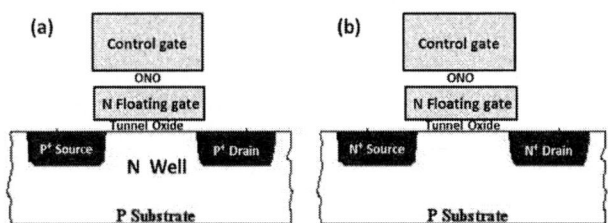

Fig.1.Cross sections of P-channel (a)and N-channel Flash memory(b).

58-nm node P-channel Flash memory found ILD void during the process development, as illustrated in Fig.2 (a). During the program and erase operation, high voltage bias needed between wordline to drain and source contacts to form large gate current which make sure the sensing operation faster and more robust to the noise, drain contacts are shared by two bits and connected by a common metal line named bitline, the programing time usually 60us and voltage bias more than 10V, this stress much higher and CT to control gate will be short together if the ILD void existed, as shown in Fig.2 (b), which induced the bit cell fail or potential risk of reliability. Actually there is one special test for this case which named Icc_ERS during CP test, the detail test condition as Fig.2(c) showed, high voltage bias exist between CT and control gate.

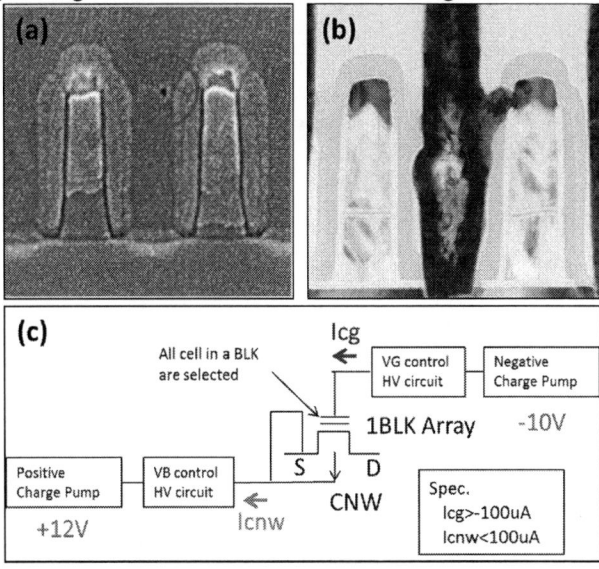

Fig.2.ILD void before (a) and post (b) P/E operation, (c) Icc-ERS test condition.

One recommended solution is to decrease the control gate polysilicon thickness to decrease the aspect ratio of the poly space and increase the ILD gap filling window, this also causes the polysilicon thickness of the logic region to be reduced because of the polysilicon in the memory array and logic region deposited at the same time for the traditional fabricating flow. But for the IO device, high break down voltage needed to meet the flash array area requirement, traditional solution to archive high BV is increasing LDD Energy, which need gate poly thickness is thicker enough to avoid punch through, so the poly thickness in the peripheral area around 2000A and cannot decreased at the same time.

In this work, we try to fabricate a new dual poly structure with low aspect ratio in array area by process optimization. Through in detail analysis in the manufacture process, physical structure, as well as CP and reliability performance of the memory cells, we proposed a novel dual poly structure to decrease cell space's aspect ratio which much enlarge the ILD gap filling window and independently to the peripheral performance.

FABRICATE FLOW DETAILS

The process flow diagram of the stack-gate NOV memory cell from ONO deposition loop to poly etch loop is illustrated in Fig. 3, process A and process B show the manufacturing process flow of the traditional and the new structure, respectively, the special processes for the novel dual poly structure are shown in green. These are also the most important manufacturing process for the recommended new structure.

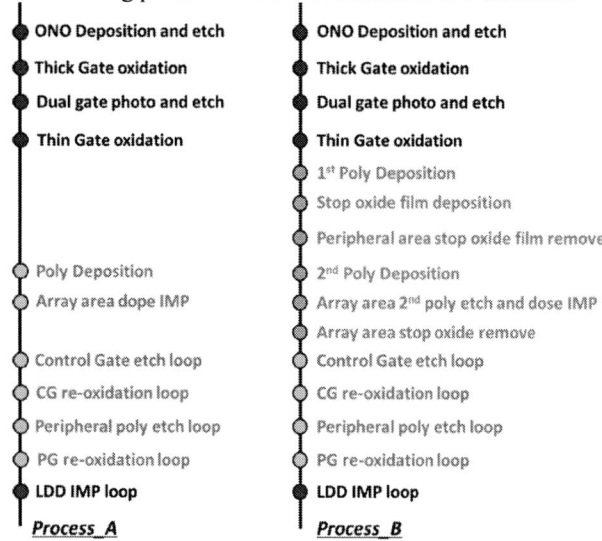

Fig.3. Fabricating flow detail information. Process_A is baseline flow and Process_B is dual poly structure flow.

The specific fabricating process of dual poly structure for floating gate type memory at 58-nm technology node is introduced in detail in the following sections. There are total 6 loops are employed before the gate etching and the corresponding structure diagram showed in Fig. 4.

[1] Deposit the 1st polysilicon on the gate oxide film which finally decide the control gate thickness of the stacked structure and can be adjusted freely within a certain range, here the poly thickness select about 1200 angstrom, and this layer of polysilicon is part of the peripheral logic gate at the same time;

[2] Thermal oxide film on the 1st polysilicon formed post one simple wet clean step and thickness around 110 angstrom, this oxide film will be as the stop layer during the 2nd polysilicon etch process;

[3] Photoresist coating and development processed with the logical area clear mask, peripheral area stop oxide film removed by wet etch while the stop oxide film in the memory array region is preserved because of the protection of PR, and it's better to add descum to harden the photoresist to avoid peeling before the large amount of wet etching process , finally photoresist was removed only by wet strip which no damage to the logic polysilicon gate surface;

[4] Deposit the 2nd polysilicon gate post wet treatment step, thickness around 800 angstrom, at the moment, about 100 angstrom oxide exist between the 1st and 2nd polysilicon at the memory array area, but 1st and 2nd polysilicon combine together and as the final gate thickness for the logic transistor device region. Note that wet etching must remove the natural oxide layer before the 2nd poly deposition.

[5] Photoresist coating and development processed with the logical area dark mask, 2nd polysilicon in the memory array area removed through dry etching technical, the oxide between two polysilicon as the stop layer during the etch process, then photoresist was removed by dry and wet strip;

[6] The remaining stop oxide layer in the memory array area was removed by wet process with 50% over etch amount.

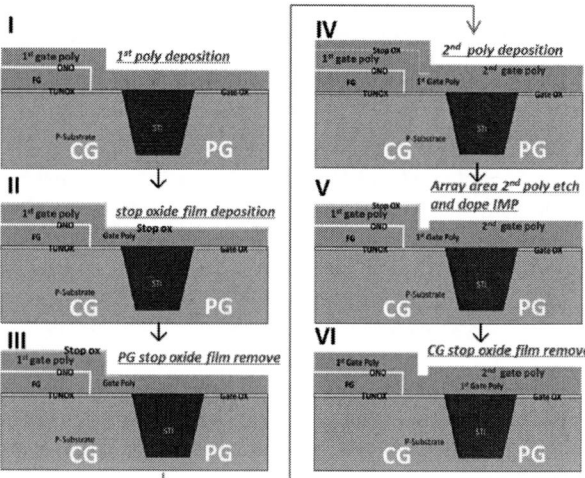

Fig.4. Dual gate poly new structure detail fabricate flow information.

At this point the dual poly films which logic and storage areas with different polysilicon thickness are formed, and the novel double-gate structures are finally formed after the gate etching process with the baseline flow.

RESULTS AND DISCUSSION

Fig.5 shows the physical structure of the floating gate stack flash array area after CT etch, the cross section along bite line direction of the baseline and the proposed dual gate poly structure in Fig.5 (a) and Fig.5 (b).The comparison of these two images shows that the total height of the new structure is about 800A lower than the baseline structure because of the 2nd polysilicon removed at the array region, and the aspect ratio of poly space is reduced from 3.2 to 2.3 for dual poly structure. The reduction of the aspect ratio of poly space greatly increases the ILD gap fill window and provides an effective method for further size reduction.

The thickness of polysilicon in the peripheral area actually no big difference for the novel dual gate poly structure to make sure that no penetration happened for the logic device. In fact, we did not make any adjustments to the peripheral logic devices of the new structure.

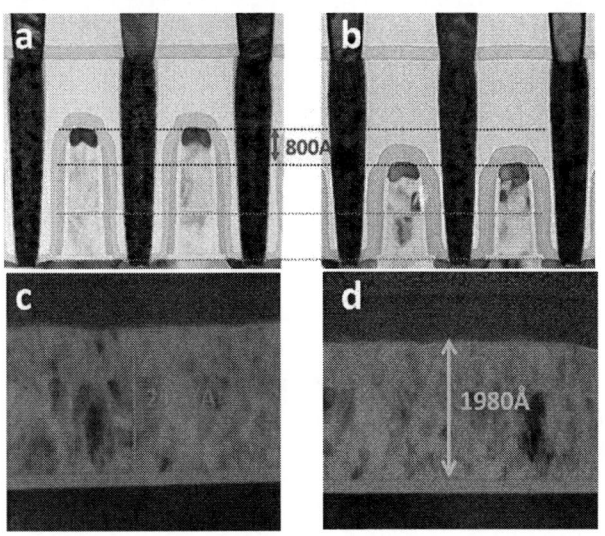

Fig.5. Cross section along the bitline post CT-ET. (a) Sample of baseline process flow and (b) sample of dual gate poly integrate flow;(c) Sample of peripheral poly thickness by baseline process (d) sample of peripheral poly thickness by dual gate poly process.

The proposed dual poly structure decreases the memory aspect ratio effectively and independently to peripheral and improves the ILD gap filling window. The yield of the new structure has been greatly improved by process optimization. Comparing the yield data of the baseline and the new structure, as shown in Fig.6a and Fig.6b, respectively, it can be seen that the failure due to the ILD void of the baseline has not been found in the novel dual gate poly structure.

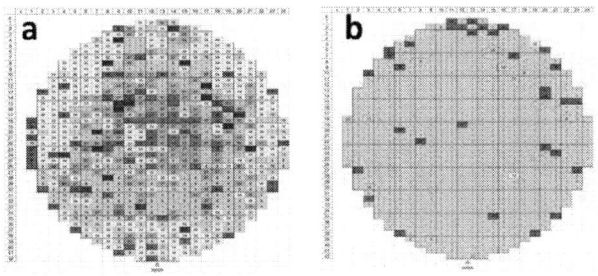

Fig.6. CP results of the BL structure (a) and dual gate poly structure (b).

The reliability performance of the new structure also was evaluated, as shown in Fig.7(b), compare with baseline performance, as shown in Fig.7(a), the new structure result is greatly improved both at room temperature and high temperature and can meet the evaluation criteria.

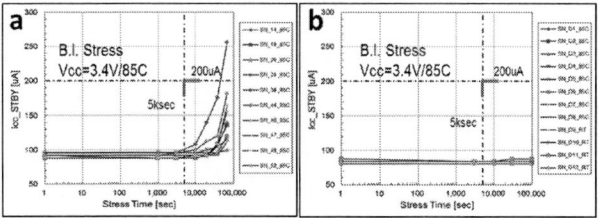

Fig.7. Stress test result of baseline (a) and dual gate poly structure (b).

CONCLUSION

In this work, we proposed a new dual poly structure. This novel structure decreases the aspect ratio of poly space in memory array area from 3.2 to 2.3 which is benefit for ILD gap filling and the ILD gap filling process window much improved. At the same time, the process flow of the novel dual poly structures no impact on the peripheral regions, so the logic device performance almost keeps no change. Good yield and reliability results are obtained by process optimization and this novel new structure is an important candidate for further scaling of stack gate type flash.

ACKNOWLEDGEMENTS

The authors would like to acknowledge the members of the 40-nm SONOS flash project team and other engineers for the support for developing this work. I am grateful to Shanghai Huali Microelectronics Corporation who gives me such a precious chance.

REFERENCES

[1] D. B. Strukov, K. K. Likharev, "Reconfigurable hybrid CMOS/nanodevice circuits for image processing," *IEEE Trans. Nanotechnology*, vol. 6, pp. 696-710, 2007.

[2] C. R. Schlottmann, P. E. Hasler, "A highly dense low power programmable analog vector-matrix multiplier: The FPAA implementation," *IEEE JETCAS*, vol. 1, pp. 403-411, 2011.

[3] S. Ramakrishnan, J. Hasler, "Vector-matrix multiply and winner-take-all as an analog classifier," *IEEE Trans. VLSI*, vol. 22, pp. 353-361, 2014.

[4] A. Kramer, "Array-based analog computation," *IEEE Micro*, vol. 16, pp. 20-29, 1996.

NEUROMORPHIC TECHNOLOGY UTILIZING NAND FLASH MEMORY CELLS

Sung-Tae Lee[1], Jong-Ho Lee[1]*

[1]Department of ECE and ISRC, Seoul National University, Seoul 151-742, Korea
*Corresponding Author's Email: jhl@snu.ac.kr

ABSTRACT

Hardware-based neural networks are expected to be a new computing breakthrough beyond conventional von Neumann architecture because of their low power operations. In this work, we introduce operation scheme of neural networks using NAND flash memory cell as analogue and binary-state synaptic devices. We report that NAND flash memory has excellent density and low error rate when compared to the results of RRAM.

INTRODUCTION

Hardware-based neural networks based on synaptic devices can reduce power consumption greatly by replacing the vector-by-matrix multiplication with a dense crossbar array of synaptic devices such as PCM [1], RRAM, and NOR flash memory. However, several problems need to be addressed before memristive crossbar arrays can be widely adopted, such as high device variability, absence of precise device models and stochastic behavior of devices [2].

As a candidate to solve the above problems, cells in Si based memories such as NOR flash memory and SRAM can be used. However, these memories are limited in density due to bit line and word line contacts in each cell device. On the other hand, NAND flash memory reduces ground wires and bit lines considerably, which allows a denser layout and greater storage capacity per chip than those devices.

The recent state-of-the-art DNN algorithms typically require enormous parameter size. As a way to accommodate this, NAND flash memory which has great advantages in cell density can be used. However, it is difficult to apply NAND flash memory composed of cell strings to neural networks due to the specificity of the string structure.

In this work, we first introduce an example of a neural network using analog synaptic weights of NAND flash memory cells and show the measured conductance response and the simulated classification accuracy [3], [4]. In addition, we introduce an example of implementing BNN (binary neural network) using NAND flash memory, and show that the Bit Error Rate and the density of NAND flash memory are superior to those of RRAM [5].

METHODS AND RESULTS

Firstly, we introduce operation scheme of neural networks using NAND flash cells as analogue synapses.

As shown in Fig.1, to use NAND flash memory cells as synaptic devices, we apply input values (voltages) to the bit-lines for the following reasons. The scheme which applies input to the bit-lines allows analogue input values satisfying weighted sum output equation, because output current is zero when the bit-line voltage is zero and bit-line current increases linearly with increasing bit-line bias in linear region. In addition, as shown in Fig. 1, by using conductance difference between a pair of adjacent cells to represent synaptic weight ($W_{ij}=G^+_{ij} - G^-_{ij}$), negative synaptic weight can be represented [1]. Note two cells in the same position in two adjacent cell strings have nearly identical device characteristics, so we can minimize the mismatch between two cells that represent one synapse.

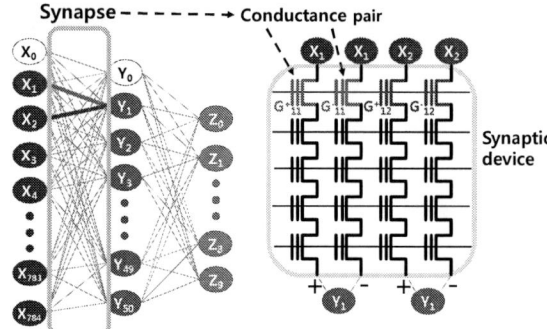

Fig. 1. 3-layer perceptron in which synapse can be implemented using NAND flash memory. Synaptic weight is encoded by the conductance difference between a pair of adjacent NAND cells.

In forward propagation, the weighted sum calculation is accomplished by simply subtracting the read current between a pair of bit lines and summing the total current using an electronic device such as a capacitor as shown in Fig. 2. By sequentially applying read pulses to the word lines, output currents for all neurons in the lth layer are sequentially generated. When the kth pulse is applied to the kth word-line, the output current flows for the kth neuron in the lth layer. During this process pass bias is applied to the unselected word-lines to read the kth cell current as shown in Fig. 2. Since the resistance of the selected cell is always much larger than that of the unselected cells with large pass bias applied, the output current primarily depends on the threshold voltage of the kth word-line cell. Then, the overall current for the kth row is summed. When the overall output current for all neurons in l^{th} layer is

978-1-7281-6559-2/20 $31.00 © 2020 IEEE

sequentially produced, the output current stored in memory is passed to all neurons in the lth layer.

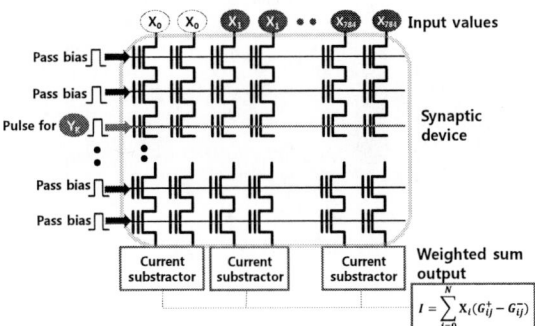

Fig. 2 Schematic of Forward propagation for producing the output current for k^{th} neuron in l^{th} layer. [3]

For comparing our synaptic device with other devices reported up to date, we use the behavior model for NAND flash cell, ideal perfect linear device, and memristive device [6]. Fig. 3 shows normalized conductance versus the number of pulses in three devices, using behavior model.

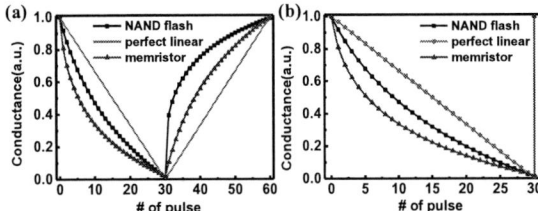

Fig. 3 (a) Bidirectional conductance responses of NAND flash, ideal perfect linear device, and memristor [6]. (b) Unidirectional conductance responses of NAND flash, ideal perfect linear device, and memristor [6].

Fig. 4 shows the classification accuracy obtained by simulation for the MNIST data set in a three-layer multi-layer perceptron network (784×50×10) using the conductance response of Fig. 3. Using bidirectional conductance response in Fig. 3 (a), the simulated accuracies for NAND flash, perfect linear, and memristor devices are 87.92 %, 94.14 % and 85.99 % respectively as shown in Fig. 4 (a). In Fig. 3 (a), the NAND flash has more linear conductance response than memristor during programming, but has more nonlinear conductance during erasing. Therefore, in bidirectional conductance case, the accuracy obtained by using NAND flash memory cells is similar to the accuracy obtained with a memristor-based synapses. Bidirectional conductance response (LTP and LTD characteristics) in NAND flash memory cells need to be improved.

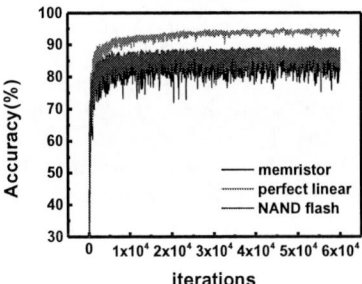

Fig. 4. Simulated classification accuracy obtained by using the unidirectional conductance response in Fig. 3 (b)

We introduce synaptic architecture for binary neural networks using NAND flash cells as binary synapses. Fig. 5 shows a novel 2T2S (two transistors and two NAND cell strings) synaptic string structure for XNOR operation. Two NAND cell strings are used for one synapse string consisting of serially connected synaptic cells with two input transistors of which two input voltages are applied to each gate. The two input transistors can be replaced by two NAND cells having gates (word-lines) isolated from each other. For each synapse consisting of adjacent two NAND cells in two cell strings, synaptic weight of +1 can be represented by the two cells of which the left cell is on-state ($V_{th,low}$) and the right cell is off-state ($V_{th,high}$). For an input value, the input value of +1 can be represented by complementary input voltages where V_{in1} is turn-on voltage (V_{on}) and V_{in2} is turn-off voltage (V_{off}). Fig. 5 (a) and (b) represent the cases when the input value is +1 and -1.

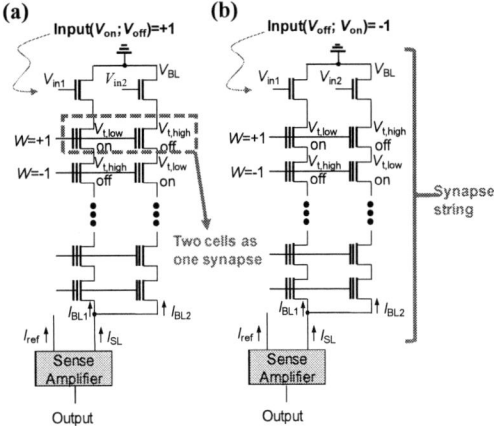

Fig. 5. 2T2S (two input transistors and two NAND strings) synapse string structure for XNOR operation. [5]

Fig. 6 compares the inference accuracy versus bit error rate for the proposed NAND flash memory synapses and reported RRAM synapses. The estimated bit error rates of NAND cells are about 4.2×10^{-8} % and 2.3×10^{-7} % when the program voltage (V_{PGM}) is 16 V and 14 V, respectively, and the estimation is based on the statistical parameters from the measurement data and the assumption of Gaussian distribution. The large on/off current ratio ($\sim7\times10^{5}$) of NAND flash cells results in sufficiently low bit-error rate. The bit-error rate of our work is much lower than those of RRAM synapses in [7], [8], while keeping much higher synapse density. Our work provides highly-reliable BNNs that do not require error correction codes (ECC), which can reduce the enormous time, energy and complex decoding circuitry required for the ECC. In addition, when program voltage is below 16V, highly reliable binary neural networks can be implemented regardless of program/erase cycling number without using conventional incremental step pulse program (ISPP) method [5]. Therefore, the time and energy can be greatly reduced compared to those of the ISPP method.

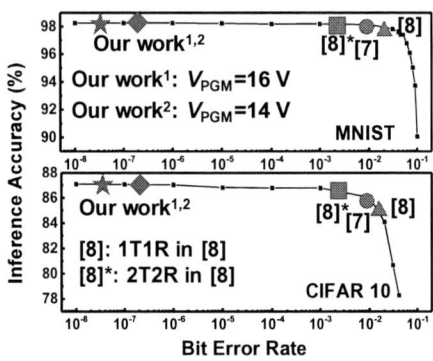

Fig. 6. Inference accuracy with bit error rate. Our work[1,2] indicate the cases when V_{PGM} are 16 V and 14 V, respectively. [5]

Although cell density is high in 2-D NAND flash, cell density can be much higher in vertical NAND (VNAND) flash than in 2-D case. Fig. 7 shows the effective area per synapse and synapse density ratio of the VNAND flash and 2T2R-based RRAM synapse with increasing number of stacks. Here, control is the area of one 2T2R-based synapse in RRAMs [7], [8]. The area occupied by the 2T2R synapse is calculated to be 24300 nm² by assuming that two 22nm FinFETs under two RRAMs determine the area of one synapse. As the number of stacks increases, the effective area of one synapse in VNAND memory becomes smaller. The synapse density of VNAND at a stack number of 128 is about ~100 times higher than that of RRAM control.

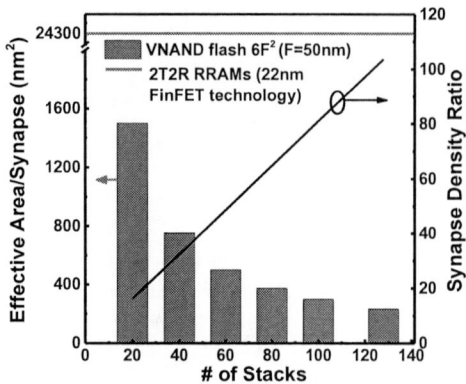

Fig. 7. Effective area per synapse and synapse density ratio with the number of stacks. Here, control is the area of one synapse in RRAMs. [5]

CONCLUSION

We have introduced a synaptic architecture based on NAND cell strings and an operation scheme for high-density and highly-reliable neural networks. The conductance response of NAND flash cell is compared with those of memristor and perfect linear device. The bit-error rate (4.2×10^{-8} %) of the proposed BNN using NAND flash memory was 4 orders of magnitude lower than that of RRAMs. In 128-stack NAND flash memory, the estimated synapse density is ~100 times higher than that of RRAMs. Thus, the proposed architecture is very promising for high-density and highly-reliable BNNs.

ACKNOWLEDGEMENTS

This work was supported by the National Research Foundation of Korea (NRF- 2016M3A7B4909604) and the Brain Korea 21 Plus Project in 2019.

REFERENCES

[1] G. W. Burr. *IEEE Trans. Electron Devices*, vol. 62, no. 11, 2015, pp. 3498–3507.

[2] P. Pouyan. *Proceeding of 5th Eur. Workshop CMOS Variability*, Oct. 2014, pp. 1–6.

[3] S. T. Lee. *Proceedings of 2018 IEEE Symposium on VLSI Technology*, Honolulu, June 18-22, 2018, pp. 169-170.

[4] S. T. Lee. *IEEE Journal of the Electron Devices Society*, vol. 7, 2019, pp. 1085-1093.

[5] S. T. Lee. *Proceedings of 2019 IEEE International Electron Devices Meeting (IEDM)*, San Francisco, Dec. 7-11, 2019, pp. 38.4.1-38.4.4.

[6] S. H. Jo. *Nano Letters*, vol. 10, no. 4, 2010, pp. 1297–1301.

[7] S. Yu. *Proceedings of 2016 IEEE International Electron Devices Meeting (IEDM)*, 2016.

[8] M. Bocquet. *Proceedings of 2016 IEEE International Electron Devices Meeting (IEDM)*, 2018.

Investigation of Multi-Level Properties of TaO$_x$-based Memristive Devices and Optimized Programming Scheme for On-Line Training

Teng Zhang[1], Yingming Lu[1], Caidie Cheng[1], Ke Yang[1], Liying Xu[1], Qingxi Duan[1], Zhaokun Jing[1], Keqin Liu[1], Rui Yuan[1], Yuchao Yang[1,2,3*], and Ru Huang[1,2,3*]

[1]Key Laboratory of Microelectronic Devices and Circuits (MOE), Department of Micro/nanoelectronics, Peking University, Beijing 100871, China
[2]Academy for Artificial Intelligence, Peking University, Beijing 100871, China
[3]Frontiers Science Center for Nano-optoelectronics, Peking University, Beijing 100871, China
Email: yuchaoyang@pku.edu.cn, ruhuang@pku.edu.cn

ABSTRACT

In this paper, we present the multi-level cell characteristics of TaO$_x$ based memristive devices as synaptic elements for neuromorphic computing. We utilized various programming conditions to modulate the conductance states with different set/reset voltage amplitude or width. Our results reveal that for LTP and LTD process, the dominant factors for write error are nonlinearity and fluctuation respectively. Furthermore, we proposed an error-aware programming scheme, thereby enabling a high accuracy for weight updating when used in ANN applications.

INTRODUCTION

Neuromorphic computing [1] has been recently considered as an alternative architecture to conventional von Neumann architecture, which performs sequential data processing. The neuromorphic computing is based on neural network algorithm which enables parallel processing on account of large connectivity between processing units. To realize such systems, artificial synapse [2], which plays an important role in the neuromorphic system, should be established using solid-state devices.

Memristive devices [3] are nowadays considered as leading candidates for artificial synapses due to their intrinsic characteristics such as excellent size scalability, fast switching speed, and low energy consumption. Therefore, in this work, we perform a thorough investigation on the multi-level property of a TaO$_x$-based memristive devices [4] by controlling various programming schemes. In long term potentiation (LTP) process, the conductance write error is mainly dominated by the non-linearity and in the long term depression (LTD) process [5], the conductance write error is mainly dominated by the conductance fluctuation. In addition, we defined two regions for the conductance and adopted an optimized programming scheme for better write accuracy. The optimized scheme can help the device to perform better in online learning when it is used as artificial synapse.

EXPERIMENTAL

The TaO$_x$ based memristive devices were fabricated on SiO$_2$ substrates with effective size of 2×2 um^2. First, reactive ion etching was used to create a 100 nm deep trench on SiO$_2$, then 10 nm Ti and 100 nm Ta was deposited by sputtering while keeping the same patterned photoresist. After lift-off, 400 °C annealing was carried out in oxygen ambient for 15 min to form about 11 nm thick TaO$_x$. 80 nm Pt top electrodes were then patterned by photolithography and lift-off. Electrical measurements were performed using Agilent B1500A semiconductor parameter analyzer. Unless otherwise specified, all tests were conducted under the condition that the top electrode Pt was grounded and the signals were applied to the Ta bottom electrode.

RESULTS AND DISCUSSION

Fig. 1 shows 100 consecutive DC current-voltage (I-V) sweeps (gray) of the TaO$_x$ based memristive device after forming process (inset), and the black line represents the averaged I-V curve. An initial forming process is required to trigger the fresh memristive device to effective bipolar resistance switching. After that, the set (reset) of the device was achieved by a positive (negative) voltage sweep on the Ta bottom electrode, in agreement with the fact that there is descending concentration of oxygen vacancies from the Ta/TaO$_x$ interface to the Pt/TaO$_x$ interface after annealing. The transition from the high-resistance state (HRS) to the low-resistance state (LRS) occurred at around 0.5 V during set process, while the gradual transition from LRS to HRS was observed during reset process. This result was attributed to the different switching mechanisms involved in each transition. Oxygen vacancies at the Ta/TaO$_x$ interface begin migrating toward the top electrode (TE) under positive bias, the conducting filament (CF) then forms with a positive feedback. Otherwise, when oxygen vacancies are driven from the CF by negative bias, a negative feedback exists and hence the gap between the CF and the electrode is steadily broadened, resulting in a gradually increased resistance [6][7]. Compliance current (CC) is set by the semiconductor parameter analyzer to prevent excessive current overshooting during set process.

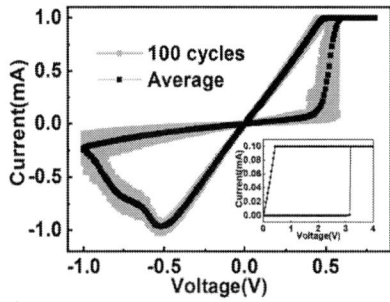

Figure 1. 100 consecutive DC I-V sweeps of the device with the inset of the forming curve.

Fig. 2(b) Further demonstrated that the resistance can be modulated by CC levels which determine the conduction current through the device during set process. Larger CC allows for the formation of thicker (multiple) CF(s) in the device, resulting in multilevel cell characteristics [8]. A number of resistance states can also be successfully accessed in reset process by changing the stop voltage as shown in Fig. 2(a). the multilevel cell characteristics ensure that the device can be used as a synaptic device [9].

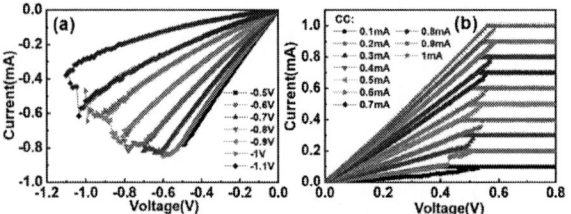

Figure 2. Multilevel characteristics achieved by different CC (b) in the set process and different stop voltage (a) in reset process.

The set event under weak pulse condition was primarily determined by probability, owing to the filamentary nature of the TaO_x based memristive device. As shown in Fig. 3(a), the higher probability from 10 cells was available for using either larger set voltage at a fixed pulse width, or the longer pulse width with a fixed set voltage, respectively. The successful switching was defined that after 250 consecutive set pulses was applied, the conductance of the device can reach 1 mS. The final conductance of the device after 250 consecutive set pulses keeps rising as the pulse width or pulse amplitude increases as shown in Fig. 3(b).

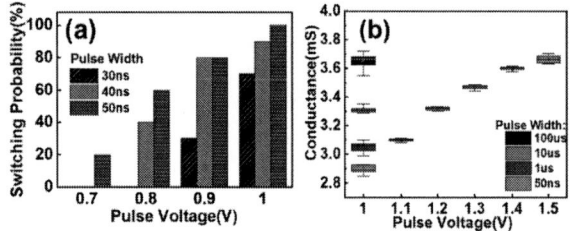

Figure 3. (a) Switching probability of the device under different pulse condition. (b) The dependence of final conductance with pulse condition.

Fig. 4 further shows the LTP and LTD characteristics of the same device using consecutive identical pulses with various pulse conditions. The results show that for LTP, the change of conductance shows a rapid change at the beginning and quickly tends to saturate. The change is relatively flat for LTD process, but the switch window is significantly different as the voltage amplitude changes.

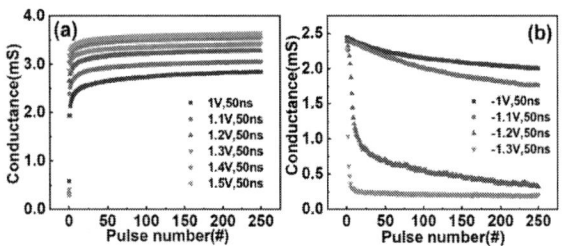

Figure 4. (a) LTP under varied pulse voltages with fixed pulse width. (b) LTD under varied pulse voltages with fixed pulse width.

We further investigated the write accuracy without verify method using 50 full LTP and LTD cycles obtained from the same device as shown in Fig. 5(a). It can be seen roughly from the curves that there are severe nonlinearity and fewer conductive states in LTP compared with LTD, while the LTD process suffers badly with conductance fluctuations. As shown in Fig. 5(b), the write accuracy using LTP branch increase rapidly as the target conductance increase, while the write accuracy using LTD branch keeps at a medium level. The cross point conductance is about 1.77 mS. This result implies that we can divide the conductance into two regions, as shown in Fig. 5. For target conductance in region I, the write accuracy using LTD branch is higher while for target conductance in region II, it is better using LTP branch to write which can lead to more accurate results.

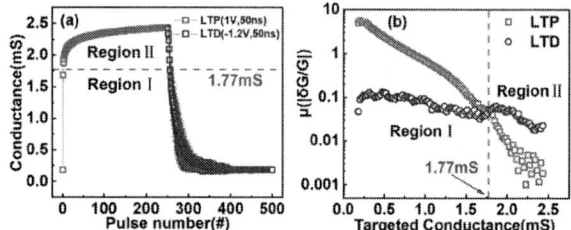

Figure 5. (a) 50 consecutive LTP (1V, 50ns) and LTD (-1.2V, 50ns). (b) The write error using LTP branch and LTD branch.

While used as artificial synapses in neural networks, the conductance of memristive devices represents for the synaptic weights connecting pre-neurons and post-neurons. As shown in Fig. 6, in the feed forward stage, each neuron computes weighted sum of its inputs, which can be implemented using memristive devices based crossbar structure [10], and applies a nonlinear activation function. While in the feed back stage, every synaptic device receives deviation which is, the ΔG in memristive neural system. To achieve more precise conductance changes, we proposed a weight update scheme as illustrated in Fig. 6(b). The difference of the proposed scheme and traditional training lies in that in this scheme, before applying switching pulse, we first judge whether the update is trying to increase/decrease conductance within region I/region II, if so, we first apply a full set/reset to force the conductance reach LRS/HRS, and then, switching pulses are applied. As we have discussed above, the proposed method avoids adjusting the conductance within region I/region II with large errors.

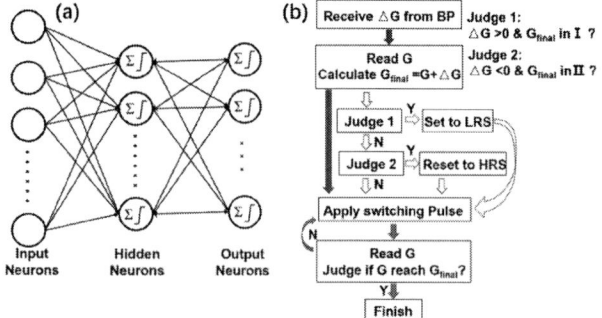

Figure 6 (a) Schematic diagram of a two-layer neural network using back perceptron algorithm. (b) Flow chart of weight update in situ training process.

Based on the proposed training method, a two-layer perceptron neural network was simulated for the classification of MNIST dataset using backpropagation algorithm. For comparison, we collected 10 training processes with and without this method and obtained the averaged results. The result shows in Fig. 7(a) indicates that this method has almost the same accuracy with software-only simulation result while the blind write method has 17% drop in accuracy. Fig. 7(b) further gives the comparison of accuracy, training time, and energy consumption. It needs to be pointed out that this conditional reset method saves about 50% training time and about 37% training energy compared with traditional verify methods while only has loss of accuracy close to 0.2%. These results prove that this is an efficient solution for solving loss of network accuracy due to non-ideal characteristics of memristive devices.

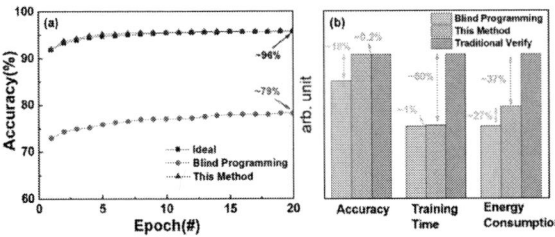

Figure 7 (a) Training process under different training methods. (b) A rough comparison of accuracy, training time, and energy consumption.

CONCLUSION

In this study, we have fabricated Pt/TaO$_x$/Ta memristive devices showing multilevel cell characteristics, which is suitable for the application as artificial synapse. we perform a thorough investigation of the multi-level property of the devices by controlling various programming schemes. Besides, an optimized programming scheme is proposed for better write accuracy when using the devices as artificial synapses for on-line training.

ACKNOWLEDGEMENTS

This work was supported by the NSFC Distinguished Young Scholar Program (Grant No. 61925401), National Key R&D Program of China (Grant No. 2017YFA0207600), the National Natural Science Foundation of China (Grant Nos. 61674006, 61927901, and 61421005), the 111 Project (B18001) and Beijing Academy of Artificial Intelligence (BAAI).

REFERENCES

[1] N. K. Upadhyay, S. Joshi, and J. J. Yang. *Sci. China. Inf. Sci.* vol. 59, 2016, pp. 061401-061426.

[2] T. Zhang, K. Yang, X. Xu, Y. Cai, Y. Yang and R. Huang. *Phys. Status. Solidi. Rapid. Res. Lett.* vol. 13, 2019, 1900029.

[3] L. O. Chua, and S. M. Kang. *Proc. IEEE.* vol. 64, 1976, pp. 209-223.

[4] M. Yin, Y. Yang, Z. Wang, T. Zhang, Y. Fang, X. Yang, Y. Cai, and R. Huang. *13th IEEE Int. Conf. on Solid-State and Integrated Circuit Technology (ICSICT)*, Hangzhou, Oct 25-28, 2016, pp. 1113–5.

[5] E. R. Kandel, J. H. Schwartz, T. M. Jessell, S. A. Siegelbaum, and A. J. Hudspeth. *Principles of Neural Science 5th edition*, McGraw-Hill, 2013.

[6] Y. Yang, P. Gao, S. Gaba, T. Chang, X. Pan, and W. D. Lu. *Nat. Commun.* vol.3, 2012, 732.

[7] Y. Yang, and R. Huang. *Nat. Electron.* vol. 1, 2018, 274.

[8] J. Woo, K. Moon, J. Song, M. Kwak, J. Park, and H. Hwang. *Trans. Electron Devices.* vol. 63, 2016, pp. 5064−5067.

[9] J. Zhu, Y. Yang, R. Jia, Z. Liang, W. Zhu, Z. U. Rehman, L. Bao, X. Zhang, Y. Cai, L. Song, and R. Huang. *Adv. Mater.* vol. 30, 2018, 1870149.

[10] C. Li, L. Han, H. Jiang, M.-H. Jang, P. Lin, Q. Wu, M. Barnell, J. J. Yang, H. L. Xin, and Q. Xia. *Nat. Commun.* vol. 8, 2017, 15666.

NEURAL SPIKE DETECTION BASED ON 1T1R MEMRISTOR

Zhengwu Liu, Jianshi Tang*, Bin Gao, He Qian and Huaqiang Wu*

Institute of Microelectronics, Beijing Innovation Center for Future Chips (ICFC),
Beijing National Research Center for Information Science and Technology (BNRist),
Tsinghua University, Beijing, China, 100084
Email: jtang@tsinghua.edu.cn; wuhq@tsinghua.edu.cn

ABSTRACT

Neural spike detection is an important step to study the working principles of the brain and construct useful neuroprosthetics for patients with neurological diseases. However, the high power consumption of conventional electronic hardware hinders the way towards highly efficient neuroprosthetics. In this paper, an energy-efficient neural spike detector based on 1T1R memristor is presented. 69.12% TPR and 19.35% FPR can be achieved when benchmarked with a conventional method. Compared with traditional neural spike detectors, our memristor-based spike detector has shown significant advantages in both power consumption (84.5 nW compared to 815 nW in conventional hardware) and data transmission bandwidth (~200× compression of raw data).

INTRODUCTION

Neuroprosthetic technology has great potential for the treatment of many neurological diseases [1]. Neural spike detection is an important step in developing neural spike-based neuroprosthetics [2]. It detects the time when a neural spike fires. Various Si-based integrated circuits for neural signal detection have been designed, fabricated and tested [2-4]. However, the real-time processing of neural signals at massive scale imposes a big challenge for those integrated systems under the constraints of energy consumption and data bandwidth [2, 5, 6].

As an emerging neuromorphic device, memristor has the advantages of compressing information in resistance change, showing low power consumption and good scalability. So it is considered as a promising candidate to solve the above-mentioned problems. In literature, there have been implementations of neural spike detection using a single memristor in one-resistor (1R) structure [6]. For practical applications where hundreds of recording sites are employed in state-of-the-art microelectrodes arrays (MEAs), a large memristor array would be needed for neural spike detection. In this case, it is very difficult to scale up 1R memristor array that suffers from sneak current paths and programming disturbances between devices. To address this issue, memristor with one-transistor-one-resistor (1T1R) structure, where the transistor serves as the selector and limits the current compliance, is a better option for large-scale spike detection. In this paper, an energy-efficient neural spike detector based on the 1T1R TiN/TaO$_x$/HfO$_y$/TiN

memristor is presented, and it achieves a True Positive Rate (TPR) of 69.12% and a False Positive Rate (FPR) of 19.35%.

METHODS

Signal flow in 1T1R memristor-based spike detection

A 1T1R memristor has three terminals to apply voltages: top electrode (TE), gate and source terminals. In the RESET process, the device is switched from low-resistance state (LRS) to high-resistance state (HRS) by applying voltage pulses, while the opposite process is called SET. **Fig. 1** shows how neural spikes are detected using a 1T1R memristor. The conditioned neural signal is applied to the source terminal of the memristor while the TE terminal is grounded and the gate terminal is biased at a constant voltage. Significant conductance change would be observed through RESET if a spike appears in the neural signal. Therefore, from the evolution of the device conductance, neural spikes can be detected.

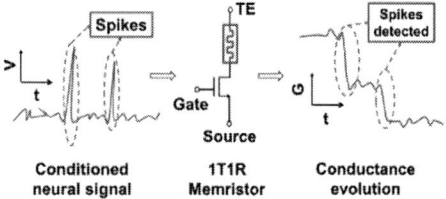

Fig.1. *Illustration of spike detection with 1T1R memristor.*

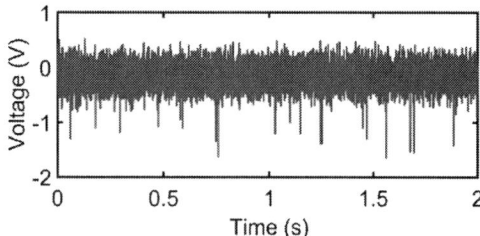

Fig.2. *A clip of neural signal in the dataset.*

Neural signal

Neural signals used in this paper are from the dataset in Ref. [6]. Extracellular electrical activities are recorded by MEAs and preprocessed by band-pass filter and amplification in analog front end. Before being applied to our 1T1R memristor, it is amplified and biased to the

978-1-7281-6559-2/20 $31.00 © 2020 IEEE

proper programming voltage range of the device. **Fig. 2** shows a clip of typical neural signal in the dataset.

RESULTS AND DISCUSSION

Analog switching behavior of 1T1R devices

The key characteristic of the memristor to achieve the presented spike detection is the analog switching behavior, where the device conductance can be tuned incrementally. The analog behavior enables the device not to change its conductance state abruptly, preventing the device from becoming invalid in the spike detection process. **Fig. 3** illustrates the evolution of the device conductance with different programming voltages. Though there is certain variation in the conductance change, it is clear that higher voltage like 1.4V leads to lower final conductance state.

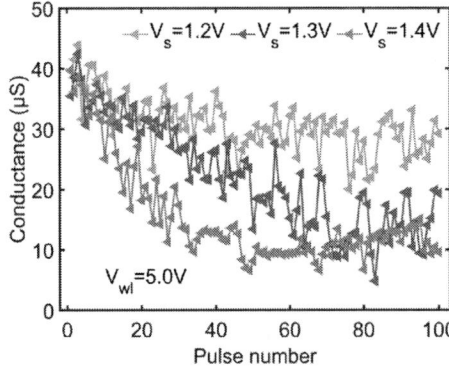

Fig. 3. The analog switching behaviors of RRAM with different RESET voltages at the source terminal.

Conductance evolution with neural signals

The conditioned neural signals are segmented into 315 clips, each of which has 200 sampling points. Signals in each clip corresponds to a length of ~16.4 ms as the sampling rate of the neural signal is 12.2 kHz. Then the neural signals are applied to the device clip by clip.

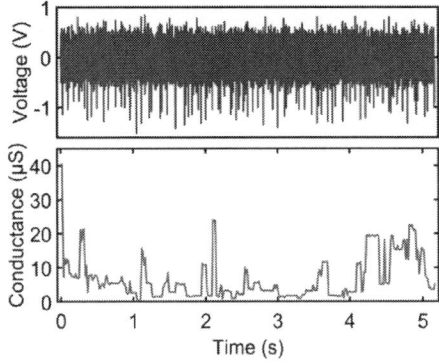

Fig. 4. Device conductance evolution with the applied neural signals.

The device conductance states are read at the beginning and the end of each clip. For each clip, 5 read operations are carried out, 3 of which are after the first three voltage pulses and the rest 2 are after the last two pulses are applied to the device. The initial conductance and the conductance values which are read after the last pulses are applied for each of the 315 clips are recorded in the conductance evolution (**Fig. 4**).

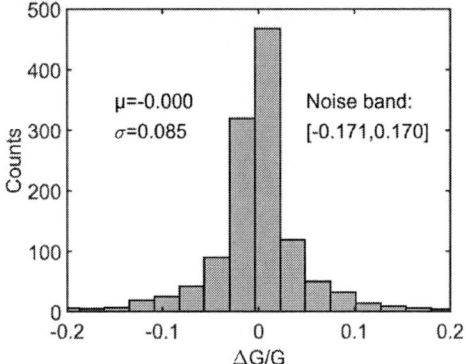

Fig. 5. Evaluation of the noise band.

The relative conductance changes ($\Delta G/G$) are then computed from the conductance evolution process, resulting in 315 data points. We need distinguish the spikes from the background noise in the neural signals using these data points, so the positive and negative thresholds for $\Delta G/G$ are needed. It is noted that there are differences between conductance states in every 5 read operations, which could be attributed to the intrinsic device noise during the programming and read process. Due to the high sample frequency of neural signals and the low spike fire rate, conductance variations for pulses applied before the read operations are considered as the background noise with lower amplitude. So the two thresholds that determine the noise band can be estimated from these read operations. The $\Delta G/G$ of every 2 adjacent read operations is calculated, resulting in $4 \times (315 - 1) = 1256$ data points for evaluating the noise band. The mean value μ and the standard deviation σ are then calculated. Based on the calculations, we use the interval $[\mu-2\sigma, \mu+2\sigma] = [-0.171, 0.170]$ as the noise band (**Fig. 5**).

Detection results

Following the above procedure, the data points outside the noise band are associated with large conductance changes that are resulted from neural spikes. In comparison, the data points within the noise band are considered as background noise in the neural signals. In this way, the clips that contain neural spikes are detected, and the results are shown in **Fig. 6**.

To demonstrate the feasibility of 1T1R memristor-based neural spike detection, we compare it with a conventional method described in Ref. [7] that can detect spikes in the same signals with high fidelity. The total numbers of detected spikes using the conventional method and the memristor method are 68 and 95 respectively. Their relative results are divided into 4 categories: True Positives (TP), False Positives (FP), False Negatives (FN) and True Negatives (TN). Here TP (TN) indicates the two methods both detect that there is a spike (no spike) in a given clip. FP indicates the memristor detects a spike while the conventional method does not. FN indicates the conventional method detects a spike while the memristor method does not. The rates of TP and FP are then used to evaluate the performance of the memristor method according to the following equations:

$$TPR=TP/(TP+FN) \qquad (1)$$
$$FPR=FP/(FP+TN) \qquad (2)$$

Using these two equations, it is calculated that the TPR and FPR of the memristor-based method are 69.12% and 19.35%, respectively.

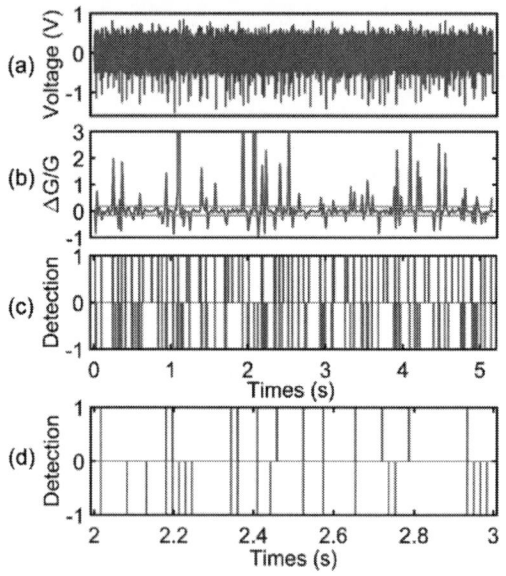

***Fig. 6**. Results of spike detection. (a) Neural signals applied to the device; (b) Relative conductance changes of the device where the green lines represent the noise band; (c) Comparison of the spike detection results for the conventional method (up) and the memristor method (down); (d) Zoom-in view of the detection results in (c).*

Finally, we evaluate the power efficiency of our memristor-based method for spike detection. Here the power consumption is calculated as 84.5 nW, which is much lower than the estimated 815 nW for a conventional hardware [4]. Furthermore, as each clip has 200 sampling points, our approach would reduce the required bandwidth for data transmission by ~200× compared with raw data.

CONCLUSION

In this work, a method based on 1T1R memristor for neural spike detection is presented. The analog switching behavior enables the memristor to successfully detect neural spikes. There are intrinsic noises during the programming process of the memristor, so the noise band based on the relative conductance changes is first evaluated. The noise band actually provides a threshold to distinguish actual spikes and background noises in neural signals. Our method achieves 69.12% TPR and 19.35% FPR when benchmarked with the conventional method. Compared with traditional neural spike detectors, the power consumption (84.5 nW vs 815 nW in conventional hardware) and the required data bandwidth (~200× compression of the raw data) of our memristor-based detector have been significantly reduced. These results show that 1T1R memristor is good candidate for neural spike detection in future neuroprosthetics.

ACKNOWLEDGEMENTS

This work is supported in part by the NSFC (61851404, 61874169, 61674089, 61674092, 61674087), MOST of China (2016YFA0201801), National Major Research Program of China (2017ZX02315001-005), Beijing Innovation Center for Future Chips, and National Young Thousand Talents Plan

REFERENCES

[1] L. R. Hochberg *et al.*, "Reach and grasp by people with tetraplegia using a neurally controlled robotic arm," *Nature,* 2012.

[2] S. Luan *et al.*, "Compact standalone platform for neural recording with real-time spike sorting and data logging," (in eng), *J Neural Eng,* 2018.

[3] A. Bonfanti *et al.*, "A low-power integrated circuit for analog spike detection and sorting in neural prosthesis systems," in *2008 IEEE Biomedical Circuits and Systems Conference (BioCAS),* 2008.

[4] S. E. Paraskevopoulou and T. G. Constandinou, "A sub-1μW neural spike-peak detection and spike-count rate encoding circuit," in *2011 IEEE Biomedical Circuits and Systems Conference (BioCAS),* 2011.

[5] J. A. Wilson and J. C. Williams, "Massively Parallel Signal Processing using the Graphics Processing Unit for Real-Time Brain-Computer Interface Feature Extraction," (in eng), *Front Neuroeng,* 2009.

[6] I. Gupta, A. Serb, A. Khiat, R. Zeitler, S. Vassanelli, and T. Prodromakis, "Real-time encoding and compression of neuronal spikes by metal-oxide memristors," *Nature Communications,* 2016.

[7] R. Q. Quiroga, Z. Nadasdy, and Y. Ben-Shaul, "Unsupervised spike detection and sorting with wavelets and superparamagnetic clustering," (in eng), *Neural Comput,* 2004.

IMPLEMENTATION OF LATERAL DIVISIVE INHIBITION BASED ON FERROELECTRIC FET WITH ULTRA-LOW HARDWARE COST FOR NEUROMORPHIC COMPUTING

Shuhan Liu, Tianyi Liu, Zhiyuan Fu, Cheng Chen, Qianqian Huang*, Ru Huang*

Key Laboratory of Microelectronic Devices and Circuits (MOE), Institute of Microelectronics
Peking University, Beijing 100871, China
*Corresponding Author's Email: ruhuang@pku.edu.cn, hqq@pku.edu.cn

ABSTRACT

In this work, a novel bio-inspired hardware design of lateral divisive inhibition is proposed and demonstrated by using only one transistor of ferroelectric FET. The proposed design is simulated based on our developed FeFET model, and is also proved to be functional in spiking neural network. The new design with ultra-low hardware cost exhibits good biological plausibility, showing its great potential for neuromorphic computing.

INTRODUCTION

Neuromorphic computing provides an attractive solution to overcome the memory wall bottleneck of von Neumann architecture by mimicking the physiology of human brain. In neuroscience, lateral inhibition [1][2], including divisive and subtractive inhibition [3], is essential for various bio-functions like associative memory. In neural networks, lateral inhibition also plays an important role in point attractor network [4] and Bayesian independent component analysis (ICA) algorithm [5]. Although there have been some approaches to the hardware implementation of lateral inhibition by using analog circuits [6], digital circuits [7] and domain wall-magnetic tunnel junction device [8], they either require high hardware cost or have poor biological plausibility. In this work, we propose a novel design of lateral divisive inhibition by using only one single ferroelectric FET (FeFET) with ultra-low hardware cost. Based on simulation using our developed FeFET model, the function of lateral divisive inhibition and leaking is demonstrated, as well as its application in spiking neural network (SNN).

DESIGN AND SIMULATION METHODS

Hardware Design of Lateral Divisive Inhibition

The biological model of lateral inhibition in [1] shows that input stimulus received by one neuron will meanwhile inhibit its competitor through inhibitory neurons. In [9], a mathematic model for leaky integrate-and-fire (LIF) neuron with lateral inhibition, including divisive and subtractive inhibition, was presented and expressed as the following equation (1).

The constant current can be used to implement subtractive inhibition [9], while the implementation for divisive inhibition is more complicated. In Eq. (1), the

$$C_{mem}\frac{dV_{mem}}{dt} = I_{PSC} + g_{leaky}(V_{rest} - V_{mem}) \\ + g_{div}(V_{rest} - V_{mem}) + I_{sub} \quad (1)$$

term of divisive inhibition has the similar form with the leaky term. However, g_{leaky} is constant, while g_{div} represents the degree of divisive inhibition and depends on both the activity level of its competitor and coupling parameter between them.

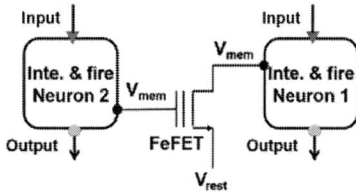

Figure 1: Hardware design of lateral divisive inhibition.

From the above model, a novel hardware design of lateral divisive inhibition is proposed as shown in Fig. 1 in this work. FeFET is a promising device in neuromorphic computing [10][11] owing to the property that its conductance can be modulated by both the gate voltage and the polarization of ferroelectric (FE) layer, and is introduced here to mimic the tunable g_{div}. According to Eq. (1), it is connected between membrane voltage (V_{mem}) and resting voltage (V_{rest}). As for the model in [1], the FeFET as an inhibitory part added to excitatory neurons is an alternative to inhibitory neurons. Besides, the V_{mem} of competitor which reflects the stimulus intensity is used as the control signal and connected to the gate of FeFET to modulate the conductance. Therefore, the above model can be satisfied and the lateral divisive inhibition can be implemented by adding only one single transistor of FeFET to a LIF neuron.

Moreover, due to the similarity between the leaky and inhibitory term in Eq. (1), this FeFET can also be functionally reused as the leaky part under some particular circumstance when the conductance of FeFET keeps constant, which largely saves the hardware cost of LIF neuron.

Simulation Method

Based on the method in [12], we have developed the

FeFET model based on the calibrated dynamic Preisach model of FE as shown in Fig. 2.

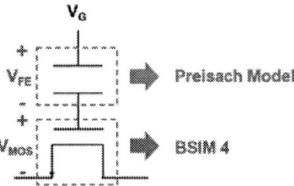

Figure 2: FeFET model based on Preisach Model and BSIM model of MOSFET.

Since the proposed design for lateral divisive inhibition can be applicable for all kinds of physical realization of LIF neuron based on Eq. (1), the typical implementation of LIF neuron in [13] is used as an example (Fig. 3). Notably a leaky transistor is no longer needed because the FeFET can be reused as alternative under certain circumstance. The gate of FeFET is connected to V_{mem} of another LIF neuron (N2) with the same configuration. For both neurons in this work, the post synaptic current (PSC) is input pulse current with 100ns period and 1ns pulse width, and the pulse amplitude for N1 PSC is 600μA while for N2 PSC is changeable.

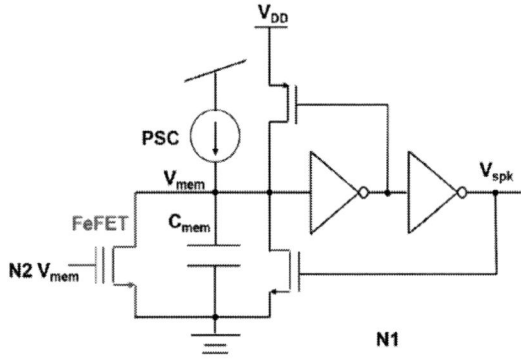

Figure 3: Circuit topology of a LIF neuron (N1) with lateral divisive inhibition. VDD is 1.5V.

RESULTS AND DISCUSSIONS
Device Characteristics

With a triangle sweeping voltage (Fig. 4 (a)), the simulated P-V loop of the standalone FE capacitor and the I-V curves of the FeFET are shown in Fig. 4 (b, c). The simulated FeFET has a large hysteresis window for the realization of lateral divisive inhibition.

Lateral Divisive Inhibition

By increasing the input pulse current of N2 in Fig. 3, which represents the stimulus intensity received by N2, the degree of lateral divisive inhibition on N1 also increases and results in lower spiking frequency of N1, as shown in Fig. 5.

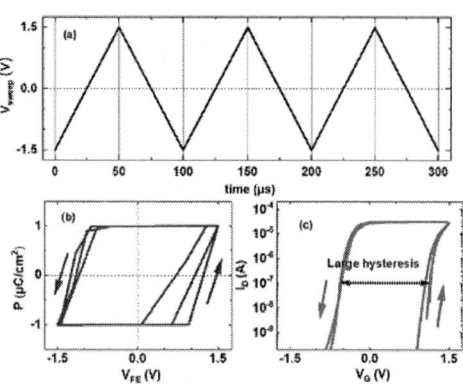

Figure 4: (a) Sweep voltage waveform, and the corresponding (b) P-V loop of the standalone FE layer and (c) I-V curve of the FeFET.

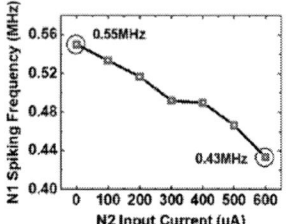

Figure 5: N1 spiking frequency as a function of N2 input pulse current.

Take two situations in Fig. 5 as examples, and the corresponding performance are shown in Fig. 6. When the input pulse current of N2 increases from 0 to 600μA, the average spiking frequency of N1 is lowered from 0.55MHz to 0.43MHz (Fig. 5). Besides, the slower integration and spiking frequency in Fig. 6 (b, d) shows that the inhibition is gradually strengthened with time due to the increased polarization of FE layer in FeFET induced by N2, which mimics the dynamic change of coupling level between neurons.

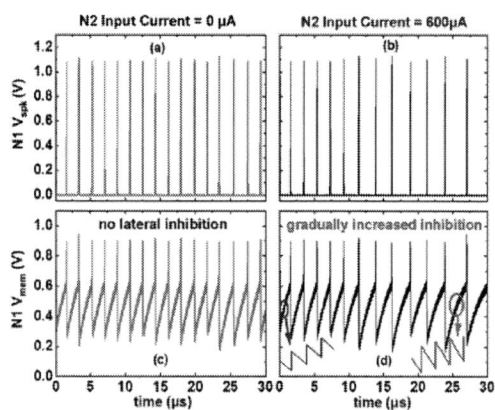

Figure 6: Comparison of V_{spk} (a, b) and V_{mem} (c, d) of N1 when input pulse current of N2 is 0µA (a, c) and 600µA (b, d).

Leaking

When there is no input stimulus to N2 and the FE layer of FeFET is preprogrammed to the positive remnant polarization (P_r) state, the FeFET can function as the leaky part in LIF neuron. Since the V_{mem} of N2 keeps at 0V (V_{rest}) without input, the conductance of FeFET keeps constant, satisfying the requirement of leaky term in Eq. (1). According to Fig. 7, when the initial preprogrammed state of FE layer is P_r, the internal gate voltage (V_{int}) of FeFET is around 0.5V, slightly higher than the threshold voltage (V_{th}) of MOSFET (0.4V), which leads to the obvious and appropriate leakage for the integration in LIF neuron (Fig. 8).

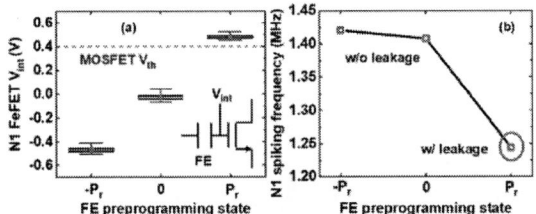

Figure 7: (a) Internal gate voltage of FeFET and (b) N1 spiking frequency with no N2 input as a function of preprogrammed polarization states of the FE layer.

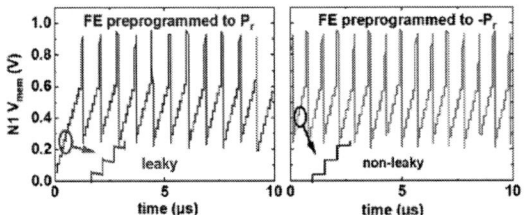

Figure 8: Comparison of V_{mem} waveform of N1 with no N2 input when the FE layer in FeFET is preprogrammed to P_r and $-P_r$ state.

Spiking Neural Network

In this work, the two-layer SNN with lateral divisive inhibition based on Eq. (1) is built and the simulation results are obtained by using MATLAB. The first layer with 10×10 neurons is fully connected to five output neurons in the second layer. Five patterns with 10×10 pixels in Fig. 9 (a) are the input to the first layer. After the training of SNN using spike-time dependent plasticity (STDP), the synaptic weights for each output neurons (Fig. 9 (b)) show specialization to each input pattern, which demonstrates that the SNN with lateral divisive inhibition has the ability to learn and recognize input patterns.

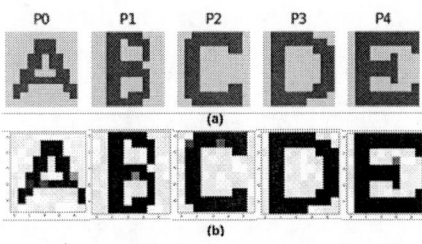

Figure 9: (a) Input pattern for training and (b) synaptic weights for each output neuron after learning.

CONCLUSION

In this work, a novel hardware design to implement lateral divisive inhibition as well as leaking in a bio-plausible way is proposed by using only one FeFET, and the bio-functions are demonstrated through simulations. Furthermore, a SNN with lateral divisive inhibition for recognition task is also demonstrated, showing its great potential for neuromorphic computing.

ACKNOWLEDGEMENTS

This work was supported by National Science and Technology Major Project (2017ZX02315001-004), NSFC (61421005, 61822401, 61851401) and 111 Project (B18001).

REFERENCES

[1] Koyama, Minoru, and Avinash Pujala. Current opinion in neurobiology 49 (2018): 69-74.

[2] Burke, Dennis A., Horacio G. Rotstein, and Veronica A. Alvarez. Neuron 96.2 (2017): 267-284.

[3] Doiron, Brent, et al. Neural computation 13.1 (2001): 227-248.

[4] Miller, Paul. F1000Research 5 (2016).

[5] Hiratani, Naoki, and Tomoki Fukai. PLoS computational biology 11.4 (2015): e1004227.

[6] Arthur, John V., and Kwabena A. Boahen. IEEE Transactions on Neural Networks 18.6 (2007): 1815-1825.

[7] Liu, Muqing, Luke R. Everson, and Chris H. Kim. 2017 IEEE Custom Integrated Circuits Conference (CICC). IEEE, 2017.

[8] Hassan, Naimul, et al. Journal of Applied Physics 124.15 (2018): 152127.

[9] Ayaz, Asli, and Frances S. Chance. Journal of neurophysiology 101.2 (2009): 958-968.

[10] Chen, Cheng, et al. 2019 Symposium on VLSI Technology. IEEE, 2019.

[11] Luo, Jin, et al. 2019 IEEE International Electron Devices Meeting (IEDM). IEEE, 2019.

[12] Ni, Kai, et al. 2018 IEEE Symposium on VLSI Technology. IEEE, 2018.

[13] Van Schaik, André. Neural networks 14.6-7 (2001): 617-628.

High Yield, Superior Quality and Reliability of IGBT and Power Devices in the Artificial Intelligence Era

Min-hwa Chi, Alicia Ding*,Kong Tjien Lim*, and Richard Chang
SiEn (QinDao) Integrated Circuits Cor., ShanDong, China 266500
*ThermoFisher Scientific, Shanghai, China 201203
Email: minhwa.chi@sienidm.com

ABSTRACT

Modern artificial intelligence and the internet-of-things (AI/IoT) drive many aspects of semiconductor technologies. Power ICs and devices need to achieve superior quality and reliability to match the high performance of AI/IoT systems. This is achieved by utilizing AI technology extensively throughout power device design and manufacturing, as well as in multi-level co-optimization, advanced process monitoring, failure analysis (FA) and 3D packaging.

INTRODUCTION

Artificial intelligence and the internet of things (AI/IoT) are becoming a significant driving force in the future of IC development. AI/IoT hardware and software [1] serve a wide range of complex functions such sensing, computing and transmission **(Figure 1 and Figure 2)**. This complex IC hardware/software must allow for fast upgrades and frequent revisions during the rapid cycles of development and manufacturing. Similarly, Power ICs and devices for high performance AI/IoT systems (e.g. data centers, smart cars, autonomous driving, robotics, industry 4.0, etc.) also need to achieve superior quality and reliability; this is possible with AI technology integrated into power devices for design/manufacturing, multi-level co-optimization, advanced process monitors, failure analysis, and 3D packaging. Thus, this paper briefly reviews those strategies, specifically for power devices (i.e. insulated-gate bipolar transistors, or IGBTs).

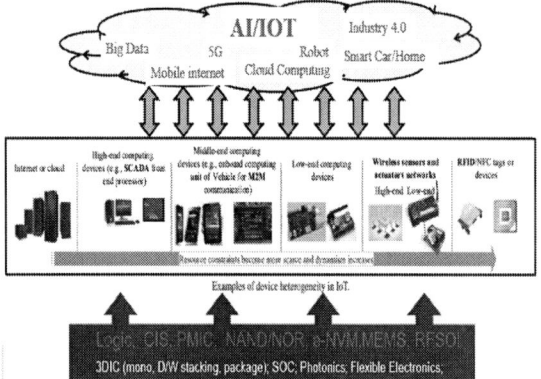

Figure 1: Fast IC development is needed for the AI/IoT era [1].

Figure 2: This complex AI chip (Nvidia 2017) is enabled by multi-level co-optimization, achieving high performance, yield, and reliability [2-3].

POWER DEVICES AND IGBTS

Si Power Devices

Silicon-based power devices (such as thyristors, TRIACs, GTOs, bipolar-junction transistors (BJTs), power MOSFETs, insulated-gate bipolar transistors (IGBTs) and integrated gate-commutated thyristors (IGCTs)) are already used pervasively, in a wide range of long-established applications. The device landscape is evolving as insulated-gate power MOSFETs and IGBTs are becoming adopted in a variety of both high and low power applications.

GTOs have traditionally been the popular choice for high-power multi-mega-watt applications, but are now becoming obsolete, being replaced by IGBTs (in low end applications) and IGCTs (in higher end applications). Furthermore, state-of-the-art IGBTs have gradually moved to high power applications, competing in some areas with IGCTs. Currently, Si-based power MOSFETs and IGBTs are the most widely used devices in power modules based on CMOS technology.

SiC and GaN Power Devices

Wide-bandgap materials, such as silicon carbide

(SiC) and gallium nitride (GaN), have been widely adopted over silicon due to their superior material properties and promise in future high-power applications [4-5] (**Figure 3**).

Material	Bandgap (eV)	Mobility (cm²/V*s)	Permittivity	Vsat (cm/s)	Critical field (V/cm)
Si	1.1	1400	11.8	1×10^7	3×10^5
GaAs	1.42	8500	12.8	2×10^7	4×10^5
4H-SiC	3.23	260	9.7	2×10^7	2.9×10^6
GaN (Bulk)	3.4	900	9	2.5×10^7	3.3×10^6
GaN (HEMT)	3.4	1800	9	2.5×10^7	3.3×10^6

Figure 3: A comparison of wide bandgap materials with silicon [4].

Generally, high-power high-voltage IGBTs have the advantages of smart power capability with control circuit integration as well as fault protection features, but they also suffer from a larger conduction drop. In comparison, new wide-band-gap materials (e.g. SiC and GaN) have a high electrical breakdown as well as high electrical and thermal conductivities, thereby creating power devices with better performance (higher voltage/power capability, higher switching frequency, lower conduction drop, higher junction temperature, and better radiation hardness). Overall, this results in increased integration of power electronics with improved efficiency and less required cooling. Currently, SiC-based power MOSFETs are expected to replace most Si-based devices in future. The power devices based on wide-bandgap materials will be widespread in future high-power applications.

IGBT

The IGBT is an integrated structure combining a power MOSFET with a bipolar transistor, which was developed circa 1980 as a superior alternative to bipolar power transistors. In the IGBT [6], the power MOSFET structure provides the base current to the integrated p-n-p bipolar power transistor, which, in turn, can modulate the conductivity of the drift region for the MOSFET structure, resulting in low forward voltage drop. In recent years, IGBTs have been applied to most power electronics applications, especially medium- and high-power equipment such as AC drive motion controls, uninterruptible power sources (UPS), and renewable energy equipment (wind/solar). Over the last 25 years, IGBTs have been leading power electronics equipment control through improvements in the IGBT chips and packaging. The evolution of IGBT structures is illustrated in **Figure 4**; a state-of-the-art IGBT with process flow and critical modules is illustrated in **Figure**

5.

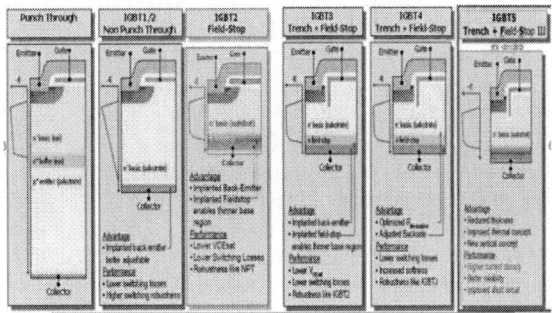

Figure 4: The evolution of early to state-of-the-art IGBT structures [7].

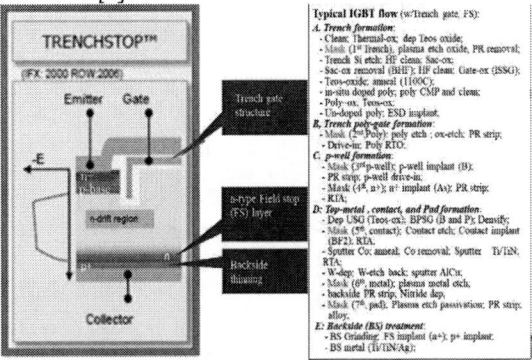

from IFX web-site

Figure 5: (a) Modern IGBTs have a set of advanced features such as trench-gate (TG), field-stop (FS), and backside thinning and metal sintering for optimized performance [7]. (b) A simplified IGBT process flow.

Power ICs for AI/IoT

The emergence of AI/IoT has caused a major change in power electronics and devices, leading to a new generation of power electronics (commonly referred to as Power Electronics 2.0) with the integration of sensors, programmable hardware, and microcontroller units (MCUs) into the power device/module (**Figure 6**) [8]. The power performance of devices and systems can be dynamically sensed, adjusted, and reconfigured for the best overall systems performance and lowest energy consumption.

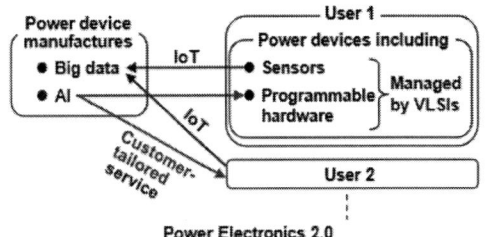

Figure 6: Illustration of new power electronics 2.0 [8] with sensors and programmable features provided by

advances in AT/IoT.

PLATFORMS ON THE CLOUD FOR DESIGN AND MANUFACTURING

Similar to very-large-scale integration (VLSI) IC design, power device and power management integrated circuits (PMICs) can be designed on technology platforms (PDK, models, fabrication data) with cloud-based electronic design automation (EDA) tools (design, simulations, technology CAD, layouts, reference flows, etc.). Prototyping in fully automated fabrication is essential for fast pace; product debug, FA, and yield enhancement are critical and are effectively accelerated by performing big-data analysis on fab data (i.e. in-line process monitors, WAT, yield data, etc.) (**Figure 7**). The massive quantity of data collected from metrology and process monitors provides a new challenge in product debug, failure analysis and yield enhancement. AI techniques are increasingly adopted for a range of analysis, including multi-level co-optimization, adjustment of process margins of IPs, failure analysis, and yield enhancement. Simulations calibrated against big data can be useful, but deep learning and judgement by AI techniques may provide better accuracy and a faster pace.

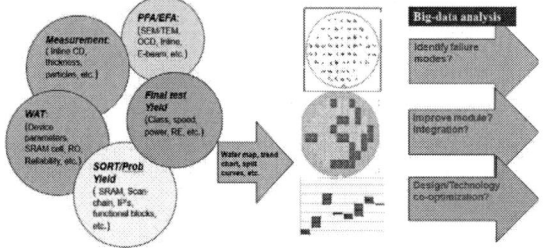

Figure 7: AI techniques are adopted in all aspects of IC design and manufacturing [9]. Big data analysis with deep learning can accelerate product debug, FA, and yield enhancement.

FAILURE ANALYSIS (FA) WITH AI TECHNOLOGY AND AUTOMATION

Yield debug and failure analysis of VLSI CMOS already employs AI techniques on the fab's large database (e.g. layout hot-spot analysis, in-line and electrical monitors, WAT data, yield, wafer map, etc.). In general, power devices (e.g. IGBTs) have additional and unique yield and reliability issues. These unique challenges, as well as a state-of-the-art FA protocol and methodology, are discussed in this section.

Simplified IGBT Process Flow and Issues

A typical IGBT cross-section is illustrated in **Figure 5** with process flow and critical modules (e.g. trench-gate formation, back-side (BS) thinning and metal sintering, as well as field-stop (FS) layer formation). The trench sidewall angle (~87−88°) and bottom corner profiles are critical for Vt and stress near the trench bottom corners, which impact the dopant diffusion and carrier mobility. The uniformity of trench CD, depth, and sidewall slope, as well as drift and FS layer thickness are all critical for an IGBT's reliability and performance. Product and reliability failures are correlated with invisible local stress and plasma damage caused by production process steps, like plasma etching and photoresist ashing, leading to aging, leakage, pre-mature breakdown of IGBTs. These process steps must induce minimal damage and to be monitored in-line.

Local Stress

Mechanical stress is a critical parameter in VLSI/power devices, altering carrier mobility and significantly impacting device performance. Currently in-line stress metrology is performed only at a monitor wafer level, with stress in active devices indirectly implied by electrical data. A new technique, called scanning surface photo-voltage microscopy (SSPVM) [10], has been used to measure mechanical stress in active Si devices with high spatial resolution. Stress analysis at the device level, such as converging electron beam diffraction (CEBD) [11], requires intensive and time-consuming crystalline analysis. The recent development of nano beam diffraction (NBD) [12] has enabled the measurement of mechanical stress in active Si devices with high spatial resolution. This is possible due to powerful, fully automated FIB sample preparation [13].

Plasma Damage

Plasma damage to the Si junction and gate/poly dielectric may occur during IGBT process steps such as trench Si plasma etching and photoresist ashing (O_2 plasma). This damage is often correlated to the drift of the cell Vt, junction leakage and early electric breakdown. Modern hot-spot detection and precise localization (**Figure 8**) as well as SEM/STEM cross-sectioning techniques [11] are effectively used as end-of-line characterization and monitoring methods (**Figure 9**). Direct photo-emission monitoring [12] has recently been demonstrated during an IGBT avalanche phenomena (only at the test wafer level). High energy hydrogen and helium ion implants [16-17] have been used to create H-related donors with a similar role to field-stop layers for optimizing electrical performance. This results in a better breakdown voltage and switching speed. An Si-rich dielectric may reduce charge traps and improve uniformity.

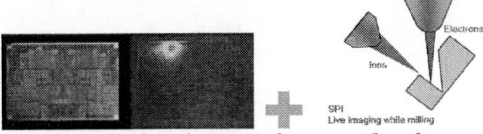

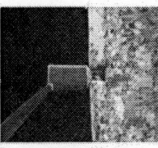

Figure 8: Modern hot-spot detection (by photoemission) and FIB/SEM cross-sectioning techniques.

AI-Assisted Failure Analysis

Electrical failure analysis (EFA) can be performed by characterization of electrical data through big-data analysis in the cloud. Primary physical failure analysis (PFA) techniques include delayering, AFM, nano-probing, SEM, and TEM. A novel delayering technique with nanoscale precision has recently been developed that uses normal xenon ion-beam milling combined with D_X chemistry (PFIB+D_X) on a plasma FIB DualBeam [18]. Site specificity, precise control, >100 µm square DUT area, nm-scale resolution and nm-scale topography have been demonstrated. This technique is now being widely used in a variety of nano-probing and memory metrology applications.

A state-of-the-art wafer fab PFA workflow adopts full automation and advanced AI techniques, as seen in **Figure 9**. Fully automated SEM/STEM sample preparation [13] can provide unattended preparation of multiple FIB/TEM samples or cross sections, leading to superior convenience and accuracy for fast product debug, failure analysis, and yield ramp. Future PFA may be performed remotely through AI/IoT technologies using robotics and 5G communication.

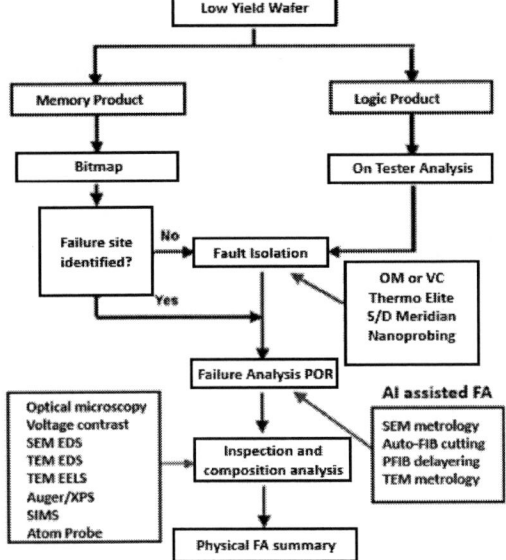

Figure 9: Workflow with fully automated SEM/STEM and unattended preparation of multiple FIB/TEM samples or cross sections. This leads to superior convenience and accuracy for fast product debug, failure analysis, and yield ramp [13].

CONCLUSIONS

The ICs for AI/IoT must have a range of capabilities and a smart integration scheme. Similarly, power ICs and devices must also achieve superior quality/reliability for high performance AI/IoT systems, which can be achieved with AI integration into design/manufacturing, multi-level co-optimization, advanced process monitors, failure analysis, and 3D packaging. Modern IGBT has critical modules of trench-gate (TG) formation, back-side (BS) thinning, and field-stop (FS) layer formation. The trench sidewall angle (~87–88°) and bottom corner profiles are critical for good performance and yield; the uniformity of trench CD/depth and sidewall slope, drift/FS layer thickness are all important for IGBT's reliability. Minimizing the invisible local stress and plasma damage from process steps can eliminate reliability failures such as aging, leakage, and pre-mature breakdown. Automated FIB-assisted NBD is suitable as a production process monitor for measuring mechanical stress in active Si devices with nano-scale resolution. Modern hot-spot detection, precise localization, and automatic SEM/STEM cross-sectioning techniques are effectively used as characterization and monitor methods. Future FA can be performed remotely through AI/IoT technologies and 5G communication.

ACKNOWLEDGEMENTS

The authors would like to thank Thermo Fisher Scientific Semiconductor Team: Paul Kirby, Alex Ilitchev, Debbora Ahlgren, Bryan Chuang, Terri Shofner, Smith Gao and Roger Alvis, for contributions and support on this paper.

REFERENCES

[1] M.A. Razzaque, et.al., IEEE Internet of Things, J. V.3, No. 1, p.70 (2016). [2] J. R. Hu , et.al., IEEE, VLSI-TSA (2018). [3] A. Sharma, et.al., IEDM, p.147 (2016). [3] J. R. Hu , et.al., IEEE, VLSI-TSA, 2018. [4] L.Spaziani and L.Lu, 30th Intl. Symp. Power Semiconductor Devices & ICs (ISPSD), p.8, 2018. [5] Bimal K. Bose, IEEE Industrial Electronics Magazine (IEM), p.7, IEM, 2009. [6] N.Iwamuro and T.Laska, IEEE ED, V.64, No.3, p.741, 2017. [7]

http:\\www.Infineon.com [8] M. Takamiya, **et.al.**, paper#1-2, ISPSD, p.29, 2017. [9] McKinsey & Company, "An Executive's Guide to AI", 2018. [10] D. Dahanayaka, et.al., 43th ISTFA, 2017. [11] Yutaka Wakayama, et.al., Jpn. J. Appl. Phys. Vol 46, 1997. [12] C. B. Vartuli1, **et.al.**, Microsc Microanal 13 (Suppl 2), 2007. [13] Anna Prokhodtseva, **et.al.**, "AutoTEM5-Fully Automated TEM Sample Preparation For Everyone", M & M, 2019. [14] Thermofisher Scientific: TAG03, June, 2019. [15] T.Matsudai1, **et.al.**, Paper# FA3, IRPS, 2016. [16] A.K.G. Wachutka, **et.al.**, 29th ISPSD, paper#HV-16, p.163, 2017. [17] S.Cheng, et.al., IIT, 2014. [18] R. Alvis, **et.al.**, ISTFA, p.393, 2015.

TRUE RANDOM NUMBER GENERATOR (TRNG) FOR SECURE COMMUNICATIONS IN THE ERA OF IOT

Zhigang Ji[1]*, James Brown[2], and Jianfu Zhang[2]

[1]National Key Laboratory of Science and Technology on Micro/Nano Fabrication, Shanghai Jiaotong University, Shanghai, 200240, P. R. China

[2] School of Engineering, Liverpool John Moores University, Liverpool, UK

*Corresponding Author's Email: zhigangji@sjtu.edu.cn

ABSTRACT

True Random number Generator (TRNG) is critical for secure communications. In this work, we explain in details regarding our recent solution on TRNG using random telegraph noise (RTN) including the benefits and the disadvantages. Security check is performed using the NIST randomness tests for both the RTN-based TRNG and various conventional pseudo random umber generator. The newly-proposed design shows excellent randomness, power consumption, low design complexity, small area and high speed, making it a suitable candidate for future cryptographically secured applications within the internet of things.

INTRODUCTION

The Internet of Things (IoT) is a fast-growing market, in which billions of devices are going to be connected. Such an increasingly connected society demands highly-secured ways to protect data, infrastructure and citizens or the fallout could be catastrophic. Implementing cryptographic security is complex and often counterintuitive. One of the most difficult challenges is creating a source of random numbers, the heart of most security systems. Random number generators are of paramount importance in these security applications since they are commonly combined with other primitives to generate encryption keys in almost all the security standards [1]. The confidentiality of the data will be under threat if weak algorithm-based pseudo RNGs (PRNG) are selected. Therefore, a hardware-based true random number generator (TRNG) is urgently required to be integrated with IoT devices for real-time protection. Apart of the security domain, the random numbers also have a wider range of applications, such as the random matrices generation in machine learning hardware [2] and rounding in stochastic computation [3]. It can be anticipated that TRNG will become one inevitable module in future IoT systems.

The standalone TRNG products emerged in the market since 2006. Currently, the major commercial players are mainly from small business such as ID Quantique in Switzerland, Entropykey in UK, TectroLabs and Comscire both in U.S. Large leading semiconductor companies, such as Intel and ARM also invest in this area. However, the products are currently only integrated into their own chips, such as Intel Ivy Bridge-EP. The hardware security market was valued at $520.3 Million in 2016 and is expected to reach $1,101.2 Million by 2022, at a CAGR of 12.87% during the forecast period. TRNG, as one of its core module, is expected to boost accordingly. Existing TRNG harvest natural noise sources in the circuit such as thermal and flicker noise [4] using various types of analog circuits, such as the resistor-amplifier-A/D converter chains or metastable elements with capacitive feedback [5].

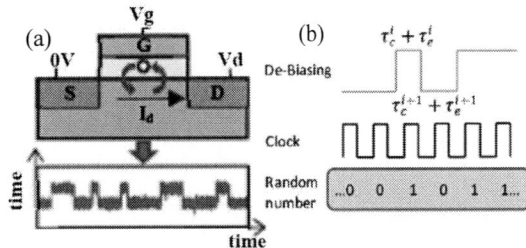

Fig. 1 Typical principle for RTN-based true random number generation. **(a)** The scholastic process of trapping and detrapping is used as the entropy source and **(b)** it can be harvested through digitization **[6]**. The post processing algorithm is usually applied.

With the aggressive scaling of CMOS technologies into nano-scale, only a handful of pre-existing traps can exist within the dielectric material of a transistor. Such pre-existing traps can communicate with the channel of the transistor by capturing and emitting electrons in a random manner, creating the so-called Random Telegraph Noise (RTN). As shown in **Fig.1**. RTN moderates the threshold voltage of the transistor. As a result, drain current exhibits two discrete levels in its magnitude, which can be readily used as a binary representation of 0 and 1 respectively. This physical phenomenon allows us to harvest entropy directly from one single nano-scaled transistor and the bulky analog circuit used in existing TRNG products can be avoided. The recent study [6] also showed that the averaged capture and emission probabilities of one RTN signal that is associated with a given trap are constant and independent of the applied voltage, which provides naturally un-biased randomness. Therefore, by repeatedly toggling after every consecutive capture and emission events, the bit stream of 0s and 1s with equal probability can be readily obtained, making it truly random. This removes the need of extra circuits for post-processing, which not only reduces the design area, but also the power consumption. In this work, we discuss

978-1-7281-6559-2/20 $31.00 © 2020 IEEE

the use of generated RTN as the entropy source for the TRNG [7]. We show that the use of generated RTN in TRNG can help tackle the challenges in reliability, design area and security, while still maintaining the low power consumption. It is expected that this proposed solution will provide the market with a TRNG that has improved reliability and security, while reducing the power consumption and cost as required in the IoT environments.

PSEUDO-RANDOM NUMBER

Conventionally, random numbers can be generated using mathematical algorithms. They are deterministic and usually named as the Pseudo-random number generators (PRNG). Such PRNGs are constructed with various non-linear relationships and some widely-used ones are Inversive congruential random number generator (ICR-RNG), linear congruential random number generator (LCG-RNG) and Primitive-root-Modulo-n random number generator (Modn1-RNG). For example, ICR-RNG starts with selecting a seed as the first number and then uses the modular multiplicative inverse to generate the next number. By repeating this procedure, a random sequence can be formed. The security relies on the non-linearity used in the algorithm. Similar to ICR-RNG, LCG-RNG makes use of the discontinuous piecewise linear equation. Although they are easy for the hardware implementation, such mathematical-based PRNGs are generally weak. Therefore in practice, the required seed is achieved by harvesting from environments such as noise, gestures etc. Apart of the pure mathematical-based PRNGs, another category of PRNG constructs the finite state machine in the hardware level. It is easy to imagine that by randomly going into a certain state, it is possible to gain the random sequences. In hardware level, such idea can be implemented with a linear-feedback shift registers. With a seed value to initialize LFSR, a random-looking pattern can be produced. This seed value typically comes from events such as user-interface interrupts, page-faults, incoming TCP/IP requests, kernel system calls and the recent sensor outputs. The seed has been known to be vulnerable to manipulation by malicious attackers, resulting in security loopholes.

TRUE RANDOM NUMBER

For demonstration purpose, the nFinFETs fabricated with a 22nm FinFET CMOS technology are used. The gate dielectric stack consists of a Hafnium oxide, a SiON interfacial layer and TiN metal gate. Our proposed solution leviages the trap from generation. The generated traps are distributed randomly in both energy within the Si bandgap and spatially across the dielectric, which controls capture and emission probability.

By stressing the transistors with the voltage higher than operating condition, the new traps can be generated within

short time through either breaking hydrogen bond or converting some kinds of precursors. These generated traps can also induce RTN, like pre-existing traps. This provides a pathway to insert RTN into nano-scaled device. Several stress modes can introduce trap generation, such as Hot Carrier Injection (HCI) and Bias Temperature instability (BTI). NBTI is uniform stress and therefore independent of device geometry. HCI stress is non-uniform and can be enhanced with shorter device channel length. It has been reported that, for advanced CMOS technologies, HCI can be more effective than BTI in generating new defects [8]. In this work, HCI is applied. Unlike conventional large-area devices, the HCI of nano-scaled devices is driven by carrier energy and carrier–carrier interaction. As a consequence, the most effective way in trap generation occurs under Vg = Vd with other two terminals grounded. The use of high temperature is to further accelerate the generation. **Fig.2a** shows the result from one nFETs at fresh, after stressing for 50s, and after recovery for 50s.

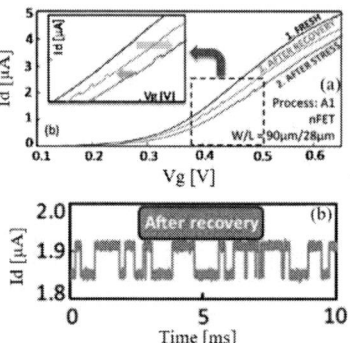

Fig. 2 (a) Id-Vg curves at fresh, after hot carrier stress and after 50s recovery. RTN generation can be clearly observed. **(b)** The RTN measurement performed on the device after recovery.

After stress, a newly generated trap can be clearly observed in the Id-Vg sweep. It is well-known that the electrical stress can not only generate new defects, it will also trigger new pre-existing traps. Unlike the generated traps, the triggered pre-existing traps is usually not stable and will disappear after the stress is removed. Therefore, we carried out the recovery at 0V for 50s on the heavily-stressed device and found that the RTN signal still exist, supporting that the true generated trap has been obtained. **Fig.2b** shows the RTN measurement on the device after recovery. In practice, this procedure should be applicable to most nFETs in one wafer. Therefore, we repeated the proposed procedure on 100 devices. The stability of the generated RTN signal was evaluated by floating the stressed nFETs at 125°C for over 6 months. The statistical summary is shown in **Fig.3a**: 81% FETs exhibit clear RTNs after applying the proposed procedure. After 6 months, RTN in 13% samples disappeared. For the rest of the 68% FETs, clear RTN can still be observed and their property did not change.

Existing TRNG design [6] assumes that the entropy source comes from the single-trap-induced RTN. However, in reality, this may not be true. Many fresh device can exhibit multi-level RTN, which is caused by the superposition of RTN signals from multiple traps. This can make the signal digitalization more difficult. The threshold scheme has been proposed to overcome this problem, however, it can introduce the design complicity. After stressing, it was also possible to generate more than one trap in a device, resulting in multi-level generated RTN. **Fig.3b** shows the proportion of devices that exhibited single or multi-level RTN. Due to its small percentage, in practice, devices showing multi-RTN can be discarded.

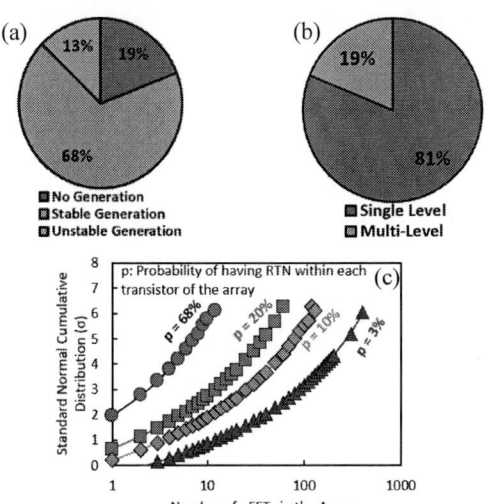

Fig. 3 (a) Percentage of different statuses for the FETs, under selected driving voltages, after applying the procedure in Fig.2. **(b)** The percentage of nFETs that showed single level and multi-level RTN after generation **(c)** The probability of finding at least one nFET with stable RTN in the nFET array.

The typical TRNG utilizes a transistor array to ensure that at least one transistor can have clear RTN that will be used as the entropy source. Based on the RTN occurrence probability in one nFET, the Monte-Carlo simulation is performed to explore the minimum number of nFETs required in the array: for a round of simulation, we generate M 1s and 0s randomly to represent the same number of transistors with or without RTN in the array. The working TRNG should have at least one '1'. By repeating this procedure for 106 times, the percentage of the working TRNG can be obtained, as shown in **Fig.3c**. It increases with the array size, because more FETs can be selected in the larger array. The minimum number of nFETs is determined by the probability of RTN occurrence of the nFET. With 68%, as we achieved through generation, 10 FETs is required to reach a yield of 6sigma.

The temporal behaviour of RTN for the pre-existing traps on fresh devices has been extensively investigated in the past few years. Apart of the normal RTN behaviour, various anomalous RTN have been observed in fresh devices due to the complex trap configurations, where the fast switching process can suddenly appear or disappear, as shown in **Fig.4a&b**. The 'quiet' phase in these anomalous RTNs cannot provide high entropy, and therefore, they are not the good entropy source for TRNG design. However, for the generated traps, it is found that all of them exhibit clear normal RTN. This suggests that the generated traps have simpler physical configurations than the pre-existing ones. Therefore, once a new trap is successfully generated, it can be used as a high-quality entropy source.

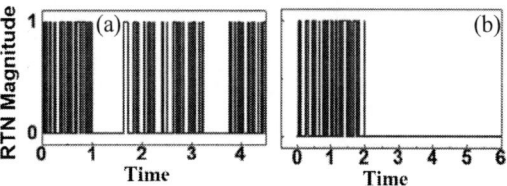

Fig. 4 Two types of anomalous RTNs from pre-existing traps on fresh devices. RTN **(a)** temporally stopped and **(b)** permanently stopped. For both cases, 'quiet' period reduces speed and security.

RTN needs to be amplified and digitalized before it can be used in TRNG design. Therefore, it will be preferable that the RTN signals are with larger magnitude. Although the average value of the RTN-induced-Vth becomes larger than the charge sheet approximation and increases with the shrinking of the device size, most of them can be still relatively small for fresh devices, because of the randomly distribution on top of the substrate [9]. On the contrary, the generation can only happen when the hot electrons injected into the dielectric. Most of the carriers located within the percolation path, and there is much higher chance for them to generation new traps on its top [10]. Therefore, the RTN magnitude from the generated traps can be larger than the pre-existing traps in fresh devices.

The speed of the RTN-based TRNG relies on the characteristic time of the trap. Although pre-existing traps can locate close to the channel/dielectric interface, most the generated traps locate away from the channel and therefore, in general, the pre-existing traps can be faster than the generated traps. The deficiency in the speed can be minimized by applying AC mode, in which both τ_c and τ_e can be adjusted by using Vg at high and low levels [7].

RANDOMNESS VALIDATION

One example of true random number generated with our entropy source is used to quantify the randomness. The generated bit stream was populated into the National Institute of Standards and Technology (NIST) randomness statistical test suite. For statistically significant results, the bit stream was divided into 56 sequences. The significance level (α) was set to 0.01 as recommended by NIST [11]. Table I tabulates the results from the NIST tests for the 15 randomness tests. In all cases, the bit stream are checked to be random.

Test ID	Name	Proportion	Pass/Fail
T1	Monobit test	99/100	Pass
T2	Frequency tests within a block	100/100	Pass
T3	Cumulative sums Test	96/100	Pass
T4	Run Test	99/100	Pass
T5	Longest Run of Ones in a block	98/100	Pass
T6	Discrete Fourier Transform Test	49/50	Pass
T7	Non-overlapping Template Matching	na	Pass
T8	Approximate Entropy Test	96/100	Pass
T9	Serial Test	100/100	Pass
T10	Binary Matrix Rank	99/100	Pass
T11	Overlapping Template Matching	8/10	Pass
T12	Universal	8/10	Pass
T13	Linear Complexity	10/10	Pass
T14	Random Excursions	n/a	Pass
T15	Random Excursions Variant	n/a	Pass

Many popular mathematical algorithms which generates pseudo-random number have been turn out to be flawed, delivering a predictable output. **Table 2** also compares the randomness from 4 different popular PRNGs including ICG, LCG, ModN1 and the PRNG generated from Security library. It is clear that none of them can pass the rigorous randomness tests. Therefore, the development of RTN-TRNG for IoT applications is essential.

TABLE II Results summary from the NIST tests for several PRNGs including ICG and LCG and ModN1. All the parameters set in the NIST test are the same as the ones used in TABLE I.

	ICG		LCG		ModN1	
	P	% pass rate	P	% pass rate	P	% pass rate
T1	0.3838	100%	0.6163	10%	0.6787	100%
T2	0.6993	100%	0.0060	81%	0.4944	99%
T3	0.3345	99%	0.0000	6%	0.5544	100%
T4	0.4944	98%	0.0902	97%	0.4012	97%
T5	0.7399	100%	0.9114	100%	0.5341	100%
T6	0.6993	100%	0.4559	98%	0.9558	100%
T7	0.0043	60%	0.0089	70%	0.7399	100%
T8	0.5228	100%	0.0000	0%	0.8935	100%
T9	0.6842	100%	0.0007	0%	0.9442	100%
T10	0.6456	100%	0.9908	100%	0.9789	100%
T11	0.3041	100%	0.0220	98%	0.1296	99%
T12	0.5749	99%	0.0097	89%	0.2133	97%
T13	0.4012	100%	0.4012	99%	0.3191	99%
T14	0.7788	100%	0.2914	100%	0.8338	100%
T15	0.4864	100%	0.2693	100%	0.8220	100%

CONCLUSION

In this work, we investigate the practical aspects of using RTN for TRNG application. We show with experimental evidence that the generated traps are suitable to be used as the entropy source. Both the advantages and disadvantages have been discussed. The use of generated RTN in TRNG can help tackle the challenges in reliability, design area and security, while still maintaining the low power consumption. It is expected that this proposed solution will provide the market with a TRNG that has improved reliability and security, while reducing the power consumption and cost as required in the IoT environments.

ACKNOWLEDGEMENTS

The authors would like to thank imec for providing the test samples.

REFERENCES

[1] D. Eastlake, J. Schiller, and S. Crocker, "Randomness Requirements for Security, " 2005.

[2] A. Krizhevsky, I. Sutskever, and G. E. Hinton, "ImageNet Classification with Deep Convolutional Neural Network," Handb. Approx. Algorithms Metaheuristics, 2007.

[3] A. S. Cassidy et al., "Real-Time Scalable Cortical Computing at 46 Giga-Synaptic OPS/Watt with ~100× Speedup in Time-to-Solution and ~100,000× Reduction in Energy-to-Solution, " Int. Conf. High Perform. Comput. Networking, Storage Anal., 2014.

[4] V. Rozic, B. Yang, W. Dehaene, I. Verbauwhede. "Highly efficient entropy extraction for true random number generators on FPGAs" in IEEE DAC, 2015.

[5] C. Tokunaga, D. Blaauw, and T. Mudge, "True Random Number Generator With a Metastability-Based Quality Control" in IEEE JSSC, 2008.

[6] A. Mohanty, K. B. Sutaria, H. Awano, T. Sato, Y. Cao, "RTN in Scaled Transistors for On-Chip Random Seed Generation" in IEEE VLSI, 2017.

[7] J. Brown, R. Gao, Z. Ji, j. Chen, J. Wu, J. Zhang. B. Zhou, Q. Shi, J. Crowford, W. Zhang, "A low-power and high-speed True Random Number Generator using generated RTN" in IEEE VLSI-T, 2018.

[8] M. Duan, J. Zhang, Z. Ji, W. Zhang, B. Kaczer and A. Asenov, "Key Issues and Solutions for Characterizing Hot Carrier Aging of Nanometer Scale nMOSFETs," IEEE Tran. Electron. Dev. 2017.

[9] S. Dongaonkar, M. D. Giles, A. Kornfeld, B. Grossnickle, J. Yoon, "Random telegraph noise (RTN) in 14nm logic technology: High volume data extraction and analysis," IEEE Symp. VLSI Tech, 2016.

[10] R. Gao, Z. Ji, S. M. Hatta, J. F. Zhang, J. Franco, B. Kaczer, W. Zhang, M. Duan, S. De Gendt, D. Linten, G. Groeseneken, J. Bi and M. Liu, "Predictive As-grown-Generation (A-G) model for BTI-induced device/circuit level variations in nanoscale technology nodes," IEEE IEDM, 2016.

[11] L. E. Bassham, A. L. Rukhin, J. Soto, J. R. Nechvatal, M. E. Smid, S. D. Leigh, M. Vangel, N. A. Heckert, D.

L. Banks, "A Statistical Test Suite for Random and Pseudorandom Number Generators for Cryptographic Applications" NIST, 2010.

YIELD ENHANCEMENT BY VIRTUAL FABRICATION: USING FAILURE BIN CLASSIFICATION, YIELD PREDICTION AND PROCESS WINDOW OPTIMIZATION TO IDENTIFY AND PREVENT PROCESS FAILURES

Qingpeng Wang, Yu De Chen, Jacky Huang, Wuping Liu, Ervin Joseph

Coventor Inc., a Lam Research Company, No.177, Bibo Rd., Pudong New Area, Shanghai, China

*Corresponding Author's Email: Qingpeng.Wang@lamresearch.com

ABSTRACT

This paper provides an example of yield enhancement using virtual fabrication. A 6 transistors based static random access memory example on 7nm node technology was used in this case study. Yield loss caused by via contact-metal edge placement error was modeled and analyzed. The results show that yield can be enhanced from 48.4% to 99.0% through process window optimization and improved specification control. We identified high resistance failure as the top failure mode in both non-optimized and optimized process models.

INTRODUCTION

Semiconductor manufacturers and practitioners are under unprecedented pressure to quickly improve yield for new semiconductor device designs, in order to deliver innovative products to market ahead of their competitors [1]. Device yield is highly dependent upon proper process targeting and variation control of fabrication steps, particularly at advanced nodes with smaller feature sizes. Large process target shifts and variability can directly lead to a hard device failure, such as an open or short on device components. A smaller process target shift or variability can introduce device performance shift which may lead to a soft fail (or performance degradation) due to mismatches between the circuit design and actual silicon performance.

Traditionally, process parameters optimization to meet yield and device specifications need gather large amount of necessary data using non-optimized wafers which is very costly [2]. Fortunately, semiconductor virtual fabrication tools (such as SEMulator3D®) are now available so that these experiments can be executed "virtually". Virtual fabrication with its multiple integrated functions can be used to understand the multi-step interactions and range for process step sensitivity to maximize yield on first lot. It accelerates the time to maximize yield and reduces the number of wafers [3].

In this paper, we demonstrate a simple example of performing yield enhancement by virtual fabrication, using failure bin classification, yield prediction and process window optimization at the 7nm technology node. We will review a specific yield loss caused by BEOL via contact and metal edge placement error.

YIELD ENHANCEMENT AND BIN CLASSIFICATION

A. Edge Placement Error and Failure Bin Analysis

Edge placement error is an important failure mode for BEOL yield loss [4]. Fig. 1 (a) shows an image of a 7nm BEOL VC and M1 top view in the logic area of the device, while Fig. 1(b) displays an image of VC overlap with M1. In Fig. 1(b), the VC overlap to the neighboring M1 window is not large. The contact resistance would be smaller if the VC overlay shifted towards the upper M1 line. Consider a simple circumstance, where M1 is split into metal A (MA) and metal B (MB), and the via contact (VC) is designed to connect with MB (see Fig. 1 (a)). Process variations of metal CD, or VC CD, or metal to VC overlay, will all make the final fabricated device function differently from the original design. Any of these process errors will lead to yield loss due to edge placement errors between vias and metal layers.

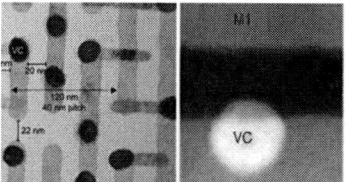

Fig.1. VC and M1 edge placement (a) Logic area, (b) VC and M1 overlap large image [5] (Courtesy: TechInsights).

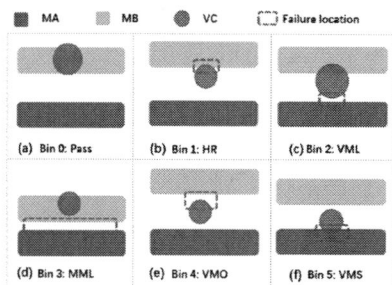

Fig.2. Bin illustration (a) Pass, (b) HR, (c) VML, (d) MML, (e) VMO, (f) VMS.

With different CD and overlay combinations, several failure bins can be defined (see Fig.2):

1. HR: VC and MB overlap area is small;
2. VML: VC to MA space is small (leakage);
3. MML: MA to MB space is small (leakage);
4. VMO: VC is disconnected from MB (open);
5. VMS: VC forms a bridge with MA (short);

978-1-7281-6559-2/20 $31.00 © 2020 IEEE

B. Structure build and calibration, Failure bin Generation and Recognition

To demonstrate the concept of yield enhancement by virtual fabrication, a 7 nm VC and M1 process was constructed virtually using a typical 6 transistors based static random access memory (6T SRAM) layout (see Fig. 3 (a)). M1 was generated by a SADP process using a drawing mandrel. For easy description and measurement, both VC and M1 are split into two groups, VA/VB and MA/MB. VA/MA is located on the mandrel while VB/MB is located on the space of the mandrel (see Fig.3 (b, c, d)) [5]. In the virtual fabrication steps, for easy analysis, VA/VB CD was set at the same value as VC CD, VA/VB had no overlay errors with each other, and no overlay existed between the mandrel and VC in the Y direction. Via CD (VCD), Mandrel CD (MCD), spacer thickness (SPT), and the mandrel to VC overlay in the X direction (MVO) are all established as process variables in the virtual fabrication model.

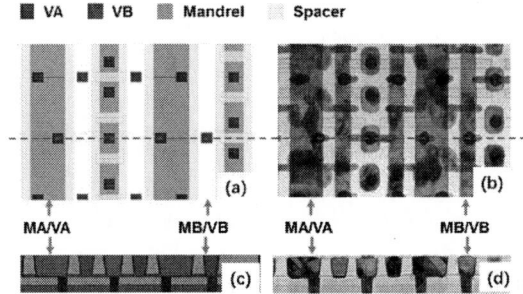

Fig.3. (a) Virtual layout, (b) top view VC/M1 on real Si, (c) cross-sectional view of virtual output structure, (d) cross-sectional view of real Si (Courtesy: TechInsights).

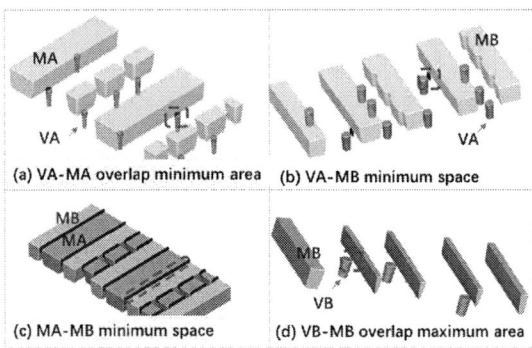

Fig.4. Virtual metrology (a), VA-MA minimum overlap area, (b) VA-MB minimum space, (c) MA-MB minimum space, (d) VB-MB maximum overlap area.

Fig. 4 shows the corresponding measurement positions on the virtual structure that were used to take measurements and classify failures into their appropriate failure bin structures. Fig. 4 (a) shows the minimum

interface area measurement between VA and MA. Fig. 4 (b) shows the minimum space measurement between VA and MB. Fig. 4 (c) shows the minimum space measurement between MA and MB. Fig. 4 (d) shows the maximum interface area between VB and MB. In order to recognize all of the failure bins in Fig.2, 7 virtual metrology were performed. These were:

1. VA-MA minimum overlap area (VAMA Amin);
2. VB-MB minimum overlap area (VBMB Amin);
3. VA-MB maximum overlap area (VAMB Amax);
4. VB-MA maximum overlap area (VBMA Amax);
5. VA-MB minimum space (VAMB Smin);
6. VB-MA minimum space (VBMA Smin);
7. MA-MB minimum space (MAMB Smin).

TABLE I. SPECIFICATION RULE FOR BIN CLASSIFICATION

	Pass	HR	VML	MML	VMO	VMS
VAMAAmin (nm²)	>280	≤280	-	-	=0	-
VBMBAmin (nm²)	>280	≤280	-	-	=0	-
VAMBAmax (nm²)	=0	-	-	-	-	>0
VBMAAmax (nm)	=0	-	-	-	-	>0
VAMBSmin (nm)	>11.2	-	≤11.2	-	-	-
VBMASmin (nm)	>11.2	-	≤11.2	-	-	-
MAMBSmin (nm)	>11.2	-	-	≤11.2	-	-
Boolean operator	AND	OR	OR	OR	OR	OR

Based upon a particular rule, the bin type could be classified using these measurement results. Table I lists the actual measurement values used to classify a virtual manufacturing "run" into a failure bin. For this concept demonstration, we set our failure bin measurement values at 70% of the nominal design values. The designed VA-MA area and MA-MB space is 400 nm² and 16 nm, so the pass specification was set to 280 nm² and 11.2 nm, respectively. For example, during a modeling run, VAMA Amin and VBMB Amin values were interrogated to see if either of these values was smaller than 280 nm², corresponding to an HR bin failure in the virtual structure.

C. Yield Prediction, Bin Count Ranking

In an actual fabrication process, process parameters such as mandrel/Via CDs and overlays are controlled within certain ranges measured against mean (or nominal) and distribution width values. For example, the MCD could be targeted within 36+/-5 nm and MVO could be targeted within 0+/-4 nm. Due to a shift in the mean value, or the normal distribution of CDs and overlays around the mean values, particular combinations of these values could produce a failed structure. SEMulator3D provides an automated method to execute a DOE, and can generate and collect user defined mean values and range width/sigma values. This data can then be used to calculate the pass rate or yield (the ratio of the pass bin to total number of runs at a specified input condition) based the collected data and our yield rule (see Table I).

978-1-7281-6559-2/20 $31.00 © 2020 IEEE

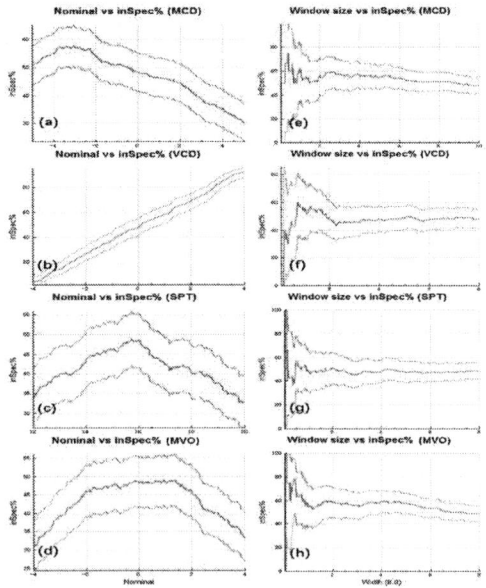

Fig.5. *Yield prediction curves (a) MCD mean, (b) VCD mean, (c) SPT, (d) MVO min, (e) MCD width, (f) VCD width, (g) SPT width, (h) MVO width;*

We executed a DOE with 3000 split virtual runs using Monte Carlo simulation. Fig. 5 (a, b, c, d) displays the yield prediction curves (in-spec %) generated using mean values of MCD, VCD, SPT and MVO that were shifted between 31 to 41nm, 16 to 24nm, 12 to 20nm and -4 to 4 nm, respectively. Fig. 5 (e, f, g, h) shows the yield prediction curves produced by varying the distribution width of MCD, VCD, SPT and MVO. Fig. 6 (a, b) shows a bin summary bar chart and the yield summary table of the bin percentages at four different input conditions. These input conditions included 1 nominal condition (with MCD=36+/-5 nm, VCD=20+/-4 nm, SPT=16+/-4 and MVO=0+/-4 nm) and 3 corner conditions with the same distribution width but varied mean shifts. The chart and table provide us with a quantified failure bin ranking at each particular input setting, useful in understanding the yield impact of specific failure modes.

D. Process window optimization

Using virtual fabrication, yield changes caused by changes in process settings can already be predicted as demonstrated. This analysis might lead to a series of additional questions. Is the predicted yield acceptable, can the nominal and width values be adjusted to gain better yield and etc? A process window optimization (PWO) function in SEMulator3D can answer these optimization questions. Table II displays a yield and process window summary for nominal, optimized, and optimized + tightened SPT width cases. It shows that the yield can be enhanced from 48.4% to 96.6% by simply optimizing the mean shift. We can further tighten the SPT width

specification from 8 to 4 nm to obtain a target yield of 99%. Fig. 7 shows the failure bin percentage Pareto plot of these 3 conditions. We can see that the failure bin percentage after PWO and specification tightening improves compared to the nominal condition. Also, the HR bin failures are the main failure mode.

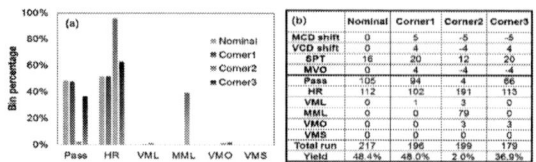

Fig. 6. *Yield status at particular MCD/VCD/MVO setting (a) Bin bar chart, (b) Yield summary;*

TABLE II. YIELD SUMMARY OF DIFFERENT INPUT CONDITIONS.

	Nominal		Optimized Meanshift		Tighten Width	
	Mean shift	Width	Mean shift	Width	Mean shift	Width
MCD	0	5	-2.7	5	-2.7	5
VCD	0	4	4	4	4	4
SPT	16	4	15.6	4	15.6	2
MVO	0	4	0	4	0	4
Yield	48.4%		96.6%		99%	

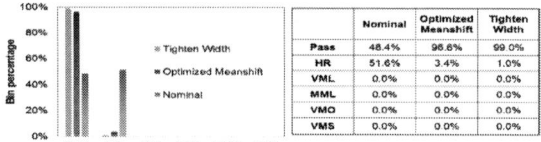

Fig.7. *Failure bin Pareto plot and summary table.*

CONCLUSION

This paper provides insight into yield enhancement through use of virtual fabrication with its multiple functions. We used a 7 nm 6T SRAM model, with edge placement error induced VC-M1 yield loss. The results showed that yield was enhanced from 48.4% to 99.0% after process window optimization and specification tightening, and that high resistance failure in this model was the primary failure mode in both non-optimized and optimized processes. Virtual fabrication can be used in a broad range of yield enhancement research to accelerate semiconductor process and technology development.

REFERENCES

[1] Moore, Gordon E. "Cramming more components onto integrated circuits." (1965): 114-117.

[2] Vincent, Benjamin, et al. "Virtual fabrication and advanced process control improve yield for SAQP process assessment with 16 nm half-pitch." Advanced Etch Technology for Nanopatterning VIII. Vol. 10963. International Society for Optics and Photonics, 2019.

[3] http://www.coventor.com/products/semulator3d

[4] Mulkens, Jan, et al. "Overlay and edge placement control strategies for the 7nm node using EUV and ArF lithography." Extreme Ultraviolet (EUV) Lithography VI. Vol. 9422. International Society for Optics and Photonics, 2015.

[5] 7nm logic tear down report from TechInsights

A Device Design for 5 nm Logic FinFET Technology

Yu Ding[1], Xin Luo[1], Enming Shang[1], Shaojian Hu[1], Shoumian Chen[1] and Yuhang Zhao[1]

[1]Shanghai Integrated Circuit Research and Development Center, No. 497 Gaosi Road,
Pudong New Area, Shanghai, P. R. China

* Corresponding Author's Email: dingyu@icrd.com.cn

ABSTRACT

In this paper, we proposed a 5 nm FINFET device, which is based on typical 5 nm logic design rules. We have performed an optimization on the process parameters and iterate through device simulation with the consideration of current process capability. Based on our preferred device architecture, we provide our process key dimensions, and simulated device DC/AC performance, and some parasitic parameters. As a part of the final evaluation, Ring Oscillator (RO) simulation result has been checked, which demonstrates that the Performance Per Area (PPA) is close to industry reference 5 nm performance.

INTRODUCTION

As MOSFET scales down, the conventional planar transistor architectures have already reached the fundamental material and process technology limits. Besides, as the size decreases, the device will suffer from the Short Channel Effect (SCE), which can result in severe leakage problem and mobility degradation, so that the effective drive current will drop. The threshold voltage (Vt) will also roll-off. Thus, a high channel doping to control the leakage current is required. However, it has major disadvantages of lower carrier mobility, higher tunneling effect, more degradation in subthreshold performance, and larger parasitic capacitance.

Therefore, the development of small devices with high performance becomes more challenging. Innovative three-dimensional (3-D) structures such as double-, triple-, fin-typed, nanosheet, and nanowire Field Effect Transistors (FET) have been of great interests. FinFET (Figure 1) is one of the most promising device structures to address short-channel effects and leakage issues in deeply nano-scale transistors. FinFET structure mitigates these problems at low channel doping conditions, which also minimizes variations of the Vt, reduces subthreshold leakage current, keeps high carrier mobility, and enhances the drive current. Thus the FinFET structure can be scaled down to 22 nm and beyond. [1-3]

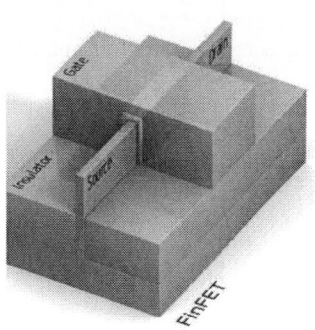

Figure 1. Schematic of FINFET Structure

In this paper, we have developed a 5 nm FINFET structure with TCAD simulation support. In details, we will propose our development procedure, introduce our process optimization (Fin loop), and present a list of key dimensions, and simulated device performance.

DEVICE DEVELOPMENT PROCEDURE

Firstly, we introduce our 5 nm development procedure shown in Figure 2. In the beginning, we set up our device targets through co-work with IMEC and material path-finding study, and then determine the architecture and process conditions. We perform the TCAD simulation with all these conditions ready and get the initial result. Next, combining the above simulation result with Middle-Of-the-Line (MOL) & Back-End-Of-The-Line (BEOL) extracted parasitic parameters, we extract the SPICE model parameters, and take the result into Ring Oscillator (RO) simulation. If the Performance Per Area (PPA) can meet our initial target, then we extract our device data and curves; if not, then as a learning circle, we will continue to optimize our simulation conditions.

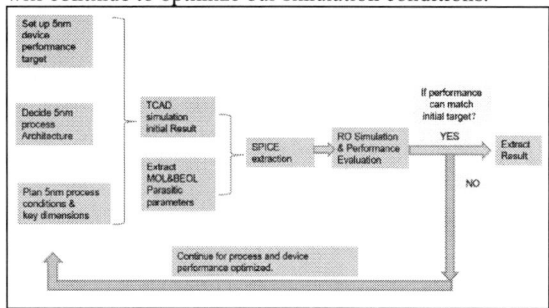

Figure 2. 5 nm device development procedure

PROCESS SIMULATION OF 5nm FINFETS

The Synopsys Sentaurus TCAD simulator was used to perform 3-D simulations, which included the density gradient (DG) solver model with quantum effects. In order to take the high doping concentration of the source and drain region, Band Gap Narrowing (BGN) model was comprised. For the accuracy of the off state current, the Band-To-Band Tunneling (BTBT) model and concentration-dependent Shockley-Read-Hall (SRH) model were considered. The mobility model used in the device simulation were Inversion and Accumulation Layer (IAL) model; Thin layer (TL) model for low-field mobility modulated as very thin Si thickness; ballistic mobility model for compensating small gate-length (below 10 nm); and Matthiessen's rule mobility model including surface acoustic phonon scattering, surface roughness scattering, and bulk mobility with doping-dependent modification effect. High Field Saturation model was also included in this simulation because the carrier drift velocity is no longer proportional to the electric field. [3]

Figure 3 illustrates the device architecture of the simulated FinFET and cross-section view of the device doping profile.

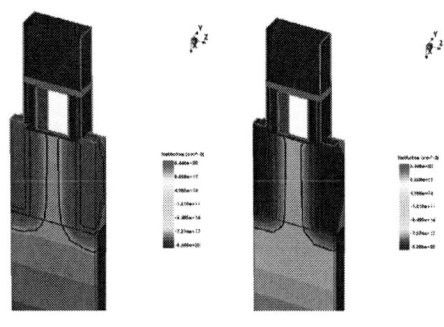

Figure 3. (a) NFET;   (b) PFET

In this simulation, a Fin pitch of 24 nm, a Poly pitch of 50 nm, and a total gate length (LG) of 19 nm have been adopted. We set the top fin-width W_{top}= 5 nm and the fin-height (FH) equal to 50-55 nm, the Fin angle=89° as a standard device; (SiO_2+HfO_2) were considered as the gate oxide materials and low K dielectric film was used as spacer material to reduce parasitic capacitance. Total spacer length of 8 nm was used. The Epitaxial doping concentration was around 1.0E21 in the Source/Drain (S/D) region. Around 15 nm contact (CT) dimension was applied. The key dimensions of different parameters for the simulated device are listed in Table I.

Process Stage	Process Description	Unit	Size
Gate Length	Gate Length	nm	19
Fin Pitch	Fin Pitch	nm	24
Poly Pitch	Poly Pitch	nm	50
TFin	Fin TOP CD	nm	5
Hfin	Fin height	nm	50-55
Afin	Fin angle	°	89°
Gate OX Thickness	Oxide/Thfo2 thickness	A	7/12
Spacer	Spacer total thickness	nm	8
	Spacer Dielectric		4.4
	Epi Trench Depth	nm	50-55
Source/Drain Epitaxy	Epi doping	cm^{-3}	~1E+21
	Epi doping (Germanium)		50%
CT	CT CD	nm	~15

Table I. Process key dimensions used in the simulation

SIMULATION RESULT

In this section, we present our simulated device DC/AC performance in Table II, which indicates that the Drain Induced Barrier Lowering (DIBL) is around ~30mV, Subthreshold Slope (SS) is around 72.7 mV/decade and 67.2 mV/decade for NMOS and PMOS respectively, which are comparable with industry reference performance and demonstrate that our short channel effect has been controlled well [4,6].

The parasitic parameters, like R_{ext}/$R_{channel}$ and capacitance are also acceptable, and further improvement is ongoing.

			NMOS	PMOS
DC Performance	Vtsat	V	0.128	0.106
	Idsat	uA/fin	73.54	94.43
	DIBL	mV	31	32
	SSSat	mV/dec	72.7	67.2
	Ieff_Ioff	uA/fin	41.63	49.86
Parasitic Parameters Extraction	Cgd0	fF/Fin	0.022	0.016
	Cgd	fF/Fin	0.039	0.04
	Rext	ohm/Fin	1261	1555
	Rchannel	ohm/Fin	1539	1880

Table II. The simulated device DC/AC performance result

Figure 4 and Figure 5 show the Id-Vd, Id-Vg, Cgg, Cgd curves for the NFET and PFET, respectively.

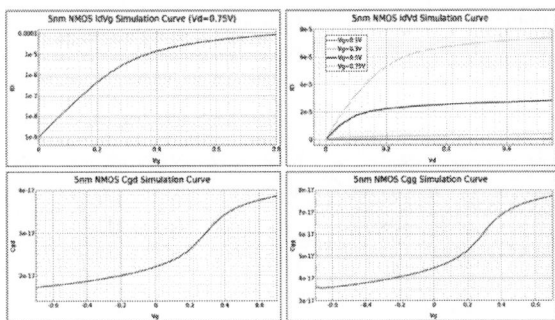

Figure 4. Characteristic curves for the NFET Device from simulation

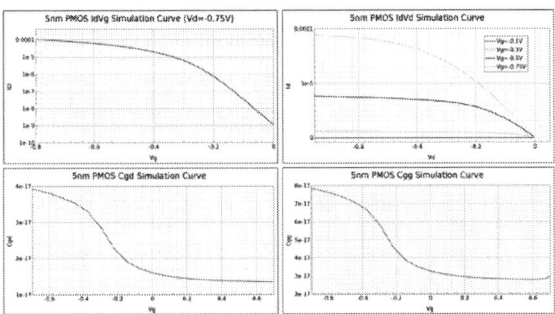

Figure 5. Characteristic curves for the PFET Device from simulation

PROCESS OPTIMIZATION

In the development process, we have done a lot of experiments by simulation. In this chapter, we focus on the analysis of fin loop process optimization.

We set the top fin width W_{top}= 5 nm, the fin-height (FH) equal to 50 nm and the gate length LG= 19 nm as a standard device. The tapered fin is tilted with an angle of 89°. Then, we have varied the parameters, such as fin height (FH), fin width (W_{top}+W_{bottom}), and only W_{top} to evaluate the impact of these parameters. Afterwards, we analyze the physical properties of the device to understand the behavior and characteristics.

In Figure 6, by increasing the fin-height from 40 nm to 60 nm, it can enhance the on-state current with off current slightly increased, the Ieff improved ~56% and ~43% for NFET and PFET respectively. This strategy is quite effective to enhance the performance, but it will be very difficult to fabrication because the aspect ratio becomes very high, not easy for production [3].

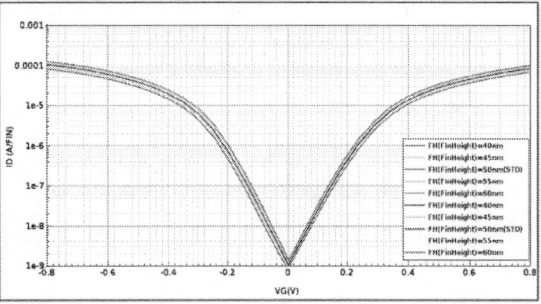

Figure 6. Transfer characteristics (in log scale) of FinFET (W_{top}=5 nm, angle=89°) with different fin heights (FHs)

The control of the fin width is essential at a given technology node because the width greatly affects the device performance. The widening of the fin can improve the saturation current. However, it may not be a good choice because a wider fin will make a larger cross section device area for source and drain epitaxy, which will induce more leakage thus may result in the severe short channel effect [3], as Figure 7(a) shows. In Figure 8, we only enlarge W_{top} (W_{bottom} is fixed), which demonstrates that the Ioff will reduce significantly. This phenomenon reversely verifies that most of the leakage current comes from the fin bottom area. And the current density distribution under off-state condition which shows in Figure 9 has also proved this behavior, so smaller W_{bottom} and straight Fin profile are preferred.

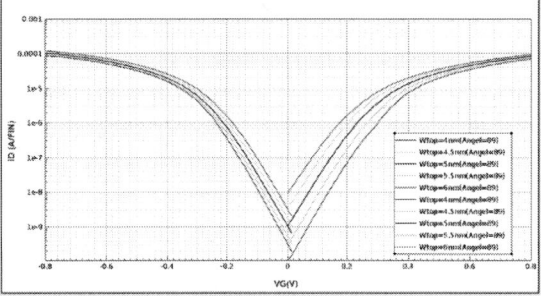

Figure 7(a). Transfer characteristics (in log scale) of FinFET (FH = 50 nm) with different fin widths (W_{top} and W_{bottom} enlarged together)

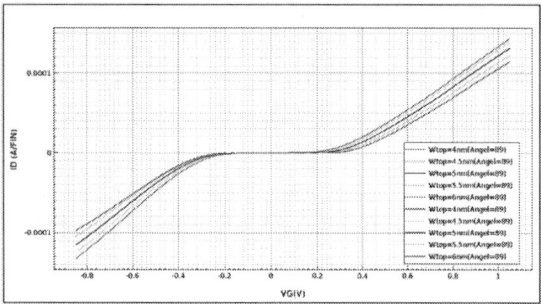

Figure 7(b). Transfer characteristics (in linear scale) of FinFET (FH = 50 nm) with different fin widths (W_{top} and W_{bottom} enlarged together)

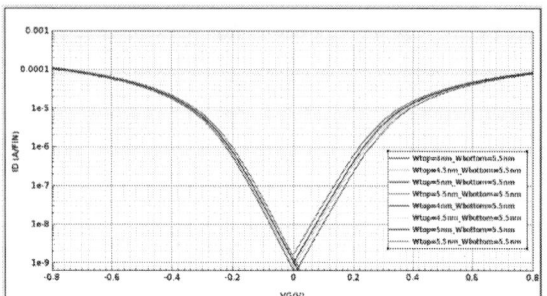

Figure 8. Transfer characteristics (in log scale) of FinFET (FH = 50 nm) with different fin top widths (fix W_{bottom})

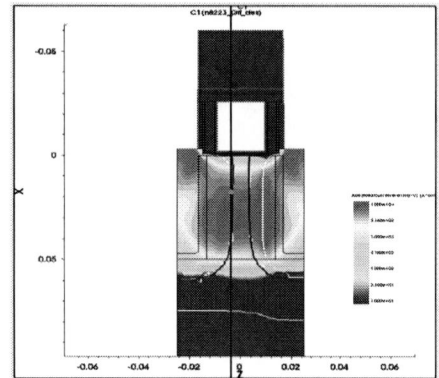

Figure 9. Current density distribution under off-state condition (VD = 0.75 V, VG=0V) with FH=50 nm and W_{top}=5 nm, angle=89°.

Considering all above, and to enhance saturation current (see Figure 7(b)), we finally set FH=50 nm, W_{top}=5 nm and 89° as a typical device fin size.

RO PPA VERIFICATION

Finally, to have a comprehensive understanding, we make the RO simulation. In Figure 10, it seems our 5 nm RO PPA performance (red line) is close or even better than

published reference 5 nm performance [5-6]. (The 5nm PPA curve of industry reference 1 (Ref 1) is calculated with which published 40% speed gain and 65% power reduction from 14 nm to 7 nm, 15% speed gain and 30% power reduction from 7 nm to 5 nm, respectively; reference 2 (Ref 2) is published only for 5 nm RO performance).

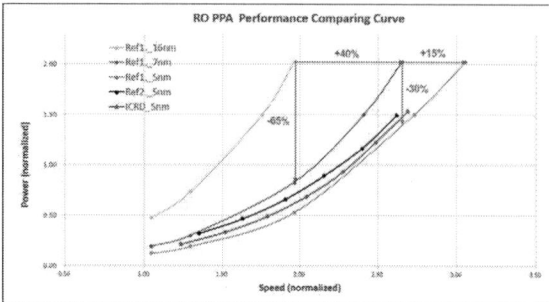

Figure 10. RO PPA performance comparison

CONCLUSION

Product innovation has been the driving force behind the semiconductor industry growth in the past years. Especially with 5G deployment starting in 2019, mobile System on Chip (SoC), High Performance Computing (HPC), and Artificial Intelligence (AI)/Data-center are propelling another period of sustained growth in the next decade. Advanced CMOS logic technology has been the key enabler for semiconductor product innovations. 5 nm technology is one of such advanced logic technology platforms. [4]

In this paper, we have performed a simulation to study a typical 5 nm FinFET device structure and performance. The result indicates that our device performance is comparable with published reference performance and the short channel effect is controlled well; our ring oscillator simulation demonstrates that the PPA is also close to industry reference. We believe that the FinFET device still has the ability to extend to 5 nm technology node with continued performance gain.

ACKNOWLEDGEMENTS

The authors would like to thank the management team and all our team members in Shanghai ICRD center, and our co-work partner IMEC's great help.

REFERENCES

[1] Colinge, Jean-Pierre, "The SOI MOSFET: From single gate to multigate." FinFETs and Other Multi-Gate Transistors. Springer, Boston, MA, 2008. 1-48 (2008).
[2] Xiong, Weize Wade, "Multigate MOSFET technology." FinFETs and Other Multi-Gate

Transistors. Springer, Boston, MA, 2008. 49-111 (2008).

[3] Kurniawan, Erry Dwi, et al. "Effect of fin shape of tapered FinFETs on the device performance in 5-nm node CMOS technology." Microelectronics Reliability 83 (2018): 254-259 (2018).

[4] Geoffrey Yeap, S.S. Lin, Y.M. Chen, et al. "5 nm CMOS Production Technology Platform featuring full-fledged EUV, and High Mobility Channel FinFETs with densest $0.021um^2$ SRAM cells for Mobile SoC and High Performance Computing Applications." 2019 IEEE International Electron Devices Meeting (IEDM). IEEE, 2019.

[5] J. Ryckaert, M. H. Na, P. Weckx, et al. "Enabling Sub-5nm CMOS Technology Scaling Thinner and Taller!" 2019 IEEE International Electron Devices Meeting (IEDM). IEEE, 2019.

[6] Shien-Yang Wu, C.Y. Lin, M.C. Chiang, et al. "A 7nm CMOS Platform Technology Featuring 4th Generation FinFET Transistors with a 0.027um2 High Density 6-T SRAM cell for Mobile SoC Applications" 2016 IEEE International Electron Devices Meeting (IEDM). IEEE, 2016.

A Study of FinFET Device Optimization and PPA Analysis at 5 nm Node

Xin Luo, Yu Ding, Enming Shang, Jie Sun, Shaojian Hu[*], Shoumian Chen and Yuhang Zhao

Shanghai IC R&D Center, 497 Gaosi Road, Zhangjiang Hi-Tech Park, Shanghai, China.

*Corresponding Author's Email: hushaojian@icrd.com.cn

ABSTRACT

Since the logic 5 nm node still uses FinFET device, it still has room for device performance improvement since its first debut in the production of 16 nm node. The goal of this paper is to investigate the FinFET device optimization and Power Performance Area (PPA) at the 5 nm node. We have simulated FinFET device electrical characteristics at Front-End-Of-the-Line (FEOL) with Technology Computer Aided Design (TCAD). We first focus on the device with different spacer thicknesses to investigate DC and AC characteristics. Then we focus on the power and speed performance of Ring Oscillator (RO) circuits based on 5 nm NMOS and PMOS devices. Detailed study has been carried out to analyze the influence of Number of fins (Nfin), Back-End-Of-the-Line (BEOL) interconnect length, and fan-out number to power and speed. We hope that our results can assist other researchers in better understanding of the 5 nm FinFET device performance.

Keywords—5 nm; FinFET; device; TCAD; PPA; ring oscillator

1. INTRODUCTION

Over the past 50 years, people have been following the development path guided by Moore's Law [1]. More and more transistors have been integrated into a single chip. Improving integration has been continuing in the Integrated Circuits (IC) industries. During this long-term period, higher speed and lower power have been continuously pursued [2-4], which has been driving IC industry to ultra-large-scale integration, fundamentally changing the ways of life: in the past ten years, people have witnessed the transformation of heavy laptops into thin and light "ultrabooks", and experienced the evolution of mobile phones from feature phones to smartphones. In not too distant future, it can be predicted that successful development of 5G baseband communication chips will enable people to connect the world through the mobile internet with lower delay time. This advantage is assumed to be of great significance for the development of smart cars, smart homes, smart supermarkets, and many other fields. Besides, WiFi chips have developed into the sixth generation nowadays, where its outstanding performance is to strengthen the network loading capacity to support more clients in the area with crowded equipment units at the same time, compared with current generation. More than this, innovations in Artificial Intelligence (AI), Data-center management, etc. can also enhance the development of microelectronics industry.

To support these innovations, advanced CMOS engineering has become a core driving force [4]. Fab industries have chosen FinFETs to replace planar CMOS devices at the 16 nm node, making System on Chip (SoC) in smartphones work faster and consume less power. FinFETs have triggered a large number of innovative scientific research in the semiconductor technology field [2-4]. Compared with normal planar CMOS devices, gate control performance can be enhanced in FinFETs and Short Channel Effect (SCE) can be reduced effectively [5]. Since the FinFETs have these advantages, device designs based on FinFETs have been adopted by the mainstream IC fabs and have continued the performance gain target of technology development from 14 nm through 7 nm nodes. As a result, recent developments in the FinFET devices have led to a continuation of the Moore's law. It is expected that the FinFETs are being extended to the 5 nm node [6-8] and beyond.

In this study, we report the 5 nm FinFET device optimization and holistic PPA analysis. We will report our study on the DC and AC characteristics of 5 nm devices with varied spacer thicknesses. And we will also present a study on the PPA performance dependence on Nfin, BEOL, and fan-out number.

2. SIMULATION METHOD

5 nm node FinFETs were simulated using Sentaurus TCAD software [9]. The Sentaurus device module has been applied to simulate DC and AC electrical performance of 5 nm node NMOS and PMOS devices. Drift-diffusion transport, poisson, and electron/hole continuity equations, have been used to calculate current. Appropriate physical mobility models and relative scattering mechanisms have been taken into account in Si, SiGe regions, and interfaces.

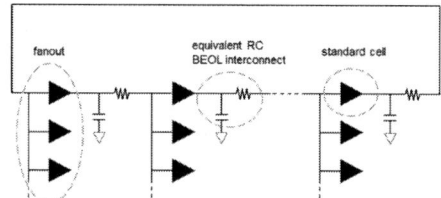

Figure 1: Schematic diagram of the Ring Oscillator (RO) circuit with 9 stages.

SPICE models of 5 nm node device has been extracted with Accelicon MBP. H-spice has been adopted to simulate the Ring Oscillator (RO) circuit. Figure 1 shows a schematic diagram of the RO circuit in which 9 stages are embedded (three stages are drawn here for simplicity). In each stage, several fan-outs (i.e. invertor quantity) are included. Unless specifically stated, two fins in each device, 3 fan-outs, and inverter cells are considered. Front-End-Of-the-Line (FEOL) performance in one stage, and equivalent RC Back-End-Of-the-Line (BEOL) interconnect between two stages are concerned while simulating Power Performance Area (PPA). The resistance and capacitance in BEOL interconnect is directly proportional to the BEOL wire length which is expressed as Contact-to-Gate Pitch (CGP) number below.

3. RESULTS AND DISCUSSION

3.1 The Influence of Spacer Thickness on 5 nm FinFET Device Characteristics

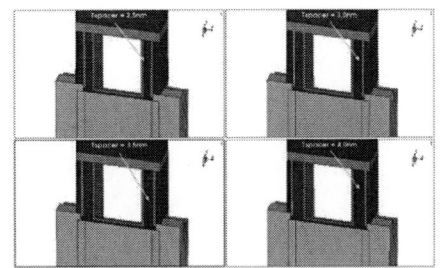

Figure 2: 5 nm FinFET device structures with increasing spacer thicknesses.

Initially, the device optimization is investigated. Devices with different low-k spacer thicknesses (T_{spacer}) have been simulated. Figure 2 shows 5 nm FinFET device structures with increasing spacer thickness.

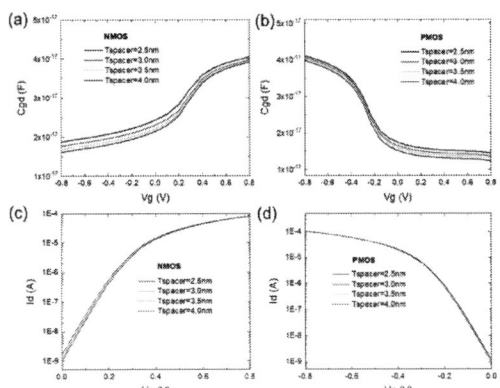

Figure 3: Dependence of C_{gd} with increasing spacer thickness for NMOS (a), and PMOS (b) vs V_g, respectively; The influence of spacer thickness on I_d-V_g curve at V_{dd}=0.75V for NMOS (c), and PMOS (d), respectively.

In Figures 3(a) and 3(b), the capacitance between gate

and drain (C_{gd}) for both NMOS and PMOS decreases while T_{spacer} increases. Figures 3(c) and 3(d) display I_d-V_g curves of NMOS and PMOS, respectively, showing that the saturation current (I_{dsat}) and off current (I_{off}) slightly declines with the change of spacer thickness. The DC characteristics does not change significantly, mainly because the drain/source side contact area is fixed in our simulation. Figure 4 shows that a slightly larger spacer thickness will be good for speed performance.

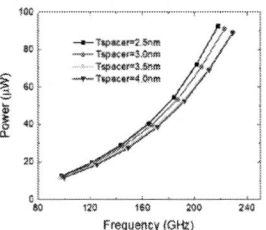

Figure 4: The relation between power and frequency with varied spacer thicknesses (interconnect length=40 CGP).

3.2 PPA Performance of Different Number of Fins and BEOL Length

As we know, the Number of fins (Nfin) in each device relates to the height of the standard cell, and the drive capability of circuits. Figure 5(a) presents the power-frequency relationship for different Nfin conditions

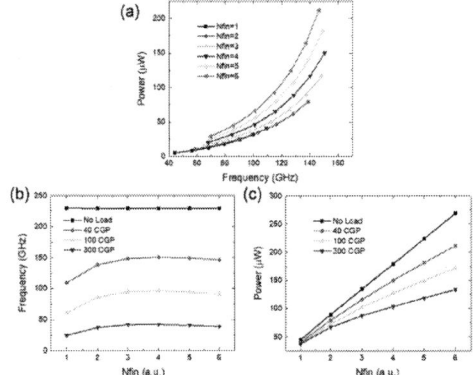

Figure 5: (a) PPA results of RO based on devices with different Nfin (interconnect length=40 CGP); (b) Frequency versus Nfin for different BEOL interconnect length; (c) The dependence of dynamic power on Nfin. (BEOL length in units of CGP)

To obtain the influence of Nfin on frequency, Figure 5(b) shows the correlation between frequency value at V_{dd}=0.75V and Nfin. In the circumstance of no BEOL load, frequency is nearly constant as device Nfin changes. This is because the delay time is proportional to CV/I, and the capacitance or current all have positive correlation with Nfin. As a result, frequency stays constant.

When the BEOL wire length increases, the frequency decreases in general for different Nfin devices. This is because the increase of BEOL load can slow down the RO circuit. While 40 CGP BEOL is added, the frequency will generally drop as Nfin decreases. This can be understood by Equations (1) [10] and (2),

$$\tau_d = \frac{CV_{dd}}{4}\left(\frac{1}{I_{onN}} + \frac{1}{I_{onP}}\right) \qquad (1)$$

$$f = \frac{1}{\tau_d} \qquad (2)$$

in which C includes both parasitic capacitance (C_{para}) and BEOL capacitance (C_{BEOL}). As Nfin decreases, C_{para} and current drops but C_{BEOL} does not change.

On the other hand, the dependence of dynamic power on Nfin is shown in Figure 5(c). The power is mainly proportional to Nfin, and the slope is notably large. The power will nearly double when Nfin increases from 1 to 2 under the 40 CGP condition for instance. Figure 5(c) also shows that more BEOL interconnect length leads to lower power. This can be explained by Equations (3) and (4) [10].

$$\text{Power} = V_{dd} * \text{average current} \qquad (3)$$

$$\text{average current} = kCV_{dd}f \qquad (4)$$

Due to the reduction of the average current, the dynamic power falls when more BEOL interconnect is added.

3.3 PPA Results of Different BEOL Capacitance

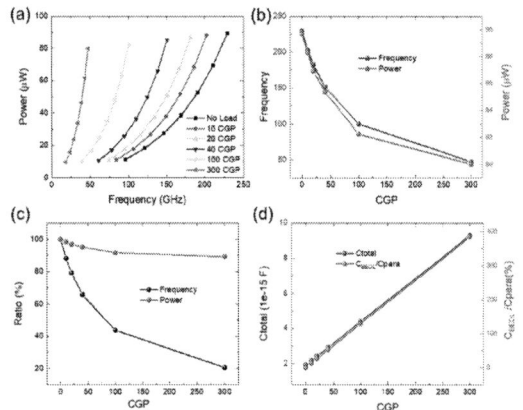

Figure 6: (a) Power vs frequency of inverter RO with different BEOL interconnect capacitance; The dependence of the (b) frequency and dynamic power, (c) frequency and power variation rate, (d) effective total capacitance, the proportion of C_{BEOL} and C_{para} on BEOL length.

In above analysis, resistance and capacitance in BEOL interconnect are all considered to study the effect of interconnect length on PPA. To study the influence of capacitance only, we simulated RO results with different interconnect capacity value (resistance in interconnect is assumed as 0). Figure 6(a) shows the power as a function

of frequency for increased interconnect capacity which is proportional to BEOL length. It can be seen in Figure 6(b) that the frequency and power all decrease with BEOL length. Figure 6(c) shows their variation ratios. The frequency drops more than power, and it is more obvious from 0 CGP to 100 CGP. This indicates that the relative importance of capacitance reduction in BEOL interconnect. The effective capacitance has been calculated from the equation (1). It is shown in Figure 6(d) that C_{BEOL} would dominate quickly when wire length increases. It also can be induced that low-k materials could be more helpful to promote IC performance [2].

3.4 The Influence of the Fan-out Number on Power and Speed

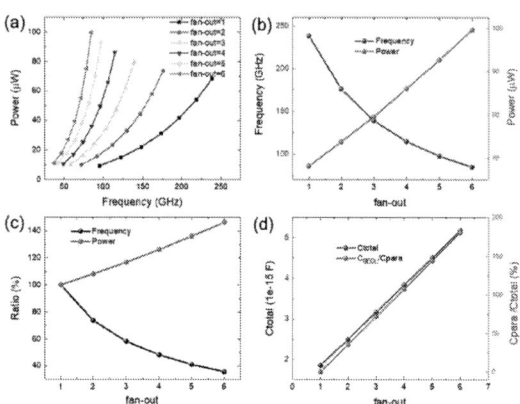

Figure 7: (a) Power vs frequency results with different fan-out number (interconnect length=40 CGP); (b) The frequency and dynamic power, (c) the frequency and dynamic power variation rate, (d) the effective total capacitance, the proportion of C_{para} and C_{total} versus the fan-out number.

The IC circuit can actually involve different number of fan-outs in one stage. Therefore, it is significant to investigate how much performance degradation it can bring when the fan-out quantity increases. The results of the power versus frequency for circuits with increased fan-out number are presented in Figure 7(a). Figure 7(b) shows that the power increases almost linearly and the frequency decreases almost inversely proportional to the fan-out number. It is normal that the increase of inverters will lead to linear power increase. The C_{para} increases because the parasitic part is multiplied with invertor number, and I_{on} can be maintained when more fan-outs are applied. According to a derivation of Equation (1), the frequency can have an inversely proportional relation with fan-out number. In Figure 7(c), the frequency and dynamic power variation rate are shown, indicating that the frequency varies more rapidly than the power. For example, the power rises 10% and the frequency falls nearly 30% when the fan-out number changes from 2 to 3. Besides, the relation of capacitance with fan-out number is shown in Figure 7(d), which is consistent with the frequency drop shown in Figure 7(b).

CONCLUSION

In summary, we have studied the 5 nm device optimization and its PPA performance. We first study the DC and AC trade-off of the 5 nm NMOS and PMOS with varied spacer thicknesses. We have found that good speed performance would need slightly larger spacer thickness. On the PPA performance, we have found that the frequency will generally drop and the power will rise as Nfin decreases with BEOL load applied. The power will increase following almost a linear relation with the added fan-out number while the frequency will decrease almost inversely proportional to the fan-out number.

ACKNOWLEDGMENTS

We thank the higher management team from Shanghai IC R&D Company for the support of this work.

REFERENCES

[1] G.E. Moore，"Cramming more components onto integrated circuits", Proceedings of the IEEE, 86(1): 82-85, 1998.

[2] M. Garcia Bardon, et al., "Dimensioning for power and performance under 10nm: The limits of FinFETs scaling", ICICDT IEEE, 2015.

[3] S.-Y. Wu, et al., "A 7nm CMOS platform technology featuring 4th generation FinFET transistors with a 0.027 um^2 high density 6-T SRAM cell for mobile SoC applications", IEEE IEDM, 2016.

[4] G. Yeap, S.S. Lin, Y.M. Chen, et al., "5nm CMOS Production Technology Platform featuring full-fledged EUV, and High Mobility Channel FinFETs with densest $0.021um^2$ SRAM cells for Mobile SoC and High Performance Computing Applications", IEEE IEDM,19-879, 2019.

[5] J.-P. Colinge, et al. "FinFETs and other multi-gate transistors", Vol. 73. New York: Springer, 2008.

[6] Y. Ding, Y. Cao, X. Luo, et al., "A Device Design for 5 nm Logic FinFET Technology", accepted by Journal of Microelectronic Manufacturing 2020.

[7] E. Shang, Y. Ding, W. Chen, et al., "The Effect of Fin Structure in 5 nm FinFET Technology", Journal of Microelectronic Manufacturing, 2, 19020405, 2019.

[8] Q. Wu, Y. Li, Y. Yang, Y. Zhao, "A Photolithography Process Design for 5 nm Logic Process Flow", accepted by Journal of Microelectronic Manufacturing 2020.

[9] Sentaurus Device User Guide I-2018.06, Synopsys Inc., Mountain View, CA, USA, 2018

[10] C. Hu, "Modern semiconductor devices for integrated circuits", Vol. 2. Prentice Hall, 2010.

A SIMPLE CURRENT TEST METHOD ON WAFER LEVEL TO PRE-VERIFY CIRCUIT FUNCTION

Jianrong Xu

Shanghai Huali Microelectronics Corporation, Shanghai, China

*Corresponding Author's Email: xujianrong@hlmc.cn

ABSTRACT

In this paper, a wafer level current testing method is presented. The proposed method gives a pre-testing on wafer level before packaging test. By a simple current testing, which can shorten the period of the process development and verification. Detailed procedure and advantages of the WAT test algorithm are given. the method is detecting the static current of the output signal (IBIAS module), which can check whether the function can be acted. Also the same data can be used for predicting the characteristics of dynamic power circuit.

INTRODUCTION

In the current advanced deep sub-micron integrated circuit manufacturing industry, people have been able to manufacture integrated circuits with relatively complex structures, high integration, and different functions. And any block integrated circuit is a single-chip module designed to complete certain electrical characteristics, and its functionality is increasingly attracting people's attention.

ILLUSTRATIONS

IBIAS module is a current generating circuit in Figure 1, power supply VDD is 1.2V, SLEEP12 and SLEPP12B as the input signal switch, (SLEEP12 high level is sleep effective, SLEEP12B low level is sleep effective) simulation results can be measured in Figure 2 IBIAS<3:0> 4uA output current. (IBIAS0=4uA, IBIAS1=4uA,IBIAS2=4uA, IBIAS3=4uA) In the existing testing technology, the electrical characteristics of the designed IP circuit are usually measured by means of a multimeter (the external pin of the chip is fixed after the package is mounted, and the tester usually uses a multimeter to test the electrical characteristics of the pins on the test board). The measurement method is mainly performed on the input, output and power pins of the device under test to determine the DC drive characteristics of the output pin of the device under test, the DC load characteristics of the input pin, and the power supply characteristics. This test method has a long cycle and requires external processing, encapsulation and other processes to return to the fab for actual testing.

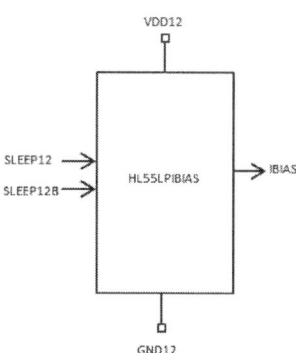

Figure 1: Schematic diagram of IBIAS module structure

Figure 2: Schematic diagram of WAT automatic tester

Agilent 4082 Figure 3 is an automatic test equipment commonly used in advanced wafer foundries. It can perform preliminary electrical parameter extraction and rapid testing. The main method is to apply voltage measurement results to the device through the SMU measurement module. With the application of the current measurement voltage, the applied voltage and current have a certain dynamic range to meet different measurement requirements. Under the premise of simulation circuit verification, it is necessary to establish an IBIAS module verification test program at the wafer level to shorten the test time.

```
v1 VDD25 0 ddv
v2 GND25 0 ssv
v3 SLEEP25B 0 ddv
v4 SLEEP25 0 ssv
v5 IBIAS[0] 0 ssv
v6 IBIAS[1] 0 ssv
v7 IBIAS[2] 0 ssv
v8 IBIAS[3] 0 ssv

.dc v1 0 ddv ddv
.op
.print i(x1.IBIAS[0]) i(x1.IBIAS[1]) i(x1.IBIAS[2]) i(x1.IBIAS[3])
```

Figure 3: IBIAS module simulation netlist

PROBLEM AND ANALYSIS

The traditional use of multimeter test methods has the disadvantages of slow speed and inaccuracy. In terms of accuracy, chip defects may cause chip defects to cause inconsistencies in the actual measured IBIAS output signals. In the measurement function, only a certain voltage and current state can be sampled and analyzed. In terms of data management, traditional multimeters can only manually record individual test data of devices, which is slow and consumes labor costs.

Agilent 4082 electrical parameter tester combined with SMU measurement module, the test results are more accurate and more efficient. Helping the establishment and circuit design of other model tests by further sampling and analysis of test programs and test results. Changed the program of data storage and management, so that the parameter data of each device before being pressed, pressure applied and pressed is very clearly stored. Moreover, the data storage method is more readable and easier to be analyzed and processed.

This wafer level verification test method is performed on the Agilent 4082 electrical parameter tester. When performing DC test, we use HP-Basic as the programming platform to apply the stress signal to the SMU module to the device. IBIAS electrical performance, complete the test.

EXPERIMENTAL DESIGN

The technical solution is based on the 55LP process platform, and a DC test program matching the simulation conditions is established in the WAT test environment to realize the actual measurement function of the IBIAS module. The implementation method is as follows in Figure 4. 1.The test environment initializes and defines the IBIAS output signal test parameter condition setting interface.2. Test the shot position on the selected wafer.3.Define the input port variables and the number of assigned pins. The order of the port variables is the same as the order defined by the port parameters.4. Connect the measurement unit SMU: FNPort(0,1)~FNPort(0,7)(SMU) where FNPort(0,9) is the ground source GNDU.5.Apply SLEEP and supply voltage and measure IBIAS current. Print out the test results and prepare to disconnect.6. Determine if the output current value is reasonable. If it is unreasonable, repeat the 2 process. If it is reasonable, continue to use the test method of scanning to analyze the current and voltage in a certain interval.

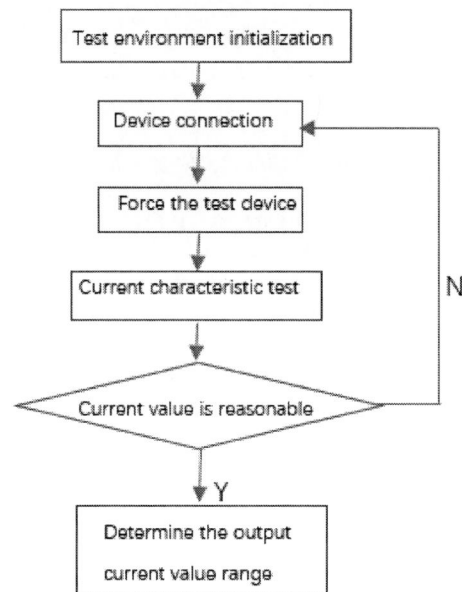

Figure 4: Schematic diagram of the test process

For the fixed voltage test of the previous IBIAS circuit, the matching between the WAT experimental results and the simulation results is high, and the difference of the four-terminal current values is less than 10% in Figure 5, thus ensuring the testability of the WAT test mode. And by applying a stress step voltage at the power supply terminal, the dynamic power supply characteristic range is determined by the output current curve. From the experimental results, the IBIAS circuit module outputs nearly 4uA current under the condition of bias voltage Vdd=0.9~1.4V, and can accurately simulate the circuit unit when Vdd=1.2V

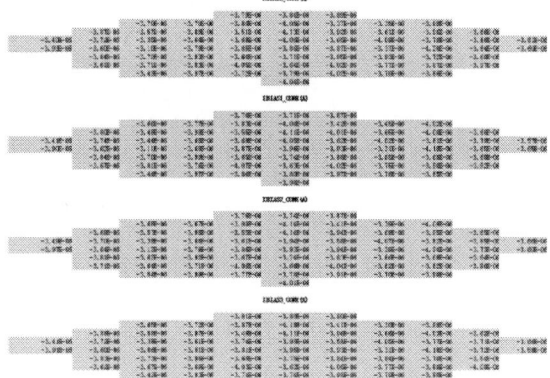

	Median	Variation
IBIAS0_CORE(A)	-3.79E-06	5.22%
IBIAS1_CORE(A)	-3.78E-06	5.61%
IBIAS2_CORE(A)	-3.77E-06	5.75%
IBIAS3_CORE(A)	-3.75E-06	6.33%

Figure 5: 1.2V device IBIAS<3:0> test distribution map

EXPERIMENTAL RESULTS

For the fixed voltage test of the previous IBIAS circuit, the matching between the WAT experimental results and the simulation results is high, and the difference of the four-terminal current values is less than 10%, thus ensuring the testability of the WAT test mode. And by applying a stress step voltage at the power supply terminal, the dynamic power supply characteristic range is determined by the output current curve. From the experimental results, the IBIAS circuit module outputs nearly 4uA current under the condition of bias voltage Vdd=0.9~1.4V, and can accurately simulate the circuit unit when Vdd=1.2V in Figure 6

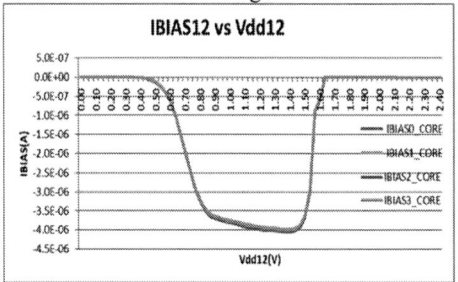

Figure 6: IBIAS signal output diagram under 1.2V device stress step voltage

CONCLUSION

This program uses the HP-Basic programming language to establish a WAT test program to implement the power supply the power supply test in the IBIAS circuit model, thereby effectively reducing test time and reducing test costs.In terms of technical effect, compared with the original test method, the new method can improve the test optimization from three aspects. From the timeliness, it takes nearly one month for the wafer to be sent from the factory to the return, and the multimeter can only be tested manually. There are drawbacks of slow speed and inaccuracy. On the contrary, WAT automatic test is more accurate and more efficient. During the experiment, all the die test time is less than 5 minutes. In terms of test functionality, the original test method can only sample and analyze a certain voltage and current state. The new method can change the test procedure of the unit model according to the actual needs of the circuit. He can scan the test mode for a certain interval. Current and voltage are analyzed for results. In the data acquisition method, the improved method can automatically output

data through the program, reducing labor costs.

The most important advantage of the WAT test method is that it can avoid the outsourcing cycle and ensure the test time is greatly reduced while ensuring the same test accuracy and physical mechanism. In the application process, the program of the tested unit model can also be changed according to the actual needs of the circuit, thereby improving work efficiency.

REFERENCES

[1] Agilent Easy_expert_manual.
[2] Array test papers
[3] Keysight Technologies 4070/4080 Series Parametric Test System---Programming Reference for Basic Users

NOVEL SEMICONDUCTOR DEVICES BASED ON SOI SUBSTRATE

K. Xiao[1], J. Liu[1], JN. Deng[1], YL. Jiang[2], WZ. Bao[2], A. Zaslavsky[3], S. Cristoloveanu[4], X. Gong[5], and J. Wan[1]*

[1]State key lab of ASIC and System, School of Information Science and Engineering, Fudan University, Shanghai, China
[2]State key lab of ASIC and System, School of Microelectronics, Fudan University, Shanghai, China
[3]Department of Physics and School of Engineering, Brown University, Providence, RI 02912, USA
[4]IMEP-LAHC, INP-Grenoble/Minatec, CS50257, Grenoble 38016, France
[5]Department of Electrical and Computer Engineering (ECE), National University of Singapore (NUS), Singapore
Email: *jingwan@fudan.edu.cn*

ABSTRACT

In this work, we review our recent studies on several novel devices built on silicon-on-insulator (SOI) substrates. The sharp-switching Z^2-FET, based on a feedback mechanism, has been demonstrated as suitable for many applications. The PISD, capable of *in-situ* photoelectron sensing, has been used as a one-transistor active pixel sensor (1T-APS). Furthermore, an SOI/MoS$_2$ heterojunction FET has been demonstrated as both a photodetector with a dynamic response spectrum and as a novel one-transistor wavelength detector (1T-WD) with an output signal sensitive to the variation of wavelength rather than intensity.

Keywords—Silicon-on-insulator (SOI), low subthreshold swing, memory, photodetector, image sensor, 2D heterojunction materials

INTRODUCTION

Silicon-on-insulator (SOI) substrates have been widely used to fabricate integrated circuits (ICs) with high density, low power consumption and high operation frequency. They possess some unique advantages, such as reduced short channel effect, low parasitic leakage and capacitance [1-3].

We review two categories of novel devices based on SOI substrates that we developed in recent years: CMOS-compatible devices and SOI-2D hybrid devices. First, zero subthreshold swing and zero impact ionization FET (Z^2-FET) and photoelectron *in-situ* sensing device (PISD) are fully CMOS-compatible and have been demonstrated in cutting-edge VLSI technology [4, 5]. They are used as sharp switches in memory and image sensing applications, with extraordinary performance compared to conventional counterparts. Second, by combining SOI and the 2D semiconductor MoS$_2$ material, hybrid devices with unique functionalities have been demonstrated, such as the interface coupled photodetector (ICPD) showing a dynamically modulated response spectrum and the one-transistor wavelength photodetector (1T-WD) that is only sensitive to the incident wavelength λ rather than intensity [6, 7].

CMOS-COMPATIBLE DEVICES

The Z^2-FET is essentially a lateral p-i-n diode with the undoped channel partially covered by the front gate, see Fig. 1(a). Thanks to its feedback mechanism, the device turns on very sharply and shows gate-controlled hysteretic behavior, see Fig. 1(b) and (c). The Z^2-FET presents subthreshold swing (SS) < 1mV/dec, much lower than the conventional MOSFET limit. The gate-controlled hysteresis can be utilized in dynamic RAM (DRAM), static RAM (SRAM) and flash memory applications [8, 9]. Figure 1(d) shows the measured waveforms of Z^2-FET used as a DRAM. It has high density, high access speed and low power consumption compared to conventional 1T-1C DRAM. There are also many other applications of Z^2-FET, such as

electro-static discharge (ESD) protection, photodetection, sharp logic switch and ion-sensitive sensor [10-12].

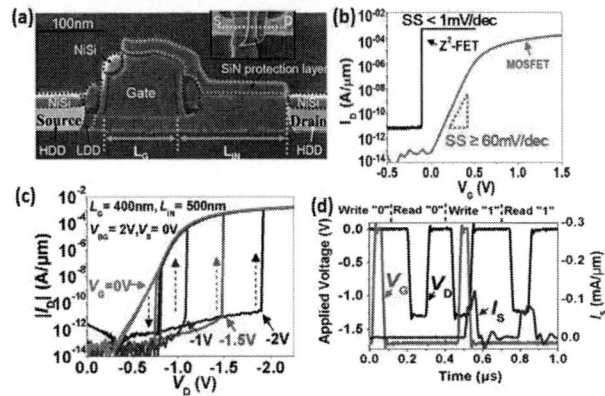

Fig. 1 (a) Z^2-FET structure with (b) I_D-V_G and (c) I_D-V_D characteristics. (c) Operating sequence of Z^2-DRAM.

A completely different SOI substrate-based device is shown in Fig. 2(a). A novel deep depletion effect in the SOI substrate is observed in the pseudo-MOS configuration, useful for photodetection application when combined with a backgate V_{BG} pulse, see Fig. 2(b) [13]. This mechanism was further elaborated into a fully CMOS-compatible PISD, schematically shown in Fig. 2(c) [5].

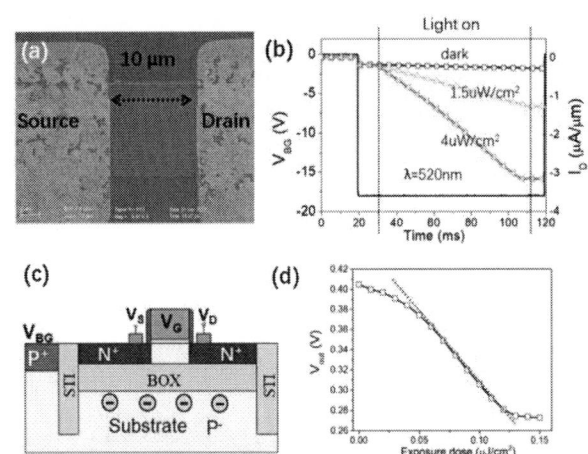

Fig. 2 (a) Original pseudo-MOS device prototype without a front gate and (b) photoelectrical characteristics. (c) Schematic view of the PISD and (d) the photoelectrical characteristics of a device fabricated in an advanced 22 nm FD-SOI process.

In the PISD, photoelectrons generated in the SOI substrate accumulate at the BOX/Si interface. The accumulated photoelectron modulates the threshold voltage (V_{th}) of the

MOSFET in the top Si channel, as illustrated in the experimental results of Fig. 2(d). The PISD is truly a one-transistor active pixel sensor (1T-APS), as it achieves photosensing, charge integration, buffer amplification and random access with only one transistor. The PISD is more compact than a conventional CMOS image sensor that combines a photodiode with three transistors.

SOI-2D MATERIAL HYBRID PHOTODETECTORS

Photodetectors with novel functionalities can be obtained by integrating 2D materials with SOI substrates. Figure 3(a) shows the SOI/MoS$_2$ heterojunction FET, where the MoS$_2$ is used as gate to control the SOI channel [6]. The device behaves as normal junction FET. However, photoelectrical measurements show that the response spectrum is tunable with the applied bias. As shown in Fig. 3(b), the increase of back-gate voltage from –9 V to zero can shift the peak of the response spectrum from the near-infrared (near-IR) to ultra-violet (UV). This is mainly attributed to the switch of photosensing junction between SOI and MoS$_2$ materials. The capability of modulating the response spectrum dynamically is attractive for multi-spectral photodetection and imaging.

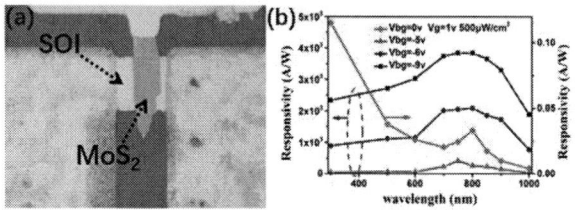

Fig. 3 (a) Heterojunction FET combining SOI and MoS$_2$, and (b) optical responsivity as a function of backgate bias V$_{BG}$.

Another interesting device obtained by combining MoS$_2$ and SOI is the one-transistor wavelength detector (1T-WD) [7]. It is essentially a MOSFET with MoS$_2$ as the top gate isolated with an HfO$_2$ dielectric from the thin 12nm Si channel, see Fig. 4(a). Both the top MoS$_2$ and the Si substrate act as photogates, but with opposite signs due to different doping polarities in the MoS$_2$ and bottom Si. As a result, the device shows ambipolar photoresponse: the zero photoresponse voltage shows a strong dependence on wavelength, but is not sensitive to the light intensity, as shown in Fig. 4(b). This wavelength detection property could be attractive for a number of applications.

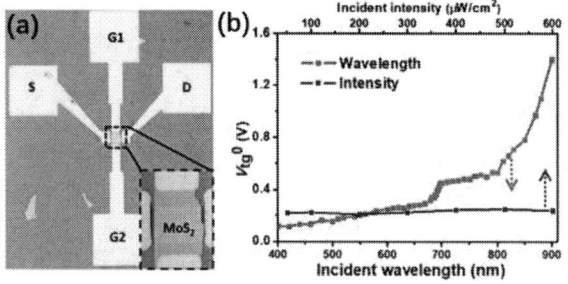

Fig. 4 (a) Structure of the 1T-WD combining SOI and MoS$_2$. (b) Photoresponse of the device under various light intensity and wavelength.

CONCLUSION

We have invented several novel devices based on SOI substrates. Of them, the Z^2-FET and PISD are fully compatible with advanced CMOS processing but are based on different device physics compared to conventional MOSFETs and exhibit high performance. On the other hand, SOI and MoS$_2$ have been combined to form hybrid devices leading to new photodetection functionalities, such as voltage-tunable response spectrum and photoresponse modulated only by incident wavelength.

ACKNOWLEDGEMENTS

The work at Fudan University is sponsored by National Natural Science Foundation of China (61904032) and the Shanghai Rising Star Program (19QA1401100).

REFERENCES

[1] R. Carter, J. Mazurier, L. Pirro, J. U. Sachse, P. Baars, J. Faul *et al.*, "22nm FDSOI technology for emerging mobile, Internet-of-Things, and RF applications," in *IEEE International Electron Devices Meeting (IEDM)*, 2016, pp. 2.2.1-2.2.4.

[2] O. Weber, E. Josse, F. Andrieu, A. Cros, E. Richard, P. Perreau *et al.*, "14nm FDSOI technology for high speed and energy efficient applications," in *Symposium on VLSI Technolog*, 2014, pp. 1-2.

[3] E. Y.-J. Kong, S. Yadav, D. Lei, Y. Kang, M. Sivan, Y. Li *et al.*, "Highly scaled strained silicon-on-insulator technology for the 5G era: Impact of geometry and annealing on strain retention and device performance of nMOSFETs," *IEEE Transactions on Electron Devices,* vol. 66, no. 5, pp. 2068-2074, 2019.

[4] J. Wan, C. Le Royer, A. Zaslavsky, and S. Cristoloveanu, "A compact capacitorless high-speed DRAM using field effect-controlled charge regeneration," *IEEE Electron Device Letters,* vol. 33, no. 2, pp. 179-181, 2012.

[5] Y.-F. Cao, M. Arsalan, J. Liu, Y.-L. Jiang, and J. Wan, "A novel one-transistor active pixel sensor with *in-situ* photoelectron sensing in 22 nm FD-SOI technology," *IEEE Electron Device Letters,* vol. 40, no. 5, pp. 738-741, 2019.

[6] J. Deng, Z. Guo, Y. Zhang, X. Cao, S. Zhang, Y. Sheng *et al.*, "MoS$_2$/silicon-on-insulator heterojunction field-effect transistor for high-performance photodetection," *IEEE Electron Device Letters,* vol. 40, no. 3, pp. 423-426, 2019.

[7] J. Deng, L. Zong, M. Zhu, F. Liao, Y. Xie, Z. Guo *et al.*, "MoS$_2$/HfO$_2$/silicon-on-insulator dual-photogating transistor with ambipolar photoresponsivity for high-resolution light wavelength detection," to appear in *Advanced Functional Materials*, 2019, doi: 10.1002/adfm.201906242, 2019.

[8] J. Wan, C. Le Royer, A. Zaslavsky, and S. Cristoloveanu, "A systematic study of the sharp-switching Z^2-FET device: From mechanism to modeling and compact memory applications," *Solid-State Electronics,* vol. 90, pp. 2-11, 2013.

[9] S.-M. Joe, H.-J. Kang, N. Choi, M. Kang, B.-G. Park, and J.-H. Lee, "Diode-type NAND flash memory cell string having super-steep switching slope based on positive feedback," *IEEE Transactions on Electron Devices,* vol. 63, no. 4, pp. 1533-1538, 2016.

[10] Y. Solaro, P. Fonteneau, C. A. Legrand, D. Marin-Cudraz, J. Passieux, P. Guyader *et al.*, "Innovative ESD protections for UTBB FD-SOI technology", *IEEE International Electron Devices Meeting*, 2013.

[11] H. E. Dirani, P. Fonteneau, Y. Solaro, P. Ferrari, and S. Cristoloveanu, "Novel FDSOI band-modulation device: Z^2-FET with dual ground planes," in *Solid-State Device Research Conference*, 2016, pp. 210-213.

[12] J. Liu, X. Cao, B. Lu, Y. Chen, A. Zaslavsky, S. Cristoloveanu *et al.*, "Dynamic coupling effect in Z^2-FET and its application for photodetection," *IEEE Journal of the Electron Devices Society,* vol. 7, pp. 846-854, 2019.

[13] M. Arsalan, X. Cao, B. Lu, Y. Chen, A. Zaslavsky, S. Cristoloveanu *et al.*, "A highly sensitive photodetector based on deep-depletion effects in SOI transistors," *IEEE SOI-3D-Subthreshold Microelectronics Technology Unified Conference (S3S)*, 2018, pp. 1-2.

Effect of Dissolved Ozone and In-Situ Wafer Cleaning on Pre-Epitaxial Deposition for Next Generation Semiconductor Devices

Ismail Kashkoush*, Darian Waugh, and Gim Chen

NAURA Akrion Inc.

6330 Hedgewood Dr Suite #150, Allentown, PA 18106

* ikashkoush@naura-akrion.com

ABSTRACT

The effect of in-situ process cleaning before epitaxial deposition was studied. The process includes using dissolved ozone to remove organics from the wafers' surface. In addition, the process was conducted in-situ without transferring the wafers from process to rinse tanks as is traditionally done. Results show that the dissolved ozone has significantly improved the yield results when compared to a process without using dissolved ozone as a surface treatment. The results also showed that dilute chemicals and *in-situ* HF/Drying are key factors required in wafer processing for successful film deposition in advanced IC manufacturing.

INTRODUCTION

The standard approach for Si surface cleaning prior to epitaxial growth processes is a high temperature, usually greater than 1050 °C, gas phase method to dissolve the native oxide along with any other contaminants on the surface of the wafer in order to prevent formation of any defects [1]. However, in the case that an advanced device with sensitive structures requires lower thermal budget treatments, a low-temperature pre-epitaxial cleaning process is needed. However, lowering the process temperature in turn causes an issue by lowering the desorption rate of SiO_2. This issue can be resolved by an HF-last process which converts the surface of the silicon wafer to a hydrogen-terminated surface, which when accomplished properly can yield a hydrophobic surface with the least defects [1,4,5,6].The pre-epitaxial cleaning of Si wafers for deposition can be approached in a wet bench at much lower temperatures than the gas phase method. The process chemicals, sequence and number of cleaning steps are becoming more critical in determining the desired end results [2,3]. The following study provides the data and process of proving that a one-step dilute in-situ-HF in the dryer is more effective than a traditional multi-tank HF-last process in a wet bench [7].

EXPERIMENTAL

All experiments were conducted on Naura-Akrion's GAMA^{TM} automated wet station which is capable of performing both a multi-tank sequence and single tank in-situ process. The silicon wafers are processed in the tool for the pre-epitaxial cleaning prior to the epitaxial growth step. Bare silicon wafers were processed with dummy oxide wafers, alternated or sandwiched, in order to simulate a situation with patterned wafers. The contamination levels from the oxide wafers on the bare wafers would be large due to the etch by-products from the oxide wafers depositing onto the bare wafers during processing. Multiple cleaning techniques were used in order to counteract the high level of contamination caused by the etching process and were compared with the conventional multi-tank method. The materials used were: a GAMA^{TM} wet bench equipped with a LuCID^{TM} dryer (HF controlled injection), KLA-Tencor SurfScan (inspected at ≥ 0.12µm), bare Si wafers with low particle counts and thermal oxide wafers. Concentrations and parameters: 100:1 HF (23 °C), 400:1 dHF (23 °C), 1:2:50 dSC1 (50°C and 800W megasonic), DIO_3 rinse (~5-10 ppm at 23 °C).

RESULTS AND DISCUSSION

The typical standard process is to use a high temperature H_2 pre-bake to desorb the native oxide on the wafers to prepare the surface for an epitaxial layer deposition. However, lower temperatures are required to ensure isothermal processing for these advanced next-generation devices [7-10]. In IC manufacturing, wafers are typically mixed with oxide wafers or the wafers are patterned, and exposed silicon is typically adjacent to oxide or nitride areas. When the wafers are exposed to HF solutions, the by-products of the etched wafers will be removed from the hydrophilic surface and be deposited on the hydrophobic surface. This deposition results in high particle counts on the exposed silicon surface. The process contained herein was created to overcome this issue.

Before proceeding with the experimental procedures, tests to ensure particle neutrality within the GAMA^{TM} wet bench were performed. The results of a conventional HF/Rinse/SC1/Rinse/Dry process yielded low particle addition even in the presence of oxide wafers; i.e. an average particle addition of – 6 (1 σ Stdev = 11, Figure 1). When only using bare silicon wafers, the conventional HF-last process yielded low particle addition as well; i.e. the average particle addition was less than 40 particles at 0.12 µm (Figure 3). In addition, post epitaxial defects were also low (~ 1.26 defects/cm^2), as shown in Figure 2 as a best case scenario.

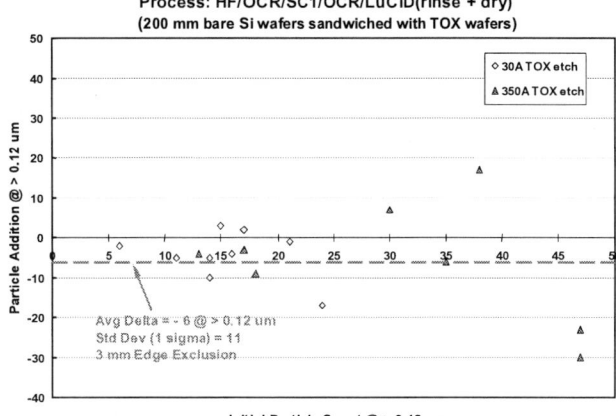

Figure 1: Particle Performance with simulated pattern oxide etch

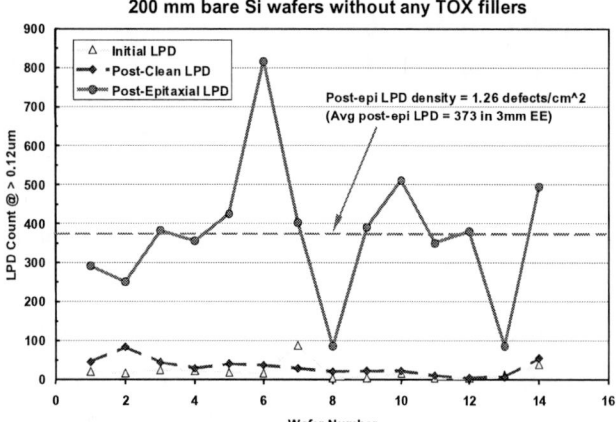

Figure 2: Post-Epitaxial LPD after conventional HF-last

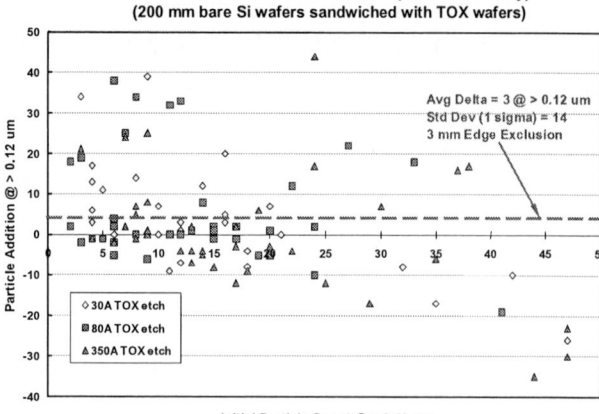

Figure 3: Particle performance with simulated pattern oxide etch (hydrophobic at the end)

Silicon wafers were sandwiched between oxide filler wafers in order to simulate patterned wafers in a typical manufacturing environment. A conventional HF-last process (SC1/Rinse/HF/Rinse/Dry) resulted in high particle counts at 0.12µm (> 1,000). The high pre-epitaxial particle counts also caused high post-epitaxial defects (> 30,000). The particulate defects are normally considered as nucleation sites of epitaxial defects during the epitaxial deposition process. Conventional methods of wafer transfer between tanks plays a significant role in increasing the deposition of silicate particles onto the silicon surface due to wafers crossing the liquid-to-air interface. To counteract the silicate deposition, two different approaches were tested.

An in-situ process was thus developed in order to prevent the wafers from crossing the liquid-to-air interface in which the contaminants reside and deposit onto the wafer surface. HF chemical injection was used in the dryer to perform the in-situ process which yielded much lower particle deposition due to wafers not crossing the liquid-to-air interface. Figures 3 and 4 show the results that the average particle adders of less than 50 particles.

An important note is that the use of ozonated rinse after HF and before going to the SC1 step is critical for eliminating any potential of metal-induced pitting on the hydrophobic surface [11,12]. As reported by Knotter [10], Fe in the SC1 can induce pitting on hydrophobic wafer surface. The oxide chemically grown in the ozonated cascade rinse (OCR) is stable and thick enough (7-10 Å, as shown in Figure 5) to protect the silicon surface from any effects of metal roughening. The post epitaxial cleaning results for the in-situ method are shown in Figure 6, and the average LPD density per wafer is about 0.89 defects/cm^2. Figure 6 also indicates that the lower the HF-last defects are, the lower the post epitaxial deposition defects will be.

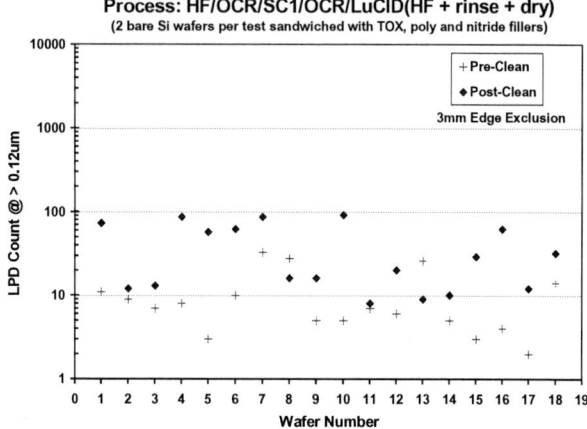

Figure 4: In-situ HF-Last with simulated pattern oxide etch

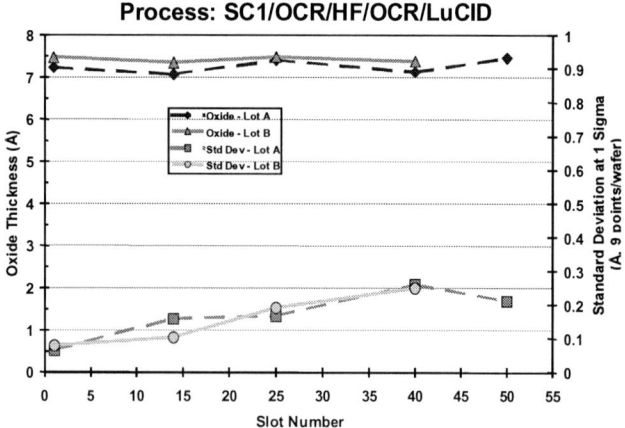

Figure 5: Oxide regrowth uniformity by ozonated rinse

The results of each of the different cleaning recipes are summarized previously [7]. The results from testing showed that the most critical step in order to achieve extremely low post epitaxial deposition defects is the in-situ process which requires no wafer transfer between steps. In order to characterize the background oxide thickness, measurements were also taken as an indicator of the oxygen content on the

978-1-7281-6559-2/20 $31.00 © 2020 IEEE

wafer surface. It is equally important to notice that the amount of oxygen content on the wafer surface could significantly increase the number of post-epitaxial defects on the wafer. The lower the oxygen content on an H-passivated surface, the lower the amount of post-epitaxial defects on the wafer surface, as shown in Figure 7.

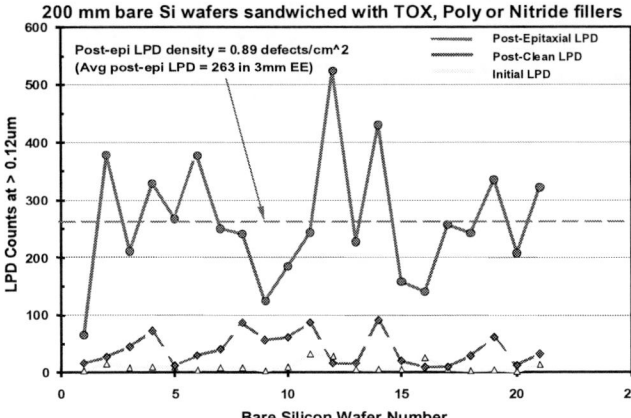

Figure 6: Post Epitaxial (900°C bake) LPD after In-situ HF-Last

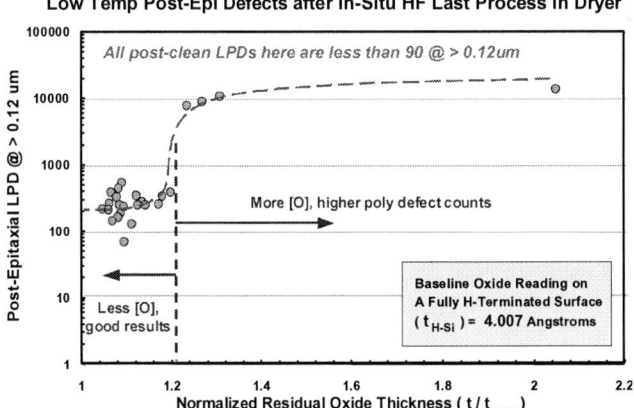

Figure 7: Low Silicate Particles and Low Residual Oxide Content on Surface are Crucial

Dissolved ozone (DIO$_3$) can be used for many different processes such as; oxide regrowth, removal of photoresist as well as organic contaminants, and pre-gate cleaning. Here ozone can be used to re-grow oxide on the wafer post HF cleaning. The oxide thickness is plotted versus time when the wafers are submerged in deionized water at various ozone concentrations at ambient temperature (Figure 8). Dissolved ozone can be used as a substitute of Sulfuric-Peroxide Mixture (SPM) to effectively remove photoresist and organics from Si surfaces [11]. In addition, dissolved ozone can also passivate the surface post HF-etch as used in this study and shown to be very useful.

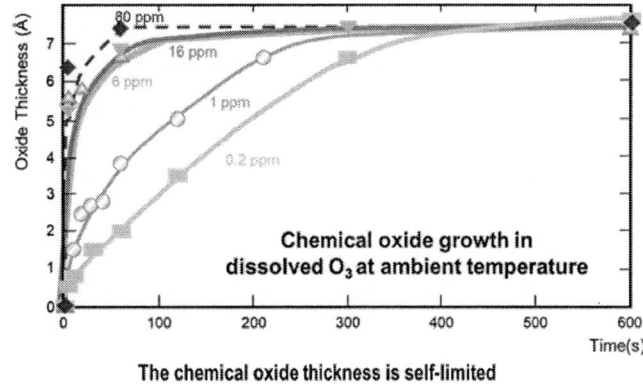

The chemical oxide thickness is self-limited by the diffusion of the oxidant species

Figure 8: Film Thickness when Wafers are Immersed in Ozonated Water at Various Concentrations and Ambient Temperature.

CONCLUSIONS

An in-situ HF-last cleaning process in the dryer was developed and used for pre-epitaxial growth. Results for the in-situ cleaning showed a significant improvement over the standard HF-last process. The reason for the in-situ process having a great impact on the elimination of particle deposition is due to the Si wafers not crossing the liquid-to-air surface between the HF-etch/rinse/dry process. The experiments conducted in the study proved that the use of dilute chemicals and the in-situ HF-etch/rinse/dry process yield lower defects than that of the standard multi-tank HF-last process. The defects after the in-situ clean are directly correlated to the post epitaxial growth defects. The lower the oxygen content and particle defects after cleaning, the lower the post epitaxial deposition defects.

REFERENCES

[1] M. Caymax, et. al. Solid State Phenomena Vols. 65-66 (1999) pp. 237-240, 1999 Scitec Publications, Switzerland.

[2] P. Besson. UCPSS '2000, Vols. 76-77 (2001) pp. 199-202.

[3] I. Kashkoush, et al. Mat. Res. Soc. Symp. Proc., Vol. 477, 1997, pp. 311-316.

[4] I. Golecki, Appl. Phys. Lett., Vol. 69 (1992) p. 1730.

[5] A. Fissel, et. al. Appl. Phys. Lett., Vol. 66 (1995), p. 3182.

[6] P. Patruno, A. Fleury, E. Andre, and F. Tardif, UCPSS '94 Proc., pp. 247-250.

[7] I. Kashkoush, et al. Elec. Soc. Clean. Symp. Proc., Vol. 26, 2001, pp. 345-351.

[8] M. Mouche., et al. UCPSS '96 Proc. Pp 269-272.

[9] S. Verhaverbeke and B. Pagliaro. Electrochem Soc. Proc. Vol. 99-36, pp. 445-451.

[10] M. Knotter and Y. Dumensil. UCPSS '2000, Vols. 76-77 (2001) pp. 255-258.

[11] J-I. Song, R. Novak, I. Kashkoush, and P. Boelen. Micro, Vol. 19, No. 1, January, 2001.

[12] C. Cowache, P. Boelen, I. Kashkoush, F. Tardif. Elec. Chem. Soc. Proc., Vols. 99-36 (2000) pp. 59-68.

FROM MICRONS TO NANOMETERS: THE IRDS AND AMC CONTROL

Chris Muller[1]*, Henry Yu[2], and David Lu[2]
[1]Muller Consulting, Lawrenceville, GA 30045 USA
[2]Purafil, Inc., Doraville, GA 30340 USA
*Corresponding Author's Email: muller-consulting@comcast.net

ABSTRACT

The Yield Enhancement Chapter of the International Roadmap for Devices and Systems (IRDS), and more specifically, the focus topics of Wafer Environment Contaminant Control and Surface Environment Contaminant Control, are responsible for identifying airborne molecular contamination (AMC) and setting guideline limits in all areas of semiconductor processing. Today AMC control is required in FEOL and BEOL operations and this control may be achieved fab-wide or at certain critical processes, potentially also at different levels for different processes.

INTRODUCTION

To those charged with establishing and maintaining the appropriate controlled environments for leading-edge semiconductor manufacturing, AMC control can appear to be a moving target. Sulfur and nitrogen oxides, ozone, and organics from outside air, as well as acids, bases, dopants and organics from sources inside the fab may all have to be considered in a successful AMC control program. But which contaminants should be targeted? What control levels should be considered? Fortunately, there is a resource that facility and process engineers can use to learn about critical issues relative to AMC and its effects on semiconductor manufacturing.

According to the International Technology Roadmap for Semiconductors (ITRS) [1] the percentage of process steps affected by nonparticulate or molecular contamination is expected to increase. The impact of AMC on wafer processing can only be expected to become more deleterious as device dimensions decrease. Pre-gate oxidation, salicidation, contact formation, DUV photolithography, EUV masks, and atomic layer deposition (ALD) have been identified as particularly sensitive production steps.

Yield Enhancement (YE).[2] Yield enhancement is represented by the functionality and reliability of integrated circuits produced on the wafer surfaces. YE for manufacturing of integrated devices addresses the improvement from R&D yield to mature yield. The YE chapter of the ITRS displays the current and future requirements for high yielding manufacturing of dynamic random-access memory (DRAM), microprocessors (MPUs), and flash memory. The YE Chapter includes the focus topic of Wafer Environmental Contamination Control (WECC) where guidelines and technology requirements for AMC can be found.

The wafer environment includes the ambient space around the wafer at all times, whether the wafers are open to the cleanroom air or stored in PODs/FOUPs. AMC needs to be controlled in front-end and back-end of line operations in semiconductor fabs. This control may be achieved fab-wide or at certain critical processes, potentially also at different levels for different processes. A summary of Yield Enhancement WECC AMC interfaces by process areas as well as target levels of ambient acids, bases, condensables, dopants, and metals for specific process steps.

The Yield Enhancement section of the ITRS, and specifically the WECC technology requirements, indicate target levels of ambient acids, bases, condensables, dopants, and metals for specific process steps. Since the 2005 edition of the ITRS, wafer environment contamination control includes the ambient space around the wafer at all times, whether the wafers are open to the ambient cleanroom air or stored in PODs/FOUPs. As the list of chemical contaminants to be controlled broadens one important WECC challenge is the accurate modeling of cleanrooms for AMC sources and distribution. This is because with the reduced air volumes many current cleanroom designs cannot dilute AMC as well as before – whether it comes from outside or inside the fab.

The ITRS takes advantage of expertise from around the world to help identify the technical challenges related to AMC and offers recommendations that can be used in establishing AMC control strategies and guidelines for advanced semiconductor device manufacturing. Requirements have been added for the control of total acids in addition to total bases in lithography applications. Subsequently, requirements have been added addressing specific acids – not just total acids (Table 1). Additional changes for AMC control that have been proposed include total acids being reduced to <20 ppbv (5 ppbv long-term), total condensable organics being reduced to <100 ppbv, and dopants being reduced to <10 pptv.

Outgassing from materials of construction in the cleanroom, wafer processing equipment, post processed wafers, and wafer environmental enclosures as well as inadequate exhaust and fugitive emissions from chemicals used in wafer processing are the main sources of AMC. In some highly congested areas or regions with poor ambient air quality, makeup air is a significant source of AMC.

Oxygen and water vapor as well as low concentration atmospheric contaminants (e.g., carbon dioxide [CO_2], ozone [O_3]) can also be considered as part of the AMC

978-1-7281-6559-2/20 $31.00 © 2020 IEEE

TABLE 1. SUMMARY OF AMC INTERFACES BY PROCESS AREA

Airborne Molecular Contaminants in Gas Phase	Short-term (long-term) Limits in pptM
Lithography (cleanroom ambient)	
Total inorganic acids (as SO_4)	2,500
Total organic acids (as SO_4)	TBD
Total bases (as NH_3)	50,000
Total condensable organics (w/ GCMS retention times $\geq$benzene, calibrated to hexadecane)	26,000
Refractory compounds (organics containing sulfur, phosphorus, silicon, calibrated to hexadecane)	100
Total surface molecular refractory condensable (SMRC) organics,	2 ng/cm^2/day
Gate/Furnace area wafer environment cleanroom/POD/FOUP ambient)	
Total metals (as copper)	1 (0.5)
Dopants (as elements B, P, As, etc.)	10
Total surface molecular condensable (SMC) organics on wafers,	2 (0.5) ng/cm^2/day
Salicidation Wafer Environment cleanroom/POD/FOUP ambient)	
Total inorganic acids (as SO4)	100 (10)
Total organic acids (as SO_4)	TBD
Exposed Copper Wafer Environment (cleanroom/POD/FOUP ambient	
Total inorganic acids (as SO_4)	500
Total organic acids (as SO_4)	TBD
Total other corrosive (oxidizing) species (as Cl_2)	1,000
Exposed Aluminum Wafer Environment (cleanroom/POD/FOUP ambient)	
Total inorganic acids (as SO_4)	500
Total organic acids (as SO_4)	TBD
Total other corrosive (oxidizing) species (as Cl_2)	1,000
Reticle Exposure (Cleanroom/POD/Box ambient)	
Total inorganic acids (as SO_4)	500 (TBD)
Total organic acids (as SO_4)	TBD
Total bases (as NH_3)	2,500 (TBD)
Total SMC organics on wafers,	0.29 ng/cm^2/day

burden. Acid vapors in the air have been linked to corrosion on wafers, as well as with the release of boron from HEPA filters. The impact of amines on deep ultraviolet (DUV) photoresists are well known examples of AMC affecting wafer processing. Hydrocarbon films of only a few monolayers may lead to loss of process control, especially for front-end processes. The impact of AMC on wafer processing can only be expected to become more deleterious with future device generations.

ITRS 2.0 [3]

In 2012, it became evident that the ecosystem of the electronic industry had undergone a major change with the proliferation of the Internet, the worldwide spread usage of wireless mobile devices like smart phones and tablets and, most of all, the Internet had evolved from the Internet of Things (IoT) to the Internet of Everything. Sensors of all kinds continue to be added to the Internet and the range of remote operations is increasing by the day allowing remote operation of tools and remote medicine as an example. To fully address these changes the process of restructuring the ITRS was initiated in 2012 and completed in 2014. The 17 International Technology Working Groups (ITWGs) were replaced in 2015 by seven Focus Teams in the new ITRS 2.0. These are:

- System Integration
- Heterogeneous Integration
- Heterogeneous Components
- Outside System Connectivity
- More Moore
- Beyond CMOS
- Factory Integration.

Yield Enhancement was now covered as part of the Factory Integration Focus Team and chapter, and more specifically the section on Wafer Environment Contamination Control was maintained

International Roadmap for Devices and Systems [4]

In 2016, the traditional process of setting an industry roadmap for semiconductors has been taken on by the Institute of Electrical and Electronic Engineers (IEEE) and expanded to include all computing. The ITRS, first published in 1965, has been reorganized as the International Roadmap for Devices and Systems (IRDS) and this new effort aims to take a broad view of the needs of computing generally. It will address road map issues that include computer systems, architectures and software as well as the chips and other components used in them.

The move comes at a time when the ITRS has lost some of its influence. The Roadmap used to point a wide variety of chip makers toward a common set of technical milestones. In recent years, however, a consolidating set of chip makers have forged their own road maps, often naming nodes in ways more driven by marketing than engineering executives.

Advanced semiconductor industry has limited ability to monitor defects on the wafer. This requires that all critical utilities and components ensure consistent quality control at all levels from manufacturing through operations. This is critical for enabling existing and future advanced semiconductor manufacturing processes.

In the new IDRS AMC guidelines will be developed by AMC experts under a reorganization of the section on Wafer Environment Contamination Control (WECC+). Specific scopes will be defined in each area such as types of gases, chemicals, and components, critical processes, and priority focus areas and limitations. Interfaces with other groups will aid in the development of new AMC guidelines. Key deliverables will include:

- Technology roadmap documents, including definition of the quality requirements, needs/risks, and potential solutions
- SEMI Standards
- Conference presentations

The goals and objectives for WECC+ under the IDRS remains similar to those of the original ITRS in that risks will be identified, and risk mitigation strategies proposed. However, with the IEEE's extensive resources collaborative experimental studies can be initiated and conducted. Communication of the risks and their mitigation strategies to the industry will be improved and to facilitate solutions SEMI standards will be updated or new standards developed. New technology trials (e.g. metrology) will be supported through benchmarking studies. All of this to continue refining the AMC requirements for advanced semiconductor manufacturing.

The Yield Enhancement (YE) Chapter of the International Roadmap for Devices and Systems (IRDS) is dedicated to ensuring that semiconductor manufacturing set up is optimized towards identifying, reducing and avoiding yield relevant defects and contamination. In the manufacture of integrated circuits, yield loss is related to a variety of sources. During processes such as implantation, etching, deposition, planarization, cleaning, lithography, etc., failures responsible for yield loss occur. A primary example of contamination and mechanisms responsible for yield loss is from AMC.

The WECC tables of the Yield Enhancement Chapter provide recommended contamination control levels which should be maintained at the interface between cleanroom environment and the part of the manufacturing equipment (minienvironments) as follows:

- AMC as measured/monitored in the cleanroom air and /or purge gas environment,
- SMC on monitoring wafers.

These values reflect the need to reduce AMC from the ambient environment as well as to keep the out-gassing emissions in the clean room environment at low level.

The YE Chapter focus topic "Surface Environment Contamination Control" (SECC) includes AMC control and crosscuts front end process technology, interconnect processes, lithography, metrology, design, process integration, test, and facility infrastructures." The SECC (formerly the ITRS's WECC) is limited to the consideration of the next two generations of the most advanced semiconductor manufacturing technologies, as defined by the production node size. It is focused on critical surfaces of the manufacturing process, including

but not limited to wafers, lithography reticles, and lithography lenses. Two typical sources of contamination include the environments of the critical surface processing and the critical system components that may shed or outgas contamination into the environment and through it to the critical surface. It is also possible that the critical surface can contribute to the contamination to the environment and subsequently contaminate the surface. SECC is intended to define (among other parameters) allowable levels of AMC that will ensure proper contamination control at the critical surfaces to reduce defects and improve yield for cost effective semiconductor manufacturing.

Even though the wavelength used in the lithographic process has not changed since the introduction of 193 nm more than 10 years ago, the integration of semiconductor devices has continued and today 14 nm node devices are in high volume and 10 nm devices in the early production stage. With this increased integration has come increased process complexity, and yield has been impacted by more than the traditional defects caused by particle contamination. This is the case where very low levels of chemical molecules, atoms and/or ions in the environment interfere with the process, equipment, device structures, parameters and/or device performance degrading yield.

With regards to AMC, the number of the chemical species in the environment has increased very rapidly as feature size decrease and new materials are integrated in the semiconductor manufacturing process. The impact of AMC on wafer processing can only be expected to become more deleterious; driven not only by decreasing device dimensions but also by the introduction of new chemistry and recipes for future technical nodes (Figure 1). As with other contamination areas, the industry's response to AMC has been reactive. A new, proactive approach is needed to deal with increased threat to device yield and performance.

Since the 1990s when the presence of amines in the environment impacted chemically amplified resists, the approach to AMC has been reactive. As with this first

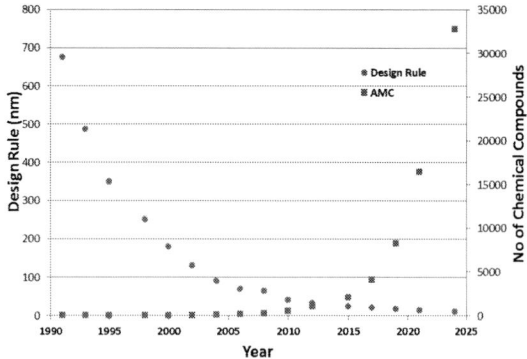

Figure 1. AMC versus design rules.

case, an investigation into a yield degradation issue reveals the presence of a chemical contaminant in the environment and a countermeasure is put in place to narrowly address the specific issue. However, this reactive approach to AMC control is no longer effective and a new approach must be developed that is based on the premise that a chemical reaction is required for an AMC contaminant to have an impact on yield. The systematic classification of these chemical reactions can be used to extrapolate and generalized from what is already known to come up with solution in a more proactive manner. The first step is the speciation and quantification of the chemical compounds in the wafer environment. Next is to understand the adsorption and deposition kinetics on the wafer. Finally, the device impact needs to be characterized.

Beginning with the 2017 IRDS and continuing today, the original classification of AMC contaminants first proposed in the ITRS is being revised. The new classification scheme is based on chemical families, their impacts on processes, equipment, and devices, and whether impacts are known or unknown, and more research is required (Table 2).

AMC CONTROL [5]

A successful AMC mitigation strategy will include the use of minienvironments like EFEM and FOUP; as well as a comprehensive chemical filtration strategy.

Chemical filters are generally applied as the first line of defense against AMC. These filters must be capable of dealing with high volumetric flow and low outgassing from filter materials themselves. Advanced sorbent materials with high specific surface area and tunable pore size structure have been developed to enhance adsorption capacity and reduce mass transfer resistance at the solid-gas interface. The porous sorbents have also been modified with either acids for the absorption of basic gases, or bases for the absorption of acidic gases. For example, strong organic bases containing ammonia and amines could be effectively removed by ion exchange resins and acid-modified sorbents through chemisorption involving proton transfer.

Development of chemical filters tailoring the removal of specific types of the micro-contaminants is still evolving and needed, though this is contingent on the identification and recognition of the key AMC species to be controlled. Recently, adsorption in combination with advanced oxidation processes has also been studied. The main advantage of such device is to prolong the operating life of a chemical filter. Adsorptive filter with photocatalytic capacity, for example, has been applied to remove volatile organic molecules and organic sulfurs with success in the testing scales. These types of chemical filters appear to be promising as they can chemically remove AMCs, though the formation of intermediate

TABLE 2. CLASSIFICATION OF AMC CONTAMINANTS BASED ON CHEMICAL FAMILIES

Chemical Family	Process	Equipment	Device		
			FEOL	MOL	BEOL
Roadmap Term.	POU	POE	POP	POP	POP
Inorganic Acids	Yes	Yes	No	Yes	Yes
Inorganic Bases	Yes	No	Yes	Yes	Yes
Carboxilic Acids	Yes	Yes	Not Known	Not Known	Yes
Organic bases	Yes	Yes	Yes	Yes	Yes
Esters	Not Known	Yes	Not Known	Not Known	Not Known
Aldehyde/Ketones	Not Known	Yes	Not Known	Not Known	Not Known
Phenolic/Alchohol	Yes	Not Known	Not Known	Not Known	Not Known
Aromatics	Not Known	Yes	No	No	No
Plastizers	Yes	No	Yes	No	No
Aliphatics	No	Yes	No	No	No
Heterocycles	Not Known	Yes	Not Known	Not Known	Not Known
Siloxanes	Not Known	Yes	Not Known	Not Known	Not Known
Fluorinated	Not Known	Yes	No	No	No
Sulfurous	Yes	Not Known	Not Known	Not Known	Not Known
Halocarbons	Not Known	Not Known	Not Known	Not Known	Not Known

products and their potential impacts remains to be studied.

SUMMARY

The percentage of process steps affected by nonparticulate or molecular contamination is expected to increase, and the impact of AMC on wafer processing can only be expected to become more deleterious as device dimensions decrease. Sensitive production steps include pre-gate oxidation, salicidation, contact formation, DUV photolithography, EUV masks, and ALD.

The proactive control of AMC in the manufacture of semiconductor devices started with the original ITRS, continued during the conversion to the IRDS, and future changes are being. A new AMC classification proposal and device defect mechanism from the IRDS can be used to systematically classify the impact of different chemical species identified in a given cleanroom environment. Lastly, new AMC countermeasures such as the use of minienvironments and the development of a comprehensive chemical filtration strategy tailoring the removal of specific types of chemical contaminants based the identification and recognition of key AMC species to be controlled is considered mandatory for leading-edge fabs.

REFERENCES

[1] International Technology Roadmap for Semiconductors, http://www.itrs2.net/2013-itrs.html
[2] International Technology Roadmap for Semiconductors, Yield Enhancement, https://www.semiconductors.org/wp-content/uploads/2018/08/2013Yield.pdf
[3] ITRS 2.0, http://www.itrs2.net/itrs-news.html
[4] International Roadmap for Devices and Systems, Institute of Electrical and Electronics Engineers, New York, https://irds.ieee.org/editions/2018
[5] C. Garza. Proactive Control of Airborne Molecular Contamination in the Manufacture of Semiconductor Devices, 2018 (unpublished).

STUDY OF SHALLOW TRENCH ISOLATION GAP FILL FOR 19NM NAND FLASH

Li Peng[*], Hongbo Li, Tiantuo Sun, Xing Gao and Qin Sun

Shanghai Huali Microelectronics Corporation, Pudong New Area, Shanghai 200433, China

pengli@hlmc.cn

ABSTRACT

Polysilazane (PSZ) curing has been introduced for 19nm NAND Flash to ensure void free Shallow Trench Isolation (STI) gap fill. PSZ film was converted into oxide mainly depending on temperature and water vapor. The high temperature PSZ curing would give rise to Si dislocation and PSZ crack. However, lowering curing temperature would lead to an insufficient conversion of PSZ film and even generate voids. As a result, wet oxidation was utilized between curing 1 and curing 2 to improve conversion rate of PSZ film. The TEM images showed good gap fill performance of PSZ curing by using low temperature/wet oxidation method.

INTRODUCTION

As the critical dimension (CD) of integrated circuit (IC) scale down, Shallow Trench Isolation (STI) gap fill becomes a big challenge due to its high aspect ratio (AR). Generally, High-density plasma chemical vapor deposition (HDP-CVD) and high aspect ratio process (HARP) are widely used for STI gap-fill at 65nm technology node. However, they become increasingly difficult to provide good gap fill for technology nodes below 28nm with aspect ratios bigger than 13:1. [1] In order to be able to fill STI geometries with these ARs, PSZ filling and curing have attracted wide attentions due to the good gap fill performance [2].

In this paper, PSZ filling and curing were used to obtain void-free STI gap fill for 19nm NAND Flash. The conversion of PSZ to oxide was mainly depending on temperature and water vapor. Thus, the influence of curing temperature, water vapor and Q-time (retention time from end time of last process to start time of next process) on conversion rate of PSZ film was investigated.

EXPERIMENT

STI structures used in this study were prepared. After the deposition of liner oxide, PSZ film was coated, and then cured to form oxide. To investigate the gap-fill performance, cross section images were obtained with Scanning Electron Microscope (SEM) and Transmission Electron Microscope (TEM).

RESULTS AND DISCUSSION

Traditional PSZ curing

In a typical STI, a trench was etched into a Si substrate, and the STI liner oxide was deposited as a conformal layer. As shown in Figure 1c, the depth of STI trench became much greater while the width became narrower, resulting in an ultra-high aspect ratio. In this study, the aspect ratio of actual structures was about 33:1. It was difficult to form a void-free gap fill using HARP.

Polysilazane (PSZ), consisting of two chemicals, precursor Per Hydro Poly Silazane (PHPS) and carrier solvent Di-n-Butyl Ether (DBE), was used to fill STI trench with ultra-high aspect ratio for 19nm NAND Flash. The void-free gap fill exhibited the good gap fill property of PSZ (Figure 1a). Then the PSZ was converted into oxide in an oxygen and water vapor (O_2/H_2O) atmosphere at an elevated temperature. During this process, N and H atoms of PSZ were replaced by O atoms to form an –Si-O-Si- network structure.

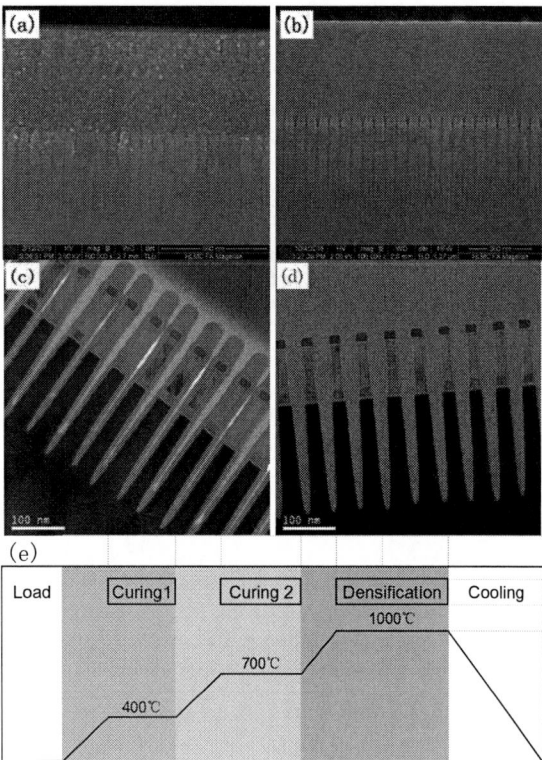

Figure 1: SEM images of (a) PSZ filling and (b) PSZ curing; TEM images of (c) STI liner oxide and (d) PSZ curing; (e) thermal curve of traditional PSZ curing.

The traditional PSZ curing mainly involved three steps: curing 1, curing 2 and densification (Figure 1e). Curing 1 was very effective to reduce impurities and realize homogeneous curing. Thus, the temperature should be as low as possible. Curing 2 was to further improve conversion rate. Besides, densification was needed to ensure wet etch rate of oxide. It could be seen that the PSZ film became denser after curing (Figure 1b). The TEM image also indicated the PSZ film was converted into oxide completely (Figure 1d). However, the high temperature process might oxidize on active area, resulting in Si consumption. Besides, it may lead to Si dislocation and PSZ crack. Thus, the temperature of PSZ curing should be decreased.

Novel PSZ curing

To decrease thermal of PSZ curing, the temperature of curing 1, curing 2 and densification was decreased to 200℃, 500℃, and 850℃. However, some voids were observed after PSZ curing, indicating insufficient conversion of PSZ film (Figure 2a). This was because PSZ film without curing would be degraded and then formed voids due to the influence of high energy ion beam under TEM.

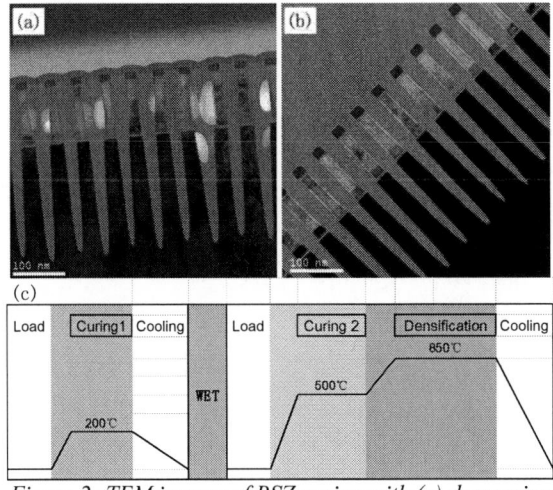

Figure 2: TEM images of PSZ curing with (a) decreasing thermal and (b) decreasing thermal and adding wet process for three times between curing 1 and curing 2; (c) thermal curve of novel PSZ curing

The conversion of PSZ to oxide is mainly depending on temperature and water vapor. To improve the conversion rate, a wet oxidation can be utilized between curing 1 and curing 2 based on decreasing thermal (Figure 2c). After curing 1, the polysilazane-based oxide was preferably treated by a wet oxidation in the presence of 70℃ water, or alternately in a Standard Clean 1 (SCI: a dilution of NH_4OH/H_2O_2) or a Standard Clean 2 (SC2: a dilution of HCl/H_2O_2) for three times. In this way, the PSZ layer could contact with water vapor adequately and then was oxidized to oxide. Besides, the SC1 and SC2 could reduce impurities. Thus, the void-free STI gap fill was observed by low temperature/wet oxidation method (Figure 2b).

Besides, the influence of water vapor on PSZ curing was investigated. As shown in Figure 3, with increasing the number of soaking in water, the frequency and size of voids were greatly reduced. This indicated that water vapor was beneficial to the conversion of PSZ film to oxide. Thus, it is necessary to be soaked in water for three times in order to ensure complete conversion of PSZ film.

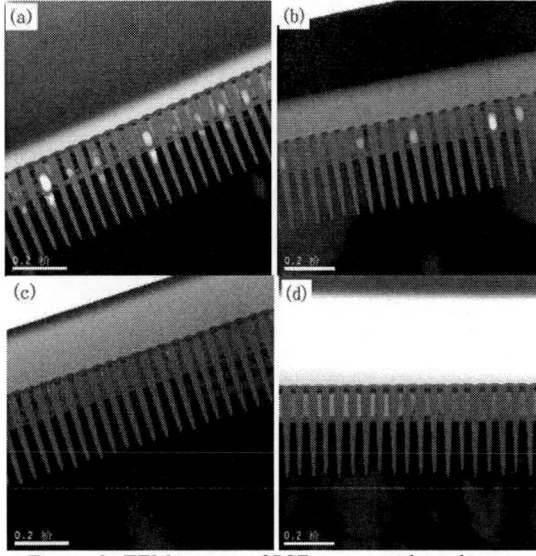

Figure 3: TEM images of PSZ curing with soaking in water for (a) zero, (b) one, (c) two, and (d) three times.

Interestingly, the conversion rate of PSZ film was also influenced by Q-time (retention time from end time of last process to start time of next process). As shown in Table 1, some samples (S1-S5) with different Q-time of novel PSZ curing were investigated. The TEM results showed that Q-time between curing 1 and wet 1 was very important for PSZ curing. This was because the surface area was preferentially converted into dense oxide, which prevented the inward diffusion of water vapor and affected the conversion of PSZ film to oxide. Thus, the Q-time between was required to less than 5 hours.

TABLE I. PSZ CURING WITH DIFFERENT Q-TIME

Q-time	S1	S2	S3	S4	S5
Curing 1 to wet 1	9 h	5 h	5 h	5 h	24 h
wet 1 to wet 2	24 h	<4 h	4 h	24 h	6 h
wet 2 to wet 3	<4 h	<4 h	<4 h	<4 h	<4 h
wet 3 to Curing 2	24 h	<4 h	24 h	24 h	24 h
TEM data	Void	OK	OK	OK	Void

CONCLUSION

PSZ filling and curing was used for STI gap fill with ultra-high aspect ratio at 19nm NAND Flash. The SEM results showed the good gap fill performance. The TEM results indicated the complete conversion of PSZ to oxide using low temperature/wet oxidation method. The conversion rate of PSZ curing was influenced by curing temperature, water vapor and Q-time.

REFERENCES

[1] Y. Bao, X. Zhou, N. Sang, et al. *China Semiconductor Technology International Conference 2015*, Shanghai, March 15-16, 2015.

[2] J. Fucsko, J. A. Smythe III, L. Li, et al. *U. S. Patent 7521378*, 2009.

MODELING METHOD OF LOCAL MISMATCH MODEL FOR MOS TRANSISTORS

Jinglun Gu

Division of Technology Development, Shanghai Huali Microelectronics Corporation,
Shanghai 201210, China
Email: gujinglun@hlmc.cn

ABSTRACT

In this paper, a MOS SPICE local mismatch model is presented, in which the temperature effect coefficient is added in the existing SPICE local mismatch model equation. The effect of temperature variation on SPICE local mismatch model of MOS devices is fully considered. This paper can accurately reflect the change of local mismatch with the change of temperature, so that the SPICE local mismatch model has a wider application range and a higher degree of coincidence with the measured data.

PREFACE

According to the classical literature, the mismatch of MOS devices is the phenomenon of random fluctuation of physical quantity of the same MOS devices that does not change with time in some manufacturing process. The final design accuracy and yield of the circuit are determined by the mismatching degree of the device under the specific process. Circuit designers need accurate MOSFET mismatch model to constrain circuit optimization design, and layout designers need corresponding design rules to reduce chip mismatch. Especially after the size of CMOS process devices enters the deep sub-micron range, the device mismatch becomes more and more serious with the size reduction, which restricts the performance of RF/Analog integrated circuits. Of course, digital circuits are not completely free from the influence of device mismatch. In the design of large-scale memory, the influence of transistor mismatch on the clock signal of sub-memory cell must be considered.

Local mismatch and global mismatch: for identical devices, local mismatch can be simply understood as parameter mismatch between devices in local area. The global mismatch is caused by the changes of parameters (such as temperature and doping concentration) on the silicon wafer.

Local mismatch is caused by two parts:

Firstly, the size of a device on a layout. The larger the area of the device, the better the matching effect. This is called the law of area.

Secondly, The distance of the device on the layout. The closer the device is on the layout, the better the matching effect. This is called the law of spacing [1].

$$\frac{\sigma_P}{P} = \frac{A_P}{\sqrt{W_{eff} \cdot L_{eff}}} + S_P \cdot D \tag{1}$$

P is an electrical parameter of the device, sigma _P is a mismatch of P, P standard deviation, Ap and Sp are area effect parameters and spacing effect parameters respectively, and D is the spacing between the two devices.

SPICE local mismatch model equation generally adds local mismatch model to some parameters in the compact model of MOS devices, such as selecting threshold voltage $vth0$, unit voltage (V), which is related to how large gate voltage Vg can be applied to the transistor to conduct and work under the given condition of leakage terminal voltage Vd.

And, carrier mobility $u0$, drift velocity of carrier under unit electric field intensity, unit is square centimeter /(volt • second), cm2/(v•s), which is related to transistor saturation current Idsat or linear current Idlin and size, and Idsat size represents the strength of transistor performance.

The traditional local mismatch model of these two parameters is as follows, which are used to adjust the voltage mismatch and the power loss distribution respectively:

$vth0$:

$$lcal\_vth0\_d\_n = (va \times gl\_1n) \times geo\_fac \\ \times mos\_local\_flag \tag{2}$$

$u0$:

$$lcal\_u0\_d\_n = (vb \times gl\_2n) \times geo\_fac \times \\ mos\_local\_flag \tag{3}$$

geo_fac is size factor, $geo\_fac = 1/\text{sqrt}(W_{eff} \times L_{eff})$, the W_{eff} and L_{eff} is equivalent width and the equivalent length, the actual width and length on the device characterization of physical, generally $L_{eff} = 1 \times scale$, $W_{eff} = (W/nf) \times scale$, L is the MOS device layout actual length, W is the MOS device layout actual width, nf is the finger number of MOS transistor. $scale$ is reduced size factor, va and vb is no units specified adjustment coefficient, va and vb can be specified for any real number.It can also be expressed by algebraic polynomials with respect to width W and length L or other dimensional parameters.

mos_local_flag is the local mismatch model identifier. This parameter is set to 1 to open the adaptation model and 0 to close the local mismatch model.

The two parameters gl_1n and gl_2n are different in

978-1-7281-6559-2/20 $31.00 © 2020 IEEE

name, but essentially the same as Normal distribution function, also known as Gaussian distribution.

Due to the random fluctuation of the same process conditions on the same MOS devices adjacent to the same module, the electrical performance of the same MOS devices adjacent to each other will be inconsistent, which is called the local mismatch of MOS tubes.

According to the classical theory, the local mismatch of MOS devices is inversely proportional to the square root of the device area.

At present, the temperature effect model has not been introduced into the local mismatch model, so the model cannot accurately reflect the changes of local mismatch with the change of temperature.

The local mismatch model without temperature coefficient can only match the local mismatch data at 25 °C, i.e. normal temperature.

As shown in figure 1~ 3, if the local mismatch model without temperature effect coefficient is used, the model cannot be consistent with the measured data of different temperatures (-40 °C, 25 °C, 125°C).

After the analysis of FIG. 1~ 3, for Idsat and Vtlin, the normal temperature mismatch models of -40°C and 125 °C are listed on both sides of the measured data, suggesting that the 125°C model should be reduced and the -40 °C model should be increased. Therefore, a monotonic function should be added to the original local mismatch model as the temperature coefficient.

The definition of electrical characteristic parameters of MOS devices is shown in table 1.

Vds is the voltage between MOS drain and source, Vgs is the voltage between gate and source, Vbs is the voltage between substrate and source, Vdd is the power supply voltage, Ids is the MOS drain current.

MODELING METHOD

Further improving the SPICE local mismatch model, the temperature effect coefficient does not work when the simulation temperature in the model is set at 25°C, and does not work when the simulation temperature in the model is set at 25°C.

The temperature effect coefficient tcoef is calculated by the following formula:

$$tcoef = 1.0 + (temper - 25.0) \times [tc1 + tc2 \times$$

Table 1 Parameters Illustration

Idsat	Vds = Vdd;Vgs = Vdd.Ids at Vbs=0(V)
Idlin	Vds = 50 (mV);Vgs = Vdd.Ids at Vbs=0(V)
Idoff	Vds = Vdd;Vgs = 0 (V).Ids at Vbs=0(V)
Vtlin	Vgs when Ids=40nA×(W/L) and Vds=50 (mV)
Vtsat	Vgs when Ids=40nA×(W/L) and Vds=Vdd

$$(temper - 25.0)]$$

$$(4)$$

temper is the simulation temperature set in the model. *temper* can automatically read the temperature set in the data document as its value. $tc1$ is the coefficient of primary term and $tc2$ is the coefficient of secondary term.

$tc1$ and $tc2$ can be specified arbitrarily.

Further improve the SPICE local mismatch model, $tc2 = 0$.

The SPICE local mismatch model is further improved. The voltage mismatch model $lcal\_vth0\_d\_n$ is calculated by the following formula: $lcal\_vth0\_d\_n = tcoef \times (va \times gl\_1n) \times geo\_fac \times mos\_local\_flag$

$$(5)$$

Parameters in this formula are declared above. va can be any real number, va can also be expressed by algebraic polynomial about width W, length L or other dimension parameters. Normally, $-2 \ll va \ll 2$.

The SPICE local mismatch model is further improved, and lcal_u0_d_n is calculated by the following formula: $lcal\_u0\_d\_n = tcoef \times (vb \times gl\_2n) \times geo\_fac \times mos\_(local\_flag)$

$$(6)$$

The method and range to use vb is the same as va.

FIGURES AND ILLUSTRATION

For 4 figures in the page 3, the horizontal coordinate is the inverse of the square root of the area of MOS devices, and the vertical coordinate is the standard deviation of the local mismatch of adjacent MOS devices. There are two electrical parameters represented, namely Idsat and Vtlin.

ACKNOWLEDGEMENTS

The author would like to thank HLMC TD1 SPICE model team for supporting this paper.

REFERENCES

[1] M. J. M. Pelgrom, A. C. J. Duinmaijer, and A. P. G. Welbers, "Matching Properties of MOS Transistors," *IEEE J. Solid-State Circuits*, vol. 24, no. 5, pp. 1433–1439, Oct. 1989.

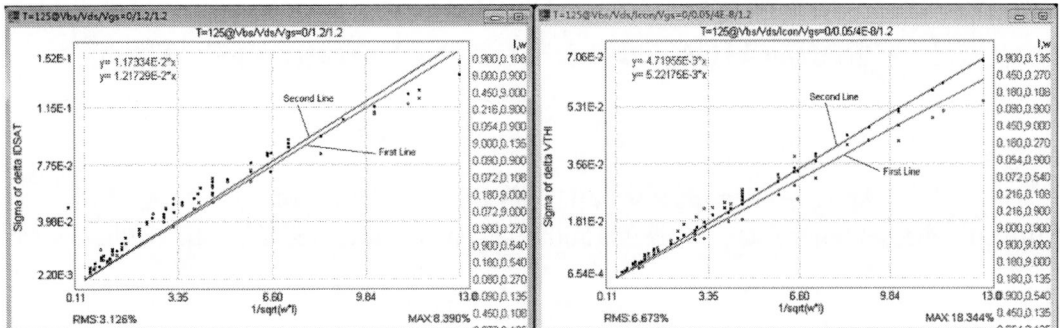

Fig.1: Schematic diagram of the local mismatch of 125°C NMOS with no temperature coefficient.

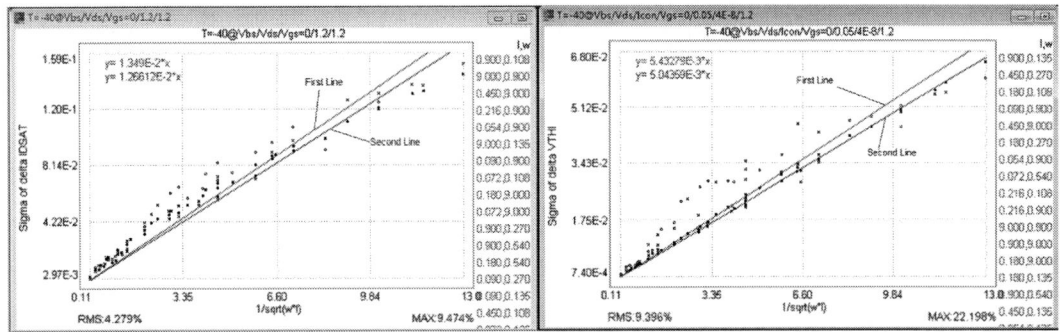

Fig.2: Schematic diagram of the local mismatch of -40°C NMOS with no temperature coefficient.

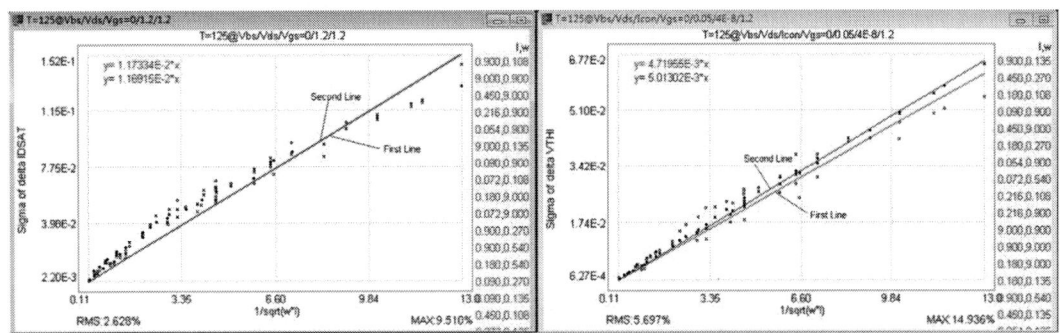

Fig.3: Schematic diagram of the local mismatch of NMOS at 125°C with the addition of linear temperature coefficient.

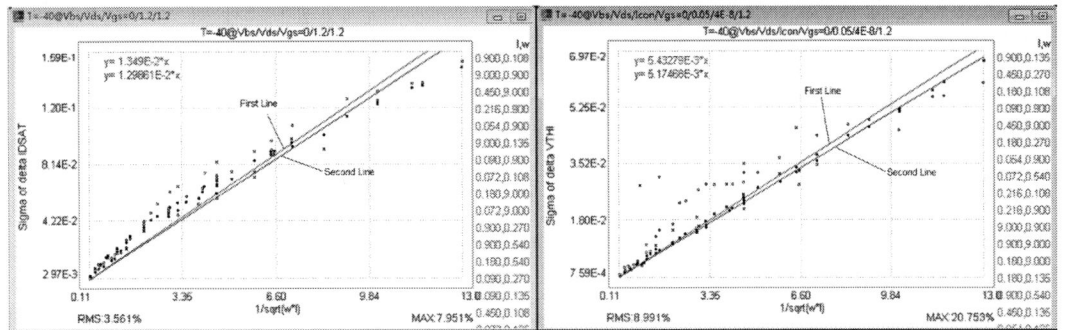

Fig.4: Schematic diagram of the local mismatch of NMOS at -40°C with the addition of linear temperature coefficient

AN APPLICATION OF LDMOS ON ESD PROTECTION

Li Wang*, Yu Chen, Hualun Chen

Huahong Semiconductor (WUXI) Limited, Wuxi, 214028, China
*Corresponding author. Tel.: +86-18921506092. E-mail addresses: Li.Wang@hhgrace.com

Biography

Li Wang Now is a technology development engineer of Huahong Semiconductor (WUXI) Limited. His research interest focuses on high voltage device, PMIC and embedded Non-volatile memory.

Abstract

This paper focuses on the application of a LDMOS ESD protection method. The application is by tuning the size of poly cover field oxide .It needn't extra process step. This way can reduce the breakdown voltage. When ESD is coming, this device can turn on first, then protect the core the Chip.

Keywords— LDMOS ; POLY cover field oxide ; ESD protection

Introduction

LDMOS is a kind of device often used in BCD process. This device can be used as switch MOS or in the analog part of the circuit. This paper is a method of applying LDMOS to power rail ESD clamp in some circuits. In use, the structure of the device is the same as that of the conventional LDMOS in the circuit, and no additional process steps are needed. In the case of fixed filed oxide size, the LDMOS breakdown voltage used as the power rail ESD clamp can be lower than the LDMOS breakdown voltage of the internal circuit only by adjusting the poly cover field size. In this way, when there is ESD on the power, power rail ESD clamp LDMOS will lead to turn on, Current through ESD LDMOS , thus playing the role of ESD protection. Of course, LDMOS used as ESD protection needs to do conventional ESD processing (s / D side does not do silicide), and this processing does not need to add additional process steps.

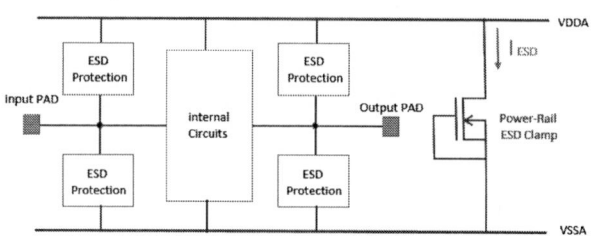

Fig.1. ESD protection circuit [1]

Device and Measurement Details

For power rail ESD clamp LDMOS, we will first define three groups of sizes: FL (filed oxide length) , PF (poly cover field oxide), PA (the distance from the poly end on the filed oxide to the drain end).

As shown in the figure below:

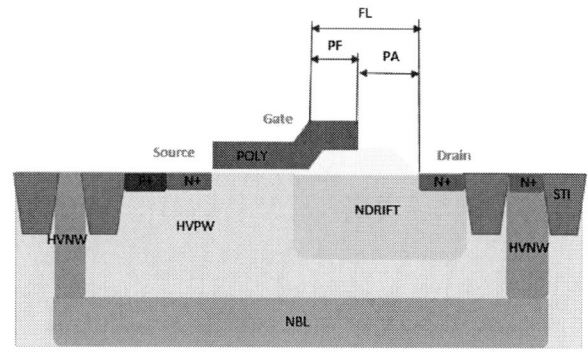

Fig.2. Schematic cross-sectional drawing of LDMOS.

For LDMOS with specific voltage, the length of FL is generally fixed, and the length of PF & PA is adjusted to make the breakdown voltage and RSP reach an optimal combination. Usually we will choose the PF & PA of breakdown voltage under the highest condition.

The following is the simulation results of LDMOS adjusting the size combination of PF & PA under the condition of fixed FL in a BCD process.

978-1-7281-6559-2/20 $31.00 © 2020 IEEE

LCH	LA	PF	PA	Rsp	BV
0.25	0.2	0.15	0.5	4.44	22.61
0.25	0.2	0.25	0.4	4.30	25.49
0.25	0.2	0.35	0.3	4.20	26.55
0.25	0.2	0.45	0.2	4.07	24.67
0.25	0.2	0.55	0.1	4.05	20.75

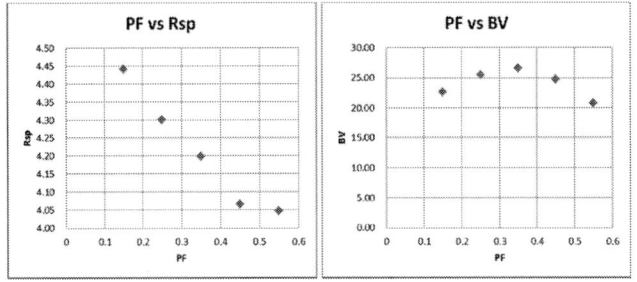

Fig.3. PF&PA the correlation between the size and BV&Rds.

breakdown voltage, the power rail ESD clamp in ESD protection can be used.

REFERENCES

[1] Prof. Ming-Dou Ker " Advanced ESD Protection Design for CMOS Circuits and Systems" 2016 ch2-4 & ch6-2.

In normal BCD process, PF & PA will be a fixed value. We can adjust the size of PF & PA through the correlation in Fig.3. In the layout stage so that the breakdown voltage can be reduced of LDMOS for ESD protection. Generally, we reduce the PA size, so as to reduce the breakdown voltage and improve the RSP of LDMOS, making the ESD performance of LDMOS better. The trigger voltage used for ESD protection LDMOS should be higher than the working voltage of the circuit, but lower than the breakdown voltage of the internal circuit, so as to play the role of ESD protection.

As shown in the figure below:

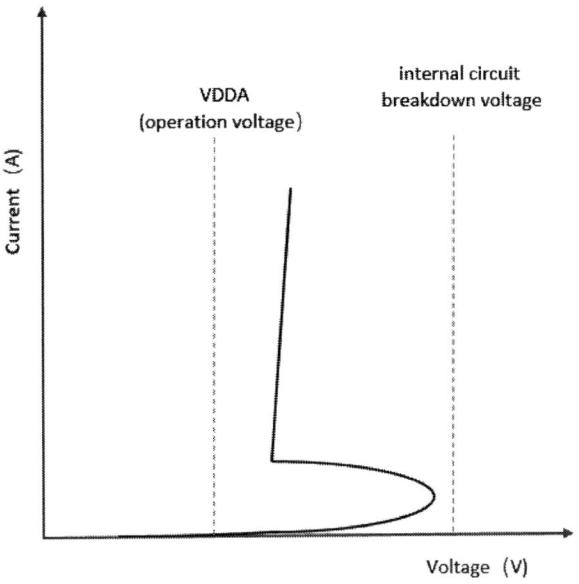

Fig.3. ESD protection LDMOS window.

Conclusions

By adjusting the proportion of PF & PA of LDMOS in BCD process, the breakdown voltage of LDMOS can be changed. By adjusting the breakdown voltage to be higher than the operation voltage and lower than the internal circuit

HV GATE OXIDE OVER-OXIDATION PROCESS OPTIMIZATION FOR SONOS 1.5T FLASH CELL

Jian Zhang[1], Wei Xiong[1], Hualun Chen[1]*

[1] Huahong Semiconductor (WUXI) Limited, Wuxi, 214028, China

Corresponding Author's Email: Zhangjian.Zhang@hhgrace.com

Biography

Jian Zhang is a technology development engineer of Huahong Semiconductor (WUXI) Limited. His research interest focuses on high voltage device, PMIC and embedded Non-volatile memory.

Abstract

In this paper, HV gate-oxide process has been optimized for solving SONOS 1.5T Flash cell Over-oxidation issue. In this type SONOS, TEOS was used as flash cell poly spacer, which has poorer ability blocking O2 diffusion during high temperature process. For this problem, traditional SONOS flash cell formation sequence was changed, N-pass HV gate oxide was preferential fabricated before flash cell. This method has demonstrated to dramatically prevent flash cell from suffering over-oxidation issue.

Keywords— Over-oxidation; SONOS; gate oxide

Introduction

Floating gate memory has been the mainstream of flash memory products such as memory stick, smart card, mobile applications etc.[1]. However, the requirements of high voltage operation and its compatibility with the logic process have been the roadblocks for developing low power and low cost applications. Fortunately, SONOS type flash memory has been involved as a potential candidate to replace the floating gate technology.

Silicon-oxide-nitride-oxide-silicon (SONOS) memory, were advantaged obtained using high density, fast erase and program time and high endurance for programing cycles electrically erasable programmable read-only memories (EEPROMs) [2]. SONOS memory has the most popular trapping type of NVM device, but also has several advantages, such as: low-power, low programming voltage and small size.

Currently, 2-Transistor (2T) structure SONOS is mainly popular in SONOS memory market, 2T SONOS composed of one memory transistor and one selective gate transistor, is widely used in high reliability and high security product such as e-passport and bank card [3]. However, the market for those products is highly competitive, small size, lower power, and operation rapidly requirements became more and more important. To solve these problems, 1.5T structure SONOS is a good substitute for overcoming those challenges.

In traditional 2T SONOS circuit, there was a Memory transistor and a Select transistor (Fig.1) in one unit, Different with traditional 2T SONOS process, 1.5T SONOS (Fig.2) have smaller size compared with 2T SONOS, and the common-source is composed of contact and doped poly. There have another difference is that the selected transistor and memory cell is fabricated by self-aligned process technology [4-5], which have less masks than 2T SONOS process. Obviously, 1.5T SONOS small size bring some advantages,

such as Lower-power, fast speed and highly reliable embedded memories, However, difficulties also occurred during it's fabricated process.

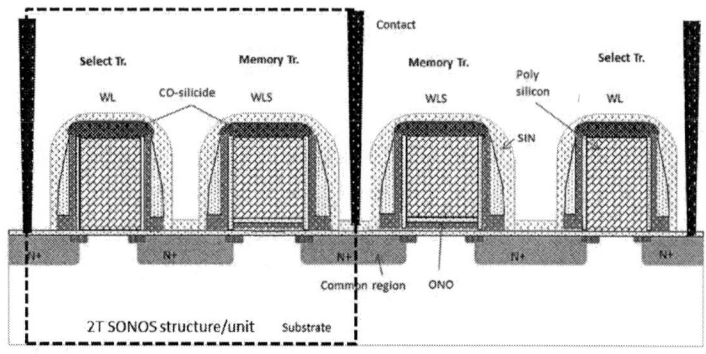

Fig.1. Typical 2T SONOS cell schematic cross-sectional.

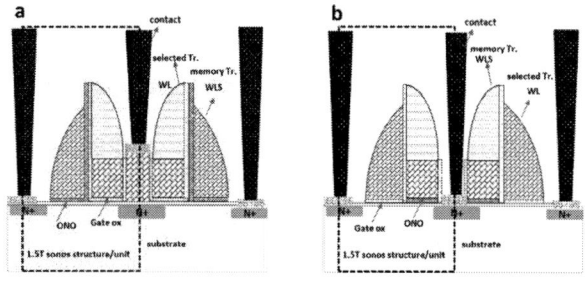

Fig.2. Typical 1.5T SONOS structure (a) n-PASS HV transistor preferential fabricated in 1.5T SONOS, (b) Memory cell preferential fabricated in 1.5T SONOS.

During SONOS Memory cell traditional preferential fabricated in 1.5T SONOS (Fig.2(b)), the cell gate was covered by TEOS(tetraethyl orthosilicate) materials(TEOS act as poly spacer), which often fabricated by CVD methods. During N-pass transistor fabricate process, the cell suffered a series of high temperature, oxygen-rich, and high humidity process. Those negative factors will influence cells performance seriously. In order to overcome these problems, traditional SONOS flash cell formation sequence was changed, N-pass HV gate oxide was preferential fabricated before flash cell. This method may prevent flash cell from suffering over-oxidation issue.

Experiment

In this paper, we describe process optimization in brief to overcome over-oxidation problems during HV gate oxide process. In Fig.3, memory cell was preferential fabricated before HV gate oxide process. ONO (Oxide-SIN-Oxide) is at the bottom of poly gate, which acted as charge trapping materials when Flash cell operation. Obviously, poly gate was

978-1-7281-6559-2/20 $31.00 © 2020 IEEE

oxidation seriously after HV gate oxide process. The interface of TEOS materials&poly gate was oxidation most serious. In the meantime, poly gate top&bottom position profile was better than poly center position (TEOS&poly interface). Assumptions was mentioned that oxygen can across TEOS and reacted with poly gate. For that reason, TEOS materials density is loosening, and poor ability to prevent oxygen gas diffusing and react with poly gate.

In order to validate assumptions and predictions quickly, we deposition 200A SIN as protect layer covered poly [6]. After a series of process steps (SONOS POLY ETCH / TEOS 250A deposition / OX ETCH / SIN 200A deposition / SIN ETCH/HV gate oxide)(Fig.4). Compared with Fig.3, it's found that poly profile have big improvements. At the spacer bottom position，oxide etch have a little oxide residue and form a little footing profile, that means SIN layer can't connect with silicon surface. During HV gate ox process, oxygen gas can diffusing and react with flash cell poly gate bottom position and silicon surface under ONO. This experiment robustly demonstrates that TEOS have poor ability for blocking oxygen gas.

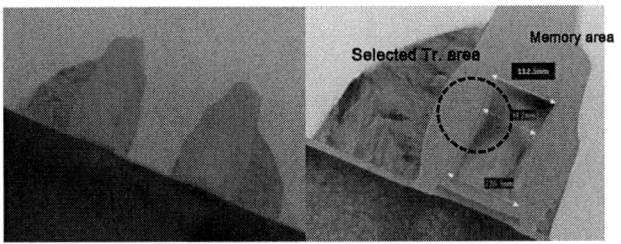

Fig.3. Typical 1.5T SONOS cross-section structure of memory cell transistor preferential fabricated in 1.5T SONOS

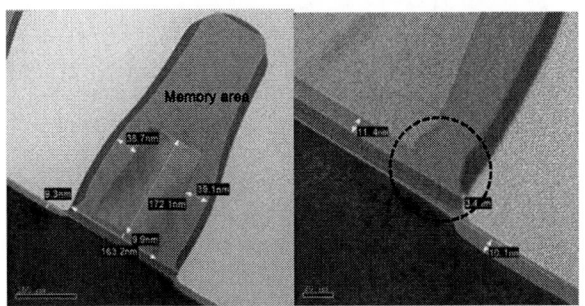

Fig.4. Typical 1.5T SONOS cross-section structure of memory cell transistor preferential fabricated in 1.5T SONOS

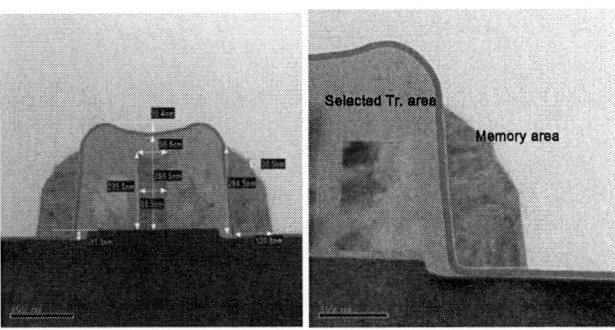

Fig.5. Typical 1.5T SONOS cross-section structure of n-PASS HV transistor preferential fabricated in 1.5T SONOS.

To overcome this problem, we changed traditional SONOS flash cell formation sequence based on Fig.3-4 results. N-pass HV transistor was preferential fabricated before flash cell (Fig.5). This type SONOS mainly process steps sequence is HV gate oxide/ HV gate poly deposition/HV gate poly etch/ TEOS spacer deposition/TEOS spacer etch/ ONO deposition/ memory cell poly deposition/ memory cell poly etch. In Fig.5, there was also have oxide footing profile after TEOS etch, but ONO was deposited after TEOS spacer etch can reduce the disadvantages. Poly cate scale and ONO thickness uniformity much better than Fig.3-4. This structure can dramatically prevent oxygen diffusing flash cell and causing over-oxidation problems.

Conclusions

In this paper, we studied 1.5T SONOS flash cell during HV gate oxide process over-oxidation problems. By experiments proving memory cell transistor preferential fabricated in 1.5T SONOS that TEOS have poor ability for blocking oxygen gas which diffusing and react with poly gate. By SONOS structure changing and process optimizing, N-pass HV transistor was preferential fabricated before flash cell. This method has demonstrated to dramatically prevent flash cell from suffering over-oxidation problems.

References

[1] J. Brewer and M. Gill, Nonvolatile Memory Technologies with Emphasis on Flash, *IEEE Press (2008)*.

[2] M. H. White and D. Adams, "Low-Voltage SONOS Nonvolatile Semiconductor Mem-ories (NVSMs)," *GOMAC 2000.*

[3] Z. Xu, W. Qian, H. Chen, W. Xiong, J. Hu, D. Liu, W. Duan, W. Kong, W.Na, and S.Zou, "Investigation and process optimization of SONOS cell's drain disturb in 2-transistor structure flash arrays," *Solid-State Electronics,* vol. 129, pp. 44-51, 2017.

[4] S.Y. Lee; Y.W. Jeon;T.J.K. Liu;D.H. Kim; D. M. Kim, "A Novel Self-Aligned 4-Bit SONOS-Type Nonvolatile Memory Cell With T-Gate and I-Shaped FinFET Structure," *IEEE Transactions on Electron Devices*, Volume: 57, P1728 – 1736, Issue: 8 , Aug. 2010.

[5] Z.Z. Xu; D.H. Liu;W.Xiong;J.Hu; Z.Q.Fang;W.T. Duan;H.L. Chen;W.S. Qian. "Investigation and three implementations for low power self-aligned 1.5-T SONOS flash device," *2018 China Semiconductor Technology International Conference (CSTIC).*

[6] W.C.Chien, H.C.Chu . "Multi-layer spacer technology for flash EEPROM," *United States Patent.*

Potential Applications of h-BN Crystals in Future ULSI

Guangyuan Lu[1]*, Yu Chen[1], Hualun Chen[1]

[1]Huahong Semiconductor (WUXI) Limited, Wuxi 214028, China

*Corresponding Author's Email: Guangyuan.Lu@hhgrace.com

ABSTRACT

Hexagonal boron nitride (h-BN) is a typical Van der Waals material possessing excellent thermal stability and chemical inertness, which has attracted great interests serving as the ideal dielectric substrates in two dimensional (2D) electronics. Here we review recent progresses of its large scale growth, summarize its corresponding applications and discuss the potentials in future ultra large scale integration (ULSI) processes.

Keywords— h-BN; potential applications.

INTRODUCTION

Among various Van der Waals stacked crystals, hexagonal boron nitride (h-BN) has attracted great interest. It possesses unique dielectric properties with a large band gap of ~6 eV [1]. Besides, h-BN exhibits remarkable intrinsic properties including high mechanical strength, thermal conductivity and excellent chemical stability. As a result, it has shown great potentials in a wide range of applications.

In this paper, we would summarize recent progresses of h-BN on both crystal fabrication and electric applications, and discuss its potential applications in future ultra large scale integration (ULSI) processes.

REVIEW OF H-BN GROWTH

When a new material gets investigated to be used in ULSI, specific requirements including crystal quality and size uniformity are critically demanded. Recently, various chemical vapor deposition (CVD) strategies have been devised to fabricate high-quality h-BN crystals. Wang *et al.* have reported the first epitaxial growth of a 10×10 cm^2 single-crystal h-BN monolayer on a low symmetry Cu(110) surface, as shown in Fig. 1(a) [2]. Wafer-scale single-crystalline h-BN monolayer films were also successfully obtained by Lee *et al.* with liquid Au selected as substrate (Fig. 1(b)). The key step is the facile rotation of circular h-BN grains on liquid Au, regulated by attractive electrostatic interaction between B and N atoms at the perimeter of each grain that eventually leads to the single-crystal growth of h-BN film [3]. For numerous electric applications, large-area continuous h-BN films with certain thickness are also highly desired. So far, efforts have been reported to grow few-layer h-BN film on various substrates including Cu, Ni, Fe, sapphire and SiO$_2$/Si [4-8]. Kim *et al.* have reported the formation of large-area h-BN film on Fe foil (Fig. 1(c)), demonstrating that the thickness of h-BN (5-15 nm) can be controlled by the cooling rate [6]. Besides, it is reported that large-area uniform h-BN film can be fabricated on rationally designed Cu-Ni alloy substrate, as shown in Fig 1(d). The thickness of h-BN films, ranging from one to eight atomic layers, can be controlled by modifying the weight of the source precursor [9]. All these efforts pave the ways to the potential usage of h-BN crystals in future ULSI.

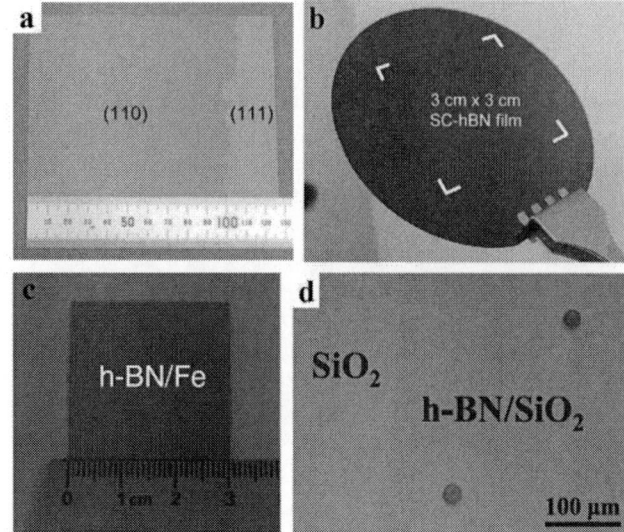

Figure 1. Recent progresses in large-area high-quality h-BN growth. (a, b) Wafer-scale single-crystal h-BN monolayer formed (a) on Cu(110) surface[2] , (b) on liquid Au and transferred on Si wafer [3]. (c,d) Large-area few-layer h-BN film formed (c) on Fe foil [6], (d) on Cu-Ni alloy and transferred on SiO$_2$/Si substrate [9].

REVIEW OF DEVICE APPLICATIONS

As a unique dielectric and transparent crystal with high in-plane thermal conductivity, h-BN typically serves as an ideal substrate or coating layer for various devices. As shown in Fig. 2(a,b), a device topology where metal electrodes are connected to a graphene layer along the 1D graphene edge could be achieved on encapsulated h-BN/graphene/h-BN structures. With h-BN serving as both dielectric substrate and protective layer, a room-temperature mobility of graphene in excess of 140000 cm^2/Vs can be achieved [10]. Recently, Choi *et al.* have improved the heat dissipation characteristics of an InGaN/GaN quantum-well (QW) green LED by using h-BN as a heat-transfer medium (Fig. 2(c,d)). After injecting current, it was found that the LED with h-BN serving as bottom layer took 6 s to reach its maximum temperature (136.1 °C), whereas the Ref-LED took specifically 11 s. After being switched off, the hBN-LED took 35 s to cool down to 37.5 °C while the Ref-LED took much longer, specifically 265 s. The results confirmed the considerable contribution of the attached h-BN to the transfer and dissipation of heat in the LED [11].

Besides, several specific devices with atomically thin h-BN serving as functional medium have also been invented. With a large band gap of ~5.9 eV, monolayer h-BN film could be a promising candidate for DUV photodetector, as shown in Fig.3 (a,b). The device gives a high on/off ratio of 10^3, and shows good reproducibility as well as high robustness [12]. Britnell *et al.* investigated the electronic properties of tunnel diodes in which h-BN acts as a barrier layer between a variety

978-1-7281-6559-2/20 $31.00 © 2020 IEEE

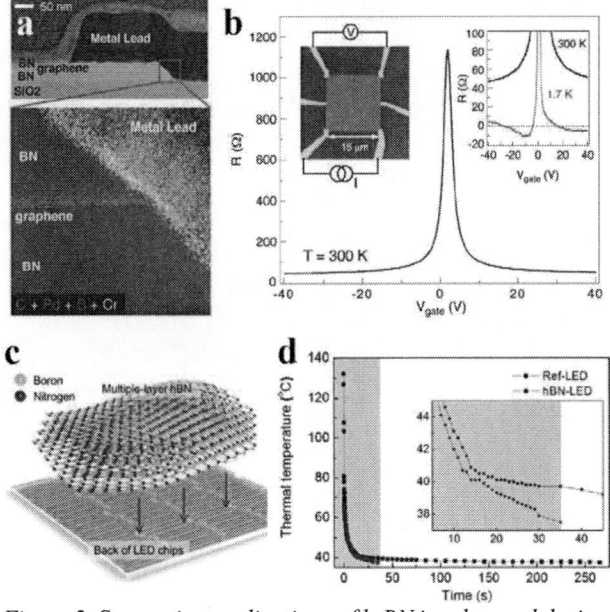

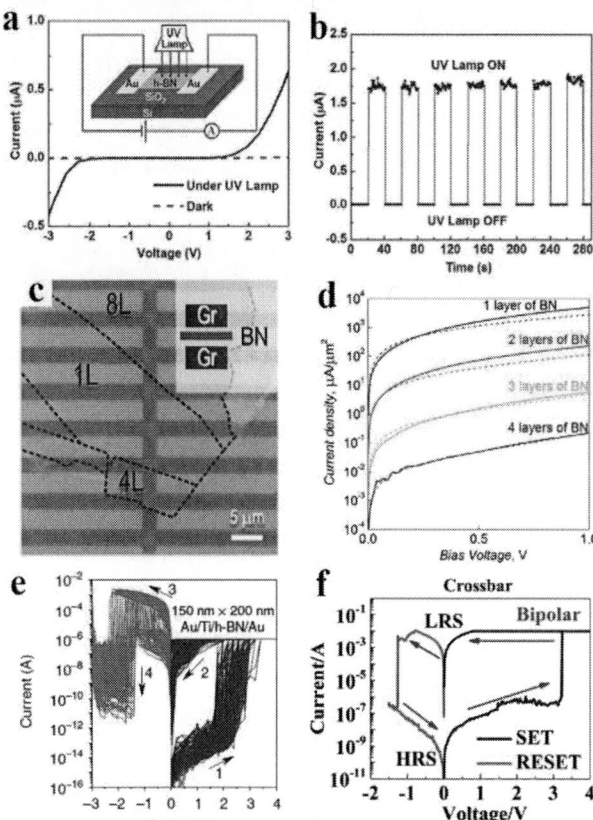

Figure 2. Supportive applications of h-BN in advanced devices. (a) STEM image showing details of the edge-contact geometry based on h-BN/graphene/h-BN structure. (b) Four-terminal resistivity measured from a 15 μm × 15 μm device with edge-contacts [10]. (c) Schematic diagram of the transfer of multiple-layer h-BN onto the back of an LED wafer. (d) The maximum temperature as a function of time after stopping the injection of current into the LED chips [11].

Figure 3. Specific applications of atomically thin h-BN. (a) I-V characteristics and (b) time-dependent current of a DUV photodetector based on monolayer h-BN [12]. (c) Optical image and (d) characteristic I-V curves for graphite/h-BN /graphite tunnel devices with different thicknesses of h-BN insulating layer [13]. (e,f) Typical I–V curves of the NVM with (e) multilayer and (f) monolayer h-BN serving as resistive switching medium [14, 15].

of different conducting materials. It is found that monolayer h-BN can act as an effective tunnel barrier and the transmission probability of the h-BN barrier decreases exponentially with the number of atomic layers, as shown in Fig.3(c, d). The current-voltage characteristics of these devices show a linear I-V dependence at low bias and an exponential dependence at higher applied voltages. The results demonstrate that atomically thin h-BN offers great potential for applications in tunnel devices and in field-effect transistors with a high carrier density in the conducting channel [13].

Recently, the resistance switching phenomena in h-BN thin film have also been investigated. Shi *et al.* proved that CVD-grown multilayer h-BN can be used as a resistive switching medium to fabricate high-performance electronic synapses. The devices can operate in a volatile or non-volatile regime, enabling the emulation of a range of synaptic-like behavior, including both short- and long term plasticity. Typically, Fig. 3(e) shows the bipolar behavior in a non-volatile memory (NVM) fabricated using 15-18 layers h-BN as switching medium [14]. While in monolayer h-BN, the nonvolatile resistance switching (NVRS) effect was also reported by Xu *et al.*, as shown in Fig. 3(f). The monolayer h-BN based atomristors showed stable bipolar and unipolar nonvolatile switching, with large on/off current ratio (up to 10^7) and fast switching speed (<15 ns) [15]. These efforts contributed greatly to the potential applications of h-BN in future flexible memories.

ON THE APPLICATION IN FUTURE ULSI

The progresses listed above indicate certain potential of h-BN crystals in future ULSI applications. With excellent

chemical inertness, wafer-scale high-quality h-BN crystals may be used as hard mask or etch stop layer in future ULSI processes. Good thermal stability and high transparence of atomically thin h-BN make it an ideal candidate for future photo detector coatings. Because of its high in-plane conductivity and insulting property, h-BN also has the potential to serve as heat-transfer and insulating medium in the integrated circuits.

Besides acting as the supportive layer in semiconductor processes, atomically thin h-BN may also be selected as the specific medium in some advanced devices in the future. For example, the high integrated tunneling transistors and memories, as some preliminary investigations are talked above. A key process for the applications of h-BN in ULSI is to make it in designed patterns. Two promising routes are suggested here, as shown in Fig.4. One is to deposit the metallic electrode in designed patterns first, then grow h-BN via CVD method. By tuning growth conditions, high-quality h-BN with controlled thickness can only form on the catalytic surfaces of designed metal electrodes. The challenging project is that such CVD process may still need a relatively high temperature (typically > 500 ℃), which confines the design of devices. Another route is to perform the traditional dry etch processes on the transferred h-BN continuous film, as shown in Fig.4 (b). For the process a commercial transfer method by plating wafer-scale h-BN thin film onto targeted substrates is being developed.

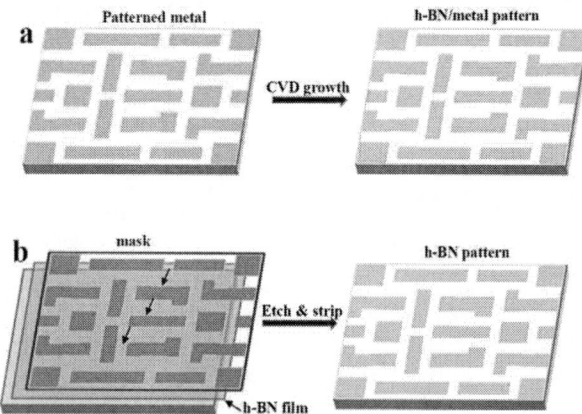

Figure 4. Schematic illustration of the two proposed routes to make h-BN in designed patterns.

SUMMARY AND OUTLOOK

Based on the various progresses obtained by worldwide research groups recently, we believe that, some specific applications of h-BN crystals in ULSI will be available in the future.

REFERENCES

[1] Y. Kubota *et al. Science* 317 (5840), 2007, pp.932–934.

[2] L. Wang *et al. Nature* 570, 2019, pp. 91–95.

[3] J. S. Lee *et al. Science* 362 (6416), 2018, pp. 817-821.

[4] L. Song *et al. Nano Lett.* 10 (8), 2010, pp. 3209–3215.

[5] A. Ismach *et al. ACS Nano* 6 (7), 2012, pp. 6378–6385.

[6] S.M. Kim *et al. Nat. Commun.* 6, 2015, 8662.

[7] A. R. Jang *et al. Nano Lett.* 16 (5), 2016, pp. 3360-3366.

[8] S. Behura *et al. J. Am. Chem. Soc.* 137 (40), 2015, pp. 13060-13065.

[9] G. Lu *et al. Mater. Lett.* 196, 2017, pp.252–255.

[10] L. Wang *et al. Science* 342 (6158), 2013, pp.614-617.

[11] I. Choi *et al. ACS Appl. Mater. Interfaces* 11, 2019, pp. 18876−18884.

[12] H.Wang *et al. Adv. Mater.* 27, 2015, pp. 8109–8115.

[13] L. Britnell *et al. Nano Lett.* 12 (3), 2012, pp. 1707-1710.

[14] Y. Shi *et al. Nat. Electron.* 1(8), 2018, 458.

[15] X. Wu *et al. Adv. Mater.* 31(15), 2019, 1806790.

IMPROVED HCI OF EMBEDDED HIGH VOLTAGE EDNMOS IN ADVANCED CMOS PROCESS

Junwen Liu*, Hualun Chen, Yu Chen

Huahong Semiconductor (WUXI) Limited, Wuxi, 214028, China

*Corresponding Author's Email: junwen.liu@hhgrace.com

ABSTRACT

HCI is always a big concern for high voltage device and normally it can be improved by device size optimization, implant energy and dosage tuning. But in some case, this is not enough, especially for the embedded high voltage EDNMOS integrated in the advanced CMOS process. In this paper, we proposed an additional oxide layer under silicide block film and demonstrated the significantly improved HCI performance of one embedded high voltage EDNMOS, which failed HCI with traditional optimization on device size and implant process.

INTRODUCTION

Per smart power and control requirement for portable devices, like power management, panel display drivers, kinds of high voltage transistors are developed, like EDMOS, LDMOS, which are easily integrated into the standard CMOS process flow. Because of high voltage operation for these devices, HCI induced device performance degradation becomes a big concern for the product reliability, which is much worse than normal CMOS HCI induced degradation. So, lots of effort were spent on the studying the mechanism of the HCI induced degradation for these high voltage transistors. Normally, we are using the substrate current as one key index for the HCI performance. Compared to normal CMOS transistor, in the curve of substrate current, there are two peaks happened for these EDMOS or LDMOS. The first peak of substrate current is similar to the normal CMOS transistor under around half Vg(gate voltage), the impacted ionization happened at the point of the junction between channel and drain; while the second peak of substrate current is induced by the Kirk effect under normal Vg(gate voltage), the impacted ionization happened near the drain side.

In this paper, we studied the HCI performance of one EDNMOS and used initial degradation of IDLIN as the quick check for the HCI reliability evaluation. And we also proposed one method to further improve the HCI performance.

EXPERIMENTAL

Figure 1 shows the cross section of the N-type EDMOS transistor used in this study. The channel length Lc is fixed value because of self-aligned P-body implant process, and the total drain extension size (Lo+Le) is also fixed for this study. This EDNMOS is typically operated under the condition of VG=3.3V and VD=8V. The device parameter IDLIN is measured under VDS=0.1V. The quick check method is that one time stressing the device with 8.8V(1.1 VD) at drain side and weeping gate voltage from 0V to 3.63V(1.1 VG) with 0.03V as a step, then using the shift of IDLIN as initial degradation for the HCI reliability evaluation.

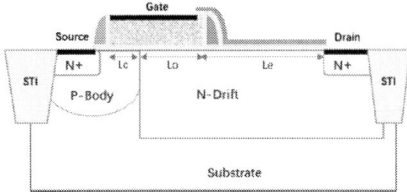

Figure 1: The cross section of N-type EDMOS transistor

RESULT AND DISCUSSION

Figure 2 shows one EDNMOS linear drain current and substrate current characteristics and also initial IDLIN degradation after one time quick check (substrate current curve measurement). The substrate current curve got two peak, one peak happened at about VG=1.6V while the second peak happened at highest VG. And the second peak value is much higher than the first peak which directly caused about 25% initial IDLIN degradation which is totally not acceptable for product functionality.

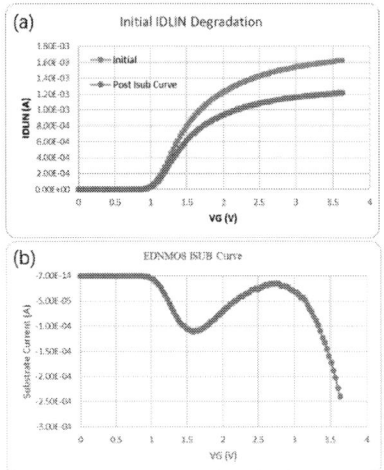

Figure 2. EDNMOS IDLIN and Substrate current. (a) Initial IDLIN degradation after one time quick check (substrate current curve measurement), (b) ISUB curve performance

A. Substrate current optimization

When the device architecture is decided like the EDNMOS used in this study, the substrate current first peak and second peak are playing like a seesaw. Figure 3 shows when we raising the first peak value of substrate current by increasing the dose of N-Drift or reducing the size of Lo, the second peak value of substrate current will go down.

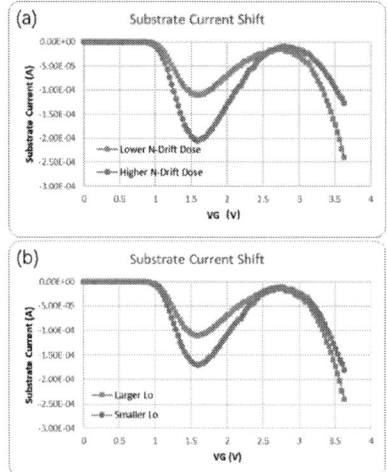

Figure 3. EDNMOS substrate current comparison. (a) Different N-Drift dose, (b) Different size Lo

Different substrate current performance will get different initial IDLIN degradation. Figure 4 shows when the second peak substrate current much higher than the first peak, the initial IDLIN degradation is the worst one. When we get the balanced first peak and second peak substrate current, the initial IDLIN degradation is almost best one.

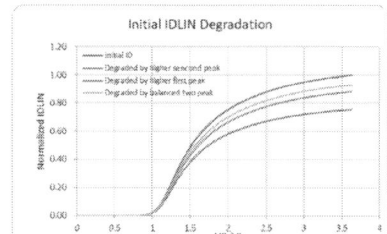

Figure 4. EDNMOS initial IDLIN degradation comparison

B. Further initial IDLIN degradation improvement

After we get the balanced first peak and second peak substrate current by optimizing N-Drift implant and the EDNMOS size Lo, the initial IDLIN degradation is still about 7%, almost cannot pass 10% degradation criteria for the standard process qualification.

Figure 5 shows the two strong ionization points, the point under poly gate is corresponding to the first peak of substrate current like normal CMOS transistor with high electrical field at the PN junction, while the another point close to the drain side is corresponding to the second peak of substrate current caused by Kirk effect.

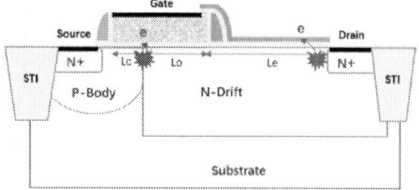

Figure 5. Two strong ionization points for HCI concern

As we know there are lots of trapping center for the silicon nitride film which is widely used to store the charge in the memory cell, especially for 3D NAND area. In our EDNMOS structure, the silicide block film is using stacked film of silicon oxide and silicon nitride which is just above drain extension area, and the silicon oxide film is quite thin, only around 50A, so the hot electrons are very easily tunneling through the silicon oxide and injecting into the silicon nitride film and then stored there, which will deplete the surface of extension area and increase the resistance of the drain extension, then it make the IDLIN degraded.

So now, we are thinking about how to block the hot carrier and minimize the damage or charge trap which can degrade the IDLIN after the substrate current is optimized. Here, we are introducing one silicon oxide film around 150A at the drain extension area before spacer formation, the film with blue color shows in Figure 6.

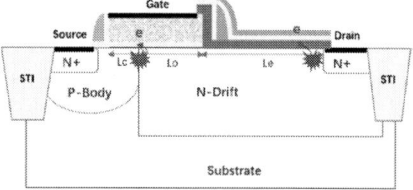

Figure 6. The EDNMOS with extra silicon oxide film

This extra silicon oxide film is processed by HTO with high quality and the hot electrons almost cannot go through this film and inject into the silicon nitride film. We minimized the trapped electrons and got less than 3% initial IDLIN degradation which shows in Figure 7.

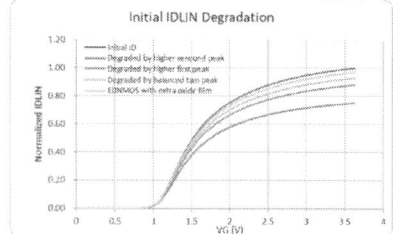

Figure 7. EDNMOS with extra HTO film initial IDLIN degradation

CONCLUSION

In this work, we are using initial IDLIN degradation as an index to quickly evaluate the hot-carrier effects on a drain extended NMOS and demonstrated a furtherly improved HCI performance with an extra silicon oxide film after substrate current optimization.

REFERENCES

[1] C. T. Kirk, "A theory of transistor cutoff frequency (fT) falloff at high current densities", IEEE Trans. Electron Devices, vol. ED-9, 1962, p. 164.

[2] A. W. Ludikhuize, "Kirk effect limitations in High Voltage IC's", Proc. ISPSD, 1994, pp.249.

[3] S. K. Lee et al., "Optimization of safe-operating-area using two peaks of body-current in submicron LDMOS transistors", Proc. ISPSD, 2001, pp. 287–290.

[4] W. F. Sun et al., "Hot-Carrier-Induced On-Resistance Degradation of n-Type Lateral DMOS Transistor With Shallow Trench Isolation for High-Side Application", IEEE Trans. Device Mater. Rel., vol. 15, no. 3, Sep. 2015, pp. 458–460.

[5] Jun Wang, Rui Li, Yemin Dong, Xin Zou, Li Shao, and W.T. Shiau, "Substrate current characterization and optimization of high voltage LDMOS transistors", Solid-State Electronics, vol. 52, 2008, pp. 886–891.

[6] F. R. Libsch and M. H. White, "Charge transport and storage of low programming voltage SONOS/MONOS memory devices", Sol. State Electron, vol. 33, 1990, p. 105.

[7] M. H. White, Y. Yang, A. Purwar, and M. L. French, "A low voltage SONOS nonvolatile semiconductor memory technology", IEEE Trans. Compon., Packag., Manufact. Technol. A, vol. 20, no. 2, Jun. 1997, pp. 190–195.

Improved Standby Leakage of Huge Volume SRAM by Thin SIN Film of STI Liner

Xiaobing Ren[1], Wei Xiong[1], and Hualun Chen[2]*

Huahong(Wuxi) Semiconductor Manufacturing Corporation, Wuxi 201203, China

*Corresponding Author's Email: Xiaobing.Ren@hhgrace.com

ABSTRACT

P+ to Pwell and N+ to Nwell leakage are the most basic leakage components in VLSI circuit and had been received many technologies to be reduced. In each technology, trade off must be made to keep low P+ to Pwell or N+ to Nwell leakage while do not degrade the other characteristic of the circuit. We found that a thin SIN layer post STI Liner OX was able to reduce the B project range at the interface of Active and STI OX, hence reduce P+ to Pwell leakage in Ultro Low Leakage Huge Volume SRAM. As a result, increased P+ diffusion resistance caused by reducing P+ implant energy can be avoided.

INTRODUCTION

The standby leakage of 6T SRAM cell is consist of subthreshold leakage, junction leakage, gate leakage and N+/P+ Leakage. The first three components are leakage of single active device. Basically, High MOS threshold voltage, graded S/D doping concentration, suitable halo implant and gate Oxide of sufficient thickness and good quality are always utilized to keep them in acceptably low level. In Ultra Low Leakage process, the sum of the three kinds of leakage will be first improved to be lower than like 1pA in a SRAM cell. Then N+/P+ leakage, namely N+ to Nwell and P+ to Pwell leakage will be considered in actual cell leakage measurement.

SRAM STANDBY LEAKAGE

MOS Leakage

Figure1 shows the MOS leakage in a 6T SRAM cell when it is in standby mode.

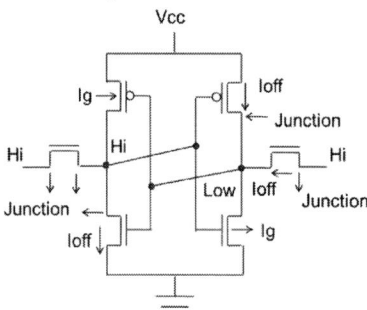

Figure 1: MOS leakage in SRAM

The sum of the leakage associated with the six MOS is:

Ig(PU+PD)+Ioff(PU+PD+PG)+Ij (PG+PD+3×PG)   (1)

Each leakage can be easily obtained by testing designed MOS structure used in SRAM.

N+ to Nwell and P+ to Pwell Leakage

The N+ to Nwell and P+ to Pwell leakage path is the subthreshold leakage of parasitic field MOS as is well known. In figure 2, If Poly gate is at Low voltage which is GND, and P+ is at Vcc, the parasitic field PMOS trends to produce subthreshold leakage.

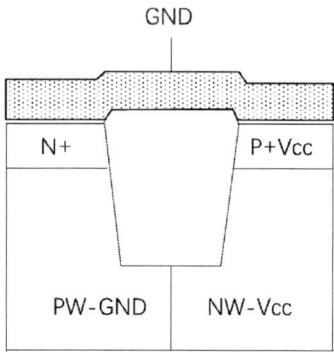

Figure 2: Parasitic MOS leakage

This leakage is a function of STI isolation and exists between adjacent cells.

SRAM TEST STRUCTURE

First a SRAM array of few cells is designed and only single MOS is connected out to be tested. MOS parameters can be measured including each leakage.

Then arrays of much more cells are given and as a small but whole SRAM circuit to be tested, especially standby leakage. The SRAM cell count of the array is determined by the cell design and the dimension of test structure.

Finally, a formal SRAM chip will accept CP test and standby leakage is as an important test bin, to be checked.

STANDBY LEAKAGE TEST RESULT

Using equation (1) we can calculate the total MOS leakage, then multiply the count of SRAM cells in array test structure. The result is used to compare with the actual test leakage of SRAM cell array.

Since equation (1) does not have the leakage of N+ to

Nwell and P+ to Pwell , the offset of calculation and actual test result is N+ to Nwell and P+ to Pwell leakage. Table 1 showed that a 1000-cell SRAM array got actual standby leakage approximately 1000 times the leakage of a single cell, which indicates that N+ to Nwell and P+ to Pwell leakage is negligible.

TABLE I.   STANDBY LEAKAGE COMPARISON

SRAM Standby Leakage		
MOS leakage Sum.	*1000Cells*	*4M SRAM*
0.97pA	1017pA	>40uA

While the test of a 4M SRAM chip showed different result. In the SRAM CP test, totally 90 4M SRAM chips in one wafer were tested and their average standby leakage is also showed in table1. If N+ to NWell and P+ to PWell leakage were still negligible in 4M SRAM chip, the standby leakage should be 4uA calculated by multiplying 1pA/cell by 4 million. The average standby leakage in table1 indicates that N+ to NWell or P+ to PWell leakage arises significantly in 4M SRAM Chip.

The 4M SRAM standby leakage is also showed by the distribution diagram in Figure 3. From the diagram we can see that the leakage is in a wide range from 4uA to 100uA and even more. A good solution to suppress the leakage will not only reduce the leakage but also improve the uniformity.

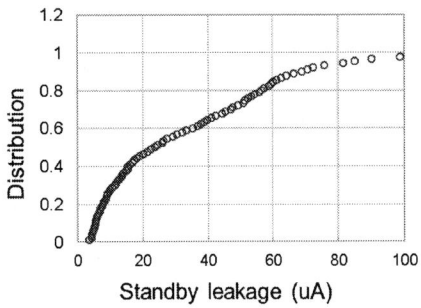

Figure 3: 4M SRAM Standby Leakage

EXPERIMENTS TO IMPROVE LEAKAGE
Reduce Active Area CD
The first split experiment is reducing width of Active area by 10%. When active area width was reduced, the STI was widen. The space of N+ to Nwell and P+ to Pwell was increased to reduce the punch through of parasitic field MOS.

The leakage comparison of reducing active area width and baseline is showed in Figure 4.

The diagram in Figure 4 shows that reducing active area CD had no effect in improving N+ to Nwell or P+ to

Pwell leakage. The reason is probably that the space widening is too little comparing to the total channel length of parasitic field MOS, because the channel length of field MOS only increased one half of the width that is reduced from active area.

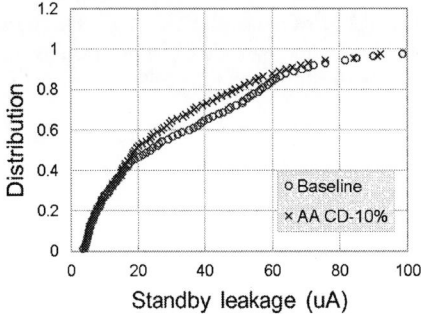

Figure 4: The leakage of Reducing Active Area CD

STI Liner SIN and Reducing P+ Implant energy
Considering dislocation at STI corner is always a leakage source, Adding a thin SIN film in diffusion process after STI lner OX is used as a split condition. The thin SIN layer is like liner SIN. The thickness of the thin SIN film is about 50-100A.

On the other hand, a lower energy P+ source/drain implant is tried to reduce P+ junction depth to increase the space of P+ to Pwell.

The result of the two split condition is showed in Figure 5. It is clear that both conditions got much better SRAM standby leakage than baseline, and the same leakage level and distribution. The results indicates that the P+ to Pwell leakage, or parasitic P field MOS leakage could be well suppressed by increasing space of P+ to Pwell and adding liner SIN.

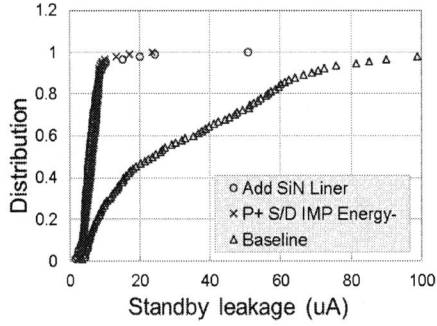

Figure 5: The leakage of Reducing p+ energy and Adding Liner SiN

THEORY EXPLAINATION
Reducing P+ implant energy will directly increase the channel length of parasitic field MOS, hence reduce subthreshold leakage. In the SRAM chip, P+ to Pwell

space is insufficient in or between cells to suppress leakage.

A thin SIN film in the STI OX act as part of the gate isolator of the parasitic field MOS, as shown in Fig.6. Because the dielectric constant of SiN is higher than Oxide, the capacitor of total isolator is higher than pure Oxide. That means the St of parasitic field MOS is reduced and subthreshold is better controlled. So the standby leakage is lowered significantly with liner SiN.

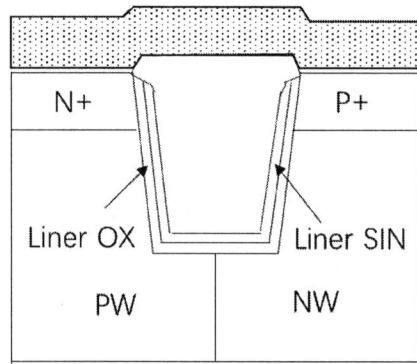

Figure 6: Liner SiN on the STI sidewall and bottom

CONCLUSION

Suppressing the leakage of P+ to Pwell can be accomplished both by increasing space of P+ to Pwell or increasing the capacitor of gate isolator of the parasitic field MOS. In the experiment of this article, it is accomplished by reducing P+ implant energy and adding liner SiN. Since reducing P+ implant energy increases the P+ diff non silicide resistor, the liner SiN layer may be a good choice for 4M SRAM standby leakage improvement.

REFERENCES

[1] Stanley Wolf. *Silicon Processing For The VLSI Era*, Volume 3, 1995, pp. 378-379.

[2] R.W.Mann, W.W. Abadeer, M.J.Breitwisch, O bula, J.S. Brown, B.C. Colwill, P.E. Cottrell, W.G.Crocco, Jr., S.S. Furkay, M.J. Hauser, T.B.Hook, D. Hoyniak, J.M. Johnson, C.H.Lam, R.D.Mih, J. Rivard, A.Moriwaki, E.Phipps, C.S.Putnam, B.A.Rainey, J.J.Toomey and M.I.Younus, "Ultrolow-power SRAM technology," *IBM J. RES&DEV. VOL.47 NO. 5/6 SEPTEMBER/NOVEMBER 2003*

[3] C.C.Wu, C.H.Diaz, B.L.Lin, S.Z.Chang,C.C.Wang, J.J.Liaw, Ch.H.Wang, K.K. Young, K.H.Lee, B.K.Liew and J.Y.C.Sun, "Ultra-low leakage 0.16um CMOS for Low-standby Power Applications, *IEDM Tech. Digest*, pp.671-674(1999)

MINIMIZED JUNCTION LEAKAGE CURRENT FOR NANOSCALE MOSFET APPLICATIONS

Wenqi Bai*, Huishan Yang, Zhisen Huang, Kunghong Lee, Shiming Wang, Zhanyuan Hu,
Huali Microelectronics Corporation, Shanghai 201203, China
*Corresponding Author's Email: baiwenqi@hlmc.cn

ABSTRACT

Ultra-low leakage current development for extreme high threshold voltage (EHVT) devices had become a challenge for the low power consumption application in advanced CMOS technology. This paper experimentally shows the reduction of leakage current by engineering Halo and LDD implant to achieve the EHVT device via 28nm high-k metal-gate (HKMG). The drain (I_{off}) and bulk (I_{offb}) leakage current exhibit 75% and 88% reduction comparing with regular sample at the same on-state current (I_{on}). The leakage current mechanisms as well as the correlation with experimental conditions have been discussed in detailed in this paper. The ultra-low leakage device has great potential for low power application.

Keywords: *28nm high-K metal-gate (HKMG), extreme high threshold voltage (EHVT), band-to-band tunneling (BTBT), on-state current (I_{on}), leakage current (I_{off}), bulk leakage (I_{offb})*

INTRODUCTION

CMOS VLSI chips are scaled continuously to obtain high transistor density, high performance and low power consumption in the past several decades. The supply voltage (V_{DD}) is also scaled down to keep the power consumption under a relatively low level. Consequently, the threshold voltage (V_{th}) has to be scaled simultaneously to maintain high drive current. Unfortunately, the scaling of V_{th} could lead to a substantial increase of leakage current. With channel length scaling to several nanometers, the short channel effect (SCE) becomes a serious problem, and thus makes it more difficult to control the leakage.

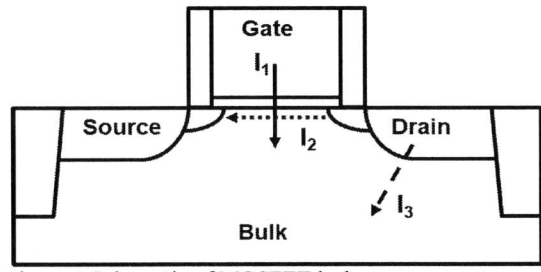

Figure.1 Schematic of MOSFET leakage

As illustrated in figure. 1, the leakage currents of a MOSFET mainly have three parts of compositions. I_1 is the gate leakage, which is significantly reduced with the introduction of HKMG technique. The gate oxide with high-k dielectric can obtain the same gate oxide capacitance with higher physical thickness. As a result, the tunneling of carrier is restrained. I_2 represents the leakage currents between source and drain, which mainly have two mechanisms, i.e. subthreshold leakage and punch-through leakage current. Subthreshold leakage current increases substantially with V_{th} scaling down. Punch-through leakage happens below the surface and becomes significant if the depletion regions of drain-bulk and source-bulk get too adjacent or merged. The leakage current (I_3) between bulk and drain mainly comes from p-n junction reverse bias current. [1-2]

In the above-mentioned, the leakage current of each MOSFET device is serious problems, and should be minimized. However, it has become a challenge to develop an EHVT device with ultra-low leakage as device keeps scaling down to several nanometers. This paper experimentally shows an EHVT NFET at 28HKMG process by significantly reducing the leakage. By tailored channel engineering, in the same I_{on}, the total and bulk leakage shows 75% and 88% reduction respectively.

EXPERIMENTS

This paper reveals EHVT NFET with low leakage current via 28nm HKMG process. The channel engineering of EHVT NFET is carried out at LDD implantation loop. The control scheme and new scheme of LDD loop are shown in figure.2.

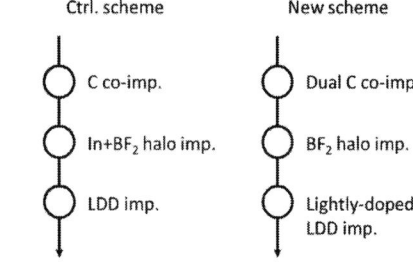

Figure.2 New scheme with channel engineering

The total dosage of BF_2 halo implant is lower than In+ BF_2 dosage, but is higher than BF_2 dosage in In+ BF_2

for comparable threshold voltage and low subthreshold leakage.

Due to higher boron elements content in new EHVT NFET via halo step, the transient enhanced diffusion (TED) should be suppressed, that use carbon co-implant to overcome TED.

I_{off}, I_{offs} and I_{offb} represent drain leakage current (total leakage), drain to source leakage current and drain to bulk leakage current.

RESULTS AND DISCUSSIONS

In figure 3, NFET with channel width of 3um is firstly analyzed, the results of V_{ts} versus I_{on} shown comparable between control and new scheme, but I_{on}/I_{off} performance is significantly improved. Comparable V_{ts}-I_{on} performance indicates that the improvement of I_{on}-I_{off} performance results from the reduction of leakage current rather than the increase of carrier mobility.

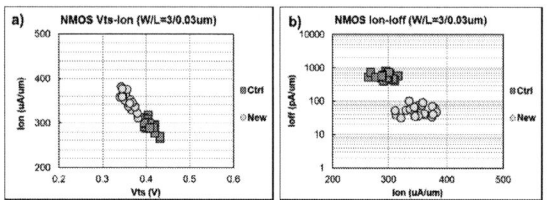

Figure.3 (a) W/L=3/0.03um N-MOSFET V_{ts} vs. I_{on} plots; (b) W/L=3/0.03um N-MOSFET I_{on} vs. I_{off} plots.

In order to study the mechanisms of leakage reduction, the plots of I_{on} versus I_{offs} and I_{offb} are shown in figure.4. The gate leakage is too low and thus not shown. The source to drain leakage, mainly subthreshold leakage in this paper, is comparable for both new and control EHVT NFET. Nevertheless, the I_{offb} decreases nearly one order. Therefore, the improvement of I_{on}-I_{off} performance can be mainly ascribed to the reduction of bulk leakage. For new EHVT NFET, the reduction of bulk leakage may come from replacing In+BF$_2$ halo implant by BF$_2$ halo implant and lightly-doped LDD implant. As pointed out by Yusuke Matsunaga et al., the Indium implant may lead to crystalline defects and thus results in large reverse-bias current [3]. Additionally, advanced CMOS process usually uses heavily doped shallow junctions and halo doping to control short channel effect (SCE). In this case, band-to-band tunneling (BTBT) dominates the reverse pn junction leakage. The BTBT tunneling current density is given by [4]:

$$J_{b-b} = A \frac{E V_{app}}{E_g^{1/2}} exp(-B \frac{E_g^{3/2}}{E})$$

where $A = \frac{\sqrt{2m^*}q^3}{4\pi^3\hbar^2}$, and $B = \frac{4\sqrt{2m^*}}{3qh}$ (1)

m^* is effective mass of electron, E_g is the energy-band gap, V_{app} is the applied reverse bias, E is the electric field at the junction, q is the electronic charge and $\hbar$ is $1/2\pi$ times

Planck's constant. Assuming a step junction, the electric field E at the junction can be expressed by:

$$E = \sqrt{\frac{2q N_a N_d (V_{app} + V_{bi})}{\varepsilon_{Si}(N_a + N_d)}}$$ (2)

Where N_a and N_d are the doping in the p and n side respectively, ε_{Si} is permittivity of silicon, k is the Boltzmann constant. As illustrated by equation (1) and (2), BTBT tunneling current can be effectively reduced by decrease electric field at the junction. Replacing In+BF2 halo implant by BF2 halo implant and lightly-doped LDD implant result in the decrease of Na and Nd, and lead to the reduction of electric field. As a result, the BTBT leakage current can be decreased.

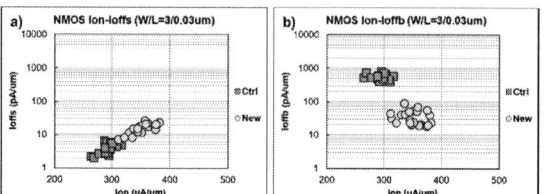

Figure.4 (a) W/L=3/0.03um N-MOSFET I_{on} vs. I_{offs} plots; (b) W/L=3/0.03um N-MOSFET I_{on} vs. I_{offb} plots.

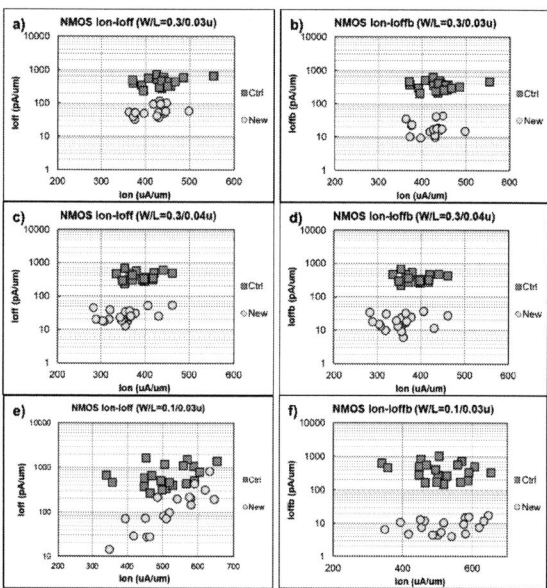

Figure.5 I_{on} vs. I_{off} and I_{on} vs. I_{offb} plots of EHVT N-MOSFET with (a) (b) W/L = 0.3/0.03um; (c) (d) W/L = 0.3/0.04um and (e) (f) W/L = 0.1/0.03um.

For high level of integration, the devices with small dimensions are widely used. Figure.5 exhibits I_{on}-I_{off} performance of new and control EHVT NFET with various dimensions. The EHVT NFET with all these dimensions shows significantly better I_{on}-I_{off} performance due to the reduction of bulk leakage. For nominal EHVT

NFET (W/L=0.3/0.03um), in the same on-state current, the total and bulk leakage current exhibit a significant reduction of 75% and 88%, respectively.

CONCLUSIONS

This paper shows the reduction of EHVT NFET leakage current fabricated by 28nm high-k metal-gate (HKMG) process. The optimization is carried out by channel engineering at LDD loop, which is very convenient to control. For EHVT NFET with various dimensions, the total leakage has been successfully decreased, which mainly results from the reduction of bulk leakage. The mechanism has also been discussed.

REFERENCES

[1]   K. Roy, S. Mukhopadhyay, and H. Mahmoodi-Meimand, Leakage current mechanisms and leakage reduction techniques in deep-sub micrometer CMOS circuits, Proc. of IEEE, vol. 91, no. 2, pp. 305-327, Feb. 2003.

[2] Singh S ,Kour B, Koushik B.K, Dasgupta.s , Leakage current reduction using modified gate replacement technique for CMOS VLSI circuit, International Conference on Communications, Devices and Intelligent Systems (CODIS), Kolkata, India, 28-29 Dec. 2012 pp.464-467.

[3] Y. Matsunaga, S. R. B. Aid, S. Matsumoto, J. Borland, and M. Tanjyo, Characterization of BF2, Ga and in dopants in Si for halo implantation, 13th Int. Workshop Junction Technol. (IWJT), 2013, pp. 74–77.

[4] Y. Taur and T. H. Ning, Fundamentals of Modern VLSI Devices. New York: Cambridge Univ. Press, 1998, ch. 2, pp. 94–95.

STUDY OF RELATED YIELD LOSS AND MECHANISM OF NOR FLASH SELF-ALIGN-SOURCE

Tian Zhi[1], Youhua Qin[1], Gu Zhen[1], Juanjuan Li[1], Qiwei Wang[1], Haoyu Chen[1]*

[1] Shanghai HuaLi Microelectronics Corporation 201203, China
*Corresponding Author's Email: tianzhi@hlmc.cn

ABSTRACT

This paper analyzed the special failure pattern in the wafer center region of 65nm NOR flash. By circuit and failure checking, confirmed the root cause is that the low read current from the big resistance induced by photo residues in self-align-source (SAS) area. The voltage non-uniformity and check-board yield failure were ascribed to the silicon dislocation in SAS active area induced by SAS loop implant. Additional anneal, and higher temperature of rapid thermal oxide, can improve these issue by repairing the dislocation. Decreasing resistance of SAS by dose can also improve yield loss corresponding to erasing cell. All above know-hows helped us to comprehend the new clue and orientation to optimize failure induced by erase failure, and provided the experience for continuous shrinkage of floating NOR flash cell.

INTRODUCTION

Both the massive demand of mobile application, toys and smart card, and development of burgeoning OLED display with more than one NOR flash to maintain the enough magnitude of light stimulated the new growth of NOR flash. NOR flash cell as shown in *Fig.1(a)* as the conventional floating gate flash cell have the advantage of mature design, verification and application environments. But other high density and low cost NOR flash request smaller cell unit and simple process integrated method for mass production, now many optimization study are executed in universal and research institute in the world. [1-4].

For floating gate NOR flash cell, shrinking mainly focused on the passive region as illustrated in *Fig.1(b)*, including space between transistors, space between poly and contact, one method to decrease these space was to use the self-alignment technology in the dimension region. Self-align floating gate is used in the early 65NOR flash process, and the area saving method for the self-align source technology, which use the exist poly gate of word line as etch stop layer to remove the oxide in the source region, and then implant high dose to decrease the resistance of source have been developed as common source as shown in *Fig.2(a),(b)*. But due to this self-align-source use the very deep and limited space for implant dopant, so any small defect in the active bottom or shallow trench isolation (STI) sidewall will increase

resistance obviously. For this common source, if the resistance increases too much, the read current will drop abruptly and induce read failure of cell array finally. This issue has become one bottle-neck of yield and performance improvement for the higher NOR flash array capability and smaller cell unit request

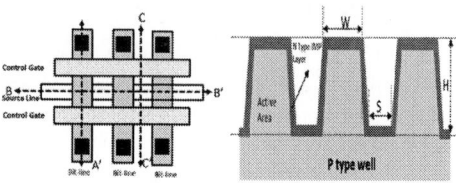

Fig.1 (a)   **Fig.1 (b)**

Fig.1 (a). Schematic of 65nm flash cell layout. Fig.1 (b).Along Fig1.(a) BB'. W demonstrates width of flash cell, S is the width of shallow trench isolation, H is the depth of shallow trench isolation.

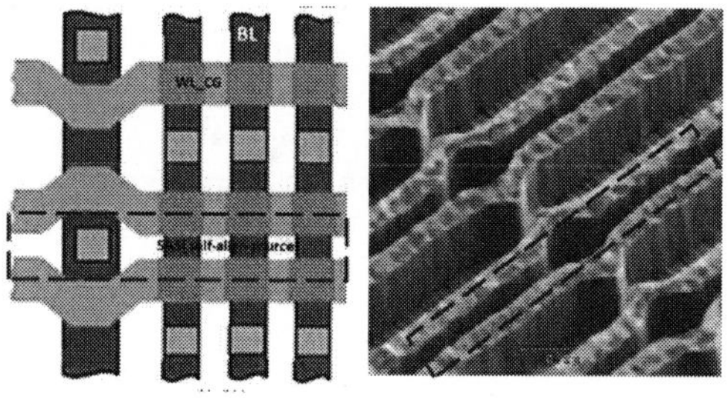

Fig.2(a)   **Fig.2 (b)**

Fig.2 (a). NOR flash top view the dotted-line region is SAS common source. Fig.2(b). NOR flash side view (tilt 45 Deg.), the blue dotted-line region is the high aspect ratio source.

RESULTS AND DISCUSSION

For the initial 65nm NOR flash integrated process development, the strip type failure along the center of the wafer with some width as shown in *Fig.3(a)* and checkboard related failure in wafer edge(*Fig.3(b)*) were serious. Base on NOR flash operation property, if one cell become over-erasing state (the threshold voltage become negative state) in one bit-line (NOR flash cell are in parallel), will induce big leakage and other cell cannot be read really.

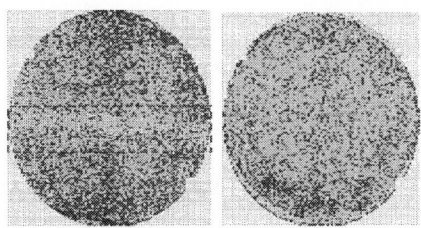

Fig.3(a)   Fig.3 (b)

Fig.3(a).NOR flash yield test results, showing the over-erase fail along the diameter region.Fig.3 (b). NOR flash yield test results, showing the improvement for over-erase fail region along the diameter direction.

To verify if the high aspect ration and high implant SAS region defect induce this special failure pattern. By adding step by step special defect scan for SAS region, found the abnormal defect after etch step. Normal region with oxide are totally removed in the shallow trench, and some damage in the control gate region as shown in *Fig.4(a)* green frame region. Abnormal region, the oxide and control profile no loss, showed the no etch for these region as shown in *Fig.4(a)* red frame region. It is the photo resist for the self-align-source that was not totally removed, blocked the following etch step. For normal defect scan, these residues in the bottom of the trench cannot be detected, and only can be found byspecial defect scanning recipe.To double confirm this issue, more wafers with same photo loop are detected, all suffered this issue. By means of stronger pre-clean recipe before photo loop, photo resist developing and stripping refined management, the wafers with improvement process showed fewer defects, as showed in *Fig.4(c)* compared with old baselin shown in *Fig.4(b)*. The yield test show about 15.0% improvement and over-erase related bin drop obviously. These results verified our failure model and improvement method.

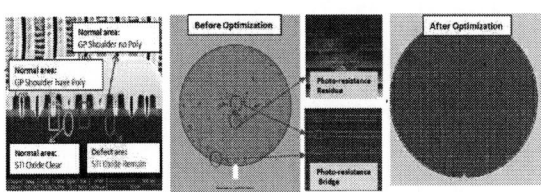

Fig.4(a)   Fig.4(b)   Fig.4 (c)

Fig.4(a).Normal region: oxide in STI removed and control-gate poly etched. Abnormal region: oxide in the STI and the intact control-gate poly. Fig.4(b).Defect scanning results before SAS Photo optimization, finding photo-resist residues. Fig.4(c).After optimizing by stronger pre-clean recipe before photo loop, photo resist developing and stripping refining, no photo-resists residue.

Base on above results, the residue or other abnormal case in self-align-source can induce the erasing related failure. So we suspect the checkboard failure suffere same issues. By analysing the implant recipe and rapid thermal temperature, the heavy element (Asenic), high energy (30KeV) and dose(3E15cm2),and just 900℃, 30 seconds for this step, suspected not enoug repair for heavy element and dose that induced the dislocation in SAS active area as shown in *Fig.5(a)*.

First group experiment: before re-oxidation, adding the long time,low tempperature anneal (700℃, 30 minutes) dislocation can improve, but cannot be eliminated as shower in *Fig.5(b)* . Incresing the anneal temperature (from 900℃ to 950℃), the related yield improved as shown in *Fig.5(c)*, suspect the silicon dislocation improve compared with that of *Fig.5(a)* of base-line condition, and of *Fig.5(b)* with 700℃ additional anneal split. The higher temperature, the less silicon dislocation in SAS active area.

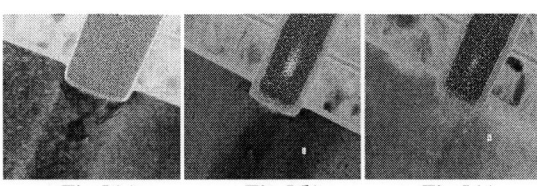

Fig.5(a)   Fig.5(b)   Fig.5(c)

Fig.5(a). TEM after self-align-source etching, silicon loss 20nm and obvious silicon dislocation. Fig5.(b). SAS TEM along the bit-line of additional anneal (700 ℃30m) before poly re-oxidation split, showing improvement compared with that of Fig.5(a) of base-line condition. Fig5.(c). SAS TEM along the bit-line of 950 ℃ poly re-oxidation split, obvious improvement compared with that of Fig.5(b) of base-line condition, and improvement compared with 700 ℃ additional anneal split.

Second group test: owed base on first group results, temperature were dominat faccotr for SAS Rs. To further confirm this, more different thickness with same temperature (baseline condition vs. condition1; condition2 vs. condition3) (-0.7%;-1.3%) showed no improve for resistance of SAS as shown in *Table 1*. For different temperature with same thickness (condition1 and condition2), increasing temperature decreases the resistance of source about 13%. It is not same with convertinal mechanism that more thermal budget, more diffusion, you can find the resistance of high thermal budget condition1 (950℃60m) is higher than low thermal budget (1000 ℃ 37.5 seconds), so the temperature dominated in this heavy element dopant. From above data, the higher temperature, the higher diffusion, the lower resistance, and more current for erasing cell, improved the over-erasing related failure for block operation.

Table1. *Different conditions for resistance of SAS.*

Split and Condition		Silicon Data		
Split	**Condition**	**Median /ohm/sqr**	**STD**	**vs. Baseline**
Baseline	950℃38Å	197.2	6.305	0%
Condition1	950℃42 Å	195.9	2.773	-0.7%
Condition2	1000℃42 Å	170.3	3.337	-14%
Condition3	1000℃46 Å	167.1	3.777	-15%
Condition4	950℃38 Å &Dose+10%	175.1	4.332	-11%
Condition5	950℃38 Å &Dose+20%	169.1	3.590	-14%
Condition6	950℃42 Å &Energy+17%	171.0	2.746	-7%

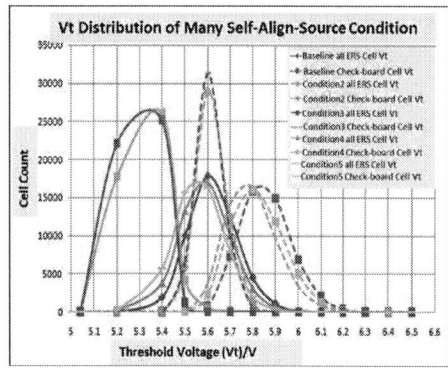

Fig.6. *Vt distributions of full erase and check-board state, high temperature re-oxidation can get more convergent distribution.*

For conventional method to decrease resistance such as higher dopant, we have tried some conditions. Even with silicond dislocation, higher dose will increase the concentration of dopant, resistance decreases as **Table 1**. For all these conditions, collecting the voltage distribution of erasing cells and checkboard cells, results showed higher temperature improve voltage distribution and failure of non-uniform erasing speed as shown in ***Fig.6***. More dose yield can improve 0.2%, and higher temperature condition yield improve 0.4%, all yield increase.

SUMMARY

By means of analysing and summarizing 65nm NOR flash special failure pattern in the center of wafer, base on operation principle of NOR flash and pephriphery design, confirmed the reason for this failure. The residu in the shallow trench after self-align-source blocked the implant dopant and induced the low erasing current for over-erasing issue. By optimizing the photo related process, improved the special failure pattern. Step by step checking, the incompleted repaire damage and silicon dislocation induced by heavy element, high energy and dose in self-align-source were main reason for this high resistance and low read current. Adding the anneal, or increasing the thermal temperature can get compact voltage distribution of erasing cells. Normal method to decrease the resistance, can improve the related failure. This paper verified the strong correlation of the resistance of SAS, silicon dislocation in SAS with final yield. So higher confidence level that by using of low resistacne can improve erasing realted yield filure. These verified methods are good inspiration and strategical methods for development of continuous shrinking cell.

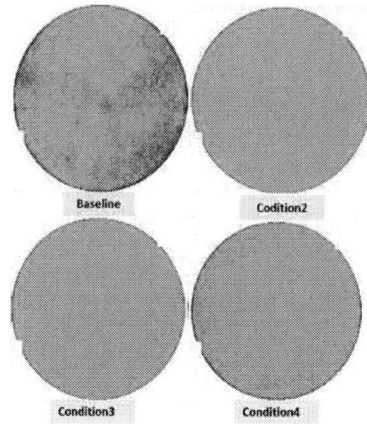

Fig.7. *Yield distribution of bin related with erase cell for baseline and three improving conditions. Increasing dose can improve the yield, and high temperature re-oxidation can improve yield obviously.*

REFERENCES

[1] R. Koval, et al, VLSI Symposium Technical Digest, pp. 204-205, 2005.

[2] Grew Atwood, "Future Directions and Challenges for ETOX Flash Memory Scaling", IEEE TRANSACTIONS ON DEVICE AND MATERIALS RELIABILITY, VOL. 4, NO.3, SEPTEMBER 2004.

[3] Stefan K. Lai, " Brief History of ETOXTM NOR Flash Memory', Journal of Nano-science and Nanotechnology Vol.12, 7597-7603,2012.

[4] Y.Huai,"Spin-transfer torque MRAM (STT_MRAM): Challenges and prospects," AAPPS Bull., vol.17,no.6,pp.34-40,Dec.2007.

[5] Nobuhiko Ito, Yoshimitsu Yamauchi, Naoki Ueda, Kaoru Yamamoto, Yasuhiro Sugita, Takitsugu Mineyama, Akira Ishihama, Kazuhiro Moritani, "A Novel Program and Read Architecture for Contact-Less Virtual Ground NOR Flash Memory for High Density Application", 2006 Symposium on VLSI Circuits Digest of Technical Papers, 1-4244-0006-6/06/ (c) 2006 IEEE.

HIGH PERFORMANCE HVNMOS DEVELOPMENT FOR ADVANCED PLANNER NAND FLASH

Juanjuan Li[*], Zhi Tian, Xiaohua Ju, Tao Liu, Shaokang Yao, Haewan Yang, and Yaoyu Chen
Shanghai Huali Microelectronics Corporation (HLMC), Shanghai 201203, China
*lijuanjuan@hlmc.cn

ABSTRACT

As the flash cell physical size is scaled down, the related down scaling of decoder and page buffer area are also a challenge for chip design. High performance N type MOS including HVN_PT (in word-line decode circuit) and HVN_PB (in page buffer circuit) for 1x-nm planner NAND flash are described in this paper. These N type MOS adopted a series of optimized structure and process integrated methods based on 2D TCAD process and device simulation, to achieve high channel, junction breakdown, and isolation voltage with smaller transistor area limited by down scaled NAND flash cell unit. Finally, these structure and process integrated methods were validated in HLMC 12-inch 1x-nm NAND process flow.

INTRODUCTION

Because of its serial architecture, high performance, high density and matured process capability, NAND flash has grown enormously, and is becoming more and more widely used by mass storage [1]. As we all known, the cost is proportional to the flash cell chip size. NAND flash scaling technology is crucial to its development. Planner NAND cell scaling has already reached below 20nm [2-3]. As both flash cell word-line and bit-line direction sizes are scaled down, the related decoder and page buffer area also need to reduce size. Because NAND flash cell need high voltage on word-line or cell P well when programing or erasing. Advanced planner NAND flash cell chip needs high performance HVN_PT and HVN_PB transistors with down scaled size, high break down voltage and high isolation voltage.

HVN_PT (in word-line decode circuit) and HVN_PB (in page buffer circuit) array sizes depend on NAND flash bit-line and word-line direction pitch size, refer to fig.1.1 [4]. If HVN_PT or HVN_PB array size larger than flash cell pitch, the total chip will be increased a lot, in that case, there's no meaning to shrink NAND flash cell. So if NAND cell shrinks to 20nm or beyond, the related HVN_PT and HVN_PB devices and the spaces between them also need to be reduced. Fig.1.2 and 1.3 show HVN_PT and HVN_PB partial array structure. Unlike to normal CMOS, these HVNMOS has additional drain/source implant mask to keep a certain distance to gate (parameter B/C in fig.1.2 and 1.3), just similar to LDMOS, in order to gain high breakdown voltage. Based on their own layout in chip, the scaling down of HVN_PT and HVN_PB is rooted in lateral shrink of HVN_PT (parameter A, B, C, D, E, F, G, H in fig.1.2), vertical shrink of HVN_PB (parameter I, J, K, L in fig.1.3), which are the most key and difficult points for HVNMOS device design.

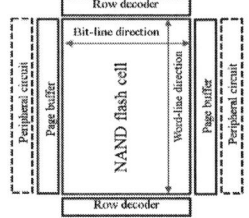

Figure 1.1: NAND Flash memory floorplan

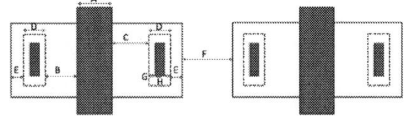

Figure 1.2: HVN_PT partial array structure

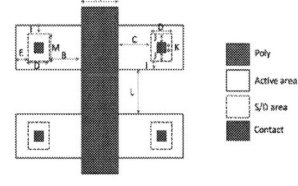

Figure 1.3: HVN_PB partial array structure

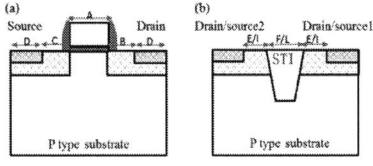

Figure 1.4: (a) HVNMOS device cross-section; (b) Isolation structure cross-section.

And also similar to LDMOS, to increase device BVDS, it's better to gain uniform surface electric filed distribution. What's more, to balance HVNMOS BVDS and isolation leakage is the challenge for the two MOS devices design. That is to balance HVNMOS transistor size and isolation size. Fig 1.4 (a) and (b) show the cross-sections of the HVNMOS device and isolation

978-1-7281-6559-2/20 $31.00 © 2020 IEEE

structure.

SPLITS AND RESULTS

The DOE (design of experiment) approach of silicon wafers verification is based on the following steps:

A. Simulating HVN_PT and HVN_PB device, with the total shrank size. Optimizing process conditions and fixing the optimal devices size.

B. Fixing the isolation structure sizes of HVN_PT and HVN_PB array area. Simulating the isolation structure with shrank size.

C. Process optimizing for both device and isolation.

D. Selection of the optimal devices and isolation structures.

It is well known that short channel effect would cause severe Vt roll off, channel leakage, HCI, et al. Here is no large room for HVN device gate length (parameter A) shrinking.

In actual applications, HVN device drain side needs high break down voltage, while source side does not need that much. In that case, unlike to normal CMOS device, HVN device could be designed as an asymmetric structure, which is keeping the same transistors total length and reducing source side space to increase drain side space. To gain the highest BVDS, LDD region at drain side needs to be fully depleted. With optimized LDD doping profile, wider drain side space, more LDD depletion space, higher BVDS. In table I, condition (a) is the drain and source symmetric structure with the same size of parameter B and C, (b) and (c) are the so designed asymmetric structure, which show more effective improvement to BVDS. Fig. 2.1, (a), (b), (c) shows the related structures electric field distributions, and (d) shows the surface electric field curve at drain break down voltage. In table II and fig. 2.2, shorter parameter E or I, lower punching through voltage and higher punching through leakage current.

It is also recommendation that, within the process control, do the best to reduce source/drain high doping area (parameter D or M), including contact (CT) size (parameter H or K) and source/drain overlap CT area (parameter G or J). It will be good both for device BVDS (larger parameter B or C) and isolation (larger parameter E or I) with shallow source/drain high doping area. Fig 2.3, (a) and (c) show the doping concentration of deep and shallow source/drain high dose area structure, (c) and (d) show the electric field curve of (a) and (c) at drain breakdown voltage. Compared with (a) in fig.2.3, (c) shows more uniform electric field at surface, and more LDD depletion width, so higher BVDS. Fig 2.4 (a) and (b) show the isolation structures current leakage of deep and shallow source/drain high dose area structure. Shallower source/drain high dose area would improve punching through leakage current for isolation structure.

HVN_PB device is much affected by narrow width effect. It is well known that narrow width transistors

subthreshold voltage roll off and hump would become worse [5]. So HVN_PB is more sensitive to the doping profile near the STI edge. Also, HVN_PB surface channel doping is much lower with narrow width, due to boron segregation or OED (oxide enhanced diffusion) [6]. To reduce HVN_PB narrow width effect, more B doping is needed at STI corner, but it will further reduce parameter E/I margin. That means HVN_PB device break down is mainly from high electric field of junction at STI corner.

TABLE I.   HVNMOS BVDS WITH DIFFERENT B AND C

Condition	TCAD simulation data		
	B/um	C/um	BVDS/V
(a)	X	X	24.1
(b)	X+0.1	X-0.1	26.7
(c)	X+0.2	X-0.2	28.3

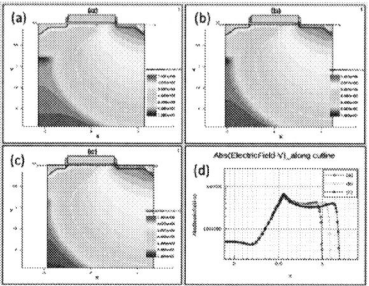

Figure 2.1: (a), (b), (c) show the related structures electric field distributions in table I, and (d) shows the surface electric field curve at drain breakdown voltage.

TABLE II.   ISOLATION BV WITH DIFFERENT E/ I

Condition	TCAD simulation data	
	E/I/um	BV/V
(a)	Z	32.3
(b)	Z-0.3	23.3
(c)	Z-0.6	16.1

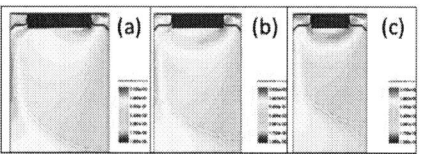

Figure 2.2: (a), (b) and (c) show the isolation structure leakage current density of different parameter E/I in table II.

Increasing implant dosage along STI trench, isolation will be improved (shrank size area in fig. 2.5), but junction BV will be worse (peripheral area when high dosage in fig. 2.5), which will also affect HVN_PB device BVDS.

Increase STI trench depth is a good way to improve isolation leakage, but would bring process difficulty. Add additional implant for only partial STI trench area is a normal way to improve isolation leakage current issue. But the cost will be increased. Another method is to add additional poly above STI, which could be compatible with normal poly process and could not increase cost. The additional poly above STI needs to be grounded or with low voltage, in that case, junction electric filed would be increased, it will reduce punching through leakage current and increase isolation BV in fig. 2.6 and fig. 2.7, but junction BV may have some impact, so needs carefully design.

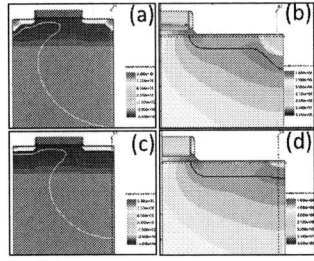

Figure 2.3: (a) and (c) show the doping concentration of deep and shallow source/drain high dose area structure respectively, (c) and (d) show the (a) and (c) electric field curve at drain breakdown voltage.

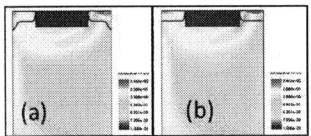

Figure 2.4: (a) and (b) show the isolation structure current leakage of deep and shallow source/drain high dose area structure.

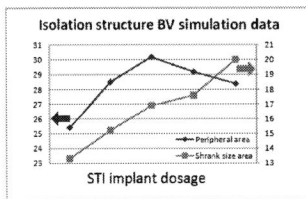

Figure 2.5: Isolation structure BV simulation data with different STI implant dosage.

Based on the above TCAD simulation study, HVN_PT, HVN_PB and isolation structures with varying shrank size are designed and processed in HLMC 12-inch 1x-nm NAND technology. Finally, we got the same results as TCAD simulation.

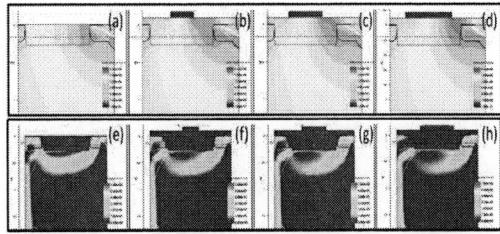

Figure 2.6: (a) shows the electric field at high isolation leakage current with baseline condition that without additional poly above STI, (b), (c) and (d) show the electric field at high isolation leakage current with different size of additional poly above STI, (e), (f), (g) and (h) show the related current density respectively.

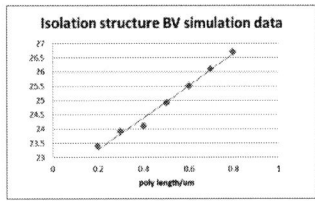

Figure 2.7: shows related isolation structure BV simulation data in Fig. 2.6.

CONCLUSIONS

We provided a thorough investigation of the performance for layout structure and process engineering for HVN_PT and HVN_PB device used in 1x-nm planner NAND flash. TCAD simulation and experiments show the N type MOS devices could achieve high channel and junction breakdown, and isolation voltage with smaller layout area limited by 1x-nm NAND unit. Finally, these structure and process integrated methods were validated in HLMC 12-inch 1x-nm NAND process flow.

REFERENCES

[1] J. Lee, S.-S. Lee, O.-S. Kwon, et al. *IEEE International Solid-State Circuits Conference*, 2003, session 16, Non-volatile memory, Paper 16.7.

[2] J. Hwang, J. Seo, Y. Lee, et al. *International Electron Devices Meeting, 2011*, pp. 9.1.1-9.1.4.

[3] A. Goda, and K. Parat. *International Electron Devices Meeting*, 2012, pp. 2.1.1-2.1.4.

[4] K.-D. Suh, B.-H. Suh, Y.-H. Lim, et al. *IEEE International Solid-State Circuits Conference 95*, February 16, 1995, pp. 128-350.

[5] P. Sallagoity, M. Ada-Hanifi, M. Paoli, et al. *IEEE Trans. Electron Devices*, vol. 43, 1996, pp. 1900-1906.

[6] W. K. Park, J. H. Lee, and G. Lim, *IEEE Electric Device Letters*, vol. 25, 2004, pp. 532-534.

DEVELOPMENT OF LOW LEAKAGE CURRENT IN EXTREME PFET DEVICE

Wenqi Bai*, Huishan Yang, Shiming Wang, Kunhong Lee, Zhisen Huang, Zhanyuan Hu
Huali Microelectronics Corporation, Shanghai 201203, China
*Corresponding Author's Email: baiwenqi@hlmc.cn

ABSTRACT

Leakage current has become a challenge to power consumption in advanced CMOS technology. This paper reveals the low leakage of extreme PFET via Halo and LDD optimization, in 28nm high-k metal-gate (HKMG). In the same on-state current (I_{ON}), the drain (I_{OFFD}) and bulk (I_{OFFB}) leakage current are reduced to ~30 pA/um (~75%) and ~10 pA/um (~88%), respectively. The mechanisms of leakage reduction have been discussed in detailed. This device has great potential for IDDQ reduction via CMOS circuits design.

Keywords: *28nm, high-K metal-gate (HKMG), on-state current (I_{ON}), drain leakage current (I_{OFFD}), bulk leakage current (I_{OFFB})*

INTRODUCTION

In recent decades, CMOS VLSI devices have been scaled down to achieve high performance and low power consumption when circuit density increased. While the geometric scaling keeps on-going, the supply voltage (V_{DD}) is also scaled down to maintain the power consumption under control. As a result, the threshold voltage (V_{th}) of MOSFET device has to be scaled simultaneously to keep high drive current and good device performance. However, the scaling of V_{th} leads to a substantial increase of leakage current, short channel device especially. The short channel effect (SCE) becomes more and more non-negligible, which further worsen this problem [1-2].

As above-mentioned, the current path of MOSFET device is illustrated in figure. 1. I_1 is the gate tunneling current; I_2 is the subthreshold current; I_3 is the punchthrough current; I_4 is the reverse PN junction current [1], while drain terminal forced as on-state voltage, others terminals forced as off-state voltage. Furthermore, I_1 can be significantly reduced with the application of HKMG technique.

The I_{OFF} has two major components between source-drain and bulk-drain path. I_2 and I_3 current show substantially increase with gate length or V_{th} scaling down. But punch-through leakage occurs below the surface that only becomes obvious when the doping of bulk between drain and source is too low.

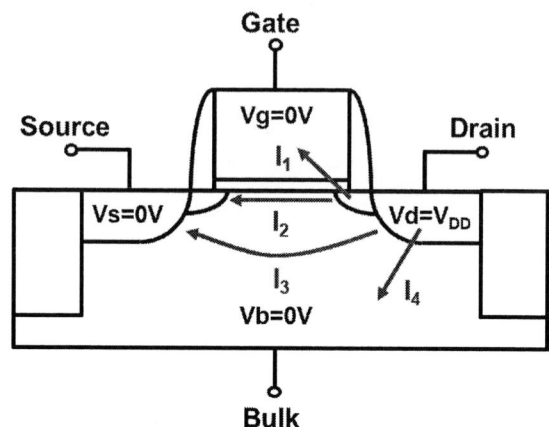

Figure.1 Schematic of MOSFET leakage

Therefore, in order to minimize the leakage current of MOSFET device, channel engineering is generally employed. This research primarily focuses on developing an extreme PFET at 28HKMG process by significantly reducing the leakage. The drain and bulk leakage are reduced to 30 and 10 pA/um respectively, by tailored channel engineering. This research reveals the low leakage current of extreme PFET for low power consumption application.

EXPERIMENTS

This paper develops extreme PFET with low leakage current at 28nm HKMG process and mainly studies short channel device. The channel engineering of extreme PFET is carried out at LDD implantation process. The conventional process of LDD loop includes Ge pre-amorphization (Ge-PAI), halo implant and LDD impurity co-implant. We tried to split halo and LDD co-implant to achieve leakage reduction of extreme PFET. Split-A, B, C and D conditions are listed in Table 1, "↑"/"↓" means the concentration of dosage increase and decrease, respectively. Besides the split of dosage, we also investigated dual step implant of halo to improve junction profiles as split-D.

The dual step halo implant has one halo implant with bigger implant tilt angle and higher implant energy, while the other one has relatively smaller implant tilt angle and lower implant energy. The total dosage of dual-halo implant is comparable with typical conditions at the same LDD.

Table 1. The details of channel engineering

Sample	Halo dosage	LDD dosage
Split-A	↑	-
Split-B	↓	↓
Split-C	↓↓	↓↓
Split-D	step1 ↑, step2 ↓	↓

The leakage current ($I_{OFF} = I_{OFFD}$) are tested at $V_{Drain} = 0.9V$ and $V_{Source} = V_{Gate} = V_{Bulk} = 0V$. I_{OFFD} means the sum of leakage current from source, bulk and gate. I_{OFFS} and I_{OFFB} describe the path of leakage from source to drain, and bulk to drain, respectively. The part of I_{OFFG} can be ignored due to it is much smaller than others. Therefore it is not discussed in detail in this paper. The on-state current (Ion) is tested at $V_{Drain} = V_{Gate} = 0.9V$ and $V_{Source} = V_{Bulk} = 0V$, in order to compare I_{ON}/I_{OFF} performance, the threshold voltage of linear (V_{th_lin}) and saturation (V_{th_sat}) are also tested.

RESULTS AND DISCUSSIONS

The halo and LDD implantation dose dependence of I_{ON}, V_{th} and I_{OFF} due to split seat is shown in figure 2 (a), (b) and (b), respectively. I_{ON} is kept at the same level to help clarify leakage path via split conditions. The difference of leakage current is very obvious. As indicated by Split-A, higher dosage of halo implant results in slight decrease of I_{OFFS} but substantial increase of I_{OFFB}, and the total leakage current is dominated by band-to-band tunneling (BTBT). Decreasing halo and LDD implant dosage lead to a significant reduction of total drain leakage current by split-B, which is mainly ascribed to the noteworthy improvement of BTBT, but I_{OFFS} increased unfortunately. This phenomenon becomes more distinct when halo and LDD implant dosage further go down. I_{OFFS} becomes the main proportion of total leakage current, and unhealthy when gate length scaling down.

By dual-halo implantation methodology, the result of I_{OFFS} and I_{OFFB} are shown the lowest value of total leakage, and without sacrificing device performance. Besides, threshold voltage is also displayed in Figure 2(b).

In further analysis, Figure 3 (a-d) show the correlation of I_{ON} and I_{OFF} by split-D. The total leakage current of Split-D is much smaller than baseline. The total leakage shows strong correlation with bulk leakage and low correlation with source to drain leakage, as indicated by Figure 3(b) and (d). It is worthy noted that the I_{OFFB} of split-D is much smaller than that of baseline at the same I_{OFFS}.

The concentration of halo and LDD implantation are

useful to control the sheet resistance (R_S) and SCE directly. In the meanwhile, BTBT may dominate the I_{OFF} from bulk

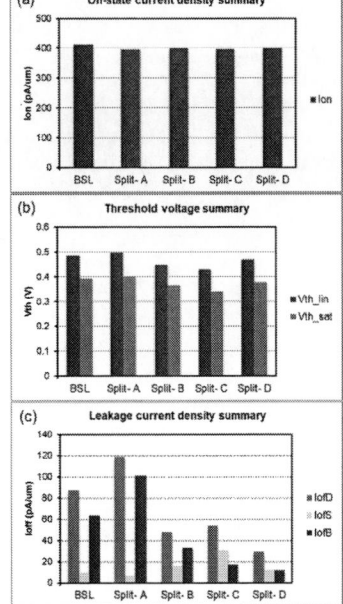

Figure.2 (a) On-state current summary; (b) Threshold voltage summary; (c) Leakage current summary.

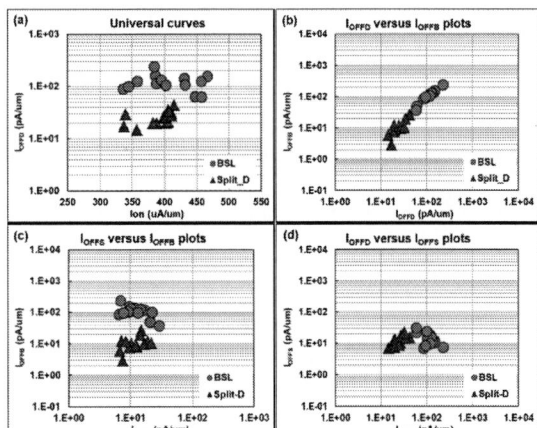

Figure.3 (a) I_{ON}-I_{OFFD} universal curve; (b) I_{OFFD} versus I_{OFFB} plot; (c) I_{OFFS} versus I_{OFFB} plot; (d) I_{OFFS} versus I_{OFFB} plot.

to drain, as shown as I_{OFFB}. The extent of geometric scaling is larger than the extent of supply voltage (V_{DD}), which results in higher electric field, and higher electric field will worsen BTBT.

The BTBT tunneling current density is given by [3]:

$$J_{b-b} = A \frac{E V_{app}}{E_g^{1/2}} exp(-B \frac{E_g^{3/2}}{E}) \qquad at$$

$$A = \frac{\sqrt{2m^*}q^3}{4\pi^3\hbar^2}, \text{ and } B = \frac{4\sqrt{2m^*}}{3q\hbar} \qquad (1)$$

Where m^* is effective mass of electron, E_g is the energy-band gap, V_{app} is the applied reverse bias, E is the electrical field at the junction, q is the electronic charge and $\hbar$ is $1/2\pi$ times Planck's constant. Assuming a step junction, the electric field E at the junction and built-in voltage across the junction V_{bi} can be expressed by:

$$E = \sqrt{\frac{2qN_aN_d(V_{app}+V_{bi})}{\varepsilon_{Si}(N_a+N_d)}} \qquad (2)$$

$$V_{bi} = \frac{kT}{q}In(\frac{N_aN_d}{n_i^2}) \qquad (3)$$

Where N_a and N_d are the doping in the p and n side respectively, ε_{Si} is permittivity of silicon, k is the Boltzmann constant, and n_i is the intrinsic carrier concentration.

The weak inversion current is given by [4]:

$$Ids = \mu_0 C_{ox} \frac{W}{L}(m\text{-}1)(\frac{kT}{q})^2 exp^{\frac{q(V_g\text{-}V_{th})}{mkT}}(1\text{-}exp^{\frac{qV_{DS}}{kT}}) \qquad \text{at}$$

$$m = 1 + \frac{3t_{ox}}{W_{dm}} \qquad (4)$$

When Vg=0V, equation (4) can be re-expressed as:

$$Ids = \mu_0 C_{ox} \frac{W}{L}(m\text{-}1)(\frac{kT}{q})^2 exp^{\frac{-qV_{th}}{mkT}}(1\text{-}exp^{\frac{qV_{DS}}{kT}}) \qquad (5)$$

Where V_{th} is the threshold voltage; C_{ox} is the gate oxide capacitance; u_0 is the zero bias mobility; m is the subthreshold swing coefficient; W_{dm} is the maximum depletion layer width, t_{ox} is the gate oxide thickness; C_{dm} is the capacitance of the depletion layer.

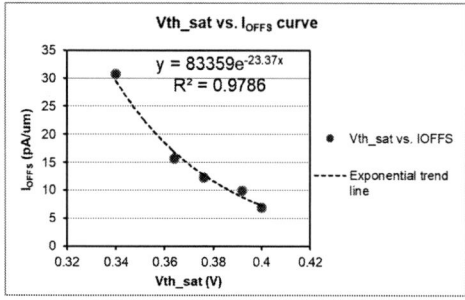

Figure.4 The sensitivity of V_{th_sat} versus I_{OFFS}.

In figure.4, it can be seen that the I_{OFFS} exhibits strong correlation with V_{th_sat}, as indicated by Equation (5). As a result, the I_{OFFS} can be reduced and adjusted by halo/LDD implantation, short channel especially. However, from equation (2) and (3), the electrical field may enlarge between drain and bulk by tuned halo and LDD

concentration, which consequently results in deterioration of junction leakage according to equation (1). Split-B demonstrates that by decreasing both N_d and N_a, total leakage is quite effective reduced, which is mainly attributed to I_{OFFB}. Nevertheless, I_{OFFB} is further reduced, but I_{OFFS} rapidly increased by split-B and C, which can be ascribed to V_{th} decreased.

As above-mentioned, the dual step of halo implantation is revealed as split-D to balance I_{OFFS} and I_{OFFB}, and the total leakage is reduced obviously. In the results, the halo implantation has more influence on channel performance that split bigger tilt angle and higher energy. This is very important methodology to reduce subthreshold leakage and keep I_{ON}. Additionally, the I_{OFFB} reduction of split-D compared to split-B indicates that the N_d at the junction between drain and bulk is also decrease while N_a remains unchanged due to the same LDD conditions.

CONCLUSIONS

This paper reveals the reduction of extreme PFET leakage current fabricated by 28nm high-k metal-gate (HKMG) process. The total leakage current has been successfully decreased by a dual step of halo implantation, which is achieved by optimizing the leakage of I_{OFFS} and I_{OFFB} simultaneously. This work provides an inspiration at low power device development.

REFERENCES

[1] K. Roy,S. Mukhopadhyay, and H. M. Meimand, Leakage current mechanisms and leakage reduction techniques in deep-sub micrometer CMOS circuits, Proc. of IEEE, vol. 91, no. 2, pp. 305-327, Feb. 2003.

[2] Singh S ,Kour B, Koushik B.K, Dasgupta.s , Leakage current reduction using modified gate replacement technique for CMOS VLSI circuit, International Conference on Communications, Devices and Intelligent Systems (CODIS), Kolkata, India, 28-29 Dec. 2012 pp.464-467.

[3] Y. Taur and T. H. Ning, Fundamentals of Modern VLSI Devices. New York: Cambridge Univ. Press, 1998, ch. 2, pp. 94–95.

[4] Fundamentals of Modern VLSI Devices. New York: Cambridge Univ. Press, 1998, ch. 3, pp. 120–128.

FABRICATION AND CHARACTERIZATION OF A NOVEL FULLY SELF-ALIGNED SPLIT-GATE SONOS MEMORY DEVICE

Zhaozhao Xu[1,], Jun Hu[1], Ning Wang[1], Kegang Zhang[1], Donghua Liu[1], Hualun Chen[1], Wensheng Qian[1]*

[1]ShanghaiHuahong Grace Semiconductor Manufacturing Corporation, Shanghai 201203, China

*Corresponding Author's E-mail: Zhaozhao.xu@hhgrace.com

ABSTRACT

A novel low-power fully self-aligned split-gate silicon-oxide-nitride-oxide-silicon (SONOS) flash memory with "memory-last" configuration has been fabricated for the first time at 90-nm node. Firstly, the fabrication sequence is introduced. Then, the operations of unit cell are presented and the transfer characteristics are also experimentally demonstrated. It was revealed that, a wider V_{TH} window can be obtained in the memory gate (MG) transistor, although it exhibits a slightly small V_{TH}. However, V_{TH} of selected gate (SG-) transistor is still too large for low-power applications. This is ascribed to the inappropriate source line (SL-) halo, arsenic implantations, and thermal budget, which induces a high threshold voltage locally at SL-side. However, it still proved that this novel split-gate SONOS memory is very promising for low-power applications.

Key Words—Split-gate; silicon-oxide-nitride-silicon (SONOS); low-power; fully self-aligned; SL-halo

INTRODUCTION

As one of popular nonvolatile memory (NVM) technologies, silicon-oxide-nitride-silicon (SONOS) has been widely used in consumer, industrial, and automotive products [1]. For example, SONOS cell is preferred for most of the financial smart IC cards because of its higher reliability compared with floating gate (FG) memory device. Extrinsic charge loss and stress induced leakage current (SILC) in FG cell [2] are the root cause of reliability, which limit its application in fields required high security. Moreover, scaling bottom oxide is also a bottleneck for FG memory technology evolution as device dimensions shrink [3, 4]. Therefore, SONOS technology has recently attracted increasing interest for various benefits. To realize the ultra-low power, both program and erase are performed by using Fowler-Nordheim (FN) mechanism. FN requires very small programming current (<1nA/cell) thus realizing low power and allowing many cells to be programmed at a time. Commercially, the greatest merit is cost effective with fewer masking layers due to its better compatibility with conventional CMOS process [5, 6].

The 1.5-T (1.5-transistor or split-gate cell) and 2-T (2-transistor cell) structure can further reduce system power by grounding the target memory gate (MG) during read operation owing to the overerase immunity, compared with stacked-gate NOR type memory. To saving chip area, split-gate structure is preferable because the common source between the selected gate (SG) and MG is removed [7]. Additionally, split-gate structure afford better suppression of short channel effect (SCE), which is ascribed to back-to-back MG and SG structure. Consequently and recently, Y. Taito et al. had demonstrated 40-nm node [8] as well as 28-nm node [9, 10] embedded split-gate MONOS. Furthermore, split-gate MONOS has been scaled 16/14-nm node by S. Tsuda et al. [11, 12]. Rather than employing the (source side injection) SSI/FN for program/erase (P/E), FN/FN is adopted to improve the reliability and power and relax the P/E procedure.

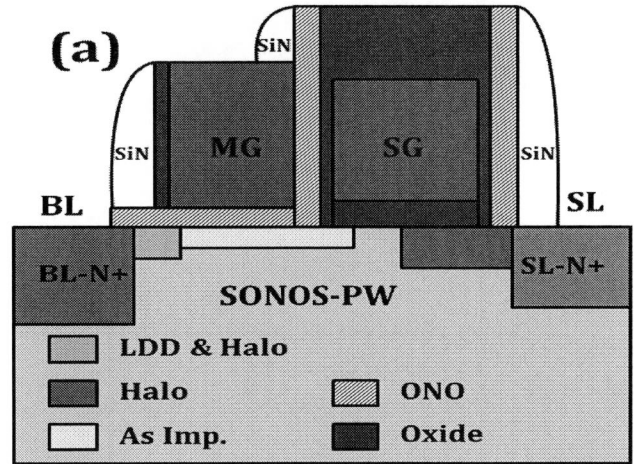

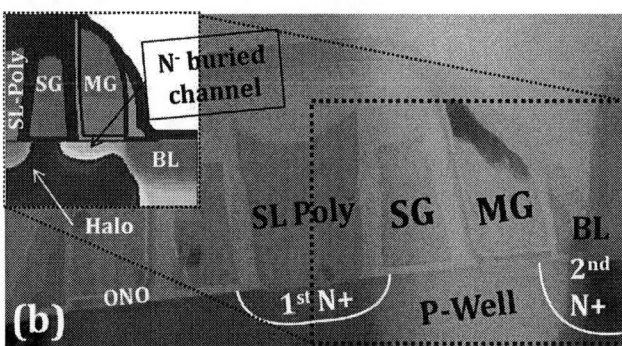

Figure 1: (a) Schematic cross-sectional drawings and (b) SEM picture of the proposed split-gate SONOS device fabricated at 90-nm node.

In this work, a novel self-aligned split-gate SONOS with FN/FN for P/E is proposed for low power applications to further reduce the chip area penalty. Firstly, the fabrication-process is presented. Secondly, the device characteristics are demonstrated and discussed. Finally, we present a conclusion about this work.

DEVICE STRUCTURE AND FABRICATION

Figure 1 shows the schematic cross-sectional view and scanning electron microscope (SEM) picture of the novel split-gate SONOS memory along the bit-line (BL) direction. This cell features a source-side halo implant to adjust threshold voltage ($V_{TH,SG}$) of SG-transistor, as illustrated in the inset of Figure 1(b). As shown, a unit cell consists of a SG-transistor and a side-wall MG-transistor. The length of SG is defined by the feature size of the given process. Therefore, the length of SG is 100nm at our 90-nm node technology. However, the length of MG is determined by the thickness of spacer polysilicon (poly) which is also 100nm in this work. Because CMOS-polysilicon is shared with SG, the length of MG is not limited by the thickness of CMOS-poly, which benefits scaling the memory size. The length of MG is merely confined by the

short channel effect of the MG-transistor. Additionally, this cell also features a self-aligned poly for source-line (SL). It reduces the SL-contact size from 0.24 μm to 0.16 μm. Smaller size of SL-polysilicon is under development and SL-poly length of 65 nm is promising according to [13]. As also show in the inset of the Fig 1(b). MG is a depletion mode transistor with N-type buried channel for low-power operation.

STI formation
SONOS & CMOS well implantation
SG/CMOS gate oxide formation
Poly and SiN deposition
SiN open and oxide deposition
Oxide spacer formation (a)
SG/CMOS Poly 1st etch
SL-Halo implant
SiN spacer formation
SL-N+ implant (b)
Oxide etch and SL-poly deposition
SL-poly CMP and Oxidation
SiN strip with wet etch
SG/CMOS poly 2nd etch
Arsenic implant (c)
PR and Gate oxide removal
ONO formation
MG-poly deposition
CMOS gate patterning
MG-poly etch/CMOS poly 3rd etch (d)
LDD, spacer, S/D formation
Salicide and contact formation
Metallization

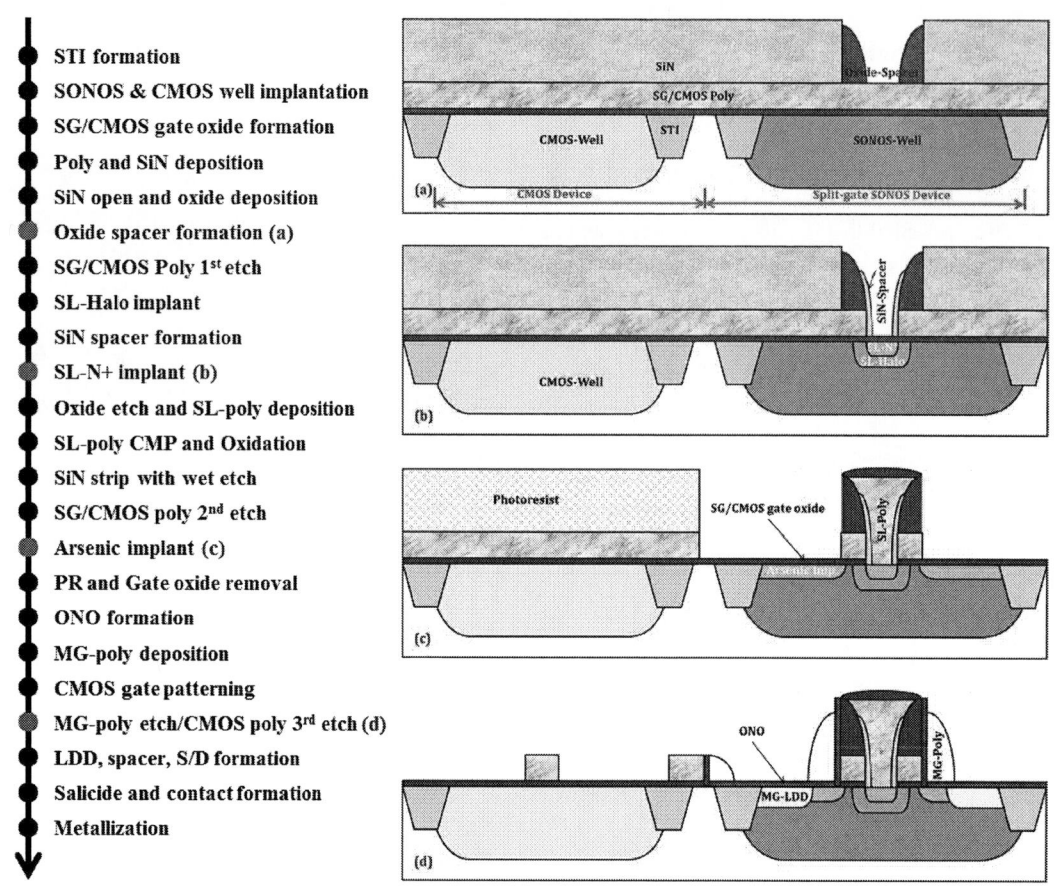

Figure 2: The fabrication sequence for the fully self-aligned split-gate SONOS device. (a) Wells formation and oxide spacer formation, (b) the 1st etch of SG/CMOS poly, SL-Halo and SL-N+ implantation, (c) SL-poly formation, growth of protecting oxide, 2nd etch of SG-poly, and MG-cell V_{TH} implant, (d) ONO deposition, MG-poly etch/CMOS poly 3rd etch.

With additional five-mask steps, this split-gate SONOS memory is integrated to our 90-nm logic process. The fabrication-process steps of the memory as well as CMOS device are illustrated in Figure 2. After the formation of shallow trench isolation (STI), both CMOS-well and SONOS-well are implanted at the beginning, followed by the growth of SG/CMOS gate oxide. Next, SG/CMOS poly and a thick silicon-nitride (SiN) film are deposited. After deposition, the total region for two SG-transistor and SL is defined in lithographic step. Then, thick tetra-ethoxy-silane (TEOS) oxide is deposited, followed by an anisotropic etch for making oxide spacers, as illustrated in Figure 2(a). Both oxide spacer and the remaining SiN are served as a hard mask to etch SG/CMOS poly for the 1st time, which is first self-alignment. After oxidation of poly side wall, self-aligned implantation of SL-Halo is performed to adjust V_{TH} of SG-transistor. Then, a high temperature oxide (HTO)/SiN (~400 A) is deposited and etched back as an insulator between the SG and SL, followed

by the SL-N+ implantation, as presented in Figure 2(b). Then, the surface oxide above the SL-N+ region is cleaned up for the SL-poly deposition. The second alignment is achieved by a chemical mechanical polishing (CMP) on the deposited SL-poly. After CMP, a protecting oxide layer on SL is thermally grown. Next, using the protecting oxide as the hard mask, 2nd etch of SG/CMOS poly is carried out while the CMOS area is masked by the photoresist, as demonstrated in Figure 2(c). The self-aligned implantation of arsenic is performed to adjust V_{TH} of MG-transistor after the removal of photoresist. Afterwards, HTO/SiN is deposited and etched back to form interpoly dielectric between the SG and MG. After surface clean, the oxide-nitride-oxide (ONO) layers are grown, followed by the deposition of polysilicon dedicated for the MG-transistor. Then, MG-poly is etched back, followed by self-aligned MG-LDD implantation. After the removal of ONO layers, the CMOS gate is patterned while the SONOS-region is blocked with the photoresist. The 3rd etch of SG/CMOS poly is performed to form the CMOS gate, as illustrated in the Figure 2(d). Finally, CMOS-LDD implantation, L-spacer, source/drain (S/D shared with BL-N+) implantation, self-aligned silicidation, and metallization processes are carried out. The fabrication steps of the conventional split gate are demonstrated in [14] and [7].

TABLE I. BIAS CONDITION DURING PROGRAMMING, ERASING, AND READING FOR BOTH SELECTED AND UNSELECTED CELLS. WLS AND WL REPRESENT THE WORD LINE OF MG-CELL AND SG-CELL, RESPECTIVELY.

Terminals	Items	Program	Erase	Read
WLS	Selected	7V	-4V	0
	Unselected	-4V	7V	0
WL	Selected	-4V	2V	2V
	Unselected	-4V	0	0
BL	Selected	-4V	7V	1V
	Unselected	1.6V	7V	1V
SL	Selected	Floating	Floating	0
	Unselected	Floating	Floating	0
P-Well	Selected	-4V	7V	0
	Unselected	-4V	7V	0

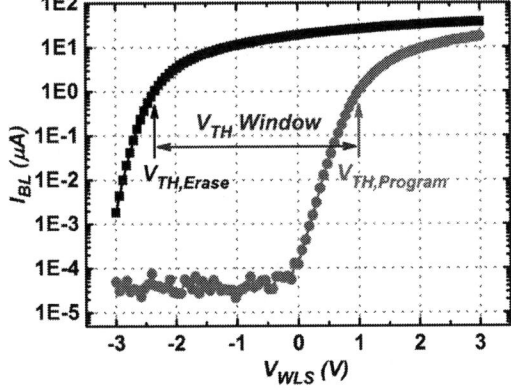

Figure 3: Measured transfer characteristics of programmed and erased MG-transistor. The time for both programming and erasing is 10 ms and the SG-transistor is highly overdriven.

RESULTS AND DISCUSSION

To enable low-power application, FN tunneling is employed for both programming and erasing. The details of unit cell operation is similar to the 2-tranisitor which are presented in [15]. As shown in Table I, bias condition of selected and unselected cells are given for programming, erasing, and reading. During the programming, a 7 V is applied to the word line of MG-cell (WLS) and both BL and p-well are biased with -4 V while keeping SL floating. It is worth noting that the word line of SG-cell is maintained to -4 V. Consequently, thicker interpoly dielectric is required for SG-MG isolation. For erasing, a block of data is erased at the same time. For this operation, -4 V is applied to WLS and both BL and p-well is biased with 7 V while still keeping SL floating. Both programming and erasing are performed with positive and negative voltage sources to decrease the operating voltage and

ease the complexity of circuit design. During the reading of data, the WL is overdriven at 2 V and 1 V is applied to the BL. Meanwhile, the WLS, SL, and the p-well are ground. It is low-power because the additional charge pump is not required for WLS in data sensing.

The measured transfer characteristics of MG-transistor are shown in the Figure 3. If the V_{TH} are defined at constant current of 1 μA, the $V_{TH,Program}$ and $V_{TH,Erase}$ is 1.02 and -2.4 V, respectively. It exhibits a large V_{TH} window of 3.42 V for low-power applications [16, 17]. However, the current $V_{TH,Program}$ and $V_{TH,Erase}$ is slightly smaller compared to our 2-Transistor SONOS cell. Further tuning is required to meet the target.

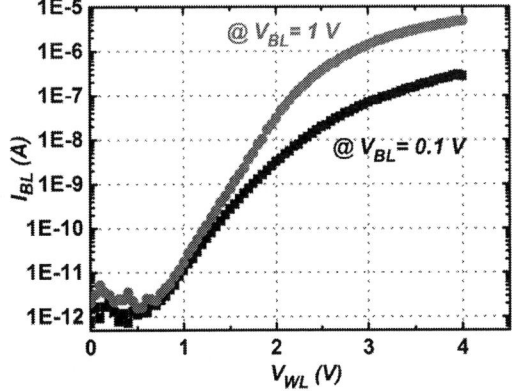

Figure 4: Measured transfer characteristics of SG-transistor at BL voltage of 0.1 and 1 V.

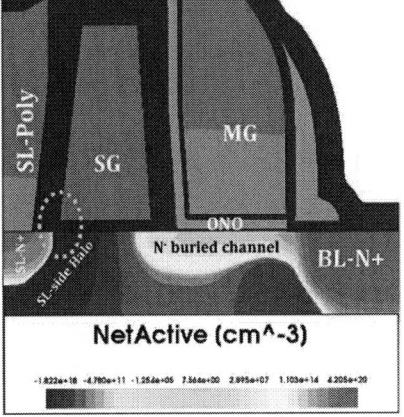

Figure 5: Doping concentration of split-gate SONOS device. Limited arsenic diffusion induces a gap (green circle) between the SL-N+ and SG-polysilicon.

The measured transfer characteristics of SG-transistor with different BL voltages are illustrated in Figure 4. Abnormal transfer curves are obtained. This effect is probably ascribed to the limited diffusion of shallow arsenic (SL-N+) due to reduced thermal budget, which induces a gap between the SL-N+ and SG-poly, as illustrated and confirmed with TCAD simulation in Figure 5. Additionally and as indicated in Figure 4, a $V_{TH,SG}$ of 2 V is extracted with a constant current of 0.1 μA with BL voltage of 1 V. It is too large according to Table I and for low-power

978-1-7281-6559-2/20 $31.00 © 2020 IEEE

applications. Adjustment of SL-side implantations is required to improve the on-state current for sensing.

CONCLUSION

To further reduce the unit cell size of SONOS memory, a novel fully self-aligned split-gate SONOS flash with Fowler-Nordheim for both program and erase has been fabricated and characterized in this work. For the first time, source-side halo implant for $V_{TH,SG}$ adjustment is experimentally demonstrated, which avoids the counter-doped MG/SG-channel. Currently, the $V_{TH,SG}$ is too large because of the inappropriate SL-halo, SL-N^+ implantations, and thermal budget, which induces a high threshold voltage locally at SL-side. Fortunately, the MG-transistor exhibits wider V_{TH} window for low-power applications and the $V_{TH,Program}$ and $V_{TH,Erase}$ is 1.02 and -2.4 V, respectively. It proves that this split-gate SONOS memory is very promising for low-power applications.

REFERENCES

[1] Y. Taito, T. Kono, M. Nakano, T. Saito, T. Ito, K. Noguchi, H. Hidaka, and T. Yamauchi, "A 28nm Embedded Split-Gate MONOS (SG-MONOS) Flash Macro for Automotive Achieving 6.4GB/s Read Throughput by 200MHz No-Wait Read Operation and 2.0MB/s Write Throughput at T_j of 170°C," *IEEE Journal of Solid-State Circuits,* vol. 51, no. 1, pp. 213-221, 2016.

[2] E. F. Runnion, S. M. Gladstone, R. S. Scott, D. J. Dumin, L. Lie, and J. C. Mitros, "Thickness dependence of stress-induced leakage currents in silicon oxide," *IEEE Transactions on Electron Devices,* vol. 44, no. 6, pp. 993-1001, 1997.

[3] M. H. White, D. A. Adams, and J. Bu, "On the go with SONOS," *IEEE Circuits and Devices Magazine,* vol. 16, no. 4, pp. 22-31, 2000.

[4] J. S. Meena, S. M. Sze, U. Chand, and T.-Y. Tseng, "Overview of emerging nonvolatile memory technologies," *Nanoscale Research Letters,* vol. 9, no. 1, pp. 526-526, 2014.

[5] Q. Lin, C. Zhao, and N. Sheng, "Hydrogen-induced program threshold voltage degradation analysis in SONOS wafer," *Solid-State Electronics,* vol. 116, pp. 60-64, 2016/02/01/, 2016.

[6] P. B. Kumar, R. Sharma, P. R. Nair, D. R. Nair, S. Kamohara, S. Mahapatra, and J. Vasi, "Mechanism of drain disturb in SONOS flash EEPROMs," *IEEE Transactions on Electron Devices,* vol. 54, no. 1, pp. 98-105, 2007.

[7] C. Charpin-Nicolle, A. de Luca, A. Persico, G. Médico, C. Tallaron, F. Aussenac, R. Kies, G. Molas, L. Masoero, O. Cueto, and B. de Salvo, "A technological and electrical study of self-aligned charge-trap split-gate memory devices," *Microelectronic Engineering,* vol. 118, pp. 15-19, 2014/04/25/, 2014.

[8] T. Kono, T. Ito, T. Tsuruda, T. Nishiyama, T. Nagasawa, T. Ogawa, Y. Kawashima, H. Hidaka, and T. Yamauchi, "40nm embedded SG-MONOS flash macros for automotive with 160MHz random access for code and endurance over 10M cycles for data." pp. 212-213.

[9] Y. Taito, M. Nakano, H. Okimoto, D. Okada, T. Ito, T. Kono, K. Noguchi, H. Hidaka, and T. Yamauchi, "A 28nm embedded SG-MONOS flash macro for automotive achieving 200MHz read operation and 2.0MB/S write throughput at Tj of 170°C." pp. 1-3.

[10] Y. Taito, T. Kono, M. Nakano, T. Saito, T. Ito, K. Noguchi, H. Hidaka, and T. Yamauchi, "A 28 nm Embedded Split-Gate MONOS (SG-MONOS) Flash Macro for Automotive Achieving 6.4 GB/s Read Throughput by 200 MHz No-Wait Read Operation and 2.0 MB/s Write Throughput at Tj of 170C," *IEEE Journal of Solid-State Circuits,* vol. 51, no. 1, pp. 213-221, 2016.

[11] S. Tsuda, Y. Kawashima, K. Sonoda, A. Yoshitomi, T. Mihara, S. Narumi, M. Inoue, S. Muranaka, T. Maruyama, T. Yamashita, Y. Yamaguchi, and D. Hisamoto, "First demonstration of FinFET split-gate MONOS for high-speed and highly-reliable embedded flash in 16/14nm-node and beyond." pp. 11.1.1-11.1.4.

[12] S. Tsuda, T. Saito, H. Nagase, Y. Kawashima, A. Yoshitomi, S. Okanishi, T. Hayashi, T. Maruyama, M. Inoue, S. Muranaka, S. Kato, T. Hagiwara, H. Saito, T. Yamaguchi, M. Kadoshima, T. Maruyama, T. Mihara, H. Yanagita, K. Sonoda, T. Yamashita, and Y. Yamaguchi, "Reliability and scalability of FinFET split-gate MONOS array with tight Vth distribution for 16/14nm-node embedded flash." pp. 19.3.1-19.3.4.

[13] L. Fang, J. Gu, B. Zhang, W. Kong, and S. Zou, "A Highly Reliable 2-Bits/Cell Split-Gate Flash Memory Cell With a New Program-Disturbs Immune Array Configuration," *IEEE Transactions on Electron Devices,* vol. 61, no. 7, pp. 2350-2356, 2014.

[14] Z. Xu, D. Liu, W. Xiong, J. Hu, Z. Fang, W. Duan, H. Chen, and W. Qian, "Investigation and three implementations for low power self-aligned 1.5-T SONOS flash device," *China Semiconductor Technology International Conference 2018, CSTIC 2018.* pp. 1-5.

[15] Z. Xu, W. Qian, H. Chen, W. Xiong, J. Hu, D. Liu, W. Duan, W. Kong, W. Na, and S. Zou, "Investigation and process optimization of SONOS cell's drain disturb in 2-transistor structure flash arrays," *Solid-State Electronics,* vol. 129, pp. 44-51, 3//, 2017.

[16] R. van Schaijk, M. Slotboom, M. van Duuren, D. Dormans, N. Akil, R. Beurze, F. Neuilly, W. Baks, A. H. Miranda, and P. G. Tello, "Low voltage and low power embedded 2T-SONOS flash memories improved by using P-type devices and high-K materials," *Solid-State Electronics,* vol. 49, no. 11, pp. 1849-1856, 2005/11/01/, 2005.

[17] M. H. White, D. A. Adams, J. R. Murray, S. Wrazien, Z. Yijie, W. Yu, B. Khan, W. Miller, and R. Mehrotra, "Characterization of scaled SONOS EEPROM memory devices for space and military systems." pp. 51-59.

INVESTIGATION AND DEMONSTRATION OF HOT CARRIER EFFECT IN LDMOS TRANSISTORS WITH ULTRA-SHALLOW TRENCH ISOLATION

Zhaozhao Xu[1,], Jun Hu[1], Ziquan Fang[1], Wenting Duan[1], Donghua Liu[1], Wensheng Qian[1]*

[1]ShanghaiHuahong Grace Semiconductor Manufacturing Corporation, Shanghai 201203, China

*Corresponding Author's E-mail: Zhaozhao.xu@hhgrace.com

ABSTRACT

A lateral double-diffused metal–oxide–semiconductor (LDMOS) field-effect transistor with ultra-shallow trench isolation (USTI) is proposed in this work. The degradation of device induced by hot-carrier injection (HCI) in n-type LDMOS transistors in the drift region is fully investigated. It revealed that improvement of the relationship between the breakdown and specific on-resistance can be obtained without sacrificing of HCI degradation. Additionally, the physical mechanism behind the results is also analyzed from the technology computer-aided-design (TCAD) simulation.

Key Words—Ultra shallow trench isolation (USTI); lateral double-diffused metal-oxide-semiconductor (LDMOS); on-resistance degradation; hot-carrier injection (HCI); Technology computer-aided-design (TCAD)

INTRODUCTION

High voltage devices like laterally double-diffused MOSFET (LDMOS) devices are widely used in high voltage, smart power technologies due to its low specific on-resistance (R_{sp}). Planar LDMOS such as shallow trench isolation (STI) [1, 2] is frequently used because the ease of co-integration with the conventional CMOS technologies [3]. Due to the high voltage operating condition, hot-carrier degradation is a key parameter and should be qualified. It is generally known that lower R_{sp} leads to the improvement of output power and efficiency of switches and drivers in mixed-signal integrated circuits [4, 5]. Hence one of the desired performances of LDMOS is achieving the lowest R_{sp} to minimize conduction losses [6] and save chip area penalty for the competitive power ICs. However, the inherent STI of CMOS technology is hardly to achieve optimal breakdown voltage (BV)-R_{sp} relationship for the low-voltage LDMOS with BV< 50V.

Consequently, an ultra-shallow trench isolation (USTI) concept [1] for further improving the tradeoff between BV and R_{sp} has been proposed for the low-voltage power LDMOS due to the rapid growing market in portable power management and automotive electronics [7, 8]. However, the HCI degradation of LDMOS with USTI is not investigated and demonstrated. Additionally, how to improve the BV-R_{sp} performance as well as minimize the HCI degradation of USTI-LDMOS is still lacked. It is well known that HCI degradation is highly related with substrate current (I_B). Therefore, the HCI effect of USTI-LDMOS is fully demonstrated and studied with the advanced TCAD tools.

This work is organized as below. Firstly, the device structure and simulation details are presented. Secondly, the impacts of channel length (L_{CH}), accumulation area size (L_A), threshold voltage (V_{TH}), and thickness of gate oxide (T_{ox}) on the HCI are demonstrated and discussed. Then, the suggestion for

improvement of BV-R_{sp} with robust suppression of HCI effect is proposed. Finally, we present a conclusion about this work.

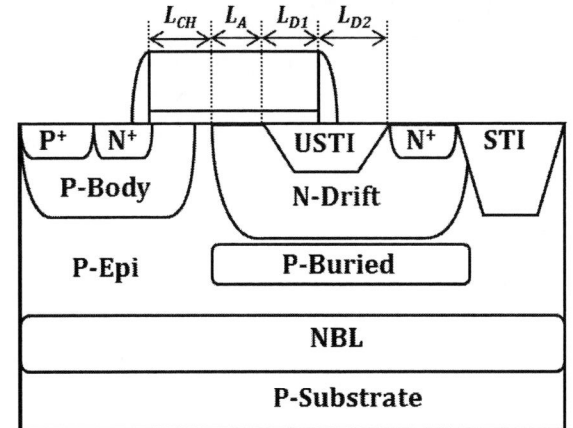

Figure 1: Schematic cross-sectional drawings of 20-V USTI-LDMOS.

DEVICE STRUCTURE AND SIMULATION

Figure 1 shows the schematic cross-sectional views of the USTI LDMOS device. As illustrated in Figure 1, the P-Epi/P-Substrate structure with N-type buried layer (NBL) is employed. Junction isolation was adopted to ease the processing. The heavily doped NBL is formed on P-Substrate by using lower energy antimony implant followed by high thermal budget. With the deep N-type well implant, NBL is peaked up and shorted to the drain through metal to isolate the device from the P-Substrate of the wafer and enable double RESURF [9]. This isolation features insensitivity of logic blocks to parasitic disturbances from the power areas [10]. As shown in the Figure 1, phosphorus is implanted into drain-side to form N-Drift region. Meanwhile, boron with high energy is co-implanted with the N-Drift to form a P-buried layer under N-Drift at the same masking step, which assists the depletion of the drift region. The channel is formed by P-Body implantation and diffusion which are both self-aligned to the gate. It further enables L_{CH} scaling. The polysilicon gate is employed as the field plate over the field oxide above drift region in USTI to release the electric field crowding effect at the polysilicon edge. As presented in Figure 1, the length of channel-side drift (accumulation region) is defined as L_A. This region is deliberately doped with heavier doping concentration to reduce the resistance. The length of drain-side drift is defined as L_{D1} and L_{D2}. The major processes of the USTI-LDMOS are detailed in [1], which is implemented in our 0.18 μm BCD (Bipolar-CMOS-DMOS) technology.

TCAD simulations are performed by Sentaurus TCAD simulation tools from Synopsys technology [11]. For the electrical simulation of LDMOS device, Shockley–Read–Hall,

Auger generation–recombination, OldSlotboom, and Okuto-Crowell avalanche model were turn on with default parameters. In addition, mobility models including PhuMob, HighFieldSaturation, Enormal are employed for device simulation with calibrated parameters. Self-heating effect is also considered when high current is conducted in this device. The main solving models are Poisson, Electron, and Hole.

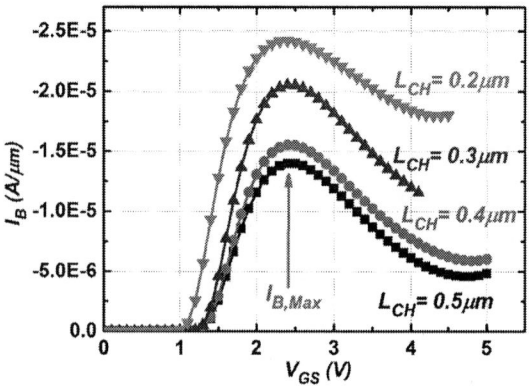

Figure 2: I_B-V_{GS} curves as function of L_{CH}.

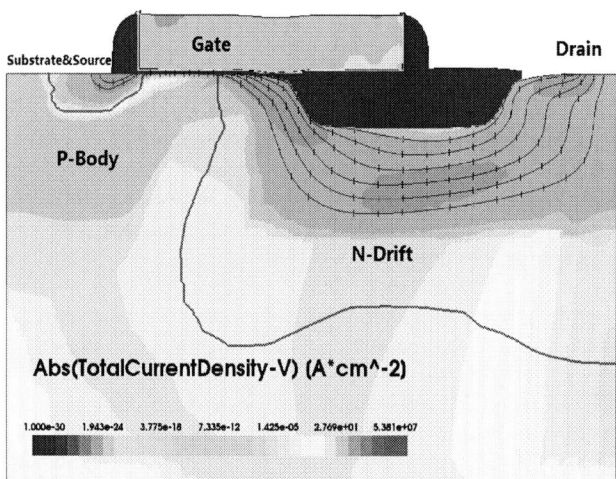

Figure 3: Current density stream trace for 20-V USTI-LDMOS.

RESULTS AND DISCUSSION

For low-voltage LDMOS, the channel resistance is comparable [12] to and even larger than the resistance of drift region. Therefore, channel length scaling is critical for further optimization of BV-R_{sp} tradeoff. As expected, significant improved BV-R_{sp} tradeoff [13] can be achieved by scaling L_{CH}. However, as illustrated in Figure 2, $I_{B,Max}$ is also obviously increased with the decrease of L_{CH}. Higher lateral E-field is induced in the channel region when L_{CH} is decreased. This enhances the impact ionization at P-Body/N-Drift junction, which induces higher I_B at first peak (about V_{GS} of 2.5V). Same as the conventional CMOS device, higher conducting current is obtained with the reduction of channel length. As shown in Figure 3, conducting current is mainly concentrated in the USTI bottom. Thus, higher current induces Kirk effect [14, 15] easily when L_{CH} is shrunk.

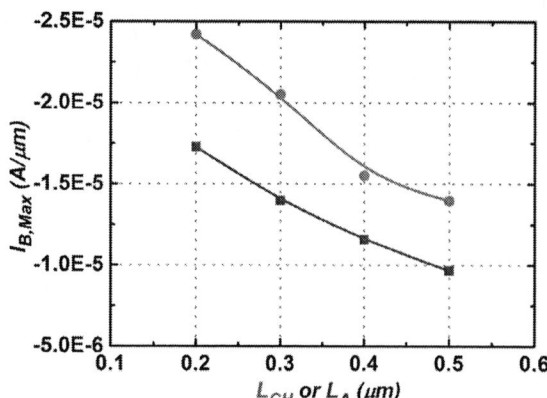

Figure 4: $I_{B,Max}$ as function of L_{CH} and L_A.

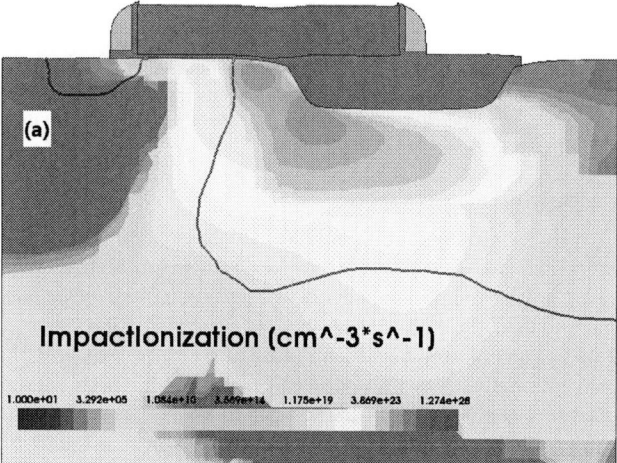

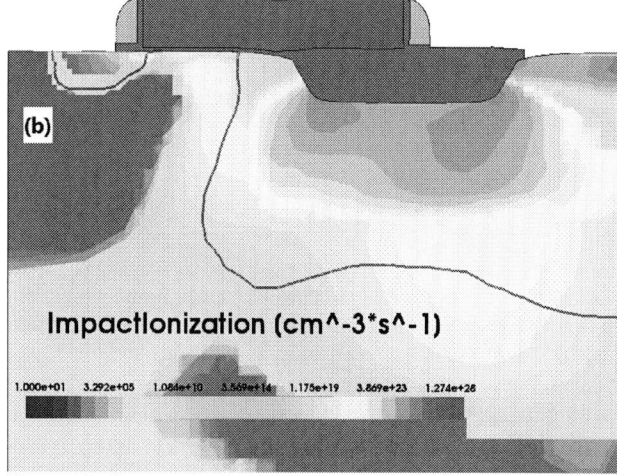

Figure 5: Impact ionization rate under (a) V_{GS}/V_{DS}= 2.5/22V and (b) V_{GS}/V_{DS}= 5.0/22V.

For best BV-R_{sp} relationship, optimal L_A should be used. L_A exhibits more complicated impact on the BV-R_{sp} tradeoff as compared with L_{CH} due to the current spreading effect in accumulation region. Additionally, it was revealed that L_A also exhibits significantly impact on the HCI degradation or $I_{B,Max}$. Figure 4 plots the I_B-V_{GS} as the function of L_A. It is manifest that

L_A also has significant effect on the $I_{B,Max}$. The longer the L_A, the lower the $I_{B,Max}$, which exhibits same trend as the L_{CH}. Figure 5 presents the impact ionization rate under $V_{GS}/V_{DS}= 2.5/22$V and $V_{GS}/V_{DS} = 5.0/22$V, respectively. It reals that currents are crowded both at the channel-side USTI bottom with $L_A \leq 0.2 \mu$m, which induces Kirk effect easily.

The substrate current as the function of T_{ox} are shown in Figure 6. Noted that the V_{GS} of first peak increases with the T_{ox} and the I_B of first peak decreases slightly. In addition, I_B of second peak is significantly reduced with the increase of T_{ox} because of higher V_{TH}. This results in lower saturated current which benefits ease of Kirk effect. According to [13], BV-R_{sp} tradeoff is hardly affected by using thicker T_{ox}. It is worth noting that thicker T_{ox} is an obstacle for scaling L_{CH}. In addition, the impact of implantation of adjusting the threshold voltage on the I_B is also studied. The conclusion is the same as the T_{ox} one and it is not shown here.

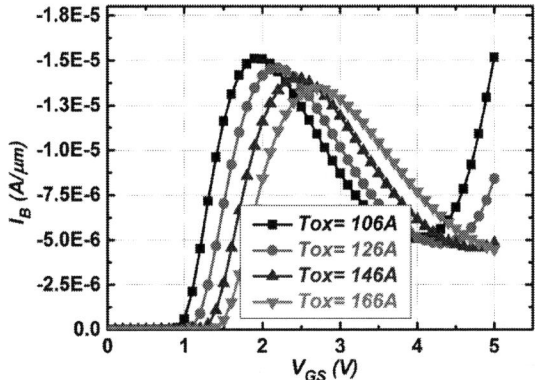

Figure 6: I_B-V_{GS} curves as function of the thickness of gate oxide.

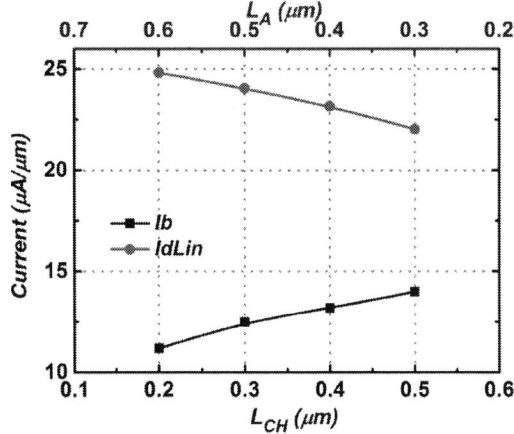

Figure 7: $I_{B,Max}$ under different L_{CH}/L_A configurations with a constant value of 0.8 μm for $L_{CH}+L_A$.

Based on above investigation, both L_{CH} and L_A exhibit significant impact on the HCI effect. To further optimize the BV-R_{sp} relationship, it is apparent that scaling L_{CH} is the most effective way [13]. Scaling L_{CH} is beneficial not only for the increasing of linear current but also for reduction of cell pitch,

which can further reduce the R_{sp} of LDMOS. With consideration to the HCI degradation, larger L_A is required for decreasing substrate current, as shown above.

Consequently, the tradeoff between L_{CH} and L_A is investigated for achieving better BV-R_{sp} relationship as well as robust suppression of HCI effect. As illustrated in Figure 7, the $I_{B,Max}$ gradually decreases with decreasing L_{CH} while keeping a constant value of 0.8 μm for $L_{CH}+L_A$. However, the linear current increases with reduction of L_{CH}, which benefits BV-R_{sp} relationship. Thus, it is reasonable inferred that better BV-R_{sp} tradeoff can be achieved without sacrificing hot-carrier resistance of device.

CONCLUSION

To further improve the performance of USTI-LDMOS, the degradation induced by hot-carrier injection is fully investigated in this paper by TCAD simulation. It was found that both channel length (L_{CH}) and accumulation length (L_A) exhibit significant impact on substrate current during stress. Consequently, optimal L_A only for best BV-R_{sp} tradeoff can not be employed when L_{CH} is shrunk. To further optimize BV-R_{sp} tradeoff with scaling L_{CH}, L_{CH}-L_A tradeoff is inevitable when higher resistance of HCI degradation is taken into consideration. Fortunately, it was revealed in this paper that better BV-R_{sp} performance can still be obtained by optimizing L_{CH}-L_A configuration and without sacrificing hot-carrier resistance.

REFERENCES

[1] F. Jin, D. Liu, J. Xing, X. Yang, J. Yang, W. Qian, W. Yue, P. Wang, M. Qiao, and B. Zhang, "Best-in-class LDMOS with ultra-shallow trench isolation and p-buried layer from 18V to 40V in 0.18μm BCD technology." pp. 295-298.

[2] T. Mori, S. Kubo, and T. Ipposhi, "A novel divided STI-based nLDMOSFET for suppressing HCI degradation under high gate bias stress." pp. 299-302.

[3] A. F. M. Alimin, H. H. Hizamul-din, S. F. W. M. Hatta, and N. Soin, "The influence of shallow trench isolation angle on hot carrier effect of STI-based LDMOS transistors." pp. 248-251.

[4] M. Imam, Z. Hossain, M. Quddus, and J. Adams, "Design and optimization of double-RESURF high-voltage lateral devices for a manufacturable process," *Electron Devices IEEE Transactions on,* vol. 50, no. 7, pp. 1697-1700, 2003.

[5] S. Sharma, T. Letavic, Y. Shi, A. Loiseau, J. E. Monaghan, N. Feilchenfeld, R. Phelps, C. Lamothe, D. Cook, and J. Dunn, "Planar dual gate oxide LDMOS structures in 180nm power management technology." pp. 405-408.

[6] B. Duan, Y. Yang, B. Zhang, and X. Hong, "Folded-Accumulation LDMOST: New Power MOS Transistor With Very Low Specific On-Resistance," *IEEE Electron Device Letters,* vol. 30, no. 12, pp. 1329-1331, 2009.

[7] M. N. Chil, O. S. Yang, D. Ke, K. J. Mo, L. Kun, M. Tiong, R. V. Purakh, and R. Nair, "Advanced 300mm 130nm BCD technology from 5V to 85V with Deep-Trench Isolation." pp. 403-406.

[8] H. Cha, K. Lee, J. Lee, and T. Lee, "0.18μm 100V-rated BCD with large area power LDMOS with ultra-low effective specific resistance." pp. 423-426.

[9] R. Zhu, V. Khemka, A. Bose, and T. Roggenbauer, "Stepped-Drift LDMOSFET: A Novel Drift Region Engineered Device for Advanced Smart Power Technologies," *Power Semiconductor Devices & Ics .ispsd .ieee International Symposium on*, pp. 1-4, 2006.

[10] R. Rudolf, C. Wagner, L. O'Riain, K. H. Gebhardt, B. Kuhn-Heinrich, B. V. Ehrenwall, A. V. Ehrenwall, M. Strasser, M. Stecher, and U. Glaser, "Automotive 130 nm smart-power-technology including embedded flash functionality," vol. 19, no. 3, pp. 20-23, 2011.

[11] U. Manual, *Sentaurus Process/Device Version G-2012.06 Synopsys*, Mountain View, CA, USA, Jun. 2012.

[12] S. Reggiani, S. Poli, M. Denison, E. Gnani, A. Gnudi, G. Baccarani, S. Pendharkar, and R. Wise, "Physics-Based Analytical Model for HCS Degradation in STI-LDMOS Transistors," *IEEE Transactions on Electron Devices,* vol. 58, no. 9, pp. 3072-3080.

[13] Z. Xu, D. Liu, J. Hu, F. Jin, X. Yang, W. Duan, W. Yue, Z. Fang, W. Qian, W. Kong, and S. Zou, "Demonstration of improvement of specific on-resistance versus breakdown voltage tradeoff for low-voltage power LDMOS," *Microelectronics Journal,* vol. 88, pp. 29-36, 2019/06/01/, 2019.

[14] V. Parthasarathy, V. Khemka, R. Zhu, and A. Bose, "SOA improvement by a double RESURF LDMOS technique in a power IC technology."

[15] A. W. Ludikhuize, "Kirk effect limitations in high voltage IC's." pp. 249-252.

A New Integration flow Study of ONO film Uniformity and Silicon recess improvement for 2T-SONOS Flash

Liqun Dong[*], Zhenghong Liu, Chris Shao, Haoyu Chen, Guanqun Huang, and Ruisheng Qi

Shanghai Huali Microelectronics Corporation, Shanghai 201203, China

*Corresponding Author's Email: Dongliqun@hlmc.cn

ABSTRACT

Non-volatile SONOS (Silicon-Oxide-Nitride-Oxide-Silicon) flash shows better application as the advantages of low program voltage, long-term data retention, good compatibility of CMOS strategy and low cost on memory device. But the device and process window also narrowed especially for cell size continue shrink. This paper applies a new integration flow study of ONO film uniformity and silicon recess improvement for 2T-SONOS Flash on 40nm Tech. Traditional non-SONOS ONO film remove method of LP device is by dry etch, but on 40nm scale, the process window is not enough due to pad oxide thinner with cell size shrink. With much study, we apply a new method that non-SONOS ONO film remove by wet, at the same time, in consideration of select gate silicon recess, we optimize Wet etch process 2 in 1. On one hand, it can enlarge ONO etch process window, optimize ONO uniformity and match production needs, on the other hand, it can improve select gate silicon recess to further reduce the risk of NiSi piping and improve yield, Furthermore, the cost can drop remarkably when use 2 in 1 Wet process particularly on FAB.

INTRODUCTION AND RESULT

Background

Non-volatile SONOS (Silicon-Oxide-Nitride-Oxide-Silicon) flash shows better application as the advantages of low program voltage, long-term data retention, good compatibility of CMOS strategy and low cost on memory device[1]. ONO technology is especially important for SONOS performance, the device and process window is narrowed with the cell size continue shrink[2]. Current 40nm LP 2.5V only SONOS pad oxide is thinner about 20A than 55nm LP SONOS, directly induced SG (select gate) ONO dry etch window being not enough, see as Figure 1, which found SIN residual with etch window lower limit and finally influence SG device performance, cause yield loss. For a mass of production, process window is critical, ONO dry etch current condition only be used as temporary delivery condition. After several rounds of experiments, it's hard for PE to resolve the problem, otherwise it will cause silicon damage with over etch time increased which is absolutely not allowed.

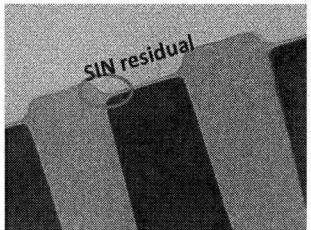

Figure 1: ONO dry Etch BL technology window lower limit wafer found SIN residual PFA result

Experiment Approach 1

According to 55HV SONOS SG (select gate) ONO remove experience, we apply the method with 40LP Eflash SG ONO removed by wet etch process which named ONO ET new, Figue 2 is the comparison fabrication process flow between ONO BL and ONO ET new flow, with ONO ET new process, First, ONO dry ET only etch SG ONO top oxide, then SG SIN removed by wet etch, and ONO CG (control gate) top oxide removed post PR strip, Gate1 pre-clean follow BL flow, check ONO ET New flow inline final film of ONO is ok, but PFA check post Cell drain ET we found bias between CG and SG is about 1 time more than BL condition with ONO ET new process, see as Figure 3 (a) and 3 (b), which will lead to silicide pipping and impact floating voltage being more close to Vneg, and finally cause SONOS inhibit cell fail and directly yield loss, see as Figure 4 (a) and 4 (b).

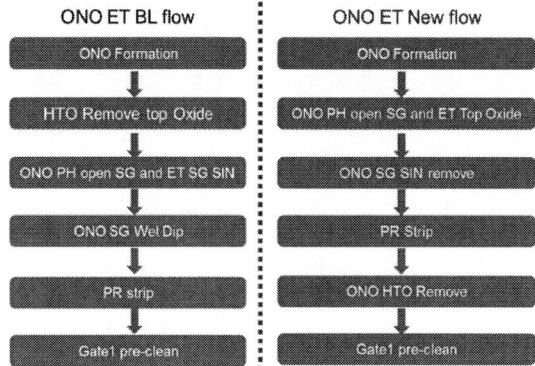

Figure 2: Fabrication process flow for the ONO etch BL and ONO ET New process.

Obviously, current method induces side effects, the

bias comes from the wet etch amount (Figure5) post PR strip after we carefully check all items with ONO ET new process, because we also need to ensure that the final thickness of ONO film match BL, which is critical for device operation and reliability, we need continue to optimize the new condition.

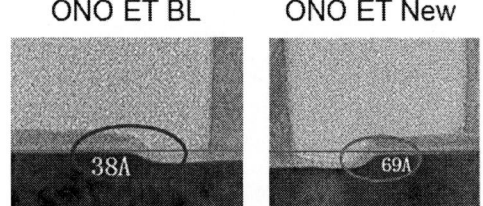

Figure 3: 40LP Eflash CG ad SG bias post Cell drain ET PFA result between ONO ET BL and ONO ET New

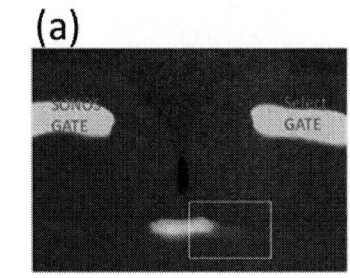

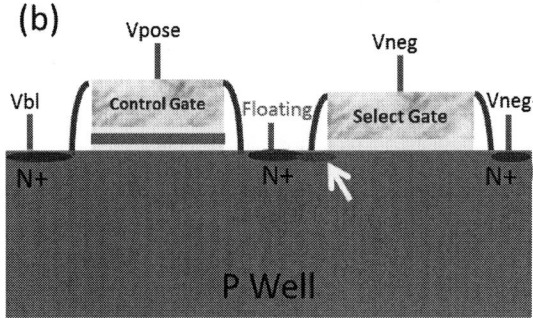

Figure 4: (a) Silicide Pipping induces SONOS Fail EFA result and (b) fail mode

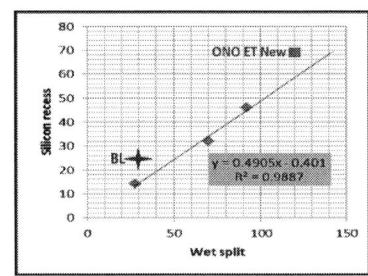

Figure 5: 40LP Eflash ONO ET new Wet split & Silicon recess correlation

Optimized Experiment Approach 2

To optimize the new condition, ONO deposition reduction, ONO CG HTO remove and Gate1 pre-clean 2 in 1 method which is named ONO New process is applied,. After several rounds of experimental data collection, we confirmed the new film thickness of ONO deposition reduction and 2 in 1 wet etch amount, at the same, we adjusted the stephight to match BL, specific process see as Figure6, 6 (a) is the ONO remove BL flow, 6 (b) is the ONO remove new process. F1 is the Logic film pattern and EFF is the SONOS film pattern, BL flow ONO Wet dip make F1 pattern loss and ONO new process have no this step, meaning that F1 pattern has no loss at ONO loop with ONO new process, so we adjust the stepheight (SH for short) at AA (active area) loop post STI CMP, because F1 pattern and EFF pattern are both clear, when F1 generates loss by wet etch, the EFF pattern will loss at the same time, so the TUN (SONOS Vt imp area) wet dip amount need corresponding reduction, so that the F1 pattern and EFF pattern stepheight respectively match BL condition.

Figure 7(a)-7(c) is the inline match data after ONO new process tuning, we can see that Figure 7(a) shows ONO final film thickness match BL and ONO film uniformity is better than BL, Figure 7(b) and 7(c) shows F1 and EFF pattern SH match BL. Figure 8 shows 40LP Eflash CG ad SG bias post Cell drain ET PFA result with ONO New process, which improve about 30% than BL, to some extent, it reduces the chance of silicide pipping and yield loss.

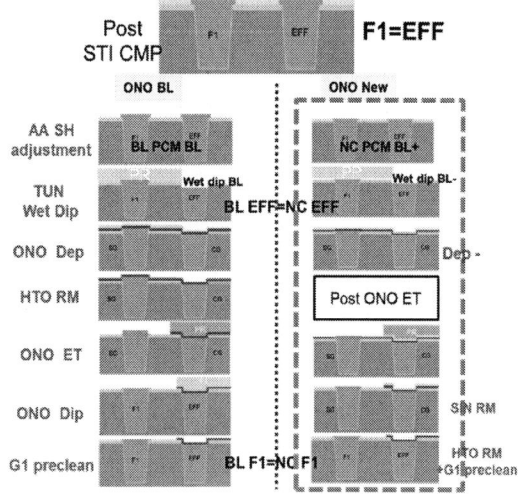

Figure 6: New studied fabrication process flow for the SONOS SG ONO remove, 6 (a) is the ONO remove BL flow, 6 (b) is the ONO remove new process

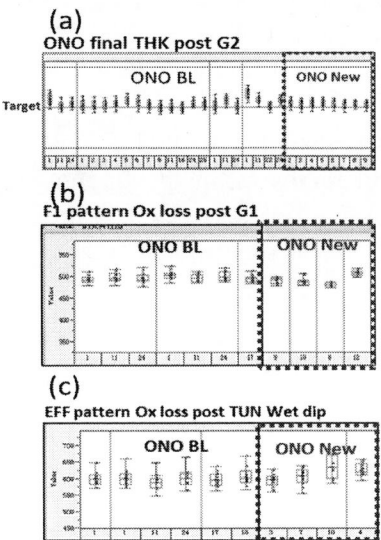

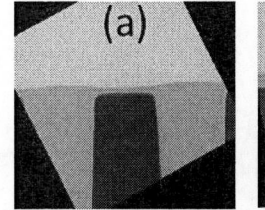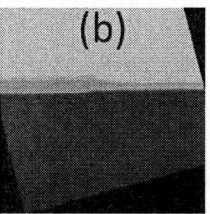

Figure 10: 10 (a) is the ONO dry ET window upper limit PFA on SG, 10 (b) is the ONO dry ET window upper limit PFA on AA

Figure 7: 40LP Eflash inline match data after ONO new process tuning, 7 (a) is the ONO final THK comparison between ONO BL and ONO new process , 7 (b) is the F1 pattern oxide loss post G1 between ONO BL and ONO new process, and 7(c) the EFF pattern oxide loss post G1between ONO BL and ONO new process

Figure 8: 40LP Eflash CG ad SG bias post Cell drain ET PFA result with ONO New process

For ONO dry etch window improvement, we check ONO new process after ONO dry etch and window upper limit PFA, see as Figure 9 and Figure 10. Figure9 shows ONO dry etch SIN have enough loss so window lower limit is ok, window upper limit PFA Figure 10 (a) and 10 (b) shows that there is still SIN remaining, means it will not induce silicon damage.

Figure 9: ONO New Process Post ONO Dry ET

CONCLUSION

For 40LP Eflash ONO ET window not enough condition, with much study, we apply a new method that non-SONOS ONO film removed by dry etch and wet etch, at the same time, in consideration of select gate silicon recess, we optimize wet etch process 2 in 1. On one hand, it can enlarge ONO etch process window, optimize ONO uniformity and match production needs, on the other hand, it can improve select gate silicon recess, make the bias between control gate and select gate improve about 30% than BL condition, can further reduce the risk of NiSi piping and reduce yield loss, Furthermore, with the BL flow and new process flow we applied comparison, our new process flow can save one step by use wet etch 2 in 1 method, which the cost can drop remarkably particularly on FAB production.

ACKNOWLEDGEMENTS

I would like to express my gratitude to all people who helped and guided me during the writing of this paper, especially my mentor at work, also a good team player Liuzhenghong engineer. In addition, I should be grateful to Shanghai Huali Microelectronics Corporation who gives me such a precious chance.

REFERENCES

[1] Joe E. Brewer and Manuzur Gill. *Nonvolatile memory technologies with emphasis on Flash*, 2008.
[2] ITRS 2013, the 2013 Edition of the International Technology Roadmap for Semiconductors, on line: http://www.itrs.net/.

STUDY OF GIDL IMPROVEMENT FOR 2T-SONOS FLASH

Zhenghong Liu*, Liqun Dong, Ruisheng Qi, Shugang Dai, Guanqun Huang, Haoyu Chen, Chris Shao

Shanghai Huali Microelectronics Corporation, Shanghai 201203, China
*Corresponding author. Tel.: +86-17717386771. E-mail addresses: liuzhenghong@hlmc.cn

ABSTRACT

The improvement of Gate induced drain leakage (GIDL) is studied in 2T SONOS (silicon-oxide-nitride-oxide-silicon) nonvolatile memory. High GIDL current from the select gate (SG) introduce inhibit disturb to the neighbor SONOS gate. It is found that these leakage bits impact the overall yield and reliability. In this paper, the variation trend of GIDL leakage with LDD dopant dose, energy and tilt is investigated in detail Results show that GIDL leakage is effectively decreased through increasing tilt and energy or decreasing the dose amount of select gate LDD IMP step. In addition, GIDL leakage also decreased by changing SG LDD dopant step from post poly re-oxidation to post spacer1 etch which increased the space of SG gate to drain. The proposed condition improves GIDL current by one order of magnitude; yield and Vt window are also greatly increased.

INTRODUCTION

2-Transistor (2T) SONOS which composed of one memory transistor and one selective gate transistor has been widely used in high reliability and high security devices such as e-passport, currency counter, and bank card [1] because of their better endurance, a higher radiation tolerance, no erratic bits and simpler processing in recent three decades [1-3].The single cell schematic and the side - view cut along the bitline of a 2T SONOS cell can be seen in Fig. 1(a) and 1(b). SG transistor channel on for the select cell while channel off for the unselect cell and this ensures small standby current from the Flash memory core and can be utilized for low -voltage operations where power consumption is critical[4].

The intolerable gate leakage and more reliability problems are caused as the shrinkage of the channel length.SG transistor as normal NMOS device and the SG channel length shrink greatly at 40-nm technology node. During the development, large GIDL current which induced disturb for 2T SONOS memory transistor was found, this issue becomes a latent crisis in reliability for the SONOS devices.

methods of the GIDL leakage in 2T SONOS memory. The variations of GIDL current under different dose, energy and tilt of select gate LDD dopant step are discussed. The effect of select gate LDD implantation step sequences on GIDL current is also compared.

GIDL MEASUREMENT DETAILS

The GIDL leakage generation and its influence on 2T SONOS device are explained by combining the operation principle of 2×2 array, which shown in Fig.2 (a).There are total four 2T SONOS cells, cell A is the PGM selected cell, cell B is PGM unselect cell and subject to drain stress, cell C is inhibit select cell and subject to gate stress, cell D is inhibit unselect cell and not subject to program disturbs. Operation conditions are showed in table1.

TABLE1
Operating Condition for the 2T SONOS cell

Operation	Source	P-well	WL	Bitline	WLS	Bitline+1	WLS+1
Program	1V	0V	0V	0V	7.5V	4.1V	1V
Erase	0V	0V	-2.5V	0V	-7.5V	0V	-7.5V
Read	0V	0V	2.5V	0.6V	0V	0V	0V

During cell A program operation, cell C memory cell channel on because of share the same word line with cell A and subject to gate stress, the bitline voltage of cell C is about 0.7V to keep it from being programmed, so the voltage condition of cell C during cell A PGM operation is showed in Fig.2(b). At this condition, the voltage of drain transfer to internode area between CG and SG and GIDL leakage from SG drain area generated, this can be explained by the GIDL generated principle of NMOS shown in Fig.2(c).Cell C drain current should be increased under the lower SG Vt condition because of GIDL leakage, which can be seen in Fig.2 (d).

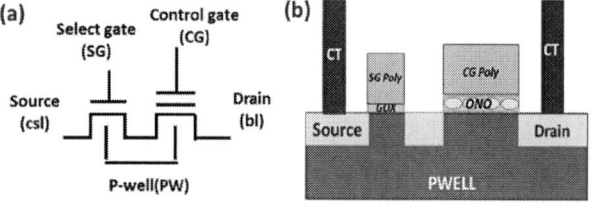

Figure1. 2T single cell schematic (a) and the side - view cut along the bitline(b).

This paper studies the causes, effects and monitoring

978-1-7281-6559-2/20 $31.00 © 2020 IEEE

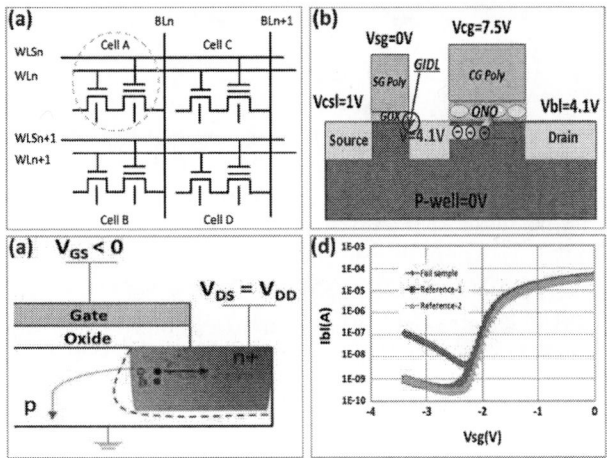

Fig.2. (a)2T 2×2 array schematic, During the program operation, cell A is the PGM selected cell. Cell B is PGM unselect cell and subject to drain stress. Cell C is inhibiting select cell and subject to gate stress. Cell D is inhibiting unselect cell and not subject to program disturbs. (b)Cell C voltage condition during Cell A PGM operating.(c)Cross section of n-MOSFET in accumulation at high VDS: electrons can tunnel the drain forming GIDL current.(d)The bitline current of 2T SONOS cell under different SG Vt and blue dots show higher GIDL current.

Based on the above analysis, the monitoring of GIDL mainly adopts two methods. One is to test EP curve for 2X2 array structure, increase the stress time under operating voltage bias and continuously test Vtp, Vte and disturb Vt which named Vinh, the curve as shown in Fig.3 (a). In order to enhance GIDL leakage, the deterioration experiment was conducted by raising the stress bias between Vbl and Vsg during EP curve test, there are two action to achieve this worse case, one is increasing Vbl while other terminals keep no change, another is just decreasing Vpw and Vsg at the same time, the result of Vinh at degradation test condition as shown in Fig.3 (b), which can be seen that the Vinh induced by GIDL leakage becomes worse as the stress bias of Vbl and Vsg increased.

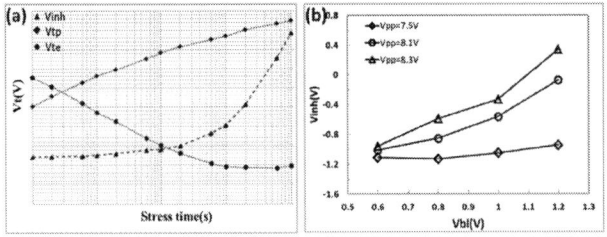

Fig.3.(a) EP curve for GIDL induced disturb.(b) Vinh shift to Vtp as the stress bias between Vbl and Vsg increased.

Another way of monitoring GIDL is to design one special testkey which contains a large number of 2T SONOS cells array. All the CG word line of these bits are connected together and drawn out by one pad, all the SG word line connected and drawn out by another pad, and as well as the bitline, source and pwell all drawn out by independent pad. The current of drain and pwell are measured at the test conditions described in Fig.2 (b) to monitor the GIDL leakage.

RESULTS AND DISCUSSION

The SG transistor of the 2T SONOS cell is constructed by the same way as the normal n-MOSFET which used in the periphery, and the principle to suppress GIDL is to reduce the overlap between the gate and the SG drain, in our work we try to improve the GIDL current from two aspects: one is from the device structure and another one is from the device adjustment such as implant tilt and energy and dose amount to decrease the dopant concentration at the overlap area.

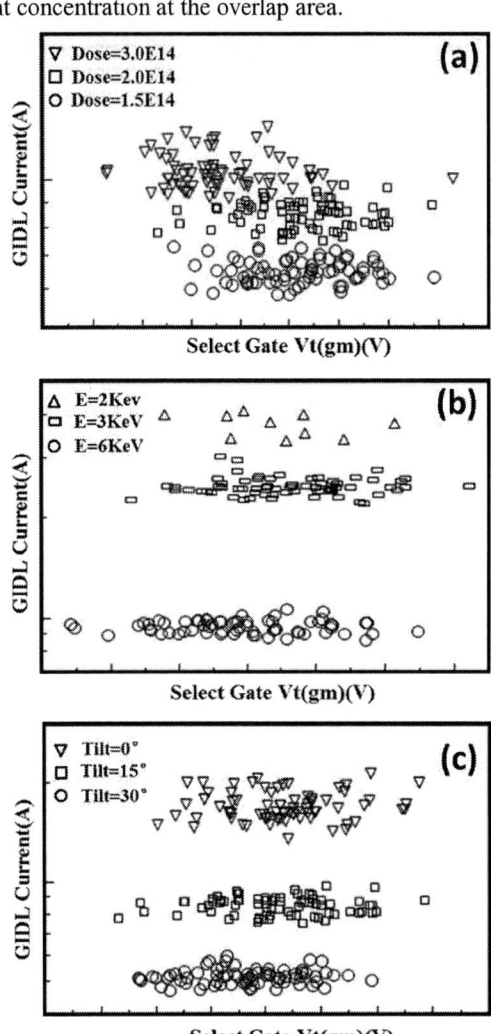

Fig.4. (a) The GIDL current under different LDD dose amount.(b) The GIDL current under different LDD implant tilt.(c) The GIDL current under different LDD implant energy.

Fig.4 shows the GIDL current under different SG LDD dose amount and energy and tilt conditions. The data shows that the GIDL leakage improved as decreasing of dose amount as well as increasing of injection energy and tilt, as Fig.4 (a), Fig.4 (b) and Fig.4(c) shown, respectively. These three directions all decreased the LDD dopant concentration of the overlap area between SG gate and drain. Of course, we cannot adjust the dose amount, the energy and the tilt indefinitely in the GIDL improvement direction because of the SG LDD implant process is share with 2T SONOS memory cell. For example, the current capacity will be decreased of the SONOS cell as decrease the dose amount, and channel leakage will be increased as the energy or tilt increased.

The another solution is to focus on structural optimization as Fig.5(a) shown, the LDD implant step change from post poly re oxidation to post spacer1 etch, the overlap between SG gate and drain was minimized because of spacer1 thickness and the GIDL was suppressed on a certain degree, which showed in Fig.5(b).

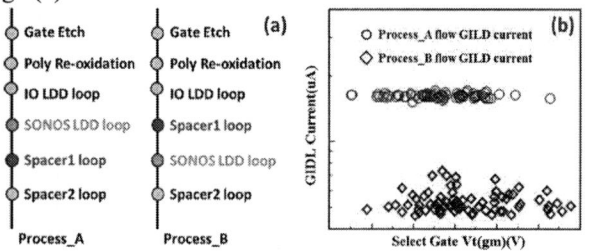

Fig.5. (a) Baseline process flow named Process_A and GIDL improvement flow named Process_B.(b)GIDL leakage compare result for Process_A and Process_B.

Fig.6 (a) and Fig.6 (b) show the comparison between the base and the improvement condition on Vinh and Vt window. It has been found that the optimized condition Vinh much improved than BL and still own ~0.7V margin post longtime stress and Vt window ~340mV increased.

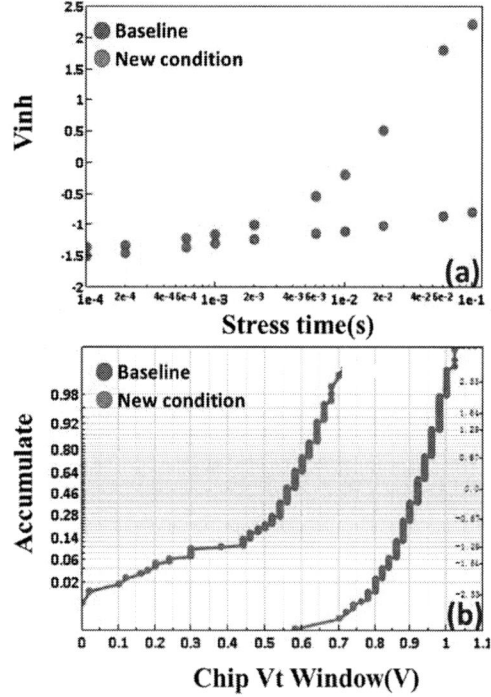

Fig.6. The comparison of Vinh (a) and Vt window (b) for baseline (red dots) and new condition (green dots).

CONCLUSION

In this work, we try to improve the GIDL leakage of the select gate transistor for 2T SONOS memory from structure and device injection condition. The variation trend of GIDL leakage with LDD dopant dose, energy and tilt is investigated in detail. The results showed that GIDL leakage is effectively decreased through increasing tilt and energy or decreasing the dose amount of select gate LDD IMP step. In addition, GIDL leakage also decreased by changing SG LDD dopant step from post poly re-oxidation to post spacer1 etch which increased the space of SG gate to drain. The proposed universal and facile condition improves GIDL current by one order of magnitude, the Vinh own ~0.7V margin even at worse test condition, the VT window about 340 mV increased.

ACKNOWLEDGEMENTS

The authors would like to acknowledge the members of the 40-nm SONOS flash project team and other engineers for the support for developing this work. I am grateful to Shanghai Huali Microelectronics Corporation who gives me such a precious chance.

REFERENCES

[1] L. Parker, R. Singh, C. N. Li, L. Walker, C. M. Hong, S. Liu, D. Farenc, K.-T. Chang, and P. Ingersoll, " Advanced Flash Devices Embedded in a 0.13 μ m CMOS Process, " *IEEE NVSM Workshop, Monterey, CA*, pp. 59–61, Aug. 2001.

[2] C. Kuo, M. Weidner, T. Toms, H. Choe, K.-M. Chang, A. Harwood, J. Jelemensky, and P. Smith, " A 512-kb Flash EEPROM Embedded in a 32-b Microcontroller, " *IEEE J. Solid – State Circuits*, Vol. 27, No. 4, pp. 574–582, Apr. 1992.

[3] K. - M. Chang , S. Cheng , and C. Kuo , " A Modular Flash EEPROM Technology for 0.8 μ m High Speed Logic Circuits, " *CICC Proc.* 13, pp.18.7.1 – 18.7.4 , May 1991.

[4] W. Liu , K. - T. Chang , C. Cavins , B. Luderman , C. Swift , K. - M. Chang , B. Morton , G. Espinor ,and S. Ledford , " A 2 - Transistor Source - Select (2TS) fl ash EEPROM for 1.8V - Only Applications," *IEEE NVSM Workshop*, Monterey, CA, pp. 4.1.1 –4.1.3 , Feb. 1997.

ONE NEW CALIBRATION STRUCTURE OF MOSFET GATE OXIDE CAPACITOR

Han Xiaojing [1]*

Shanghai Huali Microelectronics Corporation, Shanghai, China

* Email: hanxiaojing@hlmc.cn

ABSTRACT

This paper introduces a new kind of calibration structure for MOSFET gate oxide capacitance. This new calibration structure is used to remove the parasitic interconnect capacitance from the gate oxide capacitor when calculate the gate oxide capacitance. By processing measured data of gate oxide capacitance and this new capacitance calibration structure, we can get the value of gate oxide capacitance more accurately, which provides a more accurate guarantee for SPICE model and circuit design.

INTRODUCTION

Background

With the rapid development of high performance integrated circuit chips, the demand of chip process models' accuracy is becoming stronger. It turns into a hot topic how to improve the accuracy of process model in all directions.

When measuring the MOSFET gate oxide capacitance, we receive measurement data that contain part of interconnection parasitic capacitance, which make deviation of the real gate oxide capacitance [1]. In order to reduce the deviation caused by interconnection parasitic capacitance, we apply the gate oxide capacitor calibration structure. By measuring and handling with the data of standard structure and calibration structure, the parasitic effect of gate oxide can be removed and the accuracy of gate oxide capacitor can be improved.

Common Practice and its defect

At present, the widely used calibration structure of MOSFET gate oxide capacitor is a structure that based on a standard MOSFET gate oxide capacitor structure (shown in Figure 1) and removes its Active Area (AA), poly and contact from standard capacitor structure (shown in Figure 2). By measuring this kind of calibration structure, we can achieve capacitor calibration data that mostly caused of parasitic effect. Running calculation between this calibration data and the standard capacitor structure measurement data, the result of calibrated capacitance can be achieved.

However, the accuracy of the structure calibration is limited, which brings some errors to the simulation and the followed circuit design.

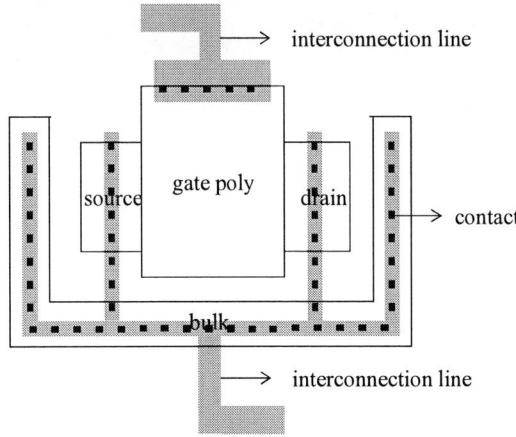

Figure 1: A standard MOSFET gate oxide capacitor structure

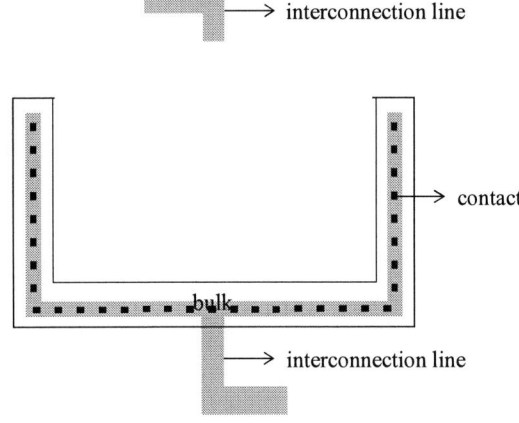

Figure 2: A widely used calibration structure of MOSFET gate oxide capacitor

CALIBRATE CAPCITANCE MORE ACCURATLY

New Calibration Structure

The technical problem to be solved is to improve the existing MOSFET gate oxide capacitance calibration structure and reduce the MOSFET gate oxide capacitance parasitic effect on measurement data, so as to meet the requirements of demanding circuit design.

In order to solve the technical problem mentioned above, a new MOSFET gate oxide capacitor calibration structure is proposed in this paper. In the existing structure, AA and gate poly are retained, and only the contact on source, drain and gate poly is removed (shown in Figure 3). Through this structure, the parasitic capacitance of gates is better removed to calibrate gate oxide capacitor, so as to improve the accuracy of capacitance measurement.

The real value of MOSFET gate oxide capacitance Cox is the measured value of standard gate oxide capacitance structure Cox_meas minus the measured value of gate oxide capacitance calibration structure Cox_cal.

$$Cox = Cox\_meas - Cox\_cal \qquad (1)$$

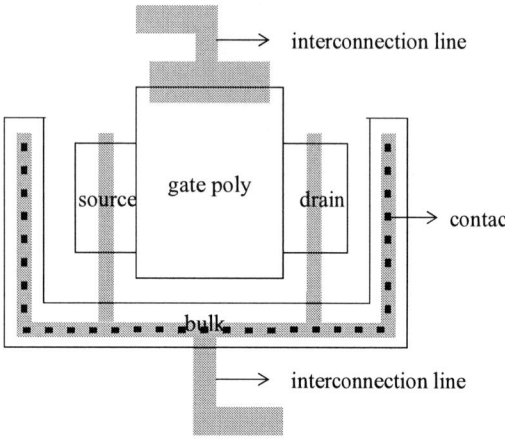

Figure 3: A new calibration structure of MOSFET gate oxide capacitor

Detailed implement methods

The common calibration structure of MOSFET gate oxide capacitance is shown in figure 5 [2]. Without adding AA or poly, the parasitic capacitance of interconnection is not completely removed, so that some error will still exist in the gate oxide capacitance value obtained [3].

Here we propose a new calibration structure based on the common MOSFET gate oxide capacitor calibration structure. We keep AA and poly structure, and only remove the source of leakage and the contact on the grid, as shown in figure 6. By testing calibration values from this structure, calibration data can be obtained which including the parasitic capacitance between source-drain interconnection line and gate, the parasitic capacitance between source-drain interconnection line and bulk, and the parasitic capacitance between grid interconnection line and bulk and so on a series of parasitic capacitance. Therefore, more comprehensive capacitance calibration data can be obtained, and MOSFET gate oxide capacitor

calibration can be done more accurately.

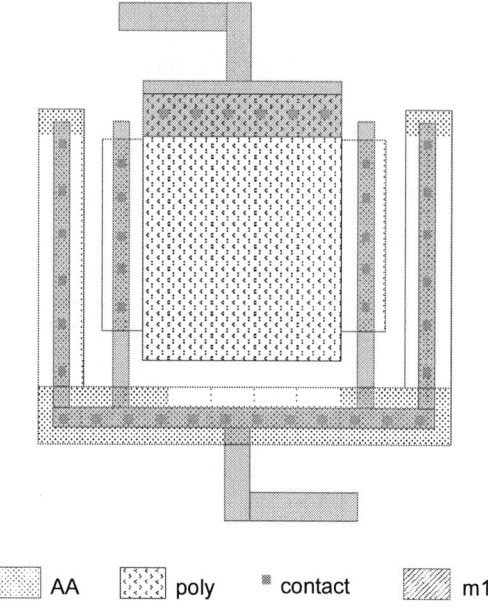

Figure 4: A standard MOSFET gate oxide capacitor layout

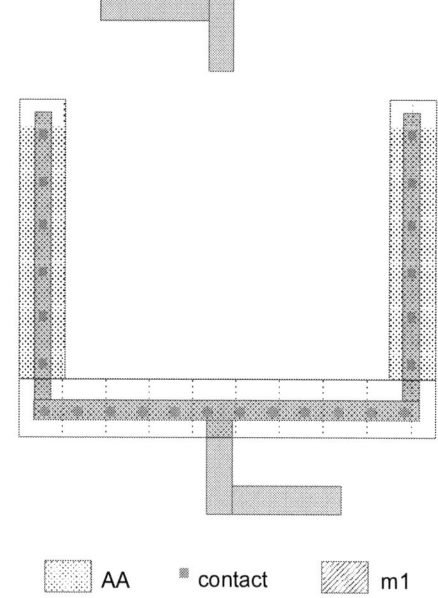

Figure 5: A common calibration structure of MOSFET gate oxide capacitor layout

[3] DAI Yusheng, WANG Guohui, GUAN Yong, WU Lifeng, LI Xiaojuan, *Degradation Analysis of Power MOSFET Parasitic Capacitance in Transient Rresponse of Switching*, China, 2014, pp. 1.

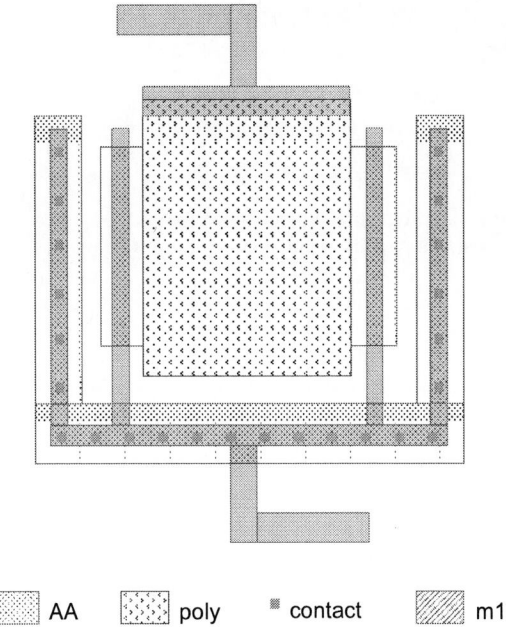

AA   poly   contact   m1

Figure 6: A new calibration structure of MOSFET gate oxide capacitor layout that keeps AA and poly

CONCLUSION

Using the calibration structure proposed in this paper, a series of parasitic capacitances, including the parasitic capacitance between source-drain interconnection line and gate, the parasitic capacitance between source-drain interconnection line and bulk, and the parasitic capacitance between grid interconnection line and bulk can be removed, so as to improve accuracy of calibration and obtain more accurate simulation model of MOSFET gate oxide capacitances.

ACKNOWLEDGEMENTS

I would like to express my sincere gratitude to all those who assisted me during the writing of this paper.

Especially, I am extremely grateful for my leader Wang wei and my teacher Peng Xingwei. The paper is not going to complete without their great inspiration and patient guide.

REFERENCES

[1] Wang Min, Wang Baotong, Ke Daoming. *Parasitic Capacitance of MOSFET with High K Gate and Short Channel [J]*, Journal of University of Science and Technology of China, vol. 10, 2013, pp. 822-829.

[2] Peng Xingwei, Wang Wei, *Yi Zhong Hu Lian Dian Zu Dian Rong Jiao Zhun Jie Gou*, CN105161487A, China, 2015.

NARROW-BAND MASK SYNTHESIS WITH SEMI-IMPLICIT DIFFERENCE

Yijiang Shen[1]* and XiaoPeng Wang[1]

[1]School of Automation, Guangdong University of Techonology, Guangzhou 510006, China
*Corresponding Author's Email: yjshen@gdut.edu.cn

ABSTRACT

In this paper, a distance level-set regularized reformulation of mask synthesis is developed to secure a simple and straightforward construction of the narrow band and provide the nonlinear diffusion term for implicit difference schemes. Subsequently, the mask update is performed only in the vicinity of the zero level set thereby reducing optimization dimensionality; moreover, the semi-implicit discretization is applied to circumvent the stability constraints enabling sufficiently large stepsize improving convergence with much less iteration numbers. Additive operator splits the mask update with respect to coordinate axes to solving multiple comparatively small scale linear systems of equations with Thomas method. Simulation results merit the superiority of the proposed approach with improved convergence by overcoming the stability constraints and reduced optimization dimensionality.

INTRODUCTION

To keep pace with the increasing reduction of the critical dimension (CD) of modern integrated circuits, the inherent distortion due to optical diffraction has to be compensated by more aggressive optical proximity correction (OPC) [1] techniques with new computational strategies.

Generally implemented iteratively, mask optimization (MO) approaches, which deliberately introducing pre-distortions on the geometrical shapes, have to process excessive calculation of aerial images and gradients of cost functions in each iteration [2], Gradient-based methods including steepest-descent (SGD) [3-5], conjugate gradient [6], augmented Lagrangian [7], level-set methods [8-11] and alike, have been extensively applied. to process the large amount of data incurred in the mask optimization (MO) procedure. Two major predicaments of the above methods arise from full-scale optimization dimensionality featuring all optimization variables that is every mask pixel in each iteration and the prohibitive time-step mandated by optimization stability for explicit difference discretization schemes

MO procedure is, by nature, non-convex with no guarantee of global minimum, therefore, optimizing every mask pixel is not always necessary. Consequently, optimization dimensionality reduction is performed in compressive-sensing (CS) based MO methods [12, 13] where only predefined sampling points are updated.

Alternatively, narrow-band based techniques [14] conduct optimization work only in the vicinity of the evolving fronts to reduce computation cost and memory usage. Furthermore, a distance-regularized level-set formulation [15, 16] is proposed to discretize the diffusion terms implicitly and the non-diffusion ones explicitly, hence 'semi-implicitly' circumvent the stability constraints. ,

We propose in this paper a combined optimization framework where reduced optimization dimensionality is achieved by localizing the optimizing variables in the neighborhood of the pattern contours while securing sufficiently large time-steps with semi-implicit difference schemes. The size of the linear system matrices with respect to coordinate axes is significantly reduced and the superiority of the proposed approach is merited by the simulation results.

NARROW-BAND BASED SEMI-IMPLICIT OPTIMZATION FRAMEWORK

According to Fourier optics and Abbe method, the aerial image formation of lithography systems can be calculated as

$$\mathbf{I_a} = \frac{1}{J_s} \sum_{(\alpha_s,\beta_s)} \mathbf{J}(\alpha_s,\beta_s) \sum_{p=x,y,z} \left\| \mathbf{H}_p^{\alpha_s\beta_s} \otimes \left(\mathbf{B}^{\alpha_s\beta_s} \odot \mathbf{M} \right) \right\|_2^2, \quad (1)$$

where $\mathbf{J} \in \mathcal{R}^{N_s \times N_s}$ and $\mathbf{M} \in \mathcal{R}^{N \times N}$ are scalar matrices representing source and mask pattern distributions, $\mathbf{J}(\alpha_s,\beta_s)$ is the intensity of the source point at (α_s,β_s), $J_s = \sum_{(\alpha_s,\beta_s)} \mathbf{J}(\alpha_s,\beta_s)$ is the sum of source intensities as a normalizing factor. $\mathbf{B}^{\alpha_s\beta_s}$ is the matrix representing the oblique incidence effect of the light rays. $\mathbf{H}_p^{\alpha_s\beta_s}$, $p = x, y, z$ are referred as the equivalent filters of the x, y, z components; $\otimes$ denotes convolution and $\odot$ is the entry-by-entry multiplication. The wafer image can be approximated by a sigmoid function to give

$$\mathbf{I} = \mathcal{T}(\mathbf{M}) = \text{sig}(\mathbf{I_a}) = \frac{1}{1+e^{-a(\mathbf{I_a} \cdot \mathbf{t_r})}}, \quad (2)$$

with a and t_r being the steepness of the sigmoid function.

Optimization Framework

From Eq. (1), we can see that MO involves deconvolution therefore generally ill-posed. One possibility is to consider a variational model by incorporating regularization terms. To this end, the distance regularized level-set (DRLS) term [17] enabling stable level-set evolution and accurate numerical computation is used to regularize the pattern fidelity cost function. Consequently, the MO is reformulated as

$$\text{minimize } \frac{1}{2}\int_\Omega \left(|\nabla\omega|-1\right)^2 d\mathbf{r}$$
$$\text{subject to } \frac{1}{2}\int_\Omega \left(\mathcal{T}(\mathbf{M})-\mathbf{I_0}\right)^2 d\mathbf{r} \qquad (3)$$

where ω is the parametric transformation $\mathbf{M}=\frac{1+\cos(\omega)}{2}$, ∇ denotes the gradient, $\mathbf{I_0}$ is the target pattern, Ω is the area of the bounded region and $\mathbf{r}$ embeds spatial coordinates (x,y). Following the derivations in [10], we arrive at the Hamilton-Jacobi equation

$$\partial\omega_t = \mu|\nabla\omega|\left(\frac{\nabla\omega}{|\nabla\omega|}\right)+|\nabla\omega|(-v(\mathbf{r},t)-\mu\Delta\omega), \qquad (4)$$

with Δ denoting Laplacian operator, t being the artificial time and $v(\mathbf{r},t)$ being the normal velocity function defined as

$$v = -\frac{\operatorname{asin}(\omega)}{2J_s}\odot\sum_{(\alpha_s,\beta_s)}\mathbf{J}(\alpha_s,\beta_s)\sum_{p=x,y,z}\mathrm{R}\big[(\boldsymbol{B}^{\alpha_s\beta_s})^*\odot$$
$$\left(\big(\mathbf{H}_p^{\alpha_s\beta_s}\big)^{*\diamond}\otimes\left\{\mathbf{E}_p^{\alpha_s\beta_s}\odot(\mathbf{1}-\mathbf{I_0})\odot\mathbf{I}\odot(\mathbf{1}-\mathbf{I})\right\}\big)\big], (5)$$

with $*$ being the conjugate operation, $\diamond$ flipping the matrix in both up-down and right-left directions, $\mathbf{1}\in\mathcal{R}^{N\times N}$ being the all-ones matrix and $\mathbf{E}_p^{\alpha_s\beta_s}=\mathbf{H}_p^{\alpha_s\beta_s}\otimes\boldsymbol{B}^{\alpha_s\beta_s}\odot\mathbf{M}$.

Semi-implicit Difference

We discretize t as times $t_k = k\tau, k=0,1,2,\cdots$ with τ being the step size and order ω lexicographically as a vector $\in\mathcal{R}^{N_2\times 1}$, which hereby remain denoted as ω whenever there is no ambiguity. To overcome the stability constraint of prohibitive step size when explicit discretization schemes are applied, we discretize the first term (diffusion term) in Eq. (4) implicitly and the second term explicitly to give the semi-implicit formulation the MO problem in vector-matrix notation as

$$\omega^{k+1}=\omega^k+\tau|\nabla\omega^k|g^k+\tau\sum_{\mathbf{r}\in(x,y)}A_\mathbf{r}(\omega^k)\omega^{k+1}, \quad (6)$$

where $g=-v(\mathbf{r},t)-\mu\Delta\omega$ as the 'balloon force', $A_\mathbf{r}$ represents the interaction in the $\mathbf{r}$ direction, with the elements $a_{ij\mathbf{r}}(\omega^k)$ of $A_\mathbf{r}(\omega^k)$ calculated as

$$a_{ij\mathbf{r}}(\omega^k)=\begin{cases}|\omega_i^k|\frac{2\mu}{|\omega_i^k|+|\omega_j^k|} & (j\in\mathcal{N}_\mathbf{r}(i))\\ -\sum_{j\in\mathcal{N}_\mathbf{r}(i)}|\omega_i^k|\frac{2\mu}{|\omega_i^k|+|\omega_j^k|} & (j=i),\\ 0 & \text{else}\end{cases} \quad (7)$$

with $\mathcal{N}_\mathbf{r}(i)$ being the 4-neighbors of the pixel i with respect to the $\mathbf{r}$ coordinates.

However, the solution of ω^{k+1} in Eq. (6) cannot be determined directly from the scheme, but requires solving the linear system of equations

$$\omega^{k+1}=\left(I-\tau\sum_{\mathbf{r}\in(x,y)}A_\mathbf{r}(\omega^k)\right)^{-1}(\omega^k+\tau g^k), \quad (8)$$

where I denotes unit matrix. Equation (8) is a very large sparse system, yet with large bandwidth because the pixels in the i_{th} row cannot be ordered to bound the non-vanishing entries diagonally within the positions of $[i,i-2]$ and $[i,i+2]$. Solving Eq. (8) with Gaussian elimination or iterative methods lead to immense computation and storage effort, therefore, we resort to its

additive operator splitting (AOS) variant

$$\omega^{k+1}=\frac{1}{2}\sum_{\mathbf{r}\in(x,y)}\left(I-2\tau A_\mathbf{r}(\omega^k)\right)^{-1}(\omega^k+\tau g^k), \quad (9)$$

offering one important advantage: The operators $I-2\tau A_\mathbf{r}(\omega^k)$ lead to strictly diagonal dominant triangle systems which can be solved very efficiently by the Thomas algorithm.

Narrow-band Implementation

It is observed that optimizing every mask pixel is not always necessary. Therefore, we perform a simple but effective edge-based narrow band implementation. In each iteration, we only update the edge pixels which are labeled as wherever $|\nabla\omega|>e_{thr}$, with $e_{thr}>0$ being a predefined threshold. The edge pixels to be updated at the beginning and the end of the optimization are displayed in Fig. 1(a) and (b) respectively.

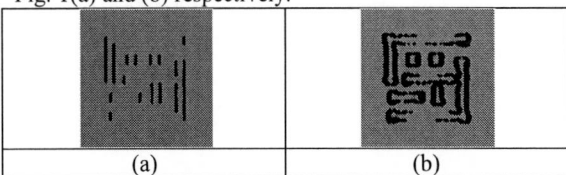

Figure 1: Edge pixels to be updated at (a) the beginning and (b) the end of the optimization, respectively..

NUMERICAL RESULTS

Numerical simulations are performed on a immersion lithographic imaging system with wavelength $\lambda=193nm$, $NA=1.35$, spatial resolution $\delta x=\delta y=4nm$, the threshold of the sigmoid function being $a=80$ and $t_r=0.85$. The system is illuminated by a partially coherent source $\mathbf{J}$ with partial coherent factor $\sigma_{in}=0.6$ and $\sigma_{out}=0.9$. We also define the pattern error (PE) as the square of the $\mathcal{L}_2$ norm of the difference between the target pattern $\mathbf{I_0}$ and the resist image $\mathbf{I}$.

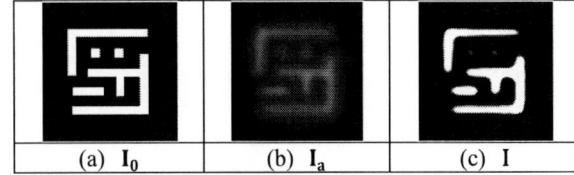

Figure 2: Lithographic simulation with (a) the target pattern $\mathbf{I_0}$ illuminated by $\mathbf{J}$. (b) The aerial image $\mathbf{I_a}$ and (c) the printed resist image $\mathbf{I}$ with PE 2866.

Figure 3 compares lithographic imaging performance of simulations with the SGD method in row (a) where the step size is bounded by the Courant-Friedrichs-Lewy (CFL) condition, the semi-implicit approach [15] in row (b) and the proposed narrow-band semi-implicit approach in row (c) which both apply step size 1.5. Pattern fidelities of the resist images are improved to PE 551, 569 563 when synthesized masks derived from the SGD, the semi-implicit approach and the proposed approach are

illuminated by the source **J**

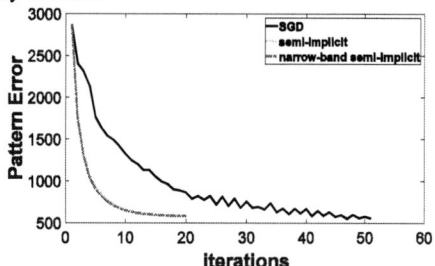

Figure 3: Lithographic simulation with synthesized masks. Rows (a), (b) and (c) use synthesized masks with the SGD method, the semi-implicit and the proposed narrow-band semi-implicit approach, respectively.

Convergence performance with respect to iteration numbers for the simulations by the SGD, the semi-implicit approach and the proposed approach is depicted in Fig. 4. Semi-implicit approaches with $\tau = 1.5$ show significant improved convergence than the SGD method with an average τ of 0.1596 . The proposed narrow-band implementation of the semi-implicit approach slightly outperforms the full-scale one [15], although not obvious, refuting the necessity of optimizing all pattern pixels. It is also noticed from Fig 5 that the edge pixel number is much smaller than the full scale optimization size $N^2 = 66049$, therefore the size of the linear systems to be solved is greatly reduced.

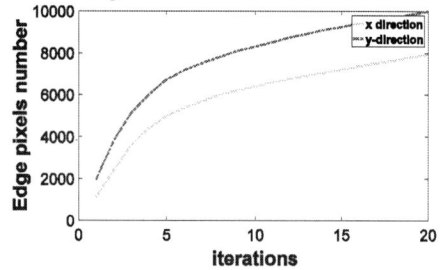

Figure 4: Convergence performance for the simulations in Fig. 3.

Figure 5: Edge pixel number with respect to iteration.

CONCLUSION

We have presented a fast computation strategy for mask synthesis in optical lithography. The proposed approach overcomes the stability constraint of prohibitive step size by explicit discretization schemes, while reducing the optimization dimensionality with a simple and effective edge-based narrow band implementation. The computation efficiency is rooted in solving one-dimensional linear systems with Thomas algorithm to update only the edge pixels in the mask patterns. .

ACKNOWLEDGEMENTS

his paper is partially supported by Natural Science Foundation of China (61875041), Natural Science Foundation of Guangdong Province, China (2016A030313709).

REFERENCES

[1] A. K. Wong, *Optical Imaging in Projection Microlithography*. Washington: SPIE, 2005.

[2] X. Ma and G. R. Arce, *Computational Lithography*, John Wiley and Sons, 2010.

[3] Y. Peng, J. Zhang, Y. Wang and Y. Zu, *IEEE. T. Image. Process*, vol. 20, 2011, pp. 2856-2864.

[4] X. Ma, C. Han, L. Dong and G. R. Arce, *J. Opt. Soc Am.*, vol. 30, 2013, pp. 112-123.

[5] N. Jia and E. Lam, *Opt. Express.*, vol. 19, 2011, pp. 19384-19398.

[6] W. Lv, S. Liu, Q. Xia, X. Wu, Y. Shen and E. Lam, *J. Vac. Sci. Tchnol. B*, vol. 31, 2013, pp. 041605-13.

[7] J. Li and E. Lam, *Opt. Express.*, vol. 22, 2014, pp. 9471.

[8] Y. Shen, N. Wong and E. Lam, *Opt. Express.*, vol. 17, 2009, pp. 23690-23701.

[9] Y. Shen, N. Wong and E. Lam, *Opt. Express.*, vol. 19, 2011, pp. 5511-5521.

[10] Y. Shen, *Opt. Express.*, vol. 25, 2017, pp. 21755.

[11] Y. Shen, *Opt. Express.*, vol. 26, 2018, pp. 10065-10078.

[12] X. Ma, D. Shi, Z. Wang, Y. Li, and G. R. Arce, *Opt. Express.*, vol. 25, 2017, pp. 7131-7149.

[13] X. Ma, Z. Wang, Y. Li, G. R. Arce, L. Dong and J. Garcia-Frias, *Opt. Express.*, vol. 26, 2018, pp. 14479-14498.

[14] Y. Shen, *Opt. Express.*, vol. 26, 2018, pp. 10065-10078.

[15] Y. Shen, F. Peng and Z. Zhang, *Opt. Express.*, vol. 27, 2019, pp. 1520-1528.

[16] Y. Shen, F. Peng and Z. Zhang, *Opt. Express.*, vol. 27, 2019, pp. 29659-29668.

[17] C. Li, C. Xu, C. Gui and M. Fox, *IEEE. T. Image. Process*, vol. 19, 2010, pp. 3243.

WAFER EDGE PEELING DEFECT MECHANISM ANALYSIS AND REDUCTION IN IMD PROCESS

Ya li Feng, Qi Liang Ni, Xiao fang GU, Guang zhi He, Hao Guo
Shanghai Huali Microelectronics Corporation, 201909
*Shanghai, China*Corresponding Author's Email: <u>fengyali@hlmc.cn</u>

ABSTRACT

An innovative model of the mechanism for peeling defect in inter metal dielectric(IMD) process induced by poor adhesion between metal and oxide film on wafer bevel is presented. Peeling defect is inspected by dark field inspection (DFI) tool of KLA. Scanning electron microscopy (SEM) and Transmission electron microscope (TEM) are used to study the stack and composition of wafer edge film. Peeling defect source will be discussed in this paper, and the tests demonstrated metal film and oxide film's growth ability varied at wafer bevel, that induce metal film accumulate approach to wafer bottom bevel, when the thick oxide film of next layer deposition, they cannot bond firming, the films peel off with the mechanical transfer or high temperature. Solution of to avoid peeling defects with bevel clean also was demonstrated as well.

Keywords—wafer edge peeling; film's growth ability; bevel clean; IMD process;

INTRODUCTION

Wafer bevel include top bevel, apex and bottom bevel [1],as showed Fig.1,with chip size continually shrinks as semiconductor advanced develop, film stack on wafer edge becomes more and more complex, but which cannot be etched because of the process design, that will bring more and more new defect types. Based on that, Wafer edge is becoming more and more important as it can be sources of defect can result in significant yield loss in semiconductor manufacturing. For this reason, wafer edge's study is becoming more and more important.

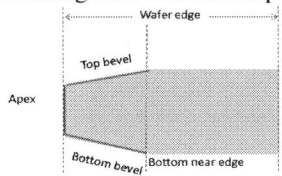

Fig.1.Wafer edge structure introduction

Surface particle induced by wafer edge peeling was reported about the stresses which tend to cause peeling in the biomaterial between interfaces of the layers [2]. In lithographic process, peeling correlate with BARC accumulate was also studied in present paper [3].

Peeling defect will induce major yield loss in BEOL (back end of flow), so it's urgently to find a solution of defect reduction. It was reported with bevel polish a clearing of wafer edge regions on wafer bevel for non-opened structures, which showed a significant drop in defect count on wafer edge region [4].But the wafer edge polish is a difficult process, which can easily lead to wafer breakage.

As of now, the current research on the mechanism of wafer peeling defect in IMD process has not been clearly stated. There is no solution that cost low and process easily but obvious improvement on wafer edge defects was reported.

Particle defect as Fig.2(a-b)in back end of process occurred in MHM(metal hard mask film) deposition process at 300~400℃ temperature with a special map and random distribution. The defects concentrated on the edge of the wafer sometime, and the defect size is larger than 1um,which will induce pattern fail after etch process Fig.2(c), finally formed Cu residue after Chemical mechanical polish(CMP)process Fig.2(d),that is the major yield killer. Collecting this type defect of each metal layer, it was found that the number of defects gradually deteriorated with the increase of the metal layer, and Fig.3 shows this correspondence. The mechanism of this type defect will be discussed in this paper.

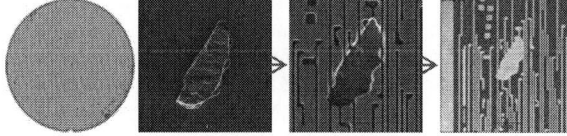

Fig.2(a)defect map Fig.2(b) Film deposition Fig.2(c) After etch process Fig.2(d) After CMP
Fig.2 Defect transferred from step by step

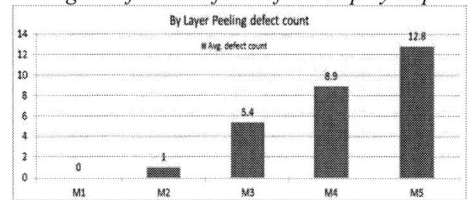

Fig.3. Statistics on the number of defects in different metal layers

EXPERIMENT

For finding the root cause of the particle, this paper proposed two modes for test. One is the film deposition tool condition worse; the other one is the source came from wafer self. Dark field inspection (DFI) tool of KLA was used for defect scan. Film deposition tool do enhance monitor and mechanical transfer test. Wafer edge was also reviewed by SEM (scanning electron microscopy).

The film stack was analyzed by TEM and SEM, mechanism of peeling in Inter metal dielectric (IMD) process will be discussed in this paper.

RESULTS AND DISCUSSION

The defect count wasn't added and no similar map produces before and after transmission to demonstrate that the defect wasn't machine's mechanical transfer induced. It showed the defect source was mostly come from wafer-self especially from wafer edge. In response to this suspicion, the SEM was used to review the wafer edge. As shown in Fig.4, peeling defects were found at the bottom bevel of the wafer. Some have been completely peeled off, the section can be seen after peeling off, and some have not been completely peeled off.

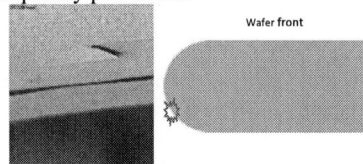

Fig.3 Wafer edge peeling source at bottom bevel

Furthermore, in this study, TEM analysis was performed on the peeling defects of the wafer, and the SEM was used for wafer bevel analysis (as shown in Fig.5).

Wafer was chose as sample which processed as Fig.5 (b) showed. It can be seen from the TEM results that the thickness of large particle defects on the wafer match with process film thickness. From EDX and film thickness analysis, it can be known from the film deposition process sequence that the film peeled off between the interface of PEOX with Ti/Ta film and IMD film. But the mechanism was not clear.

Aiming at clear the root cause of peeling mechanism, SEM analysis on the wafer bevel also be used. Wafer bevel was further inspected, and the film growth at bevel from the wafer top bevel to the bottom bevel was confirmed (Fig. 7).

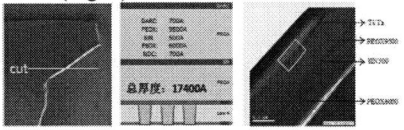

Fig.5(a) defect Fig.5(b)sample wafer as processed Fig.5(c)TEM result of defect
Fig5.TEM analysis of Defect &film stack

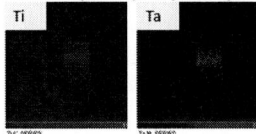

Fig6.EDX analysis of Defect

At top bevel of wafer edge, and (IMD) films accumulate here. At this position 1, the etching ability is almost zero, each layer of film exists well, and the interfaces of different films are clear. With the film accumulates and becomes thicker and thicker, at position 2 of the bottom bevel, the TEOS and low-k film of M3 began to fracture. At position 3 of the bottom bevel, fractures of the M3 and M4 Low-k film were successively found. From position 4, no accumulation of metal dielectric film has been seen. This is because the reactive gas cannot reach this position during the CVD process.

During the film deposition process, the reaction gas can enter the back of the wafer through the gap between the wafer edge and the equipment hardware. In addition to the front side of the wafer, a thin film was grown on the bevel and back end of the wafer. As the reaction gas

Concentration gradually decreases from the front to the bevel and then to the back end, the corresponding film thickness also gradually becomes thinner. At the wafer edge, the thin film uniformity on the bevel and the back end of the wafer was poor, and the film thickness gradually accumulates during the wafer production process, and continuously with the device hardware during the transfer process, which often becomes a potential source of defects.

Since the IMD films accumulate more and thicker with layer and layer deposition, there will be a tendency that the peeling defects gradually deteriorate with the increase of the metal layer. Through the inspection and analysis of more wafer edge data, it was found that the metal dielectric layer tears as shown in Fig.7(c), the EDX analysis between the fracture layers is Ti and Ta.

Based on the analysis of the above data, the root cause of the peeling defect is that the stress between the metal and the oxide film is not matched, and the fracture occurred in the high temperature process. The contact space between the wafer edge and the machine's hardware becomes smaller and smaller with the film becomes thicker and thicker, the contact space between the wafer edge and the machine's hardware becomes smaller and smaller. Constant contact and collision during the transmission process will cause the film to peel off and produce defects.

Fig.7 (a). Post M2 CMP Fig.7 (b). Post M3 Film DEP Fig.7(c). Post M3MHM Film DEP

DEFECT REDUCTION INTRODUCTION

Since the defect comes from the peeling off the thin film on the bottom bevel of the wafer edge, propose one method to reduce the defect by cleaning the films deposited on wafer edge before thick IMD film deposited. One method is using backside clean to clear the films, which can cover the bottom bevel area. The cleaning equipment is a single-wafer cleaning equipment (model DV-38) produced by LAM Company. The chemical cleaning solutions used are HNO3 and HF for Ti easily dissolves in HF and generates titanate, and N2 spin drying method is used. HF and HNO3 are a common chemical solution in semiconductor production, and are mainly used for wafer backside cleaning in copper interconnect technology.

$$Ti + 6HF \rightarrow H_2[TiF_6] + 2H_2 \tag{1}$$

The improvement of defects after wafer backside clean is shown in Fig.8: the average number of defects without wafer backside cleaning is 6 and the average number of defects after wafer backside cleaning is 1, an improvement of 83%.

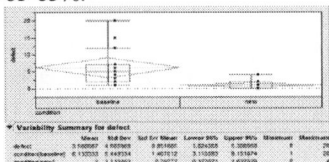

Fig.8 defect statistics under different cleaning conditions

CONCLUSION

This paper studied the mechanism of wafer edge peeling defects. It is found that the film has different growth ability on the wafer edge, and the growth ability of the metal film is stronger than that of the oxide film. The stresses between the both are different, when transferred at high temperature process, which was easy to cause the film to peel off.

This study proposes a method of backside cleaning to improve wafer edge defect. The cleaning equipment is a single-wafer cleaning equipment (model DV-38) produced by LAM Company. The chemical cleaning solutions used are HNO3 and HF, and N2 spin drying method is used. With this method, the defect improvement efficiency is 83%.

REFERENCES

[1] Alexander E. Braun, Semiconductor International, Jan. 2006

[2] Thomas D. Moore, Inter Society Conference on Thermal Phenomena, May.2004

[3] X Yuan, Z Qiang, J Hao, Semiconductor Manufacturing International Corp, 2017

[4] Anngret Wieters, Peter Thieme, International Conference on Planarization/CMP Technology, October 25 -27.2007

EVALUATION OF PRE SILICIDE IMPLANT FROM LOW TEMPERATURE TO ROOM TEMPERATURE

Zhouchun[1]*， Caowenjie[1]， Chengxinhua[1]， Fangjingxun[1]

[1]Research and Development Dept., Shanghai Huali Microelectronics Corporation, Shanghai 201912, China

*Corresponding Author's Email: zhouchun@hlmc.cn

ABSTRACT

For the purpose of increasing productivity and reducing cost in pre silicide implant, cold silicide implant is evaluated to be replaced with Xe implant in room temperature, mainly resulting in shorter process time and better device performance. In this paper, a series of experiments have been performed to collect amorphous thickness/roughness, silicide thickness as well as silicide resistance. The effects of different implant recipes are investigated especially in terms of physical thickness and WAT data. The goal of experiment is to explore the possibility of implant dosage of Xe element, energy and tilt angle in order to optimize device performance and maintain comparable yield result. As a corresponding result, device local variation and yield are achieved to basic level accompanying with higher productivity.

Keywords—Xe implant; amorphous thickness; dosage; tilt angle; productivity

INTRODUCTION

As device scales down, the different implant species and surface roughness produced by implant have bigger effect on the device performance, especially the implant temperature which improve on the device performance of production is important. So pre silicide implant by low temperature has been widely used as the amorphous layer[1] in continuous advanced technology for better device performance. During cold silicide implant (temperature is below -50°), the self-healing process of the lattice will be extremely slow, and the amorphous layer has fewer lattice voids[2], thus greatly reducing the lattice defects at the End of range after annealing, as shown in fig1.

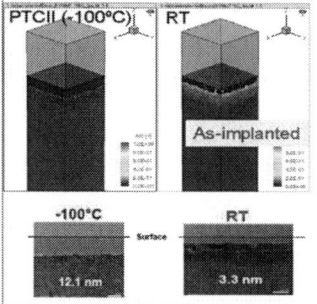

Figure 1 Amorphous layer of RT & Cold implant

But in actually, the throughput of implant in low temperature is only a quarter of the implant in room temperature, and along with a small number of defects with wafer surface, which leading to loss in yield. In addition, the production cost of cold implant is more than twice that of normal implant, containing special gas and other costs. In order to reduce cost and increase productivity, Xe implant in room temperature is being investigated as an alternative which can replace Si implant in low temperature.

In this paper, by studying the process of silicide layer at source and drain region, we found the amorphous thickness of Xe in room temperature, combine with optimizing tilt angle and dosage is similar to the thickness of cold silicide implant, result in the nickel silicide resistance comparable with baseline recipe of cold implant, contribute to yield improvement.

EXPERIMENT

Firstly, we need determine the energy, tilt angle and dosage of Xe implant in room temperature to produce an amorphous layer that is close to the thickness of basic recipe. Secondly, according to previous experimental data, establish the Xe implant split condition of the reference energy, angle and dosage, in order to explore whether there is significant difference in silicide thickness after pre silicide implant[3]. Lastly, the effects of nickel silicide resistance on silicide thickness were discussed in experiment.

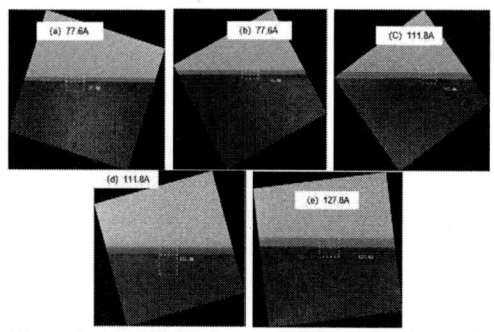

Figure 2 Amorphous layer of different implant split
(a) basic cold silicide implant; (b)Xe implant, energy=4kv; (c)Xe implant,energy=8kv;(d)Xe implant,energy=8kv,

tilt=7°; (e) Xe implant, energy=8kv, dosage double

RESULT AND DISCUSSIONS

A. The influence of dosage, angle and energy for amorphous layer

The basic amorphous thickness of cold silicide implant which used in production is about 77.6A as shown in fig2(a). For finding out the condition that can match this benchmark, the different energy of Xe implant are doped in the same Si substrate respectively, fortunately fig2 (b) shows that the thickness of Xe implant energy at 4kv is similar to that of the basic cold implant recipe.

When the energy of Xe implant increases, the gap of amorphous layer become larger, as shown in fig2(c), the thickness of Xe implant energy at 8kv is thicker about 44% than the benchmark. Therefore, there is no linear relationship between implant energy and amorphous layer[4].

On the other hand, as shown in fig2(d), it can be seen that the change of implant angle (tilt angle is changed from 0° to 7°) has no obvious effect on the thickness of amorphous layer at same implant energy, both of them are all about 111.8A. However, this does not mean that the angle has no effect on device performance, which needs to be further verified by production. Fig2(e) shows that the thickness only increases about 10 percent after Xe implant dose doubled, so by comparing the change in energy and dose, it can be concluded that the amorphous layer is more sensitive to the variation in implant energy.

B. The silicide thickness of the optimized implant energy, dosage and angle

During next experiment of pattern wafer, silicide thickness after NiSi annealing will be collected by implant energy split condition. Fig3(a) shows that silicide thickness in the bottom of trench at SRAM area is 273A approximately after pre silicide implant in low temperature and NiSi annealing. At the same optimized dosage and angle of Xe implant, silicide thickness is grown on silicide layer at SRAM area after annealing, but different implant energy form different thickness as shown in fig3(b/c/d), combine Xe implant energy 4kV with optimized dosage and angle, the thickness 268A is realized shown in fig3(c), which is closest to the reference value.

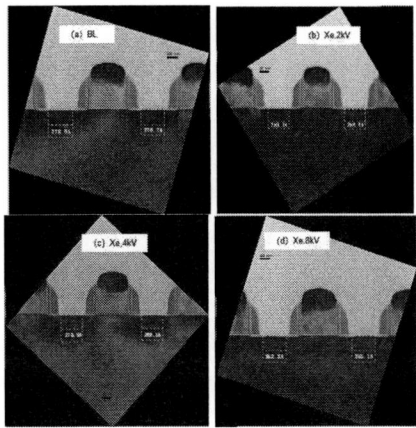

Figure 3 Silicide thickness of different implant energy

(a) basic cold silicide implant; (b) Xe implant, energy=2kv; (c) Xe implant, energy=4kv; (d) Xe implant, energy=8kv

C. Comparison of nickel silicide resistance and silicide thickness

For further studying the process of silicide layer, in order to explore whether there is strong correlation in nickel silicide resistance and silicide thickness, the related experiment has been performed on other production[5]. Under six different implant conditions, total six pattern wafers will be doped in Si and Xe species not only, but also be doped in Ge species which often used as co-implant for amorphous layer.

Si energy=2kV,tilt=0°,co H M P						25
Xe, energy=2kV,tilt=0°					24	
Xe, energy=2kV,tilt=7°				23		
Ge, energy=2kV,tilt=0°			22			
Ge, energy=2kV,tilt=7°		21				
Si energy=2kV,tilt=0°	20					
Wafer ID	#20	#21	#22	#23	#24	#25
Silicide THK	164.6	162.3	161.9	158.9	158.8	168.1
NiSi RS	28.3	29	28.9	30.9	31	27

Figure 4 Silicide THK&RS Data

After measuring the silicide thickness of all wafers, the electrical characteristic of the source/drain film with different implant conditions has been tested, then silicide resistance is obtained as shown in fig5. It can be seen that there is obvious linear relationship between thickness and resistance[6], the slope is calculated to be about 4ohm per 10A. This provides a reliable basis for adjusting parameters which related to thickness.

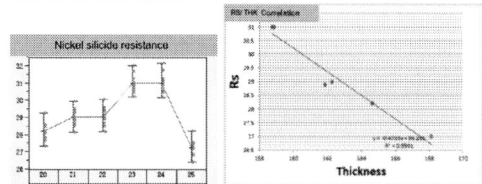

Figure 5 Silicide THK&RS Correlation Graph

In the half to half split test of production, through the optimized Xe implant instead of cold silicide implant, a large productivity increase of 200% can be obtained for silicide layer compared with traditional cold implant as shown in fig6. The silicide resistance of Xe implant split wafers is comparable with baseline, and other core device performance are achieved to basic level accompanying with higher productivity.

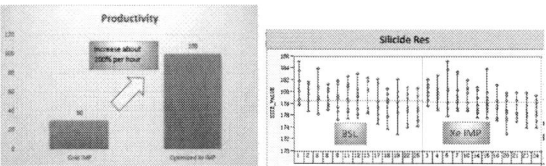

Figure 6 Productivity & silicide resistance with Xe implant

CONCLUSION

In this paper, through the introduction of pre silicide implant in traditional silicide layer, comparing different implant conditions and optimizing implant energy, angle and dosage, we finally got a reliable Xe implant recipe for amorphous layer. The Xe implant can prevent the defect of cold implant due to hardware limits, and has a shorter process time result in the increase of the productivity, which is conducive to advanced process product yield improvement.

REFERENCE

[1] Rinus T. P. Lee ; Li-Tao Yang , IEEE. 25, 39 (2000). Nickel-Silicide: Carbon Contact Technology for N-Channel MOSFETs With Silicon–Carbon Source/Drain, Volume: 29 , Issue: 1 , PP: 89 – 92, Jan. 2008,

[2] S. H. Yeong, Journal of the Electrochemical Society, v 155, n7, p H508-12(2008).

[3] Tanjun，Chenyongyue, Huangqiuming，Hongjiaqi，Yanqiang，Zhouhaifeng，Fangjingxun, CSTIC2012, Improved Ge doped Cap layer for embedded SiGe epitaxial growth in 28 nm CMOS technology, Shanghai, 201203

[4] F. H. Cioldin ; J. A. Diniz ; A. R. Vaz ; G. A. Calligaris ; L. P. Cardoso ; I. Doi , IEEE, Study of the phase transitions of Nickel Platinum Silicide obtained by sputtering and rapid thermal processing, 28 Aug.-1 Sept. 2017

[5] S. H. Yeong, Journal of the Electrochemical Society, v 155, n7, p H508-12(2008).

[6] S. H. Yeong, Materials Science and Engineering B: Solid-State Materials for Advanced Technology, v 154-155, n 1-3, p 43-48 (2008).

FRAGMENTATION OF SQUARE PATTERN MASK WITH SMALL CORNER-TO-CORNER SPACE

Yu Shirui, Chen Yanpeng, Wang Dan, Deng Guogui, Hu Yidan

Shanghai Huali Integrated Circuit Corpotation
Shanghai, China
chenyanpeng@hlmc.cn

ABSTRACT

Hole layer mask with small corner-to-corner space is usually been limited by mask rule check in OPC. For 28nm and below node, via or contact layer square pattern edges need not fragmentation in general. This paper investigates fragmentations of hole layer square mask with small corner-to-corner space to make contour critical dimension on target. The potential risks of fragmentations and limit conditions of fragments movement are also been discussed. Comparison of square pattern mask with or without fragments ADI results is also been studied in this paper.

Keywords—OPC; fragmentation; Via; Contact

INTRODUCTION

In node 28nm or below, the corner to corner space is usually smaller than projecting space in via and contact layer in design rule. Because there is corner chop effect in the corner in hole layers, and the contour of via or contact is usually a cycle, so the spaces of holes in the diagonal direction on wafer are bigger than the design value[1].

For hole layers, draw layer is sized at first to get the target layer in general. The critical dimension (CD) of the target layer equals to after developing inspection (ADI) CD. So the corner to corner space between two holes shrinks more than the projecting space as the Pythagorean Theorem.

When correcting the mask layer of via based on the OPC model (MBOPC), the edges of square via usually are not divided as the CD of square via is quite small especially in node 28nm or below. For rectangle via, the width of the long side is about 2~3 times bigger than the short side, via long sides are usually divided fragments in MBOPC[2]. Based on the exposure condition, the mask CD is bigger than the target CD in general, which makes the space between two corners of via mask could be smaller than target layer.

For the square via which has the smallest corner to corner space, in MBOPC procedure, the movement of the mask edges may be limited by MRC, although the MRC spec of corner to corner space is smaller than the projecting space. If the size of the mask can not be big enough base on the OPC model simulation, the contour of via or the ADI CD will not on target accordingly. For this situation, fragmentations of square via

mask edges can be put into use.

EXPERIMENT

A section in via layout where contains the pattern that via mask is limited by MRC in the corner to corner direction is clipped (figure 1). The corner to corner spaces of Via1~Via5 are the minimum value in the design rule, while Via6 aren't

Draw layer is sized up to get the target layer. Then sub-resolution assist feature (SRAF) is added in order to assist the contour on target and increase the depth of focus (DOF).

Segmentation of square via mask still exists several potential risks because the OPC model does not contain the complex via mask pattern, so the model may not be able to simulate the same contour with the real wafer result. During the MBOPC procedure, two groups of experiments are performed. One is to divide the square via mask edges into fragments, while the other group is without segment.

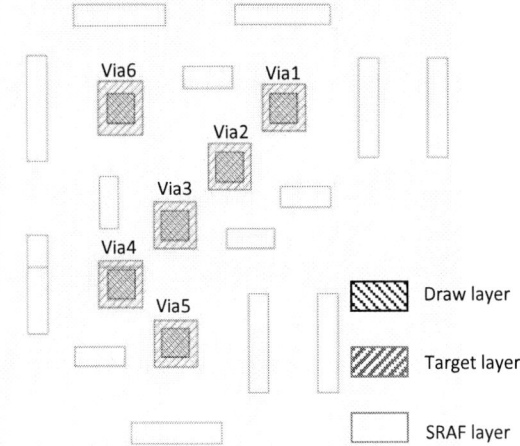

Figure 1: Layout clip of minimum corner to corner space square via

For the first group, via mask edges are cut into two segments at most as the square via width is small enough already. What's more, the maximum value of the gap between the two segments in the same edge of via mask is been limited for purpose to reduce the risk of inconformity between the wafer shape and simulation contour.

For the second group, fragments are not divided on the edges of via mask. So the movement of mask edges would be limited by MRC in the corner to corner direction in our

978-1-7281-6559-2/20 $31.00 © 2020 IEEE 117

expectation.

Then the final via mask is got based on the OPC model, and get the simulation contour of the two groups of via mask layer. The differences between the contour and the target of the two groups are compared. The two groups of mask patterns are also exposed on wafer to measure the CD data and review the images of the two experiment groups.

RESULT AND DISCUSSION

The mask layers of both experiment groups are calculated out based on the OPC model, as shown in figure 2. fig2.(a) is the mask layer with fragments, while fig2.(b) is the mask layer without fragment. The two groups contain the same draw, target and SRAF layer. The corner to corner spaces between the via mask layer of Via1~Via5 in both groups, by measurement, are the minimum value of MRC. The mask edge in figure fig2.(a) is divided into two segments at most and the distance two segments is quite small.

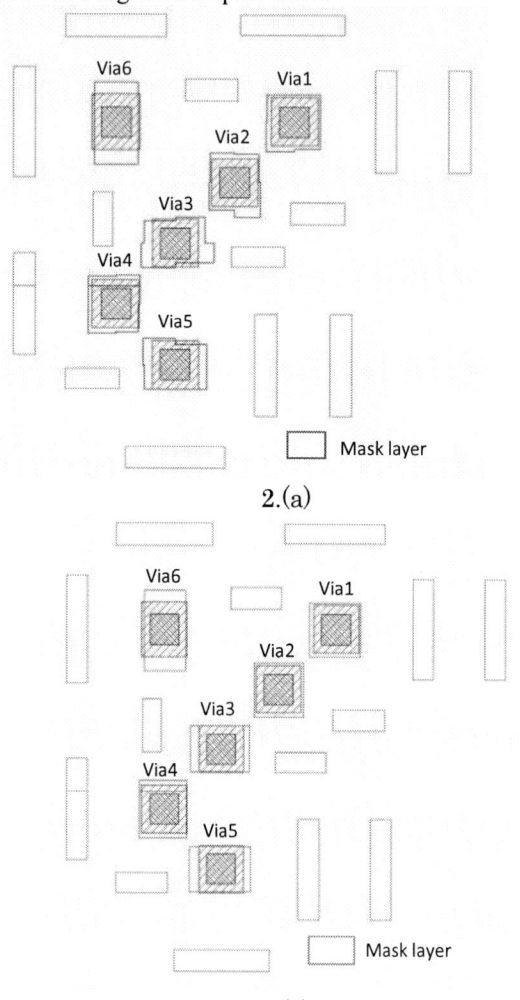

2.(a)

2.(b)

Figure 2: Layout clip of minimum corner to corner space square via (a) mask layer with fragments (b) mask layer without fragment

The pattern of two groups are added same SRAF. SRAF is quite important for via especially for node 28nm or below. SRAF condition contains SRAF width, spaces to main feature and spaces between SRAF. But the capability of SRAF to assist contour on target may be confined if the mask edges are

limited by MRC. So segmentation of mask layer may be a reasonable way to make contours on target.

The simulation results are shown in figure 3, fig3.(a) is the contour layer of mask with fragments, while fig3.(b) is the contour layer without fragment. In fig3.(a), the mask layer with fragments, contour of every via is exactly on target, while the contour of mask without fragment in fig3.(b) is not all on target. The contour of Via3 in fig3.(b) is not on target, the edge placement error (EPE) of the right edge is about -2nm.

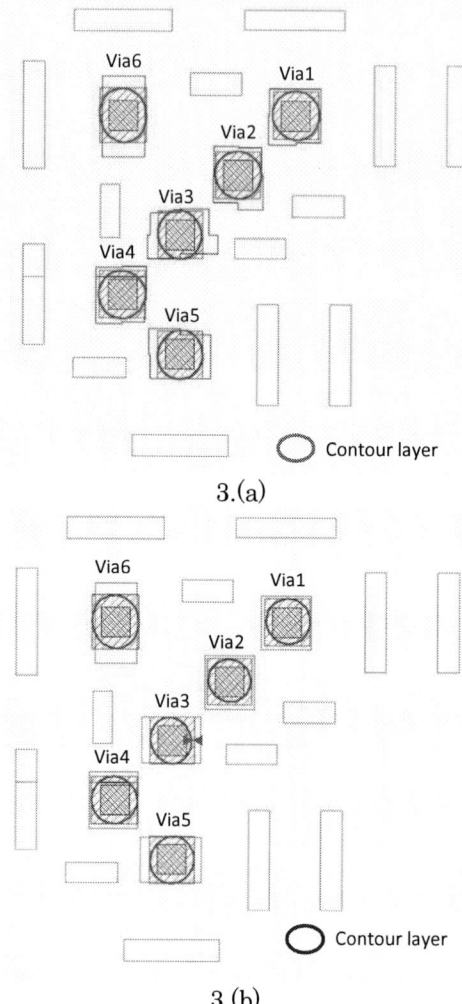

3.(a)

3.(b)

Figure 3: Layout clip of minimum corner to corner space square via (a) mask layer with fragments (b) mask layer without fragment

But the potential risks of fragmentation of mask layer have to be taken into account, because the data for OPC model fitting do not contain complex patterns with fragments. Usually square and rectangle patterns are only been used to fit the via OPC model. So the simulation contour of the mask with fragments by using the OPC model may be not very accurate. Wafer data can help us verify the simulation results.

Focus energy matrix (FEM) wafer is exposed to check CD data and images of Via3 in best focus and defocus conditions at best energy. The CDSEM images and CD data show that the nominal condition (best energy and best focus) CD of the mask with fragments is on target and the roundness of holes is acceptable. The target CD of Via3 in X-direction is 70.5nm.

The CD of mask without fragment is smaller than the target value. Process window condition CD verification is also been checked. Best energy condition defocus CD and image, to measure the DOF, spec is 63.5nm. DOF of mask with fragments is about 105nm, while mask without fragment is only 60nm. Reasonable segmentation of square via mask can improve DOF observably.

What should be considered for square via mask fragmentation is the quantity and extension of fragments should be limited strictly, so that OPC model can cover the mask pattern quite well.

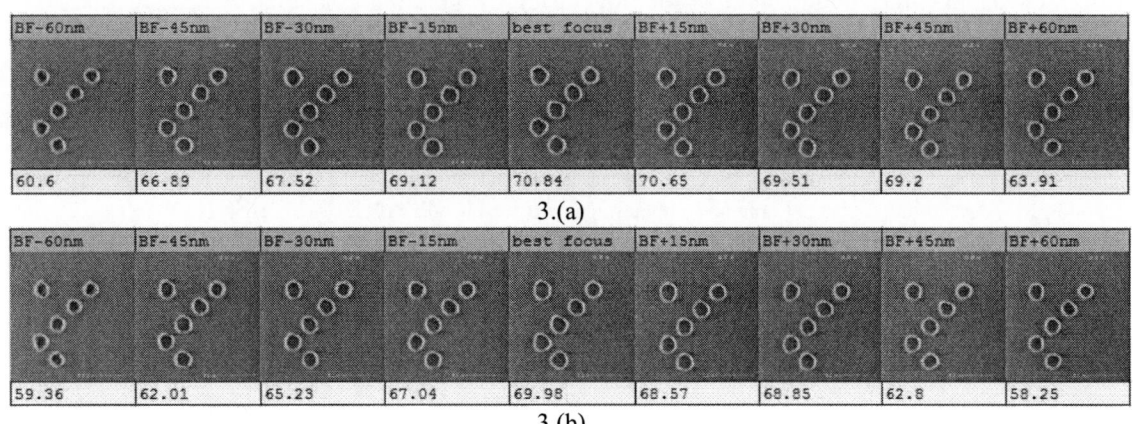

3.(a)

3.(b)

Figure 4: CDSEM data and images of Via3 (a) mask layer with fragments (b) mask layer without fragment

CONCLUSION

For via OPC, the main method to improve DOF is sizing up target CD and optimizing the addition of SRAF. For some conditions that sizing up target layer or optimizing SRAF can not make contour layer on target or improve DOF, fragmentation of mask edge can be considered. The extension of each fragment should be limited in a reasonable value. the contour can be on target, and the DOF can be enough.

ACKNOWLEDGMENTS

The authors would like to thank HLMC TD2OPC team colleagues for providing needed support.

REFERENCES

[1] Quan Chen , Shirui Yu , Zhibiao Mao , Yu Zhang , Bin Gao , Yanpeng Chen , Albert Pang , "Mask model analysis and its application in 28 OPC modeling", IEEE CSTIC, 2015.

[2] Yonghua Zhang , Lisong Dong , Yayi Wei , Haoru Hu , IEEE CSTIC, 2019.

STUDY OF LOW PINCH-OFF VOLTAGE JFET IN 500V HIGH VOLTAGE PROCESS

Wenting Duan, Ziquan Fang, Wensheng Qian

HuaHong Grace Semiconductor Manufacturing Corporation, Shanghai 201206, China

*Corresponding Author's Email: Wenting.Duan@hhgrace.com

ABSTRACT

This article presents a variety of small size JFET structures of low pinch-off voltage in 500V high voltage platform. The low pinch-off voltage is achieved by adopting low concentration channel area and implanting the opposite conductivity type impurity into the deep channel region, and the JFET current is improved by increasing the concentration of the part of channel region. Finally, the low pinch-off voltage and high current JFET device is achieved without mask adding.

INTRODUCTION

High-voltage JFET(Junction Field Effect Transistor) integrated on the BCD500V process platform is often used as an important part of the driving circuit, and traditional high-voltage JFET is typically composed of the drift region of drain side and JFET body area, which is shown in Fig. 1. The drift region of drain side mainly ensures high breakdown voltage, consisting of DNW (Deep N-type well) and a layer of anti-type implant; The body area is used to control on-current and pinch-off voltage of JFET. The channel in body area consists of N-type impurity of DNW, which is pinched off by the PW (P-type well) and Psub(P-type substrate) on both sides of the DNW. when Gate PW is applied an reverse bias, the DNW begins to deplete until the channel is pinched off [1] [2].

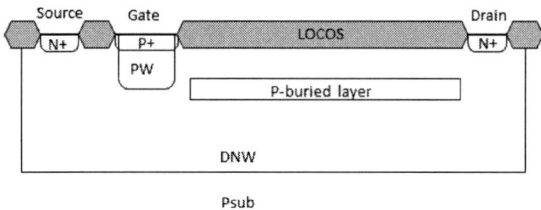

Figure 1: traditional 500V JFET structure

The HV JFET device of this structure has a high pinch-off voltage (more than 20V) due to the depth of DNW, the wide channel is not easily pinched off [3] [4]. In this paper, the pinch-off voltage is reduced by shortening the width of the channel, and the JFET current is increased by increasing the concentration of the local body area.

Unlike the high-voltage JFET on the original BCD500V process platform, this paper presents the small-size JFET structure. Due to the drift region size reduction, its breakdown voltage is low. The pinch-off voltage is the gate to source voltage when the channel is pinched off., and the pinched off voltage must be lower than the breakdown voltage of junction. In order to realize the small size, low pinch-off voltage JFET device, this paper makes the following attempt.

RESULTS AND DISCUSSION

1. After the size of the JFET is reduced, using the original pinch-off structure PW (P-type well)-DNW (deep N-type well) -Psub (P-type leader), as shown in Figure 1, the JFET device channel is DNW, The N+(the heavy doping N-type area) area at both ends of the channel are the source and drain of JFET, the PW at surface and the Psub below the channel pinch off to the channel, and the PW is picked up by P+(the heavy doping P-type area), which is the Gate of JFET. The structure pinch-off voltage is the same as the original high-voltage JFET, which is more than 20V due to the wider channel, as shown in Fig.3.

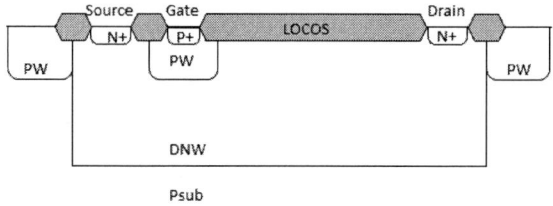

Figure 2: small size JFET structure1

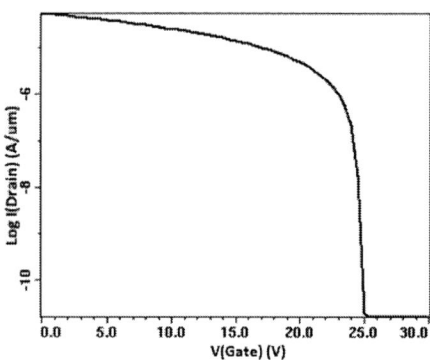

Figure 3: pinch off voltage curve in simulation

2. Shortening the width of the channel is attempted to reduce the JFET pinch-off voltage, by replacing DNW with NW on the basis of structure 1, and the PW is removed, as shown in Figure 4. The JFET channel of this structure is changed from DNW to NW, and the width of the channel is significantly shortened. The simulation

shows that the channel of the structure cannot be pinched off, that is, the channel cannot be completely depleted because of the high concentration of NW. Figure 5 is a plot of the device simulation depletion region, gray as depletion region, and green as none depletion region. When the Vgs(gate to source voltage) is -20V, the channel is still not completely depleted.

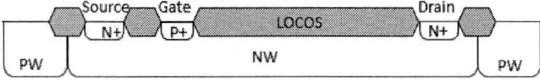

Figure 4: small size JFET structure2

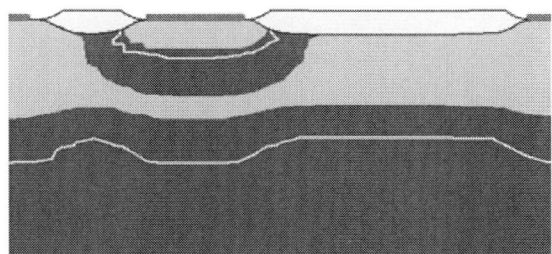

Figure 5: depletion region of JFET in simulation

3. Shorten the width of the channel with another method is also attempted, as shown in Figure 6. P-type buried layer implant which is already in the process platform is adopted in the DNW area and picked up by the PW of two sides. P-buried implant divides The DNW into two parts, the JFET channel is the upper part of the DNW, which is pinched off by P-buried implant and P+ Gate. Although pinch-off voltage of this structure is significantly reduced and the simulated pinch-off voltage is 3.5V, as shown in Figure 7, its on-current is low because of the low DNW concentration.

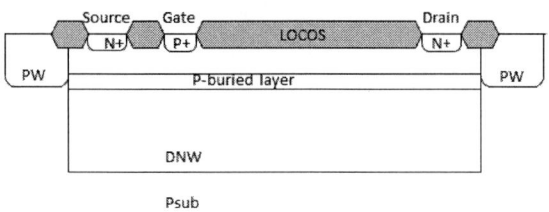

Figure 6: small size JFET structure3

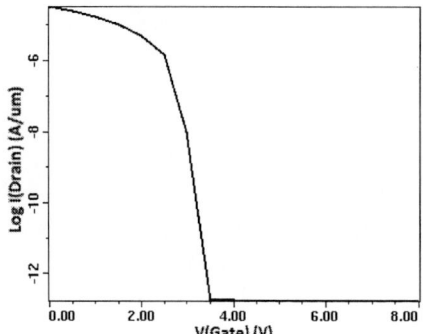

Figure 7: pinch off voltage curve in simulation

4. In order to get the pinch-off voltage down and on-current high, NW is added under P+ gate in structure3, as shown in Figure 8. In this structure, not only low pinch-off voltage is achieved, but also the on-current is improved. At the same time, the pinch-off voltage is adjustable by the NW horizontal size, the pinch-off voltage increases with NW size increasing, as shown in Figure 9. Meanwhile, the JFET on-current is also affected by the NW size, the on-current increases with NW horizontal size increasing. After NW implant adding, the JFET on-current of structure4 increases by 3 to 4 orders than that of structure 3, as shown in Table 1.

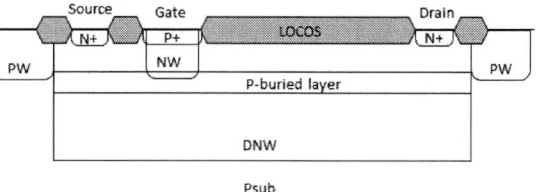

Figure 8: small size JFET structure4

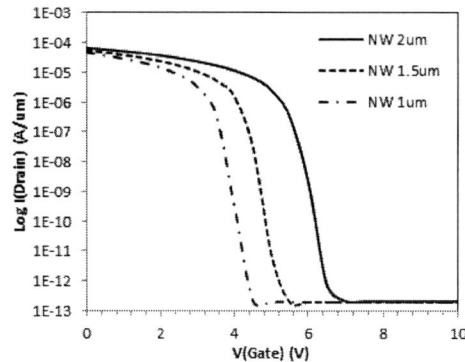

Figure 9: pinch off voltage curve in simulation

978-1-7281-6559-2/20 $31.00 © 2020 IEEE

Table1: Vp & Id

	structure3	structure4		
	DNW	NW 2um	NW 1.5um	NW 1um
Vp(V)	3.5V	6.5V	5.5V	4.5V
Id(A/um)	9.3E-09	2.2E-05	1.0E-05	3.0E-06

CONCLUSION

In this paper, the small size and low pinch-off voltage JFET devices of various structures are studied, the low pinch-off voltage of JFET devices are realized by narrowing the width of the channel, and the JFET on-current is increased by increasing the channel concentration. Finally, the JFET device of low pinch-off voltage which is adjustable is obtained.

ACKNOWLEDGEMENTS

The authors would like to thank all members of device group and high voltage group of HHgrace for their great support on this work.

REFERENCES

[1]Kun Mao, Ming Qiao, WenTong Zhang, et al. A 700V narrow channel nJFET with low pinch-off voltage and suppressed drain-induced barrier lowering effect[J]. Superlattices & Microstructures, 2014, 75(32):576-585.

[2]Nidhi, Karuna, Ker, Ming-Dou. A CMOS-Process-Compatible Low-Voltage Junction-FET With Adjustable Pinch-Off Voltage[J]. IEEE Transactions on Electron Devices:1-8.

[3]Chorng-Wei Liaw,, Leaf Yeh,, Ming-Jang Lin, et al. Pinch-Off Voltage-Adjustable High-Voltage Junction Field-Effect Transistor[J]. IEEE Electron Device Letters, 28(8):737-739.

[4]Liu Yong, Tang Zhaohuan, Wang Zhikuan, et al. Design and application of a depletion-mode NJFET in a high-voltage BiCMOS process[J]. Journal of Semiconductors, 2010, 31(8):084006.

STUDY OF MOSFET IDVG CURVE DOUBLE HUMP EFFECT

Jun Hu[1] ,Zhaozhao Xu,Wenting Duan,Ziquan Fang,Donghua Liu,Wensheng Qian[1],

[1]HuaHong Grace Semiconductor Manufacturing Corporation, Shanghai 201206, China
*Corresponding Author's Email: jun.hu@hhgrace.com

ABSTRACT

In the traditional CMOS manufacturing process, we often use the IV curve to evaluate the characteristics of the transistor, and sometimes the IdVg curves of the transistor will appear double hump , especially for the NMOS.

This paper analyzes mechanism of the double hump phenomenon of the IdVg curve. There are two main causes, one is due to the segregation effect of impurities, and the other is due to the manufacturing process of STI. This article also shares ways to improve this phenomenon.

INTRODUCTION

In current MOS and other planar processes, different devices are isolated by field oxide. Before the 0.25um process came out, they used LOCOS (Local Oxidation of Silicon) isolation technology. After the 0.25um process came out, they were changed. Using STI technology.

In a typical device with LOCOS process, a high-temperature process causes a transition region between a thin gate oxide layer and a thick oxide layer to form a bird's beak-like structure, which causes the effective or electrical width of the device to be smaller than the designed channel width.

When the channel length of the MOS field-effect transistor is constant, reducing the channel width W will also significantly affect the electrical characteristics of the device. It is generally believed that when the channel width W is small enough to be comparable to the thickness of the channel depletion layer, a phenomenon in which a decrease in the channel width causes an increase in Vt is called a narrow width effect.

This phenomenon can be explained by the surface depletion layer expanding on the side of the channel edge. Figure 1 shows a cross-sectional view of the channel width direction of a MOS field effect transistor. Where Tox is the thickness of the field oxide layer, and X is the thickness of the depletion layer under the field oxide layer. The electric lines on both sides of the channel are not perpendicular to the surface of the silicon substrate, causing the depletion layer to expand to the side, and there is an additional space charge region on both sides of the channel width. These additional charges are controlled by the gate voltage. This increases the total amount of charge in the depletion layer. Therefore the Vt voltage is increased compared to when it is not extended.

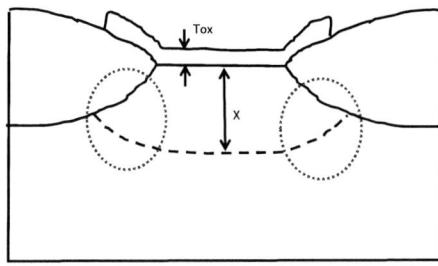

Fig 1: Depletion region with LOCOS process

The STI process is different. The space charge region is limited to the active region, and the voltage applied to the field oxygen region depletes the charge in the active region, which will widen the depletion region near the edge of the active region. In this case, Vt will be lower than in the absence of this effect.

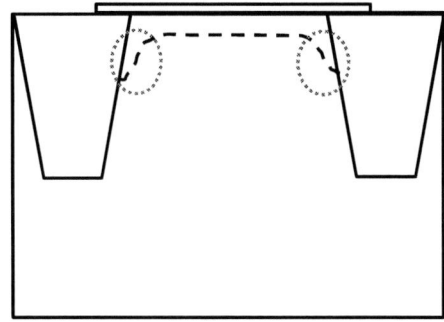

Fig2: Depletion region with STI process

The above is the trend and mechanism of Vt change in the case of the COS active area width change in the LOCOS process and the STI process. This effect can be ignored in devices with wide channel widths, and the Vt voltage does not change with the channel width. When the channel width W can be compared with the surface depletion layer width The situation will be more significant.

RESULTS AND DISCUSSIONS

In the CMOS manufacturing process, the IdVg curve of the transistor will show a double peak phenomenon, especially for NMOS, which is more obvious. The main reasons are as follows.

One is due to the different solid solubility of impurities in silicon and silicon dioxide. The device with a thermal process of growing the gate oxide layer after well implantation and Vt implantation. Due to the large thermal budget of this thermal process, in the process, the width of the channel of the device, the junction of silicon and silicon dioxide, because the solid solubility of the impurities in silicon and oxide is different, the impurity boron diffuses from the silicon into the oxide, thereby reducing the concentration in the well in the channel width direction. So from the width direction of the transistor, there are two parasitic transistors at the edge of the park, and the threshold voltage of these two transistors is lower than the transistor itself. Shown on the device's IdVg curve, a double-peak phenomenon appears, and some transistors are turned on in advance.

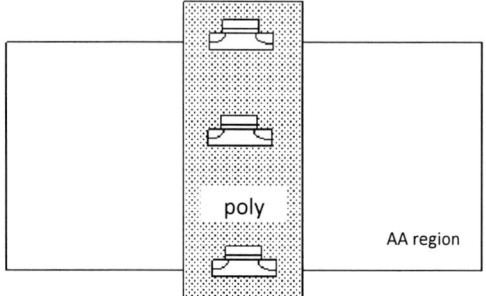

Fig3: NMOS layout structure

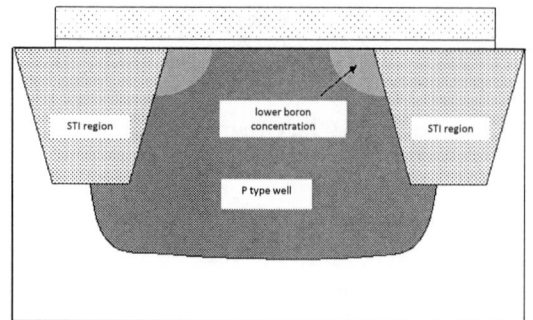

Fig4: NMOS doping profile by width direction

As shown in FIG. 3, viewed from the transistor width direction, the device is formed by a plurality of transistors connected in parallel. In the conventional process flow, the impurity distribution in the width direction of the transistor is shown in FIG. 4. The boron concentration near the shallow trench isolation region is lower than the middle part of the gate, which causes the parasitic transistor to turn on in advance, and the IdVg curve has a double hump phenomenon.

The double-peak phenomenon of the IdVg curve of the NMOS transistor caused by the difference in the solid solubility of impurities in silicon and oxide can be solved by changing the order of well implantation and gate oxidation.

1. Growth of gate oxide and polysilicon;

2. Perform well implantation, and boron ion implantation for NMOS;

3. Perform polysilicon implantation and, for NMOS, phosphorus ion implantation; this implantation can be performed on the same lithographic plate as well implantation.

4. Perform the next steps in transistor manufacturing.

With the above method, well implantation has undergone fewer thermal processes, especially the thermal process of growing the gate oxide layer, reducing the diffusion of boron from the silicon into the silicon oxide, making the impurity concentration in the well more uniform, which can improve the transistor, especially the NMOS. Double hump phenomenon of IdVg curve

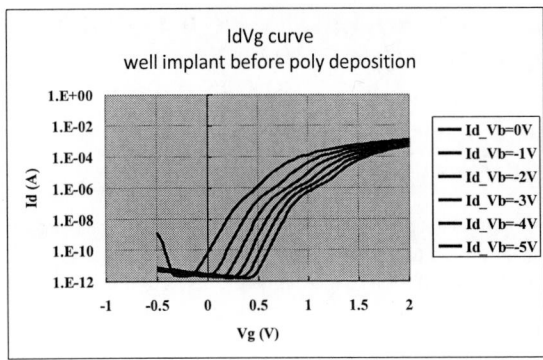

Fig5: NMOS IdVg curve with double hump

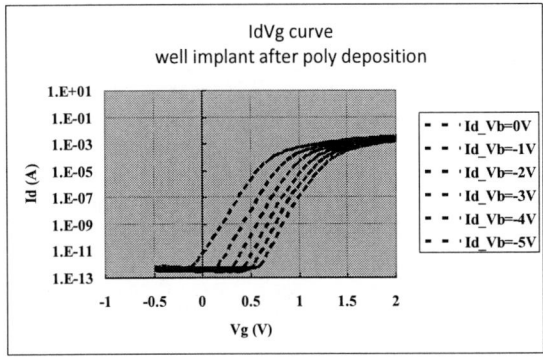

Fig6: NMOS IdVg curve without double hump

Figure 5 is the IdVg curve of the device under normal conditions, and the double hump phenomenon is more serious. Figure 6 is the IdVg curve after changing the well implant and gate oxide thermal process. The double hump phenomenon has been significantly improved.

Second, it is caused by related processes of shallow trench isolation. In the width direction of the device, at the junction of the active region and the shallow trench isolation, the thickness of the gate oxide layer of the device is lower than that of the gate oxide layer in the middle of the gate. Therefore, the threshold voltage of the two parasitic transistors in the width direction is lower than that of the transistor itself, and it appears on the IdVg curve of the device, and a double hump phenomenon appears, and some transistors are turned on in advance.

Third, because the oxide stress at the STI corner is relatively large, the loss will be faster during the etching process, so it is easy to sink down to form a divot, and the poly dep here is also easy to sink in, so

the gate electric field will be stronger here. As a result, the channel here is turned on in advance.

The fourth is that before the well implantation, the screen oxide at the STI corner will be thinner than the flat active area, resulting in less Vt implantation than the flat active area. Therefore, the Vt of the parasitic transistor here is relatively small, and the device leads in advance.

Usually, the screen oxide before the well implant can be removed, so there is no thickness inconsistency, so the effect of Vt variation of the parasitic tube can be reduced.

ACKNOWLEDGEMENTS

The authors would like to thank all members of device group and PI group of HHgrace for their great support on this work.

REFERENCES

[1] Donald A.Neamen Semiconductor Physics and Devices Basic Principles, Fourth Edition P461.

EFFECT OF IMPLANT BEAM CURRENT ON RESISTANCE OF BF2 IMPLANTED POLYSILICON

Lichao Zong[1], Chunling Liu[1], Xingjie Wang[1], Liming Chen[1]*

[1]Shanghai Huahong Grace Semiconductor Manufacturing Corporation

Corresponding Author's Email: lichao.zong@hhnec.com

ABSTRACT

The effect of different injection beams (3ma, 5ma and 7ma) on the square resistance of polysilicon was studied by implanting BF_2 with GSD200 (an energy of 30 Kev and dose of $2E^{15}$ under 1000℃, 30S rapid thermal annealing). The experimental results showed that the higher beam current would result in the lower resistance of polysilicon. The higher implant beam current will lead to more damage in polysilicon which will result in bigger poly grain size after thermal annealing, the bigger grain size will make more carriers in grain boundary and the resistance of polysilicon decreases accordingly.

Key words: Polysilicon, Resistance, Implant, BF2, Beam current

INTRODUCTION

Polysilicon film is widely used in LSI and VLSI, the heavily doped polysilicon is commonly used as gate and interconnecting material for MOS, the lightly doped polysilicon is commonly used as load resistance in the cell circuit of memory. The studies of implanted polysilicon had been reported several times [1]. The polysilicon resistance will be impacted by some factors, such as implant energy, dosage, angle, Plasma flood system: PFS (PFS is used for neutralize positive charge during implant process) and pressure compensation system: PCOM (PCOM is used for vacuum system which can lead to implant dosage difference). In this work, we studied the effect of BF_2 implant beam current on polysilicon RS, the higher implant beam current will make lower polysilicon RS.

EXPERIMENT

(1) Grow 0.5um SiO_2 on P type substrate

(2) 250nm polysilicon was deposited by LPCVD (temperature is 600℃, pressure is 0.8T, time is 60mins).

(3) Implant BF_2, energy is 65kev, dosage is $2e^{15}/cm^2$, tilt is 7° to reduce small channel effect,

Here we designed implant beam current splits (3ma, 5ma, 7ma) to study the influence on polysilicon resistance, wafer must be located on a cooling plate, cooling is necessary because the ion implantation process adds appreciable energy to the silicon substrate .

(4) Rapid thermal annealed at 1100℃ for 30 sec a $N_2+1\%O_2$ ambient to activate impurity ion.

The polysilicon RS was measured by four probes method, and use Thermal Wave (TW) to detect the surface damage. Thermal Wave (TW) is a method to measure the variation of amount of surface damage. The TW signal has a direct correlation to the damage caused by the implanted ions. The Thermal Wave Signal is produced by two lasers. A 1 MHz argon(Ar) laser is used to cause an excitation within the wafer substrate. The argon laser is called the PUMP LASER. It's energy imparted to the silicon surface and focused in the center of the HeNe laser spot laser leaves a signature which can be read by the helium neon (HeNe) laser. This excitation is then carried along the reflected HeNe path and is read by the Thermal Wave detector. In previous studies, when the implant energy was limited in 10-100kev, dosage was limited in $10e^{11}$-$10e^{16}$ (ions/cm^2), it was shown that both dose and energy are proportional to damage, the higher energy or dosage, the worse damage will happened on polysilicon surface.

The polysilicon grain size under different implant beam current was observed by scanning electron microscope (SEM) tool.

EXPERIMENT RESULT

Polysilicon RS, TW signal and Polysilicon surface profile were checked with different implant beam current of BF_2.

TABLE I.   BEAM CURRENT VS. RS/ TW/ GRAIN SIZE

	Implant beam current		
	3ma	*5ma*	*7ma*
RS (Ω .CM)	302.9	298.5	295.5
TW Unit	22425	22612	22785
Grain Size (Å)	800	900	1000

The experiment result showed that implant beam current will impact polysilicon RS obviously (see Figure 1).

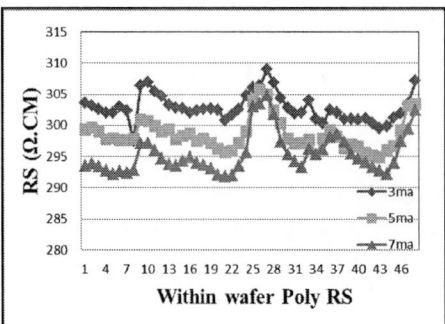

Figure 1: poly RS Vs implant beam current

Polysilicon RS will be lower with higher beam current, Poly RS with 7mA beam current is lower about 8% than with 3mA beam current (see TABLE1).

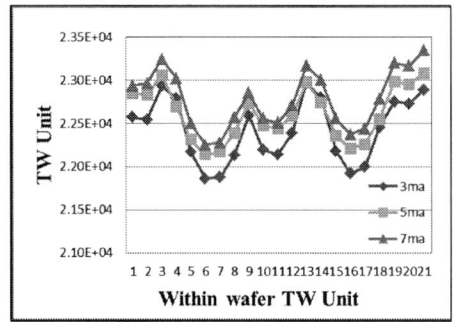

Figure 2: Poly TW signal by implant beam current

FIG2 showed that TW signal value for different implant beam current with the same implant energy and dosage, 7ma beam current will get more serious damage on polysilicon surface。

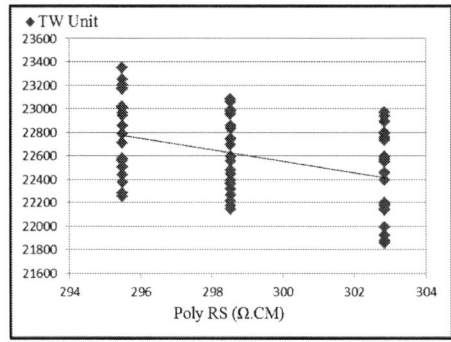

Figure 3: Poly RS correlation to TW signal

FIG3 showed polysilicon RS correlation to TW signal, it seems polysilicon RS will impacted by the surface damage, the worse surface damage leads to the lower RS.

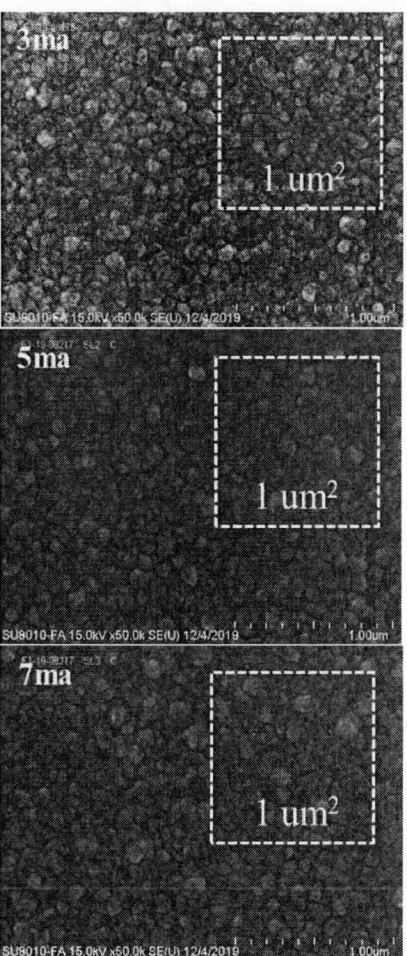

Figure 4:Poly surface profile by implant beam current

FIG4 is the SEM image for polysilicon surface under different beam current (3ma, 5ma, 7ma), the result showed that there was no obviously difference for surface profile, the flatness and compactness are good for all the samples, but the polysilicon grain size seems different with different beam current, 7ma implanted polysilicon grain size is ~1000 Å (the grain size was counted by the grain amount (n) in unit area: $1um^2$, $D=2/\sqrt{n\pi}$), 5ma grain size is ~900 Å and 3ma grain size is ~800 Å. It means that the grain size will increase with the implant beam current after RTA, that is, the polysilicon structure will be recovered more enough for higher implant beam current after RTA.

RESULT DISCUSSION

The polysilicon conductivity depends on 2 factors [2, 3] carrier density and carrier mobility, the carrier density mainly depends on implant dosage besides the temperature, while the carrier mobility is impacted by crystal structure and the defects count trapped in the boundary of polysilicon. Under the same implant dosage ($2e^{15}/cm^2$), the carrier density is almost the same, but the carrier mobility will be impacted by the grain density (the number of grains per unit volume) of polysilicon, the grain size variation will make the defects count trapped in the boundary of polysilicon different, and the carrier mobility shift accordingly.

The grain size of polysilicon varies with the implant beam current, it should be caused by injection molecular effect for BF_2 which has higher atomic mass, when BF_2 implant into polysilicon, the spatial-temporal correlativity of B and F is fade away by the implant energy dissipate gradually [4]. In the damage area, the spatial-temporal correlativity will lead to multi-body collision effect, it means BF_2 implant will lead to more damage on polysilicon surface compared to a single atom injection. The more serious damage is accompanied by greater local warming on the surface of polysilicon, the local warming on polysilicon will make some minor lattice damage restoring and some grain boundary defects recovering. Adequate recovery of polysilicon may result in larger grain size, according to the barrier model [5] of electrical conductance in the polysilicon, the polysilicon RS is positively to grain amount per unit volume (n), which means it is negatively to grain size (1), so the larger grain size will lead to polysilicon RS lower.

$$R_\square = K \frac{nk}{qSA^*T} e^{\frac{E_B}{kT}} \qquad (1)$$

A^* is effective Richardson constant, T is absolute temperature, E_B is interface barrier height, k is Boltzmann's constant, q is electron charge, n is grain amount per unit volume and it's negatively to grain size, S is sectional area of polysilicon RS, K is constant coefficient for grain size and polysilicon RS. From this formula, we can see that with other conditions unchanged, larger grain size will make lower polysilicon RS.

So, the lower polysilicon RS caused by higher implant current is mainly because of the larger grain size caused by higher beam current.

CONCLUSION

In ion implant process, the polysilicon RS depends on many factors such as implant energy, implant dosage, beam current, temperature and etc. acting together and the interrelationship between them is complex. In this paper, polysilicon RS of BF_2 implanted with 65kev, $2e^{15}/cm^2$ at room temperature were studied as a function of implant current. The beam current varied from 3ma to 7ma and we found the polysilicon RS is negatively correlation with beam current, the TW data after implant and grain size is positively correlation with beam current. This result may has some significance to adjust BF_2 implanted polysilicon RS in the production process.

ACKNOWLEDGEMENTS

Thanks very much for the guidance given by my supervisor Xingjie Wang and Chunling Liu.

REFERENCES

[1] Jialu. Liu, Yanqing. Zhang, Jianjun. Li, *Chin. Phys. Soc.*, 46, 8(1997).
[2] Qinwen Fan, *Electronics and Packaging.*, vol. 2-5, 2002, pp.15-16, 38.
[3] T. I. Kamins, *J. Appl. Phys.*, 42, 4357 (1971).
[4] G. Baccarani and B. Ricco, *J. Appl. Phys.*, 49, 5565 (1978)
[5] Jingping. Xu, Yuehui. Yu, Zhaolian. Peng, Tao. Chen, *Microelectronics.*, vol. 23-5, 1993, pp. 32-35.

Deep Power Down Leakage Study Caused by Poly L-shape Pattern

Chong Huang[1]*, Ming Zhang[1], Fangce Sun[1], Steam Cao[2], Guanghua Yang[2], Susanna zheng[2]

[1] Department of Process Integration, Shanghai Huahong Grace Semiconductor Manufacturing Corporation, Shanghai 201203, China

*Corresponding Author's Email: Chong.Huang@HHGrace.com

Abstract

Deep power down leakage is a very key requirement for IC chip, especially for the MCU chips with battery power supply. In this paper, we studied the deep power down mode chip leakage caused pocket shadowing effect due to poly L-shape layout. And we tried to optimize the layout and process to reduce the leakage.

Keywords— L-shape poly layout, deep power down leakage, LDD shadowing effect

Introduction

With the rapid development of electronic products, electronic products were used in every area of our life. And more and more MCU were used in IOT or wireless electronic products. In order to get as longer working time as within one power bank, the low leakage MCU chips were required, especially the ultra-low leakage at standby working mode. Normally, it requested less than 3uA level.

HHGrace were one of the makers of MCU chip with embedded SST super flash. During we developed the 0.11um technology embedded flash MCU products, we faced the leakage issue. There some signals:

1, all the products suffered leakage issue used the same logic stand cell. And the leakage was occurred in the logic area with the same stand cell.

2, the leakage showed random distribution on the whole wafer as Fig1.

3, the leakage distribution was from 3uA-150A as Fig2. For the high leakage samples, we cannot find the hot spot by EFA (EMMI/OBIRCH).

4, the leakage failure ratio showed strong correlation with the stand cell logic gate counts. The more stand cell logic gate counts, the higher leakage failure ratio.

The pocket implant shadow effect of L-shape poly layout was suspected [1],[2]. We designed 3 cycles of DOE to study the leakage issue and try to find the solution.

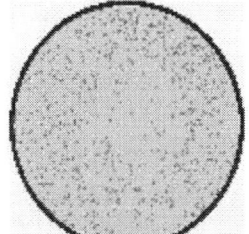

Fig1. leakage distribution map on the wafer

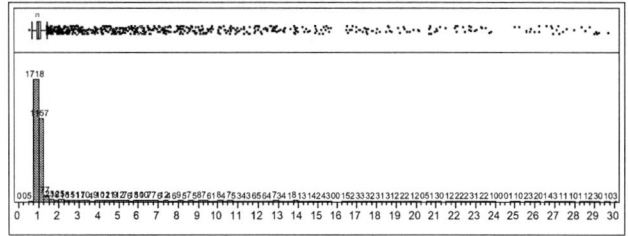

Fig2, leakage level distrubition

EXPERIMENT

In the 1st cycle of DOE split, we did device corner/Poly CD FEM/Poly AA matrix/Pocket IMP dosage split/pocket IMP direction split as following table1.

Condition		1	2	3	4	5	6	7	8	9	10
Device split	POR	v			v	v	v	v	v	v	v
	FF (N/P run 8% faster)		v								
	SS (N/P run 8% slower)			v							
Gate Photo	POR	v	v	v				v	v	v	v
	FEM				v						
	X direction AA					v					
	Y direction AA						v				
Pocket IMP	POR	v	v	v	v	v	v				
	Dosage -10%							v			
	Dosage +10%								v		
	Dosage +20%									v	
	IMP direction from 4 direction to 8 direction										v

Table1 1st cycle of DOE split

RESULTS

In the 1st cycle of learning, we have 3 groups experiment.

1, The 1st group is the logic stand cell device split, we did device POR/FF/SS corner split. And the results showed that when the device ran faster, the whole wafer leakage level was higher and the media level is 2uA and the distribution was wider (Fig3). When the device ran slower, the whole wafer leakage level was lower and the media level is 1.3uA and the distribution was narrow (Fig5). But whatever the FF corner or the SS corner, all the conditions suffered more than 10% high leakage failure ratio. (>2.5uA counted as fail samples.)

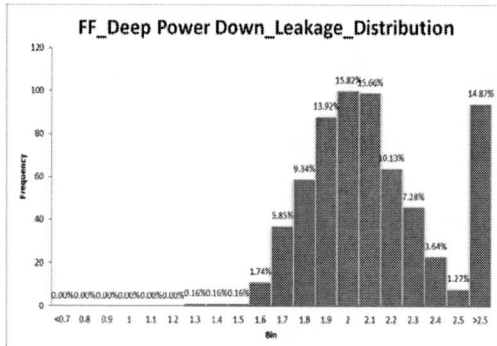

Fig3, FF corner leakage distribution

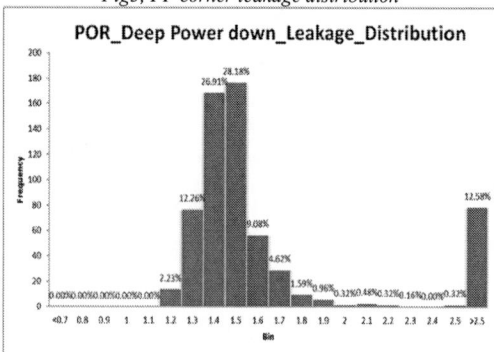

Fig4, POR leakage distribution

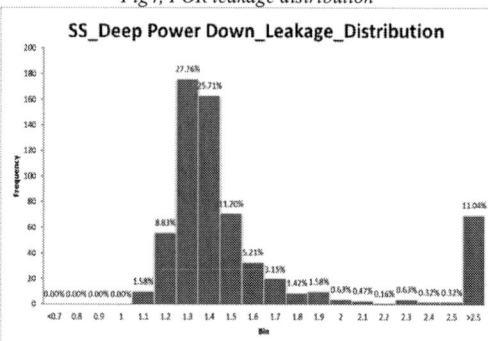

Fig5, SS corner leakage distribution

2, The 2nd group is the poly photo related experiment split. The poly FEM showed no strong CD correlation with the leakage issue. But the poly photo AA(overlay) matrix showed bigger AA would results worse leakage issue (Fig6 & 7).

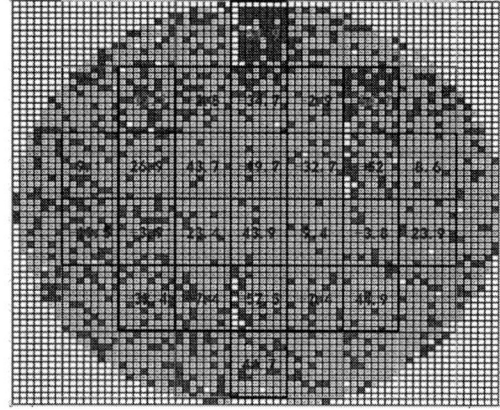

Fig6, X direction AA matrix

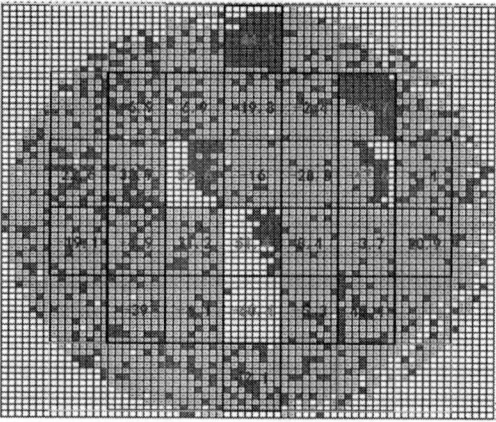

Fig7, Y direction AA matrix

3, The 3rd group is the pocket IMP related experiment split. This was channel leakage split[3]. The pocket IMP showed obviously grouping. The dosage -10% condition showed worst leakage failure ratio. Both dosage +20% and 8 directions showed better leakage failure ratio (Fig8).

This split result told us that this was channel leakage.

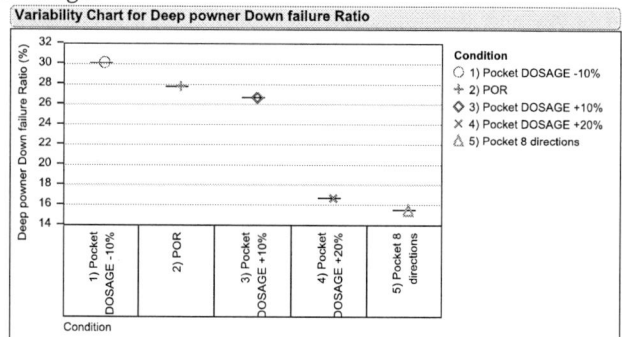

Fig8, Pocket implantation split deep power down leakage failure ratio (%)

MODELING DISCUSSION

Based on the experiment1 results, we highly suspect small device channel leakage. And one leakage model was built up, we call it as poly L-shape shadowing effect for pocket IMP.

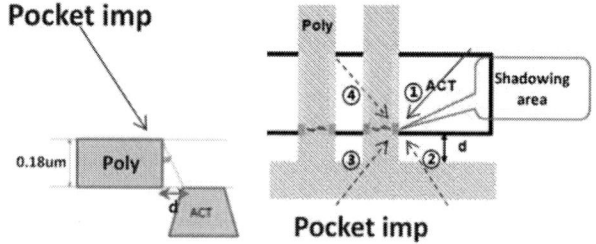

Fig9, Pocket implant L-shape shadowing effect

There is poly shadowing effect due to there is the 0.18um thickness poly on the wafer during the pocket IMP. The poly line likes a wall and it blocks 3 directions (②③④) pocket IMP. There only 1 direction (①) pocket IMP can implant into the channel (Fig9). So the shadowing area pocket IMP dosage is not enough. That why the pocket dosage +20% and 8 directions of experiment1 showed better leakage performance.

And distance "d" of field poly to Act is the most key parameter.

978-1-7281-6559-2/20 $31.00 © 2020 IEEE   130

How does the distance "d" of L-shape shadowing impact the device channel leakage? We designed one e-test test key with different d and Act width as Fig10 (d1<d1.5<d2<d3, Act width1< Act width2 Act width3 Act width4).

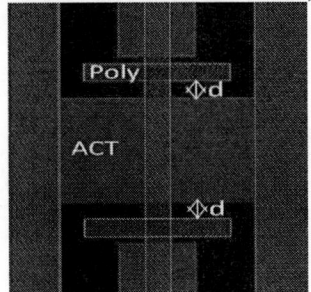

Fig10

The E-test data showed L-shape shadowing does impact the channel pocket implant. The device Ioff/Ids/Vt was obviously impacted when the device become narrow and short. When the device channel width became bigger than width3, the L-shape shadowing impact became minor (Fig 11,12,13).

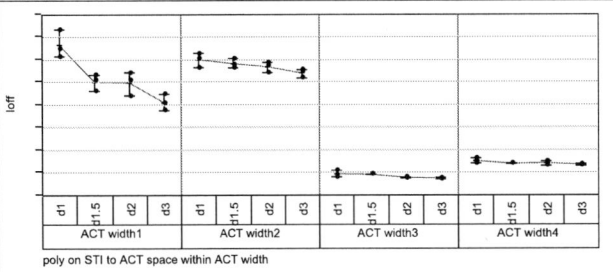

Fig11, L-shape different d and Act width e-test Ioff comparison

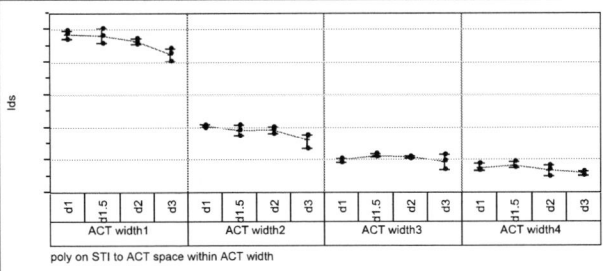

Fig12, L-shape different d and Act width e-test Ids comparison

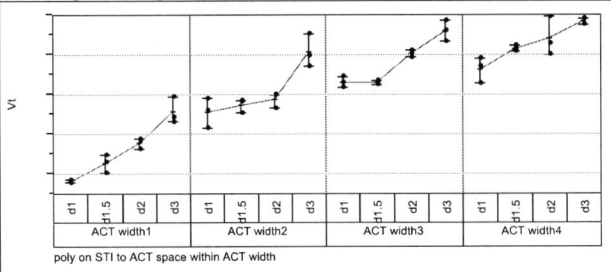

Fig13, L-shape different d and Act width e-test Vt comparison

IMPROVEMENT ACTION

We found all the products suffered high leakage failure ratio using the same logic stand cell library. And this library used very high frequency "d1" for the L-shape poly pattern. Other products did not suffered leakage issue used other library which used high frequency "d2" (d2 > d1).

So we designed one new mask. This new mask keeps most of the pattern but dig out a block at the L-shape poly area to enlarge the d1 (Fig14). And run 2nd cycle of DOE lot as table2.

Condition		1	2	3	4	5	6	7	8	9	10
GATE Photo	POR Mask	v	v	v	v						
	New Mask					v	v	v	v	v	v
Pocket IMP	POR	v	v				v				v
	IMP 8 direction					v			v		
	IMP 8 direction & Dosage+10%			v	v			v	v		

Table2, 2nd cycle of DOE split table

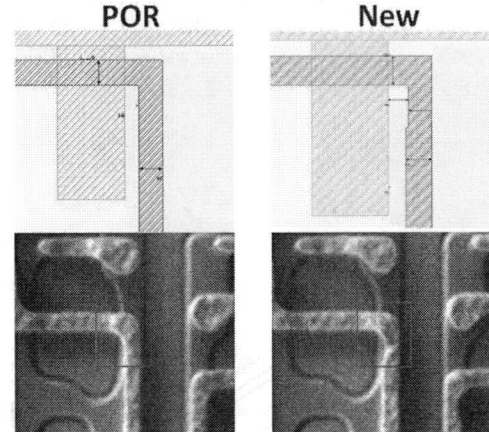

Fig14, L-shape layout optimization

IMPROVEMENT ACTION RESULTS

The experiment result was very positive. The d1 was enlarged as the SEM images (Fig14). And the deep power down leakage failure ratio was improved from 20% to less than 2% (Fig15). The media value of the leakage was improved and the distribution was much improved (Fig16).

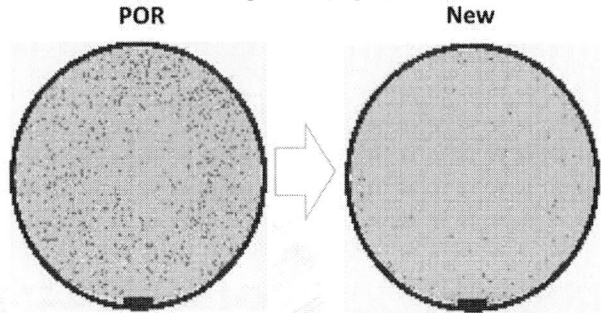

Fig15, POR and new mask deep power down leakage failure comparison

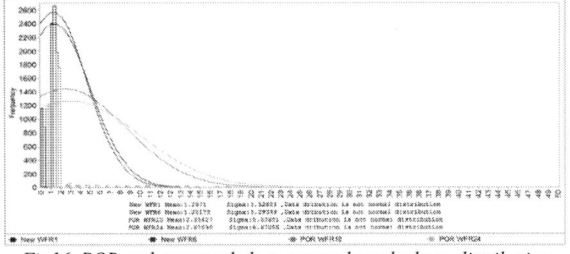

Fig16, POR and new mask deep power down leakage distribution comparison

ACKNOWLEDGEMENTS

In this paper, one device leakage model caused by poly layout was studied. We can conclude that:

1, The L-shape poly layout will result pocket implant shadowing effect.

2, The L-shape shadowing effect will result channel leakage and impact the whole chip leakage performance.

3, The "d" of the L-shape poly layout is the key effect.

4, Wide channel width will reduce the L-shape shadowing

effect.

5, Pocket implant optimization or poly layout optimization can reduce the L-shape shadowing effect.

REFERENCES

[1] K. Yoneda IEEE DOI:10.1109/IWJT.2002.1225190
[2] D. Lenoble IEEE DOI: 10.1109/IIT.2002.1257934
[3] T.Hori et al., IEDM'94 Tech.Dig.,P.75

VIRTUAL SOURCE FOR AN ODD MATHIEU-GAUSS BEAM AND COMPARE OF THE FUNCTIONAL IMAGES OF THE ODD AND EVEN

Xuxin Qi[1*], Xuxin Qi[1]

[1]School of Science, Harbin Institute of Technology, Harbin 150000, China

*Corresponding Author's Email: qixuxin@hlmc.cn

ABSTRACT

Based on the beam superposition, we identify a group of virtual sources for generating a 2n+2-th order Mathieu-Gauss beam with odd modes. The integral and differential representations for the Mathieu-Gauss wave are derived,which, in some conditions, yields the corresponding paraxial Mathieu-Gauss beam.Based on the same theory, we can get the integral and differential representations and the functional images of the 2n+1-th order Mathieu-Gauss beam with odd modes, the 2n-th order Mathieu-Gauss beam with even modes and the 2n+1-th order Mathieu-Gauss beam with even modes. Then compare the four kinds of functions.

INTRODUCTION

The conception of nondiffracting beams[1,2] was put forward by Durnin from Rochester University in 1987. Bessel beams, as one of the most important singular beams, are one of nondiffracting beams. On the basis of the nature of nondiffracting beams, the research team of Gutiérrez–Vega J C. enumerated four types of fundamental mode solutions to the paraxial wave equation, corresponding to four types of nondiffracting beams. These fundamental mode solutions are cosine beams, parabolic beams , Mathieu beams, and Bessel beams, they are obtained by solving the paraxial wave equation in circular,parabolic, elliptic, and space rectangular coordinates, respectively[3]. In the theory of nonparaxial, the virtual source technology was put forward to solve Helmholtz equation[4,5,6].

We mainly research Mathieu-Gauss beam , which is an approximate nondiffracting beam. Since, we identify a group of virtual sources for generating a 2n+2-th order Mathieu-Gauss beam with odd modes. The integral and differential representations for the Mathieu-Gauss wave are derived, which,in some conditions, yields the corresponding paraxial Mathieu-Gauss beam.Based on the same theory, we can get the integral and differential representations and the functional images of the 2n+1-th order Mathieu-Gauss beam with odd modes, the 2n-th order Mathieu-Gauss beam with even modes and the 2n+1-th order Mathieu-Gauss beam with even modes. Then compare the four kinds of functions and find the connection of these functional images.

VIRTUAL SOURCE FOR AN ODD MATHIEU-GAUSS BEAM

In order to get the odd Mathieu-Gauss beam,we suppose that $U(p,y,z)$ (n,j=0,1,2…) is a monochromatic scalar wave function that describes a 2nth-order odd Mathieu–Gauss wave propagating along the positive z-axis in circular cylindrical coordinates,when z=0,

$$U_{2n+2}(p,y,z=0)=A(q)(-1)^{j+1}sin[(2j+2)y]J_{2j+2}(By)exp(-y^2/w_0^2) \ (j=0,1,2…;A(q)=2j+2…2n+2) \quad (1)$$

in circular cylindrical coordinates, several circular loops of current strength $S_{ex}(2j+2)$, which situated in $z=z_{ex}$, and the radius is $p=p_{ex}$, in the space of z>0,we can get the 2n+2th-order odd Mathieu-Gauss beam.Supposing the source beam equation is

$$U_{2n+2,2j+2}(p,y,z)=E_{2n+2,2j+2}(y,z)sin[(2j+2)y] \quad (2)$$

Under the tradition of Eq.(1), Eq.(2) is satisfied with nonhomogeneous Helmholtz equation

$$(d^2/dp^2+d/pdp-(2j+2)^2/p^2+d^2/dz^2+k^2)U_{2n+2,2j+2}(p,z)=-A(q)Sex[d(p-p_{ex})/p]d(z-z_{ex}) \ (A(q)=2j+2…2n+2) \quad (3)$$

By use of Bessel transform pair

$$E_{2n+2,2j+2}(p,z)=J_{2j+2}(pp)U_{2n+2,2j+2} \ (pp)pdp \ (j=0,1,2…) \quad (4)$$

$$U_{2n+2,2j+2} \ (pz)=J_{2j+2}(pp) \ E_{2n+2,2j+2}(p,z) \ pdp \ (j=0,1,2…) \quad (5)$$

According to Eqs.(3) (4) (5)

$$E_{2n+2,2j+2}(p,z)=A(q)pdpJ_{2j+2}(pp)J_{2j+2}(pp_{ex})Sex(2j+2)^i/2gexp[i \ g(z-z_{ex})] \ (j=0,1…;A(q)=2j+2…2n+2;p=0,1…) \quad (6)$$

For Re(z−z$_{ex}$)>0, where $g=(k^2-p^2)^{1/2}$, if $p^2<< k^2$, g is expanded for small p^2, $g=k-p^2/2k$ retain the leading term for the amplitude factor and the first two terms, $k-p^2/2k$ for the phase factor

$$E_{2n+2,2j+2}(p,z)=A(q)exp(ikz)[iSex(2j+2)/2kexp(-ikz_{ex})] \ pdp \ J_{2j+2}(pp)J_{2j+2}(pp_{ex})exp[-ip^2/2k(z-z_{ex})] \ (j=0,1…;A(q)=2j+2…2n+2;p=0,1…) \quad (7)$$

Yield

$$E_{2n+2,2j+2}(p,z)=A(q)exp(ikz)[iSex(2j+2)/2kexp(-ikz_{ex})]exp(-3j+1)\Pi i(-ik/z-z_{ex})exp[ikp_{ex}/2(z-z_{ex})]exp[ikp^2/2k(z-z_{ex})] \ J_{2j+2}(-kp_{ex}p/z-z_{ex}) \ (j=0,1…;A(q)=2j+2…2n+2;p=0,1…) \quad (8)$$

For z>0 , in order to get Mathieu-Gauss beam and satisfy z=0at its boundary ,in the paraxial approximation the input distribution is assumed to be

$$E_{2n+2}(p,0)=exp(r)exp(-p^2/w_0^2)J_{2j+2}(Bp) \quad (9)$$

$$z_{ex}=ikw_0^2/2=ia \quad (10)$$

$$p_{ex}=iBw_0^2/2=ib \quad (11)$$

$$Sex(2j+2)=(-1)^{j+1}(-2ia)exp(-ka)exp[3(j+1)\Pi i]exp(-Bw_0^2/4) \tag{12}$$

When z_{ex}, p_{ex}, Sex are substituted in Eqs.(6) (8) yield

$$E_{2n+2,2j+2}(p,z)=A(q)(-1)^{j+1}aexp(-ka)exp[3(j+1)\Pi i]exp(-Bw^2/4)pdpJ_{2j+2}(pp)J_{2j+2}(pib)exp[ig(z-ia)/g] \tag{13}$$

$$E_{2n+2,2j+2,p}(p,z)=A(q)(-1)^{j+1}exp[ikz/(1+iz/a)]exp(-B^2w_0^2/4)exp[B^2w^2/4(1+iz/a)]exp[-p^2/w_0^2(1+iz/a)]J_{2j+2}(Bp/1+iz/a) \tag{14}$$

We can get the exact equation of 2n+2th-order odd Mathieu-Guass beam

$$E_{2n+2,2j+2}(p,z)=A(q)(-1)^{j+1}aexp(-ka)exp[3(j+1)\Pi i]exp(-B^2w^2/4)pdpJ_{2j+2}(pp)J_{2j+2}(pib)exp[ig(z-ia)]/g$$
$$sin[(2j+2)$$
$$y_{ex}] \quad (j=0,1\ldots;A(q)=2j+2\ldots2n+2;p=0,1\ldots) \tag{15}$$

Establish the differential equation of nonhomogeneous Helmholtz

$$(d^2/dp^2+d/pdp-(2j+2)^2/p^2+d^2/dz^2+k^2)G(p,z)=-Sex(2j+2)$$
$$[d(p-p_{ex})/2\Pi p]d(z-z_{ex}) \tag{16}$$

Is given by

$$G(p,z)=Sex(2j+2)exp(ikR/4\Pi R) \tag{17}$$

$$R=[(p-p_{ex})^2+(z-z_{ex})^2]^{1/2} \tag{18}$$

Both sides of Eq.(13)are multiplied by $sin[(2j+2)y_{ex}$, and integrated with respect to y_{ex} from 0 to 2Π, when z_{ex}, p_{ex}, Sex are substituted in, we can obtain the equation of 2n+2th-order odd Mathieu-Gauss beam

$$E_{2n+2,2j+2}(p,y,z)=A(q)(-1)^{j+1}(ia/2\Pi)exp(-ka)exp[3(j+1)\Pi i]exp(-B^2w_0^2/4)dy_{ex}sin[(2j+2)y_{ex}]exp(ikR)/R$$
$$(j=0,1\ldots;A(q)=2j+2\ldots2n+2; y_{ex}=0\ldots2\Pi) \tag{19}$$

Where

$$R=[p^2-b^2-i2pbsin(y_{ex}-y)+(z-ia)^2]^{1/2} \tag{20}$$

According to Helmholtz equation,for we can obtain the follow equation

$$E_{2n+2,2j+2}(p=0,z)=A(q)(-1)^{j+1}aexp(-ka)exp[3(j+1)\Pi i]exp(-B^2w^2/4)pdpJ_{2j+2}(pib)exp[ig(z-ia)]/g$$
$$(j=0,1\ldots;A(q)=2j+2\ldots2n+2;p=0,1\ldots) \tag{21}$$

COMPARE OF THE FUNCTIONAL IMAGES OF THE ODD AND EVEN

Based on the same theory,we can obtain the equation of 2n+1th-order odd Mathieu-Gauss beam

$$uc_{2n+1}(p,y,q,z,t)=2\Pi exp(ik_zz-iwt)A(q)(i)^{2j+1}sin[(2j+2)y]J_{2j}(ktp) \quad (j=0,1\ldots;A(q)=2j+2\ldots2n+2) \tag{22}$$

the exact equation of 2n+1th-order odd M athieu-Guass beam

$$E_{2n+2,2j+2}(p,z)=A(q)(-1)^{2j+1}aexp(-ka)exp[3(j+1)\Pi i/2]exp(-B^2w^2/4)pdpJ_{2j+2}(pib)exp[ig(z-ia)]/g\ sin[(2j+2)y]$$
$$(j=0,1\ldots;A(q)=2j+2\ldots2n+2;p=0,1\ldots) \tag{23}$$

The differential equation of 2n+1th-order odd Mathieu-Gauss beam is

$$E_{2n+2,2j+2}(p,y,z)=A(q)(-1)^{2j+1}ia/\Pi exp(-ka)\ exp[3(j+1)\Pi i/2]exp(-B^2w^2/4)dy_{ex}\ sin[(2j+2)y_{ex}]\ exp(ikR)/R$$
$$(j=0,1\ldots;A(q)=2j+2\ldots2n+2; y_{ex}=0\ldots2\Pi) \tag{24}$$

Where

$$R=[p^2-b^2-i2pbsin(y_{ex}-y)+(z-ia)^2]^{1/2} \tag{25}$$

The equation of 2nth-order even Mathieu-Gauss beam is

$$uc_{2n}(p,y,q,z,t)=2\Pi exp(ik_zz-iwt)A(q)(-i)^jcos(2jy_{ex})J_{2j}(Bp)$$
$$(j=0,1\ldots;A(q)=2j+1\ldots2n+1) \tag{26}$$

the exact equation of 2nth-order even Mathieu-Guass beam is

$$E_{2n,2j}(p,z)=A(q)(-1)^jaexp(-ka)exp(3j\Pi i)exp(-B^2w^2/4)pdpJ_{2j}(pib)exp[ig(z-ia)]/g\ cos(2jy)$$
$$(j=0,1\ldots; A(q)=2j\ldots2n; p=0,1\ldots) \tag{27}$$

The differential equation of 2nth-order even Mathieu-Gauss beam is

$$E_{2n,2j}(p,y,z)=-A(q)(-1)^jia/\Pi exp(-ka)\ exp(3j\Pi i)exp(-B^2w_0^2/4)dy_{ex}\ cos(2jy_{ex})\ exp(ikR)/R$$
$$(j=0,1\ldots;A(q)=2j\ldots2n; y_{ex}=0\ldots2\Pi) \tag{28}$$

Where

$$R=[p^2-b^2-i2pbcos(y_{ex}-y)+(z-ia)^2]^{1/2} \tag{29}$$

Then compare with the Eqs. (15), (23) and (27), we can get some conclusions as follows

$$E_{2n+2,2j+2}(p,z)=A(q)(-1)^{j+1}aexp(-ka)exp[3(j+1)\Pi i]exp(-B^2w^2/4)pdpJ_{2j+2}(pib)exp[ig(z-ia)]/g\ sin[(2j+2)y_{ex}]=$$
$$E_{2n+2,2j+2}(p,z)=-A(q)(-1)^jaexp(-ka)exp(3j\Pi i)\ exp(3\Pi i)exp(-B^2w^2/4)pdpJ_{2j+2}(pb)J_{2j+2}(pib)\ exp[ig(z-ia)]/g\ sin[(2j+2)y_{ex}] \tag{30}$$

CONCLUSION

1. For odd Mathieu-Gauss beam,when the functional image of 2n+1th-order move parallel towards the left to1/2 unit,which is in conformity with the functional image of 2n+2th-order if j is even number.

2. For even Mathieu-Gauss beam,when the functional image of 2n+1th-order move parallel towards the right to1/2 unit,which is in conformity with the functional image of 2nth-order if j is odd number.

3. For j is even number,when the functional image of 2nth-order even Mathieu-Guass beam move parallel towards the right to $\Pi/4y$, unit,which is in conformity with the monotone section of the functional image of 2n+2th-order odd Mathieu-Gauss beam if j is odd number.

4. For j is odd number,when the functional image of 2nth-order even Mathieu-Guass beam move parallel towards the right to $\Pi/4y$, unit,which is in conformity with the monotone section of the functional image of

2n+2th-order odd Mathieu-Gauss beam if j is even number.

5. when the functional image of 2n+1th-order even Mathieu-Guass beam move parallel towards the right to $\Pi/4y$, unit,which is in conformity with the monotone section of the functional image of 2n+1th-order odd Mathieu-Gauss beam.

ACKNOWLEDGEMENTS

Thanks to my teachers, classmates and collages who have helped me. Without your help, I would not be what I am today. Thanks for your help in my work and study, and I will keep working hard.

REFERENCES

[1] Durnin J, Miceli Jr J J, Eberly J H. Diffraction-free beams[J]. Physical Review Letters, 1987, 58(15): 1499.

[2] Durnin J. Exact solutions for nondiffracting beams. I. The scalar theory[J]. JOSA A, 1987, 4(4): 651-654.

[3] Gutiérrez-Vega J C, Iturbe-Castillo M D, Chávez-Cerda S. Alternative formulation for invariant optical fields: Mathieu beams[J]. Optics letters, 2000, 25(20): 1493-1495.

[4] G. A. Deschamps, Electron. Lett. 7, 684 (1971).

[5] L. B. Felsen, J. Opt. Soc. Am. 66, 751 (1976).

[6] S. Y. Shin and L. B. Felsen, J. Opt. Soc. Am. 67, 699 (1977).

[7] A.Chafiq,Z.Hricha,A.Belafhal,Opt.Soc.Commum.253 (2005)223.

[8] M.Couture and P.A.Belanger,Phys.Rev.A24,355-359(1981).

[9] S. R.Seshadri.Virtual source for the Bessel-Gauss beam.Optical letters,June 15,2002.

DETECTION OF ELECTRICAL DEFECTS BY DISTINGUISH METHODOLOGY USING AN ADVANCED E-BEAM INSPECTION SYSTEM

Shanshan Chen*, Hunglin Chen, Yin Long, Fengjia Pan, and Wang Kai

Shanghai Huali Microelectronics Corporation, Shanghai 201314, China

*Corresponding Author's Email: chenshanshan@hlmc.cn

ABSTRACT

With critical dimension shrinks during semiconductor process development, E-beam inspection (EBI) technique has play a vital role in detecting inline electrical defect by voltage contrast (VC). This study we introduce three different defect monitoring for 28nm process. The first is Cell to Cell inspection, which relies on comparing gray level differences between the defect site and adjacent sites or a reference image. However, while pixel size and grey level differences are small enough that defect is not easy to be detected as device shrink beyond, hot spot and die to database (D2DB) inspection that can help to distinguish true defects from a large amount of false alarm defects. These inspections provide timely and high efficiency feedback for health of line monitoring and yield improvement.

INTRODUCTION

With increasing processing capacity needs for high-volume manufacturing, multiple tool sets are required for in-line process. Optic-based inspections such as Bright Field Inspection (BFI) and Dark Field Inspection (DFI) only use laser or light to illuminate interesting regions and detect scattered or reflected light to capture physical defect which is major on the surface, which are undergoing challenges to penetrate inside metal beyond skin depth. [1] Therefore, EBI can capture electrical failure such as device leakage, contact open and circuit short by voltage contrast effect. [2]

Cell to cell inspection is a widespread detection method to capture defective optical, SEM or patch image in inspecting location and then comparing with the periodic location's reference image in real time.

A focus has been on detection failures in locations, which has been predicted to be trouble spots with pattern simulation tools. T. Luoh et al.[3] emphasized that hot spot inspection with small 1536x1536 inspection field under 7nm pixel resolution shows no misalignment issue as compared to leap & scan in the large inspection field 6144x6144 under pixel 40nm. For Hot spot, EBI can inspect the sites with exactly same (x, y) coordinates in each die due to its high precision alignment performance. T. Kitamura et al. [4] in 2005 first presented that D2DB inspection methodology can compare the design layout with the patterned images on the wafer by using the supernova computer system. The recording process can be executed as bellow: the first step is to get the high quality images at the care areas by study condition parameters and then do the alignment between images and design layout; The second step is to do the pixel counting according the image grayscale differences, then transfer the image contour into GDSII format. The final step is to gauge and report the contour profile error between image contour and design target. [5-11].

In this work, we present a classification methodology comparison of Cell to Cell and Die to Database. Cell-Cell inspection relies on the differentiation of gray level between the defect site and adjacent sites or a reference image. While pixel size and grey level differences are too small to compare as device shrink to 28 nm and beyond, it is not easy to detect defects. By overlaying the E-beam defect location onto the design layout file, D2DB E-beam inspection that helps to distinguish true defects from a large amount of false alarm defects.

RESULT AND DISCUSSION

Cell to Cell inspection

We usually only sweep SRAM area due to production capacity by using EBI. Typical SRAM are built with a conventional six transistor cell. Two inverters are used to form a latch consisting of four transistors (2NMOS, 2PMOS). Two more NMOS transistors serve as transfer gates to isolate or connect the cell to the two bit-lines. The transfer gates are controlled by a wordline, which allow the state of the latch to be read, or changed, and the same wordline is used to control both transfer gates of any given cell. These particular contacts are FET node contacts, while they merge, it result in single cell failures. Fig. 1 (a) and (b) shows contact structure image of 6T SRAM. (c) shows the basic layout of SRAM and Fig. 2. shows the inspection mode.

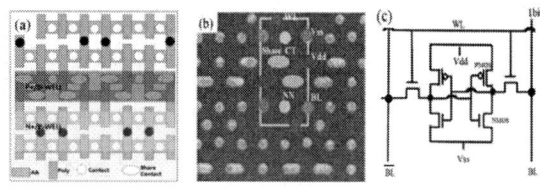

Figure 1: (a) (b) contact structure image of 6T SRAM; (c) the basic layout of SRAM

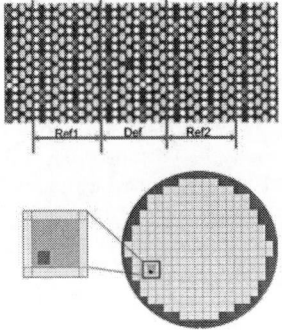

Figure 2: Cell-Cell inspection mode

Contact Defect Open Analysis

Tungsten (W) is commonly used to fill the tiny contact holes to form conducting plugs that connect metal wires to silicide on active areas of silicon such as source/drain and polysilicon gate electrodes. In order to interpret the resistive open result, a basic discussion of electron beam inspection detection mechanism is provided. Voltage Contrast (VC) inspection is always used to examine electrical health of product-like SRAM structures. Electrically open circuits result in dark VC have some different root causes such as incomplete hard mask opening, hollow strap connection, deep trench capacitors in the wafer substrate, and so on. The moderate landing energy (LE) of the primary electron beam can be controlled by the total bias between electron gun and wafer. While more electrons leave the surface than reach it, thus inducing positive charge on the surface, which is called "positive mode". Case study from inline etch process, the depth-width ratio of contact void would be influenced by the etching rate and time. For different process window, EBI can monitor the defect situation easily as shown in Figure3. We utilize focus ion beam (FIB) technique to prepare for transmission electron microscope (TEM) cross-section sample in order to analyze root cause of the electrical failure. The TEM image shows that the bottom of the CT is under etching. In order to reduce the Dark VC, contact etch process recipe optimization was conducted. By applying the optimized etching recipe, we were able to eliminate the CT open on production lot, and significantly improve electrical test failure as shown in Fig. 3.

After the split of photo, etch and CMP, the CD of contact is in its target, which is conducive of tungsten filling, and defect removed, as shown in Fig. 4.

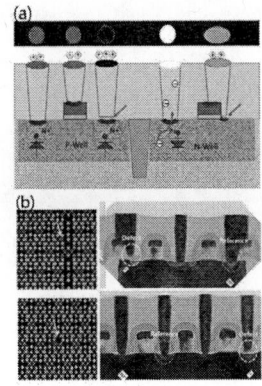

Figure 3: (a) Structure and (b) Cross-Sectional TEM image of CT open

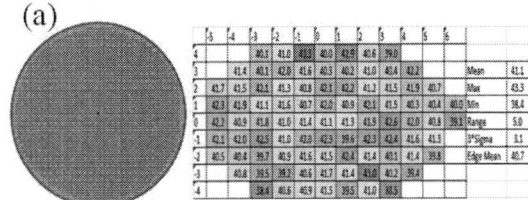

Figure 4: Wafer Map and Contact CD Map

Hot spot inspection

While the failures in locations has been predicted to be trouble spots with pattern simulation tools, EBI is particularly expert in this by using hot spot mode, and is also being able to inspect full chips for a small number of reticle fields to search for additional patterning problem areas. For example, we can frame part of the area in logic to inspect while we judge which maybe objective. The cause of hot spot maybe various, such as shorter or longer film deposition times for CVD or etch process, the partial etch of via, or resist type or parameters. So while defect is caused due to overlay offset, it may be necessary to adjust the OPC keyword or order new mask[9-11].

For the BEOL, via partial etch or mask offset, which may lead to metal poor connection, are key concerns for a recent technology particularly for the all in one layer, becoming a challenging process window. Metal pinching is best detected after metal liner whereas Via open or partial open maybe be best detected after etch.

Figure 5 shows the defect breaks in the trench dielectric which generally occur right or left below the Dielectric layer, and cannot be detected optically. In order to enhance the defect signal, we use a back scattered electron (BSE) mode to detect, which require high enough landing energy to penetrate the metal layer to reach the target defect and still has enough energy to bounce back to the sample surface and reach the detector. This mode enhance the brightness of copper, and allowing challenge defect to more easily be detected as shown by Fig. 5[9,10,11].

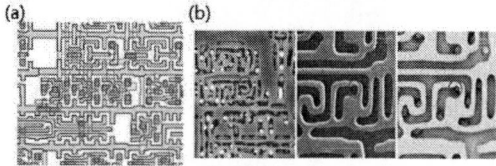

Figure 5: Hot spot inspection in Logic

Die to database inspection

A Die to Database (D2DB) system called Supernova has been developed for the Hermes Microvision Inc. eScan500 E-beam inspection system. The system is capable of recording the pixelated image of a very large area for a wafer while just ticking in the running recipe. In the production progress, lot is impossible to wait long time at each step site, so it is very important to collect the entire wafer image with supernova combine with parallel computer.

D2DB inspection not only can help engineer to seek the contact/via hole diameter process window verification but also can get the CD uniformity distribution. Just like other image processing tasks, Supernova D2DB image processing is an idle application for parallel processing. Firstly, since each region can be processed independently, which help parallel processing not that difficult to implement. Secondly, the data analysis is time intensive so a method for collecting the data like this is worth the expense. The D2DB Image Process Flow is as shown in the Fig. 6.

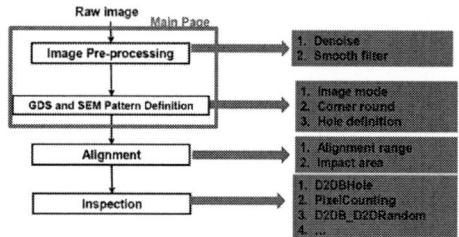

Figure 6: The process of D2DB flow

In addition, the system provides multiple outputs. One basic output is a wafer or die map showing the locations of the defects or the number of defects per die. A second output is a defect list which contains the die coordinates, type and other information for each defect along with an image. An engineer can visit the defect locations in the design and simulation result by way of the list, such as the image of Fig. 7. In other words, for D2DB inspection, a function called layer binning, which can classify defects into bins associated with layers defined in the hierarchical structure of GDS[5-11].

CONCLUSION

In this study, we illustrate different application by EBI which not only just distinguishes leakage and open issue in SRAM but truly provides more information of defect detection analysis during inline process. Three methodologies were used to illustrate this point. The first instance showed that scanning SRAM by Cell-Cell results in a much stronger VC signal was able to detect the connection of the bottom of contact. The second example shows that scanning logic area by hotspot makes detection of via partial etch in logic easily. The third suggest that for D2DB technique, Superior resolution from e-beam inspection together with capability of handling GDS information provides sufficient defect detection sensitivities and automatic defect binning capabilities. These are few examples demonstrate that as device shrink beyond, electron beam inspection has become increasingly important for integrated circuit manufacturing process monitoring and characterization.

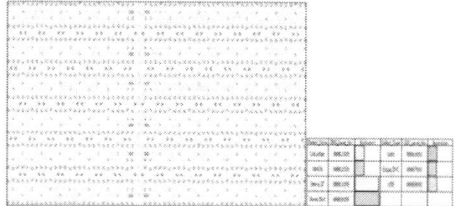

Figure 7: The process of D2DB flow

REFERENCE

[1] O. D. Oliver et al., Proc. of SPIE 2012, Vol. 8324,.
[2] S. Halle et al., SEMICON 2013, Taiwan.
[3] Tuung Luoh, et al. (2013).
[4] Tadashi Kitamura et al., Proc. of SPIE, San Jose, California, USA, 5756 (2005).
[5] Kunal N. Taravade et al., Proc. of SPIE vol. 5256, 2003.
[6] Hyunjo Yang et al., Proc. of SPIE vol. 5375, 2004.
[7] Hyunjo Yang et al., Proc. of SPIE vol. 5752, 2005.
[8] Tadashi Kitamura et al., Proc. of SPIE vol. 5756, 2005.
[9] O. D. Oliver et al., ASMC 2013, p 296-300.
[10] Ling-Wuu Yang et al., 2018, p 1-4.
[11] Philippe Leray et al., Proc. of SPIE Vol. 9778 977800-7.

STUDY OF ULTRA-HIGH VOLTAGE BCD PROCESS WITH GATE OXIDE THINNING

Donghua Liu[1], Wenting Duan[1], Ziquan Fang[1], Wensheng Qian[1]*

[1] HuaHong Grace Semiconductor Manufacturing Corporation, Shanghai 201206, China

*Corresponding Author's Email: donghua.liu@hhgrace.com

ABSTRACT

High voltage gate driver ICs, which have the advantages of short response time, low power consumption, high integration and high reliability, are widely used in motor drives, automotive electronics, electronic ballasts, switching mode power supplies and other fields.

Due to the applied voltage usually above 500V, the gate oxide thickness of the device is thick, usually higher than 400A. This paper investigates the possibility of using a thinned gate oxide thickness in an ultra-high voltage BCD process. With a thinner gate oxide, lower threshold voltage, higher saturation current and lower on-resistance can be achieved. But the side-effect is that the breakdown voltage of the device will decrease. This paper also studies how to improve the breakdown voltage of the device after using thinner gate oxide, and positive progress has been made.

INTRODUCTION

Paper 01:

1 Introduction

The bipolar CMOS DMOS(BCD)process，which was introduced twenty years ago，is an important mixed technology that allows the integration

On a single chip of bipolar linear，CMOS logic，and double diffusion MOSFET power functions. In recent years, the multiple discrete components in high voltage IC applications have been reduced through their integration into a BCD process. The integration of discrete elements provide improved performance, increased functionality, enhanced reliability, and compact solutions for applications in consumer, automotive, motor control, switched mode power supplies and medical use[1].

High voltage gate driver ICs, which have the advantages of short response time, low power consumption, high integration and high reliability, are widely used in motor drives, automotive electronics, electronic ballasts, switching mode power supplies and other fields[2]. High voltage LDMOS is used as level shift transistors and high voltage isolation region from high side to low side devices which usually account for a small proportion of the total area of high voltage gate driver ICs. The conventional high voltage gate drive ICs are mostly based on the thick epitaxial technology and have slightly larger feature size, resulting large area for high side and low side CMOS circuits.

Energy-saving lamps: In the prior art, two power MOSFETs and control ICs for oscillation have been made in a plastic package, making the circuit simpler. Almost half of the electricity consumed in the United States is in lighting and motors, so energy conservation in lighting is very important. It is estimated that each energy-saving lamp can save $ 30 to $ 70 per year. In a situation where the market for bipolar transistors is shrinking, China's market is growing rapidly due to the need for energy-saving lamps.

Automotive electronics: In modern cars, in order to improve engine performance and provide a safer and more comfortable environment, many parts require semiconductor devices (the power devices in each car can reach as many as 200). Such as engine control unit: fuel injection controller, ignition control, transmission, etc.; Comfortable applicable components: power steering wheel, power window, power door lock, lighting control, air conditioning, audio, etc.; in addition to safety requirements. So China's potential automotive market provides a lot of room for the development of power semiconductors[3].

In the current high-voltage process, the gate oxide thickness is 430A. In theory, the gate oxide thickness has great effect on Vt and Idsat.

The following formula shows that Id and Vd have a parabolic relationship.

$$I_{Dsat} = \beta \left[(V_{GS} - V_T)V_{Dsat} - \frac{1}{2}V_{Dsat}^2 \right] = \frac{1}{2}\beta \left(V_{GS} - V_T \right)^2$$

$$V_{Dsat} = V_{GS} - V_T$$

$$\beta = \frac{Z}{L}\mu_p C_{ox}$$

DEVICE STRUCTURE

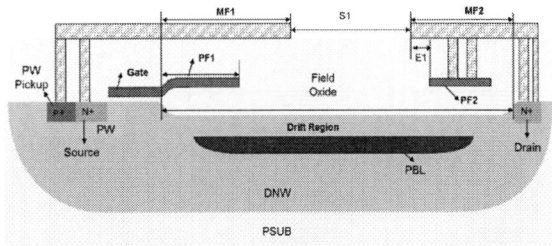

Figure 1: Schematic cross section of ultra-high voltage NLDMOS

The schematic cross section of 500V NLDMOS device is showed in Figure 1. When drain terminal is biased at operation voltage (Vd), the voltages sustaining in lateral direction and vertical direction are the same. The breakdown voltage of both directions needs to be higher than the operation voltage and at least have 10% margin. Fig.1 shows that the drift region of NLDMOS embraces deep n-type well (DNW) and p-type buried layer (PBL). The vertical bias of Vd is mostly dropped on the PN diode formed by DNW and p-type substrate (PSUB) in the vertical direction. Its breakdown is higher than 1000V and can be used in the application of 500V NLDMOS. The drift region with DNW and PBL is designed considering RESURF effect. Drift region is fully depleted and sustains the lateral voltage when drain is biased. The doping condition of the drift region is determinate due to the technology platform and will not be discussed here. P-type well (PW) and p-type heavily doped region (P+) form the channel region. N-type heavily doped region (N+) forms the source and drain. Above the silicon, there are poly gate, polysilicon field plate 1 (PF1, closed to source side), polysilicon field plate 2 (PF2, closed to drain side), metal field plate 1 (MF1, source side), and metal field plate 2 (MF2, drain side). S1 is the space between MF1 and MF2. E1 represents the extension size of MF2 over PF2[4-6].

RESULT AND DISCUSION

Low Voltage Device

When gate oxide was thinned from 430A to 155A, the characteristics of the device changed greatly. As shown in Table 1. The first is that under the same channel length, the threshold voltage Vt can be adjusted to about 0.82 volts to ensure that the leakage of the device is kept within specifications. The biggest change in the device is that the saturation drive current has nearly doubled, from 299uA / um to 663uA / um; the second is the breakdown voltage, which has been reduced from 15.1 volts to 10.9 volts as shown in figure 2.

Table I. PBL Enlarged 1um to Channel Side

Lch/um	1	0.8	0.8
Gox/A	430	430	155
Vt/V	1	0.964	0.821
Idsat/uA/um	NA	299	663
BV/V	15.1	15.1	10.6

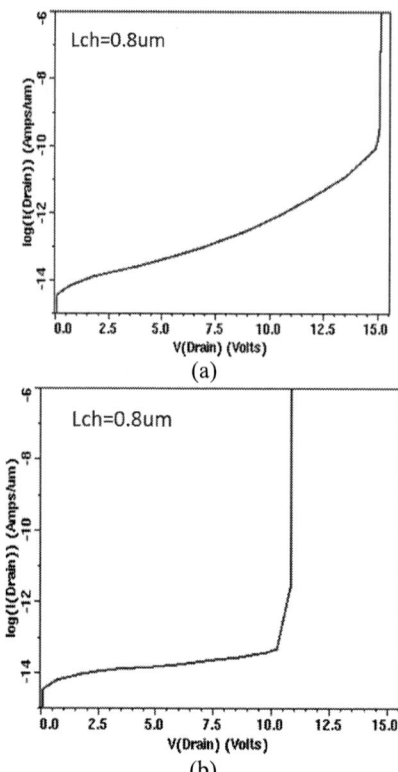

(a)

(b)

Figure 2: Breakdown voltage MOSFET with 430A and 155A gate oxide

Figure 3 shows the impact ionization distribution of devices with two structures when breakdown occurs. The left and right figures correspond to 430A and 155A, respectively. It can be seen that device breakdown occurs in the LDD region under the polysilicon gate.

The breakdown voltage of this device is finally tuned back to16V as shown in figure 4.

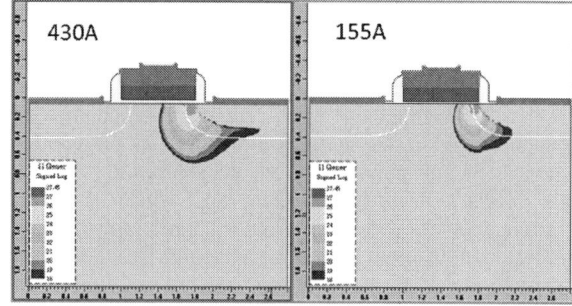

Figure 3: Impact ionization distribution of MOSFET with 430A and 155A gate oxide

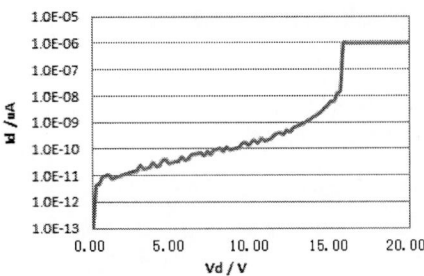

Figure 4: Breakdown voltage curve.

Ultrahigh Voltage Device

For ultra-high voltage devices, when the gate oxide is thinned, the first problem encountered is that the threshold voltage is reduced and the leakage current becomes large. However, the breakdown voltage of the device is not affected, as shown in figure 5.

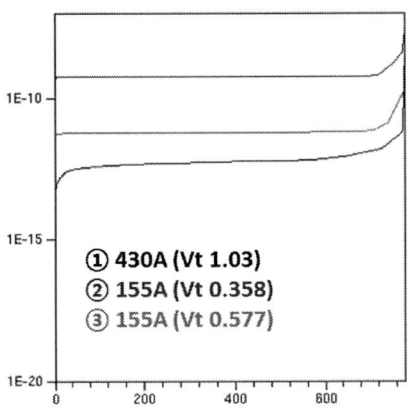

Figure 5: Breakdown voltage ultrahigh voltage device with 430A and 155A gate oxide.

Figure 6 shows the lateral electric field distribution in the breakdown of ultra-high voltage devices. From the figure we can see that the thinning of the gate oxide results in a stronger electric field at the bird's beak. Figure 7 is the longitudinal electric field distribution at the bird's beak, which further confirms the electric field change at that place.

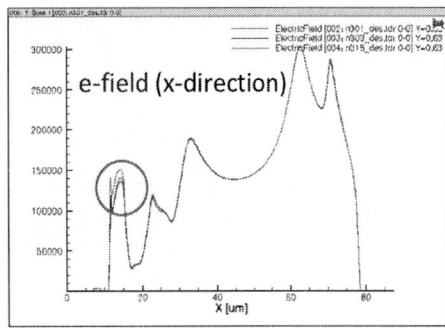

Figure 6: Electric field distribution of x direction.

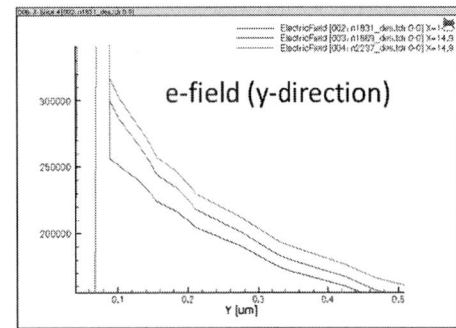

Figure 7: Electric field distribution of y direction.

Figure 8 shows a method for further optimizing the breakdown voltage of an ultra-high voltage device. Increasing or decreasing the concentration of PBL can effectively affect the breakdown voltage of the device. The simulation results show that the breakdown voltage is increased from 705 volts to 805V.

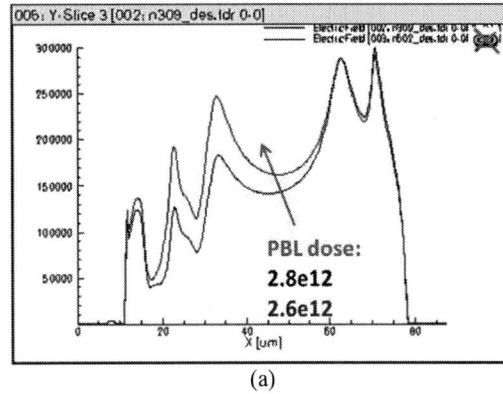

(a)

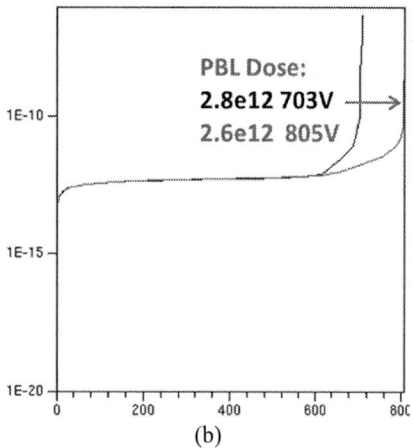

(b)

Figure 8: Electric field distribution (a) and breakdown voltage (b) of ultrahigh voltage device with different PBL dose.

CONCLUSIONS

In the ultra-high voltage BCD process, the gate oxide industry of the device uses a relatively thick 430A. This article first uses 155A thin gate oxide in this process. And through the final process optimization, the negative effects of thinning of gate oxide have been overcome, so that the device of this process platform has improved the current drive area, the breakdown voltage will not decrease, and the customer's chip area will eventually be reduced.

ACKNOWLEDGEMENTS

The authors would like to thank all members of device group and high voltage group of HHgrace for their great support on this work.

REFERENCES

[1] Qiao Ming, et al. *A High Voltage BCD Process Using Thin Epitaxial Technology* [J]. Chinese Journal of Semiconductors, 2007, V28 p1742-1747

[2] H. H. Wang, et al. A 0.35μm 600V Ultra-Thin Epitaxial BCD Technology for High Voltage Gate Driver IC[C], ISPSD, 2018

[3] K. Mao, M. Qiao, L. Jiang. *A 0.35um 700 V BCD Technology With Self-isolated and Non-isolated Ultra-low Specific On-resistance, DB-Nldmos.* IEEE ISPSD, 2013, pp. 11-13

[4] D. H. Liu, W. T. Duan, W. S. Qian. *A Novel 25V PLDMOS Design in 700V BCD Process [C].* IEEE Conference Publications of CSTIC 2017, 2017.

[5] W. T. Duan, D. H. Liu, et al. *Study of Ultra High Voltage 500V NLDMOS with Aggressive Design of Drift Region* [C]. IEEE Conference Publications of CSTIC 2019.

[6] D. H. Liu, X. M. Xu, , et al. *The Investigation of Field Plate Design in 500V High Voltage NLDMOS [J].* Advances in Condensed Matter Physics, 2015, pp. 1-6

THE METHOD OF IMPROVING ALD SICN FILM UNIFORMITY

Yanxia Hao*, Junlong Kang, Jun Yin, Yinshuai Wang, Guangyu Nie

Xinhua Cheng, Jingxun Fang

*Shanghai Huali Integrated Circuit Corporation, Shanghai 201210, China

6 Liangteng Rd., Pudong district, Shanghai, China

haoyanxia@hlmc.cn

ABSTRACT

With the development of integrated circuits, smaller field effect transistor size has higher device sensitivity on spacer thickness. Furnace film ALD (Atomic layer deposition method) SICN has been widely used as offset spacer due to its high anti-Phosphoric acid corrosion and good uniformity. In this paper, a series of process parameters such as boat rotate speed, gas flow are optimized to achieve acceptable with in wafer and wafer to wafer thickness uniformity. The formation mechanisms and the proper solutions are discussed as well.

Keywords—uniformity, boat rotate speed, gas flow

INTRODUCTION

In the logic 28/14nm CMOS technology node, low-k spacer thickness plays an important role in defining the IMP distance between LDD and source/drain. [1] As field effect transistor size getting smaller and smaller, the film uniformity with in wafer and wafer to wafer now becomes especially crucial in the semiconductor manufacturing. Compared to the Single-wafer job mode of chamber, furnace can realize the batch operation by processing several productions once. [2]. However, Considering of the setting rule of the temperature zone and gas injector in different position, the temperature and gas flow between each wafer are different, which results in the difference in thickness and uniformity with in wafer and wafer to wafer.. [3] Although the Atom Layer Deposition (ALD) way has the advantage of Monolayer atomic deposition, furnace inherent characteristics as side leading to the uneven of the wafer, which limits its development al. [4]

In order to get better device performance, increasing the film uniformity with in wafer and wafer to wafer is demanded. We studied the boat rotate speed and gas flow influence in the wafer uniformity in this paper to achieve acceptable structure. The formation mechanisms were also studied in detail.

EXPERIMENT AND DISCUSSION

boat rotate speed

Huali low-k spacer process uses the 300mm vertical furnace tube as shown in Fig.1. Atomic layer deposition method mainly depends on the sub recipe to achieve atomic thickness deposition, which theoretically deposits the thickness of an atomic layer per cycle. The thickness is controlled by reaction time and gas flow of the sub recipe. [5] During the reaction, the boat keeps a constant speed as shown in Fig.1.

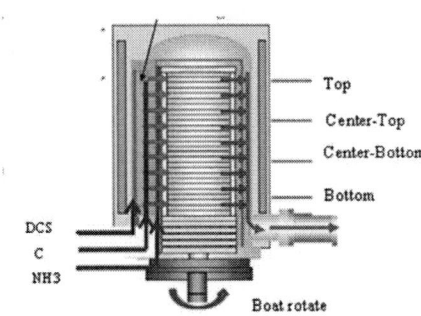

Fig.1: furnace tube

We found that when the boat rotate speed was 2r/min, a special map with a fixed direction appeared on the wafer from furnace top to the bottom end as shown in Fig. 2.

It was probably caused by the condition that when the gas flow entered the furnace tube, the position of the special map always appeared at the same time. According to this theory, we adjusted the speed of the boat rotate speed from 2r/min to 2.2r/min. After tuning the boat rotate speed we found that when the sub recipe time is not the times of the boat rotate speed, the gas flow would no longer enter from the same position, thus solving the problem of special map and realizing the concentric circle structure as shown in Fig.2.

The structure of concentric circles not only solves the problem of special map, but also makes good preparations for the low-k spacer etch. The good uniformity of the wafer provides a good structure for device performance.

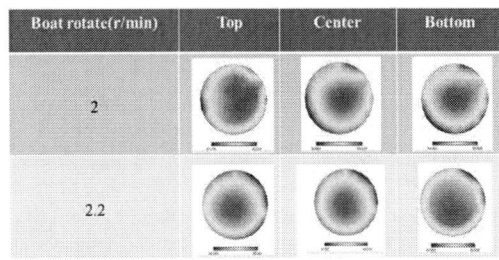

Fig.2: Monitor wafer map of different zone of tube

gas flow

According to the discussion above, we know that the gas flow plays an important role with in wafer uniformity. Reaction gases such as DCS, NH_3 and C source enter the tube through different channels. Due to the inherent property of the furnace tube that the gas flow enters from the side into the middle which results in the uneven of the wafer. [5]. The range of the same wafer is also larger.

In this experiment, the influence of the carrier N_2 content of the DCS on the range of wafer was studied. The specific experimental scheme is shown in Table 1 below. According to the gas reaction principle, DCS would first deposit a layer on the wafer surface. [6] Thus, the content of DCS is an important factor that determines the wafer thickness. We intend to increase the partial pressure of the DCS by increasing the carrier N_2 gas flow of DCS.When the DCS enter into the tube, the gas can enter the middle of the wafer more because of the increased gas pressure, so that we can achieve the good uniformity of the wafer edge and the middle thickness

Step	N2 PRG1	SI FLOW
BSL	3000	3000
DCS-N2	5000	5000

Table 1:A LD process parameters

In the BL condition, the gas content of DCS carrier N_2 is 3000sccm.It can be seen that the thickness of edge of the wafer is thick than that in the middle as shown in Fig.3. In this experiment, we increased the gas content of DCS Carrier N_2 to 5000sccm.

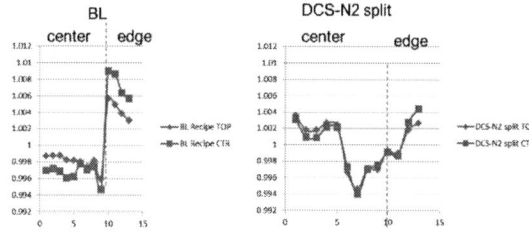

Fig.3 : Monitor wafer range

It can be seen that after increasing the gas content of DCS carrier N_2, the edge of control wafer became thinner compared with BL. The range of the whole wafer became smaller, and range became better, as shown in Fig.3 This is because that the Carrier N_2 in Si flow plays the role of carrying the DCS reaction gas into the surface of the wafer. After increasing the Carrier flow content of N_2, the total capacity of DCS- N_2 increases, the total pressure of gas increase, which increasing the activation of the DCS in furnace tube from the wafer edge to the middle. Thus,

Compared to BL condition, the wafer edge become thinner.

Fig.4 shows the TEM of the low-k spacer on the pattern wafers.Fig.4(a) is the thickness of the middle position of the wafer after adjusting the gas flow, while Fig.4(b) is the thickness of the edge position. It shows that the center and the edge thickness of the wafer is the same, which indicates that gas flow can affect the within wafer uniformity. The good uniformity of the spacer thickness is conducive to 28 nm process product yield improvement.

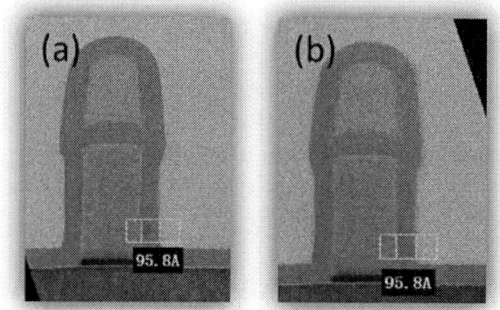

Fig. 4: The TEM images of wafer center and edge

CONCLUSION

In this paper, problems and solutions for with in wafer and wafer to wafer thickness uniformity are investigated. The mechanism is that the setting rule of the temperature zone and gas injector in different position results in the difference in thickness and uniformity. It is found that boat rotate speed, gas flow can effectively tune to get good wafer uniformity

ACKNOWLEDGEMENTS

The authors would like to acknowledge HLMC failure analysis department for the supports of TEM microscopy.

REFERENCES

[1] A H M Kamal, "Improving the yield of deep submicrometer CMOS processes by controlling the grain size of poly-Si gate through post deposition rapid thermal anneal," IEEE Trans Semicond Manuf, vol. 15, no. 4, pp. 552–554, Nov. 2002.

[2] Y. Taur and T. H. Ning. Fundamentals of Modern VLSI Devices, 2nd edition. Cambridge University Press, Cambridge, UK. p. 176, 2009.

[3] T. Hori. A 0.1-µm CMOS technology with tilt-implanted punchthrough stopper (TIPS). IEEE IEDM Tech. Dig., p. 75, 1994.

[4] H. Chen, J. Zhao, C. S. Teng, et al. Submicron large-angle-tilt implanted drain technology for mixed-signal applications. IEEE IEDM Tech. Dig., p.

91, 1994.

[5] Jianqin Gao, Jun Tan, Haifeng Zhou, Jingxun Fang, and Albert Pang, *CSTIC2015*, Shanghai, March 15-16, 2015

[6] J.M. Hartmann, V. Loup, G. Rolland, P. Holliger, F. Laugier, C. Vannuffel, M.N. Semeria *Journal of Crystal Growth* 236 (2002) 10–20

MAGNETOELECTRIC MEMORY CELL BASED ON MICROSIZED FeGa FILMS ON FERROELECTRIC 50BZT–50BCT FILMS

Zhi Tao[1] , Yemei Han[1,] , Xianming Ren[1] , Hui Li[1] , Fang Wang[1] ,and Kailiang Zhang[1,*]*
[1]Tianjin University of Technology, Tianjin 300384, China
[2]Dept.School of Electronic and Information Engineering, Tianjin Key Laboratory of Film Electronic & Communication Devices, Tianjin, China
* Corresponding Author's E-mail: Yemei Han:hanym@tju.edu.cn , Kailiang Zhang: kailiang_zhang@163.com

ABSTRACT

Magnetoelectric memories exhibit the advantages of FRAMs and MRAMs. We report the fabrication of a magnetoelectric memory cell consists of microsized Fe70Ga30 films on ferroelectric $50[Ba(Zr_{0.2}Ti_{0.8})O_3]–50(Ba_{0.7}Ca_{0.3}TiO_3)$ (50BZT–50BCT) films on Pt/Ti/SiO$_2$/Si substrates. The 50BZT–50BCT films were sputtered by RF magnetron sputtering, ferroelectric hysteresis loop and displacement voltage butterfly loop were characterized which suggest that 50BZT–50BCT thin films exhibit good ferroelectric and piezoelectric properties. $Fe_{70}Ga_{30}$ films were grown on 50BZT–50BCT films by DC magnetron sputtering deposition combined with photolithography and results indicate that $50BZT–50BCT/Fe_{70}Ga_{30}$ composite films have good soft magnetic properties with a coercive field of 50 Gauss at room temperature. The microstructure of the composite films were characterized by X-ray diffraction. The current voltage behaviors of the microsized Fe70Ga30 films were measured, the current voltage curves could be modulated effectively by applying bias electric field, and multilevel resistance states could be induced, and thus a magnetoelectric memory device was obtained which uses electric field as writing field and the resistance of the ferromagnetic layer as the media.

Keywords—Magnetoelectric memory;50BZT-50BCT/FeGa films; ferroelectric properties; piezoelectricity properties magnetic properties.

INTRODUCTION

In the multiferrous magnetoelectric materials, the change of magnetoresistance makes a single memory unit realize two logical states, which can significantly improve the storage density. At the same time, combining the advantages of ferroelectric and magnetic random memory, it is expected to prepare a new type of nonvolatile random memory for electric writing and magnetic reading, namely, magnetoelectric random memory(MeRAM)[1]. The coupling between ferroelectric and ferromagnetic properties can realize the conversion between electricity and magnetism, the information writing process can be realized by electric field and the reading process can be realized by magnetic head[2, 3].

It has been reported that the composite films consist of $Fe_{70}Ga_{30}$ (abbreviated as FeGa) grown on $50[Ba(Zr_{0.2}Ti_{0.8})O_3]–50(Ba_{0.7}Ca_{0.3}TiO_3)$ (abbreviated as 50BZT–50BCT) exhibit bias voltage dependent resistance states. The magnetoelectricity properties of the composite films prepared by 50BZT–50BCT films are similar to those based on PZT family, such as $Pb(Zr,Ti)O_3(PZT)$, $Pb(Mg_{1/3}Nb_{2/3})O_3-PbTiO_3(PMN-PT)$ and $Pb(Zn_{1/3}Nb_{2/3})O_3-PbTiO_3(PZN-PT)$[4, 5]. In this work, we prepare composite film based on microsized $Fe_{70}Ga_{30}$ grown on 50BZT–50BCT films, we discuss the current voltage (IV) characteristics of the $50BZT–50BCT/Fe_{70}Ga_{30}$ composite films modulated by the applied bias voltage, and we believe that a magnetoelectric memory is realized in the medium of the change of the ferromagnetic layer resistance.

EXPERIMENTAL PROCEDURE

50BZT-50BCT films were deposited on Pt/Ti/SiO$_2$/Si substrates by RF magnetron sputtering. The power of RF magnetron sputtering is 50W, the distance between the target and the substrate is 7.0cm. The pressure of 2.0Pa, the Ar:O$_2$ ratio of 30:20, the substrate temperature of 500°C and sputtering time of 2h are used. The films were annealed at 700°C for 30 minutes. Microsize FeGa (50um*200um) films were grown on 50BZT-50BCT films by photolithography with DC magnetron sputtering, the power of DC magnetron sputtering is 50W, the distance between the target and the substrate is 6.5cm with the pressure of 1.0Pa. Fig. 1 (a) gives the image of FeGa films growing on 50BZT-50BCT films. The thickness of 50BZT-50BCT films is about 100nm and FeGa films is about 120nm measured by profilometer (Veeco, Dektak 150). The microstructures of the films were characterized by X-ray diffraction (XRD, Rigaku, DMAX-2500). The composition of 50BZT-50BCT was determined by SEM(Verios 460L), Ferroelectric test system (Precision Multiferroic II) and Vibrating sample magnetometer (VSM, Lake Shore 7404) was used to analyze the ferroelectric and the magnetic properties. Fig. 1 (b) is a schematic of the cross-section for the composite films. The resistance states of FeGa films on top of

50BZT-50BCT films was characterized by using Agilent B1500A semiconductor device analyzer at room temperature. In order to investigate the current voltage characteristics of 50BZT-50BCT/FeGa composite films as functions of bias voltage, during the measurements, the IV characteristics of the composite films were measured by applying positive bias voltage on the bottom electrode Pt and the top FeGa films is grounded.

RESULTS AND DISCUSSION

The particle size and density of the films have great influence on the properties of piezoelectric materials[6]. Fig. 2 shows the SEM micrograph and element mapping of 50BZT-50BCT films, the micrograph indicates that the film shows uniformly distributed grains, the particle size is relatively uniform, and the element number distribution is relatively uniform, which conforms to the distribution characteristics of 50BZT-50BCT films.

In order to characterize the electrical properties of 50BZT-50BCT films, the ferroelectric hysteresis loop and piezoelectric loop are obtained in Fig. 3. The piezoelectric curve shows good piezoelectric performance, with the piezoelectric displacement is 64Å, which is similar to previous reports[7, 8]. Note that, The asymmetry behavior of displacement voltage butterfly loop may be caused by the difference of work function between the top (the probe) and bottom (Pt) electrodes[9]. The ferroelectric hysteresis loop shows that the films exhibit good ferroelectric performance.

The XRD patterns for 50BZT-50BCT/FeGa composite films are shown in Fig. 4 (a) . It is obvious that the orientation of 50BZT-50BCT films is similar to that of BaTiO$_3$-like perovskite structure in (110) and (111) and the orientation of ferromagnetic FeGa films is (110). There are no other heterophases along with ferroelectrics and ferromagnetic phases in 50BZT-50BCT/FeGa composite films. Fig. 4 (b) gives the magnetic hysteresis loop of 50BZT-50BCT/FeGa composite films, we can see from the hysteresis loop that the coercive field of the composite films is about 50Guass, which suggests that the composite films we prepared possess good soft magnetic properties[10].

Fig. 5 (a) exhibits the IV curves under different bias voltages in the range from 0 to 0.5V. When the bias voltage is not applied to the film, the current of the film increases from 0 to 25mA during the scanning voltage from 0 to 1V. When 0.5V bias is applied to the film, the current of the film increases to 50mA. The residual current is also increased from 2.26mA to 11.3mA in proportion as the bias voltage is increased from 0V to 0.5V. The slope and intercept of the IV curve change with the change of bias voltage, which may be caused by the uneven charge distribution caused by the applied bias voltage.

Fig. 5 (b) exhibited the resistance characteristic curves under different bias voltage. When bias voltage is not applied, the resistance is stable at about 40Ω; when bias voltage is 0.2V, the resistance is stable at about 31.4Ω; when bias voltage is about 0.5V, the resistance is about 20Ω. The IV characteristics of 50BZT-50BCT/FeGa composite films are effectively modulated by the applied bias voltage, and induce the change of resistance states for FeGa films. Different resistances represent different states of memory, and could also represent "0" and "1" of memory information. Applying a bias electric field, the current voltage curve of the films can be effectively modulated, and the multi-level resistance states of the films can be induced. Thus, a magnetoelectric memory with the electric field as the input and the ferromagnetic layer resistance as the medium is obtained.

CONCLUSIONS

In summary, composite films of 50BZT-50BCT/FeGa were fabricated. 50BZT-50BCT films shows a good piezoelectric curve with a displacement of 64Å. The modulation of IV behaviors is successfully realized for 50BZT-50BCT/FeGa composite films by applying bias voltage, the multi-level resistance states of ferromagnetic phase of 50BZT-50BCT/FeGa composite films is induced, which provides an important basis for the realization of magnetoelectric memory with electric field as input and ferromagnetic layer resistance as medium.

ILLUSTRATIONS

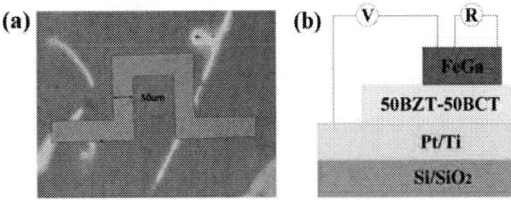

Fig. 1 (a) 50μm×200μm FeGa films grown on 50BZT-50BCT films; (b) schematic of the cross-section for 50BZT-50BCT/FeGa composite films.

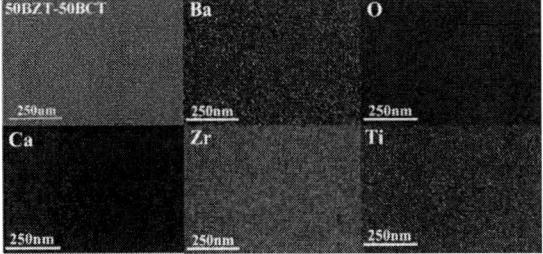

Fig. 2 SEM micrograph and Element mappin of 50BZT-50BCT films.

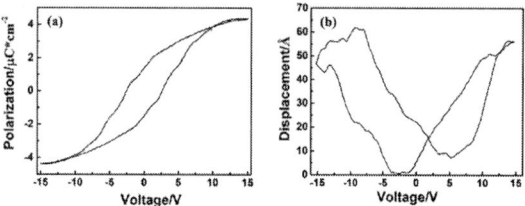

Fig. 3 (a) ferroelectric hysteresis loop; (b) piezoelectric curve of 50BZT-50BCT films.

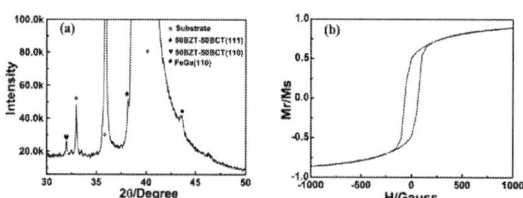

Fig. 4 (a) XRD patterns for 50BZT-50BCT/FeGa composite films; (b) magnetic hysteresis loop of 50BZT-50BCT/FeGa composite films.

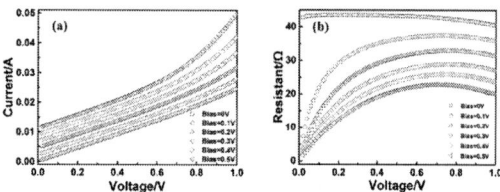

Fig. 5 (a)the IV curves; (b) the resistance voltage curves under different bias of 50BZT-50BCT/FeGa composite films.

ACKNOWLEDGEMENT

This work was supported by National Natural Science Foundation of China (Grant Nos. 51502204, 61404091, 61274113, 61804108 and 51502203), the National Key Research and Development Program of China (Grant No.2017YFB0405600) and Natural Science Foundation of Tianjin City (Grant Nos. Grant no. 18JCYBJC85700, 18JCZDJC30500 and 17JCYBJC16100).

REFERENCES

[1]   J.F. Scott. Data storage - Multiferroic memories. Nat. Mater., vol. 6, 2007, pp.256-257.

[2]   M. Bibes and A. Barthelemy. Multiferroics: Towards a magnetoelectric memory. Nat. Mater., vol. 7, 2008, pp. 425-426.

[3]   J.M. Hu, Z. Li, J. Wang, and C.W. Nan. Electric-field control of strain-mediated magnetoelectric random access memory. J. Appl. Phys., vol. 107, 2010, pp. 093912.

[4]   J.H. Gao, D.H. Xue, Y. Wang, D. Wang, L.X. Zhang, H.J. Wu, S.W. Guo, H.X. Bao, C. Zhou, W.F. Liu, S. Hou, G. Xiao, and X.B. Ren. Microstructure basis for strong piezoelectricity in Pb-free Ba(Zr0.2Ti0.8)O-3-(Ba0.7Ca0.3)TiO3 ceramics. Appl. Phys. Lett., vol. 99, 2011, pp. 092901.

[5]   H. Palneedi, V. Annapureddy, H.Y. Lee, J.J. Choi, S.Y. Choi, S.Y. Chung, S.J.L. Kang, and J.H. Ryu. Strong and anisotropic magnetoelectricity in composites of magnetostrictive Ni and solid-state grown lead-free piezoelectric BZT-BCT single crystals. J. Asian Ceram. Soc., vol. 5, 2017, pp. 36-41.

[6]   J.P. Praveen, K. Kumar, A.R. James, T. Karthik, S. Asthana, and D. Das. Large piezoelectric strain observed in sol-gel derived BZT-BCT ceramics, Curr. Appl. Phys., vol. 14, 2014, pp. 396-402.

[7]   J.G. Wu, D.Q. Xiao, and J.G. Zhu. Potassium-Sodium Niobate Lead-Free Piezoelectric Materials: Past, Present, and Future of Phase Boundaries. Chem. Rev., vol. 115, 2015, pp. 2559-2595.

[8]   Z. Tao, F. Che, Y.M. Han, F. Wang, Z.C. Yang, W. Qi, Y. Wu, and K.L. Zhang. Out-of-plane and In-plane piezoelectric behaviors of Ba(Zr0.2Ti0.8)O3 -0.5(Ba0.7Ca0.3TiO3) thin films. Prog. Nat. Sci., vol. 27, 2017, pp. 664-668.

[9]   F. Yan, T.J. Zhu, M.O. Lai, and L. Lu. Effect of bottom electrodes on nanoscale switching characteristics and piezoelectric response in polycrystalline BiFeO3 thin films. J. Appl. Phys., vol. 110, 2011, pp. 084102.

[10]   Y.M. Han, F. Che, Z. Tao, F. Wang, and K.L. Zhang. Electric field induced modulation of transport characteristics in multiferroic BZT-BCT/FeCo thin films. J. Mater. Sci.-Mater. Electron., vol. 29, 2018, pp. 4786-4790.

TaO$_x$ synapse array based on ion profile engineering for high accuracy neuromorpic computing

Jingjing Yang[1,2], Jiadi Zhu[2], Bingjie Dang[2], Teng Zhang[2], Qingxi Duan[2], Liying Xu[2], Keqin Liu[2], Zhiting Lin[1,*], Ru Huang[2,*], and Yuchao Yang[2,*]

[1]School of Electronics and information Engineering, Anhui University, Hefei 230601, CHINA
[2]Department of Micro/nanoelectronics, Peking University, Beijing 100871, CHINA
[*]E-mail: yuchaoyang@pku.edu.cn, ztlin@ahu.edu.cn, ruhuang@pku.edu.cn

ABSTRACT

Resistive random-access memory is a promising candidate for high-density, low-power neuromorphic computing. Here, we demonstrate a thermally oxidized TaO$_x$ RRAM array with built-in oxygen concentration gradient, which ensures good linearity and symmetry and consequently better inference accuracy on both MNIST and CIFAR10 datasets after training with perceptron, LeNet-5 and ResNet-18 networks, thus illustrating the potentiation of applying RRAMs for high-density, high-accuracy, large-scale neuromorphic computing.

INTRODUCTION

Neuromorphic devices have attracted great attention in recent years, and are believed to be a possible solution to overcome the von Neumann bottleneck [1-4]. Among all the neuromorphic devices, oxide based resistive random-access memory (RRAM) [5] stands out with its strong capability for high-density, low-power integration. However, the linearity of oxide-based RRAM hardly meets up the requirement for high accuracy neuromorphic computing, and could hinder its further applications [6]. Various novel structures and novel materials have been proposed recently, trying to improve the linearity and symmetry, such as bilayer structure [7,8] and organic materials [9], requiring relative complex fabrication process or CMOS-incompatible material systems. Moreover, previous work mainly utilizes perceptron network and MNIST dataset to demonstrate the effect of artificial neural network training [10]. It will be of practical significance to assess the device performance using more complex network models and training datesets.

In this work, we demonstrate a thermally oxidized tantalum oxide (TaO$_x$) RRAM [11] array. With simple device structure and facile fabricate process, it exhibits good uniformity, gradual switching behavior and outstanding linearity and symmetry. Furthermore, the inference accuracy of thermally oxidized TaO$_x$ RRAM is also compared with that of RRAM with sputtered TaO$_x$ layer and the RRAM with ideal linearity. Two kinds of datasets, i.e. MNIST [12] and CIFAR10 [13] are used with three different artificial neural network structures

[14-16] to comprehensively evaluate the impact of device non-linearity on the inference accuracy. These results demonstrate the capability of realizing high-density, high-accuracy neural network with RRAMs, and also pave the way for large-scale integration of neuromorphic devices.

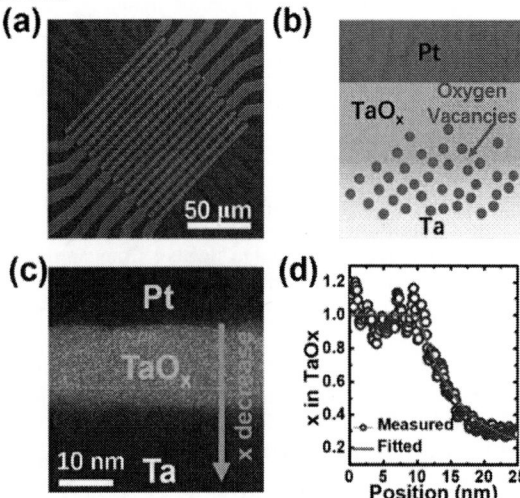

Fig.1. (a) SEM image of a 16×16 RRAM crossbar array. (b) Schematic of the cross-section of a TaO$_x$ based RRAM device, a huge amount of oxygen vacancies is generated near the Ta electrode. (c) TEM image of the cross-section of a TaO$_x$ based RRAM device, and (d) concentration of oxygen in TaO$_x$ from EDS linescan. A built-in exponential gradient in oxygen concentration can be observed ~10 nm from the Pt/TaO$_x$ interface.

DEVICE FABRICATION

Shown in Fig. 1a and b, the devices in RRAM array has a Pt/TaO$_x$/Ta structure. The bottom electrode (BE) was fabricated on Si substrates with 300 nm thermal oxide. Reactive ion etching (RIE) was used to create a trench in SiO$_2$ (~100 nm), followed by Ta/Ti deposition (100/10 nm) and lift-off processes. The Ta BE was then subjected to oxygen annealing at 400°C, forming ~ 10 nm TaO$_x$, before 80 nm Pt top electrode (TE) was deposited. The resultant devices have a crossbar structure and are 5×5 μm^2 in size. Pt/TaO$_x$/Pt devices are used for

978-1-7281-6559-2/20 $31.00 © 2020 IEEE

comparison in this study, where the TaO_x layer was formed by sputtering. Electrical measurements were performed using Agilent B1500A semiconductor parameter analyzer. The microstructure and composition of the devices were studied by transmission electron microscopy (TEM) and energy dispersive x-ray spectroscopy (EDS), respectively.

RESULTS AND DISCUSSION

Fig. 1c shows the high-resolution TEM image of an as-fabricated $Pt/TaO_x/Ta$ device, in which there exists a built-in oxygen ion concentration gradient between the TaO_x layer and the Ta BE. Such concentration gradient can be analyzed in detail by EDS line scan. Fig. 1d clearly reveals a descending concentration of oxygen ions starting 10 nm from the TE to the BE side, which is consistent with the oxidation process and can be described with an exponential decay function. As the oxidation of Ta started from the surface and gradually moves inwards, the region close to the TE has the highest oxygen concentration due to the longer exposure to the oxygen atmosphere, leading to the constantly increase of oxygen vacancy towards the TaO_x/Ta interface, as illustrated in Fig. 1b.

Such profile of oxygen ions distribution contributes to the good uniformity of the DC performance of the thermally oxidized TaO_x based device. Compared with the DC performance of RRAM with 10 nm sputtered TaO_x layer, such device possesses better uniformity, which can be shown by the 20 consecutive current-voltage (*I-V*) sweeps in Fig. 2a, b. Apart from the uniformity, gradual switching behavior is also achieved in this device due to the gradual ion concentration.

The weight modulation characteristics are also studied

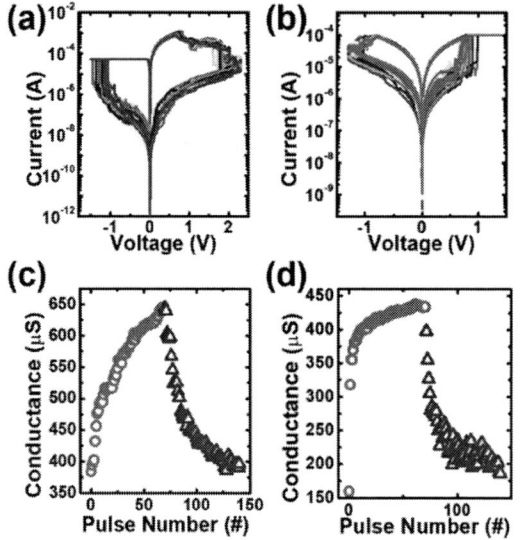

Fig.2. Comparison of DC characteristics and conduction modulation process of TaO$_x$ RRAM fabricated by thermal

oxidization and sputtering. 20 consecutive I-V sweeps of (a) thermally oxidized and (b) sputtered TaO_x RRAM. The pulse parameters used in the measurements were (c) potentiation: -1.7V, 3.5ms, depression: +1.9V, 3.5ms, and (d) potentiation: +0.82V, 10μs, depression: -0.95V, 10μs.

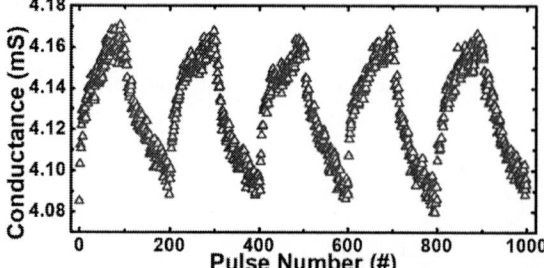

Fig.3. Cycle-to-cycle variation of the weight modulation process of the thermal oxidized TaOx RRAM. (potentiation: -1.7V, 3.5ms, depression: +1.9V, 3.5ms)

and sputtering (Fig. 2d). Significant improvement in linearity can be observed in thermally oxidized TaO_x based RRAM compared with the sputtered device, since the gradual oxygen ion concentration profile could contribute to gradual formation process of the conductive filament and thus leading to the more incremental and linear change in conductance. The cycle-to-cycle uniformity of the device based on thermally oxidized TaO_x layer is shown in Fig. 3, where good stability can be observed. Accordingly, the thermally oxidized TaO_x RRAM demonstrates robust analog switching behavior and can be a good candidate for high-accuracy neuromorphic computing.

Simulation of the training and inference processes with artificial neural networks are carried out to demonstrate the impact of non-linearity on inference accuracy. Three different network models, i.e. 3-layer perceptron [14], LeNet with 5 convolutional layers (LeNet-5) [15], and residual network [16] with 18 convolutional layers (ResNet-18) are used to benchmark the effect of non-linearity on different network structures. Two different datasets, namely, the MNIST and CIFAR10 are used to evaluate the accuracy dependence on datasets. The learning rate for the ideal linearity in all these networks is fixed at 0.001.

As shown in Fig. 4, the influence of non-linearity strongly affects the inference accuracy. With all the networks and datasets we used, higher accuracy is always obtained with better device linearity, since the non-linearity has abrupt weight change in a region and slowed down dramatically in the rest of the dynamic range, which would make it difficult to locate the global minimum during the training process. The RRAM with thermally oxidized (TO) TaO_x layer demonstrates accuracy comparable with that of devices with ideal linearity, and is significantly higher than that of device with sputtered (SP) TaO_x layer under all the circumstances. This result also shows the fact that more complex neural network may require better device linearity (see Fig. 4a and b, or c and d), as the effect of non-linearity would accumulate with the increased number of convolutional layers. Moreover, it is also revealed that the inference accuracy is also dependent on network models and datasets. As is implied from Fig. 4b and c, with the same LeNet-5 structure, the inference accuracy on CIFAR10 is significantly lower than that on MNIST. More importantly, the inference accuracy of deeper and complex networks and datasets, shows a stronger dependence on device linearity, e.g. ResNet-18 on CIFAR10 (Fig. 4d), therefore stressing the significance of careful ion profile engineering in tuning the weight modulation characteristics.

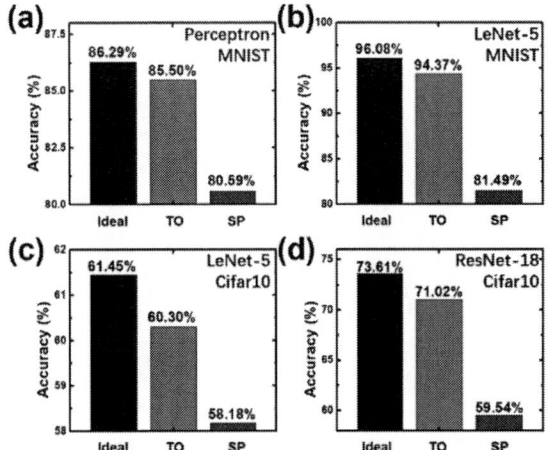

Fig.4. Comparison of the inference accuracy when different devices, network structures and datasets are used. (a) a 3-layer perceptron trained with different linearities, namely, ideal linearity, linearity of device based on thermal oxidized (TO) TaO_x layer, and linearity of device based on sputtered (SP) TaO_x layer, on MNIST dataset. (b) LeNet-5 trained with different linearities on MNIST dataset. (c) LeNet-5 trained with different linearities on CIFAR10 dataset. (d) ResNet-18 trained with different linearities on CIFAR10 dataset.

CONCLUSION

In this paper, we demonstrate a thermally oxidized TaO_x RRAM array with built-in oxygen concentration gradient which ensures good linearity and symmetry. The inference accuracy of the thermally oxidized TaO_x RRAM is evaluated after training with 3-layer perceptron, LeNet-5 and ResNet-18 on both MNIST and CIFAR10 datasets, showing better inference accuracy in thermally oxidized TaO_x RRAM compared with sputtered TaO_x RRAMs. It shows the advantage of careful ion profile engineering in building high-performance neuromorphic systems.

ACKNOWLEDGEMENTS

This work was supported by the National Key R&D Program of China (2017YFA0207600), the National Natural Science Foundation of China (61925401, 61674006, 61927901, 61421005), and Beijing Academy of Artificial Intelligence (BAAI). We acknowledge Y.-F. Y. in helping building up the simulation environment.

REFERENCES

[1] Q. Xia, and J. J. Yang, "Memristive crossbar arrays for brain-inspired computing." *Nat. Mater.*, vol. 18, 2019, pp.309-323.

[2] J. Zhu, Y. Yang, R. Jia, Z. Liang, W. Zhu, Z. U. Rehman, L. Bao, X. Zhang, Y. Cai, L. Song, and R. Huang, "Ion gated synaptic transistors based on 2D van der Waals crystal with tunable diffusive dynamics." *Adv. Mater.*, vol. 30, 2018, pp.1800195.

[3] X. Zhu, D. Li, X. Liang, W. D. Lu, "Ionic modulation and ionic coupling effects in MoS2 devices for neuromorphic computing." *Nat. Mater.*, vol. 18, 2019, pp.141.

[4] C. Cheng, Y. Li, T. Zhang, Y. Fang, J. Zhu, K. Liu, L. Xu, Y. Cai, X. Yan, Y. Yang, and R. Huang, "Bipolar to unipolar mode transition and imitation of metaplasticity in oxide based memristors with enhanced ionic conductivity." *J. Appl. Phys.*, vol. 124, 2018, pp.152103.

[5] D. B. Strukov, G. S. Snider, D. R. Stewart, R. S. Williams, "The missing memristor found." *Nature*, vol. 453, 2008, pp.80-83.

[6] C. Sung, A. Padovani, B. Beltrando, D. Lee, M. Kwak, S. Lim, L. Larcher, V. D. Marca, and H. Hwang, "Investigation of I-V Linearity in TaOx-based RRAM Device for Neuromorphic Applications." *J. Electr. Device Soc.*, vol. 7, 2019, pp.404-408.

[7] Z. Wang, M. Yin, T. Zhang, Y. Cai, Y. Wang, Y. Yang, and R. Huang, "Engineering incremental resistive switching in TaOx based memristors for brain-inspired computing." *Nanoscale*, vol. 8, 2016, pp.14015-14022.

[8] Q. Duan, L. Xu, J. Zhu, X. Sun, Y. Yang, and R. Huang, "Resistive switching and synaptic plasticity in HfO2-based memristors with single-layer and bilayer structures." *China Semiconductor Technology International Conference (CSTIC)*, Shanghai, China, Mar. 11, 2018, pp.1-3.

[9] Y. van de Burgt, E. Lubberman, E. J. Fuller, S. T. Keene, G. C. Faria, S. Agarwal, M. J. Marinella, A. A. Talin, A. Salleo, "A non-volatile organic electrochemical devices as a low-voltage artificial synapse for neuromorphic computing." *Nat. Mater.*, vol. 16, 2017, pp.414.

[10] Y. Yang, M. Yin, Z. Yu, Z. Wang, T. Zhang, Y. Cai, W. D. Lu, and R. Huang, "Multifunctional Nanoionic Devices Enabling Simultaneous Heterosynaptic Plasticity and Efficient In-Memory Boolean Logic." *Adv. Electr. Mater.*, vol. 3, 2017, pp.1700032.

[11] M. Yin, Y. Yang, Z. Wang, T. Zhang, Y. Fang, X. Yang, Y. Cai, and R. Huang, "TaOx based memristors with recessed bottom electrodes and built-in ion concentration gradient as electroni synapses." *13th IEEE International Conference on Solid-State and Integrated Circuit Technology (ICSICT)*, Hangzhou, China, Oct. 25-28, 2016, pp. 17081875.

[12] Y. LeCun, C. Cortes, C. J. C. Burges, "The MNIST Database", [Online] http://yann.lecun.com/exdb/mnist/

[13] A. Krizhevsky, "The Cifar10 Dataset", [Online] https://www.cs.toronto.edu/~kriz/cifar.html

[14] F. Rosenblatt, "The perceptron: A probabilistic model for information storage and organization in the brain." *Psychological Review*, vol 65, 1958, pp. 386-408.

[15] Y. LeCun, L. Bottou, Y. Bengio, and P. Haffner, "Gradient-based learning applied to document recognition," Proc. IEEE, vol. 86, 1998, pp. 2278–2324.

[16] K. He, X. Zhang, S. Ren, and J. Sun, "Deep Residual Learning for Image Recognition." [Online] https://arxiv.org/pdf/1512.03385.pdf

THE CAUSATION AND IMPROVEMENT OF ONE TYPE OF PARTICLES OCCURRING IN BATCH-CLEAN TOOL

Jing Ye, Jun Gao, Nan Lin, Jun Liu

Shanghai Huahong Grace Semiconductor Manufacturing Corporation

Shanghai, China

*Corresponding Author's Email: Jing.Ye@hhgrace.com

ABSTRACT

This article mainly expound the causation and improvement of one type of particles occurring in Batch-clean tool. The defect chart trend up was found since a certain time. According to many experiments, we found the particle occurring in the specific type Bath-clean tool. NH_4OH in SC1 and HCl in SC2 react to NH_4Cl when the exhaust of acid and alkali is not balance for there is no shutter between SC1 and SC2 of the Bath-clean tool. The solid NH_4Cl dispersed on the surface of the wafers and enlarged by next film. The defect case was improved by installing shutter between HQDR2 and SC2 to isolate the atmosphere of SC1 and SC2.

Key words: Particle，SC1，SC2，NH_4Cl

INTRODUCTION

It is well known that WET clean is the most important process to remove particle coming from each step in integrated circuit (IC) manufacturing [1]. As the requirements for increased device performance and reliability have become more stringent in the era of VLSI (Very Large Scale Integration) and ULSI (Ultra Large Scale Integration) silicon circuit technology, techniques to avoid contamination and processes to generate very clean wafer surfaces have become critically important [2]. The famous RCA-1 clean (sometimes called "standard celan-1, SC1") developed by Werner Kern at RCA laboratories, is a procedure for removing organic residue and films from silicon wafers. The decontamination works of SC1 based on sequential oxidative desorption.

SC1 is the mixture of $5H_2O:1H_2O_2:1NH_4OH$. And RCA-2 clean (SC2) is often used the mixture of $6H_2O:1H_2O_2:1HCl$ to further clean the surface [1].

It is very detrimental to the yield of chip if the WET clean process become the particle source, especially for the CD size is smaller and smaller and the scale of integration circuit is larger and larger. This article mainly expound one WET batch- clean tool becoming particle source after the condition changed, and the improvement method of the particle issue.

METHODOLOGY

As shown in Figure 1, the defect chart trend up since a certain time by means of the defect monitor. As shown in Figure 2, SEM photo show the defect was small particle and could be enlarged by the next film. The defect mainly concentrated in the lower right corner as the defect map shown in Figure 3.

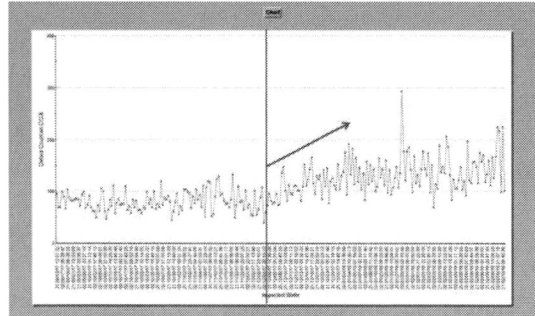

Figure 1: defect trend chart

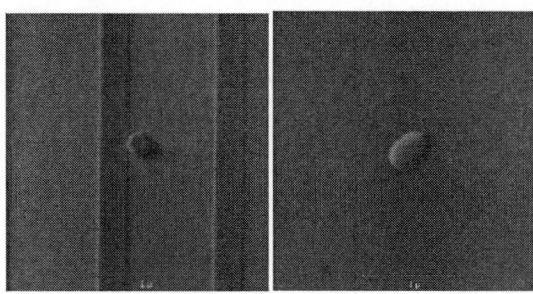

Figure 2: SEM photos of the particle before and after film deposition

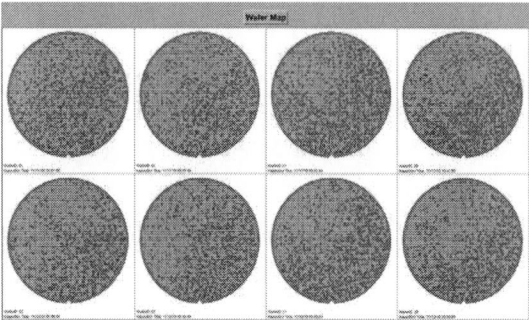

Figure 3: Defect map

The one specific batch-clean tool was suspected to result the defect issue according to the common tool comparison and the notch orientation change experiment. As shown in Figure 4, the special map changed along with the notch orientation of the specific batch-clean tool from 2 o'clock to 12 o'clock. WET engineers cleaned and changed the filter of the specific tool immediately, but it didn't work.

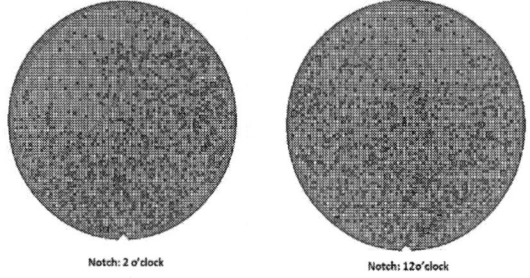

Figure 4: Notch orientation change experiment

The time point of the defect chart trend up is

suspected to be related to the setup of another new WET machine. The new machine resulted in less exhaust margin due to using the common power pipeline, so the exhaust volume of the specific batch-clean tool was insufficient, and then resulted in the particle issue. There was no shutter between the SC1 and SC2 of the specific tool as shown in Figure 5 by comparing with other batch-clean tools. The NH_4OH in SC1 and HCl in SC2 react to solid NH_4Cl as in shown in (1), and NH_4Cl dispersed on the surface of the wafers especially at the top half if the wafers if the acid-base unbalance.

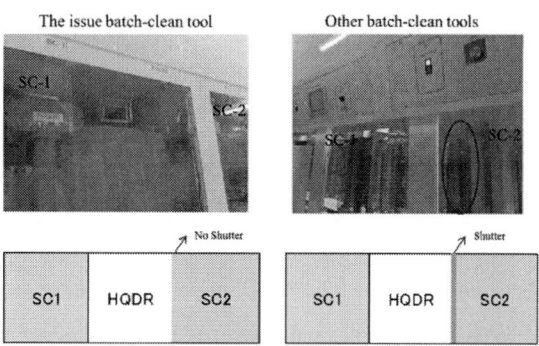

Figure 5: Difference of the issue tool and other tools

$$NH_4OH+HCl=NH_4Cl+H_2O \qquad (1)$$

RESULT & DISCUSSION

The particle issue can be solved by preventing the particle generating fundamentally, now that the root cause of the particle issue was found.

The defect case was totally resolved after installing a shutter between HQDR2 and SC2 as shown in Figure 6, to isolate the atmosphere of SC1 and SC2. The improvement effect is very obvious as shown in the Figure 7.

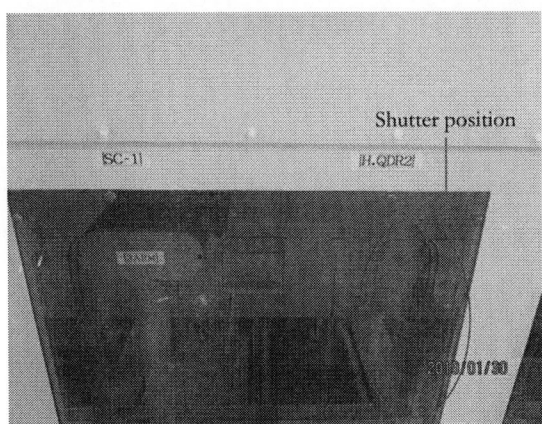

Figure 6:   Shutter position

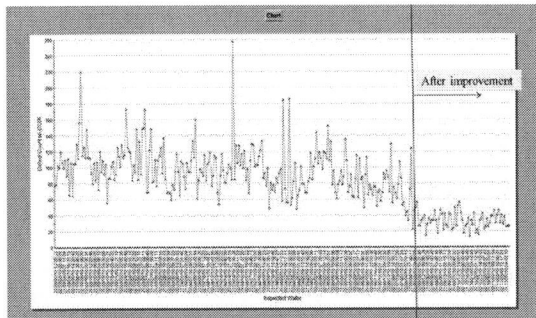

Figure 7: Defect trend chart

SUMMARY

WET clean tool was the most important to remove particles coming from each step in semiconductor manufacturing, but it may be particle source under certain conditions. By finding out the root cause, the effective improvement method can be found. This paper conducted an in-depth investigation and verification on the root cause of the particle generation, and finally found the fundamental solution.

REFERENCES

[1] W. Kern, Handbook of semiconductor wafer cleaning technology, Technology and Applications, Noyes Publication, 1999:274

[2] T Ohmi, Evolution of silicon wafer cleaning technology, Electrochem Soc, 2002:30

IMPACT OF NANOPILLAR-TYPE ELECTRODE ON HFO$_X$ -BASED RRAM PERFORMANCE

Baotong Zhang[1,2], Xiaokang Li[2], Yuancheng Yang[2], Haixia Li[2], Ru Huang[2], Ming Li[2] and Peimin Lu[1*].*

[1] College of Physics and Information Engineering,
Fuzhou University, Fuzhou 350116, China

[2] Key Laboratory of Microelectronic Devices and Circuits, Institute of Microelectronics,
Peking University, Beijing 100871, China

Email: liming.ime@pku.edu.cn, lpm@fzu.edu.cn

ABSTRACT

In this work, the performance of HfOx-based RRAM with 30nm nanopillar-type electrode was investigated. Experiment results show that the novel device has lower operation voltages and higher resistance ratio than conventional flat electrode RRAM. The underlying physical mechanism is attributed to the enhanced electric field by the nanopillar electrode. This research will provide a valuable guidance for future scaling of oxide-based RRAM.

INTRODUCTION

Metal oxide resistive random access memory (RRAM) has a great potential to be the next-generation nonvolatile memory (NVM) due to its promising properties such as low power consumption, fast switching speed, long-time retention, excellent scalability, etc. [1]. The forming voltage increases rapidly with the continuous scaling down of RRAM device, which limits it to be used in the low power application [2]. One of the effective ways to adjust forming voltage is altering the device structure. By the change of the electrode structure, we can manipulate electric field in RRAM to optimize switching voltages [3-4]. However, most of research focuses on the conducting bridge RAM (CBRAM) [5-6], and there is little research on how to reduce operating voltages on small-sized oxide-based RRAMs via the nanometer-scale modified electrode with tip-enhanced electric field.

In this paper, the Pt/HfOx/Ti/TiN RRAM devices (0.3*0.3μm^2) with 30nm nanopillar electrode were fabricated and the effect of electric field enhancement on nanometer-scale HfOx-based RRAM was demonstrated. Compared with the flat electrode device, the forming voltage of novel device is reduced by 20% and set/reset voltage has been decreased. In addition, the resistance ratio is improved in the same operation.

EXPERIMENT

All devices studied in this work were fabricated on SiO$_2$/Si substrates. Firstly, 100nm Poly-Si was deposited by LPCVD, and the Poly-Si nanopillars with the 30nm diameters were patterned by e-beam lithography and followed by reactive ion etching. Then a 5nm Ti adhesion and 20nm Pt film layer were deposited by e-beam evaporation and adopted as the bottom electrodes after the lift-off process. Subsequently, a 200nm thick oxide was deposited as the isolation layer and the 300*300nm contact holes were formed in the same layer. A 5nm HfOx layer was then deposited by ALD and served as the switching layer, and the Ti / TiN (5nm/20nm) stack layers as oxygen getting layer and the top electrode respectively were further deposited by sputtering. The conventional flat electrode RRAM were also fabricated as the control group. Fig. 1 shows the schematic diagram and optical microscopy photograph of nanopillar type RRAM and the SEM image of nanopillar.

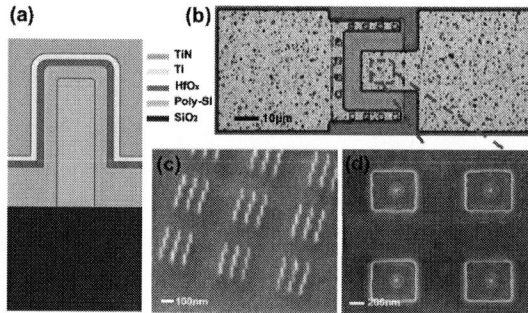

Figure 1: (a) Schematic diagram and (b) optical microscopy photograph of the fabricated Pt/HfOx/Ti/TiN RRAM. SEM images of (c) Poly-Si nanopillar array with HSQ and (d) nanopillar electrode after contact hole etching.

RESULTS AND DISCUSSION

For simplicity, NE and FE are referred to nanopillar type electrode memory device and flat type electrode memory device, respectively. Fig. 2 shows the typical switching I-V curves of NE and FE.

In order to study the effect of different electrode structures on the RRAM characteristics. Firstly, we measured the forming voltage distribution of 30 PE and NE devices by DC sweep operation. The voltage ramp from 0V to 6V was applied on electrodes and compliance currents was set to 100μA. The results shown in Fig. 3. As can be seen from the box plot, the median value of the forming voltage of NE (3.72V) is about 20% lower than that of FE (4.74V).

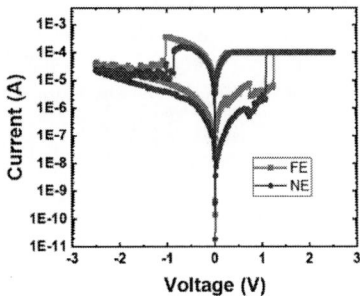

Figure 2: Typical I-V curves of FE and NE. The stop voltage is -2.5V to 2.5V and compliance currents is 100µA.

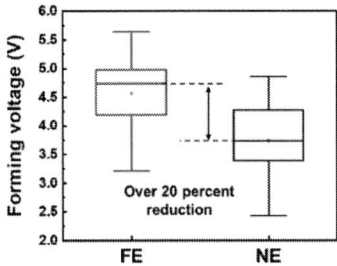

Figure 3: Forming voltage statistical results obtained by DC sweep operation for 30 devices of FE and NE.

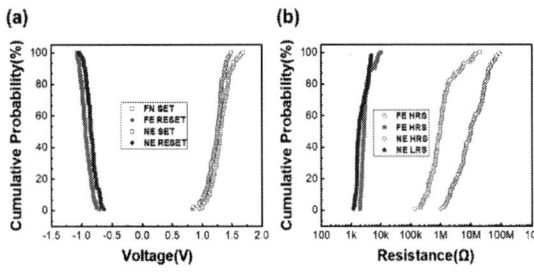

Figure 4: (a) Set/reset voltage and (b) resistance distribution for two different electrode structures during 100 continuous DC sweep cycles.

Then, we measured the set/reset voltages and resistance distribution of two samples for 100 continuous DC sweep cycles. The stop voltage of set/reset voltage is 2.5V/-2.5V respectively and read voltage is 0.1V. As shown in Fig .4, the median value of set/reset voltage of NE is lower about 0.1V than that of FE. Besides, the same median value of LRS are about 2.5K for NE and FE, while NE can achieve higher HRS (~20MΩ), about one order larger than the HRS (~2MΩ) of FE. As a result, NE can be

obtained higher on/off ratio (~10^4) compared with the normal FE (~10^3).

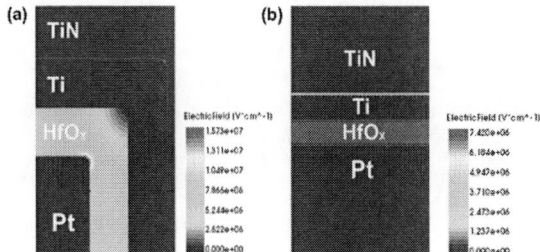

Figure 5: Electric field of the top area of (a) NE and (b) FE simulated by Synopsys TCAD Sentaurus Tools.

To better understand the mechanism behind the improved switching characteristics of Nanopillar-type RRAM, the simulation of electric field is carried with Synopsys TCAD Sentaurus. Fig. 5 shows the cross-section profile of simulated electric field along the direction from top electrode to bottom substrate. The simulation results show that the nanopillar top edge has higher electric field density, which benefits to the generation and dissolution of oxygen vacancies [7-8]. Therefore, the improvement of the operation voltages of nanopillar-type RRAM can be achieved.

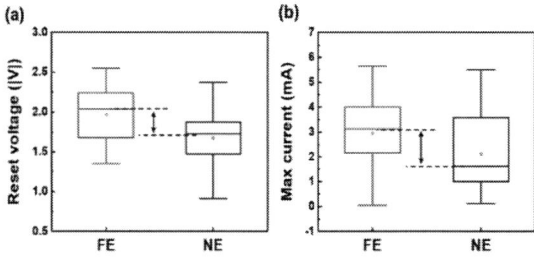

Figure 6: Statistic of (a) reset voltage and (b) its corresponding maximum current of NE and FE.

As for the higher HRS/LRS ratio, we studied the first reset process just after forming, and compared the max reset current and reset voltage of two type of devices. Fig. 6 (a) and (b) shows the statistic of reset voltage and its corresponding maximum reset current of NE and FE respectively. It can be found that both the median value of maximum current and reset voltage of FE are larger than that of NE, which means that the dissolution of filaments in FE devices is more difficult than in NE ones because the latter benefits from the enhanced electric field around the tip of nanopillar. On the other hand, the nanopillar electrode can form an O-vacancy concentrated path for the filament formation while the flat electrode results in more random distributed O-vacancy paths for the filament or even filament cluster formation. Especially, if the filament

978-1-7281-6559-2/20 $31.00 © 2020 IEEE

cluster is formed, the reset process is difficult to fuse off all the filaments so as to leave possibly leaky paths at HRS. In NE device, however, the filament is only formed at the preferred path; it is also easy to be burned off by the reverse electric field. With the enhanced electric field by NE, the forming voltage and the operation power can be possibly reduced, which is very promising for future scaling of RRAM.

CONCLUSION

In this paper, we reported an HfOx based RRAM with 30nm nanopillar type electrode. Attributing to the local tip-enhanced field, the improvement of switching voltage and HRS/LRS ratio can be realized. The nanopillar-type RRAM is expected to show the potential in low power extremely scaled memory application.

ACKNOWLEDGEMENTS

This work was supported in part by National Key Research and Development Plan (Grant No.2016YFA0200504), National Science and Technology Major Project (Grant No. 2017ZX02315001-004), National Natural Science Foundation of China (Grant No. 61421005), and 111 Project (Grant No. B18001).

REFERENCES

[1] H. S. P. Wong et al., "Metal–Oxide RRAM," in Proceedings of the IEEE, vol. 100, no. 6, pp.1951-1970, June 2012.

[2] P.S Chen et al. "Impacts of device architecture and low current operation on resistive switching of HfOx nanoscale devices." in Microelectronic Engineering, vol. 105, pp. 40-45, May 2013.

[3] H. Y. Lee et al., "Low-Power and Nanosecond Switching in Robust Hafnium Oxide Resistive Memory With a Thin Ti Cap," in IEEE Electron Device Letters, vol. 31, no. 1, pp. 44-46, Jan. 2010.

[4] K. Chuang et al., "Effects of Electric Fields on the Switching Properties Improvements of RRAM Device With a Field-Enhanced Elevated-Film-Stack Structure," in IEEE Journal of the Electron Devices Society, vol. 6, pp. 622-626, 2018.

[5] Y. Huang, W. Tsai, C. Chou, C. Wan, C. Hsiao and H. Cheng, "High-Performance Programmable Metallization Cell Memory With the Pyramid-Structured Electrode," in IEEE Electron Device Letters, vol. 34, no. 10, pp. 1244-1246, Oct. 2013.

[6] Shin, Keun‐Young, et al. "Controllable Formation of Nanofilaments in Resistive Memories via Tip‐Enhanced Electric Fields." Advanced Electronic Materials, vol. 2, no. 10, 1600233 Oct 2016.

[7] X. Guan, S. Yu and H. -. P. Wong, "On the Switching Parameter Variation of Metal-Oxide RRAM—Part I:

Physical Modeling and Simulation Methodology," in IEEE Transactions on Electron Devices, vol. 59, no. 4, pp. 1172-1182, April 2012.

[8] A. Padovani, L. Larcher, O. Pirrotta, L. Vandelli and G. Bersuker, "Microscopic Modeling of HfOx RRAM Operations: From Forming to Switching," in IEEE Transactions on Electron Devices, vol. 62, no. 6, pp. 1998-2006, June 2015.

IMPROVEMENT ON ELECTRONIC CHARACTERISTICS OF TAOX/TIOX DUAL-LAYER STRUCTURE RESIATIVE MEMORY

Yu She, Honggang Pan, Fang Wang*, Chuang Li, Zhenzhong Zhang, Yemei Han, Kailiang Zhang*
Tianjin Key Laboratory of Film Electronic & Communication Devices
School of Electrical & Electronic Engineering, Tianjin University of Technology, Tianjin 300384, China.
*Corresponding Author's Email: fwang75@163.com (Fang Wang);kailiang_zhang@163.com (Kailiang Zhang) and 86-22-60214229

ABSTRACT

With the rapid development of modern information, resistance random access memory (RRAM) has been a continuous research hotspot because of its large storage capacity, high reading speed, and excellent low-power performance. In this paper, in order to improve switching ratio of single-layer TiOx structure and uniformity of device, the double-layer TaOx/TiOx-based RRAM was fabricated and characterized by a semiconductor device analyzer (Agilent B1500A). The results showed that the uniformity was improved, and the window was enlarged from 33 to 100 by adding the TaOx layer in ITO/TiOx/TiN structure. And the devices with double-layer TaOx/TiOx had lower operating voltage and more consistent SET/RESET voltage distribution compared with the single-layer TiOx structure. In addition, the position of TaOx insertion layer was further discussed and it was found that the insertion layer could effectively increase the oxygen vacancy concentration in the resistive layer. A possible switching mechanism of TaOx/TiOx-based RRAM was inferred. It was believed that TaOx/TiOx double-layer structure can reduce the randomness of the filament's formation/fracture position to some extent, thereby making the uniformity better.

INTRODUCTION

RRAM with its high memory density, low power consumption, fast reading speed, long retention time, high density integration and other advantages have become an attractive flash replacer [1-3]. In recent years, the phenomenon of resistance switching has been found in perovskite oxides of complex metal oxides (such as SrTiO and SrZrO), and then in binary metal oxides (such as NiO, TiO2, HfO2 and TaOx). TiOx is one of the earliest materials used to explore resistive storage. Due to its multi-resistive states that can be reacted with electrodes in the preparation process, TiOx-RRAM has become a research hotspot [4]. At the same time, the poor uniformity of TiOx-RRAM has been an obstacle to be resolved. In recent years, in order to improve the performance of RRAM, a TiO_2-based RRAM device with very simple

$ITO/TiO_2/Pt$ structure was fabricated and shown high stable retention characteristic for over 104 seconds with a resistance ratio (R_{HRS}/R_{LRS}) of around 2 orders. Further, its average values of V_{set} and V_{reset} were reduced to 0.6V and −0.5V, which was remarkable in low operation voltage application. However, the uniformity of the structure cannot meet the needs of practical application [5]. To solve this problem, in this work, the TaOx layer was inserted in ITO/TiOx/TiN device. Furthermore, the effects of different insertion positions on uniformity and window were discussed. Finally, the possible resistance switching mechanism was inferred.

EXPERIMENT

The ITO/TiOx/TiN devices were fabricated on Si substrate and the base vacuum of the chamber was higher than 5×10^{-4} Pa during film deposition. A 50nm TiN as bottom electrode (BE) was deposited on Si wafer by reactive magnetron sputtering with Ti target under the condition of 90W power. Subsequently a 10nm TiOx as the switching layers was deposited on the BE by magnetron sputtering with TiO_2 ceramic target under the condition of 180W radio frequency (RF). The sputtering pressure was 0.8Pa and the oxygen partial pressure was fixed at 10%. Respectively 5nm intercalated TaOx were deposited on the BE compare with on the TiOx layer. Finally, an ITO top electrode (TE) of about 200nm was deposited with a metal mask 300μm in diameter and an ITO ceramic target by 150W RF magnetron sputtering(The sputtering pressure was 0.4 Pa.) in an Ar environment.

RESULTS AND DISCUSSIO

In this work, the ITO/TiOx/TiN, ITO/TiOx/TaOx/TiN and ITO/TaOx/TiOx/TiN structures are referred to as T1 (type 1), T2 (type 2) and T3 (type 3), respectively, for convenience. Figure 1 (a), (b), (c) showed the typical DC I–V characteristics of T1, T2 and T3, respectively, for 6 switching cycles.

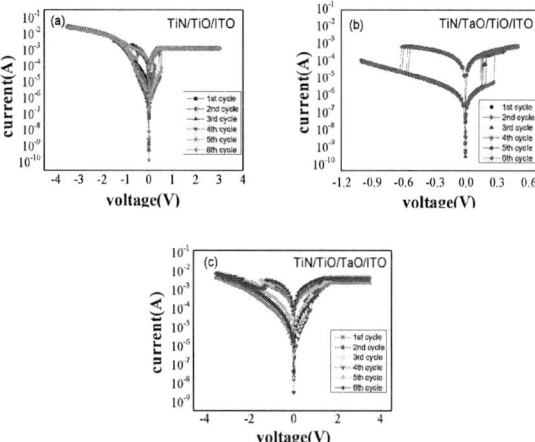

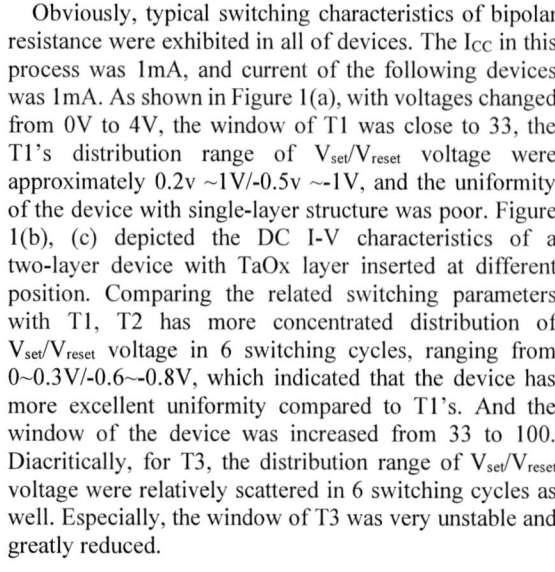

Figure 1: The typical DC I-V characteristic of (a) T1 (b)T2 (c)T3 for 6 switching cycles

Obviously, typical switching characteristics of bipolar resistance were exhibited in all of devices. The I_{CC} in this process was 1mA, and current of the following devices was 1mA. As shown in Figure 1(a), with voltages changed from 0V to 4V, the window of T1 was close to 33, the T1's distribution range of V_{set}/V_{reset} voltage were approximately 0.2v ~1V/-0.5v ~-1V, and the uniformity of the device with single-layer structure was poor. Figure 1(b), (c) depicted the DC I-V characteristics of a two-layer device with TaOx layer inserted at different position. Comparing the related switching parameters with T1, T2 has more concentrated distribution of V_{set}/V_{reset} voltage in 6 switching cycles, ranging from 0~0.3V/-0.6~-0.8V, which indicated that the device has more excellent uniformity compared to T1's. And the window of the device was increased from 33 to 100. Diacritically, for T3, the distribution range of V_{set}/V_{reset} voltage were relatively scattered in 6 switching cycles as well. Especially, the window of T3 was very unstable and greatly reduced.

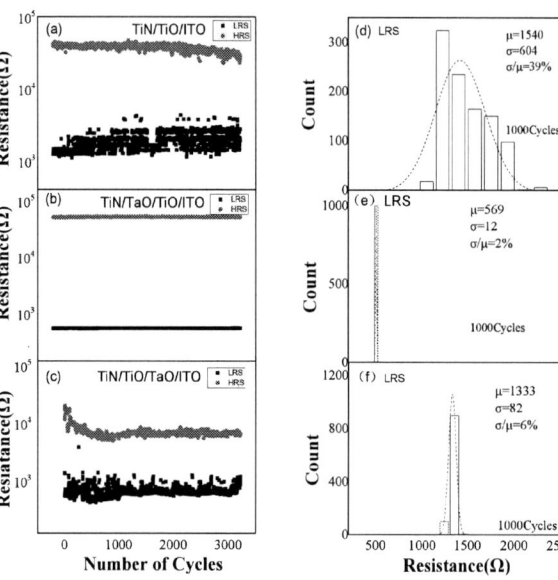

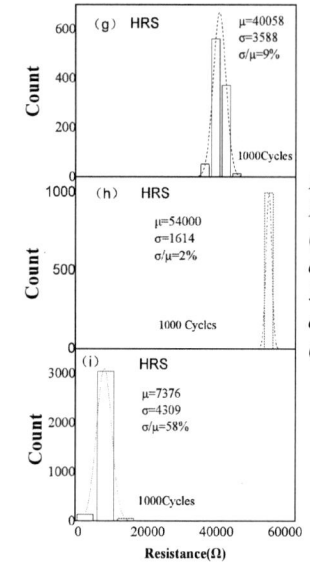

Figure 2: The endurance DC test of (a) T1 (b) T2 (c) T3; resistance value distribution from the 3000 cycles of the endurance test of LRS (d)-(f); HRS (g)-(i);

The endurance test results for the three devices up to 3000 cycles with 1mA I_{CC} were shown in Figures 2 (a)-(c). The window of T1 was about 33 within 3000 cycles. Obviously, HRS decreased while the LRS fluctuated greatly, thus the uniformity of T1 was poor. T2 showed a good uniformity over 3000 cycle that HRS/LRS were hardly any obvious fluctuation. HRS remained at around 56000 Ω and LRS remained at around 560 Ω. However, compared with T2, T3 could not achieve the improvement of uniformity. The window of T3 sharply decreased within 3000 cycles, both HRS and LRS with a large fluctuation, and finally the window attenuates to about 6(<10) near 3000. Therefore, the uniformity of T3 was not

ideal. Furthermore, a statistical summary of the R_{LRS} and R_{HRS}, respectively, of the corresponding devices were shown in *Figures 2* (d)-(i). From the fitting of the distributions according to the Gaussian function, the mean values (μ) and standard deviations (σ) that were shown in each graph. The coefficient of variation (σ/μ) corresponds to the uniformity of the devices. That is, the larger the coefficient of variation is, the worse the uniformity [6]. The coefficients of variation, R_{LRS}/R_{HRS}, for T1, T2 and T3 were 39%/9%, 2%/2% and 6%/15%, respectively. Along with the variations in the mean values (μ) R_{LRS} and R_{HRS} values as discussed above, the minimum coefficient of variation (σ/μ) of both R_{LRS} and R_{HRS} were achieved from T2, and excellent endurance with high uniformity was only achievable of T2. An important result can be determined from Figure 2. That is, the TaOx insertion layer can effectively improve the uniformity of the single-layer device. However, due to the asymmetric ITO/TiOx/TiN structure, the placement of TaOx would also affect the performance of the device. Therefore, in this work, the insertion of TaOx to the bottom interface has successfully improved the uniformity and switching ratio. Overall, ITO/TiOx/TaOx/TiN device showed the best endurance and switching uniformity. It is meaningful for future research nonvolatile memory.

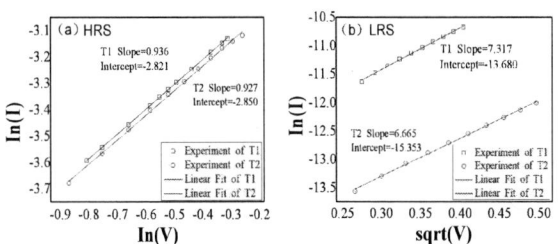

Figure 3: (a) (b) Analysis of the LRS/HRS conduction mechanism of T1/T2

To further investigate the cause of the excellent uniformity of T2 contract T1, the electrical conduction mechanisms in the samples for both R_{LRS} and R_{HRS} were fitted based on the I − V measurements, respectively. Figure 3 (a) describes linear fitting of ln (I) versus In (V) for the LRS. It can be obtained that the slopes of the T1 and T2 are 0.936 and 0.927 (approximately equal to 1). This indicates that the conduction mechanism in the LRS may be dominated by Ohmic [7]. The experimental data can be well fitted by straight lines for both T1 and T2 in the HRS from (b), which implies that the conduction mechanism for the HRS in the above samples was dominated by Schottky emission theory. Besides, the conduction mechanism for the LRS/HRS does not change after introduction of the TaOx layer. Therefore, one possible switching mechanism was inferred. In T1, the resistance switching was mainly determined by the formation and fracture of the oxygen vacancies conductive filament. In T2, the TaOx layer was inserted to obtain oxygen ions from the original TiOx layer, resulting in the increase of oxygen vacancy defects in TiOx layer. When the positive voltage was applied, the oxygen vacancy concentration in TiOx layer was higher, that is, it was easier to form oxygen vacancy conductive filaments and the resulting conductive filaments were more stable, the operating voltage of the TiOx layer was reduced at the same time. When applying negative voltage, due to the Ti-O binding energ slightly larger than Ta-O, energy that was applied to the voltage on TaOx can drive oxygen ions back on TiOx layer, which made the V_{reset} was relatively low, so the conductive filament fracture faster and more even.

CONCLUSION

In this paper, the effect of the insertion of TaOx layer on uniformity in a single-layer ITO/TiOx/TiN was investigated. The results showed that the insertion position of the TaOx layer would affect the performance of the device. When the TaOx layer was inserted into the top interface, the uniformity and endurance deteriorated, but inserted into the bottom interface, the uniformity improved and the window enlarged from 33 to 100. Compared with the single-layer TiOx structure, the double-layer structure had lower operating voltage and more consistent Vset/Vreset distribution, which coefficient of variation (σ/μ) reducing to 2%. We inferred that the HRS/LRS switching process were related to the formation and fracture process of oxygen vacancy conductive filament. The double-layer TaOx/TiOx structure captures oxygen ions from TiOx layer, increasing oxygen vacancy concentration, moreover, the TaOx layer as a series resistor, effectively the uniformity was improved, and operating voltage reduced. I believe that our work would lay a solid foundation for the future improving the uniformity of RRAM.

ACKNOWLEDGEMENTS

This work was supported by Natural Science Foundation of Tianjin City (Grant Nos.18JCZDJC30500,17JCYBJ C16 100 and 17JCZDJC31700) , National Key Research and Development Program of China (Grant No.2017YFB0405600) and National Natural Science Foundation of China (Grant Nos.61404091, 61274113, 61505144, 51502203 and 51502204).

REFERENCES

[1] Hongzhi Z, Kailiang Z, Fang W, et al. *Semiconductor Technology International Conference (CSTIC)*, 2015 China.

[2] Kailiang Z, Kuo S, Fang W, et al. *IEEE Electron Device Letters*, 2015, 36.10: 1018-1020.

[3] Kumar D, Aluguri R, Chand U, et al. *Nanotechnology*, 2018, 29.12: 125202.

[4] Sun J, Wang H, Song F, et al. *Small*, 2018, 14(27): 1800945.

[5] Shi-Xiang Chen, Sheng-Po Chang*,z, Shoou-Jinn Chang, et al. *ECS J. Solid State Sci. Technol* 2018, 7(7): 3183-3188.

[6] Yichuan Wang, Yu Yan, Chen Wang, et al. *Appl. Phys. Lett*, 2018, 113: 072902.

[7] Shi-Jian W, Wang F, Zhi-Chao Z, et al. *Chinese Physics B*, 2018, 27(8): 087701.

A NOVEL GATE ARCHITECTURE DESIGN IN STI BASED LDMOS

Ziquan Fang[1]*, Zhaozhao Xu[1], Wensheng Qian[1]

[1] HuaHong Grace Semiconductor Manufacturing Corporation, Shanghai 201206, China

* Corresponding Author's Email: Ziquan.Fang@hhgrace.com

ABSTRACT

Conventional STI based LDMOS devices always have an extended gate which performs as a poly plate on top of the STI. In this paper, we have proposed a novel LDMOS gate architecture with no overlap between gate and STI. Instead, the contacts landing on the STI perform as plate to obtain high off-state breakdown voltage (offBV). The novel LDMOS is compatible with CMOS process, and the offBV can be above 20V without additional drift implant mask. TCAD simulation is used to explain the underlying physics of the proposed novel architecture.

INTRODUCTION

Nowadays the Lateral Diffused MOS (LDMOS) are widely used in high voltage integrated circuits (HVICs). Classical researches of LDMOS are focusing on BV & R_{on} trade-off, on-state characteristics and safe-operating-area (SOA) improvement.

Conventional LDMOS technology is compatible with complementary CMOS process. Normally the doping concentration of CMOS well implant is more than $1e13/um^2$ to avoid latch-up effect. However, this doping level is too high for LDMOS drift region to be depleted, the off-state BV is gating at N/P well junction which is less than 14V. In order to obtain higher BV, additional implant mask layer with lighter doping is needed to define the drift region of LDMOS device.

In this paper, a novel LDMOS structure is presented, the drift region is shared with CMOS well and ldd implant, and the gate architecture is redesigned accordingly to ensure the LDMOS BV above 20V[1][2]. Therefore, in some circuit design cases that require a device with around 20V off-state BV, this novel LDMOS structure can be used without additional mask or cost.

DEVICE STRUCTURE

Fig.1(a) shows the cross sections of conventional STI based N-LDMOS. The drift region consists of L_A (drift and channel overlap), L_{PF} (poly gate and STI overlap) and L_{PA} (STI without gate overlap).

Fig.1(b) shows the novel LDMOS structure discussed in this paper[3][4][5]. L_{PF} is the space between poly gate and STI so that the Nldd can be implanted under the channel to form L_A region (self-align Nldd implant). L_{PA} is the width of STI, the contact landing on the L_{PA} region is connected with poly gate to perform as a plate. D_A is the space between poly gate and NW, so that the device BV is no more gating at the N/P well junction.

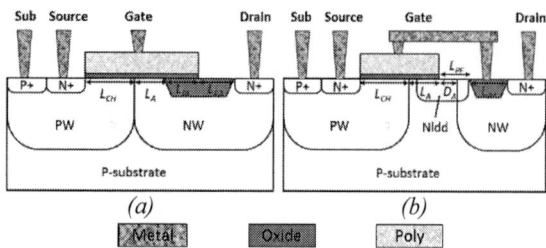

Fig.1. Cross-section of (a) conventional STI based LDMOS (b) novel LDMOS gate architecture

Fig.2 shows the process flow of the novel LDMOS by TCAD simulation. The simulation is based on a 90nm low power platform with gate operation voltage (V_g) of 5V. It can be seen in Fig.2(a) that the L_A region has not been formed after gate formation and it is formed after Nldd implant shown in Fig.2(b).

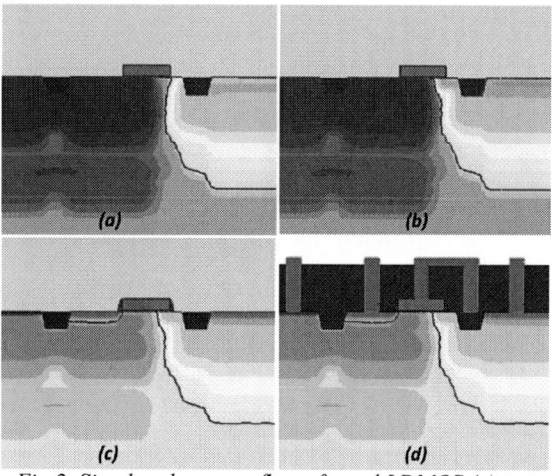

Fig.2. Simulated process flow of novel LDMOS (a)post gate formation (b) post Nldd implant (c) poly SD implant (d) post contact & metal formation

Fig.3(a) shows the simulated 2D device doping contour of conventional LDMOS (L_{CH}=0.4, L_A=0.2, L_{PF}=0.4, L_{PA}=0.2um), which drift region is shared with CMOS NW implant condition. Fig.3(b) is novel LDMOS (L_{CH}=0.4, L_A=0.2, L_{PF}=0.2, L_{PA}=0.4, D_A=0.2um) doping contour, which drift region is shared with CMOS NW and Nldd condition. Device characteristics comparison will be displayed in the following parts.

978-1-7281-6559-2/20 $31.00 © 2020 IEEE

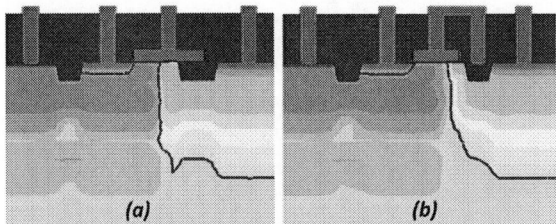

Fig.3. Simulated 2D device doping contour of (a) conventional STI based LDMOS (b) novel LDMOS

RESULTS AND DISCUSSION

BV and I_dV_d performance of the two device structures are compared on Fig.4. The BV of novel LDMOS is ~8V higher than conventional LDMOS, while the Id_{sat} is ~8% slower. It can be explained by considering the depletion width and impact-ionization (II) rate shown in Fig.5.

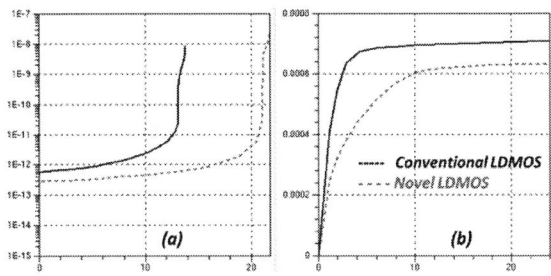

Fig.4. Comparison of (a) BV and (b) I_dV_d curve for conventional and novel LDMOS structure

The white line in Fig.5 indicated the depletion region boundary when BV occurs. In Fig.5(a), the drift region is hard to deplete due to heavy NW concentration, which cause low BV (~13V) at PW/NW junction. The severe II rate between depletion region (area in red color) results in early device breakdown. In Fig.5(b), it is obvious that the depletion width is wider, as there is 0.4um space between PW/NW, therefore higher BV (~21V) can be obtained.

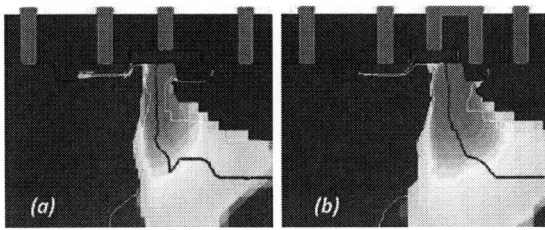

Fig.5. Simulated impact-ionization (II) rate contour and depletion width of (a) conventional LDMOS and (b) novel LDMOS

L_A variation in novel LDMOS structure is studied in Table I. As the L_A region is formed by Nldd self-align implant, the effective L_A length is more determined by Nldd implant energy and dose than affected by L_A size.

Therefore, the device BV and Id_{sat} performance almost not affected with L_A size varies from 0.1um to 0.3um. However, Id_{sat} starts to drop with L_A longer than 0.3um due to channel length increases with longer L_A.

Table II shows the L_{PF} size impact to device. With L_{PF} size varies from 0.1um to 0.4um, BV drops around 7V, while Id_{sat} remain the same. Fig.6(a)(b) show the II rate contour and depletion boundary of L_{PF}=0.1um and 0.4um accordingly. It can be seen that the L_{PF} region cannot be fully depleted when L_{PF}=0.4um, therefore II location transfer from PW/NW junction under STI to silicon surface under poly edge which cause the BV drop.

TABLE I.   PERFORMANCE COMPARISON WITH VARIES L_A SIZE FOR NOVEL LDMOS STRUCTURE

LA size (um) PF/DA/PA=0.2/0.2/0.4	0.1	0.2	0.3	0.4
BV$_{DS,off}$ (V_{GS}=0V), V	18.5	20.7	18	17.5
Id$_{sat}$ (V_{GS}/V_{DS}= 5V/20V), uA/um	645	634	638	617

TABLE II.   PERFORMANCE COMPARISON WITH VARIES L_{PF} SIZE FOR NOVEL LDMOS STRUCTURE

PF/DA size (um) LA/PA=0.3/0.4	0.1	0.2	0.3	0.4
BV$_{DS,off}$ (V_{GS}=0V), V	21.03	18.1	14.4	14.45
Id$_{sat}$ (V_{GS}/V_{DS}= 5V/20V), uA/um	630	638	626	626

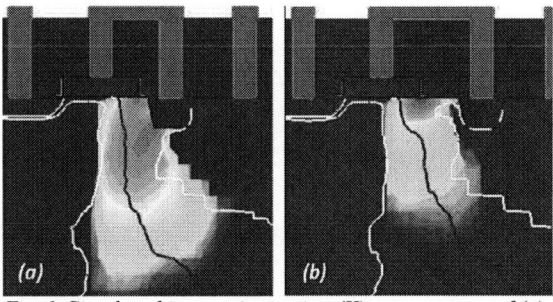

Fig.6. Simulated impact-ionization (II) rate contour of (a) L_{PF}=D_A=0.1um (b) L_{PF}=D_A=0.4um

D_A size determines the NW relative location to STI. In Table III, when D_A=0.4 (L_{PF}=0.2um), Id_{sat} drops more than 20%.

TABLE III.   PERFORMANCE COMPARISON WITH VARIES D_A SIZE FOR NOVEL LDMOS STRUCTURE

DA size (um) LA/PF/PA=0.2/0.2/0.4	0.1	0.2	0.3	0.4
BV$_{DS,off}$ (V_{GS}=0V), V	13.3	20.75	24.3	27.2
Id$_{sat}$ (V_{GS}/V_{DS}= 5V/20V), uA/um	650	634	602	502

I_dV_d curve in Fig.7(b) shows severe quasi-saturation effect with increasing D_A. Fig.8 is the simulated current density contour comparison of D_A=0.1um and 0.4um. it is obvious that the current path is limited near the left bottom of STI, the narrow current path leads to the quasi-saturation effect.

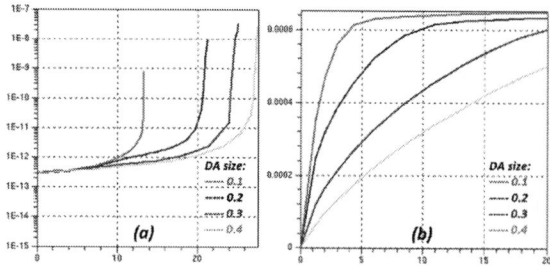

Fig.7. Comparison of (a) BV and (b) I_dV_d curve with varies DA size for novel LDMOS structure

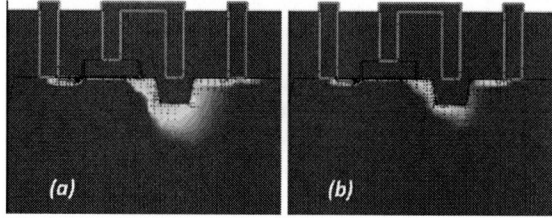

Fig.8. Simulated total current density contour with V_g=5V, V_d=10V of (a) D_A=0.1um (b) D_A=0.4um

CONCLUSION

A novel LDMOS device structure is introduced in this paper. The device BV and Id_{sat} performance are studied by TCAD simulation. The novel LDMOS can provide >20V BV with shared NW/PW condition with CMOS device. However, the device size needs to be carefully design to obtain best device performance.

ACKNOWLEDGEMENTS

The authors would like to thank all members of device group and process integration group of HHgrace for their great support on this work.

REFERENCES

[1] S. Teja, M. Bhoir and N. R. Mohapatra, "Split-gate architecture for higher breakdown voltage in STI based LDMOS transistors," 2017 International Conference on Electron Devices and Solid-State Circuits (EDSSC), Hsinchu, 2017, pp. 1-2.

[2] D. Muller et al., "High-Performance 15-V Novel LDMOS Transistor Architecture in a 0.25-um BiCMOS Process for RF-Power Applications," in IEEE Transactions on Electron Devices, vol. 54, no. 4, pp. 861-868, April 2007.

[3] D. Muller, A. Giry, D. Pache, J. Mourier, B. Szelag and A. Monroy, "Architecture optimization of an N-channel LDMOS device dedicated to RF-power application,"

Proceedings. ISPSD '05. The 17th International Symposium on Power Semiconductor Devices and ICs, 2005., Santa Barbara, CA, 2005, pp. 159-162. doi: 10.1109/ISPSD. 2005.1487975

[4] K. Na, K. Baek, G. Lee and Y. Kim, "High-Voltage LDMOS Transistor with Split-Gate Structure for Improved Electrical Performance," in IEEE Transactions on Electron Devices, vol. 60, no. 10, pp. 3515-3520, Oct. 2013.

[5] H. Liu et al., "A novel high-voltage LDMOS with shielding-contact structure for HCl SOA enhancement," 2017 29th International Symposium on Power Semiconductor Devices and IC's (ISPSD), Sapporo, 2017, pp. 311-314.

978-1-7281-6559-2/20 $31.00 © 2020 IEEE

IMPACT OF CIRCUIT LIMIT AND DEVICE NOISE ON RRAM BASED CONDITIONAL GENERATIVE ADVERSARIAL NETWORK

Shengyu Bao[1], Zongwei Wang[1,2,*], Tianyi Liu[1], Daqin Chen[1], Yimao Cai[1,3,*] and Ru Huang[1,2]

[1]Institute of Microelectronics, Peking University, Beijing 100871, China

[2]Key Laboratory of Microelectronic Devices and Circuits, Peking University, Beijing 100871, China

[3]Frontiers science center for nano-optoelectronics, Peking University, Beijing, 100871, P. R. China

*Corresponding Author's Email: {wangzongwei, caiyimao}@pku.edu.cn

ABSTRACT

In this work, a Conditional Generative Adversarial Network (CGAN) [1] is demonstrated based on the Resistive Random Access Memory (RRAM). During training, the read noise of RRAM is utilized as a random bias source to enrich the diversity of the generator in CGAN. Further, we evaluate the impact of both read noise (RRAM as weight storage cell) and the resolution of the AD/DA circuit on the performance of CGAN through a comprehensive simulation.

Keywords—CGAN; RRAM; Read Noise(key words)

INTRODUCTION

Generative Adversarial Networks (GAN) have emerged as a state-of-the-art approach to train generative models and are successfully implemented in generating realistic/stylish images and semi-supervised learning. Although GAN succeeds in generating new analogous patterns for a designated dataset, the types of examples are randomly generated without labels. To generate images that belong to a given type, conditional GAN is developed, allowing image generation to comply with a conditional class label. Recently, it is proposed that a novel GAN with a complementary generator is needed for high-performance semi-supervised learning [2]. This structure requires additional randomness to diversify the generator and change the distribution of the generated image.

The RRAM-based neuromorphic computing system (NCS) is a promising approach for neural network acceleration. In recent years, many researches have shown the great potential of RRAM to improve energy efficiency as well as the performance when implementing various algorithms, such as convolutional neural network (CNN), recurrent neural network (RNN) and GAN [3]. Nevertheless, the inherent non-ideal effects (static and dynamic variations) remain critical issues. As a result, neural network accelerators may suffer significant accuracy degradation when non-ideal effects are ignored [4].

On the other hand, the intrinsic randomness of non-ideal effects can be utilized to obtain a novel GAN, which may be suitable for high-performance semi-supervised learning. In this work, we proposed a

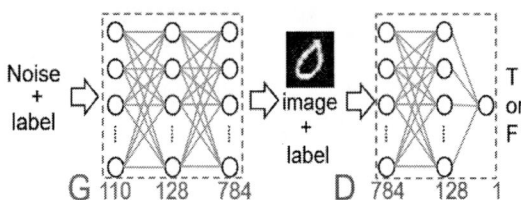

Figure 1: Illustration of the CGAN. The G of CGAN is a 110*128*784 MLP while the D is a 784*128*1 MLP.

CGAN system employing the RRAM read noise as a random input bias. Further, we investigated the impact of RRAM intrinsic non-ideal effects and circuit limits on the performance of CGAN.

CGAN

As shown in Figure 1, CGAN has two competing parts: one is Generator (G), and the other is Discriminator (D). Both are traditional neural networks, such as Multi-Layer Perceptron (MLP) and CNN.

During the training process, G is trained to take random noise and label as input and generate fake data, which is similar to the real data; D is trained to distinguish real/fake data and check whether the label and the data matches. The training process ends when the competition between G and D achieves Nash Equilibrium.

RESULTS AND DISCUSSION

The structure of the proposed CGAN system is shown in Figure 2. The RRAM array in Figure 2 (a) is used to generate the input noise for CGAN. In this case, we take advantage of the RRAM non-ideal effect to form a noise signal generator and create a complementary generator in CGAN.

As shown in Figure 2 (b), the G in CGAN is a multi-layer perceptron (MLP) based on two RRAM crossbar arrays. Our CGAN is pre-trained with MNIST data, thus can generate fake MNIST images. The quality of these fake images reflects the quality of our CGAN system.

The impact of DAC resolution on the quality of CGAN generated images is investigated. We assess the quality of the generated images by using the accuracy of a pre-trained MLP classifying the generated images. As shown in Figure 3, the accuracy of classifying generated

978-1-7281-6559-2/20 $31.00 © 2020 IEEE

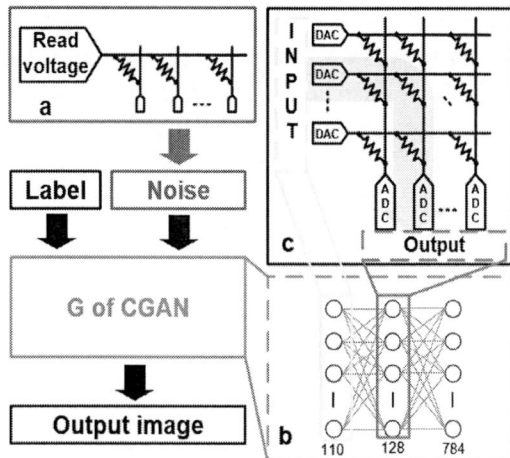

Figure 2: Illustration of G in simulation, (a) is a 1*100 RRAM array providing the input noise of G; (b) shows the structure of G, which is a 110*128*784 MLP; (c) shows the mapping of the first layer of G on a 110*128 RRAM-crossbar, and the second layer is mapped on a 128*784 RRAM-crossbar.

image increases with the resolution of DAC, and the saturation point is reached when the resolution of DAC is 5-bit. The result shows that 5-bit resolution is enough in this case

The impact of the standard deviation of read noise is studied with 5-bit DAC resolution. As shown in Figure 4, the quality of the generated images gets worse while the standard deviation of RRAM (as the weight storage) read noise increases. The quality of the generated image drops quickly when the standard deviation of RRAM read noise

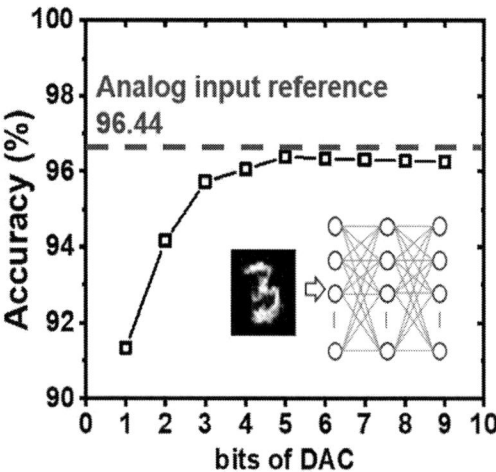

Figure 3: Influence of DAC resolution on accuracy of pre-trained MLP classifying RRAM-CGAN generated MNIST image.

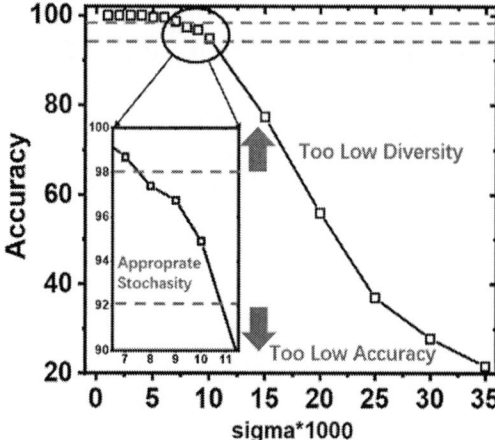

Figure 4: Influence of read noise on accuracy of pre-trained MLP classifying RRAM-CGAN generated MNIST image.

is higher than 0.01.

However, there is a tradeoff between the diversity and quality of the generated image, because our noise provider and image generator share the same kind of RRAM. As shown in Figure 4, the diversity of the generated image is poor when the accuracy of classifying generated image is higher than 98%. It is important to balance the quality and diversity of the generated image.

The proposed implementation of CGAN might be helpful in shaping semi-supervised learning. As shown in Figure 5, the generated images with high quality and diversity are mixed with an extracted dataset to train an MLP classifier to check whether they can improve the performance of the classifier. Ten thousand generated images are mixed into the original MNIST training data to train an MLP classifier as an experimental group; while another MLP only trained by original MNIST training data is used as a control group. The accuracy of the experimental group and control group classifying original

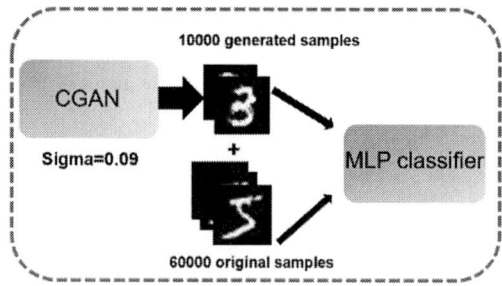

Figure 5: 10 thousand generated samples are mixed with original MNIST samples to train an MLP classifier, the quality of the classifier is assessed by recognizing original MNIST testing data.

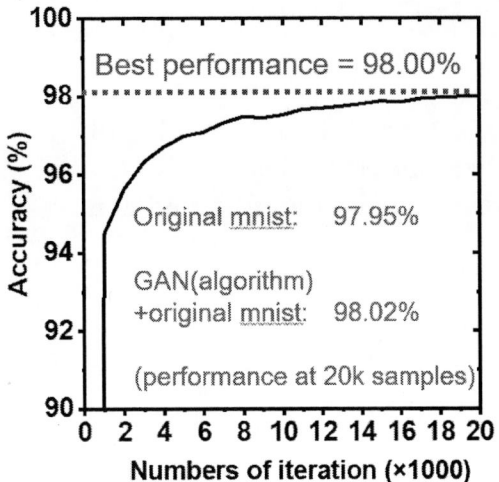

Figure 6: accuracy of MLP in experimental group (trained with CGAN generated images and original MNIST training data) classifying original MNIST testing data

MNIST testing data is shown in Figure 6. After 20 thousand training iterations, the accuracy of the experimental group reaches the saturation point of 98.00%, which is higher than the control group 97.95%. The result proves that the generated images improve the training process, which means the generated images enrich the diversity of the training data with high quality.

In addition, the quality of the images generated by our system is compared with images generated by an ideal one, which gets an accuracy of 98.02% in training MLP classifier as shown in Figure 6. This result indicates that the read noise (in the synaptic array) of the generator leads to a lower quality of generated images.

CONCLUSION

We demonstrated an RRAM-based Conditional Generative Adversarial Nets. The impact of digital circuit resolution and RRAM read noise on the diversity and quality of the CGAN is investigated through simulation. The potential of proposed CGAN for semi-supervised learning is envisioned.

ACKNOWLEDGEMENTS

This work was supported in part by the National Key Research and Development Project under grant No. 2018YFB1107701, in part by the National Natural Science Foundation of China under grant No. 61834001, No. 61904003, No. 61421005, and in part by the "111" Project under grant No. B18001. Z. W. acknowledges the support from China Postdoctoral Science Foundation (No. 2019M650340).

REFERENCES

[1] M. Mirza. *arXiv*: 1411.1784 [cs.LG], 2014
[2] Z. Dai. *Neural Information Processing Systems, NIPS 2017.*
[3] Z. Wang. *Nanoscale*, vol. 8, pp. 14015-14022, 2016.
[4] J. Kang. *IEEE International Electron Devices Meeting (IEDM), 2017.*

IMPLEMENTATION OF GRAPH CONVOLUTION NETWORK BASED ON ANALOG RRAM

Daqin Chen[1], Zongwei Wang[1,2,], Shengyu Bao[1], Yimao Cai[1,3], Ru Huang[1,2]*

[1]Institute of Microelectronics, Peking University, Beijing 100871, China

[2]Key Laboratory of Microelectronic Devices and Circuits, Peking University, Beijing 100871, China

[3]Frontiers science center for nano-optoelectronics, Peking University, Beijing, 100871, P. R. China

*Corresponding Authors' Email: {wangzongwei, caiyimao}@pku.edu.cn

ABSTRACT

In this work, the implementation of Graph Convolutional Network (GCN) based on resistive switching memory is demonstrated through simulation. After training, the RRAM-based GCN can process a semi-supervised graph classification task. Further, the impacts of read noises and circuit bit-precision on the performance of GCN are analyzed. Results show the proposed GCN can reach high accuracy when bit-precision$\geq$4-bit. Moreover, read noise can severely affect accuracy.

INTRODUCTION

Benefiting from the rapid development of computing hardware, deep neural networks (DNNs) has achieved significant success in various applications such as computer vision and natural language processing (NLP). Data on social networks, biological gene networks and log of telecommunication networks can be structured with graphs, which represent both objects and their relationships. These data usually present irregular and interdisciplinary features. As a result, neural networks aimed at processing graphs have been widely investigated. One of the main difficulties with graph neural networks (GNN) is the inefficiency of parallelism and intensive data-exchange when shuffling large amount of parameters between memories and processors. The use of graph convolutional networks (GCN) improves the parallelization while the data shuffling between memory and processor remains power/time-consuming.

Resistive random access memory (RRAM) has emerged as one of the most promising synaptic devices for neuromorphic computing applications [1-3]. With a crossbar array structure, RRAM can accelerate the time/energy-consuming operation (e.g., multiply-and-accumulate (MAC)) in neural network computing in the analog domain with great parallelism. Various data-intensive computing algorithms, such as convolutional neural network (CNN) and recurrent neural network (RNN), have been successfully demonstrated based on RRAM array with improved energy efficiency [4, 5].

Therefore, the attempt to build RRAM-based GCN is a promising approach to improve energy efficiency in graph processing.

In this paper, we demonstrate the RRAM based GCN by simulating the task of classifying a karate club network. For the first time, we realized classifying a karate club network in a RRAM-based GCN.

GRAPH CONVOLUTIONAL NETWORK

One typical GNN architecture is graph convolutional networks, of which filter parameters are usually shared across all locations in the graph [7]. The goal of the model is to learn a function of signals on a graph $G = (V, E)$, which takes X and A as input. Feature matrix X summarizes a feature description xi for every node i in a $N \times D$ form. N represents number of nodes and D represents number of input features. Adjacency matrix A represents a description of the graph structure in matrix form. Graph G produces a node-level output Z. Z is an $N \times F$ feature matrix, where F is the number of output features per node. Graph-level outputs can be modeled by introducing pooling operation [7].

As an example, let's consider a simple form of a layer-wise propagation rule (1):

$$f(H(l), A) = \sigma(AH(l)W(l)) \qquad (1)$$

where $W(l)$ is a weight matrix for the l-th neural network layer, $\sigma(\cdot)$ is a non-linear activation function such as ReLU, and $H(0)$ represents feature matrix X. As shown in Figure 1, this is a simple but representative model in GCN. We demonstrate a simple GCN model on a well-known graph dataset: Zachary's

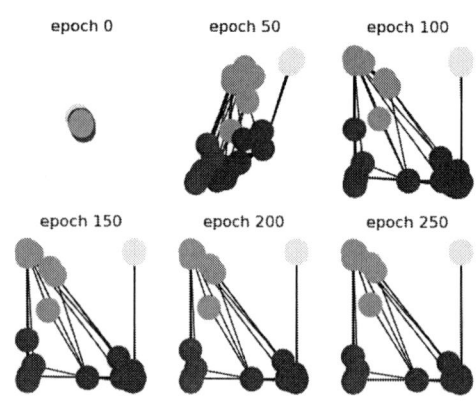

Figure 2: Dynamics for 250 training iterations with a single label per class. Labeled nodes are highlighted.

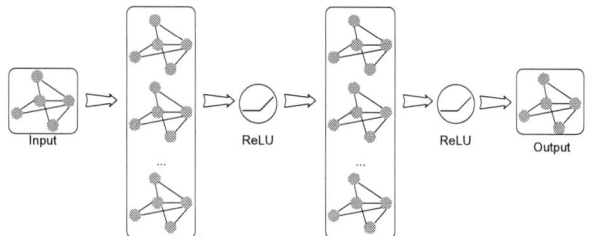

Figure 1: Multi-layer Graph Convolutional Network

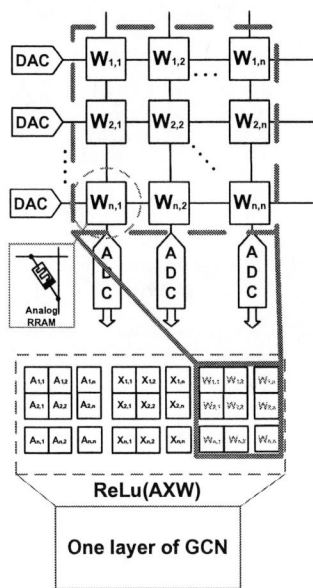

Figure 3: Mapping of the GCN to analog RRAM array

karate club network. Labels are utilized to train fully distinguishable and parameterized models. For each class, one node is labeled and trained for a few iterations. As shown in Figure 2, the 3-layer GCN model manages to divide the community linearly, giving only one labeled example for each category.

DEMONSTRATION OF RRAM-based GCN

As shown in Figure 3, each layer in GCN can be described as multiplications of three matrices in a queue. A layer of GCN consists of an adjacency matrix A, a feature matrix X, and a weight matrix W. The calculation process can be described as equation (2):

$$F = ReLu(A \square X \square W) \qquad (2)$$

In our system, Matrixes A and W are mapped into the analog RRAM crossbar arrays while registers are used to store the matrix X. As the input of the GCN, Matrix A is constant for a certain problem and can be reused in each layer. To implement equation (2), $A \square X$ is calculated by multiplying Matrix A by each column vector of X. In each clock cycle, Matrix A multiplies a column vector of matrix X, and the results are stored back into the registers. Therefore, $A \square X$ is calculated after clock cycles running equally to the amount of the column vectors. Similarly, each row vector of the registers multiplies matrix W to obtain the final results. A row vector of $(A \square X)$ multiplies matrix W, and the registers are updated by the calculated results. The result of one layer will be obtained after clock cycles running equally to the amount of the row vectors.

We also investigated the impact of read noise and circuit bit-precision on the inference procedure of GCN. The effects of RRAM read noise on classification are investigated via simulation. Figure 4 shows the whole framework of our GCN. The GCN contains 3 layers, the first layer has a 34×4 weight matrix $W1$, corresponding to a 34×4 RRAM array; the second layer is a 4×4 weight matrix $W2$, and the third layer contains a 4×2 weight matrix $W3$. The 34×34 RRAM array which represents the adjacency matrix A is shared among all layers. The performance of GCN is mainly evaluated by the accuracy.

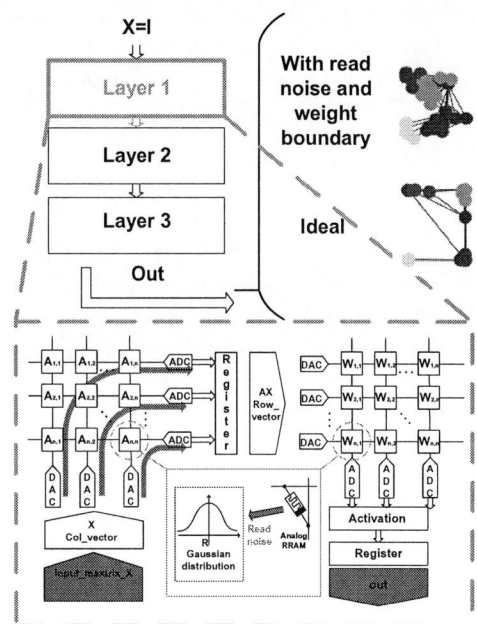

Figure 4: Whole framework of simulated GCN net

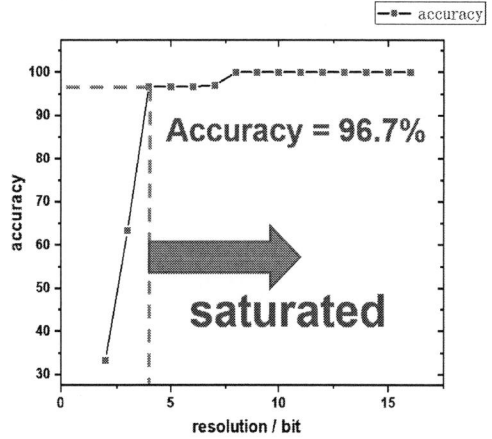

Figure 5: The impact of the A/D resolution on accuracy.

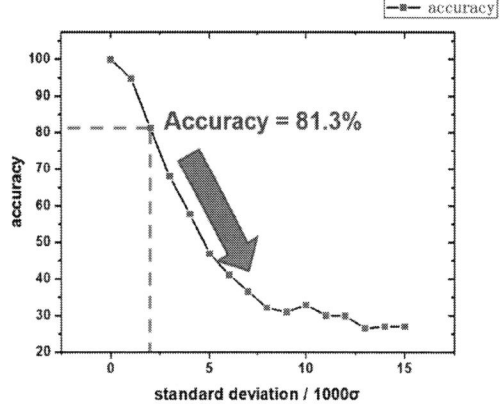

Figure 6: The impact of the read noise on accuracy.

Table 1 shows the impact of different read noises and bit-precision on accuracy. For A/D bit-precision over 4-bit, the accuracy of GCN nearly reaches saturation points, as shown in Figure 5, which means the 4-bit resolution is enough for our GCN system. However, read noise can severely affect accuracy. As shown in Figure 6, the accuracy of our GCN system decreases dramatically as the standard deviation of read noise increases. When the standard deviation of read noise is 0.001, the accuracy can be kept to 94.8%. The GCN system can't properly work if the standard deviation of read noise gets higher. In the inference mode, the read noise affects the calculation results in every MAC operation. What's worse, the impact of the read noise may accumulate from layer to layer, thus significantly impairing the accuracy.

Table 1 The impact of the read noise and A/D resolution on accuracy

Standard deviation σ	Accuracy	A/D resolution /bit	Accuracy
0.000	100%	2	33.3%
0.001	94.8%	3	63.3%
0.002	81.3%	4	96.7%
0.003	68.0%	5	96.7%
0.004	57.7%	6	96.7%
0.005	46.9%	7	97%
0.006	41.1%	8	100%

CONCLUSION

Key achievements: 1) To the best of our knowledge, this is the first time GCN was demonstrated based on analog RRAM; 2) The impact of read noises and A/D bit-precision on inference of RRAM-based GCN was investigated. In inference mode, results show that the proposed GCN can reach high accuracy with 4-bit A/D precision but is vulnerable to large read noise

ACKNOWLEDGEMENTS

This work was supported in part by the National Key Research and Development Project under grant No. 2018YFB1107701, in part by the National Natural Science Foundation of China under grant No. 61834001, No. 61904003, No. 61421005, and in part by the "111" Project under grant No. B18001. Z. W. acknowledges the support from China Postdoctoral Science Foundation (No. 2019M650340).

REFERENCES

[1] Z. Wang et al., "Fully memristive neural networks for pattern classification with unsupervised learning," Nat. Electron., vol. 1, no. 2, pp. 137–145, 2018.

[2] Z. Wang, M. Yin, T. Zhang, Y. Cai, Y. Wang, Y. Yang, R. Huang, Engineering incremental resistive switching in TaO_x based memristors for brain-inspired computing, Nanoscale, vol. 8, pp. 14015-14022, 2016

[3] W. Wu et al., "A Methodology to Improve Linearity of Analog RRAM for Neuromorphic Computing," Symp. VLSI Technology, pp. 3–4, 2017.

[4] H. Wu et al., "Device and circuit optimization of RRAM for Neuromorphic computing," IEEE International Electron Devices Meeting (IEDM), pp. 274–277, 2017.

[5] P. M. Sheridan, F. Cai, C. Du, W. Ma, Z. Zhang, and W. D. Lu, "Sparse coding with memristor networks," Nat. Nanotechnol., vol. 12, no. 8, pp. 784–789, 2017.

[6] J. Kang, et al, Time-Dependent Variability in RRAM-based Analog Neuromorphic System for Pattern Recognition, IEEE International Electron Devices Meeting (IEDM), 2017

[7] David K. Duvenaud, Dougal Maclaurin, Jorge Iparraguirre, Rafael Bombarell, Timothy Hirzel, Alan´ Aspuru-Guzik, and Ryan P. Adams. Convolutional networks on graphs for learning molecular fingerprints. In Advances in neural information processing systems (NIPS), pp. 2224–2232, 2015.

[8] Thomas N Kipf and Max Welling. Semi-supervised classification with graph convolutional networks. arXiv preprint arXiv:1609.02907, 2016

TCAD SIMULATION ON RANDOM TELEGRAPHY NOISE AND GRAIN-INDUCED FLUCTUATION OF 3D NAND CELL TRANSISITORS

Shijie Hu[1,2], Ming Li[2,3], Ru Huang[1,2*]*

[1]Shenzhen Graduate School, Peking University, Shenzhen 518055, CHINA
[2]Key Laboratory of Microelectronic Devices and Circuits (MOE), Institute of Microelectronics, Peking University, Beijing 100871, CHINA
[3]Frontiers Science Center for Nano-optoelectronics, Peking University, Beijing 100871, CHINA
Phone: 86-10-62765929, *E-mail: liming.ime@pku.edu.cn; ruhuang@pku.edu.cn

BIOGRAPHY

Shijie Hu is now pursuing the Master's degree at Institute of Microelectronics, Peking University, Beijing, China. His research interests include 3D NAND modeling and mechanism of random telegraph noise (RTN) effect in 3D NAND flash.

ABSTRACT

In this work, a TCAD simulation platform was set up to study the real poly-channel modeling, trap-induced noise and random grain doping in 3D NAND cell transistor. The random telegraph noise and size dependence was simulated and analyzed. Simulation results show that the RTN and grain size change have greater influence on threshold voltage (V_T) and channel current (I_D) in 3D NAND than doping fluctuation. It is shown that the instability caused by random doping cannot be ignored, too.

Keywords—3D NAND; random telegraph noise; variation; TCAD

INTRODUCTION

A 3D stacked NAND flash memory, featuring a vertical poly-Si channel and charge trap storage, has become the major technology for high density mass storage. However, the device-to-device variation caused by oxide traps, grains size and random grain doping will induce threshold voltage and channel current fluctuations and thus read/write failure. A single oxide trap in the nanometer-scale channel can induce great read current and threshold voltage change by capturing and emission of single electron. The use of poly-Si as channel material also gives rise to additional fluctuations due to the grain size variation and the high-density defects at grain boundaries. To accurately simulate the randomness of 3D NAND cell transistors, the challenges are mainly originated from the modeling of poly grains in the channel and the capturing/emission behaviors in the oxide traps and the boundary defects. The commercial TCAD tool, i.e. Synopsys Sentaurus, has not implemented the functions to generate the real poly grains and the transient response to single electron trapping and detrapping.

To do so, we have set up a 3D simulation platform to construct the random poly channel model and implement the electrical response to the single oxide trap. With the platform, the variability induced by grain size, oxide traps and random doping can be simulated simultaneously for the first time.

TCAD SIMULATION PLATFORM

Because Synopsys Sentaurus does not provide random geometric structure construction method, to build up a stochastic poly-Si channel, Matlab was used to generate the geometric structures of poly-Si grains with 3-D Voronoi

algorithm. Voronoi algorithm is a geometry structure randomly generate method. In two-dimensional Voronoi algorithm, random dots are generated as the initial seeds according to the designated average grain size. The dots are then connected to form the minimum triangles in which no other dot exists. The polygon region enclosed by the perpendicular lines of all the sides of triangles forms the poly-Si grain. Different from the 2D Voronoi method, the perpendicular lines should be replaced by mid-vertical planes to generate the grains in 3D Voronoi algorithm. After converting data types in the third-part tool, the geometric structures were converted into the entity models which can be recognized by Synopsys Sentaurus.

Fig. 1(a) shows the final model of 3D NAND cell transistors. Oxide filler was used to form a full deplete channel to decrease the off-state current. In O-N-O layer, tunneling oxide was set to 3 nm, charge trapping layer and block oxide layer were both set to 6 nm. Cell-to-cell space was 30 nm, and the length of single cell was 50 nm. Here we only simulated three cells to save the time consumption. Actually, the operation of cells are influenced by the nearest cells more than others, so the simple three-cell model is quite reasonable to present the nature of 3D NAND transistor. Traps are defined in tunneling oxide and near the interface.

In the real poly-Si channel, grains are usually not stacked on each other without any gap, whereas, there is about 0.5~2 nm spacing between them filled by amorphous silicon, as shown in Fig.1(b). Due to the randomness of grain shapes and sizes, the thickness of a-Si are also randomized.

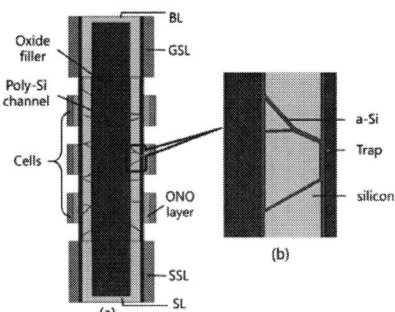

Fig. 1 (a) 3-D NAND TCAD plot for the randomly generated poly-Si grain configuration. (b) Poly-Si channel of 3-D NAND. Grains are separated by amorphous silicon and traps are locaed in oxide.

SIMULATION RESULTS

RTN results from the oxide trap capturing/emission of electrons from/to channel. At nanometer scale, both local electric field and electron density fluctuation will induce RTN in theory. But once channel doped, even very slightly, influence of one or several electrons on the electron density change would be very limited because the channel length of 3D NAND is so long that the number of electrons excited from dopants is much larger than the trapped electrons. Consequently, we focus more on the change of local electric field, which is the major reason for the change of local electron density and mobility.

Firstly, in order to investigate the influence of static electron trapping on the V_T and I_D fluctuation, we put a single trap in the tunneling oxide layer about 0.5 nm close to channel and set it as trapped or detrapped in turn. Fig. 2 (a) show the electric field distribution at channel surface with single electron trapped. It's shown that the local electric field around the trap decreased about four orders of magnitude once if the trap was filled by one electron. Under the same condition, as shown in Fig. 2 (b), the local electron density changed significantly in more extended area than the electric field was influenced. Besides, we can find that the electron density variation due to the single charge trapping strongly depends on the grain size which can be regarded as one of the main fluctuation sources in the polysilicon channel NAND device.

On other hand, the boundary gap between grains is also taken into account in this simulation. A certain gap was generated by shrinking the grains proportionally and filled by amorphous silicon. The grain boundary (GB) width in real polysilicon is randomly, which will greatly affect the average mobility. Since the grain boundary width is usually less than 2nm, we considered the thin-layer mobility model proposed by Susanna et al. in simulation. To compare the impact of different device model settings, Fig. 3 (a) shows different trans-characteristics results with different models with average grain size fixed as 31nm, where black line is simulated under condition of constant doping, without trap and grain boundary, red line including a single trap in tunneling oxide without boundary gap and blue line including random doping condition without trap in tunneling oxide(??). Random doping concentrations in polysilicon grains were generated according to normal distribution with average value μ=1e15 and standard deviation σ=0.005, and the doping concentration in the amorphous silicon was set as average value. Results show that only incorporating traps in tunneling oxide will cause about 18mV V_T shift. The incorporation of grain boundary gap and random doping in simulation will caused 14mV V_T shift. Fig. 3 (b) compares the trans-characteristics for average grain size Φ = 38 nm, 34nm and Φ =31nm, respectively. It's shown that drain current increases with Φ increasing but the sub-threshold current varies quite randomly. It indicates that the influence of grain size fluctuation may be more severe in the sub-threshold region.

Furthermore, we also simulated the dynamic drain current disturbance in time domain to respond the trapping and detrapping processes. To do that, firstly, electron capture rate ($C_{c,phonon}$) for the phonon-assisted transition be calculated by:

$$C_{c,phonon} = \frac{\sqrt{m_t m_0^3 k^3 T_n^3} g_c}{h^3 \sqrt{\gamma}} v_T S \omega \left[\frac{a(S-l)^2}{S} + 1 - \right.$$
$$\left. a \right] \exp\left[\frac{\Delta E}{2KT} + \gamma - s(1 + 2f_B) \right] \left(\frac{z}{1+\gamma} \right)^l \exp\left(\frac{E_F - E_C}{kT_n} \right) \frac{|\Psi(Z_0)|^2}{|\Psi(0)|^2} \quad (1)$$

where v_T is trap interaction volume and is related to trap position, S is Huang–Rhys factor, l is the number of

the phonons emitted in the transition, $f_B = [\exp\left(\frac{\hbar\omega}{kT}\right) - 1]^{-1}$, $z=2S\sqrt{f_B(f_B + 1)}$, $\gamma=\sqrt{l^2 + z^2}$, $\Delta E=E_c + \frac{3kT_n}{2} - E_{trap}$. Wave function ratio and phonons energy are different between trapping and detrapping. Then, probability distribution function of trap emission rate after electron been captured was generated by matlab according to the exponential distribution proposed by Tibor Grasser shown in Fig.4 (a), $g(\tau)=\frac{\tau}{\tau_e}\exp(-\frac{\tau}{\tau_e})$, where τ_e is the emission time constant and is the inverse of $C_{c,phonon}$. Finally, along with TCAD result, the time sequence of trapping and detrapping states was converted into drain current sequence along time. As shown in Fig. 4 (b), the RTN sequence was obtained for Φ = 31 nm and constant doping concentration of 1e15 cm^{-3}. With this method, the impact of grain size, grain doping concentration and the location of oxide trap on RTN can be easily investigated.

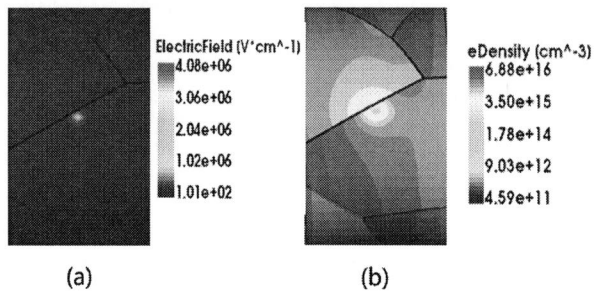

Fig. 2 (a) Channel electric field distribution, which influenced by single oxide electronic trap. (b) Channel .electric density distribution under same condition as (a).

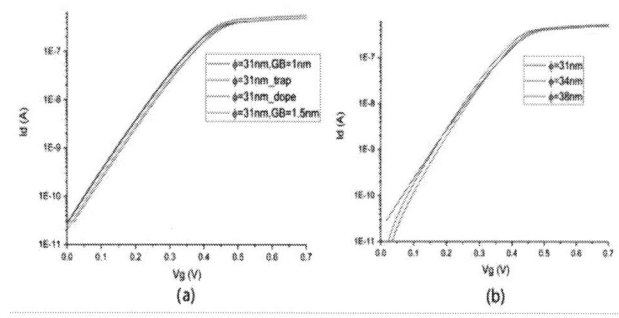

Fig. 3 (a) Different simulation condition. Black line is under constant doping, no trap and GB=1nm conditions. red line is same as black line but with single trap in tunneling oxide near channel side. Blue line is same as black line but with grains doping randomly. Green line is same as black line but GB=1.5nm. (b) Model with average grain size Φ = 38 nm, 34nm and Φ =31nm.

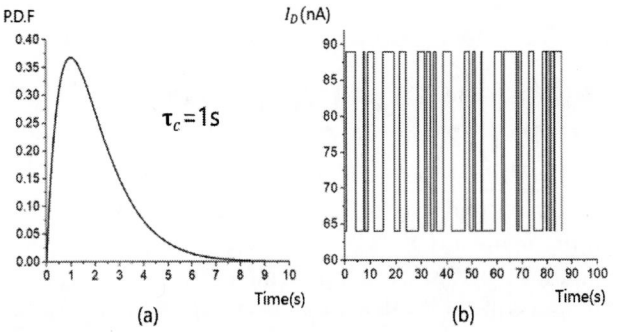

Fig. 4 (a) Trap capture electron probability distribution function varies with time when capture time constant $\tau_c = 1s$. (b) Channel current varies with time when $\tau_c = \tau_e = 1s$

CONCLUSION

In this paper, we have set up a simulation platform integrating the commercial TCAD tool and matlab tool for the random fluctuation and RTN simulation in 3-D NAND Flash cell transistors. By 3D Voronoi method, a real poly-Si channel model was built up to reflect the nature of grain boundaries. A time sequence of trapping and detrapping states was generated according to Tibor Grasser E/C model so as to combine the dynamic RTN with different static random fluctuation sources.

ACKNOWLEDGE

This work was supported in part by National Key Research and Development Plan (Grant No.2016YFA0200504), National Science and Technology Major Project (Grant No. 2017ZX02315001-004), National Natural Science Foundation of China (Grant No. 61421005), 111 Project (Grant No. B18001) and Cooperative Project from Western Digital Corp.

References

[1] R. Guerrieri, P. Ciampolini, A. Gnudi, M. Rudan and G. Baccarani, *Numerical simulation of polycrystalline-Silicon MOSFET's* in *IEEE* Transactions on Electron Devices, vol. 33, no. 8, pp. 1201-1206, Aug. 1986.

[2] G. Nicosia et al., *Impact of temperature on the amplitude of RTN fluctuations in 3-D NAND flash cells* 2017 IEEE International Electron Devices Meeting (IEDM), San Francisco, CA, 2017, pp. 21.3.1-21.3.4.

[3] C. Monzio Compagnoni, A. Goda, A. S. Spinelli, P. Feeley, A. L. Lacaita and A. Visconti, *Reviewing the Evolution of the NAND Flash Technology*, in Proceedings of the IEEE, vol. 105, no. 9, pp. 1609-1633, Sept. 2017.

[4] Grasser, Tibor . *Stochastic charge trapping in oxides: From random telegraph noise to bias temperature instabilities.* Microelectronics Reliability 52.1(2012):39-70.

[5] N. C. -. Lu, L. Gerzberg, Chih-Yuan Lu and J. D. Meindl, *Modeling and optimization of monolithic polycrystalline silicon resistors*, in *IEEE* Transactions on Electron Devices, vol. 28, no. 7, pp. 818-830, July 1981.

[6] D. Resnati *et al.*, *Characterization and Modeling of Temperature Effects in 3-D NAND Flash Arrays—Part I: Polysilicon-Induced Variability*, in *IEEE* Transactions on Electron Devices, vol. 65, no. 8, pp. 3199-3206, Aug. 2018.

[7] Sentaurus Device User Guide, Synopsys, Zurich, Switzerland, 2016.

[8] A. Ghetti, C. Monzio Compagnoni, A. S. Spinelli and A. Visconti, *Comprehensive Analysis of Random Telegraph Noise Instability and Its Scaling in Deca–Nanometer Flash Memories*, in *IEEE* Transactions on Electron Devices, vol. 56, no. 8, pp. 1746-1752, Aug. 2009.

A PHYSICAL CURRENT MODEL FOR MULTI-FINGER GATE TUNNELING FET WITH SCHOTTKY JUNCTION

Yimei Li[1], Jin Luo[1], Qianqian Huang[1,2]*, Xia An[1], Le Ye[1,2] and Ru Huang[1,2]*

[1]Key Laboratory of Microelectronic Devices and Circuits (MOE), Institute of Microelectronics
Peking University, Beijing 100871, China
[2]Peking University Information Technology Institute (Tianjin Binhai)
*Corresponding Author's Email: hqq@pku.edu.cn, ruhuang@pku.edu.cn

ABSTRACT

Compared with conventional tunneling field-effect transistor (TFET), a novel multi-finger gate tunneling FET with Schottky junction (mFSB-TFET) can effectively obtain the steeper subthreshold slope due to tunneling electric field coupling effect, the higher I_{on} due to the introduced multi-finger gate and Schottky junction, while keeping the ultra-low I_{off} simultaneously. To facilitate the circuit simulation by using this new device, in this paper, we establish a physical model of the channel surface potential in both gate handle region and gate finger region, and then further calculate the tunneling current by considering tunneling electric field coupling effect and the Schottky current by considering the equivalent barrier reduction respectively. The model results with different structure parameters agree well with the Sentautus TCAD simulation results, which shows the validity of this model.

INTRODUCTION

Tunneling field-effect transistor (TFET) operated by band-to-band tunneling (BTBT) mechanism has attracted much attention for ultra-low power application [1], because it can break the fundamental subthreshold slope (SS) limitation of MOSFET (60mV/dec at 300K) and has ultra-low off-state current for Si device. However, for conventional Si-based TFET, the low on-state current restricted by poor tunneling probability is the main challenge. In order to overcome the disadvantage of TFET, material engineering [2], structure engineering [3] and mechanism engineering [4] have been widely utilized. Through mechanism engineering, we proposed and fabricated a new multi-finger gate tunneling FET with Schottky junction (mFSB-TFET), which has the feature of Schottky junction, multi-finger gate and the electric field coupling effect between the fingers for the I_{on} enhancement and the SS improvement meanwhile maintaining the ultra-low I_{off} by self-depleted effect [5]. The CMOS-compatibly fabricated mFSB-TFET with optimized source pocket shows the record minimum SS of 29 mV/dec at 300K as well as large I_{on}/I_{off} ratio (~10^8) at 0.6 V [6], which demonstrates the great potential for ultra-low power circuit application. To evaluate the circuit performance based on mFSB-TFET, the accurate analytic current model of device is prerequisite.

In this work, we establish the current model based on surface potential for mFSB-TFET for the first time. Based on the insight of mFSB-TFET physics, the physical model of the surface potential of channel in gate handle region and gate finger region is established firstly. Based on the modeled surface potential, the BTBT current component and the Schottky current component models are developed respectively. The total current model can be finally obtained. The impacts of different structural parameters are also investigated, which shows good agreement with the simulation results from Sentaurus TCAD.

THE STRUCTURE OF MFSB-TFET AND CURRENT COMPONENTS

In this work, an n-type mFSB-TFET on silicon substrate is studied. Compared with conventional TFET, mFSB-TFET has a multi-finger gate structure, which separates p+ source and adds additional side-tunnel junctions. The other side of multi-finger gate away from the drain is designed as Schottky junction, as shown in Figure 1 (a) (b). The total drain current of mFSB-TFET consists of three components (Figure 1 (b)): (1) the I_{BTBTh} component from the tunnel junctions along gate handle region; (2) the I_{BTBTf} component from the tunnel junctions along Schot

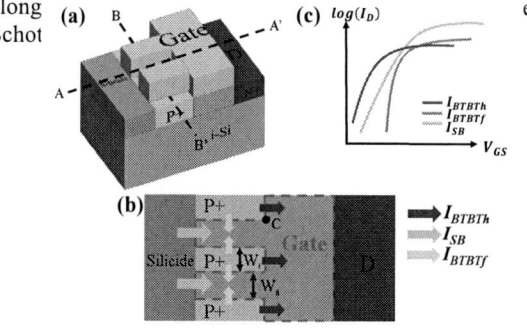

Figure 1: (a) Schematic view of mFSB-TFET with multi-finger gate structure; (b) top view and current components of mFSB-TFET; (c) diagrams of I_{BTBTh}, I_{BTBTf} and I_{SB} current components.

For the mFSB-TFET without source pocket at the tunnel junctions, due to the introduced self-depleted effect

978-1-7281-6559-2/20 $31.00 © 2020 IEEE

under the multi-finger gate, the channel under the gate finger region obtains the higher conduction band, which needs the larger gate voltage compared with that in gate handle region when the tunnel junctions turn on. Therefore, I_{BTBTf} component is not dominant compared with I_{BTBTh} in the subthreshold state. On the other hand, due to the Schottky junction, I_{SB} component is dominant in the on-state which enhances the on-current remarkably compared with conventional TFET. As a result, the I_{BTBTf} component can be ignored for the drain current modeling of mFSB-TFET without pocket. Figure 1 (c) illustrates the magnitude and turn-on gate voltage of three current components. What's more, the electric field coupling effect between fingers in mFSB-TFET could boost the tunneling electric field in source tunnel junction, which results in the steeper SS compared with T-gate Schottky Barrier tunneling FET (TSB-TFET) [4]. Due to the self-depleted effect, for the Schottky junction, the Schottky leakage current can be effectively decreased due to the higher effective Schottky barrier height. It can be seen that the key of mFSB-TFET modeling is to establish the surface potential model of channel under the gate handle region and gate finger region and then calculate the tunneling current and Schottky current respectively.

MODEL DERIVATION

Surface Potential Modeling

In the gate handle region, the surface potential model of channel is similar to the conventional TFET surface potential model, which has been proposed in our previous work [7]. The surface potential of channel in the gate handle region along AA' direction as shown in Figure 1 (a) is defined as φ_{ch} and the formula of φ_{ch} can be presented as follow [7]

$$\varphi_{ch} = F + \frac{kT}{q} \left\{ \ln \frac{q}{kT} \left[\frac{kT}{q} + \frac{\sqrt{F}}{\sqrt{F}+\gamma}(V_{GS}-V_{FB}-F) \right] \right.$$
$$\left. + \frac{1}{2}\left[\frac{F}{(\sqrt{F}+\gamma)^2} - \frac{\gamma(F-2)}{2(\sqrt{F}+\gamma)^3} \right](V_{GS}-V_{FB}-F)^2 \right\} \quad (1)$$

where F is the contiguous function that has been calculated in our previous work, γ is body factor [7], T is absolute temperature and V_{FB} is flat band voltage. The model results of φ_{ch} are in good agreement with the simulation results as shown in Figure 2 (a).

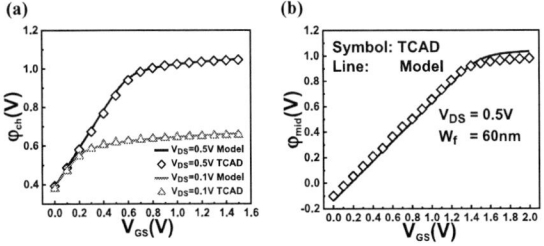

Figure 2: The model and simulation results of (a) surface potential φ_{ch} along gate handle region with different voltages in mFSB-TFET; (b) surface potential φ_{mid}.

In the gate finger region, the solution of surface potential modeling of channel in self-depleted region is similar to that in striped gate region of junction-modulated TFET (JTFET) [8]. The surface potential of channel in the gate finger region along BB' direction as shown in Figure 1 (a) is defined as φ_s, which is described in our previous work [8]. However, the value of surface potential in the middle point of gate finger regions along BB' direction, defined as φ_{mid}, is the most important for the I_{BTBTh} and I_{SB} modeling. φ_{mid} can be obtained from φ_s and expressed as

$$\varphi_{mid} = (V_{GS} - V_{FB} - \frac{qN_{ch}}{\varepsilon_{si}}\lambda^2) + (\varphi_{ch} - V_{GS} + V_{FB} + \frac{qN_{ch}}{\varepsilon_{si}}\lambda^2) \cdot \cosh(\frac{\frac{W_f}{2} - y_{sc}}{\lambda})$$

$$-\alpha \frac{(V_{GS} - V_{FB} - \frac{qN_{ch}}{\varepsilon_{si}}\lambda^2)}{\lambda} \cdot \sinh(\frac{\frac{W_f}{2} - y_{sc}}{\lambda}) \cdot \frac{W_f}{2} \quad (2)$$

$$\lambda = \sqrt{\varepsilon_{si} t_{sieff} t_{ox} / \varepsilon_{ox}} \quad (3)$$

where N_{ch} is substrate doping concentration, t_{sieff} is effective junction depth in Si-substrate, y_{sc} is the width of depleted region in gate finger area [8] and W_f is defined as shown in Figure 1 (b). The modeled results of φ_{mid} are in good agreement with the simulation results as shown in Figure 2 (b)

Current Modeling

The multi-finger gate structure can induce larger tunneling area and introduce tunneling electric field coupling effect thus increase I_{BTBTh} component. The tunneling electric field in gate handle region is defined as E_H and the tunneling electric field in gate finger region is defined as E_F, which can be calculated as follow

$$E_H = \frac{\varphi_{ch}}{x_{sc}} \qquad E_F = \frac{\varphi_{mid}}{y_{sc}} \quad (4)$$

where x_{sc} is the width of depleted region in gate handle area [7]. The enhanced tunneling electric field at the gate finger corner, such as point C as shown in Figure 1 (b), is affected by E_H and E_F, which can be calculated by the following formula

$$E_{coup} = \sqrt{E_H^2 + E_F^2} \quad (5)$$

Therefore, the average tunneling electric field along the tunnel junctions in the gate handle region can be expressed as follow

$$\overline{E_H} = \frac{2E_{coup} y_{sc} + E_H (W_s - 2y_{sc})}{W} \quad (6)$$

where W_s is defined as shown in Figure 1 (b). With the average of tunneling electric field along gate handle region, the tunneling electric field coupling effect is introduced to I_{BTBTh} component modeling which can be calculated by the following formula

$$I_{BTBTh} = (N+1)E_g t_{sieff} W_s \frac{A_{Kane}}{B_{Kane}} \cdot (\frac{\overline{E_H}}{E_H W_{t,\min}})^2 \cdot \exp\left[-B_{Kane} \left(\frac{E_g}{q}\right)^{\frac{1}{2}} \frac{E_H W_{t,\min}}{\overline{E_H}} \right] \quad (7)$$

where N is the finger number, A_{Kane} and B_{Kane} are constants of the Kane's parameters for BTBT probability [7]. $W_{t,min}$ is the minimum tunneling width [7].

For I_{SB} component, the thermal emission current and the Schottky tunneling current are modeled separately in our previous work [9]. In order to simplify the calculation complexity and improve convergence, in this work, the Schottky tunneling effect is effectively regard as Schottky barrier height lowering. The equivalent barrier reduction $\Delta\varphi_{tunnel}$ is formulated by [10]

$$\Delta\varphi_{tunnel} = \left[\frac{2q^3 N_{ch}}{\varepsilon_r\varepsilon_0}\right]^{\frac{1}{2}} x_c \quad (8)$$

where x_c is critical barrier thickness. From Figure 3 (a) it can be seen that when φ_{mid} is larger than φ_{iBn} which is the intrinsic Schottky barrier height, the effective barrier height φ_{Bn_eff} equals to φ_{mid}. On the other hand, when φ_{mid} is smaller than φ_{iBn} as shown in Figure 3 (b), the effective barrier height φ_{Bn_eff} will be (φ_{iBn}-$\Delta\varphi_{tunnel}$).

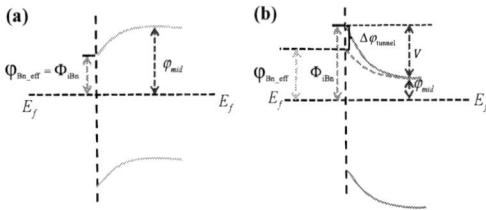

Figure 3: The energy band diagram of Schottky junction and equivalent barrier reduction (a) φ_{mid} is larger than φ_{iBn}; (b) φ_{mid} is smaller than φ_{iBn}

Therefore, with the help of contiguous function, the φ_{Bn_eff} can be expressed as follow

$$\varphi_{Bn\_eff} = \frac{1}{2}\cdot[\varphi_{iBn} + (\frac{E_g}{q} - \varphi_{mid} + V_{S0}) + \sqrt{[\varphi_{iBn} - (\frac{E_g}{q} - \varphi_{mid} + V_{S0})]^2 + \delta^2} \quad (9)$$

where V_{s0} is the surface potential in source region and δ is a smoothing factor. The I_{SB} component is given by [10]

$$I_{SB} = A^* T^2 N J_{sieff} W_f \exp(-\frac{q\varphi_{Bn\_eff}}{kT})\cdot[1-\exp(-(\frac{qV}{kT})] \quad (10)$$

where A^* is Richardson constant, W_f is defined as shown in Figure 1 (b) and V is the applied gate voltage.

For I_{total} of mFSB-TFET, it can be finally obtained by combining the calculated I_{BTBTh} compoment and I_{SB} component as follow

$$I_{total} = I_{BTBTh} + I_{SB} \quad (11)$$

Model Results and Discussion

In this paper, the simulations of transfer current characteristics of mFSB-TFET are carried out using the Synopsys Sentaurus TCAD devices simulation tools. The dynamic Nonlocal band-to-band tunnel model is adopted in device simulation and calibrated based on the experimental results of silicon-based TFET device characteristics.

As shown in Figure 4 (a) (b), the model results of I_{SB} component and I_{total} agree well with simulation results.

In order to verify the validity of mFSB-TFET model, the impact of different structure parameters (t_{ox} and W_f) is also studied and the results of model and simulation are compared. As shown in Figure 5, the model well predicts the devices performance with different parameters.

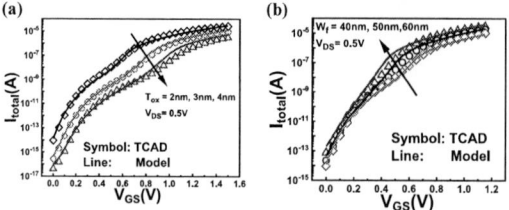

Figure 4: The model and simulation results of (a) I_{SB} component; (b) I_{total} with different V_{DS} (the finger width is 40nm).

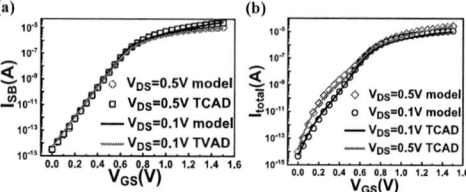

Figure 5: The model and simulation results of I_{total} with different structure parameters (a) t_{ox} and (b) W_f.

CONCLUSION

In this paper, we establish the physical compact model of mFSB-TFET based on the surface potential of channel in gate handle region and gate finger region. The tunneling current component model is developed with tunneling electric field coupling effect and Schottky current component model is developed with equivalent barrier reduction effect. The results of total current model are in good agreement with the simulation results which are strongly supported by TCAD.

ACKNOWLEDGEMENTS

This work was supported by National Key R&D Program of China (2018YFB2202801), NSFC (61851401, 61421005, 61822401, 61604006) and 111 Project (B18001).

REFERENCES

[1] M. Ionescu, et al. *Nature*, vol. 479, pp. 329-337, 2011.
[2] F.Mayer, et al. *IEDM*, pp. 1-5, 2008
[3] H. Riel, et al. *IEDM*, pp.391-394, 2012
[4] Q.Huang, et al. *IEDM*, pp. 16.2.1-16.2.4, 2011
[5] R. Huang, et al. *Nanotechnology*, vol. 25, no. 50, pp. 505201, 2014
[6] Q.Huang, et al. *IEDM*, pp.13.3.1-13.3.4, 2014
[7] C. Wu, et al. *IEEE-TED*, vol. 61, no. 8, pp. 2690-2696, 2014
[8] Zhu. Lv, et al. *CSTIC*, 2018
[9] J. Luo, et al. *ICSICT*, pp. 1-3, 2018
[10] Sze S.M., *Physics of Semiconductor Devices*, 2007

ORIGIN OF STEEP SUBTHRESHOLD SWING WITHIN THE LOW DRAIN CURRENT RANGE IN NEGATIVE CAPACITANCE FIELD EFFECT TRANSISTOR

Chang Su, Qianqian Huang[*], Mengxuan Yang, Liang Chen, Zhongxin Liang, Ru Huang[*]

Key Laboratory of Microelectronic Devices and Circuits (MOE), Institute of Microelectronics,
Peking University, Beijing 100871, CHINA
*Corresponding Author's Email: hqq@pku.edu.cn; ruhuang@pku.edu.cn

ABSTRACT

Negative capacitance FET (NCFET) can achieve the steeper subthreshold swing (SS) than conventional MOSFET for the reduction of supply voltage V_{DD}, while many experimental results indicate that NCFETs show the steeper SS only within a limited drain current range. In this work, the physical origin and parameter impacts of drain current range corresponding to steeper SS in NCFET are investigated, based on the modeling of domain switching dynamics of ferroelectric (FE). It is found that the larger remnant polarization is beneficial for the steeper SS with the larger drain current range. Besides, simulation results indicate that only within the limited window of sweeping rate and FE switching time, the drain current range can be enlarged for the steeper switching in NCFET.

Keywords—steep subthreshold swing; domain switching dynamics; drain current range

INTRODUCTION

The power consumption becomes one of the most serious problems as the scaling of MOSFET [1] due to its subthreshold swing (SS) limitation at room temperature. The ferroelectric-based negative capacitance FET (NCFET) has attracted much attention due to its capability of sub-60 SS and is regarded as one of the candidates for low power applications [2]. However, many experimental results show that the steep SS of NCFET can only be observed when the drain current is small [3]-[4]. Therefore, it is crucial to investigate how NCFET can realize the steeper SS with the larger drain current range to further decrease the supply voltage V_{DD}.

In our previous theory [5], we have experimentally verified that the physical origin of NC effect is related to the domain switching dynamics of FE. Further investigation of device design from the microscopic physical perspective is crucial for the understanding and optimization of NCFET.

In this paper, the phenomenon of steep SS within a limited drain current range in NCFET is analyzed based on the FE dynamic model. The dynamic behaviors of both ferroelectric materials and sweeping voltage strongly influence the SS and the drain current range. Moreover, in order to further improve the performance of NCFET, the impacts of the critical parameters of FE material are investigated.

ANALYSIS OF STEEP SS AND ITS DRAIN CURRENT RANGE

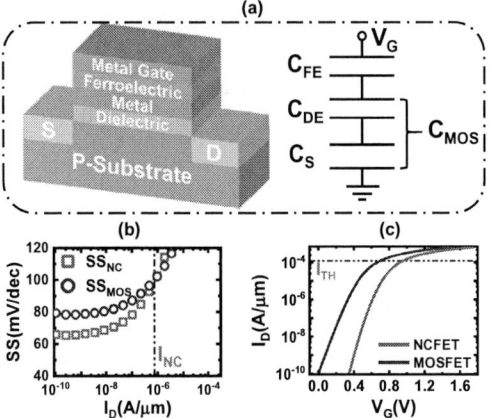

Figure 1: (a) Schematic structure and equivalent circuit of NCFET; Simulated (b) subthreshold swing (SS) of NCFET and baseline MOSFET; (c) transfer curves of NCFET and baseline MOSFET.

The schematic structure of NCFET and equivalent circuit are shown in Figure 1(a), where FE and dielectric (DE) are separated by an internal metal layer. The device simulation of NCFET in this work is based on Kolmogorov-Avrami-Ishibashi (KAI) equation integrated with baseline MOSFET model, as described in our previous work [5]. Compared with baseline MOSFET, it's noteworthy that the steeper SS only exists over a few decades of drain current range in NCFET (Figure 1(b)). Owing to the fact that NC effect usually occurs when drain current is small, it is more necessary to investigate the condition when NC effect disappears and try to obtain the steeper SS within the larger drain current range for the further power consumption reduction. Consequently, I_{NC} is defined as the larger drain current when SS_{NC} equals to SS_{MOS}. Moreover, as shown in Figure 1(c), I_{TH} represents the threshold current, and the corresponding threshold channel charge is defined as Q_{TH}.

As shown in Fig. 2, based on our previous theory in [5], the increment of FE polarization charge consists of the following two parts: time-induced and voltage-induced

978-1-7281-6559-2/20 $31.00 © 2020 IEEE

increments. Moreover, it is found that the time-induced increment of polarization charge is the origin of the emergence of NC effect [5]. For a standalone FE capacitor, during the forward sweeping, the polarization switching includes four successive stages, including nucleation, forward growth, sideways expansion and coalescence [6], in which the third and fourth stages are much slower than previous stages [7]. For NCFET, the reverse polarization charges begin to switch under the forward sweeping. The domain wall moves faster and faster with the increasing time-induced increment of polarization charge. When this increment exceeds the charge increment of MOS capacitance (C_{MOS}) in response to the total gate voltage [5], NC effect occurs.

$$dP = \frac{\partial P}{\partial t} \cdot dt + \frac{\partial P}{\partial V_{FE}} \cdot dV_{FE}$$

time-induced increment of polarization charge

voltage-induced increment of polarization charge

(1) dynamic polarization matching [5]:

$$\frac{\partial P}{\partial t} \cdot dt > C_{MOS} \cdot dV_G$$

(2) NC effect ($dP/dV_{FE} < 0$) is equivalent to:

$$\frac{\partial P}{\partial t} \cdot dt > \left| \frac{\partial P}{\partial V_{FE}} \cdot dV_{FE} \right| \quad \text{under forward sweeping}$$

Figure 2: dynamic polarization matching condition for NC effect and its equivalent form

Since the domain wall motion is fast and the time-induced increment is larger in the nucleation and forward growth stages, NC effect for steeper SS may occur during these two stages more possibly. However, the related polarization charge of FE during these two stages is usually small. Therefore, assuming that channel charge of underlying MOS structure is equal to the polarization charge of FE layer in NCFET, the steep SS is usually observed when the drain current is relatively small.

Once NC effect occurs, the voltage across FE (V_{FE}) decreases, leading to the negative voltage-induced polarization increment, which results in the larger required time-induced increment for the steep SS. In the meanwhile, the less V_{FE} contributes to the less polarization lag [8], which indicates that polarization charge can respond to changes of V_{FE}, leading to the smaller related time-induced polarization increment. In addition, during the process of polarization switching, domain wall moves slower and slower after forward growth, indicating the time-induced increment is decreasing. Therefore, NC phenomenon will disappear when the time-induced polarization increment is less than the voltage-induced polarization decrement. At this time, when the related

polarization charge is larger than Q_{TH}, the NCFET can be more likely to achieve steep SS in nearly the whole subthreshold state, which is beneficial for the operation voltage and power consumption reduction.

PARAMETER IMPACTS OF FE MATERIALS AND SWEEPING VOLTAGE

As mentioned above, domain switching dynamics is crucial to the emergence and disappearance of steep SS and its drain current range in NCFET. Therefore, the impacts of FE switching time and sweeping rate are simulated.

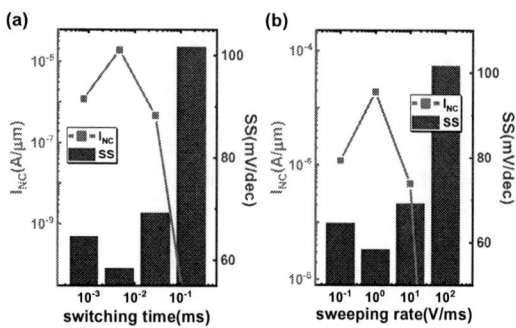

Figure 3: Simulated I_{NC} and SS respectively for (a) different switching time and (b) different sweeping rate. The left axis and the right axis represent I_{NC} and SS respectively in a plot.

The switching time of ferroelectric material is related to the domain wall motion, which reflects the speed of polarization switching. Too long switching time indicates that FE cannot respond to the forward sweeping voltage, causing negligible reversed polarization charge and the time-induced increment. Therefore, in this case, the NC effect is difficult to appear and the SS of NCFET becomes worse. Too short switching time indicates that the polarization changes instantaneously with V_{FE}, and thus the dominant voltage-induced increment may lead to the positive capacitance ($dP/dV_{FE} > 0$) instead of negative capacitance. In other words, only within a limited window of FE switching time, SS_{NC} can be smaller than SS_{MOS}, and there exists an optimal switching time to enlarge the time-induced increment for the larger I_{NC} and the better SS (Figure 3 (a)).

Similarly, with the fixed switching time of FE, subthreshold behaviors of NCFET under different voltage sweeping rate are different (Figure 3 (b)). Since domain switching dynamics depends on the relative differences between intrinsic response of FE and extrinsic sweeping voltage, the large sweeping rate situation is similar to the long switching time situation, and vice versa. It is naturally to understand there also exists an optimal sweeping rate.

978-1-7281-6559-2/20 $31.00 © 2020 IEEE

For the impacts of other material parameters, on the one hand, the remnant polarization (P_r) plays a significant role in increasing I_{NC} and decreasing SS (Figure 4 (a)). The larger P_r leads to the larger time-induced increment and the larger corresponding polarization charge with the same dynamic behaviors, leading to the larger I_{NC} and the smaller SS. Besides, it is beneficial for the emergence of NC effect. Moreover, from Fig.4a, it is shown that the I_{NC} also strongly depends on P_r, which indicates that changing FE remnant polarization by stress engineering [9] can be effective for the device optimization from the material perspective.

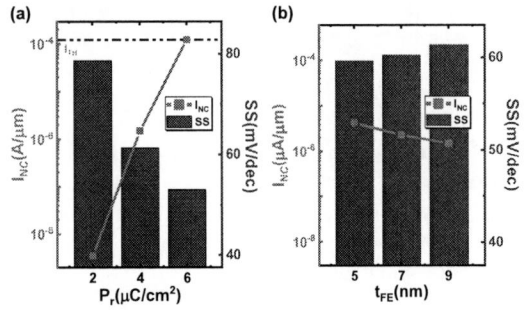

Figure 4: Simulated I_{NC} and SS respectively for (a) different remnant polarization (P_r) and (b) different ferroelectric thickness (t_{FE}). The left axis and the right axis represent I_{NC} and SS respectively in a plot.

On the other hand, the impact of FE thickness (t_{FE}) is simulated (Figure 4 (b)). In order to exclude the influence by P_r, P_r is set as a constant during the simulation. The thinner FE suggests the shorter time for nucleation and forward growth stages qualitatively, giving rise to the larger time-induced increment of polarization charge. However, since the speed of sideways expansion is immune to t_{FE} and the lateral size is too small to affect the dynamic behaviors of FE, the impact of t_{FE} is nearly negligible compared with P_r. It is also noteworthy that polarization charge per unit area would not be influenced by the variation of thickness, which indicates that I_{NC} is independent of t_{FE} especially when NC effect disappears during the sideways expansion and coalescence process. It is proven that FE thickness can hardly influence I_{NC} from Figure 4 (b). In spite of these negligible improvement by decreasing FE thickness, ultrathin FE layer faces difficulties in fabrication, high degree of crystallization, ferroelectric stability [10] and so on. In practical, the thicker sub-10nm doped HfO_2 film usually causes the distinctly increased P_r [11], leading to the better SS within the larger drain current range, thus it is feasible for the NCFET improvement by optimizing the design of HfO_2-based ferroelectric materials.

CONCLUSION

In this paper, the steep SS and its corresponding drain current range in NCFET are physically analyzed based on domain switching dynamics of FE. It is shown that there exist optimal switching time and sweeping rate for NCFET to realize steep SS within a large drain current range. In order to further increase the current range, for doped HfO_2 film, it is essential to enlarge the remnant polarization for the larger time-induced increment of polarization charge.

ACKNOWLEDGEMENTS

This work was supported by National Key R&D Program of China (2018YFB2202800), NSFC (61822401, 61421005, 61851401) and 111 Project (B18001).

REFERENCES

[1] S. Borkar. "Design Challenges of Technology Scaling." IEEE Micro, July-August 1999, pp. 23-29.

[2] S. Salahuddin, S. Datta. "Use of negative capacitance to provide voltage amplification for low power nanoscale devices." Nano Letters, vol. 8, 2008, pp. 405-410.

[3] M. H. Lee, et al., "Physical thickness 1.xnm ferroelectric HfZrOx negative capacitance FETs", IEEE Inter. Electron Devices Meeting (IEDM), San Francisco, CA, USA, Dec. 2016. pp. 12.1.1-12.1.4.

[4] P. Sharma, et al., "Impact of total and partial dipole switching on the switching slope of gate-last negative capacitance FETs with ferroelectric hafnium zirconium oxide gate stack," Symposium on VLSI Technology, Kyoto, 2017, pp. T154-T155.

[5] H. Wang, et al., "New Insights into the Physical Origin of Negative Capacitance and Hysteresis in NCFETs," IEEE Inter. Electron Devices Meeting (IEDM), San Francisco, CA, USA, Dec. 1-5, 2018. pp. 31.1.1-31.1.4.

[6] D. Zhao, et al., "Switching dynamics in ferroelectric P (VDF-TrFE) thin films." Physical Review B, vol. 92, 2015, pp. 214115.

[7] C. T. Nelson, et al., "Domain dynamics during ferroelectric switching," Science, vol. 334, 2011, pp. 968-971.

[8] S. L. Miller, et al., "Device modeling of ferroelectric capacitors," Journal of Applied Physics, vol. 68, 1990, pp. 6463-6471.

[9] Kanno. I, et al., "Thermodynamic study of c-axis-oriented epitaxial $Pb(Zr,Ti)O_3$ thin films", Physical Review B, vol. 69, 2004, pp. 064103.

[10] H. W. Park, et al., "Modeling of negative capacitance in ferroelectric thin films", Advanced Materials, vol. 31, 2019, pp. 1805266.

[11] X. Tian, et al., "Evolution of ferroelectric HfO_2 in ultrathin region down to 3 nm," Applied Physics Letters, vol. 112, 2018, pp. 102902.

CIRCUIT RELIABILITY EVALUATION OF APPROXIMATE COMPUTING

Yuwei Zhang[1,2], Zuodong Zhang[2], Zhe Zhang[2], Jiayang Zhang[2], Runsheng Wang[2],*
Zhiting Ling[1], Ru Huang[2]

[1]School of Electronics and Information Engineering, Anhui University, Hefei 23060, China
[2]Institute of Microelectronic, Peking University, Beijing 100871, China
[*]E-mail: r.wang@pku.edu.cn

ABSTRACT

With the downscaling of CMOS technology, the reliability issue has become inevitable. Even in fault-tolerant applications, the effects of timing violation caused by the transistor aging are unacceptable. In the past, the traditional method to solve timing violation is to add a wide frequency guardband, which will lead to the decrease of speed. In this paper, the reliability of the approximate computing is evaluated in an image processing applications. And a method to improve circuit reliability is proposed. The results show that this method can turn critical path timing errors, which are difficult to predict and seriously hurt circuit reliability, into deterministic logic simplification errors, and thus the actual function of the circuit is not affected.

Keywords—Approximate Computing; Circuit Reliability; Static Timing Analysis; Transistor Aging; Negative Bias Temperature Instability (NBTI).

INTRODUCTION

Due to the transistor aging, the circuit is facing reliability issues during its expected lifetime [1,2]. Among all the transistor degradation effects, the negative bias temperature instability (NBTI) has the most serious impact on the circuit [7], which is manifested as the increase of V_{TH}. Therefore, the circuit delay increases, and the timing constraint does not satisfy.

Generally, a frequency guardband is added to mitigate the effects of aging [2]. But this method inevitably leads to a loss of efficiency. Several methods have been proposed to reduce the frequency guardband. For example, adding aging control (AC) gate to reduce degradation is proposed in [2], and Ref.[3] creates an aging-aware cell library to obtain the required frequency guardband. However, these methods cannot eliminate the frequency guardband completely, and have additional overheads.

On the other hand, approximate computing is an emerging design technique that improves circuit performance while saving area and power, by allowing the calculated results to have a certain acceptable error [8]. This new computing paradigm also provides possibilities for solving the reliability issue of circuits from a new perspective.

In this paper, the approximate computing is applied to enhance the reliability of circuit. The limited precision loss is proposed to trade for reduced delay of the circuit. The proposed method is verified in practical applications of image compression.

RELIABILITY-ENHANCED DESIGN FLOW

We proposed a reliability analysis method based on critical path for very large-scale circuits. The proposed flow is shown in Figure 1.

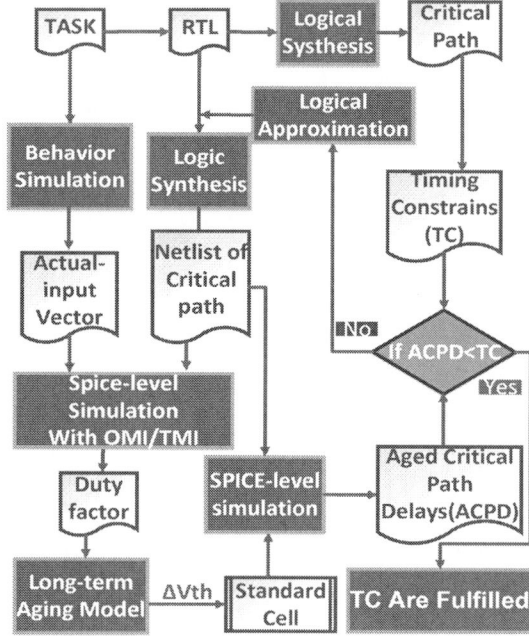

Figure 1: Design flow for reliability enhancement based on critical paths

Fresh simulation

Regardless of degradation, we performed static timing analysis on traditional binary circuits to find the critical path and timing constraint.
Reliability evaluation

First, we obtain the input vector using behavior level simulation. Combined with the netlist of critical path generated by logic synthesis, the duty factor of each transistor is calculated by SPICE-level simulation. By using the aging model proposed in our previous work [4], the corresponding ΔV_{TH} is obtained and added to the transistors of each standard cell, followed with the SPICE-level simulation to calculate the delay after degrading.

Logic approximation

If the degraded delay does not satisfy the time constraint, a timing violation occurs. We make a logical approximation of the circuit and continue with reliability evaluation until time constraints are met.

COMPARISON OF CIRCUIT RELIABILITY IN PRACTICAL APPLICATIONS

In order to evaluate the reliability of an approximate computing circuit. We take Discrete Cosine Transform (DCT) and Inverse Discrete Cosine Transform (IDCT) applications as examples. Since the critical path of the application is the adder. What used here is a 16-bit adder of a Ripple Carry Adder (RCA) structure. In the approximation flow, the full adder of the lower significant bit of the adder is replaced by an approximate adder cell until the timing constraint is met.

Approximate adder

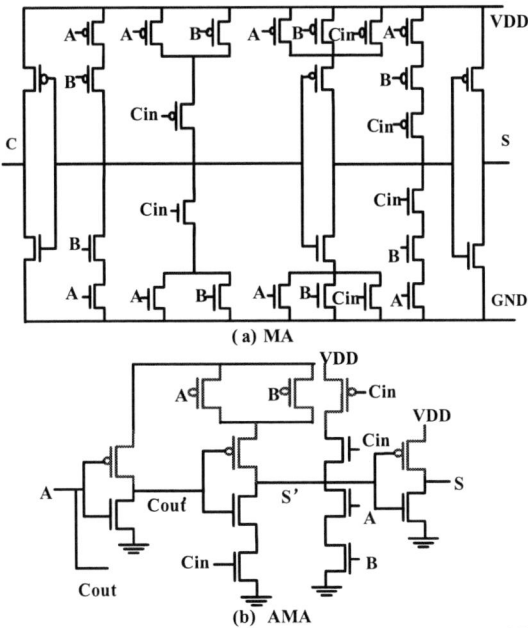

Figure 2: (a) Mirror adder, (b) Approximate mirror adder.

The approximate full adder adopted here is an approximate adder proposed in [5], which is simplified to the approximate mirror adder (AMA) according to the original mirror adder (MA). As shown in Figure 2, the carry output C_{out} is equal to the input signal A. Therefore, the delay of the one-stage adder is subtracted with each replacement of the one-bit adder unit, and in the approximate adder, aging does not change the delay.

Simulation results and discussion

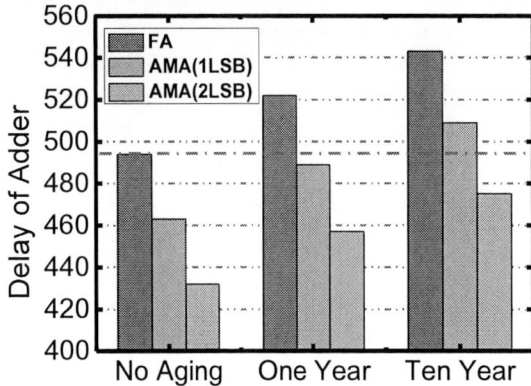

Figure 3: The dependence of the delay of the adder with different level of degradation for FA, AMA (1LSB), AMA (2LSB).

The simulation results are shown in Figure 3. After one year of degradation, the traditional 16-bit adder appears timing violation. However, by using our approximate computing method, we only need to replace the full adder on the two least significant bits with an approximate adder to ensure that the circuit will still meet the time limit after ten years of aging.

Performance evaluation

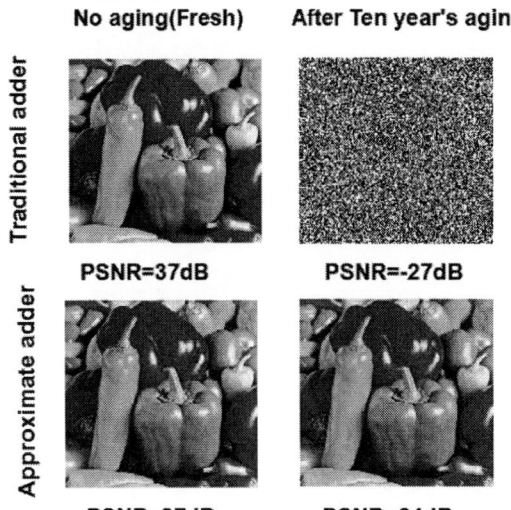

Figure 4: Evaluation of the impact of aging-induced errors of traditional (upper right) and approximate adder (bottom right) on image output quality in the image process application.

In the application of image compression, the timing violation caused by the aging of the circuit will seriously affect the image quality. As shown in Figure 4, after ten years of aging, the image has been completely unacceptable. However, using the approximate adder we proposed, peak signal-to-noise ratio (PSNR) only decreased by 3dB after ten years of aging of the circuit.

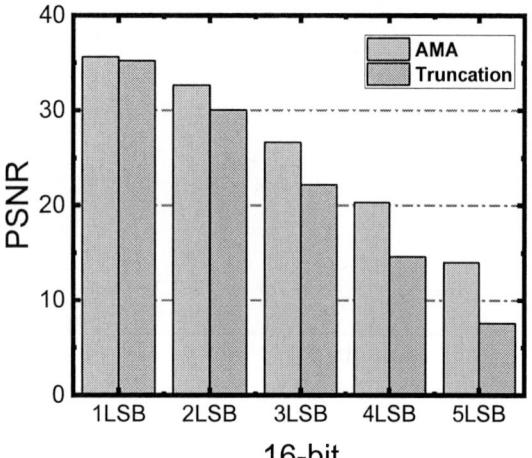

Figure 5: Output quality comparisons between AMA and truncation.

We further compare the AMA method with the method proposed in [9], which performs precision reduction by truncating the least significant bits (LSBs). In terms of latency, the AMA method is consistent with the results of truncation method. But in terms of image output quality, the AMA method is further improved. As shown in Figure 5, the value of PSNR obtained by the AMA method is always larger than that obtained by truncation method. And with the increase of LSBs, the benefit will become larger.

CONCLUSION

In this paper, a method of approximation is introduced into the reliability-enhanced design, in which limited precision loss is proposed to trade for reduced delay of the circuit, so that the frequency guardband is completely eliminated, which indicates that this approximation based method is an alternative for the reliability-enhanced circuit design.

ACKNOWLEDGEMENTS

This work was partly supported by NSFC (61522402 and 61421005) and the 111 Project (B18001)

REFERENCE

[1] J. Keane and C. H. Kim, "Transistor aging," IEEE Spectrum, 2011.

[2] S. Arasu, M. Nourani, J. M. Carulli, and V.K. Reddy, "Controlling aging in timing-critical paths," IEEE D&T, vol. 33, no. 4, pp.82-91, 2016.

[3] H. Amrouch, B. Khaleghi, A. Gerstlauer, and J. Henkel, "Reliability aware design to suppress aging," in DAC, 2016.

[4] S.Guo, et al. "Towards reliability-aware circuit design in nanoscale FinFET technology: new-generation aging model and circuit reliability simulator." Proceedings of the 36th International Conference on Computer-Aided Design. IEEE Press, 2017.

[5] V. Gupta, D. Mohapatra, A. Raghunathan, and K. Roy. "Low-power digital signal processing using approximate adders," IEEE Trans. CAD, 32(1):124-137, 2013.

[6] N. Banerjee, G. Karakonstantis, and K. Roy, "Process variation tolerant low power DCT architecture," in Proc. Design, Automat. Test Eur., 2007, pp.1-6.

[7] Cao, Yu, et al. "Cross-layer modeling and simulation of circuit reliability." IEEE Transactions on Computer-Aided Design of Integrated Circuits and Systems 33.1 (2013): 8-23.

[8] Jiang, Honglan, Jie Han, and Fabrizio Lombardi. "A comparative review and evaluation of approximate adders." Proceedings of the 25th edition on Great Lakes Symposium on VLSI. ACM, 2015.

[9] Amrouch, Hussam, et al. "Towards aging-induced approximations." 2017 54th ACM/EDAC/IEEE Design Automation Conference (DAC). IEEE, 2017.

DEVICE MODELING AND APPLICATION SIMULATION OF FERROELECTRIC-FETS WITH DYNAMIC MULTI-DOMAIN BEHAVIOR

Zhiyuan Fu, Cheng Chen, Jin Luo, Qianqian Huang[*], Ru Huang[*]

Key Laboratory of Microelectronic Devices and Circuits (MOE), Institute of Microelectronics,
Peking University, Beijing 100871, CHINA
Phone: 86-10-62768703, [*]E-mail: hqq@pku.edu.cn; ruhuang@pku.edu.cn

ABSTRACT

In this work, a ferroelectric-FET (FeFET) model based on multi-domain Preisach theory is developed. Instead of the single-domain Landau-Khalatnikvo (L-K) and tanh based model, the Preisach model with dynamic module is established to accurately model the dynamic behavior of multi-domain FeFET. Moreover, a FeFET-based neuron structure is presented and simulated based on the developed device model, showing the promising potential of FeFET for neuromorphic computing.

Keywords: FeFET model; Preisach theory; FeFET neuron

INTRODUCTION

In recent years, Ferroelectric-FET (FeFET) with non-volatile property has attracted great attention for many applications including memory and neuromorphic computing applications [1][2]. According to the multi-domain theory of ferroelectric material, polarization states of ferroelectric will not only depend on the applied voltage but also on its history [1]. In addition, the domain switching in ferroelectric material is not instantaneous, and shows the dynamic behavior. To evaluate the device and circuit performance based on FeFET, accurate current model of FeFET is perquisite. However, the current models of FeFET are mainly based on the single domain Landau-Khalatnikvo (L-K) model or the tanh function for ferroelectric materials [3][4], which cannot simulate the dynamic behavior and multi-domain effect in FeFET.

In this work, we develop a FeFET current model based on multi-domain Preisach theory with the dynamic module considered. Moreover, based on the established model, a FeFET neuron structure based on FeFET is presented and simulated to show the potential of FeFET in neuromorphic devices.

MODELING FRAMEWORK

The developed FeFET model in this work is composed of two components as shown in Fig. 1.

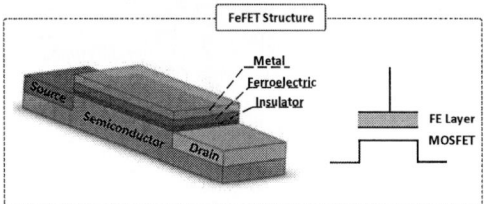

Fig. 1 Ferroelectric FET structure. FeFET model is composed of two parts, FE layer and conventional MOSFET.

The BSIM model is used for the MOSFET component of FeFET [5] in order to obtain the relationship between charge and internal gate voltage $Q_{\mathrm{MOS}}(V_{\mathrm{MOS}})$. The model of ferroelectric (FE) layer is described by the multi-domain Preisach theory with a RC-like dynamic module, through which the charge and the voltage relationship of FE layer $Q_{\mathrm{Fe}}(V_{\mathrm{Fe}}(t))$ in FeFET can be obtained. Based on the above relationships, the FeFET can be modeled as follows.

Since the FE layer is stacking on the internal gate of MOSFET in FeFET, the following two equations should be satisfied.

$$\begin{cases} Q_{\mathrm{Fe}} = Q_{\mathrm{MOS}} \\ V_{\mathrm{G}} = V_{\mathrm{Fe}} + V_{\mathrm{MOS}} \end{cases} \quad (1)$$

The equation above can be solved by iteration method as shown in Fig. 2. For each given gate voltage V_G, assuming the voltage drop on MOSFET component is V_{MOS}, then Q_{MOS} can be calculated based on BSIM model. By following the Eq.1, voltage drop on FE layer V_{Fe} can be calculated according to $V_{\mathrm{Fe}} = V_{\mathrm{G}} - V_{\mathrm{MOS}}$, and then Q_{Fe} can be obtained based on the developed multi-domain Preisach model of FE layer. Further by comparing Q_{Fe} and Q_{MOS}, if their difference is smaller than the error tolerance, denoting the assumed V_{MOS} and calculated $V_{\mathrm{Fe}}, Q_{\mathrm{Fe}}, Q_{\mathrm{MOS}}$ are correct; If not, assume another V_{MOS} until the results match. For each given V_G there is V_{MOS} that could satisfy the aforementioned modeling framework, and finally $V_{\mathrm{G}} - I_{\mathrm{D}}$ relationship of FeFET can be developed.

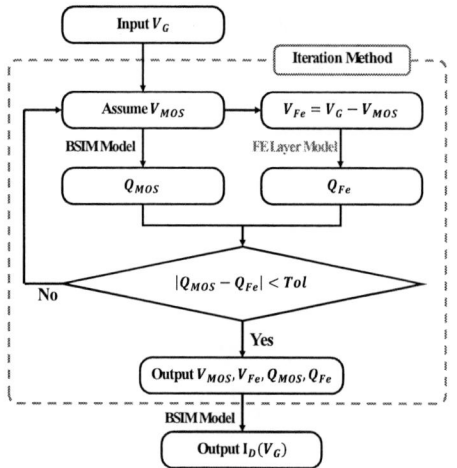

Fig. 2 Iteration method for solving relationship between gate voltage and drain current of FeFET.

MODELING of FERROELECTRIC FET

As mentioned above, key of FeFET modeling is the modeling of FE layer. Total charge of FE layer is the sum of spontaneous polarization and dielectric polarization. For spontaneous polarization, according to multi-domain Preisach theory, FE layer can be modeled as many domains with two different stable polarization states, normalized as +1 and -1, standing for the states of polarization up and down. Each domain has two coercive fields. Both the coercive fields of domains ($-\alpha_i$ and β_i) follow 2-D normal distribution (Fig. 3(b)). Spontaneous polarization of each domain can be influenced by the coercive fields, previous polarization state and external electric field. For each domain, when external electric field is smaller than coercive field α_i, the domain will switch to polarization down state instantly; when electric field is larger than coercive field β_i, domain will switch to polarization up state, and polarization state would maintain unchanged as previous state when electric field is between α_i and β_i as shown in Eq.2 and Fig. 3(a).

$$P_i(t) = \begin{cases} +1 & (E(t) \leq \alpha_i) \\ P_i(t - \Delta t) & (\alpha_i < E(t) < \beta_i) \\ -1 & (E(t) \geq \beta_i) \end{cases} \quad (2)$$

The spontaneous polarization of FE layer is calculated as the sum of polarization of domains.

As for dielectric polarization charge of FE layer, it can be calculated by external electric field E_{ext} and dielectric constant ε_{Fe} of the layer. The equation $P_{DE} = \varepsilon_{Fe}E_{ext}$ should be satisfied, where $\varepsilon_{Fe} = \varepsilon_0 \chi_e$ shown in Fig. 3(c).

For the dynamic switching behavior of ferroelectric material, a RC-like dynamic module is introduced into the above static Preisach model for FE layer in this work. The applied voltage $V_{Fe}(t)$ is converted into an introduced effective voltage $V_{eff}(t)$ for the non-instantaneous time

response of the FE layer [4]. The $V_{eff}(t)$ is calculated by Eq. (3) .

$$\frac{dV_{eff}}{dt} = \tau(V_{Fe} - V_{eff}) \quad (3)$$

Where τ is the time constant of domain switching. The $V_{eff}(t)$ and $V_{Fe}(t)$ are illustrated in Fig. 3(d), and the simulation results are shown in Fig. 4(a). It can be seen that compared with V_{Fe}, the peak voltage of V_{eff} is decreased and a lagging effect is shown, reflecting the dynamic behavior in FE layer.

Based on aforementioned FE layer model, P-V response for given input voltage is simulated as shown in Fig. 4. When sweeping voltage is relatively high, P-V loop shows saturation characteristic; and when the sweeping voltage is relatively low (point "b" to "c"), an unsaturated minor loop will appear. At point "b" and "c", when the sweeping voltage change sweeping direction, P-V loops also demonstrate dynamic behavior of FE layer.

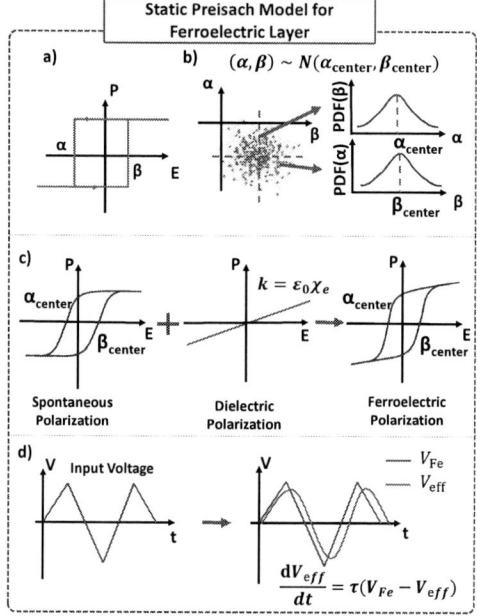

Fig. 3 (a) Single domain P-E loop, α and β indicate coercive field of one domain. (b) Multi-domain effect and classic P-E loop of FE layer with normal-distributed coercive field. (c) Total polarization is the sum of spontaneous polarization and dielectric polarization. (d) Dynamic module. FE layer will react to $V_{eff}(t)$ instead of $V_{Fe}(t)$ to simulate dynamic polarization switching behavior.

Therefore, with the MOSFET BSIM model and the aforementioned dynamic Preisach model of FE layer, the FeFET model can be finally developed. The simulation results based on the FeFET model are shown in Fig. 5, showing that the FeFET have two different V_T with different gate voltage sweeping direction.

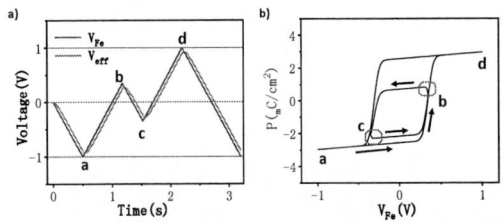

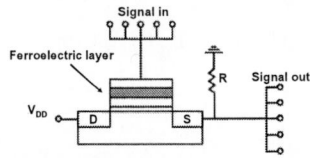

Fig. 6 FeFET-based neuron structure. The gate will integrate input pulses acting as dendrites, and source will give output pulse as axons.

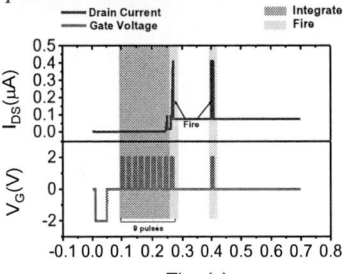

Fig. 4 (a) Simulation result of dynamic module, the peak voltage of V_{eff} is smaller than V_{Fe}, and time lagging effect. (b) P-V loop of ferroelectric capacitor simulated with the voltage input of (a), the saturation loop and unsaturation loop are shown. Point b and c reflect the dynamic behavior of ferroelectric.

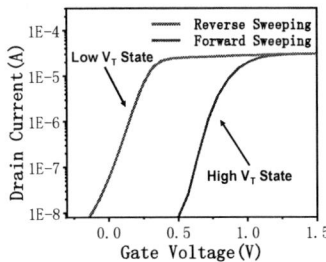

Fig. 5 FeFET simulation result with $V_{DD}=1V$. Different sweeping direction shows different V_T state.

FEFET-BASED NEURON AND SIMULATION RESULTS

CMOS based implementation for neuron structures usually requires high hardware cost [6]. In this work, we present a FeFET-based neuron structure with low hardware cost, showing the potential of FeFET to build neuron devices. Simulation results of FeFET neuron is based on the aforementioned FeFET model.

'Leaky Integrate and Fire (LIF)' function is the most basic neuron behavior. Neuron can integrate input pulses from dendrites, which give rise to the potential of neuron. When potential reaches threshold, the neuron will 'fire' with output spike delivered to axon, and 'leaky' stands for spontaneous protential drop between input pulses.

The proposed FeFET-based neuron structure is shown in Fig.6. Gate of n-type FeFET can integrate input pulses and source is designed to give output pulse. Polarization of FE layer in the FeFET will accumulate gradually with the input pulses, and change the device threshold voltage.

The simulation result of FeFET neuron is shown in Fig. 7, the FE model is calibrated with experimental results in [7]. It is shown that after 9 input pluses of 2V, the neuron will fire when V_{DD} is 1.5V. However, since the FE layer of FeFET usually has relatively long retention time, there is no obvious potential drop after the fire. To realize the leaky effect, accelerated polarization degradation is required for the FE layer design of FeFET [8][9].

Fig. 7 Simulation results of FeFET-based neuron structure. V_{DD} is 1.5V and V_G is 2V for the input pulses, with initialize pulse -2V/50ms. The neuron will fire after 9 pulses.

CONCLUSION

In this work, we develop a FeFET model based on dynamic multi-domain Preisach theory. In contrast to single-domain L-K model and tanh approaches, non-instantaneous behavior is realized in this model. Moreover, we present and simulate a FeFET neuron structure based on developed model, showing the potential of FeFET for neuromorphic computing, especially for neuron devices.

ACKNOWLEDGEMENTS

This work was supported by National Science and Technology Major Project (2017ZX02315001-004), NSFC (61851401, 61421005, 61822401) and 111 Project (B18001)

REFERENCES

[1] Mulaosmanovic, H., et al. *2017 VLSI*. IEEE, pp. T176-T177, 2017.

[2] Jerry, Matthew, et al. *Journal of Physics D: Applied Physics* 51.43 (2018): 434001.

[3] Aziz, Ahmedullah, et al. *IEEE Electron Device Letters* 37.6 (2016): 805-808. C

[4] Ni, Kai, et al. *2018 VLSI*. IEEE, pp. 131-132, 2018.

[5] Sheu, Bing J., et al. *IEEE Journal of Solid-State Circuits* 22.4 (1987): 558-566.

[6] Chu, Myonglae, et al. *IEEE Transactions on Industrial Electronics* 62.4 (2014): 2410-2419.

[7] Wang, Huimin, et al. *2018 IEDM*. IEEE, pp. 31.1.1-31.1.4, 2018.

[8] Chen, C., et al. *2019 VLSI*. IEEE, pp. T136-T137,

2019.

[9] Luo, J., et al. *2019 IEDM*. IEEE, pp. 6.4.1-6.4.4, 2019

A NOVEL ELECTRICAL ISOLATION SOLUTION FOR TUNNEL FET INTEGRATION

Ting Li[1,2], Qianqian Huang[2,3]*, Le Ye[2,3], Yuan Zhong[2], Mengxuan Yang[2], Yiqing Li[2], Yimei Li[2], Zhongxin Liang[2], and Ru Huang[2]*

[1] School of Electronic Information Engineering, Anhui University, Hefei 230601, China
[2] Key Laboratory of Microelectronic Devices and Circuits (MOE), Institute of Microelectronics, Peking University, Beijing 100871, China
[3] Peking University Information Technology Institute (Tianjin Binhai)
* E-mail: hqq@pku.edu.cn; ruhuang@pku.edu.cn

ABSTRACT

In this work, for bulk tunnel field-effect transistors (TFET), the electrical isolation solutions between neighboring devices for TFET integration are investigated. To suppress the leakage current of the P-type doped regions through the P-type substrate, a new effective isolation method is proposed and verified via simulation. The simulation shows the leakage current can be reduced from 10^{-5} A/μm to 10^{-13} A/μm. The solution is beneficial for TFETs to keep its advantages for ultra-low power applications such as implantable medical devices and Internet of Things.

Keywords — leakage current; isolation; static power; N⁻ WELL-PWELL

INTRODUCTION

The tunnel field-effect transistors (TFET) is a gated P-I-N structure operated by band-to-band tunneling mechanism. Since the off-state leakage current of TFET is the reverse-biased P-I-N junction current, compared with the MOSFET, the TFET has a much lower off-state current, which leads to much lower static power [1]. The off-state leakage current in the silicon-based TFET is usually in the level of 10^{-13} A/μm [2-3]. For the P-I-N structure of TFET on bulk substrate, the traditional twin-well processes for CMOS will increase the source-to-drain direct tunneling current.

To suppress the direct tunneling current, TFET devices are needed to fabricate on the high-resistivity bulk Si substrate without twin-well processes. However, it is found that the P-type doping region of every two devices will form a leakage path between the P-type substrate. This leakage current will in turn deteriorate the off-state current and static power and may even make a simple inverter consisting of two TFET devices cannot operate properly [4].

Therefore, an effective leakage isolation solution for TFET integration is indispensable. In our previous work, we have proposed that implanting a reasonable dose and energy of phosphorus to form the N⁻ WELL can introduce a reverse-biased p-n junction with the P-type substrate for the leakage isolation [4]. However, the dose and energy of the phosphorus implantation as well as the distance between N⁻ WELL need to be precisely controlled to ensure the formation of reverse p-n junction, otherwise it may introduce a new leakage path between N-doped regions through the N⁻ WELL when the p-type substrate between the two N⁻ WELL is fully depleted.

To solve this problem, this work designs a new isolation by further introducing PWELL between the two adjacent N⁻ WELL for device isolation. The PWELL process is compatible with CMOS well process and can be formed by implanting boron. For the new isolation method, the process requirement about the doping concentration and energy as well as the distance between N⁻ WELL can be more relaxed without increasing the manufacturing process complexity.

LEAKAGE PATH AND SIMULATION SETUP

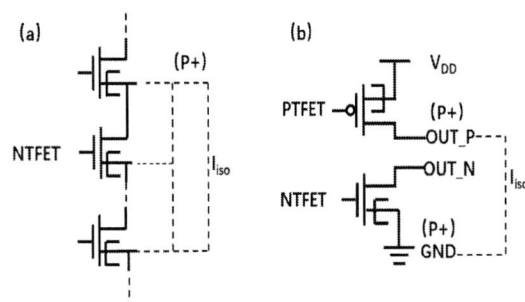

Fig. 1 (a) The leakage paths between different NTFETs; (b) The leakage path between the adjacent NTFET and PTFET. I_{ISO} is the leakage current.

As shown in Fig. 1(a), we take the NTFET transistor as an example. It will form several leakage paths between every two P-type doped regions through the P-type substrates when there is no well processes for isolation. In this work, to be simplified, the leakage current between the adjacent NTFET and PTFET, as shown in Fig. 1(b), are simulated and analyzed.

In this paper, all simulations are carried out by using Synopsys TCAD Sentaurus tools [5]. The gate voltage is fixed to be 0 V and VDD is set to be 1.5 V. The source and drain of NTFET and PTFET are implanted and formed by using the 3D process simulation. For any two TFETs, the leakage current of the P-type doped regions through the P-type substrate, and the leakage current of N-type doped regions through the N⁻ WELL are simulated and extracted. The I_{OUT_P} and I_{GND} represent the leakage currents of the P-type doped regions through the P-type substrate, while the I_{OUT_N} and I_{VDD} represent the leakage current of N-type doped regions through the N⁻ WELL.

LEAKAGE WITHOUT ISOLATION SOLUTION

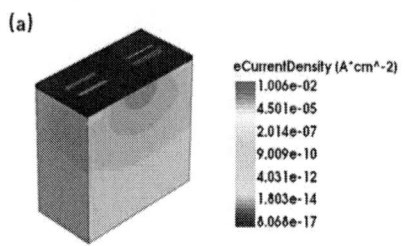

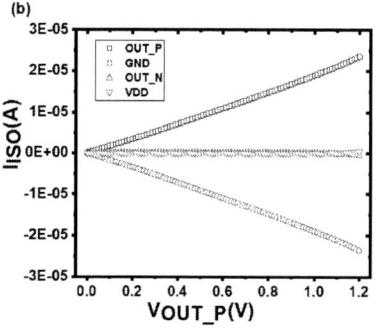

Fig. 2 (a) The electron current density profile when there is no leakage isolation ;(b) The currents of different electrodes as a function of the voltage of OUT_P when there is no leakage isolation.

In this part, the leakage current without isolation solution are theoretically analyzed and simulated. As shown in Fig. 2a, when there is no isolation solution, the electron current density profile shows that there is a leakage path between the two TFET devices. The extracted results as shown in Fig.2b show that the I_{OUT_N} and I_{VDD} are very low, while the I_{OUT_P} and I_{GND} can reach to the 10^{-5} A/μm, which is much larger than the off-state leakage of TFET. Fig.2(b) also shows that the I_{OUT_P} and I_{GND} increase with the increase of V_{OUT_P}. This is because

as V_{OUT_P} increased, the electric potential difference between OUT_P and GND is increased. At this time, the leakage current is mainly from P-type doped regions through the P-type substrate. It indicates that an effective isolation solution is necessary for TFET integration.

THE PROPOSED ISOLATION SOLUTIONS
(a) N⁻WELL-PSUB

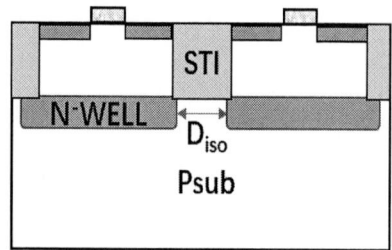

Fig. 3 The schematic isolation solution of N⁻ WELL -PSUB. D_{ISO} is the distance between two N⁻ WELL.

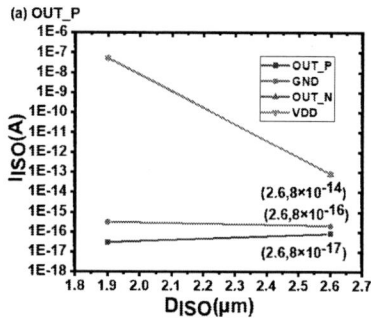

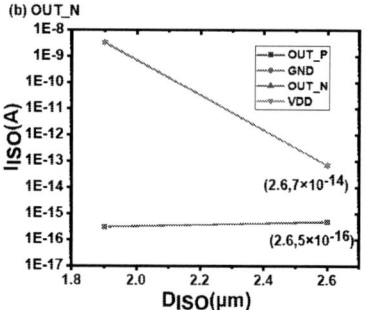

Fig. 4 The currents of different electrodes as a function of the distance between two adjacent N⁻ WELL in the case of: (a) the leakage of OUT_P; (b) the leakage of OUT_N.

The N⁻ WELL-PSUB isolation solution proposed in [4] is to implant a reasonable dose and energy of phosphorus to form the N⁻ WELL in the device area region through the P-type substrate. The schematic isolation solution of N⁻ WELL-PSUB is shown in Fig.3

978-1-7281-6559-2/20 $31.00 © 2020 IEEE   188

In order to observe whether the N⁻ WELL-PSUB isolation is effective, the current of each electrode is observed and compared with the intrinsic off-state leakage current (10^{-13} - 10^{-14} A/μm) of TFET as the voltage of OUT_P increases. The gate voltage and VDD is the same as the situation when no isolation solution is adopted. Meanwhile, the leakage current of N-type doped regions through the N⁻ WELL as the voltage of OUT_N increases also needs to be observed.

As shown in Fig.4, as the value of D_{ISO} decreases, the values of I_{OUT_N} and I_{VDD} become much larger, the values of I_{OUT_P} and I_{GND} are almost constant. The results indicate that the leakage current of two N-type doped regions through the N⁻ WELL is introduced.

(b) N⁻ WELL -PWELL

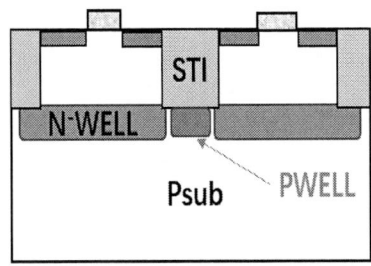

Fig. 5 The schematic isolation solution of N⁻ WELL -PWELL

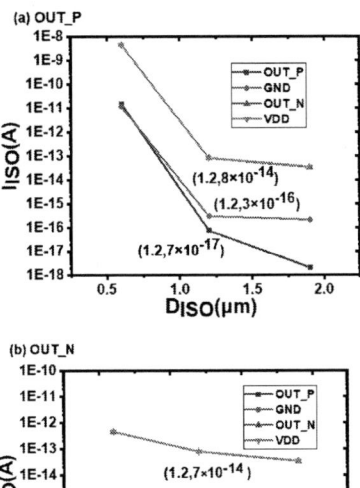

Fig. 6 The currents of different electrodes as a function of the distance between two adjacent N⁻ WELL in the case of: (a) the leakage of OUT_P; (b) the leakage of OUT_N.

The observed new leakage path between N-type doped regions as shown in Fig.4 is much larger than the off-state leakage of TFET and may increase as the D_{ISO} decreases, resulting in an invalid isolation when the D_{ISO} is too small. In order to solve the above problem, we propose a new isolation solution of N⁻ WELL-PWELL as shown in Fig.5.

Fig. 6 shows the current of different electrodes for the new isolation solution. Since the PWELL is formed by implanting a much higher dose and energy of boron than p-type substrate, the PWELL is more difficult to be depleted by the N⁻ WELL than the p-type substrate. By comparing Fig.4 and Fig.6, it can be seen that the N⁻ WELL interval can be reduced by more than two times with the introduction of PWELL, indicating this type of isolation solution is a very effective solution for the TFET integration.

CONCLUSION

In this paper, for TFET integration on the bulk substrate, the leakage current problems between any two devices are studied. To solve the problem, a new N⁻ WELL -PWELL isolation solution is proposed and verified by simulation. This solution ensures an effectively electrical isolation while considering the requirements of the area as well as the implantation process.

ACKNOWLEDGEMENTS

This work was supported by National Key R&D Program of China (2018YFB2202800), NSFC (61421005, 61851401, 61822401) and 111 Project (B18001).

REFERENCES

[1] Appenzeller, J. (2004-01-01). "Band-to-Band Tunneling in Carbon Nanotube Field-Effect Transistors". Physical Review Letters. 93 (19). doi:10.1103/PhysRevLett.93.196805

[2] Q. Huang, et al. "A novel Si tunnel FET with 36mV/dec subthreshold slope based on junction depleted-modulation through striped gate configuration." IEEE Inter. Electron Devices Meeting (IEDM), pp.187-190, 2012.

[3] Shinji Migita, et al. "Experimental Demonstration of Temperature Stability of Si- Tunnel FET over Si-MOSFET." IEEE Silicon Nanoelectronics Workshop(SNW), pp. 7-8,2012.

[4] Q. Huang, et al., "First Foundry Platform of Complementary Tunnel-FETs in CMOS Baseline Technology for Ultralow-Power IoT Applications: Manufacturability, Variability and Technology." IEEE Inter. Electron Devices Meeting (IEDM), , pp.22.2.1-22.2.4,2015.

[5] Synopsys Sentaurus TCAD Ver. J-2017.09, Synopsys, Inc., https://www.synopsys.com/home.aspx

THE FACTORS THAT INFLENCE THE EFFECTIVE MOBILITY IN 5 NM PMOS FINFET DESIGN

Enming Shang[1]*, Xin Luo[1], Yu Ding, Shaojian Hu[1], Shoumian Chen[1] and Yuhang Zhao[1]

[1]Shanghai IC R&D Center 497 Gaosi Road, Zhangjiang Hi-Tech Park, Shanghai 201210, China

*Corresponding Author's E-mail: shangenming@icrd.com.cn

ABSTRACT

In the 5 nm FinFET design, more induced stress can bring charge mobility improvement and device performance. There are several influence factors on stress and mobility in the process design. Substrate orientation and channel direction combination is a early considered factor. The source/drain epitaxy (S/D epi) plays an important role in channel stress formation. The two steps in S/D epi, seed and bulk epitaxy processes play a different role in stress formation and mobility improvement. Fin structure is also an important factor in process design. We have found that a fin height of 50 nm for a 5 nm PMOS FinFET device can provide maximum hole mobility.

INTRODUCTION

According to Moore's Law, the CMOS manufacturing design rules have been continuously shrinking from a few microns to the current a few nanometers in the past decades . Since the first FinFET at 22 nm technology node made by Intel Corporation, 16/14 nm FinFET logic technology have been developed the TSMC and Samsung sequentially [1-3]. As the device evolves to 5 nm technology node and beyond, more physical effect, such as, ballistic transport, quantum confinement effect must be considered because of shorter gate length and smaller critical dimensions [4-6], will affect device performance besides Short Channel Effect (SCE). In order to reduce SCE and improve the device performance, the stress applied in the channel become more and more important. However, as the critical dimension shrinks to several nanometers, it is very difficult to apply stress to the channel, especially in the p-type FinFETs. To find an effective method to apply stress and improve the mobility of channel material is an urgent task. As we know, the hole carriers in device transport for p-type MOSFET have lower mobility then electrons. In this work, the effective mobility of holes for PMOS with different influence factors have been studied. It shows that a FinFET at substrate orientation/channel direction combination (100)/<110> is the best combination, which will supply highest channel stress and hole mobility. S/D epi in PMOS stress formation is very significant. The Germanium fraction in seed epitaxy and Phosphorous doping in bulk epitaxy plays an important role in improving the hole mobility. Finally, the fin structure and channel length influence on the mobility have been studied.

SIMULATION RESULTS AND DISCUSSION

In this work, we focus on the 5 nm technology node FinFET stress and effective mobility. According to public information [7], key dimensions for a typical 5 nm FinFET are listed in Table 1.

TABLE 1. KEY PARAMETERS AND THEIR DIMENSIONS OF 5 NM FINFET DEVICE STRUCTURE

Parameter	Critical Dimension (nm)
Fin pitch	24 nm
Gate pitch	50 nm
Fin width	5 nm
Gate length	19 nm
Fin height	50 nm
Source/Drain epi height	50 nm
Channel doping	2×10^{15} cm^{-3}
Substrate initial doping	2×10^{15} cm^{-3}

The FinFET process is simulated by Sentaurus TCAD^{TM} software [8].

The effective gate field induced by depletion and inversion charge density can affect the surface roughness scattering on the surface of channel and adjacent materials, which are the intrinsic mechanics that affect the channel mobility. The effective gate field E_{eff} is descripted by[9],

$$E_{eff} = (q/\varepsilon_{Si})(N_{dep} + \eta \, N_{inv}) \qquad (1)$$

where $\eta=1/3$ in PMOS[10]. And the depletion charge density per unit area N_{dep}, is given by Equation (2)

$$N_{dep}=(4\varepsilon_{Si}*k_b *T *\ln(N_{sub}/n_i)*N_{sub}/q^2)^{1/2} \qquad (2)$$

$$N_{inv}=\int_{\infty}^{Vg} C_{gc}(V_g)\,dV_g \qquad (3)$$

The inversion charge density per unit area N_{inv} is given by Equation (3). Effective mobility u_{eff} is associated with the inversion layer charge density N_{inv} and conductivity of device $g_d(V_d)$, described in Equation (4).

$$\mu_{eff}=L\times g_d(V_d)/(q\times W_{eff}\times N_{inv}) \qquad (4)$$

Where $g_d(V_d)$ is the conductivity under the 50 mV drain bias voltage in this work, and $W_{eff} =2H_{fin}+W_{fin}$ represents the effective channel width of the FinFET device.

Since different substrate orientations and channel crystallographic direction combination will exert different channel stress in MOSFET devices, it is found that the combination (100)/<110> supplies the highest channel stress along the channel and the best hole mobility among all other combinations, as shown in Figure 1. Next combination (100)/(100) gives the second best channel stress and mobility. It suggests that (100) substrate orientation is suitable for MOSFET devices. In the following simulation, the data are based on the best combination (110)/<110>

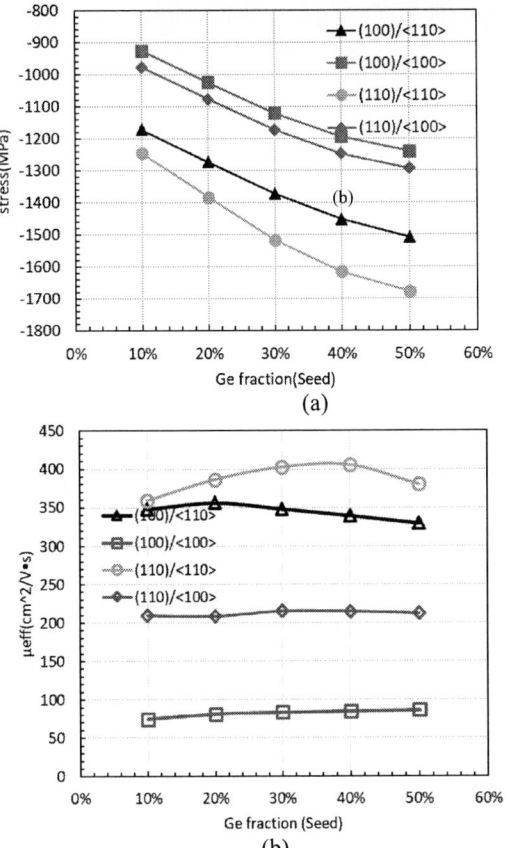

Figure 1: Germanium fraction in seed epitaxy vs. (a) stress in channel and (b) effective mobility with different substrate orientation/channel direction combination.

S/D epi have two steps consist of the seed and bulk epitaxy processes (Figure 2). To introduce compressive stress, the Silicon Germanium (SiGe) S/D epi is applied to the PMOS FinFETs because of the larger lattice constant of Germanium than to that of Silicon. The seed Germanium fraction in SiGe is usually attributed to have more contribution to the channel stress and mobility. As shown in Figure 1(a), a 40% stress improvement can be achieved as the Germanium fraction in SiGe seed increases from 10% to 50%. However, the hole mobility does not perform a monotonous increase like the stress, yielding a maximum value at 40% Germanium fraction. It can be explained that Germanium crystal constant is larger than that of Silicon, the Coulomb scattering and phonon scattering effects in the surface between Seed and channel become larger due to the mismatch of two

lattice constants at their interface, inducing effective hole mobility decrease.

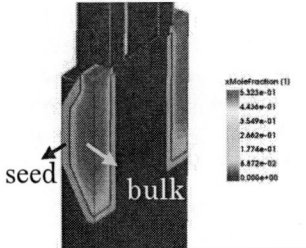

Figure 2: Germanium fraction distribution diagram showing bulk and seed epitaxial layers in the FinFET

Conventionally, Germanium fraction in the seed is set to below 30% in order to avoid excessive stress release and interface distortion. SiGe bulk epitaxy has higher Germanium fraction (>50%) for contributing a larger compressive stress and hole mobility.

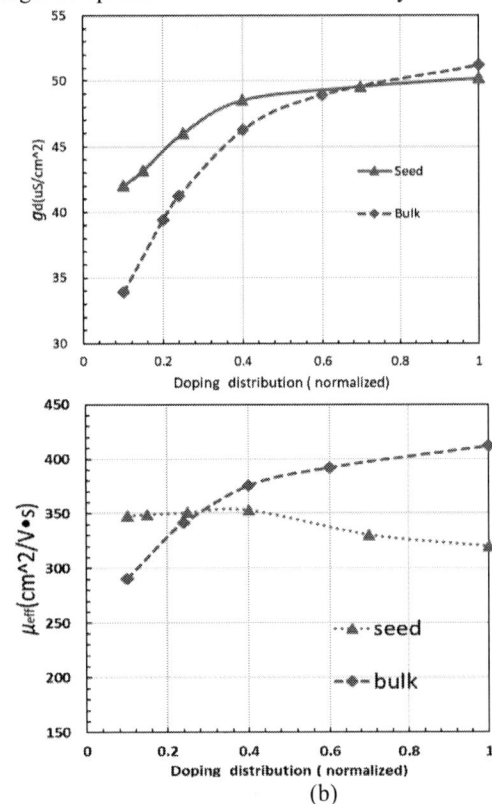

Figure 3. (a) g_d and (b) μ_{eff} vs. doping distribution of seed and bulk S/D epi

At same time, the p-type doping in two S/D epi sections have difference influence on the effective mobility μ_{eff}. The phosphorous doping in seed contribute little influence on the μ_{eff}. As shown in Figure 3, an increase of phosphorous doping in bulk epitaxy will give a significant boost to the effective hole mobility since heavy doping in bulk epitaxy can lower the resistance, or increase the conductivity g_d

(Figure 3(a)) of the S/D junctions, while heavy doping in seed can introduce more mismatch between the channel and seed interface, which can offsets the S/D resistance reduction. While the inversion charge density N_{inv} is not affected by the S/D epi because N_{inv} is mainly affected by initial substrate doping and gate field. As a result, the increase of doping in the bulk will boost the effective hole mobility but the increase of doping in the seed will almost not.

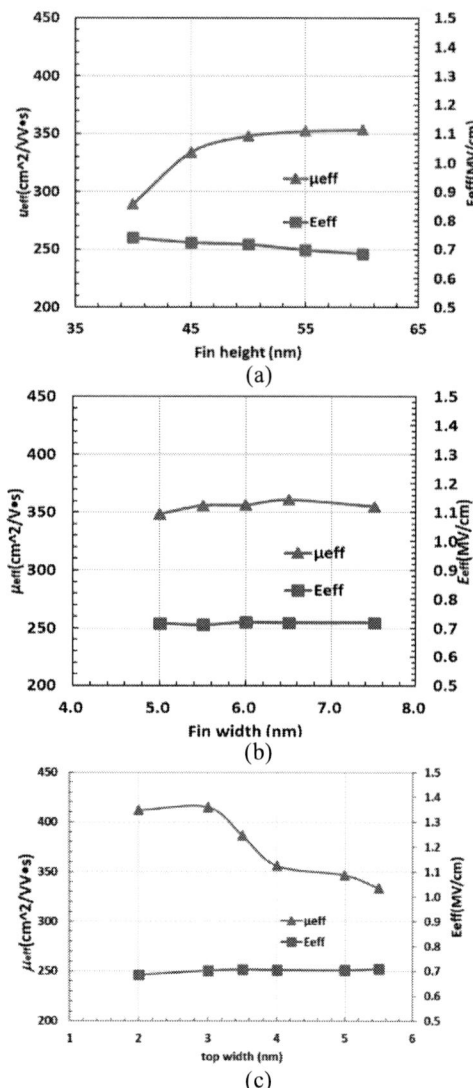

Figure 4: E_{eff} and μ_{eff} vs fin structure variations

In a 5 nm FinFET device design, the Critical Dimension (CD) of device is very small and become more significant to device performance, such as the fin height, width and channel length etc. . When the fin height increases from 40 nm to 60 nm, the effective hole mobility will also increase, but will saturate at around 350 cm^2/V• s (50 nm fin height) as shown in Figure 4(a)[11]. Equation (4) tells us that the effective mobility is influenced by conductivity and inversion

layer charge density. When the fin height increases, both two factors in Equation (4), the g_d and N_{inv} will increase simultaneously. More N_{inv} is caused by the increase of gate to channel capacitance C_{gc}, which is closed related to the fin height. However, the gate control ability for the whole fin becomes weak if the fin height is larger than one critical value (found to around 50 nm). As a consequence, the conductivity of device grows slowly and the mobility reaches a saturation value. The fin width does not affect mobility as shown in Figure 4(b). As the top of fin width increases from 2 nm to 5.5 nm when the bottom fin is fixed at 6 nm, the mobility decrease by about 20%. The wider of fin top will reduce the gate control ability for channel inversion layer, as shown in Figure 4(c). By the way, Figures 4(a)-4(c) show that effective gate field E_{eff} of inversion layer is not very sensitive to the above fin structure variation.

CONCLUSION

We have studied the PMOS stress and effective mobility affecting factors. The (100)/<110> substrate orientation /channel direction are the best combination for compressive stress formation and hole mobility improvement. We have found that heavier p-type doping in bulk S/D epi can provide higher hole mobility of device and there seems to be an optimized Germanium fraction (~40%) in seed for the mobility consideration. The fin structure influence on hole mobility are also studied and we have found that the 50 nm fin height is optimum and can provide a saturation hole mobility values as the fin height increases.

ACKNOWLEDGEMENTS

We would like to thank Shanghai ICRD colleagues and higher management team for the support of this work.

REFERENCES
[1] S. Natarajan, et.al, *IEEE, IEDM* **14**, 2014, pp. 71-74.
[2] S. Y. Wu, et. al, *IEEE, IEDM* **13**, 2013 pp.224-227.
[3] S. Y. Wu et al., *IEEE, IEDM* **4,** pp. 48-51 (2014).
[4] Y. K. Choi, et al., 2001, pp.85-86.
[5] J. B. Roldán, et. al, *IEEE, TED* **57,** pp. 2925-2933, (2010).
[6] V. P. Trivedi, et.al, 2005, pp. 579-582.
[7] 中国半导体论坛（*China Semiconductor Forum*).
[8] *Sentaurus Process User Guide* I-2018.06, *Synopsys Inc.*, Mountain View, CA, USA, (2018).
[9] S. Takagi, et.al, *IEEE, IEDM*, 1994. pp. 2357-2362
[10] S. Takagi, et.al, IEEE, IEDM, 1994. pp. 2362-2368
[11] E. Shang et.al, *Journal of Microelectronic Manufacturing* 2019, pp. 19020405-19020409

Optimization of embedded SiGe process to enhance PFET performance on 28nm low power platform

Wei Liu, Haibo Lei, Xuejiao Wang
Technology Development Department
Shanghai Huali Integrated Circuit Corporation
Shanghai, China
Wei Liu@hlmc.cn

Abstract—This paper presents a new SiGe profile of 28nm CMOS technology using conventional poly gate and SiON gate dielectric (Poly/SiON) with best-in-the-class 27nm pFET transistor. PFET Drive current of 431 µA/µm at off current $7.5×10^{-10}$A/µm were achieved at Vd = -1.05V, which performance is 12% higher than standard SiGe structure. TCAD simulation reveals that compressive stress intensity of modified SiGe is ~3% higher than that of standard SiGe. The effective mobility curves are obtained by split CV method, the mobility peak value of modified SiGe is also higher. This reveals compressive stress induced in the channel and decreasing parasitic resistance in SD region by modified SiGe structure are shown to be the major source of the observed performance enhancement. This research about pFET performance boosting through SiGe profile modification has given an optimized direction for mass production in 28nm platform.

Keywords- pFET, 28 nm technology node, SiGe, stress, split CV

I.   INTRODUCTION

As CMOS technology is scaled toward 28nm and below, the conventional poly/SiON technology still remains as a very viable and competitive technology choice, due to its lower cost and the wealth of manufacturing experiences accumulated [1-3]. In order to enhance driving current, especially for pMOSFET, strain engineering has become an attractive and increasingly important technique for mass production [4-7].

For pMOS devices, a uniaxial compressive stress in the channel is induced by the selective epitaxial growth (SEG) of $Si_{1-x}Ge_x$ in recessed source/drain (S/D) regions. In order to boost the transistor performance, it is crucial to achieve an optimum Ge contents and SiGe shape in the SiGe region, which yields a hole mobility and a drive current enhancement [8]. Many research such as different composition of $Si_{1-x}Ge_x$ [9], U-shape SiGe, Σ-shaped SiGe [10], refilled SiGe [11] have been widely developed to improve compressive strain in channel. Furthermore, the use of embedded SiGe reduces the S/D resistance is also a good choice for performance boosting [8, 12].

In this paper, we present a state-of-the-art 28nm Poly/SiON technology with developed modified SiGe stress control to improve CMOS pFET performance, including the following technologies: 1) mobility improvement by optimizing Σ-shaped SiGe-Source/Drain shape for pFET; 2) Reduction of parasitic resistance by embedded B-doped SiGe-SD. We successfully demonstrated high performance

with 27nm gate length pFETs. The DC electronic parameter and CV measurements were performed using an Agilent 4082 semiconductor parameter analyzer and an Agilent E4980A Multifrequency LCR Meter. Compressive stress with different SiGe structure have been investigated by TCAD simulation along gate channel direction at short channel-length pFET device (W/L = 2.7/0.027 µm/µm). The effective hole mobility behavior in long-channel pFET (W/L = 2.7/0.9 µm/µm) have been studied with Split CV technology.

II.   DEVICE FABRICATION AND DEVICE STRUCTURE

The pFET devices described in this research were fabricated following a state of the art 28nm low power platform flow, which incorporates eSiGe process as stressor in the S/D. The pFET devices in this study were fabricated on <110> wafer surfaces. Continuously, the AA process, STI process, well implant, gate poly stack formation, spacer formation, eSiGe formation, LDD implant, SD implant, thermal, NiSi and contact formation etc. were step by step processed. Detailed pFET SiGe-SD fabrication is as following: First, hard mask Si_3N_4 deposition to protect poly gate stack. Second, KrF lithography and etch to define a self-aligned sigma-shaped trench in the pFET SD region. Then, multiple-layer eSiGe structure was deposited in the trench by selective epitaxy. Finally, Si_3N_4 hard mask remove. The cross-section profile and top view of pFET is shown in Fig. 1.

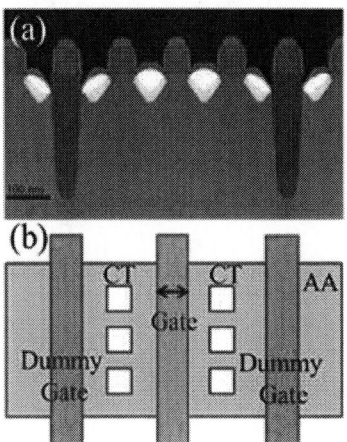

Fig 1. SiGe pMOSFET: (a) schematic cross-section after SiGe formation of HAADF XTEM image and (b) layout top view.

Main factors of pFET performance in the eSiGe profile are trench depth D, tip depth and eSiGe tip distance to channel T2G. In this paper, we optimized profile through trench depth from 700Å to 600Å, tip depth modified from 230Å to 160Å, T2G from 20 to 5 Å; meanwhile keep same SiGe total volume. Fig. 2(a)-(d) show the standard eSiGe (trench 700Å, up) and the optimized eSiGe (down) profile picture in which the optimized eSiGe profile have smaller trench depth and tip depth value.

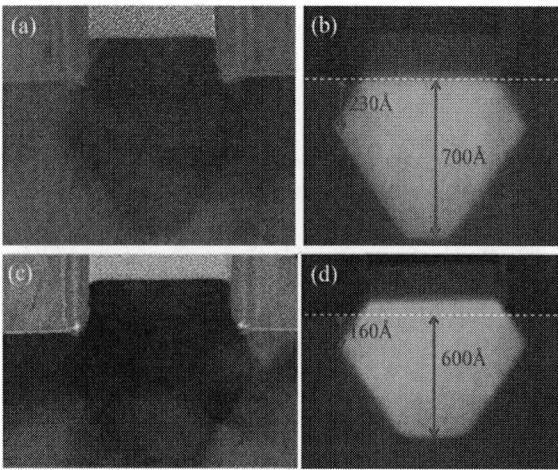

Fig. 2 TEM and HAADF XTEM images of Lmask = 27nm PFET. (a) & (b) baseline SiGe recipe: trench depth 700Å; (c) & (d) modified SiGe recipe: trench depth 600Å.

Furthermore, we increased Boron content in the cap of SiGe profile to reduce resistance of ohmic contact, as illustrated in Fig. 3. From SIMS result, Boron content increased from 4.4×10^{20} to 1.5×10^{21} after ion implantation and subsequent anneals. In this paper, SiGe T700 stands for standard SiGe profile, SiGe T600 stands for optimized SiGe profile, respectively.

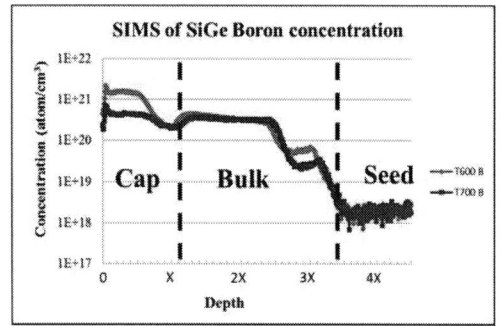

Fig. 3 Boron concentration of eSiGe by Secondary Ion Mass Spectroscopy (SIMS)

III. ELECTRICAL CHARICTERIZATION

A. Basic electrical performance

The electrical properties of pFET are listed in Table I (tox_inv = 24.4 nm, and VDD= -1.05V). On/Off current (Ion/Ioff) is exacted at Vg = -1.05/0V and Vd= -1.05V. As table I show, pFET with optimized SiGe T600 saturation threshold voltage (Vtsat) have shift ~44mV, with I_{ON}/I_{OFF} 431 µA/µm/7.5×10^{-10}A/µm, which performance is better than standard SiGe T700 with 435 µA/µm/2.3×10^{-9} A/µm. Normalized Rsd of SiGe T600 is about 5 ohm/sqr lower. Vt local variation improved by 2mV, which indicated threshold voltage distribution of modified SiGe T600 for pFET is more uniform.

Table 1. Electrical performance of PFET with 27nm gate length

Parameters	L_G=27 nm PFET	
	SiGe T700	SiGe T600
Vthlin (V)	0.394	0.418
Vthsat (V)	0.225	0.269
ION (µA/µm)	435	431
IOFF (A/µm)	2.3×10^{-9}	7.5×10^{-10}
Rsd (ohm/sqr)	95	90
Vt local variation (mV)	26.96	24.97

Fig. 4 shows I_{ON}/I_{OFF} and I_{ON}/Vtsat plot at -1.05V operating voltage (V_{dd}). A significant device performance improvement (~12%) in optimized SiGe T600 than that in standard SiGe T700 is observed in Fig. 4 (a) (red triangle vs. blue circle symbols). Fig. 4 (b) shows comparisons of I_{on}-Vtsat curve, about 7% Idsat improvement is obtained with the optimized SiGe T600.

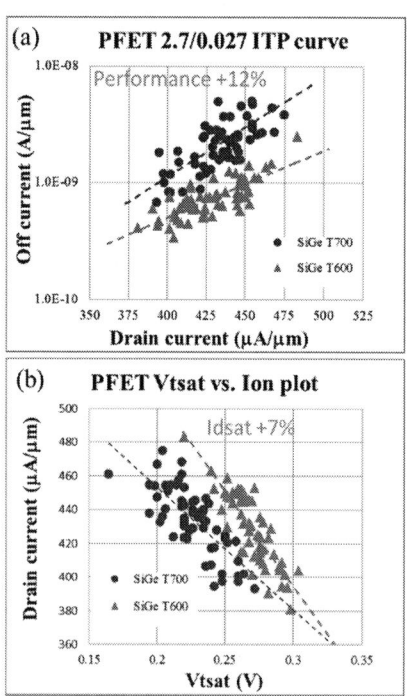

Fig. 4 I_{on} vs. I_{off} and Vtsat vs. Ion characteristics of 27nm minimum gate length pMOSFETs at V_{dd} = -1.05V.

978-1-7281-6559-2/20 $31.00 © 2020 IEEE

B. Simulation of SiGe profile and stress

In order to reveal the channel stress difference of SiGe T700 and T600, TCAD simulation has been researched. Fig. 5 has shown two stress contours with SiGe T700 and SiGe T600 conditions. Point of origin 0 along channel length direction corresponds to the center of the channel. The compressive stress (marked by blue) enriches mainly in SiGe S/D and along channel surface region. From the SiGe region towards the inner channel, stress content decreases gradually with the distance. Comparison of lateral stress along the channel length direction, compressive stress from SiGe T600 is much obviously stronger than that of SiGe T700.

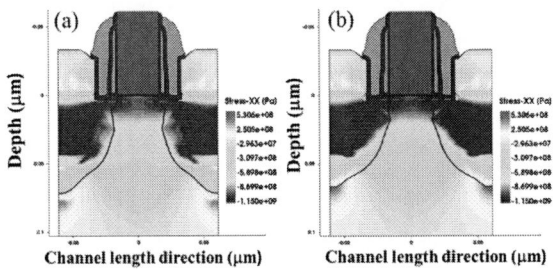

Fig. 5 pFET structures with the corresponding compressive stress distributions of (a) standard SiGe T700 and (b) the optimized SiGe T600.

To quantify the stress along the channel length direction, further stress intensity with different SiGe along the channel (red line marked in Fig. 5) is measured, as indicated by the blue and red line in Fig. 6, respectively. Stress intensity is plotted as a function of distance from channel length. The result shows stress intensity decreases from gate edge to center. At the center of gate, compressive stress intensity of SiGe T600 has a relatively higher value (~-1040 MPa) than that of SiGe T700 (~-1010 MPa). From the TCAD result, reducing the trench depth and tip to gate will bring high-stress region closer to the channel, and will lead to a higher channel stress. Strengthening the channel stress is attributed to the excellent I_{on}-I_{off} performance, as indicted in Fig. 4.

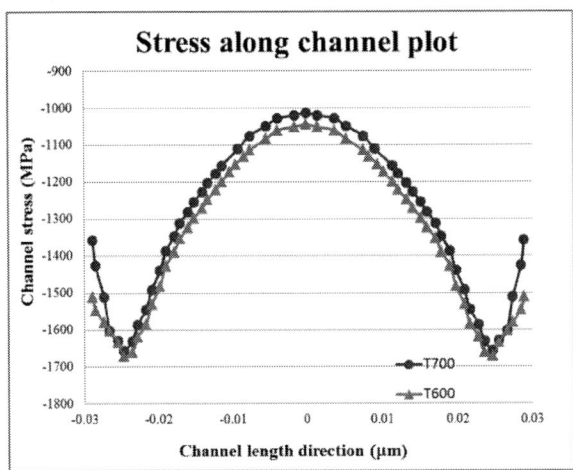

Fig. 6 Channel Stress contribution is plotted as a function of distance along channel length direction.

C. Split C-V technique for mobility exaction

Effective mobility (μ_{eff}) extraction for the pFET devices was based on split C-V method [13-19]. μ_{eff} is thus defined as:

$$\mu_{eff} = \frac{I_{DS}}{V_{DS}} \cdot \frac{1}{Q_n} \cdot \frac{L}{W} \qquad \text{EQ. 1}$$

Where L is gate length, W is active area (AA) width.

Q_n is obtained by integration of the accurate gate to channel capacitance data (C_{gc}).

$$Q_n(V_g) = \int_{V_0}^{V_g} C_{gc}(V_g) \, dV_g \qquad \text{EQ. 2}$$

Where Vg is the gate voltage, and V_0 is the gate voltage at which $Q_n = 0$. The I_{DS} is achieved by integration of the channel conductance (G_m).

$$I_{DS} = \int_{V_0}^{V_g} G_m(V_g) dV_g \qquad \text{EQ. 3}$$

Hence, substitution Q_n and I_{ds} of equation 2 and 3 in equation 1, μ_{eff} can be expressed as followed:

$$\mu_{eff} = \frac{L}{W \cdot V_{DS}} \cdot \frac{\int_{V_0}^{V_g} G_m(V_g) dV_g}{\int_{V_0}^{V_g} C_{gc}(V_g) dV_g} \qquad \text{EQ. 4}$$

This technique relies on the computation of the inversion charge (Q_n), the channel conductance (G_m).

G_m vs. V_g curves of pFET with standard SiGe T700 is presented in Fig. 7, extracted from $I_d V_g$ curve ($V_d = -0.05V$).

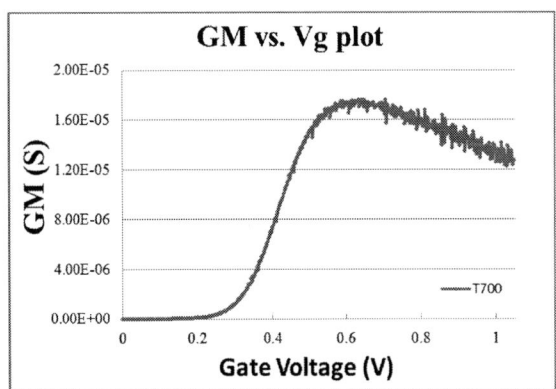

Fig. 7 Transconductance Gm vs. Vg curves of pFETs with different SiGe measured at DC (with delay times: 2 sec). L = 0.9 μm, W = 2.7 μm, Vd = -0.05 mV.

The gate to channel capacitance C_{gc} is measured in the device W/L = 2.7/0.9 μm/μm at frequency 100 KHz and plotted as a function of gate bias for pFET with standard SiGe T600 (Fig. 8). After removing pad parasitic capacitances by de-embedding structures, the final capacitance represents the intrinsic gate to effective channel capacitance C_{ch} is achieved and can thus be integrated from accumulation to strong inversion to obtain the inversion charge (Q_n) curve based on EQ. 2 (as Fig. 8).

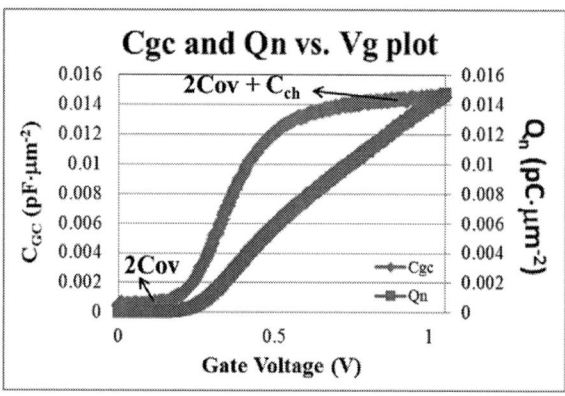

Fig. 8 C_{GC} and Q_n versus V_g

Based on the above curves, a μ_{eff} calculation using a simple split C-V model was performed. The hole effective mobilities shown in Fig. 9 of SiGe T700 and T600. With Gate bias increased, μ_{eff} gradually increased to reach a peak value and then slightly decreased. This degradation under higher gate bias can be attributed to a combination of coulomb and phonon scattering mechanisms [13]. Comparison of μ_{eff} of SiGe T700 and T600, the maximum peak value of SiGe T600 is about 5% higher than that of SiGe T700. Combine with TCAD simulation result (Fig. 5 and 6), the hole mobility in the channel region induced by the compressive stress in the modified SiGe SD structure cause the pFET performance enhancement, which is consistent with Fig. 4. Therefore, the SiGe T600 is more effective to enhance hole mobility for pFET, which is a good choice for pFET performance enhancement in mass production.

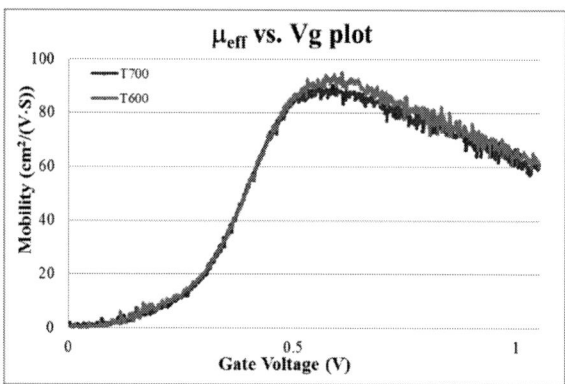

Fig. 9 Effective mobility as a function of gate voltage, extracted by standard split C-V method (EQ. 4).

IV. CONCLUSIONS

We described high performance 27nm pMOSFET with modified SiGe profile for 28nm technology node. Drive current of 431 μA/μm at off current 7.5×10^{-10}A/μm for 27 nm pMOS were achieved at Vd = -1.05V, device performance is 12% higher and the hole effective mobility is

~5% higher than standard SiGe structure. The increasing compressive stress induced in the channel and decreasing parasitic resistance in SD region by modified Σ–shaped SiGe structure are the major sources of the observed performance enhancement. This research about pFET performance boosting has given an optimized direction for mass production in 28nm platform.

ACKNOWLEDGMENT

The authors would like to thank the support from Cuiqin Xu, Qiaozhi Zhu, Jing Shi, Peng Gou, Haitao Wang, Xiaojun Zhou, Tong Zhao, Peng Zhang, Xin Guo, Yongwang Zhang, Shenlong Xuan, Yinghua Liu, Xubin Jing, Tianpeng Guan, Ming Tian, Yu Zhang and Albert Peng.

REFERENCES

[1] Wu, S. Y., Liaw, J. J., Lin, C. Y., Chiang, M. C., Yang, C. K., Cheng, J. Y. et al. (2009, June). A highly manufacturable 28nm cmos low power platform technology with fully functional 64mb sram using dual/tripe gate oxide process. In 2009 Symposium on VLSI Technology (pp. 210-211). IEEE.

[2] Chakravarthi, S., Chidambaram, P. R., Machala, C. F., & Mansoori, M. (2006). Dopant diffusion modeling for heteroepitaxial SiGe/Si devices. Journal of Vacuum Science & Technology B: Microelectronics and Nanometer Structures Processing, Measurement, and Phenomena, 24(2), 608-612.

[3] Liang, C. W., Chen, M. T., Jenq, J. S., Lien, W. Y., Huang, C. C., Lin, Y. S., Chou, P. Y. et al. (2011, June). A 28nm poly/SiON CMOS technology for low-power SoC applications. In 2011 Symposium on VLSI Technology-Digest of Technical Papers (pp. 38-39). IEEE.

[4] Luo, Z., Chong, Y. F., Kim, J., Rovedo, N., Greene, B., Panda, S., Davis, R. et al. (2005, December). Design of high performance PFETs with strained Si channel and laser anneal. In IEEE InternationalElectron Devices Meeting, 2005. IEDM Technical Digest. (pp. 489-492). IEEE.

[5] T. Ghani, A 90-nm high volume manufacturing logic technology featuring novel 45-nm gate length strained silicon CMOS transistors, 2003. IEDM Tech. Dig.. (pp. 978-980).

[6] Chidambaram, P. R., Smith, B. A., Hall, L. H., Bu, H., Chakravarthi, S., Kim, Y. et al. (2004, June). 35% drive current improvement from recessed-SiGe drain extensions on 37 nm gate length PMOS. In Digest of Technical Papers. 2004 Symposium on VLSI Technology, 2004. (pp. 48-49). IEEE.

[7] Chang, W. T., & Lin, Y. S. (2013). Performance Dependence on width-To-length ratio of Si Cap/SiGe channel MOSFETS. IEEE Transactions on Electron Devices, 60(11), 3663-3668.

[8] Washington, L., Nouri, F., Thirupapuliyur, S., Eneman, G., Verheyen, P., Moroz, V., Smith, L., Xu, X. et al. (2006, Jun). pMOSFET with 200% mobility enhancement induced by multiple stressors, 2006. IEEE Electron Device Lett.. (pp. 511–513).

[9] Eneman, G., Verheyen, P., Rooyackers, R., Nouri, F., Washington, L., Schreutelkamp, R. et al. (2006). Scalability of the $Si_{1-x}Ge_x$ Source/Drain Technology for the 45-nm Technology Node and Beyond. IEEE Transactions on Electron Devices, 53(7), 1647-1656.

[10] Ohta, H., Kim, Y., Shimamune, Y., Sakuma, T., Hatada, A., Katakami, A. et al. (2005, December). High performance 30 nm gate bulk CMOS for 45 nm node with Σ-shaped SiGe-SD. In IEEE InternationalElectron Devices Meeting, 2005. IEDM Technical Digest. (pp. 4-pp). IEEE.

[11] Yang, H. C., Li, C. W., Liao, W. S., Du, C. K., Wang, M. C., Yang, J. M. et al. (2013, January). The enhancement of MOSFET electric performance through strain engineering by refilled SiGe as Source and Drain. In 2013 IEEE 5th International Nanoelectronics Conference (INEC) (pp. 251-253). IEEE.

[12] Arden, W. (2006). Future semiconductor material requirements and innovations as projected in the ITRS 2005 roadmap. Materials Science and Engineering: B, 134(2-3), 104-108.

[13] Kilchytska, V., Lederer, D., Simon, P., Collaert, N., Raskin, J. P., & Flandre, D. (2005, October). Revised split C-V technique for mobility investigation in advanced devices. In 2005 IEEE International SOI Conference Proceedings (pp. 110-111). IEEE.

[14] Ramos, J., Severi, S., Augendre, E., Kerner, C., Chiarella, T., Nackaerts, A. et al. (2006, September). Effective mobility extraction based on a split RF CV method for short-channel FinFETs. In 2006 European Solid-State Device Research Conference (pp. 363-366). IEEE.

[15] Akkez, I. B., Cros, A., Fenouillet-Beranger, C., Boeuf, F., Rafhay, Q., Balestra, F., & Ghibaudo, G. (2012, September). New parameter extraction method based on split CV for FDSOI MOSFETs. In 2012 Proceedings of the European Solid-State Device Research Conference (ESSDERC) (pp. 217-220). IEEE.

[16] Pirro, L., Ionica, I., Ghibaudo, G., & Cristoloveanu, S. (2014, March). Split-CV for pseudo-MOSFET characterization: Experimental setups and associated parameter extraction methods. In 2014 International Conference on Microelectronic Test Structures (ICMTS) (pp. 14-19). IEEE. "Parameters extraction in SiGelSi pMOSFETs using split CV technique"

[17] Soussou, A., Leroux, C., Rideau, D., Toffoli, A., Romano, G., Tavernier, C. et al. (2013, March). Parameters extraction in SiGe/Si pMOSFETs using split CV technique. In 2013 14th International Conference on Ultimate Integration on Silicon (ULIS) (pp. 41-44). IEEE.

[18] Romanjek, K., Andrieu, F., Ernst, T., & Ghibaudo, G. (2004). Improved split CV method for effective mobility extraction in sub-0.1-μm Si MOSFETs. IEEE Electron Device Letters, 25(8), 583-585.

SRAM and Single Device Isolation analysis in FinFET Technology

Yijun Zhang, Li Tan, Yu Li

Technology R&D Center, SMIC

18 Zhangjiang Road, Pudong New Area, Shanghai, P.R. China 201203

E-mail: YiJun_Zhang@smics.com

ABSTRACT

It has been recognized that SRAM memory is an ideal vehicle for defect monitoring and yield improvement during process development because of its highly structured architecture and simplified approach to memory bitmapping [1]. As SRAM transistors are aggressively scaled down, isolation leakage become an important reliability concern shallow trench isolation (STI) continues to scale down. This isolation leakage has to be considered to improve SRAM yield and performance. In FinFET technology, SRAM isolation is mainly dominated by NW CD and STI depth, while single device isolation is mainly impacted by AA to AA space and AA to well space. In this paper, NAA-NAA/NAA-NW/PAA-PAA/PAA-PW four types of isolation are tested to study the NW CD impact to SRAM isolation. STI depth +10 nm and STI depth +20 nm two splits are designed to research its impact to SRAM AA-well isolation. At last, AA to AA space and AA to well space influence to single device is analyzed.

Keywords: Isolation, SRAM, Single Device.

INTRODUCTION

SRAM is the main feedback for process improvement and yield learning. It is because the SRAM high density and small feature size make its yield be very sensitive to process variation and the failing bit cells can be precisely localized by functional test for physical failure analysis [2-3]. As the dimension shrink, SRAM failures caused by isolation leakage become easier to happen, especially in FinFET technology. Isolation technique is used to obtain the electrical signature of the transistors of SRAM cell. The technique consists of physical isolation and electrical isolation. The purpose of physical isolation is to isolate the interested SRAM cell from the array, and the electrical isolation can electrically separate the n-channel (NAA) and p-channel (PAA) transistors of the cell through probing with bias voltage scheme [4]. Physical isolation is performed by cutting the bit line and bit line#. For electrical isolation, well implant and STI technique is the main method to separate NAA and PAA. As shown in Fig. 1, there is mainly four types of isolation in SRAM cell, including NAA-NAA, PAA-PAA, NAA-NW and PAA-PW.

In this paper, NW CD and STI depth split are designed to research their impact to SRAM isolation. Besides, AA to AA space and AA to well space split are designed to research their impact to single device.

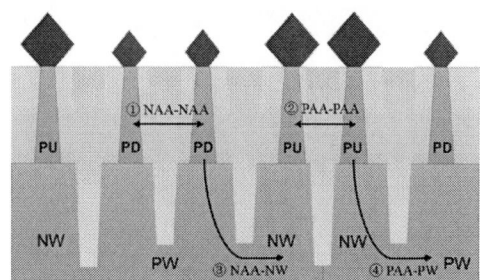

Fig. 1: Illustration of SRAM ISO

EXPERIMENTAL

The MOSFET samples in this paper were fabricated by FinFET process. Schematic of SRAM isolation test condition is shown in Fig. 2. To approach the SRAM real work status, NW is connected to Vdd and PW is connected to 0 voltage (GND). NW CD and STI depth are two key parameter to impact SRAM isolation: NW CD +5/ NW CD -5 (nm) and STI depth +10/STI depth +20 (nm) are design to test.

(a) NAA-NAA test schematic

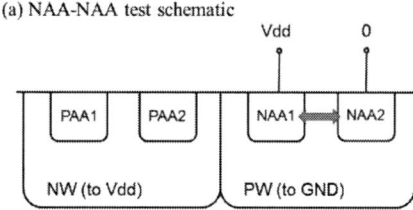

Fig. 2(a): Schematic of NAA-NAA test condition

(b) PAA-PAA test schematic

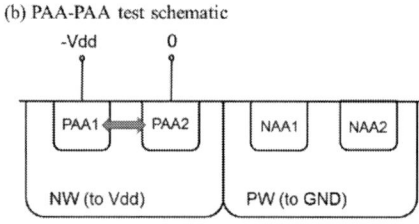

Fig. 2(b): Schematic of PAA-PAA test condition

978-1-7281-6559-2/20 $31.00 © 2020 IEEE

(c) NAA-NW test schematic

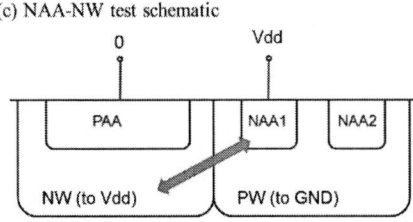

Fig. 2(c): Schematic of NAA-NW test condition

(d) PAA-PW test schematic

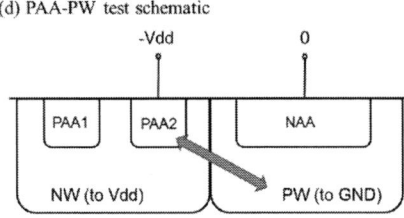

Fig. 2(d): Schematic of PAA-PW test condition

RESULT AND DISCUSSION

I. NW CD impact to SRAM isolation

NW CD split impact to SRAM isolation is displayed in Fig. 3. In modern CMOS processes, wells are formed using high-energy ion implantation that require a thick photoresist layer to mask the well implants [5]. Well implant and Source/Drain implant region will change along NW CD split. According to Fig. 2c and Fig. 2d, NAA-NW isolation increased while PAA-PW isolation decreased when NW CD become larger, which can be explained by metal boundary effect (MBE). During NP boundary shift from PAA side to NAA side, N plus implant will be more close to NW implant while P plus implant will be far away from PW region, which lead to more diffusion between NAA and NW. NAA-NAA isolation nearly keep the same during NW CD change, while PAA-PAA isolation show the same trend with PAA-PW isolation. It is suspected that PAA-PAA isolation contained PAA-PW isolation.

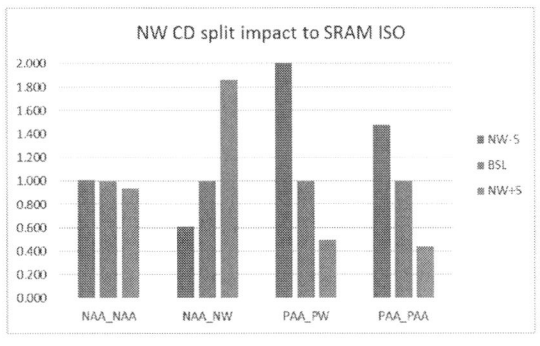

Fig. 3: NW CD split impact to SRAM ISO

II. STI depth impact to SRAM isolation

As shown in Fig. 1, NW and PW is separated by STI cut, STI depth show an obvious impact to SRAM isolation in FinFET technology [6]. STI depth +10 nm and STI depth +20 nm two splits are designed to research the impact of STI depth to SRAM AA-well isolation. According to Si data (Fig. 4), at different STI depth, P plus to PW isolation show no obvious change, while N plus to NW isolation will become better when STI depth become larger. According to TCAD simulation results (Fig. 5), when STI depth increase 20nm, current density on the Fin surface from PU to PD obviously smaller than baseline. This phenomena is not hard to explain: when STI depth + 20 nm, it is harder for electronic or carrier ion to diffuse from one side to another side. So STI depth controlling is a new challenge in SRAM FinFET technology.

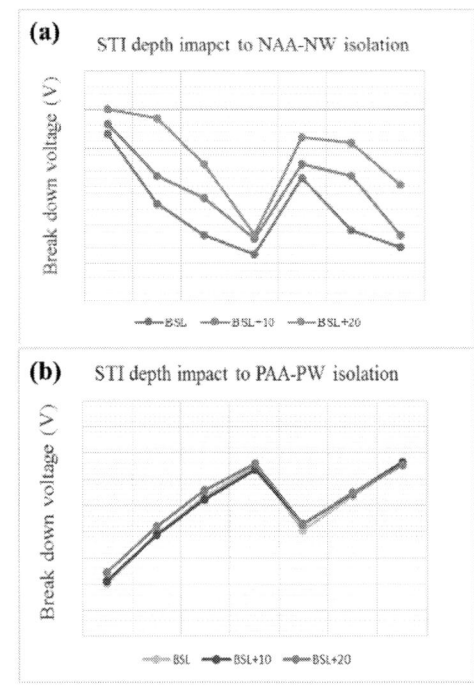

Fig. 4: STI depth impact to SRAM isolation: (a) NAA-NW; (b) PAA-PW.

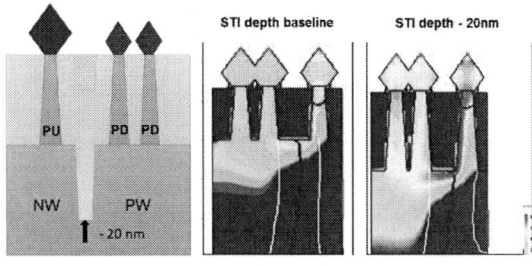

Fig. 5: TCAD simulation of STI depth impact to SRAM ISO

III. AA to AA space and AA to well space impact to single device isolation

AA to AA space and AA to well space are two key parameters in layout design, N plus to N plus and P plus to Plus isolation of single device at different AA space is researched and results displayed in Fig. 6. At X direction (along the Fin), both N plus to N plus isolation and P plus to P plus isolation become better as the AA space increase. At Y direction (along the gate), AA space changing show no obvious impact to isolation. Which means that along the Fin direction, isolation will be more sensitive to AA space, which can be explained by LOD effect (Length of Oxide Definition) [7]. When LOD distance (AA space) increase, carrier diffusion length will be longer and leakage will be smaller, such lead to better isolation.

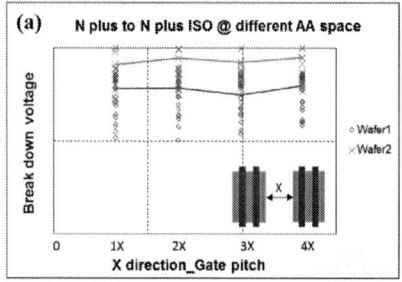

Fig. 6(a): N plus to N plus ISO at X direction

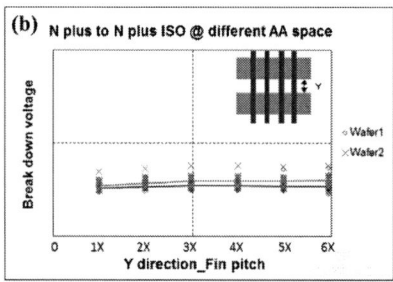

Fig. 6(b): N plus to N plus ISO at Y direction

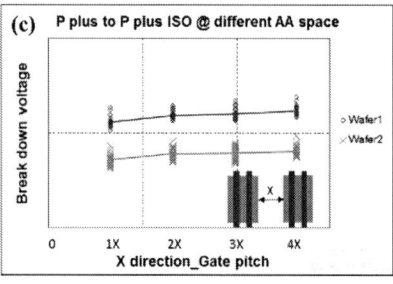

Fig. 6(c): P plus to P plus ISO at X direction

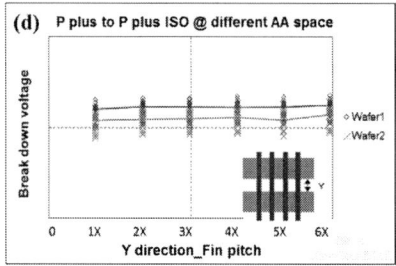

Fig. 6(d): P plus to P plus ISO at Y direction

N plus to NW and P plus to PW isolation of single device at different AA to well space is shown in Fig. 7. For N plus to NW isolation, break down voltage become better when NAA to NW space increase in both X direction and Y direction. While for P plus to PW isolation, break down voltage nearly keep the same when PAA to PW space change. As NW implant dosage is always larger than PW implant, it is not hard to explain that NAA-NW isolation is more sensitive to AA-well space.

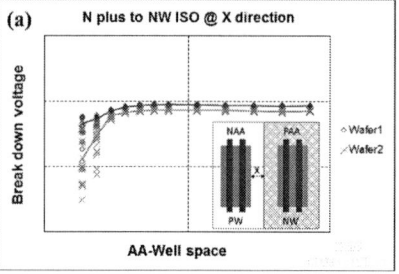

Fig. 7(a): N plus to NW ISO at X direction

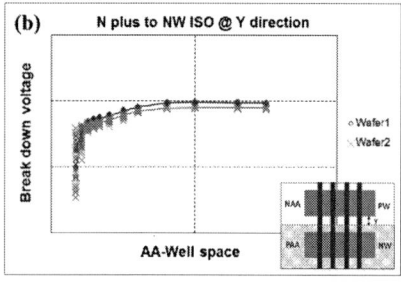

Fig. 7(b): N plus to NW ISO at Y direction

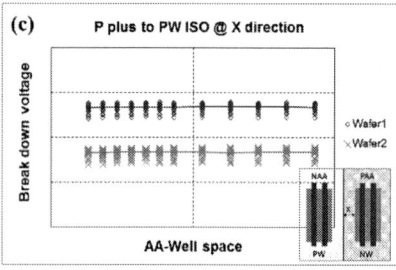

Fig. 7(c): P plus to PW ISO at X direction

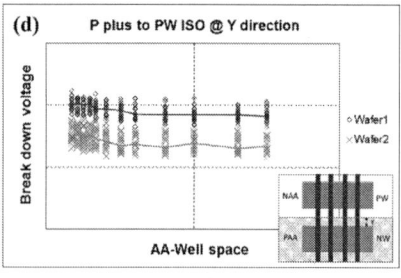

Fig. 7(d): P plus to PW ISO at Y direction

SUMMARY

NW CD and STI depth influence to SRAM isolation is studied. When NW CD become larger, both PAA-PAA and PAA-PW isolation reduced, while NAA-NW isolation increased, which can be explained by metal boundary effect. STI depth show an obvious impact to SRAM isolation in FinFET technology. According to Si data and TCAD simulation results, STI depth increase will improve AA to well isolation efficiently. AA to AA space and AA to well space also influence the isolation. Isolation along the Fin direction will be more sensitive to the AA space, while isolation in gate direction nearly keep the same when AA to AA space change. N plus to NW isolation become better when AA to well space increase, while P plus to PW isolation show no obvious change, which can be ascribed to the dosage difference between NW and PW implant.

REFERENCE

[1] Hung-sung Lin, Wen-tung Chang, Chun-lin Chen, et al. 2006 13th International Symposium on the Physical and Failure Analysis of Integrated Circuits, pp. 63-66, 2006.

[2] D. Maji, P. J. Liao, Y. H. Lee, et al. 2013 IEEE International Reliability Physics Symposium (IRPS), pp. 3E1.1-3E1.5, 2013.

[3] May Yang, J. H. Li, Smith Gao, et al. 2014 12th IEEE International Conference on Solid-State and Integrated Circuit Technology (ICSICT), 2014.

[4] Yit-Wooi Lim, Teong-San Yeoh, 1998 IEEE International Conference on Semiconductor Electronics, PP. 64-69, 1998.

[5] John V. Faricelli, IEEE Custom Integrated Circuits Conference 2010, pp. 1-8, 2010.

[6] Bryant. A, Hansch. W, Mii. T, proceedings of 1994 IEEE International Electron Devices Meeting, pp. 671-674, 1994.

[7] Bob Peddenpohl, Max Otrokov, Jeremy Wells, 2018 IEEE International Conference on Microelectronic Test Structures (ICMTS), pp. 31-34, 2018.

ETCH MODEL BASED ON MACHINE LEARNING

Rui Chen[1], Haoru Hu[1], Xiaoting Li[2], Ying Chen[1], Xiaojing Su[1], Lisong Dong[1], Lei Qu[2], Chen Li[1], Jiang Yan[2], Yayi Wei[1]*

[1]Key Laboratory of Microelectronics Devices and Integrated Technology, Institute of Microelectronics, Chinese Academy of Sciences, Beijing 100029, China
[2] North China University of Technology, Beijing 100144, China
*Corresponding Email: weiyayi@ime.ac.cn

ABSTRACT

This paper introduces a machine learning based etch bias prediction model and demonstrates its capability of extracting the features of the one-dimension and two-dimension layout post etch and predicting of etch bias. For 1D pattern, this model achieves absolute error less than ±2nm and average relative error below 10%. For 2D layout we achieved average absolute error less than 4nm with a standard deviation of 3nm, and average relative error of 15%. These results are promising for the developing the model-based etch proximity correction model towards the high volume manufacturing.

INTRODUCTION

The Moore's law drives the semiconductor industry development and aggressive node scaling. To meet the emerging patterning challenges, computational lithography employs Source Mask Co-Optimization (SMO) and Optical Proximity Correction (OPC) to improve the patterning fidelity. The etch process following lithography transfers the trench or hole patterns from photo-resist or hard mask into the substrate, thus is the key process step to determine the final patterning profile. It can be inferred that etch and lithography process jointly determine the final critical dimension (CD) of the substrate pattern which is critical to the high performance integrated circuit manufacturing.

The state of the art technologies mainly emphasis on compensating the light proximity effect as the CD has pushed to the mother nature limit of existing scanner light source. However etch process also suffers from various undesired effects, such as loading effect, line-end shorting, etc., which distort the post-etch profile or CD from post-lithography and downgrades the circuit performance. The CD discrepancy between post-etch and post lithography is described as the etch bias which is simply calculated by subtracting post-etch CD from post-litho CD (ADI-AEI). This phenomenon is inevitable in practical manufacturing and the industry has taken steps to compensate it, the manner is called etch proximity effect correction (EPC).

The industry widely adopts the rule-based (RB) etch proximity correction by implementing an empirical etch bias table in SMO and OPC while worked well for previous nodes in the past decade. However in the advanced node the layout consists of extremely scaled and complicated patterns, making it unrealistic for etch bias table to cover all scenarios well. As a result the post-etch CD with RB compensation could exhibit larger discrepancies from design than before.

In order to mitigate the abovementioned issues, model-based (MB) etch bias compensation is highly desired. Due to the sophisticated source of etch proximity effect, various physical models have been developed since the early 1980s [1-3]. Rather than physical simulation, it is more practical and efficiency to take advantage of machine learning to capture the etch effects in the complicated layout and predict etch bias, as the machine learning has already been adopted in OPC and manufacturing related applications [3-11].

Although not too much, there have been some notable reported progress on the machine learning based etch model, Seongbo et al. have developed a novel machine learning based EPC by taking segments of interest together with its surroundings as the geometric and optical parameters as input of the artificial neural network (ANN) to predict the etch bias [4]. Byungwhan et al. constructed a neural network to predict the silicon oxynitride (SiON) etching [5]. Haoru et al. demonstrated an etch bias prediction model for one-dimensional (1D) layout by using error back prorogation (BP) based neural network [3].

In this paper, we first briefly review our previous work in the 1D layout model, then we introduce the progress on the etch bias prediction model for two-dimensional layout. Finally we discuss the model performance optimization, demonstrate the etch bias predicting application followed by conclusion.

1D ETCH BIAS PREDICTION

In our previously published study, we built a BP based predicative etch bias algorithm flow that enables the etch bias prediction in 1D line-space structures with various width and space.

In this model, firstly the dataset was established by extracting features directly from the 1D layout before and after etching, following that the BP neural network model was trained given the extracted dataset. Finally the trained model was used to predict the etch bias given another set of extracted dataset and in addition to that the model was optimized based on the R^2 evaluation criterion in order to achieve the best performance. As shown in Figure 1, the inset illustrates the etch bias data extracting form the measurement of ADI and AEI, and the predicted data exhibits similar trend comparing to the true data.

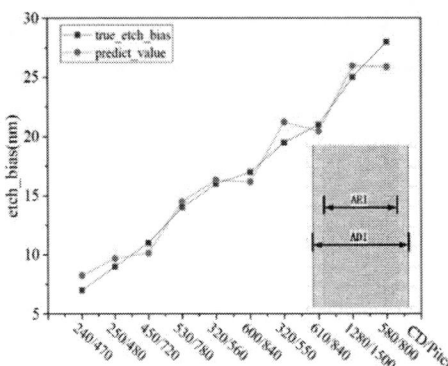

Figure 1: Comparison between BP neural network prediction value and actual etching bias value [3]

As shown in Figure 2, the absolute errors are within ±2nm with less than 10% relative error, demonstrating a good predicting capability.

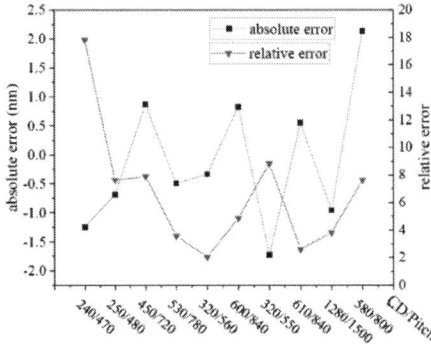

Figure 2: Absolute error and relative error distribution in terms of CD/Pitch [3]

2D ETCH BIAS PREDICTION
Work Flow

The 2D layout etch bias prediction model consists of two phases, the training phase and testing phase as shown in the Figure 3.

As the first step of the training phase, we employ the concentric circle area sample (CCAS) method which includes the parameterization of a segment using local pattern densities [4] to extract the features from the training layout and build the training dataset. Following that, we downsample the data by principle component analysis (PCA) and divide each sample in the dataset into N clusters according to similarities. Finally the model is

trained based on each individual cluster using ANN as illustrated in Figure 4, it's noteworthy that the training phase requires the training for etch bias prediction neural network, as well as the training for clustering model.

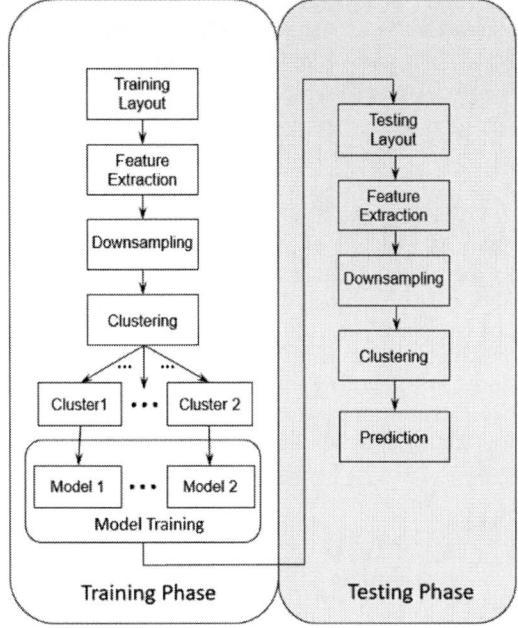

Figure 3: Two-dimensional etch bias prediction model work flow

In the testing phase, similar to the training phase, we first extract features from the test layout by using the CCAS manner and establish the eigenvectors representing the surroundings of the etch pattern center. Secondly the dataset is downsampled. Lastly the trained clustering model divides the eigenvectors of the testing layout, and picks the corresponding neural network to predict the etch bias.

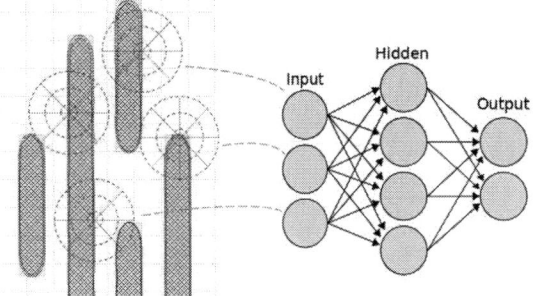

Figure 4: The ANN used in the model to predict etch bias

Model Optimization

With the introduction of the 2D etch bias prediction

model work flow, it can be referred that the *K*-means manner based layout eigenvector analysis is a key step to establish a precise model. Therefore the eigenvector clustering is critical.

In order to make sure that the clustering number of the input layer vectors is optimal, this study employs the Calinski-Harabasz Index (CH) to evaluate the classification result as shown in the following equation.

$$CH = \frac{SS_B}{SS_W} \times \frac{N-k}{k-1} \qquad (1)$$

Where k is the number of clusters, N is the total number of observations (data points), SS_B is the overall between-cluster variation, SS_W is the overall within-cluster variation. The higher the CH index, the better the clustering. As shown in Figure 5, the vertical axis denotes the CH index value while the horizontal axis represents the clustering number. It can be inferred that when the clustering number is eight, the CH index is maximum, and we take this result for the best clustering result.

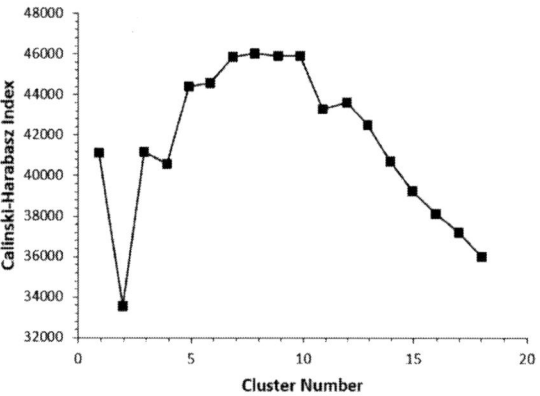

Figure 5: Calinski-Harabasz Index's dependence on classification cluster number

By dividing the input eigenvector into eight clusters, it's needed to train the corresponding neural network on each individual cluster. In this paper we only introduce the neural network optimization flow of single cluster. We randomly select one cluster and explore the impact of various activation functions on the R^2 index. As shown in Figure 6, when the activation function type is tanh, R^2 reaches maximum. Therefore we select tanh activation function type for the hidden layer nodes.

In addition, in order to determine the optimal hidden layer node number for each cluster, we incased the hidden layer node number from 10 to 35 and observed their impact on the R^2. As illustrated in Figure 7, R^2 reaches peak while the hidden layer node number hits 32.

Therefore the optimal model for this cluster consists of 26 hidden layer nodes.

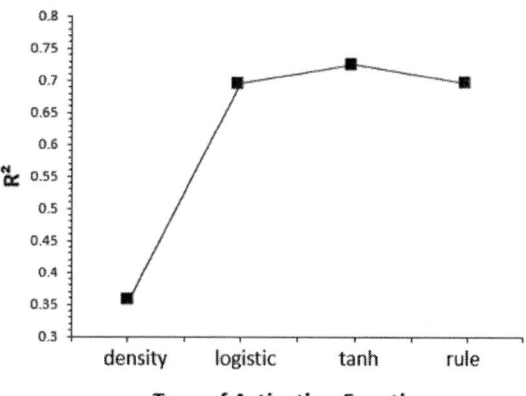

Figure 6: Impact of node's activation function on R^2

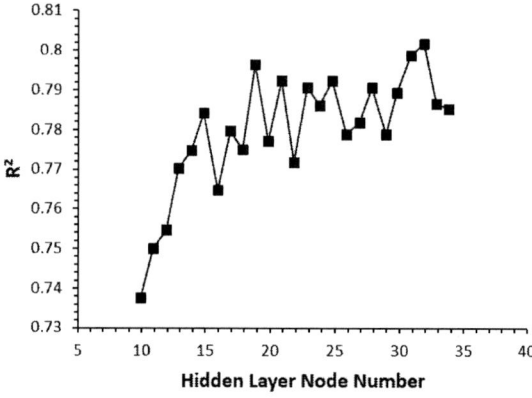

Figure 7: Impact of hidden layer node number on R^2

Result Analysis

In order to verify the effectiveness of the etch bias prediction, this study randomly takes 50 sampling points to verify the model performance. As shown in Figure 8, by comparing the predicted and measured etch bias, they do exhibit the similar trend, in which each individual predicted value does not differ too much from the corresponding measured value, excluding few exceptions showing relatively larger gap.

By doing a simple subtraction we achieve the absolute error of the 2D etch bias prediction. As plotted in Figure 9, except a few extra-ordinary pints, most portion of the absolute error stays within ±7.5nm. The calculated mean value of these errors is around 3.9nm with a standard deviation of 3nm. As can be inferred in the abovementioned analysis, the overall relative prediction error is calculated to be about 15%.

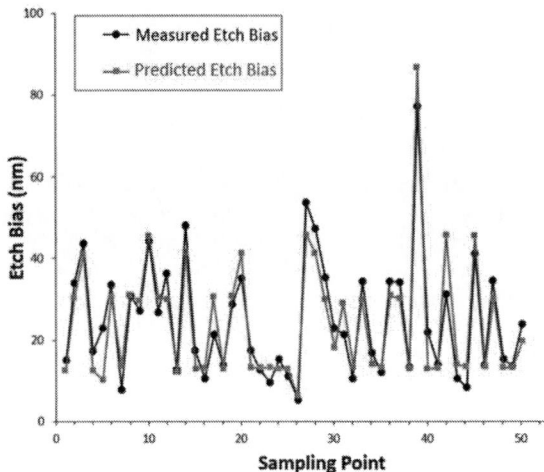

Figure 8: Comparison of predicted and measured etch bias in 2D layout

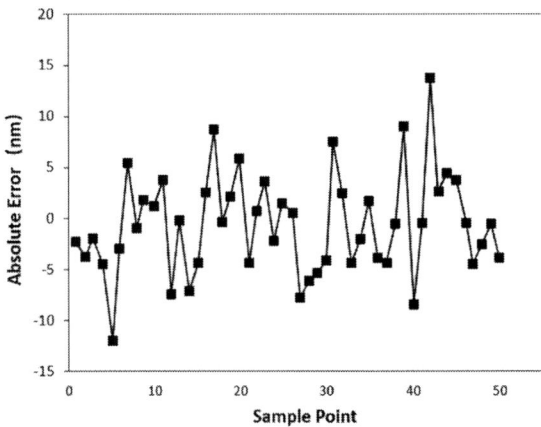

Figure 9: Absolute error of etch bias prediction in 2D layout

CONCLUSION

In this paper, we have demonstrated a machine learning based etch bias prediction model for 2D patterns. In this study, the environment dependent features are sampled using the CCAS method. As well in order to improve the prediction accuracy, PCA and K-means clustering are employed. In the 2D pattern etch bias prediction test we have achieved encouraging result, the model successfully predicts the similar trend as the measured value, the calculated average absolute error is less than 4nm with 3nm standard deviation, and average relative error of 15%. These result opens new window for the development of precise, fast and economy EPC applications in the high volume manufacturing.

ACKNOWLEDGEMENTS

This work was supported by the National Natural Science Foundation of China 61874002.

REFERENCES

[1] Y. Wei and R. Brianard, *Advanced Processes for 193-nm Immersion Lithography*, SPIE PRESS, Bellingham, Washington, 2009.

[2] S. Sato, K. Ozawa, F. Uesawa, *Proc. SPIE*, 2006. pp. vol. 6155, 2006, pp. 615504–1.

[3] H. Hu, L. Dong, Y. Wei and Y. Zhang, *Proceedings of China Semiconductor Technology International Conference (CSTIC)*, Shanghai, 2019.

[4] S. Shim, and Y. Shin, *IEEE Trans. Semicon. Manuf.*, vol. 30(1), 2017, pp. 1–7.

[5] B. Kim and B. T. Lee, *Journal of Applied Physics*, vol. 98, 2005, pp. 034912 1–6.

[6] S. Shang, Y. Granik and M. Niehoff, et al., *Proceedings of the SPIE*, vol. 6730, 2007, pp. 67302G–1.

[7] R. Chen, G. Lobb, A. Clancy, B. Morgenfeld and S. Pal, *Proc. SPIE*, vol. 10146, 2017, pp. 101461D 1-7.

[8] Y. Chen, Y. Lin, T. Gai, Y. Su, Y. Wei, and D. Z. Pan, *Proceedings of the 24th Asia and South Pacific Design Automation Conference*, 2019, pp. 420–425.

[9] H. Yang, S. Li, Y. Ma, B. Yu, and E. F. Young, *55th ACM/ESDA/IEEE Design Automation Conference (DAC)*, 2018, pp. 1–6.

[10] X. Ma, B. Wu, Z. Song, S. Jiang, and Y. Li, *J. of Micro/Nanolithography, MEMS, and MOEMS*, vol.13(4), 2014, pp. 043007 0–12.

[11] X. Ma, S. Jiang, J. Wang, B. Wu, Z. Song, and Y. Li, *Micro. Eng.* vol. 168, 2017, pp. 15–26.

AI COMPUTATIONAL LITHOGRAPHY

Xuelong Shi*, Yuhang Zhao, Shoumian Chen, Chen Li
Shanghai IC Research and Development Center, Shanghai, China 201206
*Corresponding Author's Email: shixuelong@icrd.com.cn

ABSTRACT

Machine learning based computational lithography is intended to accelerate the speed of the solutions significantly. There are three critical aspects of AI computational lithography: (1). The feature vector design, (2). The approximate mapping function construction, (3). The model training scheme. Approximate mapping function construction can be realized using forward neural network architecture in theory, model training is an art with the help of mathematical understanding, while feature vector design must achieve optimal resolution, sufficiency and efficiency simultaneously. To pave the way of successful AI computational lithography implementation, we have designed physics based optimal feature vector for AI computational lithography. By combining this feature vector design method with deep neural network architecture, a universal machine learning based computational lithography framework can be established.

INTRODUCTION

Computational lithography has been playing a critical role of enabling the Moore's law in semiconductor industry. With the semiconductor industry progressing from 7nm node to 5nm node and beyond, the gap between lithography process control capability and lithography process window is growing rapidly. To bridge the gap, computational lithography community has invented key technologies to enlarge lithography process window significantly in the past, those technologies include assist features (scattering bars) and source-mask co-optimization. At present, assist feature placement remains at sub-optimal level in terms of performance, the full benefit of assist features must come from placement scheme derived from inverse lithography technology (ILT). ILT algorithms do exist to achieve optimal mask solution; however, current algorithms are still unable to offer full chip solution with acceptable speed without GPU arrays. As a compromise, ILT engine is only invoked for lithography hotspots fixing in a hybrid implementation scheme. Such a hybrid implementation scheme, of course, serves practical purpose in some sense with careful boundary segments engineering, but it is theoretically non-ideal. To overcome the difficulties, the computational lithography community has made attempts to seek solutions by employing ever-maturing machine learning technologies. In essence, machine learning based ILT mask solution can be viewed as constructing a nonlinear mapping function between the target design and the ILT

solution mathematically, while feed-forward multilayer neural network architecture has been proven to possess such a capability.[1,2] In semiconductor industry, the feed-forward neural network structure was first applied to electron beam lithography and mask making proximity effect correction.[3,4] The technique was then applied to model based OPC, assist feature placement and inverse lithography solution (ILT).[5-14] Machine learning techniques have also been applied to hot spot detection as a classification problem. [15-18] Great progress has been made in the area of machine learning based computational lithography using the deep convolution neural network architecture (DCNN).[13,17] Convolution neural network (CNN) has been applied successfully in computer vision area and it has be proven to possess the capability of object recognition and classification with excellent accuracy through training on a large amount of data. The reason behind of the success lies in the architecture of the CNN, it contains automatic feature extraction layers followed by fully connected feed-forward neural network layers. However, feature vectors extracted from DCNN for ILT lack of intuitive physical interpretation, they cannot address the critical questions regarding feature vector design, the resolution, the sufficiency, and the effectiveness. Therefore, it is still in great desire for machine learning based computational lithography to develop a method of designing physics based feature vectors.

FEATURE VECTOR DESIGN FOR MACHINE LEARNING BASED COMPUTATIONAL LITHOGRAPHY

Machine learning based ILT can be generally stated as: *For a given ADI target layer and a fixed optimal mask generation mechanism (ILT algorithm), there should exist a unique mapping function between ADI target data and ILT data, as shown in Fig.1*

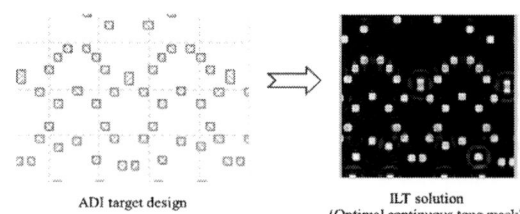

ADI target design   ILT solution
(Optimal continuous tone mask)

Figure 1: Mapping from ADI target to ILT image

Mathematically, it can be expressed as

ILT patterns = F (ADI target patterns)   (1)

Equation (1) is not a point-to-point mapping function, instead, it is a function-to-function mapping. To construct such a mapping function F successfully, two important questions must be addressed first. One is the universal method of constructing approximation function, and the other is the optimal feature vector design that can be used to describe the local characteristics of the ADI target patterns within a range. As to the approximate function construction, it has been proven mathematically that any nonlinear function can be approximated by

$$F = \sum_{l=0}^{l=m} \omega_l f(\sum_{k=0}^{k=n} v_{kl} x_k)   (2)$$

Where $\{x_k\}$ is the feature vector, and f is a nonlinear function often termed as activation function, which must meet certain requirements for equation (2) to be valid. Equation (2) can be represented graphically by a feed-forward neural network with one-hidden layer. Although a neural network with one hidden layer has the capacity of approximating any nonlinear function, in practice, it is often found that multilayer neural network offers better efficiency in approximating a function. As to the feature vector design, different representations or schemes in describing the neighboring environment around a point often have dramatic difference in terms of resolution, sufficiency and efficiency; careful engineering is demanded based on the problem at hand.

Now, let's turn to the critical question, that is, how to design an optimal feature vector for ILT? To begin with, let's look at Fig2.

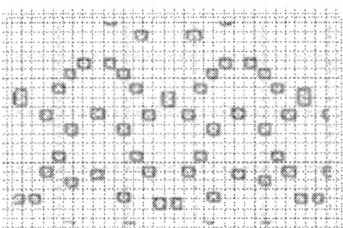

Figure 2: Divide the neighboring environment into cells

The value of a pixel in the continuous tone mask from ILT is fully determined by its neighboring environment within a finite range, say 1um per side. *The first question is how many degrees of freedom of the neighboring environment around a point have.* The theoretical answer is: *the degree of freedom of the neighboring environment is infinite.* Therefore, a complete description of the neighboring environment is an impossible task. However, there is an error tolerance margin in practice; this renders us to the possibility of describing the neighboring environment

using finite degrees of freedom. For example, one can partition a pattern into small cells as shown in Fig. 2, the cell dimension is related to the resolution of the lithography process condition $k \cdot \lambda / NA(1+\sigma_{max})$. For a typical immersion lithography process, the cell size is around 15nm to 20nm. If one considers 1000nm range each side for neighboring environment, and uses geometry weighting in each cell to describe the neighboring environment, then the feature vector length will be around 100^2. Obviously, such a simple scheme of generating the feature vector in describing a point's neighboring environment for computational lithography is inefficient. Intuitively, not every cell has the same influence on the point of interest, on average, the closer the cell is to the point of interest, the more important the cell is. Incremental concentric square sampling[5], incremental concentric circle area sampling[11], polar Fourier transform[12] have all been proposed to construct feature vectors for computational lithography. Unlike the feature vectors from CNN training, those feature vectors do not address the feature vectors' effectiveness and sufficiency. CNN can extract optimal features by training large amount of data. To extract features from CNN with 2um spatial range taken into account, the number of convolution layer and pooling layer pairs should be around 6 or 7 if 3x3 convolution kernels and 2x2 pooling layers are to be used, with each convolution layer having more than 8 kernels.[15] This network structure constitutes a quite complex network for extracting features by training. *The features extracted from CNN are optimal to some extent; however, the optimality depends on training data set and the network design itself.* To seek optimal feature vectors in a more general sense, unique characteristics of computational lithography problems at hand must be explored. As we all know, all lithography problems are derived from aerial images, furthermore, *all aerial images derived from a given imaging condition and polygons following a given set of design rules constitutes a special class of functions. Therefore, optimal feature vector design for computational lithography must be related to optimal and efficient representation of aerial images of the class at hand.* Based on this principle, we have designed our feature vector.

MACHINE LEARNING BASED INVERSE LITHOGRAPHY (ILT) AND RESULTS

To explore the machine learning based ILT feasibility, we have used our physics derived optimal feature vector as the input for deep neural network (DNN) with five to six hidden layers. Due to very nonlinear nature of the problem (growing assist features), we have expanded our feature vector to have elements around 150-200. In the model training, we have used about sixty images with each image having 143x143=20449 pixels (each pixel is 14nm in size). The training of the neural network model needs to include training samples and verification samples, and they are selected from part of the target patterns selected from the design target patterns, which is the periphery area of a SRAM design. The pattern selection strategy is the same as that for OPC model calibration and SMO. Generally, many training target images contain a large area, in which information for model training is repeated. If we include these areas into model training completely, the training samples can be biased initially. To avoid the initial sampling unbalance issue, we first identify the areas, in which information for model training is highly repeated, and we only select a small portion of the area into model training, and the other areas are fully used for model training, as shown in Fig. 3.

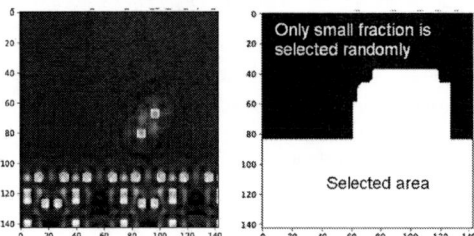

Figure 3: Sampling area selection

In model training, each feature vector element is centered and normalized by itself before feeding into neural network model. ILT rigorous solutions from same design patterns are used as training targets after being normalized by its maximum value. Besides the tricks commonly used in training a deep neural network model such as batch normalization and learning rate scheduling, we introduced dynamic and adaptive sampling weighting to guide the training process and to achieve better model. The training process is divided into two phases. In the first phase, stochastic gradient descent (SDG) is used with gradually increasing batch size; in the second phase, we used Adam optimizer to train all samples as a single batch. With our current feature vector length and DNN architecture, good machine learning based ILT solutions can be achieved with about 15X speed up factor, Fig. 4

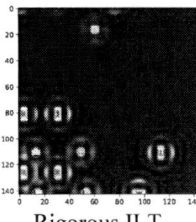 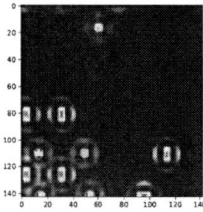

Rigorous ILT   Machine learning ILT

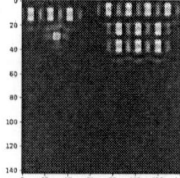

Rigorous ILT   Machine learning ILT

Figure 4: Rigorous ILT vs. Machine learning ILT

CONCLUSIONS

With paralleling computing architecture and intelligent machine learning algorithms, machine learning is rapidly gaining applications in a wide range of fields, including computational lithography. Due to high accuracy requirement, machine learning based computational lithography is not intended to replace current computational lithography techniques entirely; rather, it provides a mean of achieving optimal computational lithography solution on a full chip scale with acceptable speed. One of the critical aspects in machine learning based computational lithography is the design of feature vector. In our study, *we have developed a physics based method of designing optimal feature vector for computational lithography. In combining this feature vector design method with deep neural network architecture, a universal machine learning based lithography framework is established.* What left for a particular machine learning based computational lithography problem is to experiment with {feature vector size + DNN design} to achieve a balanced solution between speed and accuracy. The results of our neural network are very encouraging.

REFERENCES

1. Cybenko, G., "Approximation by superposition of a sigmoidal function.", *Mathematics of Control. Signals and Systems.* 2, (1989), 303-314.
2. Hornik, Kurt, "Approximation Capabilities of Multilayer Feedforward Networks.", *Neural Networks*, Vol. 4, (1991), 251-257.
3. R. Frye, E. Rietman, and K. Cummings, "Neural network proximity effect corrections for electron beam lithography.", *Proc. IEEE Int. Conference on Systems, Man and Cybernetics*, (1990), pp 704-706
4. P. Jedrasik, "Neural networks application for OPC (optical proximity correction) in mask making.", *Microelectronic Engineering*, Vol. 30, no. 1-4, (1996).
5. Gu A. and Zakhor A., "Optical proximity correction with linear regression.", *IEEE Trans. Semicond. Manuf.* Vol. 21, (2008), 263-71.
6. R. Luo, "Optical proximity correction using a multilayer perceptron neural network." *Journal of Optics.* Vol. 15, no. 7, (2013).

7. Jia N, Lam E.Y., "Machine learning for inverse lithography: using stochastic gradient descent for robust photomask synthesis.", *Journal of Optics*, vol. 12. (2010).

8. W.C. Huang et al., "Intelligent model-based OPC", *Proc. SPIE* 6154, (2006), 1065-1073.

9. Xu Ma, Bingliang Wu, Zhiyang Song, Shangliang Jiang, and Yanqiu Li, "Fast pixel-based optical proximity correction based on nonparametric kernel regression.", *J. Mirco/Nanolith MEMS MOEMS*, 13(4), 043007, (Oct-Dec 2014).

10. P. Gao, A. Gu, and A. Zakhor, "Optical proximity correction with principal component regression.", *Proc. SPIE 6924*, 69243N (2009).

11. Tetsuaki Matsunawa, Bei Yu and David Pan, "Optical proximity correction with hierarchical Bayes model.", *Proc. SPIE 9426*, (2015).

12. Suhyeong Choi, Seongbo Shim, Youngsoo Shin, "Machine learning (ML)-guided OPC using basis functions of polar Fourier transform.", Proc. SPIE 9780, (2017).

13. Shibing Wang et al., "Efficient full-chip SRAF placement using machine learning for best accuracy and improved consistency.", Proc. SPIE 10587, (2018).

14. Song Lan, Jun Liu, Yumin Wang, Ke Zhao, Jiangwei Li, "Deep learning assisted fast mask optimization.", Proc. SPIE 10587, (2018).

15. Haoyu Yang, Luyang Luo, Jing Su, Chenxi Lin, and Bei Yu, "Imbalance Aware Lithography Hotspot Detection: A Deep Learning Approach.", Proc. SPIE 10148, (2017).

16. Yibo Lin, Xiaoqing Xu, Jiaojiao Ou, David Pan, "Machine Learning for Mask/Wafer Hotspot Detection and Mask Synthesis", Proc. SPIE 10451, (2017).

17. Tetsuaki Matsunawa, Shigeki Nojima, and Toshiya Kotani, "Automatic Layout Feature Extraction for Lithography Hotspot Detection Based on Deep Neural Network.", Proc. SPEI 9781, (2016).

18. Yiwei Yang, Zheng Shi, Litian Sun, Ye Chen, Zhijuan Hu, "A kernel-Based DfM Model for Process from Layout to Wafer", Proc. SPIE 7641, (2010).

ACCURATE MASK MODEL APPROACHES FOR WAFER HOT SPOT PREPDITION AND VERIFICAITION

Young Ham*, Colbert Lu[a], HJ Lee[a], Mohamed Ramadan, Michael Green, and Chris Progler

R&D, Photronics Inc., Boise, Idaho 83717, US

[a]Photronics DNP Mask Corporation (PDMC) Inc., Hsin-Chu City, Taiwan, R.O.C.

*Email : yham@photronics.com

ABSTRACT

Hotspots often lead to unexpected critical dimension (CD) behavior, degradation of process window and ultimately impact wafer yield. The most important aspects of the patterning process in device manufacturing are to calibrate the OPC model, correct for process window (PW) limiting hotspots, and correct defect issues. Despite advanced techniques for low k1 lithography such as optical proximity correction, source mask optimization, multiple patterning, and EUV lithography, PW is still narrowing and it is hard to estimate the impacts of hot spots as device features shrink at an accelerated rate from node to node.

In this paper, we will demonstrate the limitation of traditional hotspot detection technology and introduce a practical lithography hotspot identification method using a mask process model. Mask model-based hotspot verification is a better approach to precisely identify lithography hotspots and will provide the information needed to improve hotspots lithographic performance. Hot spot detection and correction are under development with the goal of minimizing process window impact of weak points. As standard operating procedure (SOP) of our advanced mask characterization and optimization (AMCO), contour-based prediction is the key part of hot spot modeling in order to match wafer process results. AMCO is a series of techniques designed to enhance wafer performance by harmonizing the mask and wafer processes. Our contour based approaches show good feasibility to predict hot spots in lithography.

INTRODUCTION

Weak points in the design that have lower process latitude than other key features still adversely affect wafer process window performance. Efforts to overcome the narrow process window of weak points include extensions of current techniques such as more process-aware OPC. In the mask manufacturing process, mask process correction (MPC) is becoming a popular technique to reduce mask process-induced pattern fidelity limitations and to provide the best wafer patterning performance. Ultimately, regardless of enhancement techniques like MPC, there will always be some disparity between printed mask features and the design. The purpose of AMCO is to bridge the gap between mask process variation and its wafer process effect to achieve optimal wafer process window performance.

Essentially, AMCO is the study and analysis of hot spot or weak patterns between wafer manufacturers and the mask supplier. Usually, hot spots must be monitored to find CD issues that cause lower process window and defects during wafer patterning. The typical solutions are to optimize the OPC pattern by using bias, scattering bars and pattern patches. During this process, we find that an accurate simulation model or predictive model is required in the wafer process. Also, we are interested in how the mask impacts the wafer patterning, especially on hot spots and how it finally influences the process window.

AMCO is the methodology we developed to evaluate these lithography aspects and adapt a model. From previous experience, we found that using the OPCed design patterns for lithography simulation is not accurate for predicting wafer hot spot and process window performance. A SEM-based mask contour model is a better method to predict and verify weak patterning areas and hot spots more accurately as it accounts for mask patterning capability.

Hot spots and correspondence in the wafer process

The cornerstones of stable lithographic performance are wider process latitude and lower CD tolerances. An anchor pattern with larger process window is used to setup lithographic conditions. In the process, we monitor the stability of the anchor feature. However, some features other than the anchor cannot be resolved in the given process condition regardless of depth of focus (DOF) of the target CD. We thus define those patterns as hot spots or weak points. Those patterns are detected in defect inspection using DOF and exposure latitude (EL) matrices and will be shown in the process window as failure points. Generally, these structures are optimized with OPC treatment or a new OPC model.

In simulation, we tune the model to predict hot spot performance in the aerial image and/or in post-develop if a resist model is available. Figure 1 is an example of the AMCO approach to find the root cause that generates hot spots in our experiment. If we use design layouts for CD analysis, there will be a mismatch in CD and PW between designs and mask patterns. The purpose of this approach is to find an accurate model for hot spot analysis.

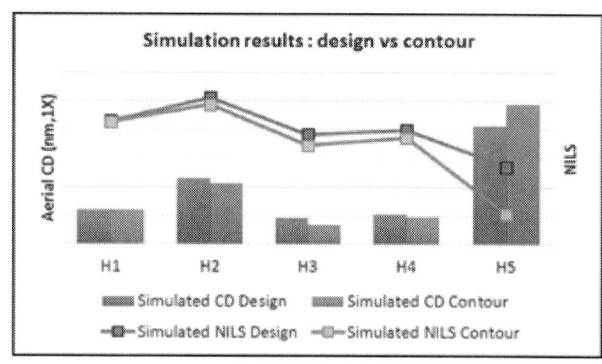

Figure 1: CD differences between design CD and mask SEM contour CD in wafer process.

Hot spot prediction and accuracy in lithography

If the OPC model is accurate enough to predict the lithography process in CD targeting, pattern fidelity and iso-dense (ID) bias, we can simulate the process and CD performance by using an adapted model such as a resist model. If we define that when the hot spot is off target in the optimum process window, we can evaluate the CD variation to induce a pattern bridge, shortening or necking. Often the predictive model is the same as the lithography result or shows a similar trend, but different from our expected result. Model accuracy depends on correlation between the simulated model and the actual process result. To check model accuracy relative to the wafer result, we use real hot spot pattern structures in 14nm level logic.

A resist model based simulation is more accurate than a simple aerial image model. The difference between the two models is that aerial only considers intensity distribution at the resist surface and the others incorporate intensity and photo-acid generator (PAG) diffusivity predicting wafer results after develop. In addition, when we apply a mask contour model in simulation instead of post-OPC design structures, the accuracy of the results is further increased. In this paper, we evaluate and compare simulated results based on design (post OPC), mask contours, single Gaussian fitting mask model and a more complex mask process model. Our goal is to improve modeling accuracy and correlation with wafer results. We also simulate using a resist aerial image model to match the lithography process as the data trend of the standard aerial image model is unable to predict hot spot trends in various structures.

EXPERIMENTAL AND RESULTS

In the experimental, we utilize a test chip that includes hot spot arrays for an in-depth study of mask impact on wafer process margin and CD as well as hot spot analysis. The test chip is used to characterize wafer hot spots and their behavior relative to mask fidelity. We select a few different design arrays to show bridges, shortening, necking and CD difference in the wafer patterning process. The test chip is evaluated for resist CD trend and process window. We compare the simulation result by the various models to the wafer CD result. For the mask process, we obtain the mask process model by simple MPC and use it for compensation in the hot spot structures. We added pre-MPC and post MPC structures in the test chip to verify accuracy for hot spot verification. Ultimately, we consider accuracy and correlation to lithographic performance of various hotspots.

Comparison of hot spot performance predictability of Gaussian fitting, SEM contour and MPCed design models

Figure 1 shows the comparison of simulation results from a SEM contour of the mask vs the original OPCed layout. The CD difference in simulated wafer process between the two is clear. The actual wafer CD trend correlates much better to mask SEM contour rather than OPCed layout. That small difference in CD or energy level is the main reason that CD and PW of hot spot structures is misunderstood. We demonstrated that corner rounding on the CD feature is affected by both mask process and design scaling. Since mask pattern fidelity varies with the mask making process, we concluded that contour-based metrology is essential for hot spot detection and accurate wafer process prediction.

As another approach, we applied a simple single Gaussian fitting formula to the post OPC design as a mask model. We evaluated hot spot structures for process issues such as necking, bridging, and shortening. While more accurate than post OPC design based simulation, the Gaussian mask model is not quite as correlative with wafer results as the mask contour based simulation. It is known that single Gaussian fitting is not perfectly accurate as a mask model. This is largely due to it not accounting for e-beam and mask process variations such as loading and fogging effects. In order to obtain a high accuracy simulation, there must be consistency in the data trend regardless of design structure being simulated. Due to limitations in mask writing and other process steps, final mask fidelity and bias does not exactly match post-OPC design. For this reason, simulation using the post-OPC design cannot predict the behavior of hot spot patterns accurately.

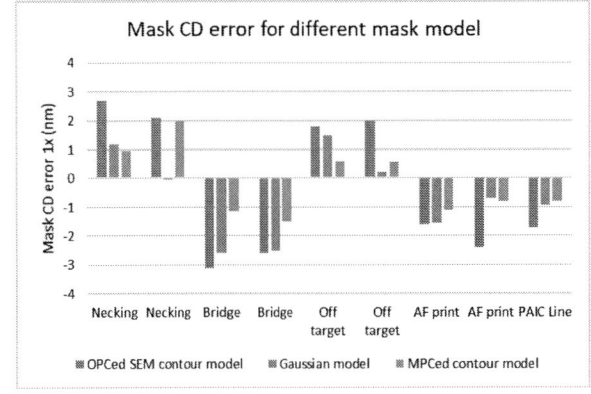

Figure 2: CD differences of MPCed contour and Gaussian contour model vs SEM contour,

The simulation results of various models are shown in Figure 2. The CD of the single Gaussian-fitting model is close to the CD of the SEM contour image in the simple structures. These structures show a small CD difference in the 1nm range (1X). However, more complex structures show 2~3nm differences on wafer scale (8~12nm on mask scale). To characterize those differences, we measured of the pull back of the SEM contour from the post OPC design layout. We found the worst case CD difference to be ~10nm for line and space and ~20nm for round structures. From our hot spot structure and model study, we concluded that mask resolution capability is a limitation in resolving OPCed structures. The hot spot with extreme OPC structures has 2~3nm CD difference on wafer and is an obstacle to utilizing the mask model for hot spot prediction and verification. The other problem is to control stable CD target with enough PW in all patterning areas. This depends on ID bias and linearity in the mask process and will effect hot spot generation. Correcting 1~2nm CD difference is necessary for compensation to remove hot spot impact on the lithography process. For the MPCed contour model, we can get close to zero CD offset. The results in Figure 2 show model accuracy for hot spot patterns. Here, the MPCed contour model the best choice for predicting CD error of hot spots.

Mask process model and wafer verification

In the paper, we applied a simple mask process model to reduce hot spot prediction error across different CD targets. The mask process model was developed using a simple MPC model fitting with 1480 metrology points to characterize feature size and patterning issues. As a result, we achieved 0.186nm edge placement error (EPE) for adapting structures. The MPC contour is more realistic than the design and other simple mask models. Our goal for the MPC model is to correct 1~2nm CD tolerance in the hot spot areas and for the contour applied by MPC to be accurate in prediction of hot spots.

Figure 3 shows the wafer CD target error result of the wafer with and without MPC for 8 hotspot structures. Mask CD results showed the difference between the revised mask CD vs the original to be about ~1nm. Therefore, based on the magnitude of the wafer CD off-target, we found that most of the hotspots were improved in 1nm to 3nm. However, some patterns were not obviously improved. The likely reason is due to process and measurement variation which is in 0.5nm to 1nm range. Finally, we verified that the mask process model generated by simple MPC methods corrected the edge-to-edge error by about 4nm CD on the mask. This correction reflected on the wafer CD and process by about 3~4nm.

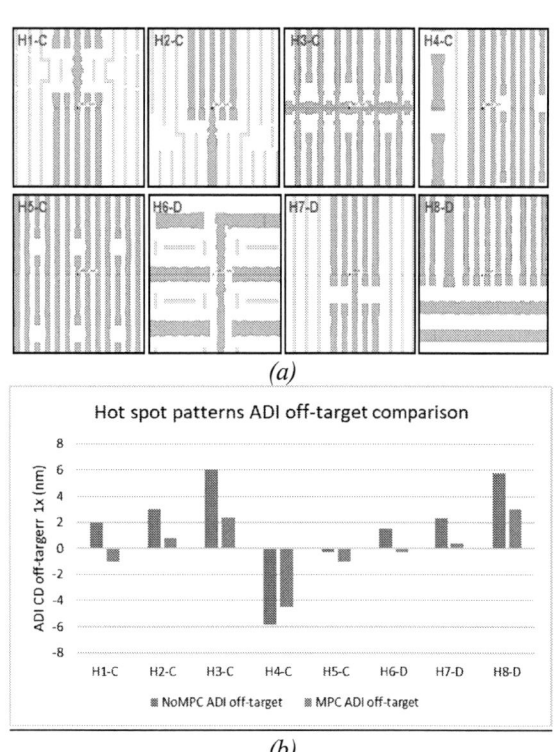

(b)

Figure 3: ADI CD offset comparison, (a) Hot spot patterns, (b) MPC vs no MPC wafer CD offset

As a result, we have concluded that root cause of inaccuracy of hot spot models come from mask writing and process limitations, mask pattern's full back from OPCed structures, and boundary effects in SMO and MPC model inaccuracy. Figure 4 shows EPE between the MPC model and hot spot area. Each MPC models distribution of EPE is ~4nm for hot spots. For the full chip area, the distribution of 1nm error is 87%, 2nm is 11% and 3nm is 2%. To reduce wafer offset obtain accurate prediction, we will need to optimize the MPC model, contour image and exposure condition (ie SMO and dose control).

The purpose of this experimental is to create highly accurate models for hot spot prediction and verification based on accuracy and correlation with wafer results. We have concluded that the biggest error in wafer hotspot is from the difference between the OPCed layout and the mask pattern. In order to obtain more accurate wafer prediction, we must extract a better contour image from MPC model.

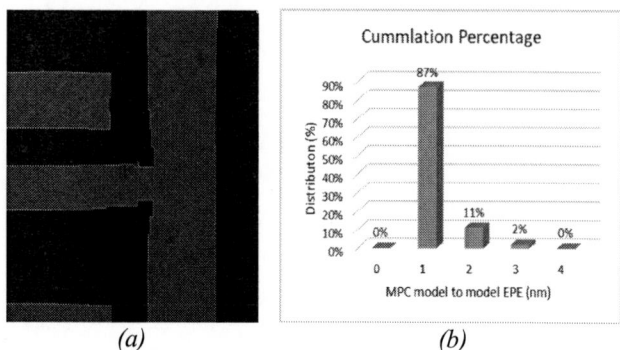

(a)   *(b)*

Figure 4: CD difference by different MPC models in tot spot area (a) Hot spot pattern, (b) EPE in MPC models

SUMMARY

In this study, mask models were compared with the manufacturing model using resist aerial images that were obtained from the OPCed layout. The single Gaussian-fitting model showed better correlation in simple test structures, but did not match with wafer results perfectly. This was due to inaccurate contour and CD target in the conditions of 10% EL and 50nm DOF. The Gaussian fit contour was relatively close to the real SEM image contour created by the mask process. If we apply a more accurate model relative to the mask processes, it is a more realistic representation of the mask SEM contour. We analyzed those approaches and demonstrated that a 1~3 nm CD difference among contour models exists in specific hot spots with extremely complex OPC structures. This difference comes from model inaccuracy and mask process limitations. For model matching, MPC contour is the preferred method to compensate CD differences in a hot spot pattern. Our contour based simulation results show good predictability of hot spots. From our study, we verified our mask contour models correlated to wafer CD trends and applied the best method to correct CD difference between OPCed layout and SEM contour. Finally, we showed positive direction in hot spot area and proposed technical motivation to enhance mask and wafer lithography.

REFERENCES

[1] Young Ham, Yohan Choi, Michael Green, Mohamad Ramadan, and Chris Progler, Proc. of IEEE 2019, April 2019
[2] Yohan Choi, Michael Green, Young Cho, Young Ham, Howard Lin, et al, Proc. of SPIE, Vol. 10148, March, 2017
[3] Yohan Choi, Michael Green, Jeff McMurran, Young Ham, Howard Lin, et al, Proc. of SPIE, Photomask technology 2016, Vol. 9985, Oct. 2016

A Simulation Study for Typical Design Rule Patterns and Stochastic Printing Failures in a 5 nm Logic Process with EUV Lithography

Yanli Li, Qiang Wu, Yuhang Zhao

Shanghai IC R&D Center

497 Gaosi Road, Zhangjiang Hi-Tech Park, Shanghai 201210

Contact: liyanli@icrd.com.cn

ABSTRACT

We have done a simulation study for typical 5 nm logic design rule patterns with a self-developed aerial image simulator based on the Rigorous Coupled Wave Analysis (RCWA) algorithm and the Abbe imaging routine. Generally speaking, critical structures in a lithography process are semi-dense patterns, the array edge structures, line end to line end structures, line end to perpendicular line structures (under 2D design rules), the minimum area structures, the bi-line, tri-line, ..., etc. Compared to those from the 193 nm immersion process, the behaviors for the above structures are different. For example, the minimum area of the EUV photolithographic process is found to be significantly larger than the minimum area of 193 nm immersion process.

In our simulation, we have kept aware of the stochastics impact due to drastically reduced number of photons absorbed compared to the DUV process, the criteria used for various structures of image contrast are tightened. For example, in EUV lithography, the minimum Exposure Latitude (EL) for the gate layer, the metal layer, and the line end pattern have been raised to, respectively, >18%, 18%, and 13%. We have studied the stochastic printing failures of dense patterns and 2D structures, for example, line end to line end structures, line end to perpendicular line structures. Compared to 1D dense patterns, 2D structures may have lower failure probabilities with the same dimension. In contrast to dense line/space patterns, for example, the line end to line end, and the line end to perpendicular line structures have a larger area on both sides of the line ends, which is easier for developer to flow in and out so that they are relatively not easy to form bridge defects. In case of the reverse tone structures, the trench end to trench end, and the trench end to perpendicular trench structures have a large, connected under-exposed area (dark tone area) which is more robust to etch.

We have also taken account of the influence of EL on LWR and Critical Dimension (CD) on stochastic printing failures. We will present the results of our work and our explanations.

1. INTRODUCTION

The semiconductor manufacturing field has been continually growing with linewidth shrinking. When the Logic process shrinks to 7 nm node, the use of Extreme Ultra-Violet (EUV) wavelength in the lithography process becomes available. EUV is a potential method considered for exposure down to 36 nm pitch and possibly below. Shrinking to 5 nm logic node, the arrangement of transistor is denser and it becomes too complicated for 193 nm immersion multiple patterning since it may require more than 4 exposures for a single cut layer which can fit well within one single EUV exposure. Therefore, the 5 nm node is to be expected as the first node that will accept EUV lithography on a large scale. Due to the Chief Ray Angle at Object space (CRAO) of incidence with 6° in EUV lithography, there will be pattern shift resulting from shadowing effect [1].

Figure 1 displays a typical multilayer mask structure for EUV, which is used by our simulation. The multilayer film stack consists of a 2.5 nm Ru protection layer and a 60 nm TaN absorber and a reflecting layer made of 40 pairs of alternating 4.2 nm silicon and 2.8 nm molybdenum thin films. The reflecting layer is placed below the absorber. Due to the thick mask stack, the invalidity of periodical boundary condition as a result of the non-vertical illumination CRAO, the reflective optical beam path, and the need for multiple illumination pupil locations, if we use the Finite Difference Time Domain (FDTD) algorithm, we need to use Perfect Matching Layer (PML) boundary condition and use multiple periods to approximate the dense patterns. A simple estimate indicates that the simulation work can be more than 300 times that of 193 nm immersion and 1000 times more for 1D and 2D, respectively, In other words, the algorithm of FDTD is too slow to be used for EUV simulation of 2D patterns. In our paper, the result of typical 5 nm logic design rule has been simulated with the self-developed aerial image simulator and the Abbe imaging method based on the much faster algorithm of Rigorous Coupled Wave Analysis (RCWA) instead of FDTD. Our model has been calibrated with wafer exposure data from several PhotoResists (PR) under collaboration with IMEC. We have also matched the FDTD 1D results with our 1D results based on RCWA.

In our simulation program of RCWA, 35 orders in each of X and Y dimensions have been used. There have been two angles that describe the 3D direction: the polar angle θ of 6° and the azimuthal angle φ, which depends on the slit position, which spans 26 mm in the X-direction and has a 1.6 mm width with an angle span of ±18.6° around the optical axis [2].

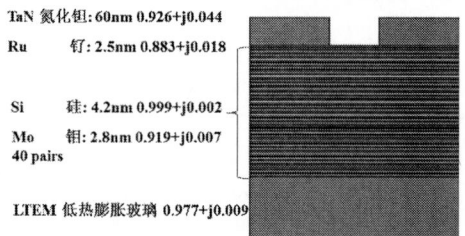

Figure 1. A mask cross section structure we use for EUV simulation [5].

2. RESULTS AND DISCUSSION

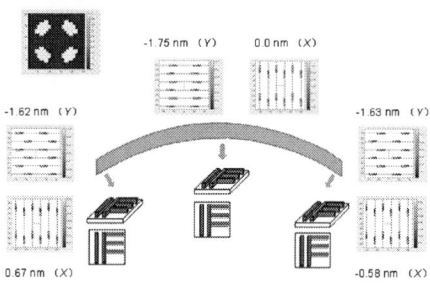

Figure 2. Simulated EUV shadowing effect with a 32 nm pitch line and space pattern. The illumination condition is a Quasar 45° 0.9-0.5 partial coherence factor and 0.33 NA [3].

The illumination condition is a Quasar 45° with 0.9-0.5 partial coherence factor and 0.33 NA. The PR is a typical 30 nm thick Chemically Amplified photoResist (CAR). The shadowing effect at different slit positions, 0, -13 mm, and 13 mm is shown in Figure 2. The anchoring pitch is 32 nm with After-Developing-Inspection (ADI) Critical Dimension (CD) of 16 nm [3]. The Exposure Latitude (EL) is set to ≥ 18% [4]. At the center of the exposure slit, the pattern shift along -Y direction results from the shadowing effect is about 1.75 nm for pattern oriented along X direction. With the slit position located away from the slit center, the Patterns oriented either along X or Y direction will have shifts.

Table I(a) and Table I(b) show the simulation results of typical design rule with 2D patterns. Simulated EL and CD and recommended EL and CD are displayed in these two tables. The illumination condition is a Quasar 35° with 0.9-0.7 partial coherence factor and 0.33 NA. The PR is a typical 30 nm thick CAR. For Table I(a), the patterns

are, from right to left, minimum isolated area on one side of wider line, short bar on one side of wider line, staggered Tip-to-Tip (TtT), TtT on one side of dense pattern, TtT, TtT within dense pattern, Tip-to-Line (TtL). For Table I(b), the patterns are, from right to left, right line of isolated bi-lines (the left line is represented by a hollow bar), left line of isolated bi-lines (the right line is represented by a hollow bar), right line of isolated tri-lines (the left two lines are isolated bi-lines (the left line is represented by a hollow bar), center line of isolated tri-lines (the two edge lines are represented by hollow bars), left line of isolated tri-lines (the right two lines are represented by hollow bars), minimum isolated area on one side of dense line/space, isolated short bar on one side of dense line/space.

Table I(a) The simulation data of typical design rule 2D patterns [5].

Metal Required EL	13%	13%	13%	13%	13%	13%	13%
Simulation EL	17.36%	14.34%	16.31%	15.1%	13.1%	15.61%	14.39%
Simulation Contour CD (nm)	20.11	19.78	19.85	20.30	22.4	18.05	17.32
Recommended CD (nm)	20	20	20	20	22	18	17.5

Table I(b) The simulation data of typical design rule 2D patterns [5].

Metal Required EL	13%	13%	13%	13%	13%	13%	13%
Simulation EL	14.39%	14.27%	16.04%	19.35%	15.88%	16.87%	16.96%
Simulation Contour CD (nm)	18.03	17.65	17.99	17.90	17.80	18.02	17.98
Recommended CD (nm)	18	17.5	18	18	18	18	18

Recalling in 193 nm immersion lithography process, according to the setting of typical design rules, the EL for the 2D patterns, such as TtL and TtT, is 10%, which is smaller than the EL for the most dense space and line patterns, say ≥13% for the metal layer and ≥18% for the gate layer [6]. While for the 2D patterns in EUV lithography process, taking stochastics effect (will discuss in detail later) into account，EL for the dense space and line patterns are recommended to be at least 18%, at the same time, EL for the 2D patterns must be enhanced to ≥ 13%. The simulation data of several 2D patterns are discussed as follows. Appropriate Optical Proximity Correction (OPC) which is not shown here has been added in order to obtain acceptable process window data.

2.1 The Simulation of Typical Design Rule patterns with EUV Photolithographic Process

A typical mask of TtL pattern is depicted in Figure 3. The main pattern pitch is 36 nm and trench CD on mask is

17.6 nm with the upper TtL gap CD of 15.7 nm and the lower TtL gap CD of 14.2 nm and the trench CD along the X direction of 16.3 nm (at line cut ④ position) . From the simulation results, line cut ① is used as the anchor point with ADI CD of 18 nm, taking EUV stochastics impact into account, EL must be ≥ 18% [4]. ③ is the TtL line cut which EL must be ≥ 13%. ④ is the line of TtL which EL must be ≥18% [4]. Line cuts ② and ⑤ are the trench width near the tip, which is used as a monitor of OPC. Our simulation indicates that the CD and EL at all line cut positions satisfy our target requirement with TtL CD close to target of 20 nm.

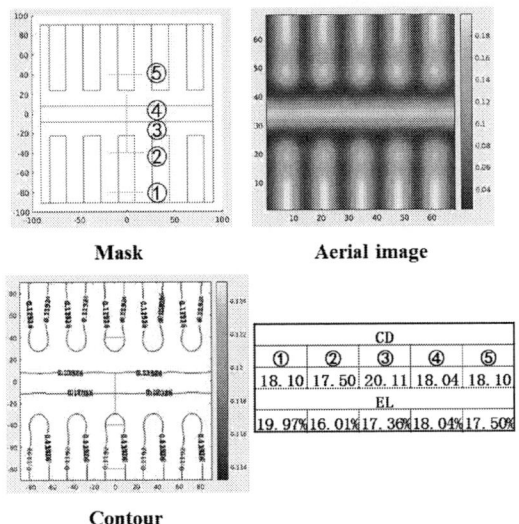

Figure 3. The mask and simulation results of TtL pattern.

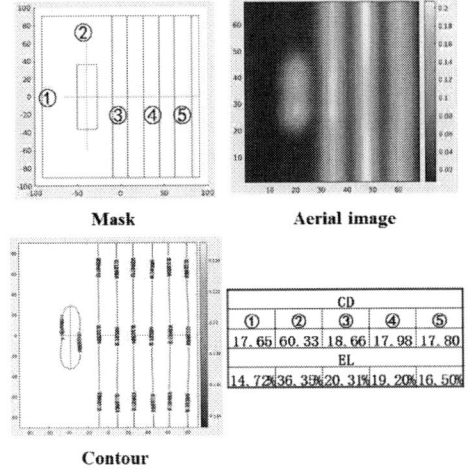

Figure 4. The mask and simulation results of minimum isolated area on one side of dense line/space.

Shown in Figure 4 is another situation. The mask CD

of the isolated area is 23 nm along X direction (at line cut ① position) and 72 nm along Y direction (at line cut ② position) and the dense trench next to the isolated area has a CD of 18 nm (at line cut ③ and ④ positions) and the right trench has a CD of 20 nm (at line cut ⑤ position). In order to achieve an EL more than 13% of the isolated trench, we have experimented with different minimum area and found it has to be 5 times that of minimum pixel squared, i.e. 5×18×18 nm², with 17.5 nm contour CD at line cut ① position.

2.2 Increased Impact of Aberration in EUV Lithography

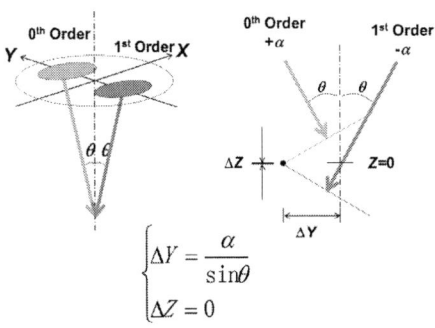

$$\begin{cases} \Delta Y = \dfrac{\alpha}{\sin\theta} \\ \Delta Z = 0 \end{cases}$$

Figure 5. Schematic diagram of coma induced pattern shift as a function of illumination pupil position [5].

In Figure 5, it is explained that the pattern shift caused by aberration is inversely proportional to NA. Since the NA of EUV is 0.33, which is much smaller than 1.35 for 193 nm immersion, the horizontal position of pattern in EUV process is more sensitive to odd aberration and thus the aberration from 0.33NA EUV exposure tools needs be controlled to a tighter specification. In the following sections, the pattern shift resulting from Transverse Coma aberration of 2D patterns, as mentioned in Tables I(a) and I(b), is discussed.

Figure 6 shows the transverse aberration (Coma) in X- (Z7) or Y-(Z8) direction of 0.2 nm root-mean-square (rms) and 0.7 nm rms for various 2D patterns [5]. The OVL is expected to be controlled to within ± 2.5 nm in 5 nm Logic Process. When Z8 is 0.2 nm rms shown in Figure 6(a), the pattern shift for typical 2D patterns contributing to OVL budget is in the range of -1~+1 nm, which is reasonable. These pattern shifts of bi-lines and tri-lines are all in the range of 0.6 nm under typical Z7 aberration of 0.2 nm rms and 0.7 nm rms. The pattern shifts of two special patterns owing to the Coma aberration will be described in detail in the following section.

The main pattern pitch is 36 nm and mask trench CD is 17.6 nm with the upper TtL gap CD of 15.7 nm and the lower TtL gap CD of 14.2 nm. For TtT pattern shown in Figure 7, a transverse Coma in the Y-direction of 0.7 nm rms is found to result in a pattern shift and cause a distortion of the tips: necking on the "a" side (at line cut ③ position).

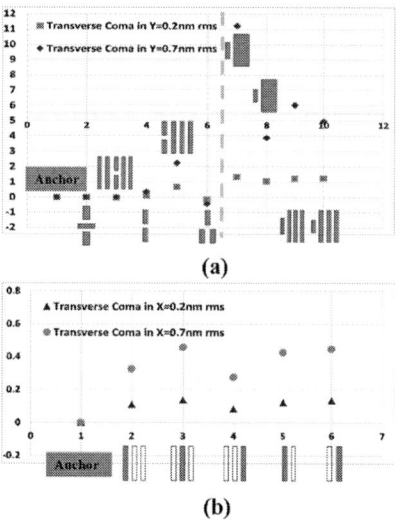

(a)

(b)

Figure 6. Pattern shift caused by aberration of 0.2 nm rms and 0.7 nm rms,(a): Transverse Coma in Y (Z8) (b): Transverse Coma in X (Z7)[5]

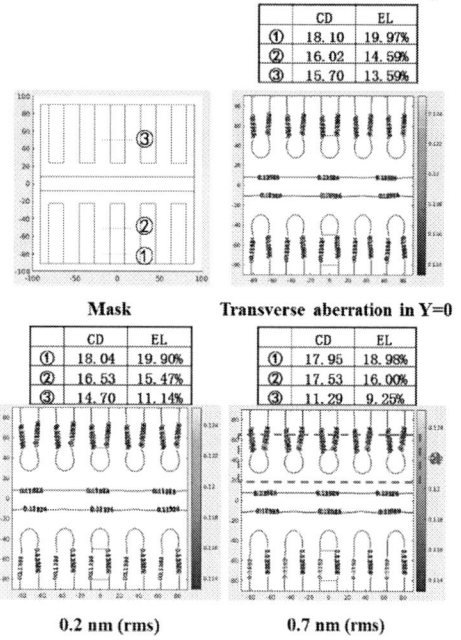

Figure 7. Pattern shift caused by Transverse aberration in Y of 0.2 nm rms and 0.7 nm rms for TtL pattern.

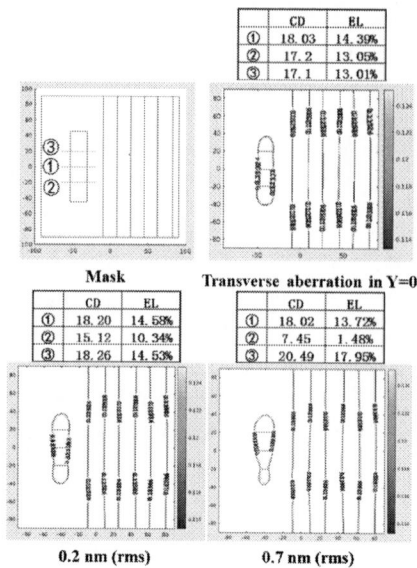

Figure 8. Pattern shift caused by Transverse aberration in Y for short bar on one side of dense line/space.

As shown in Figure 8, the mask CD of isolated area is 22.5 nm along the X direction (at line cut ① position) and 90 nm along the Y direction (at line cut ② position) and the dense trench next to the isolated area has a CD of 18 nm (at line cut ③ and ④ positions) and the right trench has a CD of 20 nm(at line cut ⑤ position). For isolated patterns, the aberration can not only cause the pattern shift, but also result in significant distortion, especially in the case when the aberration is 0.7 nm rms.

From above simulation results, the rms aberration for EUV lithography with 0.33NA needs to be kept under 0.2 nm [5] while the rms aberration for 193 nm immersion lithography are usually kept under 1 nm [7]. All above aberration effects, including in EL, CD, and pattern shape fidelity must be compensated by OPC.

3. Process Setting Optimization and Stochastic Printing Failures Analysis

3.1 Process setting Optimization for Line Width Roughness

In order to obtain better device performance, the Line Width Roughness (LWR) can be optimized by adjusting the thickness of PR and exposure energy, as shown in Figure 9. We have used a NA of 0.33 and 0.819-0.546 Dipole 70° partial coherence setting with a typical CAR photoresist. The pitch of the 1D pattern along the X direction is 40.5 nm and the ADI CD is 20 nm. The LWR is the unbiased LWR here. Thicker PR setting has better unbiased LWR from the simulation results. When the PR thickness becomes thicker by 10 nm, the LWR is roughly equivalent to the adding of 10 mJ/cm^2 in exposure energy.

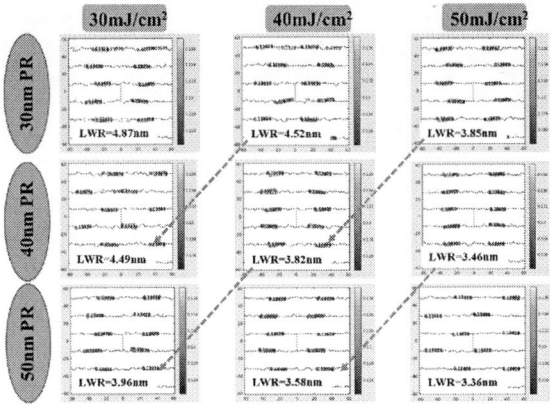

Figure 9. The process setting optimization for LWR.

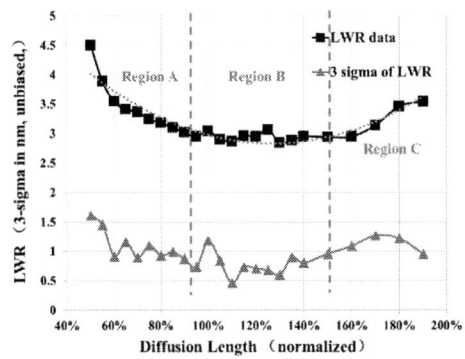

Figure 10. LWR v.s. Diffusion Length of Pitch 40.5 nm and ADI CD 20 nm

We found that the LWR has a correlation with the effective photoacid diffusion length (we use "diffusion length" later) which roughly behaves like a quadratic function. We can split the diffusion length axis into three regions, as shown in Figure 10. The exposure energy is set at 54 mJ/cm^2 with the PR thickness of 30 nm. In region A, when the diffusion length is too small, the averaging of the photon absorption stochastics is not significant, which results in larger LWR. With the increase of diffusion length, the LWR will be reduced due to the diffusion averaging we just discussed and finally achieved a stable state, shown in region B. In region C, when the diffusion length continue to increase, the LWR becomes worse because smaller EL caused by excessive diffusion will

affect LWR near the threshold.

In other words, there is a minimum EL that is required for keeping LWR small enough.

3.2 Defect density Analysis

We have collected larger than 0.16 million 2D TtL and TtT patterns for three different CDs: 15, 17, and 20 nm, respectively. For TtL and TtT, the main pattern pitch is 36 nm and ADI contour CD is 18 nm.

We have used a NA of 0.33 and 0.9-0.7 Quasar 35° partial coherence with the PR thickness of 30 nm and exposure energy of 54 mJ/cm^2 and the same diffusion length. From the simulation results, the distributions of TtL and TtT CD satisfy Gaussian distribution and the 20 nm TtL CD distribution is used as an example shown in Figure 11.

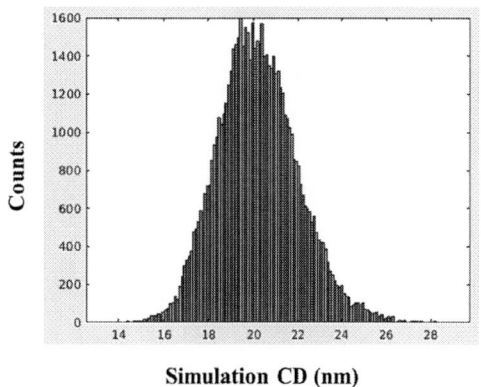

Figure 11. The Gaussian distribution of TtL CD.

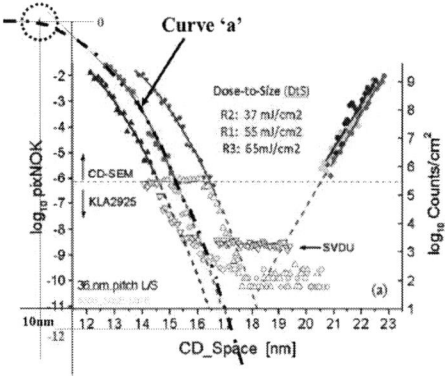

Figure 12. Defect density as a function of linewidth and pitch. The black dashed line ("Curve 'a'") is a simulated curve done by our self-developed photon absorption stochastics model [8]

Since rough edges will generate a distribution (Gaussian) of dimensions, or measured CDs at different patterns, or at different parts of a single pattern. Defect will emerge at location where the dimension is below certain limit. Therefore, we can use the error function which is the definite integral of the Gaussian function to get the population that is bound to have defects to fit the actual measured defect density at a given CD.

A defect study is shown in Figure 12[8]. The black dashed line is the simulation result on the side of micro-bridges from our simulation and we can obtain an LWR value 2.7 nm from the fit. At the same time, the minimum CD which has 100% defect density from the fit is 10 ~ 11 nm for 1D patterns as pointed by a dotted circle. From Figure 12, we can also find that the pitch of 36 nm may have zero CD window if we extrapolate the defect density to 10^{-12}.

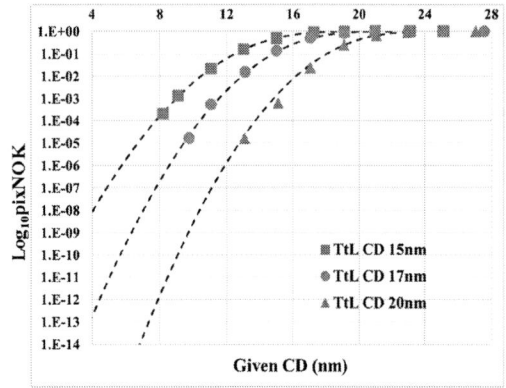

Figure 13. Defect density as a function of ADI CD for TtL pattern.

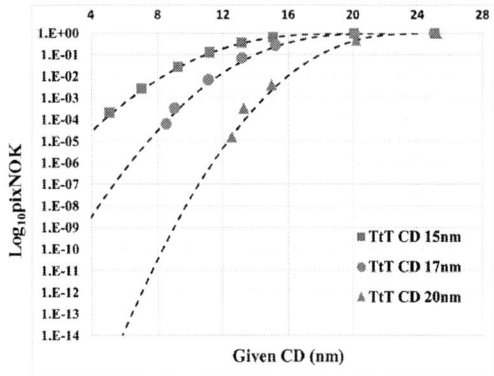

Figure 14. Defect density as a function of ADI CD for TtT pattern.

We have plotted the TtL and TtT integrated probability (vertical axis) from zero CD to a given CD (horizontal axis), shown in Figure 12 and Figure 13, respectively. Similarly, the six groups of simulation data are fit by the same error function. The probability of stochastic printing failures increases when the CD decreases from 20 nm to 17, and to 15 nm. When the same level of defectivity density from TtL and TtT needs to be achieved compared to the 1D patterns, the minimum CD that have 100% probability in causing defect can be found to be ~8 nm from our simulation, which is smaller than 10~ 11 nm of 1D patterns. The above phenomenon is reasonable and we will explain in detail as follows.

Firstly, the density of TtL or TtT pattern is much less than that of the 1D patterns with about two orders of magnitude.

Secondly, although the simulated defect density is high, after the developing and etch processes, the defect density can be substantially reduced. This is explained as follows.

The line end to line end, and the line end to perpendicular line structures have larger clear areas on both sides of the line ends, after some developing, it becomes easier for the developer and rinse water to flow into the area between the line ends or the line end to the perpendicular line so that it is easier for the partially developed PR to be removed compared to the dense lines and spaces. In case of the dark field, the trench end to trench end, and the trench end to perpendicular trench structures have large, connected under-exposed areas and can generate more polymers that may deposit on the potential broken line defect between the tips or tip to line.

4. Conclusion

We have studied the typical 5 nm logic FinFET design rule patterns with EUV process by a self-developed simulation program. What we have found is as follows. The shadowing effect results in pattern shift which is about 0.5~2 nm. The minimum EL for the metal layer and the tip-to-tip patterns are recommended to be, respectively, ≥18%, 13%. The minimum area of trench pattern is about 5 times that of minimum pixel squared, i.e. 5×18×18 nm2. Generally speaking, rms aberration in EUV lithography needs to be kept under 0.2 nm. We can fit the actual measurement data from IMEC with an self-developed stochastics model, from which we can get the minimum CD of different patterns which can cause 100% defectivity density or forecast the tendency of stochastic printing failures with the changing of CD. In addition, the model suggests that for each design rule pattern, there exists an optimized effective photoacid diffusion length together with EL and LWR.

ACKNOWLEDGEMENTS

We thank the higher management team from Shanghai IC R&D Company for the support of this work.

REFERENCES

[1] D. Civay, E. Hosler, V. Chauhan, T. Guha Neogi, L. Smith, D. Pritchard, "EUV telecentricity and shadowing errors impact on process margins", Proc. SPIE, 9422, 94220Z, 2015.

[2] Renzo Capelli, Anthony Garetto, Krister Magnusson, Thomas Scherubl,
"Scanner arc illumination and impact on EUV photomasks and scanner imaging", Proc. SPIE 9231, 923109, 2014.

[3] Qiang Wu, Yanli Li, Yushu Yang, and Shoumian Chen "A Study of Image Contrast, Stochastic Defectivity, and Optical Proximity Effect in EUV Photolithographic Process under Typical 5 nm Logic Design Rules", accepted by Institute of Electrical and Electronics Engineers 2020, accepted for publication.

[4] Qiang Wu, Yanli Li, Yushu Yang, Yuhang Zhao, "A Photolithography Process Design for 5 nm Logic Process Flow", accepted by Journal of Microelectronic Manufacturing 2020, accepted for publication.

[5] Yanli Li, Qiang Wu, Yushu Yang, Shoumian Chen, "Simulation Study for Typical Design Rule Patterns in 5 nm Logic Process with EUV Photolithographic Process", accepted by Journal of Microelectronic Manufacturing 2020, accepted for publication.

[6] Qiang Wu, "The Variables and Invariants in the Evolution of Logic Optical Lithography Process", Journal of Microelectronic Manufacturing 2, Issue 1: 19020101, 2019.

[7] Hironori Ikezawa, Yasuhiro Ohmura, Tomoyuki Matsuyama, Yusaku Uehara, Toshiro Ishiyama, "A Hyper-NA Projection Lens for ArF Immersion Exposure Tool", Proc. SPIE 6154, 615421, 2006.

[8] P. De v Bisschop, E. Hendrickx, "Stochastic printing failures in EUV lithography", Proc. SPIE 10957, 109570E, 2019.

A STUDY OF IMAGE CONTRAST, STOCHASTIC DEFECTIVITY, AND OPTICAL PROXIMITY EFFECT IN EUV PHOTOLITHOGRAPHIC PROCESS UNDER TYPICAL 5 NM LOGIC DESIGN RULES

Qiang Wu, Yanli Li, Yushu Yang, and Shoumian Chen

Shanghai IC R&D Center
497 Gaosi Road, Zhangjiang Hi-Tech Park, Shanghai 201210, PR China
Contact: wuqiang@icrd.com.cn

ABSTRACT

The introduction of Extremely Ultra-Violet (EUV) lithography in the photolithographic process can simplify process flow at 7 nm or more advanced technology nodes, which includes good linewidth and overlay budget control and reduction of hard mask layers. In a typical 5 nm logic process, the Contact-Poly Pitch (CPP) is 44-50 nm, the Minimum Metal Pitch (MPP) is 30-32 nm. And the overlay budget is estimated to be 2.5 nm (On Product Overlay, OPO). We have studied the process window of the 5 nm lithographic process with a self-developed RCWA algorithm based EUV simulation program and will present our results on process window and defectivity.

INTRODUCTION

Photolithography has been the driving force for the continuous design rule shrink. Early lithography used contact-proximity printing, which has a limit in spatial resolution around 2 μm for practical photoresist thickness. The design rule shrink to below 1 μm must be attributed to the introduction of projection imaging system, which has more generous Depth of Focus (DoF) to accommodate usable photoresist thickness. From the first Perkin Elmer's scanning projection printer with a NA of 0.167 in 1973 to the modern 193 nm immersion scanner with a NA of 1.35 (2007), optical resolution has been improved from a few microns to 38 nm [1]. With the introduction of the Extremely Ultra-Violet (EUV) lithography, the resolution can be further extended to 18 nm. One big advantage of the optical imaging is pattern replication with massive parallelism, around 10^{11}-10^{12} pixels, as contrasted by the scanning beam methods, such as the scanning electron beam methods even with multi-beam system [2]. There is also the self-assembly method for pattern formation, but the defectivity is still too high [3].

I ROADMAP OF OPTICAL LITHOGRAPHY FOR LOGIC PROCESSES

Starting from the 0.25 μm technology node in the mid-1990, photolithography has been using imaging projection to replicate design patterns from a mask to the wafers. Due to diffraction, the resolution of any imaging system is finite and given by the Rayleigh criterion below for one dimensional (1D) lines and spaces,

$$\Delta X = 0.5 \frac{\lambda}{NA} \tag{1}$$

where ΔX denotes the minimally resolvable distance, λ is the wavelength of light, and NA is the numerical aperture. In Figure 1, two diffraction-limited trenches are separated by 0.5λ/NA and the sum of the intensities are added showing a center dip about 20% lower in intensity compared to the peaks. The minimally resolvable distance defined here has an imaging contrast around 10~11%, defined by,

$$\text{Contrast} = \frac{I_{\max} - I_{\min}}{I_{\max} + I_{\min}} \tag{2}$$

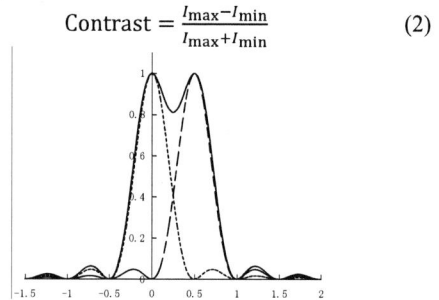

Figure 1: Resolution limit described by Rayleigh criterion: X-axis unit: λ/NA.

In photolithography, there is a similar equation that describes the minimally printable linewidth, as given by,

$$\Delta X = k_1 \frac{\lambda}{NA} \tag{3}$$

where ΔX denotes the minimum linewidth, or Half Pitch (HP), and the k_1 is the factor that is dependent on the type of illumination with a minimum value of 0.25. In imaging lithography, one of the most important parameters that describes the quality of imaging is imaging contrast. It can vary from 100% under ideal situation to a level that barely prints an image. In the case of Rayleigh limit we just discussed, the contrast limit is around 10~11%. The contrast for a given imaging system is usually described by the Modulation Transfer Function (MTF), as depicted in Figure 2. Figure 2 indicates that it is difficult to make patterns with all pitches all at very high imaging contrast.

978-1-7281-6559-2/20 $31.00 © 2020 IEEE

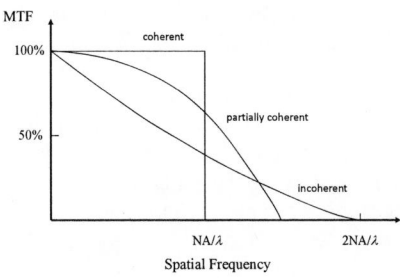

Figure 2: Modulation Transfer Function (MTF) of an ideal lens (Schematic).

Besides imaging considerations, photoresists also play a critical role in the realization of final imaging contrast. Photoresists have evolved from the initial Novolak-DNQ type i-line photoresist for the generations before 0.25 μm to the revolutionized Chemically Amplified photoResist (CAR) invented by IBM in the 1980's [4], with which the resolution and energy sensitivity have been fundamentally improved. One key to the imaging with CAR is the effective photoacid diffusion length. The diffusion can bring chemical amplification but at the cost of some imaging contrast. In 2004, it can be measured to 1~2 nm accuracy [5], which enabled quantification of its effect to the imaging, as described by Equation (4). A list of the effective photoacid diffusion length values for various technology nodes is shown in Table I.

TABLE I.   EFFECTIVE PHOTOACID DIFFUSION LENGTH FOR 250 NM TO 5 NM FOR CHEMICALLY AMPLIFIED PHOTORESISTS (CAR)

Logic Technology Node (nm)	Photoresist Type	Effective Photoacid Diffusion Length (nm)
250	248 nm CAR	70-100
180	248 nm CAR	40-70
130	248 nm CAR	20-40
90	193 nm CAR	25-40
65	193 nm CAR	17-25
45/40	193 nm CAR	10-15
32/28	193 nm CAR	5-10
20/14	193 nm CAR	5
	193 nm CAR, NTD	10
10/7	193 nm CAR	5
	193 nm CAR, NTD	5-10
7/5	13.5 nm EUV	<=5

$$I_D(x, y, z) =$$

$$\left(\frac{1}{a\sqrt{2\pi}}\right)^3 \iiint_{-\infty}^{+\infty} I_0(x', y', z') e^{-\frac{(x-x')^2}{2a^2} - \frac{(y-y')^2}{2a^2} - \frac{(z-z')^2}{2a^2}} dx' dy' dz'$$

(4)

In Equation (4), I represents the aerial image intensity and I_D the intensity with photoacid diffusion, and a is the diffusion length.

We have done a process window optimization simulation based on typical design rules (e.g., from wikichip.org) on the gate layer lithography with typical process conditions, such as the use of 6% phase shifting mask starting at 180 nm generation node, the use of 193 nm immersion lithography starting at 45 nm node, the use of X/Y polarization starting at 32/28 nm node, the use of Self-Aligned Double Patterning (SADP) process starting at the 10 nm node, etc. and we have summarized our result in Table II, as follows. In Table II, the Exposure Latitude (EL) numbers are mostly above 18% (to ±10% linewidth tolerance) and the Mask Error Factor (MEF) simulated also showing a trend: mostly below 1.5.

TABLE II.   EXPOSURE LATITUDE, MASK ERROR FACTOR, AND DOF SIMULATION FOR THE GATE LAYER FROM 250 NM THROUGH 5 NM TECHNOLOGY NODES UNDER TYPICAL PROCESS CONDITIONS

Logic Tech Node (nm)	Mask Type	Photoresist Thickness (nm)	CD (nm)	Pitch (nm)	Wavelength (nm)	Photoacid diffusion length (nm)	Illumination condition	Exposure Latitude, EL (%)	Mask Error Factor, MEF	Depth of Focus, DoF (nm)
250	Binary	700	250	500	248	70	0.55NA conv	19.3	1.47	500~600
180	6% PSM	500	180	430	248	60	0.65NA annular	17.7	1.39	450
130	6% PSM	400	150	310	248	30	0.70NA annular	18.9	1.66	350
90	6% PSM	300	120	240	193	30	0.70NA annular	19.7	1.56	350
65	6% PSM	220	90	210	193	25	0.85NA annular	18.6	1.51	250
45	6% PSM	200	90	180	193i	15	1.1NA annular	22.5	1.51	150
28	6% PSM	100	55	118	193i w XY pol	5	1.35NA weak DP	21.5	1.4	80
20	6% PSM	90	45	90	193i w XY pol	5	1.35NA weak DP/SMO	22.6	1.45	60
14	6% PSM	90	39	78/84	193i w XY pol	5	1.35NA strong DP/SMO	13.7/2 3.1 @ pitch 78/84	2.8 /1.47	60
10	6% PSM	90	33	66	193i SADP w XY pol	5	1.35NA weak DP/SMO	16.2/2 4.3 @ pitch 90/13 2	3.1 /1.1	60
7	6% PSM	90	27	54	193i SADP w XY pol	5	1.35NA weak DP/SMO	19.9/2 5.4 @ pitch 90/10 8	2.3 /1.22	55
5	6% PSM	90	25	50	193i SAQP w XY pol	5	1.35NA weak DP/SMO	>18% @ pitch 100	<1.5 @ pitch 100	50

We have also done a simulation study for the typical Back-End-Of-the-Line (BEOL) metal layer. The simulation result is shown in Table III. In the BEOL, the design rules have evolved from the bi-directional design to the unidirectional design starting at the 10 nm technology node. And the NTD process has been first adopted at the 14 nm node for the contact, via, and the metal layers. The data show that the EL from 250 nm node has witnessed a falling trend until 28 nm, staying at ~ 13%. The MEF follows a rising trend until 28 nm, staying at ~3.5.

TABLE III.   EXPOSURE LATITUDE, MASK ERROR FACTOR, AND DOF SIMULATION FOR THE METAL LAYER FROM 250 NM THROUGH 5 NM TECHNOLOGY NODES UNDER TYPICAL PROCESS CONDITIONS

Logic Tech Node (nm)	Mask Type	Photoresist Thickness (nm)	CD (nm)	Pitch (nm)	Wavelength (nm)	Photoacid diffusion length (nm)	Illumination condition	Exposure Latitude, EL (%)	Mask Error Factor, MEF	Depth of Focus, DoF (nm)
250	Binary	1000	320	640	248	70	0.55NA conv	29.3	1.03	600~800
180	6% PSM	600	230	460	248	40	0.65NA annular	18.1	1.85	600
130	6% PSM	400	160	340	248	25	0.70NA annular	19.8	450	450
90	6% PSM	300	120	240	193	25	0.70NA annular	16.9	2	450
65	6% PSM	200	90	180	193	17.5	0.85NA annular	13.4	2.85	300
45	6% PSM	180	80	160	193i	15	1.1NA annular	14.9	2.63	200
28	6% PSM	90	45	90	193i w XY pol	5	1.35NA weak DP	12.6	3.2	80–100
20	OMOG	90	32	64 (SE pitch 90)	193i w XY pol	5	1.35NA weak DP/SMO	12.6	3.2	60–80
14	6% PSM	70, NTD	32	64 (SE pitch 90)	193i w XY pol	10	1.35NA strong DP/SMO	13.98 w PDB	3.58	60–80
10	6% PSM	70, NTD	22	44 (SAL ELE)	193i SADP w XY pol	7	1.35NA weak DP/SMO	12.55	3.35	60–75
7	OMOG	70, NTD	20	40 (SAL ELE)	193i SADP w XY pol	7	1.35NA weak DP/SMO	12.7	3.5	55–70
5	EUV	50, EUV CAR	16	30–32 (EUV SALELE)	13.5 nm EUV SALELE	4	0.33NA Quasar	18% ~20%	1.55	55

II LIMIT OF EUV PHOTOLITHOGRAPHY PROCESS

Current EUV exposure tools have a NA of 0.33, which theoretically has a resolution limit of 0.5λ/NA, or 20.5 nm. It is believed that the practical resolution is 26 nm pitch for lines and spaces under unidirectional designs and 36 nm pitch for patterns that have both orientations. Shown in Figures 3(a) and 3(b) is a trench width through pitch simulation result with a minimum pitch of 32 nm for a mask width of 18 nm and an After-Developing-Inspection (ADI) width of 16 nm. The simulation is done with 60 nm TaN absorber, a 2.5 nm Ru protection layer under the absorber, and an equivalent 80 alternating layer of 4.2 nm Si and 2.8 nm Mo mirror with Finite-Difference Time-Domain (FDTD) algorithm calibrated with wafer data. The illumination condition is 0.33NA, 0.8-0.3 dipole 90° and a SRAF width of 12 nm with a SRAF pitch of 32 nm. A typical CAR photoresist model is used and we have found that, at the minimum pitch, an EL of 19% and a MEF <1.5 can be obtained. Note that in EUV, due to the defectivity originating from photon absorbing stochastics, the practical minimum pitch needs to be relaxed to about 36~40 nm [6-7].

Shown in Figure 4 is a process window simulation with a typical CAR resist at 27 nm pitch under a dipole illumination with 54 mj/cm^2 showing an EL of 15.3% and a LWR (unbiased) of 3.73 nm. The model has included a description on the stochastics in photon absorption. The simulation has assumed an underlayer which contributes about 30% more EUV absorption. This LWR is similar to that of 193 nm photoresist, but the EUV process prints much smaller linewidth which makes the LWR appear relatively bigger.

For CAR photoresist, the effective photoacid diffusion length is usually smaller than 5 nm. We have found that acid diffusion can smooth out some non-uniformity from LWR. However, longer diffusion will reduce the EL, which will make LWR worse. The two mechanisms compete with each other and a compromise will live in the maintenance of acceptable EL while maximizing photon absorption or smoothing. Shown in Figure 5 is a plot of the simulated (unbiased) LWR as a function of normalized (to the maximum acceptable number) effective acid diffusion length under an exposure dose of 54 mj/cm^2 and the average EL is 12.3%. Figure 5 indicates that about 75% of the normalized diffusion length is the optimized number, which we use to yield an EL of 15.3%. Shown in Figure 6 is a plot of simulated LWR as a function of EL under a fixed photoacid diffusion length (75% against a normalized scale), which indicates that once the effective acid diffusion length is determined, the LWR will become lower if the EL is getting higher.

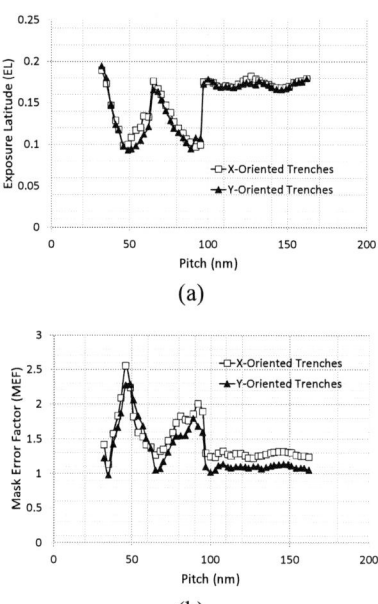

Figure 3: Simulated process window data for 1D trench width through pitch: (a) EL and (b) MEF at both X and Y orientations; The pattern is located in the center of the curved slit.

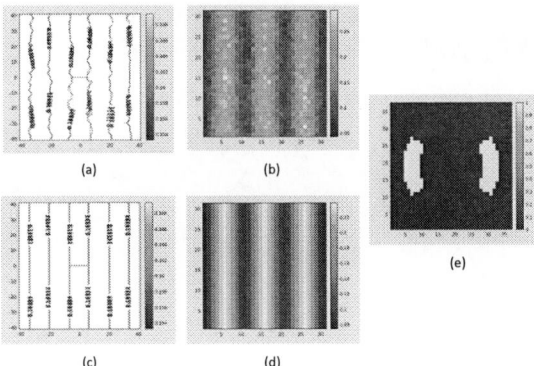

(a)   (b)

(e)

(c)   (d)

Figure 4: Simulated top down image profile of a 27 nm pitch dense trenches with dipole illumination (shown in right). The pattern is located in the center of the curved slit. (a) Image profile with stochastics, (b) image intensity with stochastics, (c) image profile, (d) image intensity, (e) illumination condition on pupil.

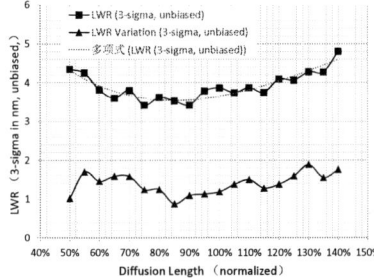

Figure 5: Simulated LWR as a function of the effective photoacid diffusion length. The pattern is a 27 nm dense lines and spaces oriented along the Y direction and the illumination is a dipole with 0.33 NA. Each LWR data point is an average of 20 single measurement, with each calculated from about 30 linewidth simulations. The pattern is located in the center of the curved slit.

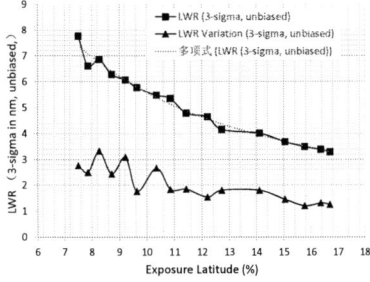

Figure 6: Simulated LWR as a function of EL. The pattern is a 27 nm dense lines and spaces oriented along the Y direction and the illumination is varying from a narrow dipole (DP X 20°) to annular with 0.33 NA under an exposure dose of 54 mj/cm^2. Other conditions are the same to those in Figure 5.

III TYPICAL 5 NM LOGIC DESIGN

RULES AND LITHOGRAPHY PROCESS DESIGN WITH CONSIDERATION OF OPTICAL PROXIMITY EFFECT

Starting from the 7 nm technology node, the EUV has been adopted and in the 5 nm node, the EUV has been adopted to cover most cut, contact, and metal layers. Shown in Table IV is a list of three critical pitches for the 10, 7, and 5 nm logic design rules, respectively.

In the Front-End-Of-the-Line (FEOL), at the fin and gate layers, the 193 nm immersion lithography is still used in combination with self-aligned quadruple patterning (SAQP) and SADP processes, respectively. It has been shown that in SAQP process, the LWR after 2nd mandrel strip and etch can improve from an initial 7.6 nm after lithography to about 2 nm [8].

For layers that are not sensitive to the pattern edge roughness, such as the metal and cut layers, can adopt EUV lithography. For hole pitch around 50 nm, the use of EUV can save about 4~6 exposures in 193 nm immersion. Because of the impact of stochastics, the metal 30~32 nm pitch cannot be accomplished by single EUV exposure (limited to 36~40 nm pitch). Instead, the industry adopts self-aligned litho-etch litho etch (SALELE) process.

TABLE IV.   TYPICAL DIMENSIONS FOR CRITICAL LAYERS UNDER 5 NM LOGIC DESIGN RULES

Logic Tech Node (nm)	10	7	5
Fin Pitch	33~42	27~30	22.5~24
Contact to Poly Pitch	66~68	54-56	44~50
Metal Pitch	44~48	36~40	30~32

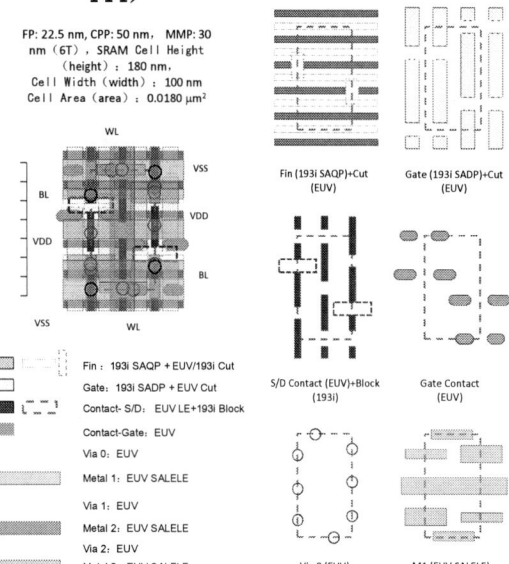

Figure 7: A tentative SRAM layout [6].

In Figure 7, a tentative SRAM cell with a Fin Pitch (FP) of 22.5 nm, a contact-to-poly pitch (CPP) of 50 nm, and a Minimum Metal Pitch (MMP) of 30 nm is shown.

According to the last section, once the photo-lithographic process has been optimized for the consideration of stochastics, it needs to be further adjusted to accommodate all pitches allowed by the design rules to satisfy the regular process window requirements in EL, DoF, and MEF. Shown in Figures 8 is a line through pitch simulation result for the 193 nm immersion layers (for fin and gate). Shown in Figures 9 and 10 are EUV trench and hole width through pitch simulation results. All simulation is under typical 5 nm logic design rules and near the optimized process conditions.

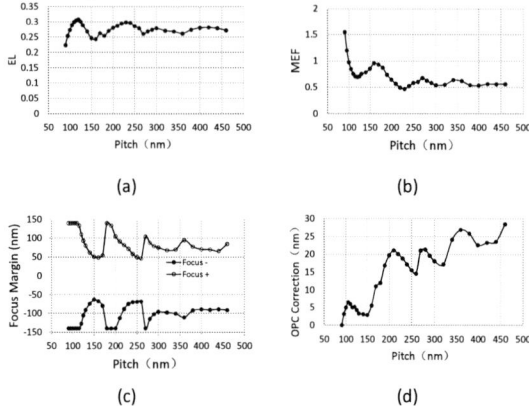

Figure 8: 45 nm ADI linewidth through pitch simulation under 193 nm immersion. (a) EL; (b) MEF; (c) Focus Margin; (d) Optical Proximity Correction (OPC). Illumination condition: 1.35NA, 0.9-0.7 Dipole 90°, X/Y polarization, and 6% PSM mask.

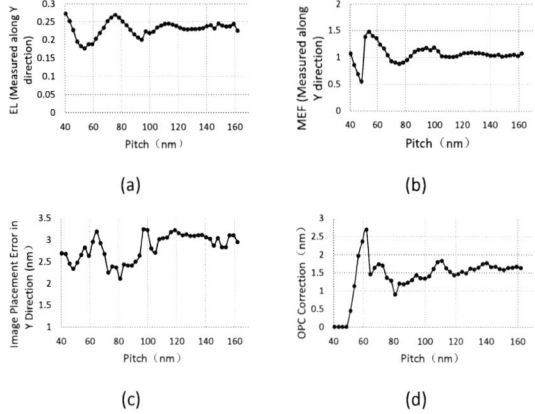

Figure 9: 22.5 nm ADI trench width along the X direction through pitch simulation with EUV lithography: (a) EL; (b) MEF; (c) Image Placement Error; (d) OPC. Pattern

location: slit center. Illumination condition: 0.33NA, 0.9-0.3 Quasar 45° with a 30 nm CAR photoresist model

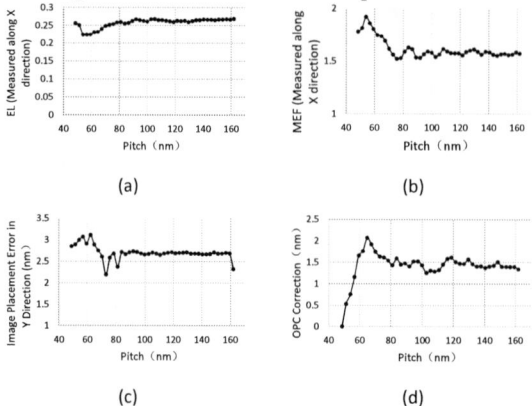

Figure 10: 26 nm ADI hole width through pitch simulation with EUV lithography. (a) EL; (b) MEF; (c) Image Placement Error; (d) OPC Correction. Pattern location: slit center. Illumination condition: 0.33NA, 0.8-0.4 Annular with a 30 nm CAR photoresist model.

IV DISCUSSION OF OPTICAL PROXIMITY EFFECT IN EUV LITHOGRAPHIC PROCESS

Compared to 193 nm immersion lithography, the amount of Optical Proximity Correction (OPC) is usually less than 3 nm, or less than 10~15% of the linewidth, as depicted in Figures 9 and 10. This is due to that the EUV processes have relatively larger k_1 values, defined in Equation (3). For example, for the trench layer that is described by Figure 9, the k_1 value is,

$$k_1 = \frac{\Delta X \times NA}{\lambda} = \frac{p NA}{2\lambda} = \frac{40.5 \times 0.33}{2 \times 13.5} = 0.495 \quad (5)$$

where p represents the minimum pitch and the ΔX denotes linewidth and is defined as the HP, or $p/2$. In DUV lithography process, the OPC is not very significant when the k_1 value is greater than 0.4. For the 193 nm immersion lithography that is used for the FEOL fin and gate processes described in Figure 8, the k_1 value is given by,

$$k_1 = \frac{\Delta X \times NA}{\lambda} = \frac{p NA}{2\lambda} = \frac{90 \times 1.35}{2 \times 193} = 0.315 \quad (6)$$

which is much smaller than that of EUV and 0.4! As we can find that the maximum OPC from the dense pitch of 90 nm to the isolated pitch of 460 nm is 28.5 nm, or 63% the linewidth.

In EUV, due to existence of Chief-Ray-Angle-at-Object (CRAO) of 6° in illumination and a ring-shaped exposure slit [9], there are other sources of patterning error that need to be taken care of by OPC, which are the horizontal-vertical linewidth bias (H-V bias), the shadowing effect caused pattern shift, the mask 3D scattering effect caused DoF reduction, which are described by simulations shown in Figures 11, 12, and 13.

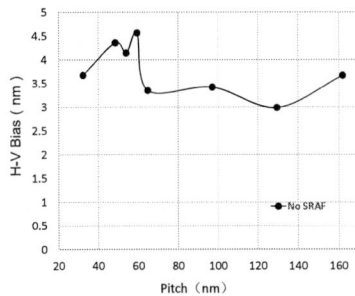

Figure 11: Simulated EUV H-V bias for line and space through pitch pattern. Illumination condition: 0.33NA, Quasar 45° 0.9-0.2 with a 30 nm CAR resist model.

Beside the shadowing effect, the small NA (0.33) used in EUV can subject the imaging result to be more sensitive to aberration, which is described in Figure 14. Figure 14 shows that Coma in Y can cause pattern shifts which is inversely proportional to NA. A comparison between 193 nm immersion and the current 0.33 NA EUV lithography is shown in Table V, indicating that a 0.2 nm rms Coma can cause 1.8 nm maximum pattern shift in a 0.33 NA EUV system, while the much larger 1 nm rms can only cause a maximum pattern shift of 3.2 nm in 193 nm immersion.

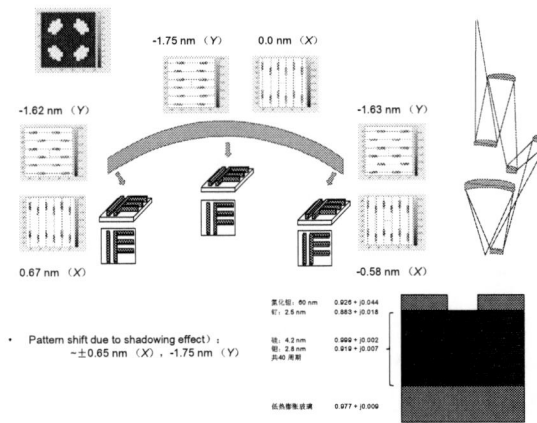

Figure 12: Simulated EUV shadowing effect with a 32 nm pitch line and space pattern. Illumination condition: 0.33NA, Quasar 45° 0.9-0.5 with a 30 nm CAR resist model.

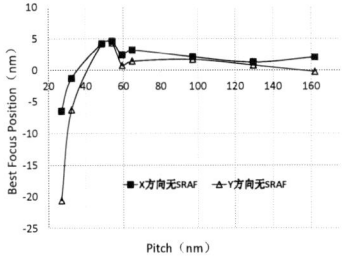

Figure 13: Simulated best focus shift for X and Y oriented line and space through pitch pattern with no SRAF placed. Illumination condition: 0.33NA, Quasar 45° 0.9-0.2 with a 30 nm CAR resist model.

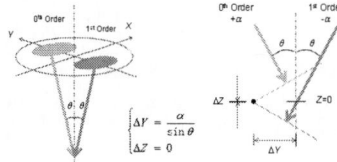

Figure 14: Schematic diagram of coma induced pattern shift as a function of illumination pupil positions [6].

TABLE V. COMPARISON BETWEEN 193 NM IMMERSION LITHOGRAPHY AND EUV LITHOGRAPHY PROCESSES

Type	193 nm Immersion	0.33NA EUV
Min Pitch (nm)	90	36
CD (nm)	45	18
Typical Tech Node (nm)	28	5
On-Product-Overlay (nm)	7	2.5
rms Aberration Control (nm)	1	0.2
Max Pattern Shift due to Max coma (nm)	3.2	1.8

CONCLUSIONS

We have studied image contrast, stochastic defectivity, and optical proximity effect in EUV photolithographic process under typical 5 nm logic design rules. The FEOL fin and gate processes will continue to adopt 193 nm immersion with, respectively, self-aligned quadruple and double patterning techniques. EUV will be applied to most of the cut, contact and metal layers. The EUV photon absorption stochastic effect in EUV photoresist will limit the minimum pitch that can be used for single exposure, which is about 36~40 nm for lines and spaces and 48~50 nm for the hole layers.

ACKNOWLEDGMENTS

We would like to thank Shanghai ICRD colleagues and higher management team for the support of this work.

REFERENCES

[1] http://www.lithoguru.com/scientist/litho_history/Kato_Litho_History.pdf.

[2] M.J. Wieland et al., "MAPPER: High throughput maskless lithography", Proc. SPIE 7637, 76370F, 2010.

[3] R. Gronheid, I. Pollentier, "Addressing the challenges of directed self assembly implementation", Litho extensions symposium Miami, 2011.

[4] H. Ito and G.C. Willson, "Chemical Amplification in the Design of Dry Developing Resist Materials," Polym. Eng. and Sci. 23, 1012, 1983.

[5] Q. Wu, S. Halle, and Z. Zhao, "The Effect of the Effective Resist Diffusion Length to the

Photolithography at 65 and 45 nm Nodes, A Study with Simple and Accurate Analytical Equations", Proc. SPIE 5377, 1510, 2004.

[6] Q. Wu, Y. Li, Y. Yang, and Y. Zhao "A Photolithography Process Design for 5 nm Logic Process Flow", Journal of Microelectronic Manufacturing 2, 19020408, 2019.

[7] P. De Bisschop, E. Hendrickx, "Stochastic printing failures in EUV lithography", Proc. SPIE 10957, 109570E, 2019.

[8] L. Sun, X. Zhang, S. Levi et al., "Line Edge Roughness Frequency Analysis for SAQP process", Proc. SPIE 9780, 97801S, 2016.

[9] R. Capelli, A. Garetto, K. Magnusson, and T. Scherübl, "Scanner arc illumination and impact on EUV photomasks and scanner imaging", Proc. SPIE 9231, 923109, 2014.

THE TOPOGRAPHY EFFECT ON THE LITHOGRAPHY PATTERNING CONTROL FOR IMPLATATION LAYERS

Dongyu Xu[1]*, Dingshuo Luo[1], Zhihong Wang[1], Wenzhan Zhou[1], Zhanyuan Hu[1]

[1]Shanghai Huali Integrated Circuit Cooperation, No. 6 Liangteng Road
Shanghai 201620, People's Republic of China
*Corresponding Author's Email: xudongyu@hlmc.cn

ABSTRACT

Lithography patterning is controlled by aerial images and photoresist behaviors. The projected aerial images could be changed by the reflectivity of material and topography. The photoresist process can be affected by baking, development, and local thickness changes. In this paper, we report a study of the optical contribution and the photoresist local/global loading effect from topography, showing a substantial influence on CD (critical dimension) control. Therefore the consideration of photoresist thickness and the refection from underlayers is a must when patterns locate on a complicated topographical environment.

INTRODUCTION

As the technology nodes are getting more advanced, the chip size is shirking, the CD control becomes more critical [1]. To promote the accuracy and uniformity of CD, advanced tools with shorter wavelength and large NA, elaborate OPC patterns, planarization and antireflection chemicals are introduced into the lithography patterning.

There are a bunch of issues that could affect the CD, such as underlying filmstack thickness variations, resist and bottom antireflection coating thickness uniformity. Among these issues one nonnegligible aspect is the complicated topography from underlayers, such as long-range height variation resulting from deposition or chemical mechanical polish, or short-range variation from particular pattern design. The non-uniform distribution on wafer tends to cause focus drift, reflection, photoresist thickness variation [2]. Especially when some advanced devices are induced into semiconductor fabrication, for example FinFET (fin field-effect transistor) and GAA (gate all around) technology [3]. The topography issue becomes more severe due to their complex spatial structures.

Here we aim to understand how to troubleshoot the unexpected CD variation, and qualificationally analyze the contribution from the underlying topography is more like the aerial images or the photoresist variation. Through simulation and experiments, we report a study of the optical contribution and the photoresist thickness from topography, showing a substantial influence on CD control. Furthermore a simple correction flow has been set up for the patterns in chip through the correlation between normalized CDs and the underlying pattern density.

RESULTS AND DISCUSSIONS

CD Measurement Results

To understand the contribution from optical effect and photoresist local thickness independently, three patterns were carefully designed and studied. Their layouts are shown in Figure 1. The pattern I is a dense array on a large AA (active area). The pattern II is an identical dense array as well but on a small AA, surrounded by some dummy gates. The pattern III is a semi-trench on AA and gates.

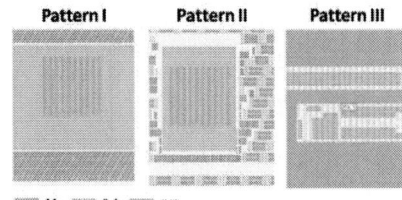

Figure 1: The local environment of selected patterns.

Their CDs both on bare silicon wafer and structured wafer were collected by CD-SEM (critical dimension scanning electron microscope). The CD-SEM images and the average CDs of these selected patterns are shown in Figure 1. The CDs of these 3 patterns are very close to the target on a bare silicon wafer. However, when exposed at a structured wafer, only pattern I is on target. The other 2 patterns are far away from the target, a near 20 nm bias cannot be neglected.

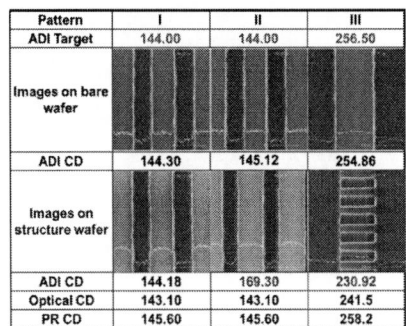

Pattern	I	II	III
ADI Target	144.00	144.00	256.50
Images on bare wafer			
ADI CD	144.30	145.12	254.86
Images on structure wafer			
ADI CD	144.18	169.30	230.92
Optical CD	143.10	143.10	241.5
PR CD	145.60	145.60	258.2

Figure 2: The CD-SEM images of patterns on a bare silicon wafer and on a structured wafer.

The optical and photoresist CDs simulated by SLITHO (Sentaurus Lithography) cannot explain the error. To investigate the reason of unexpected CD bias, the local photoresist thickness of patterns on unexposed wafers was measured by SEM (scanning electron microscope). The cross-section SEM images are shown in Figure 3a, where the local photoresist thicknesses are 180 nm (I), 193 nm (II), 243 nm (III) respectively. These dramatic differences of photoresist thickness can be attributed to the photoresist viscoelasticity, resulting from the topographical environment at their specific location.

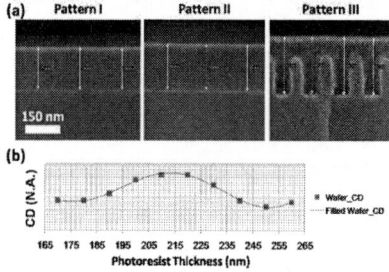

Figure 3: The photoresist thickness of different patterns at specific location (a). The CD swing curve in the resist trench width with resist thickness (b).

Due to the photoresist reflectivity is correlating with its thickness. A CD swing curve in the resist trench width with resist thickness is measured and fitted in Figure 3b. If we do the simple calculation we could find the period is near 71.7 nm base on equation (1), which is in accord with the as-measured results.

$$Period = \lambda/2n_2 \qquad (1)$$

Where λ is the vacuum wavelength of light (248 nm for KrF) and n_2 is the photoresist reflectivity (1.75). The unexpected CDs bias can be accounted for the different local photoresist thickness preliminarily, resulting from the photoresist mobility shown in Figure 4.

Figure 4: The mechanism scheme for CD variation, due to a local photoresist thickness difference.

The Photoresist Loading Effect

To study the photoresist loading effect, preceding a straightforward photoresist calibration is a must to catch the optical model and the photoresist behavior. In this paper, both FEM (focus and energy metrics) and nominal conditions of a series through-pitch patterns were measured to calibrate the model by SLITHO.

Despite the local environment difference (Figure 1),

pattern I and II are designed identically. Because the local photoresist thickness is investigated clearly, so this factor will be used to simulate the photoresist cross-section images as shown in Figure 5. The simulated CDs of pattern I and II are in accord with the CDs on structured wafer, indicating our photoresist model is accurate. The CD of pattern II is near 20 nm larger than pattern I for both optical and photoresist contours. It can be explained that more reflecting light is out of photoresist when the photoresist thickness is increasing from 180 nm to 193 nm.

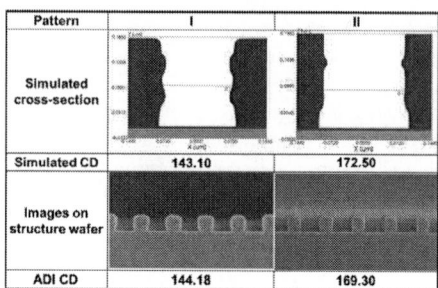

Figure 5: The simulated cross-section images and the CD-SEM cross-section images of pattern I and II.

Topographical Reflection Effect

To further understand the contribution from underlying topography, the pattern III on a structured wafer and on a bare wafer is simulated as shown in Figure 6. It is well-matched with the CD measured by CD-SEM. If the model is just simplified at a planar, there is a bias caused by aerial image error (~ 15 nm).

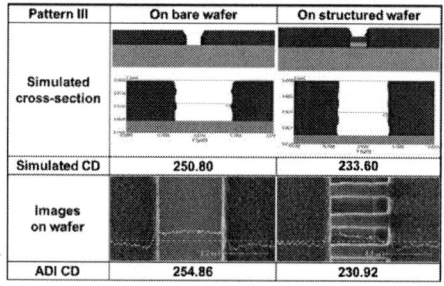

Figure 6: The simulated cross-section images and the CD-SEM cross-section images of pattern III on bare silicon wafer and structured wafer, respectively.

Therefore the exposure intensity of pattern III on a structured wafer and on a bare silicon wafer is shown in Figure 7. With the underlying topography, the standing wave phenomenon has been effectively reduced, confirming the decreased illuminating intensity. That is due to the light reflection from underlying gates disrupt the original standing wave on a bare silicon wafer, and followed by a destructive interference.

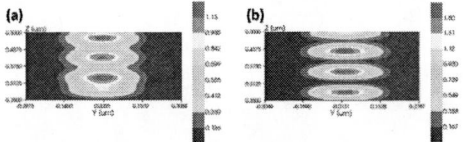

Figure 7: The exposure intensity of pattern III on a structured wafer (a) and on a bare silicon wafer (b).

A Straightforward CD Correction Flow

According to the discussion above, there is a bias between the CD on structured wafer and expected. To compensate that, an empirical resist spin-on model is used to estimate the local photoresist thickness. Here the pattern density is defined as the area of gate layer (underlying layer) divided by total area (2 * 2 μm). Therefore, both the photoresist thickness and the normalized CD for specific pattern serve as the function of pattern density as shown in Figure 8.

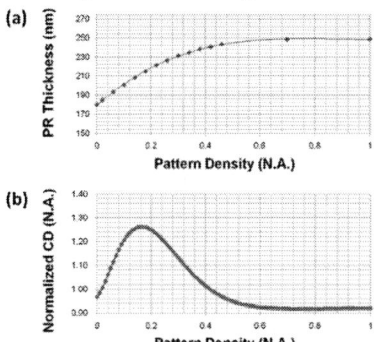

Figure 8: The correlation between pattern density and underlying pattern density (a) or the normalized CDs (b).

According to the discussion above, the topography model is superior to the planar model. To verify this conclusion, a series of weak patterns are used evaluate the accuracy of simulation as shown in Figure 9. These patterns in a chip are selected to reflect the real situation of process. We can tell with the topographic model in hand the predicted CDs are very close the CDs on a structured wafer.

Pattern	IV	V	VI	VII	VIII
Planar model	234.48	204.16	288.69	202.80	242.23
Topographic model	211.42	183.76	271.63	183.22	223.21
Images on structure wafer					
ADI CD	209.21	179.81	273.11	178.20	220.25

Figure 9: The predicted CD with optimized/non-optimized model and their CD-SEM images on a structured wafer of some weak patterns.

CONCLUSIONS

In this paper, we report a study of the optical contribution and the photoresist local/global loading effect from topography, showing a substantial influence on CD (critical dimension) control. Therefore the consideration of photoresist thickness and the reflection from underlayers is a must when patterns locate on a complicated topographical environment. Furthermore a straightforward CD correction flow has been set up for patterns in chip through the correlation between normalized CDs and the underlying pattern density.

This solution is used for a two-dimensional MOSFET (Metal-Oxide -Semiconductor Field Effect Transistor) process at 28/22 nm nodes here, which also shows potential at more advanced nodes such as FinFET or GAA due to their topography is more complicated.

ACKNOWLEDGEMENTS

D.X. particularly thanks the insightful discussion with Angmar Li from ASML.

REFERENCES

[1] S. G. Hansen. *Proceedings of SPIE*, 2002, vol. 4690, pp. 366-380.

[2] J. Kim, H. Bak, Y. Sohn, et al. *Proceedings of SPIE*, 2001, vol. 4346, pp. 982-993.

[3] N. Loubet, T. Hook, P. Montanini, et al. *2017 Symposium on VLSI Technology*, 2017, Kyoto, June 5-8, 2017, pp. T230-T231.

HIGH SPEED WAFER GEOMETRY ON SILICON WAFERS USING WAVE FRONT PHASE IMAGING FOR INLINE METROLOGY

J.M. Trujillo-Sevilla, J.M. Ramos-Rodríguez, J. Gaudestad

Wooptix

La Laguna, Tenerife, Spain

Presenting Author Jan Gaudestad: jangaudestad@wooptix.com , +1-415-684-3384

ABSTRACT

In this paper we introduce a new metrology technique for measuring wafer geometry on silicon wafers. Wafer geometry will be critical for the next generation integrated circuits (IC) for improvements in lithography overlay and to measure Nanotopography (NT) and roughness in conjunction with Chemical Mechanical Polishing (CMP). Wave Front Phase Imaging (WFPI) has high lateral resolution and is sensitive enough to measure NT and roughness on a silicon wafer by simply acquiring a single image snapshot of the entire wafer. WFPI is achieved by measuring the reflected light intensity from monochromatic uncoherent light at two different planes along the optical path with the same field of view.

We show that the lateral resolution in the current system is 24μm though it can be pushed to less than 5μm by simply adding more pixels to the image sensor. Also, we show that the amplitude resolution limit is 0.3nm. First, 3 mirrors simulating a 50mm blank wafer with a known geometry was used to compare WFPI to the industry standard chromatic confocal microscopy. Then, a 2-inch wafer was measured while laying it on a flat sample holder without chucking it and NT and roughness was revealed by applying a double Gaussian high pass filter to the global topography data. The exposure time was 0.1 seconds and the time to analyze the data was just under 2 seconds while processing 4.34 million topography data points.

INTRODUCTION

The geometry of unpatterned silicon wafers used as substrates for integrated circuits (IC) manufacturing is critical for process control and ultimately for device yield. Wafer geometry has many characteristics that have been classified based on spatial wavelength (λ_s) and amplitude (Z height resolution) (Fig. 1.) [1]. Nanotopography (NT) is defined as height variations with amplitudes in the tens of nm at the wafer surface and within λ_s in the range of 200μm–20mm. Beyond NT lies roughness, with amplitude in the single digit nanometer to sub-nanometer and λ_s in the range of tens of nanometers to microns. Shape, flatness, and to certain degree also NT has typically been measured using optical techniques such as KLA-Tencor's WaferSight 2. However, this system has not been capable of measuring roughness due to its poor spatial resolution [1], [2].

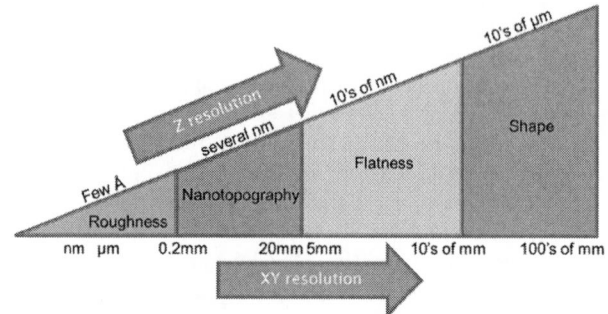

Figure 1. Definition of wafer geometry

Several tools are available today with its own advantages and disadvantages, however all tools, except for Atomic Force Microscopy (AFM), struggle to measure the small amplitude, few nm to sub nanometer, caused by roughness [3], [4]. Thus, no single optical tool can cover the whole range of topographical variations leaving AFM as the only candidate for measuring roughness, however the general problem with AFM is its small imaging area coupled with slow speed [5]. We show in our work that WFPI is the first optical solution to measure deeper into the roughness than any other optical metrology systems available today by applying a filter to the global topography depth map.

DESCRIPTION OF WFPI

The working principle of WFPI is based on registering the intensity distribution at two different optical planes (measurement planes). The intensity distribution is recorded by a conventional imaging sensor. The wave front phase is defined as the surface perpendicular to the direction of propagation of the light rays. The sensor assumes geometrical propagation of light, and in this regime the light can be considered as a collection of light rays which bends according to Snell's law and reflects on a surface keeping its angle with respect to the surface normal. The reflected beam will carry the wave front phase, which value is proportional to the surface height map. In our case, we are using a collimated red (λ = 650nm) light beam reflects onto the surface, the reflection angle of each ray is exactly two times the angle of the surface normal [6].

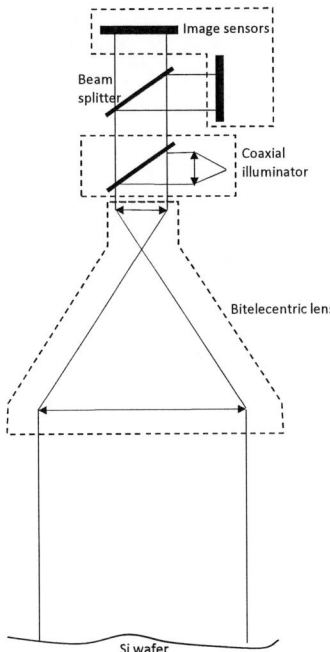

Figure 2. Typical configurations of WFPI for silicon wafer metrology in a manufacturing mode [6].

The reflected light beam, which carries the wave front phase information of the sample surface profile is de-magnified by a telecentric lens. For image acquisition, two paired imaging sensors were used, each one placed at a different optical plane allowing the pair of images to be acquired at the same time at different optical paths from the sample being measured. With this setup, a single image snapshot can collect wafer topography data of the entire wafer with the same number of pixels as being present in the imaging sensor [7].

The specifications of the camera setup used is summarized in Table 1.

Table 1. Specifications of the demo system for 50mm diameter sample size.

Lens f#	7.5
Image Sensor	2680 × 4024 (10.8MP)
Data points (50mm round)	≈ 4.34 million
Pixel Size (50mm round image)	24μm

LATERAL RESOLUTION

As it has been shown that WFPI can measure amplitudes in the sub nanometer regime [6] the main advantage of WFPI is its capacity to produce high lateral resolution topography maps of silicon wafers. Getting an accurate value of the maximum lateral resolution is more difficult for WFPI than in conventional imaging systems, as in conventional systems the modulation transfer function (MTF) is obtained measuring the contrast of several line groups at different resolutions. In WFPI, on the other hand, the wave front phase map must first be calculated, and then one can measure the different contrast values for resolution values in a known target.

From a theoretical point of view, WFPI is considered as a coherent optical system, and so, its transfer function consists of a flat response of unity value up to its cutoff frequency [9]. The cutoff frequency value depends on the criterion chosen. Here we have chosen Abbe's criterion because it is more restrictive than other popular criterions such as Rayleigh or Sparrow [10]:

$$f_c = \frac{1}{N\lambda} \tag{1}$$

Applying Equation 1 using the F# at 7.5 and using a red (650nm) LED external light setup one gets that the theoretical resolution limitation of this system is 4.875μm. The resolution limitation is then the pixel size rather than the optical limitation.

For the measurement of the actual frequency response of the prototype we used a USAF-1951 test target. This test target is made by a glass substrate with a deposition of chrome that forms line pairs with frequencies up to 512 cycles per mm (cy/mm), one cycle being 2 pixels side-by-side with different contrast ratio. Since the chrome deposition layer creates surface topography, it can be used to measure the performance of WFPI in terms of lateral resolution. The process consisted of placing the test target in the WFPI measurement system and run a single measurement of an area that contains line groups with spatial frequencies in a range valid for this study [7].

Due to the two cameras and the magnification of the telecentric lens arrangement used, the pixel size in object space is 24 μm, which translated to frequency that is equivalent to approximately 20 cy/mm ($1mm/(2 \times 24\mu m) \approx 20$). A phase image of the test target was acquired in an area in that covered frequencies from 1 cy/mm to 20 cy/mm. The measured WFPI was used to calculate the phase information and measure the local contrast. In Figure 4, the results are summarized and plotted against the theoretical contrast limitation calculated by applying Equation 1 and getting a cutoff frequency of 103 cy/mm ($1mm/(2 \times 4.875\mu m) \approx 102.56$). However, with the current pixel size (24μm), the system can only measure up to 20 cy/mm.

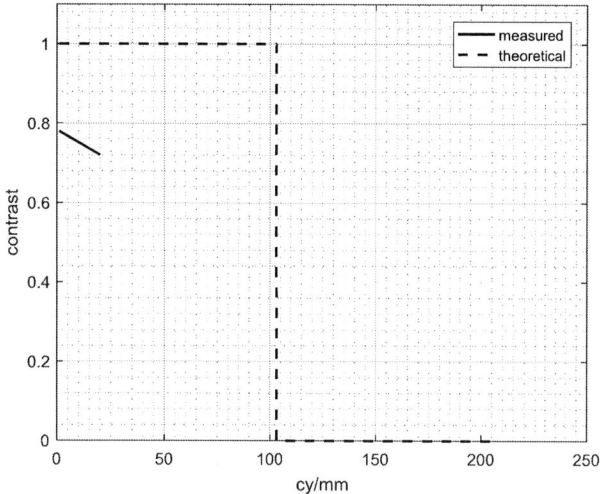

Figure 3. Comparison of measured and theoretical contrast response versus lateral resolution.

It can be observed a large difference between the predicted and the actual frequency responses. It is important to note that the theoretical behavior is only valid with an optical system which is completely free of aberrations; any wave front error in the lenses of the instrument contributes to a poorer frequency response. However, looking at the slope, one can expect this to continue close to linearly, indicating that by increasing the number of pixels by using a higher resolution camera setup, one can reach the optical resolution limitation of the system (4.875µm) [7].

Z-HEIGHT RESOLUTION – AMPLITUDE

A key aspect in determining the minimum amplitude that can be measured by a system is the noise level. The noise level for AFM is about 0.1 nm and the noise level for an optical profiler is on the order of 0.4 nm. However, both techniques struggle with speed and a very small field of view [5].

In WFPI the noise level is mostly dependent on the pixel size and the signal-to-noise-ratio (SNR in dB) of the image sensor being used. A noise simulation was done using the pixel size 6µm square (which is the size of the pixels used in the 50mm WFPI system). The noise is given in nano meter root-mean-square (RMS) with its standard deviation (σ). The noise level is summarized in Figure 4.

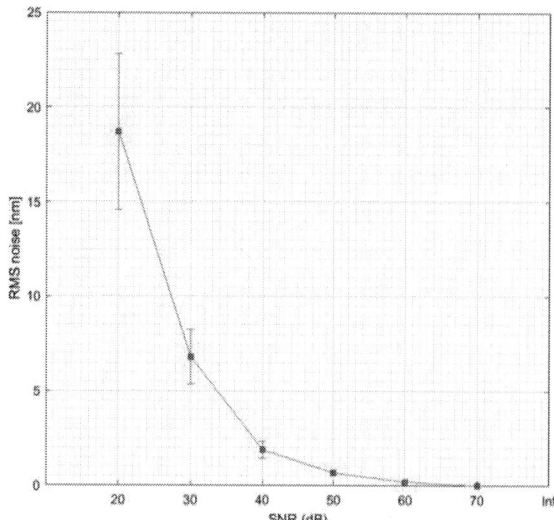

Figure 4. Noise plotted with its RMS noise (nm), standard deviation (nm) and signal-to-noise-ratio (dB)

The noise in the camera image sensor was 62dB, which translated to RMS noise level at 0.3nm with standard deviation of 0.1nm. Combined with a 24µm lateral resolution, WFPI is a good candidate among optical techniques for measuring into the roughness regime.

COMPARISON BETWEEN WFPI AND CHROMATIC CONFOCAL MICROSCOPY

WFPI was tested against commercial chromatic confocal microscopes. These instruments are laboratory reference tools in the field of wafer geometry metrology, they exhibit height resolutions as good as 10 nm and a maximum lateral resolution of 5 µm. The main drawback of these instruments is that they need raster scanning to measure an extended surface such as silicon wafers, requiring several minutes even for smaller sample sizes, to acquire the data.

We have used commercial reference mirrors with a known warpage (Table 2) as target samples to compare the performance of our instrument against two commercial chromatic confocal microscopes [11, 12].

Table 2. Specifications for the 3 mirrors that were used. All 3 mirrors were 50.8mm in diameter with aluminum coating.

#	Type	Company	Model	Curv r (m)	warpage (µm)
1	Flat	Edmund Optics	69-249	∞	0
2	Concave	Lambda	PCCM-5006B-10000	10	29.40
3	Concave	Lambda	PCCM-5006B-3000	3	98.01

The first set of data was done on a flat mirror which means there is no warpage no warpage (radius of the bow is infinite ∞) (see Figure 5 below).

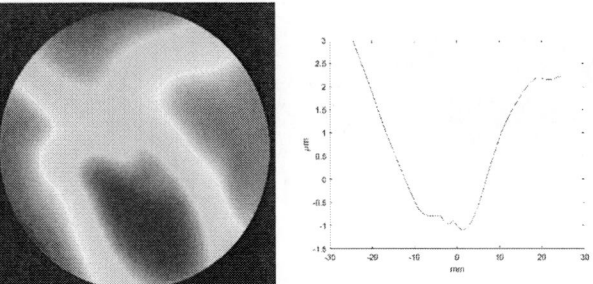

Wave Front Phase Imaging: 100ms

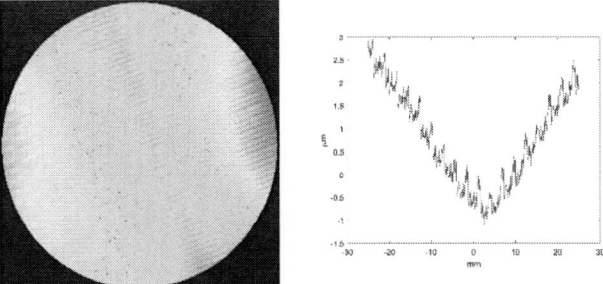

Chromatic confocal microscopy 1: 2 min [11]

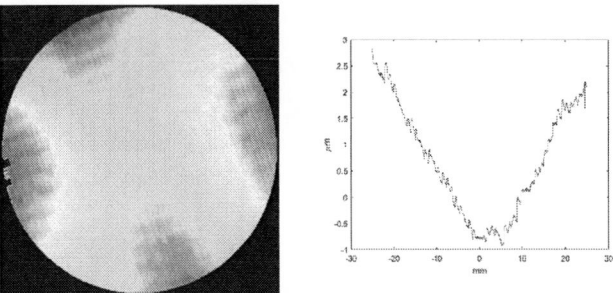

Chromatic confocal microscopy 2: 11 min [12]

Figure 5. Comparison of reference mirror measurement using three instruments. All data shown with a false color image and a line scan through the center of the mirror sample. Top: WFPI snapshot. Middle: chromatic confocal at higher speed raster scanning. Lower: chromatic confocal with lower raster scanning speed.

Some warpage is to be expected, even on a flat surface, due to temperature changes among other factors. In all three data sets taken, one can clearly see that there is some warpage in the single digit micron range. This is still within specs of these samples and would also be according to Semi wafer specifications [13]. It is also clear that when a raster scanning system acquires data faster, the raster scanning noise increases. No raster scanning noise can be seen when using WFPI.

The second set of samples were acquired on a sample with 29.4μm bow between the center low point and the edge high point (radius of the bow is 10m) (see Figure 6 below).

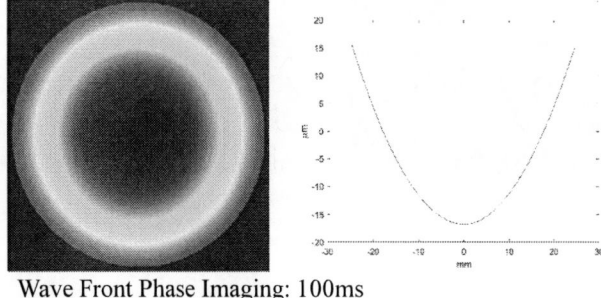

Wave Front Phase Imaging: 100ms

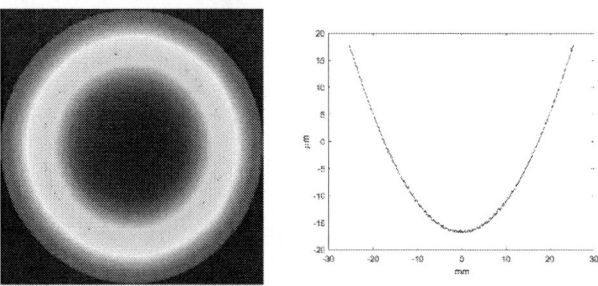

Chromatic confocal microscopy 1: 2 min [11]

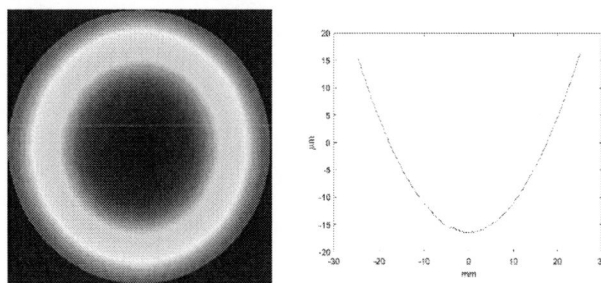

Chromatic confocal microscopy 2: 11 min [12]

Figure 6. Comparison of reference mirror measurement using three instruments. All data shown with a false color image and a line scan through the center of the mirror sample. Top: WFPI snapshot. Middle: chromatic confocal at higher speed raster scanning. Lower: chromatic confocal with lower raster scanning speed.

Due to the bow of the mirrors, it's harder to notice the raster scanning noise, however the noise is still present but more visible in the faster scanning speed data sets (Middle).

The last and third sample was a sample with warpage of 98.01μm from center low point to edge high point (radius of the bow is 3m, smaller radius gives higher warpage and large height difference between lowest and highest point) (see Figure 7 below).

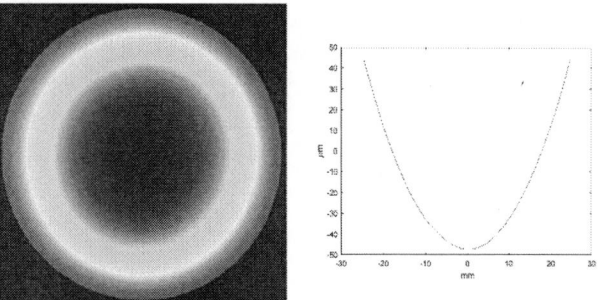

Wave Front Phase Imaging: 100ms

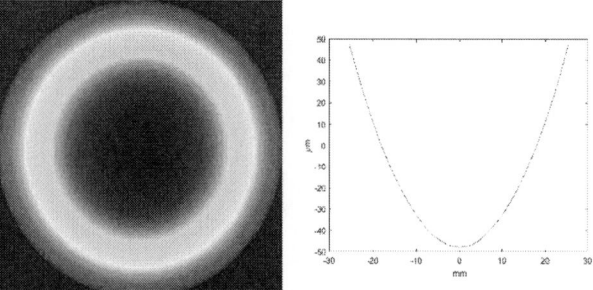

Chromatic confocal microscopy 1: 2 min [11]

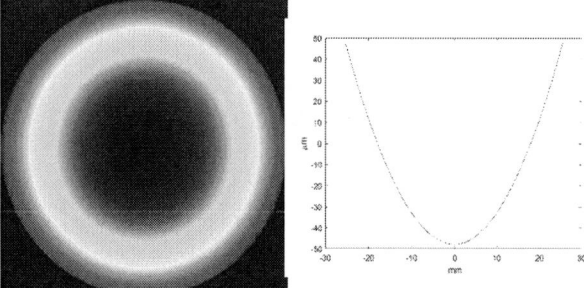

Chromatic confocal microscopy 2: 11 min [12]

Figure 7. Comparison of reference mirror measurement using three instruments. All data shown with a false color image and a line scan through the center of the mirror sample. Top: WFPI snapshot. Middle: chromatic confocal at higher speed raster scanning. Lower: chromatic confocal with lower raster scanning speed.

The data comparison demonstrates a strong correlation between WFPI and chromatic confocal microscopy.

WFPI OF BLANK SILICON WAFER ON FLAT SAMPLE HOLDER

A 2-inch blank silicon wafers were used for WFPI measurements while laying on a flat wafer holder without using vacuum leaving gravity as the only force working on the wafer. Since the sample does not move during the measurement no clamping or vacuum is required.

WFPI, using exposure time of 0.1 seconds, acquired a topography image from which a global depth map was generated and displayed in a color map (Fig. 8 Left) and a

3D map (Fig. 8 Right). A double Gaussian high pass filter was applied on the global topography map of the wafer to remove low frequency spatial resolution which revealed roughness with a spatial resolution equal to the pixel size (24μm) (Fig. 8 Bottom).

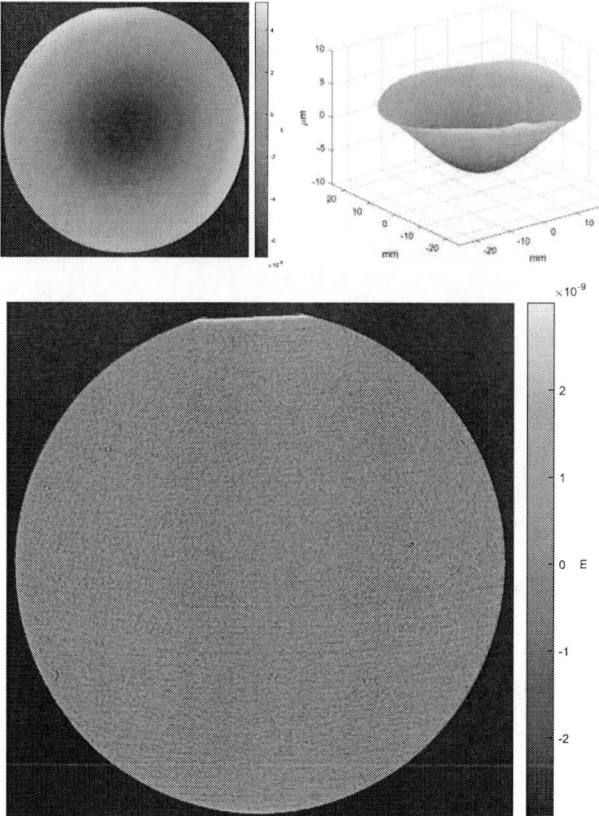

Figure 8. Upper images show the depth map and the lower image shows the high-pass filtered depth map.

The data collected was analyzed through the unique PC algorithm for WFPI to calculate the wave front phase and the associated amplitude (Z height). The topography data was then filtered using a double Gaussian to reveal the higher spatial frequencies clearly showing the low amplitude depth map associated with NT and roughness stemming from wafer polishing. The total time for image acquisition was about 0.1 seconds and the analysis algorithm took just over 2 seconds to run.

CONCLUSION

WFPI has shown to have great correlation with existing global wafer geometry techniques used by the semiconductor industry in both the lab and the fab. It also shows the noise improvements by not have any moving parts involved in the silicon wafer data acquisition.

WFPI was also proven to reveal NT and roughness by applying a high pass filter on the global topography data to reveal the higher spatial frequencies. With lateral resolution of 24µm and an amplitude sensitivity at 0.3nm while collecting data on the entire wafer in a single image snapshot. With an exposure time of 0.1 seconds with an added 2 seconds for data analysis, WFPI has the potential to become the only wafer geometry technique capable of revealing roughness at a speed fast enough to satisfy the extreme demands required by high volume silicon device manufacturing in wafer fabs.

The current WFPI system is made for samples with diameter of max 50mm. However, with different optics one can easily make a system that works on a full 300mm diameter sample, which is currently being built that will collect more than 10 million data points on a 300mm wafer with the same data acquisition time.

REFERENCES

[1] Pradeep Vukkadala, Kevin T. Turner and Jaydeep K. Sinhaa, "Impact of Wafer Geometry on CMP for Advanced Nodes", Journal of The Electrochemical Society (ECS), 2011

[2] SEMI M43-1109, "Guide for reporting wafer nanotopography," www.semi.org (Oct. 2009).

[3] T. A. Brunner, Y. Zhou, Ch. W. Wong, B Morgenfeld, G. Leino, S. Mahajan, Sunit Mahajan, "Patterned wafer geometry (PWG) metrology for improving process-induced overlay and focus problems", Proc. SPIE 9780, Optical Microlithography XXIX, 97800W (15 March 2016).

[4] V.Brouzet, V.Gredy, F.Chenevas-Paule, K.Le-Chao, D. Guiheux, A.Laurent, V.Coutellier and D.Le-Cunff, "Full Wafer Stress Metrology for Dielectric Film Characterization: Use Case", Advanced Semiconductor Manufacturing Conference (ASMC), 2019.

[5] C.Beitia, M. Abdel, M. Cordeau1, S.Godny, S.Petitgrand, D.Alliata, "Optical profilometry and AFM measurements comparison on low amplitude deterministic surfaces", Advanced Semiconductor Manufacturing Conference (ASMC), 2019.

[6] Juan Trujillo, Jose Manuel Ramos Rodrigues, Jan Gaudestad, "Wave Front Phase Imaging of wafer warpage", International Wafer Level Packaging Conference (IWLPC) San Jose, CA, USA (Oct. 2018)

[7] J.M. Trujillo-Sevilla, O. Casanova-Gonzaleza, S. Bonaque-González, J. Gaudestad, J.M. Rodríguez-Ramos "High-resolution wave front phase sensor for silicon wafer metrology", SPIE Photonics West, San Francisco CA, USA (Feb. 2019).

[8] J.M. Trujillo-Sevilla, J.M. Ramos-Rodríguez, J. Gaudestad, "Wave Front Phase Imaging of Wafer Warpage, Advanced new metrology technique for blank incoming wafers", Advanced Semiconductor Manufacturing Conference (ASMC), 2019.

[9] Joseph W. Goodman, "Introduction to Fourier optics". Roberts and Company Publishers (2005)

[10] Harding, Kevin, ed. "Handbook of optical dimensional metrology" CRC Press, 2013.

[11] https://www.cybertechnologies.com/en/products/ct-300/

[12] https://nanovea.com/profilometers/

[13] SEMI M1-0918 "SPECIFICATION FOR POLISHED SINGLE CRYSTAL SILICON WAFERS" (November 2017)

HOW TO IMPROVE 'CHEMICAL STOCHASTIC' IN EUV LITHOGRAPHY ?

Toru Fujimori

Electronic Materials Research Laboratories, R&D Management Headquarters

FUJIFILM Corporation

4000 Kawashiri, Yoshida-Cho, Haibara-Gun, Shizuoka, 421-0396, Japan

toru.fujimori@fujifilm.com

ABSTRACT

Extreme ultraviolet (EUV) lithography is almost ready for realize 7nm generation manufacturing and beyond. A key factor for the realization of EUV lithography is the choices of EUV resist materials that are capable of resolving below 15nm half pitch with high sensitivity. However, the performances of EUV resist materials are still not enough for the true HVM requirements. One critical issue is 'Chemical stochastic', which will be become 'defectivity'.

We report herein how to improve 'Chemical Stochastic'.

INTRODUCTION

EUV lithography is most promising process for alternative to 193 nm immersion multiple lithography, realizing 7 nm generation node manufacturing and beyond. With recent rapid progress on source power improvement [1], process and material explorations are more and more accelerated to achieve HVM requirements. Therefore, a key factor for the realization of EUV lithography is the choice of EUV resist material that is capable of resolving below 15-nm half pitch with high sensitivity [2]. However, the performance of EUV resist is still not enough for the true HVM requirements, even by using the qualified EUV resist materials. One critical issue is 'stochastic issues', which will be become 'defectivity' [3].

Previously, the stochastic issue was basically considered from low photon number from EUV source, which means 'photon shot noise'. However, the stochastic issue was not only from them but also from EUV resist materials, called 'chemical stochastic' [4].

The present study aims to clarify how to improve 'Chemical Stochastic'. Accordingly, we investigated the reduce of photon shot noise effect and also reducing 'Chemical Stochastic' by using new designed EUV resist materials.

RESULTS AND DISCUSSION

1. Concept of the reducing stochastic effect

In the past, speaking of the stochastic issue was basically considered from low photon number from EUV source, which means 'photon shot noise'. It was still critical concerning point of the stochastic, even with recent progress on source power improvement. Also, several researchers aiming high EUV absorption system were reported to improve 'photon shot noise' stochastic [5-8], but material design and its lithographic performance still need detailed investigation.

The other hand, EUV resist materials location randomness in the film was also concerning to be stochastic to become defectivity.

2 (two) kinds of stochastic factor were focused to improve the lithographic performance, especially the defectivity.

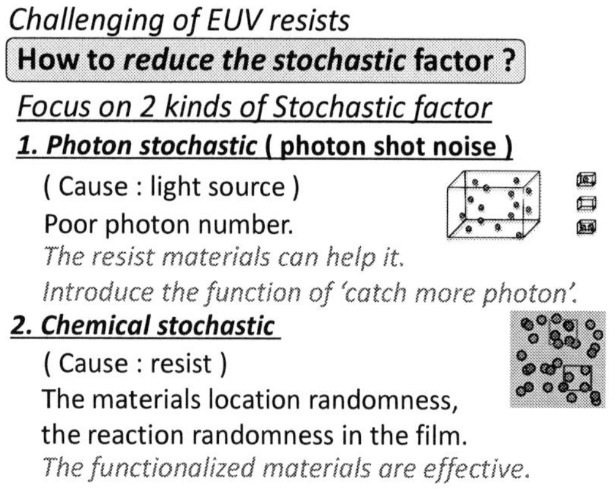

Challenging of EUV resists

How to *reduce the stochastic* factor ?

Focus on 2 kinds of Stochastic factor

1. Photon stochastic (photon shot noise)

(Cause : light source)

Poor photon number.

The resist materials can help it.

Introduce the function of 'catch more photon'.

2. Chemical stochastic

(Cause : resist)

The materials location randomness, the reaction randomness in the film.

The functionalized materials are effective.

Figure 1: The concept and focus points of improvement of the lithographic performance.

The stochastic factor, which is basically same as the source of randomness, observed at every process steps.

The 1st step was resist materials coating procedure, which observed materials location randomness. The 2nd step was exposure procedure, which observed photon randomness due to their low photon numbers. The 3rd step was PEB (Post Exposure Bake), which observed reaction site/area randomness. The final step was development, which observed dissolving randomness with developer.

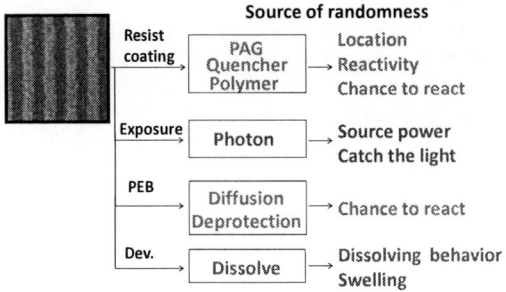

Figure 2: The source of randomness consideration.

2. The results of photon stochastic noise reducing by using 'Organic high EUV absorption materials'

Photon shot noise effect on EUV lithography was well-known issue and various studies have emphasized its impact on LWR performance [9]. As well as LWR, the photon shot noise effect on defectivity have to be considered to obtain pattern quality satisfying HVM requirements.

We designed 'Organic high EUV absorption materials' to be able to catch photon more efficiency from the source. For lithographic performance evaluation of them, we synthesized new polymer using high absorption monomer unit [10]. High absorption resist showed 20% dose reduction with keeping its LWR value. Besides that, nano-bridge is clearly decreased on the high absorption resist.

Organic high-EUV absorption CAR

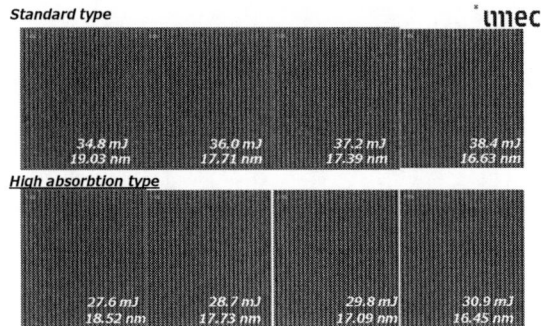

High Abs. resist showed *better bridging performance*.

Figure 3: The lithographic performance of 'Organic high EUV absorption materials', which showed excellent performance.

3. The results of 'Chemical Stochastic' reducing by using the novel functionalized materials

In the past, speaking of the stochastic issue was photon shot noise; however, it was not only from them but also from EUV resist materials themselves, called 'chemical stochastic'.

For 7 nm node application, traditional chemically amplified resist (CAR) system is first candidate because of its well-studied property through KrF and ArF manufacturing process for many years, e.g. stability, metal contamination and post-litho process compatibility. A traditional CAR material is mainly composed from polymers, PAGs (photo acid generator) and quenchers. The loading amount of the materials was almost 10: 1: 0.1. In case, the materials location locality could be observed in the film. For instance, there are PAG rich area and / or Quencher rich area in the film (polymer matrix). PAG rich area observed higher react with polymer and Quencher rich area observed lower react with polymer. Consequently, the solubility difference part might be formed in the film, polymer matrix.

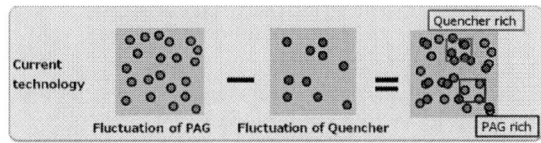

Figure 4: The image figure of 'chemical stochastic'.

We designed the novel functionalized materials, which were connection unit introduced materials, interaction unit introduced materials, and / or bounded materials.

These novel functionalized materials introduced photo resist can be avoided the aggregation of each chemical.

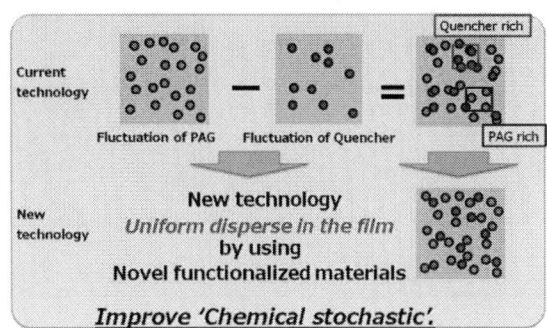

Figure 5: The concept of novel designed functionalized materials to reduce location randomness in the film.

As a result, new resist, which introduced new designed materials, observed excellent lithographic performance. The previous generation pitch (CD size), which means 18 nm CD size, showed not so different between current technology's one and new technology's one. However, in the real EUV generation, like 14nm CD size, it looks totally different. There was lots of kind of defectivity, like pinching, bridging and collapsing, in 14 nm by using current technology. This is the most critical issue for realize EUV lithography HVM. It will be expected to resolve to the issue by using new technology, which described in this study.

Lithographic performance

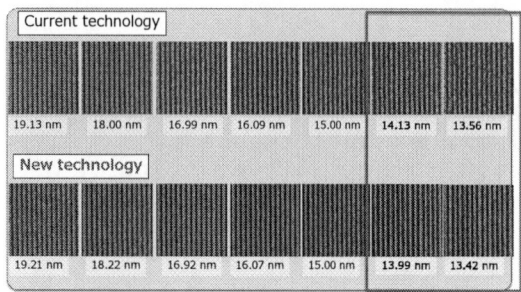

New technology, reducing 'Chemical Stochastic', showed better performance especially at smaller CD region.

Figure 6: The lithographic performance with new resist, which by using new technology to reduce the location randomness.

CONCLUSION

This study hereby showed excellent improving the lithographic performance due to reduce the stochastic issue.

'Organic EUV high absorption materials' and new technology and materials to reduce 'Chemical Stochastic' were expected to apply the real EUV lithography HVM.

ACKNOWLEDGEMENTS

The full-field EUV exposures were done on the ASML NXE:3300 scanner at imec, as part of the imec Advanced Lithography Program. We would like to thank Danilo De Simone and Geert Vandenberghe for their support with the experiments at imec and helpful discussions.

REFERENCES

[1] A. A. Schafgans, D. J. Brown, I. V. Fomenkov, Y. Tao, M. Purvis, S. I. Rokitski, G. O. Vaschenko, R. J. Rafac, D. C. Brandt, *Proc. SPIE **10143**, 1014311 (2017).*

[2] H. Furutani, M. Shirakawa, W. Nihashi, K. Sakita, H. Oka, M. Fujita, T. Omatsu, T. Tsuchihashi, N. Fujimaki, and T. Fujimori, *J. Photopolym. Sci. Technol. (2018).*

[3] P. De Bisschop, J. Van de Kerkhove, J. Mailfert, A. Vaglio Pret, J. Biafore, *Proc. SPIE **9048**, 904809 (2014).*

[4] T. Fujimori, *International symposium on extreme ultraviolet lithography 2019.*

[5] H. Tsubaki, W. Nihashi, T. Fujimori, T. Tsuchihashi, K. Yamamoto, T. Goto, *Proc. SPIE **9776**, 977608 (2016).*

[6] J. Jiang, D. D. Simone, G. Vandenberghe, *Proc. SPIE **10146**, 101460A (2017).*

[7] S. Higashino, A. Saeki, K. Okamoto, S. Tagawa, T. Kozawa, *J. Phys. Chem. A 114, 8069-8074 (2010).*

[8] O. Ongayi, M. Christianson, M. Meyer, S. Coley, D. Valeri, A. Kwok, M. Wangner, J. Cameron, J. Thackerey, *Proc. SPIE **8322**, 83220T (2012).*

[9] S. Bhattarai, W. Chao, S. Aloni, A. R. Neureuther, P. Naulleau, *Proc. SPIE **9422**, 942209 (2015).*

[10] T. Fujimori, *International symposium on extreme ultraviolet lithography 2018.*

IMPACTS OF RTP PYROMETER OFFSETS ON WAFER OVERLAY RESIDUE

Jian Lv [1*], Qing Wang [1], Yetao Lu [1], Xing Gao [1] and Qin Sun [1]

[1] Shanghai Huali Microelectronics Corporation, Shanghai 201203, China

*Corresponding Author's Email: lvjian@hlmc.cn

ABSTRACT

Throughout the history of rapid thermal process (RTP), the solution to wafer distortion has been a focused issue of researchers. RTP featuring high temperatures, fast ramp rates and high strain rates always results in the deformation of the silicon wafer substrate, which in turn causes lithographic overlay errors. This problem has become more and more serious in recent years as wafer size keeps increasing and device geometry continues shrinking. Even though a number of events have been documented, an investigation of more subtle correlation between overlay failure and RTP is in urgent requirement. Herein, the effects of RTP process conditions on overlay accuracy in the fabrication of 55 nm e-flash memory gate dielectric were studied. Results from this investigation demonstrated that pyrometer offsets at wafer edge influence overlay residue significantly. The overlay vectors shrink with the decrease of pyrometer offset delta between probe seven and probe six. Moreover, this offset delta had to be controlled between 2.5 °C to 4 °C in order to ensure the oxide thickness uniformity and tolerable overlay residue.

INTRODUCTION

As the device geometries and critical dimensions in semiconductor Integrated Circuit (IC) fabrication continue shrinking, RTP process gradually replaced long-time furnace steps during past decades. Nowadays, RTP process has been widely used in high temperature silicon oxidation, silicidation and implant dopant activation [1-3]. However, this process always induces wafer distortion due to the thermal non-uniformity across the wafer, especially at the wafer edge location. In the subsequent masking step, the RTP induced wafer distortion will result in severe overlay residue [4, 5]. Temperature non-uniformity across the whole wafer has been widely accepted as the root cause for wafer distortion [6]. Although RTP tool has been equipped with multi-zones tuning system to optimize the temperature uniformity, the overlay requirement of device has been advancing at higher speed due to the increase of wafer size and pattern densities [7, 8]. Up to today, one of the major tasks in RTP process is to minimize the overlay residue while achieving the best film thickness uniformity.

Herein this work, we describe some study aimed to solve the RTP induced overlay residue issue during the development and manufacture of 55nm e-flash memory device. As previously reported, the peak temperature, soak time, gas pressure of RTP process and many other factors would affect overlay residue. We will focus on a few aspects of RTP process and evaluate their effects on overlay residue. Also, we give a simple but effective method to balance the RTP induced overlay residue and oxide thickness uniformity.

EXPERIMENTS

A 55 nm e-Flash memory device lot was adopted as the vehicle to investigate the effects of RTP process on photolithography overlay residue. Herein, the experiment was conducted on a 300 mm ISSG system. During ISSG oxidation, wafers were loaded on a ceramic edge ring, rotating at a speed of ~200 rpm. The ISSG chamber was heated by 409 tungsten halogen lamps located on the front side of chamber. The temperature was measured by seven pyrometers located at the backside of the wafer. Herein, the ISSG oxidation temperature was 1000 °C. With such ISSG system, we could optimize the oxide thickness uniformity within wafer by tuning the temperature of different wafer zones through the adjustment of seven, independent, recipe specific, "temperature offsets". The seven offsets were applied to modify the temperatures of seven zones of wafers, which would affect the oxide growth rate and the deformation of wafers. Herein, we focused on adjusting the temperature offsets of probe seven and probe six. The effects of offset delta between probe seven and probe six on thickness uniformity and overlay residue were investigated carefully.

RESULTS AND DISCUSSION

Prior to performing investigations on overlay residue issue, the correlations between oxide thickness uniformity and offsets had to be established firstly. During our experiments, the offset delta between probe seven and probe six (T7-T6) was adjusted from -0.5 °C to 5.5 °C. Fig. 1a shows the oxide thickness full map of the wafer run at T7-T6 = -0.5 °C. It can be founded that the oxide at the wafer edge was obviously thinner than inner oxide. Thus we could conclude that the temperature of the wafer's edge is too low when the offset delta equals to -0.5 °C. When the value of offset delta increases to 4 °C, the thickness of wafer edge oxide became close to wafer inner oxide. The increase of the temperature of wafer edge detected by pyrometer seven accelerated the growth of wafer edge oxide, and as a result, the oxide thickness uniformity within wafer was improved.

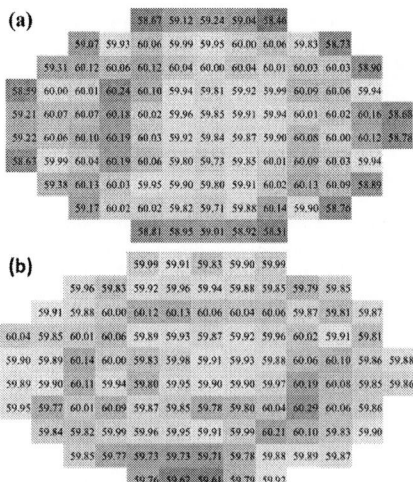

Figure 1: The oxide thickness full map of the wafer run at (a) T7-T6 = -0.5 ℃; (b) T7-T6 = 4 ℃.

Table 1 shows the correlations between oxide thickness uniformity and offset delta. At first, the range of oxide thickness within the whole wafer was 1.78A when offset delta (T7-T6) was -0.5 °C, which meant the oxide thickness uniformity was unacceptable. However, the range of oxide thickness within the whole wafer gradually decreased from 1.78A to 0.69A while the offset delta increased from -0.5 °C to 4 °C. It seems that the edge of wafers loaded on ceramic edge rings are colder than wafers' inner area, which could be attributed to the thermal transmission between wafers and edge rings. The increase of T7 offset decreased the temperature variation between wafer edge and wafer inner area, and thus, the thickness uniformity of oxide film within wafer was optimized. However, the range of oxide thickness began to increase reversely when offset delta exceeded 4.5 °C. So the offset delta should be kept within 4.5 °C in order to optimize the film thickness uniformity.

TABLE I. THE OXIDE FILM THICKNESS RANGES OF WAFERS RUN AT DIFFERENT OFFSET DELTAS.

T7-T6 (°C)	MAX (A)	MIN (A)	FM Range (A)
-0.5	60.24	58.46	1.78
0.5	60.22	58.82	1.40
1.5	60.25	59.15	1.10
2.5	60.05	59.19	0.86
3.5	60.15	59.41	0.74
4	60.29	59.61	0.69
4.5	60.28	59.58	0.70
5.5	60.71	59.65	1.06

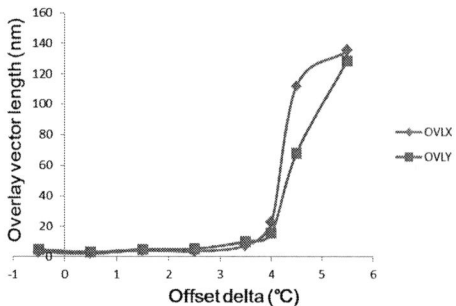

Figure 2: The overlay vectors length of each tested wafer run at different temperature offsets.

Fig 2 shows the overlay vectors measured from each tested wafer. It could be found that the overlay vectors at both X direction and Y direction were less than 20 nm while the values of offset delta were within 3.5 °C. However, the overlay vectors dramatically increased when the offset delta exceeded 4 °C. We observed that the overlay vectors length at both directions exceeded 120nm when offset value was 5.5 °C, which would cause severe misalignment at a subsequent masking step. The data also showed that negative offset value yielded a good result. We believed that the relatively low edge temperatures helped to relieve thermal stress during ISSG process.

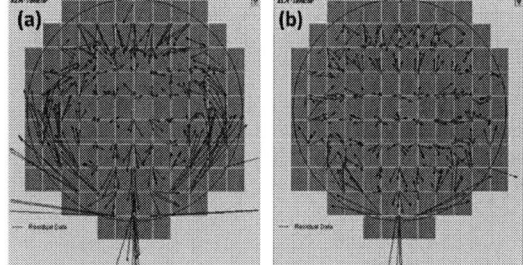

Figure 3: Overlay residue maps of wafers when offset delta value was (a) 4 ℃ and (b) 0.5 ℃.

From above data, we could conclude that the offset delta should be controlled within 3.5 °C in order to minimize the overlay vectors length. The best result was obtained when offset delta was 0.5 °C. When offset delta exceeded 4 °C, the overlay vectors began to increase rapidly. Fig 3 showed that the overlay residue of wafers were quite sensitive to offset delta value. A small change of offset delta value would cause obvious variations of overlay residue map. The two overlay maps revealed a significant improvement of the overlay residue when offset delta decreased from 4 °C to 0.5 °C, especially the overlay vectors at the wafer edge.

CONCLUSION

In summary, we systematically investigated the

effects of offset delta between probe seven and probe six on oxide thickness uniformity and overlay residue. It was found that the overlay vectors length would increase dramatically when the offset delta value exceeded 4°C. The overlay vectors of wafers at both X direction and Y direction were lower than 20 nm if the offset delta was under 4 °C. On the other hand, the offset delta should be controlled above 2.5 °C in order to keep the range of oxide thickness within wafer under 1A. Otherwise, the oxide thickness uniformity was unacceptable. Herein, we found that the offset delta between probe seven and probe six had to be controlled between 2.5 °C to 4 °C to meet the requirements of thickness uniformity and overlay residue.

ACKNOWLEDGEMENTS

The authors wish to thank Shanghai Huali Microelectronics Corporation for the support of this work.

REFERENCES

[1] R. Kakoschke. *Mat. Res. Soc. Proc.*, vol. 224, 1991, pp. 159- 170.

[2] J. C. Gelpey, K. Elliott, D. Camm, S. McCoy, J. Ross, D. F. Downey and E. A. Arevalo. *Electrochem. Soc. Proc.*, vol. 11, 2002, pp. 313-324.

[3] W. Skorupa, R.A. Anwand, W. Anwand, M. Voelskow, T. Gebel, D. F. Downey and E. A. Arevalo. *Mater. Sci. Eng. B.*, vol. 114, 2004, pp. 358-361.

[4] J. F. Buller et al., *IEEE. T, Semiconduct. M.*, vol. 9, 1996, pp. 108-114.

[5] R. Deaton and H. Z. Massoud. *J. Appl. Phys.*, vol. 70, 1991, pp. 3588-3592.

[6] M. Akatsuka et al., *J. Electr. Soci.*, vol. 146, 1999, pp. 2683-2688.

[7] J. F. Shepard, W. A. Muth and S. Macnish. *Proc. 18th IEEE Conf. on Advanced Thermal Processing of Semiconductors.*, RTP2010.

[8] J. P Hebb and K. F. Jensen. *IEEE. T, Semiconduct. M.*, vol. 2, 1998, pp. 99-107.

APPLICATIONS OF SPARSE AND COMPACT RESIST MODELING IN ADVANCED NODE IMPLANT LAYER

Shirui Yu, Mudan Wang[*], Yiqun Tan, Juan Wei, Renyang Meng

Technology Development, Shanghai Huali Integrated Circuit Corporation, Shanghai, Chinacompact

*Corresponding Author's Email: wangmudan@hlmc.cn

ABSTRACT

OPC model, basically consisting of optical and resist model, has been regarded as an effective and popular resolution enhancement technique. With the layout of integrated circuit to more advanced technology node and more complicated design, resist model has been evolving from traditional sparse simulation model to compact simulation model due to powerful computation function. As the pattern design of implant layer is relatively simpler compared with AA/PO, sparse VT5 model and compact CM1 model can still coexist for use with respective typical characteristic. Sparse VT5 model consume less runtime due to simpler function calculation. Compact CM1 model exhibits higher accuracy due to taking more items into account. A suitable model has to achieve good balance between the model accuracy and real runtime during mask tape-out. VT5 and CM1 are both calibrated in advanced node (28/14 nm) implant resist modeling, and systematic performance comparison between the two models are made from methodology, accuracy, stability and simulation runtime based on a design layout. CM1 model shows obvious advantages in lower node implant layer resist modeling. It provides good guidance for other implant layers' resist modeling with thick photoresist.

Keywords—resist modelling; sparse model; compact model; advance node; implant layer

1. INTRODUCTION

OPC model is a modeling and simulation technique of actual lithography process which consists of sophisticated optical system and chemical reactions occurred in resist. Accordingly, OPC model usually divides into optical and resist models two part. An optical model is a physical and optical process of a lithographic exposure system, which models several constituents of the system such as illumination source, optical system, kernel parameters, image parameters, pupil parameters and film stack. A resist model is a chemical process that express series of complicated photochemical reactions in resist after exposure. A resist model uses the aerial image, the output of the optical model, as the input to predict the threshold at which edges print. VT5 model calculates the wafer print image threshold as a function of four image parameter terms, such as slope, max intensity (I_{max}), min intensity (I_{min}), and factor, plus one or more convolution kernels. It generates the resist model by optimizing the function's polynominal coefficients to best fit the empirical measurements from a test chip. CM1 model takes a combination of up to 16 terms into account to describe the exposure threshold of the resist. It adds the consideration of physically- or empirically-based terms, such as curvature, acid/base concentration, diffusion of acid/base and so on. The difference between VT5 and CM1 is that the VT5 methodology uses a variable threshold printing test at each layout location, however, CM1 methodology fits to a single global threshold.

As the IC industry critical dimensions decrease to 28 nm and beyond, OPC model has developed from traditional variable threshold models such as VTR, VTRE and VT5 to compact model (CM1) for most key layer. However, VT5 is still used in some implant (IMP) layers such as well layer. In this paper, VT5 and CM1 are both applied to calibrate in 28nm implant resist modeling, and the performance comparison between the two models are made from methodology, accuracy, stability and simulation runtime based on a design layout. CM1 model shows obvious advantages in advanced note implant resist modeling. It provides guide for advanced node implant layer resist modeling with thick photoresist effect.

2. PERFORMANCE COMPARISON

2.1 METHODOLOGY

VT5 model uses a relatively simple mathematical function to predict the image threshold at which edges print. Figure 1 illustrates the physical meaning of few image parameter terms. VT5 is site based resist model. At each given site the aerial image is used to determine the value of each term. The common threshold of VT5 model can be calculated by the following formula:

$$T_V = f(I_{max}, S_{slope}, F_{actor}, I_{min}, I_{slope}, D_1, \ldots, D_n) = \sum c_j \cdot \prod_i Param_{i,j}^{power_{i,j}}$$

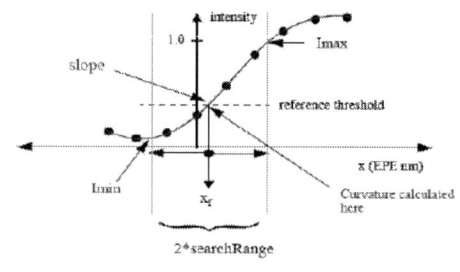

(default searchRange = .5*lamda/NA)

Figure 1: Common image parameter items of VT5 model

CM1 model is grid based. It uses a single threshold to describe the exposure threshold of the resist based on resist surface profile. The resist surface is generated by performing a summation of modified aerial image surface by a variety of kernels, such as Neutralization, Diffusion, Acid/Base concentration, Imin, Imax, Slope, Curvature and G-L convolution. It is regarded as an analogue of SOCS, namely, the sum of coherent system approximation of the TCC function. The threshold of CM1 model can be expressed by the following equation:

$$T = \sum_{i=0} c_i M_i(I) \quad (T = const)$$

$$M = \left((\nabla^k I_{+b})^n \otimes G_{s,p} \right)^{\frac{1}{n}}$$

Where Gs,p is a mixture of Gauss and Laguerre equations, s is the diffusion length, p is the order of the equation, b represents acid or base, n is also the order of equation.

Figure 2 simply illustrates series of items involved in CM1 modeling and their respective mathematical expressions and physical meanings. AI is the base term in CM1 model. It will be taken as the input of each items of Gauss-Laguerre equations and finally obtain the resist image after integrative convolution contribution of these items. There are many available model forms in CM1. The difference lies in the number of calculation items. The more the item is, the more accurate the model is. As a result, longer run time model simulation requires. In reality, a good compromise should be balanced between accuracy and model simulation speed during product tape-out period. Therefore, a suitable model form should be adopted carefully

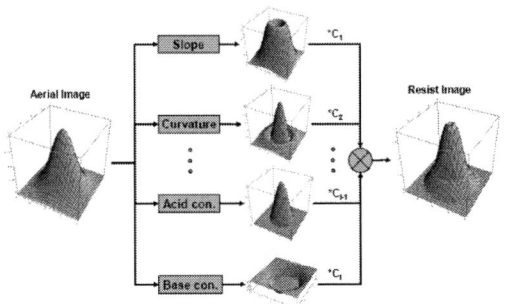

Figure 2: A simple illustration of compact CM1 modeling item.

2.2 ACCURACY

An OPC model generation process mainly includes two parts. One part is the CD data collection from wafer via high-magnification CDSEM measuring. Another part is the model calibration using the measured CD of selected test mask pattern printed on wafer by photolithography. Taken the technology node attributes and practical situation occurred in logic chip into consideration, we select following pattern structures to collect model data.

Table 1 shows the data structure of 28 implant model. Three kinds of test pattern structures are counted in 28 implant model calibration data. The ratio of one-dimension (1D) pattern to two-dimension (2D) pattern is 2:1. The percentage of each structure occupied in total data points is also listed in table 1. Specification, also simplified as spec, has been regarded as one crucial criterion to evaluate the accuracy of model calibration. There is some difference about the spec between VT5 and CM1. For VT5, the model spec refers to the error per edge (epe) between simulation CD and measured CD. While for CM1, the model spec refers to the CD error between simulation value and measured value. As a result, the CM1 model spec is twice of that of VT5, as shown in table 1.

Table 1 CDSEM Data structure of 28 implant model

Dimension	Structure	Percentage (%)	Model Spec	
			VT5	CM1
1D	Linearity	6	±6.5	±12
	Line Proximity	32		
	Space Proximity	30		
2D	Linearity	8	±13	±24
	Line Proximity	12		
	Space Proximity	12		

Figure 3 shows the model calibration results of VT5. Each color represents one specific pattern structure. Most data points are well controlled in the model spec range except few points. The few OOS (out of spec) points are marked in circle, as shown in figure 3. These points contain the small CD of Iso Line, large CD of Dense Line and large pitch of proximity patterns, such as Line Pad, Many Space, 3 Space and Space Pad, as well as some 2D patterns. The model fitting error at the main part of each structure group basically locates close to zero. It means good model consistency and prediction between measurement value and simulation value.

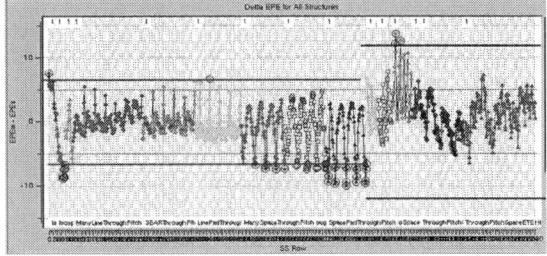

Figure 3: The fitting result of sparse VT5 model

Same model data was applied to calibration using CM1 model. Figure 4 shows the corresponding model fitting result. The structure plotted in figure 4 keeps the same as that of VT5. The whole trend is similar to that of VT5 fitting result. A common phenomenon was observed in both model types, that is the simulation error rise extremely high at large pitch for 1D proximity structure patterns. It probably dues to long proximity range effect. For CM1 model, however, those OOS points occurred in VT5 model are all disappeared and all model data points are well controlled in spec. It can be concluded that CM1 model plays an important role in the improvement of long proximity range effect.

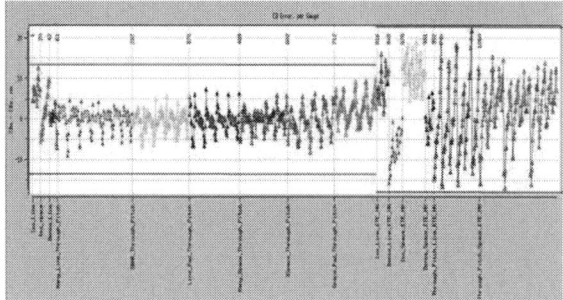

Figure 4: The fitting result of compact CM1 model.

2.3 STABILITY

Model stability is another very important factor in practical application. According to accumulated industry experience, some good practices can help to improve model stability. For VT5 model, the parameter eigenvalue can be used to judge model stability. A suitable value is between 0.01 and 0.001. Smaller may cause overfitting, while larger may results in underfitting. Another way to check the model stability is to check the threshold coefficients graph in the analyze fitness dialog box.

CM1/AI difference, the difference between resist and optical image intensity, is widely used to evaluate the stability of CM1 model. At current 28 and more advanced note, the recommended value is 10%, which means the maximum acceptable difference. If the value is much larger than 10%, the model has a high probability of producing artificial extra printing features or false holes in the contour. In addition, accompanying setting of 1.5 for threshold tolerance field is recommended as a pair of setting. The final values of above mentioned parameters for VT5 and CM1 are all within the recommended value range, indicating good model stability.

2.4 RUNTIME

The ultimate application of OPC model is used to correct mask. What we care about not only is model accuracy and stability, real run time for model simulation is also of great importance during product mask tape-out.

Especially for implant layers, as there are many different implant layers sharing a same model in common, long run time will influence the tape-out schedule and even the whole product manufacture. A complete OPC process containing five steps, namely, TDOPC, TDOPC verify, MBOPC, OPCV and MRC. Figure 5 shows the run time percentage of each step occupied in whole OPC process. Among them, only MBOPC and OPCV will use OPC model for simulation. The total run time of the two steps comes to one third of the whole OPC running time, which cannot be ignored. A run time test was conducted based on a test chip. Table 2 lists the run time test results of MBOPC and OPCV in VT5 and CM1, respectively. The run time of MBOPC and OPCV for CM1 is higher than VT5. It mainly ascribes to the advanced and sophisticated mathematical calculation of Gauss-Laguerre convolution equation in CM1 model.

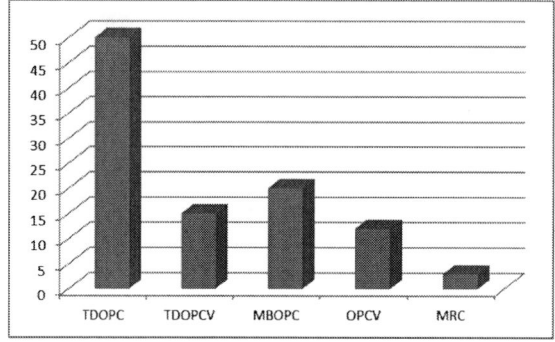

Figure 5: The run time percentage of each step in whole OPC process

Table 2. Run time comparison between VT5 and CM1

Item	Model-based OPC		OPC verify	
Model type	VT5	CM1	VT5	CM1
CPU count	260		260	
Run time(h)	0.5	0.8	0.5	1.2
Percentage(%)	9	14	10	18

3. CONCLUSION

In this paper, we have investigated the difference between two common resist model sparse VT5 and compact CM1 from several important aspects, such as methodology, model accuracy and stability, as well as real model simulation run time. Compared with VT5, CM1 takes more items into consideration and adopts more advanced and complicated Gauss-Laguerre equation. As a result, compact CM1 resist model displays better result with all data points within spec and accordingly it requires longer run time in return. It was found that CM1 model plays an important role in the improvement of long proximity range effect in 28 and more advanced node implant case. Compact CM1 model was regarded as a better one choice considering the balance between model accuracy, prediction and runtime. Further model verify on wafer level is still on going to confirm the conclusion.

REFERENCES

[1] Quan Chen, Shirui Yu, Zhibiao Mao, Yu Zhang, Bin Gao, Yanpeng Chen, Albert Pang. Mask model analysis and its application in 28 OPC modeling, 2015 China Semiconductor Technology International Conference, 2015, 1- 3.

[2] Yaojun Du . The comparison of the effectiveness of model-based SRAFs and rule-based SRAFs, 2016 China Semiconductor Technology International Conference (CSTIC) , 2016, 1 - 7.

[3] Yiqun Tan, Weiwei Wu, Quan Chen, Shirui Yu. Application of resist profile model and resist-etch model in solving 28nm metal resist toploss, 2017 China Semiconductor Technology International Conference (CSTIC), 2017, 1 − 3.

[4] Shanhu Shen, Chunlei Xie, Zheng Shi, Xiaolang Yan. OPC model calibration based on circle-sampling theorem, 2006 8th International Conference on Solid-State and Integrated Circuit Technology Proceedings, 2006.

STUDY OF ALIGNMENT & OVERLAY STRATEGY IN 14 NM LITHOGRAPHY PROCESS

Lulu Lai*, Rui Qian, Biqiu Liu, Xiaobo Guo, Cong Zhang, Jun Huang, Yu Zhang J
Shanghai Huali Integrated Circuit Corporation, Shanghai, China
*Corresponding Author's Email: lailulu@hlmc.cn

ABSTRACT

A more accurate and precise control of overlay performance in lithography process is required as design rule shrinks. Overlay performance is mainly determined by alignment and overlay measurement process, of which alignment and overlay marks play an important role. SADP (Self-aligned double patterning) process becomes widely adopted to realize half pitch of original design for 14nm technology node and beyond. The alignment and overlay marks formed by SADP process differ from traditional ones, which should be well designed to better comply with process condition and reduce the pattern loading effect induced by CMP and ETCH process, and eventually improve overlay performance. In this paper, the alignment behavior of different alignment marks formed via SADP process is investigated. On the other side, the overlay performance of segmented overlay marks is designed and compared with traditional ones to reveal the effect of segmentation on improving the overlay measurement precision and accuracy.

INTRODUCTION

For advanced integrated circuit manufacturing, overlay performance in lithography process has been an extremely critical factor. Overlay performance is greatly influenced by scanner alignment and overlay measurement process, which are achieved with the assistance of the corresponding marks. Alignment mark and overlay mark, which are laid on scribe line under most circumstances, are of the most key factor for overlay control. It's known that alignment mark and overlay mark with diverse design respond differently to the same process condition [1]. The accurate alignment process of scanner guarantees the overlay performance between previous and current layers; on the other hand, precise overlay measurement can well reflect the real overlay and assure correct feedback to the scanner.

Immersion tools with wavelength of 193nm cannot fulfill the design demanded, therefore SADP Process, which employs single lithography process followed by self-aligned spacer and etch [2], is used to achieve pitch doubling in 14 nm node and beyond. The original alignment and overlay marks with micrometer size will be converted into fence like structures with tens of nanometers in width, which is likely to peel in the subsequent processes and even become defect source [3]. Since overlay measurement precision is highly dependent on the mark contrast for image-based overlay measurement, those fence-like marks are considered unable to generate robust contrast. Besides, even though alignment process is diffraction based, the signal formed by fence-like structures is also considered too weak to assurance alignment accuracy. Therefore, mark segmentation of SADP process is essential to get robust signal and contrast for health alignment and overlay measurement process, respectively. For current layers, especially with small design rules, segmented marks are expected to gain better process compatibility than non-segmented marks. Mark segmentation should also be considered to match the scanner illumination to make sure the formation of desired profile [4].

In this paper, the alignment marks produced by SADP process are evaluated and the optimal mark is selected to guarantee the accuracy of wafer alignment process and improve overlay performance. What's more, the segmentation split of overlay mark is investigated to reveal the effect of segmentation on precise overlay measurement.

EXPERIMENT

The pre-layer overlay mark and alignment mark was formed by SADP process so that mark segmentation is required to avoid fen-like structure and form mark with robust contrast and signal. The core layer of SADP process and inner part of the mark were both exposed by ASML NXT1965i. The overlay measurement was carried out by KLA-Tencor Archer-600 and the sampling was executed with 9 sites per shot and 13 shots per wafer, and total 117 overlay targets were measured per wafer.

Alignment mark design

The alignment marks (AM) chose for evaluation is labeled as AM1, AM2 and AM3, illustrated in Figure1. The marks have been segmented to avoid fence like structures, which will greatly weaken alignment signal, and the segmentation follows minimum design rule. Mark segmentation direction follows main pattern direction. Optical proximity correction is applied to the alignment marks to avoid pattern collapse or bridge defects.

Overlay mark design

Image based overlay mark (AIM) is used and mark segmentation is applied. The pre-layer mark, which is always segmented, has fine CD/pitch following minimum design rule, while coarse CD/pitch varies according to our mark design.

The definition of coarse and fine CD/pitch is shown in

Figure2. The segmentation of current layer also follows its minimum design rule and the segmentation direction of both pre-layer and current layer follow main pattern direction. Optical proximity correction is applied to the overlay marks as well.

TABLE I.   OVERLAY MARK LAYOUT SUMMARY

Mark No.	T1	T2	T3	T4	T5
Current layer Segmentation	No	No	No	Yes	Yes
Coarse&Fine Pitch/ nm	1300	1700	2200	1300	1700

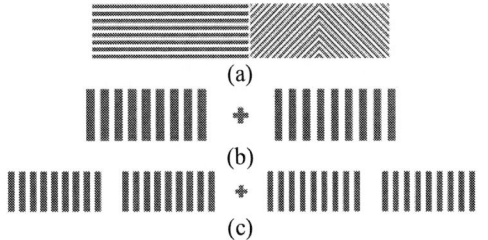

Figure 1: Alignment mark evaluated in this paper: (a) AM1 combination mark; (b) AM2; (c) AM3

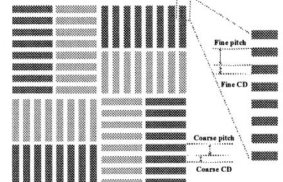

Figure 2: Coarse and fine CD/pitch illustration of AIM mark

Five overlay marks with top ranking simulation results were chose to be placed on the scribe line (Table 1), including segmented and non-segmented current layer marks while previous layer marks keep segmented. For a certain overlay mark in this paper, the inner mark and outer mark share same coarse pitch and close CD (CD may not be exactly the same because of different fine CD/pitch). The coarse CD is close to half of the pitch size.

RESULT AND DISCUSSION
Alignment mark evaluation

Diverse alignment mark design has different response to current process, and in this paper AM1/AM2/AM3 combination marks are chosen as candidate to evaluate and the optimal alignment strategy is defined.

Figure 3 shows two key parameters of coarse alignment, wafer quality (WQ) and multiple correlation coefficients (MCC) under different illumination. AM3 outperformed to AM2 and AM1 in WQ and MCC. For mark AM3, WQ performance sequence is FIR > Red > Green > NIR. Fine alignment signal cannot be acquired by

AM1 combination mark even with its WQ and MCC of coarse alignment meet the criteria. It's inferred that other factors defined by scanner recipe or machine constant failed to pass the corresponding criteria. Hence, this mark is not suggested to use as coarse alignment mark for smash mark.

For fine alignment, AM3_FIR_5th exhibited best performance with the combination of high WQ and small WQ standard deviation (StdDev), along with comparable residue overlay performance indicator (ROPI), shown in Figure 4. AM1_Red_1st has WQ a little bit higher than AM3_FIR_5th while the standard deviation is larger as well. Therefore, considering WQ, WQ StdDev, MCC and ROPI of both coarse and fine alignment signal, AM3_FIR_5th is suggested to be used as alignment strategy for current layer.

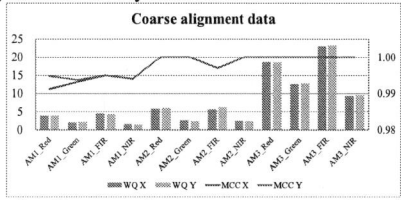

Figure 3: Coarse alignment wafer quality (WQ) & multiple correlation coefficient (MCC) of various marks AM1, AM2 and AM3

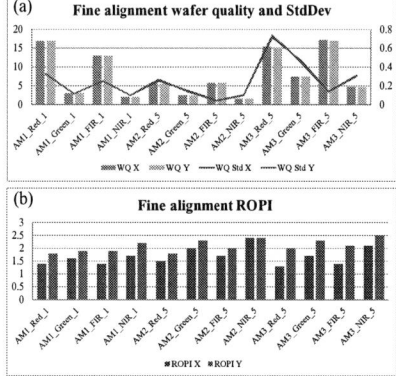

Figure 4: Fine alignment wafer quality & wafer quality StdDev (a) and ROPI (b) of various marks AM1, AM2 and AM3

Limited to mark design of AM2 and AM3, which have large empty area in the mark, are prone to be affected by process that have pattern loading effect such as CMP and ETCH. Alignment mark should present sufficient manufacturability and robustness to survive various processes and be able to provide accurate signals. Otherwise mark damage may lead to alignment fail and wafer reject. AM1 mark has evenly distributed trench and line area in the mark, so it is less affected by pattern loading effect. Besides, AM1 fine alignment mark also shows better WQ and MCC with low ROPI performance.

978-1-7281-6559-2/20 $31.00 © 2020 IEEE

Therefore, AM1 fine alignment mark is also considered as candidate for fine alignment, nevertheless, coarse alignment mark for AM1 mark should be carefully selected and evaluated.

Overlay mark evaluation

Overlay performance of five selected marks with segmentation and CD/pitch split was compared and the corresponding data including overlay raw data, residue, TIS (tool induced shift) and Qmerit (an indication of mark quality) is shown in Figure 5. The result suggests that segmented marks T4 and T5 with larger TIS and Qmerit exhibited clearly worse overlay performance, compared with non-segmented marks T1-T3. It's inferred that it's not beneficial for current layer to be segmented, and it's may be resulted from the relatively weak contrast, compared with non-segmented marks.

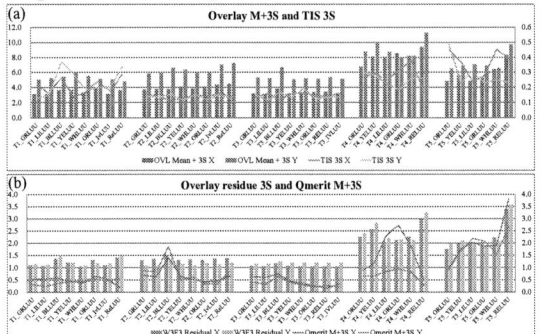

Figure 5: Overlay M+3S & TIS 3S (a), and residue 3S & Qmerit M+3S (b) comparison among segmented and non-segmented mark with different coarse CD/Pitch design

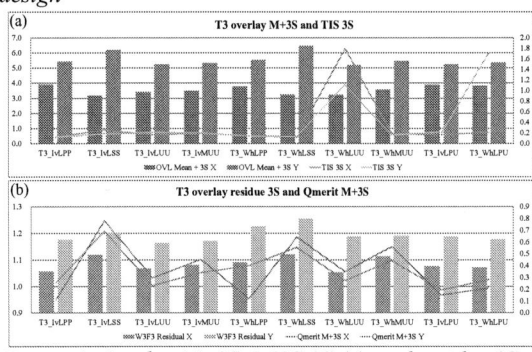

Figure 6: Overlay M+3S & TIS 3S (a), and residue 3S & Qmerit M+3S (b) comparison of T3 with different illumination/NA/polarization condition

T3 mark, with largest CD/pitch among three non-segmented marks, shows most stable TIS 3S and Qmerit M+3S performance through variable illumination/NA/polarization, while there is a jump in the chart of T1's TIS 3S and T2's Qmerit M+3S. What's more, T3 also shows best overlay raw and residue performance. It's hence inferred that relatively larger coarse CD/pitch is helpful for stable overlay performance for current layer being evaluated, which need to be further investigated.

T3 mark was further evaluated to determine the optimal measurement condition. Figure 6 shows through various measurement conditions. Under ivory/low NA/P-polarized condition T3 mark exhibits lowest TIS and Qmerit, which is chose as golden measurement condition to get accurate and precise overlay measurement result.

CONCLUSIONS

The alignment and overlay marks for SADP process are well designed and evaluated in this paper. Segmentation for alignment mark is required for SADP process to avoid fence-like structures and obtain robust alignment signal for accurate scanner alignment process, which in return will guarantee precise and stable overlay performance. Alignment mark AM3 with segmentation following minimum design rule shows best alignment signal compared with AM1 and AM2 mark. AM1 mark, which also shows good fine alignment performance, can be selected as second candidate since it's not prone to be affected by pattern loading effect brought by CMP and ETCH process. Segmentation is also required for outer overlay marks formed by SADP process, while it's not a necessity for inner overlay marks of current layer. The mark design of previous and current layer affects the measurement accuracy and precision. The study suggests that non-segmented mark for current layer can get more precise overlay measurement result with low TIS and Qmerit for image-based overlay metrology. What's more, non-segmented mark with largest CD/pitch exhibited most stable TIS 3S and Qmerit M+3S through variable illumination/NA/polarization, which may give a hint for future mark design.

ACKNOWLEDGEMENTS

We appreciate great supports from colleagues of lithography group of TD2 in Shanghai Huali Microelectronics Corporation and KLA-Tencor Corporation.

REFERENCES

[1] C.W. Yeh, C.T. Huang, K. Lin, C.H. Huang, E. Yang, T.H. Yang, K.C. Chen and C.Y. Lu. *Proc. of SPIE,* vol. 8324, 2012, 832431.

[2] H.L. Hsu, E.C. Lio, C. Chen, J.H. Chang, S.S. Lee, S. Buhl, M. Gutsch, P. Lomtscher, M. Freitag, B, Habets and R. Liu. *Proc. of SPIE,* vol. 10585, 2018, 105851H.

[3] L.Ye, H. Hu and W.M. He. *2016 CSTIC,* 2016, pp. 1-4.

[4] C.C. Huang, C.T. Huang, A. Golotsvan, D. Tien, C.F. Chiu, C.Y. Huang, W.B. Wu and C.L. Shih. *Proc. of SPIE,* vol. 7971, 2011, 79712B.

LITHO PROCESS OPTIMIZATION TO IMPROVE OVERLAY MEASUREMENT IN THICK PR LAYER

Jiantao Wang*, Jun Yu, Yuming Sun, Cong Zhang, Xiaobo Guo, Biqiu Liu, Song Gao, Shuo Liu, Jun Huang, Yu Zhang

Shanghai Huali Integrated Circuit Corporation

*Corresponding Author's Email: wangjiantao@hlmc.cn

ABSTRACT

Overlay (OVL) is a key index in lithography, which will determine product quality, and its measurement is affected by many factors, such as measurement tool, measurement strategy, OVL mark and STI CMP, etc. For some layers used thick Photoresist(PR), OVL results are out of control due to worse PR profile, especially for asymmetric profile which will cause inaccurate mark signal reading, so it is difficult to guarantee to get actual OVL performance. In this paper, we take some of investigations to optimize OVL measurement accuracy of layer with thick PR, and the corresponding mechanism is also analyzed.

INTRODUCTION

In semiconductor fabrication, Lithography is one of the most important processes, and overlay (OVL) is a key index of current layer and pre-layer alignment, which will impact product yield. Mark damage or bad mark profile will cause inaccurate signal reading during measurement, and false results. Generally, bad mark profile usually appeared on the layers with thick resist.

A thick resist with thickness varied from 3 to 5um is usually used for block layer of high dose implant. As experienced, too thick resist will cause the pattern asymmetric after lithography exposure and measurement error. The mechanism of PR asymmetric had been studied to solve this issue previously. S.I.Yet etc. think it is caused by solvent evaporation from photoresist film, and the OVL mark profile is found closely related to the pattern density at surroundings. It can be improved by increased post-apply bake temperature [1]. This hypothesis can explain the phenomenon reasonably, but they ignore the effect with the different PR type and thickness, and also not reveal influence of other temperature steps process. Zhu Liang etc. and Mike Adel etc. also do some efforts to solve this measurement issue based on the OVL mark design optimization [2, 3]. This will bring great challenge to mark OPC and design.

In this paper, we studied the correlation of PR profile asymmetry with temperature of photo process step and pattern density in design layout comprehensively. We found PR baking temperature affected PR profile, and pattern density, outgassing model would act as a coordinating role for PR profile. Based on experiment

results, increasing soft bake temperature up to 110℃ or removing hard bake can get better PR profile and accuracy OVL measurement.

EXPERIMENTS

Materials and Machine

PR with 4um thickness, I-line tool, including the track tool with TEL Pro Z and scanner with ASML XT400, OVL tool of Archer600, CDSEM tool of Hitachi CG5000, and cross section SEM cut tool is field emission scanning electron microscope (FESEM)

Method

Wafers were prepared by soft bake temperature split, and temperature range from 90-110 ℃ with its step of 5℃, meanwhile, split of removing hard bake is also studied. PEB delay 20min (add PEB delay time) experiment is taken to verify the outgassing model.

RESULTS

OVL Measurement Results

Figure1 shows the OVL measurement results, including residual data and Qmerit data. Residual is belongs to uncorrection component of OVL measurement results which are usually used to character best OVL performance in theory by OVL model. Qmerit is an indicator of mark symmetry. The OVL measurement is more accuracy as Qmerit values are smaller. Figure 1 (a) and (b) shows the residual of OVL in condition of PAB 110℃ are smaller than that of PAB 90℃. Note that the map has the same scale, and the arrow length on the map represents the OVL value. Figure 1 (c) and (d) results show the Qmerit data of PAB 110℃ are much smaller than that of PAB 90℃. It proves the mark PR profile is improved apparently as the temperature up to 110℃.

PR Profile

PR profile can be seen from top view image. If PR profile is asymmetry, the image will show large white slope from top view. Figure 2 shows the OVL mark PR profile with different conditions from top view using CDSEM measurement. From top view images, as increase of the soft bake temperature, the white slope becomes smaller, that is, the symmetry of PR profile became better. When soft bake is increased to 110℃, the best PR profile is got based on top view images, as shown in Figure2(e). Compared to the lower soft bake temperature condition, split of combining removing hard bake after PR coating and adding PEB delay time after exposure will also have a degree improvement of PR profile through the images, shown as Figure2(h). But these two conditions still do not achieve comparative PR profile as soft bake of 110℃. This result reveals that the soft bake temperature indeed have a great influence on the symmetry of PR profile.

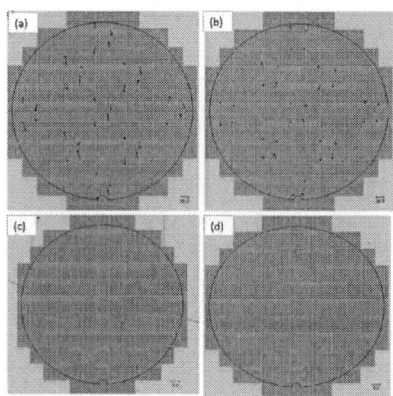

Figure1: OVL measurement results, including residual data and Qmerit data. (a) PAB 90 ℃ residual, (b) PAB 110 ℃ residual, (c) PAB 90 ℃ Qmerit, (d) PAB 110 ℃ Qmerit

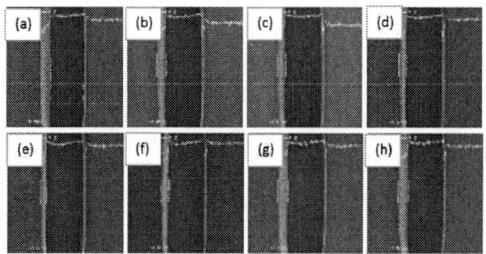

Figure2: OVL mark PR profile for different conditions from top view using CDSEM measurement. (a) PAB 90 ℃, (b) PAB 95 ℃, (c) PAB 100 ℃, (d) PAB105 ℃, (e) PAB 110 ℃, (f) removing hard bake, (g) PEB delay 90s, (h) PEB delay 20min.

In order to verify the PR profile, cross section SEM cut is taken. Figure 3 shows the cross section SEM cut images of PAB 90℃ and PAB 110℃. From the images, PAB 90℃ has bad PR profile, while there is a great improvement on PR profile in the condition of PAB 110℃. The cross section SEM cut results match with top view CDSEM results, as shown in Figure 2(a) and 2(e). It also proves the correlation of top view white slope with PR profile. Meanwhile, the OVL results are also consistent with the SEM cut results. From the SEM cut image, PR

thicknesses are almost equal even if temperature changes, it will meet the requirement of implant block.

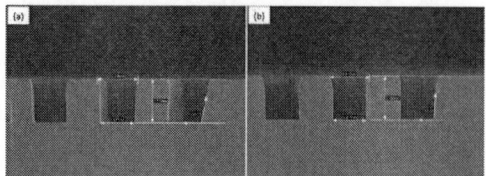

Figure3: cross section SEM cut images of different PAB conditions using FESEM. (a) PAB 90 ℃, (b) PAB 110 ℃.

Process Window Comparison

Based on mass production consideration, temperature change is a big process change, and baseline process change may influence chips quality. Process window is an important index of Lithography, mainly including DoF and EL. Besides, critical dimension uniformity (CDU) is another key index for change. Table 1 showed the process window and CDU comparison results. From the data, the two conditions have the comparable process window and CDU performance. Here, the percentage of CDU on target is about 2.1% and 2.7%, The performance of PAB 90℃ and PAB 110℃ are comparable. It means that will not affect the Lithography process window as PAB temperature tuned to 110℃. So the change is safe.

Table1 Process window and CDU comparison for different PAB conditions

Index	PAB 90℃	PAB 110℃
DoF	0.9um	0.9um
EL	15%	15%
CDU (3sigma)	64.1nm	83.3nm

DISCUSSION

From the above results comparison, it shows PR profile is related to the soft bake temperature. Soft bake temperature increased can greatly improve PR symmetry, while the mechanism is always unclear. The mechanism will be discussed as below.

Solvent Evaporation

The solvent in thick PR is continually evaporated at all of Lithography process, and it will cause PR shrink. If one side of the pattern evaporates more solvents in comparison with other sides, PR will suffer more shrink, resulting in tilt to some extent between both sides, the white slope are found at top view images. While increasing of soft bake temperature will increase evaporation of solvent in PR, and the solvent evaporation is lessen at the following photography process. So the PR profile tilt is great improved to reduce effect of OVL measurement, induced by asymmetry mark profile. Figure 4 shows the solvent evaporation diagram. The model based on solvent evaporation can explain phenomenon of PR asymmetry improvement by tuning temperature, but this method has limitation of pattern profile improvement. If the pattern loading very high, such as very dense pattern and ISO pattern, the solvent evaporation speed and amount will have big difference. In those cases, the model based on solvent evaporation cannot fully explain those results.

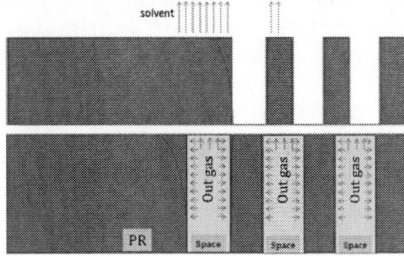

Figure4: the solvent evaporation diagram

Pattern Density

As mentioned above, when pattern is in large loading effect, the behavior of PR shrink will be decided by the surrounding environment of mark pattern, as shown in Figure 4. If the area of PR region is large, the large PR region evaporation amount of solvents will be much larger than smaller PR region, then the PR shrink will be more for the large PR area. The result will result in PR profile tilted to large PR side, as shown in Figure 2 and Figure 3.

In order to verify this results, different pattern loading mask was used to expose. Figure 5 shows mark top view images, and find the OVL mark layout with different pattern density. These two layouts are used to prepare wafers to investigate OVL marks, and the result showed larger PR area difference surrounding the OVL mark, the mark profile asymmetry is worse. For example, Figure 5(a) S1 showed the right region has large area PR, while the left smaller, the result is the OVL mark PR showed asymmetry seriously. For Figure 5(b), the OVL mark are surrounded uniform PR area, the OVL mark images showed PR symmetry good. From the OVL mark image can get the conclusion of Figure 5(a) OVL measurement will be abnormal, the actual OVL measurement matched the expected results. While adding PEB delay times split cannot be explained well by this model.

Outgassing Model

As shown in Figure 2(h), PEB delay 20min can also improve PR symmetry a little, but the improvement is limited, it may have no any improvement for OVL measurement. This phenomenon had been mentioned by S.I.Yet[1], exposure of resist induces photochemical reaction and generates N_2 gas in the exposure region, N_2 gas will push PR, then caused PR asymmetry. While different pattern density also affect the N_2 amount.

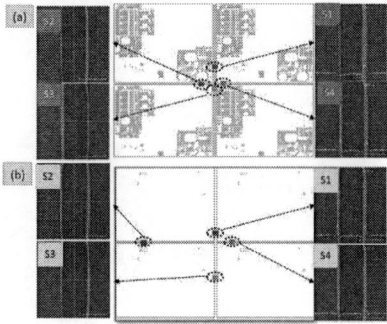

Figure5: the OVL mark layout surrounding different pattern density and mark top view images, the color area is the light transmission region (a) transmission rate 40%, (b) transmission rate 0.4%

Comprehensive Effects

From all above analysis, not any theory can fully explain the PR asymmetry phenomenon alone, while the phenomenon can be explained by model of the joint effects.

CONCLUSION

We study the relationship of PR profile symmetry with both temperature of photography process and pattern density in design layout comprehensively. We find that baking temperature affects the PR profile, and pattern density also affects it. Outgassing model will act as a coordinating role. Through experiments, the PR asymmetry is improved significantly seen at SEM images, and OVL measurement also becomes precise.

ACKNOWLEDGEMENTS

The authors would like to acknowledge all HLMC TD2 Litho colleagues, especially thanks the works of Billy Yu, the paper modification by Dr. Biqiu Liu and HLMC failure analysis department for the supports of SEM microscopy.

REFERENCES

[1] S.I.Yet, E.C. Goh, Faith Lim, A.E. Ling, B.C. Lee, Y.K. Ng, W.B. Sheu, et al., ICSE2006 Proc. 2006, pp861-865.

[2] Zhu Liang, Li Jie, Zhou Congshu, Gu Yili, and Yang Huayue, Journal of Semiconductors, Vol. 30, No. 6, 2009, pp 066002-1-006002-5.

[3] Mike Adel, Mark Ghinovker, Boris Golovanevsky, Pavel Izikson, Elyakim Kassel, Dan Yaffe, Alfred M. Bruckstein, Roman Goldenberg, Yossi Rubner, and Michael Rudzsky. IEEE Transactions on Semiconductor Manufacturing, Vol. 17, No. 2, 2004, pp166-178.

EFFECTS OF ELECTRON BEAM ON PHOTO RESIST SHRINKAGE AND CRITICAL DIMENSION IN SEM MEASUREMENT

Yuyang Bian[1]*, Hongxu Sun[1], Lipeng Wang[1], Xijun Guan[1], Xiaobo Guo[1], Biqiu Liu[1], Cong Zhang[1], Jun Huang[1], Yu Zhang[1]

[1]Shanghai Huali Integrated Circuit Corporation

Pudong District, Shanghai, China

bianyuyang@hlmc.cn

ABSTRACT

Scanning electron microscope (SEM) measurement is the dominant method to obtain critical dimension (CD) in lithography. In SEM measurement, electron beam exposed on photo resist (PR) will affect PR physical-chemically and cause PR shrinkage inevitably. This phenomenon is well-known in lithography and also studied by many researchers. In this paper, we reviewed previous research results and investigated the PR shrinkage effects on CD measurement of 193nm immersion photo resist in 14nm node. With patterns sizing down, it is important to choose an appropriate voltage, current and frame to avoid large amount shrinkage in the measurement. A 30nm size CD shrinks more than 10% until getting saturation. Different patterns and CD sizes possess different through pitch shrinkage performance, which means PR shrinkage will also impact OPC model setup.

Keywords—lithography; photo resist; SEM; electron beam; shrinkage; LER; PSD

INTRODUCTION

193nm photoresists (PR) suffer significant shrinkage under electron beam (e-beam) irradiation during scanning electron microscope (SEM) measurements [1]. With lithographic generation upgrade, these phenomena become more important, and will impact critical dimension (CD) results significantly. It is essential to investigate the CD trend under e-beam exposure. It benefits lithography process setup in new generations.

Photoresists shrinkage is a physical and chemical interaction between e-beam and resist polymer. Su et al. considered shrinkage was caused by resist cross-linking or other chemical reactions, while high e-beam dose increased the depth of electron penetration [2]. Habermas et al. used e-beam to cure resist sample and found the disappearance of FTIR peaks corresponding to carbonyl region, which reveals that shrinkage may be caused by the decomposition of carbonyl to carbon monoxide and dioxide [3]. Akerman et al. also showed a positive correlation between the content of carbonyl in resist and CD shrinkage [4]. From the point of phenomenology view, Bunday et al. investigated shrinkage trend by fitting CD shrinkage curve with dose or measurement times [5]. Hence, the no shrink 0th exposed CD can be calculated by extrapolation. He et al. studied the CD x and y direction shrinkage trends in 2D features in the TV or slow scan mode in which the e-beam will scan from top to down of the sample in order [6]. Due to the charging effect of the scanned area, e-beam will irradiate to the wafer in oblique angles in y direction and make more shrinkage.

PR shrinkage depends on several factors, involving voltage, current, materials, etc. In this paper, shrinkage performance of different line sizes, mainly for 14nm node, was studied by changing measurement recipe conditions. The through pitch CD shrinkage was also investigated.

EXPERIMENTAL

Commercial 193nm immersion PR, also with SOC (spin-on-carbon) and Si-ARC (silicon anti-reflection coating) materials were used. Wafers used in these experiments were with designed film stacks. Fig.1. demonstrates the film stacks of wafers used in the following experiments. Each wafer was prepared in the best exposure energy and focus condition, which is verified by FEM (focus energy matrix). The wafers were coated and then exposed by TEL Clean Track LITHIUS-Pro Z (Tokyo Electron LTD.) and ASML NXT1980i (ASML Holding N.V).

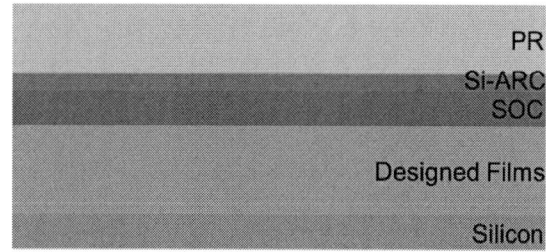

Figure 1: Film stacks of the wafers used in the experiments

Hitachi CG5000 (Hitachi High-Technologies Corporation) was used in CD measurement. Each condition used in paper with different voltages (300, 500 and 800V), currents (5, 6 and 8pA) and modes (HR: high resolution; HDOF: high depth of focus) was calibrated before measurement.

RESULTS AND DISCUSSION

A. Effect of dose, voltage and pattern size on CD shrinkage

Voltage and current are important factors for PR shrinkage. Meanwhile, dose is introduced to demonstrate the level of sample irradiation by e-beam. Dose can be calculated by formula (1), in which I, t and A represent current, time and inspected area, respectively. Dose signifies the amount of charges in unit area. However, the value can only be gotten in a fully automated measurement mode.

$$D=It/A \qquad (1)$$

Fig.2 shows shrinkage value with variation of CD sizes, voltage and dose. It clearly demonstrates that with the increase of dose, CD shrinkage increases. When dose reaches a certain level, shrinkage saturated. The shrinkage rate is fast on the first several scan times, and then trends down. Shrinkage of small size CD (~30nm) saturated at dose of 500μC/cm2, while large size CD (~70nm) can hardly get saturated at dose above 600μC/cm2. For small line size, e-beam can easily penetrate into the materials and consume the reactants, then shrinkage will saturate. But for large size line, the products of the reaction covered the not shrunk PR line. High voltage strengthens the penetration ability of electrons. As we seen, because of further consuming of the remaining PR, the saturated shrinkage value of large size CD increases with increase of voltage. High voltage will accelerate the shrinkage of PR, especially for large size CD.

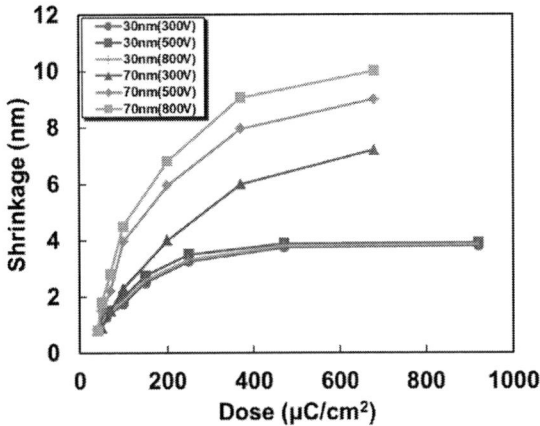

Figure 2. Normalized CD shrinkage curve with increase of dose

A 30nm size CD shrinks more than 10% until getting saturation. It should be noted that different materials have different shrinkage curve. With patterns sizing down, it is important to choose an appropriate voltage, current and frame to avoid large amount of shrinkage in the measurement. However, lower voltage or current and less scan frame will inevitable induce image blurring and reduce measurement precisions.

B. Effect of shrinkage on through pitch patterns measurements

Due to shrinkage performance depending on CD sizes, the shrinkage trends of through pitch patterns both for 1D and 2D patterns are future investigated by tuning measurement doses. Fixing voltage, current and scan mode, only frames are changed from 24 to 64 to represent the variation of dose. Fig.3 (a)-(c) show the dense patterns used in the measurements, in which line (a) with width of ~160nm and pitch of 220nm, line (b) with width of ~80nm and pitch of 140nm and dot pattern (c) with width of ~50nm in x direction, width of ~100nm in y direction. Fig.3 (d)-(f) are the corresponding isolation patterns to (a)-(c), respectively.

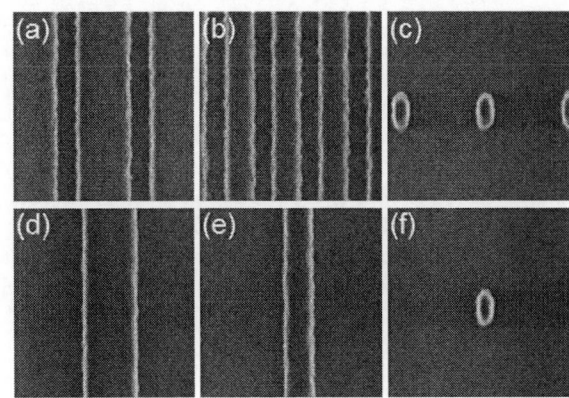

Figure 3. (a) and (b) are dense lines with CD of ~160nm and ~80nm, (c) is dense dot pattern. (d)-(f) are the corresponding isolation patterns

Adjusting frame from 24 to 64, the surface of sample will be irradiated by e-beam for longer time and PR width will shrink more. Fig.4 displays the shrinkage of CD by calculating the difference of CD width of frame 24 and 64. Shrinkages are fitting through linear curve, as shown in formula (2), in which, a and b represent the slope and intercept.

$$Bias = a \times Pitch + b \qquad (2)$$

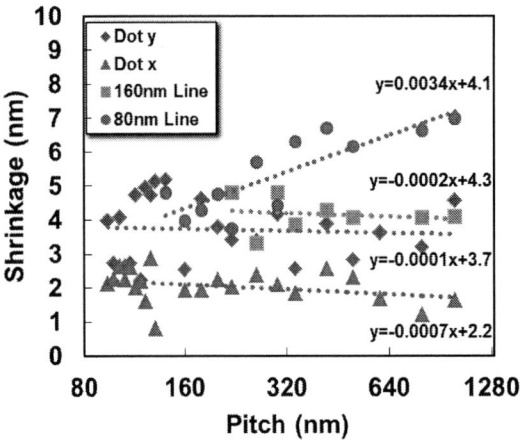

Figure 4. CD shrinkage curve with increase of pitch for different patterns

For 1D dense line of ~160nm, through pitch shrinkage value is almost a constant (a=-0.0002). Nevertheless, for small CD size of ~80nm, shrinkage increases with increase of pitch (a=0.0034). It implies shrinkage is not a fixed value with variation of pitch for a given measurement condition. It also depends on CD sizes. From Fig.3 (a) and (d), we can see line edge roughness (LER) of dense lines is larger than isolated pattern. The same appearance can also be observed from Fig.2 (b) and (e). These results are in agreement with He et al. results that sidewall of isolated line suffers more backscattering electrons striking which leads more PR shrinkage [6].

Dot patterns, as shown in Fig.3 (c) and (f), are common used in 14nm for space cutting. From Fig.4, we can see that the

978-1-7281-6559-2/20 $31.00 © 2020 IEEE   253

shrinkage of dot pattern both in x and y directions are a constant. Generally, e-beam scans from left to right, and goes from top to down in SEM measurements,. The scanned area accumulated large amount electrons which will repulse e-beam in y directions. It results in the difference of shrinkage in x and y direction.

Different patterns and CD sizes have different through pitch shrinkage performance, which means PR shrinkage will also impact OPC model setup. Optimizing CD-SEM recipe not only benefit for measurement accuracy and precision but also for optical behavior interpretation.

C. Effect of shrinkage on LER characterization

From Fig.3, the difference of edge roughness for dense and isolated line is observed. PSD (power spectral density) method was applied for LER analysis. LER of dense line was obviously worse than isolated line. The overall LER of dense line is 4.5nm, and that for isolated line is 2.6nm. It confirmed that under same measurement setting, isolated line suffers more shrinkage.

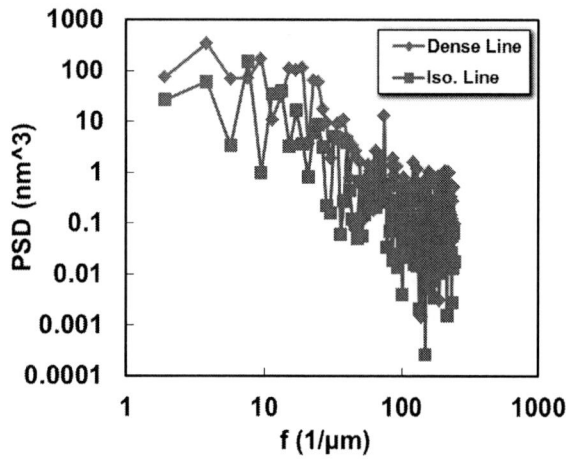

Figure 5. PSD curves of dense and isolated line under same measurement recipe setting

PSD curve can be divided into 3 zones, of which low frequency (LF) zone is strong correlation with EPE (edge placement error) performance, middle frequency (MF) zone is mainly dominated by PR materials, and high frequency (HF) zone responds to measurement error. Due to the e-beam curing effects to line edges, HF zone of isolated line is much smaller than dense line. The difference of MF is partial contributed by the changing of PR constituents caused by e-beam irradiation, which can be used in the further study of the underlying shrinkage mechanism.

CONCLUSION

SEM measurement is the dominant method to obtain CD in lithography. Shrinkage is well-known in lithography, and can hardly be prevent during SEM measurement. In this paper, we studied shrinkage behaviors for different line sizes, patterns and pitches. The shrinkage rate is fast on the first several scan times, and then trends down. A 30nm size CD shrinks more than 10% after get saturated. Different patterns and CD sizes possess different through pitch shrinkage performance. With

patterns sizing down, it is important to choose an appropriate voltage, current and frame to avoid large amount shrinkage in the measurement. Optimizing CD-SEM recipe not only benefit for measurement accuracy and precision but also for optical behavior interpretation.

ACKNOWLEDGEMENTS

We appreciate great supports from colleagues of lithography group of TD2 in Shanghai Huali Microelectronics Corporation and Hitachi Corporation.

REFERENCES

[1] A. Horiba1, M. Yasuda1, H. Kawata, M. Okada, S. Matsui and Y. Hirai, "Impact of resist shrinkage and its correction in nanoimprint lithography," Jpn. J. Appl. Phys, Vol. 51, 06FJ06, 2012.

[2] B. Su, G. Eytan and A. R. Romano, "193-nm photoresist shrinkage after electron-beam exposure," Proc. SPIE, Vol. 4344, pp. 695-706, 2001.

[3] A. Habermas, D. Hong, M. F. Ross and W. R. Livesay, "193-nm CD shrinkage under SEM: modeling the mechanism," Proc. SPIE, Vol. 4689, pp. 92-101, 2002.

[4] L. Akerman, G. Eytan, R. Uchida, S. Fujimura and T. Mimura, "Examination of possible primary mechanisms for 193nm resist shrinkage," Proc. SPIE, Vol. 5752, pp. 744-754, 2005.

[5] B. Bunday, A. Cordes, J. Allgair, E. Piscanib, B. J. Riceb, Y. Avitanc, R. Peltinovc and O. Adan, "Phenomenology of ArF photoresist shrinkage trends," International Symposium on Semiconductor Manufacturing (ISSM), pp. 3-6, 2008.

[6] W. He, X. Shi, H. Hu and Q. Wu, "The characterization of photoresist shrinkage difference in X-Y directions with CDSEM metrology," 2016 China Semiconductor Technology International Conference (CSTIC), pp. 1-4, Shanghai, 2016.

ENLARGE PROCESS WINDOW OF BSI IN DTI LOOP ： A NOVEL OPC APPROACH TO ADD SRAF

Qiao Yanhui[1]*, Li Baoxuan[2], Wan Dan[2], Chen Yanpeng[2], and Yu Shirui[2]
[1]Huali Integrated Circuit Corporation, Shanghai, 201314, China
* E-mail: qiaoyanhui@hlmc.cn

ABSTRACT

Compared with conventional Front-Side illuminated (FSI) CMOS image sensor (CIS), the photodiode of back-side illuminated (BSI) has a higher photoconductivity due to its special structure. In order to erase the effect of photoelectron scattering between different photosensitive units, the Deep Trench Isolation (DTI) layer was used to isolate it. However, the main figure of DTI layer is relatively isolated. Under KrF exposure condition, the process window is insufficient. In this paper, the influence of adding different sub-resolution assist features on the process window is discussed.

INTRODUCTION

The demand for portable mobile devices such as mobile phones with higher camera quality and richer application scenarios has provided a huge driving force for the development of backside illuminated CMOS image sensor (BSI-CIS). It has the natural advantages of less photon loss, less impact by light incidence Angle, and flexible wiring, which is related to different structural sequence compared with traditional front-side illuminated CMOS (FSI) [1]. Unfortunately, some issues perplex BSI, such as spectral crosstalk, photon crosstalk, electronic crosstalk, and so on. However, by introducing grid Deep Trench Isolation (DTI) layer formed by metal-insulator-silicon structure can eliminate the photon and electron crosstalk caused by photons and electrons penetrating pixels [2].

However, the after developing inspection (ADI) critical dimension (CD) located in grid cross is much larger than design on the wafer，because of the corner rounding effect. In general, the grid-shaped DTI with the same design size will be specially Optical Proximity Correction (OPC) processed to reduce the cross CD in order to obtain larger pass area and more optical information to improve the camera quality. Under the same lithographic exposure conditions, the process window decreases as the cross CD decreases, it is an effective way to optimize the process window to add sub-resolution assist features (SFAF) to mask. In this paper, Mentor caliber software simulation was used to study the influence of different SRAF on the process window. PVband is the difference between the maximum and minimum CD of contour simulated under different doses and focus conditions, which can reflect the sensitivity of cross CD to different exposure conditions and is the main representation method for evaluating the size of process window..

EXPERIMENT

In this experiment, a DTI grid-shaped structure of 3um*5um was clipped, in figure 1 its cell morphology was shown, the pixel pitch is 1.1um approximately. Then the line CD was sized up by 26, 32, 36 and 42nm respectively, as shown in figure 2, this is also the case without adding SRAF.

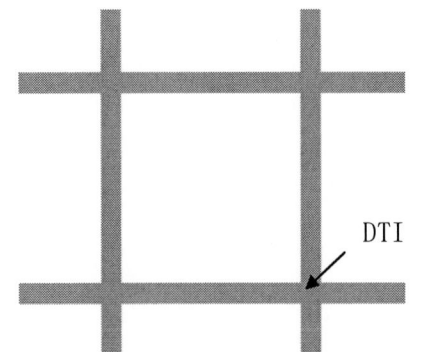

Figure 1: The morphology of grid-shaped DTI unit

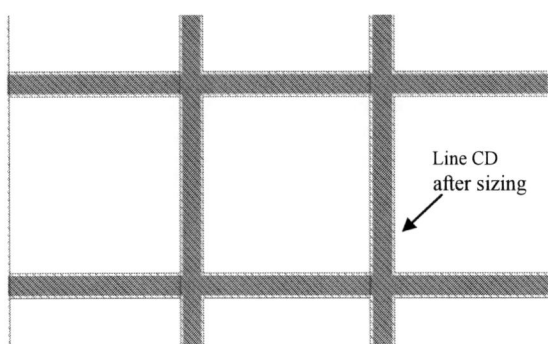

Figure 2: DTI after sizing up & without SRAF

Then, conventional SRAF and novel SRAF were added to the graphs of different line CDs by rule based OPC, and both conventional SRAF and novel SRAF are 80nm wide and 170nm away from the main pattern, the morphologies of after adding SRAFs are shown in figure 3a and 3b respectively. Finally, the same MRC limit was set for different graphics with different line CDs and different SRAFs in the recipe, and different masks were generated after running model-based OPC.

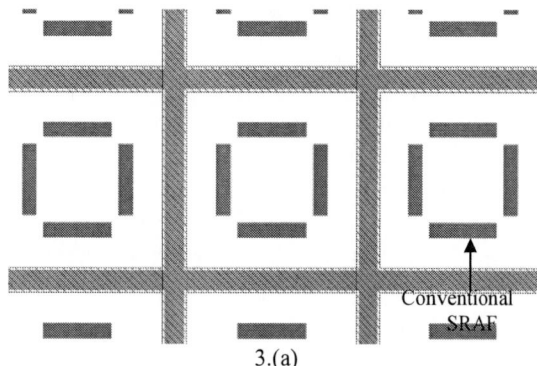

3.(a)

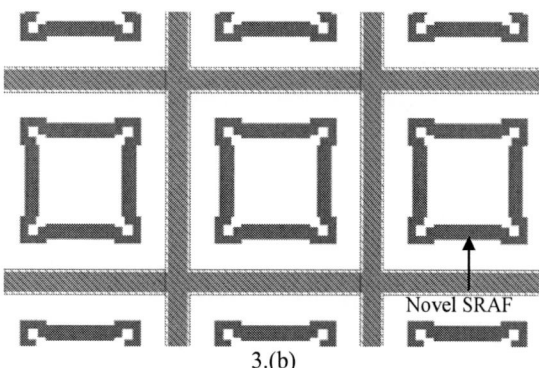

3.(b)

Figure 3: DTI with conventional SRAF (a) and with novel SRAF (b)

RESULT AND DISCUSSION

We simulated the masks we obtained before under the nominal condition and under the condition of PW, where PW conditions were the matrix of focus { -150, 0, +150 } nm and dose { 0.96, 1 ,1.04 }, and measured the simulated contour corner & line of PVBand under PW condition and contour corner CD under nominal condition.

TABLE I
The influence of different SRAF on the corner PVband
Unit (nm)

Line CD	No SRAF	Conventional SRAF	Novel SRAF
CD+26	29.7	29	26.9
CD+32	24	25.5	22.6
CD+36	22.6	21.9	19.8
CD+42	20.5	19.1	17.7

Table 1 shows the influence of different SRAF addition methods on Corner PVBand under the condition of the same pitch and different line CD; table 2 shows the influence of different SRAF addition methods on line PVBand under the condition of the same pitch and different line CD; table 3 shows the different masks corner contour CD distribution under nominal condition.

TABLE II
The influence of different SRAF on the line PVband
Unit (nm)

Line CD	No SRAF	Conventional SRAF	Novel SRAF
CD+26	14	14	12.5
CD+32	12.5	13	12
CD+36	12	12.5	11.5
CD+42	11.5	11.5	10.5

TABLE III
Corner CD under nominal condition
Unit (nm)

Line CD	No SRAF	Conventional SRAF	Novel SRAF
CD+26	246.8	248.2	241.8
CD+32	258.8	258.1	256.7
CD+36	265.9	264.5	265.2
CD+42	274.4	275.1	272.9

As can be seen from the above tables, both the corner PVBand and line PVBand will decrease with the increase of line CD under the same SRAF addition method. However, a larger line CD will lead to a larger corner CD, and a larger corner CD will reduce the amount of photons entering and affect the imaging quality eventually. Therefore, If DOF can guarantee the yield rate, a smaller line CD should be selected. Compared with without SRAF, conventional SRAF has little effect on the PVBand of corner& line, this is because an SRAF with a length of only 400nm is added to the space with a length of almost 1um, while the SRAF with a too short length has little effect on the light intensity at the corner and line. But the novel SRAF decreases PVBand of corner & line by an average of 10% and 7% respectively. It thanks to two aspects, on the one hand the conventional SRAF is lengthened with a narrower SRAF, the total SRAF length is longer, on the other hand the narrower SRAF become more close with the main pattern corner, causing the pitch of SRAF and main pattern corner equals the pitch of SRAF and main pattern line, the grating interference effect is strengthened.

CONCLUSION

For grid-shaped DTI layer, when line is expanded to the same CD, compared with no SRAF and conventional SRAF, the novel SRAF addition method can effectively improve the process window at both line and corner with the almost same contour corner CD.

REFERENCES

[1] [1]. Z. C. Hsiao et al., "TSV-less BSI-CIS wafer-level package and stacked CIS module" IEEE, 2013

[2] [2]. Tournier et al., "Pixel-to-Pixel Isolation by Deep Trench Technology: Application to CMOS Image Sensor," IISW, 2011.

DEVELOPMENT OF 90 NM & 5 NM HIGH RESOLUTION ADVANCED LITHOGRAPHIC PATTERNING MATERIALS

*Xuemiao Li, Zhenyu Yang, and Hai Deng ***

School of Microelectronics, State Key Laboratory of Molecular Engineering of Polymers，Fudan University，Shanghai 200433, China
* Corresponding Author's Email: haideng@fudan.edu.cn

ABSTRACT

Directed Self-assembly is a chemical assisted patterning technology, which has attracted a great deal of interest due to its potential high resolution and low cost[1-4]. Researchers around the world discovered some sub-5 nm resolution DSA materials with high χ value, which normally requires high annealing temperature or long annealing time. As modern semiconductor manufacturing litho process normally requires hot baking time less than 3 min to ensure cost-effective throughput, fast annealing BCPs thus are essential for future DSA patterning technology. Also our group focus on the development of advanced materials for ArF lithography and we are trying to contribute to China own advanced patterning material industry.

INTRODUCTION

Patterning technology is a key enabled technology for semiconductor industry. Japan-based suppliers dominate ArF resist market share more than 90% worldwide and 100% in China. From 90 nm to 10 nm node technologies, semiconductor manufacturers in China are all depending on Japan-made photo resists.

Recently, 7 nm to 5 nm node (12~10 nm LS) technologies entered into HVM in TSMC and Samsung, which are also using Japanese EUV and ArF photo resists. By developing 193 nm lithographic materials, we are trying to contribute to China own advanced patterning material industry.

For 3 nm node (8 nm LS) or below, DSA lithography is a strong candidate competing with EUVL. Researchers around the world discovered some sub-5 nm resolution DSA materials with high χ value, which normally requires high annealing temperature or long annealing time. As the production of integrated circuits demands high-speed processing at mild temperature, the desirable time for hot baking or a thermal annealing process should be less than 3 min. Therefore, fast self-assembly DSA materials are essential for future 3 nm node patterning technology.

In our study, we designed and synthesized a family of fluorine-containing BCPs for DSA application [5,6]. The finest half pitch of these BCPs is 5 nm or less, with thermal annealing time up to 1 min at 80 °C, which is the record fast DSA patterning material reported so far.

In order to obtain high resolution DSA material, high χ (Flory–Huggins interaction parameter) value is necessary. Our resulted BCPs demonstrated high χ ranged from 0.2 to 0.6, which displayed resolution up to 5 nm half pitch by SAXS. However, the thermal annealing time and temperature are varied even with similar χ value of different BCPs. We therefore carried out investigation of the driving factors for the annealing speed or kinetics by comparing chemical structure and the segment mobility of each block in the BCPs.

EXPERIMENT AND DISCUSSION

Synthesis of block copolymers and characterization

Each block copolymer was synthesized according to the following procedure. The flask was dried under vacuum and purged with argon 3 times. 25-35 mL of THF and 1-5 mL of styrene were transferred to a flask loaded with dry LiCl (5 eq. of the initiator). The flask was cooled to −80 °C - −85 °C, and 0.5 mL of sec-butyllithium (1 M in hexane) was injected. After 30 min of polymerization, styryl anions were capped with 0.1 mL of DPE. Then 1.5 mL pentadecafluorooctyl methacrylate monomer was added into the system and polymerized for 30 min, and then terminated with 0.5 mL of degassed methanol. The solution was precipitated with methanol twice and the resulting white powder was dried in vacuum.

This procedure was for the synthesis of polystyrene-b-poly(fluoromethacrylate) ,and the synthesis of BCPs with other monomers were similar to this route.

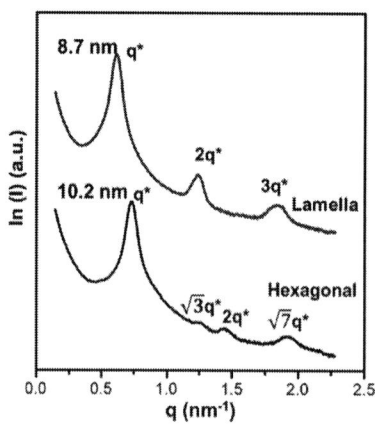

Figure 1. Bulk morphologies at room temperature, after thermal annealing of typical lamella structure (blue line) and hexagonal structure (black line) respectively, which were defined by SAXS measurement. Both of them showed 5 nm resolution.

^{1}H-NMR spectra were acquired on a 400 MHz AVANCE III instrument using CDCl$_3$ as solvent and TMS as internal standard. From the ^{1}H-NMR spectra we can identify the characteristic peak of each component.

GPC measurements were used to calculate the molecular weight and PDI. (Polydispersity index) Narrowly distributed polystyrene samples were used as calibration standards. All the block copolymers show a narrow distribution up to 1.13 during this synthesis process.

By controlling the amount of the initiator and the ratio of monomers, we can precisely alter the pitch dimension and morphology. For the hexagonally packed cylindrical morphology (Hex): L$_0$ is the center-to-center spacing and for the lamellar morphology (Lam): L$_0$ is the domain spacing (full-pitch).

Sample preparation and thermal annealing process

For the sample preparation for SAXS measurement, a 5 wt% BCP solution (in toluene) was drop-cast on the silicon wafer and then dried in vacuum oven at room temperature for 2 h. The resulting samples were annealed on an 80 °C hot plate, usually for 1 min, and then quenched on a chill plate at 0 °C.

AFM image of block copolymer thin film

Atomic force microscopy (AFM) images of the directed self-assembly pattern in Si templates were measured by Brucker Dimension Fast Scan in Peak Force mode. AFM images in Figure 2 further revealed the domain spacing and morphology of the self-assembled pattern. The finger print pattern showed 5 nm resolution and the height difference between two blocks are shown in this image. The film thickness of the BCP was ~23 nm, as measured by ellipsometry.

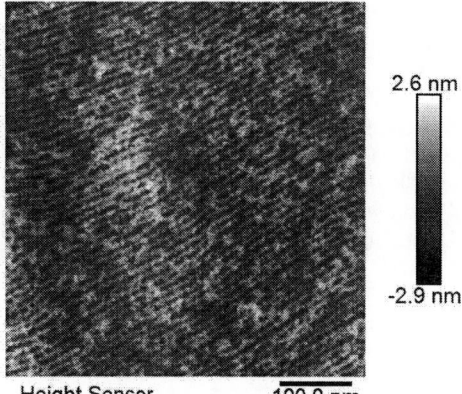

Figure 2. AFM image of block copolymer thin film (23 nm) on Si wafer after thermal annealing.

Cross-sectional FESEM images

A 45°-tilted cross-sectional SEM image of the thin film showed that a long-range-ordered nanopattern formed during the thermal annealing process. The line pattern showed sub-5 nm resolution after moving the "skin layer" by CF$_4$ plasma, which is shown in Figure 3.

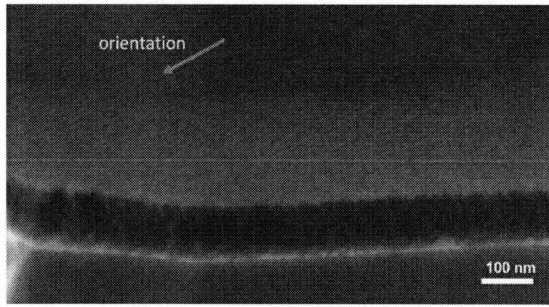

Figure 3. SEM image (45° tilted) of block copolymer thin film on Si wafer after thermal annealing.

Calculation of χ-parameter of block copolymer via SAXS Analysis

The χ value of block copolymer was estimated by Leibler's mean-field theory using the random phase approximation of the absolute intensity from SAXS. [7,8]

SAXS data and fitlines for disordered block copolymer at various temperatures are calculated. The best-fit lines were obtained by changing χ as one of the adjustable parameters using Leibler's mean-field theory. Four parameters, including the statistical segment lengths of both polymers and fitting constants K and χ, were optimized to fit the SAXS pattern at a certain temperature.

Our resulted BCPs demonstrated high χ ranged from 0.2 to 0.6, which due to the different monomer involved and these BCPs displayed resolution up to 5 nm half pitch by SAXS.

CONCLUSIONS

In this study, we designed and synthesized a family of fluorine-containing BCPs for DSA application. The finest half pitch of these BCPs is 5 nm or less, with both lamellar and hexagonal structure. The resulted BCPs demonstrated high χ ranged from 0.2 to 0.6, which displayed resolution up to 5 nm half pitch by SAXS, AFM and SEM. The small domain size combined with their rapid self assembly makes them promising as high throughput materials for next generation lithographic patterning technology.

ACKNOWLEDGEMENTS

This work was supported financially from the Shanghai Science and Technology Committee (18511104900), Fudan University (IDH1717041) and the Ministry of Science and Technology of China (2016YFA0203302). The authors also acknowledge experimental support from the State Key Laboratory of Molecular Engineering of Polymers and the Nano-fabrication Laboratory of Fudan University.

REFERENCES

[1] Kim, H.-C.; Park, S.-M.; Hinsberg, W. D. *Chem. Rev.* 2010, 110, 146.

[2] Jeong, S.-J.; Kim, J. E.; Moon, H.-S.; Kim, B. H.; Kim, S. O. *Nano Lett.* 2009, 9, 2300.

[3] Kennemur, J. G.; Yao, L.; Bates, F. S.; Hillmyer, M. A. *Macromolecules* 2014, 47, 1411.

[4] Bates, C. M.; Maher, M. J.; Janes, D. W.; Ellison, C. J.; Willson, C. G. *Macromolecules* 2014, 47, 2.

[5] Wang, C. X.; Li, X. M.; Deng, H. ACS Macro Letters, 2019, 8, 368.

[6] Li, X. M.; Li, J.; Wang, C. X.; Liu, Y. Y.; Deng, H. Journal of Materials Chemistry C, 2019, 7, 2535.

[7] Leibler, L. Macromolecules, 1980, 13, 1602.

[8] Sakamoto, N.; Hashimoto, T. Macromolecules, 1995, 28, 6825.

THE LAW THAT GUIDES THE DEVELOPMENT OF PHOTOLITHOGRAPHY TECHNOLOGY AND THE METHODOLOGY IN THE DESIGN OF PHOTOLITHOGRAPHIC PROCESS

Qiang Wu, Yanli Li, Yushu Yang, Shoumian Chen, Yuhang Zhao
Shanghai IC R&D Center
497 Gaosi Road, Zhangjiang Hi-Tech Park, Shanghai 201210, PR China
Contact: wuqiang@icrd.com.cn

ABSTRACT

Photolithography has been one of the key enabling technologies that continue to support the shrink of semiconductor manufacturing design rules. This technology started from 1 to 1 proximity or contact replication from a mask pattern to wafer image to the current large imaging projection based circuit pattern transfer [1]. The underlying principle for the fast development of the photolithography technology is that the replication process is through light propagation, which can process billions of patterns in parallel. For example, for modern 193 nm immersion process, the minimum pixel size is around 45 nm at a minimum pitch of 90 nm. For a full exposure shot spanning 26 mm by 33 mm, there are a total of 4.2×10^{11} pixels. If we use ASML NXT1980i exposure tool with a throughput of 275 wafer per hour, it only takes about 13 seconds for one 12 inch wafer exposure or about 160 ms for each exposure shot. In EUV, the parallelism will be higher by more than a factor of 4. This paper will summarize key process window performance parameters, such as imaging contrast/Exposure Latitude (EL), Mask Error Factor (MEF), Depth of Focus (DoF), linewidth uniformity, etc. from typical logic 0.25 μm, 0.18 μm, 0.13 μm, 90 nm, 65 nm, 45 nm, 28 nm, 20 nm, 16/14 nm, 10 nm, 7 nm, and 5 nm technology nodes and key enabling photolithography technologies that have been adopted in time to continually support technology advancement, such as anti-reflection coating, phase shifting mask, chemically amplified photoresist, polarization imaging, optical proximity correction, etc. This summary will result in a law that guides the continuous development of the photolithographic process for generations.

INTRODUCTION

The technology nodes and their year of production from 0.25 μm through the final 1 nm in shown in Figure 1 with nodes at 5 nm or beyond plotted as estimated [2]. It basically follows the Moore's law as a straight line in the logarithms plot. This linewidth shrink originates from the demand for continuous advancement of integrated chip manufacturing technology for Performance, Power, Area, and Cost (PPAC).

Starting at the 32/28 nm logic technology node, the Front-End-Of-the-Line (FEOL) Active Area (AA) and the gate layers have adopted unidirectional design rule for improved lithography performance, mostly in Critical Dimension Uniformity (CDU), Line Edge Roughness (LER), and LineWidth Roughness (LWR), which are related to the imaging contrast. Since larger LER and LWR may cause increased variability in the device parameters, such as the threshold voltage, the leakage current, and performance [3-5], in the FEOL, a key to the photolithography process is the imaging contrast.

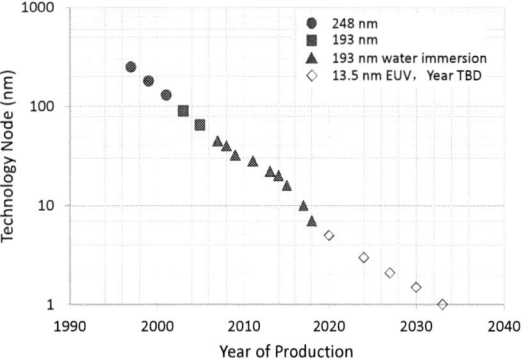

Figure 1: Technology node, year of production, and Exposure Wavelength [5].

I LIHEWIDTH UNIFORMITY REQUIREMENT AND ERROR BUDGET BREAKDOWN

To satisfy the PPAC requirement, in photolithography, there is the final gate CDU requirement, which is displayed in Figure 2 and Table I [2]. Note that the ratio between the CDU and the physical gate length is around 10%, which is nearly constant throughout the generations.

Shown in Figure 3 is a plot of CDU requirement and the mask CDU contribution, which is mask CDU (1X) multiplied by the Mask Error Factor (MEF), or the Mask Error Enhancement Factor (MEEF) determined by the photolithographic process. The data show that the contribution of mask CDU to the total gate CDU is about 50%. The rest of the CDU budget will be split among other sources, such as photolithography process, the etch process, the Optical Proximity Correction (OPC) accuracy, etc.

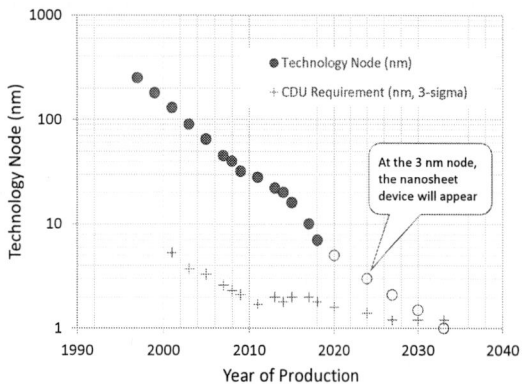

Figure 2: Gate CDU requirement, Technology node, and year [2].

TABLE I. GATE PITCH, CDU, PHYSICAL GATE LENGTH, AND CDU RATIOS FOR LOGIC TECHNOLOGY NODES FROM 130 NM TO 1 NM (DATA ARE OBTAINED FROM ITRS AND IRDS ROADMAPS AND FROM AUTHOR'S EXPERIENCE, 5 NM TO 1 NM ARE ESTIMATED FROM 2017 IRDS ROADMAPS [2])

Tech Node (nm)	Gate Pitch (nm)	Gate CDU Requirement (nm)	Ratio: Gate CDU/Gate Half Pitch	Physical Gate Length (nm)	Ratio: Gate CDU/Physical Gate Length
130	310	5.3	3.4%	65	8.2%
90	240	3.7	3.1%	45	8.2%
65	210	3.3	3.1%	32	10.3%
45	180	2.6	2.9%	25	10.4%
40	162	2.3	2.8%	23	10.0%
32	130	2.1	3.2%	20	10.5%
28	118	1.7	2.9%	16	10.6%
22	90	2	4.4%	20	10.0%
20	90	1.8	4.0%	18	10.0%
16/14	87	2	4.6%	17	11.8%
10	66	2	6.1%	20	10.0%
7	54	1.8	6.7%	18	10.0%
5	50	1.6	6.4%	16	10.0%
3	42	1.4	6.7%	14	10.0%
2.1	32	1.2	7.5%	12	10.0%
1.5	32	1.2	7.5%	12	10.0%
1	32	1.2	7.5%	12	10.0%

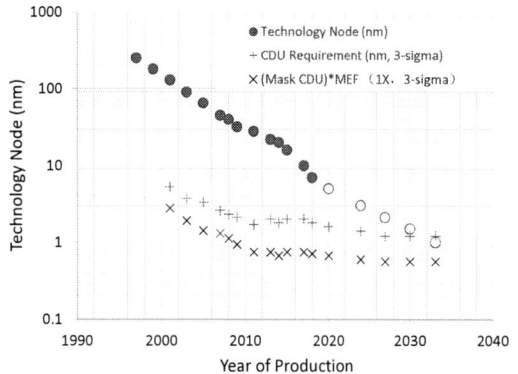

Figure 3: CDU, CDU breakdown, Mask CDU, Technology node, and year. The contribution of Mask CDU to the total gate CDU is about 0.5 [2].

II PARAMETERS THAT

CHARACTERIZE RESOLUTION LIMIT AND PHOTOLITHOGRAPHY PROCESS OPTIMIZATION

In projection imaging, the most important function that is used to characterize the lens performance is the Modulation Transfer Function (MTF), depicted by Figure 4. This plot indicates that it is not possible to provide high modulation, or contrast for all pitches with the minimum pitch being $0.5\lambda/NA$ (spatial frequency $2NA/\lambda$). In photolithographic processes, there are 3 basic performance indicators that are used: the Exposure Latitude (EL), the Depth of Focus (DoF), and the MEF. The EL is defined as the ratio between the range of exposure energy and the best exposure energy to within ±10% linewidth tolerance, and is related to the imaging contrast through Equation (1) for 50% duty ratio patterns near the minimum pitch.

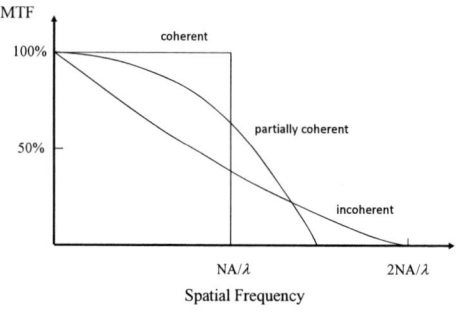

Figure 4: Modulation Transfer Function (MTF) versus illumination conditions with different coherence.

$$EL = 10\,\pi \times \text{imaging contrast}\,(\%) \qquad (1)$$

Here we show an example of a 20 nm gate process with 90 nm minimum pitch and 45 nm as the After-Developing-Inspection (ADI) linewidth, or CD. Since the 90 nm pitch uses unidirectional design along the Y direction, we use dipole X illumination condition for enhanced contrast. The dipole condition has an outer ring of 0.9 and inner ring of 0.7 in the illumination pupil. The pole angles that we experiment with are 35°, 60°, 90°, and 120° as depicted in Figure 5. Shown in Figures 6(a) through 6(d) are the simulated lithographic process windows of EL, MEF, DoF, and OPC, respectively as a function of the pole angle. The data indicate that as the pole angle increases, the EL at the minimum pitch (90 nm) will drop. The pole angle of 90° has an EL of 22% (for all pitches and the limit is 31.4% given by Equation (1)), a MEF ≤1.5, and a DoF 109.5 nm which are the best condition among the 4. Note that the minimum EL for the gate layer is around 18%, which will be explained later.

Besides one-dimensional (1D) situation, we need to review the printing in the two-dimensions (2D) as well, e.g., the line end to line end condition for the chip area consideration. Shown in Figures 7(a) through 7(c) are the simulation of a group of line end to line end (Tip-to-tip, or gap) structures as a function of the 4 dipole conditions we have just discussed. We have used the dense lines and spaces at the 90 nm pitch as our anchoring position (shown in Figure 7(a) as the "Energy Anchoring Line Cut" position) where the mask CD is 40 nm and ADI CD is 45 nm as before.

Figure 7(b) shows the 2D imaging result of the 4 dipole conditions. From the left to the right, we vary the gap mask CD to make sure that the EL of the gap is more than 10%, a requirement from experience which is less demanding than the 18% for the gate width. Shown in Figure 7(c) is a plot of the EL versus Gap CD on wafer (gap ADI CD). The dashed line is the minimum EL of 10% requirement, the cross points between the curves and the dashed line is the minimum ADI CD of the gap with EL equals to 10%. The data indicate that the minimum ADI gap CD will increase when the pole angle reduces. Although the gap CD for the 35°, 60°, and 90° look similar, the tip profile for the 35° and 60° are more pointy, which means that the area at the tip position become less as well, which will increase the tip to tip separation after plasma etch process. The optimization of pole angle will impact the final PPAC.

Besides the illumination condition optimization, we also need to optimize the photoresist process conditions. Ever since the introduction of Chemically Amplified photoResist (CAR) at the 250 nm technology node [6], a jump in imaging resolution have been obtained with more vertical cross section profile and lower exposure dose. However, too much diffusion will damage EL. Since 2004, the photoacid diffusion length has been first measured to 1~2 nanometer accuracy [7], the impact to the EL can be accurately quantified. For each technology node, an appropriate diffusion length is needed. Besides CAR, many other technologies have been introduced during the past 20 years, as shown in Figure 8. To name a few: starting at the 14 nm, 10 nm node, self-aligned processes have been introduced in the FEOL and BEOL, respectively, and starting at the 5 nm node, EUV technology has been implemented to large extent in the cut, contact/via, and metal layers.

sigma, and inner sigma.

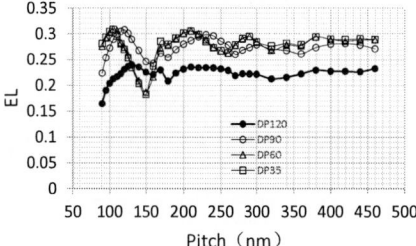

(a)

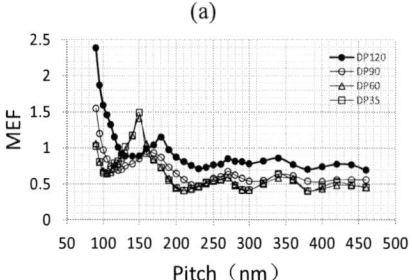

(b)

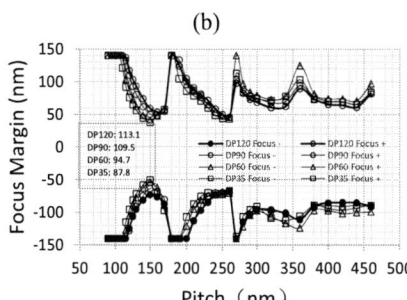

(c)

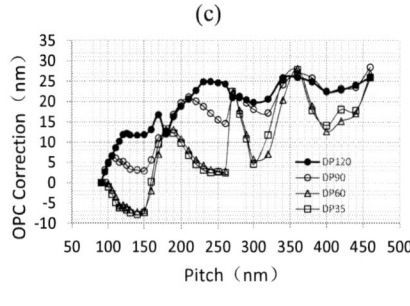

(d)

Figure 6: (a) EL, (b) MEF, (c) Focus Margin, and (d) OPC correction through pitch from different dipole (DP) illumination conditions: DP120°X, DP90°X, DP60°X, and DP35°X. The NA is 1.35 and the annulus setting is 0.9 outer sigma and 0.7 inner sigma with X/Y polarization setting and a typical CAR photoresist model with 90 nm thickness.

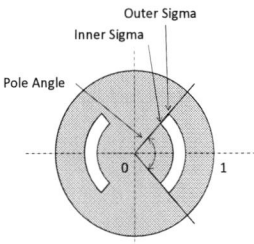

Figure 5: Schematic of dipole X illumination condition in the illumination pupil with definitions of pole angle, outer

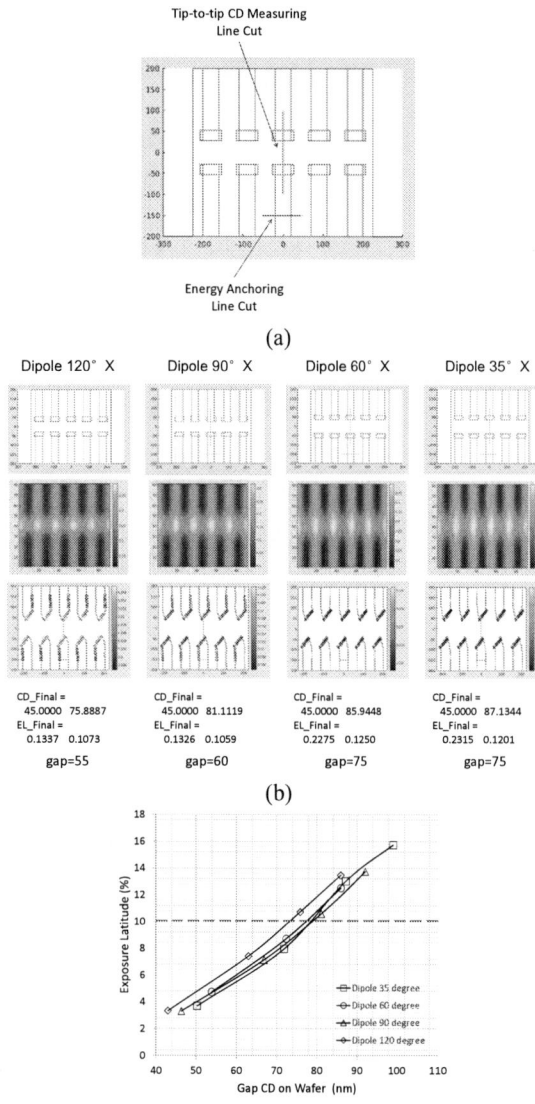

(a)

Dipole 120° X   Dipole 90° X   Dipole 60° X   Dipole 35° X

CD_Final =
45.0000 75.8887
EL_Final =
0.1337 0.1073

gap=55

CD_Final =
45.0000 81.1119
EL_Final =
0.1326 0.1059

gap=60

CD_Final =
45.0000 85.9448
EL_Final =
0.2275 0.1250

gap=75

CD_Final =
45.0000 87.1344
EL_Final =
0.2315 0.1201

gap=75

(b)

(c)

Figure 7: Typical tip-to-tip (gap) simulation at 90 nm pitch under dipole 120°X, 90°X, 60°X, and 35°X illumination conditions. The NA is 1.35 and the annulus setting is 0.9 outer sigma and 0.7 inner sigma with X/Y polarization setting and a typical CAR photoresist model with 90 nm thickness: (a) Line cut position description; (b) Mask patterns and aerial image intensity plots; (c) tip-to-tip EL vs ADI CD plot. The dashed line at EL of 10% represents the minimally accepted EL for the gap. The gap width is measured with exposure energy anchored at the dense lines and spaces described in (a).

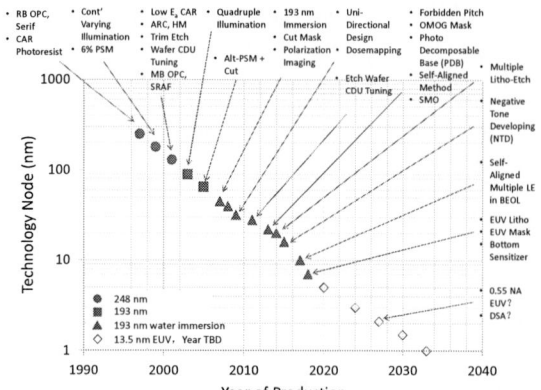

Figure 8: Key Technologies vs year of use. Acronyms: RBOPC: Rule-Based OPC; MBOPC: Model-Based OPC; ARC: Anti-Reflective Coating; HM: Hard Mask; Alt-PSM: Alternating Phase Shifting Mask; OMOG: Opaque Molybdenum silicide On Glass.

III   ROADMAPS   FOR   THE PHOTOLITHOGRAPHY   PROCESS DEVELOPMENT INCLUDING EUV

As mentioned in section I, the semiconductor integrated circuit has been evolving for more than 20 years since the 0.25 μm generation. There is a roadmap that guides the evolution of the design, device, process, equipment, material, and EDA tools. Here in the paper, we only list the requirement for photolithography process, which originates from the requirement of CDU and the chip area.

A recommended photolithography method for various technology nodes is summarized in Table II. Shown in Figures 9, 10, and 11 are the roadmap plots for EL, MEF, and LER/LWR from the 250 nm technology node to the now 7 nm technology node and the future 5 nm, 3 nm, 2.1 nm, 1.5 nm, and 1nm nodes.

TABLE II.   PHOTOLITHOGRAPHY METHOD RECOMMENDED FOR TECHNOLOGY NODES FROM 250 NM TO THE CURRENT 7 NM, AND ESTIMATED FOR THE FUTURE 5, 3, 2.1, 1.5, AND 1 NM.

Tech Node (nm)	Gate Pitch (nm)	Gate Lithography Method	Metal Pitch (nm)	Metal Lithography Method
250	500	248 nm	640	248 nm
180	430	248 nm	460	248 nm
130	310	248 nm	340	248 nm
90	240	193 nm	240	193 nm
65	210	193 nm	180	193 nm
45	180	193 nm immersion	160	193 nm immersion
40	162	193 nm immersion	100	193 nm immersion
32	130	193 nm immersion	90	193 nm immersion
28	118	193 nm immersion	90	193 nm immersion
22	90	193 nm immersion	80	193 nm immersion
20	90	193 nm immersion	64	193 nm immersion LELE
16/14	87	193 nm immersion SADP	64	193 nm immersion LELE
10	66	193 nm immersion SADP	44	193 nm immersion SALELE
7	54	193 nm immersion SADP	40	193 nm immersion SALELE
5	50	193 nm immersion SADP	30	0.33 NA EUV SALELE
3	42	193 nm immersion SADP	24	0.33 NA EUV SALELE
2.1	32	193 nm immersion SAQP	18	0.55 NA EUV SALELE
1.5	32	193 nm immersion SAQP	14	0.55 NA EUV SALELE
1	32	193 nm immersion SAQP	14	0.55 NA EUV SALELE

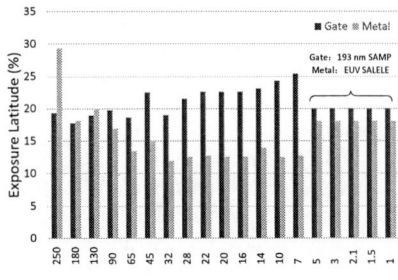

Figure 9: EL Roadmaps

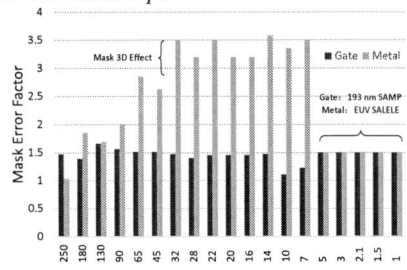

Figure 10: MEF Roadmaps.

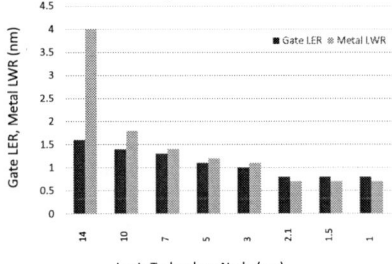

Figure 11: LER/LWR Roadmaps

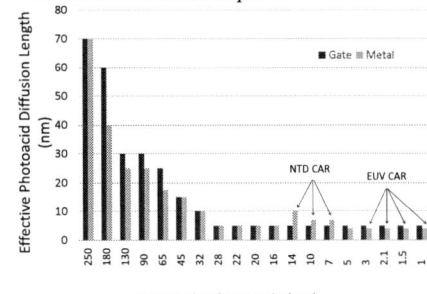

Figure 12: Roadmap for the Effective Photoacid Diffusion Length

The future nodes are estimated from the available sources, such as the performance of the current EUV exposure tools and the EUV photoresists, and the 2017 IRDS roadmaps that summarize current progress in the device research and process development in general [2].

Figure 12 shows the roadmap of photoacid diffusion length for chemically amplified photoresist. Note that the diffusion length drops sharply from the initial 70 nm to the current 5 nm for the PTD 193 nm immersion process and about 7~10 nm for the NTD 193 nm immersion process and about <5 nm for the EUV process.

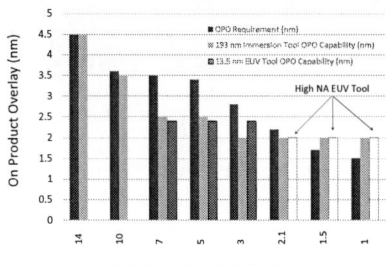

Figure 13: Overlay Roadmaps

TABLE III. DEPTH OF FOCUS REQUIREMENT ROADMAP FROM 250 NM TO THE CURRENT 7 NM, AND ESTIMATED FOR THE FUTURE 5, 3, 2.1, 1.5, AND 1 NM.

Tech Node (nm)	Gate Pitch (nm)	Gate Lithography Method	Depth of Focus (nm)	Metal Pitch (nm)	Metal Lithography Method	Depth of Focus (nm)
250	500	248 nm	500-600	640	248 nm	600-800
180	430	248 nm	450	460	248 nm	600
130	310	248 nm	350	340	248 nm	450
90	240	193 nm	350	240	193 nm	450
65	210	193 nm	250	180	193 nm	300
45	180	193 nm immersion	150	160	193 nm immersion	200
40	162	193 nm immersion	150	100	193 nm immersion	200
32	130	193 nm immersion	80	90	193 nm immersion	80-100
28	118	193 nm immersion	80	90	193 nm immersion	80-100
22	90	193 nm immersion	60	80	193 nm immersion	60-80
20	90	193 nm immersion	60	64	193 nm immersion LELE	60-80
16/14	87	193 nm immersion SADP	60	64	193 nm immersion LELE	60-80
10	66	193 nm immersion SADP	60	44	193 nm immersion SALELE	60-75
7	54	193 nm immersion SADP	55	40	193 nm immersion SALELE	55-70
5	50	193 nm immersion SADP	50	30	0.33 NA EUV SALELE	55
3	42	193 nm immersion SADP	50	24	0.33 NA EUV SALELE	55
2.1	32	193 nm immersion SAQP	50	18	0.55 NA EUV SALELE	30-40
1.5	32	193 nm immersion SAQP	50	14	0.55 NA EUV SALELE	30-40
1	32	193 nm immersion SAQP	50	14	0.55 NA EUV SALELE	30-40

Shown in Figure 13 is the roadmap of another important parameter besides imaging, the overlay, which is largely dependent on the exposure tool development in positioning accuracy. It also depend on the processes such as thermal processes (e.g., rapid thermal annealing), which may cause wafer warpage, the etch process, and the Chemical Mechanical Planarization (CMP) which may distort the overlay and alignment marks into rotational like distribution on wafer. Due to the use of self-aligned process starting at the 14 nm process, the requirement has been relaxed which seems to keep exposure tool performance within renewed requirement.

Besides the above parameters, there is also the DoF. Although this is mostly determined by the exposure tool performance in focusing stability and wafer leveling accuracy, the thickness of photoresist also play an important part. Since the photoresist is used as an etch mask, thicker photoresist can prevent etch damage to the protected area on wafer, and thus provide more etch process window. But, it will reduce imaging DoF, which will cause more CD variation. A DoF requirement

978-1-7281-6559-2/20 $31.00 © 2020 IEEE

roadmap based on experience and simulation is displayed in Table III.

IV CHARACTERISTICS OF EUV LITHOGRAPHY TECHNOLOGY

The introduction of EUV can simplify patterning process with fewer masks. However, due to the all-reflective design, there will be an off-axis angle in the illumination Chief Ray Angle at Object (CRAO). The angle is set to 6° in the current mainstream tools, like the ASML NXE3XXX series. And the all reflective surfaces have to use high-low multilayer design due to most materials don't reflect EUV very much. This will cause imaging shadowing effect, Horizontal-Vertical CD bias (H-V Bias), mask 3D effect [8-9]. The use of 13.5 nm higher energy photons will cause photon absorption stochastics which limits the line space minimum pitch to be around 36~40 nm and 48~50 nm for holes [8-10]. In addition, the use of small NA 0.33 will cause more aberration sensitivity [8-9].

VI THE METHODOLOGY IN THE DESIGN OF PHOTOLITHOGRAPHIC PROCESS

1. Choose photoresists that have balanced properties for both photolithography and etch processes: for example, the NTD process provides a reduction of etch bias while sacrificing some imaging contrast, or EL and MEF near the minimum pitch.
2. Optimize photoresist thickness, PEB temperature, BARC thickness, substrate reflectivity, and etch bias.
3. Keep minimum EL and max MEF requirements at the minimum pitches and as much as possible for other pitches and 2D patterns with illumination condition optimization (may use Source Mask co-Optimization, SMO), mask bias optimization (for photoresist type, MEF, EL, and CD through pitch consideration).
4. Optimize photoresist process in EUV lithography, minimize impact from stochastics while maintaining acceptable exposure energy and EL. Manage aberration impact in the exposure tools and the mask 3D effect, maintain acceptable DoF.
5. Use self-aligned multiple patterning processes for maximum gain in overlay and LWR/LER and with unidirectional design rules.

CONCLUSIONS

We have studied the evolution of photolithography process and process requirement from 0.25 μm to the current 7 nm technology nodes and provide an estimate of the future 5, 3, 2.2, 1.5, and 1 nm technology nodes. We have studied the CDU requirement and CDU requirement for the mask. We have analyzed the evolution of key lithography parameters, such as the EL, MEF, and LWR/LER, overlay, and DoF. We have also done a study of the evolution of the effective photoacid diffusion length and provide an estimate for the future developments in EUV photoresist. Finally, we have mentioned distinctive characteristics in the EUV lithography.

As a result of this study, we have summarized information from exposure tools, photoresists, and process information etc. into a guideline on the setup of photolithography and have provide an outlook for future generations with EUV lithography process.

ACKNOWLEDGMENT

We would like to thank Shanghai ICRD colleagues and higher management team for the support of this work.

REFERENCES

[1] http://www.lithoguru.com/scientist/litho_history/Kato_Litho_History.pdf.
[2] 2001 ITRS Roadmap, 2005 ITRS Roadmap, 2013 ITRS Roadmap, 2017 IRDS Roadmap.
[3] G.F. Lorusso et al., "Impact of Line Width Roughness on Device Performance", Proc. SPIE 6152, 61520W, 2006.
[4] C. Gustin, L.H.A. Leunissen, A. Mercha, S. Decoutere, and G. Lorusso, "Impact of line width roughness on the matching performances of next-generation devices", Thin Solid Films 516, 3690, 2008.
[5] R.Wang, X. Jiang, T. Yu, J. Fan, J. Chen, D. Z. Pan, and R. Huang, "Investigations on Line-Edge Roughness (LER) and Line-Width Roughness (LWR) in Nanoscale CMOS Technology: Part II–Experimental Results and Impacts on Device Variability", IEEE Trans. electron Devices 60, 3676, 2013.
[6] H. Ito and G.C. Willson, "Chemical Amplification in the Design of Dry Developing Resist Materials," Polym. Eng._and_Sci. 23, 1012, 1983.
[7] Q. Wu, S. Halle, and Z. Zhao, "The Effect of the Effective Resist Diffusion Length to the Photolithography at 65 and 45 nm Nodes, A Study with Simple and Accurate Analytical Equations", Proc. SPIE 5377, 1510, 2004.
[8] Q. Wu, Y. Li, Y. Yang, and Y. Zhao "A Photolithography Process Design for 5 nm Logic Process Flow", Journal of Microelectronic Manufacturing 2, 19020408, 2019.
[9] Yanli Li, Qiang Wu, and Yuhang Zhao, "Simulation Study for Typical Design Rule Patterns and Stochastic printing failures in 5 nm Logic Process with EUV Photolithographic Process", submitted to Proc. CSTIC 2020.
[10] P. De Bisschop, E. Hendrickx, "Stochastic printing failures in EUV lithography", Proc. SPIE 10957, 109570E, 2019.

14NM FIN SADP PATTERNING PROCESSES AND INTEGRATION

Chunyan Yi，Ming Li，Yongjian Lou, Weijun Wang, Zhunhua Liu，Xiaoqiang Zhou，Wen Xu, Ying Zhang

Shanghai IC&RD Center, shanghai 201210,China

Email: yichunyan@icrd.com.cn

ABSTRACT

This paper describes the development of a Self-Aligned Double Patterning (SADP) scheme for advanced technology node 14 nm Fin. The process development focus on profile tuning and pitch walking reduction. This study has achieved a tunable fully integrated SAPD process flow, which has met all key specs for 14nm Fin patterning. This study established an technical foundation which can be used in other key SADP applications and a platform for extending our patterning process capability to SAQP(Self-Aligned Quadruple Patterning) for advanced technology nodes.

INTRODUCTION

As integrated circuits line width continually shrinks, at 14nm tech node, some critical dimensions are out of the traditional lithographic limitation. SADP(Self-Aligned Double Patterning) technique successfully fills in the gap between 193nm immersion lithography and EUV lithography, moreover, tt also serves as the foundation for 7nm FinFET SAQP process. Because of its process complexity, there is no domestic semiconductor tools used in mass production of 14nm SADP yet. Based on mass production criterion, Shanghai IC&RD Centre developed a integrated 14nm Fin patterning scheme including SADP and Fin etch. All etches are performed with NAURA NMC612D ICP etcher.

SADP patterning process flow is shown in figure1:

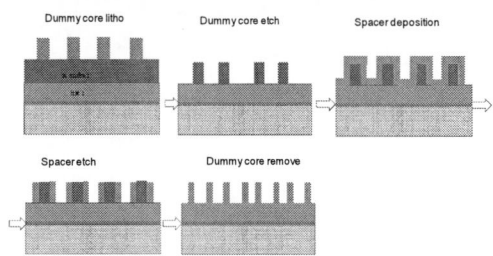

Figure 1 SADP patterning process flow

In above scheme, the dummy core material is amorphous silicon, which has better LWR performance comparing to Poly core, also this material is compatible with current mainstream CMOS process. Spacer film is ALD oxide, which has higher etch selectivity.

EXPERIMENTS

FIN+STI etch process development and optimization

Fin profile is the most important parameter for device performance. The most challenging index is pitch walking which is inherent for SADP technique. The final fin space CD could be different between two adjacent fin structures [1-5]. This pitch walking could induce STI depth micro-loading. Meanwhile we also need control within wafer STI depth macro-loading effect. Figure 2 is an example of serous pitch walking induced STI depth loading.

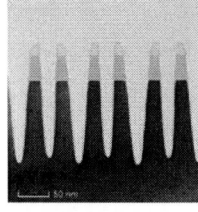

Figure 2 bad pitch walking example of fin etch

Fin profile tuning

In order to get an ideal smoothing fin profile, we used multi-steps fin etch recipe. This fin etch recipe including HM etch, fin etch, STI etch and other transition steps for polymer clean and deposition. Good Fin profile is achieved through adjustment of these step's etch and sidewall polymer deposition.

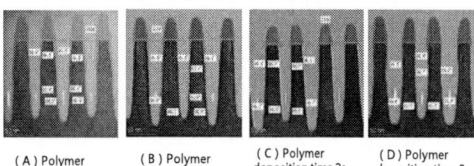

(A) Polymer deposition time 6s   (B) Polymer deposition time 4s   (C) Polymer deposition time 2s   (D) Polymer deposition time 0s

Figure 3 14nm Fin profile tuning TEM results

Figure 3 a shows a fin profile tuning trend. Through adjusting polymer deposition time, fin profile changes from double slope (figure3A) to smooth tapered (figure3B), then to smooth bowing (figure3D). Above results showed that we can find a balance point (figure3C) to make profile vertical and smoothing, when polymer deposition time is 2s we got the best fin profile. Even though above fin etch profile meets integration target, it's has serious STI depth micro-loading due to pitch walking.

Fin Pitch walking optimization

Figure 4 illustrates CD transfer procedure of SADP technique. Spacer CD is determined by ALD deposition thickness and core remove step's CD bias. As illustrated in figure4, a, b and c represent space CDs between fins, spl and spr represent left and right spacer CD respectively. In ideal case:

spl=spr

a=core CD

b= core pitch-2spl- core CD

c= core CD

Pitch walking=|a-b|=2 core CD+2spl- core pitch

If core pitch is 96nm，then：

pitch walking=2core CD+2spl- 96

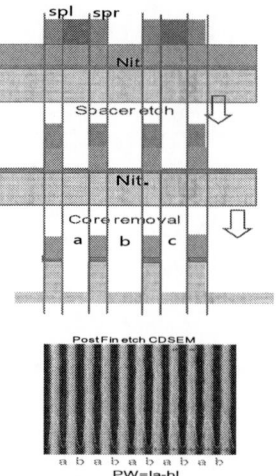

Figure 4 SADP double patterning CD transfer process

Based on above equations, if spacer CD is fixed, we can adjust core CD to get a minimum pitch walking. But in actual process, there are too many factors impacting pitch walking, such as core profile, core AEI CD measurement method, core removal etch step induced spacer CD bias etc. [2], all these factors need large amount of experiments to find their complex interaction. In order to solve STI depth loading issue caused by   pitch walking in short time, we kept spacer deposition thickness at 200A, collected full wafer Space CD after fin etch, calculated each die's pitch walking value, finally correlated this fin space CD with core AEICD die by die(shown as figure 5). Through this method we got an optimal core CD value which can make STI depth loading meeting integration target.

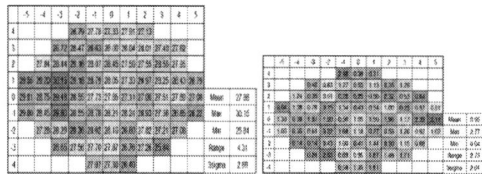

Figure5 (A) Fin core AEICD with CDSEM algorithm A （B）calculated fin pitch walking value

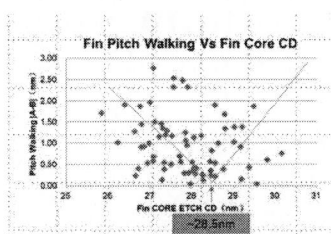

Figure 6. Core CD VS pitch walking with CDSEM algorithm A

According to figure 6, we targeted core CD at 28.5nm and carried out fin etch test on real wafer, the fin etch TEM result as showed in figure 7.

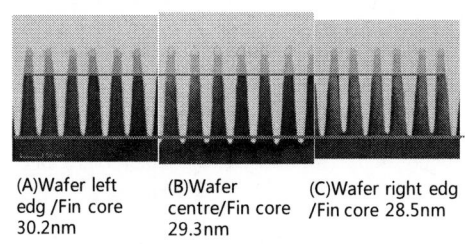

(A)Wafer left edg /Fin core 30.2nm   (B)Wafer centre/Fin core 29.3nm   (C)Wafer right edg /Fin core 28.5nm

Figure 7. Algorithm A STI depth loading results

Base on CDSEM algorithm A, the calculated pitch walking doesn't match with real data（figure7C）. The actual TEM results showed that, when core CD is 28.5nm, pitch walking induced STI depth loading (77A) is the worst among those three core CDs. So we optimized CD-SEM algorithm, figure 8 is based on new algorithm B, the simulated best fin core CD is 30.2nm.

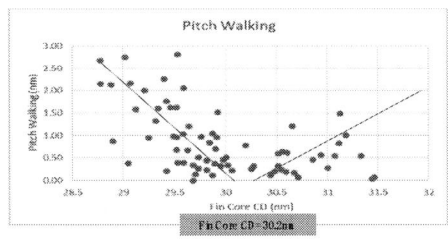

Figure 8 Simulated core CD VS pitch walking based on CDSEM algorithm B

Base on above algorithm B's results we carried out real wafer tests to verify it, test results are shown is figure 9.

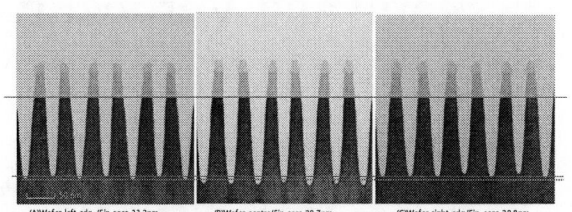

(A)Wafer left edg /Fin core 33.2nm   (B)Wafer centre/Fin core 29.7nm   (C)Wafer right edg/Fin core 28.8nm

Figure 9 algorithm B STI depth loading results

Fin core AEICD	STI depth microloading
33.2nm	27A
29.7nm	11A
28.9nm	37A

When Fin core CD is 29.7nm (figure 9B), the STI depth loading is the best (11A), this data is quite match with simulated value.

CONCLUSION

In this paper we introduced a 14nm FinFET fin patterning scheme and the main challenges in each of those important process steps and provided solutions. The final fin etch results showed that we got a mass production satisfied integration scheme. This work also demonstrated that NAURA's advanced ICP etcher's capability on 14nm node fin patterning and etch.

ACKNOWLEDGMENT

The authors gratefully acknowledge the support of dry etch group of NAURA for their contribution to this work. This research was also supported by ICRD's advanced module group.

REFERENCES

[1] Min-hwa Chi, USA. Challenges in Manufacturing FinFET at 20nm node and beyond, 2012

[2] Sylvain Baudot,and Sofiane Guissi et.al, "N7 FinFET Self-Aligned Quadruple Patterning Modeling", IEEE , 2018.

[3] M. LaPedus, "Top Five Design and Manufacturing Challenges at 20nm", Mar. 21, 2012.
http://semimd.com/blog/2012/03/21/
top-five-design-and-manufacturing-challenges-at-20nm

[4] K. J. Kanarik, G. Kamarthy, and R.A. Gottscho "Plasma etch challenges for FinFET transistors", Solid State Technology, V.55, No.3, April 2012.

Towards Microstructures with Ultrahigh Aspect-Ratio and Verticality in Deep Silicon Etching

Yuanwei Lin
Department of Semicondutor Etching
NAURA Technology Group Co., Ltd.
Beijing, P. R. China
Email address: yuanweilin@pku.edu.cn
ORCID: 0000-0002-5293-1777

Abstract—Devices containing deep silicon trenches with higher aspect ratio and higher verticality could achieve better performance, such as higher carrier mobility in trench gate field effect transistors, lower on-resistance in power switching devices, larger capacitance in silicon capacitors, and so on. In this work, we demonstrate a deep silicon trench structure with aspect ratio of >65, depth of >100 μm and high perpendicularity of 90°±0.1°. This is realized through optimization of the process recipe, which could control the balance between deposition and etching. The high aspect ratio silicon trench with high verticality has significance to advancing the field of silicon device fabrication.

Keywords—Deep silicon etching, Bosch process, Silicon microstructures, Aspect ratio, Verticality

I. INTRODUCTION

Though two-dimensional materials and III-V group materials are thought to be promising in the fabrication of next-generation electronic devices, silicon-based devices are still the mainstream of the modern semiconductor industry, and tremendous efforts have been made to perpetuate the vitality of silicon. Introducing deep trench structure into silicon devices is one of the approaches to ameliorate the device performance. For instance, carrier mobility reduction caused by channel electron scattering and undesirable short channel effect could be both inhibited when trench gate is utilized to substitute traditional planar gate [1, 2]. Super junction is another example for device performance improvement by deep microstructures [3, 4], which could effectively reduce the on-resistance/higher doping concentration and increase the switching rate/shorten depletion region. For the reason that the driving current in super junction devices with planar structure is low, vertical super junction is generally adopted, where the fabrication technology is developed from multiple epitaxial process to deep trench process. As for discrete devices that originally require trench etching, they are also developing toward high aspect ratio and high verticality. Silicon capacitors are typical devices that requires deep trench structures, which could increase the use efficient of device space and increase the capacitance [5-7]. The higher verticality makes it more consistent between the practical device and the theoretical model, which better meets the design demands. Therefore, the development of semiconductor discrete devices has witnessed increasing demands on deep trench etching.

Owing to the high aspect ratio and high verticality in the deep silicon trench, conventional wet etching with isotropic reaction is impossible to realize, whereas anisotropic dry etching should be used. Dry etching is a low temperature plasma based technique that uses ionized gas to etch the wafer by physical bombardment and chemical reaction [8-

10]. For deep silicon etching, it becomes more difficult as the etching depth increases, where a V-shaped topography is eventually formed. This is mainly due to two reasons: First, the reactive plasma or etchant is hard to enter the deep microstructure, inhibiting the etching at the bottom of the microstructure; second, the product of the etching formed at the bottom is also difficult to carry away from the reaction system, suppressing etching according to chemical equilibrium. Optimizing the hardware of the etching apparatus might partially address the issue with high costs. Thus, to obtain deep silicon microstructures with high aspect ratio and high verticality in practical way has been the direction of both the scientific and industrial community.

In this study, through optimization of the process recipe, a deep silicon trench structure with aspect ratio of >65, depth of >100 μm and high perpendicularity of 90°±0.1° is achieved. The balance between deposition and etching is controlled by introducing ramping to both the bias power and the etching time. The sidewall of the trench is smooth with scallop size of less than 20 nm, which would not damage the electrical performance of the device. The silicon trench with high aspect ratio and high verticality has significance to advancing the field of silicon device fabrication.

II. EXPERIMENTAL SECTION

Commercially available 8 inch silicon wafer with (100) face upward was used in this study. SiO_2 hard mask (HM) with thickness of ~0.6 μm was grown on the silicon wafer by PECVD using a NMC EPEE 550. The HM was opened by using Ar and CF_4 plasma in a GDE chamber of a NMC MASE 200 etcher after defined by photo-lithographically. As shown in Fig.1a, the remaining photoresist (PR) is ~2.4 μm thick. The deep dry silicon etching was then performed in a HSE chamber of a NMC MASE 200 etcher. The scanning electron microscope (SEM) images were taken by using FEI Inspect and Hitachi SU8000.

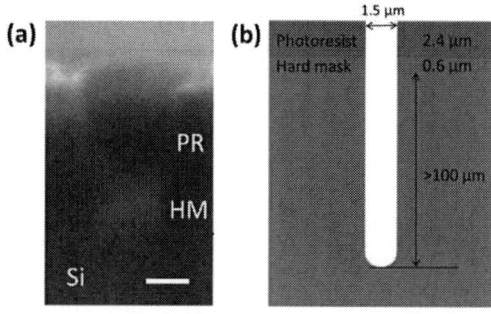

Fig. 1.   (a) Cross-sectional SEM image of the wafer before etching. Scale bar: 1 μm. (b) Schamatic of the cross-sectional view of the deep silicon trench.

III. RESULTS AND DISCUSSION

Considering the plasma has higher mean free path at lower chamber pressure, it is feasible to obtain deeper microstructures by simply lowering the chamber pressure. However, extremely low chamber pressure would cause troubles in evacuation capability and ignition of radiofrequency (RF). Many hardware upgrades are thus needed to address these issues, such as increasing the pumping rate of the vacuum pump, or improving the RF system.

When compared with hardware upgrade, optimization of the Bosch process recipe is a more convenient method. As shown in Fig. 1b, the etchant needs to enter the trench with depth of >100 μm. Bias electrode of RF is the key part to make the plasma generated by the source electrode move downward. Thereby, increasing the bias power could guarantee more plasma to enter the bottom of the deep trench. On the other hand, extending the single-step etching time could promote the products at the bottom of the deep trench to move away from the chemical reaction system. Taking these two aspects into consideration, we design a combined ramping technique as following:

$$P = P_{\text{initial}} + (P_{\text{final}} - P_{\text{initial}})*n/n_{\text{total}} \qquad (1)$$

$$t = t_{\text{initial}} + (t_{\text{final}} - t_{\text{initial}})*n/n_{\text{total}} \qquad (2)$$

where P, t, n, n_{total}, P_{initial}, P_{final}, t_{initial} and t_{final} are defined to be bias power, etching time at present, cycle number at present, Bosch total cycle number, initial bias power, final bias power, initial etching time and final etching time, respectively. The method could be seen by flow chart mode in Fig. 2. Definitely, the final bias power should be larger than the initial bias power, and so is the situation in the etching time.

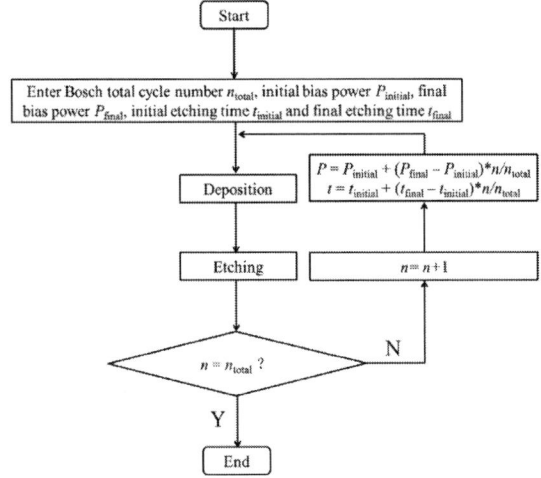

Fig. 2. Flow chart of the optimized Bosch process.

By using this optimized process technique shown above, a deep trench of silicon with depth of more than 100 μm could be obtained with critical dimension (CD) width of 1.5 μm, as illustrated in Fig. 3.

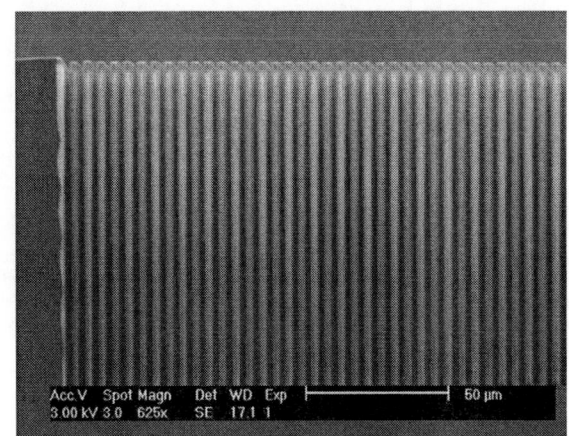

Fig. 3. Cross-sectional SEM image of the deep silicon trench.

The optimized Bosch process recipe is shown in TABLE I.

TABLE I. MODIFIED BOSCH RECIPE

Step	Recipe					
	Chamber pressure	Source power	C₄F₈ flow	SF₆ flow	Bias power	Time
deposition	35 mTorr	1800 W	150 sccm	0	1 W	0.7 s
etching	50 mTorr	2200 W	0	200 sccm	55 W ~ 70 W[a]	1.2 s ~ 1.5 s[b]

[a.] Ramping of bias power.

[b.] Ramping of etching time.

The statistics on the etching profile is shown in TABLE II.

TABLE II. ETCHING PROFILE

Depth	Designed CD	Top CD	Middle CD	Bottom CD	Sidewall angle
125 μm	1.50 μm	1.36 μm	1.36 μm	1.47 μm	~89.95°

Due to the alternate deposition and etching steps in Bosch process, undesirable scallop structure would be introduced making the sidewall rougher than that by using non-Bosch process. Though the ramping of etching time is utilized in the modified Bosch process shown in TABLE I, the scallop size of the deep trench is less than 20 nm (Fig. 4).

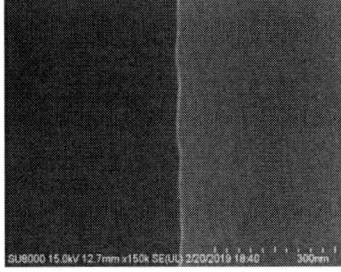

Fig. 4. SEM image showing the scallop size of the deep trench.

It is noted that porous silicon structure could be fabricated by illuminated electrochemical etching with aspect-ratio larger than 100 [11, 12]. However, the sidewall of the microstructure obtained by electrochemical method is rather rough, which hamper the practical application to electronic devices [13, 14]. In addition, Tang *et al* [15] reported ultra-deep reactive ion etching of high aspect-ratio in thick silicon through ramped-parameter process with relatively large trench width variation from top to bottom, whereas the CD width variation in this work is within 0.2 μm (i.e. better sidewall angle).

Otherwise, control experiments showed that damage and grass appeared in the bottom of the trench if the ramping of bias power and etching time is eliminated (Fig. 5). To avoid the abnormal profile, the depth would be decreased to ~80 μm (with aspect ratio of ~50). These control results prove the ramping technique is necessary to obtain ultrahigh aspect-ratio and verticality in deep silicon etching.

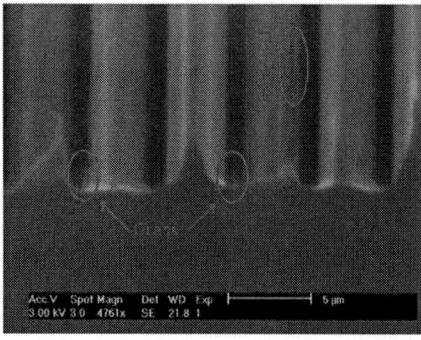

Fig. 5. SEM image showing the damage and grass in the control experiment without ramping of bias power and etching time.

The deeper of the silicon trench, the harder for the reactive plasma or etchant to enter the deep microstructure. So is the situation for the product of the etching formed at the bottom to carry away from the reaction system. That's why a gradient ramping of the bias power and etching time is required rather than a maximum of them.

IV. CONCLUSION

In summary, through optimization of the process recipe, we demonstrate a deep silicon trench structure with aspect ratio of >65, depth of >100 μm and high perpendicularity of 90°±0.1°. The ramping of both the bias power and the etching time is introduced to control the balance between deposition and etching. The smooth silicon trench with high aspect ratio and high verticality would have potential applications in power chip, micro-electro-mechanical systems (MEMS), and advanced packaging.

ACKNOWLEDGMENT

The author thanks Ms. Yi Pang from NAURA Technology Group Co., Ltd. and Ms. Jie Chen from 58th Institute of China Electronic Technology Group Co., Ltd. for beneficial discussions on this work.

REFERENCES

[1] B. Han, H. Takamizawa, Y. Shimizu, K. Inoue, Y. Nagai, F. Yano, Y. Kunimune, M. Inoue, and A. Nishida, "Phosphorus and boron diffusion paths in polycrystalline silicon gate of a trench-type three-dimensional metal-oxide-semiconductor field effect transistor investigated by atom probe tomography," *Appl. Phys. Lett.* vol. 107, pp. 023506 , 2015.

[2] W. Wang, R. Ning, and Z. John Shen, "Numerical study of PN-doped poly-silicon shield gate trench MOSFET with reduced output capacitance," *IEEE Electron Device Lett.*, vol. 38, pp.1055-1058, August 2017.

[3] Sameh G. Nassif-Khalil, and C. Andre T. Salama, "Super-Junction LDMOST on a silicon-on-sapphire substrate," *IEEE Trans. Electron Devices,* vol. 50, pp.1385-1391, May 2003.

[4] F. D. Bauer, "The super junction bipolar transistor: a new silicon power device concept for ultra low loss switching applications at medium to high voltages," *Solid State Electron.* vol. 48, pp. 705–714, 2004.

[5] H. Hu, C. Zhu, Y. F. Lu, M. F. Li, B. J. Cho, and W. K. Choi, "A high performance MIM capacitor using HfO_2 dielectrics," *IEEE Electron Device Lett.*, vol. 23, pp.514-516, September 2002.

[6] F. Roozeboom J. H. Klootwijk, J. F. C. Verhoeven, F. C. van den Heuvel, W. Dekkers, S. B. S. Heil, J. L. van Hemmen, M. C. M. van de Sanden, W. M. M. Kessels, F. Le Cornec, L. Guiraud, D. Chevrie, C. Bunel, F. Murray, H. D. Kim and D. Blin "ALD options for Si-integrated ultrahigh-density decoupling capacitors in pore and trench designs," *ECS Trans.*, vol. 3, pp. 173-181, May 2007.

[7] J. H. Klootwijk, K. B. Jinesh, W. Dekkers, J. F. Verhoeven, F. C. van den Heuvel, H.-D. Kim, D. Blin, M. A. Verheijen, R. G. R. Weemaes, M. Kaiser, J. J. M. Ruigrok, and F. Roozeboom, "Ultrahigh capacitance density for multiple ALD-grown MIM capacitor stacks in 3-D silicon," *IEEE Electron Device Lett.*, vol. 29, pp.740-742, July 2008.

[8] F. Lärmer, and A. Schilp, Patents DE 4241045, US 5501893 and EP 625285, 1996.

[9] B. Wu, A. Kumar, and S. Pamarthy, "High aspect ratio silicon etch: A review," *J. Appl. Phys.* vol. 108, pp. 051101, 2010.

[10] Y. Lin, R. Yuan, X. Zhang, Z. Chen, H. Zhang, Z. Su, S. Guo, X. Wang, and C. Wang, "Deep dry etching of silicon with scallop size uniformly larger than 300 nm," *Silicon*, vol. 11, pp. 651-658, 2019.

[11] T. Geppert, S. L. Schweizer, U. Gosele, R. B. Wehrspohn, "Deep trench etching in macroporous silicon," *Appl. Phys. A*, vol. 84, pp. 237-242, 2006.

[12] E. V. Astrova, G. V. Fedulova, and E. V. Guschina, "Formation of 2D photonic crystal bars by simultaneous photoelectrochemical etching of trenches and macropores in silicon," *Semiconductors*, vol. 44, pp. 1617-1623, 2010.

[13] V. Lehmann, and H. Foll, "Formation mechanism and properties of electrochemically etched trenches in n-type silicon," *J. Electrochem. Soc.*, vol. 137, pp. 653-659, 1990.

[14] H. Foll, "Properties of silicon-electrolyte junctions and their application to silicon characterization," *Appl. Phys. A*, vol. 53, pp. 8-19, 1991.

[15] Y. Tang, A. Sandoughsaz, K. J. Owen, K. Najafi, "Ultra deep reactive ion etching of high aspect-ratio and thick silicon using a ramped-parameter process," *J. Microelectromech. S.*, vol. 27, pp. 686-697, 2018.

THE SOLUTION OF AIO-ET VIA OPEN AND PROCESS WINDOW IMPROVEMENT

Baichun Zhang*, Jianguo Yang, Lei Sun，Quanbo Li，Jun Huang，Yu Zhang
Shanghai Huali Integrated Circuit Corporation, Shanghai 200120, China
*Corresponding Author's Email: zhangbaichun@hlmc.cn

BIOGRAPHY

Baichun Zhang got the master degree in Shanghai University, Shanghai, China, in 2017. The same year he joined the Huali Integrated Circuit Corporation where he is engaged in etch process development of advanced copper/low-k interconnects for ultra large scale integration.

ABSTRACT

With the development of VLSI (Very Large Scale Integrated Circuits) manufacture, the continuous shrinkage of line width critical dimensions should meet high performance and high integrated level of device requirement. More and more new materials and new processes were introduced to meet the process requirement of critical layer. This paper investigates the solutions of the via open and the window improvement of All in One etch (AIO-ET). In this paper, Transmission Electron Microscope (TEM), Wafer Acceptance Test (WAT) and Chip Probing (CP) were performed to evaluate the etch process.

Keywords—via open; window improvement; AIO-ET

INTRODUCTION

Integrated circuits have dramatically changed our life. In 1960s, integrated circuit chips were insignificant. While the technology has come a long way in terms of complexity and practicality. Nowadays, every family in developed country has at least hundreds of integrated circuit chips, so the integrated circuit chip has been beginning the most important scientific and technological revolution in human history. Integrated circuit is the foundation of computer technology, and it also promotes the development of related technologies, such as software industry and network. Almost all products in the information era originate from integrated circuit technology [1].

With the development of the integrated circuit, in order to further reduce the resistive capacitance delay effect of BEOL (Back End of Line), the dual Damascus process and ultra-low-k material (BDII) have been widely used. In terms of reducing the influence of plasma on the dielectric constant of ultra-low dielectric materials and overcoming the challenges of lithography with the reduction of key dimensions, the MHM (Matel Hard Mask) layer has shown great advantages in AIO etch. Meanwhile, compared with the traditional VFTL (via-first-trench-last) process, AIO etch has the traditional craft incomparable advantages in facet profile control, etch selection ratio and line edge roughness.

With the introduction of ultra-low dielectric materials, the reliability problem of BEOL becomes more and more difficult. This is mainly due to low mechanical properties, low thermal conductivity, poor breakdown resistance and poor adhesion between the substrates and the upper dielectric. This poses a great challenge to each process module in the following process, so a good process integration becomes especially important. It also puts forward higher requirements for etching technology [2-3]. This paper mainly describes the research on Via Open through the low-power chip developed by Huali Integrated Circuit Corporation.

EXPERIMENT

The integrated (AIO: all-in-one) etching process is shown in figure 1. The black box is an integrated (AIO) etching process. The so-called integrated via etch, photoresist strip and trench etch are completed in the same process.

The manufacturing process involved in this paper is based on the low power chip technology of Huali Integrated Circuit Corporation. The dual Damascus process is used, and the dielectric material is BDII (Black Diamond II), which is the ultra-low dielectric material with a dielectric constant of 2.55.

In this paper, the experiments are all performed on Vigus LK II. TEL Vigus LK II is a typical capacitance coupled plasma etching chamber and the superposition of electrode on a DC source. The electrodes have two different frequencies: one frequency is 13 MHz, which is called LF (low frequency); while the other frequency is 40 MHz, which is called HF. The DC source is designed to reduce the content of fluoride free radicals in the plasma in the process chamber and the ion bombardment energy on the surface of the wafer, so as to improve the etching selection ratio, reduce the critical size, and reduce plasma damage. LF is a low frequency, mainly used to control plasma energy; while HF is a high frequency, mainly used to control plasma concentration. The SEM images are obtained by the machine CG5000 CD-SEM of Hitachi company, while the TEM images are got from the machine JEM-2100F of Japanese electronics company.

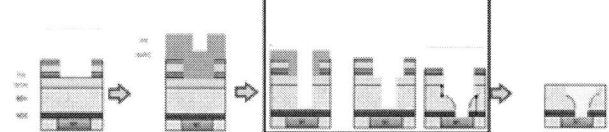

Figure 1: schematic diagram of AIO-ET

The wafer was found yield loss problem after AIO etch process on TEL Vigus LK II as illustrated in the following figure 2. The TEM image (figure 3) and EDS analysis result (figure 4) showed that Via Open occurred between the metal. Suspected AIO-ET window margin.

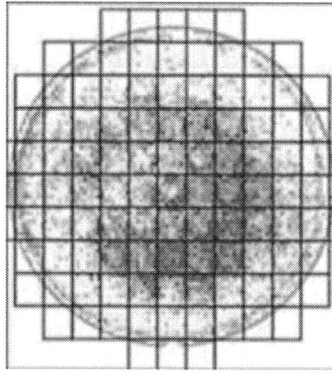

Figure 2: Via Open Map in Wafer

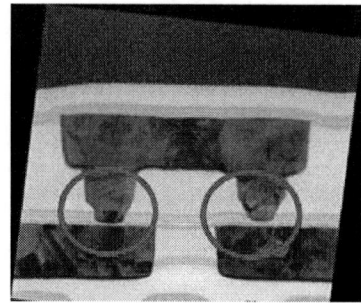

Figure 3: Schematic Diagram of Via Open after AIO-ET

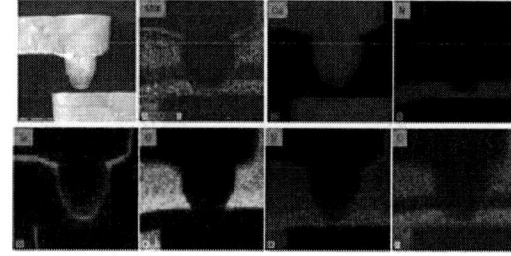

Figure 4: EDS analysis results

Results and Discussion

During the process of recipe optimization, in addition to conventional measurement, WAT (Wafer Acceptance Test) and MA (Misalignment) tests can also well represent the wafer performance after etching. For the case of Via Open in AIO-ET, according to the changes of morphology between each step as shown in figure 5, the linear relationship between PV time and MA have been investigated (figure 6). The figure 6 shows clearly that MA Window increases with the increase of PV time. However, with the increase of the PV time, Via CD decreases (as show in figure 7), which will inevitably lead to the rise of RC.

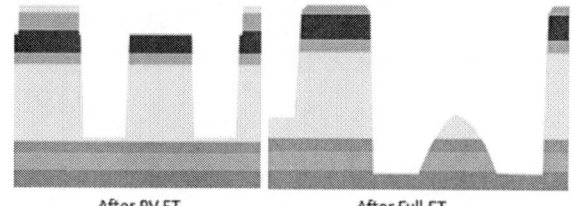

Figure 5: schematic diagram of Film changes during etching

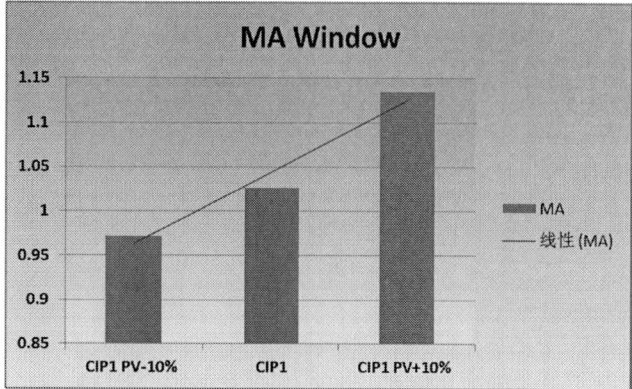

Figure 6: The relationship between PV time and MA

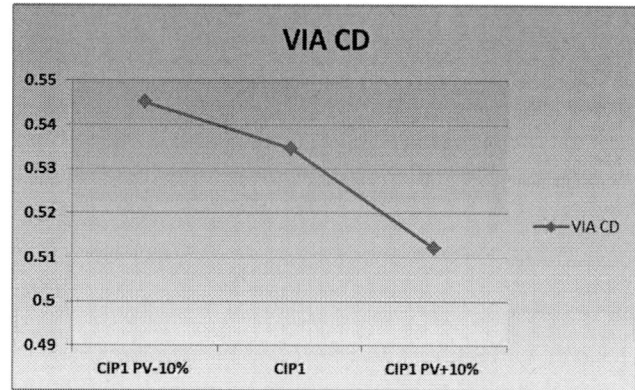

Figure 7: The relationship between PV time and Via CD

From the above experiment, it can be seen that the increase of PV time can enlarge MA Window, but RC will be sacrificed. In order to match the MA and RC, we balance them by adjusting the temperature. From figure 8 below, we can know that RC increases gradually with the increase of the PV time at the same temperature, which is consistent with the conclusion above. The reason for the increase of RC is the decrease of Via CD. Besides, from the study of the relationship between RC and PV time and temperature, it can also be seen that RC can well match the target when PV increases by 10% and temperature increases 3 degree. At the same time the wafer's uniformity is the best (as shown in figure 9).

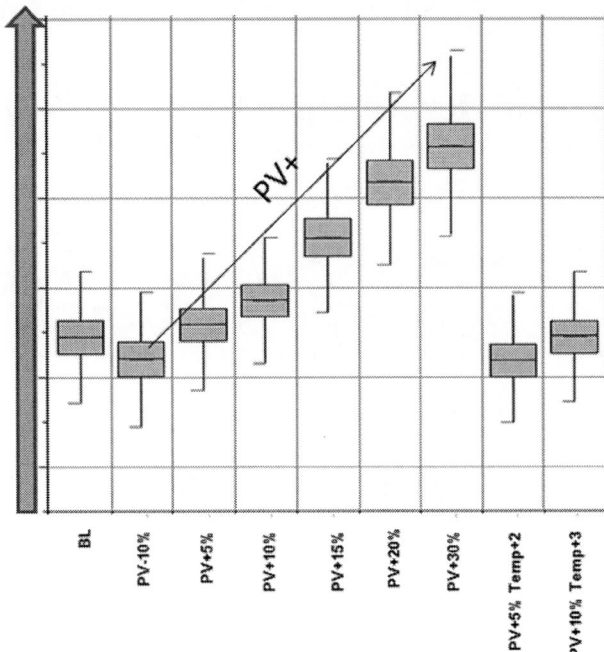

Figure 8: The relationship between PV time and temperature and RC

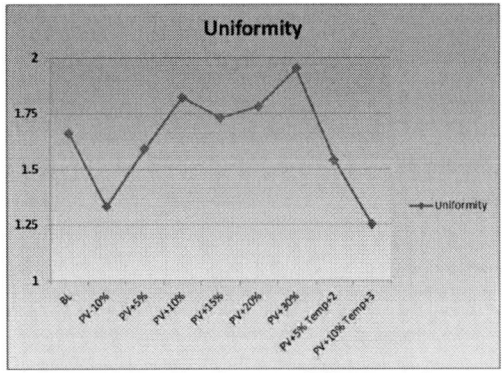

Figure 9: The relationship between PV time and temperature and uniformity

CONCLUSION

This paper mainly describes the process problems during the development of low power chip technology in Huali Integrated Circuit Corporation. Combining with process equipment and process characteristics, the direction and method are given to solve these process problems. The experimental results show that the problem of Via Open can be effectively solved by increasing PV time and adjusting temperature.

References

[1] H. Xiao. *Semiconductor manufacturing technology and introduction*, SPIE Press，2013．

[2] J.Q. Zhou, M.D. Hu, H.Y. Zhang, D.J. Wang, X.P. Wang, C.L. Zhang, X.H. Song, S.M. Chang and K.F. Lee. *CSTIC*, 2012

[3] M.D. Hu, J.Q. Zhou, D.J. Wang, C.L. Zhang, J. Huang, X.P. Wang, H.Y. Zhang, Z. Mo, T. Shindo and L.H. Chen, *CSTIC*, 2013

From the above experimental results, it can be obtained that at PV+10% and temperature + 3 °C, RC reaches target, and CD has the best uniformity; at the same time, when PV time increase, MA Window will trend up too. Therefore, the TEM was verified under this condition (as figure 10). The image showing Via Open has been solved.

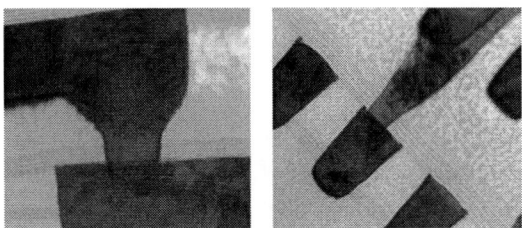

Figure 10: TEM morphology under new conditions

THE SOLUTION OF CONTACT OPEN AND SHORT

Renhui Xu*, Jie Zhang, Jianguo Yang, Lei Sun, Quanbo Li,Jun Huang and Yu Zhang
Shanghai Huali Integrated Circuit Corporation (HLIC),
Shanghai 200120, China
*Corresponding Author's Email: xurenhui@hlmc.cn

Biography

Renhui Xu received the B.S. degree in Beijing University of Aeronautics and Astronautics, Bei Jing, China, in 2005. In 2017, he joined the Technology Development Deapartment, Huali Microelectronics Corporation, Shanghai, China, where he is engaged in etch process development of advanced copper/low-k interconnects for ultralarge scale integration.

Abstract

With the development of VLSI (Very Large Scale Integrated Circuits) manufacture, the line width CD (Critical Dimensions) shrink continuously in order to meet the high performance and high integrated level of device reqirement. More and more new material and process step were introduced to meet the process requirement of critical layer. With the development of integrated circuit technology, the device density become more and more large. The CD of CT (contact) hole needs to be greatly reduced. Such a large CD shrink in contact etch process leads to some challenges to ensure free CT open and other problems.

There is a strong negative correlation between CT CD and CT open (DVC). But large CD will reduce MA (miss alignment) window, which leads CT short to poly gate easily. The difficulty of CT etch process development on TEL LK2 is the tradeoff between CD, CT open and CT short. Some factors were found that how to improve CT open/short window during process development. In this study, we provide solutions to the problems during the development and optimize the CT etch process.

Keywords—CT etch; Open short; MA window; DVC; CT-open

Introduction

Since CMOS technology moved to 40nm node and beyond, both patterning and non-patterning process will encounter big challenges, and the high aspect ratio contact etch process becomes more an more cntical to yield Improvement [1,2]. It is well known that contact etch has to use richer polymer to balance CD shrink and contact open. However, due to smaller contact hole, etch byproduct cannot be removed completely by the following single wafer wet clean process, the remained fluoride coupled with the moisture absorbed from the air will react with silicide to form one thin film on the silicide surface with high resistance as shown in Figure 1, and/or generated the byproduct to prevent the following metal gap-fill , thus leading to yield loss. Contact etch with the embedded and post etch treatment process could greatly benefit the byproduct removal and the wet clean process window.

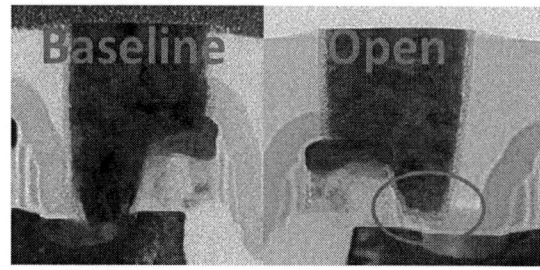

Figure 1: Contact open

Contact to poly short also induce yield loss. It may induce by overlay shift, CD shift, profile bowling or CESL (contact etch stop layer) damage such as Figure 2. CESL damage may be induced by sidewall polymer. MA window can be used as characterization for contact to poly short, as shown in Figure 3.

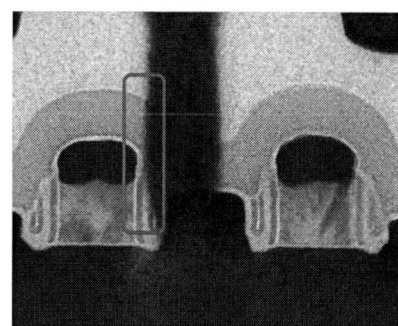

Figure 2: Contact to poly short

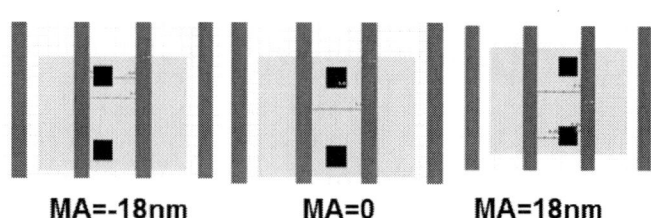

Figure 3: MA window

Experiment I

For Contact open, more SIN etching amount obviously improve DVC performance as Figure 4. Enough over etch is necessary for a robust recipe, but over etching amount need to control within safe level. Too much oxide over etching amount contact open window enough, but side effect is bowling profile and degrade MA window.

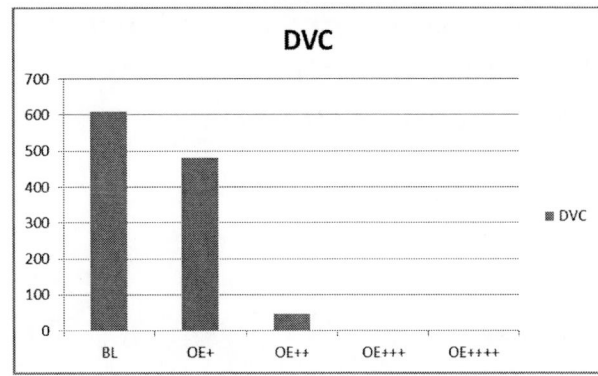

Figure 4: Over etch experiment

Experiment III

Too much oxide and SIN over etching can induce profile bowling and contact to poly short. CESL damage by sidewall polymer also degrade contact to poly short window.

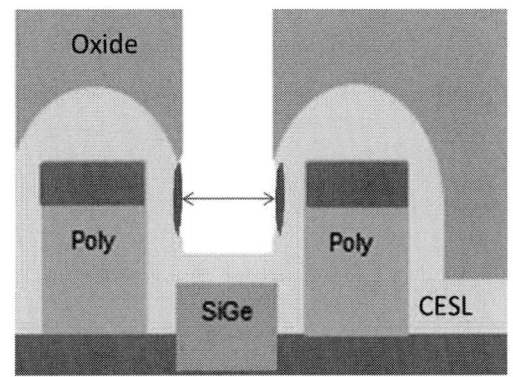

Figure 5: Bowling profile and sidewall CESL damage

Reduce oxide over etching amount can avoid bowling profile and enlarge MA window which can improve CT to poly open window.

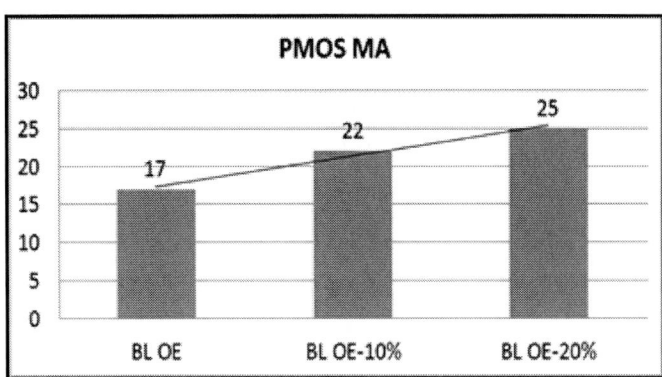

Figure 6: MA vs OE

Experiment III

Add in-situ treatment step post hardmask remove can cleare sidewall polyer and avoild sidewall CESL damage. As shown in Figure 7, we did an experiment: reducing treatment process time, MA obvious down grade a level.

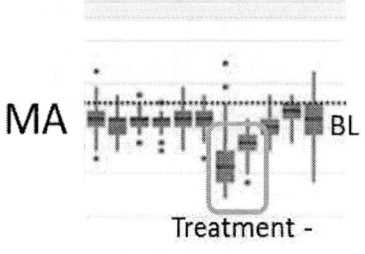

Figure 7: Treatment time- experiment

Conclusion

For contact open solusion, enough over etching amount is nesscessary. Over etching amount need control within safe level. More over etching amount induce bowling profile and degrade contact to poly short window. Control over etching amount is very important to balance contact open and contact to poly short. Another solution for improving contact to poly short: add in-situ treatment step post hardmask remove .

References

[1] M. Chen, et al. *Electron Devices*, VoI.57, 2010.
[2] S. Demuynck, et al. *Semiconductor Manufacturing*, 2007, ISSM 2007.

OPTIMIZED WORK FUNCTION METAL LAYER DAMAGE EFFECT IN METAL GATE BARC ETCH PROCESS BY ICP ETCH SYSTEM

Kai Qian[1]*, Shaoxiong Liu[1], Lian Lu[1], Quanbo Li[1], Jun Huang[1], Yu Zhang[1]

[1]Shanghai Huali Integrated Circuit Corporation, Shanghai City, China

*Corresponding Author's Email: qiankai@hlmc.cn

ABSTRACT

Metal gate BARC open process is considered to be more and more critical in semiconductor manufacturing in and beyond 14nm technical node, due to its important effect on WAT result through the work function metal layer damage side-effect. In this work, metal gate BARC open process is performed in ICP etch system. The pattern loading of dense/isolated area is critical for process health. The effects of work function metal layer damage are studied systemically to setup ideal process. Partial etch profile to control pressure impact is studied in detail. The effect mechanism is also discussed through polymer passivation on Fin. It is proposed that reasonable over etch window is very important to eliminate Work Function Metal damage and BARC residue in ICP etch system.

INTRODUCTION

As the semiconductor device technology advances to 16/14nm node and beyond, FinFET is considered as main stream technology due to its advantage of higher transistor performance, larger density and lower leakage. In FinFET technology node, the replacement metal gate is also introduced for its lower gate leakage, thinner Tox and faster electron mobility. [1][2] To control varied device Vt values, work function metal layer as TiN/TaN are used in metal gate film stack. [3] For multiple Vt tuning of metal gate in 16/14nm FinFET node, BARC open is critical as a mask for multiple TiN thickness in device tuning.

In this work, metal gate BARC open process is investigate in ICP (Inductively Coupled Plasma) etcher equipment. The key requirements for BARC open in metal gate loop are no BARC residue and no TiN/TaN damage. For no BARC residue concern, enough etch amount is need. Meanwhile, for no TiN/TaN damage, the balance of polymer deposition on fin in short space CD area should be taken into account. Therefore, the effects of work function metal layer damage need to be fine control. The etch window is studied in detailed. Finally, the effect mechanism of polymer deposition is also discussed, which will evidently impact the health status of metal gate BARC open process.

EXPERIMENT

All the dry etching experiments were performed in the same types of 300mm commercial etcher of ICP system with tunable ESC (Electrostatic Chuck) and TCCT (Transformer Coupled Capacitive Tuning). PR and BARC were coated in the trench of Poly which is already removed. Cross section pictures of PR/BARC and work function metal are performed using SEM (Scanning Electron Microscope), TEM (Transmission Electron Microscope) and EDX (Energy Dispersive X-Ray Spectroscopy). Test wafer condition is as blow: Wafer diameter size is 300mm. Film stack is patterned PR/BARC as Figure 1 (a).

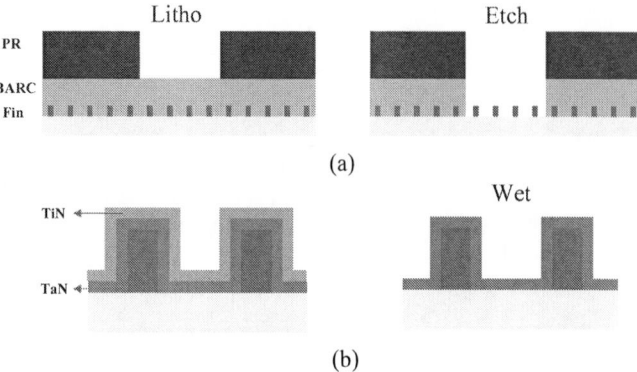

Figure 1: (a) MG BARC Open Process; (b) TiN wet remove process post MG BARC open

RESULT AND DISCUSSION

A. Pattern loading

In this process film stack is PR/BARC, etch process is etching BARC in the pattern defined by Litho, as shown in Fin.1 (a). Total BARC etch without residue is required for the following TiN wet remove process. Otherwise, BARC residue will block wet process and then cause TiN residue finally. In this situation, the pattern loading among dense and isolated area becomes critical for the health evaluation of BARC etch and the incoming BARC thickness check become necessary. As shown in Fin.2(a, b), the incoming BARC in isolated pattern is ~400A thicker than that in dense pattern. This loading will result in BARC residue in isolated pattern while dense pattern is fully etched, as shown in Fin.2(c, d). In Fin.2(c), TiN residue is found post wet TiN remove process, due to not enough BARC etch amount. A thick TiN layer is observed clearly on TaN layer. In order to cover this residue, enough etch amount need to be added in OE and BARC residue will clear in isolated area. In Fin.2(d), only TaN layer is found. The incoming BARC thickness is directed related with BARC residue. Detailed check in all pattern to clarify enough etch amount is necessary for a health BARC etch process.

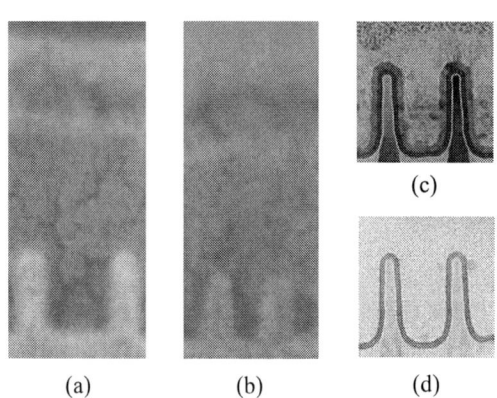

Figure 2: (a) Dense area incoming BARC profile; (b) Isolated area incoming BARC profile; (c) BARC residue in isolated area; (d) BARC clear in isolated area.

B. TiN Damage

It is known that plasmas induced damage will result in uncontrolled pattern-dependent etch rate variation and physical damage of etching pattern, and cause device damage and etch process failure finally. [4, 5] As BARC etch lands on TiN/TaN on fin, much OE amount may impact the quality of TiN and finally impact device WAT performance. The BARC etch process design should take both enough etch amount and TiN damage into account. Therefore, the BARC etch process is designed as three etch step as low pressure, high pressure, high pressure and high temperature. In low pressure step, etch amount is controlled to located above fin ~300A, as shown in Fin.3(a). The plasma pressure is directly related with bombardment of plasma to substrate. The lower pressure exhibits the higher bombardment. To avoid possible TiN damage on fin, low pressure step should not touch fin top for safety's sake. Meanwhile, in order to control CD, the pressure of first step is not allowed to be too high. The higher pressure exhibited the higher isotropic etch, resulting in higher space CD. Afterwards, the other step will etch BARC clearly without residue, as shown in Fin.3(b). Higher pressure transition is aiming for total BARC clearance with higher isotropic etch, which benefit the residue clearance.

High pressure high temperature step is designed as over etch step with higher etch rate and lower polymer deposition. Too much polymer deposition is generally considered as residue source. So lower polymer deposition is good for BARC residue clearance. But it should be noticed that polymer deposition plays two parts in BARC etch: one as fin protection, one as space CD modification knob. That means polymer deposition should balance between fin protection and BARC residue. Enough high pressure high temperature OE with lower polymer deposition will benefit BARC residue clearance, resulting in good TiN/TaN/Fin structure post TiN wet remove, as shown in Fin.3(C) by TEM and EDX characterization. But too much OE with lower polymer deposition will cause TiN layer damage just post dry etch and finally TaN damage post wet TiN remove, as shown in Fin.3(d). TiN damage will conjoin with bottom TaN damage and impact device WAT performance. Therefore, no TiN damage is a criterion for BARC open process. The OE step window should consider both BARC residue clearance and TiN damage at the same time.

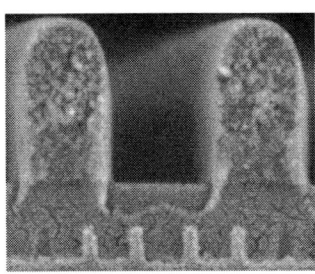 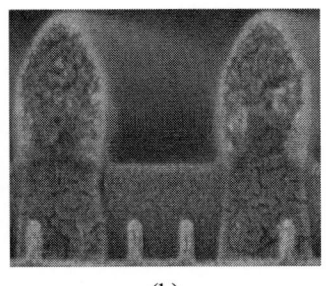

(a)　　　　　　　　　　(b)

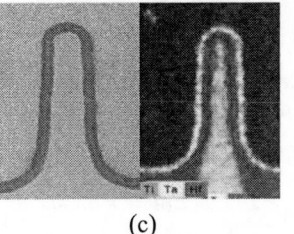 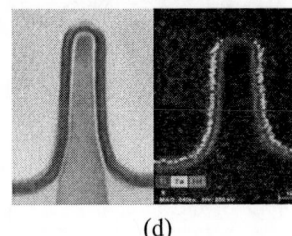

(c)　　　　　　　　　　(d)

Figure 3: (a) Partial BARC etch Profile; (b) Full BARC etch profile; (c) TEM and EDX profile of full OE; (d) TEM and EDX profile of too much OE.

C. BARC Profile

In order to define exact space CD, vertical BARC profile is need for inline long-time process maintenance. The pressure, polymer deposition and Bias.voltage can impact the BARC profile. A sidewall concave BARC profile is easily observed if process condition is not improved, as shown in Fig4.(a). The isotropic effect of pressure, sidewall protection of polymer deposition and plasma flux directivity of Bias.voltage can be used as efficient tuning knobs for BARC profile. After systematic design of three factor, a nearly vertical BARC profile can be obtain, as shown in Fig4.(b). Meanwhile, it is important to note that Bias.voltage is the most effective knob. But TiN damage also need to double check if higher Bias.voltage is used.

(a)　　　　　　　　　　(d)

Figure 4: (a) Side wall concave BARC profile; (b) Nearly vertical BARC profile.

CONCLUSIONS

In this work, Metal gate BARC etch have been investigated in detail. BARC etch loop is described by schematic diagram for easy understanding. For a health BARC etch process, pattern loading of dense/isolated area, reasonable over etch amount to balance TiN damage and BARC residue clearance, vertical BARC profile are the main issue. Depending on yield and WAT test, metal gate BARC etch process can be further improved after overall consideration.

ACKNOWLEDGMENTS

Thanks for all the authors for great work and collaboration. Thanks for FA (failure analysis) for TEM and EDX image support.

REFERENCES

[1] S. Senturia. Proceedings of Transducers2003, Boston, June 8-12, 2003, pp. 10-15.

[1] W. Wu, M. Chan. *IEEE Electron Device Letters*, Vol. 27,

2006, pp. 68-70.

[2] R. A. Wachnik, S. Lee, L. H. Pan, N. Lu, H. Li, R. Bingert, M. Randall, S. Springer, C. Putnam. *Proceedings of the IEEE 2013 Custom Integrated Circuits Conference*, DOI: 10.1109/CICC.2013.6658494.

[3] J. Xu, A. Wang, J. He, X. Jing, Z. Zhang, B. Zhang. *2017 China Semiconductor Technology International Conference (CSTIC)*, DOI: 10.1109/CSTIC.2017.7919796.

[4] Y. Karzhavin, W. Wu. *1998 3rd International Symposium on Plasma Process-Induced Damage (Cat. No.98EX100)*, DOI: 10.1109/PPID.1998.725579.

[5] A.T. Krishnan, V. Reddy, S. Krishnan. *International Electron Devices Meeting. Technical Digest (Cat. No.01CH37224)*, DOI: 10.1109/IEDM.2001.979650.

Analysis of Linewidth Uniformity and Line Edge/Width Roughness in a 5 nm Logic SAQP Process

Bowen Wang[1], Yushu Yang[1], Yibo Wang[2], Yuning Zhu[2], Yongjian Luo[2], Qiang Wu[1], Weihao Lin[1], Zhunhua Liu[1], Yanli Li[1], Qingqing Wu[1], Shoumian Chen[1], Ying Zhang[2]

1-Shanghai IC R&D Center
497 Gaosi Road, Zhangjiang Hi-Tech park, Shanghai 201210, China
2-Beijing NAURA Microelectronics Equipment Co., Ltd.
No. 8 Wenchang Avenue, Beijing Economic-Technological Development Area, Beijing 100176, China
Tel:021-60860986*3115 E-mail:wangbowen@icrd.com.cn

Biography

Graduated in 2016 from Northwest University with B.S. and M.A. from Department of Optical Science and Engineering, Fudan University in 2019, joint Shanghai IC R&D Center as a lithography engineer from 2019. Has published three SCI academic papers during graduate study period.

Abstract

As the semiconductor manufacturing design rules have been continually shrinking, the requirement in linewidth uniformity, linewidth roughness, line edge roughness, and overlay has been rising to stringent levels [1]. In a FinFET process, the fins are made with 193 nm immersion lithography together with the Self-Aligned Quadruple Patterning (SAQP) techniques. The linewidth uniformity, line edge roughness (LER) and linewidth roughness (LWR) are critical limiting factors in the process of SAQP as LER does not scale down with the dimensions of the devices. Three times the root mean square (3σ) is the most common LER characterization parameter. In this paper, we have conducted a SAQP process study on a typical film stack based on a 5 nm logic process flow. We will make the SAQP patterns with domestically made etching machine and Atomic Layer Deposition (ALD) tools and will characterize all etch steps within the mandrel 1 and spacer 1 of SAQP process in terms of variations. We will benchmark the measured data against required targets in linewidth uniformity and line edge/width roughness. Moreover, we will present and analyze the result of our study.

Keywords—5 nm; SAQP; 193 nm immersion; linewidth roughness; line edge roughness; FinFET

Introduction

Besides EUV, the exposure tool that has the highest resolution is the 193nm immersion scanner, which has a fundamental printing limit of 38 nm (half-pitch). However, device performance requirement continues to push for smaller feature sizes, which are deeply below the 38 nm resolution limit [2]. Various multi-patterning techniques have been developed

to satisfy the device needs, among which the Self-Aligned Quadruple Patterning (SAQP) technique can not only produce better linewidth roughness (LWR)/line edge roughness (LER), but also provide a better overlay owning to the self-aligned nature. This process flow involves sidewall spacers as a means in doubling the printed line density [3]. The self-aligned scheme in SAQP greatly reduces the overlay issue compared with other multi-patterning techniques [4]. However, the linewidth uniformity and LER/LWR are still challenging. We should do our best to reduce the LWR/ LER and improve the uniformity of linewidth as which may induce the variability in the threshold voltage, leakage, and drive current in fin field-effect transistor (FinFET).

In a 5 nm FinFET logic process, the targeted fin pitch is a 22.5~25 nm, which requires careful adjustment of the initial lithographic patterning process, the two mandrel and spacer etch process as well as the two spacer deposition process parameters to keep the linewidth uniformity and avoid a systematic pitch variation (pitch walking) [5, 6]. Firstly, we will improve the uniformity and profile of mandrel 1 and spacer 1. Unequal spaces between the neighboring spacer 1s will lead to undesired variability for mandrel 2 and spacer 2 etch or deposition steps. In order to minimize pitch walking, we optimize the bias power, etching gas composition, and plasma pressure in the mandrel 1 and spacer 1 etching processes. We will benchmark the measured data against required targets in linewidth uniformity, line edge/width roughness, and pattern position variation in mandrel 1 and spacer 1 etch processes of 5 nm FinFET logic process. We will analyze the result and present the TEM cross section images in the next section.

Experiment

For the 5 nm FinFET SAQP process, we have designed a patterning film stack: a bottom layer of silicon, a PE CVD nitride cap, followed by an α-Si layer to serve as a sacrificial layer 2, dielectric cap 2 and, another layer of PE CVD nitride and α-Si to serve as a sacrificial layer 1 and dielectric cap 1, and finally, APF ("Advanced patterning Film". or a type of CVD carbon hardmask material) cap. A schematic of the patterning film stack is shown in Figure 1.

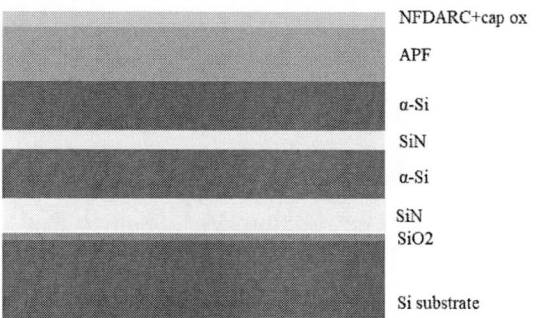

Figure 1. Schematic structure of the patterning film stack for a 5 nm FinFET SAQP. NFDARC: nitrogen free dielectric anti-reflection coating.

Figure 2 shows all the individual etch or deposition process steps of our 5 nm FinFET SAQP process. The initial pattern is made with 193 nm immersion lithography. The domestically made etching machine used in 5 nm FinFET logic process is NMC612D ICP etcher from NAURA and the Atomic Layer Deposition (ALD) tools is FT-300T from Piotech. A Hitachi CG6300 is used in Critical Dimension measuring Scanning Electron Microscope (CD-SEM) metrology.

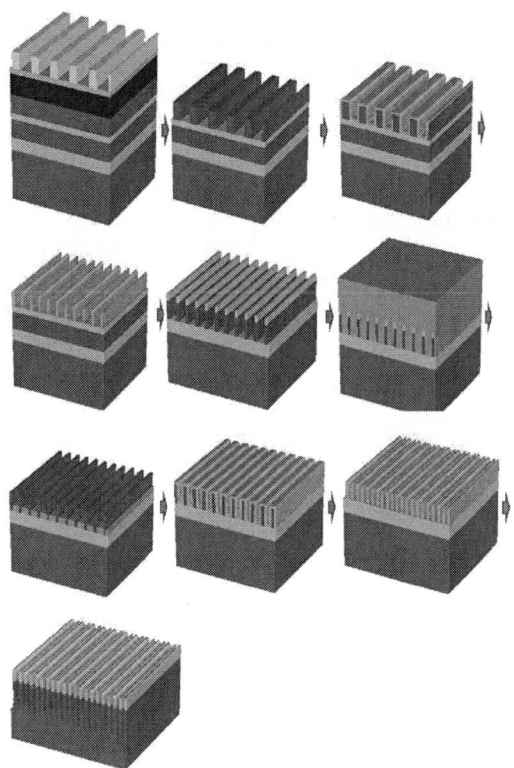

Figure 2. Schematic structure of the individual etch or deposition steps in the 5 nm FinFET SAQP process. From beginning to end: Lithography, mandrel 1 etch, spacer 1 deposition, spacer 1 etch back/mandrel 1 removal, mandrel 2 etch, SOC deposition, SOC etch back/SiN stop layer removal/SOC removal, spacer 2 deposition, spacer 2 etch back/mandrel 2 removal, Fin etch.

We have obtained that the pitch walking is at its minimum (near 0) when the mandrel 1 CD is 35.5 nm from the Figure 3.

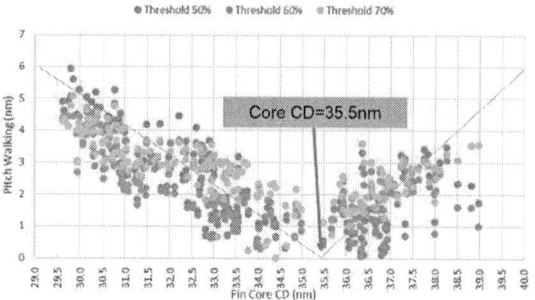

Figure 3. The relationship between the mandrel 1 AEI CD and pitch walking.

Based on the target mandrel 1 CD, we have chosen an appropriate etch recipe for the first run (R01). The After Etch Inspection (AEI) CD, CD uniformity (CDU, in 3σ), and LWR/LER, measured from the CG6300 CDSEM, are 39.00 nm, 1.73 nm, 3.10 nm and 1.90 nm, respectively. Before running the mandrel 1 etch process, we have set mandrel 1 etch specifications to 36.00 nm, 1.50 nm, 3.50 nm and 2.50 nm, respectively. The AEI mandrel image from the CDSEM is presented in Figure 4. Most indicators have met the specifications, except for the CD in the mandrel 1 etch from R01. It's easy to shrink the mandrel 1 CD by extending the photoResist Hardening (RH) or Hard Mask (HM) trim etch time in mandrel 1 etch process.

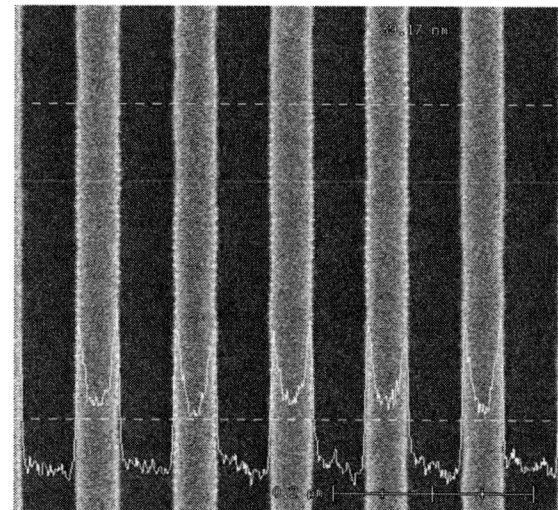

Figure 4. The CD-SEM image of mandrel 1 after mandrel 1 etch process.

In addition to AEI, CDU, and LWR/LER, the difference between Bottom CD (BCD) and Top CD (TCD) in the mandrel 1 profile is also an important indicator to characterize the profile of the mandrel. In order to observe the cross section image of the mandrel 1, we did the TEM analysis assisted by the integrated Service Technology (iST) as shown in Figure 5. By using PCI5 software for the characteristizing of the BCD and TCD, the difference between BCD and TCD (ΔCD) is 3.20 nm, which is unfortunately beyond the specification of 2.00 nm.

Figure 5. The TEM cross section image of mandrel 1 after R01 mandrel 1 etch process.

In order to avoid pitch walking caused by larger ΔCD, we increased SF$_6$ flow rate and increase the bias power. As shown in Figure 6. The ΔCD in Figure 6 is about 1.10 nm and within the specification. However, we can see that there is some notching exist near the top of polysilicon mandrel 1 and slight necking at the top of mandrel 1. At the same time, there is some footing present at the bottom of mandrel 1.

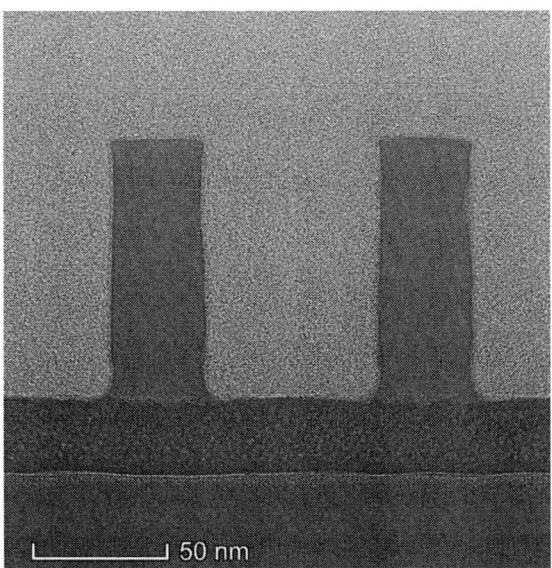

Figure 6. The TEM cross section image of mandrel 1 after R07 mandrel 1 etch process.

In response to the above issues, we have once again increased O$_2$ and SF$_6$ flow and adjusted the etching time and plasma pressure in Main Etch (ME) process, at the same time we increase the bias power in Over Etch (OE) step. The increase of O$_2$ flow can produce polymer to protect the top half of polysilicon, the increase of SF$_6$ flow and bias power is to eliminate the footing. The TEM cross section image is shown in Figure 7. In Figure 7, we have shown that the notching in R07 (Figure 6) has been solved and the necking has also been

improved while the footing still present and requires more efforts.

Figure 7. The TEM cross section image of mandrel 1 after R11 mandrel 1 etch process.

After solving most of the problems in mandrel 1 etch process, we have also done an optimization of spacer 1 etch process. Before spacer 1 etch process, an oxide layer needs to be deposited through ALD by FT-300T. Figure 8 is the TEM cross section image after ALD process.

Figure 8. The TEM cross section image after ALD spacer 1 deposition process.

In spacer 1 etch process, we tried the adjustment of the bias power and etching gas composition in ME process, and C$_4$F$_6$ flow in OE step. Those changes have resulted in a satisfactory ΔCD and effective height of spacer 1. At the same time, it is unfortunately that these changes have not brought satisfactory results about footing in spacer 1. As shown in Figure 9, the only problem that needs to be solved is footing by observing the spacer 1 TEM cross section image from R05 spacer 1 partial etch process.

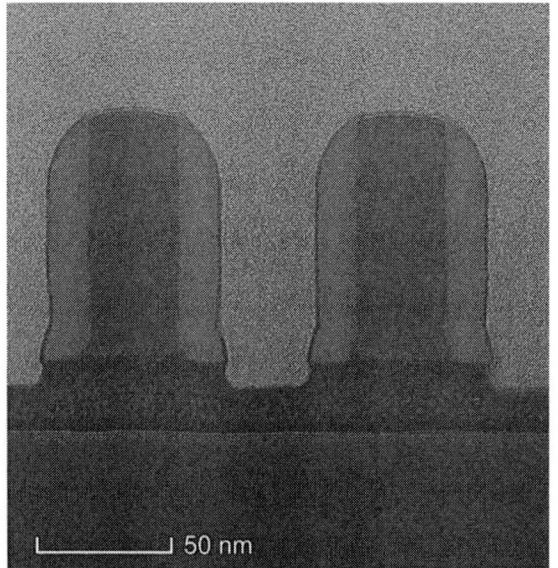

Figure 9. The TEM cross section image of mandrel 1 after R05 spacer1 partial etch process.

As shown in Figure 10, there are large tilt in the cross section image of spacer 1. We believe that this tilt is caused by the stress release during the preparation of the TEM samples. At the same time, we also need to reduce the thickness difference of the nitride layer on both sides of the spacer 1, otherwise it may cause pitch walking in the subsequent mandrel 2 etch process, because different nitride layer thicknesses will cause different amounts of lateral etching during the transfer process.

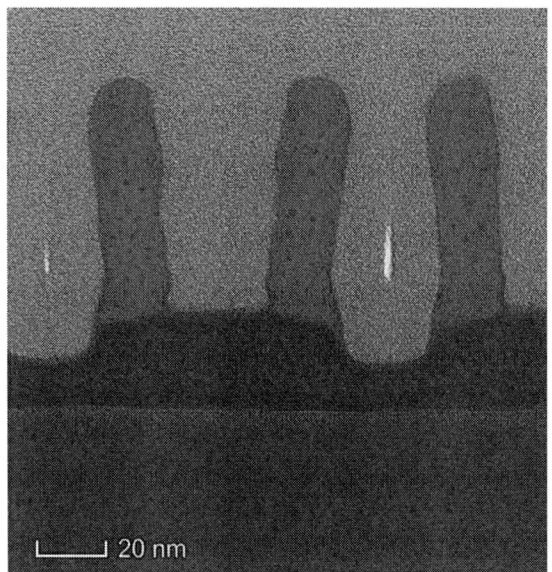

Figure 10. The TEM cross section image of mandrel 2 after R07 spacer1 etch process.

Conclusion

We have performed 5 nm Fin SAQP process study and compared our process data against required targets in AEI CD, linewidth uniformity, line edge/width roughness and ΔCD. The required targets of ΔCD, linewidth uniformity, and line edge/width roughness in 5 nm FinFET SAQP mandrel 1 and spacer 1 etch process are 35.50 nm, 1.50 nm, 3.50 nm/2.50 nm and 2.00 nm, respectively. The measured data from CD-SEM are 35.50 nm, 0.96 nm, 3.10 nm/1.90 nm and 1.10 nm, respectively. In summary, all the indicators show that our 5 nm FinFET SAQP mandrel 1 and spacer 1 etch processes meet the required targets. As future work, we will continue to develop the SAQP process.

Acknowledgment

The authors would like to thank the Etch II BU Eastern China Site at Beijing NAURA Microelectronics Equipment Co., Ltd. and all the members of the Advanced Device Material Lab at Shanghai IC R&D Center, including numerous engineers from spacer deposition. We will also give special thanks to the lithography team and the etch team at the Technology Development Division of Shanghai IC R&D Center for the great collaboration.

References

[1] 2017 IRDS roadmap

[2] C. Bencher1 et al., "22nm Half-Pitch Patterning by CVD Spacer Self Alignment DoublePatterning (SADP)", Proc. SPIE Vol. 6924, 69244E,(2008)

[3] W. Jung et al., "Patterning with Amorphous Carbon Spacer for Expanding the Resolution Limit of Current Lithography Tool", Proc. SPIE Vol. 6520, 65201C, (2007)

[4] L. Sun, X. Zhang et al., "Line Edge Roughness Frequency Analysis for SAQP process", Proc. SPIE Vol. 9780 97801S-1, (2016)

[5] S. Guissi, J. Ervin, S. Baudot et al., "N7 SAQP Pitch Walk & Fin Variability Modeling", imec partner technical week.

[6] Fang F , Herrera P , Kagalwala T , et al. "SAQP pitch walk metrology using single target metrology"[C]// SPIE Advanced Lithography. International Society for Optics and Photonics, 2017.

METAL HARD MASK OPEN PROCESS WINDOW ENHANCED BY INSERTION OF POLYMER DEPOSITION

Li Fei Sun[,1], Qing Peng Wang[1], Ji Hong Zhang[1], Lei Sun[3], Andrew Li[2] and Yu Shan Chi[1]*

[1]Lam Research Corporation, Shanghai 201210, China
[2]Lam Research Corporation, Fremont, CA 94538, USA
[3]Shanghai Huali Integrated Circuit Corporation (HLIC), Shanghai 201206, China
*Corresponding Author's Email: Lifei.Sun@lamresearch.com

ABSTRACT

Insufficient BARC remaining after TiN etch can result in the CD and profile shift and residue defect issue in MHM open process. In this work, CH_4 deposition is utilized to gain BARC remaining and two approaches for the insertion of polymer deposition were systematically investigated both by simulation and experiment. It is shown that deposition step inserted before BARC etch is more effective for improving the BARC remaining thickness and profile after MHM open. "Visibility deposition" model of the SEMulator3D® semiconductor modeling is utilized to well explain the different behavior of these two approaches and reveal that the high AR and vertical profile of the patterns result in the selective polymer deposition, thus contribute to the BARC remaining improvement. In addition, the process tuning window is also explored for TiN CD and profile control.

Keywords—MHM open; TiN etch; BARC remaining; polymer deposition; visibility deposition

INTRODUCTION

Titanium nitride (TiN) shows many unique physical properties, such as high melting point and hardness, good electrical conductivity and excellent stability, thus has been widely utilized for many applications in sub-micron semiconductor integrated circuit (IC) [1]. TiN is usually served as a diffusion barrier between metal layer and SiO_2 to avoid multilevel metallization [2]. As the dimension continuously scaling down, the poly-Si gate depletion and high sheet resistance become a serious concern, thus poly-Si gate is replaced by using high-k/metal gate to enable further scaling. W/TiN is proved to be one of the most promising candidates, with TiN served as work function metal [3]. Besides, precise patterning from designed layout to real features is getting more and more challenging as device dimension shrinks. Due to the effective low-k damage reduction, line roughness improvement and dual damascene facet control, Metal Hard Mask (MHM) for low-k trench etch application has been widely used in back-end-of-line (BEOL) since 45 nm node [4]. As a trench etch mask, MHM etch process becomes a key step, for any loading effect at the mask open step, particularly the critical dimension (CD) bias loading, will be transferred into the subsequent trench etch step. Therefore, the CD control is very critical at MHM

open etch process, which is highly dependent on the remaining and profile control of its mask layer, namely photoresist (PR) and bottom anti-reflective coating (BARC). Under such a concern, to keep enough BARC remaining after TiN etching becomes a key goal at MHM open application to ensure enough process window. Additionally, it is found that there is a strong correlation between the BARC remaining and Ti and Si based residue defects for MHM open application, indicating that insufficient mask may induce the sputtering of corners of SiO_2 cap layer and the underneath TiN layer. Therefore, keeping adequate BARC remaining is important in MHM open process.

In this work, gaining BARC remaining for MHM etch process has been systematically investigated. Methane (CH_4) deposition step is added into the MHM etch sequence to increase the mask thickness. Two approaches for CH_4 deposition sequence were demonstrated and compared: one is to deposit polymer before BARC etch step and the other is to deposit after BARC etch. Both simulation and experiment results show that small modification of the insertion sequence of deposition step makes a big difference. Polymer deposition inserted before BARC etch is more effective to improve the BARC remaining thickness and profile, thus shows better control of TiN CD and profile, and enhance MHM etch process window. "Visibility deposition" model is utilized to reveal that the pattern aspect ratio (AR) plays a critical role in the selective deposition on the patterns. To explore the process window, the CH_4 deposition time, breakthrough (BT) and BARC step etch time are optimized for BARC remaining improvement and TiN CD control.

RESULTS AND DISCUSSION

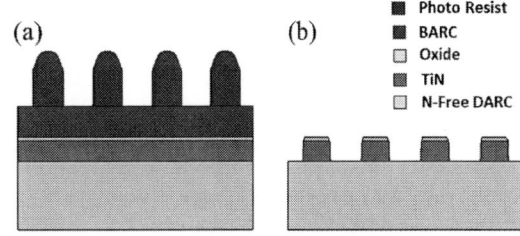

Figure 1: The schematics of MHM open process. (a) Pre

MHM open film stack; (b) Post MHM open film stack.

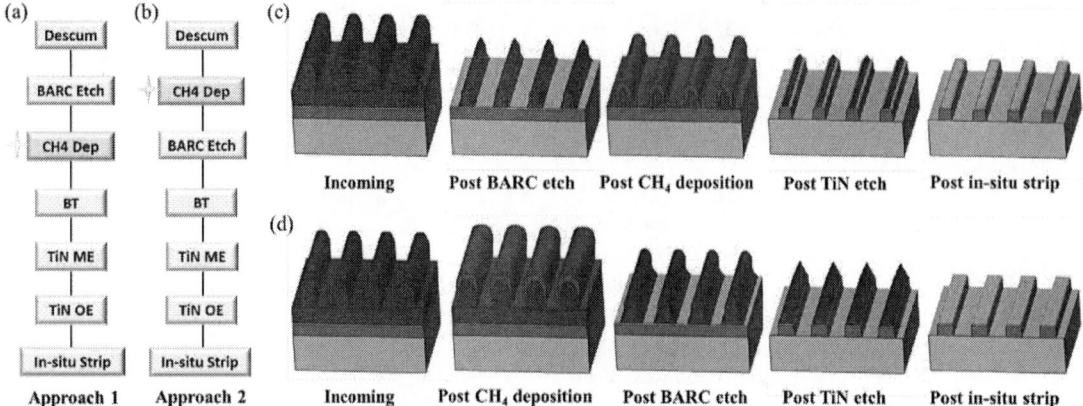

Figure 2: (a-b) The schematics of MHM etch sequence for (a) Approach 1 and (b) Approach 2, respectively; (c-d) By step 3D simulation schematics of MHM open process by using (c) Approach 1 and (d)Approach 2, respectively.

MHM open is one of the metal etch applications, which patterns the hard mask for the low-k trench etch application. The film stacks of MHM open are shown in Figure 1, PR and BARC films act as the mask during TiN etching. The cap oxide is inserted and particularly designed for process rework, which require an extra breakthrough (BT) step. To break through the oxide layer, fluorine-based chemistry will be used, which usually shows relatively low selectivity to PR (PR/Ox >1.6) and will consume additional PR at this step. For TiN etch step, chlorine-based chemistry is utilized, which will consume more PR with lower selectivity to PR (PR/Ox >3). In that case, the PR and BARC will be mostly consumed after these two etch steps, especially at the shoulders. It is often found during the Fab production that the BARC remaining is less than 200A at the top, and hardly any remaining at the shoulders, even worse at the pattern weak points. As it is known to all, the CD and profile of the mask will be transferred into the subsequent films. In this regard, the lack of BARC remaining contributes to the shift of the CD, breaking down of fine CD lines at weak points. Moreover, the insufficient BARC remaining is very likely to enhance residue defects.

To well protect the top corner of TiN and reduce the residue defects, additional deposition step is introduced into the MHM open etch flow. There are two options for the insertion point as shown in Figure 2, one is to insert the deposition step after BARC etch (Figure 2a), the other one is in front of BARC etch step (Figure 2e). Coventor SEMulation3D® semiconductor virtual fabrication platform is utilized to simulate the etch process of MHM open for both approaches. According to the simulation results, with the same starting patterns and simulation setting for deposition and etching, different polymer insertion sequence will make a significant difference on the final BARC remaining thickness and TiN CD and

profile. As shown in Figure 2c, the CH_4 deposition following the BARC etch only gain a little more BARC remaining after TiN etch for approach 1. The final TiN CD is out of specification and TiN top rounding shows due to the lack of BARC mask at the shoulders. However, while the CH_4 deposition step is moved forward to be in front of the BARC etch step, the BARC remaining is distinctly increased and the final CD and profile of TiN meet the specification (Figure 2d). To further verify the significant improvement of BARC remaining with polymer deposition, and compare these two deposition sequences, MHM etching with the above two approaches is conducted in Lam Research Versys® Metal M chamber. Cross-sectional secondary electron microscopy (XSEM) is utilized to image the lateral profile of the patterns and to identify the film thickness. As shown in Figure 3a, with the polymer deposition after BARC etch, hardly any BARC remaining at the top of TiN after MHM open. With in-situ strip, the TiN CD is smaller than the specification with top rounding (Figure 3b). In contract, inserting CH_4 deposition in front of BARC etching can obviously gain

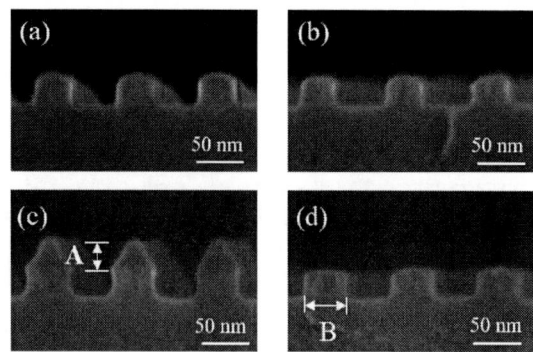

Figure 3: XSEM images of (a) Post TiN etch and (b) Post

Strip (Approach 1); (c) Post TiN etch and (d) Post Strip   (Approach 2).   A: BARC Remaining; B: TiN bottom CD.

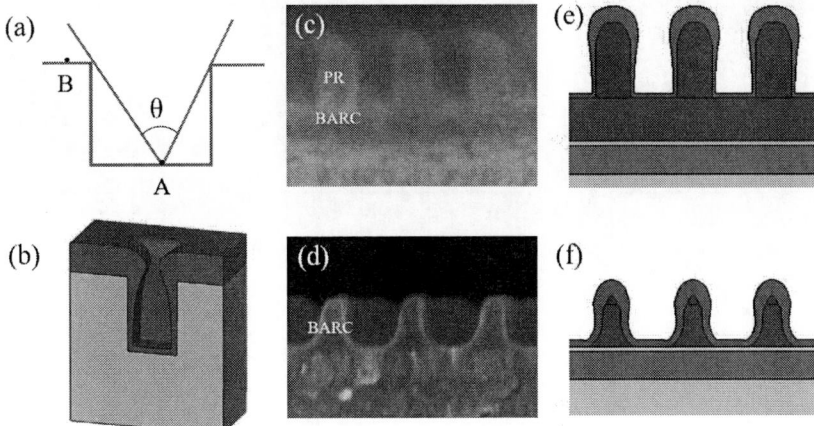

Figure 4: Visibility deposition model. (a) The 2D schematic of visibility in the trench; (b) 3D simulation schematic for visibility deposition; (c) XSEM image of incoming pattern profile; (d) XSEM image of pattern profile after BARC etch; (e) Simulation schematic of the pattern cross-section profile after polymer deposition before BARC etch (Approach 2); (f) Simulation schematic of the pattern cross-section profile after polymer deposition before BARC etch (Approach 1).

the BARC remaining, and the top corners of TiN are well covered and protected by BARC layer during TiN etch process (Figure 3c), thus the profile and CD of TiN patterns can be well controlled (Figure 3d). These experimental results are very comparable to the virtual fabrication results (Figure 2c and 2d).

To further reveal the key point which results in such a big difference between the two approaches, "Visibility Deposit" model of the SEMulator3D® is also utilized to well simulate the deposition process and clearly demonstrate that the profile and AR of the incoming patterns plays a critical role on the selective polymer deposition on the patterns [5]. "Visibility Deposit" is one of the models for deposition simulation, with which the deposition rate at each point on the wafer is a function of the visibility to a directional source of each point on the wafer surface [6]. For instance, as depicted in two-dimension (2D) in Figure 4a, a point A at the bottom of a trench will only see a fraction of the hemisphere of "sky" (angle θ), from which particles can impinge on its surface. In contrast, point B on the top surface will see the full 180 degrees of 2D sky [6]. Therefore, with visibility deposition, material deposition rate on the top of the features is obviously faster than that in the trench (Figure 4b), and the gap will get larger with the increasing of AR. For Approach 2, polymer deposits onto the incoming PR patterns after development, as shown in Figure 4c, the PR profile is relatively vertical. Whereas, after BARC etching for Approach 1, the pattern CD become smaller and more tapered than incoming PR patterns (Figure 4d). After polymer deposition by using Approach 2, selective deposition is exhibited under the action of "Visibility

Deposit" mode, polymer mainly deposits on the top of the patterns while only a little polymer deposits into the trench (Figure 4e). Therefore, mask thickness is effectively gained after deposition with Approach 2. For the deposition by using Approach 1, with relatively small AR and tapered profile, the deposition is more like conformal deposition with the polymer thickness deposited on the top only slightly thicker than that in the trench (Figure 4f). Such a thick polymer layer in the trench need to be etched by the subsequent BT step, which will take extra time and induce the isotropic consuming of the BARC film. As a result, there is hardly any BARC film gained after TiN etch, which has been verified by both simulation and experiment results (Figure 2c and 3a).

After proving that Approach 2 is more promising to gain the BARC remaining for MHM open, we turn to check the process window for polymer deposition and TiN CD and BARC remaining control. The incoming PR patterns were treated with 15s and 30s CH_4 deposition, respectively. As shown in Figure 5b, after 15s CH_4 deposition, polymer deposits both at the top and in the trench of the PR patterns. The height and width of PR patterns are both enlarged, thus increases the pattern AR. Continuing to increase the polymer deposition time to 30s, polymer tends to mostly deposit on the top of the patterns and there is only a little deposition in the trench (Figure 5c). The deposition rate at the top of patterns is calculated to be about 7 A/s, which is ~3 times of that in the trench (Figure 5d). In addition to the deposition step, the BT and BARC etch time are also optimized for tuning the TiN CD and BARC remaining. As shown in Figure 5e, TiN CD only decreases slightly with increasing BT time, while the BT time obviously correlates to the BARC remaining. To clear up the oxide cap layer and keep enough BARC

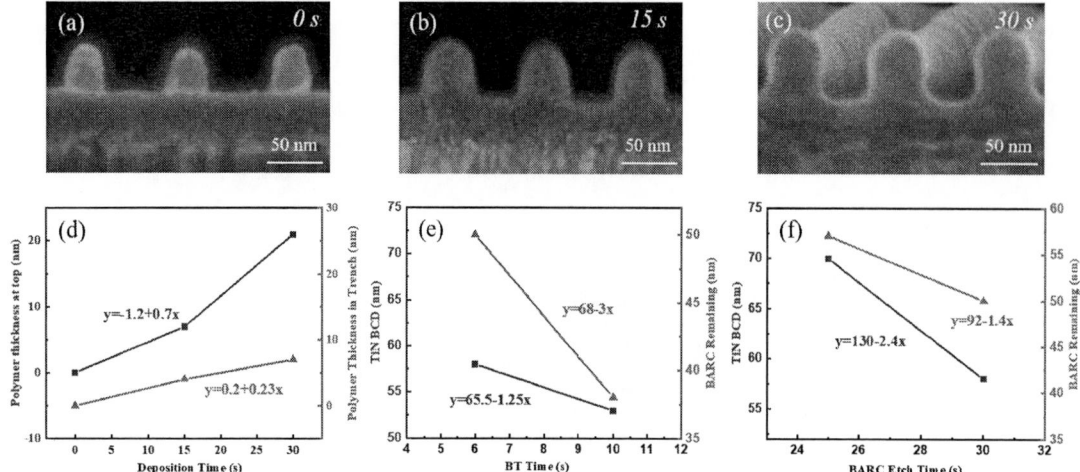

Figure 5: Process window check. XSEM images of pattern profile (a) before CH_4 polymer deposition, (b) after 15s CH_4 polymer deposition and (c) after 30s CH_4 polymer deposition. (d) Polymer thickness increased with CH_4 deposition time both at the top and trench of patterns. (e-f) TiN bottom CD and BARC remaining tuning by (e) BT etch time and (f) BARC etch time.

remaining, BT time need to be well controlled. Under the guarantee for enough BARC remaining, BT time should be less than 10 s (Figure 5e). On the other hand, TiN CD is more sensitive to the BARC etch time. To get TiN CD on target, 45nm for instance, the BARC etch time should be set about 35 s, which can also ensure enough BARC remaining for it only decreases slightly with the prolonged BARC etch time.

In conclusion, we proposed the strategy to insert additional polymer deposition step into MHM open process to gain the BARC remaining for better TiN CD and profile control. We systematically compared two polymer deposition insertion sequences and found high AR pattern benefits to the selective polymer deposition with the effect of visibility deposition, thus effectively increases the BARC remaining after TiN etching. Finally, it was also found that BT time mainly affects the BARC remaining and the BARC etch time can be used to tune the TiN CD. Our findings thus not only reveal the mechanism of selective deposition, but also provide an innovative strategy to enhance process window for mask open.

REFERENCES

[1] F. Fracassi, R. d'Agostino, R. Lamendola, and I. Mangieri, *J. Vac. Sci. Technol. A*, Vol. 13, No. 2, 1995, pp. 335-342.

[2] Wan Lee, Chang Hee Han, Ji-Soo Park, and Jin Won Park, *Journal of The Electrochemical Society,* Vol. 148, No. 3, 2001, pp. G95-G98.

[3] Elke Erben, Klaus Hempel and Dina Triyoso, IntechOpen, 2018, pp. 28-40, DOI: 10.5772/intechopen.78335.

[4] D. J. Wang, M. D. Hu, J. Q. Zhou, C. L. Zhang, X. P. Wang, and H. Y. Zhang, *ECS Transactions*, Vol. 52, No. 1, 2013, pp. 317-323.

[5] http://www.coventor.com/products/semulator3d

[6] Coventor SEMulator3D documentation

APPLICATION OF A BEVEL ETCH PROCESS FOR IMPROVING PARTICLE PERFORMANCE IN CMOS IMAGE SENSOR MANUFACTURE

Yiling Sun1*, Jihong Zhang1, Yu Jiang1, Fulong Qiao2, Keqiang He2, Zhigang Zhang2, Kang Huang2, Yushan

[1]Lam Research Corporation

[2]Huali Micro Electronic Semiconductor Corporation Ltd.

Shanghai City, China

*Corresponding Author's Email: Lonzo.Sun@lamresearch.com

ABSTRACT

Due to the continuing improvements of CMOS image sensor (CIS) technology, the back side-illumination (BSI) structure was involved to overcome optical characteristics deterioration in smaller pixels. In BSI structure, bonding loop is necessary to adhere two wafers. However, multiple dielectric and metal layer on the bevel of the device wafer possibly to be the source of defect as it is so loose that peeling particles will fall on the wafer after trimming step and bubble defect will formed at last. The main objective of the work is to improve peeling particle performance. After insertion of bevel etch, the number of peeling particles was reduced to less than 5ea compared with more than 300ea without bevel etch. Ultimately, no more bubble defect was found due to peeling particles clear off. What's more, we also discuss the mechanism of peeling particles and bubble defect forming. And the roles of different components of the gas mixtures in the bevel etch process.

INTRODUCTION

The resolution increase of CIS has attracted much attention in the past several decades owing to consumer application like mobile phone, digital still camera, camcorder and other mobile devices [1-3]. To promote sensor resolution, CIS technology has migrated from frontside illumination (FSI) to BSI duo to pixel size shrinks down. Unlike FSI, BSI structure has no metal layer above the pixel. So, it can avoid photon absorption and reflection of metal layer. As a result, BSI will improve quantum effect, fill factor and crosstalk [4-5]. During the BSI sensor manufacture, wafer bonding loop was involved as the device wafer is so thin that need another wafer to carry it. However, there will be multiple nonuniform like photoresist, dielectric and metal layers on the bevel of device wafer which will be the source of particles. Because during the bonding loop, device wafer need trim the edge which will cut the films and release the peeling particles from the multiple bevel layer. In this case, the fall-on peeling particles will bring bubble defect after bonding two wafers without bevel etch.

In this paper, we found oxygen-related peeling particles after trimming step on the edge of wafer and macroscopic bubble defect formed after bonding. To improve the defect performance, we study a highly

effective plasma dry etch solution with mixed halogen gas on the Lam Research tool Coronus HP®. With the insertion of bevel etching in the bonding loop, we found multiple nonuniform defect-source layer clear up and almost no peeling particles any more. Beyond that, we have built up the model of peeling and bubble defect forming.

METHODOLOGY

Bonding loop started with finishing frontside of device wafer. Namely most logic-related part of the device has completed at this time.

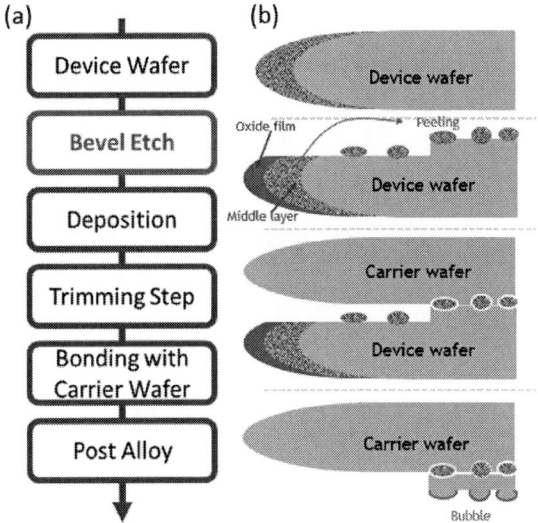

Figure 1: (a) Bonding loop flow w/ or w/o bevel plasma etch. (b) The schematic diagram of peeling and bubble defect forming

In the conventional approach, the device wafer will deposit oxide layer at first. But is this study, for the purpose of cleaning the peeling particles source, the 300mm device wafer will receive the plasma dry bevel etch before deposition. Continuous and uniform plasma in the Coronus HP® chamber will clean up double sides of the particles source film on the wafer bevel area from the gas mixture of NF3 (70 sccm), CF4 (300 sccm) for 1 min with a pressure of 3000 mT. After this plasma etch, the peeling source films on the bevel area will removed.

Then the wafer will send to the plasma enhanced chemical vapor deposition chamber to coat 25kÅ oxide layer. This film is aiming to bond the device and carrier wafer together with the interaction of van der Waals force.

Before bonding with carrier wafer, trimming step will be introduced to cover wafer different thick profile on the edge. About 0-8mm distance of the edge wafer was trimmed in the process. What's more, some clean step will be inserted as there are some particles generated during trimming step with the broken of film structure.

Next, to enhance van der Waals force, bonding step will heat the carrier wafer at the temperature of 600 °C for the 10min.

At last, adhere these two wafers together and alloy them to bond tightly. The flow of bonding loop with bevel etch was schematically shown in Fig. 1a.

RESULTS AND DISCUSSION

In the conventional approach, there is no plasma dry bevel etch after trimming step, we found hundreds of peeling particles on the edge of the wafer by optical scanning, as illustrated in Fig. 2a. Scanning electron microscope characterization (SEM) exhibits there are kinds of peeling particles shape, and the size of the particles is 1-2 μm(Fig. 3), and the energy dispersive spectrometer result shows most composition of the peeling is Ti, W and Si. Based on the EDS data, we have reason to believe that the source of peeling particles is not from bonding loop due to there is no metal-related process in it.

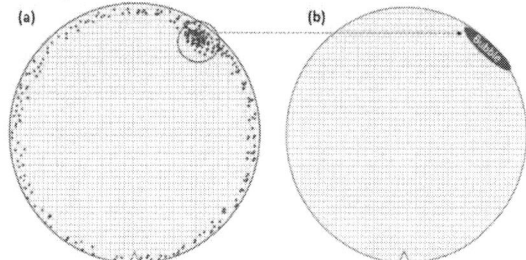

Figure 2. (a) The schematic diagram of peeling particles within wafer map. (b) The peeling particles accumulation area will bring macroscopic bubble defect after bonding loop.

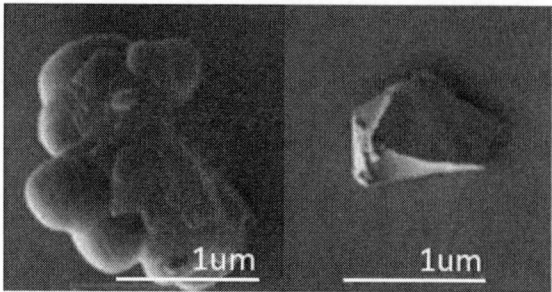

Figure 3. SEM picture of peeling particles.

After the total bonding loop finished, we found the bubble defect on the edge with macroscopic size, as illustrated in Fig. 4a. Compared with the peeling distribution on the wafer, it can be found that the map of accumulative peeling particles matches well with the bubble defect location.

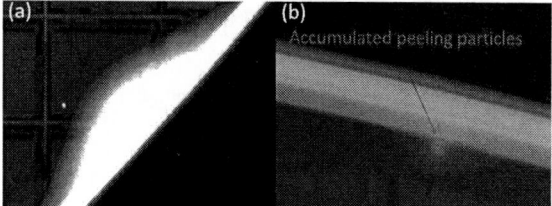

Figure 4. (a) Macroscopic bubble defect on the edge of the wafer. (b) Accumulated peeling particles found under the bubble defect

In order to obtain the cause and effect between peeling particles and bubble defect, more detailed analysis was carried out, the low magnification of SEM pictures shows accumulative peeling particles under the bubble defect, as illustrated in Fig. 4b.

Based on these data, it revealed that the source of peeling particles is from frontside loop of the device wafer as there is no metal deposition or any metal induced process during bonding loop. The root cause of the bubble defect is accumulative particles on the wafer edge. By now we have built up the mechanism of the peeling particles induce the bubble defect (Fig. 1b), but stronger evidence is needed to make it clear that where is the peeling source.

To further investigate mechanism of peeling particles formed, the middle layer adhesiveness characteristics were measured. On the wafer bevel, it's hard to accurate estimate the detailed films thickness and composition ration on this area, we call these multiple dielectric and metal films, formed during frontside loop, as middle layer in this paper. Fig. 5 is SEM picture is the morphology of oxide layer and middle layer after trimming step on the wafer bevel area. From the picture, it is not difficulty to find that the adhesiveness of middle layer is loose. As we know, the oxide film is impacted. The large adhesiveness variation of these two films that next to each other makes it is easy to release the peeling particles after trimming step.

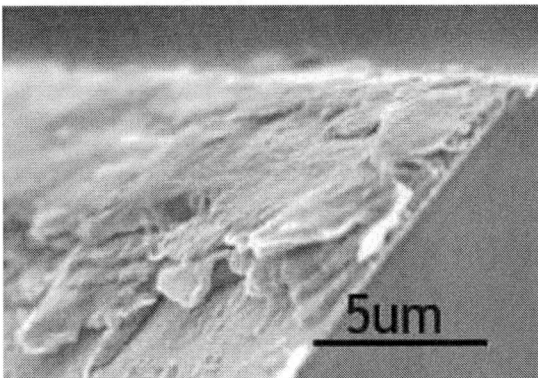

Figure 5. Middle layer on the bevel area

From these results it was clearly established that the source of peeling particles is middle layer which formed during the frontside loop, it must be in the back end of the line as there is a large proportion of metal composition. The middle layer is coherent itself as there is no peeling particles on the incoming wafer. But after oxide film deposition and trimming step, the gap of adhesiveness between two layers makes middle layer act as the peeling particles source.

In order to improve the particle performance, a highly effective plasma dry etch was inserted into the bonding loop. Consider the middle layer is formed during frontside loop, we etched the incoming device wafer before any other step on the Lam Research tool Coronus HP® chamber. This tool is specially designed for bevel etch with strict controlling of etch distance. The mixture of halogen chemistry was involved to remove the middle layer on the both sides of the wafer. The halogen plasma atmosphere has the ability to react with the W, Ti and Si, forming WFx, TiFx and SiFx respectively. And these gaseous by product will flush out at last. With the help of active halogen ions and enough over etch, the middle layer on the bevel of the wafer was clear up as expected.

After insertion of bevel etch, almost no peeling particle was found (<5ea). Compared with conventional approach particle performance(>300ea), it is big improvement. With no peeling particle, there is no more bubble defect inevitably as the root cause of the bubble is particles accumulation.

CONCLUSIONS

In summary, we exhibited the mechanism of peeling particle forming. And the correlation model of peeling particles and bubble defect was built up. We also have demonstrated that the particle performance of CIS during bonding loop can be improved by insertion a highly effective plasma dry etch on the bevel of the wafer. Less than 5ea peeling particles performance was achieved on the wafer during the bonding loop through clear up the particle source layer. This technology also permits improvement in performance of wafer yield due to reduction of bubble defect in the bonding loop.

ACKNOWLEDGEMENTS

Authors would like to thank the etch process group in the Technical Development Department of Huali Microelectronic Ltd. for enabling the experiments to be conducted and valuable technical discussions. Further we would like to acknowledge Guibin Wu and Yushan Chi from Lam for knowledge sharing and technical discussion. Coronus is a registered trademark of Lam Research Corporation.

REFERENCES

[1] M. Furumiya, H. Ohkubo, Y. Muramatsu, S. Kurosawa, F. Okamoto, Y. Fujimoto, and Y. Nakashiba, "High sensitivity and no-crosstalk pixel technology for embedded CMOS Image Sensor," IEEE Trans. Electron Devices, vol. 48, pp. 2221-2227, October. 2001.

[2] S. Chieh, G. Agranov, H. Tian, C. Baron, H.-W. Lee, and R. Madurawe, "Challenges and opportunities in small pixel development for novel CMOS image sensors," in Proc. IEEE VLSI-TSA, pp. 1-2, Apr. 2013

[3] R. Fontaine, "The evolution of pixel structures for consumer-grade image sensors," IEEE Trans. Semicond. Manuf., vol. 26, pp. 11-16, February. 2013

[4] B. Vereecke, C. Cavaco, K. De Munck, L. Haspeslagh, K. Minoglou, D. Sabuncuoglu, K. Tack, and H. Osman, "A Platform for Backside Illuminated CMOS Image Sensors for UV and Visible Applications," Solid State Devices Conference, Tsukuba 2014, pp990-991

[5] G. Taverni, D. Moeys , C. Li, C. Cavaco, V. Motsnyi, D. Bello, and T. Delbruck, "Front and Back Illuminated Dynamic and Active Pixel Vision Sensors Comparison," IEEE Transactions on Circuits and Systems, vol. 65, pp. 677-681

SILICON WAFER THINNING PROCESS BY DRY ETCHING WITH LOW ROUGHNESS AND HIGH UNIFORMITY

Zihan Dong[1], Renzhi Yuan[1], and Yuanwei Lin[1*]*

[1]Department of Semiconductor Etching, NAURA Technology Group Co., Ltd., Beijing 100176, China

*Corresponding Author's Email: dongzihan@naura.com; linyuanwei@naura.com

ABSTRACT

To overcome the issues of surface roughness, morphology and damage in the wafer thinning process, dry etching has been introduced to improve the surface quality of the ultra-thin wafers after mechanical grinding. At present, only non-Bosch process is reported for wafer thinning to our best knowledge. In this work, we demonstrate a time-multiplexed alternating thinning (TMAT) process for 12-inch silicon wafer thinning with higher uniformity and lower roughness compared with that of non-Bosch process. This method is operating convenient and might help obtain ultra-thin wafers with better quality in the semiconductor industry.

INTRODUCTION

As a fundamental material, silicon wafer has been widely used in modern electronic industry, such as integrated circuits (ICs), micro-electro-mechanical System (MEMS) and advanced packaging (AP). Since the demand for smaller and thinner chips/packages increases, manufacturers are finding their ways to reduce thickness of standard wafers. How to control the surface uniformity and roughness of the ultra-thin wafer is a fundamental challenge in wafer thinning process, especially for the wafers with large size (12-inch). Generally, the conventional method for wafer thinning is mechanical grinding and polishing [1-5]. Mechanical grinding could effectively reduce the thickness of the wafer in a few minutes but cause the surface damage (i.e., crystalline defects/dislocations and microcracks) that induces residual stress on the wafer surface [6-9]. To overcome these issues, dry etching process was introduced, which could remove the damage layer and reduce surface residual stress after the mechanical grinding [10]. However, it is difficult to obtain low roughness and high uniformity through current dry etching technology when the thinning thickness is larger than 20 μm. Herein we demonstrate a dry etching silicon wafer thinning process with high uniformity and low roughness, where the physical bombardment was reduced and the chemical reaction was enhanced by the time-multiplexed process.

EXPERIMENTAL

Our process of wafer thinning is similar to Bosch process [11], mainly including three steps, deposition step, gas exchange step and etch step, which is a time-multiplexed alternating thinning (TMAT) process. The details of TMAT process are shown in Figure 1. Step 1: Inject the depositional gas, turn on the RF power and deposit a thin film on the silicon wafer. Step 2: Turn off the power and the depositional gas, then inject the chemical etching gas. Step 3: Turn on the RF power and etch the silicon. Step 4: Turn off the power and inject the depositional gas. Step 5: It is judged whether the current cycle number of the Step 4 is equal to the total number of cycles, and if so, the process ends. Otherwise, the current cycle number is incremented by 1, and the process returns to the Step 1.

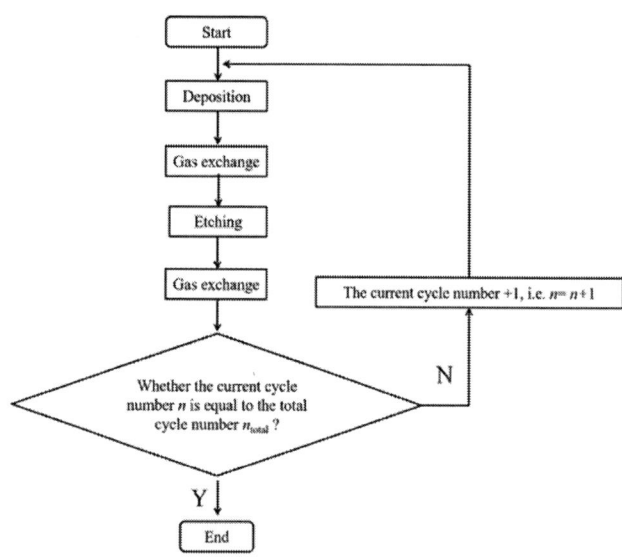

Figure 1. Flow chart of TMAT Process.

To perform the process mentioned above, the 300 mm silicon wafer was thinned in a HSE chamber [12]. The atomic force microscope (AFM) and thickness measurement were taken on a Brucker Dimension EDGE and a micrometer caliper, respectively.

RESULTS AND DISCUSSIONS

The recipe of TMAT process is shown in TABLE 1, and the control recipe of conventional non-Bosch process is shown in TABLE 2.

TABLE 1 Recipe parameters used in this work except for the cycle number

Step	Pressure (mTorr)	Source power (W)	Bias power (W)	Duty	SF_6 flow (sccm)	C_4F_8 flow (sccm)	BCl_3 flow (sccm)	Time*(s)
Deposition	45	1800	1	50%	1	400	50	1.5
Gas exchange	45	1	1	50%	1200	1	1	2
Etching	45	3000	200	50%	1200	1	1	1.5
Gas exchange	45	1	1	50%	1	400	50	2

*Each step time can be increased or decreased as appropriate according to the actual situation.

TABLE 2 Recipe parameters used in the control experiment

Process	Pressure (mTorr)	Source power (W)	Bias power (W)	Duty	SF_6 flow (sccm)	C_4F_8 flow(sccm)	BCl_3 flow (sccm)	Time*(s)
Etching	45	3000	200	50%	1200	0	0	300

*Etching time selects different values depending on the thickness of the thinning.

The results show that when the thickness of the thinning increases, the surface roughness becomes better (Figure 2 and Figure 3). The wafer thickness uniformity is controlled to be below 5%, and it does not cause much deterioration with thinning thickness (from 30 μm to 100 μm), as shown in TABLE 3. But the surface roughness and uniformity of single step dry etching process cannot be controlled without deposition step, as shown in Figure 4 and TABLE 4.

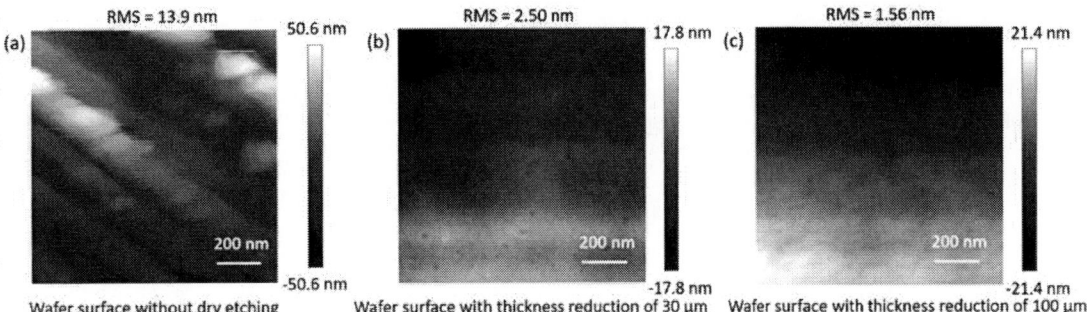

Figure 2 AFM images of the wafer surface by the TMAT process with different thickness reductions.

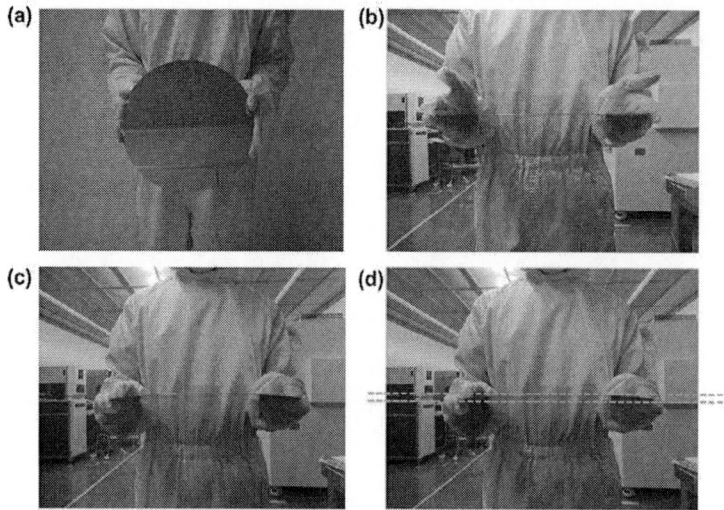

Figure 3 Thinned wafer photos by the TMAT process.

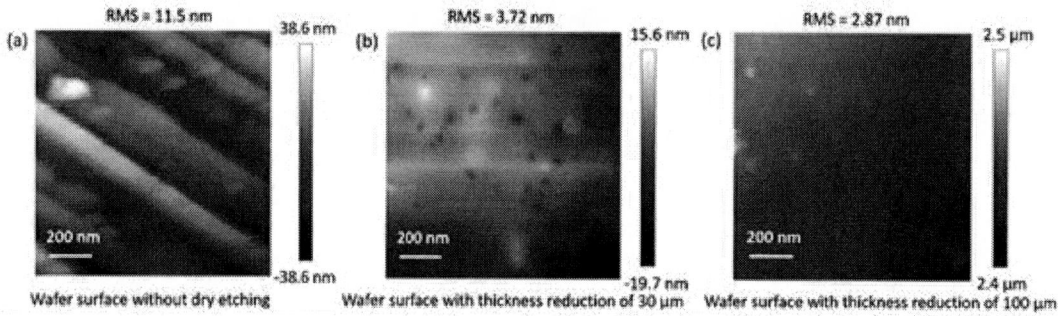

Figure 4 AFM images of wafer surface by the conventional single step dry etching process with different thickness reductions.

TABLE 3 Wafer thinning uniformity by the TMAT process

Target thinning thickness (μm)	Min thickness (μm)	Max thickness (μm)	Uniformity
30	29	31	3.33%
100	97	105	3.96%

TABLE 4 Wafer thinning uniformity by the conventional single step dry etching process

Target thinning thickness (μm)	Min thickness (μm)	Max thickness (μm)	Uniformity
30	28	31	5.08%
100	97	116	8.92%

Compared with conventional plasma etching process, the wafer thinning process reported herein has a deposition step before the etching step. During the etching step, the deposited film could withstand the strong ion physics bombardment at the beginning of the etching, which could further reduce damage to wafer. When the deposited film is removed, the process is mainly based on chemical etching. The chemical etching has good isotropic etching to achieve better roughness and uniformity, as shown Figure 5.

The lifetime of ions within physical bombardment is shorter than that of free radicals within chemical reaction, therefore only a thicker deposited film is needed, which could increase the proportion of isotropic chemical etching.

In addition，the gas exchange step is introduced between the deposition and etching step. Its function is to stabilize the gas atmosphere in the chamber and reduce the generation of particles.

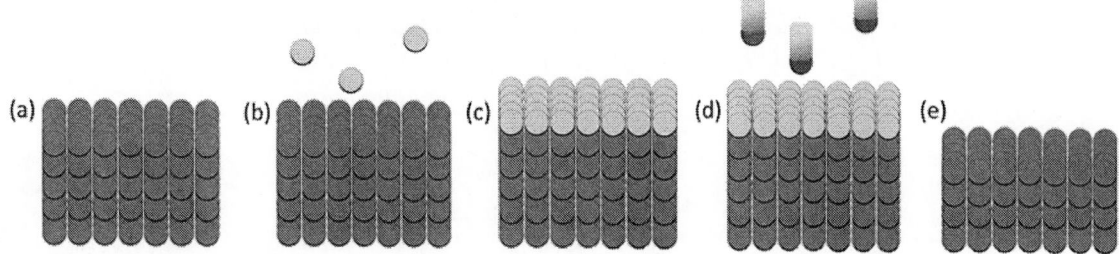

Figure 5 Schematic diagram of surface roughness control mechanism in the TMAT process.

Though TMAM process is very similar to Bosch process, there is still a big difference between the two methods. First, in terms of technical methods, the chemical composition of the deposited film in TMAM process is not limited to fluorocarbon polymers, but any chemical composition that can passivate the surface of the silicon, and does not need to consider the influence of etching depth and etching selectivity on the process parameters. Second, in terms of mechanism, Bosch process utilizes the fluorocarbon polymer to protect sidewall, thereby achieves anisotropy etching for high aspect ratio silicon etch. Conversely, the deposited film in TMAT process is used to withstand the strong ion physics bombardment at the beginning of the etching, which could achieve isotropic etching as much as possible to facilitate wafer thinning and surface polishing uniformly.

SUMMARY

In conclusion, we have demonstrated a time-multiplexed alternating thinning (TMAT) process for 12-inch silicon wafer thinning with 100 μm thickness reduction. Compared with the conventional single step dry etching process, TMAT process could facilitate wafer thinning and surface polishing with high uniformity and low roughness. Though TMAT process is very similar with Bosch process, these two methods are significantly different in terms of technical methods and mechanisms. TMAT process is easy to thin a greater thickness wafer while controlling its uniformity and surface roughness, and might help the industrial community obtain ultra-thin wafers with better quality.

REFERENCES

[1] Yoshida S, Nagai O. Wafer grinding method: U.S. Patent 7,462,094[P]. 2008-12-9.

[2] Hirata K, Nishino Y, Morikazu H, et al. Wafer thinning method: U.S. Patent Application 10/319,593[P]. 2019-6-11.

[3] Karlsrud C E, Van Woerkom A G, Odagiri S, et al. Wafer polishing method and apparatus: U.S. Patent 5,329,732[P]. 1994-7-19.

[4] Pei Z J, Strasbaugh A. Fine grinding of silicon wafers: designed experiments[J]. International Journal of Machine Tools and Manufacture, 2002, 42(3): 395-404.

[5] Pei Z J, Fisher G R, Liu J. Grinding of silicon wafers: a review from historical perspectives[J]. International Journal of Machine Tools and Manufacture, 2008, 48(12-13): 1297-1307.

[6] Pei Z J, Billingsley S R, Miura S. Grinding induced subsurface cracks in silicon wafers[J]. International Journal of Machine Tools and Manufacture, 1999, 39(7): 1103-1116.

[7] Chen L Q, Zhang X, Zhang T Y, et al. Micro-Raman spectral analysis of the subsurface damage layer in machined silicon wafers[J]. Journal of Materials Research, 2000, 15(7): 1441-1444.

[8] Chen J, De Wolf I. Study of damage and stress induced by backgrinding in Si wafers[J]. Semiconductor science and technology, 2003, 18(4): 261.

[9] Zhou L, Tian Y B, Huang H, et al. A study on the diamond grinding of ultra-thin silicon wafers[J]. Proceedings of the Institution of Mechanical Engineers, Part B: Journal of Engineering Manufacture, 2012, 226(1): 66-75.

[10] McLellan N, Fan N, Liu S, et al. Effects of wafer thinning condition on the roughness, morphology and fracture strength of silicon die[J]. Journal of Electronic Packaging, 2004, 126(1): 110-114.

[11] Laermer F, Schilp A. Method of anisotropically etching silicon: U.S. Patent 5,501,893[P]. 1996-3-26.

[12] Cui Y, Jian S, Chen C, et al. Uniformity improvement of deep silicon cavities fabricated by plasma etching with 12-inch wafer level[J]. Journal of Micromechanics and Microengineering, 2019, 29(10): 105010.

IMPROVEMENT RESEARCH OF ROUND CONVEX RESIDUE IN DUAL GATE LAYER

Mingguang Hang*, Lili Jia, Fang Li, Jun Huang, Wenyan Liu
Shanghai Huali Integrated Circuit Manufacturing Corporation, Shanghai 200433, China
*Mingguang Hang's Email: hangmingguang@hlmc.cn

ABSTRACT

With the development of integrated circuit technology, the application of new materials and new process in integrated circuit process also brings new challenges. This paper reported some improvement research for round convex residue (oxide residue) in dual gate layer which may result in low yield. Improvement research include change wet clean process condition before thick gate oxidation, add wet clean process post thick gate oxidation, change lithography process conditions and change thick gate growth mode. The results show that oxide residue can be effectively removed by these methods, and which can improve yield about 20%.

INTRODUCTION

Dual gate oxidation is the important process during integrated circuit manufacturing. Forming different thickness gate oxide layers by partial wet etch and re-oxidation. Thick gate oxide layer is work for high voltage device layer, and thin gate oxide layer is work for low voltage device layer, to achieve the needs of different device in the same process.

Otherwise, defect issue is a huge problem that plagues a lot of engineers. Defect on wafers are divided into many different types such as particle [1], residue, scratch and so on. And defect sources and formation mechanism are also diverse. These defects will kill the yield of wafers to varying degrees. In this paper, we studied the formation mechanism and solution of oxide residue in dual gate layer which can kill the yield above 20%, and the access to resolve oxide residue mainly include the following 3 points: add wet clean process post thick gate oxidation, change lithography process conditions and change thick gate growth mode.

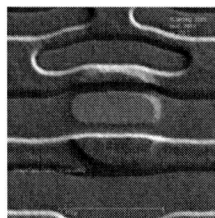

Figure1: the image of oxide residue

RESULTS AND DISCUSSION

1.the formation mechanism of oxide residue

In dual gate loop process, after thick gate oxidation and dual gate photo, dual gate wet etch and strip, it can be found some round convex defect which can be proved is oxide residue by SEM and TEM.

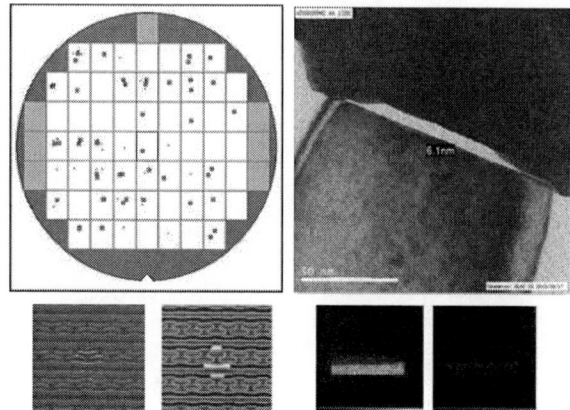

Figure2: oxide residue map, SEM image and TEM image

Based on experimental data and material properties, we further study the formation mechanism of oxide residue. At present, thick gate oxidation using wet oxygen oxidation process, which is using water vapor and silicon to react to generate silicon dioxide[2] and hydrogen. The chemical reaction equation as below:

$$Si(solid)+2H_2O(vapor) \rightarrow SiO_2(solid)+H_2(gas)$$

During the reaction, H • and H_2O react to form $-OH$. SiO_2 film is easy to absorb H_2O and $-OH$ due to its hydrophilic nature. In the subsequent lithography process, the soluble part of the photoresist will be melted to produce H^+ during exposure, and then removed by developer (alkaline chemical). However, if SiO_2 film contains $-OH$, $-OH$ and H^+ will react to form H_2O, which will prevent photoresist removal. And this residue will prevent the following up SiO2 film removal, eventually formed oxide residue.

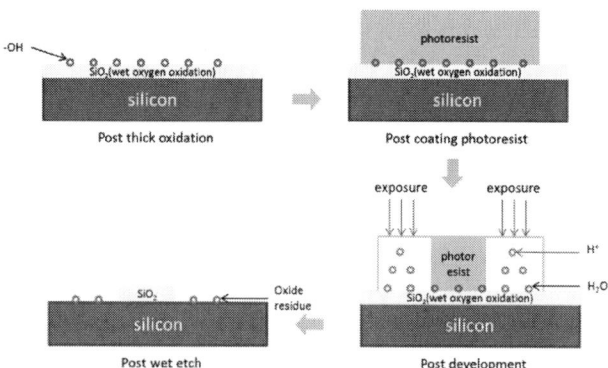

Figure3: the formation mechanism of oxide residue

2.Research on wet clean to improve oxide residue

Wet clean is a method of removing defects by cleaning the surface of wafers with a variety of chemicals. In this case, for improving oxide residue issue, SC1 and SPM have been used to clean wafer surface after thick gate oxidation, and inspect defect after wet etch and photoresist strip. The experiment result shows that both SC1 and SPM clean wafer can improve oxide residue performance, and use SC1 works better than SPM, the number of defects can be reduced to zero.

TABLE I. IMPROVEMENT OF OXIDE RESIDUE BY WET CLEAN

Condition	Total defect count	Oxide residue count
No clean	283	74
SC1 clean	96	0
	69	0
SPM clean	142	0
	74	9

For further analyze the mechanism, Qcept detection technology has been used to inspect the charge of wafer after wet clean. The experiment result shows that there are large amounts of negative charges on the surface of wafer after thick gate oxidation, and the number of charges can be greatly reduced after SC1 clean and SPM clean, but SC1 has better cleaning effect. Accordingly, there are lots of chemical residue with negative charges (–OH) on the wafer after thick gate oxidation, which may result in the formation of oxide residue. These residue with negative charges (–OH) can be removed by SC1 or SPM clean, and then form to the oxide residue will be prevented successfully.

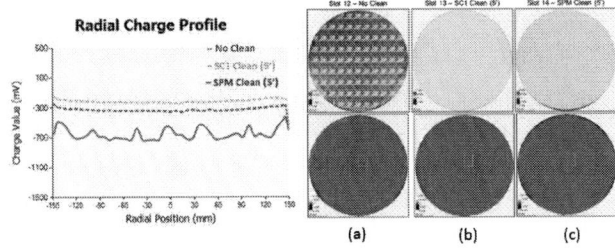

Figure4: charge and defect distribution on wafer, (a) no clean, (b) SC1 clean, (c) SPM clean

3.change thick gate growth mode to improve oxide residue

Dual gate oxide layers are generally obtained by thermal growth. Thermal growth methods can be divided into dry oxygen oxidation and wet oxygen oxidation. Dry oxygen oxidation is that expose the silicon to a high temperature atmosphere and high purity oxygen without water vapor to complete the growth of the oxide layer, the chemical reaction equation as below:

$$Si(solid)+O_2 \rightarrow SiO_2(solid)$$

When water vapor is involved in the reaction, it is called wet oxygen oxidation, the chemical reaction equation as below:

$$Si(solid)+2H_2O(vapor) \rightarrow SiO_2(solid)+H_2(gas)$$

In this case, we changed the growth mode of gate oxide layer from wet oxygen oxidation to dry oxygen oxidation, which can suppress oxide residue from root cause by reduce the generation of H_2.

TABLE II. IMPROVEMENT OF OXIDE RESIDUE BY CHANGE GATE OXIDE GROWTH MODE

condition	Dry oxygen oxidation			Wet oxygen oxidation		
Oxide residue count	0	0	0	13	9	25

Although the use of dry oxygen oxidation to grow the gate oxide layer can reduce the generation of oxide residue, in actual production, not only the oxidation rate of dry oxygen oxidation is slow, but also the slow oxygen diffusion rate and low solubility which will limit the growth thickness of the oxide layer.

4.change lithography condition to improve oxide residue

During the integrated circuit manufacturing, the main role of photoresist is to protect the part covered by photoresist during etching or ion implantation. In dual gate process, the protection object of the photoresist is silicon dioxide. Due to the surface of silicon dioxide is hydrophilic, and the photoresist is hydrophobic, the surface of silicon dioxide is easily to absorb water molecules in the air. It is easily to cause detachment of silicon dioxide and photoresist. On the other hand, there are large numbers of –OH on the surface of silicon dioxide, which may cause photoresist poison if it contacts with photoresist directly. Both of the above two reasons will cause to the formation of oxide residue. Therefore, coating an organic layer (HMDS) [3] on the wafer before coating photoresist to increase the surface hydrophobicity of silicon dioxide, which can enhance the adhesion of silicon dioxide and photoresist and avoid photoresist poisoning.

TABLE III. IMPROVEMENT OF OXIDE RESIDUE BY CHANGE LITHOGRAPHY CONDITION

Condition	Coating HMDS		No HMDS	
Oxide residue count	4	6	22	25

Experiment result proved that chemical method is more superior than physical method on improving defect issues.

CONCLUSION

In summary, we have conducted a series of experimental studies and verified their effects, and finally concluded three effective solutions as following. 1. Add SC1 clean post thick gate oxidation; 2. Thick gate oxide growth mode changed to dry oxygen oxidation; 3. Coating an organic layer (HMDS) on the wafer before coating photoresist. Due to the limitations of dry oxygen oxidation in actual production, we eventually adopted the condition of add SC1 clean post thick gate oxidation to effectively eliminate oxide residue, and which can improve yield about 20%.

REFERENCES

[1] L. J. Duan, H. H. Au etc., A New Mechanism of Poly-silicon Crater Defect Induced from Al Tiny Particle Charging Effect during Water Rinse in Oxide Patterning Process, South Lake Tahoe, CA, USA, 16 Oct.-19 Sept. 2006.

[2] N. S. Pshchelko, E. G. Vodkaylo etc., Technological features of silicon dioxide electret manufacture, St. Petersburg, Russia, 1-3 Feb. 2017.

[3] K. Thodkar, C. Schonenberger etc., Characterization of HMDS treated CVD grapheme, Ottawa, ON, Canada, 10-15 July 2016.

ADVANTAGE TIMELY ENERGIZED BUBBLE OSCILLATION MEGASONIC NANO-SPRAY METHOD TO ELIMINATE SURFACE PARTICLE DEFECT IN LIGHTLY DOPED DRAIN 28NM

Hong Li [1], Fang Li[1]*, Wenyan Liu[1], Jun Huang[1], Yu Zhang[1]
Wenjun Wang[2], Xiaoyan Zhang[2], Ting Yao[2], David H. Wang[2]
[1]Huali Microelectronics Corporation
[2]ACM Research (Shanghai), Inc.
[1]No.6 Liang Teng Rd., Pudong New Area, Shanghai, 201314, China
[2]No.1690 Cai Lun Rd., Pudong New Area, Shanghai, 200120, China
*Corresponding Author's Email: lifang@hlmc.cn

ABSTRACT

After lightly Doped Drain (LDD) photoresist wet strip, the random surface particles were often detected by wafer inspection, especially in the 28 nm IC technology and beyond. Sensitive pattern structures were very easily damaged by traditional violent clean, therefore it was more difficult to remove random defects completely within LDD loop. In this study, an advantage Timely Energized Bubble Oscillation Megasonic Nano-spray (TEBO Nano-spray) wet clean method was introduced to remove unwanted surface particle defect within LDD loop in a single wafer clean tool. Remarkable particle remove efficiency (PRE) was as high as 90 % with pattern damage free by using the new wet clean method. At the same time the device reliability was qualified. The cleaning condition and related substrate loss were also taken into consideration.

Keywords—LDD, Megasonic Nano-spray clean, Single wafer cleaning, PRE

INTRODUCTION

Wet cleaning technology plays a critical role in semiconductor manufacturing industry, and takes up around 20~30 % of whole process. As evident by the critical dimension continuous shrinking and moving to larger diameter substrates, there are many additional requirements related wet cleaning exposed. The surface particle within LDD loop is a major problem for front end of line (FEOL) process. One of the reasons is the pattern itself is deep and wide as feature sizes shrink, and fragile pattern structure easily damaged, furthermore the final wafer accept test (WAT) and chip probe (CP) was also correlated directly to substrate losses in this area.

Although particle removing and yield enhance has been widely discussed in the literature, most of the cases were based on the Nano-spray or mega clean tools [1-3]. Traditionally, single Nano-spray wafer clean can overcome most of surface particle and defect problem with the help of SC1 (SC1 compositions of NH4OH with H2O2 and H2O) and SC2 (SC2 compositions of HCL with H2O2 and H2O) [4]. However, tiny particle inside pattern were still difficult to remove, in addition to what easily damage the pattern structures under micro Nano-spray. For better particle removal, a study Timely Energized Bubble Oscillation Megasonic Technology (TEBO) was reported, which can provides stable control of bubble cavitation, without pattern damage at the different modes [5,6]. However, the TEBO wet clean related process parameter and actual performance are less mentioned so far.

Accordingly, we employed a simple DOE (design of experiment) to screen the effect of surface particle within LDD loop. The effects of the TEBO mega, the process mode, and the chemical time and concentration were investigated. Based on the analysis, an advantage new method of Timely Energized Bubble Oscillation Megasonic Nano-spray method was proposed and particle removal efficiency (PRE) test was performed. Meanwhile, the cleaning condition qualified corresponding mechanism is also discussed.

EXPERIMENT

A CMOS process was carried out with P-type 300mm silicon wafers within LDD loop. By tuning TEBO parameters (like mega, mode, power, etc.) and cleaning chemical condition, different defect performance was found. All DOE split condition and defect level are shown in Table 1. In this paper, Scanning Electron Microscope (SEM) was used to detect defect condition after cleaning processes, and the silicon Substrates film thickness was measured using SFX200 (KLA-Tencor, USA) before and after process.

In addition, the WAT and Yield are also judged by professional test.

RESULTS AND DISCUSSION

PRE test

According to the wet condition tuning table 1, TEBO mega cleaning has significant difference of PRE with TEBO mode and power. The same mode with different power showed different properties as shown in wet condition 1, 2 and 3 with mode B. The higher the energy (from 60 w to 80 w) by mode B, the better PRE was got. Meanwhile, the mega power is a trade-off between PRE and pattern damage risk. As shown in condition 1 with mode B, PRE as higher as > 90 %，but suffer pattern damage, which is exposed as in Fig. 1. Under the same mega power, different modes also show different properties. In the test of condition 2, only mode B no pattern damage occurred and got a best PRE of 88 %. Meanwhile, as condition 3 shows, when the mega power migrates to 60 w and below, a very low PRE was got, it suggested the mega power can't generate enough cavitation which applied mechanical force uneasy to remove particle from patterned structure wafer.

TABLE.1 WET CONDITION TUNING TABLE

Test table	Mega Power (watts)	TEBO Mode			SC1 time(s)	SC1 Con.	Nano spray
		A	B	C			
1 MHz TEBO process window							
1	80	DMG	>90%	DMG	I	1	No
2	70	DMG	88%	DMG	I	1	No
3	60	<70%	<70%	<70%	I	1	No
2 MHz TEBO process window							
4	100	DMG	>80% remain PA>500nm	DMG	I	1	No
5	100	DMG	>90%	DMG	I	1	YES
6	100	>80%	>90%	>80%	II	2	YES
7	100	<80%	<80%	<80%	III	2	YES

Notes: The SC1 time I>II>III; The SC1 con. 1>2

Fig.1 The SEM picture of pattern collapsed structure

Due to poor and margin PRE by 1 MHz, a 2MHz cleaning system was fabricated and researched. As a result, 2 MHz TEBO without pattern damage window is larger than 1 MHz, and the biggest mega power setting window near twice than 1Mhz. Under the high operating frequency of 2 MHz, the displacement at the end of the waveguide is as small as 1/2 of a 1 MHz type, which generate smaller bubble with lower pattern damage risk. With the wet condition 4, mode B by the power 100 w, get PRE of > 80 %. Fig. 2 shows the removed and remained particle defect with best condition 4, moreover, < 500 nm defect remove efficiency near 100 % and > 500nm ~ 1um remove ability lower. Non-optimized TEBO mega clean condition can't remove the surface PA absolutely, due to the van der Waals forces acting between the substrate and the PA of size (500 nm ~ 1um) increase, which can't be removed effectively. Therefore, a TEBO Megasonic Nano-spray cleaning systems have been first proposed and fabricated. By combine the TEBO mega and Nano-spray clean, condition 5 & 6 show much better cleaning efficiency that leads to remarkable big Particle size defect reduction.

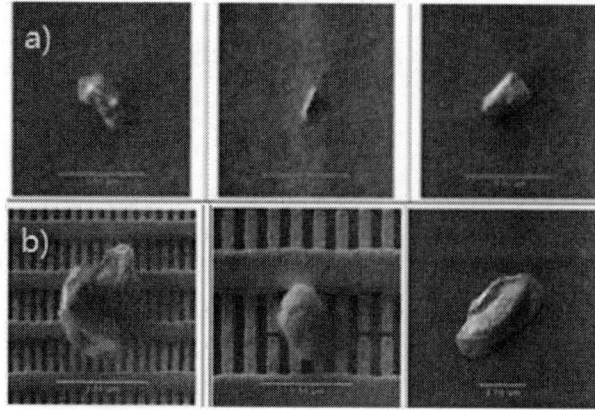

Fig.2 The SEM picture of particle :a）removed particle; b）remained particle

Substrate films loss test

For the assessment of substrate loss cleaning ability, KLA tests were performed after 10 times wet process. AS shown in table 2, By shorter SC1 rinse time and the lower concentration, the bias between baseline and TEBO cleaning get shorter, where the SC1 concentration changed from 1 to 2, and SC1 time degreased from I to III got a ~0.2 nm CD and 0.26A thickness bias with baseline, respectively. Considering PRE and Substrate films loss results, the condition 6 was the best one.

TABLE.2 THE SUBSTRATE FILMS LOSS OF DIFFERENT WET CONDITION

Wet condition	Item	Pre	Post	Loss	Loss /10	Bias /10
Baseline *10	THK(A)	27.48	21.84	5.63	0.56	NA
	CDP1(nm)	59.20	55.60	3.60	0.36	NA
	CDP2(nm)	62.00	58.10	3.90	0.39	NA
Condition 5 *10	THK(A)	27.43	15.38	12.05	1.21	0.56
	CDP1(nm)	59.10	51.90	7.20	0.72	0.36
	CDP2(nm)	61.90	54.40	7.50	0.75	0.36
Condition 6*10	THK(A)	27.54	19.35	8.19	0.82	0.26
	CDP1(nm)	59.20	53.70	5.50	0.55	0.19
	CDP2(nm)	62.10	56.10	6.00	0.60	0.21
Condition 7*10	THK(A)	28.19	20.64	7.55	0.76	0.19
	CDP1(nm)	59.10	53.90	5.20	0.52	0.16
	CDP2(nm)	61.70	56.20	5.50	0.55	0.16

WAT and CP test

To verify the improved TEBO Meagonic-Nanospray wet cleaning, the go-to process condition 6 was applied to check WAT and CP yield. A comparable WAT parameter is observed between BL and TEBO clean. In addition, the CP yield showed ~0.1% improvement. The result was further verified of defect reduction with optimized TEBO clean.

CONCLUSION

In this work, an advantage Timely Energized Bubble Oscillation Measonic-Nanspary method was designed and fabricated to eliminate surface particle in Lightly Doped Drain 28 nm. By tuning the TEBO parameters and cleaning chemical (like TEBO frequency、mode and power，SC1 concentration、SC1 time), the PRE window and substrate films loss was got, where we obtain a PRE as higher as > 90 % without pattern damage and acceptable substrate about ~ 0.2 nm CD and 0.26 A thickness bias with baseline, respectively. At the end, CP yield shows ~ 0.1 % improvement. Considering the results, it is thought that the TEBO mega Nano-spray method can be applicable to LDD loop or patterned structure with improved PRE, which gives us possibility of lower pattern damage and higher effectiveness to semiconductor cleaning processes.

REFERENCES

[1] Hilscher, D.F. Jaeger, D. , Dewan, C. , Brodsky, M. Single Wafer Cleaning Lessons in Advanced

[2] I. S. Park et al., "Meeting the critical challenges for 65 nm and beyond using a single wafer processing with novel megasonics and drying technologies," *ECS Trans.*, vol. 1, no. 3,

[3] H. Kim, Y. Lee, and E. Lim, "Design and fabrication of a horn-type megasonic waveguide for nanoparticle cleaning," *IEEE Trans. Semicond. Manuf.*, vol. 26, no. 2, pp. 221–225, May 2013

[4] W. Kern, A. D. Puotinen, "Cleaning solutions based on hydrogen peroxide for use in silicon semiconductor technology", *RCA Review*, vol. 31, pp. 187-206, 1970.

[5] David H. Wang, Yue Ma, Fuping Chen, Liangzhi Xie, et al., Pyo Leem, Geunmin Choi, "Remove of fine particle using SAPS Technology and Function water", *SEMICON Korea 2013 Conference,* Febury 8, 2012.

[6] David H. Wang, a, Fuping Chen, Xiaoyan Zhang et al., "Damage-Free Cleaning of Advanced Structure Using Timely Energized Bubble Oscillation Megasonic Technology", *Solid State Phenomena,* Vol. 282, pp 64-72, 2018

OPTIMIZATION OF 28NM SIGE SIGMA SHAPE TRENCH DEPTH LOADING EFFECT

Lili Jia, Fang Li*, Wenyan Liu, , Jun Huang
Shanghai Huali Integrated Circuit Manufacturing Corporation,
No. 6 Liang Teng Rd., Pudong New Area, Shanghai, 201315, P.R.China
*Corresponding Author's Email: lifang@hlmc.cn

ABSTRACT

Since CMOS technology moved to 28nm node and beyond, selective epitaxial embedded SiGe (e-SiGe) is widely used for Source/Drain for introducing compressive strain to PMOS channel to improve the hole mobility. Studies have shown, the depth loading (or range) of sigma-shaped silicon trench at different opening CD area has influence on device performance, and device performance can be significantly improved by depth loading reduction. In this paper, influence of chuck speed of TMAH process to improve sigma-shaped silicon trench depth loading effect was investigated. It was shown that higher chuck speed process mode has more superiority than lower chuck speed mode. The depth range can be reduced 10A with such high chuck speed mode. And no negative impact was observed on sigma-shaped trench physical profile.

INTRODUCTION

As the critical dimensions of semiconductor device continuously shrink and circuit operating frequency keeps increasing, higher drive current is required to increase circuit speed. Applying stress to induce appropriate strain in the channel region has become a critical method for enhancing the performance of CMOS transistor[1]. Selective Embedded SiGe(B) (e-SiGe) in Source/Drain(S/D) region which see as a very promising technique is extensively introduced for improving the recent short channel p-MOS performance, since the induced compressive strain within the channel area enhance hole mobility[2~4]. Normally, a major wet etchant Tetramethyl- ammonium hydroxide (TMAH) and one step WET Etch process mode is used to realize the sigma shape formation. Studies have shown, the depth loading (or range) of sigma- shaped silicon trench within wafer has an influence on device performance, and better device performance needs smaller depth loading.

This study focuses on the correlation between TMAH process chuck speed and the depth loading of sigma-shaped silicon trench at different opening CD area, discovering higher chuck speed mode can optimize the trench depth range to improve the final device performance.

EXPERIMENT

Mono-crystalline silicon has a diamond structure with many crystal plane(Fig.1). TMAH has different etch rate at different crystal orientation. Etch rate at <100> crystal plane was much quicker than <111>.the mechanism is related to atomic density of crystallographic orientation, and this character make silicon formed sigma-shaped recess.

After the formation of main spacer and S/D on 300mm pattern wafer, the plasma-based silicon etch was utilized to shape a U-trench silicon recess within bulk silicon substrate. Then WET TMAH etch process was followed up to achieve the desired sigma shaped recess. The schematic diagram of U-shaped and sigma-shaped silicon recess is shown in Fig.2. The parameter TU and TS means trench depth of U and sigma-shaped respectively, and the depth @different opening CD is measured by TEM, and nova OCD is another measurement method to analyze the proformence.Fig.3 is single wafer WET TMAH etch process mode. In this mode, wafer is high-speed rotation on the spin base, chemical liquid is sprayed from the chemical nozzle and flow to the surface of wafer, meanwhile, the nozzle arm scan at the wafer surface with certain speed. In this work, different chuck speed mode was discussed.

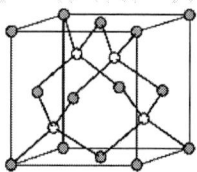

Figure1. The crystalline structure of silicon

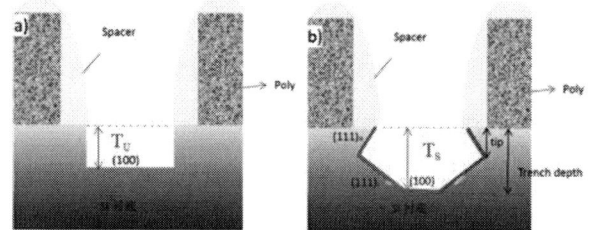

Figure2. a）U-trench shape; b）sigma-shaped trench

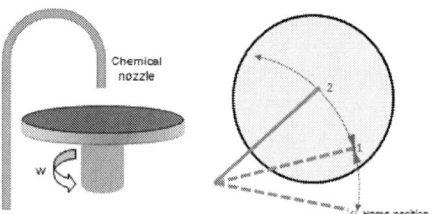

Figure3.Single wafer WET Etch

RESULTS AND DISCUSSION

In this work, U-trench depth(TU) was fixed and controlled depth loading within certain limits(<20A), and used dilute hydrofluoric acid(DHF) to remove native oxide on silicon surface. Then TMAH etch was closely follow up to achieve sigma-shaped trench recess. Several test splits were investigated with different chuck speed .

Fig.4 and Fig.5 shows the sigma-shaped trench physical profile and measured the trench depth(Ts), the trench loading of baseline condition is nearly 23.6A.Increasing chuck speed, the trench loading can reduce to 16.85A.

Fig.6 shows the sigma-shaped trench depth(TS) fullmap distribution within wafer(All data is minus the target trench depth value). As shown in the figure, increasing chuck speed had no impact on OCD pad.

CONCLUSION

In this paper, the depth loading of sigma-shaped silicon trench @different opening CD area was investigated. Experiment results indicate different chuck speed mode of TMAH process will induce the range variation. Higher chuck speed mode has smaller range than lower chuck speed mode. From increasing the chuck speed, depth range can reduced by 71%, and has no impact on sigma-shaped trench physical profile.

REFERENCES

[1] S.V. Vanderbroek, E.F. Crabbe, B.S. Meyerson, D.L. Harame, P.J. Restle, J.M.C. Stork, et al., *IEEE Trans. Elec. Dev.*, vol. 41, 1994, pp. 92-101.

[2] S.-L. Zhang, *Microelectron. Eng.*, vol. 70, 2003，pp. 174-185.

[3] S.E. Thompson, G.Y. Sun, Y.S. Choi, and T. Nishida, *IEEE Trans. Elec.* Dev., vol. 53, 2006, pp. 1010-1020.

[4] Ohta, H. Kim, Y. Shimamune, Y. Sakuma, T. Hatada, A. Katakami, A. Soeda, T.Kawamura, K. Kokura, H. Morioka, H. Watanabe, T. Hayami, J.O.Y. Ogura, J.Tajima, M. Mori, T. Tamura, N. Kojima, M. Hashimoto, K., *Electron Devices Meeting*, p. 240, (2005).

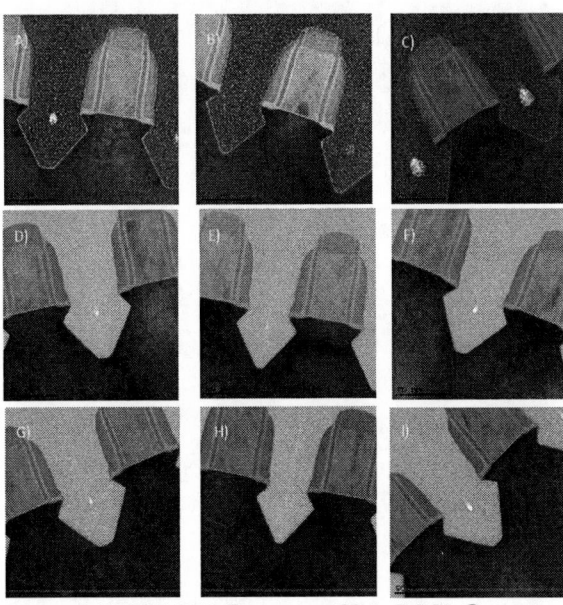

Figure4. TEM(A BSL @opening CD1; B BSL @opening CD2; C BSL @opening CD3; D Speed1 @opening CD1; E Speed1 @opening CD2; F Speed1 @opening CD3; G Speed2 @opening CD1; H Speed2 @opening CD2; I Speed2 @opening CD3)

Split	BL	Speed1	Speed2
Range(A)	23.6	18	16.85

Figure5. Trench depth range at different opening CD with different chuck speed measured by TEM

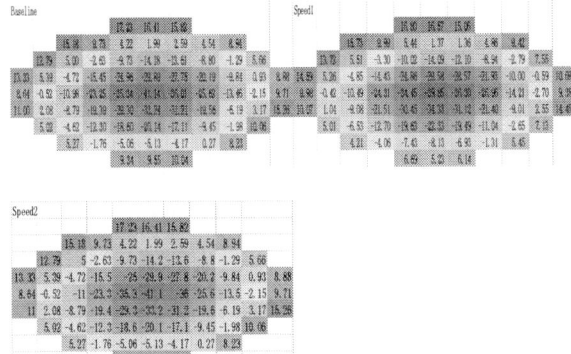

Figure6. Trench depth distribution with different chuck speed

WELL CD CONTROL AND VERTICAL PROFILE BARC ETCH DEVELOPMENT AND RELATED THEORY RESEARCH

Jiang Linpeng *; Zhu Yizheng; Lu Lian; Li Quanbo; Huang Jun; Zhang Yu

Shanghai Huali Integrated Circuit Corporation Shanghai City, China

* Corresponding Author's Email: jianglinpeng@hlmc.cn

ABSTRACT

The BARC as a lower cost structure material is widely used in IC manufacture. For 14nm technology node，it is used to determined ion implantation area. However, the ideal BRAC profile is hard to achieve since its soft material characteristic. This deeply restricts its application. In our study, the idealized vertical BARC profile is obtained by variety of BARC profile learning on ICP etcher, with the physical structure evaluated by SEM. In addition, the analysis of radicals and ions processing on the BARC etching and related profile shaped mechanism is proposed. The result induced PR profile plays very important roles in BARC profile develop.

Keywords—BARC Etch ; vertical profile ; Etching mechanism

INTRODUCTION

In more than 20 nm technical node, single photoresist (PR) structure is used to determined ion implantation region. Continuously with the development of integrated circuit, the critic dimension (CD) is reduced, also the device structure shift from two-dimensional design to three-dimensional FinFET structure. In order to cover all such changes, PR thickness should be substantially added. But it would directly induce PR residual high, PR profile taper and some other side effects.

For alleviating such problems, a good structure of basal antireflective material BARC (Bottom Anti-Reflection Coating) were introduced, and the PR/BARC design was used to undertake ion implantation region limit above complex substrate. Lithography combining with dry etch is used to replace previous single lithography process.

As well know, the formation of vertical BARC profile would take a crucial role on maintain boundary of implantation. For etch, BARC profiles are affected by the chemical composition of side wall surface, etch rate, the ion scattering angle distribution and the specific surfaces exposed to the plasma. Therefore, the evaluation of the profile is extremely essential to understand. Also the key influencing factors should be explored. In our work, variety of BARC profile is achieved by chemical gas and TCP power tuning. SEM is used to examine the BARC Profile formation.

EXPERIMENT

The Scheme of BARC ETCH process is described as Fig.1. the film stack is PR/BARC/Substrate, the PR was exposure firstly (Fig.1.(A)), then the BARC is etched by inductively coupled plasma (ICP) source etcher (Fig.1.(B)).

The various BARC profile was achieved by chemical gas and power tuning. The physical structure is evaluated by SEM.

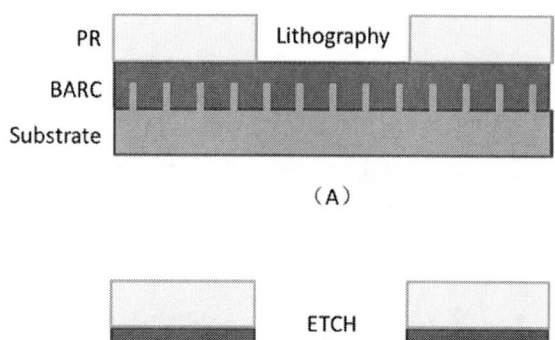

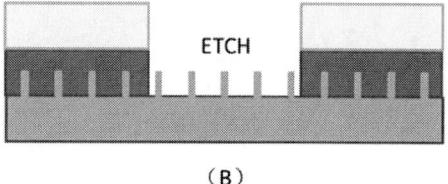

Fig.1. (A) PR/BARC/Substrate Film stack and PR was exposure firstly by lithography; (B) Barc profile formed by etch

RESULT AND DISCUSSION

As shown in Fig.2., we gain three profile samples by ICP etch. From Fig.2.(A), there are a lot of PR accumulate in the sidewall, which caused PR bowing Profile, BARC wall slight footing. In Fig.2 (B), less PR sidewall stacking makes PR profile and BARC profile more vertically. InFig.2.(C), side PR loss, and with PR sidewall stacking become further weaken, BARC profile displays notching near PR surface.

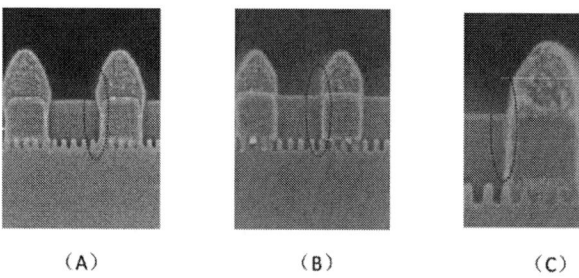

Fig.2. (A) Bowing PR Profile brings out worse taper Barc profile; (B) Vertical PR profile results in ideal Barc Profile and (C) Too much loss of PR induces Barc notching.

Base on experimental results, BARC profile shows greatly influenced by PR profile. More sidewall PR stacking will obstruct the plasma touch the BARC bottom (as Fig.3.(A))，

978-1-7281-6559-2/20 $31.00 © 2020 IEEE

which induces unacceptable BARC footing. On the contrary, too little sidewall PR stacking shield give rise to too much BARC top loss (as Fig.3.(C)). Only the vertical PR profile can bring out straight BARC profile and contribute to well CD control (as Fig.3.(B)).

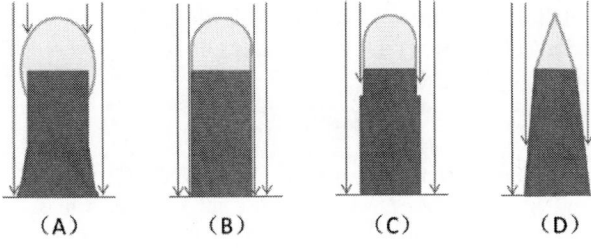

Fig.3. Barc etch theoretical model on (A) Bowing PR Profile; (B) Vertical PR profile and (C)Too much loss of PR Profile (D) Hat shape PR Profile

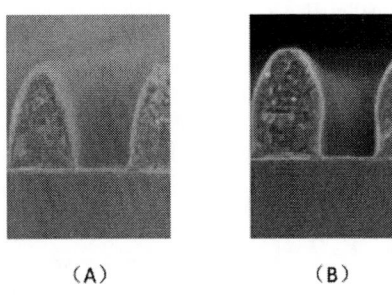

Fig.4. (A) Hat shape PR Profile bring worse footing Barc profile; (B) Vertical PR profile shows ideal Barc Profile

Furthermore, we apply our proposed mechanism to explain other related case as Fig.4.. The hat shape PR profile was gain by high bias bombardment. In such etching situation, BRAC shows taper profile as Fig.4. (A)., too less PR will cause plasma hit the BARC sidewall forming the taper profile as Fig.3.(D) . However, more PR shield can restrain the sidewall loss (as Fig.4.(B)), and built the better vertical profile. In conclusion, PR profile can define the BARC plasma bombarding condition in advance. It will be an effective means to adjust BARC profile.

In addition, the selection of etch chemical species also plays an important role on BARC profile formation. Since H_2 plasma etch displays few substrate plasma damage, it is chosen to be main organic-film etch gas. Its lower boiled byproduct C_xH_y compounds (-161.2 ℃)[1] makes H_2 plasma also perform well sidewall thrust ability. Therefore, the rational application of H_2 plasma will beneficial to side wall control.

CONCLUSIONS

In this work, BARC vertical profile was achieved with well CD control. The related straight BARC profile formation mechanism is studied, which indicates that the PR profile shows greatly influence on BARC profile. Bowing and fewer shields PR will induce BARC footing and taper. The suitable PR structure is an essential step for BARC etching. Also, such mechanism can be extended to other soft film stacking etching.

ACKNOWLEDGMENTS

Thanks for all the authors for great work and collaboration. Thanks for FA (failure analysis) for SEM and EDX image support.

REFERENCES

[1] H. Nagai, S. Takashima, M. Hiramatsu, M. Hori and T. Goto, J. Appl. Phys. 91, 2615 (2002)
[2] H. Lee, B. Kwon, Y. Park, Journal of the Korean Physical Society, 56,1441 (2010)
[3] S. Junking, Y. HeeJoung, S.YeolMun, Transactions on Semiconductor Manufacturing, 20, 150(2007)
[4] T. Tsutsumi, Y. Fukunaga, K. Ishikawa, K. Takeda, Transactions on Semiconductor Manufacturing, 28, 515 (2015)
[5] K. Wang, J. H. Zhang1, Y. S. Chi, Z. H. Ji, H. Y. Pei, Y Z Zhu, Q. B. Li, J. Huang, China Semiconductor Technology International Conference (CSTIC), 25(2016)

IMPACT OF REWORK PROCESS TO ETCH BIAS AND THE CORRESPONDING SOLUTION

Pengkai Xu, Penggang Han, Wenyan Sun, Sen Wu, Bin Zhao, Yi Wang, Fulong Qiao*
ShangHai HuaLi Microelectronics Corporation-Technology Development Division
Pudong New Area , Shanghai, P. R. China
*Corresponding Author's Email: qiaofulong@hlmc.cn

ABSTRACT

Etch bias is a key parameter to determine AEICD performance, and the photo rework is a routine method if photo process abnormal. Dependence of etch bias on photo rework is firstly found and studied in this work. By theoretical analysis and experimental verification, the model of the this dependence is proposed, which mainly attributes to the new dense OX film left below Barc after rework and the high selectivity of Barc to OX for Barc open step. Under the above two conditions, part of plasma convert its etch direction from vertical to horizontal during the OX open period, which attributes to trimming Barc CD, and thus changing etch bias. This dependence gets weaker when transmission rate increases due to macro-loading effect. Besides, it also gets stronger for dense pattern than for iso pattern, which attributes to micro-loading effect. While this dependence may induce AEICD and thus the WAT shift, solution to weaken the side effect of etch bias shift is proposed.

1: INTRODUCTION

With the development of technology, Ultra Large-Scale Integrate (ULSI) microelectronic is now becoming more and more important role in our daily life. ULSI is composed of humorous nm-scale micro structures, like gate, contact and via etc., and its device performance shows strong dependence on the CD (Critical Dimension) of these structures. Moreover, since the size of these structures is directly determined by etch process, acceptable ULSI production requires optimized etch process.

For etch process, etch-bias is an key parameter to determine the CD of the micro structure, like AA CD, Poly gate CD, Contact Diameter, etc.[1, 2, 3]. It is defined as the difference between AEICD (After Etch Inspection Critical Dimension) and ADICD (After Developing Inspection Critical Dimension), as shown in equation. (1).

$$\text{Etch Bias} = \text{AEICD} - \text{ADICD} \qquad (1)$$

Its shift induces CD shift and thus device performance, like threshold voltage, RC delay, shift. Therefore, understanding of etch-bias dependence on others factors is necessary for predicting device performance.

Etch process is used to remove multi-layer films including main etch material film, like poly, silicon, inter layer dielectric (ILD) etc. and mask layer film, like APF, Barc, PR etc.. Etch bias is mainly determined by mask-layer etch process[4 , 5]. Therefore, factors influence mask-layer etch will also influence etch bias. For example, with STI step height increase, Barc (Bottom Anti-Reflective Coating) etch time will be decreased and thus etch bias increases. With polymer gas increased in mask layer etch step, more polymer will be formed on mask layer side wall, and thus increases etch bias. Compared with that of dense pattern, etch bias of iso pattern is smaller, due to micro-loading effect in mask etch steps. Moreover, with transmission rate increase, etch bias may also be decreased, due to macro-loading effect.

While the above influence can be understood easily, the dependence of etch-bias on photo rework has not been proposed before, although rework often happen in ULSI mass production. Such study is necessary because it can provide us not only useful information for rework flow design and selection, but also useful experience for etch recipe development. Furthermore, transmission rate of different products normally shows significant difference, and the multiple design pattern is also widely used even for the same layer of the same product. Therefore, investigation into the variation of this dependence with factors like transmission rate and pattern density is also important for predicting the etch bias shift, and necessary for compensation solution.

In this work, etch bias dependence on rework is studied, based on analyzing 55nm and 65nm FEOL etch processes. Moreover, the variation of this dependence with transmission rate and pattern density is also studied, based on analyzing etch characteristic. Finally, the solution to compensate this etch bias shift, and thus avoid AEICD and WAT shift, is also proposed.

2: EXPERIMENT INTRODUCTION

2.1 Etch and rework process introduction

Etch bias is generated in etch process. To understand the etch bias shift reason easily, introduction to etch process is necessary. Base on incoming etch film, etch process can be divided into 2 main step groups: mask open step group (like PR and BARC etch step) and main material remove step group (like poly and silicon etch step). While mask open steps mainly determine etch bias, main material remove steps determines parameters like profile and final film stack, etc..

Base on mask layer material difference, etch process mainly includes 3 approaches: PR approach, APF approach and Hard Mask approach. While PR approach takes PR and Barc as etch mask layer, APF approach & Hard Mask approach take APF and OX or SIN as etch mask layer, respectively.

Etch process is designed base on its film stack. Taking PR approach poly ET process as an example. its etch film is shown in Fig. 1a. It can be seen that etch film is mainly composed of 3 layers: poly, Barc & PR, in which Barc & PR are mask film, poly is the main material film. Its etch process is mainly composed of the following steps:

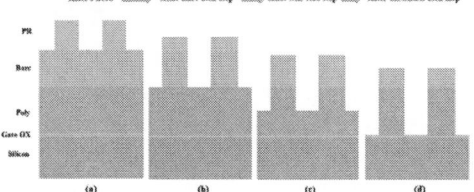

Fig. 1: 65nm Nor-Flash GP ET Film stack after (a): GP Photo, (b): Barc etch step, (c): ME etch step and (d): SL and SLOE etch step

(1): Barc etch step: like Fig. 1b shows, this step works as mask layer open step, and plays main role in etch bias determination;

(2): Poly main etch (ME) ET step: like Fig. 1c shows, this step has high etch rate to all kinds of materials, like OX, SiN, Barc & Poly;

(3): SL ET step: like Fig. 1d show, this etch step stops on gate OX, and thus requires high selectivity to gate OX. Otherwise, too much OX loss or even AA damage may happen, which further induces the following implant process shift.

(4): OE (Over Etch) step: this step also has high selectivity to gate OX. Suitable OE amount is necessary to avoid poly residue or worse poly profile like undercut.

In addition to the etch process, rework flow is also selected by taking etch film stack into consideration, as summarized in table I. It can be seen that, while dry strip is widely used in PR approach and Hard Mask appoach rework flow, only wet trip is used for APF approach rework flow. This is because dry strip may induce APF poping and thus contaminate dry strip tool. As for hard mask mode, rework flow mainly composed of steps including dry strip, wet strip, and scrubber. etc. As for PR mode etch process mentioned above, its rework flow is only composed of dry strip and wet strip steps. This is because some high aspect ratio (Height/CD) pattern has already been formed before this process, and wet scrubber in rework flow may induce defect like pattern collapse and thus prohibited.

In rework flow, while dry strip play main role in PR and Barc removing, wet strip process is mainly used for removing the remaining Barc. Wet strip process should be chosen carefully. Taking Poly Etch-4 rework flow as an example, its wet strip should not use OX remove fluid, since gate structure has already been formed beofore, and

the OX remove fluid may induce thinner gate oxide and thus impact the following implant step.

Table I: Etch Process rework procedure statistics of different etch mode

Etch approach	Etch Process	Etch film stack (Top to bottom)	Rework Flow
APF approach	AA Etch-1	PR/Barc/OX/DARC/APF/OX/Si	Wet Strip
	Poly Etch-1	PR/Barc /OX/DARC/APF/ Poly	
HM approach	AA Etch-2	PR/Barc/SiN/OX/ Silicon	Dry Strip, Wet Strip, Scrubber et. al.
	Poly Etch-2	PR/Barc/OX/Poly	
	Poly Etch-3	PR/Barc/SIN/Poly	
PR approach	AA Etch-3	PR/Barc/SiN/OX/ Si	Dry Strip, Wet strip
	Poly Etch-4	PR/Barc/Poly/OX/ Si	

2.2 Etch bias calculation introduction

Base on equation (1), this work calculated etch bias of every measurement site for those wafers both ADI and AEI CD were obtained. For rework lot, the ADICD after rework is used for bias calculation. For every wafer, the above ADI and AEI CD in both wafer center and wafer edge were measured. Consequently, the etch bias in both wafer center and wafer edge is also obtained.

2.3 Etch chamber introduction

In this work, we mainly focus on FEOL (front end of line) etch process. Nearly all of these etch process are finished in Lam tool, by Kiyo chamber or Kiyo 45 chamber. For these chambers, TCP mode is used for plasma generation. This mode is widely used in AA and Poly etch processes. As for the plasma concentration in chamber, RF power can be adjusted to meet requirement.

Pressure of above chambers can be adjusted between 3 and 100 mT, which is normally adjusted for optimizing residence time distribution of reactant, products and by products. Reaction temperature of the above chamber is optimized by adjusting E-Chuck (electronic chuck) temperature. Specifically, both inner and outer E-Chuck temperature can be adjusted separately to meet etch temperature requirement in both wafer center and wafer edge, respectively.

3: RESULTS AND DISCUSSION

3.1 Root cause of etch bias dependence on rework

Influence of rework on current poly etch processes is shown in Fig. 2. It can be seen that, Poly Etch-4 process etch bias gets smaller for those rework lots. Fig. 2a also shows that this bias shift are the same for those lots rework 1 time and more than 1 times. However, for other poly etch processes, like Poly Etch-1 process and Poly Etch-2 processes, they does not show the above variation.

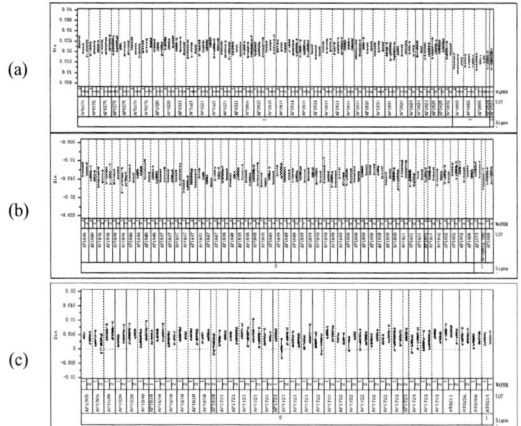

Fig. 2: Rework influence on Poly ET Bias for (a) Poly Etch-4 process, (b): Poly Etch-2 process; (c): Poly Etch-1 process, where, signal 0 means no-rework lot, while signal 1 means rework lot with 1 time and signal 2 (in 3.1a) means rework lot with 2 times

As described in sec. 2.1, poly etch-4 process rework flow is composed of dry strip and wet strip steps. Dry strip step adds new OX film above poly if poly is exposed to strip chamber directly [6], just as poly etch-4 process rework does. Then, the following wet strip process in the above rework flow cannot remove this OX. Therefore, the above rework process left a new dense OX above poly. As there is no dense OX film above poly for non-rework lot, this change means poly etch-4 process film stack has been changed after rework. Besides, this new film is directly formed below the mask layer, and thus may influence mask layer etch process. Therefore, this change is highly suspected to induce etch bias shift.

Experiment was designed and carried out to verify the above suspect. Results shown in Fig. 3. Firstly, only rework dry strip process is added before Barc coating step. With this step added, a new dense OX is formed above poly, which is similar as that of rework lot. Results show that its etch bias is nearly the same as that of rework lot. Then, another experiment with OX remove step added between this dry strip process and Barc coating step done. The results show that its etch bias is nearly the same as that of non-rework lot. This proves that it is the OX above poly, which is left after rework, induces etch bias shift for poly etch-4 process.

Base on the above results, suspect model is proposed to understand this bias shift, like Fig. 4 shows. In poly etch-4 process, Barc etch step has high selectivity for Barc to OX, but low selectivity for Barc to poly. As a new dense OX is left above poly for the rework lots, plasma etch

along vertical direction is weakened when it touches this OX film, and part of it transfers to horizontal direction, which trim Barc CD. After OX breakthrough, this trim process stops. As for the non-rework lots, no such trim process happens since no such OX above poly. Consequently, compared with non-rework lot, rework lot suffers trim effect in Barc etch step, and thus show low etch bias.

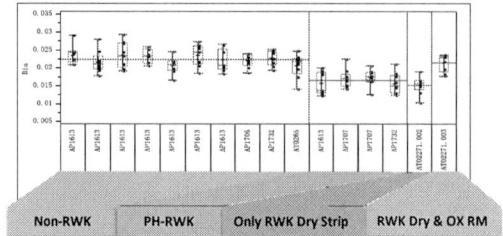

Fig. 3: Experiment result for eliminating rework influence on etch bias

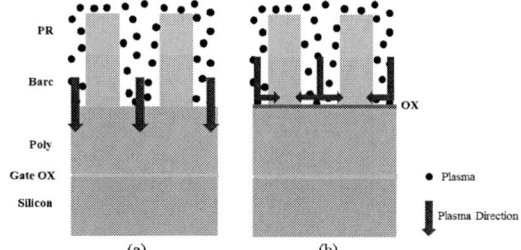

Fig. 4: Plasma direction for (a) non-rework lot and (b) rework lot

Base on the above model, it is easier to understand that rework induced trim process only happens during the OX open period. Longer open period induces larger etch bias shift, while shorter open period induces small etch bias shift. For lot rework more than 1 times, OX thickness above poly keeps nearly unchanged after 1st rework. This means OX open period remains unchanged even if more than 1 times rework is done. Therefore, lots rework more than 1 times show same etch bias shift performance as those rework 1 time.

With respect to other FEOL etch process mentioned in table II, their etch process and the rework flow do not meet the above two requirements (Table 3.2), and thus do not show the above dependence.

3.2 Variation of etch bias dependence on rework

For poly etch-4 process etch bias shift, its variation with transmission rate can be obtained is shown in Fig. 5a.It can be seen that the bias shift decreases with

Table II: OX film formation after rework and Barc etch selectivity to oxide for different AA and Poly etch processes

ET Process	AA Etch-1	AA Etch-2	AA Etch-3	Poly Etch-1	Poly Etch-2	Poly Etch-3	Poly Etch-4
New OX	N	N	N	N	N	N	Y
High OX selectivity	N	N	N	N	N	N	Y

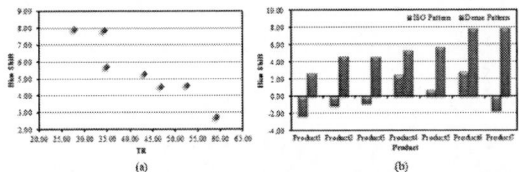

Fig. 5: Variation of etch bias dependence on rework with (a): transmission rate and (b): pattern density.

transmission rate (Fig. 5a). Besides, this bias shift is also more obvious for dense pattern than for iso pattern, as Fig. 5b shows.

With transmission rate increase, larger photo resist open area is opened to etch chamber in etch process, meaning more poly gate is exposed to chamber. While chamber plasma keeps unchanged, this increased poly gate number leading to decreased plasma amount around every one of them, and thus weakens trim effect induced by rework. Consequently, etch bias shift decreases with transmission rate. Besides, this result also nearly show linear correlation between etch bias shift amount and transmission rate when transmission rate increases from 25% to 60%. Their correlation is:

$$S=-0.1567*T+12.204 \qquad (2)$$

Where S is etch bias shift amount (with nm unit) and T is transmission rate.

With respect to the etch bias shift difference between dense and iso patterns, possible reason is given to the micro loading effect. For dense pattern, it is harder for plasma to diffuse into the opened area due to high diffusion resistance. However, as for iso pattern, plasma diffusion resistance is low due to the large opened area. Therefore, etch rate in dense pattern is lower than that in iso pattern. Base on the above analysis, OX break through time is longer for dense pattern than for iso pattern. Consequently, the above trim effect induced by rework is stronger for dense pattern than for iso pattern, and thus induces low etch bias for dense pattern.

3.3 Solution to avoid AEICD & WAT shift due to etch bias shift

While AEICD gets smaller for rework lots, WAT items, like Idsat (saturation current), may get worse (Fig. 6). As discussed in Sec. 2.1, OX remove fluid should be prohibited in poly etch-4 process rework. Besides, its dry strip step is necessary for removing the thick PR & BARC above poly. Otherwise, PR residue may be left after rework, which further influences etch process. Therefore, the etch bias shift post rework can not be avoid for this etch process. Other compensation solution to this etch bias shift issue is necessary for rework lot keeping normal device performance .

One solution that increases ADICD for rework lots is employed to compensate this bias shift. Base on the bias shift correlation with transmission rate of different product (equation 2), different ADICD compensation value is used.

The results show that, AEICD nearly returns to the baseline performance, like Fig. 7 shows.

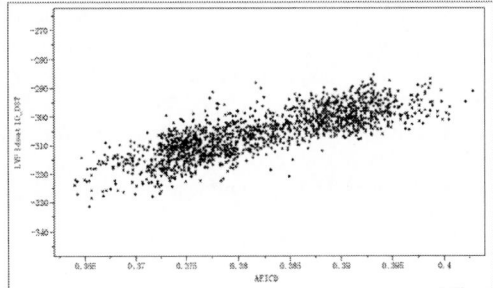

Fig. 6: Correlation between Idsat and AEICD of poly etch-4

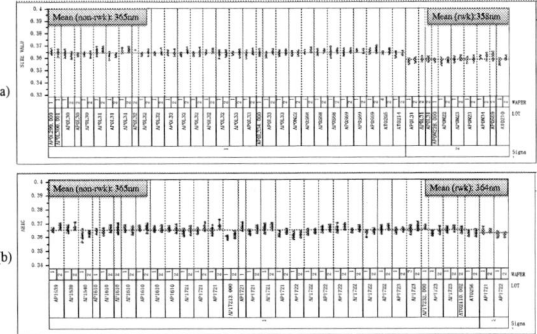

Fig. 7: Poly etch-4 AEICD for non-rework and rework lot under (a): no ADICD compensation & (b): ADICD compensation conditions.

4: CONCLUSION

Dependence of etch bias on rework is studied in this work. It was found that only poly etch-4 process show its etch bias dependence on rework flow in all the FEOL etch processes proposed in this work. This bias shift issue attributes to two factors as: a new OX film is left above poly after rework, and the Barc etch step has high selectivity to OX, which induces enhanced trim effect during Barc open step. Since other etch processes does not meet the above two conditions, their etch bias is independent of rework.

By decreasing plasma amount around poly, like increasing transmission rate does, the above dependence gets weakened. By extending OX open time in Barc etch step, like Barc etch for dense pattern, the above dependence gets stronger. Correlation of etch bias shift amount with transmission rate is proposed to predict the etch bias shift of new product in new tape out pilot run process. To avoid AEICD and corresponding WAT shift, caused by this etch bias change, ADICD compensation is done for rework lots, which has been proved to be effective.

ACKNOWLEDGEMENTS

In this work, we thanks for support from HLMC TD1 etch, litho and PIE. Besides, support from HLMC E3 etch is also sincerely appreciated.

REFERENCES

[1]: Y. H. Huang, S. S. Du, H. Y. Zhang, et. al., *65nm poly gate etch challenges and solutions*, 9th International Conference on Solid-State and Integrated-Circuit Technology. 2008, pp. 1166-1169;

[2]: Y. Karzhavin, *Shallow trench isolation etch process for 0.2um trench capacitor DRAM technology*, 10th Advanced Semiconductor Manufacturing Conference and Workshop, 1999, pp. 239-254;

[3]: T. T. Quinn, S. Johnston, R. Lindquist. *Contact etch in the LAM 4520XL using standard CF4/CHF3 chemistry.* Advanced Semiconductor Manufacturing Conference and Workshop. Theme-Innovative Approaches to Growth in the Semiconductor Industry, 1996, pp. 185.

[4]: A. Hamouda, *Efficient etch bias compensation techniques for accurate on-wafer patterning*, Spie Advanced Lithography, 2015 , 9427 (2) : pp. 129-132;

[5]: D.A. Williams, A.R. Locander, T. Herrera, et. al. *Improvements in polysilicon etch bias and transistor gate control with module level APC methodologies*, IEEE Transactions on Semiconductor Manufacturing, 2005 , 18 (4) : pp. 522-527

[6]: K. Han, S. Luo, O. Escorcia, et. al.. *Evaluation of plasma strip induced substrate damage*, Solid State Phenomena, 2009, 145-146: pp. 249-252

Rectangular suspended single crystal Si nanowire with (001) planes and <001> direction developed via TMAH wet chemical etching

Shuang Sun, Baotong Zhang, Yuancheng Yang, Xia An, Xiaoyan Xu, Ru Huang and Ming Li[*]

Key Laboratory of Microelectronic Devices and Circuits, Institute of Microelectronics,Peking University, Beijing 100871, China

* Email: liming.ime@pku.edu.cn

Abstract

In this study, a kind of rectangular suspended single crystal Si nanowire with (001) planes and along <001> direction is developed via a CMOS-compatible top-down scheme. In this scheme, the nanowires are formed by anisotropic etching of TMAH on different silicon crystallography orientations. By designing the initial orientations of hard mask patterns, the rectangular suspended silicon nanowires can be successfully fabricated without any sacrificial epitaxial layers. Due to the damage-free process and the high mobility on (001) planes, this scheme will provide a high-quality channel for the future gate-all-around silicon transistor technology.

Introduction

Today's semiconductor manufacturing industry under the guidance of Moore Law is rapidly developing. Therefore, preparation of ultra-short channel devices with high integrated circuit performance and low power consumption will become the focus of the semiconductor manufacturing industry in the future. However, as the technology node towards 3 nm, the researchers found that due to the short-channel effect, high off-state current, DIBL effect and other problems have all emerged. In order to solve the gate-control problems, a large number of new structure semiconductor devices began to emerge. Such as double gate FET, FinFet, gate all around nanowire MOSFET and so on. The multi-gate structure devices, especially the gate all around structure devices, are able to strengthen the control ability of the gate to the channel and suppress the short channel effect in the meantime[1-7].

However, the channel carrier mobility degradation is another critical issue in the driving current of the extremely scaled SNWTs, especially the devices approaching atomic size. The fundamental reasons for mobility degradation come from high electric field, phonon scattering and especially the surface scattering due to the non-defined surface orientations. As a result of top-down scheme, plasma etching and lithography fluctuation both are unavoidable to cause very severe lattice damage and roughness, which are the major scattering centers of carriers under high-field transport. To overcome the mobility degradation in SNWTs, there have been developed several ways including strain technology[8][9], non-Si high mobility channel materials[10-18], and preferred crystal surface technology[19]. Among them, one of the best promising schemes is to fabricate suspended silicon nanowires with preferred surface orientations such as (001) family. The process challenges for such structure are how to avoid plasma damage and the extraordinary process costs such as sacrificial epitaxy.

In this study, we proposed to fabricate the suspended nanowires with anisotropic etching of TMAH. With special layout design, the nanowires can be formed with four planes all belong to (001) family and the channel direction along <001>.

Experiments

The overall process, sketched in Fig.1, involves the following steps: first, deposit 5nm SiO_2 and 50nm poly-Si on (001) Si wafer and follow electron beam lithography to define the spacing of nanowires and the sidewall crystal planes. Then the poly Si is etched and followed by 20 nm Si_3N_4 deposition. Si_3N_4 is etched by RIE to form the inner spacer to shrink the spacing between nanowires. After that, 25 wt% TMAH wet chemical etching was carried out at 35°C for 5min. The portion between the trenches forms suspended nanowire. Finally, H_3PO_4 acid remove the Si_3N_4, HF acid remove the SiO_2.layer.

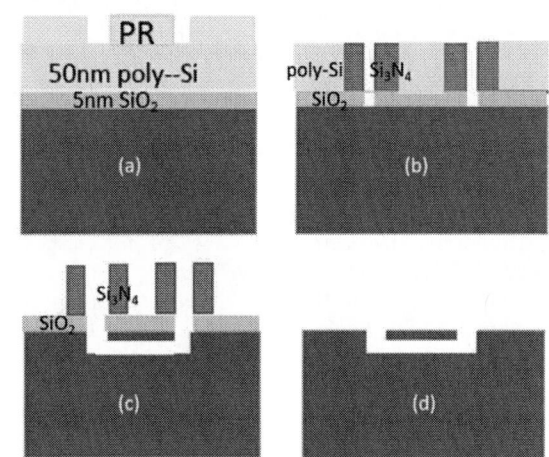

Figure1: Rectangular suspended single crystal Si nanowire with (001) planes and <001> direction preparation process

The Si (100) crystal plane has the lowest atomic arrangement density and the (111) crystal plane has the largest atomic arrangement density. Therefore, when TMAH is used to wet etch silicon microstructures, the precise silicon microstructure can be obtained due to the different etching rates of TMAH on each Si crystal face.

During the process, some of the side surfaces of the rectangular trench will have some high Miller index crystal planes with a high etching rate and disappear as the reaction continues, and wet etching will make some random Si defects expose at different locations[20]. When the defects are exposed, TMAH quickly hollows out the Si below the defects. In addition, due to the large etch window in the layout, the (100) bottom surface will erode before the two (111) planes meet. As a result, it forms suspended nanowire channels surrounded by (001) lattice plane all around. The dimension of the suspended nanowire is determined by the trench space. It's worth noting that, the orientation of Si, doping type, dopant concentration, temperature of TMAH and stirring speed all affect the forming speed of suspended single crystal Si nanowire channel.

Results and Discussion

The plane view SEM picture is shown in Figure 2. The specific crystal faces have been marked.

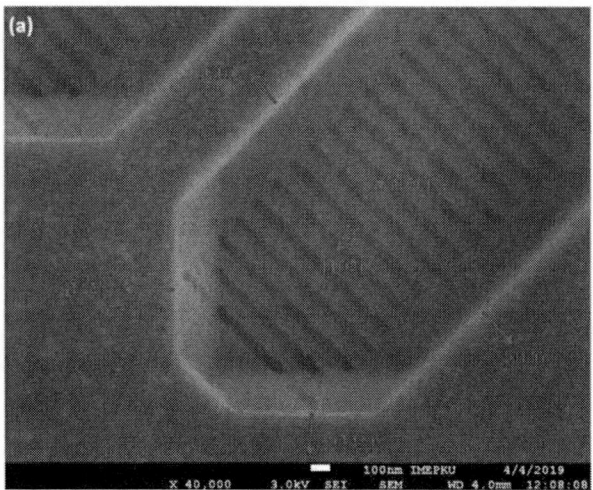

Figure 2: The plane view SEM picture and specific crystal faces

The anisotropy of single crystal Si is a prerequisite for the precise microfabrication of TMAH wet chemical etching. We also observe the aero-plane view of the suspended nanowires in Figure 3 (a). In order to increase the sample conductivity, we deposit 30 nm Ti layer by electron beam evaporation and Pt on Si nanowires. We can see the sidewalls of the suspended nanowires are highly steep and have a low surface roughness. From the observed cross-sectional view SEM picture in Figure 3 (b), we can see due to the stress introduced by the 30nm Ti layer, the specific surface of the nanowire is in a concave shape. It can be inferred that the cross-section of the nanowire structure should be rectangular combined with plane view SEM picture before depositing 30 nm Ti conducting layer. The lattice planes around the rectangular nanowires are all belong to the (001) crystal family.

It is noting that the electron mobility is the highest in the Si (001) lattice plane family, so this method can effectively improve the electron mobility in nanowire transistors. Besides, because of the suspended structure, it's convenient to fabricate the gate-all-around MOSFET subsequently.

The holes in the pattern are caused by the wet etching of TMAH and over-etching of H_3PO_4. The number of holes can be reduced by adding ammonium persulfate or isopropanol to TMAH or by precisely controlling the time of H_3PO_4 acid etching the Si_3N_4.

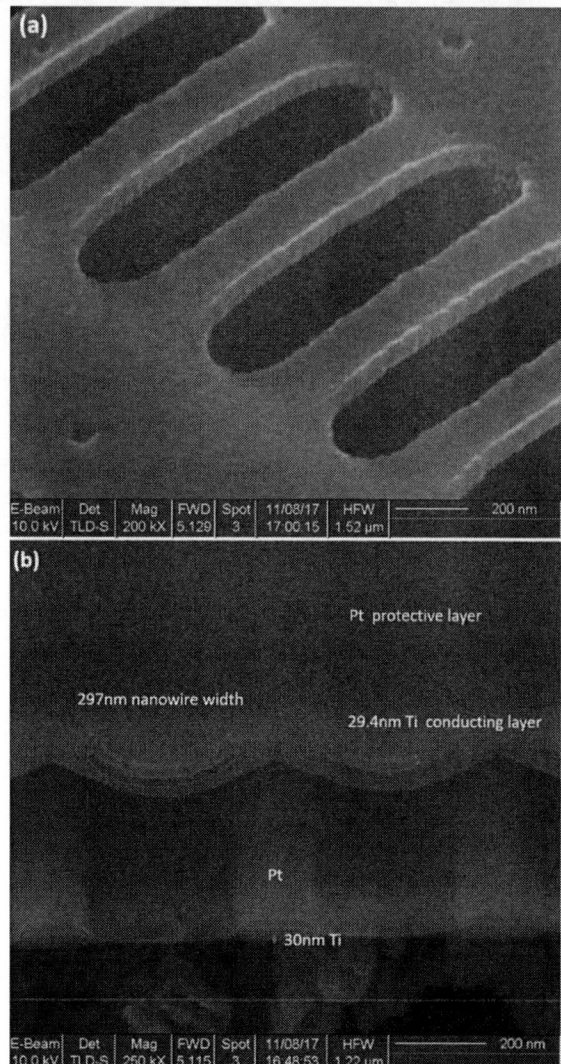

Figure 3: The aero-plane view HRSEM picture after depositing 30nm Ti conducting layer (a); The material name of each layer has been marked in the figure. We can find that due to the stress introduced by the 30nm Ti layer and Pt protective layer, the specific surface of the nanowire is in a concave shape (b).

We can also find that as the origin trench width increases, the diameter of formed nanowire channel first decreases and then increases as shown in Figure 4. It can be attributed to the enhanced capillary force as the trench size decreases. As a result, the smaller trench width, the faster TMAH enters the trench, but the less total amount of solution that can be accommodated. When the size of the trench is small, the volume of TMAH that can be accommodated in the trench is a major factor affecting the line width of the nanowire. The larger the groove width, the larger the volume of TMAH that can be accommodated, and the more lateral corrosion of Si, the smaller the diameter of the formed nanowire. As the trench width increases, the siphon phenomenon will be caused by the infiltration of TMAH. When the corrosion time of TMAH is insufficient, the siphon phenomenon will suppress the TMAH solution enter the trench. Therefore, the amount of lateral etching of TMAH on Si decreases and thus the nanowire diameter increases.

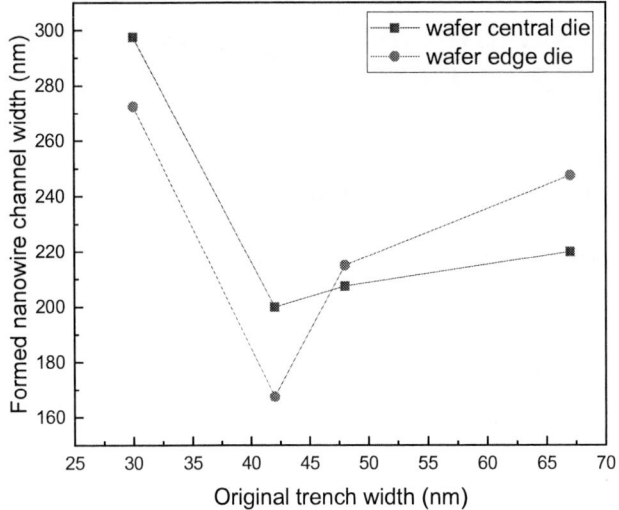

Figure 4: The relationship between the formed nanowire channel width and the original trench width. The black line present wafer's central die, the red line present the wafer edge die.

In the following research, we will focus on decreasing the nanowire diameter through reducing the initial design of the trench space in layout and adjusting TMAH concentration, wet etching time and etching temperature.

Summary

In this work, we report an epi-free top-down scheme to prepare suspended rectangular single crystal Si nanowires with (001) planes and along <001> channel direction, for the first time. This technique will provide very clean interface and thus improve electron mobility for the future sub 3nm node gate-all-around nanowire transistor technology.

Acknowledge

This work was supported in part by National Key Research and Development Plan (Grant No.2016YFA0200504), National Science and Technology Major Project (Grant No. 2017ZX02315001-004), National Natural Science Foundation of China (Grant No. 61421005), and 111 Project (Grant No. B18001).

References

[1] Ferain I et al. *"Multigate transistors as the future of classical metal-oxide semiconductor field-effect transistors."* Nature 2011;479:310–6.

[2] Bangsaruntip S et al. *"Density scaling with gate-all-around silicon nanowire MOSFETs for the 10 nm node and beyond."* IEDM Tech Dig 2013:526–9.

[3] Bangsaruntip S et al. *"High performance and highly uniform gate-all-around silicon nanowire MOSFETs with wire size dependent scaling."* IEDM Tech Dig 2009:297–300.

[4] Suk SD et al. *"Investigation of nanowire size dependency on TSNWFET."* IEDM Tech Dig 2007:1129–31.

[5] D. Hisamoto, W.-C. Lee, J. Kedzierski, H. Takeuchi, K. Asano, C. Kuo, E. Anderson, T.-J. King, J. Bokor, and C. Hu, *"FinFET-a self-aligned double-gate MOSFET scalable to 20 nm,"* IEEE Trans. Electron Devices, vol. 47, no. 12, pp. 2320–2325, Dec. 2000.

[6] J. P. Colinge, *"Multi-gate SOI MOSFETs,"* Microelectron. Eng., vol. 84, nos. 9–10, pp. 2071–2076, Sep./Oct. 2007.

[7] J.-S. Yoon, T. Rim, J. Kim, M. Meyyappan, C.-K. Baek, and Y.-H. Jeong, *"Vertical gate-all-around junctionless nanowire transistors with asymmetric diameters and underlap lengths,"* J. Appl. Phys., vol. 105, no. 10, Sep. 2014, Art. no. 102105.

[8] Auth C，Cappellani A，Chun J S，et al, *"45 nm High-k - metal gate strain- enhanced transistors"* VLSL,pp 128-129(2008)

[9] Yang H S，Malik R，Narasimha S，et al, *"Dual stress liner for high performance sub-45 nm SOI CMOS manufacturing"* IEDM, pp1075－1077(2004).

[10] Liaows，Liawy G，Tang M C,et al, *"PMOS hole mobility enhancement through SiGe conductive channel and highly compressive ILD- SiN$_x$ stressing layer"*, IEEE Electron Devices Letters, pp 86-88(2008).

[11] Kanno H, Sadoh T and Miyao M. *"Electrical and structural properties of poly-SiGe film formed by pulsed-laser annealing".* Journal of Applied Physics, pp 72-74(2014).

[12] Mahajan R, Gautam D K, *"Analytical study of effect of channel doping on threshold voltage of metal gate high K SiGe MOSFET"*,Silicon ,Vol. 10, no 1, pp 85-90 2018

[13] Poorvasha S and Lakshmi B, *"Investigation and statistical modeling of InAs-based double gate tunnel FETs for RT performance enhancement"*,Journal of Semiconductors, vol. 39, no. 5,pp. 054001, 2018

[14] Schwierz F *"Graphene transistors"*, Nature nanotechnology, Vol. 5, pp. 487-496, 2010

[15] Thingujam T, Luisier M, Han S J, Tulevski G, Breslin C M et al, *"sub 10nm carbon nanotube transistor"* Nano letters, Vol 12, pp. 758-762, 2012

[16] Kamata Y, *"High K/Ge MOSFETs for future nanoelectronics"*, Materialstody, vol. 11, no 1-2, pp.30-38, 2008

[17] Claeys C L and Simoen E, *"Ge based technologies: From materials to devices"*, Elsevier, 2007

[18] Norton D P, Budai J D and Chisholm M F, *"Hydrogen-assisted pulesd laser deposition of (001) GeO2 on (001) Ge"*, Appl.Phys. Lett, vol. 76, no.13, PP. 1677-1679, 2000

[19] Yang M，Chan K，Shil，et al, *" Hybrid- orientation technology (HOT): opportunities and challenges"*. IEEE Transactions on Electron Devices，pp 965－978 (2006).

[20] Steinsland, E, Finstad, T, Hanneborg A, *"Etch rates of (100), (111) and (110) single-crystal silicon in TMAH measured in situ by laser reflectance interferometry"*, Sensors and Actuators a-physical pp 73-80(2012)

EFFECTIVE LITHOGRAPHY LEVELING IMPROVEMENT WAS ACHIEVED BY RETAINING WAFER BACK-SURFACE NITRIDE DURING A NOVEL SMT NITRIDE REMOVE PROCESS

Weiwei Ma[1]*, Chao Sun[1], Xiaolin Xu[1], and Wei Zhou[1]

[1]Shanghai Huali Integrated Circuit Corporation, Shanghai 200433, China
*Corresponding Author's Email: maweiwei@hlmc.cn

ABSTRACT

The shrinkage of critical dimension requires more delicate lithography leveling; otherwise, it will easily leads to various pattern failures. Worse wafer back-surface flatness caused by repetitive BEOL backside clean leads to worse lithography leveling and defocuses follow. BEOL lithography leveling performance of a 28 nm node wafer is significantly improved by retaining its back-surface nitride during a novel SMT nitride remove process. The thin back-surface nitride layer could resist the corrosion of backside clean chemicals and maintain wafer back-surface flatness in a more preferable level. In this study a conventional one-step batch SMT nitride remove wet process is replaced by two continuous steps. The first one is a single wafer front-side wet process, and the second one is a conventional batch one. The via-5 leveling data demonstrate a 60% to 80% improvement of leveling range from novel two-step SMT nitride remove wafers. These wafers are defocus free, while the baseline wafers still suffer various kinds of defocus defects. Average CP yield improvement is about 0.5%.

INTRODUCTION

The lithography (Litho) depth of focus (DOF) and overlay (OVL) tolerance margin reduce with the shrink of the critical dimension (CD). The importance of wafer backside clean (BSC) increases as the CD shrinking to 2x node and beyond. Without BSC the defectivity of wafer front side will increase due to upper wafers shad off particles to the wafers underneath [1]. Moreover, particles attached to wafer backside alone can also have a significant impact on wafer flatness and lithography related defects [2]. However, BSC itself also has drawbacks especially when its chemicals have a relatively high etch rate to the back-surface of wafer. High etch rate will result in a distinct uneven film loss of the back-surface, because the area blocked by defects will inevitably loss less film or even have no loss [3]. One kind of BSC chemicals is a mixture of nitric acid (HNO_3) and hydrofluoric acid (HF), which has a high etch rate to silicon. In our case, poly-silicon is the back-surface film when wafer achieves BEOL. Thus, the back-surface flatness will get worse and worse during repetitive BSCs. And defocuses with special patterns happen.

Therefore, we choose to alter wafer back-surface during SMT nitride removing. A thin nitride film formed through atomic layer deposition during spacer 2 deposition is retained to ensure a preferable flatness throughout the whole BEOL.

The reasons why we choose to obtain the desired nitride film during SMT removing are as follow. Firstly, high temperature thermal processes have finished before SMT, thus the backside nitride film will have the minimum impact on device during subsequent thermal processes. Secondly, there is no more nitride removing process after SMT removing so that this film can be kept permanently in theory.

EXPERIMENT

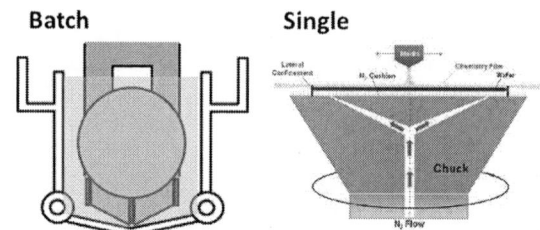

Figure 1. Batch and Single Nitride Remove Processes

The process difference between batch nitride removing and single nitride removing was shown in figure 1. During batch process the whole wafer was immersed in phosphoric acid solution and as a consequence the nitride layers on both wafer sides were etched. Moreover, in order to make sure that the nitride layer on the front surface was removed thoroughly extremely high amount of chemicals were used so that even the nitride on wafer back surface was cleared away. While during single process only the front surface of the wafer was flushed by phosphoric acid solution and the nitride on the back surface of the wafer was retained.

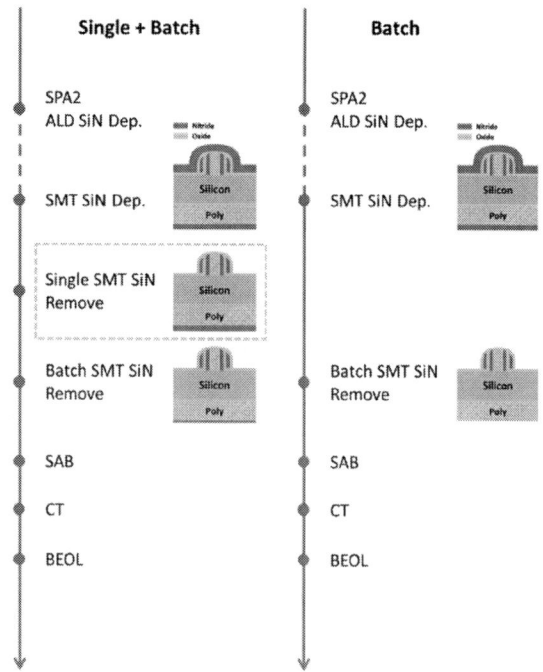

Figure 2. Two SMT Nitride Remove Process Flows

978-1-7281-6559-2/20 $31.00 © 2020 IEEE

In our experiment, a batch nitride removing process with much less process time was followed after a single one to ensure the thoroughly removal of front surface nitride. Process flow differences were illustrated in figure 2. The conventional one-step batch SMT nitride remove wet process was replaced by a novel two-step one. During the first single process only front-side nitride was removed. The second batch process ensures that the front-side nitride was removed thoroughly and the back-surface nitride was retained with a proper thickness. The chemicals used in both single and batch processes were conventional phosphoric acid (H_3PO_4). All wafers were prepared based on HLMC 28 nm low power platform.

RESULTS AND DISCUSSION

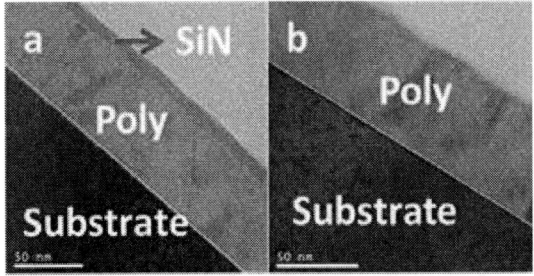

Figure 3. Back Surface Film Stack after Two Different SMT Removing Processes a. Single+Batch SMT remove b. Batch SMT remove

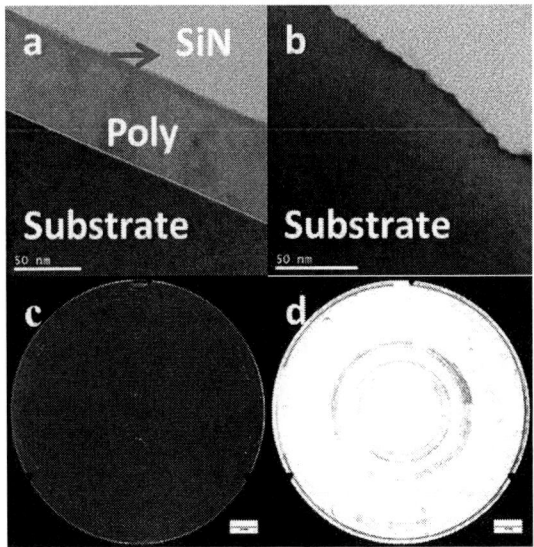

Figure 4. Back Surface Film Stack and Image after Final WAT Test a.c. Single+Batch SMT remove b.d. Batch SMT remove

Film stacks of wafer back-surface after two different SMT nitride remove process were illustrated in figure 3. These TEM images demonstrated that this novel two step single+batch (S&B) SMT nitride remove process retained a thin silicon nitride to the wafer back-surface successfully. Moreover, the back-surface flatness was also improved by this new thin layer. And this flatness improvement would benefit BEOL litho leveling remarkably.

After final WAT (wafer accept test), we check the back-surface condition and film stack once more. Wafer final backside inspection images, as shown in figure 4, confirmed that wafers processed by this novel two step single+batch (S&B) SMT nitride remove managed to keep a thin silicon nitride layer to the wafer back-surface throughout the whole BEOL processes. The TEM images also demonstrated a significant back-surface flatness improvement. While the wafer processed by conventional batch SMT nitride remove had much worse back-surface flatness. As could be seen from figure 4d, poly and gate oxide layer were completely eroded leaving a rough surface of silicon substrate, which should have had a bad effect on BEOL lithograph process.

BEOL metal-x leveling maps were shown in figure 5 along with defocus maps and yield maps. Figure 5a was a leveling map of a wafer processed by S&B SMT nitride remove and figure 5b were of wafers processed by conventional batch one. It's obvious that an S&B wafer had a remarkable (60% to 80%) reduction on leveling range. Moreover, an S&B wafer was leveling hot spots free. A strong correlation between litho leveling hot spots and defocus defects was also demonstrated. And defocus defects would finally result in yield loss (0.5% in average).

Namely, a 0.5% yield improvement could be achieved by replacing conventional batch SMT nitride remove with a novel two-step one. Further, this back-surface nitride layer also benefits OQA (outgoing quality control) visual inspection.

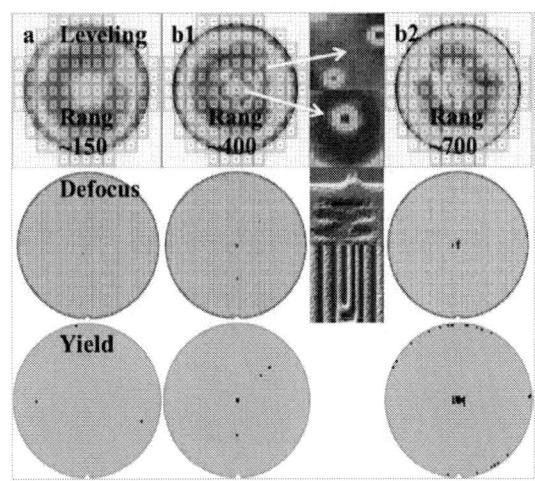

Figure 5. Wafer Leveling, Defocus and Yield Map a. Single+Batch SMT remove b1 and b2 Batch SMT remove

CONCLUSION

A novel two-step SMT nitride remove process was proposed. By simply replacing the conventional batch SMT remove process with the novel two-step one a thin ALD nitride film was retained on the wafer back-surface. And this thin nitride film protected the wafer back-surface throughout the whole BEOL processes including repetitive BSCs. 60% to 80% reduction of BEOL litho leveling range was achieved from S&B wafers. No defocus defect was detected from S&B wafers and yield improvement was about 0.5% in average.

ACKNOWLEDGEMENTS

I would like to appreciate senior engineer Chao Sun, Xiaolin Xu, Wei Zhou for their support and advice on this experiment.

REFERENCES

[1] "Yield Improvement in 2x Node Technology by Introducing Backside Cleaning," ASMC, 2015.

[2] "Wafer Backside Cleaning for Defect Reduction and Litho Hot Spots Mitigation," ASMC, 2018.

[3] "Backside Wafer Damage Induced Wafer Front Side Defect and Yield Impact" ASMC, 2007.

IMPROVED SELECTIVE SILICON NITRIDE ETCH FOR ADVANCED LOGIC AND MEMORY APPLICATIONS

Chien-Pin Sherman Hsu

Avantor

No. 38-1, Tai-Yuan St., Chu-Bei, Hsinchu, Taiwan

sherman.hsu@avantorsciences.com

ABSTRACT

Mechanisms of H_3PO_4-based selective etch of silicon nitride (Si_3N_4) vs. silicon oxide (SiO_2) and oxide regrowth challenges in 3D NAND stack are reviewed. Effective additives have been developed to fortify phosphoric acid etchants and deliver up to 7.5-fold selectivity enhancement over standard H_3PO_4. These additives are expected to give equal or better oxide regrowth control. They also provide built-in bath seasoning effects for more stable silicon nitride etch rate and process reproducibility.

INTRODUCTION

Selective Etch of silicon nitride (Si_3N_4) vs. silicon oxide (SiO_2) has wide applications in semiconductor IC and memory fabrication processes, including recent growth in the development of vertical NAND (V-NAND) flash memory devices [1]. Hot phosphoric acid (H_3PO_4) has been a preferred chemistry for the selective etch of Si_3N_4 [2-3]. However, considerable interest in improving H_3PO_4-based etch chemistry has emerged and are described as follows: 1) Si_3N_4 vs. SiO_2 selectivity enhancement, 2) oxide regrowth control or suppression, and 3) improved bath and process stability with more effective seasoning method or agents [4].

This study reports the development of new fortified H_3PO_4-based formulations to provide enhanced Si_3N_4 etch selectivity, unique oxide regrowth suppression agents and alternative silicon bath seasoning additives. Their beneficial effects have also been successfully demonstrated.

EXPERIMENTAL

Wet clean experiments were conducted in beakers and the SEZ (LAM) SP304 Single Spin Tool. Film thicknesses were measured with X-ray Fluorescence (XRF), ellipsometry or 4-point probe. Removal of PR films or residues was examined with optical microscopy or High Resolution Scanning Electron Microscopy (HR-SEM).

RESULTS AND DISCUSSION

Si_3N_4 Selective Etch and Oxide Regrowth Mechanism

Si_3N_4 etch by 85% phosphoric acid proceeds through a well-established reaction and generates ammonium phosphate and silicic acid (H_2SiO_3) (Eq. 1). Silicic acid may be converted to orthosilicic acid, $Si(OH)_4$, as a hydrated form of H_2SiO_3 (Eq. 2). While water solubility of SiO_2 or Si_3N_4 is relatively low, any such dissolutions also produce H_2SiO_3 (Eq. 3) or $Si(OH)_4$ (Eq. 4), which are expected to contribute to oxide regrowth by adding more SiO_2 moieties to the existing silicon oxide framework.

The addition of silicon-containing additives such as silicic acid or TEOS, have been proposed to improve Si_3N_4-to-SiO_2 selectivity. However, their effects on oxide regrowth need to be carefully assessed.

$$3\ Si_3N_4 + 27\ H_2O + 4\ H_3PO_4 \rightarrow$$
$$4\ (NH_4)_3PO_4 + 9\ H_2SiO_3 \tag{1}$$

$$H_2SiO_3 + H_2O \leftrightarrow Si(OH)_4 \tag{2}$$

$$SiO_2 + H_2O \leftrightarrow H_2SiO_3 \tag{3}$$

$$Si_3N_4 + 12\ H_2O \rightarrow 4\ NH_3 + 3\ Si(OH)_4 \tag{4}$$

Development of Improved H_3PO_4-based Silicon Nitride Etch Chemistry

A series of unique additives have been developed to deliver the following benefits (Table I):

1) Additive A1 – It serves as a silicon oxide (SiO_2) etch suppressor to enhance nitride-to-oxide selectivity. This unique additive also provides built-in "bath seasoning" without using sacrificial silicon nitride wafers. In contrast to silicic acid or TEOS types of additives, the A1 additive is not expected to contribute to oxide regrowth.

2) Additive A2 - It functions as a silicon nitride etch promoter, and has very limited effects on silicon oxide etch rate increases compared to other additives, such as hydrofluoric acid (HF) [5-6].

3) Combination of Additive A1 and A2 – Use of dual additives has demonstrated almost 2X Si_3N_4-to-SiO_2 selectivity enhancement compared to a standard H_3PO_4 etchant. Optimization of various A1 contents at a fixed A2 level further boosts selectivity by 7.5X compared to H_3PO_4 (Fig. 1).

TABLE I. ETCH RATES AND SELECTIVITY OF SELECTED FORTIFIED Si_3N_4 ETCH CHEMISTRIES (AT 160°C)

Chemical Description[1]	Additive Type	Si_3N_4 Etch Rate Å/min	SiO_2 (Thermal Oxide) Etch Rate Å/min	Si_3N_4-to-SiO_2 Selectivity
H_3PO_4	None	43.04	1.34	32
H_3PO_4 + A1	A1, Oxide Etch Suppressor	48.57	1.33	37
H_3PO_4 + A2	A2, Nitride Etch Promoter	142.68	3.12	46
H_3PO_4 + A1 + A2	A1, Oxide Etch Suppressor + A2, Nitride Etch Promoter	143.79	2.44	59

[1] H_3PO_4 - 85% phosphoric acid in water

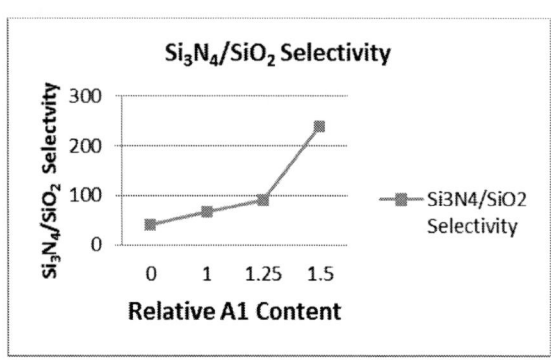

Fig. 1: Additive A1 effects on Si_3N_4-to-SiO_2 selectivity for H_3PO_4-based etchant at 160°C

REFERENCES

[1] M. Lapedus, *Semiconductor Engineering, Knowledge Center*, August 16, 2018.

[2] W. Van Gelder and V. E. Hauser, *J. Electrochem. Soc,* **114**, pp. 869-872, 1967.

[3] L. Liu, I. Kashkoush. G. Chen and C. Murphy, *212th Electrochem. Soc. Meeting,* Paper #1016, October 7-12, 2007.

[4] T. Kim, C. Son, T. Park and S. Lim, *Surface Preparation and Cleaning Conference (SPCC)*, Paper 02-02, April 2-3, 2019.

[5] D. Seo, J. S. Bae, E. Oh, S. Kim and S. Lim,, *Microelectronic Eng.* **118**, pp. 66-71, 2014.

[6] L. Liu, et al., J. Phys: *Condens. Matter* **28**, 094014, 2016.

QUASI-ATOMIC LAYER ETCHING TECHNOLOGY FOR HIGH UNIFORMITY ETCHING APPLICATIONS

Y. Zhang, J. Chong, C. Wang, Q. Xie, D. Li

NAURA Technology Group Co., Ltd.

ABSTRACT

This paper illustrated Quasi-Atomic Layer Etching (Q-ALE) process, based on Inductively Coupled Plasma (ICP) etching technology. Q-ALE process could solve several conventional plasma etching issues, for instance, Aspect Ratio Dependent Etching (ARDE) effect. Furthermore, Q-ALE process could achieve relatively high etching uniformity with low surface roughness and low etching damage for silicon and compound semiconductors etching applications.

Key Words: ALE, GaN, HEMT

INTRODUCTION

Continuous plasma etching has some non-ideal properties, such as Aspect Ratio Dependent Etching (ARDE). However, the Atomic Layer Etching (ALE) shows several advantages, including excellent uniformity, aspect ratio independence and surface smoothing effect. In addition, low etching damage could be realized using ALE.

The typical ALE consists of several stages, shown in Figure 1. Stage A is the surface modification step. Stage B is the removal step. The modification and removal gaseous state materials in Stage A and B separately, depend on the material to be etched. Between Stage A and B, gases were purged and switched.

EXPERIMENTS

The ALE tool were equipped with a Radio Frequency Induction Coupled Plasma (ICP) source, pumped by a turbo molecular pump, backed by a dry pump. Gases are metered by Mass Flow Controllers (MFC) in a gas box and then switched by the fast switching valves. The Electro-Static Chuck (ESC) temperature is controlled by a chiller sub-system, and the wafer is electrostatically clamped with Helium gas injected under the wafer for heat transfer. The ALE tool is run with a computer and Programmable Logic Controller (PLC) as well for automatic recipe sequence operation.

The wafers with polycrystalline silicon and silicon dioxide layers on monocrystalline silicon substrate wafers were utilized for silicon ALE experiments, while the multiple Gallium Nitride (GaN) layers with a number of buffer layers on monocrystalline silicon substrate wafers for GaN ALE experiments.

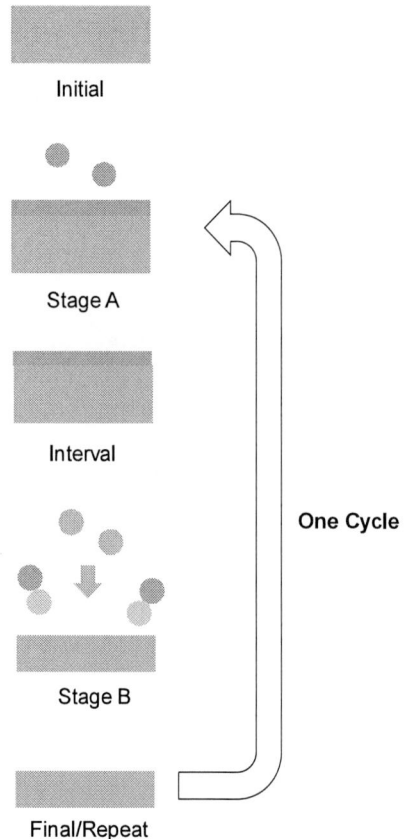

Figure 1: The mechanism of ALE

The wafer to wafer experiments shows etch depth uniformity less than ±2% measured by Filmetrics F60, with the goodness of fit large than 0.99 (Figure 3).

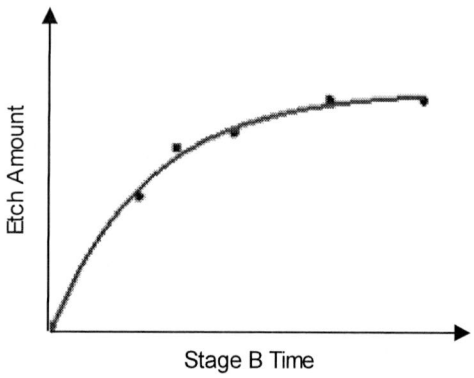

Figure 2: Measurements showing self-limiting effect

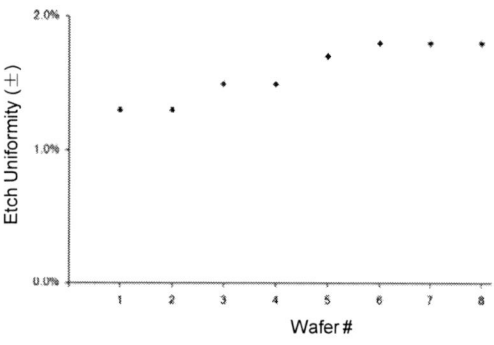

Figure 3: Wafer to Wafer etch depth uniformity

RESULTS

The relationship between etch amount per cycle and the Stage B time is demonstrated in Figure 2, showing self-limiting characteristic of the removal step, which is the critical feature of ALE compared to continuous plasma etching.

The measured GaN self-limiting etching amount per cycle is about 0.5nm/cycle, while the thickness of the etched GaN layer is about 0.5nm, measured by Thermo Fisher Scientific FEI Focused Ion Beam Transmission Electron Microscope (FIB-TEM) before etching.

DISCUSSION

For both silicon and GaN ALE, Argon was chosen in Stage B while chlorination gas in Stage A, e.g. Cl_2 and BCl_3. The main difference between silicon and GaN ALE is the energy of Argon ion due to the surface bonding energy difference of silicon and GaN.

The GaN ALE could be applied to the manufacture of the High Electron Mobility Transistor (HEMT) because of the low etching damage consequence, which would bring little harmful influence to the 2-Dimensional Electron Gas (2DEG) of the HEMT.

CONCLUSION

ALE tool were developed base on ICP technology, with gases fast switching capability. Silicon and GaN ALE experiments were implemented and the process recipes were optimized. The measurement results illustrate the self-limiting characteristic of ALE.

ACKNOWLEDGEMENTS

The authors would like to thank the enormous amount of help and collaboration with the colleagues of NAURA and all the partners.

REFERENCES

[1] C. Huard et al., J. Phys. D: Appl. Phys. 51 (2018) 155201
[2] K. Ishikawa et al., Jpn. J. Appl. Phys. 56, 06HA02 (2017)
[3] S. Athavale et al., J. Vac. Sci. Technol. B 14(6), 1996
[4] W. Lu et al., IEDM 2018
[5] J. DuMont et al., ACS Appl. Mater. Interfaces 2017, 9, 10296−10307

XPS ANALYSIS OF GALLIUM NITRIDE FILM AFTER O2/BCL3 DIGITAL ETCH

Jiale Tang[1], Yongjie Hu[1], Yudong Zhang[1], Zhiqiang Gu[2], Dongchen Che[2], Dongdong Hu[2], Lu Chen[2], Kaidong Xu[1,2]*, Shiwei Zhuang[1]*

[1] School of Physics and Electronic Engineering, Jiangsu Normal University, Shanghai Road 101, Xuzhou, China

[2] Leuven Instruments Co. Ltd (Jiangsu), Liaohexi Road 8, Xuzhou, China

*Corresponding Author's Email: zhuangshiwei@jsnu.edu.cn

ABSTRACT

Oxygen as modification gas for the first step, boron trichloride for the etch step was used to sequentially remove GaN from sapphire for 20 cycles. The relationship between oxidation and etching parameters with etching depth of each cycle was investigated. For 9 s etching time per cycle, the etch rate increased from 0.39 to 1.04 nm/cycle by increasing oxidation time from 0 to 20 s, and remained fairly constant with higher oxidation time revealing a self-limiting oxidation. The surface chemistry of GaN was investigated by X-ray photoelectron spectroscopy after ICP oxidation. A comprehensive analysis of oxygen chemisorption on GaN films with different oxidation time and possible mechanisms were discussed.

INTRODUCTION

Wide bandgap AlGaN/GaN high electron mobility transistors (HEMTs) are emerging as excellent candidates for RF/microwave power amplifiers (PAs) because of their high-power handling capabilities [1, 2]. Their demonstrated low-noise and high breakdown properties indicate their potential for protection-circuit-free low-noise amplifiers (LNAs). With the wide bandgap of III-nitride materials, AlGaN/GaN HEMTs also offers promising possibilities for high temperature digital circuit applications. So far, the focus has been on improving the performance of depletion-mode (D-mode) AlGaN/GaN HEMTs [3-6].

Dry plasma etching techniques are useful for fabricating GaN devices. Plasma etching of GaN uses primarily chlorine-containing plasma to produce volatile $GaCl_3$ [7]. Various chlorine-containing plasma is formed using Cl_2 and BCl_3 gases in etch chambers [8, 9]. GaN is also etched by low energy electron enhanced etching in H_2 plasma [10]. Another digital etching technique is proposed, which uses a dry plasma step and a wet etching step, but the disadvantage of this technique is that the throughput is too low [11]. A semi-self-limiting process for etching GaN under Ar and Cl_2 plasma using a two-step cycle method was proposed, but no voltage device data was provided. The semi-self-limiting processed for etching GaN using a two-step cycle method of O_2 and BCl_3 plasma was

presented recently and the voltage device data was provided [12] but what happened on the GaN films base on O_2 plasmas are still unclear. This work revealed that the etch step did not remove oxide completely per cycle. Surface composition of the ICP oxidized GaN was analyzed using X-ray photoelectron spectroscopy after the plasma etching with an additional oxidation step. In this study, deconvolution of the peaks was modelled using Gaussian and Lorentzian line shapes and a Shirley-type background. X-ray photoelectron spectroscopy (XPS) has been used to characterize the formation of oxide bond by O_2 exposure and oxygen ion bombardment on GaN surface [13-15]. XPS Ga(3d) CL (core level) and O (1s) CL shifted to higher bong energy from native oxidation to ICP oxidation, simultaneously, the change of O (1s) CL with different oxidation indicated the growth of multiple types of oxide on GaN films in ICP oxidation.

EXPERIMENTAL

In this work, GaN was grown on a sapphire substrate by MOCVD (Metal Organic Chemical Vapor Deposition). Firstly, the sample was washed in acetone and isopropanol for 5 mins respectively, then was soaked in deionized water for about 10 mins before oxidized by ICP. The GaN film was approximately 100 nm thick, using Si3N4 as the hard mask which was deposited by PECVD 3116 from Jiangsu Leuven Instruments Co.Ltd. The ICP equipment we used for etching is LICP 2127 from Jiangsu Leuven Instruments Co. Ltd. The coil frequency was operated of 13.56 MHz. In the oxidation step, the O_2 flow rate was 100 sccm, the chamber pressure was 80 mTorr, the source power was 900 W, the bias power was 30 W and the oxidation time was 10, 20, 30, 60, and 90 sec per cycle. In the etch step, the BCl3 flow rate, the chamber pressure, the source power and the bias power was 70 sccm, 15 mTorr, 300 W and 10 W respectively. The etch time was 9, 14, 19 and 24 secs per cycle. The etch depth and etched profiles were characterized by HITACHI UHR FE-SEM SU8200/8220. The GaN was oxidized in situ chamber by an additional oxidation step after sequentially remove GaN from sapphire for 20 cycles while the digital etch process was stopped and the surface chemical states were characterized using THERMO ESCALAB 250Xi X-ray photoelectron spectroscopy.

RESULT AND DISUSSIONS

The GaN etching morphologies with O_2/BCl_3 measured with SEM was shown in Figure 1. The etch depth was 25.8 nm, oxygen as a modifying gas in the first step for 60 s per cycle and boron trichloride in the etching step, by sequentially remove GaN from sapphire for 20 cycles. The etching rate is calculated as 1.29 nm per cycle. Oxygen plasma oxidizes GaN to form oxide and then boron trichloride plasma to remove the oxide. After etching, there is a 180nm SI_3N_4 as a hard mask to block further etching.

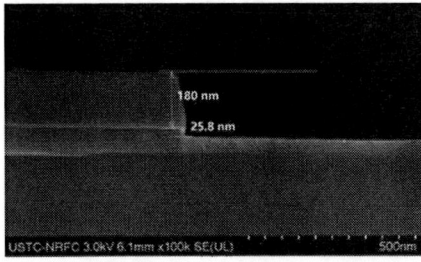

Fig. 1. SEM image of GaN after 20 etch cycles.

In order to verify the self-limitation of the process, the relationship between the oxidation time and the etch rate and the etch depth of per cycle was studied, and the oxidation time was set to 0, 10, 20, 30, 60 and 90 s per cycle, and the etch time was fixed for 9 s and the result is shown in Figure 2. The result shows the etch rate increase with oxidation time up to 10 seconds of O_2 plasma per cycle. After 15 seconds, the etch rate remains fairly ~1 nm per cycle with increasing oxygen time. The etch depth would no longer increase by prolong the oxidation time, which indicated that the oxidation step was a self-limiting process.

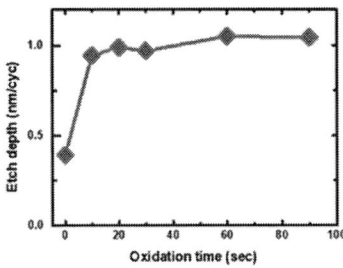

Fig. 2. Etch depth per cycle of different oxidation time (0, 10, 20, 30, 60 and 90 s

To confirm the self-limit of the etch step, the relationship of etch time per cycle and etch rate or etch depth per cycle was investigated. The etch time was set at 9 14 19 and 24 s with fixed oxidation time at 9 s. All the results are shown in Fig.3. In Fig.3, the etching depth per cycle increased by prolong the etching time, the etch rate per cycle increased instead of decreased. The reason why this happened is that the oxide on GaN has not been completely removed or the etch rate should decrease. If all the oxidation reaction layers are removed during the etch step of each cycle, the etch rate should be about 0.35 nm, as shown in FIG. 2. However, the high etch rate shown in FIG.3 indicated the thickness of the oxidized reaction layer was more than 1 nm and the top of the oxide may have a higher density. The saturated etch depth of GaN remained fairly constant at ~1nm shown in Fig.2. Similar to ALD, the implementation of ALE has long been based on a series of independent, self-limiting surface reactions to replace the complex plasma-surface interactions of steady-state plasma etching. The etch step should be self-stopped after removing all the modified layer [16]. Considering all phenomena of Figure 2 and Figure 3, there is no self-limiting during the etch step, so the process is semi-self-limiting.

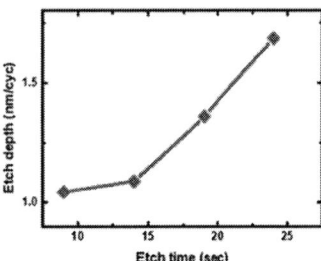

Fig. 3. Etch depth per cycle of different etch time (9, 14, 19 and 24 s)

To investigate the chemical state and the interaction of chemisorbed oxygen with the surface of the films, XPS was used to measure chemical state on GaN surface of different samples. FIG.4 (a)displays the XPS Ga(3d) CL spectra of sample 1, 2, and 3 observed at bonding energy positions of 19.9 eV 21.4eV and 21.0 eV respectively. The peaks were deconvoluted into two major components attributed to Ga-O and Ga-N bonding [13, 17-19]. The Ga (3d) CL shifted to higher BE (bond energy) values and FWHM (full width a half maximum) increases from native oxidation to ICP oxidation shown in the FIG.4 (from b to c) , indicating that the contribution to surface oxide increased (from b to c) and the binding energy of Ga(3d) increased. The shift of Ga(3d) CL towards higher BE values, and increment in FWHM reveals a higher amount of surface oxide. However, the opposite happened between simple 2 and simple3 which may be caused by the change of density of Ga_XO_Y film, high density Ga_XO_Y may prevent further oxidation of GaN. The change dense of structure may decrease the etch rate [20]. The oxide layer is difficult to be etched during the etch step and much oxide is left on the GaN surface, which is difficult to be further oxidized during the next cycle due to the barrier of the oxide film which retained from last cycle.

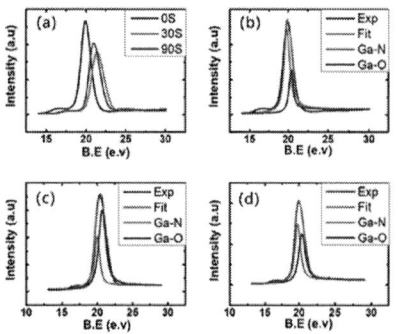

Fig. 4. (a): XPS spectra of GaN films with their native oxide, 30s oxidized by ICP, and 90s oxidized by ICP. Deconvoluted Ga (3d) XPS spectra of sample 1 (b), 2 (c) and 3(d) displaying various components.

The XPS O (1s) CL and de-convoluted XPS O (1s) CL are shown in FIG.5(a). The shift of O (1s) towards to higher and increment in FWHW from native oxide to ICP oxide indicated the contribution of surface oxide increased (from simple b to c and d) and the binding energy of O (1s) increased. The de-convoluted XPS O (1s) CL are showed in FIG.5 (b), (c) and (d) indicated multiple types of oxide growth with ICP oxidation. However, the opposite happened from simple 2 to 3 which shift towards to lower BE values may result from the change of Ga_XO_Y film's structure is dense. The GaN oxidation follow simple reaction below

$$GaN + O_2 \rightarrow Ga_XO_Y + NOx.$$

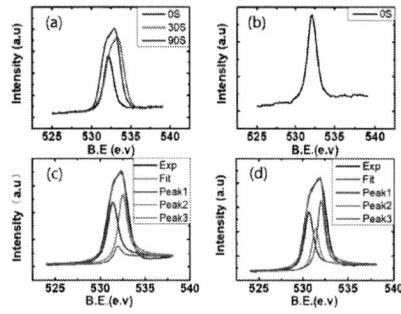

Fig. 5. (a): XPS (a) spectra of GaN film with their native oxide, 30s oxidized by ICP, and 90s oxidized by ICP.

De-convoluted O (1s) XPS spectra of sample 1 (b), 2 (c) and 3(d) displaying various components.

The nitrogen removal from the GaN film and exchange reaction with incoming oxygen [20]. The GaN film may adsorb multiple types NOx during ICP oxidation (from b to c and d). The GaN was etched ~20nm with 20 cycles. The etch depth was ~1 nm per cycle when the etch time was 9 s. The etch depth was increased by prolonging etch time. As shown in Fig.3 trend of etching depth per cycle increased as extending the etch time indicated that the thickness of the oxidized film thicker than 1nm. However, the trend of etch rate should decrease when increasing the etch time from 14 s to 24 s for the etch rate of GaN is much lower when the oxide layer is completely removed. As depicted in Fig.6, the patterned GaN was oxidized in the oxidation step, and the Ga_XO_Y film was formed on the GaN which is thicker than 1nm. The Ga_XO_Y was etched by BCl3 plasma with low bias power, and thin Ga_XO_Y film was removed. However, in etch step the Ga_XO_Y was not completely removed Ga_XO_Y film, and the remaining Ga_XO_Y film was oxidized again, therefore, the density of Ga_XO_Y film may become higher.

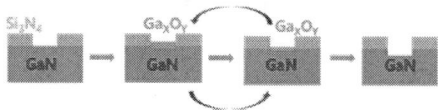

Fig. 6. Schematic illustrations of GaN etch.

This is a quasi-atomic layer etching process. The etch step in each cycle does not completely remove the oxide layer on the GaN. Prolonging the oxidation time in ICP will increase Ga_XO_Y's dense of structure. As depicted in FIG.7 a complexity chemical states on the GaN surface during the oxidized in the ICP. Initially, Ga_XO_Y is formed on the GaN film and multiple types of oxide was measured by X-ray photoelectron spectroscopy which may result from NO_X [20]. The nitrogen was removed from the GaN film and react with oxygen and formed NO_X adsorbed on the GaN surface [20]. The density of gallium oxide becomes higher and forms oxide film as the barrier for subsequent etch.

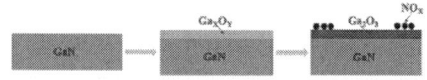

Fig. 7. Schematic illustrations of oxidized the GaN

CONCLUSIONS

A semi-self-limiting processed for etching GaN using a two-step cycle method of O2 and BCl3 plasma was investigated. It was found that, etching rate will no longer be affected with increased oxidation time indicated oxidation step is a self-limiting process. However, the etching depth and the etching rate increased when prolong the etching time in etching step indicated that the process is semi-self-limiting. During the etching step the oxide was not completely removed which result in the increase of etching rate when prolong the etching time. The contribution of surface oxide and the binding energy was higher than the native oxide compared to oxide of ICP. But the opposite happened during the further ICP oxidation which may be due to a higher density of gallium oxide. Multiple types of oxide were detected on GaN surface with ICP oxidation. The density of Ga_XO_Y becomes higher when prolong oxidation time in ICP which may decrease etching rate and oxidation time. The nitrogen removal from the GaN film and exchange reaction with incoming oxygen. This study may reveal that an atomic layer etch control may not depend on atomic layer modified thickness on the surface for the etching rate remained fairly constant at ~1nm in this work but the thickness of the oxide layer was more than 1 nm. This work will enable a deeper understanding of the etching process and changes in the surface chemical states at critical oxidation steps in GaN etch.

ACKNOWLEDGEMENTS

This work was supported by the National Foreign Experts Bureau High-end Foreign Experts Project (Grant No. G20190114003), the Key Research and Development Program of Jiangsu Province (Grant No. BE2018063), the Natural Science Research Projects of Colleges and Universities in Jiangsu Province (19KJD140002) and the Scientific Research Program for Doctoral Teachers of JSNU (Grant No. 9212218113).

REFERENCES

[1] X. Shi, S. Posysaev and M. Huttula. J. Small., vol. 22, 2018, pp. 45-26

[2] Z. Shi, Y. Zhang, B and Wu. X. Cai., J Applied physics letter vol. 2 ,2013, pp. 161-101.

[3] V. Kumar, W. Lu, R. Schwindt, A. Kuliev, G. Simin, J. Yang, M.A. Khan, I. J IEEE Electron Device Letters. Adesida, vol.23 2002, pp. 455-457.

[4] W. Saito, Y. Takada, M. Kuraguchi, K. Tsuda, I. Omura, T. Ogura, H.J. IEEE Transactions on Electron Devices., vol. 50 ,2003, 2528-2531.

[5] J. Johnson, E. Piner, A. Vescan, R. Therrien, P. Rajagopal, J. Roberts, J. Brown, S. Singhal, K. J. IEEE Electron Device Letters., vol.25 ,2004, pp. 459-461.

[6] Y.-F. Wu, A. Saxler, M. Moore, R. Smith, S. Sheppard, P. Chavarkar, T. Wisleder, U. Mishra, P.J IEEE Electron Device Letters., vol.25 ,2004, pp. 117-119.

[7] S. Pearton, J. Zolper, R. Shul, F.J.J.o.a.p. Ren. J. IEEE Electron Device Letters., vol.86, 1999, pp. 1-78.

[8] M. Lin, Z. Fan, Z. Ma, L. Allen, H.J.A.P.L. Morkoc, J. Applied Physics Letters., vol.64 ,1994, pp. 887-888.

[9] R. Shul, G. McClellan, S. Casalnuovo, D. Rieger, S. Pearton, C. Constantine, C. Barratt, R. Karlicek Jr, C. Tran, M.J.A.p.l. Schurman, J Applied Physics Letters., vol.69 ,1996, pp. 1119-1121.

[10] H. Gillis, D. Choutov, K. Martin, S. Pearton, C.J.J.o.t.E.S. Abernathy, J. Journal of The Electrochemical Society., vol.143 ,1996, pp. 25-1.

[11] R.E. Leoni, J. Proceedings IEEE Lester Eastman Conference on High Performance Devices at University of Delaware, Newark, Delaware, August 6-7, 2003, pp.8-9

[12] S.D. Burnham, K. Boutros, P. Hashimoto, C. Butler, D.W. Wong, M. Hu, M.J.p.s.s.c. Micovic, physics solid state., vol.7 ,2010, pp. 210-201.

[13] S. Wolter, B. Luther, D. Waltemyer, C. Önneby, S.E. Mohney, R.J.A.P.L. Molnar, J. Applied Physics Letters., vol.70, 1997, pp. 2156-2158.

[14] S. Wolter, J. DeLucca, S.E. Mohney, R. Kern, C.J.T.S.F. Kuo, J. Thin Solid Films. Vol. 371 ,2000, pp.153-160.

[15] M. Grodzicki, P. Mazur, S. Zuber, J. Brona, A.J.A.s.s. Ciszewski, J. Applied Surface Science., vol.304 ,2014, pp. 20-23.

[16] G. Oehrlein, D. Metzler, C.J.E.J.o.S.S.S. Li, Technology, J. Solid State Science., vol. 4, 2015, pp.50-41.

[17] M. Mishra, T. Krishna, P. Rastogi, N. Aggarwal, A.K.S. Chauhan, L. Goswami, G.J.M.F. Gupta, J. Materials Focus vol.3 ,2014, pp. 218-223.

[18] M. Mishra, S.T. Krishna, N. Aggarwal, S. Vihari, A.K.S. Chauhan, G.J.S.o.A.M. Gupta, J. vol.7 ,2015, pp.546-551.

[19] V. Thakur, S.J.A.S.S. Shivaprasad, J. Science of Advanced Materials, vol. 327 ,2015, pp. 389-393.

[20] T. Yamada, J. Ito, R. Asahara, K. Watanabe, M. Nozaki, S. Nakazawa, Y. Anda, M. Ishida, T. Ueda, A.J.J.o.A.P. Yoshigoe, J. Journal of Applied Physics., vol.121, 2017, pp. 35-303.

5 NM FIN SAQP PROCESS DEVELOPMENT AND KEY PROCESS CHALLENGE DISCUSSION

Yushu Yang*, Bowen Wang[1], Qiang Wu[1], Yanli Li[1], Yibo Wang[2], Yuning Zhu[2], Yongjian Luo[2], Weihao Lin[1], Qingqing Wu[1], Jianjun Zhu[1], Shoumian Chen[1], Ying Zhang[2]

[1]Shanghai IC R&D Center, Shanghai 201210, China
[2]Beijing NAURA Microelectronics Equipment Co., Ltd., Beijing 100176, China
*Corresponding Author's Email: yangyushu@icrd.com.cn

ABSTRACT

When CMOS technologies entered nanometer scales, FinFET has become one of the most promising devices because of its superior electrical characteristics. The 5 nm FinFET logic process is the cutting-edge technology currently being developed by the world's leading foundries. With the shrinkage in size, the usage of various multiple patterning methods (e.g., Self-Aligned Double Patterning, SADP, or Self-Aligned Quadruple Patterning, SAQP, Litho-Etch-Litho-Etch, LELE, 2D cut) becomes more and more frequent.

In this study, we will briefly introduce 5 nm logic key layer process approach with EUV photolithography technology and, as an example, present in detail the 5 nm Fin patterning process with a Fin pitch of 24 nm based on the SAQP patterning method. Key process challenges are also discussed such as Critical Dimension Uniformity (CDU) and pitch walking. Finally, we proposed the Module Technical Specification (MTS) of 5 nm Fin SAQP key process as a reference. Moreover, we co-work with NAURA to develop 5 nm Fin SAQP etch processes on domestic made etcher tool NAURA NMC612D with very good initial results.

INTRODUCTION

Since the CMOS technology entered nanometer era, FinFET has become the mainstream technology of device scaling. In a typical 5 nm logic process, the Fin pitch is 22~27 nm, the Contact-Poly Pitch (CPP) is 48~55 nm, and the Minimum Metal Pitch (MPP) is around 30~36 nm [1]. With the continuous pattern pitch shrinking, especially into the 5 nm node, Extreme Ultra-Violet (EUV) lithography is a promising technology considered for exposure of 36 nm pitch and below. It can produce small patterns with single exposure and simplify the process flow significantly. So it is considered that 5 nm logic technology node is the first node that will adopt Extremely Ultra-Violet (EUV) lithography on a large scale.

Based on extensive investigation and research such as EUV lithography simulation, we provided 5 nm key layer process design rule and process table including pitch and minimal size data, process approach, and mask count summary with ArF immersion lithography and EUV lithography (Table I). Our proposed process approach is marked in beige. Due to the fact that the Front-End-Of-the-Line (FEOL) layers are related to the CMOS devices, which have a very stringent requirement on the Line Edge Roughness (LER)/ LineWidth Roughness (LWR), and because SADP or SAQP process can smooth out the LER/LWR significantly, Fin and gate patterning prefer to use ArF immersion lithography by SAQP or SADP approach. Other layers, such as various cut layers can use EUV lithography. EUV lithography will reduce mask count significantly and simplify the process flow to reduce process variation such as edge placement error induced by double or quadruple patterning overlay variation.

TABLE I. 5 NM KEY LAYER PROCESS DESIGN RULE AND PROCESS APPROACH

Layer	Shape	Pitch(nm)	Minimal size(nm)	193i	193i mask count	EUV	EUV mask count	Remark
Fin	L/S (H)	24	5	SAQP	1	SADP	1	
Fin cut (Horizontal)	Trench	48	24	LE2	2	SE	1	
Fin cut (Vertical)	Trench	100	50	SE	1	SE	1	
Gate	L/S(V)	50	20	SADP	1	SE	1	
Gate cut	Trench	48	24	LE2	2	SE	1	
MDA	Trench	50	14	LE6	6	SE	2	193i: MDA LE2, Cut LE4; EUV: MDA SE(EUV), Cut SE (193i)
MDG	Hole	25×48	43.5×24	LE4	4	SE	1	MDG: 72T 25nm
V0	Hole	50	16	LE4	4	SE	1	
M1	L/S(H)	32	16	SAQP+LE4	5	SALELE	2	193i: M1 SAQP, Cut LE4; T2T:25nm
Total					26		11	

▨ Proposed process approach

Based on the 5 nm key layer process design rule and process approach assumption, we did 3D process simulation to check process window and feasibility with Synopsys software Sentaurus Process Explorer (Figure 1). For Fin patterning process, we adopted SAQP patterning approach with 193 nm ArF immersion lithography which will be described in detail later. And the proposed process approach will be a reference for the process development of 5 nm logic technology.

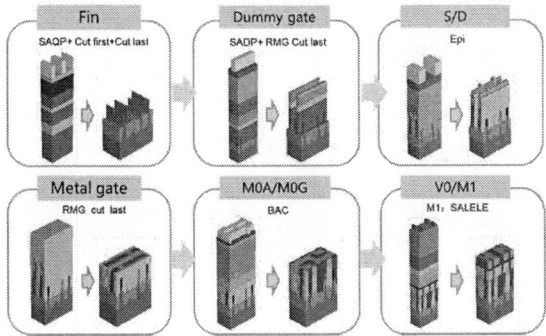

Figure 1. 5nm key layer process simulation

5 NM FIN SAQP PATTERNING PROCESS

Fin is the one of the most important structures with the minimum pitch size in the chip which is related to device performance closely. According to our 5 nm design rule and process approach for Fin structure, we setup a 5 nm Fin SAQP process flow to define Fin features with 193 nm immersion lithography technology. The original lithography pattern pitch is 96 nm and final Fin pitch is 24 nm after quadruple pattering processes. The mandrel material is amorphous Si which has better LER/LWR performance comparing to traditional poly material. The spacer material is Atomic Layer Deposition (ALD) SiO_2, which has high etch selectivity to Si and Si_3N_4 (about 1:6) and good step coverage performance for high aspect ratio feature. We choose Si_3N_4 as etch stop layer for mandrel or spacer etch. The brief film structure and process flow is shown in Figure 2. And all the etch processes are implemented by NAURA ICP etcher tool NMC612D.

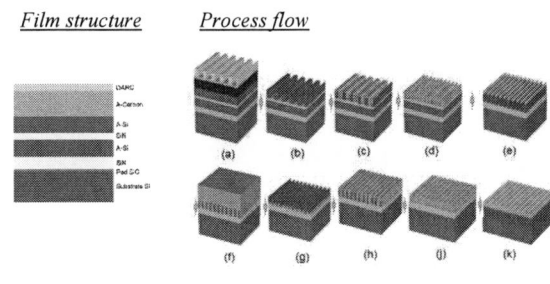

Figure 2. 5 nm Fin SAQP film structure and process flow (a) litho patterning (b) mandrel 1 etch (c) ALD SiO_2 deposition (d) Spacer 1 etch and core removal (e) Mandrel 2 etch (f) SOC deposition (g) SOC etch back and Si_3N_4 removal (h) ALD SiO_2 deposition (j) Spacer 2 etch and core removal (k) final Fin etch

KEY PROCESS CHALLENGE DISCUSSION

1) CD Uniformity

CDU is a key metric which affects post Fin pitch walking and etch depth loading [2-3]. Across-wafer CDU requirement becomes more stringent for 5 nm Fin SAQP process due to smaller pattern size and multiple pattern transfer. In our SAQP process, CDU control includes two parts: mandrel 1 etch CDU control, ALD SiO_2 deposition uniformity control.

Mandrel 1 etch CDU improvement mainly depends on litho Dose Mapping, or DOMA (a function designed by ASML through exposure dose adjustment for individual fields and some limited adjustment within fields) compensation on the specific area of the pattern. Figure 3 shows that the mandrel 1 etch CDU can be improved from 3σ=1.73 nm to 0.87 nm.

Before DOMA

	-5	-4	-3	-2	-1	0	1	2	3	4	5		
4					36.49	36.47	36.65						
3			36.60	36.63	37.26	37.37	36.93	36.96	36.77				
2		37.37	37.10	37.04	37.63	37.35	36.71	36.70	36.82	36.31			
1	36.96	37.53	37.59	37.51	37.17	36.73	36.56	36.70	37.11	36.88	36.04		
0	37.11	37.99	37.88	37.17	37.18	36.41	36.42	36.36	36.55	36.39	36.16	Mean	37.0
-1	37.77	37.96	37.12	37.61	37.49	37.28	37.28	36.70	37.34	36.74	35.62	Max	38.15
-2		37.22	37.77	38.15	37.59	36.90	36.55	36.69	36.23	37.00		Min	35.17
-3			37.47	37.57	37.65	36.99	36.79	36.86	36.49			Range	2.98
-4					36.58	36.12	35.17					3sigma	1.73

Post DOMA

	-5	-4	-3	-2	-1	0	1	2	3	4	5		
4					38.36	38.40	38.57						
3			38.14	38.33	38.03	38.64	38.18	38.30	38.31				
2		38.40	38.48	38.45	38.00	38.20	38.13	38.52	37.91	38.15			
1	38.21	38.25	38.50	37.80	38.19	38.55	38.35	38.35	38.39	38.03	38.89		
0	37.98	38.52	37.66	38.32	37.84	37.98	38.43	38.43	36.16	38.97	38.49	Mean	38.31
-1	37.98	38.58	38.17	37.79	38.59	38.21	38.51	37.91	38.31	38.29	38.38	MAX	39.02
-2		38.54	37.73	38.45	37.96	38.60	38.27	39.02	38.14	38.57		MIN	37.73
-3			38.29	38.45	38.58	38.76	37.78	38.34	38.17			Range	1.29
-4					38.98	38.55	38.65					3sigma	0.67

Figure 3. Mandrel 1 etch CDU improvement by litho DOMA compensation

ALD SiO_2 deposition on amorphous Si and nitride use domestic made Plasma Enhanced ALD (PEALD) tool Piotech FT-300T. The blanket wafer deposition thickness variation 3σ is 2.36 Å and spacer CDU on pattern wafer 3σ is 0.42 nm, which can meet our spec of 0.5 nm. (Figure 4)

Blanket wafer thickness U%   *Pattern wafer ALD SiO CDU*

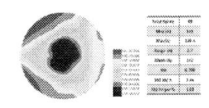
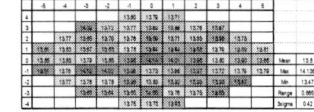

Figure 4. ALD SiO_2 uniformity

2) Pitch walking

The pitch walking is defined as the difference between the largest and smallest space between the fins: [4-5]

$$\text{Pitch walking} = \max(\alpha,\beta,\gamma) - \min(\alpha,\beta,\gamma) \qquad (1)$$

where α, β, γ are space between the fins from different origins after SAQP.

Unbalanced space between Fins will affect device performance as well as subsequent etch and deposition processes such as etch depth loading or Fin bending effect. The pitch walking evaluation and improvement of SAQP process is much more complex than previous SADP process due to its multiple pattern transfer. Figure 5 depicts SAQP pattern transfer procedure under ideal condition.

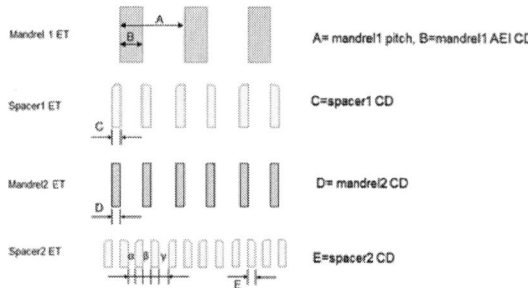

Figure 5. SAQP pattern transfer procedure

In ideal case, we can calculate the Fin spaces α, β, γ can be obtained from the following equations:

$$\alpha = B - 2 \times E \tag{2}$$

$$\beta = D \tag{3}$$

$$\gamma = A - B - 2 \times D - 2 \times E \tag{4}$$

where A, B, D, E are dimensions defined in Figure 5. In order to minimize pitch walking, we need to follow an order in parameter adjustment to get the optimized result: firstly, adjust spacer 1 CD C to change mandrel 2 CD D to make space β on target. Because mandrel 1 pitch A is fixed and spacer 2 CD E is related to Fin CD directly, both of them cannot be adjusted randomly. Secondly, adjust mandrel 1 CD B to balance α and γ on target.

5 NM FIN SAQP KEY PROCESS MTS

Based on extensive investigation and simulation, we provided 5 nm Fin SAQP key process MTS table as below.

TABLE II.   5NM FIN SAQP KEY PROCESS MTS TABLE

Layer	Inline Spec						TEM Spec											
	CD (nm)	CDU (2σ)(nm)	Depth (A)	Depth Uniformity (3σ)(A)	LER (2σ)(nm)	Others	Top CD (nm)	Bottom CD (nm)	BCD-TCD (nm)	Sidewall angle (°)	I/D CD loading (nm)	Dense depth (A)	ISO depth (A)	I/D depth loading (A)	Profile	Substrate loss (A)	Fin hard mask remaining (A)	Others
Mandrel1 Litho	1.0					3												
ALD SiO deposition		0.5																
Mandrel1 ET	36+/-1	1.5			2.5		36+/-1	36-/-1	<2	89+	<1				Smooth Footing <1nm	SiN loss <30A	APF shoulder >300A	
Spacer1 RT	14+/-0.5	0.6			2		14+/-0.5	CD Loss <1nm		>89+	<0.7				Footing <2nm	SiN loss <60A	Top no remaining	Spacer shoulder height >800A
Mandrel1T ET	12+/-0.7	0.6			2		12+/-0.7	12+/-0.7	<2	>89+	<0.7				Smooth Footing <1nm	SiN loss <30A		
ROC ET back																Bottom SiN no loss	Top a-Si loss <100A	
Spacer2 ET	12+/-0.5	0.5			1.5		12+/-0.5	CD Loss <1nm			<0.7				Footing <1.5nm	SiN loss <60A		Spacer shoulder height >800A
Fin ET	9+/-0.5	0.5	80		1.5	pitch walking <1	9+/-0.8	9+/-1		89+ (Above 600A)	<0.7	1200 +/-90	1900 +/-60	<50	Smooth		Top 870 remaining >100A	

CONCLUSION

In conclusion, we have introduced an initial study on the key process loops for a 5 nm logic process flow with EUV lithography, according to typical design rules. And we have setup a 5 nm Fin patterning process flow with a Fin pitch of 24 nm based on SAQP patterning method, in which the etch process is developed with NAURA co-operation on the ICP etcher tool NMC612D. Some key process challenges are discussed such as CDU and pitch walking of SAQP. Finally, we provide key process MTS of our 5 nm Fin SAQP as a reference.

ACKNOWLEDGEMENTS

The authors will very much acknowledge the support of dry etch group of NAURA for their contribution to this project. And thank Shanghai ICRD colleagues and higher management team for the support of this work.

REFERENCES

[1] Q. Wu, Y. Li, Y. Yang, and Y. Zhao "A Photolithography Process Design for 5 nm Logic Process Flow", Journal of Microelectronic Manufacturing 2, 19020408, 2019.

[2] Min-hwa Chi, USA. "Challenges in Manufacturing FinFET at 20nm node and beyond", 2012

[3] Changwoo Lee, "Across-wafer sub-1 nm critical dimension uniformity control by etch tool correction", SPIE 10963_109630G_2019

[4] Sylvain Baudot, and Sofiane Guissi et.al, "N7 FinFET Self-Aligned Quadruple Patterning Modeling", IEEE , 2018.

[5] Taher Kagalwala, Alok Vaid, Sridhar Mahendrakar et.al, "Measuring self-aligned quadruple patterning pitch walking with scatterometry-based metrology utilizing virtual reference", J. Micro/Nanolith. MEMS MOEMS 15(4), 044004 (Oct–Dec 2016)

TITANIUM SILICIDE ANNEAL PROCESS RESEARCH FOR 14NM FINFET TECHNOLOGY

Lan Jiang[*], Yan Gui, Yaoting Shen
Shanghai Huali Integrated Circuit Corporation
6 Liangteng Rd., Pudong district, Shanghai, China
*Corresponding Author's Email: jianglan@hlmc.cn

ABSTRACT

The contact resistance between silicides and Si/SiGe substrates plays more and more important role in both the intrinsic device resistance and parasitic external resistance at 14nm FinFET technology and beyond. Titanium silicide (TiSix) is commonly used as S/D contact material in 14nm nodes because of its good thermal stability. In this study, soak anneal and millisecond anneal (MSA) as different titanium silicide (TiSix) formation processes are investigated in 14nm node both on blanket and pattern wafers. The sheet resistance is measured by four-point probe method. Also contact resistance of silicide on pattern wafer is obtained by wafer acceptance test (WAT).

INTRODUCTION

For the 55/40/28nm technology node, Ni(Pt)Si silicide was widely used as the contact material in planar devices because of its low thermal budget, low resistivity, and low Si consumption [1]. However, the fast diffusion of Ni lead to several defects such as nickel pipes or spiking defects, which are the source of leakage current [2]. Compared with traditional contact material, titanium silicide (TiSix) based materials show better thermal stability. Si instead of Ti is the dominant diffusing species for the formation of TiSix [3]. With the continuous dimensional downscaling of CMOS devices into 14nm node and beyond, the shrinking of contact area at source/drain regions raises serious concerns of high metal/semiconductor contact resistance. Previously, Shujuan Mao et al. [4] reported that TiSix/n-Si contact resistivity can be reduced by proper Ge precontact amorphization implantation (PAI). The ultralow contact resistivity was also achieved according to moderate thermal budget [5]. Soak anneal and millisecond anneal (MSA) is quite different rapid thermal processing (RTP) to form TiSix phase.

In this work, soak and millisecond anneal (MSA) as different TiSix formation processes were investigated in 14nm node both on blanket and pattern wafers. The influence of soak anneal and millisecond anneal on sheet resistance were studied in detail.

EXPERIMENT

The 300nm p-type Si (100) wafers with 10-12Ω•cm resistivity are used as the starting material. To simulate PMOS device condition, blanket wafer are implanted by boron (B), and followed by dopant activation using spike annealing at 920°C. After that, all wafers undergo the same process including precontact amorphization implantation (PAI) and Ti/TiN deposition. Physical Vapor Deposited (PVD) and Atomic Layer Deposited (ALD) are used for deposition of 150A Ti and 50A TiN respectively. Then, the wafers were splited into soak anneal and milliseconds anneal (MSA) process conditions to form titanium silicide. The blanket wafers are processed with soak anneal method annealed at different temperatures of 500°C, 550, 575, 600, 625, 675°C for 15s. The other wafers by millisecond anneal (MSA) are heat treated with a dwell time of 1000us. The experiment process flow for the blanket wafer is shown in Fig. 1. On the blanket wafers, the sheet resistance is measured by four-point probe method.

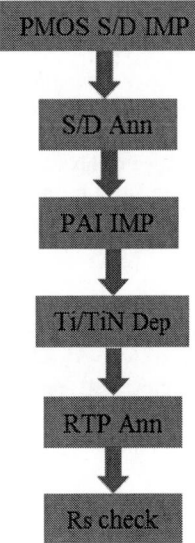

Figure1: Process Flow for the Blanket Wafer

On 14nm pattern wafers, all wafers undergo the same process except for the rapid thermal processing (RTP) which is soak anneal and millisecond anneal. Also contact resistances of silicide on pattern wafer were obtained by wafer acceptance test (WAT).

RESULT DISCUSSION

The sheet resistance of blanket wafers by using millisecond anneal is shown in Fig.2. Specifically, dwell time of MSA is just 1000us.With rising annealing temperature, the sheet resistance initially increases and then decreases, reaches a maximum value at 900°C. There are three stages in Ti/Si reaction: firstly, α-TiSi alloy is formed at a low annealing temperature; secondly, TiSix phases are transformed into C49 phase with increasing temperature; further, C49 phase changed to C54 as the activation energy is enough [6]. Among of them, C49 phase with a higher resistivity is about 60-70μΩ•cm and C54 phase is about 15-20μΩ•cm [7].Hence, it can be speculated that α-TiSi alloy and C49 phase are formed when temperature below 900°C. With further increasing of MSA temperature such as 950°C, it may force the phase transformation from C49 to C54.

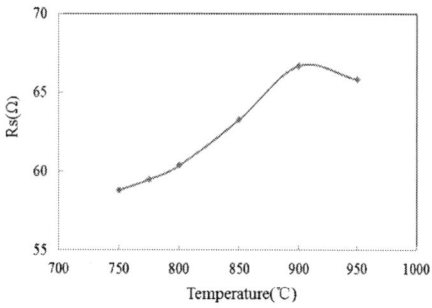

Figure2: Sheet Resistance of the Blanket Wafers Milliseconds Annealed at Different Temperatures

The sheet resistance of blanket wafers which using soak anneal is shown in Fig.3. It can be seen that the sheet resistance is reduced with the temperature increasing. Higher temperature means higher activation energy. As increasing temperature, diffusion of Si became faster and consumed the Ti rapidly. Then, it was related to the phase change from C49 phase with a higher resistivity to C54 phase with a lower resistivity. This behavior was in good agreement with the previously reported [8]. Combined with the sheet resistance of millisecond anneal (MSA) process, it could be found that thermal budget of soak anneal was higher than millisecond anneal. Although the process temperature of soak anneal was lower, the dwell time was longer than millisecond anneal. Above all, the resistance of TiSix was seriously affected by thermal budget.

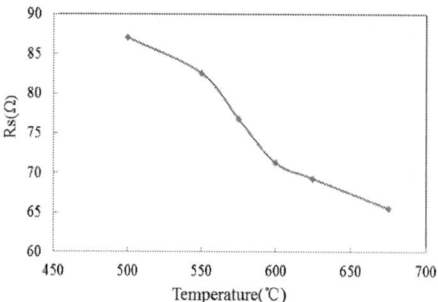

Figure3: Sheet Resistance of the Blanket Wafers Soak Annealed at Different Temperatures

On 14nm pattern wafers, the temperature of soak anneal and millisecond anneal (MSA) are 575°C and 800°C respectively. According to WAT results, contact resistances of soak anneal and millisecond anneal (MSA) are about 90 ohm*um and 62 ohm*um. Milliseconds anneal shows significant improvement on contact resistance.

CONCLUSION

With increasing temperature, the sheet resistance of blanket wafers using millisecond anneal with a dwell time of 1000us initially increased and then decreased. The sheet resistance of blanket wafers which using soak anneal with a dwell time of 15s was reduced when temperature raised. The resistance of TiSix was seriously affected by thermal budget. On 14nm pattern wafers, Milliseconds anneal shows significant improvement on contact resistance.

REFERENCES

[1] H. Iwai *et al.*, "NiSi salicide technology for scaled CMOS," *Microelectron. Eng.*, vol. 60, 2002, pp. 157-169.

[2] C. Lavoie *et al.*, "Towards implementation of a nickel silicide process for CMOS technologies," *Microelectron. Eng.*, vol. 70, 2003, pp. 144-157.

[3] H. Yu *et al.*, "Ultralow-resistivity CMOS contact scheme with precontact amorphization plus Ti (germano-) silicidation," *IEEE Symp. VLSI Technol.*, 2016, pp. 1-2.

[4] Shujuan Mao *et al.*, "Impact of Ge preamorphization implantation on both the formation of ultrathin TiSix and the specific contact resistivity in TiSix/n-Si Contacts," *IEEE Trans. Electron Devices*, vol. 65, 2018, pp.4490-4496.

[5] H. Yu et al., "Titanium silicide on Si:P with precontact amorphization implantation treatment: Contact resistivity approaching 1×10-9 Ohmcm2," IEEE Trans. Electron Devices, vol. 63, 2016, pp. 4632-4641.

[6] M. H. Wang and L. J. Chen, "Phase formation in the

interfacial reactions of ultrahigh vacuum deposited titanium thin films on (111)Si," *J. Appl. Phys.*, vol. 71, 1992,pp. 5918-5925.

[7] R. W. Mann et al.,"Silicides and local interconnections for high-performance VLSI applications," *IBM J. Res. Develop.*, vol. 39, 1995, pp. 403-417.

[8] E. H. Lim et al., "Monitoring of TiSi Formation on narrow polycrystalline silicon lines using raman spectroscopy," IEEE Trans. Electron Devices, vol. 19, 1998, pp. 171-173.

SURFACE SMOOTHING AND ROUGHENING EFFECTS OF HIGH-K DIELECTRIC MATERIALS DEPOSITED BY ATOMIC LAYER DEPOSITION AND THEIR SIGNIFICANCE FOR MIM CAPACITORS USED IN DRAM TECHNOLOGY PART II

W.S. Lau

Zhejiang University, Department of Information Science and Electronic Engineering,
No. 38 Zheda Road, Hangzhou 310027, People's Republic of China
Email: liuweicheng@zju.edu.cn

ABSTRACT

Previously, the author suggested that the atomic layer deposition (ALD) of an amorphous high-k dielectric thin film has a surface smoothing effect on a rough surface. In this paper, the author points out that for ALD high-k dielectric materials which tend to be polycrystalline, the situation is different. When the film is very thin, it can be amorphous with a surface smoothing effect; when the film is thicker than a critical thickness, it can be polycrystalline with a surface roughening effect. An asymmetry in interfacial roughness will lead to an asymmetry in the top and bottom Schottky barrier heights, resulting in I-V polarity asymmetry. The significance of this theory on the leakage current mechanism of ZAZ MIM capacitors used in DRAM technology will be explained.

INTRODUCTION

MIM capacitor structures based on high-k dielectric materials deposited by atomic layer deposition (ALD) are used in dynamic random access memory (DRAM) ICs. The physics underlying the leakage current vs. voltage (I-V) characteristics is still a hot research topic. For example, the I-V characteristics appear to be asymmetrical even for an apparently symmetrical MIM capacitor structure. Previously, Lau et al. [1]-[4] pointed out that the deposition of an amorphous high-k dielectric thin film, for example Al_2O_3, by atomic layer deposition (ALD) on the surface of a metal film with nano-sized roughness has a "surface smoothing effect", resulting in an asymmetry in interfacial roughness such that the asymmetric I-V characteristics of symmetrical MIM capacitors can be explained. ALD of titanium oxide (TiO_2), zirconium oxide (ZrO_2) and hafnium oxide (HfO_2) is different from ALD of Al_2O_3. If ALD is done in an intermediate temperature (IT) range such that there is a threshold thickness $t_{threshold}$. When the thickness of the film $t_{high-k} < t_{threshold}$ the film is amorphous [5]. The author's theory is that the "surface smoothing effect" exists for the ALD of an amorphous film. When the thickness of the film $t_{high-k} > t_{threshold}$ the film is polycrystalline. The author's theory is that the "surface roughening effect" exists for the ALD of a polycrystalline film [5].

THEORY

In this paper, the author will explain in more depth how the surface smoothing and roughening effects can influence the I-V characteristics by making use of the electron Smoluchowski effect [6]. The effective Schottky barrier height of the top TiN-ZAZ interface was higher than that of the bottom TiN-ZAZ interface when there was surface smoothing effect. Conversely, the effective Schottky barrier height of the top TiN-ZAZ interface was smaller than that of the bottom TiN-ZAZ interface when there was surface roughening effect. The leakage current mechanism of ZAZ is either Poole-Frenkel (P-F) or Schottky emission (SE). Knebel et al. [7] suggested that for their 7.5 nm TiN/ZAZ/TiN MIM capacitors, the leakage current mechanism for both polarities of bias voltage is P-F, which is a bulk effect, the electron Smoluchowski effect cannot be important because it is basically an interfacial effect. If the suggestion by Knebel et al. [7] is true, the ALD surface smoothing and roughening effects will have no effect on the leakage current. In addition, the I-V characteristics should be symmetrical for both polarities of bias voltage. However, actual experimental data show that the I-V characteristics are not symmetrical for any measurement temperature.

The author believes that the leakage current due to SE in MIM capacitors is quite frequently misinterpreted as leakage current due to P-F. Many authors may identify the leakage current mechanism in ZAZ MIM capacitors as P-F because of excessive faith in a theory regarding the image force dielectric constant which has to be modified. This modification of theory will be discussed in detail in another paper also submitted to CSTIC 2020 [8]. According to the current theory (Theory X1), the image force dielectric constant is equal to the square of the refractive index, which is about 4 for a high-k dielectric. When the image force dielectric constant is equal to about 1 for SE and about 4 for P-F, many authors will take the value of "about 1" to be unphysical and then take P-F as the leakage current mechanism; however, the author's opinion is that the value of "about 1" is reasonable according to his theory (Theory L) and the leakage current mechanism is actually SE [8]. Table I shows the effect of switching from Theory X1 to Theory L on the refractive index and the image force dielectric constant using data from Knebel et al. [7]. Knebel et al. interpreted the leakage current mechanism as P-F for both polarities based on Theory X1. If this is true, the leakage current should be symmetrical for both polarities of bias voltage. However, this is obviously not true. Instead, the author will interprete the leakage current mechanism as SE for both polarities based on Theory L. The effect of switching from Theory X1 to Theory L is shown in Table I. The author points out the leakage current mechanism is more easily by examining the 2 curves of logI vs $V^{1/2}$ plotted together for both polarities of bias voltage in the same figure instead of just examining 1 curve of logI vs $V^{1/2}$ for only one polarity of bias voltage [8].

The other problem is the extraction of the Schottky barrier height by I-V measurements at various temperatures (IVT). The author points out that this IVT approach is problematic if the leakage current mechanism is SE modified by tunneling [8]. Only a relative measure of the Schottky barrier height by examining the leakage current can be done for such a situation [8].

Table I Effect of switching from Theory X1 to Theory L

ZAZ thickness (Bias polarity)	7.5 nm (TE+ve)	7.5 nm (TE-ve)
Refractive index n (Leakage Mechanism)	2.6 (P-F)	3.2 (P-F)
Refractive index n (Leakage Mechanism)	1.3 (SE)	1.6 (SE)
Image force dielectric constant n^2 (Leakage Mechanism)	1.69 (SE)	2.56 (SE)

APPLICATION OF THE THEORY

The above theory can be applied to symmetrical $ZrO_2/Al_2O_3/ZrO_2$ (ZAZ) MIM capacitors with TiN used for both top and bottom electrodes used in DRAM technology [9]. Fig. 1 shows ZAZ MIM capacitors influenced by the surface smoothing and roughening effects. Besides the ZAZ film thickness, the heat treatment (PDA/PMA: post deposition anneal / post metallization anneal) after ZAZ deposition is also important.

For 4.6 nm ZAZ MIM capacitors with no strong heat treatment after ZAZ deposition by ALD, there is a surface smoothing effect. Asymmetry in roughness will lead to an asymmetry in Schottky barrier height because of the Smoluchowski effect.

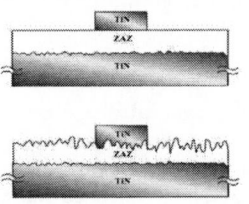

Figure 1: TiN/ZAZ/TiN MIM capacitors influenced by the surface smoothing (top) and roughening (bottom) effects for ZAZ deposited by ALD. When the thickness of the ZAZ film $t_{high-k} < t_{threshold}$, there are 2 cases: (a) surface smoothing will be observed for no or weak heat treatment after ZAZ deposition and (b) surface roughening will be observed for strong heat treatment after ZAZ deposition. When the thickness of the ZAZ film $t_{high-k} > t_{threshold}$, there is only 1 case: (c) surface roughening will be observed independent of the heat treatment after ZAZ deposition.

According to the author's theory and analysis, 4.6 nm ZAZ MIM capacitors behave like two back-to-back Schottky diodes D1 and D2 with two different Schottky barrier heights due to the Smoluchowski effect in the higher electric field region, as shown in Fig. 2. Experimental data for 4.6 nm ZAZ capacitor with TiN top and bottom electrodes at 303 K from An et al. [10] support the interpretation as 2 back-to-back diodes, as shown in Fig. 2.

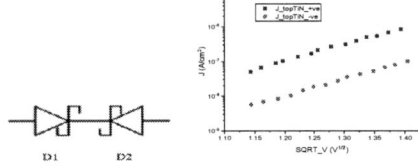

Figure 2: (Left) The leakage current vs. electric field curves of 4.6 nm TiN/ZAZ/TiN MIM capacitor at the high field region can be represented by two back-to-back Schottky diodes D1 and D2 (SE) with two different Schottky barrier heights due to the surface smoothing and Smoluchowski effects even though both top and bottom electrodes are TiN. (Right) Data for 4.6 nm TiN/ZAZ/TiN capacitor at 303 K support the above interpretation.

According to the author's theory and analysis, 4.6 nm ZAZ MIM capacitors behave like two back-to-back Schottky diodes D1 and D2 with two different Schottky barrier heights due to the Smoluchowski effect in the higher electric field region. However, in the medium field region, 4.6 nm ZAZ MIM capacitors behave like one Schottky diode in series with a nonlinear resistor, resulting in P-F leakage for one polarity and SE leakage for the other polarity. Experimental data for 4.6 nm ZAZ capacitor with TiN top and bottom electrodes at 343 K from An et al. [9] support this interpretation, as shown in Fig. 3. Fig. 2 shows only the high field region at 303 K without the medium field region. Actually, the P-F leakage mechanism also has some effect at 303 K in the medium field region; however, the existence of P-F leakage is more obvious at 343 K than at 303 K. The above theory originated from the author's extended unified Schottky-Poole-Frenkel theory [10].

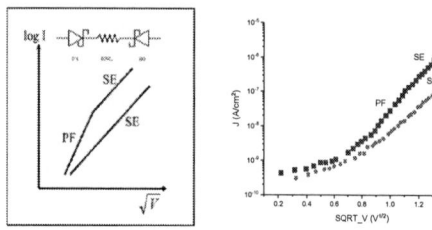

Figure 3: (Left) The leakage current vs. electric field curves of ZAZ MIM capacitor used for DRAM at the high field region can be represented by two back-to-back Schottky diodes D1 and D2 (SE) with two different Schottky barrier heights in series with a nonlinear resistor (P-F). (Right) Data for 4.6 nm ZAZ capacitor with TiN top and bottom electrodes at 343 K support the interpretation as 2 back-to-back diodes (SE) in series with a nonlinear resistor (P-F).

Further analysis by the author shows that ZAZ MIM capacitors behave like a simple linear resistor in the lower electric field region, resulting in symmetric Ohmic leakage current, as shown in Fig, 4.

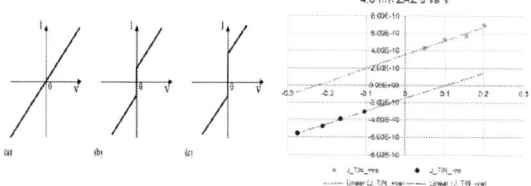

Figure 4: (Left) The leakage current vs. electric field curves of ZAZ MIM capacitor used for DRAM at the low field region is just symmetric Ohmic conduction and can be represented by a simple linear resistor. However, there may be some measurement artifacts (e.g. a jump at 0 V) such that the actual curve (Right) looks like Curve (c) (Left). Data for 4.6 nm ZAZ capacitor with TiN top and bottom electrodes at 343 K support the above interpretation.

CONCLUSION

The surface smoothing and roughening effects together with the Smoluchowski effect are important for the understanding of the I-V characteristics of symmetrical ZAZ MIM capacitors used in DRAM technology. For extremely thin ZAZ MIM capacitor (e.g. 4.6 nm ZAZ) without subsequent strong heat treatment after ALD, there is only surface smoothing. The leakage current mechanism cannot be P-F for both polarities of bias voltage according to I-V symmetry considerations. A new theory is required [8]. With the help of this new theory [8], the leakage current mechanism is more likely to be SE and ZAZ MIM capacitors behave like two back-to-back Schottky diodes with two different Schottky barrier heights in the higher electric field region. In the medium field region, the P-F mechanism can be important for positive bias applied to the top electrode. Further analysis by the author shows that ZAZ MIM capacitors behave like a simple linear resistor in the lower electric field region. The above theory originated from the author's extended unified Schottky-Poole-Frenkel theory [10].

REFERENCES

[1] W.S. Lau, J. Zhang, X. Wan, H. Wong, J.K. Luo and Y. Xu, ECS Trans., vol. 60(1), pp. 527–531, 2014.

[2] W.S. Lau, J. Zhang, X. Wan, J.K. Luo, Y. Xu and H. Wong, AIP Advances, vol. 4, article no. 027120, 2014.

[3] W.S. Lau, L. Du, D.Q. Yu, X. Wang, H. Wong and Y. Xu, ECS J. Solid State Sci. Technol., vol. 6, pp. N111-N116, 2017.

[4] W.S. Lau, L. Du and H. Wong, "Difficulty involved to observe the surface smoothing effect of an amorphous high-k dielectric thin film deposited by atomic layer deposition on a metastable metal film," CSTIC 2019 (China Semiconductor Technology International Conference, Shanghai, 2019, IEEE), pp. 1-3, 2019.

[5] W.S. Lau, "Surface smoothing and roughening effects of high-k dielectric materials deposited by atomic layer deposition and their significance for MIM capacitors used in DRAM technology," CSTIC 2019 (China Semiconductor Technology International Conference, Shanghai, 2019, IEEE), pp. 1-3, 2019.

[6] W.S. Lau, "The application of the Smoluchowski effect to explain the current-voltage characteristics of high-k MIM capacitors," CSTIC 2017 (China Semiconductor Technology International Conference, Shanghai, 2017, IEEE), pp. 1-3, 2017.

[7] S. Knebel, U. Schroeder, D. Zhou, T. Mikolajick and G. Krautheim, IEEE Transactions on Device and Materials Reliability, vol. 14, pp. 154-160. 2014.

[8] W.S. Lau, "Some key modifications of theory required to understand the leakage current mechanisms for MIM capacitors used in DRAM technology," accepted for presentation in CSTIC 2020.

[9] C.H. An, W. Lee, S.H. Kim, C.J. Cho, D.-G. Kim, D.S. Kwon, S.T. Cho, S.H. Cha, J.I. Lim, W. Jeon and C.S. Hwang, Phys. Status Solidi RRL, vol. 13 (2019), article no. 1800454.

[10] W.S. Lau, ECS J. Solid State Sci. Technol., vol. 1, pp. N139-N148, 2012.

FDSOI SIGE MORPHOLOGY OPTIMIZATION ON BOUNDARY OF AA AND STI

Jiaqi Hong, Qiuming Huang, Qiang Yan, Jun Tan, Yongyue Chen, Haifeng Zhou, Jingxun Fang*

Research and Development Dept., Shanghai Huali Integrated Circuit Corporation, Shanghai 201314, China

*Corresponding Author's Email: hongjiaqi@hlmc.cn

ABSTRACT

As short channel effect turns into a major constraint for traditional bulk silicon devices when CMOS technology scales down to 22nm and beyond, fully depleted silicon on insulator (FDSOI) devices becomes more popular in integrated circuit manufacture. The selective SiGe epitaxial is applied for FDSOI PMOSFET as raised source and drain to enhance hole mobility, whereas a defect, SiGe shrinkage, that SiGe non-growth on boundary of STI and AA can kill device yield. In this paper, an optimized Siconi pre-clean process and a novel Si-nucleation layer are proposed to improve the SiGe shrinkage, thereby boosting the device performance.

INTRODUCTION

As the critical dimension of IC devices getting smaller and smaller, the short channel effects (SCE) become a major restriction for conventional bulk devices since CMOS technology scales down to 22nm and beyond. Fortunately, the fully-depleted silicon-on-insulator (FDSOI) devices has demonstrated excellent SCE suppression with its rather thin surface silicon thickness and simple isolation structures. In addition, planar FDSOI show advantages in highly-efficient power management, back gate bias control, high-speed circuits as well as small parasitic capacitance [1-3].

Stress technology of SiGe is widely used since 28nm node to boost PMOS device performances through hole mobility enhancement [4]. In this paper, selective SiGe epitaxy was applied in FDSOI raised source-drain (RSD) to improve saturation current of PMOS and reduce the silicide contact resistance.

In fabrication, the initial surface silicon of FDSOI wafer is about only 10nm, while several processes in front of SiGe epitaxy, including oxidation, etching, cleaning, etc. will cause a certain amount of silicon loss, especially on the boundary of STI and AA, resulting no SiGe growth. This defect, called SiGe shrinkage, may kill device yield. In this paper, an optimized Siconi pre-clean process and a novel Si-nucleation layer are proposed to improve the SiGe shrinkage, thereby boosting the device performance.

EXPERIMENT AND ANALYSIS

Fig. 1 shows a TEM HAADF cross-section view of a PFET S/D along the gate direction with Width=3μm. It can be seen that a defect occurs on the boundary of STI

and AA, where there's no SiGe growth, which we call it "SiGe shrinkage". This defect will be detrimental to the subsequent silicide CT connection process, and in the worst case, CT puncture can happen, as shown in Fig. 2, which will directly kill device yield.

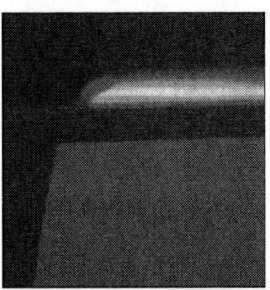

Fig. 1: TEM HAADF cross-section view of a PFET S/D

Fig. 2: TEM cross-section view of CT puncture

Experiments were carried out on 300mm patterned wafers, where the Buried oxide thickness is 20nm, and the initial surface silicon thickness is 10nm. The front-end process flow before RSD is briefly described in Fig. 3(a). The processes in PMOS RSD loop are shown in Fig. 3(b), where SiGe will selectively grow only on exposed silicon surface. The SiGe S/D contains three layers: a SiGe seed layer with Ge% around 20%, a SiGe bulk layer with Ge% around 35%, and followed by a Si cap layer on top.

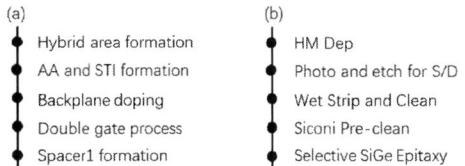

(a)	(b)
Hybrid area formation	HM Dep
AA and STI formation	Photo and etch for S/D
Backplane doping	Wet Strip and Clean
Double gate process	Siconi Pre-clean
Spacer1 formation	Selective SiGe Epitaxy

Fig. 3: (a) front-end process flow before RSD; (b) processes in PMOS RSD loop

Usually, A dry plasma chemical pre-clean technology (Siconi) within the epitaxy equipment is applied first to form a defect-free Si surface for better followed SiGe growth, which will cause some additional Si loss. Fig. 4 shows the remaining surface Si before SiGe Siconi, it can be seen that the remained Si on the boundary of STI and AA is significantly less than elsewhere. Therefore, Si on the boundary of STI and AA after Siconi process may be insufficient for following SiGe growth. On the other hand, as partition check result shown in Fig. 5, it is found that SiGe shrinkage on the boundary of STI and AA appears at the seed layer, so the followed bulk and cap layers are even less able to grow.

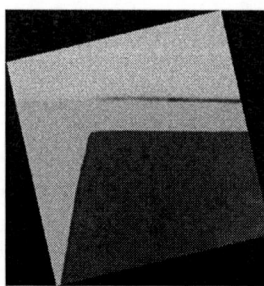

Fig. 4: remaining surface Si before SiGe Siconi

Fig. 5: SiGe shrinkage at seed layer

Based on the above analysis, solutions are proposed to improve the SiGe shrinkage. On one hand, the Siconi pre-clean process was optimized to prevent excessive surface Si loss; on the other hand, a novel step of about 3nm pure silicon growth was added before SiGe seed growth. This pure Si layer, which we called it "Si-nucleation" layer, can increase the surface Si thickness to compensate the previous surface Si loss. In addition, this novel "Si-nucleation" layer was well Si-latticed to form a defect-free Si surface, which is beneficial to the followed SiGe seed layer growth.

Through Design of Experiments (DOE), a well-grown PFET S/D was shown in Fig. 6, and no SiGe shrinkage defect was found on the boundary of STI and AA. The well-grown PMOS RSD increased S/D dimensions to reduce the S/D resistance and increase saturation current, thereby improving device performance.

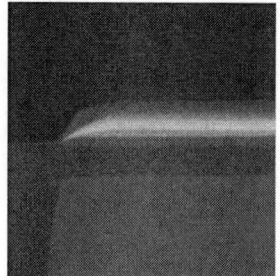

Fig. 6: a well-grown PFET S/D

CONCLUSION

In summary, an optimized Siconi pre-clean process and a novel Si-nucleation layer are proposed to optimize SiGe morphology on boundary of STI and AA. A well-grown PFET S/D was archived by DOE to improve device performance.

REFERENCES

[1] Josef Watts, et. al, RF-pFET in Fully Depleted SOI demonstrates 420 GHz FT, *RFIC*, 2017, pp.84-87

[2] R. Berthelon, et. al, A novel dual isolation scheme for stress and back-bias maximum efficiency in FDSOI Technology, *IEDM*, 2016, pp.468-471

[3] D.H. Triyoso, et. al, Extending HKMG scaling on CMOS with FDSOI: Advantages and integration challenges, *ICICDT*, 2016, pp.1-4

[4] Yiqun Liu, et. al, Conformal SiGe selective epitaxial growth for advanced CMOS technology, *CSTIC*, 2018, pp. 1-3

OPTIMIZATION OF IMPERFECT MORPHOLOGY FOR SELECTIVE EPITAXIAL SIGE GROWTH

Yongyue Chen*, Qiang Yan, Jiaqi Hong, Qiuming Huang, Jun Tan, Haifeng Zhou, Jingxun Fang

Research and Development Dept., Shanghai Huali Integrated Circuit Corporation, Shanghai 201314, China

*Corresponding Author's Email: chenyongyue@hlmc.cn

ABSTRACT

With advanced devices continuously scaling down, embedded SiGe in source and drain regions has been demonstrated to improve p-MOSFET device performance through hole mobility enhancement. However, as germanium and in-situ boron concentrations are required to increase simultaneously, the pattern loading effect of selective epitaxial growth of SiGe:B in recess sigma structure is enhanced after processes. Compared with logical areas, the SRAM areas show the imperfect morphology of SiGe:B growth, which becomes the major killer of the yield. In this paper, we investigate the effect of annealed SiN films as hard-mask layer and films of various thicknesses on the formation of SiGe growth in recess sigma structure. The formation mechanism of the imperfect morphology is analyzed, and the selective epitaxial SiGe growth is optimized to improve electrical properties.

INTRODUCTION

With the development of CMOS technology, transistors are suffering from short channel effect (SCE), mobility loss and large access resistance [1-3]. Embedded epitaxial SiGe materials at the source/drain (S/D) areas have been shown to boost the hole mobility through the local strain of p-MOSFET device silicon technology. SiGe epitaxial defects such as tiny particle, dislocation and imperfect morphology degrade the performance of the device. Imperfect morphology of the SiGe layer could cause detrimental problems in current leakage by degrading drivability due to lack of strain in the channel [4]. A high quality epitaxial SiGe layer is necessary to minimize these defects to improve the device performance, and thus promote the yield.

Micro-loading effect is easy to be found in the epitaxial growth of SiGe, because of different pattern density for SRAM and logical areas [3]. As shown in Figure 1, the SiGe imperfect morphology emerges in the SRAM area of the wafer with smaller width and incomplete edge formation. The total raised height is lower than the normal level. The SiGe bulk layer cannot fill the recess sigma trench with the inversed shallow arc profile, which affect the operation of device performance. Increasing the SiGe deposition time can fill the SRAM area completely, but will cause PMOS area overfilled. Thus, choose a suitable method to suppress this

phenomena is particularly important.

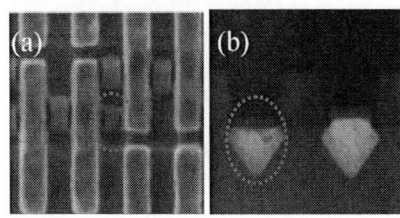

Figure 1: SiGe imperfect morphology in Σ-shape trench: (a) top view; (b) cross section view

In this paper, by studying the process of epitaxial SiGe layer at source and drain region, we found the hard-mask thickness and annealing process played a significant role on SiGe epitaxial morphology. The mechanism of the imperfect morphology was analyzed, contribute to the device performance and yield.

EXPERIMENT

The experiments were carried out on 12 inches patterned wafers following the advanced planar CMOS integration scheme.

The process flow in the SiGe loop is briefly described in Figure 2. After dummy gate patterning, the spacer fabricated with oxide and nitride by ALD process. Pre-clean and HCD SiN deposition followed as hard-mask layer. Then the NMOS area was covered up by lithography, and the PMOS source/drain area was etched to Σ-shape trench by reactive ion etch (RIE) and wet process. After photoresist strip and wet clean, the in-situ SiCoNi process was utilized to remove oxide before selective SiGe epitaxy process.

SiGe epitaxial films were selectively embedded on recess PMOS source and drain areas. The SiGe stressor contains a SiGe seed layer with low Ge concentration and a SiGe bulk layer with higher Ge concentration serving as the main stress layer. A set of Si cap layer on the SiGe forming ohmic contact with following NiSi process to reduce resistance.

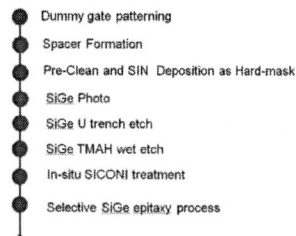

- Dummy gate patterning
- Spacer Formation
- Pre-Clean and SiN Deposition as Hard-mask
- SiGe Photo
- SiGe U trench etch
- SiGe TMAH wet etch
- In-situ SICONI treatment
- Selective SiGe epitaxy process

Figure 2: Brief description of SiGe loop process flow in advanced CMOS integration scheme.

RESULT AND DISCUSSIONS

As shown in Figure 3 (a), SiGe imperfect morphology counts occurred significant decrease from thousands of counts to several dozens with increasing hard-mask thickness from reference thickness to ref.+30 A. This indicates that thick hard-mask layer is a practical method to suppress defect formation.

The other approach is to optimize hard-mask quality by using post annealing process as shown in Figure 3(b). Post annealing of hard-mask layer further reduce the number of defect up from hundreds to a few counts.

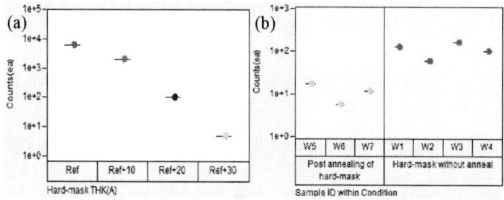

Figure 3: Effect of (a) different hard-masker thickness and (b) post annealing of hard-masker on the formation of SiGe imperfect morphology.

Figure 4 shows the schematic diagram of process-of-reference (POR) hard-mask, post annealing and thicker hard-mask layer for the formation of SiGe Σ-shape trench. As the SRAM structure is more density than PMOS area, SiGe growth rate at SRAM area is much more sensitive to surface open ratio trench than PMOS area at the same temperature. With thicker hard-mask deposited, the sidewall spacer is thicker, leading to the smaller open CD and shallower trench depth, which result in the decrease of imperfect morphology. For another approach of post annealing hard-mask layer, make membranous more dense to resist the loss of the RIE and TMAH process, also lead to reduce the open CD and shallower trench, which is advantageous to the trench filling.

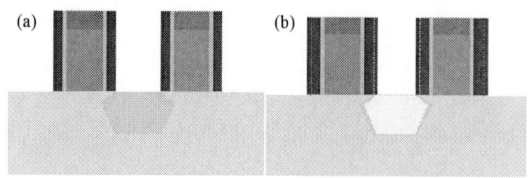

Figure 4: Schematic diagram of (a) POR hard-mask and (b) post annealing or thicker hard-mask layer for the formation of SiGe Σ-shape trench

SiGe imperfect morphology provides the extra path of current leakage from source/ drain areas to substrate. The Isoff performance of device could be strongly improved by suppressing SiGe imperfect morphology. As shown in figure 5, compared to the condition of POR, the counts of outliers above the acceptable line of I_{soff} are only 50% with thicker hard-mask and 25% with additional post annealing process, respectively.

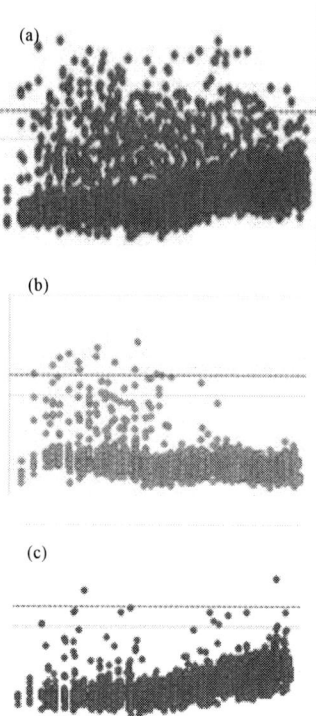

Figure 5: I_{soff} performance of device versus imperfect morphology of (a) POR hard-mask; (b) thicker hard-mask layer; (c) post annealing of hard-mask layer

SUMMARY

The effect of annealed SiN films as hard-mask layer and films of various thicknesses on the formation of SiGe growth in Σ-shape trench was analyzed. The formation mechanism on the imperfect morphology of selective epitaxial SiGe growth is studied to improve electrical properties. A high performance p-MOSFET with a SiGe source/drain was successfully achieved by suitable hard-mask thickness and post annealing process, and results in excellent I_{soff} performance.

ACKNOWLEDGEMENTS

The authors would like to acknowledge HLMC failure analysis department for the supports of TEM microscopy.

REFERENCES

[1] Huojin Tu, Yonggen He，Youfeng He，Jialei Liu，Lan Jin，Guohui Cai，Yu Liu， Yujian Huang, *CSTIC 2016*，March 13-14, 2016, pp. 1-3

[2] Y. S. Kim, Y. Shimamune, M. Fukuda*, A. Katakami, A. Hatada, K. Kawamura*, H. Ohta, T. Sakuma, Y.Hayami, H. Morioka*, J. Ogura*, T. Minami*, N. Tamura, T. Mori*, M. Kojima*, K. Sukegawa, K.Hashimoto*, M.Miyajima*, S. Satoh, and T. Sugii, in *IEDM Tech. Dig.*, Dec. 2006, pp. 871–874.

[3] Jianqin Gao, Jun Tan, Haifeng Zhou, Jingxun Fang, and Albert Pang, *CSTIC2015*, Shanghai, March 15-16, 2015

[4] Yiqun Liu*, Lan Jin, Kunshan Song, Qiong Wu, Youfeng He, Yonggen He, *CSTIC2018*, Shanghai, March 11-12, 2018, pp. 1-3

MECHANICALLY STABLE ULTRA-LOW-K DIELECTRIC AND AIR-GAP TECHNOLOGY

*Clarissa Prawoto, Ying Xiao and Mansun Chan**
Department of Electronic and Computer Engineering,
the Hong Kong University of Science and Technology, Hong Kong SAR, China
*Corresponding Author's Email: mchan@ust.hk

ABSTRACT

This paper described two approaches using structured voids in the dielectric among the interconnect metals to achieve low interlayer and intralayer capacitance while maintaining sufficient mechanical strength to withstand the CMP process. The first approach is to use vertically aligned voids and experimental results show that it can be used to achieve very high porosity with much stronger mechanical strength than conventional structures. To further reduce the intralayer dielectric constant, air-gap technology with large void-to-solid ratio has been proposed. The fabrication method and measurement results are presented.

INTRODUCTION

With the continuous technology scaling process, the performance bottleneck of integrated circuits has shifted from the active devices to the interconnect [1]. The main objective in interconnect technology research and development is to reduce the metal wire resistance and inter/intralayer dielectric capacitance. The most popular way to reduce the interconnect capacitance is by using low dielectric constant material with porous structure [2][3]. However, the presence of voids reduces the mechanical strength of the dielectric for the subsequent Chemical Mechanical Polishing (CMP) process [4][5]. In our previous work, we have demonstrated that by structuring the geometry of the voids can achieve much higher porosity in both the interlayer and intralayer dielectrics without compromising the mechanical strength [6][7]. Furthermore, the use of air-gap in between metal wires can be used to achieved even more reduction in intralayer capacitance. In this presentation, an overview of the status in using structured voids to achieve low dielectric constant is given among the tradeoffs between porosity and mechanical strength. The fabrication process, geometrical design, material properties, electrical measurement and mechanical deformation under external forces of the dielectric material will be presented.

ALIGNED CYLINRICAL VOIDS

The general schematic structures of porous dielectric materials are given in Figure 1. Without loss of generality, silicon dioxide is assumed to be the solid material in this work. Conventional approach to form voids using porogens has a structure similar to Figure 1(a) which consists of spherical voids randomly placed inside the dielectric. Such structure can easily lead to cracks when the voids are connected. On the other hand, the voids can also be structured as shown in Figure 1(b) with columnar pores. Mechanically, such structure is more stable and higher porosity can be achieved with the same mechanical strength. A possible fabrication process using vertically synthesized carbon nanotube (CNT) to as a template to form the dielectric with columnar voids is summarized in Figure 2. More process details with experimental demonstration for CNT synthesis under a CMOS compatible environment and can be found in [6][7].

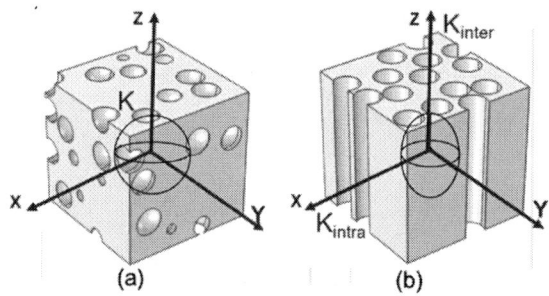

Figure 1: Schematic of (a) a generic porous dielectric structure with existing technology formed by using porogens; and (b) a porous dielectric material with vertically aligned voids

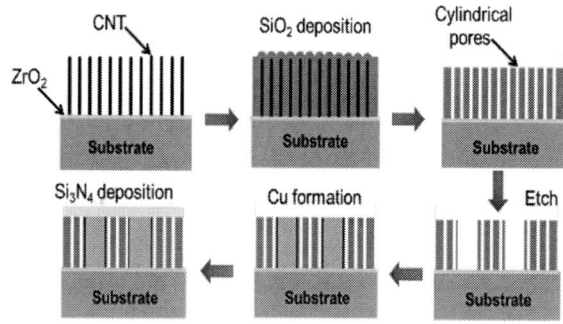

Figure 2: A possible fabrication process using a vertically synthesized carbon nanotube as a template to form the vertically aligned voids.

The dielectric with vertically aligned voids can be used as both interlayer and intralayer dielectrics. When it is used as an interlayer dielectric, capacitance between two different layers is formed by a parallel combination of voids and the dielectric material. When the dielectric is placed between two metal lines in the same layer as an intralayer dielectric, the capacitance is formed by a more interleave structure of voids and the dielectric material. Compared with the interlayer case, we consider it closer to series combination between the voids and the dielectric material. As a result, the equivalent dielectric constants for the interlayer capacitance and intralayer capacitance are different and the material becomes anisotropic. From our experimental results, the interlayer and intralayer capacitance are measured to be 1.97 and 1.75 respectively. And the distribution of the dielectric constants is shown in Figure 3. With the process of technology scaling, intralayer capacitance is becoming the dominant component for delay and the lower intralayer capacitance in the dielectric with vertically aligned voids is thus more advantageous from this perspective.

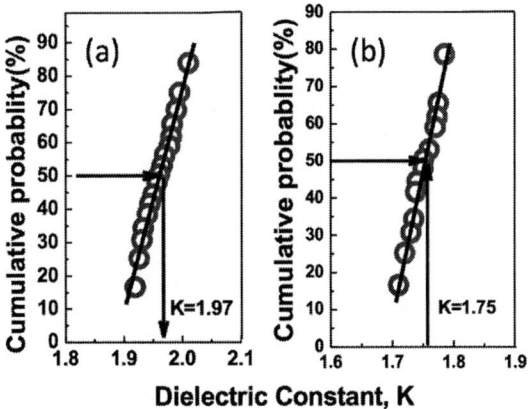

Figure 3: A possible fabrication process using a vertically synthesized carbon nanotube as a template to form the vertically aligned voids.

The mechanical strength of the dielectric is measured by the nano-indentation method. A stylus is used to apply a force to create an indentation. The depth of the indentation is measured as a function of the applied force during loading and release. The force versus displacement data can be directly translated into the elastic modules as shown in Figure 4. And the measured elastic modules are found to be around 15.8GPa. This value is higher than the required mechanical strength for the CMP process [8]. A comparison with other reported works on low-k dielectric is shown in Figure 5 [8-12], indicating the structured void approach can achieve the lowest dielectric constant without compromising mechanical strength.

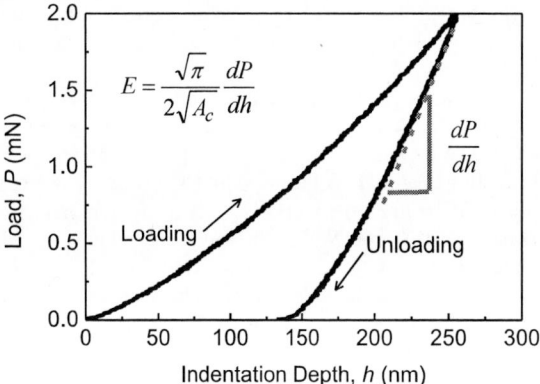

Figure 4: Elastic modulus measurement by nano-indentation experiment. The projected indentation area, A_c, was about 1.25 μm^2.

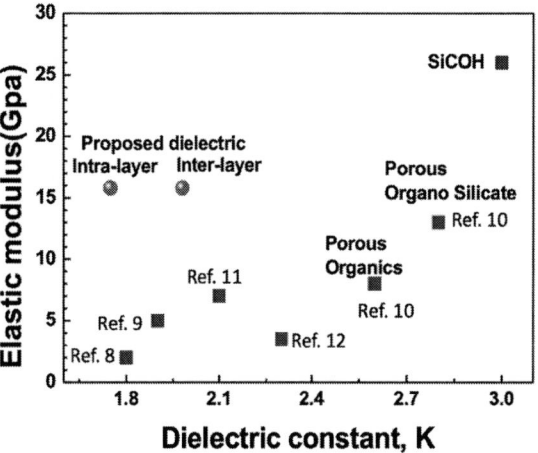

Figure 5: Comparison of reported k-value and elastic modules of reported low-k dielectrics.

2-D MATERIAL ASSISTED AIR-GAP

It has been pointed out that intralayer capacitance has become the dominant components contributing to the circuit delay. While the porous dielectric with vertically aligned voids can achieve a low effective dielectric constant, it is still limited by the low achievable porosity. Ideally, the space between two metal wire can just be filled with air by forming an airgap. That will result in the ultimate dielectric constant close to 1.0 [3]. A simple way to form intralayer airgaps is by using non-conformal dielectric deposition that eventually seal the top opening [13]. The resulting air-gaps typically have the shape of a teardrop and is only applicable for air-gaps with small dimensions. Another common method to form air-gap is to use some sacrificial materials as a template and the material is allowed to escape through a porous capping layer [14]. While the process is more flexible, the mechanical strength is compromised due to the porous sealing material [15].

To improve the mechanical strength of the airgap structure as well as allowing the formation of larger airgaps, a new approach to cover the airgap with 2-D insulating material in the form of 2-D sheets has been proposed in our earlier work [16]. By using transferred 2-D hexagonal Boron Nitride (h-BN) as a sealing material, large airgaps with high mechanical strength can be fabricated. The detail fabrication process can be found in [16] and the key steps are summarized in Figure 6 after the patterning of the metal lines and CMP process.

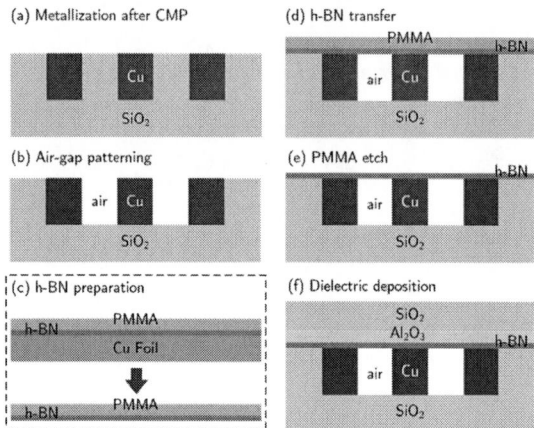

Figure 6: Process flow of the proposed air-gap technology. (a) Initial condition in a typical metallization process. (b) Patterning of air-cavities by lithography and dielectric etch. (c) h-BN grown using CVD on copper foil, PMMA coating and copper etch. (d) Transfer of h-BN and PMMA stack to interconnect structures. (e) PMMA etch by O2 plasma. (f) Additional capping layer and dielectric layer deposition by ALD and PECVD.

While airgap can theoretically give an effective k-value equal to 1.0, the actual reduction is actually not as dramatic due to a number of factors. First of all, the metal lines cannot be placed directly next to the airgap due to reliability considerations. A liner is needed on the two side of the metal for protection purpose, and the liners reduced the size of the airgap. The effectiveness of the airgap is thus dependent on the ratio of width of the airgap to the distance between the two metal lines as shown in Figure 7(a). That means the approach is less effective when the separation between the two metal lines is small. Furthermore, the fringing capacitances as shown in Figure 7 also reduce the effectiveness of the airgap by adding a parallel capacitance component in between the two metal lines. The resulting effective k-value is still far from one. The simulated effective dielectric constant and effective

capacitance reduction compared with the case without using airgap is shown in Figure 7(b). It is observed that the effectiveness of the airgap technology increases with larger metal line spacing. Therefore, the formation of large airgap in the upper interconnect layer is an attractive approach to reduce the intralayer dielectric constant.

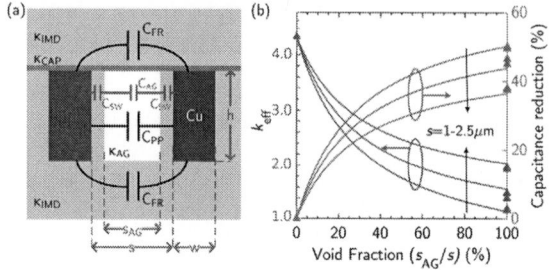

Figure 7: (a) components contribution go the intralayer capacitance (b) k_{eff} and the amount of capacitance reduction as a function of airgap dimension (s_{AG}) and spacing between the metal lines (s).

To evaluate the mechanical strength of the structure after introducing airgap, the nanoindentation method is again used. The setup is shown in Figure 8(a). The young modules actually depend on the size of the airgaps, metal line geometry and the thickness of the capping layer before the upper layer of interlayer dielectric is deposited. The results presented in figure is for airgap with a metal spacing of 2μm and the supporting material in the range of 1-2.5μm. Experimental data shows that the dimension of the supporting material is not as important as the spacing of the metal lines. To further strengthen the airgap structure, a thin layer of aluminium oxide can be deposited on top of the h-BN layer. With all the measures in place, a large young module of 30-65GPa is obtained. It provides sufficient margin for the subsequent CMP process after the airgap formation.

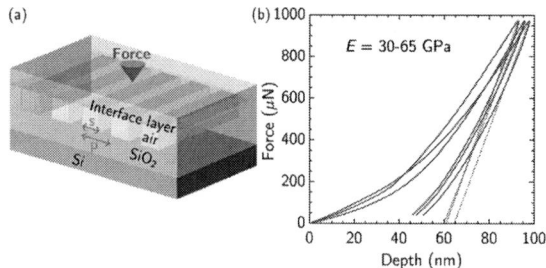

Figure 8: (a) Schematic of the nanoindentation measurement setup and (b) the measured force-displacement curves and the extracted Young's modulus.

CONCLUSION

Through experimental demonstration, it has been shown that using structured voids can achieve low effective dielectric constant with high porosity without significantly compromising the mechanical strength. The porous dielectric with vertically aligned voids can be used in both interlayer and intralayer dielectrics to achieve an effective dielectric constant below 2.0. And the structure is found to be more effective in reducing the intralayer dielectric constant. To achieve ultimate reduction of intralayer dielectric constant, airgap technology can be used. And with the assistant of 2-D material such as hexagonal Boron Nitride, the process to form airgap can be simplified. Ultimately, airgap can be used for the intralayer dielectric constant reduction and porous dielectric with vertically aligned voids can be used as the interlayer dielectric.

While the approach looks promising, the use of structured voids to reduce the *k*-value of the backend dielectrics is still in the stage of proof-of-concept. A lot of reliability related studies are necessary to transfer the technology to production.

ACKNOWLEDGEMENTS

This work is supported by the Guangdong Basic and Applied Basic Research Foundation under grant number GDSTI19EG20. We would also like to acknowledge the support of the Nanosystem Fabrication Facility at the HKUST in supporting all the experimental work.

REFERENCES

[1] M. T. Bohr, Proceedings of International Electron Devices Meeting, Washington DC, USA, Dec. 1995. doi: 10.1109/IEDM.1995 .499187.

[2] D. Sylvester and C. M. Wu, Proceedings of IEEE," vol. 89, May. 2001. doi: 10.1109/5.929648.

[3] ITRS 2.0, 2015. [Online]: http://www.itrs2.net.

[4] K. Maex, M. R. Baklanov, D. Shamiryan, F. lacopi, S. H. Brongersma, and Z. S. Yanovitskaya, Journal of Applied Physics, vol. 93, no. 11, pp. 8793–8841, Jun. 2003. doi: 10.1063/1.1567460.

[5] D. J. Michalak, J. M. Blackwell, J. M. Torres, A. Sengupta, L. E. Kreno, J. S. Clarke, and D. Pantuso, Journal of. Materials Research, vol. 30, no. 22, pp. 3363–3385, Nov. 2015. doi: 10.1557/jmr.2015.313.

[6] S. Raju, S. Li, C.J. Zhou and M. Chan, IEEE Electron Device Letters, vol.37, pp.1493-1496, Nov. 2016. doi: 10.1109/LED.2016.2609099.

[7] Y. Xiao, Z. Ma, C. Parawoto, C. Zhou and M. Chan, "Ultra-low-k Dielectric with Structured Pores for Interconnect Delay Reduction", unpublished

[8] S. Kondo, B. U. Yoon, S. Tokitoh, K. Misawa, S. Sone, H. J. Shin, N. Ohashi, and N. Kobayashi, Proceedings of International Electron Devices Meeting, pp. 6.4.1–6.4.4, Dec. 2003, doi: 10.1109/IEDM.2003.1269187.

[9] W. Volksen, R. D. Miller, and G. Dubois, Chemical Review, vol. 110, no. 01, pp. 56-110, 2010. doi: 10.1021/cr9002819.

[10] S. Chikaki, K. Kinoshital, T. Nakayama, K. Kohmura, H. Tanaka,M. Hirakawa, E. Sodal, Y. Seino, N. Hata, T. Kikkawa, and S. Saitol, Proceedings of International Electron Devices Meeting, Dec. 2007, pp. 969–972, doi: 10.1109/IEDM.2007.4419115.

[11] N. Oda, S. Chikaki, T. Kubota, S. Nakao, K. Tomioka, E. Soda,N. Nakamura, A. Gawase, J. Nogawa, Y. Kawashima, R. Hayashi, and S. Saito, IEEE Trans. Electron Devices, vol. 57, no. 11, pp. 2821–2830, Nov. 2010, doi: 10.1109/ TED.2010.2066568.

[12] S. Eslava, L. Zhang, S. Esconjauregui, J. Yang, K. Vanstreels, M. R. Baklanov, and E. Saiz, Chemical Materials, vol. 25, no. 1, pp. 27–33, 2013, doi: 10.1021/cm302610z.

[13] S. Natarajan et al., 2014 IEEE International Electron Devices Meeting, Dec 2014, pp. 3.7.1–3.7.3

[14] P. A. Kohl, D. M. Bhusari, M. Wedlake, C. Case, F. P. Klemens, J. Miner, B.-C. Lee, R. J. Gutmann, and R. Shick, IEEE Electron Device Lett., vol. 21, no. 12, pp. 557–559, Dec 2000

[15] X. Zhang, S.-K. Ryu, R. Huang, P. S. Ho. J. Liu, D. Toma, Proc. Future Fab International, pp. 81–87, 2008

[16] C. Prawoto, Z. Ma, Y. Xiao, S. Raju, C. Zhou, and M. Chan, IEEE Electron Device Letters, Vol. 40, No. 11, Nov. 2019, pp. 1876-1879

SOME KEY MODIFICATIONS OF THEORY REQUIRED TO UNDERSTAND THE LEAKAGE CURRENT MECHANISMS FOR MIM CAPACITORS USED IN DRAM TECHNOLOGY

W.S. Lau

Zhejiang University, Department of Information Science and Electronic Engineering,
No. 38 Zheda Road, Hangzhou 310027, People's Republic of China
Email: liuweicheng@zju.edu.cn

ABSTRACT

Two important key modifications of theory of leakage current mechanisms for MIM capacitors will be proposed. The first modification is the proposal of a new unified theory for the image force dielectric constant used in the Schottky emission and Poole-Frenkel equations. The second modification is that when the leakage current mechanism is Schottky emission modified by tunneling, a different approach has to be used to evaluate the Schottky barrier height. They are important, for example, for 4.6 nm ZAZ MIM capacitors used in DRAM technology.

INTRODUCTION

MIM capacitor structures based on high-k dielectric materials deposited by atomic layer deposition (ALD) are used, for example, in dynamic random access memory (DRAM) ICs. However, the leakage current vs. voltage (I-V) curve is still not well understood. In 2014, Knebel et al. [1] published their study on the leakage current mechanism of TiN/ZrO$_2$/Al$_2$O$_3$/ZrO$_2$/TiN (TiN/ZAZ/TiN) MIM capacitors. They proposed Poole-Frenkel emission as the leakage current mechanism for both polarities of bias voltage. If this theory is true, the I-V characteristics of TiN/ZAZ/TiN MIM capacitors should be symmetrical. In reality, the I-V characteristics are obviously asymmetrical. Previously, the author [2] applied his theory of surface smoothing and roughening of ALD ZrO$_2$ to explain the asymmetry of the I-V characteristics of TiN/ZAZ/TiN MIM capacitors. In this paper, the author will try to explain the deeper reasons why Knebel et al. could not identify the correct leakage current mechanism.

THEORY

Scientists have been trying to use the theories of Schottky emission (SE) and Poole-Frenkel (P-F) emission theories to explain the leakage current of MIM capacitors. In the equations for SE and P-F, there is a parameter known as the "image force dielectric constant" ε_{if}. The SE image force dielectric constant ε_{if_SE} is usually considered to be equal to the PF image force dielectric constant ε_{if_PF}.

The current dominant theory (Theory X1) regarding the image force dielectric constant is that $\varepsilon_{if} = n^2$, where n is the refractive index of the insulator measured in the visible light range. The author would like to point out that Theory X1 is an imperfect theory. Knebel et al. [1] had excessive faith in this Theory X1 such that they believed that the leakage current mechanism in TiN/ZAZ/TiN MIM capacitors is P-F for both polarities of bias voltage. The author believes that leakage current mechanism is more likely SE. In fact, Wu et al. [3] had a similar opinion. The author would like to point out that there exists a theory (Theory X0) older than Theory X1; for example, in 1953, Krömer [4] pointed out that that $\varepsilon_{if} = 1$ for germanium. In 1964, Sze et al. [5] pointed out that that $\varepsilon_{if} = n^2$ for silicon; after 1964, Theory X1 became the dominant theory whereas Theory X0 became forgotten. In this paper, the author would like to unify both Theory X0 and Theory X1 into a more generalized theory (Theory L).

In this new Theory L, Theory X0 and Theory X1 are not mutually exclusive; instead, they are 2 parts of the same unified theory, which includes (a) $\varepsilon_{if} = n^2$ approximately, (b) $n^2 > \varepsilon_{if} > 1$ and (c) $\varepsilon_{if} = 1$ approximately. For a high-k dielectric, the refractive index is usually about 2; Theory L points out that ε_{if} can have a value from slightly bigger than 4 to slightly smaller than 1. Fig. 1 shows the dielectric constant as a function of frequency. At the upper frequency range, there is a frequency range A with $\varepsilon = n^2$ and there is another frequency range B with $\varepsilon_{if} = 1$. Theory L just means the image force dielectric constant can correspond to the frequency range A + B. Internal photoemission of Al on HfO$_2$ shows that the image force dielectric constant is about 0.6. This is not acceptable for the old Theory X1 but acceptable for the new Theory L.

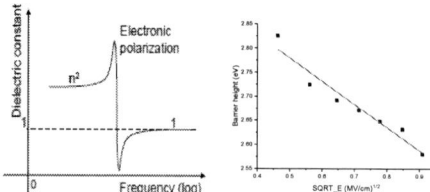

Figure 1: (Left) The dielectric constant as a function of frequency at the higher frequency range showing that the image force dielectric constant can be smaller than 1. (Right) The barrier height measured by internal

photoemission as a function of the square root of the electric field for Al on HfO$_2$. The image force dielectric constant calculated from the slope is about 0.6.

After adopting the new Theory L, the method to distinguish SE and P-F has to be modified. The author noticed that this can be done for MIM capacitors by examining the 2 curves of logI vs $V^{1/2}$ plotted together for both polarities of bias voltage in the same figure instead of just examining 1 curve of logI vs $V^{1/2}$ for only one polarity of bias voltage. The only key assumption is that $\varepsilon_{if\_SE\_top} = \varepsilon_{if\_SE\_bottom} = \varepsilon_{if\_PF}$. Theory X1 will not be used. For example, two different but parallel lines imply SE for both polarities, as shown in Fig. 2(c). According to the author's experience, P-F for both polarities, as shown in Fig. 2(a) is a very rare case for MIM capacitors with an ultrathin insulator; however, Mechanism B I-V symmetry [6] may be observed for leaky samples. The majority of experimental observations are similar to Fig. 2(b) or Fig. 2(c).

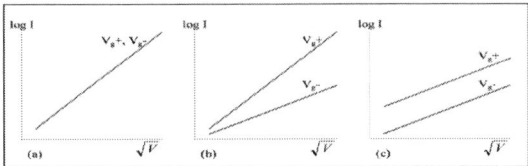

Figure 2: 3 diagrams of 2 curves of log I vs $V^{1/2}$ plotted together for both polarities of the bias voltage V. (a) P-F for both polarities, (b) P-F for positive bias and SE for negative bias and (c) SE for both polarities.

For MIM capacitors used in DRAM technology, continuous scaling leads to a thinner and thinner high-k dielectric. P-F effect is basically a bulk effect. For an extremely thin high-k dielectric, SE or tunneling will be the more likely the leakage current mechanism compared to the P-F effect. SE is usually strongly thermally activated with a Schottky barrier height energy E_{SB}. Tunneling at most only has a weak temperature dependence. E_{SB} is usually smaller for an insulator with a smaller bandgap Eg; conversely, E_{SB} is usually larger for an insulator with a larger bandgap Eg. In general, SE is observed for insulators with small Eg (e.g. Ta$_2$O$_5$); E_{SB} can be measured by evaluating the thermal activation energy during I-V-T measurement. Tunneling is observed for insulators with large Eg (e.g. SiO$_2$); E_{SB} cannot be measured by evaluating the thermal activation energy during I-V-T measurement. Actually, Schottky emission and tunneling occur simultaneously; however, the percentage of tunneling current tends to become bigger when E_{SB} becomes bigger. In order to reduce leakage current, Eg has to be larger. However, insulators with large Eg tends to have low dielectric constant. Thus it is necessary to have a high-k dielectric whose Eg is not too small or too large. For example, ZrO$_2$ is such a high-k dielectric. Then a question will arise: what will happen for an insulator with Eg not too small or too large. In this paper, the author would like to propose a new leakage current mechanism "Schottky emission modified by tunneling (SE-tunneling)". As shown in Fig. 3, the thermal activation energy E_{IVT} measured by I-V-T is plotted against E_{SB}. In the SE range, $E_{IVT} = E_{SB}$. When E_{SB} increases, E_{IVT} will also increase. In the tunneling range, $E_{IVT} = 0$ approximately. In this paper, the author would like to suggest a new SE-tunneling leakage current mechanism, which is a hybrid of SE and tunneling. In the SE-tunneling range, when E_{SB} increases, E_{IVT} can decrease. For this new SE-tunneling mechanism, an experimentalist can only get a relative measure of E_{SB} by directly examining the leakage current; a smaller leakage current implies higher E_{SB} and vice versa. This crude method actually also applies to the SE range and also to the tunneling range. Let the Schottky barrier height measured by the crude I-V method be E_{IV}. When $E_{IV\_1} > E_{IV\_2}$ and $E_{IVT\_1} > E_{IVT\_2}$, this is the normal case of Schottky emission. When $E_{IV\_1} > E_{IV\_2}$ but $E_{IVT\_1} < E_{IVT\_2}$, this is an indication of the presence of "modification by tunneling". Case 1 and Case 2 can be 2 measurements on the same sample for 2 opposite polarities of applied voltage.

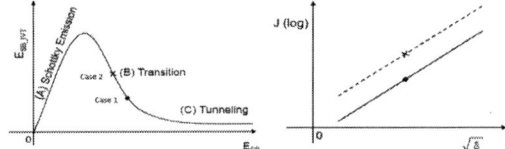

Figure 3: (Left) E_{SB_IVT} vs. E_{SB} according to the author's hypothesis. In Region (A), $E_{SB\_1} > E_{SB\_2}$ and $E_{SB\_IVT\_1} > E_{SB\_IVT\_2}$ for the normal case of Schottky emission. In Region (B), $E_{SB\_1} > E_{SB\_2}$ but $E_{SB\_IVT\_1} < E_{SB\_IVT\_2}$ for Schottky emission modified by tunneling. (Right) A simple plot of J vs $E^{1/2}$ can give a more reasonable idea regarding the Schottky barrier height.

EXPERIMENTAL SUPPORT

The image force dielectric constant can be measured by "internal photoemission (IPE)". The IPE barrier height can be measured as a function of the applied electric field. The IPE barrier height plotted as a function of the square root of the applied electric field yields a slope which can be used to calculate the image force dielectric constant. The author has analyzed a lot of existing experimental data published in the literature on Si, GaAs, SiO$_2$, Al$_2$O$_3$, ZrO$_2$, HfO$_2$, SrTiO$_3$, GeO$_2$, etc. There exist some experimental data which show up some sort of "screening effect"; the IPE barrier height is insensitive to the change of the applied electric field. If the experiment data showing up "screening effect" are ignored, 3 cases can be observed: (a) $\varepsilon_{if} = n^2$ approximately, (b) $n^2 > \varepsilon_{if} > 1$ and (c) $\varepsilon_{if} = 1$ approximately. Thus ε_{if} can have a range from less than 1 to slightly bigger than n^2. Fig. 1 (right) provides experimental data for the case of $\varepsilon_{if} < 1$.

In 1969, Pennebaker published experimental results on $Cu/SrTiO_3/Au$ MIM capacitor [7]. The leakage current was obviously different for the two polarities of bias voltage applied to the top Cu electrode. Pennebaker believed that the leakage current mechanism was P-F instead of SE because PF will yield a refractive index n=2.8 (n^2=7.84) whereas SE will yield a refractive index n=0.8 (n^2=0.64). Many other scientists believed that the leakage current in $SrTiO_3$ MIM capacitors is due to SE but could not point out what was wrong in Pennebaker's analysis. The author made his analysis of Pennebaker's experimental results and found that $\log I$ vs $V^{1/2}$ curves for both polarities looked like Fig. 2(c) and so the leakage current of the $Cu/SrTiO_3/Au$ MIM capacitor was due to SE for both polarities. The asymmetry in leakage current was obvious due to the use of 2 different metals for the top and bottom electrodes. Pennebaker's problem came from his strong faith in Theory X1; in this particular case, Theory X0 is the more correct theory. Furthermore, the internal photoemission experimental data from Hikita et al. [8] showed that the image force dielectric constant of $SrTiO_3$ is about 1; this also supported Theory X0. Similarly, in 2019, An et al. published experimental results on 4.6 nm TiN/ZAZ/TiN MIM capacitors [9]. The leakage current was obviously different for the two polarities of bias voltage applied to the top TiN electrode. An et al. believed that the leakage current mechanism was P-F for both polarities instead of SE because SE will yield a refractive index n=0.75 (n^2=0.57) and a number of 0.57 is unphysical. However, according to the author's new Theory L, 0.57 is reasonable. The I-V characteristics for both polarities appeared to be similar to Fig. 2(c) and the leakage current was due to SE in the higher field region for both polarities at a temperature close to RT. The problem of An et al. came from their strong faith in Theory X1; in this particular case, Theory X0 is the more correct theory. If the top TiN was replaced by Ru, An et al. found that the leakage current was very greatly reduced; this observation was more readily explained by SE than by P-F [9], as shown in Fig. 4. In another paper also submitted to CSTIC 2020 [10], the significance of the extended range of ε_{if} will be much more obvious; in addition, the author will point out that the leakage current mechanism in ZAZ MIM capacitors used in DRAM technology can be the SE-tunneling mechanism proposed by the author.

CONCLUSION

In this paper, the author would like to propose 2 key modifications of theory in order to understand the leakage current mechanism in ZAZ MIM capacitors used in DRAM technology. Besides a new Theory L for the image force dielectric constant, the author proposes a new leakage current mechanism which is quite similar to the traditional Schottky emission mechanism in terms of dependence on voltage and temperature; however, the Schottky barrier height cannot be extracted by the traditional method used for the traditional Schottky emission mechanism. The above theory is a further development of the author's extended unified Schottky-Poole-Frenkel theory [11].

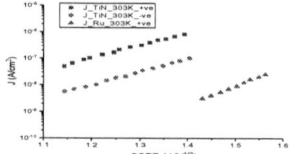

Figure 4: A semi-log plot of J vs $V^{1/2}$ plotted for 4.6 nm ZAZMIM capacitors with TiN (TE+ve), TiN (TE-ve) and Ru (TE-ve) based on data extracted from An et al. 2019 [9]. Please note the similarity to Fig. 2(c), indicating SE for all 3 curves.

REFERENCES

[1] S. Knebel, U. Schroeder, D. Zhou, T. Mikolajick and G. Krautheim, IEEE Transactions on Device and Materials Reliability, vol. 14, pp. 154-160. 2014.

[2] W.S. Lau, "Surface smoothing and roughening effects of high-k dielectric materials deposited by atomic layer deposition and their significance for MIM capacitors used in DRAM technology," CSTIC 2019 (China Semiconductor Technology International Conference, Shanghai, 2019, IEEE), pp. 1-3, 2019.

[3] Y.-H. Wu, C.-K. Kao, B.-Y. Chen, Y.-S. Lin, M.-Y. Li and H.-C. Wu, Appl. Phys. Lett., vol. 93 (2008), article no. 033511.

[4] H. Krömer, Z. Physik, vol. 134 (1953), pp. 435-449. (In German)

[5] S.M. Sze, C.R. Crowell and D. Kahng, J. Appl. Phys., vol. 35, no. 8 (Aug. 1964), pp. 2534-2536.

[6] W.S. Lau, ECS Trans., vol. 45 (3) (2012), pp. 151-158.

[7] W.B. Pennebaker, IBM J. Res. Develop., vol. 13, pp. 686-695, 1969.

[8] Y. Hikita, M. Kawamura, C. Bell and H.Y. Hwang, Appl. Phys. Lett., vol. 98 (2011), article no. 192103.

[9] C.H. An, W. Lee, S.H. Kim, C.J. Cho, D.-G. Kim, D.S. Kwon, S.T. Cho, S.H. Cha, J.I. Lim, W. Jeon and C.S. Hwang, Phys. Status Solidi RRL, vol. 13 (2019), article no. 1800454.

[10] W.S. Lau, "Surface smoothing and roughening effects of high-k dielectric materials deposited by atomic layer deposition and their significance for MIM capacitors used in DRAM technology Part II," accepted for presentation in CSTIC 2020.

[11] W.S. Lau, ECS J. Solid State Sci. Technol., vol. 1, pp. N139-N148, 2012.

THIN FILM PROCESSES: ABATEMENT OF WASTE GASES FROM PLASMA ASSISTED MATERIAL PROCESSES

Christopher P. Jones

Edwards Ltd, Clevedon, North Somerset, BS21 6TH, United Kingdom

chris.jones@edwardsvacuum.com

ABSTRACT

This paper describes the challenges associated with the management of particulate matter (PM) that are generated during the abatement of waste gases from plasma-assisted thin-film semiconductor processes. We describe operational challenges associated with the PM management within the fab and mitigating strategies. From an environmental protection viewpoint, we describe the health risks associated with the inhalation of fine PM and the standards that are currently adopted both by central and local governments in China.

INTRODUCTION

Plasma enhanced chemical vapor deposition (PECVD) is one of several methods used for the preparation of semiconductor thin films. Silicon oxynitride (SiN_xO_y) films can be deposited using PECVD at low temperatures (<400C). The process utilizes gases such as ammonia (NH_3), silane (SiH_4), nitrogen (N_2), and nitrous oxide (N_2O). The flow and composition of the gas mixture varies during deposition. Reactions occur in the chamber and exhaust gas flows and compositions will vary from mixtures that may be silane and ammonia-rich to those that will be nitrous oxide rich. Silane is a pyrophoric gas, ammonia is both flammable and ecotoxic, nitrous oxide is both a greenhouse and oxidizing gas. Point of use (POU) abatement is desirable to manage the risks associated with such gases in a fab. A layout (Figure 1) includes the tool, the fore-line connection to the vacuum pump that then exhausts to the abatement, that in turn vents to a facility duct.

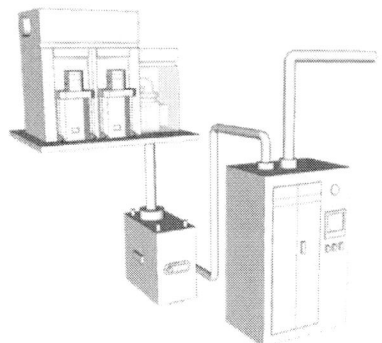

Figure 1: Schematic of a tool, vacuum pump, and abatement.

The chamber is contaminated with nitride and oxynitride solids during deposition. It is cleaned with fluorine gas that is often generated by passing the greenhouse gas nitrogen trifluoride through a plasma. The waste gases from the chamber clean include unused nitrogen trifluoride, silicon tetrafluoride, and fluorine – all are toxic and corrosive.

ABATEMENT OF GASES.

Overview

The process gases are pumped to the abatement, as illustrated in Figure 2, are oxidized in a high-temperature zone. Rapid quenching both reduces the temperature of the gas and removes some of the water-soluble gases and PM. An integral wet scrubber then removes the remaining water-soluble gases and some of the PM.

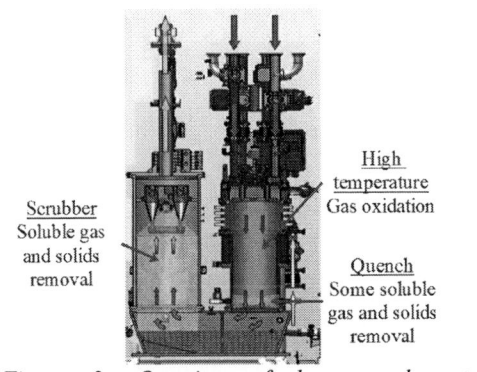

Scrubber
Soluble gas and solids removal

High temperature
Gas oxidation

Quench
Some soluble gas and solids removal

Figure 2: Overview of burner-washer type abatement.

The oxidation reaction mechanisms are complex, but the overall reactions are summarized by simple equations.

$$SiH_{4(g)}+O_{2(g)}=SiO_{2(s)} \qquad (1)$$

$$NH_{3(g)}+O_{2(g)}=N_{2(g)}+H_2O_{(l)} \qquad (2)$$

$$NH_{3(g)}+H_2O_{(l)}=NH_{3(aq)} \qquad (3)$$

$$4N_2O_{(g)}+CH_{4(g)}=4N_{2(g)}+CO_{2(g)}+2H_2O_{(g)} \qquad (4)$$

$$2NF_{3(g)}+1.5O_{2(g)}+1.5CH_{4(g)}=N_{2(g)}+6HF_{(aq)}+1.5CO_{2(g)} \quad (5)$$

$$2F_{2(g)}+CH_{4(g)}+O_{2(g)}=CO_{2(g)}+4HF_{(aq)} \qquad (6)$$

$$SiF_{4(g)}+CH_{4(g)}+2O_{2(g)}=SiO_{2(s)}+4HF_{(aq)}+CO_{2(g)} \qquad (7)$$

Both nitric oxide and nitrogen dioxide are produced from side reactions of the oxidation of the nitrogen bearing gases (ammonia, nitrous oxide, and nitrogen trifuoride). The nitrogen oxides (NOx) are regulated pollutants in all jurisdictions. The reaction of fluorine and methane can yield carbon tetrafluoride [1], a potent greenhouse gas. It

978-1-7281-6559-2/20 $31.00 © 2020 IEEE

is important for such side reactions to be minimized, but this is for the abatement user to discuss with the supplier.

Challenges with solids - facility

The oxidation of silane yields silica dust. The particle size distribution of the dust from a POU abatement system treating silane has been measured using an Electrical low-pressure impactor (ELPI). The data is provided in Figure 3 and illustrates that most of the solids are less than 1μm. Environmental scientists refer to particle size categories as PM 10 and PM 2.5, meaning particles less than 10 μm and 2.5 μm respectively. Most of the PM leaving the abatement would be classified as PM 2.5.

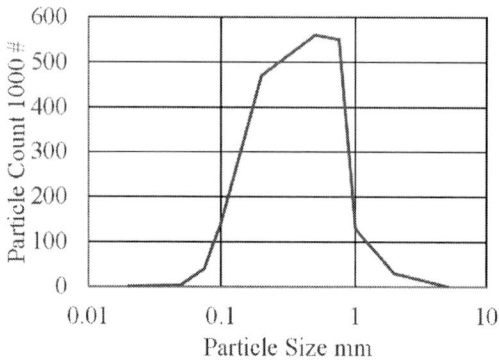

Figure 3: Particle size distribution for silica dust leaving abatement after oxidizing silane.

Damp powder leaving the abatement can accumulate and block the ducts as is illustrated in Figure 4.

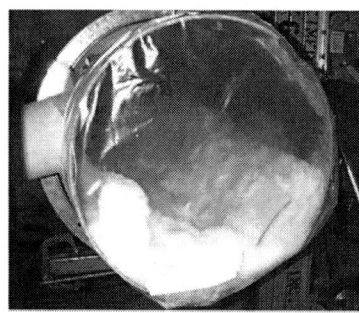

Figure 4: Facility duct blocked with silica.

Blocked ducts lead to factory downtime, hazards to personnel who clean the duct, and may reduce wafer yield if such contamination spreads within a fab.

Challenges with solids – environmental compliance and health impacts.

The health of employees working in fabs and people living nearby such operations is a high priority for society. Hazards associated with inhaling particles increase with decreasing particle size. Suspended particles are potentially hazardous because they can enter the body through the airways and lungs. Small PM 2.5 and the smaller PM 1 particles can get into fine lung structures, leading to reduced lung function. Ultrafine PM 0.1 can penetrate even further and may lead to cancers and cardiovascular failure through inflammation. The releases of particulate from fabs are covered in national and local regulations (Table 1). Note that the local regulations are often more stringent than the national regulations. There are also national regulations that target ambient breathable air. The part of the regulations that are concerned with PM is provided in (Table 2).

Table 1: National and regional regulations for particulate matter release from 40m stack in Grade II area.

	China National	Beijing
Particulates releases permitted for Grade II site and 40m stack.	GB16297 60mg/m³ 21kg/h	DB11-501-2017 10mg/m³ 8.8kg/h
	Guangdong DB44-27-2001 60mg/m³ 18kg/h	Shanghai DB31-933-2015 30mg/m³ 0.8kg/h
	Chongching DB50-418-2016 20mg/m³ 7kg/h	

Table 2: National ambient standard for breathable total suspend solids (TSP), PM10, PM2.5.

GB3095-2012	24 hr Average (μg/m³)		Annual Arithmetic Mean (μg/m³)	
Standard	Class I	Class II	Class I	Class II
TSP	120	300	80	200
PM 10	50	150	40	70
PM 2.5	35	75	15	35
Class I - national parks and equivalent				
Class II - commercial areas and equivalent				

The levels for breathable particulate (e.g. TSP 24 hr Class II average 300μg/m³) are much lower than for stack discharges from fabs (e.g. GB16297 Grade II 60mg/m³) TSP 24 hr Class II average 300μg/m³). The difference reflects that stack emissions are dispersed and diluted before interacting with the general population.

Amorphous silica particles produced during the abatement of silane have much less impact on lung function than the crystalline silica particles more often encountered in the mining and building industries. Silica particles may have acids or other components adsorbed on the particle surface and, as such, constitute a substantially greater health risk than the pure oxide. Other particulate oxides associated with the abatement of gases semiconductor production also represent serious health challenges. Beijing has been proactive in regulating such solid materials (Table 3).

Table 3: Solid compounds that may be generated as waste products from semiconductor thin film manufacturing. (Beijing regulations.)

Solid Compounds (Beijing)	Limit mg/m^3
Reactive fluoride (NH_4F, SiO_2-HF)	3
Zirconium compounds	8
Cobalt compounds	8
Molybdenum compounds	8
Nickel compounds	0.2
Arsenic compounds	0.5
Tantalum compounds	8
Antimony compounds	8
Copper compounds	8
Tungsten compounds	8
Selenium compounds	8
Titanium dioxide	8

Challenges with solids – measurement and mitigation. Accurate measurements of particulates are essential not only to ensure compliance with regulatory limits but also to monitor emissions and develop effective POU abatement solutions. Small particles, PM 2.5 and less, can be challenging to measure accurately and repeatably. Any measurement technique must:

- Use methods that are internationally recognized
- Be portable (easily move from system to system)
- Be easy to install and operate within the fab/sub-fab operating environment and able to test both point-of-use (POU) or end-of-pipe (EOP) equipment
- Follow best practices for sampling – in duct sampling is best and any sample lines must be kept short
- Measure PM10, PM 2.5 and TSP
- Measure concentration, mass flow, and composition

In recently published work, Shoo-Nan Li and his colleagues at the Industrial Technology Research Institute (ITRI) described a measurement methodology.

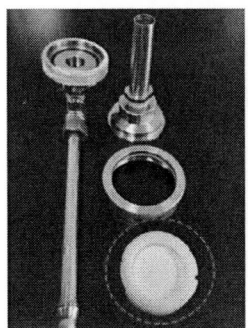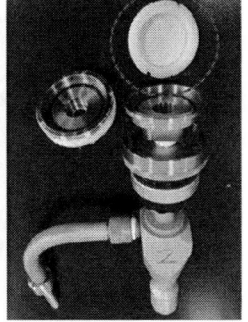

Figure 5: TSP and PM2.5/PM10 measuring equipment.

This equipment (Figure 5) was used to measure particulate releases from an Edwards Atlas Helios system that was abating both hydrogen and silane. A wet electrostatic precipitator (WESP) was evaluated during these tests for the removal of the fine particulate released by the abatement. By imposing a high voltage between electrodes (Figure 6) it is possible to remove the dust particles from the waste gas into the wastewater stream.

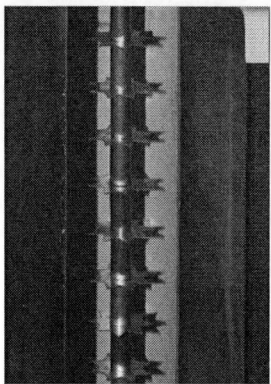

Figure 6: WESP electrodes.

The operation of the abatement and the WESP ensured that over 99% of the silica from the abatement process was recovered into the wastewater from the abatement and WESP. It is possible to use other technologies such as filters and venturi scrubbers, to remove particulate from the exhaust of the POU abatement unit.

CONCLUSIONS

It is important to understand the challenges associated with particulate matter that is generated during the abatement of gases that are used in thin-film processes. Blocked ducts cause facility downtime, lost productivity, and rectification of these problems may impact on yield. Environmental considerations are also important as we better understand the impact of pollution on society. Many different types of particulate may be generated by the fab. Accurate and reliable measurements along with effective mitigation strategies are important.

ACKNOWLEDGMENTS

The author wishes to thank Dr. Shou-Nan Li and his colleagues at ITRI for the provision of photographs and test data.

REFERENCES

[1] L. Beu, S. Raoux Y.C. Chang, M.R. Czerniak, F. Illuzzi, T. Kitagawa, D. Ottinger, and N. Parasyuk, *Intergovernmental Panel on Climate Change 2019 Refinement to the 2006, Volume 3, Chapter 6 Electronic Industries Emissions*, Intergovernmental Panel on Climate Change 2019,

STUDY OF INFLUENCE OF STI PROFILE ON HARP GAP-FILLING PERFORMANCE

Kai Wang[1]*, Zhigang Zhang[1], Ping Wang[1], Lingzhi Xu[1], Shenzhou Lu[1], Andy Tan[1], Zhenjie Qiao[1],
Kang Huang[1], Qimeng Wang[1], Duo Shan[1], Fan Zhang[1], Chang Fu[1], Zhaoyuan Zhao[1], Qin Sun[1]
[1]Shanghai Huali Microelectronics Corporation, Shanghai, China
*Corresponding Author's Email: wangkai_td1@hlmc.cn

ABSTRACT

HARP gap-filling performance is related with trench profile. In this paper, the influence of STI morphology on HARP gap-filling performance is studied. Both the slight undercut between top SiN and active area (AA), and top SiN CD, don't show impact on HARP gap-filling performance for the observed range. The side wall angle is key factor. For side wall angle of 89°, we have proved from both theory and experiment, that even 0.5° reduction of side wall angle can greatly reduce the STI void density.

INTRODUCTION

HARP gap filling is a widely used for shallow trench isolation (STI) and pre-metal dielectric (PMD)[1-4]. During the shrinkage of Nor flash, the size of STI is decreasing and the aspect ratio is increasing, which is challenging for HARP gap-filling. HARP gap-filling performance is strongly related with the STI profile. In this paper, the influence of STI morphology on HARP gap-filling performance is discussed.

RESULTS AND DISCUSSTION

Firstly we studied the impact of slight undercut between top SiN and active area (AA), which is formed during wet etch STI liner, on HARP gap-filling performance. During HARP process, the process chemical drops down to the wafer surface. The undercut is like a shield that the chemical is maybe not easy to enter into. Voids between undercut may form after HARP process. In order to check the actual influence, a sample with slight undercut was deposited with thin HARP film and was taken XTEM images. From the TEM images in Figure 1, before wafer is deposited with HARP, there is small undercut in the red circle. After the wafer is deposited with thin HARP film, the undercut are well filled with HARP film and no more undercut observed. This shows that slight undercuts don't introduce void. Since HARP process chemical is TEOS, it can mobile for some distance after deposited on sample surface. The slight undercut can be filled by the moving TEOS molecular.

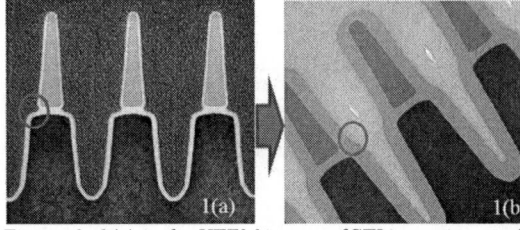

Figure 1: 1(a) is the XTEM image of STI incoming profile, on which the undercut is shown using the red circle. 1(b) is the XTEM image after a thin HARP film deposited. No undercut shown in red circle.

Secondly we studied the impact of SiN CD on HARP gap-filling performance. During the shrinkage of NOR flash, the SiN CD is also getting narrow. During HARP process, wafer was placed on heater with temperature around 540℃, while the temperature of top showerhead is below 100℃. A temperature gradient exists between wafer surface and wafer backside. Thus the temperature of top SiN should be lower than bottom of STI. Since the HARP deposition rate is very sensitive to temperature, the HARP deposition rate on top SiN should be faster than bottom of STI and results in the overhang. A narrower SiN may deteriorate this phenomenon. In order to verify the SiN CD influence on overhang, a set of samples with the same trench angle but different SiN CD were grown with very thin HARP oxide. TEM images were taken to check the overhang shown in Figure 2. No overhang for the all different samples. Above results show no impact of SiN CD on overhang for the observed SiN CD.

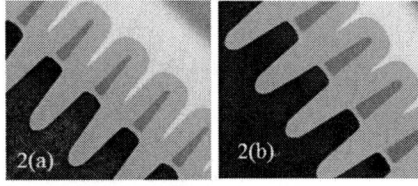
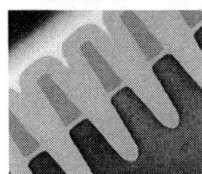

Figure 2: 2(a), (b), (c) show the XTEM image of samples with small, middle, large SiN CD, respectively, deposited

978-1-7281-6559-2/20 $31.00 © 2020 IEEE

with the same HARP recipe.

Thirdly we studied the impact of a slight side wall angle improvement on HARP gap-filling performance. For HARP process, a smaller side wall angle can make HARP gap-filling better. For the large side wall angle case (>88°), how about the HARP gap-filling performance if side wall angle is decreased by only 0.5°? A schematic of HARP partial growth is drawn in figure 3. After the side wall is deposited with HARP thickness of a, and no overhang exists during HARP growth, y and x have a relationship of

$$y/x = 1/cos(\theta) \qquad (1)$$

where θ is the STI side wall angle. That means the deposition rate along y direction is $1/cos(\theta)$ times larger than x direction. For different side wall angle, y/x is calculated and drawn in Figure 4. As θ is 89°, y/x is 57.3. As θ is 88.5°, y/x is reduced to 38.2. That means the deposition rate along y direction is much lowered and the SiO_2 density is much more densified. As etchant wet etch from sample top, the possibility of void is much more reduced. In order to verify this, two samples with side wall angles of 89.1° and 88.5° are processed by the same HARP recipe. Then the two samples were annealed, CMP polished. Finally they were wet etch from top to check void density by plan SEM shown in Figure 5. For sample with side wall angle of 89.1°, there are many voids found on surface. For sample with side wall angle of 88.5°, the voids density are reduced a lot. Thus the minor improvement of side wall angle can also reduce the void density. The side wall angle plays an very important role in reducing void density.

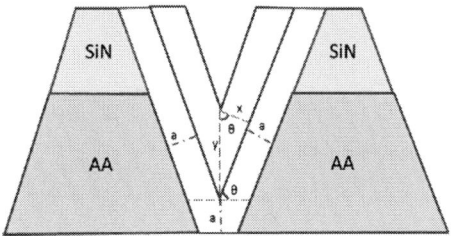

Figure 3: Schmatic of HARP partial growth without overhang

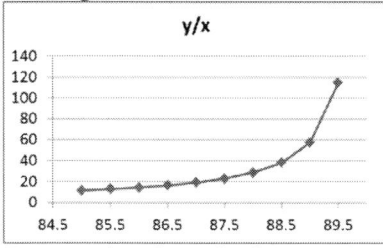

Figure 4: the relationship between y/x and side wall angle

Figure 5: 5(a) is the cross section of STI with side wall angle of 89.1°. 5(b) is the corresponding plan view SEM image for checking the void. 5(c) is the cross section of STI with side wall angle of 88.5°. 5(d) is the corresponding plan view SEM image for checking the void.

CONCLUSION

From above results, the side wall angle reduction of only 0.5° can significantly improve the gap filling performance, and reduce the void density, while the slight undercut and SiN CD has no effect for the observed case.

REFERENCES

[1] H. Liu, L. Wong, W. Lu, ZG. Sun, I. Bangun, YP. Shen, YP Wu, Z. Chen, M. S.Zhou, T. Chu, R. Leong, A. Jain, C. Ching, K. Raguputhi, H. Whitesell, J. Kasim, Z.Shen, Advanced STI Technology with Void Free Gap-fill and Superior Device Performance for 45nm Devices." ISTC 2007, Shang Hai, 2007

[2] H. Liu, W. Lu, Z.G. Sun, L. Goh, B. Zuo, V. Ho, M. S. Zhou, L. C. Hsia, "The Application of HARP for PMD gap-fill for 65nm Technology Node and Below", 23rd Advanced Metallization Conference, U. S. A., San Diego, 2006.

[3] Y.W. Teh, J. Sudijono, C. Ching, S. Venkataraman, A. Jain, "A Novel High-Stress PMD Film to Improve Device Performance for sub-65nm COMS Manufacturing", in 2006 MRS Spring Meeting,

[4] A. Al-Bayati, L. Washington, L.Q. Xia, Z. Yuan, M. Balseanu, M. Kawaguchi, F.Nouri, R. Arghavani, "Processes for Inducing Strain in CMOS Channels", Nanochip Technology Journal, 6-11, Issue 2 2005,

A NOVEL METHODOLOGY TO MONITOR WAFER PLACEMENT SHIFT IN LASER SPIKE ANNEAL

Yan Gui*, Lan Jiang, Yaoting Shen, Qingwei Dong, Kecheng Chen, Xinhua Cheng, Jingxun Fang

Shanghai Huali Integrated Circuit Corporation, Shanghai 201314, China

*Corresponding Author's Email: guiyan@hlmc.cn

ABSTRACT

For advanced technology node, laser anneal is widely used for ultra-shallow and low resistivity junctions which are needed to suppress short-channel effects and improve device performance. However, if wafer is not put in proper position, laser anneal will lead to PP source drain photo overlay residual abnormal, and even cause wafer broken. In this paper, a novel methodology is evaluated to earlier alert wafer placement shift in laser anneal which including measuring first laser scan region Rs fluctuation range control to detect the abnormal temperature spot combining with monitor camera installation, by this method wafer broken ratio is significantly reduced and productivity is improved eventually.

Keywords—laser anneal, placement shift, fluctuation range control

INTRODUCTION

With the development of large-scale integrated circuit technology, the device dimensions are continuously scaling down, and shallower p-n junction depth and lower sheet resistance are essential to control short-channel effect and improve device performance [1]. Laser generated by gas excitation is used to form ultra-shallow junction in CMOS fabrication, and lower dopant diffusion and higher dopant activation are achieved [2]. In the actual production process, if wafer is not placed in proper position, laser annealing will lead to PP source drain photo overlay residual abnormal, and even cause wafer broken. Through the current offline monitor method, the laser annealing handoff issue cannot be accurately detected, which leads to a long process of issue discovery and unnecessary losses. In view of this problem, the present invention provides a method for off-line monitoring of the shift of wafer position, that is, adding the first region fluctuation range control in Rs monitoring and monitor camera installation, so as to reduce the wafer broken rate and improve the product yield.

EXPERIMENT

The N type wafers were implanted 5KeV 2E15 boron ions, then the wafers were divided into at least three zones to scan laser, the first zone at the top of the wafer, the second zone in the middle of the wafer, and the third zone at the bottom of the wafer, as shown in Fig.1. Laser annealing is carried out through Rofin (CO_2 mix gas) laser beam spot scanning, temperature is 1100~1200℃, laser Dwell time is 300~600µs. The laser beam starts forward arc scan from the first zone to the second zone (as shown in Fig.2), then reverse arc scan is performed on the third zone. In this work, the implanted offline wafers and inline wafers were all placed in two positions to check the influence of wafer placement, position (1): wafer top area leave 0.5~1mm for thermal diffusion, position (2): wafer did not leave any space for heat diffusion, as shown in Fig.3, and the offline Rs (121dots contour map, EE 3mm) and inline PP1 overlay were detected.

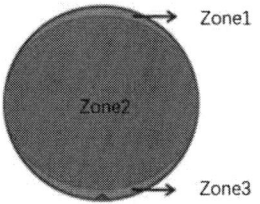

Fig.1. Division of laser annealing working area

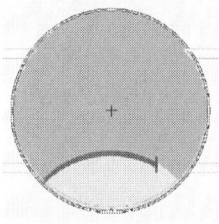

Fig. 2. Laser spike anneal process scan mode

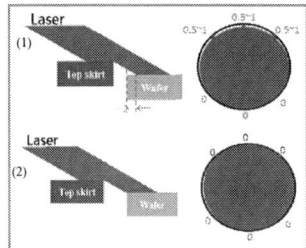

Fig.3. The scheme of wafer placement

RESULTS AND DISCUSSION

The Rs mean value and Rs uniformity of the wafers placed in two positions were detected, and this is the daily monitor content of tool maintain, as we can see the chart below, different placement did not show obviously different.

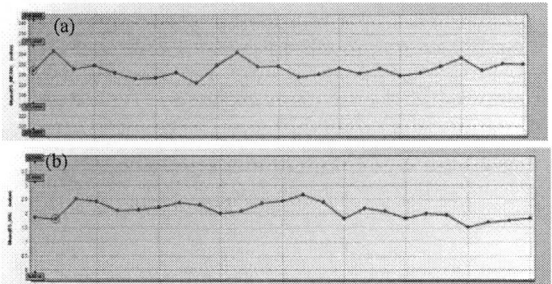

Fig.4. (a) Rs monitor mean chart;(b) Rs monitor U% chart

The offline Rs maps were also plotted, as shown in Fig.5, and the Rs map of the wafer placed in position (1) is quite uniform (Fig.5(a)), but the Rs map of the wafer placed in position (2) shows the top area Rs is relatively low (Fig.5(b)), that means wafer top area temperature is overheated, and it has a very high possibility to cause wafer broken, especially for production wafer. And production wafers were detected with PP1 overlay residual abnormal, and the abnormal map was shown as Fig.6(a)., and compared with PP1 overlay normal wafer, the abnormal wafer top area residue is about 12mm, as shown in Fig.6(b).

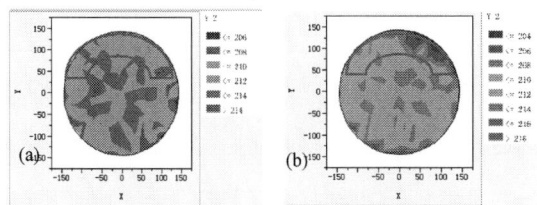

Fig.5. Rs monitor map

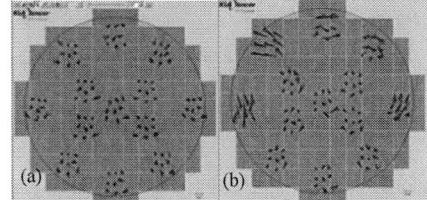

Fig.6. (a) PP1 overlay residual abnormal map;
(b) PP1 overlay residual normal map

In order to quantificat wafer top area over-heated extent, the Rs gap between mean value and top scan was monitored, as shown in Fig.7.We assume if the gap is over $4\,\Omega$ (Rs sensitivity $\sim 1\Omega/°C$), the wafer top edge is over heated, and has a risk of PP1 overlay residual issue or even wafer broken.

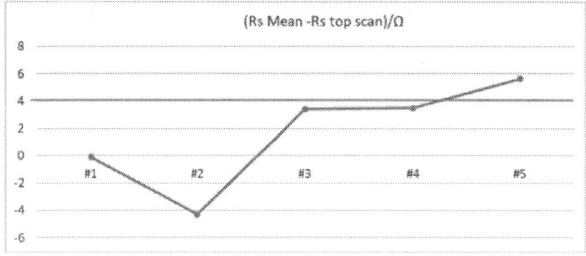

Fig.7. Rs gap between mean value and top scan

At the same time, to timely detected the wafer position shift, we install a special monitor camera to monitor the distance between top skirt and wafer edge, as shown in Fig.8. As the measured distance and wafer top area margin are inverse ratio (as shown in Fig.3), according to the measured distance, wafer position can be adjusted.

Fig.8. Monitor the distance between wafer and top skirt

We combine the zone1 Rs fluctuation range control and monitor camera installation to monitor wafer placement, and wafer broken ratio is marked falling, as shown in Fig.9.

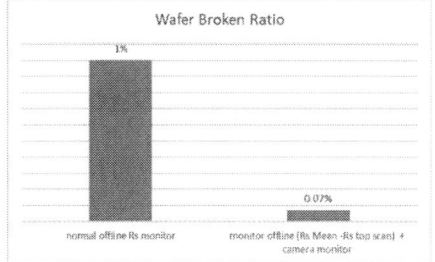

Fig.9. Wafer broken ratio comparison

CONCLUSION

We monitor offline wafer position by both Rs gap control and wafer top scan margin control, and this method can effectively alert wafer placement shift, to avoid PP source drain photo overlay residual abnormal and reduce wafer broken ratio.

REFERENCES

[1] A. Shima, Y. Wang, S. Talwar and A.Hiraiwa., Symp. *VLSI Tech. Dig.*, (2004),174

[2] J. Venturini, *Advanced Thermal Processing of Semiconductors*, RTP 2005,IEEE (2005), 7

Investigation and characterization of silicon concentration in N-free anti-reflective layer films

Luhang Shen, Jiepeng Zhou, Chunwen Liu, Haixia Li, Yiqi Gong, Yu Bao, Jingxun Fang,

Shanghai Huali Integrated Circuit Corporation

021-61871212, shenluhang@hlmc.cn

ABSTRACT

In the 28 nm high voltage flow, the N-free anti-reflective layer (NFARL) taking the place of silicon oxide as Poly hard mask (HM) film can achieve more vertical profile for Poly. However, there is (Ni, Pt)Si impurity generated on the top of the NFARL film after nickel silicide process. In this paper, the effects of silicon (Si) concentration on the characteristic, Poly profile and purity of NFARL films were systematically investigated by experimental techniques. With increasing of Si concentration, the refractive index n and extinction coefficient k of films increase monotonically. The transmission electron microscopy (TEM) measurements indicate (Ni, Pt)Si impurity can't be formed on the top of the NFARL film for the lowest Si concentration, because there is no C-Si bond in the film.

Key words: NFARL, silicon concentration, n, k, (Ni, Pt)Si impurity

INTRODUCTION

With the decrease of integrated circuit size, silicon oxide as poly HM film has widely utilized in semiconductor manufacturing. However, the vertical profile of Poly can't be achieved after Poly etches in our 28nm high voltage process. In order to improve Poly profile, NFARL, a class of Poly HM films, has been taken into consideration due to its optical properties and high selectivity. In this paper, we introduce different Si concentration in the NFARL films to investigate intrinsic optical properties of SiOC film and Poly profile by a plasma enhanced chemical vapor deposition

(PECVD) technique. For the form of NFARL film, the stoichiometry of the reaction as follows.

$$SiH_4 + CO_2 + HE \xrightarrow{RF\ 400°C} SiOC\ (NFARL) +$$

by-products

The different Si concentration in the NFARL films can be obtained by controlling the reaction gas flow of SiH_4 and CO_2. These results clearly revealed that the Si concentration can play an important role in achieving vertical Poly profile and pure NFARL in the subsequent nickel silicide process. The formation mechanism of (Ni, Pt)Si impurity is also discussed.

EXPERIMENT

The NFARL films with a thickness of about 70 nm were deposited by PECVD tool of LAM RESEARCH. All the films were grown in an atmosphere of pure SiH_4, CO_2 and HE gas flow. The 32.5, 32.6, 33.1 and 34.1 at. % Si concentrations in the NFARL films were fabricated by controlling the gas flow of SiH_4 and CO_2. The Si concentration and Valence state of the NFARL films were determined by X-ray photoelectron spectroscopy (XPS). The refractive index n and extinction coefficient k of all films were measured using the KLA-Tencor tool. The Poly profile and impurity component were characterized by Transmission Electron Microscope (TEM) and Energy Dispersion Spectrum (EDS).

RESULT AND DISCUSSION

The dependence of the refractive index n and extinction coefficient k at wavelength of 193 nm of the SiOC films on the Si concentration in the films was shown in Fig. 1. With increasing the Si concentration, the observed C concentrations are 0.96, 0.97, 0.99, 1.55 at. %, respectively. It is clear that both the refractive index n and extinction coefficient k gradually increase with the increase of Si concentration, which is ascribed to the increase of sp^2-hybridized carbon atoms.

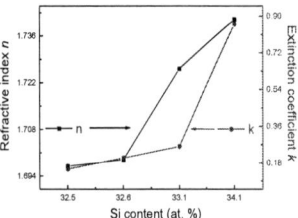

Fig.1 Refractive index and extinction coefficient vs Si concentration in the films

Fig. 2 shows the core level XPS spectra of Si 2p and C 1s with different Si concentration in the NFARL films. The XPS spectra of Si 2p for all films display one peak located at 99.5-99.7 eV, which is associated with the binding energy of Si-Si bond in bulk Si [1]. Therefore, it is obvious that this is the characteristic peak of substrate Si. For the highest Si concentration in the film, the asymmetric Si 2p spectrum can be divided into two characteristic peaks located at 101.9 and 103.5 eV using Gaussian fitting, respectively. The higher binding energy peak can be attributed to Si-O bond; the other one peak corresponded to Si-C bond in SiOC [2]. For others Si concentration, the Si 2p spectra show the characteristic peak of Si-O bond located at about 104.4 eV [2]. The observed energies for C 1s are located at 284.2 eV and 285.9 eV, which match with the binding energy of C-Si bond and C-O bond [3], respectively. It is indicate that the existence of carbon in the film will change from C-Si bond to C-O bond when the concentration of Si less than 34.1 at. %.

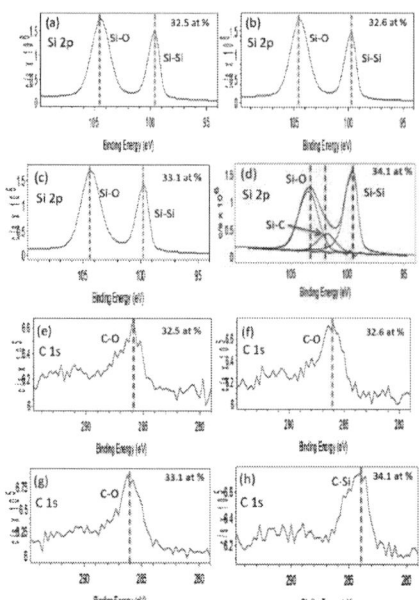

Fig. 2 the XPS spectra including Si 2p (a)-(d) and C 1s (e)-(h) with different Si concentration in the films

Fig. 3 (a) and (b) show the Poly profile for silicon oxide and the highest Si concentration of the NFARL as Poly HM after Poly LEC etch respectively. It is clear that the middle of Poly had necked when silicon oxide as Poly HM. For the highest Si concentration of the NFARL taking place of silicon oxide as Poly HM, the observed Poly profile become more vertical, and the Poly HM profile also has improved.

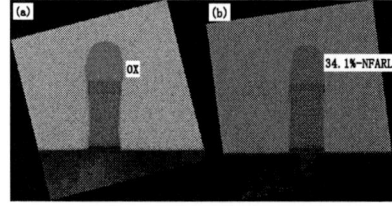

Fig. 3 TEM cross section of silicon oxide (a) and NFARL with 34.1 at. % Si concentration (b) as Poly HM

The highest Si concentration of the NFARL as Poly HM can improve the Poly profile. However, the top of the NFARL film has appeared black impurity after nickel silicide process, as shown in Fig. 4. In order to investigate the impurity component, the EDS measurement of impurity was carried out. It is obvious that the black impurity contains nickel and platinum. According to XPS results, there is lower binding energy of C-Si bond in the NFARL film with the highest Si concentration. The C-Si bond was broken, then the reaction between Si and (Ni, Pt) produces (Ni, Pt)Si in the nickel silicide process.

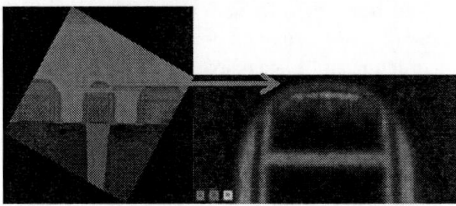

Fig. 4 TEM and EDS of NMOS (34.1 at. % Si concentration) after nickel silicide process

In order to remove the (Ni, Pt)Si impurity on the top of the NFARL film, the lowest Si concentration has been taken into consideration. According to TEM results, it is clear that the Poly profile can still keep vertical, and there is no (Ni, Pt)Si impurity on the top of the NFARL film, as shown in Fig. 5. One possible reason for the disappearance of (Ni, Pt)Si impurity is that the C-Si bond change to C-O bond with the decrease of Si concentration.

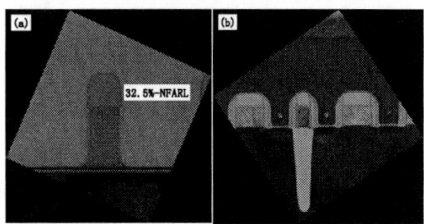

Fig. 5 TEM cross section of NFARL with 32.5 at. % Si concentration as Poly HM

CONCLUSION

In summary, the different Si concentration in the NFARL films were deposited by PECVD technique to investigate the refractive index n, extinction coefficient k and Poly profile. With increasing the Si concentration, both the n and k of films gradually increase. One possible reason is that the increase of sp^2-hybridized carbon atoms. The XPS results show that the highest Si concentration of the NFARL film has C-Si bond, but others Si concentration of films have C-O bond. The NFARL as Poly HM can obviously

improve the Poly profile. By comparing the TEM results of highest and lowest Si concentration after nickel silicide process, the existence of C-Si bond can cause (Ni, Pt)Si impurity in the highest Si concentration of the NFARL film.

ACKNOWLEDGEMENTS

The authors would like to acknowledge HLMC failure analysis department for the supports of TEM/EDS microscopy.

REFERENCES

[1] K. Prabhakaran, Y. Kobayashi and T. Ogino. Chemically prepared oxides on Si(001): an XPS study. *Surface Science*. 1993, pp. 239-244.

[2] Hongjiao Lin, Hejun Li, Tiyuan Wang, Qingliang Shen, Xiaohong Shi and Tao Feng. Influence of temperature and oxygen on the growth of large-scale SiC nanowires. The Royal Society of Chemistry. 2019, pp. 1-8.

[3] Efraín Ochoa-Martíneza, Mercedes Gabása, Laura Barrutiab, Amaia Pesquerac, Alba Centenoc, Santiago Palancoa, Amaia Zurutuzac, Carlos Algorab. Determination of refractive index and extinction coefficient of standard production CVD-graphene. 2014, pp. 1-10.

THE INVESTIGATION OF DOMESTIC MACHINES LARGE-SCALE PRODUCTION IN SOAK ANNEAL PROCESS

Shen yaoting*, Jiang lan, Gui yan, Dong qingwei, Chen kechen, Cheng xinhua, Fang jingxun
Shanghai Huali Integrated Circuit Corporation, Shanghai 201210, China
*Corresponding Author's Email:shenyaoting@hlmc.cn

ABSTRACT

In soak and spike anneal process, applied materials vantage family tools are widely used in 28/40nm and beyond. The domestic machines of this kind, being still on the stage of development, have a great potentiality in the semiconductor industry. As a domestic vendor, Mattson has introduced Helios XP for rapid thermal process(RTP).In this paper, the temperature profiles about soak anneal process have been collected on Mattson Helios XP. Besides, sheet resistance, oxide thickness, device data and yield also have been investigated both on blanket wafer and pattern wafer. The result shows that the performances of Mattson Helios XP are comparable with applied materials' vantage tool. Therefore, Helios XP as domestic machine can be applied in soak anneal at 28/40nm node for the large-scale industrial production.

INTRODUCTION

As further progress in electronic technology becomes increasingly dependent on success in rapid development cycles that include both materials innovations and changes in CMOS device architecture [1], Rapid thermal processing (RTP) has become a key process for manufacturing advanced semiconductor devices [2], with a wide variety of applications including oxidation, implant annealing and silicide formation. The thermal budget and accurate temperature control is the key to the rapid thermal processing (RTP) process [3]. It will impact the diffusion and activation rate of implanted Ion and so on. AMAT' Vantage family tools occupy the annealing market with their strong temperature control. Mattson is a domestic factory. Helios XP is their rapid thermal processing (RTP) tool. The picture was shown in Figure 1.

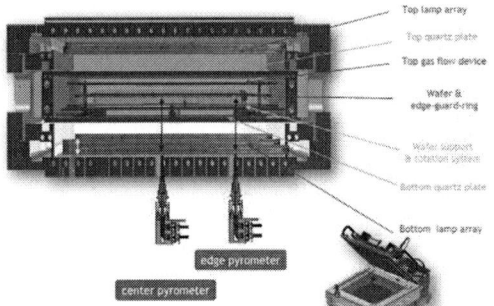

Figure1: The picture of Helios XP chamber

It has 26 halogen lamps at the top and bottom for heating. Compare with the heating on one side, double side heating can reduce pattern loading effect. The uniformity of thermal on device wafer will better.

Each chamber has three dual head Pyrometers (DHP) for collecting light from radiation and reflection, two pyrometers were located at the center and edge of the wafer.one pyrometer is used to measure in-situ radiation at wavelengths of 2.3 μm and 2.7 μm. The temperature control system converts the received optical signal into temperature feedback to the lamp tube. The voltage of lamp tube will be adjusted automatically by system at100Hz frequency.

EXPERIMENTAL

Figure.2 summarizes the items of the experimental plan for this study.

300mm diameter, n-type (100) prime silicon wafers of 8-12ohm resistivity were cleaned before implanted, the wafers were doped by implants of 1013 B/cm2 at 3KeV. Besides the Boron implants, some wafers were cleaned in SCI and SC2 solutions to remove native silicon dioxide.

New blanket wafer was prepared for collecting temperature profile on AMAT Radiance plus& Mattson Helios XP. Blanket wafers were annealed for 30s at 1000℃ in an ambient with 20slm N2, we get the real time temperature from the pyrometers. Then draw the temperature profile to compare the trend whether match.

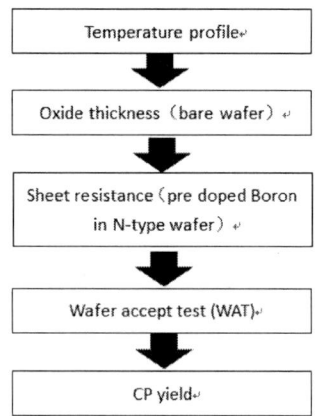

Figure2: Summary of the experimental plan

Secondly, wafers which have been remove native oxide were annealed for 60s at 1100℃ in an ambient with

20slm O2, total 8 times in 3-day marathon ,the oxide thickness is measured by KLA-Tencor thickness measurement tool. Pre doped B wafers were annealed for 30s at 920℃ in an ambient with 24slm N2 , the sheet resistance (RS) is measured by four-point probe.

Finally, on 28nm pattern wafers, we spilt two conditions on anneal process, AMAT's tool and Mattson's tool. Then WAT and CP were measured after wafer completed the whole process flow.

RESULT AND DISCUSSION

First, figure3 shows the soak anneal temperature profile on different chamber. It was observed that Helios XP's temperature profile match with baseline chamber (AMAT's radiance plus) at 750℃~1000℃ area. It was process key area that means thermal budget.

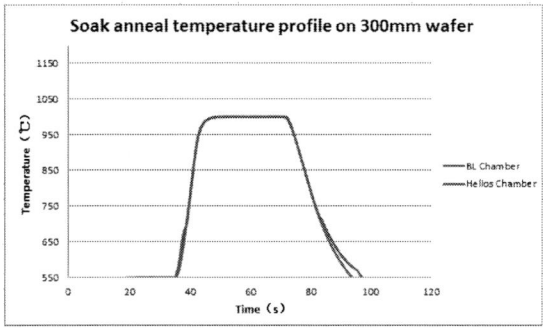

Figure3: Soak anneal temperature profile

Next, figure 4 shows the repeatability and uniformity results for marathon tests of rapid thermal oxidation, as performed on Mattson Helios XP. This data obtained through 3-day marathon testing on 300 mm wafers, shows an 8-repeatability equivalent to 1 ℃ temperature variation. Within wafer uniformity was 0.25%. It was comparable with AMAT's radiance plus.

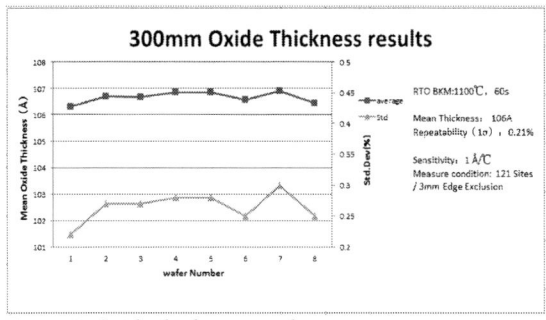

Figure4: Oxide thickness results

Figure5 shows the sheet resistance performance, RS's sensitivity is -2Ω/℃ in this implant condition, the process results demonstrated a variation that a total range of only

2C within wafer. Helios XP has the ability to do the RS uniformity as well as Radiance plus.

Uniformity of Sheet resistance
Implant condition:3 Kev 1E13 B/cm2

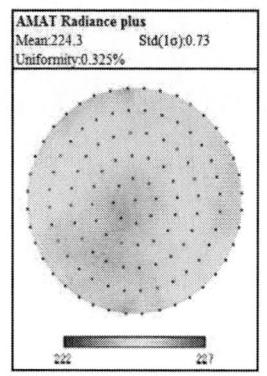
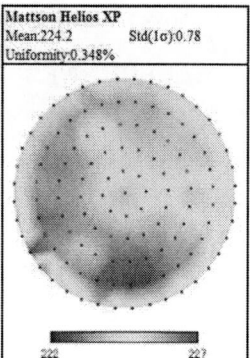

Figure5: sheet resistance results

Figure6 shows the wafer accept test (WAT) results of different process chamber. Final WAT shows all device parameters could hit design target. Helios XP performance is comparable with baseline tool.

Figure.7 shows that Circuit Probe (CP) yield of Helios XP maintains the same high level as baseline chamber (Radiance plus).

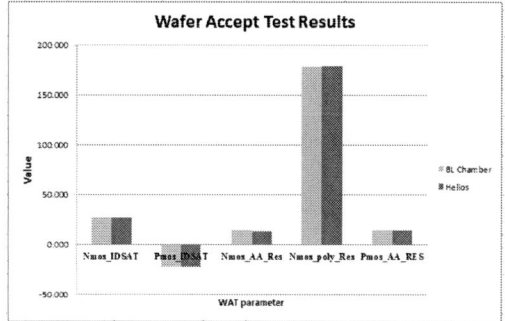

Figure6: Wafer accept test results

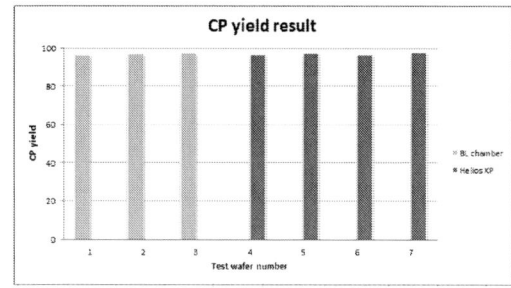

Figure7: CP results

CONCLUSION

In this work, we explored domestic machine whether

had mass production capacity in soak anneal process. Base on all temperature profile/oxide thickness/sheet resistance/WAT/CP yield data, the results showed that the performance of Helios XP was comparable with Radiance plus, Helios XP had qualification for soak anneal process in large-scale production.

REFERENCES

[1] R.B. MacKnight, P.J. Timans,S.-P. Tay *RTP applications and technology options for the sub-45 nm nodes* Mattson Technology 47131 Bayside Parkway Fremont, CA 94538, U. S. A

[2] Brad Mattson, Paul Timans, *the future of RTP, a technology that can change the IC fab industry*, Mattson Technology, 2800 Bayview Drive, Fremont, CA 94538 U. S. A

[3] P.J.Timans, Y.Z.Hu, Y.Lee, *optimization of diffusion, activation and damage annealing in millisecond annealing* Mattson Technology, Inc., Fremont, CA 94538, U. S. A

BEOL Cu Gap-fill Performance Improvement for 14nm Technology Node

Zhaoqin Zeng*, Bao Yu, Yanpeng Cao, Xingkun Xue, Jianhua Xu, Yanyan Zhang, Xiaofang Wang, Jingxun Fang, Yu Zhang

Shanghai Huali Integrated Circuit Corporation, Shanghai 201314, China

*Corresponding Author's Email: zengzhaoqin@hlmc.cn

ABSTRACT

The continuous scaling in the logic device towards 14nm node dimensions has presented numerous engineering and manufacturing challenges in BEOL Cu interconnection. Small process marginality between the CuMn seed deposition and Cu plating processes is the primary source of voids in Cu interconnects. A chemical vapor deposited (CVD) cobalt liner was used as an enhance layer to replace conventional PVD Ta in 14nm technology node. This paper analyze cobalt property and present how Cu seed thickness (THK), Deposition combine Etch (DCE) ratio, AC bias power impact on defect and top opening of trench. Experiment result shown the defect between dense trench line area and SRAM area are quite tradeoff. Trench area requests bigger top opening, while SRAM area requests more Cu step coverage on sidewall. Defect mechanism was also detail discussed. Optimized parameters were finally presented and the defect level was significantly reduced. For more challenge structure wafer, A L9 Taguchi DOE was designed and the result show ECP MW current increase combine with start earlier can further benefit defect. Defect count almost free. The benefit was thought comes from Cu better nucleation capability at ECP process.

INTRODUCTION

Cu-based interconnection has been widely used in the IC industry because of its high resistance to electro-migration and low resistivity. A typical Cu metallization consists of a Cu seed layer physical vapor deposited on the PVD Ta/TaN barrier, followed by electroplating Cu. Advanced 14 nm metal lines are significantly thinner in cross-sectional area than previous generation devices. Smaller dimension of the metal line limits thicker seed Cu deposition inside metal line by conventional PVD technologies, and degrades Cu gap-fill performances.

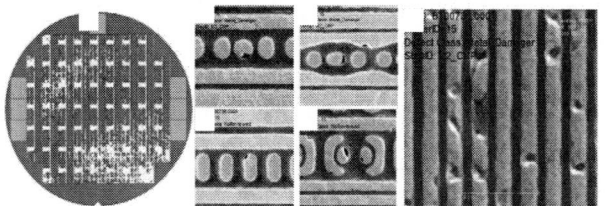

Fig.1 Inline damaged metal defect map and image post CMP step

As Fig.1, serious damaged metal defect was found both in SRAM area and dense trench line area when CuMn seed deposition and Cu plating processes was applied on 64nm pitch Cu wire interconnects wafers. To overcome Cu void issue, many researchers investigated CVD-liners such as CVD-Co and CVD-Ru due to its outstanding gap-fill performance by excellent Cu wettability [1][2][3] [4].

In this work, we use Cobalt as the seed enhancement layer

instead of PVD Ta layer. Investigate trench overhang (top opening post Barrier & Seed Dep) behaviors under different condition. Balance top opening and Cu continuity. Optimize ECP entry parameter. Combine with inline defect result, the gap-fill improvement solutions, root causes as well as the mechanism of metal damaged defect in 64nm pitch Cu wire interconnects for the 14nm technology node are discussed.

EXPERIMENT

Cu lines were fabricated via a 300mm dual damascene process in Ultra low-k dielectrics with 64nm pitch. TaN based barriers were deposited by physical vapor deposition (PVD). Co liner was deposited by chemical vapor deposition (CVD). A PVD 2step Cu seed and Cu electroplating were applied for Cu fill with 40g/L Cu VMS, followed by chemical-mechanical polishing (CMP) and Co Cap process. Gap-fill was evaluated post CMP process by a KLA-Tencor inspection tool, SEM (Scanning Electron Microscope) review was used to obtain defect type based on the scanned map and location. CD SEM was used to collect top opening data before and post barrier and seed deposition. TEM (Transmission Electron Microscope) and EDS (Energy Dispersion Spectrum) was used to analysis metal film step coverage.

RESULTS AND DISCUSSION

Cu agglomeration was studied on below 3 film stacks. The film stacks are Cu40A/Co30A/TaN30A, Cu40A/Ta30A/TaN30A, Cu40A/TaN60A. After thermal anneal (350C, 90s), No Cu agglomeration was found when Co was introduced between TaN and Cu layer, while Ta and TaN substrate wafers were found serious Cu agglomeration as Fig.2a, 2b and 2c. EDX data revealed no copper signal on the de-lamination area, but strong copper signal on small Cu agglomeration islands as Fig.2e &2f, which indicate cobalt under Cu layer help to prevent Cu agglomerating.

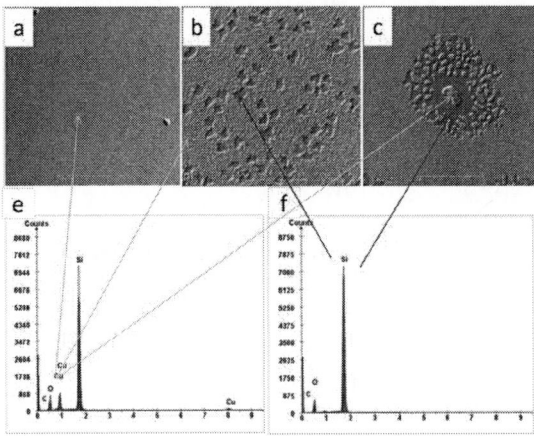

Fig.2 Top view of Cu continuity with (a) Co substrate,(b) Ta substrate, (c) TaN substrate. And its EDX data(e &f)

For gap-fill study, TaN film was deposited by PVD after Etch process, followed by Co liner, which was deposited by CVD technology. TEM elemental cross-section views of TaN barrier film and Co liner was shown in Fig.3 respectively. Co liner step coverage is very good.

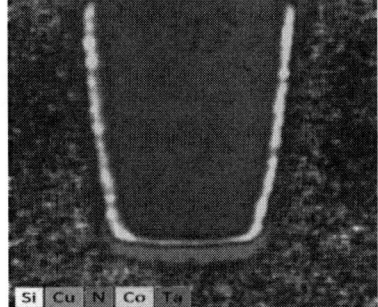

Fig.3 EDS of TaN and Co film

Cu seed thickness was split from 1x to 7x. Metal damaged defect was quite tradeoff between SRAM area and Dense trench line area as Fig.4. SRAM prefer thicker Cu seed and defect start getting worse if Cu seed less than 4x as Fig.4a. Too thin Cu seed might cause discontinuous Cu around the sidewall and result in damaged metal defect. Inversely, dense trench line was more prefer thinner Cu seed as Fig.4b. This behavior can be well explained by the trench top opening trend as Fig.7. The thinner of Cu seed, the bigger top opening of trench, which was benefit for the subsequent electroplating process (ECP). However, further decrease Cu seed thickness in dense line area, serious damaged metal defect was also found, which may due to discontinuous Cu layer in trench sidewall under extreme thin Cu seed thickness. So thickness within 3x~4x help wafer achieved best defect performance on dense line area. As there is no Cu thickness overlay area to achieve best defect performance both for SRAM and Dense line area, more actions need to be taken.

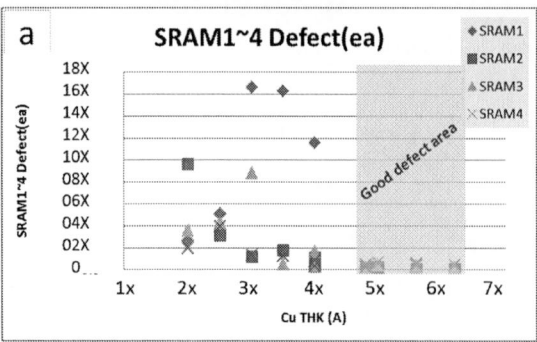

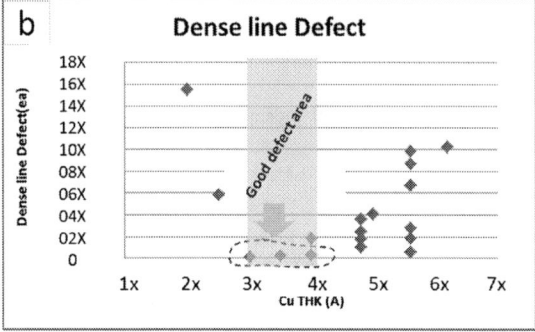

Fig.4a Defect on SRAM area VS Cu seed thickness, Fab.4b Defect on Dense trench line area VS Cu seed thickness.

Cu seed process includes 2 steps. 1st step with very low AC bias power is called "Deposition only". 2nd step is called "Deposition combines Etch (DCE)", which usually use high an AC bias power to achiever good sidewall step coverage. Defect performance with different bias power in DCE step was shown in Fig.5. With higher AC bias power, defect both on SRAM and dense line area were significantly improved. Dense line improvement was due to the bigger top opening when AC bias power was high as Fig.8a. SRAM defect reduction was considered Cu step coverage on sidewall was improved by higher AC bias power, which result in more re-sputter effect from bottom.

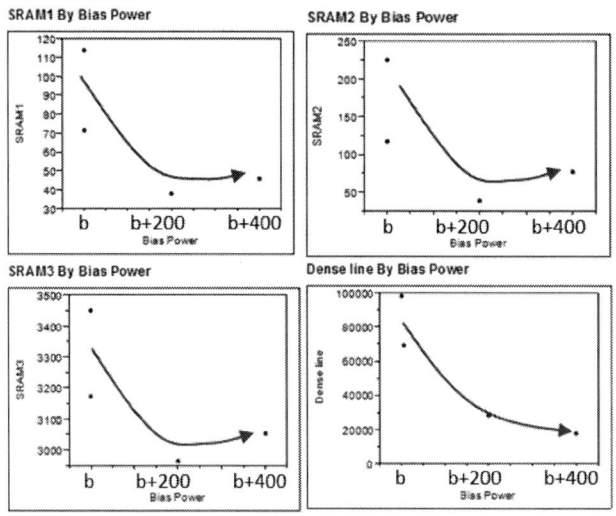

Fig.5 Defect VS AC bias power

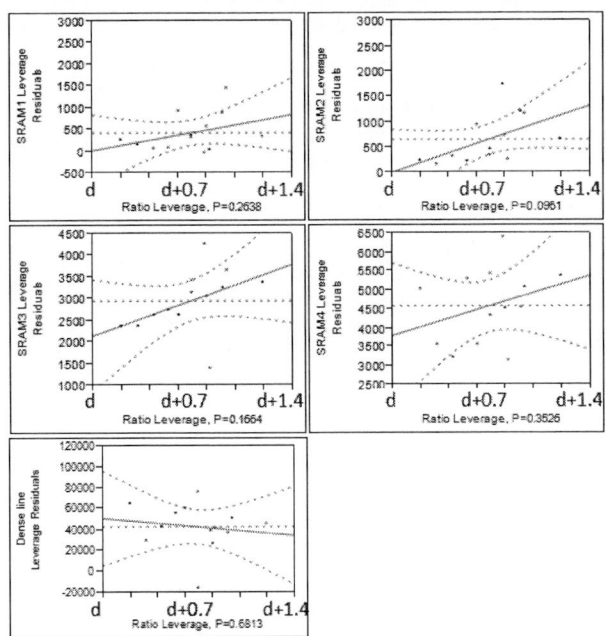

Fig.6 Defect VS DCE Ratio

To investigate DCE effect on SRAM / Dense line area defect. DCE ratio was defined as *1st step time / 2nd step time* ratio. Defect in all SRAM area were found significantly increased as DCE ratio increase, while defect in dense trench line area has opposite trend as Fig.6. DCE ratio decrease means DCE effect increase, which makes more re-sputter to sidewall and improve Cu continuity, finally result in better defect on SRAM area. However, DCE ratio decrease also induces top opening reduction as Fig.8b, and finally result in defect in dense line get worse.

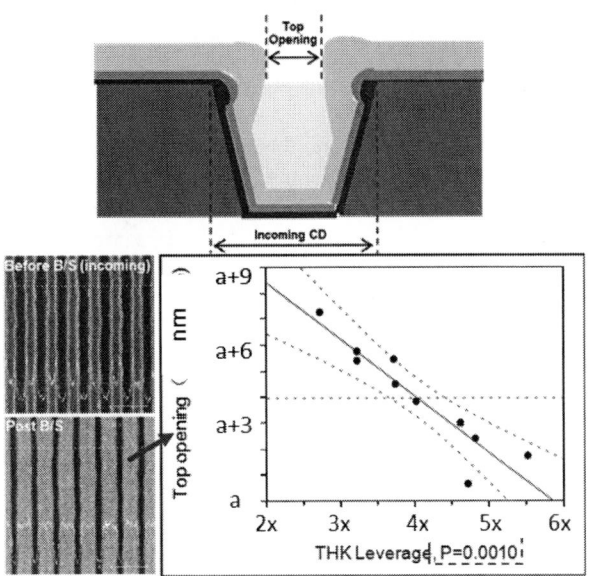

Fig.7 Top opening VS Cu seed THK

Base on above experiment result, we can concluded that the weak point in SRAM area is Cu continuity, this may due to the height of SRAM is higher than that of Dense trench line. Inversely, the weak point in Dense trench line is top opening. As the good defect

area between SRAM and Dense line is almost no overlap as Fig. 4a and Fig. 4b, Cu seed thickness was selected nearby their boundary. 4x~5x Cu seed thickness was selected as next experiment. DCE ratio split from d+0.3 to d-0.3. AC bias power was selected b+400w as higher than this value might cause seed discontinue in chamfer of Via chain base on 28nm process development experience as Fig. 9.

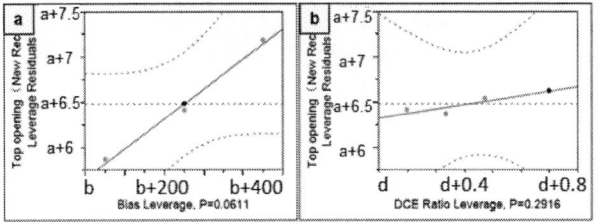

Fig.8a Top opening VS AC bias power, Fab.8b Top opening VS DCE ratio

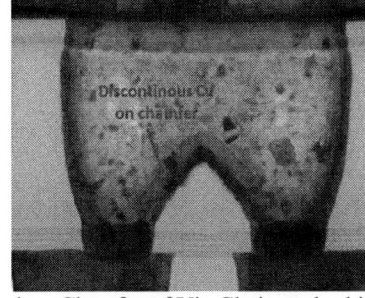

Fig.9 Cu seed on Chamfer of Via Chain under higher AC bias power.

Further experiment shown 5x Cu and a-0.3 DCE ratio achieve best defect performance as Fig.10. Either SRAM defect, or dense line defect are less than 100ea.

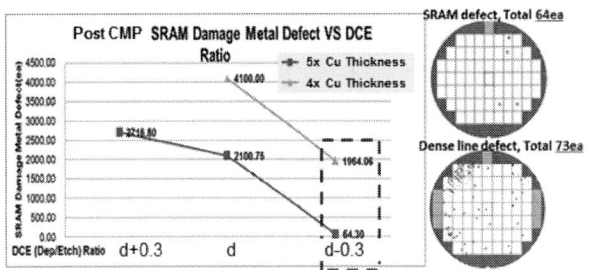

Fig.10 SRAM defect with different condition and its maps.

However, when this best condition applied to TVO product (include more minimum design rule structure), total damaged defect count was spiked up and an obvious special defect map (Let top or right top) was occurred as Fig.11. The defect location quite matches with ECP entry spot as Fig.12. Suspect gap fill widow not enough and result in damaged metal defect in ECP process weak point location (Entry spot).

978-1-7281-6559-2/20 $31.00 © 2020 IEEE   363

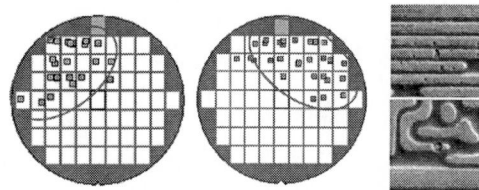

Fig.11 Defect map and image of TV0 product.

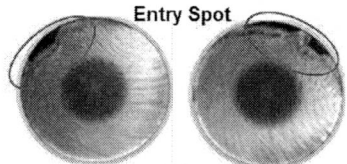

Fig.12 Haze Map of ECP entry location.

Table1. L9 DOE of gap-fill window enlargement

Wafer	Cu Seed	ECP Parameter		
		MW Current	MW Time	Triger Time
#01	1	1	1	1
#02	1	2	2	2
#03	1	3	3	3
#04	2	1	2	3
#05	2	2	3	1
#06	2	3	1	2
#07	3	1	3	2
#08	3	2	1	3
#09	3	3	2	1

A L9 Taguchi DOE was designed for gap-fill window improvement. 3 ECP factors of entry process are included, which will affect seed protection and Cu nucleation. Detail as Table1. Experiment result show 3/3/3/2 is best condition as Fig. 13. Confirm result show defect count <10ea as Fig. 14. the defect improvement is benefitted by MW current increase combine with start earlier as Fig. 15. The mechanism can be well explained by the paper [5]. An increase of plating current produced smaller nuclei and higher nuclei population density, which is very necessary when Cu seed is extreme thin.

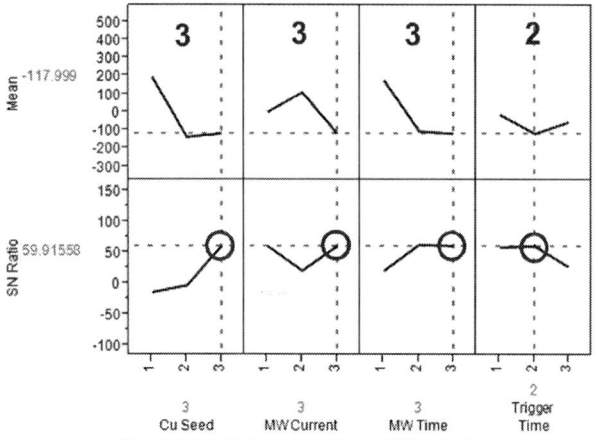

Fig.13 DOE Result of gap-fill window

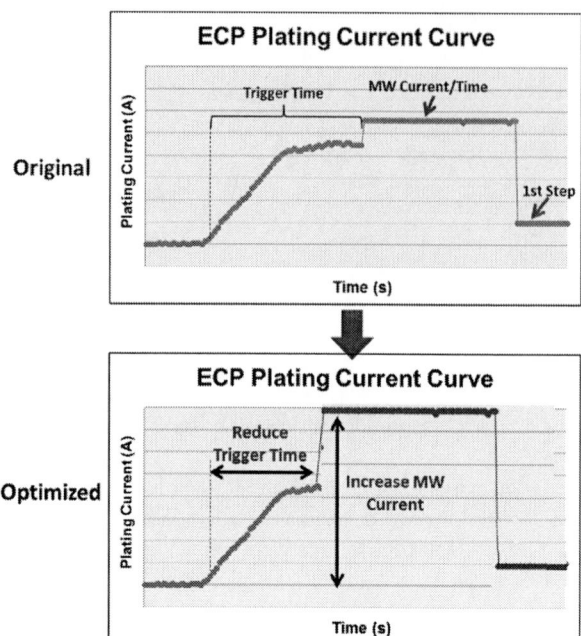

Fig.14 Confirm result of DOE good predict condition

Fig.15 ECP plating current curve optimization.

CONCLUSION

Cobalt was used as the seed enhancement layer instead of PVD Ta layer to improve Cu continuity. Defect between dense trench line area and SRAM area are quite tradeoff. Trench area requests bigger top opening, while SRAM area requests more Cu step coverage on sidewall. Thicker Cu seed thickness, Higher AC bias power, Smaller DCE ratio are benefit for the defect in SRAM area. Inverse, thinner Cu seed, higher DCE ratio help keep bigger top opening of trench, finally result in good defect in Dense trench line area. Further experiment result shown defect on both SRAM and dense line area will be significantly reduced if wafer under 5x thickness, d-0.3 DCE ratio and b+400 AC bias power condition.

For more challenge structure wafer, A L9 Taguchi DOE was designed and the result show ECP MW current increase combine with start earlier can further benefit defect. Defect count almost free. The benefit was thought comes from Cu better nucleation capability at ECP process.

ACKNOWLEDGEMENTS

The authors would like to acknowledge HLMC FA department for the supports of TEM/EDS analysis and PIE, YE team for data collection and technical discussion.

REFERENCES

[1] C.C.Yang, T.Spooner, S.Ponoth, K.Chanda, A. Simon, C, Lavoie, et al., "Physical, Electrical, and Reliability

Characterization of Ru for Cu Interconnects", IITC2006. pp187-189

[2] F.Gstrein, R.Akolkar, S.Balakrishnan, B.Boyanov, M.Kobrinsky, A.Schmitz and M.Hussein, et al., "Reliability of Cu Interconnects with Gap Fill-Enabling Ruthenium Liners", AMC2008.pp13-14

[3] S.C.Seo, C. C. Yang, C.K.Hu, A.Kerber, D.Horak, K.Petrillo, et al.,"Copper Contact Metallization using Ru-baseedd Barrier Liners for 45nm and Beyond", AMC 2008. pp31-32

[4] Hung Yi Huang, C. H. Hsieh, S. M. Jeng, H. J. Tao, Min Cao, et al., "A New Enhancement Layer to Improve Copper Interconnect Performance"

[5] Darko Grujicic, Batric Pesic, "Electrodeposition of copper: the nucleation mechanisms"

OPTIMIZATION ON DEPOSITION OF ALUMINUM NITRIDE BY PULSED DIRECT CURRENT REACTIVE MAGNETRON SPUTTERING

Yu-Pu Yang[1], Te-Yun Lu[1], Song-Ho Wang[1], Hsueh-Er Chang[1], Peter j. Wang[2], Walter Lai[2], Yiin-Kuen Fuh[1,] and Tomi T. Li[1]*

[1] Department of Mechanical Engineering, National Central University, Taoyuan City 32001, Taiwan
[2] Delta Electronics In., Taoyuan City 32063, Taiwan, China
*Corresponding Author's Email: michaelfuh@gmail.com

ABSTRACT

In this study, pulsed dc reactive sputtering of aluminum nitride (AlN) thin films was investigated. The aluminum nitride thin films were deposited on Si (100) using a reactive direct current (DC) unbalanced magnetron sputtering system. The DC reactive sputtering was used in sputtering the aluminum targets in a mixture of argon (Ar) and nitrogen (N_2) plasma. Processes of aluminum target sputtering were carried out in an atmosphere of a mixture of Ar and N_2. However, pulsed DC reactive sputtering of aluminum targets was carried out at total pressures with N_2:Ar ratios from 7:30 to 45:15. In-situ optical emission spectrometry (OES) was applied to obtain the optimal deposition rate and the highest sputtering yield from the effects of flow nitrogen/argon (N_2:Ar) ratio and pulse frequency on OES intensity. Thus, we have compared Fourier-transform infrared spectroscopy (FTIR) spectra and X-ray diffraction (XRD) patterns of AlN films deposited on Si (100) by DC reactive sputtering with an Al target in the mixture of Ar and N_2. FTIR and XRD investigated the quality of the films and the preferred orientation.

Keywords—Sputtering; Physical vapor deposition (PVD); Aluminum Nitride (AlN); Pulsed DC; X-ray Diffraction(XRD); Fourier-transform infrared spectroscopy (FTIR); Power Generator

INTRODUCTION

Aluminum nitride (AlN) is an III-V compound hard coating material with wide bandgap, favorable thermal conductivity [1], superior anti-corrosive [2], and anti-wear [3] properties. Additionally, AlN is one of the materials in several piezoelectric applications for surface acoustic wave (SAW) devices [4]. Aluminum nitride was typically produced by sintering processes [5], chemical vapor deposition (CVD), or various techniques like pulsed laser deposition [6], molecular beam epitaxy [7]. Physical vapor deposition (PVD) such as DC reactive sputtering has used for the deposition of AlN [8] with various parameters including reactive gas and its flow rate, distance from target to substrate distance, substrate temperature, and pressures during sputtering [9]. Nevertheless, the dielectric compound material to be deposited may fall back on the target during reactive sputtering of the Al

target such that electrically insulates the target by inhibiting local sputtering and causing an arc [10]. Intermittent sputtering techniques of radiofrequency (RF) sputtering has been used [11]. However, RF power supplies are expensive with matching issues, and deposition rates are low since the cycle is reduced [12]. Recently, the deposition of dense coatings of oxides and nitrides by DC reactive sputtering [13] and is a well-developed deposition technique for coatings [14]. Moreover, the pulsed magnetron sputtering [15] is a promising alternative for efficient synthesis. AlN films at lower temperatures have excellent adherence towards substrate [16].

EXPERIMENTAL AND DISCUSSION

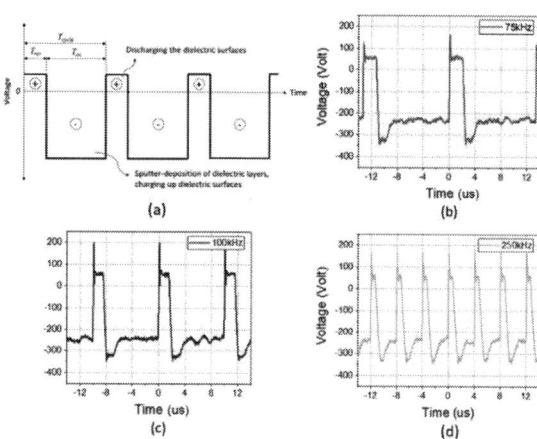

Fig.1 (a) Schematic representation of the voltage sequence applied to asymmetrical bipolar-pulsed DC sputtering of dielectrics. Oscilloscope trace of the cathode voltage waveform when operating in asymmetric bipolar pulsed mode at the waveform of (b) 75 kHz, (c) 100 kHz and (d) 250 kHz (85% duty).

Fig. 1 schematically shows a typical target voltage waveform sequence used in asymmetric bipolar pulsed DC sputtering. These particulates are involved in the films which affect the uniformity and the quality of the deposited films. The critical parameters are the pulse frequency, duty factor, and reverse voltage. Pulsed DC power should generate a rectangular waveform in Fig. 1 (a). The wave can be clearly seen in the asymmetric bipolar pulsed mode in Fig. 1 (b) 75 kHz, (c) 100 kHz and (d) 250 kHz (85% duty) applications and an oscilloscope trace of the target voltage wave. This oscillatory feature is not related to the plasma, but it is the nature of the pulsed DC power supply. A voltage spike above (b) +120 to +200 V was observed when the target voltage was changed from negative to positive. After this fast ringing, the reverse voltage flattened and remained at +50 V. The target voltage reaches up to -325 V on polarity reversal and then became constant at around -240V. Table I shows process parameters via the N_2 and Ar source gas in 300W and 5mtorr.

Table. 1. The reactive sputtering process parameters for aluminum nitride thin films

Parameters	Value	
Power	300 W	
Pressure	5 mtorr	
Distance	8 cm	
Substrate temperature	Room temperature	
Gas flow rate	Part 1	Part 2
(1) N_2	0-10 sccm	7, 45 sccm
(2) Ar	15 sccm	30, 15 sccm
Frequency	250 kHz	75, 100, 250 kHz
Duty cycle	85 %	85 %

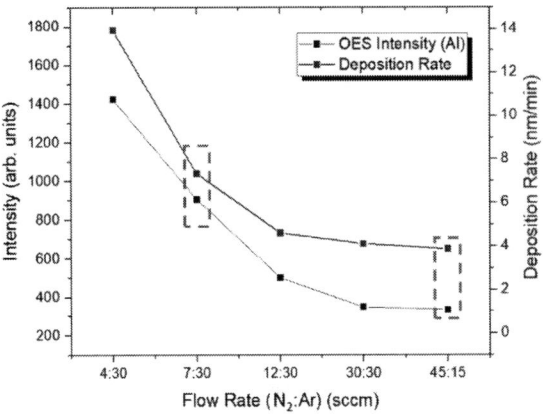

Fig. 2 Intensity of the spectral lines of excited aluminum species Al compared with a deposition rate of AlN films at different flow rates.

Fig. 2 shows the effect of the flow ratio of nitrogen/argon (N_2:Ar) on aluminum OES intensity was measured for obtaining deposition rate and the highest sputtering yield. The max value of the Al line intensity has been discerned at N_2:Ar as 4:30 by the experimental results. Discovering optimal deposition parameters must be carried out on Al spectral evolution. The Al intensity decreases markedly until increasing the N_2:Ar to above 30:30 from which the intensity of aluminum is kept at a constant near to 300. The change can be divided into two parts, named the transition (4:30, 7:30, 12:30) and compound (30:30, 45:15) sputtering regions, respectively. Increase in N_2:Ar continuously with deposition rate decreases mainly due to the increased proportion of N^+ or N_2^+ ions with lowered transfer momentum to the target, such then decreased the energy of the sputtered atoms. OES results further show that the transition region is sputtering at N_2:Ar as 7:30 (blue dotted rectangle) and the compound region is sputtering at N_2:Ar as 45:15(red dotted rectangle). In order to compare the AlN films grown in the transition and compound regions, 7:30 and 45:15 are chosen as N_2:Ar to study the effects of pulse frequency to the sputtered AlN.

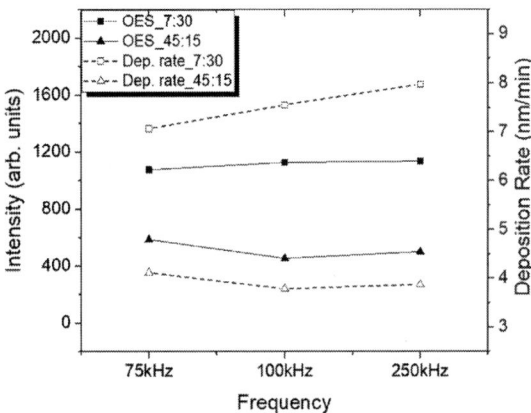

Fig. 3 The intensity of the spectral lines of Al compared with deposition rates of AlN films at different pulse frequencies.

Fig. 3 shows the intensity of the spectral lines of excited aluminum species Al compared with a deposition rate of AlN films at different pulse frequencies. We can clearly see that the deposition rate and Al intensity of the deposition process at N_2:Ar=7:30 are higher than at N_2:Ar=45:15. However, the effect of pulse frequency is not obvious. The deposition rate of the AlN films is approximately 7.5 nm/min (N_2:Ar=7:30) and 2.5 nm/min (N_2:Ar=45:15). Then increasing the pulse frequency from 75 kHz to 250 kHz, the deposition rate is increased from 7.06 nm to 7.97 nm (N_2:Ar=7:30) and 3.78 nm to 4.11 nm (N_2:Ar=45:15). The results show that the pulse frequency affects the deposition rate of the AlN film, but the effect on the Al line intensity is not obvious.

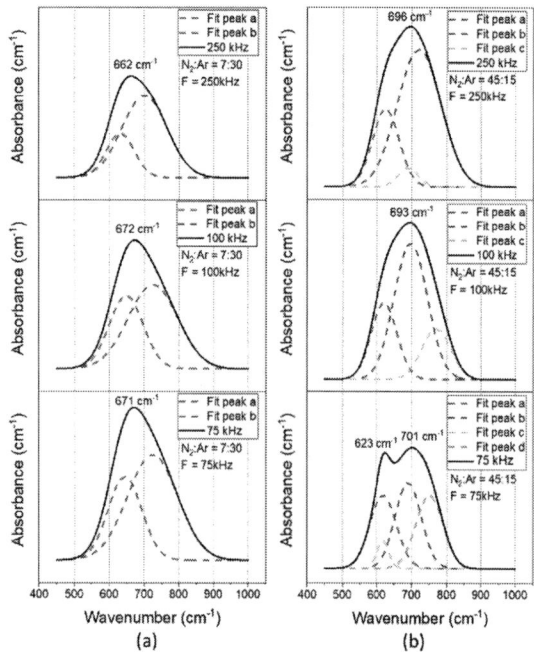

Fig. 4 Fourier transform infrared spectroscopy (FTIR) spectra of AlN films deposited under different sputtering conditions: (a) N_2: Ar=7:30 and (b) N_2: Ar=45:15

Fig. 4 shows the FTIR spectra of AlN films deposited under different N_2:Ar as 7:30, 45:15 and different pulse frequency of 75, 100 and 250 kHz, respectively. The spectra were obtained with a strong absorbance peak at 662~672 cm^{-1} by subtracting the background of the Si(100) substrate in Fig. 4(a). We observed energy shifts in the peaks a and b that were correlated with residual stress. However, a broader peak centered around 701 cm^{-1} under the films deposited at N_2:Ar as 45:15. Due to the formation of nitrogen AlN, the other peaks, namely c and d could possibly be inside the deposited AlN films. This additional residual stress leads to a shift in the FTIR absorption peaks namely peaks c and d could possibly be due to residue stress inside the deposited AlN films. We inferred that energy shifts in the peak were correlated with residual stress due to the fact that residual stress distorts the crystal unit cell, producing variations of the peak energy of IR absorption bands.

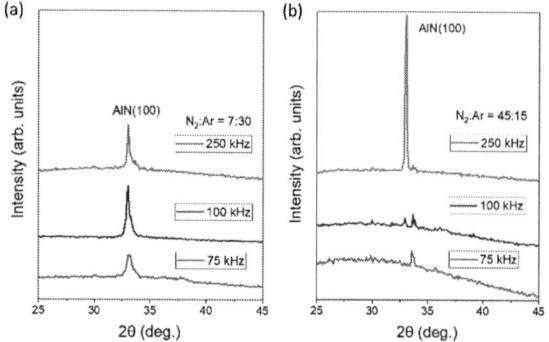

Fig. 5. XRD patterns of AlN thin films sputtered at N_2:Ar as (a) 7:30 and (b) 45:15.

Fig. 5 shows the XRD spectra at various frequencies (75 kHz, 100 kHz and 250 kHz). At N_2:Ar as 7:30, sharp AlN (1 0 0) peaks were observed in each sample, which means better crystallinity. However, At N_2:Ar as 45:15, the XRD results didn't show visible diffraction peak at the pulse frequency of 75 and 100 kHz, indicating the amorphous films. As shown in Fig. 5(a), XRD results further show that N_2:Ar=7:30 is the more suitable for AlN thin film growth.

CONCLUSION

In this paper, the deposition of AlN films was studied by pulsed DC reactive sputtering in a mixture of Ar and N_2 plasma. The FTIR spectra were obtained with a strong absorbance peak at around 696 cm^{-1} in optimal process parameters of the gas ratio of N_2:Ar=45:15 and the sputtering frequency of 250kHz. Peak shift can be correlated with residual stress, which is a parameter of importance in the deposition of AlN thin film that should be well controlled. In the XRD figure, with the same parameters, the peak intensity of AlN(100) is much more prominent. We can infer that the higher crystallinity induced from pulsed DC sputtering might become more prominent at a higher frequency. Therefore, this study based on the experimental design is explored allowing fast optimization of film deposition by means of pulsed DC sputtering process.

ACKNOWLEDGMENTS

This study was financially supported by Delta Electronics, Inc. Taiwan and Department of Mechanical Engineering, National Central University, Taiwan.

REFERENCES

[1] B. J. Pong et al., Journal of Applied Physics, vol. 83, 1998, pp. 5992-5996.
[2] H. Hoche, S. Groß, T. Troßmann, J. Schmidt, and M. Oechsner, Surface and Coatings Technology, vol. 228,

2013, pp. S336-S341.

[3] P. Vissutipitukul and T. Aizawa, Wear, vol. 259, 2005,pp. 482-489.

[4] H. Morkoç, S. Strite, G. B. Gao, M. E. Lin, B. Sverdlov, and M. Burns, Journal of Applied Physics, vol. 76, 1994, pp. 1363-1398.

[5] J.-Y. Qiu, Y. Hotta, K. Watari, K. Mitsuishi, and M. Yamazaki, Journal of the European Ceramic Society, vol. 26, 2006, pp. 385-390.

[6] M. Ishihara, K. Yamamoto, F. Kokai, and Y. Koga, Japanese Journal of Applied Physics, vol. 40, 2001, pp. 2413-2416.

[7] X. Wang and A. Yoshikawa, Progress in Crystal Growth and Characterization of Materials, vol. 48-49, 2004, pp. 42-103.

[8] M. A. Auger, L. Vázquez, M. Jergel, O. Sánchez, and J. M. Albella, Surface and Coatings Technology, vol. 180-181, 2004, pp. 140-144.

[9] G. David A and S. Ismat Shah, "Handbook of Thin Film Process Technology," ed: CRC Press, 2018.

[10] W. D. Sproul, Vacuum, vol. 51, 1998, pp. 641-646.

[11] I. C. Oliveira, K. G. Grigorov, H. S. Maciel, M. Massi, and C. Otani, Vacuum, vol. 75, 2004, pp. 331-338.

[12] J. L. Vossen and W. Kern, Thin Film, 1991.

[13] M. Benegra, D. G. Lamas, M. E. Fernández de Rapp, N. Mingolo, A. O. Kunrath, and R. M. Souza, Thin Solid Films, vol. 494, 2006, pp. 146-150.

[14] P. J. Kelly and R. D. Arnell, Vacuum, vol. 16, 1998, pp. 2858-2869.

[15] V. Dimitrova, D. Manova, T. Paskova, T. Uzunov, N. Ivanov, and D. Dechev, Vacuum, vol. 51, 1998, pp. 161-164.

[16] J.L. Vossen, and W. Kern (Eds.), Thin Film Processes, Academic Press, New York, 1978.

STUDY OF PREB PROCESS IN FDSOI

Yang Song*, Zhanhai Yang, Xia Tang, Yanfei Ma, Feng Niu, Changfeng Wang

Huali Microelectronics Corporation, Shanghai 201203, China

*Corresponding Author's Email: songyang@hlmc.cn

ABSTRACT

In this paper, photo resist etch back (PREB) process is studied for 22nm node HKMG FDSOI technology. Prior to dummy poly removal (DPR) process, PREB process is introduced in order to overcome pattern loading induced challenge --- the wide and narrow poly gate structure need to be opened separately in order to make sure the dummy gate within the whole chip can be removed thoroughly. Advantages and challenges of this PREB technique will be explained and studied.

INTRODUCTION

As the CMOS technology reaches 22nm and beyond, FDSOI devices are proposed for ultra-low power and IoT (Internet of Things) market. The ability to better control of short channel effect and the interest of using back-bias makes it considerable attractive for many applications. [1] High K metal gate (HKMG) technique can also be implemented to further improve the device performance, which makes it possible to adjust the NMOS and PMOS threshold voltage separately. In this work, HKMG FDSOI transistors with 22nm gate lengths are fabricated with high K first and metal gate last technology, as is shown in *Figure 1*. However, this HKMG approach requires more complex processes, of which one critical step is the dummy poly removal (DPR) process.[2] The DPR process always starts with a dry plasma etch step. However, one of the most significant challenges is the pattern loading effect due to the plasma etch selectivity --- the hardmask layer of large poly area is more difficult to be removed as compared with the narrow poly area.

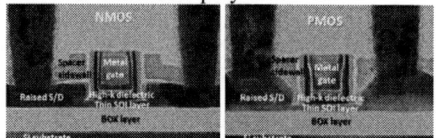

Figure 1: TEM pictures of FDSOI devices

In this work, photo resist etch back (PREB) process is introduced to open the hard mask of large poly area firstly and then followed by the etching back of photo resist in the rest areas. This new added process may increase the process complexity by adding one more mask to process flow. But the advantage is also obvious, which is able to eliminate the pattern loading effect. Meanwhile, it also put challenges on the process control, as the etching back process needs to be well controlled in order to obtain conformal dummy gate profile without damage on source and drain NiSi structure.

EXPERIMENT

The FDSOI devices are processed on 300mm (110) SOI wafers with BOX thickness of 20nm. In this work, we propose the PREB scheme, which consists of four main steps, as is shown in *Figure 2*: first, use photo process to open the PR in large poly area; second, PR etch back first step (EB1), in order to remove the hardmask oxide in the large poly area; third, hardmask etch back within the left areas and stop on spacer nitride (EB2); Finally, H_3PO_4 will be used to remove the PREB spacer nitride.

PREB Photo
Etch back first step
Etch back second step
Spacer wet removal

Figure 2: PREB scheme;

It should be noticed during the PR coating process the coating uniformity needs to be well concerned, as is shown in *Figure 3*. It is found that the using of traditional PR material shows $\Delta d \sim 900A$ ($\Delta d = d1 - d2$) loading effect between wide-narrow poly boundary and isolated poly structure, as is shown in *Figure 3*. In this work, we investigated the impact of different coating materials and coating schemes. The material change will have a slight improvement of the coating uniformity, and needs to be further investigated.

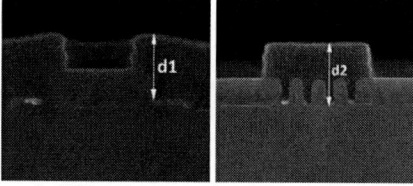

Figure 3: TEM pictures after PR coating of isolated poly (left) and wide-narrow poly boundary (right)

The gate profile is a critical concern during the PREB process. As the hardmask of poly gate usually consists of one oxide layer and one nitride layer, the EB1 and EB2 plasma etch rate need to be well controlled. If the etch rate selectivity of oxide is much higher than the nitride, horn profile will be seen, as shown in *Figure 4*. The horn profile will put challenges to subsequent DPR process, and leading to incomplete dummy poly removal. However, if the selectivity of SiN is too high, the spacer loss will be more and the spacer will be hard to protect the gate, leading to sharpened gate top, which will make subsequent metal gate filling becoming more challenged. In our work, finally, with proper etching gas chemistry, conformal dummy poly profile can be obtained.

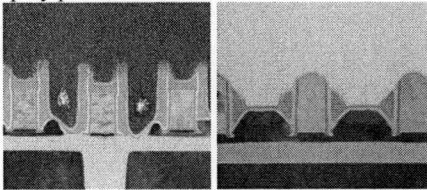

Figure 4: TEM pictures of horn profile example (left) and sharpened gate top example (right)

The PR remaining thickness is another critical parameter in PREB process. For FDSOI devices, this challenge becomes more highlighted due to its unique raised up source and drain feature --- NiSi structure becomes easier to be damaged during the PR etch back process. In this work, we investigated this issue from two aspects: one is to thin down the poly hardmask thickness; the other is to perform EB1 etch amount test to search for a proper etch condition.

Thin down the hardmask thickness has two benefits for PREB process, the first one is to ensure more PR remaining in large active area, and the other is to suppress the horn formation, as thinner hardmask comes with less pattern loading effect.

If we fix the etching gas chemistry, the plasma dry etch amount of EB1 also put significant influence on the device profile. If the etch amount is too high, in large active areas, the PR remain amount will not be enough to protect the NiSi structure on source and drain in subsequent EB2 process, resulting in large source drain resistance in final WAT test. But if the etch amount is too low, in dense/iso boundary areas, there will be PR remain left on poly gate, leading to the hardmask hardly removed in EB2 step, as is shown in *Figure 5*.

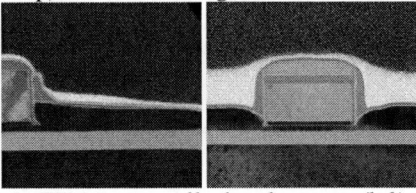

Figure 5: TEM pictures of high etch amount (left) and low etch amount (right)

In our work, we studied the PREB EB1 etch amount versus etch time. The correlation is almost linear, as is shown in Fig. 4. Thus the etch time should be well controlled in a specific range, in order to make sure there is no PR remain left on poly, but considerable amount of PR remain on source and drain NiSi structure.

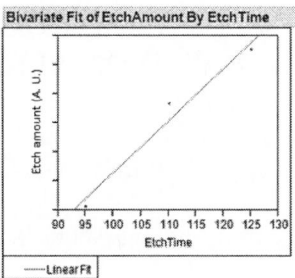

Figure 6: Correlation of etch amount vs etch time

CONCLUSION

Pattern-dependent non-uniformities are investigated in this work focusing on the pattern loading effect in 22nm node FDSOI. In our work, PREB process is introduced in order to overcome this loading issue. The PREB photo and etch back process are investigated. Finally, with optimized PREB condition, conformal dummy gate profile without damage on source and drain NiSi structure can be obtained.

ACKNOWLEDGEMENTS

The authors wish to express their gratitude to litho and etch colleagues at HLMC for their kind support. Special thanks to Duanquan Liao for his expertise, feedback and advice during the whole work.

REFERENCES

[1] C. Fenouillet-Beranger *et al.*, *ESSDERC 2008 - 38th European Solid-State Device Research Conference*, Edinburgh, 2008, pp. 206-209.

[2] F. Li, Q. Han and H. Zhang, "A study of photo-resist strip at dummy poly gate removal process," *2016 China Semiconductor Technology International Conference (CSTIC)*, Shanghai, 2016, pp. 1-3.

IDT STRUCTURE OPTIMIZATION DESIGN
BASED ON ALN/SI SUBSTRATE FOR SAW DEVICES

Kaixuan Li, Fang Wang*, Shuo Yan, Meng Deng, Huanhuan Di, Wei Li, Kailiang Zhang*

Tianjin Key Laboratory of Film Electronic & Communication Devices

School of Electrical & Electronic Engineering, Tianjin University of Technology, Tianjin 300384,

China

*Corresponding Author's Email: fwang75@163.com; kailiang_zhang@163.com

ABSTRACT

In this work, aluminum nitride (AlN) film with good piezoelectric properties was grown on the silicon (Si) substrate as piezoelectric layer, and the properties of surface acoustic wave (SAW) devices with different interdigital transducer (IDT) structures were researched by using Rectangle function, Hanning function and Kaiser function. MATLAB and e-LINE plus software were used to generate layout files quickly and accurately. Devices with 300nm finger width were tested at room temperature and the results indicated that devices with Kaiser function structure show better resonant waveforms, the center frequency was up to 4.94GHz, the inhibition degree of sidelobe increased obviously to 43.53dB, and insertion loss was -5.87dB. This work play an active role in the design and research of high performance surface acoustic wave devices.

INTRODUCTION

With the wide application of SAW devices in the field of communication systems, bio-sensing and microfluidic applications, how to improve the performance of devices had attracted extensive attention [1-2]. As for the design of SAW devices should be based on the requirements of center frequency and insertion loss of the device, the substrate material and the IDT material which were easy to be realized in the process should be selected, and then the structure of the IDT was optimized according to the technical parameters of the device [3].

IDT was the core part of SAW device, the restraint of sidelobe of its frequency response has always been an important research content. Using various window functions to weight the electrode of IDT was a common means to restrain sidelobe. According to the simulation analysis, out-of-band restraint of general IDT with equal length was only about 13dB. However, the effect of window function on out-of-band inhibition of IDT with apodized weighting electrode length was very significant [4-5].

In this work, the IDT structure was modified by the method of apodization with the input IDTs weighted and the output IDTs unweighted to obtain good frequency response characteristics of the devices. The structure of IDT was characterized by scanning electron microscope (SEM), and frequency response characteristics of the

devices were tested by network analyzer to analyze the influence of different window functions weighting.

EXPERIMENT

Apodized weighting SAW devices based on AlN thin films were fabricated. First, the 900nm AlN thin films were deposited on Si (100) substrates by DC magnetron sputtering with pure aluminum target (99.999%) and 220W DC power at 0.3Pa with the gas mixture of Ar/N2 (1:1) was used. After the use of MATLAB simulation, the selection of a better frequency response characteristics of the Kaiser window design and production. To optimize the design of IDT, MATLAB software was used to write the corresponding program to generate a common intermediate format (CIF) file containing the graphic information of a single device that can be recognized by e-LINE plus software. With calling the generated file in CIF format, drawing function of the software was used to complete the design of whole device. Bidirectional IDTs with Kaiser function weighting were fabricated by electron beam lithography system. Titanium was evaporated by electron beam evaporation system as the top electrode. Device schematic diagram is shown in Figure 1.

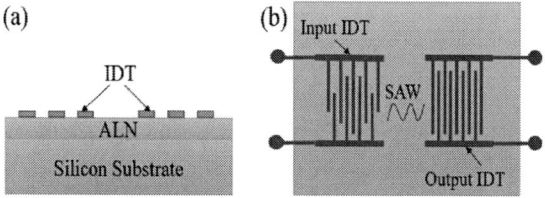

Figure 1: Schematic diagram of surface acoustic wave device, (a) Section structure of the device; (b) structural representation of IDTs

RESULTS AND DISCUSSION

It was found that AlN (002) crystal orientation has the better piezoelectric properties [6]. The XRD diffraction pattern of AlN films on Si (100) substrate are shown in Figure 2 (a). Obviously, there is a peak at 2θ of 36.20°corresponding to AlN (002).

978-1-7281-6559-2/20 $31.00 © 2020 IEEE

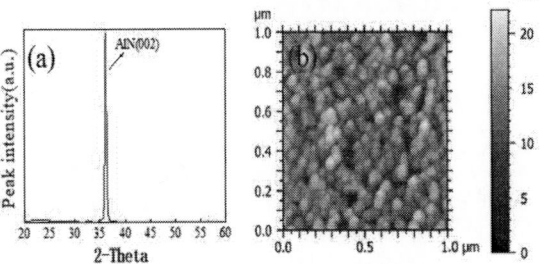

Figure 2: Test diagram of ALN, (a) XRD diffraction pattern; (b) AFM images of AlN films

As shown in Figure 2 (b), AFM (1×1 μm²) was used to characterize the surface morphology and the average root mean square (RMS) roughness of AlN films on Si (100) substrate. The AlN grains are oval, uniform and dense, with a root mean square roughness of 3.2nm, which satisfies the requirements for preparing a SAW device.

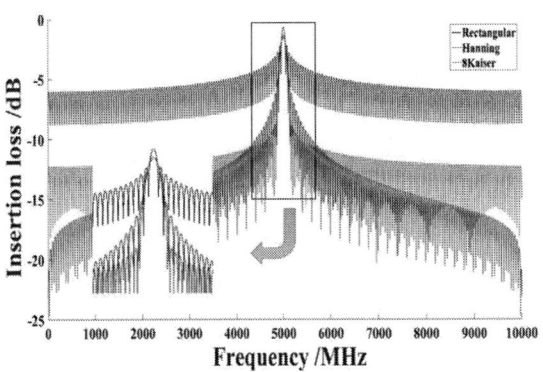

Figure 3: Rectangular function, Hanning function and Kaiser function weighted IDT frequency response simulation

Window functions were often used for apodized weighting the electrode length of IDT. Rectangular function is the most widely used window function, the default that signal without weighting is to make the signal through the Rectangular function. As shown in Figure 3, the frequency response characteristics of Rectangular function, Hanning function and Kaiser function were simulated by MATLAB. Rectangular function has a narrow main lobe width, but when weighted by the Rectangular function, the sidelobe has strong interference to the main lobe, which makes the signal resolution difficult. While, Hanning function takes an advantage of the shift characteristic of Fourier transform, and gradually reduces the height of the sidelobe by means of frequency curve superposition until the height of the sidelobe is almost eliminated, which can enhance the strength of the signal main lobe and reduce the strength of the sidelobe at

the same time. Kaiser function achieves higher sidelobe restraint performance by giving up part of the performance advantage of the main lobe width, which fully reflects the tradeoff between the main lobe and sidelobe attenuation. Therefore, the electrode length of the IDT by apodized weighting using Kaiser function is selected for further production and testing.

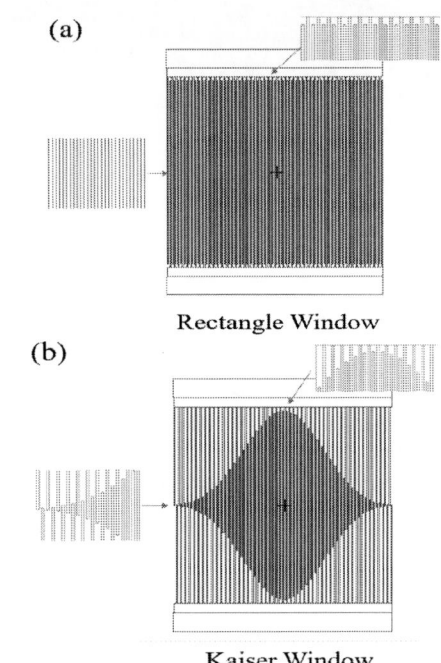

Figure 4: Derived graph of apodized weighting IDT, (a) Rectangle function; (b)Kaiser function

When given the frequency response of the transducer to design the weighting IDTs, taking the inverse Fourier transform of frequency response and getting the impulse response, the window functions are selected to truncate the impulse response to obtain a finite length impulse response. According to the correspondence between the impulse response and IDT geometric structure, the design is carried out through a CIF file using the e-LINE plus software in Figure 4.

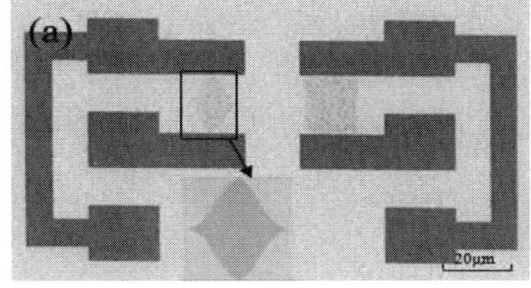

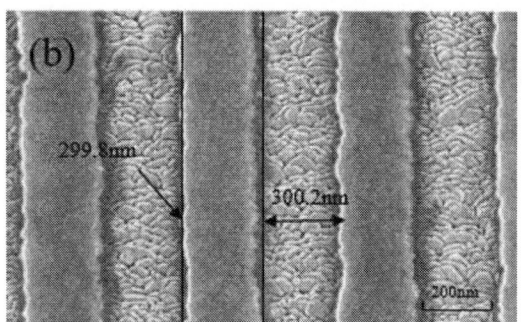

Figure 5: IDT structure diagram and width of finger and interval, (a) Device structure export diagram; (b) Width of finger and interval

It is necessary to add a multi-strip coupler (MSC) when both transmitting and receiving transducers using weighting IDTs [7]. Here we take the apodized weighting transducer as transmitting IDT and the homogeneous transducer as receiving IDT as shown in Figure 5(a). The SAW devices with interdigital width of 300nm were fabricated by electron beam lithography. SEM was used to observe the completeness of interdigital fingers made of 80nm Ti as the electrode in Figure 5(b).

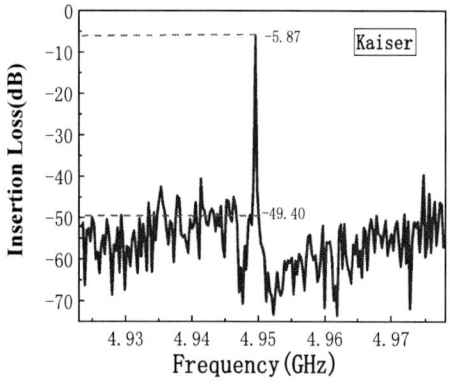

Figure 6: Frequency response test diagram of the SAW devices composed of apodized weighting IDT

In Figure 6, the frequency characteristic curve (S_{21}) of apodized weighting IDT indicates that the apodized weighting method based on Kaiser function can restrain the out-of-band response of IDT to 43.53dB, but also increase the width of the main lobe. Test result also shows that the central frequency of the prepared SAW device is 4.94GHz, and insertion loss down to -5.87dB. Width of interdigital at nanometer level can obviously increase the central frequency of device, but also increase the insertion loss to some extent.

CONCLUSION

In this work, simulation and measured results show that, the apodized weighting method based on Kaiser function can restrain the out-of-band response of IDT to 43.53dB. Compared with 13dB of general IDT with equal length, it has a great improvement. SAW device based on the AlN /Si substrate has a frequency up to 4.94GHz and insertion loss down to -5.87dB. The combination of MATLAB and e-LINE plus software can quickly and accurately generate IDT layout and improve the experimental efficiency. In addition, the apodized weighting method and fabrication process can be further researched and trial-produced to improve the frequency response characteristics of SAW devices.

ACKNOWLEDGEMENTS

This work was supported by Natural Science Foundation of Tianjin City (Grant Nos. 17JCYBJ C16 100, 18JCZDJC30500 and 17JCZDJC31700), National Key Research and Development Program of China (Grant No.2017YFB0405600) and National Natural Science Foundation of China (Grant Nos.61404091, 61274113, 61505144, 51502203 and 51502204).

REFERENCES

[1] Y.Q. Fu, J.K. Luo, N.T. Nguyen, A.J. Walton, A.J. Flewitt, et al, *Progress in Materials Science*, vol. 89, 2017, pp.31-91.
[2] KALININ V A, et al, *George Heiter Proceeding of the 2004 IEEE Radio and Wireless Conference*, New York, USA: IEEE, 2004: 187-190.
[3] Ivan D. Avramov, et al, *International Frequency Control Symposium and Exposition*, 2006 IEEE.
[4] Deboucq J, Duquennoy M, Ouaftouh M, et al. *Review of Scientific Instruments,* 2011, 82(6): 64905.
[5] Fu C, Elmazria O, Sarry F, et al. Sensors and Actuators A: Physical, 220(2014):270-280.
[6] Fang Wang, Fuliang Xiao, Dianyou Song, Lirong Qian, Yulin Feng, et al, *Microelectronic Engineering*, vol.199, 2018, pp.63-68.
[7] Ryo Nakagawaı, Ken-ya Hashimoto, et al, *Japanese Journal of Applied Physics*, 57, 07LD18 (2018).

INVESTIGATION OF FDSOI RAISED S/D FORMATION

Yanfei Ma[*], Yang Song, and Changfeng Wang

Huali Microelectronics Corporation, Shanghai, China 201203

*Corresponding Author's Email: mayanfei@hlmc.cn

ABSTRACT

In this paper, the formation of FDSOI raised source/drain (RSD) will be presented targeting for 22nm node technology. The source and drain are formed by epitaxial growth of Si or SiGe from SOI layer for NMOS or PMOS respectively. During the fabrication process, there are two major concerns regarding to the RSD formation --- the remained SOI acts as the seed for epitaxial growth, the epitaxy growth method will influence the performance of the device. In this work, we will discuss experimentally from three aspects in order to obtain sufficient SOI remain, and will propose two SiGe epitaxial schemes in order to attain uniform epitaxial profile. Advantages and disadvantages of each scheme will be investigated.

INTRODUCTION

Strain engineering has been involved in advanced CMOS generations to improve device performance, which is, to introduce tensile stress to NMOS channel and compressive strain to PMOS channel. For NMOS devices, SMT (stress memorization technique) has been widely used to transfer tensile stress into the channel, where for PMOS transistors, SiGe structure can significantly enhance the channel carriers' mobility due to germanium induced compressive strain and thus to improve the device drive current. For FDSOI devices, take PMOS as an example, two approaches has been proposed to incorporate strain into channel --- epitaxial SiGe and channel SiGe. [1] As channel SiGe approach may increase the process complexity and induce contamination problems, in this work, the epitaxial SiGe is adopted, which is, to form SiGe raised up source and drain (RSD) besides gate.

The RSD region is formed by epitaxial growth of Si or SiGe from the SOI layer for NMOS or PMOS, which puts a requirement on SOI remained thickness before the formation of source and drain. There are various etching and deposition steps can lead to significant SOI loss and exacerbate process challenges, hence it is necessary to make process changes in order to obtain sufficient SOI retains. In this work, three approaches are proposed to reduce the SOI loss --- to use a new pad oxide scheme; to introduce an extra Si-epi layer prior to source/drain formation; and to improve SOI loss uniformity by etching process tuning to indirectly reduce SOI loss. This will be elaborated experimentally in the experiment part. Second, the epitaxial growth method will also be presented. Advantages and disadvantages will be discussed.

EXPERIMENT

Three approaches will be illustrated to control the SOI loss: to reduce the SOI loss, to compensate the SOI loss, and to improve SOI loss uniformity. With sufficient SOI remains, a series of SiGe epitaxial experiments are performed and optimized to obtain uniform epitaxial profile.

Approach 1 to reduce SOI loss

Prior to the formation of the active-area (AA) for devices, the pad oxide (SiO_2) is usually thermally grown on substrate in O_2 environment, on top of the pad oxide is a layer of silicon nitride (Si_3N_4), which serves as the hardmask to define the AA pattern. The main purpose of the pad oxide under the nitride is to compensate SiN induced high tensile stress, as the thermal oxide deposited on silicon is under compressive strain. Pad oxide also plays a significant role in subsequent well implant process as the oxide thickness determines the depth of implantation.

For FDSOI devices, the formation of pad oxide inevitably consumes considerable amount of SOI, and this SOI loss makes the subsequent epitaxial process more difficult.

In this work, a novel approach has been proposed and adopted to achieve pad oxide formation, which is, to grow a thin oxide layer by ISSG (in-situ steam generation) process and then followed by ALD (atomic layer deposition) oxide process. ISSG enables growth of high quality oxide films due to the presence of reactive species (atomic O) at reduced pressure, and it also helps to reduce the intrinsic oxide defects (eg. strained Si-O bonds, Si dangling bonds etc.) compared with traditional furnace process. Post ISSG oxidation, low temperature ALD oxide is chosen for thermal budget concern. Its self-limiting surface reaction mechanism makes ALD films have many excellent properties, including atomic-level thickness control, good surface uniformity and perfect step coverage. *Figure 1* shows the comparison of ISSG combined ALD scheme and tradition furnace process, and ~20A SOI thickness gain can be obtained using this novel approach.

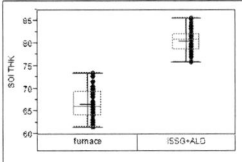

Figure 1: Comparison of SOI retains between furnace and ISSG+ALD approach

Approach 2 to reduce SOI loss

Compared with traditional FDSOI, after gate formation, we propose to introduce a thin Si epitaxial layer (5~10nm) prior to source/drain formation, in order to compensate the SOI loss consumed during the previous fabrication process. *Figure 2* shows an example of this added layer. In addition, this added layer also increases S/D size effectively both vertical and horizontally, leading to a reduction of S/D resistance and S/D to gate parasitic capacitance, thus enhances the drive current and attenuates the RC delay effect.

Figure 2: TEM cross section of addition of the Si epitaxial layer

Approach 3 to improve SOI loss uniformity

The within wafer range of SOI loss will increase after poly etch process, as is shown in Table I. The SOI thickness of edge zone is thinner than wafer center, which will cause wafer edge epitaxial growth risk and weaken the uniformity of wafer thickness map. After poly etching tuning, the within wafer SOI thickness range is improved, as is shown in Table II. The improved uniformity of SOI will be helpful for epitaxial growth.

TABLE I.   THE SOI THICKNESS VARIATION BY POLY ET

Wafer map	Before poly ET	After poly ET
SOI center-SOI edge	8A	16A

TABLE II.   THE SOI THICKNESS VARIATION BY DIFFERENT POLY ET RECIPE

Wafer map	baseline	Experiment 1	Experiment 2
SOI center-SOI edge	16A	12A	10A

There are two methods for SiGe epitaxial growth: firstly, epitaxial growth with seed layer, as is shown in *Figure 3*. This epitaxial growth structure is formed by seed, bulk & cap layer. The seed layer with low doping Ge% is also necessary to serve as a strain buffer layer in order to form high quality and high Ge% bulk SiGe stressor in the subsequent step [2]. Simultaneously, this seed layer can also act as transition layer to reduce defects on silicon interfacial bonding. The Si cap above SiGe bulk is used to protect SiGe from damage by wet process with H3PO4. Secondly, epitaxial growth without seed

layer, as is shown in *Figure 4*. This epitaxial growth structure is formed by bulk & Si cap layer. This structure has better epitaxy profile and uniformity, but the interface states between SiGe bulk and SOI layer may become worse.

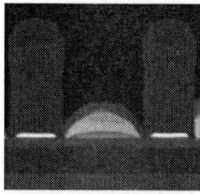

Figure 3: TEM cross section of SiGe epitaxy with seed layer

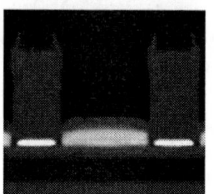

Figure 4: TEM cross section of SiGe epitaxy without seed layer

CONCLUSION

In this work, we proposed three approaches to improve the SOI remaining thickness before SiGe epitaxial growth: to use a new pad oxide scheme by ISSG and ALD oxide deposition; to introduce a thin Si epitaxial layer prior to source/drain formation and to control the SOI loss uniformity during poly etch step. In addition, we also investigated two epitaxial growth methods. Advantages of each method are also analyzed --- SiGe epitaxial growth without seed layer shows better profile uniformity, and it can be adopted if the stability of devices is of more concerns; whereas SiGe epitaxial growth with seed layer method is able to achieve better interfacial bonding, thus can be adopted if the device performance is of more concerns.

ACKNOWLEDGEMENTS

The authors wish to express their gratitude to Jun Tan and Jiaqi Hong at HLMC for their kind support. Special thanks to Duanquan Liao for his expertise, feedback and advice during the whole work.

REFERENCES

[1] K. Cheng *et al.*, "High performance extremely thin SOI (ETSOI) hybrid CMOS with Si channel NFET and strained SiGe channel PFET," *2012 International Electron Devices Meeting*, San Francisco, CA, 2012, pp. 18.1.1-18.1.4.

[2] Ming Mao Chu and June-Hua Chou, "Physical yield improvement for SiGe Selective Epitaxial Growth fabrication process on nano scale pMOS strain

engineering," *2009 IEEE Nanotechnology Materials and Devices Conference*, Traverse City, MI, 2009, pp. 42-45.

OPTIMIZATION OF THE CD UNIFORMITY (CDU) IN SILICON OXIDE SPACER PROCESS FOR 5 NM FIN SAQP PROCESS FLOW

Qingqing Wu[1], Weihao Lin[1], Xiaoqiang Zhou[1], Jinhua Zhang[1], Jing Li[2], Leng Han[2], Jianjun Zhu[1], Yushu Yang[1], Qiang Wu[1], and Shoumian Chen[1]

[1] Shanghai IC R&D Center, 497 Gaosi Road, Zhangjiang Hi-Tech Park, Shanghai, China.

[2] Piotech Co., Ltd., 900 Shuijia Road, Hunnan District, Shenyang, Liaoning Province, China.

Corresponding Author's Email: wuqingqing@icrd.com.cn

ABSTRACT

As dimensions of semiconductor devices continue to shrink, ordinary film deposition processes, such as Chemical Vapor Deposition (CVD), Physical Vapor Deposition (PVD), cannot meet the requirement of the film uniformity and target thickness. Atomic Layer Deposition (ALD), based on surface-controlled and self-saturating adsorption reactions, where the film consists of sequential atomic layers, becomes more and more popular. To meet scaling requirements, multi-patterning solutions, like Self-Aligned Double Patterning (SADP), Self-Aligned Quadruple Patterning (SAQP) and Litho-Etch-litho-Etch (LE/LE), that utilize the already installed base of 193 nm immersion exposure tools are first adopted by the industry to not only reduce the linewidth, but also improve line edge roughness or line width roughness in 20/14 nm node and beyond, say 5 nm node. Within the framework of SAQP, the final Fin Critical Dimension Uniformity (CDU) and pitch walking is related to the profile and CDU of spacer deposition closely. In this study, we report a brief summary of ALD application in 5 nm FinFET process flow with a fin pitch of 24 nm. Meanwhile, we have demonstrated the oxide spacer deposition for 5 nm fin SAQP process, and the oxide spacer CD uniformity can be controlled to below 0.5 nm (3 sigma). Moreover, the deposition process is demonstrated on domestic made apparatus.

INTRODUCTION

Thin film deposition has always been a critical process in semiconductor industry [1-2]. Many deposition technology, such as PVD, CVD, have been developed over the past few decades for materials like amorphous silicon, silicon nitride (Si_3N_4), silicon oxide (SiO_2), Hafnium oxide (HfO_2) and others. Nevertheless, more precise and controllable deposition technology is needed as semiconductor manufacturing requirement becoming increasingly stringent. ALD relies on alternated and self-limiting reaction, which is suitable for conformal deposition with a thickness uniformity controlled at the sub-monolayer level [3-5].

Herein, we first describe the reaction mechanism of ALD, and analyze the control parameters. Then we give a brief summary of ALD application in 5 nm FinFET process flow with a fin pitch of 24 nm. Finally, we will demonstrate an oxide spacer deposition for 5 nm fin SAQP process, and the CD uniformity of the oxide spacer can be controlled to below 0.5 nm (3 sigma).

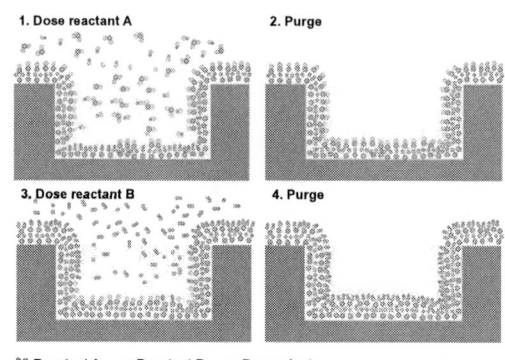

Figure 1. Schematic illustration of one ALD reaction cycle.

Atomic layer deposition is a self-limiting reaction between gaseous reactants and exposed solid surface. One full ALD cycle typically consists of four steps as schematic illustrated in Figure 1. Step 1: The first reactant A reacts with the limited functional groups on the surface in a self-terminating way. Step 2: The unreacted reactant A or gaseous by-products are purged. Step 3: Introduction of the reactant B to react with the absorbed reactant A on the surface in a self-terminating way. Step 4: The excess of reactant B and gaseous by-products are pumped away. In other words, the substrate is exposed to the reactants A and B alternatively, and sequential deposition an ABAB will follow...

In traditional ALD, namely thermal ALD, the conformality of deposition is influenced by the molar mass and reactivity of the reactants as well as partial pressures and exposure time. Plasma Enhanced Atomic Layer Deposition (PEALD) is an energy enhanced ALD, where a plasma is employed during step 3 (Figure 1) of the cyclic deposition process. Due to the use of plasma species, PEALD allows for a wider choice in reactants, substrate temperature, and process conditions, compared with thermal driven ALD [6].

With respect to PEALD, different control process parameters (e.g., pressure, plasma power, temperature, reagent flow rates) have an influence on the conformality, film composition, and deposition rate. Figures 2(a)-2(c) present the saturation curves for the Grow-Per-Cycle (GPC) as a function of the PEALD process parameters, such as precursor dose time, precursor purge time, and plasma exposure time. When using shorter precursor dosing, shown in Figure 2(d), there is no saturated chemisorption of precursor on substrate, especially for high aspect ratio structure, which will eventually bring problems, such as, lower deposition rate, and poor step coverage, and different thickness in dense and isolated areas. For shorter precursor purge time (Figure 2(e)), residual precursors remaining in the chamber will react under plasma like parasitic Plasma Enhanced Chemical Vapor Deposition (PECVD), which leads to high deposition rate and poor step coverage as well as undesirable particles and loading effect. For very short plasma exposure time, as shown in Figure 2(f), it may result in poor conversion that can bring impurities in film as well as more susceptible to wet etch. Meanwhile, the properties and conformality of PEALD film also depend on the other process parameters, such as deposition temperature, gas pressure, plasma configuration, and chamber design.

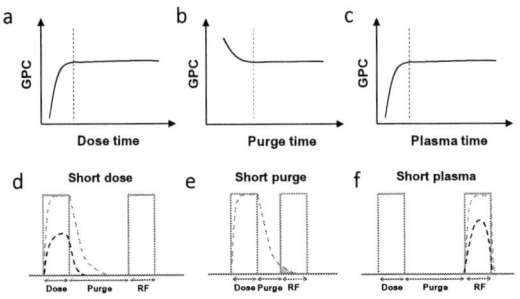

Figure 2: (a)-(c). Saturation curves for the deposition rate as a function of the ALD process parameters. (d)-(f). Schematic illustration of the effect of different process parameters on deposition.

ATOMIC LAYER DEPOSITION IN 5 NM FIN SAQP PROCESS

The miniaturization of semiconductor devices leads to the increasingly stringent requirement in film deposition. In 2007, Intel introduced the 45 nm technology including Hf-based gate dielectric film fabricated by ALD. Then ALD has been gradually used in several key process steps. Here we report a brief summary of ALD application in 5 nm FinFET process flow with a fin pitch of 24 nm (Table 1). In Front End Of the Line (FEOL), the spacers for SADP and SAQP (e.g., oxide, nitride) are all deposited by ALD to achieve a uniform thickness along the sidewall, which will be closely related to the final Fin CDU and some pitch walking. From the device consideration, in order to reduce the effective

capacitance and ensure basic mechanic strength, low k spacer is generally integrated by several low k dielectric materials, such as SiOCN, SiCN, SiBCN or SiN. And there may exist several deposition and etch steps to form the final low k spacer. In Middle Of the Line (MOL), high k material is still HfO_2, and TiN needs to be deposited by PEALD on the surface of the HKMG as a protective layer. Then, an ALD TaN process is required to prevent the diffusion of work function metals in a role of barrier layer or etch stop layer.

Table 1: A brief summary of ALD application in 5nm FinFET process flow.

Zone	Process Description	Application	Reference Thickness (Å)
FEOL	ALD oxide	SAQP spacer for fin	150 Å
	ALD SiN	SADP spacer for gate	220 Å
	ALD SiN	Offset spacer	10 Å
	ALD SiOCN	Low k spacer	35 Å
	ALD SiN	Low k spacer	15 Å
	ALD SiN	Etch stop layer	30 Å
	ALD oxide	Liner for STI	25 Å
MOL	ALD oxide	Protection layer	20 Å
	ALD HfO2	HKMG	15 Å
	ALD TiN	HKMG cap layer	10 Å
	ALD TaN	HKMG ESL	10 Å
	ALD TiN	HKMG PWF	15 Å
	ALD SiN	Etch stop layer	150 Å
	ALD TiN	Barrier Deposition	15 Å
BEOL	ALD TiO	SADP spacer for M1	150 Å

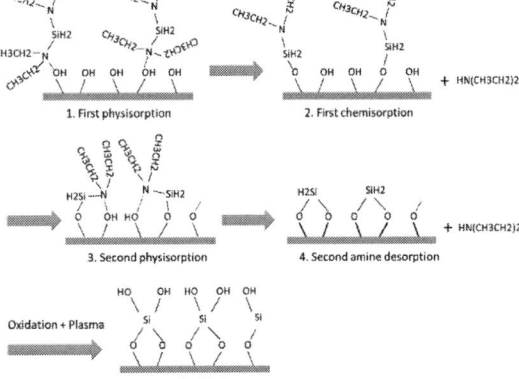

Figure 3: Reaction mechanism of PEALD oxide based on $H_2Si[N(C_2H_5)_2]_2$ as a precursor.

The reaction mechanism of oxide deposition based on $H_2Si[N(C_2H_5)_2]_2$ as a precursor is shown in Figure 3. Analogous surface reaction mechanism was studied earlier [7-8]. When the precursor $H_2Si[N(C_2H_5)_2]_2$ is introduced to the reaction chamber, it is to form hydrogen bonding between the amine of $H_2Si[N(C_2H_5)_2]_2$ and the hydroxyl of Si surface, or the initial physisorption, which is quite exothermic. Then the precursors bond to the surface, and the first amine is removed. Research has shown that the first amine desorption is entropy driven, and more

978-1-7281-6559-2/20 $31.00 © 2020 IEEE

exothermic during first amine elimination, which yields a larger population of post reaction complex on the surface. In return, it will lead to an overall faster first chemisorption and amine desorption. In the second deamination reaction, the earlier reaction will still be physical adsorption between the amino group and the hydroxyl group, and then the amine is removed to form a silicon-oxygen bond. The above reaction is done during precursor dosing, and the precursors that have not been adsorbed will be pump away during purge. In the last reaction, the surface reaction will be dominated by Oxygen radical species delivered by plasma.

We then focus on the optimization of oxide spacer in 5 nm SAQP flow. We use amorphous Si for the mandrel and silicon oxide for the spacer, nitride for the etch stop layer and 2nd hard mask. Silicon oxide spacer is deposited on amorphous Si and nitride by the Piotech FT-300T that is a 12-inch production equipment suitable for PEALD. $H_2Si[N(C_2H_5)_2]_2$ is used as the Si precursor, and Ar as carrier and O_2 as oxidant. The reaction temperature stays at 400 ℃ and the reaction pressure stays at 1.8 Torr.

RESULTS AND DISCUSSION

In order to meet the CDU requirements of 5 nm process, we carried out the SiO_2 film uniformity test on bare silicon wafers, as shown in Figure 4. After 215 deposition cycles, the mean thickness of film is 140 Å with a value of 0.65 Å/cycle. The standard 3σ shows 2.36 Å and standard 3σ percent is 1.68%, which completely meets the specification for film uniformity.

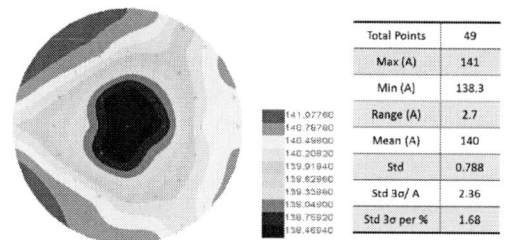

Total Points	49
Max (A)	141
Min (A)	138.3
Range (A)	2.7
Mean (A)	140
Std	0.788
Std 3σ/ A	2.36
Std 3σ per %	1.68

Figure 4: Film uniformity of PEALD SiO2 on bare wafer.

We then explore the film deposition on patterned wafer, and a schematic is depicted in Figure 5. After mandrel polysilicon etch, there will form a trench with a depth of 90 nm and a width of 60 nm. After 192 silicon oxide deposition cycles, the CDSEM shows that the mean value of oxide spacer thickness is 137.9 Å, and the 3σ shows a good value of 4.2 Å, which meets the specification of value less than 5 Å.

We then use High-Resolution Transmission Electron Microscopy (HRTEM) to further determine the uniformity of first oxide spacer (Figure 6). The cross-sectional HRTEM demonstrates a uniform film along sidewalls with no visible defect. The depression at the bottom of

sidewall may be related to the notch formed during mandrel etch. As a future improvement, the carrier gas flow and pressure will be adjusted to optimize the overall uniformity and step coverage.

Figure 5: CDSEM of before and after ALD oxide spacer.

(1) After mandrel1 etch

	-5	-4	-3	-2	-1	0	1	2	3	4	5		
4					36.49	36.47	36.65						
3			36.60	36.63	37.26	37.37	36.93	36.96	36.77				
2		37.37	37.10	37.04	37.63	37.35	36.71	36.70	36.82	36.31			
1	36.96	37.53	37.50	37.61	37.17	36.73	36.56	36.70	57.11	36.88	36.04	Maximum	38.15
0	37.11	37.99	37.86	37.17	37.16	36.41	36.42	36.38	36.55	36.39	36.16	Minimum	35.17
-1	37.77	37.96	37.12	37.61	37.49	37.28	37.28	36.78	37.34	36.74	35.82	Mean	36.95
-2		37.22	37.77	36.15	37.50	36.90	36.55	36.69	36.23	37.00		Max-Min	2.98
-3			37.47	37.57	37.86	36.99	36.79	36.86	36.49			3 Sigma	1.73
-4					36.58	36.12	35.17						

(2) After ALD oxide spacer

	-5	-4	-3	-2	-1	0	1	2	3	4	5		
4					64.09	64.05	64.07						
3			64.60	64.07	64.80	64.74	64.68	64.46	64.10				
2		64.91	64.40	64.44	65.15	64.53	64.12	64.35	64.73	63.77			
1	64.18	65.18	64.93	64.76	64.72	64.61	64.44	63.85	64.68	64.66	63.66	Maximum	65.71
0	64.81	65.64	65.46	64.86	65.10	64.68	64.43	64.28	64.15	64.19	63.48	Minimum	63.03
-1	64.79	65.47	65.31	65.66	65.40	64.74	64.59	64.71	64.77	64.32	63.40	Mean	64.54
-2		64.76	65.20	65.71	65.46	64.56	64.56	64.55	64.06	63.94		Max-Min	2.68
-3			64.67	64.84	64.94	65.06	64.31	64.35	63.50			3 Sigma	1.76
-4					63.97	63.57	63.03						

(3) ALD oxide spacer c=(b-a)/2

	-5	-4	-3	-2	-1	0	1	2	3	4	5		
4					13.8	13.8	13.7						
3			14	13.7	13.8	13.7	13.9	13.8	13.7				
2		13.8	13.7	13.7	13.8	13.6	13.7	13.8	14	13.7			
1	13.6	13.8	13.7	13.6	13.8	13.9	13.9	13.6	13.8	13.9	13.8	Maximum	14.14
0	13.9	13.8	13.8	14	14.1	14	14	13.8	14	13.9	13.7	Minimum	13.47
-1	13.5	13.8	14.1	14	14	13.7	13.7	14	13.7	13.8	13.8	Mean	13.79
-2		13.8	13.8	13.8	13.9	13.8	13.9	13.9	13.9	13.5		Max-Min	0.67
-3			13.6	13.8	13.6	14	13.8	13.7	13.8			3 Sigma	0.42
-4					13.7	13.7	13.9						

Figure 6: Cross-sectional HRTEM of first ALD oxide spacer for SAQP.

CONCLUSION

In conclusion, we introduce the reaction mechanism of ALD, and analyze the effect of control parameters on deposition rate and conformality. Then we report a brief summary of ALD applications in 5 nm FinFET process flow with a fin pitch of 24 nm. Finally, we have demonstrated the silicon oxide spacer for 5 nm fin SAQP process by PEALD. And the CDU of oxide spacer can be controlled to 0.42 nm (3 sigma). Moreover, the deposition process is demonstrated on domestic made apparatus.

ACKNOWLEDGEMENTS

This work was the collaborative project with Piotech

Co., Ltd. and NAURA Microelectronics Equipment Co., Ltd. The authors would like to thank all the members of the Advanced Device Material Lab, the ALD team at the Technology Development Division of Shanghai IC R&D Center for the great collaboration.

REFERENCES

[1] D. M. Mattox. Handbook of physical vapor deposition (PVD) processing. 2010.

[2] H. O. Pierson. Handbook of chemical vapor deposition (CVD). 1999.

[3] V. Miikkulainen, M. Leskelä, M. Ritala, R. L. Puurunen. J. Appl. Phys., 2013, 113, 021301.

[4] V. Cremers, R. Puurunen, J. Dendooven, Appl. Phys. Rev. 2019, 6, 021302.

[5] H. B. Profijt, S. E. Potts, M. vandeSanden, W. Kessels. J. Vac. Sci. Technol., 2011, 29 (5).

[6] P. L.G. Ventzek, T. IWAO, A. Ranjan. IEEE Nanotechnology Magazine. 2019, 9.

[7] M. L. O'Neill, H. R. Bowen, A. Derecskei-Kovacs, K. S. Cuthill, B. Han, M. Xiao. The electrochemical society interface, 2011, pp 33-37.

[8] G. Dingemans, C. A. A. van Helvoirt, M. C. M. van de Sanden, and W. M. M. Kessels. ECS Transactions, 2011, 35 (4), pp 191-204.

ENHANCING HIGH TEMPERATURE ADHESION PERFORMANCE VIA A RENOVATED LEADFRAME SURFACE TREATMENT

Din-Ghee Neoh[1], Tee Weikok[2], Liao Jinzhi Lois[3], Jia Wenping[4], Yee Boonhwa[5], Boon-Seong LEE[6], Wang Bisheng[7], Zhang Xi[8], Hua Younan[9], Li Xiaomin[10], Mario Strauch[11], Boon-Chye LEE[12]

[1, 6, 11, 12]Atotech (Singapore) Chemicals Pte Ltd, Buroh Street, Unit #03-01,
Surface Engineering Hub, Singapore 627563

[2,5] Sumitomo Bakelite Singapore Pte Ltd, 1 Senoko S Rd, Singapore 758069

[3,8,9,10] WinTech Nano-Technology Services Pte. Ltd.,10 Science Park Road, #03-26, The Alpha
Science Park II, Singapore 117684

[4,7] Huawei Technologies Co Ltd, Bantian Huawei Base, Longgang District, Shenzhen, China 518129.
*Corresponding Author's Email: din-ghee.neoh@atotech.com

ABSTRACT

Leadframe based micro-electronic packages are widely used in the semiconductor industry. The metal leadframe surface properties directly impact its adhesion with epoxy molding compound (EMC). During the manufacturing process and application services, the electronic packages unavoidably suffer from moisture and heat stress. When the moisture in the package volatilizes during the reflow process, its force may be strong enough to break the adhesion between the leadframe and EMC, causing interfacial delamination. In addition, the leadframe surface inevitably oxidized (especially Cu surface) under high thermal loading during the assembly process, and the thickness of metal oxides layer could rise quickly at such high temperature operation. The thick metal oxides layer may deteriorate the interfacial strength between leadframe and EMC, leading to higher tendency of package delamination after stressed

The work here evaluated an innovated leadframe surface treatment method called NEAP, with which the copper surface of leadframe was first covered with a thin layer of silver, and then the silver surface was further chemically treated to form a special oxidization film to act as an adhesion promoter for EMC. The NEAP leadframe was compared with the conventional copper leadframe (selective silver) and rough copper leadframe, in terms of its package level reliability. Under severe thermal loading conditions, most of the leadframes could not survive the reliability tests and showed package delamination. On the contrary, the NEAP leadframe show outstanding adhesion performance compared to the others. The NEAP process concept was discussed in detail. The NEAP leadframe is suggested for use in micro-electronic packages that require high thermal performance.

1. INTRODUCTION

The "substrate" based advanced micro-electronic packages are showing increasing demands in the industry

in recent years, but the conventional leadframe based micro-electronic packages remain its importance as a cost-effective solution in the packaging world. This is especially true after the emerging of micro-leadless packages such as DFN and QFN.

To continuously improve the cost efficiency, there is a trend in leadframe market whereby the leadframe strip size (length x width) has been slowly increased and evolved from 225mm x 50mm, to 250mm x 75mm, to 270mm x 90mm, and finally now to 300mm x 100mm. Many assembly challenges had been reported as a result of this change in leadframe dimension. However, the highlighted problems were more related to strip handling issues and how to prevent frame warpage. There is little information about the impact of this change to package delamination.

With higher pin counts of ICs and increased leadframe strip size, leadframe would have to undergo longer wire bonding time during assembly process, which involves high temperature operation. The new generation leadframe finish needs to survive under this harsh high temperature condition. Thus, the objective of this study is to evaluate an innovated leadframe coating which can be compatible with the harsh high temperature conditions.

Button shear test is commonly applied in the industry to assess the adhesion strength between epoxy molding compound (EMC) and leadframe. There are many factors which may affect the button shear test results, especially the preconditions before molding step. Therefore, in this paper, the relationship among preconditions, leadframe finish, button shear test results and Confocal Scanning Acoustic Microscopy (CSAM) results were also studied.

2. EFFECT OF "PRECONDITIONING" BEFORE MOLDING

In order to see the effect of preconditioning on button shear test, 6 types of leadframe finish were prepared, together with 5 types of preconditioning, respectively. 6 types of leadframe finish were listed in TABLE 1, that is copper (Cu), silver (Ag), pre-plated finish (PPF), Rough Cu, Non-Etched Adhesion Promoter (NEAP) X.2 and NEAP X.1. All the leadframes were provided by Atotech. 5 types of preconditioning were listed in TABLE 2. Most people in industry perform button shear test without any preconditioning to simulate heat loading in assembly process. In TABLE 2, set A is the control conditioning, that is without any heat treatment and plasma, which is similar to most practice in the industry. Set B is with heat treatment of 200°C, 30 min, to simulate wire bonding temp effect. Set C is with heat treatment of 200°C, 30 min and plasma using argon (Ar) gas for 5min, to simulate effect of plasma after frame undergo hear treatment. Set D is with heat treatment of 175°C, 30 min + 200°C, 30 min and plasma using argon (Ar) gas for 5min, to simulate the combination heat treatment involved in both die attach (DA) and wire bonding process. Set E is with heat treatment 175°C, 30 min + 230°C, 60 min and plasma argon (Ar) gas for 5min, to simulate the extreme heat loading conditions in the assembly process.

TABLE I. LEADFRAME SURFACE TREATMENT

#	Surface	Surface Treatment Process
1	Cu	No Adhesion Promoter (Control)
2	Ag	No Adhesion Promoter (Control)
3	PPF (Ni/Pd/Ni)	No Adhesion Promoter (Control)
4	Rough Cu	MoldPrep HMC
5	NEAP X.2	Immersion Ag + AgPrep
6	NEAP X.1	Ag Strike + AgPrep

TABLE II. PRECONDITION (BEFORE MOLDING)

Set	Heat Treatment	Plasma
A	No	No
B	200°C, 30 min	No
C	200°C, 30 min	Yes (Ar, 5 min)
D	175°C, 30 min + 200°C, 30 min	Yes (Ar, 5 min)
E	175°C, 30 min + 230°C, 60 min	Yes (Ar, 5 min)

The above 6 types of leadframe went through the 5 types of preconditioning in TABLE 2. After that, they were molded with EMC pallets with custom made molding equipment. The molded samples were post mold cured (PMC) at 200°C for 4h. Button shear test was performed on these samples after PMC with Dage 4000 Plus Bond Tester, as showed in Figure 1.

The button shear test results were showed in Figure 2. It was found that preconditioning before molding could have a significant impact to adhesion performance of Cu based surfaces, i.e. untreated Cu and Rough Cu.

For the untreated Cu, it was found that: a) Heat loading deteriorates the adhesion performance significantly; b) Under moderate heat loading conditions, Plasma could help to restore the adhesion performance; c) Under extreme heat loading conditions, Plasma could not restore the adhesion performance.

On the other hand, for the Rough Cu, it was found that: a) Rough Cu showed better heat resistance than Untreated Cu surface; b) however, at extreme heat loading conditions, Rough Cu started to showed deterioration in adhesion performance.

Compared to Cu base surface, both Ag and PPF were not so much affected by the preconditioning. However, the adhesion strength for these 2 surfaces are relatively low.

It is interesting to observe that both NEAP X.2 and NEAP X.1 showed high adhesion strength under all kinds of preconditioning.

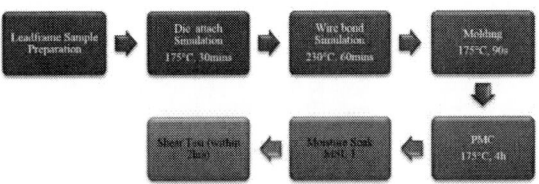

Figure 1: Test flow of button shear test.

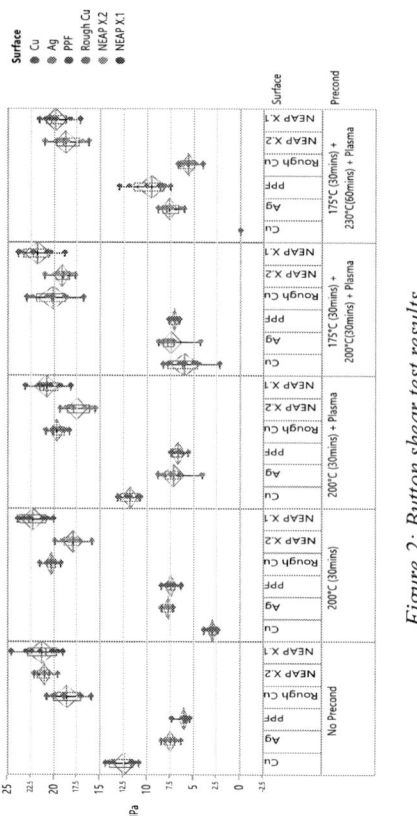

Figure 2: Button shear test results.

TABLE IV. EMC TYPES

Parameters	*EMC 1*	*EMC 2*	*EMC 3*
Filler Shape	Spherical	Spherical	Spherical
Filler Content (%)	88.0	88.0	88.0
Filler Sieving Size	55	55	55
Epoxy	MAR	MAR	MAR
Harderner	MAR	MAR	MAR
Adhesion system	A	B	C

TABLE V. DIE SIZE

#	**Type**	**Size**
1	Big	5mm x 5mm
2	Medium	4mm x 4mm
3	Small	3mm x 3mm

The 3 types of leadframe were die attached at 175°C, 30 min, followed by wire bonding (oven bake 200°C, 30 min). Plasma cleaning was applied before molding. After molding and post mold cure, the samples went through moisture sensitivity level 1 (MSL1) 1 at 85°C/85% RH, for 168hr. Three times reflow was conducted at 260°C after MSL1. CSAM were used to check all the devices after reflow. The test flow was showed in Figure 3.

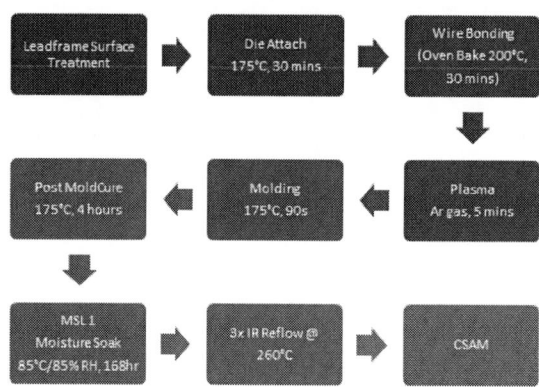

Figure 3: Delamination checking test flow of DOE 1.

Figure 4 shows the CSAM results from the impacts of leadframe types, EMC types and die size. Ag untreated and PPF untreated leadframe encountered 100% delamination, with all conditions.

Untreated Cu leadframe showed the best CSAM performance, by comparing with Ag untreated and PPF. Compared 3 different types of EMC. Besides, in terms of

3. CSAM FOR PACKAGE DELAMINATION

3.1 DOE 1

Delamination after MSL1 test is a big challenge in IC assembly house. In order to see the factors affecting the delamination, CSAM (model 301 HD2 PVA TePla) is applied to check the molded device after MSL 1 test. 3 main factors were studied, that is leadframe type, EMC type, and die size. Here 3 different leadframe types were used provided by Atotech (see Table 3). 3 types of EMC were provided by Sumitomo (see Table 4) and 3 types of die size were also provided by Sumitomo (see Table 5).

TABLE III. LEADFRAME TYPES

Test Vehicle	*Surface*	*Description*
1	Cu	QFN 10 X 10, 84L - Ring Ag
2	Ag	QFN 10 X 10, 84L - Spot Ag
3	PPF	QFN 10 X 10, 84L - Full Ni/Pd/Au

EMCs, EMC 1 showed the best results. One more finding on the die size revealed that smaller die size showed less tendency of delamination. It is because smaller die size leads to more contact area between EMC and leadframe, consequently higher adhesion strength. However, all the 3 test vehicles could not achieve zero delamination with die size of 5mm x 5mm, and 4mm x 4mm.

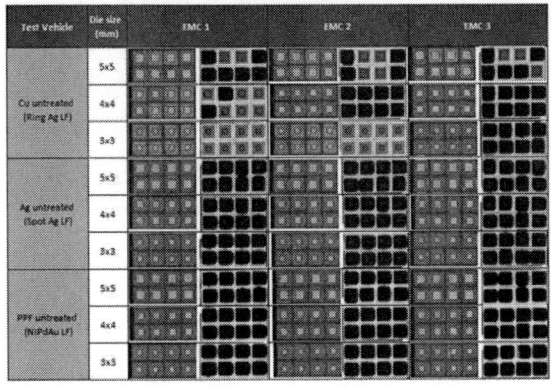

Figure 4: CSAM results of DOE 1.

3.2 DOE 2

From DOE 1, all the test vehicles could not achieve zero delamination with die size of 5mm x 5mm. To further evaluate the leadframe surface treatment impact on delamination, DOE 2 was conducted. In DOE 2, die size 5mm x 5mm and EMC 1 were selected. The test flow was the same as DOE 1. The leadframe surface treatment types were listed in Table 6.

TABLE VI.   LEADFRAME SURFACE TREATMENT TYPES

Leg	Surface	Test Vehicle	Surface Treatment Process
1	Cu	QFN 10 X 10, 84L – Ring Ag	Untreated (Control)
2	Rough Cu	QFN 10 X 10, 84L – Ring Ag	MoldPrep HMC
3	NEAP X.2	QFN 10 X 10, 84L – Spot Ag	Immersion Ag (<20nm) + AgPrep

CSAM results of DOE 2 after MSL 1 and 3x reflow were showed in Figure 5. Only untreated Cu leadframe encountered delamination. Both Rough Cu and NEAP X.2 showed zero delamination.

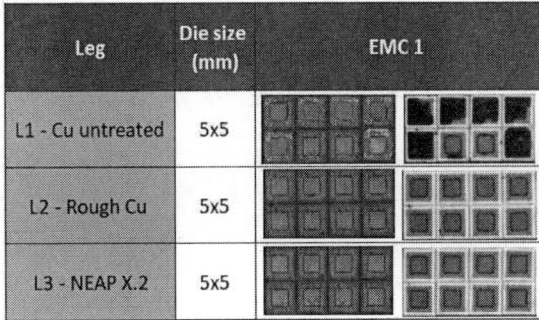

Figure 5: CSAM results of DOE 2.

3.3 DOE 3

In the earlier DOE 2, both Rough Cu and NEAP X.2 showed zero delamination. Further extreme condition was applied to differentiate the leadframe surface performance. In DOE 3, die size 5mm x 5mm and EMC 1 were selected. The leadframe surface treatment types were listed in Table 7. NEAP X.1 was added in this test. The test flow was the same as DOE 1, except wire bonding simulation conditions was increased from 200°C, 30 mins to 230°C, 60 mins which is to simulate the harsh wire bond condition in application. The test flow was showed in Figure 6.

The CSAM results of DOE 3 were showed in Figure 7. In can be seen that under the extreme wire bonding simulation conditions (i.e 230°C, 60 mins), all units in L1 showed severe delamination. NEAP X.2 was also found to show package delamination. However, both Rough Cu and NEAP X.1 was able to survive under this harsh testing conditions without any delamination.

TABLE VII.   LEADFRAME SURFACE TREATMENT TYPES

Leg	Surface	Test Vehicle	Surface Treatment Process
1	Cu	QFN 10 X 10, 84L – Ring Ag	Untreated (Control)
2	Rough Cu	QFN 10 X 10, 84L – Ring Ag	MoldPrep HMC
3	NEAP X.2	QFN 10 X 10, 84L – Spot Ag	Immersion Ag (<20nm) + AgPrep
4	NEAP X.1	QFN 10 X 10, 84L – Spot Ag	Immersion Ag ((~0.1 μm) + AgPrep

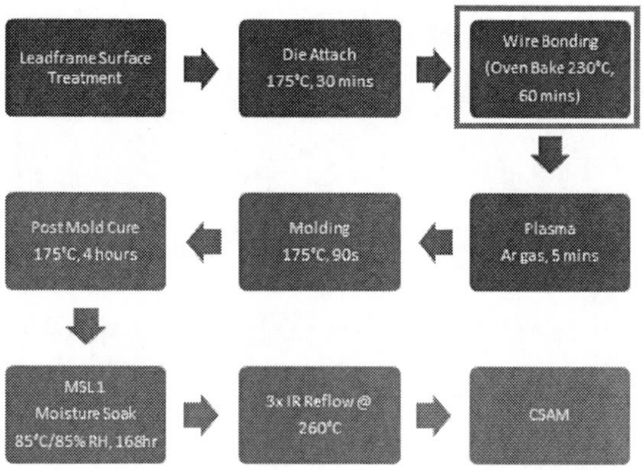

Figure 6: Delamination checking test flow of DOE 3.

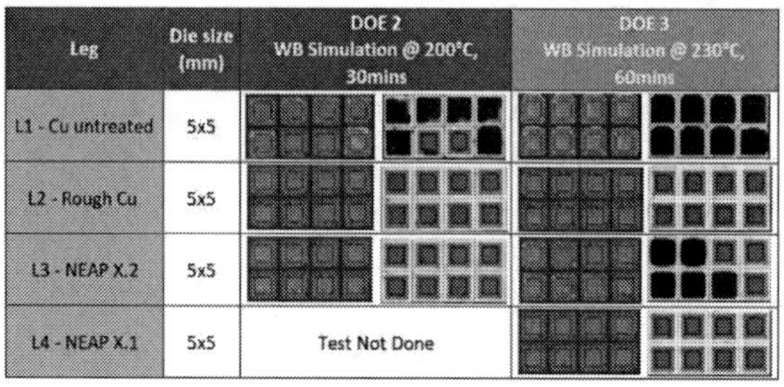

Figure 7: CSAM results of DOE 3.

3.4 DOE 2 & DOE 3 further reliability testing

Due to the high reliability of requirements for automotive industry, meeting MSL 1 with zero delamination is no longer the ultimate aims. New requirements as below are demanded by the industry:

1) MSL 1 @ 260°C + 2000 hours HTS (bake at 175°C)
2) MSL 1 @ 260°C + 2000 cycles TC (-55°C/+150°C)
3) MSL 1 @ 260°C + 96 hours Autoclave Test (121°C / 15 psi)

The samples from DOE 2 and DOE 3 with no delamination, that is Rough Cu, NEAP X.2, NEAP X.1 (see Table 8), were further subjected to high temperature stress test at 200°C up to 1000h to the compare the package reliability under this extreme test condition.

TABLE VIII. LEADFRAME SURFACE TREATMENT TYPES

Leg	*Surface*	*Test Vehicle*	*Surface Treatment Process*
2	Rough Cu	QFN 10 X 10, 84L – Ring Ag	MoldPrep HMC
3	NEAP X.2	QFN 10 X 10, 84L – Spot Ag	Immersion Ag (<20nm) + AgPrep
4	NEAP X.1	QFN 10 X 10, 84L – Spot Ag	Immersion Ag ((~0.1 µm) + AgPrep

The CSAM results were showed in Figure 8. For MSL 1 combined with high temperature stress (HTS), it was found that both NEAP X.2 and NEAP X.1 performed better than Rough Cu.

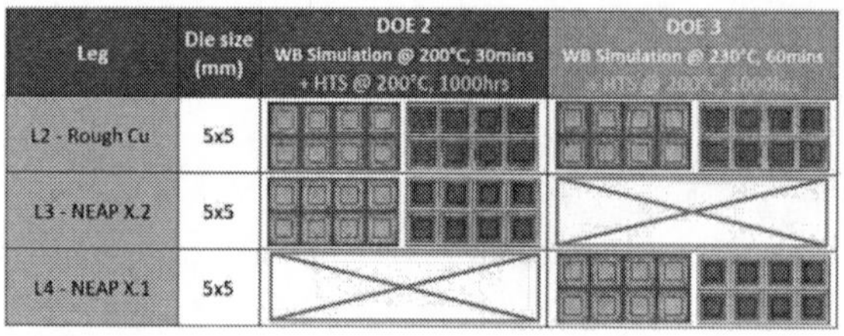

Figure 8: CSAM results of DOE 2 & DOE 3 further reliability testing.

4. CORRELATION BETWEEN BUTTON SHEAR TEST RESULTS AND MSL 1 PERFORMANCE

Effort was tried to correlate the button shear results and MSL 1 delamination performance. Figure 9 showed the button shear test results and Figure 10 showed the MSL 1 CSAM results. From Figure 9, it can be found that the button shear test results rank (from best to worst): NEAP X.1 > Rough Cu > NEAP X.2 > Ag > PPF > Cu. On the other hand, from Figure 10, the MSL 1 CSAM ranking is same as button shear test ranking, that is: NEAP X.1 = Rough Cu > NEAP X.2 > Ag = PPF = Cu.

When comparing both ranking results, it indicates that certain levels of correlations exist between button shear test results and MSL 1 delamination performance. Thus, button shear test could be a useful tool to predict the "relative" package delamination performance.

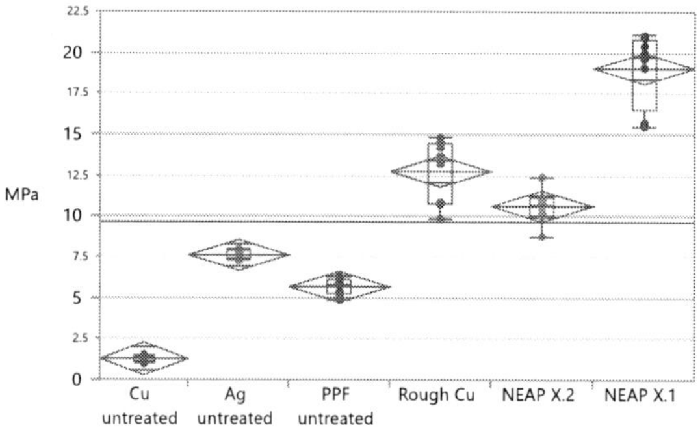

Figure 9: Button shear test results.

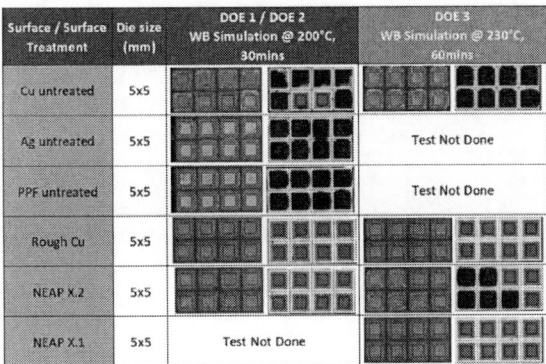

Figure 10: MSL 1 CSAM results.

5. NEAP: <u>N</u>ON-<u>E</u>TCHED <u>A</u>DHESION <u>P</u>ROMOTER

5.1 What is NEAP Process

NEAP process is a new innovative way of preparing leadframe finish to enhance the adhesion of leadframe to EMC and other polymeric materials within the IC package. NEAP involves an adhesion promoter step called "AgPrep", which is designed for silver surfaces. In order to achieve good adhesion properties, the entire leadframe surface must be silver coated before the AgPrep step.

The conventional plating line for copper leadframe with selective silver plating can be easily converted into NEAP Process, simply by replacing its silver stripping step with AgPrep step, as illustrated in Figure 11. By skipping the silver stripping step, this means the whole leadframe surface will be then covered with silver, and subsequently AgPrep is applied on this silver surface to enhance its adhesion properties.

If silver strike (silver thickness approximately 0.1μm) is used prior to selective silver plating step, the yielded process is named "NEAP X.1"; If silver immersion (silver thickness < 20nm) is used prior to selective silver plating step, the yielded process is named "NEAP X.2".

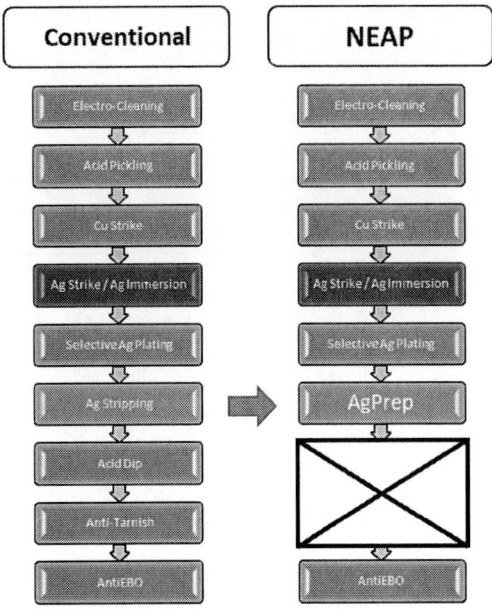

Figure 11: Converting a Conventional Plating Process for Copper Leadframe with Selective Silver to NEAP Process.

5.2 The Principle of AgPrep

The AgPrep process employs the concept of modifying the chemical state of the silver surface through an applied electrochemical potential in order to promote enhanced chemical bonding to the EMC or other polymeric materials. The modified silver surfaces show an enhanced silver oxide layer, which is later hydrated at ambient environment to form silver hydroxide. These hydroxyl groups formed on the silver surface are responsible for the interactions with the polar organic resins in the epoxy molding compound, as explained in Bolger's Acid-Base Interaction Model.

5.3 Other Benefits of NEAP Leadframes

Apart from the excellent adhesion performance, NEAP leadframes also show better compatibility to the IC assembly process:

5.3.1 Epoxy Bleed Out (EBO) Control

During die attach process, EBO is a critical problem as it could become a barrier layer that cause poor adhesion between leadframe surface and EMC or deter subsequent ground bonding on die pad (if any).

Most adhesion promoters for leadframe in the market are based on creating a metal surface of rough topography to increase the mechanical bonding with EMC or other polymeric materials. These include Rough Cu leadframes and Rough PPF (Ni/Pd/Au) leadframes. Figure 12 indicates that both Rough Cu and Rough PPF surface showed high tendency of EBO during die attach process due to strong capillary effect caused by rough surface. EBO still observed even when Anti Epoxy Bleed Out (AntiEBO) process was applied.

On the contrary, NEAP surface showed less tendency of EBO without applying AntiEBO process, and it could achieve "No Bleed" results when AntiEBO process was applied.

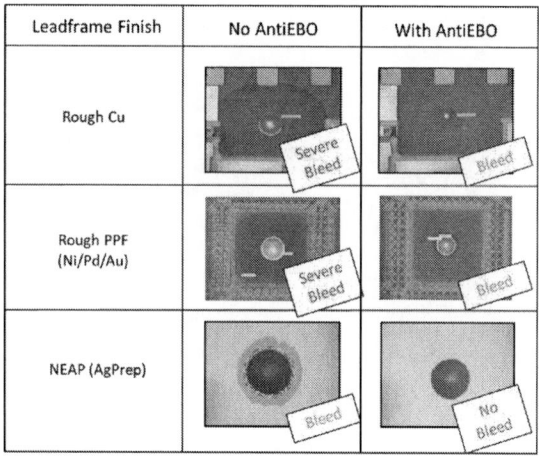

Leadframe Finish	No AntiEBO	With AntiEBO
Rough Cu	Severe Bleed	Bleed
Rough PPF (Ni/Pd/Au)	Severe Bleed	Bleed
NEAP (AgPrep)	Bleed	No Bleed

Figure 12: Comparison of Epoxy Bleed Out (EBO) on different leadframe finish (Epoxy type: Ablestik QMI 519)

5.3.2 Mold Flash Removal

After molding, excessive mold flash on leadframes has to be removed before subsequent tin plating on IC package outer leads. The chemical process used is known as deflash.

When the adhesion between leadframe and EMC becomes stronger, logically the deflash process will become more difficult. However, results revealed that despite NEAP surface exhibits very good adhesion performance, removing mold flash from NEAP surface is much easier than from Rough Cu surface (see Figure 13)

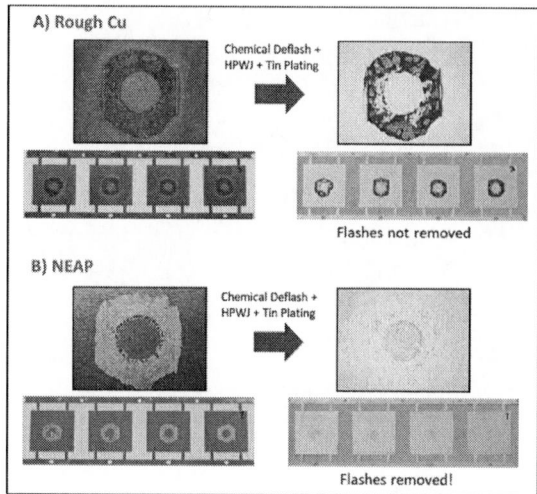

Figure 13: Deflashing comparison test Based on different leadframe finish

6. CONCLUSTIONS

IC assembly process parameters could have a direct impact to the adhesion between leadframe and EMC, especially for leadframes with Cu based finish. More severe heat loading during wire bonding process when processing "big" matrix frames could lead to severe package delamination issue.

Though, "Rough Cu" is well-known as the adhesion promoter technology in leadframe industry, the adhesion performance may deteriorate under the high temperature conditions (eg. Severe heat loading during assembly process or High Temperature Stress Test after MSL1)

As shown in all the experimental results in this paper, NEAP process demonstrates excellent and superior adhesion performance even under very harsh high temperature conditions, whilst offering some other assembly compatibility benefits compared to Rough Cu technology, such as (1) better EBO control, and (2) easy to be deflashed

NEAP process is believed to be the next generation adhesion promoter for leadframes to meet the demand of mirco-electronic packages with high reliability requirements.

ACKNOWLEDGEMENTS

The authors would like to express their sincere gratitude to the support from Huawei, Sumitomo, Atotech and Wintech-nano.

REFERENCES

[1] S. Senturia. *Proceedings of Transducers2003*, Boston, June 8-12, 2003, pp. 10-15.

[2] T. Tsuchiya, O. Tabata, J. Sakata and Y. Taga. *J. Microelectromech. Syst.*, vol. 7, 1998, pp. 106-113.

[3] R. P. Feynman, *Lectures on Physics*, Addison Wesley, 1989.

[4] K. Elissa. unpublished.

[5] R. Nicole. *J. Name Stand. Abbrev.*, in press.

[6] Bolger J.C. Acid-Base Interaction between Oxide Surfaces and Polar Organic Compound.

[7] Bolger J.C. and Michael A.S. Molecular structure and electrostatic interaction at polymer solid interfaces, Interface conversion for polymer coating.

SOLVING CMP CHALLENGES FOR CHEMICALLY STABLE MATERIALS AND 3D SHAPES

Hitoshi Morinaga

FUJIMI Incorporated, Kakamigahara 509-0109, Japan

Email: morinagah@fujimiinc.co.jp

ABSTRACT

In order to expand the use of CMP technologies into new applications, it is necessary to develop polishing technologies for a diverse range of materials and shapes. To improve the material removal rate of chemically stable materials, it is important i) to increase the abrasive particle velocity (vs. polishing object) to maximize the friction, and ii) to increase the number of working (adhered/active) particles by controlling the surface charge. To polish 3D shape precisely, design of polishing consumables (3D pad, magnetic polish slurry, polishing compound) and accurate pressure control are the keys.

INTRODUCTION

Polishing processes are required to achieve i) planarized surfaces that are ii) defect free and iii) surface roughness free; also iv) contamination free and with v) higher productivity [1]. Polishing technology has evolved since ancient times to support the development of civilization. Current CMP (Chemical Mechanical Polishing) technology, in which polishing is performed using a hard flat pad and CMP slurry, can realize the extremely flat surfaces with a surface roughness (Ra) of 0.1 nm or less. On the other hand, the applications are still limited for high purity materials or flat substrates such as semiconductor wafers. To expand CMP technologies into new applications, it is necessary to develop polishing technologies for a diverse range of materials (e.g. hard brittle materials, resins, alloys, compound materials) and shapes (Figure 1). This study will focus on CMP technologies for i) chemically stable materials and ii) 2.5-3D shapes, and discuss novel mechanisms that can help realize these challenges.

SOLVING CMP CHALLENGES FOR CHEMICALLY STABLE MATERIALS

The essential functions required for CMP are: i) material removal, ii) surface protection and iii) contamination control [1]. Material removal in CMP proceeds by the chemical action of the chemistry, the mechanical action of the abrasives, and their combined action such as particle adsorption/desorption control with additive chemistry.

Chemically stable materials such as hard brittle materials (e.g. sapphire, SiC, GaN, Si_3N_4, AlN, $LiTaO_3$, $LiNbO_3$, Ceramics) or resin are widely introduced for such application as IC package substrates, RF filters, power devices, LED and sensors. CMP is used currently or in the future for these applications. However, low removal rate of these chemically stable materials has become a challenge in improving productivity. Since the chemical action is not effective for these materials, maximizing the mechanical driving force becomes more important.

It is well-known that the material removal rate is proportional to relative velocity between the polishing object and the grinding/polishing wheel. Preston reported this experimental evidence in 1927 [2] and the law has been quoted many times in CMP. However, it is not well known that Preston's experiment was conducted with rouge-covered felts (Fe_2O_3 abrasive coated pads), unlike conditions with current CMP (Figure 2). In CMP, using a polishing pad and polishing slurry, the abrasive particle velocity is not always be the same as the polishing wheel

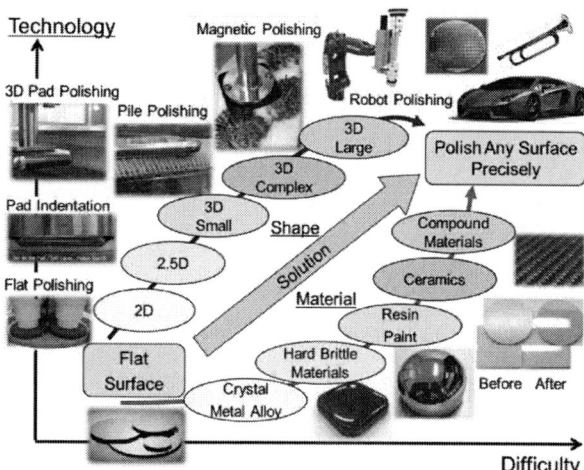

Figure 1: Technology roadmap to expand polishing fields

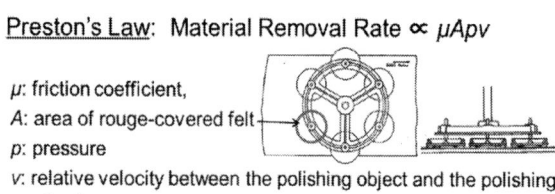

Preston's Law: Material Removal Rate $\propto \mu A p v$

µ: friction coefficient,

A: area of rouge-covered felt

p: pressure

v: relative velocity between the polishing object and the polishing wheel

Figure 2: Preston's law

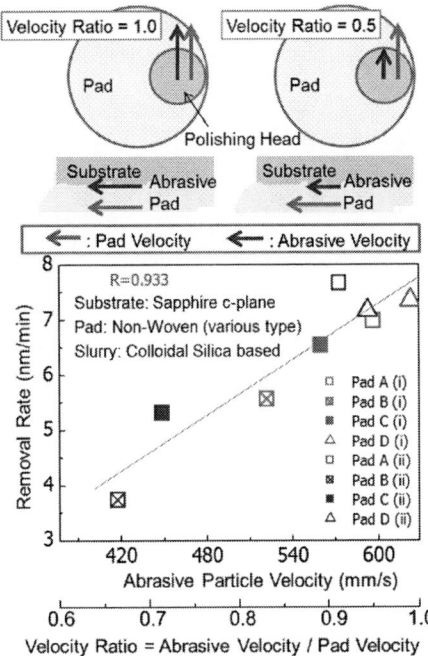

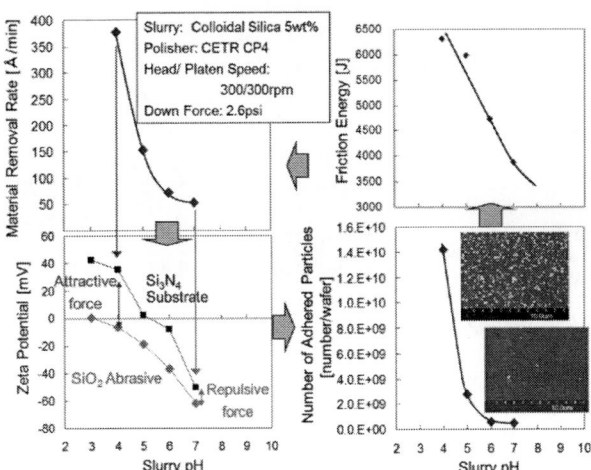

Figure 4: Impact of working particle number on Si_3N_4 removal rate in CMP

Figure 3: Impact of abrasive particle velocity (vs. substrate) during CMP on removal rate

Material Removal Rate in CMP $\propto \mu A p v_{abrasive}$

μ: friction coefficient (difficulty of rolling of abrasive particles)

A: (number of working particles) x (contact area of each particles)

p: pressure

$v_{abrasive}$: relative velocity between the polishing object and abrasive particles

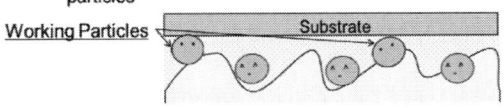

Figure 5: A model for material removal rate in CMP

(polishing pad) velocity. Then, the question arises as to which velocity is important for improving the material removal rate.

In Figure 3, using a high-speed camera and DIC (Digital Image Correlation) method, the influence of the relative velocity between the abrasive particle velocity and polishing object on the material removal rate of sapphire is analyzed [3]. The pressure and the rotation speed of the platen (the relative velocity of the polishing object and the pad) are constant, and the types of CMP slurry and pad are varied. Then, the velocity of abrasive particles in the vicinity of sapphire surface is observed. Interestingly, it was found that the material removal rate is proportional to the relative velocity between the polishing object and abrasive particles as shown in the following equation.

$$\text{Material Removal Rate in CMP} \propto p v_{abrasive} \qquad (1)$$

Here p is pressure and $v_{abrasive}$ is relative velocity between the polishing object and abrasive particles.

In order to improve the material removal rate, it is important to not only increase the relative velocity of the pad, but also to hold the abrasive firmly on the pad and increase the relative velocity of the abrasive to the polishing object to maximize the friction. In addition to the design of the polishing slurry, the design of the surface condition and density of the pad are also important in order to create such a condition.

Another question of Preston's law is the number of particles involved in polishing. Preston has reported that the material removal rate is proportional to the area of the "rouge-covered pad" (Figure 2). However, in CMP using polishing pad and slurry, the abrasive does not cover the whole surface and not all particles are working at the surface. To maximize the mechanical driving force, it is important to increase the number of working (active) particles during CMP [1]. Figure 4 shows the impact of the number of working particles on Si_3N_4 removal rate. By controlling the zeta potential of the abrasives and the substrate to induce attractive force, more abrasives can be attached to the substrate surface to increase friction, which is effective in improving the material removal rate. These novel technologies for maximizing the friction energy are extremely effective, as well as the technology of increasing the contact area and the difficulty of rolling of abrasive particles [1], to improve the removal rate of chemically stable materials. Figure 5 summarizes a model for material removal rate in CMP.

978-1-7281-6559-2/20 $31.00 © 2020 IEEE

SOLVING CMP CHALLENGES FOR 2.5D-3D SHAPES

Current CMP technologies, which utilize flat platens and hard pads, are very effective to polish planar substrates. However, these technologies are not applicable to 3D shaped objects [4, 5]. Conventional polishing methodology used for surface finishing of 3D shapes is called buffing, in which a soft buffing pad made of sponge or cotton/wool is used to hold wax-containing abrasives. Figure 6 compares the waviness (orange peel) removability in CMP using soft buffing pad and hard pad. In the case of buffing, the buff follows the uneven shape to achieve high gloss surfaces. However, it is not possible to realize precision polishing that eliminates surface micro-roughness and waviness as it also follows the surface waviness.

Despite recent advances of 3D processing technology such as CNC processing and 3D printing, the surface after processing still has surface roughness derived from cutting or grinding, and waviness derived from forming or coating. In this section, precise polishing technologies for 3D shapes which are developed to expand CMP fields will be discussed.

Pad Indentation Polishing

This method can be used for a tablet type shape (2.5D shape) that consists of a plane and a curved surface at the end (Figure 7). In this method, the polishing slurry is held by various pads (polyurethane, non-woven, suede, etc.) used for planar polishing, and a sponge layer on the back side of the pad is used in order to follow the loose three-dimensional shape of the end.

3D Pad Polishing

In this method, a pad having the same 3D shape as the shape of the polishing object is prepared in advance. Slurry flows and the pad presses against the object to render uniform pressure, while the object is rotated and polished. As a result, it is possible to eliminate the deformation of the shape after polishing, and to remove only scratches, roughness and waviness of the surface. This technology can be applied to the polishing of a 3D shaped object such as a shaft or cylindrical shape or a curve on the side of an object.. Figure 8 shows a polishing apparatus and a polishing example using this method. In the method shown in the figure, the pad is dressed regularly with a diamond dresser having the same curve as the polishing object so that the shape of the pad always mirrors the object.

Pile (Brush) Polishing

In order to polish complicated 3D shapes, it is effective to press the fiber vertically against the object to be polished and polish it like a toothbrush or a carpet (see photo in Figure 1). Fibers that are elastic and hard to cut are preferable.

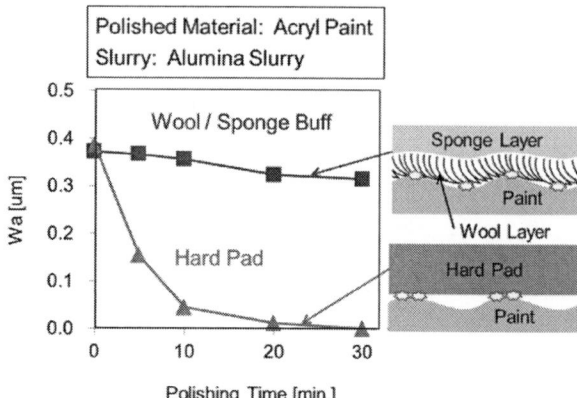

Figure 6: Waviness (orange peel) removability in CMP using soft buffing pad and hard pad

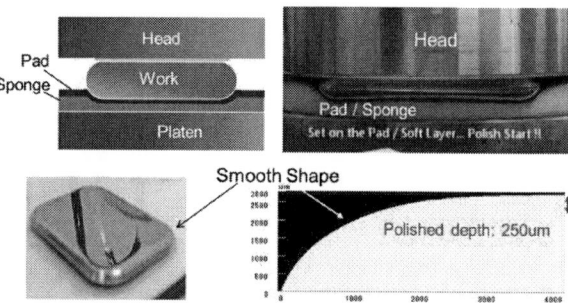

Figure 7: Pad Indentation polishing and its application (2.5D Aluminum Alloy)

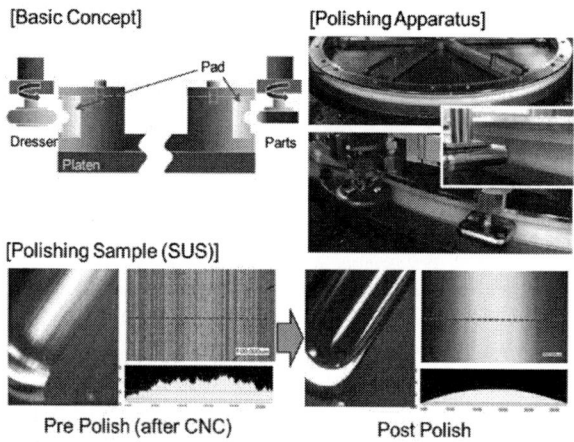

Figure 8: 3D pad polishing and its application example

Magnetic Polishing

When a magnetic force is applied to magnetic powder such as iron powder, a magnetic brush is formed along the magnetic field line. Magnetic polishing is a method of harnessing abrasives on this magnetic brush to polish. The improvement over Pile Polishing mentioned above is that broken brushes and fallen brushes are regenerated as long as there is magnetic force. In order to maintain stable polishing performance, it is important to maintain the regenerative power of magnetic brush. By adding an oxidation inhibitor for iron powder to the polishing slurry in addition to the abrasives and polishing accelerator, it is possible to suppress the oxidation of iron powder during polishing and to suppress the deterioration of the magnetic brush (Figure 9).

Robot Polishing

There are two types of robot polishing. One is a method of holding a polishing tool on a robot arm and the other is a method of holding a polishing object. The latter is mainly used for small parts. In the former, the polishing tool is held by the arm of the robot and rotated, following the contours of various sizes of 3D shapes, pressing and polishing. In this method, the various polishing methods (e.g. Pad Indentation, Magnetic Polishing) described above can be held by the robot arm and used in combination. Robots are already widely utilized in cutting and grinding application, but accurate pressure control is important when used for precision polishing. Figure 10 shows a high precision robot polishing system equipped with dual polishing head and pressure control unit. The figure also shows that orange peel of clear coat paint on automobile exterior could be eliminated with the system utilizing the polishing consumables (3D pad and polishing compound) designed for precision 3D polishing.

SUMMARY

To improve the material removal rate of chemically stable materials, it is important to increase the abrasive particle velocity (vs. polishing object) to maximize the friction, and to increase the number of working particles by controlling the surface charge. To polish 3D shapes precisely, design of polishing consumables (such as 3D pad, magnetic polish slurry, and polishing compound) and accurate pressure control are keys for solving these challenges. It is important i) to give uniform pressure to the 3D polishing objects, ii) to hold slurries on the pad for every angle, and iii) to keep shape and hardness of the pad during polishing.

ACNOWLEDGEMENTS

The author wishes to thank FUJIMI Incorporated and related companies for supporting this research, as well as all members involved in this research at FUJIMI. Further, the author thanks Prof. M.Uneda, Prof. K.Ishikawa and the

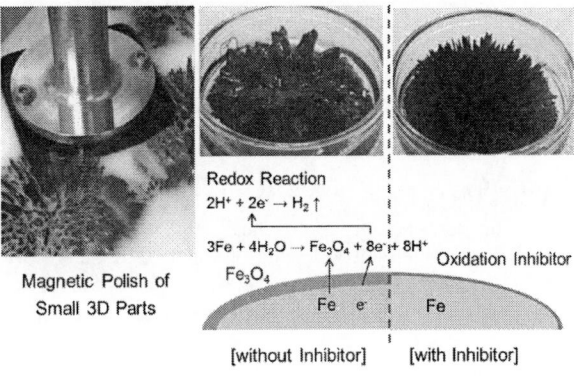

Figure 9: Effectiveness of oxidation inhibitor in Magnetic Polishing

[High Precision Robot Polishing System]

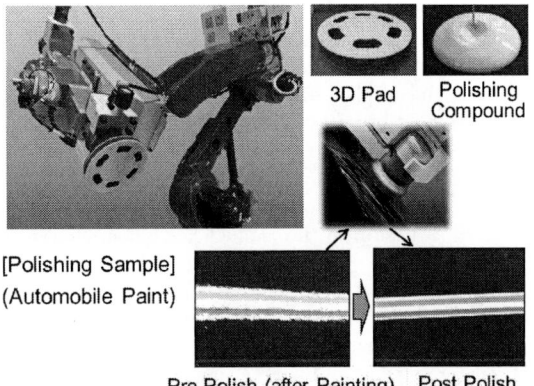

Figure 10: High precision robot polishing system equipped with pressure control unit, polishing consumables and example of automotive paint application

team of Kanazawa Institute of Technology for their great contributions to in-situ analysis of abrasive particle velocity in CMP.

REFERENCES

[1] H. Morinaga and K. Tamai. *ECS Transactions*, vol.34 (1), 2011, pp.591-596.

[2] M F. W. Preston. *Journal of the Society of Glass Technology*, vol.11, 1927, pp.214-256.

[3] M.Uneda, Y.Tomiie, K.Hotta, K.Tamai, H. Morinaga, K.Ishikawa. *Journal of the Japan Society for Precision Engineering*, vol.83 (8), 2017, pp.756-761.

[4] H.Morinaga and K.Tamai. *Journal of the Japan Society for Precision Engineering*, 2018, vol.84(3), pp.235-238.

[5] H.Morinaga. *Proceedings of ICPT 2019*, Hsinchu, September 16-18, 2019, pp15-18.

STOP ON NITRIDE SLURRY DEVELOPMENT

Shoutian Li*, Changzhen Jia, Xiaoming Ren

Anji Microelectronics Technology (Shanghai) Co. Ltd. , Shanghai 200123, China

*Email: shoutianli@anjimicro.com

ABSTRACT

The formulation approaches on ceria-based SoN (Stop on Nitride) slurry depend on the surface charge of ceria particles. For the ceria particles with positive charges, the slurry formulation will be different from the negative charged ceria. In this paper, we discuss the pros and cons of different approaches in formulating SoN slurry. Finally, we present the CMP performance results from the SoN slurry that is formulated with positive charged ceria.

INTRODUCTION

Silicon nitride (SiN) is commonly used as stop layer in manufacturing ultra large-scale integration (ULSI) microchip. The associated CMP process requires that slurry is able to planarize the topography of silicon oxide and stop on nitride (SoN) film. The most common used SoN slurry is for STI (shallow trench isolation) process. Ceria-based STI slurry was adopted for sub-90nm technology node in the production of integrated circuits with ability to stop on nitride much better than silica-based slurry [1].

Without any surface treatment, ceria particles can have isoelectric point at pH ~9, meaning the ceria particles surfaces have positive charge at pH <9. On the other hand, surface treatment with anionic polymeric molecules, for example, with polyacrylic acid results in negative charged ceria surface with isoelectric point at pH ~2 [2].

Depending on the surface charges of the ceria particles, formulating SoN slurry requires different approaches. The key component in formulating SoN slurry is the nitride inhibitor. In anionic polymer-treated ceria, the anionic polymer (AP) serves as nitride inhibitor since the anionic polymer can strongly adsorbed on nitride film surface through surface charge attraction. In untreated, positive ceria slurry, small molecules like amino acids or carboxylic acids are common nitride inhibitors [3]. The pros and cons of these two different approaches in formulating SoN slurry will be discussed. Finally, we present the CMP performance results from the SoN slurry that is formulated with positive charged ceria. This slurry shows high silicon oxide rate, high oxide to nitride selectivity, long pot and shelf life.

EXPERIMENT

Colloidal ceria with mean particle size 150nm measured by dynamic laser light scattering was used in this study. The as-received ceria has no surface treatment and zeta potential is +40mV at pH 4.5. The same colloidal ceria can be treated with anionic polymer (AP) to have a negative potential -35mV at pH 5.0. The SoN slurry with 0.4% AP-treated ceria was formulated at pH 5.0. In this slurry, AP served as effective nitride inhibitor. The SoN slurry with positive potential was formulated at pH 4.5 with 0.2% ceria loading, and a small molecule was used as nitride inhibitor.

CMP was carried in clean-room environment using Mirra polisher from Applied Materials (San Jose, CA). Blanket PTETEOS and nitride wafers were used to monitor removal rates and selectivity; patterned STI wafers were purchased from Ramco Technology Inc. (San Jose, CA). The pad was IC1010 and conditioning diamond disk was 3M A165. The Mirra polishing settings were as following: platen speed was 93rpm, carrier speed was 87rpm, slurry flow rate was 150ml/min, polishing down force varied from 1.5psi to 4psi.

RESULTS AND DISCUSSION

Blanket Removal Rates and Selectivity

Fig. 1 shows the blanket removal rates comparison between the positive ceria slurry and negative AP-treated ceria slurry. For the slurry with positive zeta potential, the oxide rates are higher but the nitride rates are much lower than the AP-treated slurry, resulting very high selectivity (>500). On the other hand, the AP-treated slurry has oxide to nitride selectivity less than 20.

It can be seen from Figure 1, the TEOS removal rates of the positive ceria slurry at 2psi are equal to the AP-treated ceria slurry at 4psi. Thus, the nitride removal rates on the patterned were compared with equal TEOS rates.

Patterned Nitride Removal Rates

Fig.2 shows the nitride removal rates on the patterned STI wafers for the AP-treated slurry and the un-treated ceria slurry with positive zeta potential. The comparison was done at the same TEOS removal rates. To maintain the same TEOS removal rates, the AP-treated slurry was polished at 4psi and the un-treated ceria slurry was polished at 2psi. For the AP-treated slurry, the nitride removal rates on the patterned wafers are similar to blanket wafers. For the un-treated ceria slurry, the nitride removal rates on the patterned wafers are significantly higher than blanket wafers but still low (~≤60A/min). On the patterned STI wafers, the two slurries have similar nitride rates.

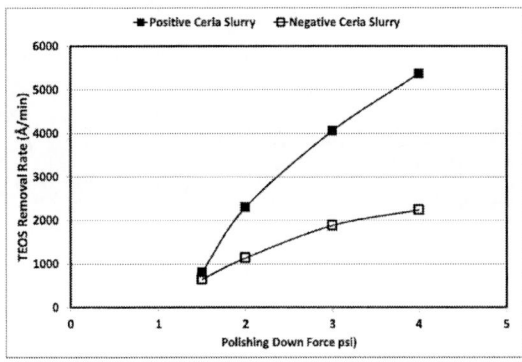

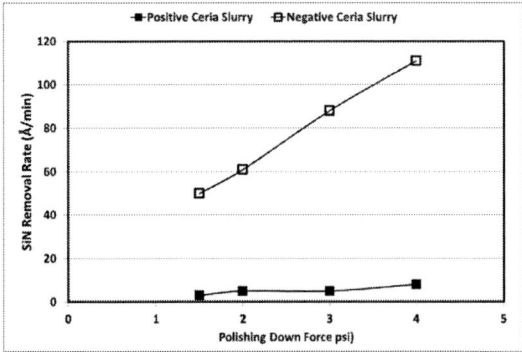

Fig. 1. the blanket removal rates comparison between the positive ceria slurry and negative AP-treated ceria slurry. (top), TEOS rates; (bottom), SiN rates.

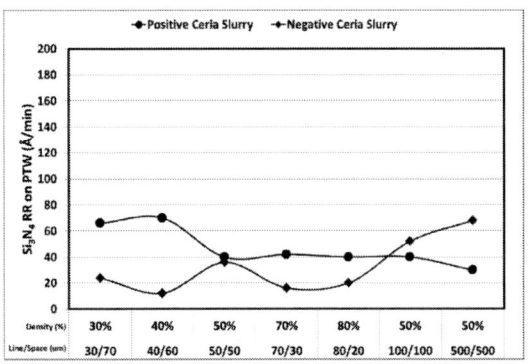

Fig. 2. the nitride removal rates on the patterned STI wafers for the AP-treated slurry (solid diamond symbols) and the un-treated ceria slurry with positive potential (solid circle symbols).

The removal rates of the positive charged ceria slurry were further studied as functions of polishing down force (DF) and platen speeds in rpm. DF varies from 1 to 3psi, and platen speeds vary from 53rpm to 127rpm. The results

are presented in Fig. 3. While the blanket TEOS rates increases from 500 A/min to > 5000 A/min, the blanket SiN rates remain very low (<30A/min).

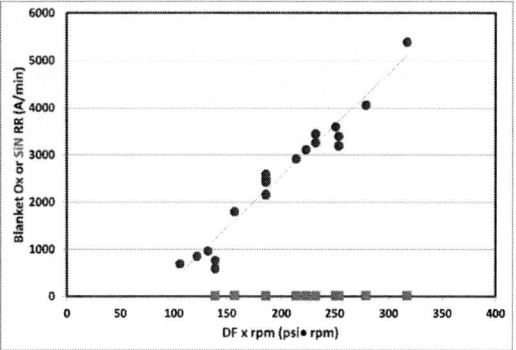

Fig. 3. Removal rates vs DF × speed of the positive charged ceria slurry

The real test how good of SoN slurry can stop on nitride lies on patterned wafers. For the positive charged ceria slurry, the SiN rates on the patterned wafers show a deflection point as a function of down force, as seen from Fig. 4. The SiN rates on the patterned wafers remain low at DF ≤ 2.5psi but exponentially increase once DF passes the threshold of 2.75psi. As long as DF ≤ 2.5psi, the SiN rates on the patterned wafers remain low regardless of the platen speeds are 53rpm or 127rpm. It should be pointed out that the SiN rates on the patterned wafer of the AP-treated ceria slurry remain low even if the DF is 4psi (see Fig. 2).

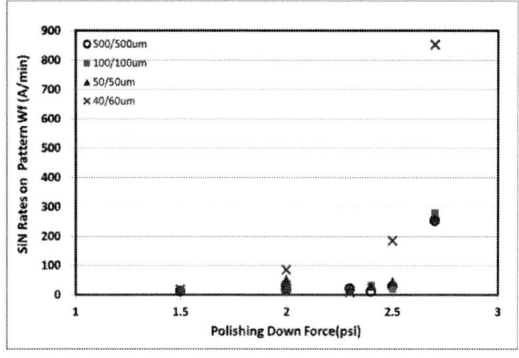

Fig.4. SiN rates on the patterned wafers of the positive charged ceria slurry

The DF depended SiN rates of the positive charged ceria slurry can be explained by the strength of the packed inhibitor film, as depicted in Fig. 5. The SiN inhibitor in the positive charged ceria slurry is small molecules. The inhibition on removal rates comes from the packed inhibitor film adsorbed on the nitride surface. On the blanket nitride wafers, the packed inhibitor film can form long range monolayer film which is difficult to push away by the polishing friction force. On the patterned wafers, the SiN is discontinued; only exist on the active area in a few μm to a few hundred μm range. The inhibitor film cannot pack well at the edge of nitride lines. During the polishing, the friction force is proportional to down force. When the friction force exceeds the inhibition film strength, the inhibition film will collapse and loss its ability to prevent nitride from abrasion. This is not the case in the AP-treated ceria slurry. When AP works as nitride inhibitor, the multiple anchor points from the polymer chains make the inhibitor film difficult to be abraded away. Even on the patterned wafers when nitride film is discrete, the strong anchor actions of the AP film can extend to side walls of the nitride as the oxide recesses at trench area. Since AP film is formed by polymer, the inhibition film is unlikely a monolayer. On the wafer scale, the polymer strength is less than the packed monolayer formed by small molecules. This leads to the observation of blanket SiN rates with the AP-treated ceria slurry is higher than the positive charged ceria slurry.

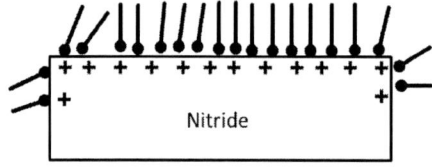

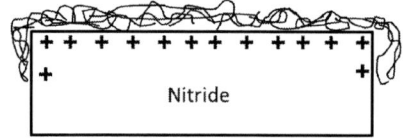

Fig. 5. The illustration of inhibitor film formed on the patterned SiN lines with small molecules (top) and polymeric AP molecules (bottom)

Patterned Nitride Removal Rates in Over Polishing

As shown in previous section, the SiN rates on the patterned wafers of the positive charged ceria slurry can keep low as long as DF ≤ 2.5psi. This slurry can withstand long over polish for low nitride loss. The results are shown in Fig. 6. The end point is defined when the oxide on the top of nitride is cleared. Any polishing passes the end point is called over polishing. Fig. 6 shows for the feature with 40um line width and 60um space (abbreviated as 40um/60um), 200s over polishing results in 150Å nitride loss, while wider lines in the features like 50um/50um,

100um/100um, 500um/500um have nitride loss less than 90Å. The results in Fig. 6 demonstrate the high performance SoN slurry can be formulated with positive charged ceria and use of small molecules as nitride inhibitor.

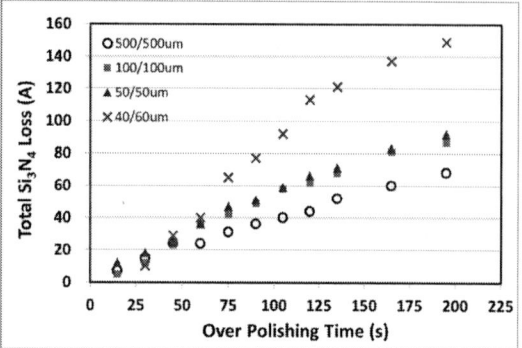

Fig. 6. SiN removal amount on pattern wafer of the positive ceria slurry vs over polishing time

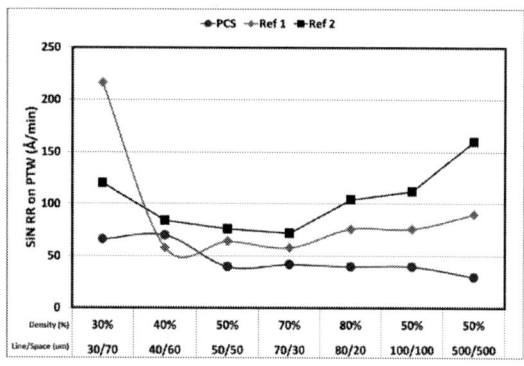

Fig.7. The comparison of the SiN removal rates on the patterned wafers.

Positive Ceria Slurry for STI

The positive ceria slurry (PCS) is formulated for a commercial STI application. This slurry is concentrated at 5× and it needs to dilute 5 times (1 part of slurry and 4 parts of water) before use. The CMP performance of PCS is compared to commercial STI slurries (Ref 1 and Ref 2) with negative zeta potential. For the fairness of the comparison, the polishing conditions were set as all of the three slurries having the same TEOS rates ~2300Å/min. Fig.7 shows the comparison of the SiN removal rates on the patterned wafers. The PCS shows lower SiN rates. Fig.8 shows the dishing comparison at 500um/500um line/space feature at different over polishing time. The end point is defined when the oxide on the top of nitride is cleared. The wide line/space 500um/500um is chosen since the dishing is typically higher at wider space features when the deformed pad can touch the bottom of the trench. PCS shows lower dishing than the reference STI slurries and dishing increase vs over polishing has the lowest slop.

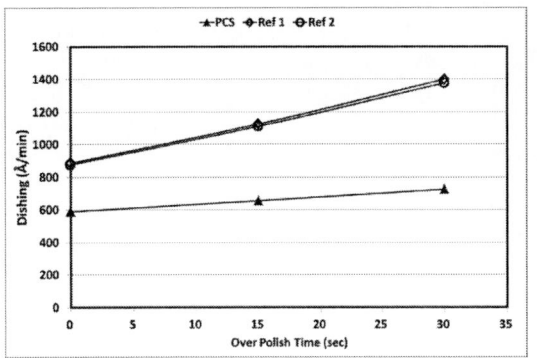

Fig. 8. The dishing comparison at 500um/500um line/space feature at different over polishing time

Pot Life and Shelf Life of PCS Slurry

Pot life is monitored over one month in the circulation through 0.5um filter and shelf life is monitored over one year. There are no changes in physical properties or CMP polishing performance during the pot life and shelf life monitoring. Given the page limit, only the data of SiN loss on the pattern wafers are shown here. Fig. 9 shows the pot life data and Fig. 10 shows the shelf life data.

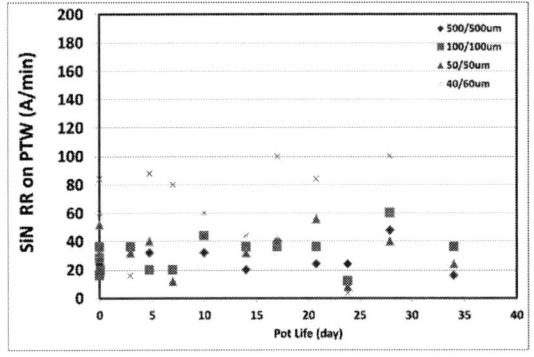

Fig. 7. The SiN removal rates on the patterned wafers of PCS during 34-day pot life test

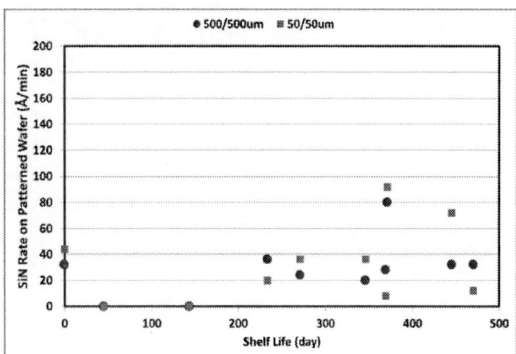

Fig. 10 The SiN removal rates on the patterned wafers of PCS during 470-day shelf life test

CONCLUSION

The SoN slurries can be formulated with the AP-treated negative charged ceria and untreated positive charged colloidal ceria. The anionic polymer in the AP-treated negative charged ceria can serve as nitride inhibitor. To formulate positive charged ceria slurry, small molecules can be used as nitride inhibitor. Using this approach, high performance SoN slurry is developed. The untreated positive charged ceria slurry shows higher oxide rates than the AP-treated negative charged ceria slurry and low nitride rates at down force less than 2.5psi.

ACKNOWLEDGEMENTS

The authors would like to acknowledge Anji Microelectronics Corporation Ltd. for the permission to publish this paper. We also would like to thank the individuals who help us in preparation of the manuscript.

REFERENCES

[1] D. Merricks, B. Santora, B. Her, C. Zedwick. *Semiconductor technology (ISTC 2008) : proceedings of the 7th International Conference on Semiconductor Technology ;*Shanghai, China, March 15 - 17, 2008
[2] C. Zedwick, D. Merricks, S. Frink, B. Santora, B. Her, *Proceedings of the CMP-MIC*, Fremont, CA, 2008.
[3] P. Carter, T. Johns, *Electrochemical and Solid-State Letters,* 8 (8) G218-G221, 2005

NOVEL ABRASION-FREE CMP TECHNOLOGY WITH HIGH PERFORMANCE POLISHING SLURRY

Chong Luo[1*], Yuling Liu[1,2*]

[1]School of Electronic Engineering, Hebei University of Technology, Tianjin 300130, China
[2] Tianjin Key Laboratory of Electronic Materials and Devices, Tianjin 300130, China
*Corresponding Author's Email: 719506927@qq.com

ABSTRACT

At present, there are many scholars who have abrasive-free chemical mechanical polishing (AFP) research, because it effectively reduces micro-scratch and grinding adsorption, micro-corrosion, etc., but has not been applied to the project so far, mainly because it has not developed a polishing slurry suitable for engineering requirements. In addition, the removal mechanism of AFP technology is not clean yet. The results of this research, the copper achieve high-performance with a novel abrasive-free polishing solution, and the mechanism is discussed.

INTRODUCTION

Copper as a multi-level interconnecting material in IC manufacturing has attractive properties owing to its good thermal and electrical properties[1] .The integration of copper interconnection can be carried out by the dual damascene technique[2]. Up to now chemical mechanical polishing (CMP) is the only way to realize Global and local planarization simultaneously[3,4]. However the number of defects after CMP could inevitably translates to yield loss of integrated circuits (ICs)[5-7].The hanging bond on the surface after CMP makes the surface with high free energy which could adsorb abrasive in the slurry and generate defects. For example, it could chemical adsorb abrasive generate polish residues, which could induce leakage current and time-dependent dielectric breakdown(TDDB), and then cause severe yield loss and circuit failure[8]. As the critical features of semiconductor devices have shrank to 14nm, the size of defects detected deceased to 0.06nm and ever more 0.03nm. It is obvious that these defects originate from abrasives (especially aggregative abrasives) included in the slurry[9-12]. Therefore abrasive-free chemical mechanical polishing which could solve the adsorption after CMP is the urgent need for development of microelectronic technique.

The advantages of AFP(abrasive-free polishing) are[13, 14]: selective for polishing materials, no need for complex post-CMP cleaning technology, simple brushing, low environmental requirements, environmental protection, low cost, and especially low defect. That is illustrated in Table I . Kondo et al.[13] applied AFP technology to the copper damascene process and achieved a very clean, damage-free and corrosion-resistant Cu surface. Since then, AFP technology has received great attention[14-16].

However it has not been applied to the project so far, mainly because it has not developed a polishing slurry suitable for engineering requirements. For example, in polishing without abrasives, the removal rate was quite low; thus, AFP could not be put to practical use. Furthermore, the removal mechanism of AFP technology is not clean yet. Accordingly, we have developed a abrasive-free slurry for Cu damascene interconnection with high performance, and the mechanism is discussed.

Table I : Comparison between traditional CMP and AFP Cu

Disadvantages of CMP	Traditional CMP	AFP	AFP advantage
Microscopic scratches	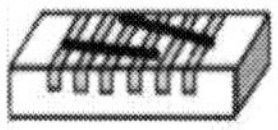	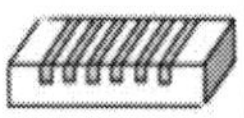	Product yield increase
Residual particles		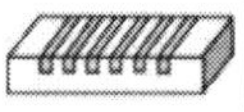	Reliability enhancement
Oxide loss	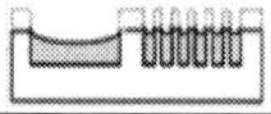	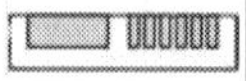	Design validity
Depression		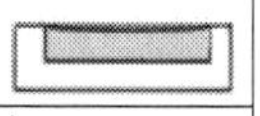	
corrosion		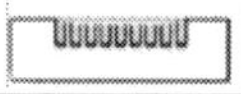	

EXPERIMENT

Abrasive-free slurry was prepared, consisted of 2.5 wt% FA/O chelating agent (Tianjin Jinling, China, pH=12.56), 0.447wt% FA/O compound surfactant, and 6 wt% H_2O_2 as the oxidant. The PH of the AFP slurry is 10.50. The FA/O chelating agent is one of macromolecular organics with many chelating rings which can react with the metal ions rapidly and the reaction products are all soluble and stable under the condition of alkaline. The structure of the FA/O chelating agent is shown in Figure 1.

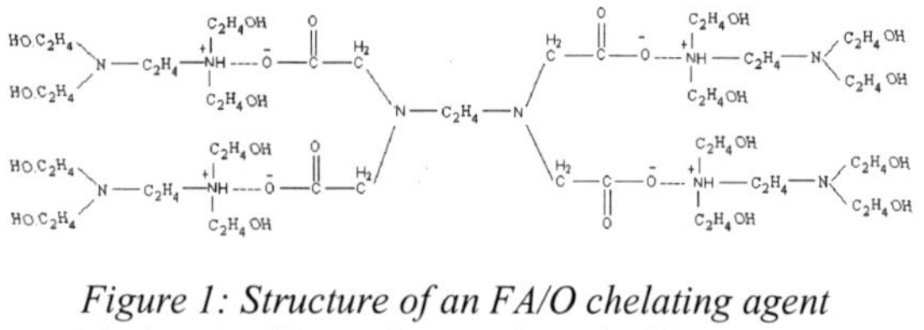

Figure 1: Structure of an FA/O chelating agent
3 inches Cu film wafer was deposited by sputtering. It

is polished on Alpsitec E460E polisher using a Dow Electronic IC1000 polishing pad with the slurry pumped on to the pad at a flow rate of 300 ml/min. The head speed/the plate speed is 87 rpm/93 rpm with an applied down force of 1.5 kPa, and the polishing time was set to 3 minutes. Prior to polishing, the pad was conditioned for 60second at 1.5 psi down force using DI water and the actual slurry, respectively.

The materials removal rate is calculated by sheet resistance at 81 points on the wafer before and after the polishing which is measured by four point probe. And wafer nonuniformity (WIWNU) of the removal rate is evaluated by the equation:

$$WIWNU=STDEV(MRR)/AVERAGE(MRR) \quad (1)$$

The etch tests are carried out in a 1000 ml glass beaker containing 800 ml of AFP slurry and 3 inches Cu blanket wafers(pure 99.99%) are dipped into it. The static Cu etch rate is calculated by the weight difference between copper before and after. That is measured by a professional electronic balance (OHAUS DISCOVERY) with a precision of 0.00001 g. The static etch rate(SER) is calculated as follows:

$$SER = \frac{\Delta m}{\rho \pi r^2 t} \quad (2)$$

Here Δm is the removal weight of copper, r is the radius of copper, and t is polishing time.

RESULTS

Table II: MRR and SER of the AFP slurry and DI water

MRR and SER results(Å/min)	
CMP using DI water slurry	59.28
CMP using AFP	3035.48
copySER using AFP slurry	75.86

Table II shows copper polish materials removal rate(MRR) and static dissolution rate(SER) of the AFP slurry and DI water. We can see that MRR by polishing using DI water is 59.28 Å/min. And cu dissolution rate in AFP slurry is 75.86 Å/min. However, MRR by polishing using AFP slurry is 3035.48 Å/min, which is one order of magnitude higher than MRR by DI water or SER by AFP slurry.

Cu film CMP MRR profile by AFP slurry is shown in Fig.2. The WIWNU=2.07%, which is lower than 3%.

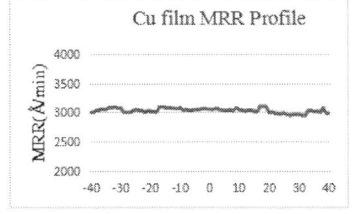

Figure 2: Cu film MRR profile by AFP slurry

DISCUSSION

How to remove abrasives and still have a higher rate is the core of realizing no abrasives. The chemical reaction in multi-level interconnecting CMP is a multiphase reaction, which takes place at the solid-liquid interface. Abrasive particles can act as mass transfer carriers, speeding up the reaction and improving the removal rate. Since AFP slurry have no abrasive, how to improve the chemical reaction rate under the condition of AFP is the key to solve the problem. At present, CMP enters the age of alkalinity from acidity, wiring from aluminum to copper[1], and its chemical reaction to copper wiring.

The chemicals in slurry alter the properties of the copper surface that comes into contact with them. This surface layer can form Cu(I,II) species (i.e. Cu_2O and CuO)[17]. Meanwhile, anions of a complexing agent react with Cu(I,II) species, forming a soluble species or insoluble salt as passivation films on the protruded region under different experimental conditions[18]. The surface layer is then removed by mechanical action of the slurry particles or by dissolution into the slurry through a convective diffusion process.

$$2Cu+H_2O_2=Cu_2O+H_2O \quad (3)$$
$$H_2O_2+Cu_2O=2CuO+H_2O \quad (4)$$

And in Alkaline solution：

$$Cu_2O+H_2O \rightarrow 2CuOH \quad (5)$$
$$CuO+H_2O \rightarrow Cu(OH)_2 \quad (6)$$

As Cu_2O, CuO and $Cu(OH)_2$ is insoluble in water, according to the insoluble material ionization balance:

$$CuOH \rightarrow Cu^+ + OH^- \quad (7)$$
$$Cu(OH)_2 \rightarrow Cu^{2+} + 2OH^- \quad (8)$$
$$[Cu^+][OH^-]=K1 \quad (9)$$
$$[Cu^{2+}][OH^-]=K2 \quad (10)$$

Where K1 and K2 is constant.

If Cu^+ and $Cu2^+$ is removed rapidly, the ionization equilibrium in equation (6) and (7) is broken. And the insoluble material will move rapidly toward the direction of ionization, accelerating the chemical reaction. Therefore, how to break the equilibrium and reduce the activation energy of reactants is the key point.

The typical mathematical model of CMP mechanical action is Preston equation[19], which is purely an empirical relation that indicates the linear dependence of the material removal rate (MRR) on the down pressure (P) and the relative velocity (V) between the wafer and the pad.

$$MRR=KPV \quad (11)$$

However the typical mathematical model of chemical action is Arrhenius's equation:

$$K_c = Ae^{-Ea/RT} \quad (12)$$

Here Kc is the chemical action rate, A, R are constants. Ea, T means activation energy and thermodynamic temperature, respectively.

In order to realize planarization, its essence is to

improve the bump and recessed area rate difference, that is, fast reaction at the bump area, slow reaction or no reaction at the recessed area, and effectively reduce WIWNU with the rate difference. As it is shown in the figure1, FA/O chelating agent is used as a polyhydroxypolyamine macromolecule with strong non-toxic activity and relatively large activation energy Ea. It do not react with copper ions in the normal state at the recessed area. But under the condition of CMP, with high pressure and high speed it reacts at the bump area, and the product is soluble. In contrast, in the recessed area, Ea is difficult to reduce and cannot response. That implements the bump and recessed area rate differential, achieves planarization.

The polish rate of recessed area

In the recessed area, the material removal rate (MRR) of Cu is only induced by the chemical dissolution:

$$MRR_{recessed}=Kc_{recessed}[Cu(\ I\ II\ )] \qquad (13)$$

Kcrecessed is the chemical reaction rate given by equation (12).

Generally, the activation energy Ea is rather considered as the minimum energy that must overcome in a chemical reaction[20]. For the complexation reaction in a static environment, there is no chemical reaction between the complexing agent and copper ion if the activation energy is high enough. Usually a chelating agent with a high molecular weight possesses the higher activation energy, and the key feature of FA/O is that it is an organic macromolecule. From the viewpoint of reactive collision theory, the higher molecular weight of the organic chelating agent, the more energy will be needed for the intense thermal motion of macromolecules to collide with copper ion. A static environment can't provide enough energy for FA/O to overwhelm this barrier to chelate metal ions, leaving Cu(I,II) species including Cu_2O, CuO and $Cu(OH)_2$ on the surface at recessed area to passivate the copper surface and limit its dissolution. In fact, surfactant molecules are adsorbed preferentially onto the surface generating an isolated layer, which makes chemical reaction more difficult[21, 22], so the material removal rate of Cu in the recessed regions is approximately equal to zero. That illustrates SER data of the AFP in table Ⅱ.

$$MRRrecessed \approx 0 \qquad (14)$$

The polish rate at bump area

In traditional CMP, the surface layer on the protruded region is removed simultaneously by mechanical action of the abrasive particles and by chemical dissolution into the slurry through a convective diffusion process, and the removal rate in bump area includes the chemical polishing rate and the abrasive removal rate:

$$MRR_{bump}=Kc_{bump}[Cu(\ I\ II\ )]+Km[Cu(\ I\ II\ )] \qquad (15)$$

Here Kc_{bump} is the chemical reaction rate given by equation (11), and Km is abrasive removal rate.

But in AFP there is no abrasive in the slurry, and the pad asperity is much softer than the abrasive particles. So the Km is approximately equal to zero, which illustrates

CMP MRR data using DI water in tableⅡ.

During the CMP process, the wafer is pressed down against a rotating platen, which holds a compliant polishing pad. The wafer moves on the pad surface with a relative velocity generated by the rotation of the carrier and the platen. The polishing pressure and the velocity can provide the mechanical energy for the chemical reaction on the protruded region[23], the distance between reactants decreases, and on the critical distance of the chemical reaction, the reaction activation energy is reduced.

$$Ea=\beta/PV \qquad (16)$$

The reaction activation energy decreases exponentially with mechanical PV in CMP, and reaction happens at the bump area. The chemical reaction rate at bump area is reflected through Kc_{bump}, so the Arrhenius's equation should be modified as:

$$K = Ae^{-\beta/PVRT} \qquad (17)$$

Here, PV is the mechanical energy provided from the applied pressure and the velocity of the pad asperities to the metal atoms of the Cu surface. The decreasing of the activation energy makes it easier for FA/O to react with copper ion, consequently accelerating the removal of Cu(I,II) species at bump area. That illustrates CMP MRR data using AFP slurry in tableⅡ.

The Schematic illustration of the chemical reaction model in AFP process is described in Fig.3. According to the above mathematical model, a high-rate difference between the bump and recessed area can achieve a high planarization as it is shown in Fig.2.

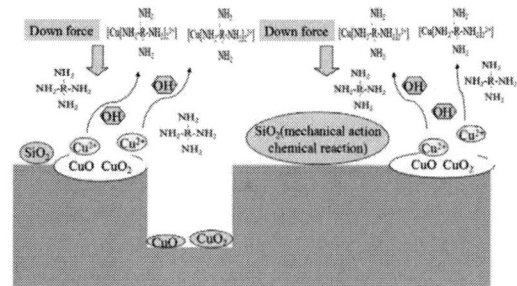

Figure3 The mechanism of the AFP process by FA/O slurry

CONCLUSIONS

Applied abrasive-free chemical mechanical polishing to the copper damascene process could achieved a very clean, damage-free Cu surface, and we developed a novel abrasive-free high-performance polishing solution using FA/O chelating agent which has high Ea. By reason of Ea could reduced during CMP, Cu MRR could higher than 3000 Å/min. In addition, it can generate high rate difference between the bump and recessed area and and effectively reduce WIWNU with the rate difference, so the

WIWNU is lower than 3%.

ACKNOWLEDGEMENTS

The authors are grateful for the support provided by the Major National Science and Technology Special Projects (No. 2016ZX02301003-004-007)

REFERENCES

[1] P.B. Zantye, A. Kumar, A.K. Sikder. *Mater. Sci. Eng. R,* vol. 45, 2004, pp. 89.

[2] Y. Gotkis, S. Guha. *J. Electron. Mater.*, vol. 30, 2001, pp. 396–399.

[3] S. Shou-mei. *Electronic Design Engineering*, vol.19, 2011, pp. 190–192.

[4] K. Asghar, et al. *Colloids and Surfaces*, vol. 497, 2016, pp. 133–145.

[5] B.T. Murph. *Proc. IEEE*, vol. 52, 1964, pp. 1537-1545.

[6] C.H. Stapper. *IBM J. Res. Dev.*, vol. 27, 1983, pp. 549-557.

[7] C. Hess. *IEEE Trans. Semicond. Manuf*, vol. 12, 1999, pp. 175-183.

[8] W.-T. Tseng, V. Devarapalli, J. Steffes, et al. *IEEE Trans. Semicond. Manuf*, vol. 26, 2013, pp. 493-499.

[9] Li Y. *Microelectronic Applications of Chemical Mechanical Planarization,* John Wiley Sons, 2007.

[10] Ring T A, Feeney P, Boldridge D, Kasthurirangan J, Li S, and Dirksen J A. *J Electrochem Soc*, vol. 154, 2007, pp. H239−H248.

[11] Seo Y J, Kim S Y, Lee W S. *Microelectron Eng* , vol. 65, 2003, pp. 371−379.

[12] Lee S I, Hwang J, Kim, Jeong H. *Microelectron Eng,* vol. 84, 2007, pp. 626−630

[13] Kondo S, Sakuma N, Homma Y, et al. *J Electrochem Soc,* vol. 147, 2000, pp. 3907-3913.

[14] Yamada Y, Konishi N, Ohashi N, et.al. *IEEE International Interconnect Technology Conference*, Burlingame, June 3-5, 2002, pp. 108-110.

[15] H. P. Amanapu, U. R. K. Lagudu, A. John-Kadaksham, S. V. Babu, and R. Teki. *ECS J. Solid State Sci. Technol.*, vol. 2, 2013, pp. 362-367.

[16] FangJ Y, TsaiM S, DaiB T, et al. *Electrochem Solid-state Lett*, vol. 8, 2005, pp: 128-130.

[17] T. Du, Y. Luo, V. Desai. *Microelectron. Eng.*, vol.71, 2004, pp. 90.

[18] S. Deshpande, S.C. Kuiry, M. Klimov. *J. Electrochem. Soc.*, vol. 151, 2004, pp: 788.

[19] Preston F W. *Journal of the Society of Glass Technology*, vol. 11, 1927, pp. 214-256. 4

[20] K.J. Laidler. *Int. Union Pure Appl. Chem.*, vol. 68, 1996, pp. 149–192.

[21] S. Pandija, D. Roy, S.V. Babu. *Microelectron. Eng.*, vol. 86, 2009, pp. 367–373.

[22] Y. Hong, V.K. Devarapalli, D. Roy, S.V. Babu. *J. Electrochem. Soc.,* Vol.154, 2007, pp. H444–H453.

[23] J. Sorooshian, D. DeNardis, L. Charns, et al.. *J. Electrochem. Soc.* Vol. 151, 2004, pp. G85–G88.

STUDY OF CERIA SETTLING IN CMP SLURRY

Li Zhang[1,2], Chuangyun Wan[1]*, Changzhen Jia[2], Chengyao Shi[2], Xiaoming Ren[2]*, Shoutian Li[2]

1.Shanghai Institute of Technology

100   Haiquan Rd, Fengxian District, Shanghai, P. R. China

1* cywan@sit.edu.cn

2．Anji Microelectronics Technology (Shanghai) Co. Ltd.

890 Bibo Rd. Pudong New District, Shanghai, P. R. China

2* xiaomingren@anjimicro.com

ABSTRACT

Colloidal stability is a basic requirement for CMP slurry to have stable performance. Due to its high density and high gravity pulling force, ceria particles in CMP slurries cannot be suspended homogenously in static state for long period of time. Improperly mixing can result in changing in CMP performance due to inhomogeneous distribution of ceria particles at different layers of slurry tank. In this work, we collect experimental data how different size of ceria particles and concentration affects the settlement and compare the ceria particles settlement data to the formulated ceria slurry. The ceria particles with larger size will settle faster than smaller size but particle concentration and additives in STI slurry has little impact on particle settlement dynamics. In the formulated ceria, the settlement of ceria particles can change the balance of abrasive to silicon oxide inhibitor ratio and has adversary effect on oxide polishing rates.

INTRODUCTION

Colloidal stability is a basic requirement for CMP slurry to have stable performance [1]. Typical silica-based CMP slurry can be colloidal stable for a long period of time, e.g., longer than a year without settling. Yet, silica-based slurries still can settle and the settling characteristics of silica-based slurries have been subjected to a few studies [2,3]. Silica has low density (2.2g/ cm^3 for fused silica and 2.65 g/ cm^3 for quartz). Thus, the gravity pulling force on silica is much less than high density ceria (7.2g/cm^3). Due to its high density and high gravity pulling force, ceria particles in CMP slurries cannot be suspended homogenously in static state for long period of time and the particles eventually settle down. Ceria slurry must be homogenously mixed well before and during use. Handling of ceria slurries is an important practical process to ensure the repeatable, stable CMP process. During CMP application, ceria slurries must maintain humongous suspension under agitation.

Many factors could affect the ceria settling. In this paper, we conduct experiments to study the colloidal ceria particles settling rates regarding to factors like particle size and particle concentration, and compare the ceria particles settlement data to the formulated ceria slurry.

EXPERIMENT

Colloidal ceria with mean particle size 84nm, 150nm and 230nm measured by dynamic laser light scattering was used in this study. The as-received ceria has no surface treatment and zeta potential is +35mV at pH 4.5. Table I shows the physical properties of colloidal ceria particles in this study.

TABLE I. PHYSICAL PROPERTIES OF COLLOIDAL CERIA PARTICLES

Ceria particles	Size (nm)	Zeta potential (mV)	pH	Conductivity (uS/cm)
Ceria-1	84 nm	+35mV	4.5	60
Ceria-2	150 nm			
Ceria-3	230nm			

In the static settling experiment, homogenously dispersed ceria abrasive particles were placed in 1-liter graduate cylinder about 35cm tall. The particles were allowed to settle without agitation. The particles were carefully drawn out with small tube through a suction method from top. About 143mL of slurry was drawn out each time, which is equal to seven 5cm-tall layers. The physical properties of each layer were measured. For simplicity, we report only two layers: top layer with depth 0-5cm and the bottom layer with depth 30-35cm.–_–A formulated STI ceria slurry was also subjected to the same settling study. The STI slurry was 5x concentrated and solid content was 1% before dilution for point-of-use.

CMP was carried out in clean-room environment using Mirra polisher from Applied Materials (San Jose, CA). Blanket PETEOS and nitride wafers were used to monitor removal rates and selectivity.

RESULTS AND DISCUSSION

Settling Dynamics of Ceria Particles

To study the dynamics of ceria settling, 1% ceria particles at pH4.5 was allowed to static settle for a period of three weeks. The solid content and particle size of different layers were monitored over settling period.

Figure 1 shows the solid content of the 84nm, 150nm, and 230nm ceria over 24 days. For simplicity, only the solid content at top and bottom layers was plotted out. The particle concentration started at 1% at time zero when the ceria slurry was homogeneously mixed. For the larger 230nm ceria particles, the solid content in the top layer (0-5cm depth) quickly became nearly zero after 5 days; while the solid content of the bottom layer continued to increase with time. For the smaller 84nm ceria particles, the difference in solid content between top layer and bottom layer is small, i.e., the top layer still had 0.77% ceria particles and bottom layers had 1.51% after 24 days. The separation in solid content between top layer and bottom layer for the 150nm ceria behaves in between 84nm and 230nm.

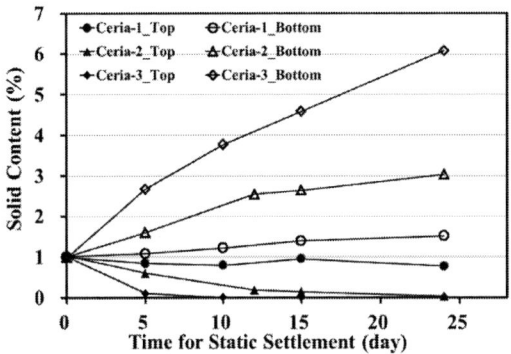

Figure 1: Solid content of different layers of 1% ceria particles at pH 4.5 over settling time

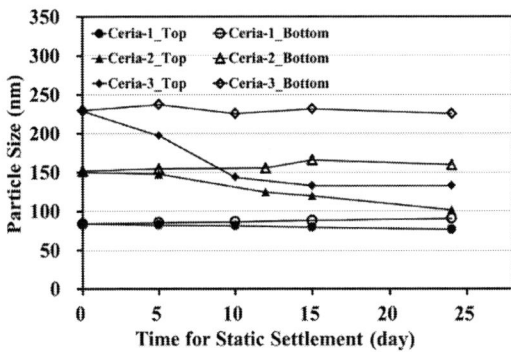

Figure 2: Particle size of 1% ceria vs settling time

For particle size vs. settlement time, the results are shown in Figure 2. In Figure 2, the particle size from top layers decreases over time and such decrease is more pronounced in larger particles than the smaller particles. Both 150nm and 230nm ceria particles show significant drop in particle size on the top layers than the 84nm particles. The particle size of the bottom layers does not change as significant as the top layers over the same

period of time.

The reason for the drop in particle size on the top layers is because the larger particles have a higher sediment rate than the smaller particles. The buoyant force exerts on a ceria particle is proportional to the volume of the ceria particle, as shown in equation (1) and the gravity pulling force is the mass force of the particle which is equal to ceria density times its volume as shown in equation (2).

$$F_b = d_0 \bullet V = d_0 \bullet \frac{4\pi}{3} r^3 \qquad (1)$$

$$F_g = d \bullet V = d \bullet \frac{4\pi}{3} r^3 \qquad (2)$$

Where F_b is the buoyant force, F_g is the gravity force, d_0 is water density, d is ceria density, r is the radius of the ceria particle.

Thus, the force difference Δ between F_g and F_b is

$$\Delta = F_g - F_b = (d-d_0) \bullet \frac{4\pi}{3} r^3 \qquad (3)$$

The net pulling force on a ceria particle is cubic power of particle size. This explains that larger particles settle faster than smaller particles.

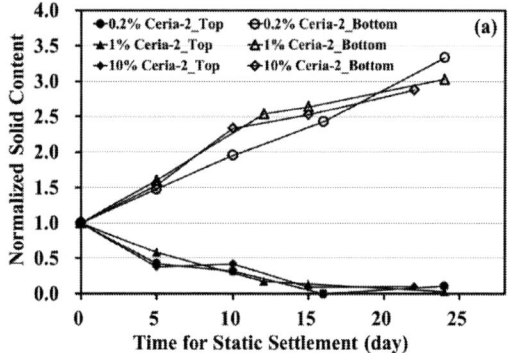

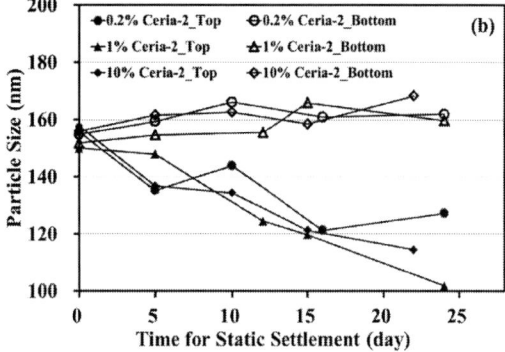

Figure 3: (a) Solid content vs settling time; (b) Particle size vs settling time

The viscosity and friction force in slurry could affect the

settling dynamic of ceria particles. To study such effect, the ceria-2 particles (150nm) at 0.2%, 1% and 10% were subjected to the settling dynamic study. Higher concentration of particles will increase solution viscosity and increase the friction force among particles. The results are shown in Figure 3. In Figure 3a, the normalized solid content was plotted vs. settling time. Figure 3b shows the mean particle size vs. settling time. Again, for simplicity, only the results from the top and bottom layers are shown in Figure 3. The results in Figure 3 show the ceria concentration has little effect on the ceria settling dynamic. This indicates the overwhelming force for the ceria settling is the gravity pulling force. The contribution to the settling dynamic from inherent viscosity or friction of ceria particles in solution is negligible.

Settling Dynamics of Ceria Particles in STI Slurry

To study the settling dynamics of formulated ceria slurry, STI slurry formulated with ceria-2 particles (150nm) at 1% solid was subjected to settling dynamics study. The formulated STI slurry has pH 4.5 and contains additives such as oxide rate booster, dishing inhibitor and nitride inhibitor.

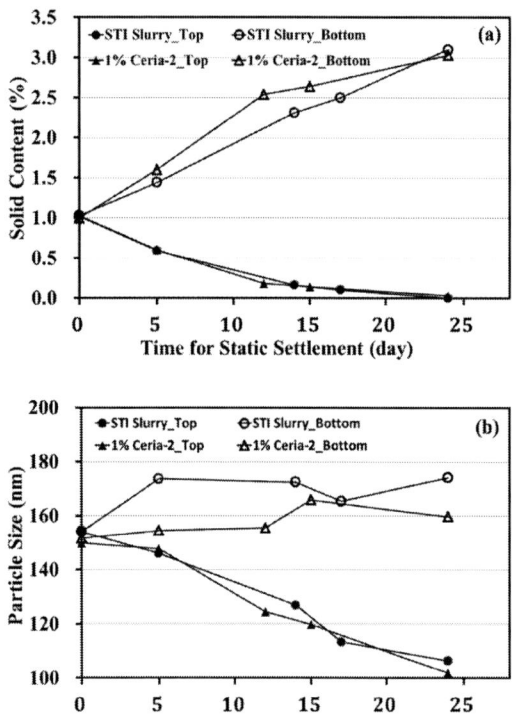

Figure 4: (a) Solid content vs settling time; (b) Particle size vs settling time of STI slurry and 1% ceria particles only

The solid content and particle size vs. settling time is

compared to 1% ceria particle alone and the results are shown in Figure 4, in which Figure 4a shows the solid content and Figure 4b shows the particle size. The results from Figure 4 show that adding chemicals (oxide booster, dishing inhibitor and nitride inhibitor) does not affect the ceria particle settling dynamics. The gravity force is the overwhelming driving force for the ceria particle settlement.

TEOS Rates of STI Slurry Affected by Ceria Settling

CMP performance of STI slurry formulated with 1% ceria-2 (150nm) was evaluated in Mirra polishing. Silicon oxide (TEOS, tetraethyl orthosilicate) and silicon nitride (SiN) removal rates were monitored. CMP was conducted in Mirra polishing. The polishing down force was set at 1.5, 2 and 2.5 psi, platen speed was 93rpm and carrier speed was 87rpm. For simplicity, only polishing rates at 2 psi of top layer and bottom layer were plotted in Figure 5.

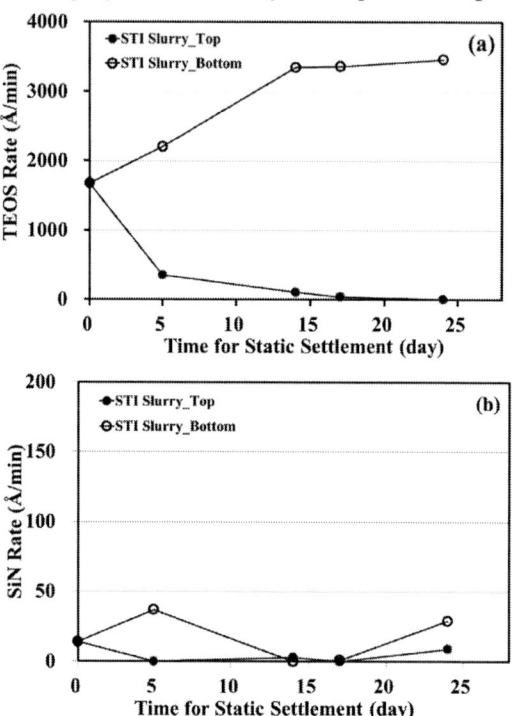

Figure 5: (a) TEOS removal rates vs settling time; (b) SiN removal rates vs settling time of STI slurry

Figure 5 shows that ceria settling has dramatic impact on TEOS removal rates but has little effect on SiN removal rates. In STI slurry, the oxide inhibitor and ceria solid have great influence on TEOS rates. Increase ceria content or reduce oxide inhibitor will increase TEOS removal rates. The SiN inhibitor added in STI slurry was excessive and the SiN removal rates remain low even when the ceria concentration is excessive (the data is not shown here).

Although ceria particles will settle in formulated slurries, proper mixing can restore homogeneity for stable removal rates. A STI slurry formulated with 1% ceria-2 (150nm) was monitored for TEOS and SiN rates over a period of more than 1 year. The STI slurry was allowed to sit on shelf without agitation. Before polishing, the STI slurry was homogeneously mixed through a pail shaker for at least 20 minutes. Then the well mixed concentrated slurry was diluted with deionized water. CMP was conducted in Mirra polishing. The polishing down force was 2.5psi, platen speed was 127rpm and carrier speed was 123rpm. Figure 6 shows the TEOS and SiN removal rates. The well mixed STI slurry shows stable TEOS and SiN removal rates more than 1 year.

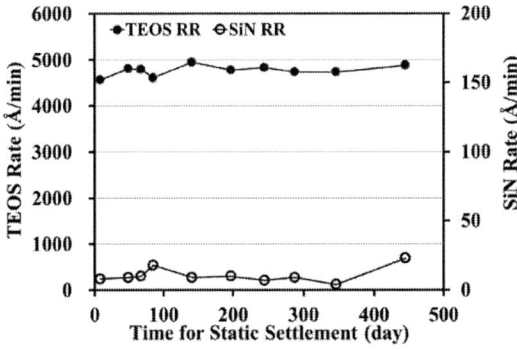

Figure 6: STI slurry's TEOS and SiN rates through shelf life

CONCLUSION

Ceria slurries must be agitated and mixed well during CMP applications since without agitation, ceria particles will settle down due to its high density and high gravity pulling force. The overwhelming driving force for the ceria particle settlement is the gravity force, which explains why ceria particle size greatly impacts on settling dynamics while other factors such as particle concentration, additives in formulated slurries have little influence. Through proper mixing, the settled ceria slurries still can offer stable removal rates and long shelf life.

ACKNOWLEDGEMENTS

The authors would like to acknowledge Anji Microelectronics Corporation Ltd. for the permission to publish this paper. We also would like to thank the individuals who help us in preparation of the manuscript.

REFERENCES

[1] A. Philipossian, et al., CMP Technology for ULSI Interconnection Tech. Prog. Presentation, Semicon West, 2000.

[2] R. K. Singh and B. R. Roberts, Proceedings of VMIC Conference, 2000 IMIC – 200/00/0545, pp. 545-547.

[3] J. Lin and W. S. Rader, CSTIC 2017, Shanghai, China, 2017.

PATTERN LOADING EFFECT OPTIMIZATION OF BEOL CU CMP IN 14NM TECHNOLOGY NODE

Lei Zhang, Yuanyuan Meng, Yi Xian,Wei Zhang, Haifeng Zhou, Jingxun Fang
Shanghai Huali Integrated Circuit Corporation
NO.6, Liangteng Rd. Pudong Shanghai, 201314, China
*Corresponding Author's Email: zhanglei_td2@hlmc.cn

ABSTRACT

To achieve the local, as well as global, planarity of the wafer surface many innovative technologies have been developed. A robust Cu chemical mechanical polishing (CMP) process with better post CMP polishing profile, smooth copper surface, tighten metal line sheet resistance (Rs) and pattern loading control has been evaluated during the Cu CMP process at 14nm and beyond. It is well known that CMP causes pattern loading of a layer to be planarized due to uneven distribution of device structures and thus reducing the effectiveness of this technology. This paper will present how to improve pattern loading and dishing control with optimized polish methodology. Experiment results shown that there is no loading between dense line area and ISO line area, and better dishing performance.

INTRODUCTION

In multilevel IC manufacturing, it's important to have a planar surface preceding the next layer to avoid topographical margin issues. In Cu planarization, the bulk copper is removed first using a high removal rate process with high selectivity slurry, followed by reducing polish down force with lower removal rate process and finished with barrier removal process. The topography after polishing such as dishing and erosion causes the increasing or variation of wiring resistance, and the sort between the wirings in upper layer. And the barrier removal process as the last polish process in Cu CMP decided the final topography and defectively performance. Poor surface topography and defect will bring some challenges in lithography and wiring for the next layer. [1-2]

One of the solutions for this pattern loading phenomenon has been the introduction of pattern fill methodology to improve the planarity of a given layer. However, dummy pattern adds capacitive load and thus, parasitic effects on both analog and digital circuits. Lower defectivity, pattern loading effect and better stability of removal rate through soft pad polishing life time have been identified as key influencing factors in copper CMP process. In order to meet the strict requirement, novel copper barrier slurry is designed and developed as a consequence. One of the biggest issue of soft pads in Cu barrier CMP process is that it tends to suffer serious removal rate trending through pad life. [3-4]

In addition, it is well known that scratch and organic residue often have a significant but negative effect on yield and the following topography. Chemical mechanical planarization become a mainstream process in semiconductor industry, it is a key technology to generate flat and smooth surface at several critical steps in the manufacturing processes. The planarization performance is influenced by topography characteristics, line/space width, pattern density, slurry chemistry, rotation speed, pad type, force/pressure, etc. However, as device continuous shrink, CMP process becomes more challengeable to achieve planarization. There exist two common issues that often occur at different pattern densities and line widths are dishing and erosion. In Cu CMP process, dishing is defined as the oxide loss relative to the level of the neighboring nitride space, and erosion refers to the nitride loss relative to the nitride level of the neighboring area. Wide trenches or open structures usually enhance the dishing issue, while dense trenches lead to more erosion. On the other hand, oxide dishing include Cu residue at next layer. [5]

In this paper an optimized polish methodology is presented that reduces the impact on RSFV while improving the pattern loading through optimizes the buffer pressure and down force. To overcome the problem faced, this pater optimized low down force and over polish condition, to balance Cu and barrier selectivity. Improve removal rate stability and adjustable selectivity for process demanded.

EXPERIMENTAL

CMP polishing equipment mainly consists of three steps in the process, as shown in figure 1. Step 1 for preliminary polishing stage, with high Cu remove rate, the main role for preliminary polishing to remove most surface Cu, real time profile closed loop control for Cu process, well control surface uniformity; Step 2 is given priority to polishing, remove the surface part of the Cu and blocking layer, with low cu remove rate, and high selectivity for endpoint signal pressure, barrier layer as polishing end layer; Step 3 as barrier polish, the oxide has high selectivity, thus forming Cu pattern.

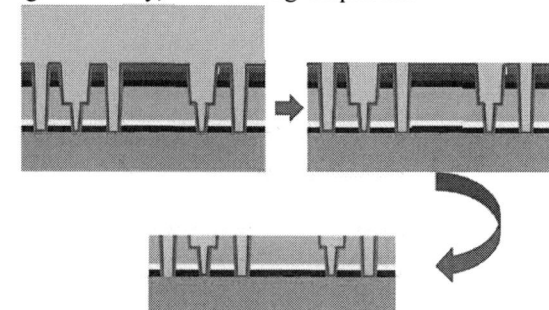

Figure 1: Cu CMP polishing process diagram

For 14nm Cu CMP, baseline condition with high down force pressure for both Cu remove step and high pressure at barrier polish step. WAT result shows Rs increase fast at large pitch area, as shows in figure 2. WAT for different width not

978-1-7281-6559-2/20 $31.00 © 2020 IEEE

follow ideal trend, thickness match with WAT, wide metal much thinner.

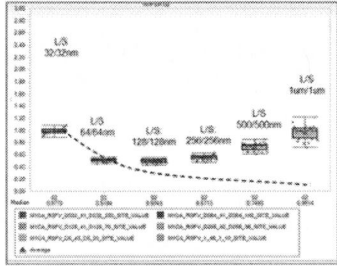

Figure 2: WAT result about Metal Rs with Different W/S(density 50%)

ISO metal shows much open, while 25%~75% density reasonable, and 100% some open. This result shows different loading performance with 14nm node process. As shown in figure 3. WAT and TEM result show that, for bigger IMD area, polish rate and Cu remove rate much faster.

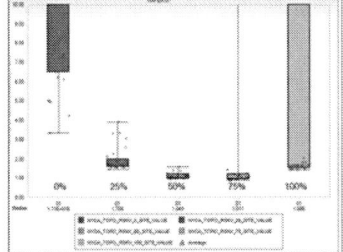

Figure 3: Metal Rs with Different Dummy Density

For 14nm node, with small line/space width, as device continuous shrink, CMP process becomes more challengeable to achieve planarization. The planarization performance is influenced by topography characteristics, line/space width, pattern density, slurry chemistry, rotation speed, pad type, force/pressure, etc.

As shown in figure 4, the result shows there is exist pattern loading at different pattern densities and line widths. Pattern loading is defined as the oxide loss relative to the level of the neighboring Cu space.

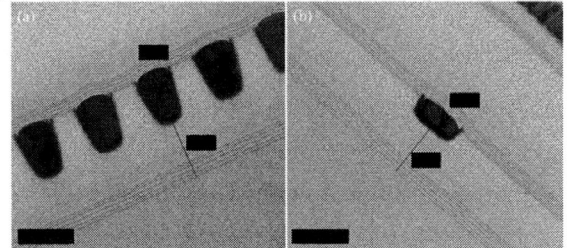

Figure 4: TEM image for different pattern densities and line widths after Cu CMP process

There exist two common issues that often occur at different pattern densities and line widths are dishing and erosion. Dishing is affected by pattern size and trench width, and erosion is affected by active pattern density. The result shows pre layer dishing include next layer suffer Cu residue defect at oxide area. The Cu residue occurred at Cu CMP process, over polish not enough.

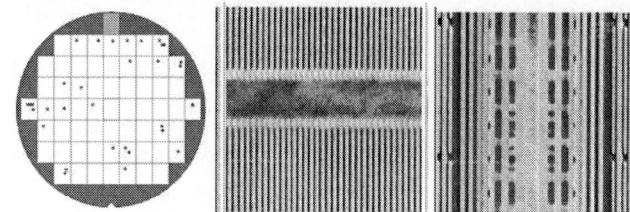

Figure 5: Cu residue defect image post Cu CMP scan result

Defect scan result shows suffer Cu residue defect issue, random map, as shown in figure 5; check layout pre layer is non-pattern oxide area. Suspect pre layer CMP oxide dishing induce Cu residue defect, as shown in figure 6.

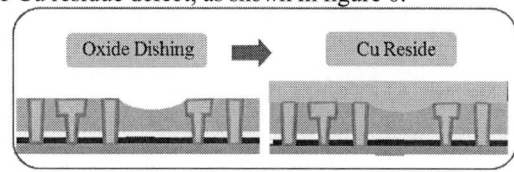

Figure 6: Pre layer Cu CMP include Cu residue diagram.

In this paper an optimized polish methodology is presented that reduces the impact on RSFV while improving the pattern loading through optimizes the buffer pressure and down force.

After via etch, and deposition of Cu. CMP process to remove most of the surface of the Cu and barrier layer. In order to balance Cu remove and oxide remove effect during Cu CMP, series of experiments were carried with various LDF, over polish and barrier polish time split.

CMP split focus on: over polish time, low down force, buffer polish force. In order to get minimal and stable dishing, CMP changed recipe parameters, such as top ring pressure and slurry dilution, polished wafers, and then measured dishing data.

RESULT AND DISCUSSION

WAT results show that, for both by width and by density, low down force condition shows better Rs performance, as shown in figure 7.

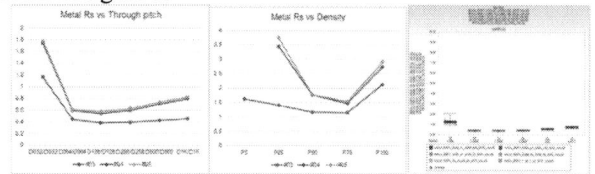

Figure 7: WAT result about Metal Rs with Different W/S with CMP new condition

Check initial dishing performance after bulk CU removal, Initial dishing is similar (35A~54A), there is no loading after LDF. Cu CMP new condition shows better WAT performance. And TEM check physical thickness has obvious improvement. Pattern loading has obvious improved.

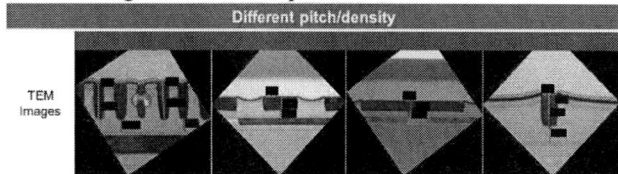

Figure 8: TEM images for different pattern densities and line widths after Cu CMP new condition

978-1-7281-6559-2/20 $31.00 © 2020 IEEE   407

And for defect scan split, a) b) are old condition, c) is improvement condition, which shows Cu residue free.

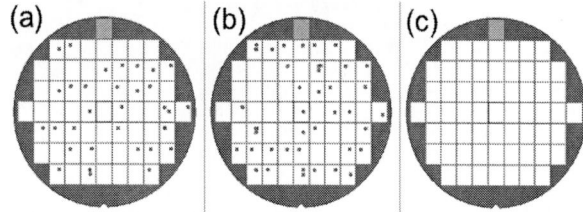

Figure 9: Defect map for different Cu CMP condition (a,b: old condition, c: new condition)

According to the results, the performance of dishing and Cu remaining thickness after CMP process is highly dependent on the pattern density characteristics. In this case, optimized pressure and over polish time can be obvious improve pattern loading. From overall experimental results, under new CMP condition, dishing performances were better and more stable and had better uniformity. And narrower structures had much lower dishing.

These data clearly show that optimized condition is a promising candidate for the dishing improvement of the pattern loading. Currently we are continuing the CMP optimization (i.e. an optimization of process times, further development with new slurries with lowered oxide remove rated and new post-CMP cleaning to further improve the electrical performance for 14nm node.

CONCLUSION

The continuous scaling in the logic device landscape towards 14nm node dimensions has presented numerous engineering and manufacturing challenges in BEOL Cu interconnection. CMP engineers always struggle with dishing or erosion performance in critical layers during new product pilot run.

In this paper, the performance of new CMP condition not only meet the difficult criteria of pattern loading but also present much more steady performance about oxide dishing .

Optimization CMP process between the down force and polish time processes is the primary source of pattern loading and dishing in Cu interconnects. Experiment result shown the defect between dense trench line area and oxide area are quite tradeoff. Trench area requests big top opening, while SRAM area requests more Cu step coverage on sidewall. Optimized parameters were finally presented and the defect level was significantly reduced

REFERENCES

[1] R. DeJule, et al. CMP Challenges Below a Quarter Micron. Semiconductor International. 1997.

[2] H. Nancy, et al. CMP process optimization for improved compatibility with advanced metal liners. IEEE Conference Publications. 2010

[3] Y.L. Hsieh, et al. Effects of BEOL Copper CMP Process on TDDB for Direct Polishing Ultra-Low K Dielectric Cu Interconnects at 28nm Technology Node and Beyond. Crown. 2013

[4] L. Zhang, et al. Study and Improvement on Tungsten Recess in CMP Process. IEEE Conference Publications. 2019

[5] H.F. Kao, et al. Dishing and Erosion Amount Prediction According Pattern Density Calculation Algorithm in 3D Design Layout. Joint Symposium. 2015

THE ADSORPTION AND REMOVAL OF CORROSION INHIBITORS DURING METAL CMP

Jin-Goo Park[1]*, Heon-Yul Ryu[2], Tae-Gon Kim[3], Nagendra Prasad Yerriboina[1],
Yutaka Wada[4], Satomi Hamada[4], and Hirokuni Hiyama[4]

[1]Department of Materials Science and Chemical Engineering, Hanyang University, Ansan, Korea.
[2]Department of Bio-Nano Technology, Hanyang University, Ansan, Korea.
[3]Department of Smart Convergence Engineering, Hanyang University, Ansan, Korea.
[4]EBARA Corporation, Fujisawa, Kanagawa, Japan.
*Corresponding Author's Email: jgpark@hanyang.ac.kr

ABSTRACT

Corrosion inhibitor plays a key role during Chemical mechanical planarization (CMP) of metal surfaces during semiconductor processing. Strong metal-inhibitor passivation formation during the CMP process and its easy removal during post-CMP cleaning are highly required. However, there are no studies available explaining this phenomenon. In this work, passivation changes of copper (Cu) and cobalt (Co) surfaces during CMP and post CMP cleaning by adsorption and removal of benzotriazole (BTA), was characterized using a new sequential electrochemical impedance spectroscopy (EIS) technique. It was found that stable Cu/Co-BTA complex (metal-inhibitor passivation) was formed when each metal surface was exposed to BTA solution. However, it was found that adsorbed BTA on Co surface could be removed just by de-ionized (DI) water rinsing while BTA on Cu surface was not removed.

INTRODUCTION

Use of a corrosion inhibitor is critical during the metal CMP process because it not only prevents corrosion defects but also contributes to global planarization during polishing. However, the presence of the corrosion inhibitor on the metal surface in the form of the metal-inhibitor complex after CMP would remain as the residual organics (not removed by DI water rinse) generating organic type defects. Generally, it is a challenging task to remove it completely during post-CMP cleaning. For example, benzotriazole (BTA) which is the most commonly used corrosion inhibitor makes the Cu surface hydrophobic and causes problems such as drying issues and poor adhesion of stacking layers apart from generating organic defects [1]. On the other side, recently, it has been started to use Co, relatively very sensitive to corrosion, for contacts and interconnects by replacing tungsten (W) and Cu. BTA is believed to be a suitable Co inhibitor for CMP [2], but no studies are available on the cleaning of Co-inhibitor complex, which is very critical as explained above for Cu. Therefore, understanding the metal-inhibitor interactions during both CMP and post-CMP process is important.

In this study, the formation and its removal of BTA passivation with Cu and Co surface were investigated using a novel sequential EIS. Using this technique, Cu/Co-BTA complex formation stability and also its removal ability by DI water rinsing was evaluated.

EXPERIMENTAL

EP (Electroplated) Cu wafers and PVD (Physical Vapor Deposition) Co wafers were cut into coupons and used for all experiments. Before EIS measurements and BTA treatment, native oxide of Cu and Co was removed by dipping in 25 mM citric acid for 2 min. EIS experiments were performed using a potentiostat (VersaSTAT3, AMETEK, USA) with a three-electrode system. Cu/Co coupon was used as a working electrode, Ag/AgCl (saturated KCl) and platinum mesh were used as a reference and counter electrodes, respectively. Sequential measurements were performed with the Cu and Co coupon by treating the electrolyte solutions accordingly. All solutions were adjusted to pH 7 by HNO_3 and KOH. 10mM BTA solution and DI water were used as an electrolyte to investigate the change of surface passivation of Cu and Co before and after BTA treatment. EIS measurements were performed at open-circuit potential (OCP) values and spectrums were obtained in the frequency range from 0.068 to 100,000 Hz with a potential amplitude of 10 mV rms. Commercial software (ZSimpWIN, EChem Software, US) was used to retrieve the electrical equivalent circuit (EEC) parameters. To measure the thickness of BTA passivation layer on Cu and Co surface, each metal surface was selectively treated with 10 mM BTA solution followed by DI water rinse by using fluidics chip and the metal-BTA passivation layer was measured by atomic force microscope (AFM, NX-20 300 mm, Park Systems, Korea). The water contact angle of Cu and Co surface was measured by a contact angle analyzer (Phoenix 300, SEO, Korea).

RESULTS AND DISCUSSION

Since EIS is a non-destructive technique, a novel approach was followed to estimate the stability of metal-BTA complex on metal surface. Figure 1 shows the comparison of surface passivation with the metal surfaces in different electrolyte conditions. The polarization resistance (R_p) in step 1 is the passivation degree of the metallic surface after native oxide removal. At step 1, Cu and Co showed similar R_p around 1-2 kohm-cm^2. After the 1st EIS measurement (step 1), the electrolyte was changed to 10 mM BTA solution and the 2nd EIS measurement was performed (step 2). As shown in Figure 1, both Cu and Co surface showed increased R_p compared to step 1. It indicates that BTA adsorbed on the surface and formed a passivation complex with Cu and Co. At step 2, Cu surface showed 2.5 times higher R_p value than Co surface implying more passivation of BTA. After the 2nd EIS measurement (step 2), the electrolyte was also changed to DI water for the 3rd EIS measurement (step 3). At this time, metal-BTA passivation showed different behavior for Cu and Co surface. Cu surface showed increased R_p value which indicates more BTA was adsorbed. It is well-known that BTA forms a stable and thick polymeric film on Cu surface after BTA treatment [3]. The increased R_p value might be due to the time between 2nd EIS measurement and the drain of the BTA solution. In the case of Co, it showed a dramatic decrease of R_p value from 69 kohm-cm^2 to 5.8 kohm-cm^2 in DI water. It means that most of the adsorbed BTA were removed from Co surface by DI water.

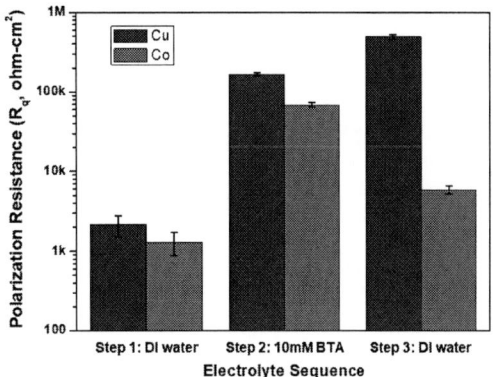

Figure 1: Polarization resistance (R_p) values of Cu and Co surfaces modeled from impedance spectra of sequential EIS measurements. Step 1 with DI water, step 2 with 10 mM BTA and step 3 with DI water.

Figure 2 shows the AFM measurement result of passivated BTA on Cu surface. As already known in many works of literature, the Cu-BTA complex layer remained after DI water rinse and AFM result showed that its thickness was around 1-2 nm. However, it was not possible to determine a boundary between Co and Co-BTA film. It might be due to the removal of BTA from Co surface by DI water rinse providing the indistinctive result.

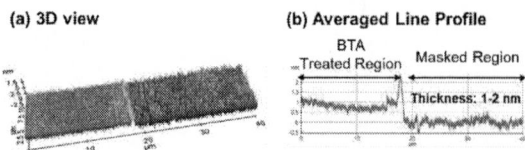

Figure 2: AFM measurement result of Cu-BTA complex film on Cu surface. (a) 3D view and (b) averaged line profile.

For further confirmation on the passivated metal-BTA layer on Cu and Co surface, contact angle of each metal surface with water before and after BTA treatment was measured because it is an effective method to analyze the adsorption of organic molecules on film surfaces qualitatively. The metallic Cu and Co showed hydrophilic surface after native oxide removal as shown in Figure 3. After BTA treatment followed by DI water rinse, Cu surface became hydrophobic showing a contact angle of 63° and Co surface showed a lower contact angle around 30° which is higher than the metallic Co surface. The reason for increased contact angle was found to be due to the residual BTA on Co surface by using an attenuated total reflectance-Fourier-transform infrared (ATR-FTIR) spectroscopy as shown in Figure 4. As-received Co showed a Co-oxygen (Co-O) vibrational mode at the wavenumber of 580 cm^{-1}. BTA treated Co surface only showed a peak at 750 cm^{-1} which is attributed to BTA and it did not show Co-O peak because BTA was treated on metallic Co surface. From this result, it could be expected that not all the BTA was removed by DI water. However, when compared with the contact angle of a thick Cu-BTA complex layer (63°), it can be inferred that BTA does not exist as a layer form on the Co surface because the contact angle measurement is sensitive to the outermost surface. Thus, the water contact angle measurement suggests that only a small amount of Co-BTA complex was remained on the Co surface with low coverage after the BTA treatment and DI water rinse.

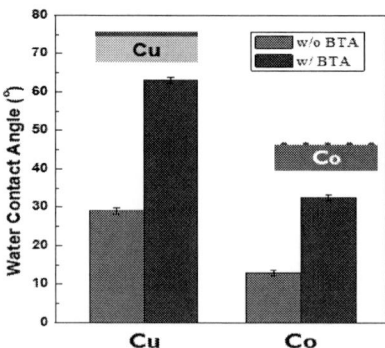

Figure 3: Contact angle of Cu and Co surface with water

before and after BTA treatment.

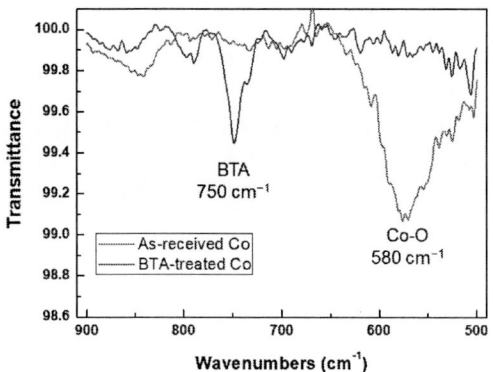

Figure 4: Contact angle of Cu and Co surface with water before and after BTA treatment.

In conclusion, BTA was adsorbed well on Cu surface and form a thick polymeric passivation film which can lead to the organic residue during post Cu CMP surface, but for Co surface, BTA showed good rinse ability by DI water even though it showed similar adsorption degree to Cu during the exposure to BTA (in-situ condition).

CONCLUSIONS

In this study, passivation stability of BTA on Cu and Co surface was characterized by using a novel sequential EIS technique. AFM and water contact angle measurements were performed to support EIS data. When each metal surfaces were exposed to BTA, passivation was enhanced due to the strong BTA adsorption and complex formation on Cu and Co surface. However, most of adsorbed BTA was removed from Co surface at ex-situ EIS measurement in DI water, while adsorbed BTA was maintained on Cu surface. In conclusion, it could be expected that Cu and Co can be protected by BTA during CMP or post-CMP cleaning process, but adsorbed BTA on Co surface can be removed easily by rinsing step. Hence the proposed 'sequential EIS technique' is a good option for evaluating not only the corrosion inhibition ability but also the removal ability of corrosion inhibitors during post metal CMP cleaning.

REFERENCES

[1] M. Ramachandran, B. J. Cho, T. Y. Kwon and J. G. Park, *Jpn. J. Appl. Phys.*, vol. 52, 2013, pp. 1-6

[2] H. S. Lu, X. Zeng, J. X. Wang, F. Chen and X. P. Qu, *J. Electrochem. Soc.*, vol. 159, 2012, pp. C383-C387

[3] B. J. Cho, S. Shima, S. Hamada and J. G. Park, *Appl. Surf. Sci.*, vol. 384, 2016, pp. 505-510.

EFFECTS OF DIFFENT INHIBITORS ON Cu-Co GALVANIC CORROSION IN POST CMP CLEANING

Xiaoqin Sun[1,2], Baimei Tan[1,2]*, Chenwei Wang[1,2]*, Mengrui Liu[1,2], Pengcheng Gao[1,2], Qi Wang[1,2], Siyu Tian[1,2]

[1]School of Electronic Information Engineering, Hebei University of Technology, Tianjin 300130, China
[2]Tianjin Key Laboratory of Electronic Materials and Devices, Tianjin, China
*Email: bmtan@hebut.edu.cn,cwtjy206@163.com

ABSTRACT

The large potential difference between copper and cobalt caused the corrosion of copper and cobalt films and the galvanic corrosion in the process of chemical mechanical planarization (CMP) and post cleaning of cobalt barrier layer. In this paper, the effects of different inhibitors on the inhibition of copper-cobalt galvanic corrosion in the cleaning solution are studied. Different concentrations of inhibitors were added to FA/O II chelating agent as cleaning agent. The inhibition effect was characterized by electrochemical workstation and the cleaning effect of particles was tested by SEM. The results show that the corrosion potential difference of Cu and Co and surface roughness can be significantly reduced with different concentration of inhibitors.

Keywords—Chemical mechanical polishing (CMP); Post-CMP cleaning; Copper interconnection; Cu-Co galvanic corrosion; Inhibitors

INTRODUCTION

With the gradual decrease of IC feature size, copper became the main wiring metal for multi-layer interconnection of integrated circuits due to its low resistivity and high resistance to electromigration. However, when the feature size is reduced to 14nm or less, there are many problems in the conventional Ta/TaN double-layer barrier structure of copper interconnect layer, such as the rapid increase of interconnect resistance and uneven deposition of copper seed layer.[1] In order to solve these problems, cobalt (Co), as a substitute for Ta liner in Ta/TaN double-layer structure, has been widely concerned.[2]

However, as the standard equilibrium potentials of Cu^{2+}/Cu and Co^{2+}/Co couples are +0.34 V (SHE) and −0.28 V(SHE), the larger potential difference will cause the separate corrosion of Cu and Co thin films and the couple corrosion in the CMP process. The Cu/Co interface is electroeroded, and cobalt dissolves as the cathode to form holes, which leads to the diffusion of copper into the dielectric layer and device failure.[3] Corrosion inhibitors are generally selected to inhibit the galvanic corrosion of Cu and Co barrier layers. [4]

In recent years, a lot of researches have been carried out on the inhibition of Cu-Co galvanic corrosion. He et al. [5] suggested a polishing slurry containing TAZ (pH=3) to reduce the corrosion current from $162\mu A/cm^2$ to $22\mu A/cm^2$, demonstrating that TAZ can effectively reduce Cu/Co galvanic corrosion under acidic conditions. K. V. Sagi et al. [6] used a polishing slurry containing oxalic acid and nicotinic acid(NA), which greatly reduced the removal rate of Cu/Co materials, while the removal rate selection ratio was about 1:1. The Cu CMP post-cleaning is a key step in the chemical-mechanical polishing process, as the copper surface is affected by a large number of particles (such as silica) and organic residues (such as inhibitors).

In this paper, the effects of different inhibitors in the inhibition of Cu-Co galvanic corrosion in cleaning solution are studied, SEM and electrochemical method were used to characterize the cleaning effect.

EXPERIMENT

Electrochemical experiments.

The galvanic corrosion current density measurement was conducted on an electrochemical workstations (Shanghai Chenhua CHI660E electrochemical workstation). Co and Cu disks (99.99% purity) were sealed in epoxy with an exposure area of 6 mm in diameter. Platinum as counter electrode and saturated calomel electrode (SCE) as reference electrode. And the working electrodes were mechanically abraded with 2000# alumina sand paper, and then rinsed with DI, dried with high purity nitrogen before all the experiments. The solution is based on 150 ppm FA/OII, including different inhibitors - NA / TAZ and their combinations. The pH value was adjusted to 9.5 by using KOH/HNO₃ .

Cleaning experiment.

3 inch diameter Cu blanket wafer were used in cleaning experiments. The purity of discs were 99.99%. In the cleaning experiment, silica abrasive particles (2.5wt%, 60nm) was coated on the copper surface to test the cleaning effect of the cleaning slurry. The wafer cleaning experiment with PVA scrub for 1 min and the cleaning solution flow is 1 L/min, then rinsed with DIW

followed by N_2 dry. The cleaning solution for the cleaning of copper surface consisted of FA/OII as chelating agent and inhibitor. The morphology of original and cleaned copper surface was characterized by scanning electron microscope (SEM).

RESULTS AND DISCUSSION
Electrochemical experiments

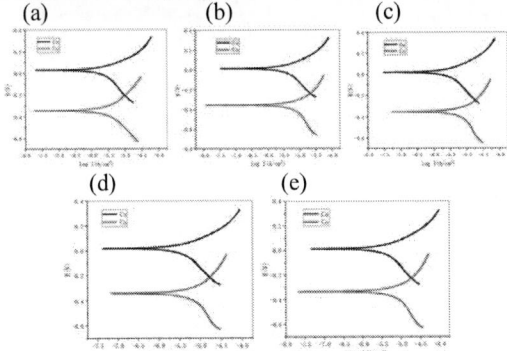

Fig 1:Tafel curves of copper and cobalt with different concentrations of NA:(a)40mM,(b)50mM,(c)60mM,(d)70 mM,(e)80mM.

Figure 1 shows the dynamic electrochemical diagram of copper and cobalt with different concentrations of NA. The corrosion voltage(E_{corr}) and potential difference(ΔE) are listed in Table 1. It can be seen from Figure 1 that with the change of NA concentration, the E_{corr} of copper and cobalt has little change. Table 1 shows that with the increase of NA concentration, the E_{corr} of copper first decreases and then increases, while that of cobalt is the opposite, first increased and then decreased, and the ΔE decreases, reaching the minimum value of 348mV at 80mM. Nicotinic acid is known to form a Co(II)-NA complex which can inhibit any Co film corrosion by forming an adsorbed surface complex at pH 10. However, the inhibition effect of Cu-Co galvanic corrosion is not obvious.

TABLE 1: Potential difference of copper and cobalt with different concentrations of NA.

Concentrations of NA (mM)	Cu E_{corr} (V)	Co E_{corr} (V)	$|\Delta E_{corr}|$ (V)
40	0.029	-0.342	0.371
50	0.014	-0.355	0.369
60	0.02	-0.348	0.368
70	0.019	-0.339	0.358
80	0.014	-0.334	0.348

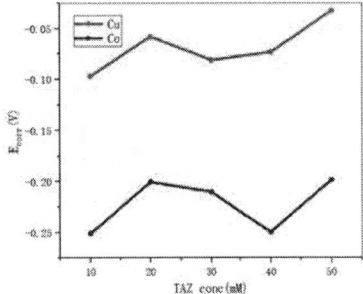

Fig 2: Corrosion potential of copper and cobalt electrodes at different concentrations of TAZ.

Figure 2 shows the E_{corr} of copper and cobalt with different concentrations of TAZ. It can be seen from Figure 2 that with the increase of TAZ concentration, the E_{corr} of copper increases first, then decreases, and then increases, while that of cobalt fluctuates obviously. The ΔE decreases, reaching the minimum value of 152mV at 30mM.

The presence of TAZ inhibits the reaction of the cathode, thus affecting the reaction of Co on the cathode surface. Due to the enhanced passivation effect of TAZ, the Co surface was covered by an insoluble Co-TAZ, which inhibited the corrosion of Co. When the concentration of the inhibitor TAZ in the solution increased further, a soluble complex was formed on the surface of cobalt. Due to the partial dissolution of the film, the corrosion potential of Co decreased.

There are two ways of TAZ passivation on copper surface, one is the growth of Cu-TAZ passivation film on copper surface, the other is the redeposition of Cu-TAZ complex on copper oxide surface. Thus TAZ as an inhibitor can reduce the corrosion potential difference between copper and cobalt, and weaken the galvanic corrosion between Cu and Co.

TABLE 2: Potential difference of copper and cobalt with different ratio of TAZ and NA.

TAZ: NA	Cu E_{corr} (V)	Co E_{corr} (V)	$|\Delta E_{corr}|$ (V)
1:1	-0.137	-0.468	0.331
2:3	-0.147	-0.391	0.244
3:2	-0.162	-0.42	0.258
1:4	-0.141	-0.471	0.33
4:1	-0.164	-0.464	0.3

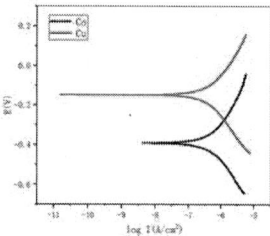

Fig 3: Tafel curves of copper and cobalt at TAZ:NA concentration ratio of 1.

It can be seen from the previous study that when the NA concentration is 80 mM, ΔE is 348mV, reaching the minimum value; when the TAZ concentration is 30 mM, ΔE is 152 mV, reaching the minimum value.

Therefore, Table 2 shows the corrosion voltage and corrosion potential difference of 80 mM NA and 30 mM TAZ mixed in different proportions with a total volume of 70 mL, and Figure 3 shows the Tafel curve when TAZ: NA is 1:1.

It can be seen from the Table that when TAZ: NA = 2:3, its ΔE reaches the minimum value of 244 mV; and after mixing, the corrosion potential difference is smaller than that when NA is used alone. Therefore, the combination of TAZ and NA can better inhibit the galvanic corrosion between copper and cobalt.

Cleaning experiment

(a)

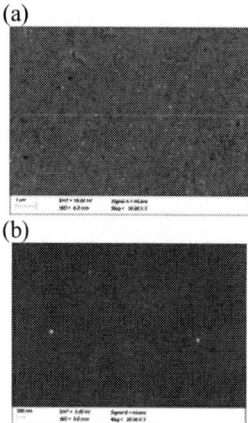

(b)

Fig 4: SEM image a)before,b)after FA/OII+TAZ cleaning of copper film contaminated by silica.

In the previous experiments, we added inhibitors of different concentrations to the alkaline cleaning solution, and found that adding inhibitors to the alkaline cleaning solution can effectively inhibit the galvanic corrosion problem introduced in the cleaning process. The experimental results show that TAZ with a concentration of 30 mM has the best inhibition effect on Copper-Cobalt galvanic corrosion, and the corrosion potential difference is 152 mV.

Therefore, through the removal of particles on the copper surface by the cleaning solution, further verify the cleaning effect of the cleaning solution. Figure 4 is SEM before and after cleaning the particles adsorbed on the copper surface with the cleaning solution mixed with FA/OII and TAZ. It can be seen from the Figure 4 that the particles on the copper surface can be cleaned better by using the cleaning solution, but there are still residues on the surface particles, so further research on the cleaning solution is still needed.

CONCLUSION

By adding different inhibitors to the alkaline polishing solution, it is found that TAZ can effectively inhibit the galvanic corrosion of copper and cobalt, while NA has a weak corrosion inhibition effect. The combination of these inhibitors can make the corrosion inhibition effect stronger than NA alone. Therefore, it can be considered that the combination of inhibitors has a good effect on corrosion inhibition. The feasibility of this method verifies the effectiveness of copper and cobalt galvanic corrosion inhibition, and provides a useful work for the next step of corrosion inhibition.

ACKNOWLEDGEMENTS

This work was supported by the Major National Science and Technology Special Projects (No. 2016ZX02301003-004-007), the Natural Science Foundation, China (No. 61704046), and the Hebei Natural Science Foundation Project (No. F2018202174). The authors thank the teachers and classmates for helpful discussions.

REFERENCES

[1] Jun Ji, Guofeng Pan, Wenqian Zhang, Yichen Du, Ping He, Ye Tian, et al. *J. ECS Journal of Solid State Science and Technology*,vol. 6, 2017,pp.813-818.

[2] Xiangzhou Li, Guofeng Pan, Chenwei Wang, Xuehai Guo, Ping He, and Yue Li. *J. ECS Journal of Solid State Science and Technology*,vol. 5,2016, pp.540-545.

[3] Simpson, D.E., C.A. Johnson, and D. Roy. *J. Journal of The Electrochemical Society*, vol. 166,2018, pp.3142-3154.

[4] Wu, J.-L. and J.-S. Fang.*J. Applied Surface Science*, vol. 477,2019,pp. 280-284.

[5] Peng He, Bingbing Wu, Shuai Shao, Tong Teng, Peng Wang, and Xin-Ping Qu. *J. ECS Journal of Solid State Science and Technology*, vol. 8,2019, pp.3075-3084.

[6] Sagi K V , Teugels L G , Van d V M H , et al. *J. ECS Journal of Solid State Science and Technology*, vol. 6,2017, pp.276-283.

MOLECULAR DYNAMICS STUDY ON SUB-NANOSCALE REMOVAL MECHANISM OF 3C-SIC IN A FIXED ABRASIVE POLISHING

Piao Zhou[1], Yongwei Zhu[1]*, Tao Sun[2]*

[1]College of Mechanical and Electrical Engineering, Nanjing University of Aeronautics and Astronautics, Nanjing 210016, China

[2]College of Chemistry and Chemical Engineering, Shanghai University of Engineering Science, Shanghai 201620, China

[1]*Corresponding Author's Email: meeywzhu@nuaa.edu.cn
[2]*Corresponding Author's Email: 04190006@sues.edu.cn

ABSTRACT

The mechanical removal mechanism of silicon carbide crystal is investigated by Molecular Dynamics (MD) simulation in a fixed abrasive polishing. Special attention is paid to the effect of the sub-nano scratching depth on the mechanical removal behavior. It was found that only the amorphous phase transition occurs in SiC. The temperature, subsurface damage depth and removal rate of SiC substrates increase with the increase of scratching depth. Furthermore, the result shows that the scratching force increases as the scratching depth increases.

Keywords—Mechanical removal; Local temperature; Amorphous phase transition; Fixed abrasive polishing; MD simulation; Sub-nanoscratch

1 INTRODUCTION

Monocrystalline silicon carbide as a ceramics material has been widely applied in LED, laser machine and semiconductor industry. It is difficult to achieve the high processing efficiency and quality for SiC[1]. Due to its higher removal efficiency relative to the a free abrasive polishing, the fixed abrasive polishing (FAP) has been widely applied in the processing of super hard and brittle materials[2, 3]. It is necessary to understand the removal mechanism of SiC wafer for further improving the removal efficiency and surface quality by using FAP technology.

SiC wafers polished by fixed abrasive have been researched by Xu[4]. It is found that the single cycle removal thickness is on the nano even sub-nanoscale for SiC substrates. The scratching depth of some abrasives is on sub-nanoscale due to its short asperity. It is difficult to investigate the removal mechanism by experiments when the scratching depth of diamond abrasives is on sub-nanoscale. MD simulation was found to be a suitable tool for studying the mechanical removal behaviors on sub-nanoscale.

The nanoindentation and nano-scratching of SiC substrates with diamond have been researched by MD simulation[5-9]. Zhang[10] conducted MD simulation of two-body and three-body contacting sliding and analyzed the removal mechanism of silicon material. They proposed that materials followed the regimes of no-wear, adhering, ploughing and cutting for two-body wear and removal. However, the three-body contact sliding divides into four stages which are no-wear, condensing, adhering and ploughing. In Noreyan's research, MD simulation was used to research the critical depth for the elastic-to-plastic transition of the cubic silicon carbide by a diamond tip[11]. It is observed that the critical pressure for elastic-to-plastic transition decreases with the increase of diamond tip width. Guo[12] investigated the removal mechanism underlying CMP process, and found that one monoatomic layer was removed when the indentation depth of diamond was 0.1nm. Furthermore, amorphous layer and dislocation was not generated during mono-layered atomic scratching process.

The removal thickness is less than 1 nm for a single cycle scratch for SiC substrates in a FAP. In order to investigate the deformation and removal mechanism of SiC substrates, the influence of sub-nano scratching depth on the structural phase transition is analyzed by MD simulation. Also, the abrading temperature and force caused by the interaction between SiC substrates and diamond abrasives are discussed.

2 SIMULATED MODEL

A MD simulation of silicon carbide substrates machined by the diamond abrasive is mentioned in a sub-nano scratching process. The mechanical removal mechanism was researched. In the model of MD simulation (Fig. 1), the substrates of moncrystalline silicon carbide presents the zinc blende structure with a lattice constant of 4.348 Å. The lattice constant of diamond abrasive is 3.75 Å. The silicon carbide substrates was comprised of 112632 atoms with a dimension of 82.612*169.572*82.612 Å^3 along x-, y-, and z-directions, respectively. The scratching depth is set as 0.6nm, 0.8nm, 1.0nm and 1.2nm. Furthermore, ABOP potential is selected for Si-Si, C-C, and Si-C interaction[13]. The canonical ensemble (nvt) and micro-canonical ensemble (nve) are applied in the relaxation and abrading process, respectively. Detailed computational parameters applied in the MD simulation are shown in Table 1.

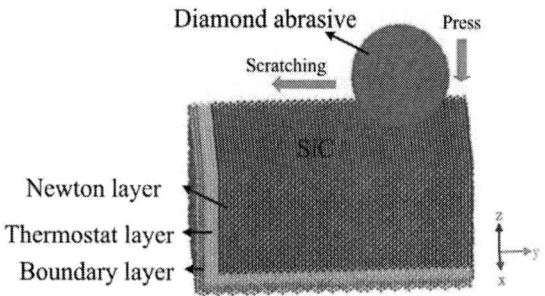

Fig. 1. Scratching process of MD simulation model

TABLE I. Computational parameters applied in the MD simulation

Configuration	scratching
Substrates	3C-SiC
Abrasive particle	Diamond abrasive
SiC dimensions	$82.612*169.572*82.612 \ \text{Å}^3$
Abrasive radius	3nm
Indentation depth	0.6nm, 0.8nm, 1.0nm,1.2nm
Scratching velocity potential function	100m/s ABOP
Timestep	0.001ps
Bulk temeprature	298K

3 DISCUSSION AND RESULTS

3.1 CRYSTAL DEFORMATION

The scratching process is proceeded as a two-body contact sliding, and is shown in Fig. 2. The red atoms are the amorphous atoms. We can see that the structural phase transition occurs of SiC substrates when the diamond abrasive is pressed into it. The removed chip piles up gradually during sub-nano scratching process. The reason producing chips is that the shear stress causes the plastic flow of materials under the interaction between diamond abrasives and SiC substrates.

It is easier to machine the SiC substrates in a ductile manner for the scratching depth on several nanometer even sub-nanometer scale. The amorphous atoms in the region of chips and structural phase transition are observed in Fig. 2 and Fig. 3. That is because the energy generating the phase transition is not sufficient to cause lattice dislocations. So, amorphous phase transition which is neither the elastic deformation nor the plastic deformation occur after two-body contact sliding on nanoscale[10].

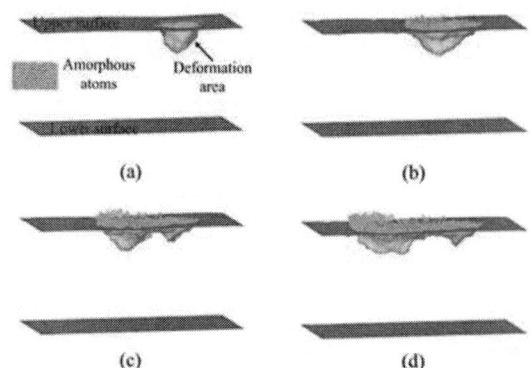

Fig. 2. The state of amorphous phase transition in different scratching distance (a) 0nm, (b) 2nm, (c) 4nm (d) 6nm

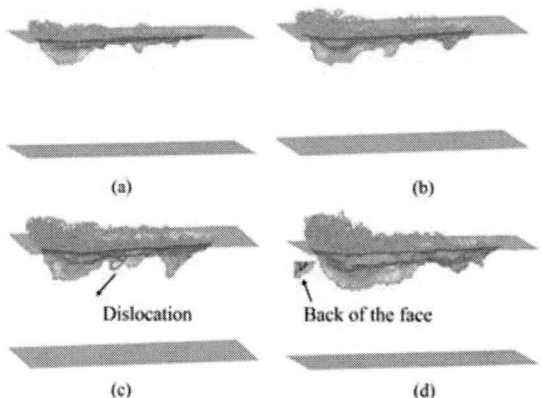

Fig. 3. The state of amorphous phase transition in different scratching depth (a) 0.6nm, (b) 0.8nm, (c) 1.0nm, (d) 1.2nm

The scratching depth is considered as a factor that affects the processing quality of fixed abrasive polishing by applying MD simulation methods. A thin amorphous layer on the surface of SiC substrates is shown in Fig. 3. It is found from Fig. 3 that the dislocation cannot be generated until the scratching depth reaches beyond 1nm. It verified that the plastic flow of amorphous atoms is the main mechanism for removal on sub-nanometer scale. The effect of scratching depth on the amorphous phase transition is depicted in Fig. 4, and shows that the amorphous atoms increase with the increase of scratching depth. The bonding force in the amorphous layer has been greatly weakened, which contributes to the surface reaction and removal in fixed abrasive polishing. It can be seen from Fig. 5 that the structure in the subsurface is broken as the substrates is scratched by the abrasive particle, and found that the depth of subsurface damage increases as the scratching depth increases. Based on the analysis above, the two-body contact sliding is important for promoting the generation of reaction layer due to the weakened effect of bonding force.

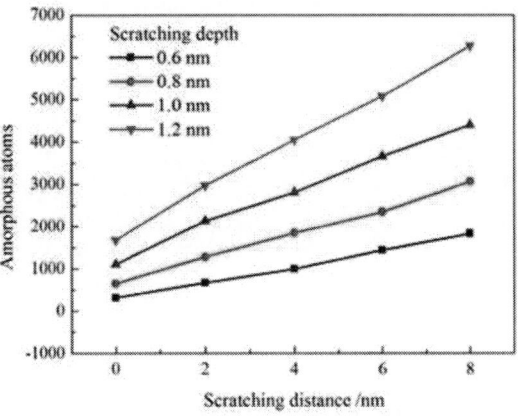

Fig. 4. The number of amorphous atoms with the scratching distance

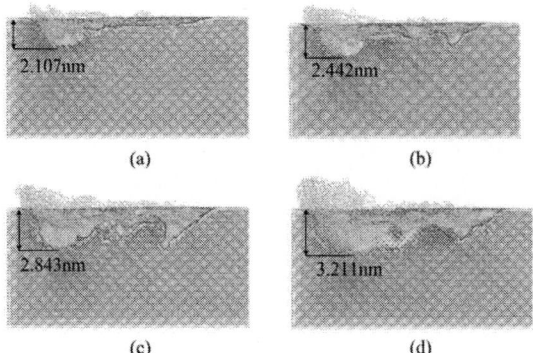

Fig. 5. The subsurface damage at different scratching depth (a) 0.6nm, (b) 0.8nm, (c) 1.0nm, (d) 1.2nm

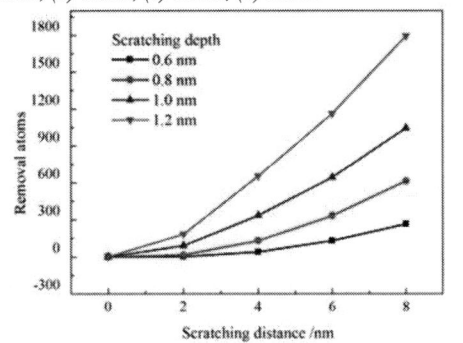

Fig. 6. The removed atoms with the scratching distance

3.2 SCRATCHING FORCES

The friction force and normal force propagate along the Y direction and Z direction, respectively, in our simulation. The effect of scratching depth (0.6nm, 0.8nm, 1.0nm, 1.2nm) on the friction property during the sub-nano abrading process is calculated by MD simulation. It is found from Fig. 7 that the frictional force and normal force both increase with the increase of scratching depth. It is due to a larger scratching depth

resulting in larger contact of substrates and diamond abrasives, which increases the resistance for sliding.

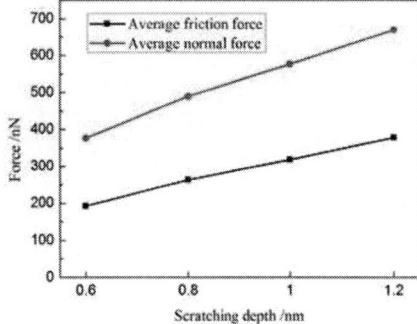

Fig 7. The average scratching force with the scratching depth

3.3 ABRADING TEMPERATURE

The formation rate of reaction layer is affected by the abrading temperature in a fixed abrasive polishing. It is necessary to study the influence of scratching depth on the abrading temperature. As can be seen from Fig. 8, the average abrading temperature caused by the interaction between diamond abrasives and SiC substrates increase with the scratching depth. More removed atoms are observed with a deeper scratching depth, and then, lead to more heat generated.

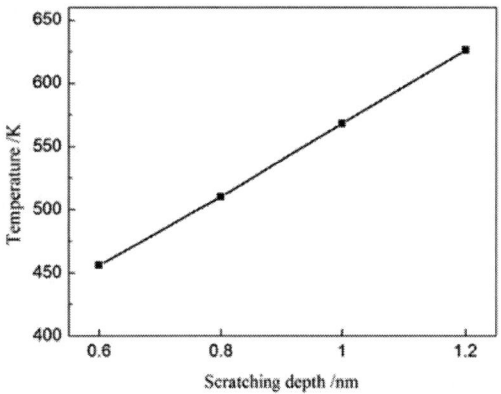

Fig. 8. The average temperature vs the scratching depth

4 CONCLUSIONS

MD simulation is used to research the mechanical removal mechanism of the SiC substrates in the FAP on sub-nanometer scale. We found that the amorphous phase transition is actively taking place during the sub-nano abrading process. However, no dislocation is observed in the substrates until the scratching depth is larger than 1nm. It is summarized that only the amorphous phase transition occurs in SiC, and amorphous atoms are removed in a form of plastic flow during sub-nano scratching process. Furthermore, it is observed that the averaged abrading temperature and

978-1-7281-6559-2/20 $31.00 © 2020 IEEE

scratching force rise with the increase of scratching depth.

ACKNOWLEDGEMENTS

This work was supported by National Natural Science Foundation of China (Grant Nos. 51675276) and Jiangsu Province Key Laboratory of Precision and Micro-manufacturing Technology.

REFERENCES

[1] Goel and Saurav. Journal of Physics D Applied Physics, vol. 47, 2014, pp. 243001.

[2] J. Li, Y. W. Zhu, D. W. Zuo, K. Lin and M. Li. Key Engineering Materials, vol. 426-427, 2010, pp. 589-592.

[3] Y. B. Tian, Z. W. Zhong and S. T. Lai. Int J. Adv Manuf Tech, vol. 68, 2013, pp. 993-1000.

[4] Q. F. Luo, J. Lu and X. P. Xu. Wear, vol. 350-351, 2016, pp. 99-106.

[5] B. Zhu, D. Zhao, Y. Tian, S. Wang, H. Zhao and J. Zhang. Mat Sci Semicon Proc, vol. 90, 2019, pp. 143-150.

[6] S. Sun, X. Peng, H. Xiang, C. Huang, B. Yang, F. Gao and T. Fu. Ceram Int, vol. 43, 2017, pp. 16313-16318.

[7] S. Z. Chavoshi and X. Luo. Materials Science and Engineering: A, vol. 654, 2016, pp. 400-417.

[8] B. Meng, D. Yuan and S. Xu. Int J. Mech Sci, vol. 151, 2019, pp. 724-732.

[9] Y. Liu, B. Li and L. Kong. Ceram Int, vol. 44, 2018, pp. 11910-11913.

[10] L. Zhang and H. Tanaka. Tribol Int, vol. 31, 1998, pp. 425-433.

[11] A. Noreyan, J. G. Amar and I. Marinescu. Materials Science and Engineering: B, vol. 117, 2005, pp. 235-240.

[12] L. Si, D. Guo, J. Luo and X. Lu. J. Appl Phys, vol. 107, 2010, pp. 64310.

[13] P. Erhart and K. Albe. Physical Review B (Condensed Matter and Materials Physics), vol. 71, 2005, pp. 35211.

ROLE OF SLURRY CHEMISTRY FOR DEFECTS REDUCTION DURING BARRIER CMP

Chenwei Wang[1,2]*; Yue Li[1,2]; Guoqiang Song[1,2]; Zhaoqing Huo[1,2]; Jia Liu[1,2]; Yuling Liu[1,2];

[1]School of Electronic Information Engineering, Hebei University of Technology, Tianjin 300130, China

[2]Tianjin Key Laboratory of Electronic Materials and Devices, Tianjin 300130, China Tianjin 300130, China

*E-mail: wangchenwei@hebut.edu.cn

ABSTRACT

In state of the art technologies, defect reduction is central to the achievement of low cost, high yield manufacturing. The defects occurred during the CMP process would lead to severe circuit failure and affect yield. In this paper, effect of slurry chemistry on surface defect during barrier CMP was studied. The experimental results showed that the complexation can effectively remove the copper residue, but would induce large dishing and erosion, if the complexation is so strong. The strong electrostatic attraction on oxide surface can improve the removal rate selectivity of OX to Cu and reduce the dishing and erosion. The dispersion effect and wetting effect can prevent the agglomeration of abrasive particles and make the copper surface hydrophilic, it can effectively reduce scratch defect during CMP.

Keywords—chemical mechanical planarization; surface defect; slurry chemistry; dishing and erosion; scratch;

INTRODUCTION

Copper chemical mechanical planarization (CMP) technology has been widely used as the essential process for giga scale integrated circuits (GLSI) applications[1-6], but there are many problems to be solved to obtain a defect-free global planarization e.g., copper residues, dishing, erosion, scratches and the various contaminations generated during copper CMP process. The copper residues remained on the wafer surface would cause short circuit. The large dishing and erosion can aggravate copper loss and lead to higher resistance , it also result in copper residue in the next layer and lead to the copper thickness in trench become more thicker. Scratches, such as micro-scratches on the wafer surface may lead to device failure and yield reduction as well as potential reliability issues.

In this study, the slurry chemistry on defect reduction during copper barrier CMP was investigated. The influence of complexation, electrostatic interaction, dispersion effect and wetting effect on defect reduction was studied, in terms of dishing and erosion, copper residues and scratches on the copper surface.

EXPERIMENTAL

Experiments were carried out on Applied Materials Reflexion LK300 mm tool (Rohm and Haas, Politex Reg Pad). 12-inch pattern wafers and 300 mm diameter copper and TEOS blanket wafers were used in the experiments. A GTSR01(KOKUSAI ELECTRIC ALPHA, Inc.) resistivity measurement was used to measure the copper film thickness. TEOS film thickness was measured by Model F-REX300X (EBARA, Inc.). 81 test points were taken in diameter of the film to calculate MRR by measuring the thickness before and after polishing.Top viewed scanning electron microscope (SEM, KLA-Tencor SEM Vision) was used to evaluate defect performance on 300mm pattern wafer. The measurement of step height was determined by Atomic Force Profiler (AFP) produced by Veeco company in America. Setting the middle position on the wafer as the test point, each group was measured three times and took the average. The slurry A to slurry F used in this experiment has the same components except complexing agent A and B, which was shown in Table 1.

Table 1 Slurry components

Item	Complexing agent A (wt%)	Complexing agent B (wt%)	H_2O_2 concentration (wt%)
Slurry A	0.1	N/A	0.1
Slurry B	0.15	N/A	0.1
Slurry C	0.3	N/A	0.1
Slurry D	0.15	1	0.1
Slurry E	0.15	3	0.1
Slurry F	0.15	4	0.1

RESULTS AND DISCUSSION

Fig.1 shows the removal rate (RR) of copper and TEOS as a function of different slurry conditions. As can be seen from Fig.1, as the concentration of complexing agent A increases (slurry A to C) from 0.1 wt% to 0.3wt%, the Cu RR was significantly increase from 152A/min to 985 A/min, while the TEOS RR was decreased from 568A/min to 426A/min. The removal rate selectivity of TEOS/Cu was reduced from 3.7 to 0.4, thus it is undesirable to add higher concentration of complexing A in the slurry to avoid large dishing and erosion occured. Because the complexing A has strong chelating chemisorption ability to Cu ions, and has little inhibition effect on TEOS layer, it can react with the copper ions which are generated during CMP, the formed complex products were dissolved in aqueous solution and then removed by mechanical abrasion of abrasives and pad,thus it can remarkably enhance the Cu RR during polishing compare with TEOS.

It can be seen in Fig.2 that the the dishing and erosion were increased as the concentration of complexing A increase, it was due to the decrease of removal rate selectivity of TEOS to Cu as the content of complexing A increase in slurry.

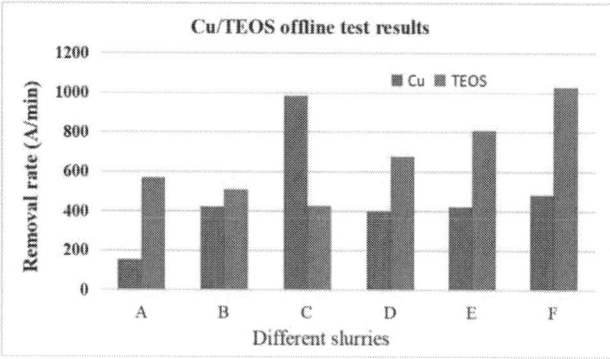

Fig.1 The removal rate of Cu and TEOS as a function of different slurries

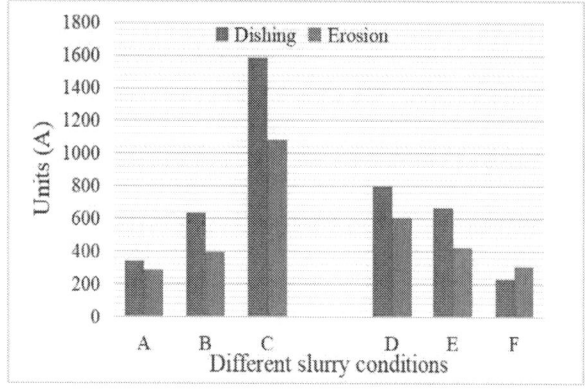

Fig.2 The dishing and erosion as a function of different slurries

TEOS is mainly composed of acidic oxides silica. During CMP, the TEOS surface may react with hydroxyl ions in the alkaline solution and form soluble silicate which will flow away with slurry. According to the literature, the zeta potentials of the surfaces played important roles in controlling the interaction forces of colloidal silica particles on polished TEOS surface and finally has a profound influence on removal rate. Normally, the surface of TEOS in water is expected to be terminated with -SiOH. At some of these sites, OH$^-$ ion in water can removal H$^+$, resulting in -SiO$^-$ termination, that reveal that the TEOS surface may end negatively charged in water suspension for basic pH, such as pH 10. The test reported that the surface charge of silica particles was more positive as the concentration of complexing B increases in the experimental range, that will reduce the electrostatic repulsion between the negative polished surface and silica particles, the enhanced attraction force between abrasives and polished substrate will be helpful for improving the removal of TEOS. However, it has small influence on Cu RR, thus the removal rate selectivity of TEOS to Cu was increased as the concentration of complexing B increase. So we can find that the dishing and erosion was decreased as the concentration of complexing B increase, as shown in fig.2.

Fig.3 shows the defect map of wafer as a function of different concentration of complexing A and complexing B. The defect counts significantly increase along with complexing A concentration increasing in a range of 0.1wt% to 0.3wt%. The SEM image shows that the defect on the wafer surface is mainly consists of copper residue when the wafer polished by using slurry A, as shown in Fig.4 (a). The result reveals that the lower concentration of complexing A is insufficient to chelating with copper ions, and the Cu RR was lower, it will result in copper residue. However, higher concentration of complexing A can cause large defect counts, it may be due to the lower removal rate selectivity of TEOS to Cu, which leading to larger dishing and erosion or copper recess during CMP, as can be seen in fig.4(c).

It can be seen from fig.3 that with the increasing of the complexing B in slurry, the total defect number was decreasing and then increasing. When the ionic strength is 3wt%, the total defect is minimum 3. From the above analysis, with the addition of complexing B, the selectivity of TEOS and copper RR is suitable at 3wt%. But if ionic strength in the slurry is too high, it may undesirably effect slurry stability. Large agglomerated particles formed by the addition of ionic salts are not easily broken under the applied down pressure and yields an enhancement in the indentation depth into the wafer surface, which causes the defects and scratches formed on the wafer surfaces during polishing.

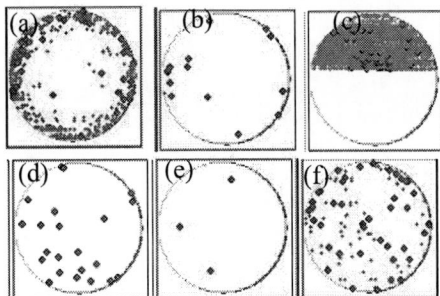

Fig.3 The defect map of copper patterned wafer after polishing by using slurry A-F

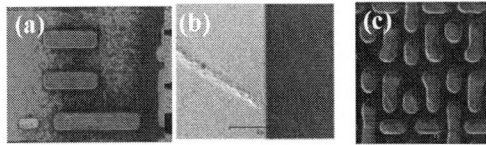

Fig.4 Typical SEM review image from defect map: (a) copper residue; (b) scratch; (c) copper recess.

CONCLUSIONS

The effect of slurry chemistry of complexing A and complexing B on copper and TEOS CMP performance on blanket wafers and pattern wafers were investigated. The results showed that complexing A has strong chelating effect with copper and little supress TEOS removal rate, it can effectively improve Cu RR. However, the dishing and erosion were increased as the concentration of complexing A increase.

the removal rate selectivity of TEOS to Cu was increased as the concentration of complexing B increase. Lower defect can be obtained by adjusting content of complexing A and B.

ACKNOWLEDGEMENTS

Financial supports for this research work is provided by Natural Science Foundation of Hebei Province (E2019202367), Major National Science and Technology Special Projects (No. 2016ZX02301003-004-007), and the Key Laboratory of Electronic Materials and Devices of Tianjin, China. Natural Science Foundation of Tianjin, China (14JCYBJC18500)

REFERENCES

[1] Chen Su, Zhang Kailiang, Song Zhitang. Advances inchemical mechanical polishing of copper in mutilevel interconnect. Semicond Technol, 2005, 30(8): 21

[2] Liu Yuling, Li Weiwei, Zhou Jianwei, et al. Microelectronics chemical technology foundation.

[3] J. Cheng et al. Material removal K. Shan, P. Zhou, J. Cai, R. Kang, K. Shi, and D. Guo, Electrogenerated chemical polishing of copper, Precision Engineering, 39, 161 (2015).

[4] Mechanism of copper chemical mechanical polishing in a periodate-based slurry,Appl. Surf. Sci., 337, 130 (2015).

[5] J. Bao, H. Shi. J. Liu, H. Huang, P. S. Ho, M. D. Goodner, M. Moinpour, andG. M. Kloster. J. Vac. Sci. Technol. B. 26. 219 (2008).

[6] O. T. Le. J-F. de Marneffe. T. Conard, I. Vaesen, H. Struyf, and G. Vereecke. J.Electrochem. Soc., 159, H208 (2012).

Role of slurry Additions on Chemical Mechanical Polishing of Cu/Ru/TEOS in H₂O₂-based Slurry

Chao Wang[1,2]; Jianwei Zhou[1,2*]; Chenwei Wang[1,2]; Xue Zhang[1,2]

[1]School of Electronic Information Engineering, Hebei University of Technology, Tianjin 300130, China
[2]Tianjin Key Laboratory of Electronic Materials and Devices, Tianjin 300130, China Tianjin 300130, China
*Email: jwzhou@hebut.edu.cn

Abstract

In this paper, the influence of a novel complex agent $(NH_4)_2SO_3$ and inhibitor 2,2'-[[(methyl-1H-benzotriazol-1-yl) methyl]imino]diethanol (TT) based on hydrogen peroxide （H_2O_2）on Cu/Ru/TEOS was studied. Since our research group has studied the relationship between Cu and Ru before, we now make further supplement. The results showed that both Ru and TEOS increased with the increase of ammonium ion(NH_4^+) concentration, and the mechanism was studied by means of electrochemistry, possibly because of the electrostatic attraction between the NH_4^+ ions and Ru, and the addition of ammonium ions may reduce the thickness of the double electric layer on the surface of TEOS. So the Ru and TEOS removal rate goes up.

Keywords—chemical mechanical polishing; ruthenium; ammonium ion；TT; removal rate

Introduction

Due to the rapid development of integrated circuits, the gradually shrinking of feature size leads to the traditional barrier of Ta and TaN bilayer for copper (Cu) interconnects encounter some restrictive problems, Such as a sharp increase in interconnects resistance，RC delay and copper seed layer non-uniformity in the back end of the line (BEOL) interconnects. In response to these problems, ruthenium (Ru) has demonstrated his outstanding excellent performance. It is not easily corroded by chemicals, and has a lower resistivity (Ru=7.6μΩ•cm, Ta=14μΩ•cm, TaN=200μΩ•cm). [1~3] Most importantly, Cu can be electroplated directly on Ru without the need for a copper seed layer. Ru is the most promising liner layer materials for the sub-10 nm technology node of BEOL.[4]

As is known to all, chemical mechanical polishing is the best way to achieve global planarization and local planarization, and can also achieve the removal of a variety of materials However, for the high hardness of Ru, it is difficult to have a high removal rate of ruthenium, which will produce more obvious dishing and erosion on the surface of the wafer. To avoid this effect, it is necessary to reduce the rate of copper as much as possible. In ruthenium barrier polishing, potassium periodate and hydrogen peroxide are usually used as oxidants. However, when potassium periodate as an oxidant, more pitting corrosion will be generated and RuO_2 porous structure will be formed. So we chose H_2O_2 as the oxidant and pH8 to avoid the production of toxic RuO_4. Q. W. Wang[5] and Jiang L [6] found that Gnd+ and K+ could be used to achieve a higher removal rate of Ru. Nevertheless, in industrial applications, high metal ion content (such as K+) and expensive price (such

as Gnd+) in slurry additive are not allowed, and hence NH_4^+ which is a low-cost non-metal ion was studied to increase the removal rate of Ru in this paper.

$(NH_4)_2SO_3$-TT was used as a novel slurry addition to improve Cu/Ru/TEOS removal rate. The effect of different concentration of ammonium on Cu/Ru/TEOS removal rate was studied. the effect of mechanism of NH4+ on removal rate of Cu/Ru/TOES were analyzed by electrochemistry and zeta potential.

Experimental

Materials—200 nm Ru film was deposited on TaN film, again by plasma-enhanced atomic layer deposition. Subsequently, 76.2 mm diameter sample was cut from the wafer with Ru/TaN layer and used in the polishing experiments.

Polishing experiments—All polishing experiments were performed on Alpsitecin-E460E polisher made in France and the polishing parameters were set as follows: the polishing pressure was 1.5 psi; the carrier and platen rotational speeds was 87/93 rpm; the fade rate of slurry was 300 mL/min and all the experiments were conducted at laboratory temperature (25±1℃).The main component of slurry was colloidal silica (mean particle size was about 85 nm. the pH nearly 9.6), hydrogen peroxide (H_2O_2, the mass fraction of 30 wt%, semiconductor grade), ammonium salt. The removal rates of Ru were determined by measuring the difference of film thickness before and after polishing by an aver333A four-point probe from Four Dimensions of company and by taking the averaged result.

Electrochemical experiment — Potentiodynamic polarization curves tests were carried out to describe the electrochemical property of the Cu and Ru samples immersed into different solutions (without silica particles) by using a Shanghai Chenhua-CHI660E electrochemical workstation with a three-electrode glass cell (100 ml volume, with a Pt counter electrode and saturated calomel electrode). Ru and Cu film electrodes (1 2 cm2) were used as a working electrode. Prior to measurement, the working electrode was wiped with 1wt% citric acid solution, followed by rinsing with deionized water for 2 minutes, then dried with high purity nitrogen.

Results and Discussion

Effect of $(NH_4)_2SO_4$ concentration and H_2O_2 concentration on removal rates of Cu/Ru/TEOS.- My group has previously studied the effects of TT on copper and Ru. Because TT is a derivative of BTA, the structure diagram is shown in Fig.1. Comparing the properties of BTA, we know that TT and Cu can form a Cu-TT complex, which can protect the surface of Cu. We can also get this from Fig.2 In the case of 1000 ppm TT, the removal rate of Cu hardly changes with the increase of

978-1-7281-6559-2/20 $31.00 © 2020 IEEE

hydrogen peroxide concentration. This is due to the protective effect of Cu-TT. At the same time, because TT has almost no effect on ruthenium, the surface of ruthenium will be oxidized first after adding hydrogen peroxide, and then the complexation of ammonium ions will cause ruthenium to be largely removed. However, because the concentration of ammonium ions is constant, the complexing ability to ruthenium is also limited, so the removal rate of Ru does not increase to a certain height. The effect of hydrogen peroxide on TEOS is not obvious.

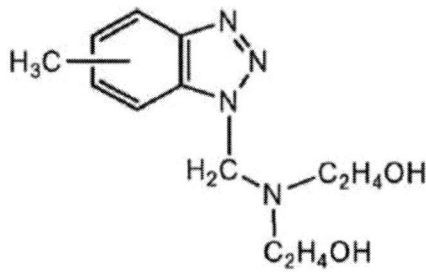

Fig.1. *The molecular structure of TT*

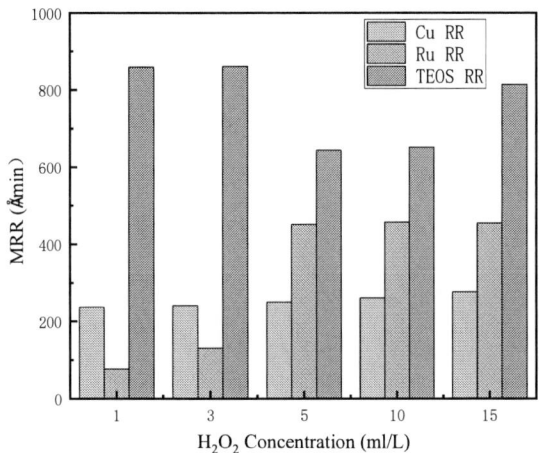

Fig.2. *Effect of different concentration of $(NH_4)_2SO_4$ on the RRs of Ru/Ru/TEOS in presence 5 wt% of silica abrasives , 1000ppm TT and 0.15 wt% H_2O_2 at pH 8.*

From Fig.3, we can see that with the increase of the ammonium ion concentration, the rate of Cu / Ru / TEOS all shows an upward trend. This is because that the ammonium ion in the solution increases with the increase of the ammonium ion concentration. It will reduce the distance between the SiO_2 particles and the surface of the material, that is, enhance the electrostatic adsorption force. This is well explained on the surface of ruthenium and TEOS, because RuO_2 on the Ru surface shows a negative charge, and the SiO_2 particles also have a negative charge. The positive charge of the root ions has a good attracting effect and promotes the improvement of the ruthenium removal rate, As shown in Figure 4. However, the removal rate does not increase with the increase of ammonium ion concentration after 60mM. This shows that the ammonium ion's ability to promote removal has reached saturation. This theory can also be applied to TEOS.

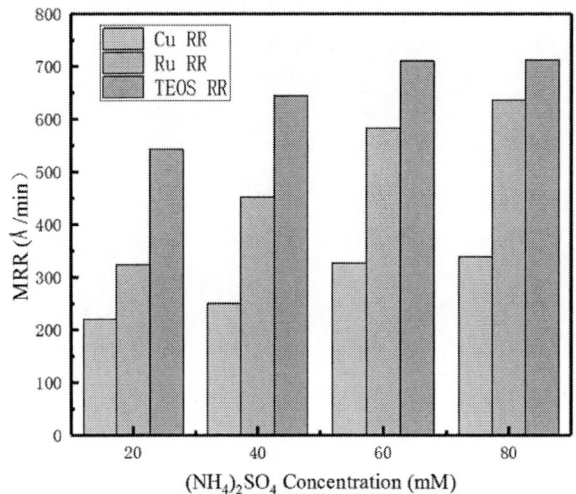

Fig.3. *Effect of different concentration of H_2O_2 on the RRs of Ru/Ru/TEOS in presence 5 wt% of silica abrasives , 1000ppm TT and 40mM $(NH_4)_2SO_4$ at pH 8.*

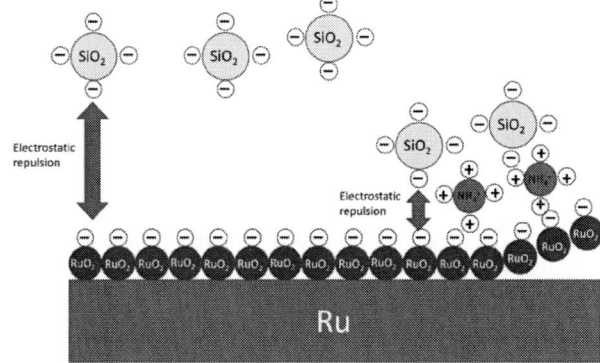

Fig.4. *Schematic diagram of electrostatic interaction between Ru and silica particles in CMP.*

Effect of TT on electrochemical behavior of Cu/Ru/ TEOS. —From Figures 5 and 6, we can clearly see that as the ammonium ion concentration increases, the corrosion current of copper gradually increases, which is in good agreement with polishing experiments. It shows that ammonium ions can form complexes with copper ions, promote the dissolution of copper and increase the polishing rate of copper. This theory can also be implemented on the curve of ruthenium. Under the conditions of different concentrations of hydrogen peroxide, we found that the corrosion current for ruthenium is the largest at a hydrogen peroxide concentration of 5ml / L, that is, a good dynamic balance is reached. For copper, the change is not very obvious, which may be due to the formation of a thick passivation layer between TT and copper, which makes it difficult for the copper to be oxidized again, and ultimately causes little change in the rate. The main reactions should be as follows.

$$Cu + H_2O_2 \rightarrow CuO + H_2O \qquad (1)$$

$$CuO + 4NH_4^+ + OH^- \rightarrow [Cu(NH_3)_4]^{2+} + 2OH^- \qquad (2)$$

$$Cu^{2+} + TT \rightarrow Cu\text{-}TT \qquad (3)$$

$$Ru + H_2O_2 \rightarrow RuO_2 + H_2O \qquad (4)$$

$$RuO_2 + 4OH^- \rightarrow RuO_4^{x-} + 2H_2O + (4\text{-}x)e^- \ (x=1 or 2) \qquad (5)$$

$$NH^{4+} + RuO_4^{x-} \rightarrow (NH4)_xRuO_4 \ (x = 1 \ or \ 2) \qquad (6)$$

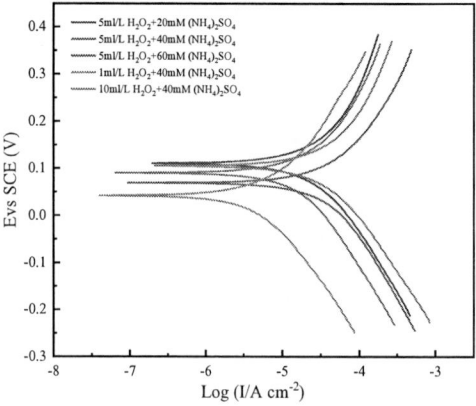

Fig.5. *potentiodynamic polarization plots of Ru in the different solution with 1000ppm TT at pH 8.0.*

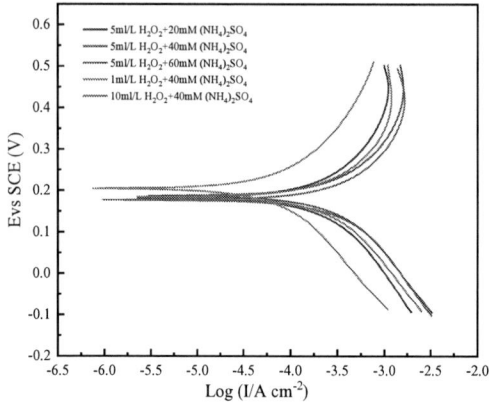

Fig.6. *potentiodynamic polarization plots of Cu in the different solution with 1000ppm TT at pH 8.0.*

Tab.1. Tafel parameters of Ru and Cu in different components solution at pH 8.

Solution system (pH=8.0)	Cu		Ru	
	E_{corr} (V)	I_{corr} (uA/cm²)	E_{corr} (V)	I_{corr} (uA/cm²)
5ml/L H₂O₂+ 20mM(NH₄)₂SO₄	0.110	3.33	0.183	41.75
5ml/L H₂O₂+ 40mM(NH₄)₂SO₄	0.090	4.48	0.177	69.25
5ml/L H₂O₂+ 60mM(NH₄)₂SO₄	0.069	6.88	0.187	59.31
1ml/L H₂O₂+ 40mM(NH₄)₂SO₄	0.042	1.00	0.204	14.39
10ml/L H₂O₂+ 40mM(NH₄)₂SO₄	0.105	6.29	0.184	50.52

Conclusions

The results show that Cu, Ru, and TEOS all increase with the increase of the ammonium ion (NH_4^+) concentration, and the mechanism is studied by electrochemical methods, which may be due to the electrostatic attraction between NH_4^+ ions and Ru, and the RuO_4^{x-} Reacts with ammonium ions to increase ruthenium removal. At the same time, the thickness of the electric double layer on the TEOS surface was reduced due to the addition of ammonium ions. Therefore, the removal rates of Ru and TEOS are improved.

Acknowledgements

Financial supports for this research work is provided by Major National Science and Technology Special Projects (No.2016ZX02301003-004-007), the Natural Science Foundation of Hebei Province, China (No. F2015202267), the Scientific Innovation Grant for Excellent Young Scientists of Hebei University of Technology under Grant No 2015007, and the Key Laboratory of Electronic Materials and Devices of Tianjin, China. Natural Science Foundation of Tianjin, China (14JCYBJC18500).

Reference

[1] S. F. Ding, S. R. Deng, H. S. Lu, Y. L. Jiang, G. P. Ru, W. Zhang, and X. P. Qu, J.Appl. Phys., 107, 103534 (2010).

[2] M C. Turk, S. E. Rock, H. P. Amanapu, L. G. Teugels, and D. Roy, ECS J. Solid StateSci. Technol., 2, 205 (2013)..

[3] H. Cui, J. H. Park, and J. G. Park, J. Electrochem. Soc., 159, 335 (2012).

[4] Cheng J , Wang B , Wang T , et al. Chemical Mechanical Polishing of Inlaid Copper Structures with Ru/Ta/TaN as Barrier/Liner Layer[J]. ECS Journal of Solid State Science and Technology, 2018, 7(11):P634-P639..

[5] Q. W.Wang, J. W. Zhou, C. W.Wang, X. H. Niu, Q. Y. Tian, and Y Xiao, Journal ofSolid State Science & Technology, 7(10),567 (2018).

[6] Jiang L , He Y , Li Y , et al. Effect of ionic strength on ruthenium CMP in H2O2-based slurries[J]. Applied Surface Science, 2014, 317:332-337..

[7] Du, Y., Wang, C., Zhou, J., Zhang, W., Ji, J., and Han, L., et al. "Effect of Guanidinium Ions on Ruthenium CMP in H2O2-Based Slurry ," Journal of Solid State Science and Technology, vol.6,pp.521–525,2017.

EFFECT OF VARIOUS SURFACTANTS ON SURFACE ROUGHNESS REDUCTION DURING COBALT "BUFF STEP" CMP

Yuanshen Cheng[1,2], Shengli Wang[1,2] *, Chenwei Wang[1,2] *, Yundian Yang [1,2]

[1]School of Electronic Information Engineering, Hebei University of Technology, Tianjin 300130, China
[2]Tianjin Key Laboratory of Electronic Materials and Devices, Tianjin 300130, China Tianjin 300130, China
*Corresponding Author's Email: shlwang@hebut.edu.cn

ABSTRACT

In this paper, effect of various surfactants on surface roughness reduction during cobalt "Buff step" chemical mechanical planarization were investigated. Various surfactants were characterized by atomic force microscope (AFM), surface tension, contact angle, viscosity, zeta potential and material removal rate. The result showed that all kinds of surfactants inhibit the removal rate of Cobalt, Solution Containing anionic surfactant conduce to obtain small surface roughness compared with others. The mechanism of reducing cobalt surface roughness by surfactant is proposed.

Keywords—Cobalt; chemical mechanical polishing; surfactant; surface roughness;

INTRODUCTION

As the feature size of integrated circuit devices continues to be scaled to 10nm down, Cobalt (Co) is the most promising metal to replace Cu because of its low resistivity at low nodes, good step coverage and thermal stability [1-3].

As we all know, if the surface roughness of cobalt is uneven, during the device conduction process, which will cause noise and leakage current increase, poor consistency electrical characteristics, and influence device frequency characteristics, such as the Resistive Capacitive (RC) delay time [4-5]. Reducing the surface roughness of Co coupon is significant to improve the circuit performance and reliability. Surfactants are widely used to reduce the surface roughness of copper in barrier polishing, but it has not been reported to reduce the surface roughness of cobalt.

In this paper, the effect of several surfactants on polishing of Cobalt buff layer were studied. The reference solution is 4wt% SiO2, 15g/L potassium tartrate, and 3ml/L 30% H2O2. Define 1# reference solution, 2# chemical A, 3# chemical B, 4# chemical C. The surfactant used in this paper are shown in the following table.

Table I. Description of chemical reagents

Chemical A	Amine oxide	$RN^+(CH_2CH_2OH)_2{\rightarrow}O)$
Chemical B	Fatty alcohol polyoxyethylene ether	$C_{12}H_{25}O.(C_2H_4O)_n$
Chemical C	Ammonium lauryl sulfate	$C_{24}H_{54}NO_4S$

EXPERIMENTAL

Polishing experiments.—All the polishing experiments were conducted on a France-E460E polisher from Alpsitec Inc and the polishing pad was POLITEX TM REG (purchased from Dow Electronic Materials, USA). The polishing experimental parameters are: 1.5 psi polishing pressure and 200 mL/min slurry flow rate，head/plate speeds is 87/93 rpm, polishing time is 1 min. Before each polishing experiment, using diamond conditioning disk conditioning pad for 15 min. When polishing were finished, the wafers were rinsed with deionized water and dried by N2 gas. The materials removal rates were determined by measuring the difference in film thickness before and after polishing using an aver333A four-point probe from Four Dimensions of company. To ensure the accuracy of the experiment, each polishing experiment was conducted three times and averaged.

Surface morphology measurement.—An Agilent 5600LS taping mode of atomic force microscopy (AFM) was used for surface roughness examination of the cobalt wafer after CMP, the detection range was 10 μm × 10 μm.

JC2000D2 POWEREACH was used to measure the contact angle and surface tension of the polishing solution on the cobalt surface, and NiComp380 DLS was used to measure the zeta potential and particle size of the polishing solution. Each group of polishing fluid was measured 3 times.

RESULTS AND DISSCUSSION

2.1 Effect of different Surfactants on Co removal rate

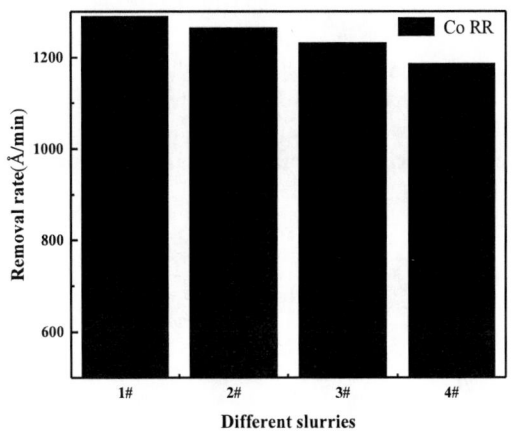

Fig.1：Effect of Different surfactants on Co RR

It can be seen from the Fig.1, with the addition of surfactants, the removal rate of Co decreased. The reason may be that the surfactant has both ends of nonpolar lipophilic group and polar hydrophilic group, in the polishing process, the nonpolar lipophilic group first contacts the surface of the material to be polished, and the polyoxyethylene chain at the hydrophilic group extends in the water, preventing the chemical reaction. That is to say, the addition of surfactants will

978-1-7281-6559-2/20 $31.00 © 2020 IEEE

weaken the chemical action and reduce the removal rate. In addition, add to surfactants increases the fluidity and lubricity of slurry on the material surface, reduces the friction coefficient between slurry and material surface, weakens the mechanical effect, and thus reduces the removal rate of material [6-7].

2.2 Effect of different surfactants on wettability of slurry

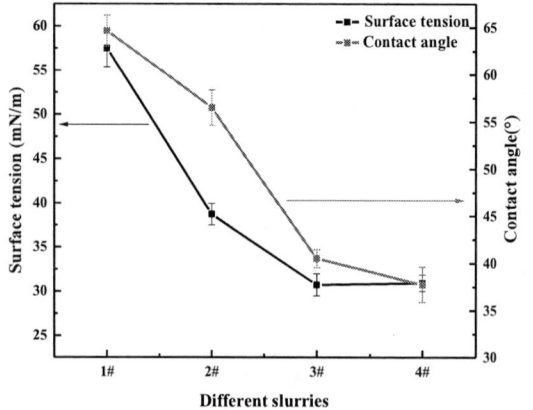

Fig.2: Surface tension and Contact angle under Different solutions

It can be seen from Figure 2 that the surface tension and contact angle of the slurry decrease with the addition of different kinds of surfactants, and the surface tension and contact angle reach the minimum value when the anionic surfactants are added.

The decrease in contact angle and surface tension indicates that the wettability of the slurry is increased and the spreading ability is enhanced, that is the hydrophilicity is enhanced. The reason is that the addition of surfactants will form micelles with outward hydrophilic groups and inward hydrophobic groups [8]. During the CMP process, these micellar will aggregate on the surface of the cobalt film and contact the cobalt surface more fully, which is more beneficial to take away the reaction products on the cobalt surface, accelerate mass transfer, and reduce the friction coefficient, which is beneficial for cleaning after polishing.

2.3 Effect of different surfactants on surface roughness

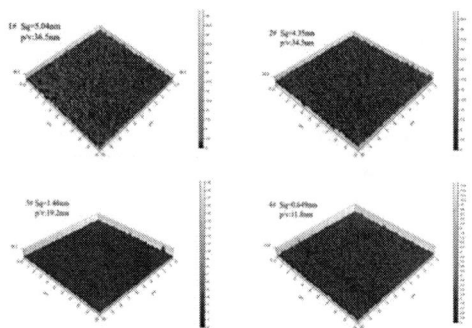

Fig.3: AFM images of Cobalt 1# reference solution, 2# reference solution +A, 3# reference solution +B, 4# reference solution +C

As shown in Figure 3, the addition of surfactants can reduce surface roughness and improve surface quality effectively. Among them, the anionic surfactant has the best effect, and the surface roughness is reduced to 0.649 nm. There

are two reasons to explain surfactants reduce surface roughness. First of all, the surfactants forms a dense passivation layer on the concave surface of the material to suppress the chemical reaction, while the mechanical pressure on the convex place is large, the chemical reaction activation energy is reduced, therefore the reaction is accelerated, and the polishing rate is accelerated. Secondly, the addition of the surfactants will accelerate the mass transfer during the polishing process, such as transfer the temperature of the concave to the convex. as the temperature of the convex increases, the thermal motion between the molecules accelerates, the number of active molecules in a unit volume increases, and the probability of effective collisions resulting in an increase in bonding, the chemical reaction rate increases, thereby accelerating the rate of cobalt in the convex. Conversely, the rate of cobalt in the concave is reduced due to the decrease in temperature [6].

Thus, the removal rate difference between concave protection and convex acceleration is used to eliminate the height difference of material surface and realize the global planarization.

2.4 Effect of different surfactants on the stability of slurry

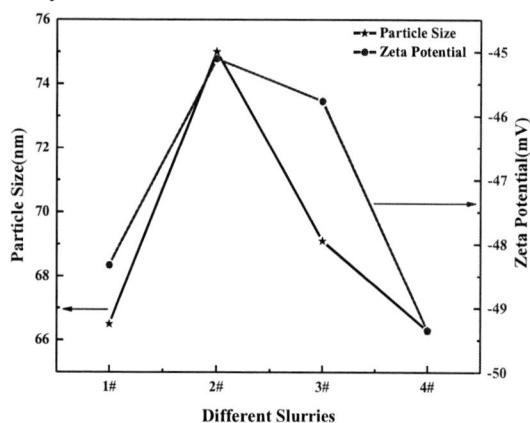

Fig.4: Effect of Different slurries on particle size and zeta potential.

Figure 4 show the effect of different slurries on particle size and zeta potential. We can see from Fig.4 that after the addition of cationic surfactant, the slurry particle size becomes larger, and the absolute value of zeta potential becomes smaller, that is to say, the adsorption of abrasive and cationic surfactant produces agglomeration, resulting in the instability of slurry. The nonionic surfactant does not exist in the form of ionic in the slurry, so its particle size/zeta potential has change hardly and the solution is relatively stable. After the addition of anionic surfactant, the anions adsorbed on the surface of nano-SiO2, resulting in the mutual repulsion of particles and good dispersion of slurry.

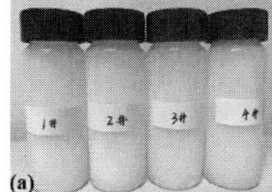
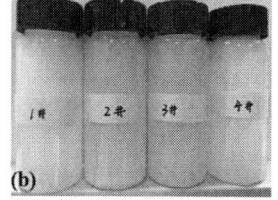

Fig.5: The slurries with different surfactant for fresh (a) and 7days (b)

It can be seen from the Fig.5, Each slurries is well dispersed when stand for 7 days except for 2#, which we can see has occurred condensation, causing precipitation at the bottom. The reason should be that the introduction of cations, as discussed above, reduces the electrostatic repulsion between SiO2 and makes the abrasive particles easier to agglomerate.

CONCLUSION

The effect of various surfactants on surface roughness reduction during cobalt "buff step" chemical mechanical planarization were investigated. The role and mechanism of different surfactants on material removal rate, surface tension, contact angle, particle size, zeta potential and after polishing surface quality were studied. The result showed that compared with other surfactants, anionic surfactants have better wettability and stability, which may be the key factor to reduce the surface roughness.

ACKNOWLEDGMENTS

Financial supports for this research work is provided by Natural Science Foundation of Hebei Province (E2019202367), Major National Science and Technology Special Projects (No. 2016ZX02301003-004-007), and the Key Laboratory of Electronic Materials and Devices of Tianjin, China. Natural Science Foundation of Tianjin, China (14JCYBJC18500).

REFERENCE

[1] X. Shi, D. E. Simpson, D. Roy, ECS Journal of Solid State Science and Technology, 4, P5058-5067 (2015).

[2] X. Shi, D. AJ. Gonyer, D. Roy - ECS Journal of Solid State Science and Technology, 5 P88-P99 (2016).

[3] X. Shi, S. E Rock, M. C. Turk, D. Roy Materials Chemistry and Physics, 136, P1027-P1367 (2012).

[4] J. Bao, H. Shi. J. Liu, H. Huang, P. S. Ho, M. D. Goodner, M. Moinpour, and G. M. Kloster. J. Vac. Sci. Technol. B. 26. 219 (2008).

[5] O. T. Le. J-F. de Marneffe. T. Conard, I. Vaesen, H. Struyf, and G. Vereecke. J.Electrochem. Soc., 159, H208 (2012).

[6] Zhang Wenqian, Liu Yuling, Wang Chenwei, Niu Xinhuan, Han Linan, Ji Jun, Du Yichen. The role of nonionic surfactants in tantalum based barrier CMP [J]. Microelectronics, 2018,48(03):421-424.\

[7] Wang yan, wang shengli, The effect of surfactant ADS on the planarization in the alkaline copper polishing slurry [J]. Semiconductor manufacturing technology, 2017) 11−0838−06.

[8] JANG S, JEONG H, YUH M, etal. Effect of surfactant on package substrate in chemical mechanical planarization [J]. Int J Precis Engineeri &Manufac-Green Technol, 2015,2(1):59-63.

MECHANISM ANALYSIS OF CHEMICAL MECHANICAL POLISHING OF 4H-SIC WAFER

Gaoyang Zhao[1,2], Aoxue Xu[1,3], Fan Xu[1,2], Daohuan Feng[1,2], WeiliLiu[1,2]*, Zhitang Song[1,2]

[1] Shanghai Institute of Microsystem and Information Technology, Shanghai, 200050, China

[2] University of Chinese Academy of Sciences, Beijing, 100049, China

* Corresponding Author's Email: rabbitlwl@mail.sim.ac.cn

ABSTRACT

As a third-generation semiconductor material, silicon carbide has excellent electron mobility and band gap, which makes it shine in power and optoelectronics applications. 4H-SiC has the highest band gap of all crystal type of SiC; chemical mechanical polishing technology is the only effective global planarization process today. In this paper, we studied the polishing rate of 4H-SiC silicon surface and carbon surface based on the chemical mechanical polishing processing. The structure of the resulting material layer was studied by TEM, and the difference between the chemical mechanical polishing of the carbon surface and the silicon surface was explained.

Key words: 4H-SiC; Surface State Analysis; Chemical Mechanical Polishing.

INTRODUCTION

With the gradual commercialization of 5G and the increasing popularity of high-speed railway in China, the application of silicon carbide power devices has attracted more and more attention. Of all crystals of silicon carbide, 4H-SiC have the highest band gap of its brothers and sisters [1]. Due to the manufacture difficulty of 4H-SiC, 4H-SiC power device leads a very expensive price on industry usage. [2] The roughness of crystal surface and interface will do harm to performance of device. [3] Moreover, since the invention of blue LED, silicon carbide has high thermal conductivity and high lattice compatibility with gallium nitride, which contributes to large-scale integration, making it an excellent LED substrate in high power applications [4]. As to fastly remove the damage layer which introduced in manufacture of silicon carbide device，the chemical mechanical polishing is taken to achieve a sub-nanometer flat surface. As the former researches shows the different plane has a quite different CMP material removal rate. The MRR of carbon plane is much higher than silicon plane. [5] A conclusion given on this: "The carbon face atoms arrangement is easier for oxidation than the silicon face atoms." But the researchers didn't give out a powerful reason for his conclusion, moreover the result is made based on 6H-SiC.

In this paper, experiment was carried out to certain the difference on the CMP processing between the silicon and carbon plane. The result will be explained by differences in hardness and thickness of the material production layer which produced in different slurry.

EXPERIMENT

Materials

We select the 4H-SiC wafer with the crystal orientation [0001] which is produced by Beijing Tianke-Heda Blu-ray Semiconductor Co., Ltd. Alumina is chosen as abrasives. HNO_3 and KOH which is 10wt% is utilized as the pH adjusting agent. KMnO4 acts as oxidant.

Experiment setting and processing

All experiments were carried out on the CETR CP-4 polisher. The polishing pressure set as 6 psi. (1psi=6894Pa). Polishing schedule is 60 minutes. Before polishing, the pad should be repaired by a diamond disk for 10 minutes under a pressure of 4 psi. The flow of slurry is 125mL/min. And the slurry is recycled with a polishing head speed at 100RPM.

After a carefully testing, the slurry component is modified as such materials: KMnO4 as an oxidant is 0.19mol/L; alumina is 5wt% as abrasives; the pH should be adjusted to 2 by using pH adjusting agent before polishing.

After polishing, the wafer needs to be ultrasonically cleaned and dried with a dust-free paper. The weighting was carried out by a using an electronic balance (3 times weighing) which is manufactured by Mettler-Toledo with a precision at 0.0001g and model ME204E.

Data processing

By using the modified slurry, we got a set of 4H-SiC different surface material removal rate data. Utilizing the weight loss, a calculation can get the Material Removal Rate based on layer thickness.

$$MRR = \frac{\Delta m}{\rho \bullet S \bullet t} \times 10^7 (nm/h) \cdot$$

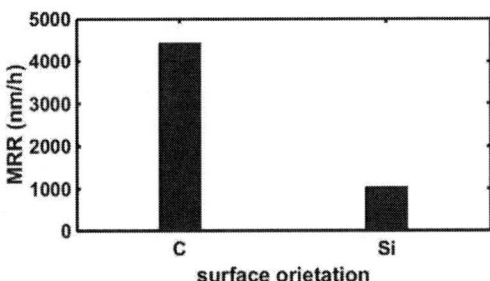

Figure 1 :The MRR differences between Silicon Surface and Carbon Surface

The carbon surface MRR is 4 times faster than silicon surface. For a more detailed explain to the difference on MRR on surface. We utilized more techniques to reveal the inner reason.

STATIC ETCHING EXPERIMENT

Giving 2 samples with 5mm*5mm square cut, immersion treatment with slurry for 1 week. And the one analyzed the silicon surface marked as S, other analyzed the carbon surface marked as C. After cleaning the samples, we use nanoindentation to study the product layer's mechanical properties.

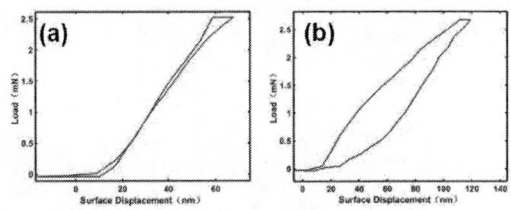

Figure 2: silicon surface (a) and carbon surface(b) static etching nanoindentation testing

As we can see, the loading and unloading curves in Figure 2 (a) are close together, indicating that less plastic material is generated on the silicon carbide silicon surface in the acid polishing solution; the loading and unloading curves in Figure 2 (b) are separated obviously, it shows that there are many plastic substances generated on the silicon carbide carbon surface in the acid polishing solution.In order to value the surface hardness more directly, we draw the hardness of silicon face and carbon face on the Figure.3 as below.

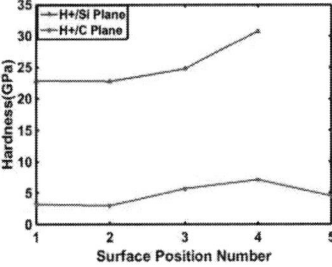

Figure.3 The Hardness of Si/C plane with the treatment of acidic slurry

As the Figure.3 shows the hardness of Si plane is much higher than C plane when they are treated with acidic slurry.

SURFACE LAYER TEM ANALYSIS

To explain the mechanical properties differences, TEM and XPS was introduced into this discussion.

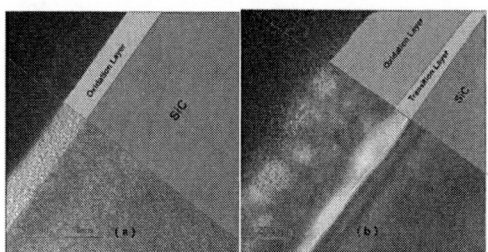

Figure 4: The HRTEM image of 4H-SiC Silicon Surface(a) and Carbon Surface(b)

The silicon surface product layer has a thickness about 3 nm. The layer is regular and thin. This is a confirmation of figure 2 which has a very thin curve displacement. The product layer has a thickness at 40~66 nm. The layer is irregular and very thick. The product layer seems parts of amorphous materials. This is a confirmation of figure.2b which has a very thick curve displacement.

SURFACE LAYER ANALYSIS

In order to acknowledge the composition of the surface layer of different surface types. The XPS was introduced into certain the surface composition.

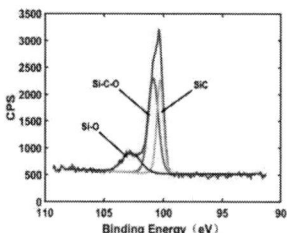

Figure 5: XPS fitting of silicon carbide silicon plane with the treatment of KMnO4 in acidic slurry (Si2p)

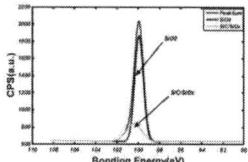

Figure 6:XPS fitting of silicon carbide carbon plane with the treatment of KMnO4 in acidic slurry (Si2p)

For the XPS Si2p image of the acid-treated SiC silicon surface, the characteristic peak positions of silicon element are three respectively 100.3eV, 101eV, and 103eV, which correspond to three valence bonds of Si-C, Si-CO, and Si-O The proportions of the three products are 18.56%, 49.10% and 32.35%. This shows that the SiC immersed in acidic slurry produces silicon carbon oxides and silicon oxides. The presence of silicon carbide also illustrates the corrosion resistance of silicon carbide.Prolonged immersion only indicates the formation of a thin oxide layer.And for the XPS Si2p image of the acid-treated SiC carbon surface, the characteristic peak positions of silicon element are two, which correspond to three valence bonds of SiO_2, and Si-Ox. This shows that the SiC immersed in acidic slurry produces silicon oxides. The product of the SiC carbon surface in the acid slurry is softer and thicker silicon oxide also explains the reason why the carbon surface MRR is higher than the silicon surface.

RESULTS

Due to the difference of product, the layer construction and mechanical property shows different. Through the nanoindentation, we got the mechanical property difference on silicon and carbon surface. For that, a simple explain to MRR difference between silicon and carbon surface can be dropped. TEM and XPS help to reveal the construction and material consistent. Abusolutely silicon oxide produced on carbon surface layer give a decrease in hardness than the silicon surface layer which remaining the silicon carbon oxide.

REFERENCES

[1]Development Status of Manufacturing Equipment for Silicon Carbide Materials and Devices[J]. Electronic Technology, 2017(4).

[2]Sheng Bozhen. Silicon Carbide Devices and Their Applications[J].Electronic Components & Applications, 2001(5):19-23.

[3]Baliga B. Fundamentals of power semiconductor devices[J]. 2010.

[4]Nakamura S, et al. Candela-class high-brightness InGaN/AlGaN double-heterostructure blue-light-emitting diodes[J]. Applied Physics Letters, 1994, 64(13): 1687-1689.

[5] Chen Guomei. Ultra-precision polishing process and mechanism of silicon carbide wafers [D]. 2017.

INVESTIGATION ON THE GALLIUM NITRIDE POLISHING PROCESS UNDER HYBRID-FIELD EFFECTS

Zhigang Dong[1], Liwei Ou[1], Kang Shi[2], Renke Kang[1*], and Dongming Guo[1]*

[1] Key Laboratory for Precision and Non-traditional Machining Technology of Ministry of Education, Dalian University of Technology, Dalian 116024, China

[2] Department of Chemistry and State Key Laboratory of Physical Chemistry of Solid Surfaces, College of Chemistry and chemical Engineering, Xiamen University, 422 Siming South Road, Xiamen, 361005, PR China

*Corresponding Author's Email: kshi@xmu.edu.cn; kangrk@dlut.edu.cn

ABSTRACT

The third-generation semiconductor material gallium nitride (GaN) has attracted more and more researchers attention due to its excellent performance. In the processing of GaN wafer, surface/subsurface damage would inevitably generate after diamond abrasive lapping. Chemical mechanical polishing (CMP) with soft abrasive is widely used to remove surface/subsurface damage and obtain the ultra-smooth wafer surface. However, traditional CMP of GaN wafer has very low material removal rate (MRR) because of the high hardness and chemical inertness of wafer. To improve GaN polishing MRR, hybrid-field effects polishing methods include photochemical mechanical polishing (PCMP) and photoelectrochemical mechanical polishing (PECMP) were proposed. Experimental results illustrated that both PCMP and PECMP methods could significantly improve MRR and surface quality. Meanwhile, the PCMP and PECMP mechanism were discussed.

INTRODUCTION

The development of semiconductor materials from the 1950s to today could be divided into three generations. The first-generation semiconductor typical materials are silicon and germanium et.al. The second are gallium arsenide and indium phosphide et.al. GaN and SiC et.al are the third-generation semiconductor materials. GaN was widely used in high temperature, high power and high efficiency devices due to its advantages of wide bandgap and high electron saturation speed [1]. As a result, GaN has been utilized in 5G communication and high-power switching devices.

However, GaN was hardly reacts with acid and base reagent at normal temperature and its Mohs hardness is greater than 9. So, it is known as one of different to machine materials. Diamond abrasive lapping was mainly used for flattening and thinning GaN wafers, but the surface/subsurface damage would generate by super hard diamond abrasive in material removal process [2]. GaN wafer surface/subsurface damage should be completely removed and smooth surface should be obtained before the wafer was prepared as substrates for further

application [3,4]. CMP with soft abrasives (eg. SiO_2 or CeO_2) could remove surface/subsurface damage and obtain ultra-smooth wafer surface and CMP technique has achieved good results in silicon wafer processing. Nevertheless, CMP MRR of GaN wafer was only 17 nm/h, and it would take about 150 hours to completely remove the damage layer [4]. The main reason for the low rate of material removal in CMP was the low oxidation rate of wafer. In another word, the oxidation of GaN wafer was the rate-determining step in whole CMP process. To improve wafer oxidation rate, strong oxidizers (eg. H_2O_2, $KMnO_4$, $K_2S_2O_8$, $OH^{\cdot-}$, $SO4^{\cdot-}$, et.al) were added into CMP slurry, but the MRR did not increase significantly [5,6].

Photochemical (PC) and photoelectrochemical (PEC) etching of GaN wafer were good candidates for high rate of material removal [7-9]. When GaN was under ultraviolet (UV) irradiation, the separated photo-excited holes (h^+) would oxide wafer effectively. In simple PC and PEC etching system, the wafer surface was not smooth enough. In addition, the distribution of surface roughness and MRR were not even on whole wafer surface after PC or PEC etching [7, 10].

In our work, PCMP and PECMP methods were proposed to polishing GaN wafer, which based on CMP and PC or PEC oxidation. In GaN PCMP and PECMP processing, high MRR and smooth surface could be achieved under hybrid-field effects [11,12].

PCMP APPARATUS AND MECHANISM

GaN wafer PCMP apparatus was shown in Figure 1a. Polishing disc glued with pad and rotated at speed ω_p. GaN wafer was fixed on the worktable and rotated at the speed ω_w. Through-holes were designed on polishing disc and pad. UV-light outlet was 40 mm above the polishing disc to ensured that the wafer surface underneath through-holes of polishing disk and pad could receive UV-light irradiation. J. Guo reported that phyllotactic distributed holes could improve the material removal uniformity in their copper substrates electrochemical mechanical polishing [13]. The through-holes were phyllotactic distributed in polishing disk and pad to improve the uniformity of the alternating hybrid-field effects of PC oxidation and mechanical polishing.

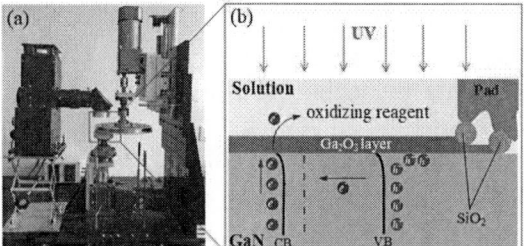

Figure 1: PCMP apparatus and mechanism

GaN PCMP mechanism was shown in Figure 1b. When GaN wafer surface was under UV illumination, photo-excited electron(e^-) would jump from the valence band (VB) to the conduction band (CB). As a result, e^--h^+ pairs would generate near wafer surface (Eq. 1). Potassium peroxydisulfate ($K_2S_2O_8$) was used as strong oxidizing reagent in PCMP. $S_2O_8^{2-}$ in polishing solution would capture photo-excited e^- and facilitate the separation of pairs (Eq. 2). After the separation of e^--h^+ pairs, the remaining h^+ would accumulate near GaN wafer surface. Then, GaN wafer surface would be oxidized to Ga_2O_3 by h^+ effectively (Eq. 3). The hardness of Ga_2O_3 was much less than GaN, so it could be more easily removed by SiO_2 abrasives. In PCMP, the hybrid-field effects of photochemical oxidation and mechanical polishing were alternately performed. Therefore, the GaN material could be continuously removed.

$$GaN \xrightarrow{h\nu} e^-(CB) + h^+(VB) \tag{1}$$

$$S_2O_8^{2-} + e^-(CB) = SO_4^{2-} + SO_4^{\bullet-}(radical) \tag{2}$$

$$2GaN + 6h^+(VB) + 3H_2O = Ga_2O_3 + 6H^+ + N_2 \tag{3}$$

PECMP APPARATUS AND MECHANISM

GaN wafer PECMP apparatus was shown in Figure 2a. PECMP apparatus has the similar structure to PCMP one. In PECMP apparatus, GaN wafer could receive the UV irradiation under through-holes, and it was wired to the anode of electrochemical workstation. The polishing disc was made of conductive stainless steel and connected to the cathode of electrochemical workstation. An electric field was applied to the wafer and polishing disc, and it could promote the separation of photo-excited e^--h^+ pairs. Remaining h^+ were forced to flow to wafer surface and oxidize GaN to Ga_2O_3 (Eq. 3). Meanwhile, e^- were driven by the electric field to the cathode and react with H^+ to form H_2 (Eq. 4).

Compared to GaN PCMP, the application of electric field in PECMP could separate photo-excited e^--h^+ pairs more efficiently. As a consequence, it does not require the addition of strong oxidizing reagent to consume e^- in PECMP solution.

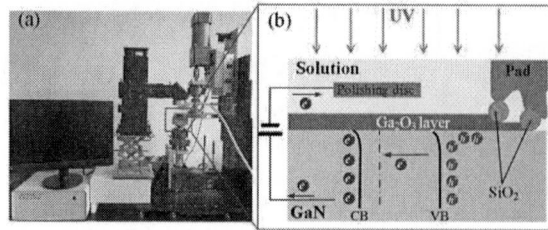

Figure 2: PECMP apparatus and mechanism

$$2H^+ + 2e^- \rightarrow H_2 \tag{4}$$

MRR AND SURFACE QUALITY IN GAN WAFER PCMP AND PECMP

Free-standing GaN wafers (n-type, unintentionally doped, 1 inch in diameter and 350 μm in thickness) were used in our experiments. The ω_p was equal to ω_w (250 rpm) and the polishing load was set to 6.5 psi. The MRR could be calculated according to Eq. 5. Figure 3 illustrated the MRR of GaN in CMP, PCMP and PECMP.

$$MRR = \frac{\Delta m}{\rho \times S \times t} \times 10^7 \, (nm/h) \tag{5}$$

Where Δm (g) represents the weight loss of wafer, and ρ represents the density of GaN (6.2 $g \cdot cm^{-3}$). S and t represent surface area of wafer (5 cm^2) and processing time (1 h), respectively.

In GaN CMP, polishing solution pH was 1.5 and it contained 0.1 M $K_2S_2O_8$ and 5 wt% SiO_2 abrasive (diameter was 25 nm), After 1-hour CMP, the MRR was only 13.4/h. In PCMP, when GaN wafer was under UV (175 $mW \cdot cm^{-2}$) irradiation, it would be photochemical oxidized, and the MRR could increase to ~235.1 nm/h. In addition, when there was no $K_2S_2O_8$ in polishing solution but the electric field was applied in PECMP, the MRR was further increased to ~966.1 nm/h (voltage between wafer and polishing disc was 2.1 V).

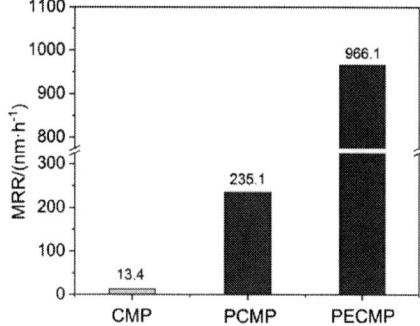

Figure 3: GaN MRR of CMP, PCMP and PECMP

According to the results of MRR shown in figure 3, we could infer that hybrid-field effects of photochemical oxidation and photoelectrochemical oxidation in GaN polishing could improve the oxidation rate and result in high MRR.

In GaN CMP, PCMP and PECMP, soft SiO_2 were selected as the abrasive. In general, CMP with soft SiO_2 abrasive could obtain ultra-smooth wafer surface and it could not introduce any new surface/subsurface damage. Figure 4 showed the GaN wafer surface roughness Ra values in different status. After diamond lapping wafers were used as-received. Scratches could be observed on wafer surface and the roughness Ra was 1.42 nm (10×10 μm^2). After CMP, PCMP and PECMP the surface roughness Ra could significantly decrease to 0.11 nm, 0.19 nm and 0.12 nm respectively. In PCMP, photo-excited e^--h^+ pairs could recombine at defect sites, and if there were insufficient mechanical polishing at these sites, nano-whiskers would be formed [11]. The electric field in PECMP could not only promote the separation of photo-excited e^--h^+ pairs, but also inhibit the recombination of e^--h^+ pairs at defect sites (dislocations and scratches et.al) efficiently. Therefore, there were not obvious nano-whiskers found on the surface in atomic force microscope measurement after PECMP [12].

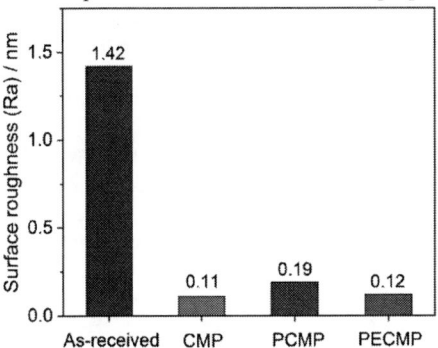

Figure 4: GaN surface roughness Ra values

In our previous work [12], cathode luminescence (CL) mapping and spectra were utilized to measure wafer damage. Experimental results indicated that only dislocations that formed in GaN material preparation process were observed on wafer surface, and the diamond lapping scratches were gradually removed. In the meantime, in CL spectra the intensity of CL signal was obviously enhanced. Hence, it could be deduced that PECMP with soft SiO_2 abrasives could remove wafer damage layer effectively and without introducing new damage.

SUMMARY

The MRR of GaN in traditional CMP was still very low and it was difficult to increase MRR through conventional ways. Two kinds of hybrid-field effects polishing methods PCMP and PECMP were discussed. (1) In PCMP, photochemical oxidation rate of GaN wafer was much faster than that of strong oxidizing reagents in CMP, and it could result in high MRR; (2) By improving the separation efficiency of photo-excited e^--h^+ pairs in PECMP, MRR could be further increased to 966.1nm/h. After both PCMP and PECMP, damage layer on wafer could be removed and better surface quality could be obtained. Next, we will continuously optimize the hybrid-field effects polishing methods of PCMP and PECMP and explore the application of these methods to other related materials.

ACKNOWLEDGEMENTS

The authors would like to acknowledge the supports by the National Natural Science Foundation of China (No. 51575085, 21972120, 91523102, 21273183).

REFERENCES

[1] C.R. Miskys, M.K. Kelly, O. Ambacher, M. Stutzmann, Freestanding GaN-substrates and devices, *Phys. status solidi C*, (2003) 1627-1650.

[2] H. Aida, H. Takeda, S. Kim, N. Aota, K. Koyama, T. Yamazaki, T. Doi, Evaluation of subsurface damage in GaN substrate induced by mechanical polishing with diamond abrasives, *Appl. Surf. Sci.*, 292(2014) 531-536.

[3] H. Aida, T. Doi, H. Takeda, H. Katakura, S. Kim, K. Koyama, T. Yamazaki, M. Ueda, Ultraprecision CMP for sapphire, GaN, and SiC for advanced optoelectronics materials, Curr. *Appl. Phys.*, 12(2012) S41-S46.

[4] H. Deng, K. Endo, K. Yamamura, Plasma-assisted polishing of gallium nitride to obtain a pit-free and atomically flat surface, Ann. CIRP. 64(2015) 531-534.

[5] X. Shi, C. Zou, G. Pan, H. Gong, L. Xu, Y. Zhou, Atomically smooth gallium nitride surface prepared by chemical mechanical polishing with $S_2O_8^{2-}$ - Fe^{2+} based slurry, *Tribol. Int.*, 110(2017) 441-450.

[6] J. Murata, A. Kubota, K. Yagi, Y. Sano, H. Hara, K. Arima, T. Okamoto, H. Mimura, K. Yamauchi, Chemical planarization of GaN using hydroxyl radicals generated on a catalyst plate in H_2O_2 solution, *J. Cryst. Growth*, 310(2008) 1637-1641.

[7] D.H. van Dorp, J.L. Weyher, M.R. Kooijman, J.J. Kelly, Photoetching Mechanisms of GaN in Alkaline $S_2O_8^{2-}$ Solution, *J. Electrochem. Soc.*, 156(2009) D371.

[8] D. Zhuang, J.H. Edgar, Wet etching of GaN, AlN, and SiC: a review, *Mat. Sci. Eng. R*, 48(2005) 1-46.

[9] J.A. Bardwell, J.B. Webb, H. Tang, J. Fraser, S. Moisa, Ultraviolet photoenhanced wet etching of GaN in $K_2S_2O_8$ solution, *J. Appl. Phys.*, 89(2001) 4142-4149.

[10] C. Youtsey, L.T. Romano, I. Adesida, Gallium nitride whiskers formed by selective photoenhanced wet etching of dislocations, *Appl. Phys. Lett.*, 73(1998) 797-799.

[11] L. Ou, Y. Wang, H. Hu, L. Zhang, Z. Dong, R. Kang, D. Guo, K. Shi, Photochemically combined mechanical polishing of N-type gallium nitride wafer in high efficiency, *Precis. Eng.*, 55(2019) 14-21.

[12] Z. Dong, L. Ou, R. Kang, H. Hu, B. Zhang, D. Guo, K. Shi, Photoelectrochemical mechanical polishing method for n-type gallium nitride, *CIRP-Annals*, 68(2019) 205-208.

[13] Z. Liu, Z. Jin, D. Wu, J. Guo, Investigation on material removal uniformity in electrochemical mechanical polishing by polishing pad with holes, ECS *J. Solid. State Sc.* 5(2019), P3047-P3052.

EFFECT OF POTASSIUM SALTS ON THE CHEMICAL MECHANICAL POLISHING EFFICIENCY OF SAPPHIRE SUBSTRATE

Yanan Lu[1,2], Xinhuan Niu[1,2*], Yaqi Cui[1,2], Xin Zhao[1,2], Zhaoqing Huo[1,2] and Chenghui Yang[1,2]*

[1] School of Electronics and Information Engineering, Hebei University of Technology, Tianjin 300130, China

[2] Tianjin Key Laboratory of Electronic Materials and Devices, Tianjin 300130, China

* 861738038@qq.com, xhniu@hebut.edu.com

ABSTRACT

Sapphire substrate is the most commonly used material in semiconductor industry for GaN-based light emitting diodes (LEDs). Chemical mechanical polishing (CMP) is one of the most effective methods to achieve atomic-scale smooth surface. The effect of potassium salts on the CMP removal rate and surface roughness of sapphire substrate was investigated. In this paper, KNO_3, KCl and $K_2S_2O_8$ were used as an additive in sapphire slurry, respectively. From the result, it is found potassium salts can significantly improve the removal rate of sapphire substrate. Meanwhile, $K_2S_2O_8$ is slightly better than the other two in the same condition. Furthermore, the removal mechanism of potassium salts for sapphire CMP was analyzed briefly.

Keywords—chemical mechanical polishing(CMP), sapphire, potassium salts , removal rate

INTRODUCTION

Sapphire (α-Al_2O_3 single crystal) has been widely applied in semiconductor device, light emitting diodes (LED), solid lasers, infrared window and precision optics due to its excellent optical, chemical and mechanical properties such as high hardness, thermal stability, chemical inertness, great electrical and dielectric properties [1]. However, because of the superior performance that the polishing of sapphire remains a challenge [2].

Chemical mechanical polishing (CMP) is becoming one of the most important technologies because of its ability to obtain super-smooth and non-damaged wafer surface [3]. In order to improve the CMP performance of sapphire substrate, the research on sapphire slurry has never stopped. Zhao et al. studied a new type chelating agent in chemical mechanical polishing of sapphire substrate. It revealed that the chelating agent can significantly enhance the surface quality of sapphire [4]. The next Zhao also researched the effect of a new-type alkaline slurry containing KNO_3 on sapphire substrate CMP removal rate. The results showed that the alkaline slurry is effective [5]. Cui et al. found the effect of chloride ions on the chemical mechanical planarization efficiency of sapphire substrate, and the improvement of removal rate by KCl was verified [6]. Furthermore, $K_2S_2O_8$ was also used as an additive in slurry in this paper. Therefore, combined with previous studies, it was

reasonable to explore the influence of potassium salts on the removal rate of sapphire.

Potassium salts were used as a new slurry additive to improve sapphire material removal rate (MRR). In this paper, the effect of different potassium salts on the removal rate of sapphire was investigated. Meanwhile, the contrast of different potassium salts in same condition was researched. Furthermore, the mechanism of the action of potassium salts was also described.

EXPERIMENTAL

Two-inch single crystal commercial c-plane sapphire wafers were used for experiments. Polishing experiments were performed on X62 S82×305-D-S single-side CMP polisher produced by Suzhou Herriot with a Suba600 polishing pad, and sapphire wafers were put in inlaid layer holes on the wax-free polishing template adsorption film.

Alkaline sapphire slurry prepared in the lab was used and nano-SiO_2 sol was selected as abrasive, whose concentration was 40wt%. The average particle size of the abrasive was 80-90nm. The surfactant volume fraction was 0.2 vol%. KOH was used as pH regulator to adjust the pH value of the slurry to 10.5. KNO_3, KCl and $K_2S_2O_8$ were added to the sapphire slurry to improve the material removal rate and reduce surface roughness of the sapphire. Each group of experiment was repeated at least three times to take the average.

Sapphire CMP process parameters were shown in Table I. Professional electronic balance was used to measure the weight of sapphire substrate before and after polishing, with a precision of 0.1mg (AUY120 ASSY). The surface topography and roughness Sq were measured by Agilent 5600LS atomic force microscopy (AFM). The mean particle size was measured by NICOMP 380ZLS laser nanoparticle size analyzer.

Table I. Sapphire CMP process parameters

Parameters	Conditions
Polishing time	30 min
Polishing head speed	50 rpm
Platen rotation speed	50 rpm
Slurry flow rate	160 mL/min
Downward pressure	0.1 Mpa
Upward pressure	0.06 Mpa

Material removal rate (MRR) is determined by the Equation (1):

$$MRR = \frac{\Delta m \times 10^4}{\rho \pi r^2 t} \qquad (1)$$

where Δm (g) is the mass loss, t is the polishing time (in the test, t=1/2h), ρ is the sapphire density ($\rho = 3.98$g/cm^3), r is the radius of sapphire substrate (r = 2.54 cm), and MRR(μm/h) is the corresponding removal rate.

RESULTS AND DISCUSSION

Effect of different potassium salts concentration on sapphire removal rates. —In order to improve the MRR of sapphire substrates, KNO_3, KCl and $K_2S_2O_8$ solution were used as an additive in sapphire slurry, respectively. The ratio of deionized water to nano-SiO_2 sol was 1:1. The additive with different concentration of 0.0wt%-0.4wt% were added to the slurry. From Figure 1 to Figure 3, it was found that MRR of c-plane sapphire increased significantly with the enhancing of additive concentration.

When the KNO_3 concentration increased from 0 to 0.1wt%, the MRR of sapphire increased sharply to 6.251μm/h, as shown in Figure 1. However, with the KNO_3 concentration continued to increase, the MRR began to show a downward trend. Therefore, the optimal KNO_3 concentration was selected at the inflexion point 0.1wt% to obtain higher MRR.

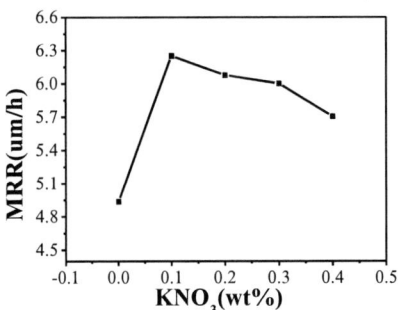

Figure 1: Effect of KNO_3 concentration on sapphire removal rate.

As shown in Figure 2, when the KCl concentration was less than 0.2wt%, the removal rate increased as the concentration increased. However, the removal rate gradually decreased when the concentration was higher than 0.2wt%. Therefore, the removal rate of the sapphire substrate was optimal when the concentration was 0.2wt%, and the maximum removal rate was 6.350μm/h.

From Figure 3, it can be seen that when the $K_2S_2O_8$ concentration was less than 0.3wt%, the removal rate increased as the concentration increased, but when the $K_2S_2O_8$ concentration was higher than 0.3wt%, the removal rate decreased as the concentration increase. In terms of the $K_2S_2O_8$ concentration, 0.3wt% was optimal. When the $K_2S_2O_8$ concentration was 0.3wt%, the removal rate was 6.549μm/h.

To sum up, adding KNO_3, KCl and $K_2S_2O_8$ can significantly improve the removal rate of sapphire substrate and $K_2S_2O_8$ was slightly better than others.

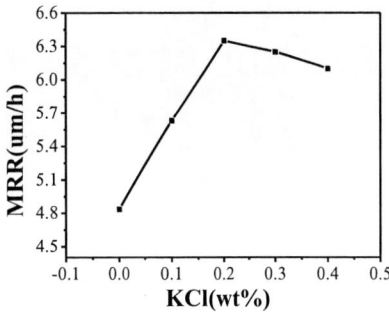

Figure 2: Effect of KCl concentration on sapphire removal rate.

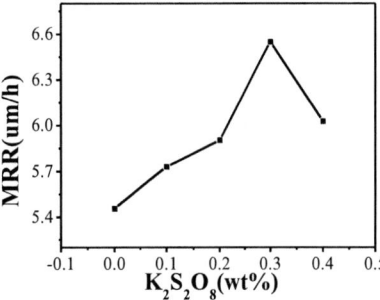

Figure 3: Effect of $K_2S_2O_8$ concentration on sapphire removal rate.

Effect of different potassium salts concentration on sapphire roughness.

After 0.1wt% KNO_3, 0.2wt% KCl and 0.3wt% $K_2S_2O_8$ slurry polishing respectively, the surface roughness was measured and good surface morphology was obtained, as shown in Figure 4. Sq of sapphires was also shown in the figure, $K_2S_2O_8$ was slightly better than others. At the same time, no scratches, corrosion pits and other defects were observed on the surface.

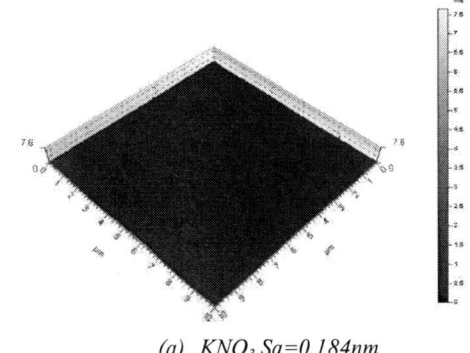

(a) KNO_3 Sq=0.184nm

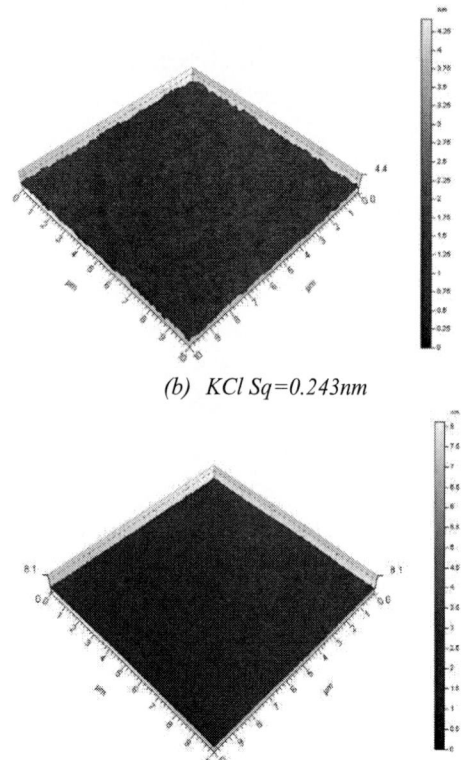

(b) KCl Sq=0.243nm

(c) $K_2S_2O_8$ Sq=0.113nm

Figure 4: The typical surface morphology and surface quality of sapphire wafers after polishing by adding different additive in sapphire slurry.

Analysis of removal mechanism of potassium salts for sapphire CMP

There are abundant hydroxyl groups on nano-SiO_2 sol surface. Under alkaline conditions hydroxyl groups were dissociated to form surface functional groups with negative charges [5]. There existed compound of SiOK, as shown in Formula (2):

$$SiOH + OH^- + K^+ \rightarrow SiOK + H_2O \qquad (2)$$

The particle size is gradually larger because SiOK particle size was bigger than SiOH.

Based on above analysis, strong electrolyte at lower concentration stabilized nano-SiO_2 sol structure. So when particle size was slightly increased and a stable slurry was maintained at medium ionic salt conditions, increase of total attractive forces between particle and wafer surface and individual particle contact area as a result of increase in particle size were found to be major factors influencing increase of material removal rates.

With increase of the concentration of potassium salts, strong electrolyte concentration compresses diffusion layer to be thinner and repulsive potential energy decreases. When attractive potential energy is more than repulsive potential energy, particles will move closer to

each other and partially agglomerate. So when particle size significantly increased and slurry was unstable at high ionic salt conditions, nonuniform contact of agglomerated particles and decreasing of number of particles led to decreased of total contact area between agglomerated particles and wafer surface, resulting in the reduction of removal rates.

Meanwhile, the following reaction occurs as well:

$$Al_2O_3+2OH^-+2K^++3H_2O\leftrightarrow2KAl(OH)_4 \qquad (3)$$

The complex compound $KAl(OH)_4$ was produced under alkaline conditions, which also speed up the CMP process.

CONCLUSION

Based on the above experimental analysis, it was found potassium salts can significantly improve the removal rate of sapphire substrate. With the concentration of potassium salts increased, MRR of sapphire increased signally. When the concentration of KNO_3, KCl and $K_2S_2O_8$ was 0.1wt%, 0.2wt%, 0.3wt% respectively, MRR reached the maximum and good surface morphology was obtained. Furthermore, it was found there existed compound of SiOK and $KAl(OH)_4$, which can improve the removal rate of sapphire substrate. All in all, potassium salts can significantly improve the CMP performance of sapphire substrate.

ACKNOWLEDGEMENTS

This work was supported by the Major National Science and Technology Special Projects (No.2016ZX023 01003-004-007), Natural Science Foundation of Tianjin, China (16JCYBJC16100, 18JCTPJC57000). The authors also thank the teachers and classmates for their helpful suggestions.

REFERENCES

[1] T. Saito, T. Hirayama, T. Yamamoto and Y. Ikuhara. *Journal of the American Ceramic Society*, vol. 88, 2005, pp. 2277-2285.

[2] X.H. Niu, Y.L. Liu, B.M. Tan, L.Y. Han and J.X. Zhang. *Transactions of Nonferrous Metals Society of China*, vol. 16, 2006, pp. 732-734.

[3] X.H. Niu, X. Zhao, D. Yin, J.C. Wang and C.W. Wang. *International Conference on Planarization /CMP Technology*. VDE, 2017, pp. 1-6.

[4] X. Zhao, X.H. Niu, J.C. Wang, D. Yin and C.H. Yao. *ECS Journal of Solid State Science and Technology*, vol. 6, 2017, pp. 618-625.

[5] X. Zhao, X.H. Niu, D. Yin, J.C. Wang and K. Zhang. *ECS Journal of Solid State Science and Technology*, vol. 7, 2018, pp. 135-141.

[6] Y.Q. Cui, X.H. Niu, J.K. Zhou, Z. Wang and R. Wang. *ECS Journal of Solid State Science and Technology*, vol. 8, 2019, pp. 488-495.

A STUDY OF CAUSES AND IMPROVING METHODS OF CHIPPING IN BSI PROCESS

Yurong Cao*, Hu Li, Zhe Feng, Zujun Ji, Zhengyuan Zhao, Jin Chen, Youfeng Xu, Xiang Peng and Feng Ji
Huali Microelectronics Corporation
Shanghai 201210, China
*Corresponding Author's Tel: <u>60570000-67501</u>, Email: <u>caoyurong@hlmc.cn</u>

ABSTRACT

The causes and improving methods of chipping were studied from CMP (Chemical Mechanical Polishing) and Trim1 (First Trimming process before bonding) perspectives. Chipping is caused by worse wafer edge bonding quality which is impacted by wafer edge pattern step-height and edge profile. Increasing CMP remove amount could reduce step-height and improve chipping performance. Bias of carrier wafer WEE (Wafer Edge Exposure) and Trim1 width should be fixed for better extreme edge bonding quality, which brings better chipping performance.

INTRODUCTION

In recent years, CMOS image sensor (CIS) market expanded very quickly and expansion rate is as high as 20% [1]. The Back-Side Illuminated (BSI) technology takes an important position in CIS market due to its downscaled pixel area and high resolution [2-4].

Image sensor has a demanding requirement for luminousness and transparency uniformity so that the control of defects is very important in BSI technology [5]. Chipping is the peeling at wafer edge of the bonded top wafer and it is a characteristic defect in BSI technology. Chipping is mainly affected by wafer edge bonding quality, which is related to the film thickness of wafer edge. Oxide CMP and Trim1 are two processes related to wafer edge film thickness in pre-bonding loop. In this paper, we studied the causes and improving methods of chipping from CMP and Trim1 perspectives separately.

RESULTS AND DISCUSSION
OX CMP perspective

After wafer bonding, CSAM (C type Scanning Acoustic Microscope) is scanned to check bonding quality. In general, the darker the color of the CSAM image, the better the bonding quality. Fig. 1a is partial of a CSAM image and Fig. 1b is an OM image of wafer before bonding. As shown in Fig. 1a, the wafer edge can be divided into three regions. The Logic2 region is white with low bonding quality, the Logic1 region is gray with medium bonding quality, and the Array region is black with good bonding quality. Logic2 region has the lowest bonding quality of these three regions. Combining the corresponding positions of these three regions in OM image in Fig. 1b, we can see that Logic2 region has a distinct chromatic aberration. The chromatic aberration of Logic2 region comes from top metal photo process of FSI wafer. Photoresist in logic2 region was completely exposed in photo

process and ETCH would ETCH out all the oxide in this area, forming a continuous step-height of about 10000A depth. Therefore, it would be very difficult for CMP planarization in logic2 region with such a large step-height, which would further influence wafer edge bonding quality and chipping result.

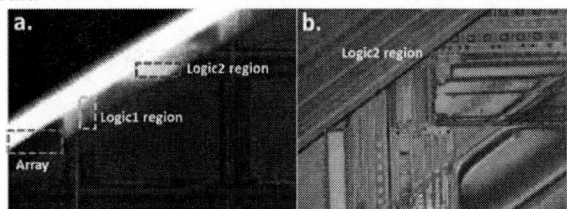

Figure 1: (a), CSAM image of bonding wafer; (b), OM image of wafer before bonding.

Wafer edge CMP down force decrease experiment was did to study the influence of CMP remove amount on the step-height elimination. As shown in Fig. 2b, edge down force decrease wafer has more bonding worse area which can be seen in CSAM image and the chipping result is worse than baseline (BL) wafer (Fig. 2a). Comparing with BL condition, CMP down force decrease leads to less remove amount and step-height elimination becomes more insufficient. And the bonding quality of wafer edge is lower and the chipping is even worse. Therefore, step-height needs to be eliminated for chipping improvement, and the only way to eliminate step-height is to increase the remove amount of CMP.

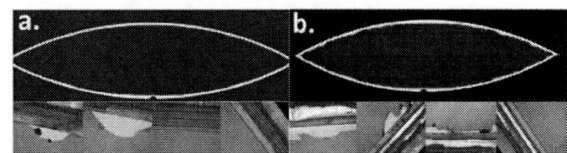

Figure 2: (a), CSAM and chipping image of BL wafer; (b), CSAM and chipping image of edge down force decreased wafer.

Remove amount increasing experiment was conducted and the remove amount was increased by 10000A and 20000A separately. Fig. 3 shows the CSAM and chipping results of wafers with increased CMP remove amount. It can be seen that there is no white area at wafer edge in CSAM image both for the wafers under this two conditions and the chipping results are also improved to < 10 ea (BL about > 100 ea). And the chipping result of 20000A remove amount increased wafer is much better. This means that the more the CMP remove amount increases, the more sufficient the step-height is eliminated and

978-1-7281-6559-2/20 $31.00 © 2020 IEEE

the better the chipping result would be.

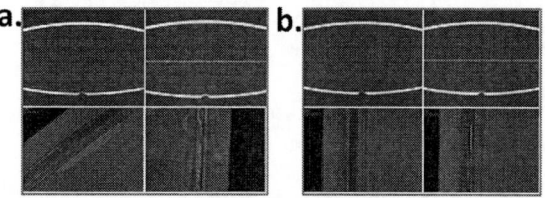

Figure 3: CSAM and chipping results of wafers with increased CMP remove amount, (a), 40K wafers; (b), 50K wafers.

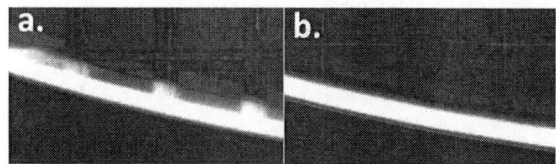

Figure 4: (a), CSAM images of BL wafer; (b), CSAM image of remove amount increased wafer.

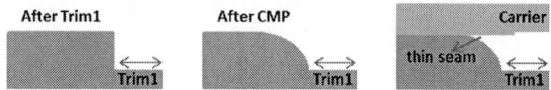

Figure 5: Wafer edge profile evolution before bonding.

Fig. 4 is the CSAM image of BL wafer and remove amount increased wafer. It can be seen that white area is not found for wafer with increased remove amount, but an even gray area can be seen at wafer edge (Fig.4b). This is caused by the faster remove rate at wafer far edge area during CMP process. We can see from Fig. 5 that wafer edge was cut out a right angle by Trim1 process and the right angle area was polished faster during CMP process. Then wafer edge would be polished into a rounded corner and a thin seam between the two bonded wafers at wafer edge appeared after bonding process, which was presented as an even gray area in CSAM image. For this thin seam, the chemical used in subsequent WET ETCH1 thinning process will get in along the seam and break the bonding surface. And this will lead to poor bonding quality and a bad chipping result. Therefore, simply increasing the CMP remove amount cannot completely improve chipping effectively, and the poor bonding quality in the thin seam area also needs to be improved.

Trim1 perspective

This paper will then discuss how to effectively eliminate the thin seam area and improve chipping result from the perspective of Trim1. It's important to note that the study on Trim1 process is based on the increase of CMP remove amount.

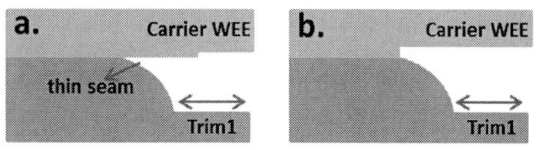

Figure 6: (a), Wafer edge structure of BL bonding wafer; (b), Wafer edge structure after adjusting carrier wafer WEE.

As we know, there also exists a photo WEE process on carrier wafer. Fig. 6a is the wafer edge structure of BL bonding wafer, and the WEE of carrier wafer is smaller than the Trim1 width of FSI wafer. To eliminate the thin seam shown in Fig. 6a, we need to adjust the WEE of carrier wafer and Trim1 width so that the carrier wafer WEE can cover the rounded corner area (Fig. 6b). In other words, the WEE of carrier wafer and the Trim1 width of FSI wafer need to be kept at a fixed bias, which is actually the width of the rounded corner in the direction of the radius. By measuring the width of the even gray area in CSAM image and checking the bonding wafer's SEM (Scanning Electron Microscope) image, we can get that the width of the rounded corner is *a*. Therefore, we suspect that chipping result will be the best when the bias is *a*. And experiment was conducted to verify the conclusions.

Table1: Trim1 width experiment split table

Item	Width(mm)	#1	#2	#3	#4	#5	#6	#7	#8	#9
Bias	a-0.3	v								
	a-0.2		v	v						
	a				v	v	v	v		
	a+0.3								v	v

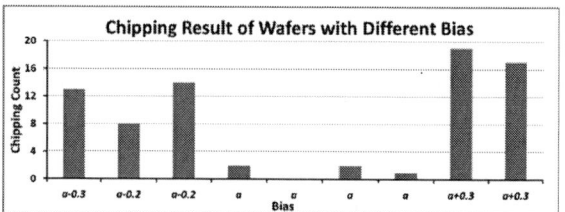

Figure 7: Chipping result of wafers with different bias.

The WEE of carrier wafer was fixed as *n* and the width of Trim1 was changed in the experiment. Then the bias variation changed from *a-0.3mm* to *a+0.3mm*, as shown in Table1. Fig. 7 shows the chipping results of these wafers with different bias. It can be seen that the chipping results are the best when the bias is *a*, only 2 ea in average. When the bias is smaller than *a* (Fig. 8a), the thin seam exists and leads to bad chipping result. And when the bias is bigger than *a* (Fig. 8c), the FSI wafer edge will be too thin and stretch out too long after thinning and peeling occurs easily, also leading to bad chipping result. Therefore, the best chipping result will be obtained when the bias is just the width of the rounded corner.

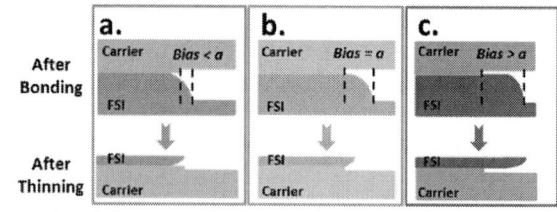

Figure 8: Chipping performance of wafers with different bias.

Furthermore, as shown in Fig. 9, when the bias is fixed, the smaller the WEE of carrier wafer, the larger the window of Trim2 (Second Trimming process after bonding) is, and the chipping result will be more stable.

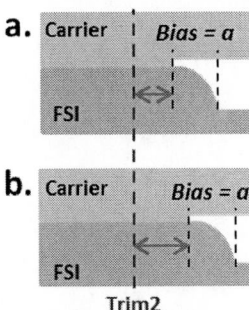

Figure 9: Trim2 window for different WEE wafers, (a), BL; (b), Smaller WEE.

CONCLUSIONS

In this paper, we quantitatively studied the causes and improving methods of chipping and the main conclusions are as follows:

1.   Chipping improvements require the elimination of step-height, and the only way to eliminate step-height is to increase the remove amount of CMP. The more the CMP remove amount increases, the more sufficient the step-height is eliminated and the better the chipping result is.

2.   However, simply increasing the CMP remove amount cannot completely improve chipping result effectively, and the poor bonding quality in the even gray area also needs to be improved. The best chipping result will be obtained when the bias of carrier wafer WEE and Trim1 width is just the width of the rounded corner.

3.   And when the bias is fixed, the smaller the width of carrier wafer WEE, the larger the window of Trim2 is, and the chipping result will be more stable.

ACKNOWLEDGEMENTS

The authors would like to thank all the members of CMP group and thank Z. Y. Zhao and C. Fu for their useful discussions.

REFERENCES

[1] R. Fontaine, "The State-of-the-Art of Mainstream CMOS Image Sensors", conf IISW conference, 2015.

[2] Y. W. Cheng, T. H. Tsai, C. H. Chou, K. C. Lee, H. C. Chen and Y. L. Hsu, "Optical performance study of BSI image sensor with stacked grid structure", IEEE, pp. 783-786, 2015.

[3] Yole Development, "More than Moore market and technology trends", report, 2015.

[4] H. Moriceau, "Low Temperature Direct Bonding Assisted by CMP and Plasma Activation", IEEE, pp. 123, 2012

[5] Z. C. Hsiao; C. T. Ko; H. H. Chang; H. C. Fu; C. W. Chiang; W. L. Tsai, "TSV-less BSI-CIS wafer-level package and stacked CIS module", IEEE, pp. 323-326, 2013.

Study on the Properties of Silica Colloid Prepared by Different Processes in Silicon Wafer CMP

Weiwei Li[*], Zhilin Zhao, Zhen Liang, Yunqian Sun

School of Electronic Information Engineering, Hebei University of Technology
Tianjin 300130, China
*Corresponding Author's Email: liweiwei@hebut.edu.cn

ABSTRACT

The properties of nano-silica colloid prepared by different processes in silicon wafer chemical mechanical polishing (CMP) were studied. The different principles of preparing nano-silica colloid by ion exchange, silica hydrolysis and hydrolytic of TEOS were analyzed respectively. The differences in structure, dispersion, density, and surface morphology were compared. Under the same CMP process parameters, silica colloids prepared by three different processes were used for polishing experiments. The results demonstrate that the hydrolysis of TEOS silica colloid is not suitable for CMP due to the network structure, which made the silica colloid as abrasive can not imply sufficient mechanical friction. The colloidal particles prepared by silica hydrolysis are denser, more uniform, with better dispersion and rough surface. The polishing rate of it is higher than that of ion exchange silica in a certain particle size range, and as the diameter increases, the growth increases.

Keywords—Colloidal silica; Ion-exchange; Silica hydrolysis; Hydrolysis of TEOS; Density; Polishing rate

INTRODUCTION

Chemical mechanical polishing (CMP) is currently recognized as the only way to achieve global flattening in semiconductor device manufacturing processes. CMP is the process of mechanical removal and chemical removal interaction with many parameters and complicated mechanisms[1,2]. Silica colloid has good stability, low hardness, simple process and low cost, as the main raw material for slurry[3], its density, surface morphology, smoothness, and other properties have a great influence on the polishing rate.

The most commonly used silica sol preparation processes in industrial production are mainly three methods: ion-exchange, silicon hydrolysis, hydrolysis or alcoholysis.

S. Sadegh Hassani et al. focused on the synthesis of nano-silica colloid via ion exchange, prepared stable nano-silica colloid with an average particle size in the range of 23-100nm[4]. M.M. Rashad et al. synthesized a spherical nano-silica colloid (10-30 nm) with good dispersion using silica fume. The wafer roughness was significantly improved after CMP[5]. Haihong Gu et al. prepared a superhydrophobic nano-silica particle from tetraethyl orthosilicate (TEOS) using a two-step synthesis route, through hydrolysis and condensation[6]. Many scholars have done a lot of detailed research on silica colloid for CMP. However, there are few reports on the differences in polishing properties of silica colloid prepared by different processes.

In this paper, the effects of silica colloid prepared by different processes on the polishing performance of silicon wafers were studied. The differences of colloidal growth mechanism in different processes, as well as the colloidal dispersion, density, morphology, and other properties were analyzed and compared. The TEM images of the silica colloid were given. The design experiment has carried out preliminary verification, showing that within a certain particle size range, the polishing properties of the three silica sols prepared by different processes have obvious differences, which provides some guidance for the optimization of the slurry.

PRINCIPLE AND ANALYSIS

a) Principle of Silicic Acid Polymerization

Silicic acid polymerization is carried out according to two mechanisms, acidic and basic, in different pH ranges. Silicic acid mainly exists in the form of $H_3SiO_4^-$ and H_4SiO_4 in slightly alkaline solution, and a silicic acid polymerization reaction occurs:

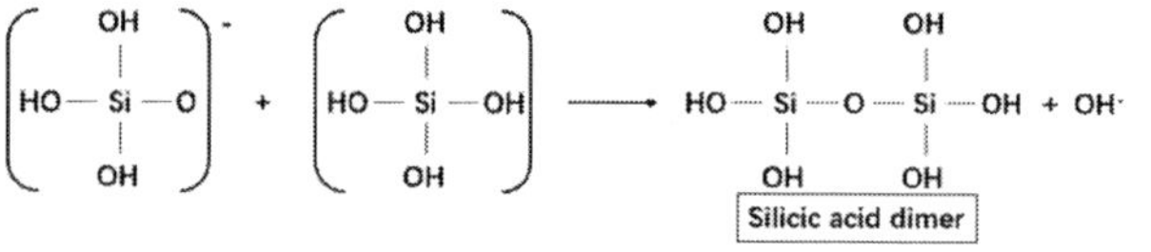

Fig1: Silicic acid polymerization

Silicic acid dimer further reacts with $H_3SiO_4^-$ to produce polysilicate such as silicic acid trimer and silicic acid tetramer. Under special requirements, a large number of active silicic acid molecules are dehydrated and polymerized to form silica particles, namely silica colloid. At the same time, the active silicic acid molecules will polymerize with the formed silica colloidal particles (parent nucleus), making the particle size continuously increase:

$$mH_2SiO_3 + nH_2SiO_3 \rightarrow (n + m)SiO_2 + (n + m)H_2O \quad (1)$$

$$nSiO_2 + H_2SiO_3 \rightarrow (n + 1)SiO_2 + H_2O \quad (2)$$

b) Principle of Ion Exchange

High-purity active silicic acid solution obtained from dilute water glass solution by ion exchange reaction. Active silicic acid smaller than 2nm form colloidal nucleus larger than 4nm, under alkaline conditions; the growth rate slows down when the particles grow to 5-10nm at room temperature, as the temperature increases, the particles continue to aggregate and grow. Therefore, adjust the pH value to 9-10, heat the polysilicate solution obtained in the previous step and left it for a while, and the active silicic acid polymerize to grow into

978-1-7281-6559-2/20 $31.00 © 2020 IEEE

the colloidal nucleus larger than 5-10nm. With this as the mother liquor, SiO_2 in polysilicate solution added at a certain rate was absorbed by the colloidal nucleus surface and grew regularly to synthesize colloidal silica with appropriate particle size and uniform distribution.

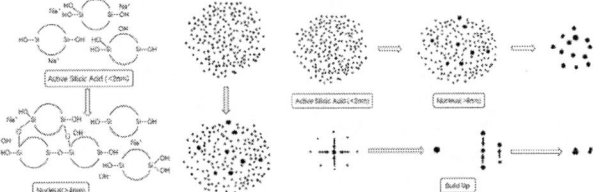

Fig2: The active silicic acid polymerize to nucleus and grow

c) Principle of Silica Hydrolysis

The silica hydrolysis uses industrial silica fume as a raw material. After activation treatment, under the condition of alkali as a catalyst and heating, the elemental silicon reacts with water generating hydrated silicic acid monomer, and the monomer is polymerized to obtain polymerized silicic acid. After the concentration of silicic acid is oversaturated, part of it undergoes dehydration condensation reaction to precipitate nucleus.

Continue to add silica fume, a large amount of active silicic acid produced by hydrolysis and silica colloidal particles (parent nucleus) undergo a hydroxyl condensation reaction to polymerize on the colloidal nucleus to achieve the increase in particle size.

$$Si + 2OH^- + H_2O \rightarrow SiO_3^- + 2H_2 \uparrow \qquad (3)$$

$$SiO_3^- + H_2O \rightarrow SiO_2 + 2OH^- \qquad (4)$$

$$Si + (2m+n)\ H_2O \rightarrow mSiO_2 \cdot nH_2O + 2mH_2 \uparrow \qquad (5)$$

Compared with the ion exchange, the silica hydrolysis colloid has different particle structures due to the different mechanisms of the two processes.

The ion exchange method is a process in which active silicic acid molecules generated after ion exchange are gathered onto the colloidal nucleus to form colloidal particles. The smaller the size, the easier it is to adsorb active silicic acid molecules on the surface. The size of the colloidal nucleus increases with the progress of the reaction, resulting in that the adsorption of active silicate molecules on the surface of the colloidal nucleus is less uniform and dense than that on the surface of colloidal nucleus with smaller size. Therefore, as the particle size of the colloidal particle increases, its density gradually becomes worse.

The silica hydrolysis method is a process of hydrated silicate monomer molecules formed by the reaction of silicon and water and then freely combine into colloidal particles. The new molecules continue to gather on these colloidal particles, and gradually increase the particle size. Therefore, the difference between the size of the colloidal particles and the size of the newly formed molecules is slower than that of the ion exchange silica colloid. And then, as the particle size increases, the difference in densification gradually increases. The mechanical friction of the particles during polishing also changes with the density of the colloid, resulting in different polishing effects.

On the other hand, during the ion exchange process, small particles regularly aggregate on the colloid nucleus to form a smooth colloidal surface. In the process of silica hydrolysis, the aggregation of new ecological silica molecules makes the surface of the particles form a "burr structure". And as the particle size increases, the difference between the two becomes more obvious. During polishing, the uniform and dense colloidal particles with rough surface will increase the mechanical friction and show a higher polishing rate.

The pictures of the silica colloid prepared by the two preparation methods are as follows:

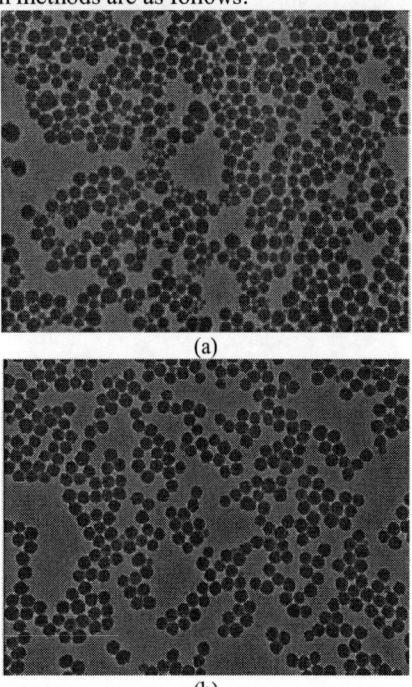

Fig3: TEM of silica colloidal abrasive
(a) Ion-exchange (b) Silica hydrolysis

d) Principle of TEOS Hydrolysis

The sol-gel method refers to a physical and chemical process in which a sol is hydrolyzed and polymerized at the same time after a metal alkoxide or a compound thereof undergoes hydrolysis and polycondensation to form a sol, and then the solute gel is solidified. Sol-gel method is usually used to prepare silica colloid with TEOS as raw material.

The hydrolysis and condensation reaction of TEOS is divided into three steps: the first step is the hydrolysis of TEOS to form a hydroxylated product (silicic acid) and the corresponding alcohol:

$$Si(OC_2H_5)_4 + 4H_2O \rightarrow Si(OH)_4 + 4C_2H_5OH \qquad (6)$$

The second step is a condensation reaction between silicic acid or silicic acid and TEOS to form a colloidal mixture:

Fig4: Condensation reaction between silicic acid and TEOS

The third step is to form a low polymer and continue to polymerize to form a three-dimensional network structure:

$$X (Si - O - Si) \rightarrow (- Si - O - Si -)_X \quad (7)$$

From the mechanism of TEOS hydrolysis, since the hydrolysis reaction can proceed at room temperature, the generated silicic acid molecules will aggregate into a nucleus in a short time to form a network structure. At the same time, part of the $-C_2H_5$ group, due to the rapid reaction and too late to detach, formed voids in the colloid, resulting in imperfect network crosslinking. The structure inside the colloidal particle is loose, resulting in ineffective mechanical friction that can cause physical removal during the polishing.

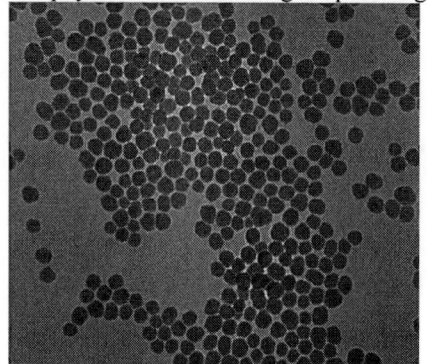

Fig5: TEM of TEOS hydrolysis silica colloidal abrasive

EXPERIMENT

Polishing experiments were carried out using the 3inch silicon wafers (100) with the following conditions: the polishing pressure 0.4MPa, the slurry flow rate 50 ml/min, and down plate rotating speed 100 rpm, every CMP process was performed for 30 min, repeat 4 times, take the average polishing rate.

The slurry is prepared from 20wt.% silica colloid prepared by ion-exchange, silica hydrolysis, and the TEOS hydrolysis process with average particle diameters of 40-50nm, 60-70nm, and 80-90nm, respectively. The results are as follows.

From the polishing rate of three different types of silica colloid in different particle size ranges, it can be seen that the TEOS hydrolysis silica colloid shows extremely low polishing rates, and the polishing rate did not change significantly with the increase of particle size.

Therefore, only the polishing rates of different particle size ranges for the silica colloid prepared by ion-exchange (1#) and silica hydrolysis (2#) are compared.

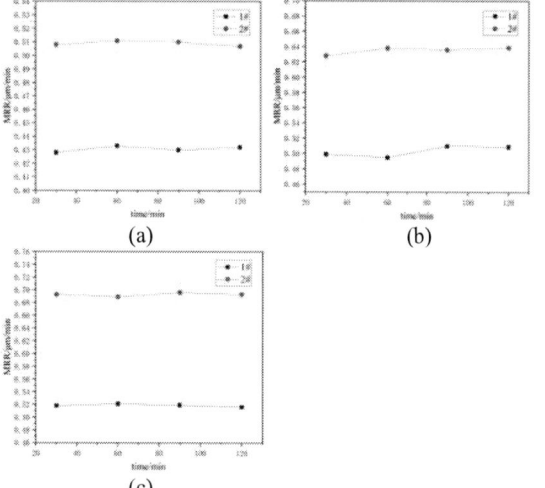

Fig7: Polishing rate of silica colloid prepared by silica hydrolysis and ion exchange in different particle size range (a) 40-50nm (b) 60-70nm (c) 80-90nm

According to the analysis of polishing rate data, when the abrasive particle size is within a certain range, the polishing rate is: silica hydrolysis > ion-exchange > TEOS hydrolysis.

The polishing rate of the silica hydrolysis abrasive is higher than that of the ion-exchange, as shown in Table 1. It is shown that as the increase of particle size, the growth of polishing rate of silica hydrolysis silica sol increase.

TABLE I
PERCENTAGE INCREASE IN POLISHING RATE

Particle Size		40-50nm	60-70nm	70-80nm
MRR (μm/min)	Ion-exchange	0.431	0.500	0.519
	Silica Hydrolysis	0.508	0.634	0.692
Increased Percentage		17.91%	26.72%	33.31%

CONCLUSION

1) Due to different growth mechanisms, resulting in different colloidal particle structures, densities, uniformities, surface morphologies, etc., the mechanical friction effect on the silicon wafer surface during polishing also changes, showing different polishing rates. When the abrasive particle size is within a certain range, the polishing rate is: silica hydrolysis > ion-exchange > TEOS hydrolysis.

2) In the process of TEOS hydrolysis to prepare silica colloid, silicic acid molecules aggregate into a nucleus in a short time, forming a network structure, making the colloidal particle structure loose, and unable to form effective mechanical friction during polishing. The silica colloid prepared by this method is not suitable for CMP.

3) In a certain particle size range, the percentage growth in polishing rate of the silica hydrolysis silica colloid is gradually increasing, compared with the ion-exchange silica colloid.

ACKNOWLEDGEMENTS

This research has been supported by Science and Technology Electro-Optical Information Security Control Laboratory Fund (No.614210701041705).

REFERENCES

[1] Park C, Kim H, Lee S, et al. The influence of abrasive size on high-pressure chemical mechanical polishing of sapphire wafer[J]. International Journal of Precision Engineering and Manufacturing-Green Technology, 2015, 2(2):157-162.

[2] Jiang L, He Y, Li Y, et al. Effect of ionic strength on ruthenium CMP in H2O2-based slurries[J]. Applied Surface Science, 2014, 317: 332-337.

[3] Lee H, Lee D, Kim M, et al. Effect of mixing ratio of non-spherical particles in colloidal silica slurry on oxide CMP[J]. International Journal of Precision Engineering and Manufacturing, 2017, 18(10): 1333-1338.

[4] Sadegh Hassani S, Rashidi A, Adinehnia M, et al. Facile and economic method for preparation of nano-colloidal Silica with controlled size and stability[J]. International Journal of Nano Dimension, 2014, 5(2): 177-185.

[5] Rashad M M, Hessien M M, Abdel-Aal E A, et al. Transformation of silica fume into chemical mechanical

polishing (CMP) nano-slurries for advanced semiconductor manufacturing[J]. Powder Technology, 2011, 205(1-3): 149-154.

[6] Gu H, Zhang Q, Gu J, et al. Facile preparation of superhydrophobic silica nanoparticles by hydrothermal-assisted sol–gel process and effects of hydrothermal time on surface modification[J]. Journal of Sol-Gel Science and Technology, 2018, 87(2): 478-485.

IMPACT OF BEVEL CONDITION ON STI CMP SCRATCH

Yuanyuan Meng*, Lei Zhang, Yibin Li, Wei Zhang, Haifeng Zhou, Jingxun Fang

Research and Development Dept., Shanghai Huali Integrated Circuit Corporation, Shanghai 201314,
*Corresponding Author's Email: mengyuanyuan@hlmc.cn

ABSTRACT

Shallow trench isolation chemical mechanical polishing (STI CMP) technology has been widely applied in the fabrication of ultra large scale integrated (ULSI). In STI CMP, the defect, topography control, thickness uniformity and so on are all so critical, especially, scratch defect is the major problem. Pad, disk, agglomerated slurry particles and incoming particles are the main sources of the tiny scratch. In this paper, we conducted a detailed study on the influence of one-step AA pull back process on the bevel region of wafer, which led to the introduction of incoming particles and ultimately led to the increase of STI CMP scratch. It was find that by adding a brush bevel process before STI CMP can reduce the scratch by 66%.

INTRODUCTION

With the development of modern integrated circuit (IC) technology, the increasing number of active components has led to a significant decrease in feature dimensions. The nanoscale dimensions of multilevel-interconnection have put forward a higher demand for the planarization [1]. Shallow trench isolation chemical mechanical polishing (STI CMP) technology has been widely applied in the planarization of IC chips. This technology possesses an excellent performance in removing unwanted topography and obtaining a flat isolation structure [2]. By using appropriate selectivity abrasive slurry and polish conditions, STI CMP can effective stop at a hard mask layer, nitride, for guaranteeing the uniformity of post layers. And it makes STI CMP a crucial process to enhance packing density and high degree of planarity.

However, in STI CMP, besides the topography control and thickness uniformity, the defect, especially, micro-scratch is always a major problem to limit its development [3]. The micro scratch defect on the wafer surface can induce the leakage current and short between gates [4]. And how to reduce the micro scratch defect is always a hot spot of STI CMP research.

As we all know, during the device manufacturing process several defects, damages and pollutions occur at wafer edge caused by film deposition, etching, annealing and wafer handling. The particles, witch easily transferred to wafer front side during semiconductor processes, are the major source to generate scratch defect. During the CMP process, these particles are easily to adhere on the surface of wafers before main polishing process. Once the particle is brought into the polish process, under the action of down force, it can generate the micro scale scratch defect.

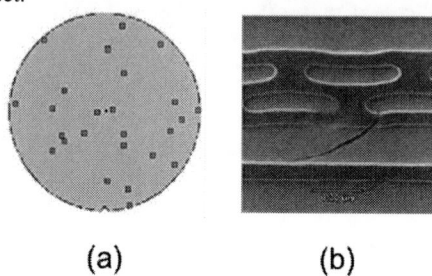

(a) **(b)**

Figure 1: Typical polishing scratches in STI CMP by AIT & SEM images taken by KLA (a) Defect map image of micro-sized scratch; (b) SEM image of micro-sized scratch

As shown in figure 1 (a) & (b), bevel can introduce particles exceeding 1 μm size, which could cause micro-scratches on the wafer surface. These kinds of scratches defect will induce the stability of the device. In order to reduce defects after STI CMP, wafer edge clearing technique was used. With this polish a clearing down to wafer substrate and a smoothing of wafer edge is feasible. Edge polishing can eliminate defects and prepare the edge for a better subsequent process yield.

EXPERIMENT

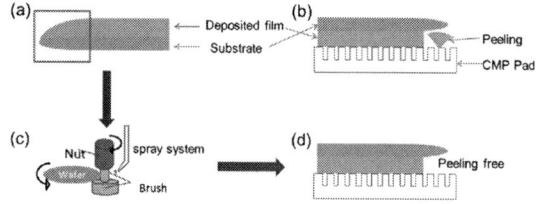

Figure 2: (a) film layer deposited on substrate; (b) particles fall off during CMP process; (c) scrubber bevel schematic diagram; (d) peeling free after scrubber bevel.

As shown in figure 2(a) and 2(b), during the device manufacturing process, the contact between the film on the edge of wafer and the substrate is often not as firm as in the middle caused by film deposition, etching, annealing and wafer handling. When the wafer was turned

978-1-7281-6559-2/20 $31.00 © 2020 IEEE

over and polished on the pad, it is easy for the weak part to fall off under the action of mechanical force and transferred to wafer front side to generate scratch. In order to solve the above problems, as shown in figure 2(c), directly adopt the machine device of wet clean, scrubber bevel was added before STI CMP in this experiment. Brush and spray device were used to clean the wafer edge, so as to reduce the source of particle generated in the polish process.

The polishing experiment was performed on the AMAT-LK system. Micro-sized scratches counts were measured by KLA-Tencor AIT after completing the nitride strip process.

RESULT AND DISCUSSION

In our experiment different film stacks at wafer edge region have to be brushed, due to the smooth surface after bevel brush the bevel inspection is more sensitive to new defects generated in subsequent process steps. As shown in figure 3(a) and 3(b), the defect map and classification in opposition to the expectations, there was no impact of bevel brush to wafer defect count visible. That means there was no increase or decrease in defect count on wafer level seen as well as no change in defect types. The film thickness as well as step height measurements after the following CMP processes indicating no impact of bevel brush to the CMP polish profiles.

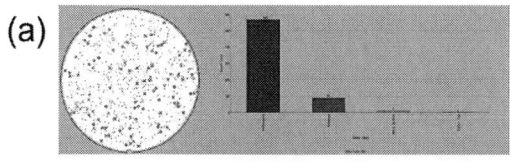

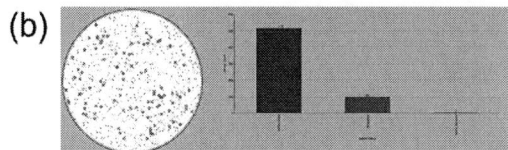

Figure 3: STI CMP defect map and classification
(a) without brush bevel clean; (b)with brush bevel clean

As shown in figure 4, for STI CMP baseline condition and the corresponding conditions of the split defect analysis data, the results show that brush bevel clean used to reduce the micro-scratch defect counts by approximately 66%, with an average of 2ea defects, effectively improve the product yield and reliability.

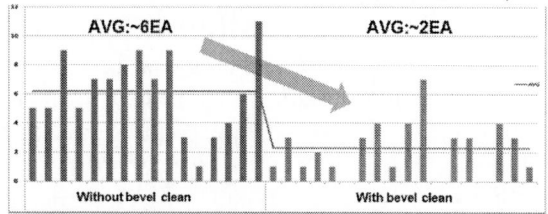

Figure 4: Defect counts observed on wafers with & without bevel clean

CONCLUSION

This paper mainly discusses the mechanism that the edge of the wafer causes scratch defects on the surface of the device in the process of manufacturing. Based on this, a method to improve the micro scale scratch defect after STI CMP is proposed, that is, a step of brush bevel clean is added before STI CMP. In the experiment, we cleaned the film layer with weak bonding on the edge of wafer in advance, so as to reduce the source of scratched particles generated in the CMP process. Experimental results showed that this method effectively reduced the number of micro scratch by 66%, significantly improved the stability of the product, and made an effective contribution to the improvement of product yield. At the same time, the process is simple and low coast, which can be effectively popularized and applied in actual mass production.

REFERENCES

[1] T. Y. Kwon, M. Ramachandran, J. G. Park. "Scratch Formation and its Mechanism in Chemical Mechanical Planarization (CMP)", Friction, 1(4) 279-305 (2013).

[2] P. Song, D. Yao, J. D. Sun. "STI CMP: Exploration of a Colloidal Silla Based Slurry System", ECS Transactions, 34(1) 113 -117 (2011).

[3] F. Bai, Z. J. Zhang, J. Wang, H. D. Wang, "Impact of Wafer Transfer Process on STI CMP Scratches", IEEE Conference Publications. 2017

[4] L. L. Hwee, S. Balakumar, S. Mahadevan, Z. M. Sheng, A. See, M. Rahman, A. Senthilkumar, "Dishing and Nitride Erosion of STI-CMP for Different Integration Schemes", Journal of Electronic Materials, 30(12) 2001.

EFFECTS OF HEAT TREATMENT IN AIR ENVIRONMENT ON THE DISPERSIVITY OF NANODIAMOND

Menggang Lu[1], Xiaoguang Guo[1], Song Yuan[1], Zhuji Jin[1], Renke Kang[1], Dongming Guo[1]*

[1] School of Mechanical Engineering, Dalian University of Technology, Dalian 116024, China
*Corresponding Author's Email: guoxg@ dlut.edu.cn

ABSTRACT

Heat treatment of nano-diamond (ND) in air can alter the ND surface chemistry and improve its dispersion. The relationship between the heat treatment temperature and ND surface groups was studied and the effects of different heat treatment conditions on ND size distribution and Zeta potential were elucidated in this paper. The results showed that suitable heat treatment could increase the number of hydrophilic groups as well as the absolute value of Zeta potential on the ND surface, and improve the dispersion and dispersion stability of ND.

INTRODUCTION

In recent years, due to its excellent physical, chemical and biological properties, nano-diamond (ND) has shown more and more promising applications in grinding and polishing, semiconductor, composite materials and biomedicine. Detonation nano-diamond (DND) and high-temperature and high-pressure nano-diamond (HTHP ND) are the most commonly used production methods.

DND has been widely discussed because of its uniform primary particle size (about 5nm). However, owing to the production method, DND is always accompanied by some lattice defects and other elemental impurities, which affects its performance [1]. HTHP ND is considered to have more uniform structure and fewer lattice defects [2]. Also, the cost of HTHP ND with larger particle size is lower. Therefore, HTHP ND is getting more and more attention.

Nevertheless, the application of ND is seriously limited due to its large primary particle size and serious agglomeration. Sanju Gupta et al. used salt beyond the solubility limit to break the agglomeration of ND with ultrasonic energy. This technique is called salt-assisted ultrasonic de-agglomeration (SAUD), and it is expected to prepare stable nano-diamond colloidal solution [3]. Stepan Stehlik et al. reduced the size of DND through the "combustion" of nano diamond itself at 520℃, then carried out high-power ultrasonic dispersion and ultracentrifugation, and finally obtained the DND with an average size of 1.4nm [4]. But after combustion and ultracentrifugation, a lot of ND will be lost. Muhammad Khan et al. used different grinding media (NH_4HCO_3, NaCl, sucrose) to ball mill DND, and investigated the effects of different grinding media on the de-agglomeration and dispersion of DND in different

solvents. The results showed that the performance of NaCl as grinding media was the best one [5]. Salt-assisted milling introduces impurities and NaCl, which will bring troubles to the subsequent separation of DND.

The heat treatment of HTHP ND in air is a green and pollution-free method. The process is relatively simple, not only avoids introducing impurities, but also removes impurities such as graphite in ND [6]. Therefore, the dispersion behavior of HTHP ND under different temperature heat treatment, and the reason for the increase of dispersion was studied in this paper.

MATERIAL AND METHODS

We used the commercial HTHP ND (supplier: Zhengzhou Zhongnan jet superabrasive Co.) with a nominal size of 250nm. The HTHP ND is annealed in air for 5h at 300℃, 400℃, 420℃, 440℃, 460℃, 480℃, 500℃, respectively.

Thermogravimetry and differential thermal analysis (TG-DTA) was used to study the thermal stability of HTHP ND raw materials in order to obtain a suitable temperature range for heat treatment.

The particle size and surface Zeta potential of the ND suspension were measured on a nanoparticle size and potentiometer (ZEN 3690, Malvern Co. UK).

X-ray diffractometer (XRD-6000, SHIMADZU Co. Japan) was used to measure the XRD pattern.

The composition and relative content of ND surface elements were analyzed by XPS. The test instrument used is an ESCALAB XI+ X-ray photoelectron spectrometer from Thermo Corporation of the UK.

The surface chemistry of HTHP ND before and after annealing was studied by Fourier transform infrared spectroscopy (FTIR). HTHP ND powder and anhydrous KBr were ground in an agate mortar with a mass ratio of 1:50-100 for 10min, pressed and then put into a Fourier transform infrared spectrometer (6700, thermofisher Co. USA) for measurement.

RESULTS AND ANALYSIS

Thermal stability analysis

The results of thermogravimetry and differential thermal analysis (TG-DTA) are shown in Figure 1. As can be seen from the TG curve, there is a slow mass loss at 500℃, and when the temperature exceeds 500℃, the sample quality drops sharply and continues to about 1200℃. The mass loss of ND reached 99.89% at about

1200℃, indicating that ND was completely oxidized. In order not to cause a large mass loss, the temperature of the heat treatment should not exceed 500℃.

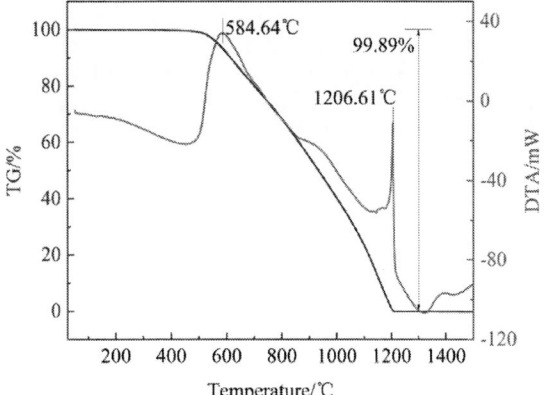

Figure 1: TG-DTA curves of HTHP ND raw material

It can be seen from the DTA curve that the HTHP ND sample is endothermic at the beginning of the heating process, that is, when the temperature is low, the desorption of adsorbed water on the ND surface and the change of some functional groups are endothermic. Obviously，there are exothermic peaks at about 580℃ and 1200℃, which indicates that the intense oxidation (combustion) of ND is an exothermic process.

Particle size and surface potential analysis

The average particle size measured by dynamic laser scattering (DLS) method is shown in Figure 2. It can be seen that after ultrasonic dispersion, the average particle size of each sample is about 250 nm. However, the lowest average particle size can be measured only when the sample is heat-treated at 420℃, which indicates that heat treatment at appropriate temperature can reduce the average particle size to some extent.

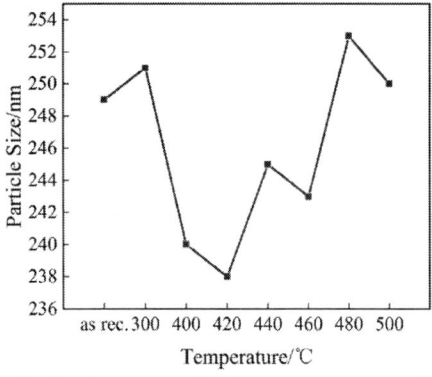

Figure 2: Particle size after heat treatment at different temperatures

As shown in Figure 3, we also measured the Zeta potential of HTHP ND surface in solution and continuously tracked the change of average particle size with time. The heat treatment at 420°C can increase the absolute value of Zeta potential of HTHP ND and enhance the electrostatic repulsion of ND in solution, which is conducive to the improvement of dispersion and dispersion stability, and keep the average particle size of ND solution at a low level for a long time.

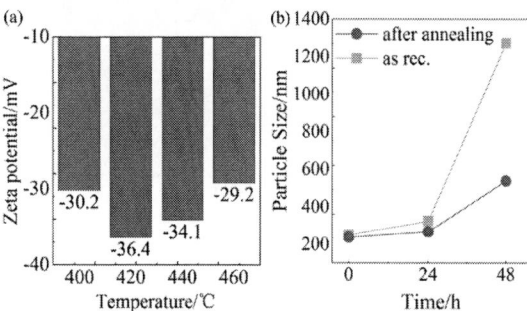

Figure 3: Zeta potential after heat treatment at different temperatures (a) and average particle size change with time (b)

Structure and surface analysis

The XRD patterns of raw materials and HTHP ND after heat treatment at 420℃ are shown in Figure 4, The diffraction patterns of the two powders showed three diffraction peaks at 2θ=44°, 75°, and 91°, corresponding to the three crystal faces of (111), (220), and (311) of ND respectively. Because HTHP ND has a large primary particle size, its XRD pattern is basically the same as that of bulk diamond. The peak is narrow and pointed, and the bottom is small, indicating that HTHP ND has good crystallinity. The XRD spectrum after heat treatment is consistent with that before treatment, which shows that the diamond structure of ND will not be changed by heat treatment at 420℃.

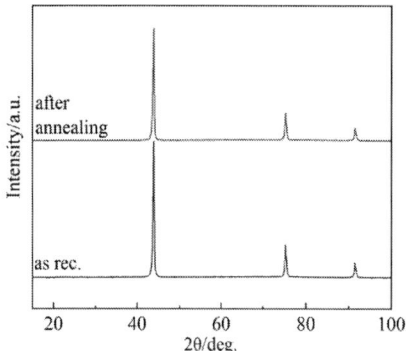

Figure 4: The XRD patterns

The results of XPS analysis is given in Table 2. According to XPS analysis, the HTHP ND sample contained 90.97 at.% C and 9.03 at.% O, after annealing in air, oxygen content increased to 9.72 at.%.

TABLE I. THE XPS RESULTS

XPS	*As rec.*	*After annealing at 420℃*
C1s	90.97 at.%	90.28 at.%
O1s	9.03 at.%	9.72 at.%

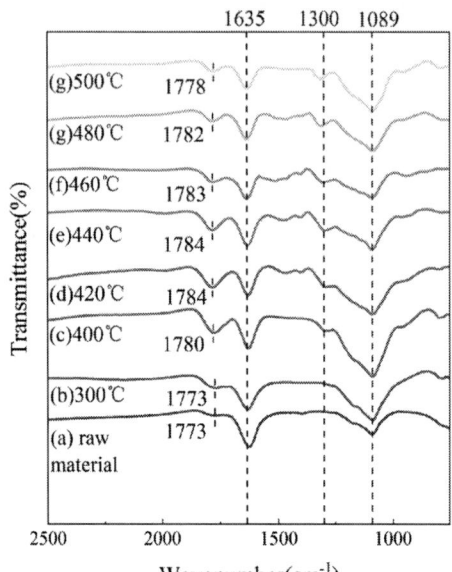

Figure 5: The FTIR spectra

The FTIR spectrum of HTHP ND after heat treatment at various temperatures are shown in Fig. 5. The spectral region of 1770-1790cm^{-1} is assigned to the stretching vibration of carbonyl C=O, and its peak intensity reaches the strongest at 420℃, and it moves from 1773cm^{-1} to 1784cm^{-1}, indicating that ketone, aldehyde, ester and other substances on ND surface are converted into carboxylic acid, anhydride or cyclic ketone. The dissociation of these groups, especially carboxyl groups, leads to negative Zeta potential, which is consistent with the measurement of the Zeta potential. The peak at 1635cm^{-1} is due to the OH bending vibration in absorbed water. The peaks at 1089cm-1 and 1300cm-1 is associated with the stretching vibration of the ether bond C-O-C, because the ether bond may form on the surface of a single ND or between ND particles, which promotes the aggregation of ND. Therefore, we can see that the strength of carbonyl peak is not much different at 400℃ and 420℃, but the average particle size is smaller at 420℃, probably because the formation of ether bond between ND particles weakens

the improvement of dispersion brought by the enhancement of carbonyl group to some extent.

CONCLUSION

By comparing the dispersion behavior and surface chemistry of HTHP ND after heat treatment at different temperatures, we found that heat treatment of HTHP ND in air at 420°C for 5 h can increase the carboxyl content of ND surface. Due to the increase of hydrophilic groups on the ND surface and the increase of the absolute value of the Zeta potential, the electrostatic repulsion is enhanced, and the dispersion and dispersion stability of HTHP ND are improved. The heat treatment at this temperature has substantially no mass loss, will not change the ND structure, and can improve the oxygen content of ND surface to a certain extent. Therefore, heat treatment in air can be used as an efficient, convenient and pollution-free method to improve the dispersion of HTHP ND.

ACKNOWLEDGEMENTS

The authors greatly appreciate the financial support of Science Fund for Creative Research Groups (No.51621064) and National Natural Science Foundation of China(No. 51575083).

REFERENCES

[1]. Woodhams, B., et al., Graphitic and oxidised high pressure high temperature (HPHT) nanodiamonds induce differential biological responses in breast cancer cell lines. Nanoscale, 2018. 10(25): p. 12169-12179.

[2]. Salava, J., et al., Influence of air annealing on the luminescence dynamics of HPHT nanodiamonds. Diamond and Related Materials, 2016. 68: p. 62-65.

[3]. Gupta, S., et al., Salt-Assisted Ultrasonicated De-Aggregation and Advanced Redox Electrochemistry of Detonation Nanodiamond. Materials, 2017. 10(11): p. 1292.

[4]. Stehlik, S., et al., High-yield fabrication and properties of 1.4 nm nanodiamonds with narrow size distribution. Scientific Reports, 2016. 6(1).

[5]. Khan, M., et al., Dispersion behavior and the influences of ball milling technique on functionalization of detonated nano-diamonds. Diamond and Related Materials, 2016. 61: p. 32-40.

[6]. Stehlik, S., et al., Size and Purity Control of HPHT Nanodiamonds down to 1 nm. The Journal of Physical Chemistry C, 2015. 119(49): p. 27708-27720.

EFFECT OF COMPLEXING AGENT IN SLURRY ON CMP PROPERTY FOR BARRIER MATERIAL COBALT

Jinsong Zuo, Fang Wang*, Kai Hu, Luguang Wang, Yujie Yuan, Kailiang Zhang*

Tianjin Key Laboratory of Film Electronic & Communication Devices, School of Electrical
&Electronic Engineering, Tianjin University of Technology, Tianjin 300384, China.
*Corresponding Author's Email: kailiang_zhang@163.com fwang75@163.com

ABSTRACT

Recently, Cobalt (Co) is focused as the barrier material for the next-generation copper interconnects technology, owing to its superiority in anti-diffusibility for barrier effects and adhesion. In this work, Chemical Mechanical Polishing (CMP) samples were prepared by Direct current (DC) magnetron sputtering, CMP tests were done by the polisher (nspire-6EC from Revasum Company), the thickness of Co films pre-CMP and post-CMP were measured by step-profiler and their surface roughness were characterized by Atomic Force Microscope (AFM). In order to optimize the CMP slurry recipe, the effects of different complexing agents (including citric acid and acetic acid) on material removal rate (MRR) and root mean square (RMS) roughness are firstly compared. And the concentration effects of complexing agents on MRR and RMS roughness were also discussed. CMP results show that the optimized MRR could reach 457.8 nm/min when 0.4 wt% acetic acid was used as complexing agent and the number of RMS roughness would decrease from 8.19 nm to 5.98 nm, which confirms that acetic acid is a more suitable complexing agent for CMP slurry of Co.

INTRODUCTION

IC (Integrated Circuit) has advantages of excellent performance and high reliability, benefitting from Cu interconnection structure [1] which is made by Damascus process and needs multiple CMP [2]–[5] processes. In the whole CMP process, the barrier layer is a critical step which directly determines the performance and yield of IC devices. With the development of microelectronics technology, TaN/Ta [6] exhibit good behavior as adhesive/barrier in the 32 nm technology of Ultra Large Scale Integration (ULSI). But feature size shrinking result in traditional barrier layer materials can't satisfied the performance of the device, when the device size comes to 14nm, innovative barrier layer materials need to be explored urgently. Co is chosen to be the barrier layer of next generation, owing to its superiority in the anti-diffusibility for barrier effects and adhesion [7]. In addition, Co has the physical properties of low resistivity ($6.3\mu\Omega{\cdot}cm$) and high thermal stability, on which Cu can be electroplated directly [8].

In this work, the effects of different complexing agents on MRR and RMS roughness have been researched.

H_2O_2 was used as oxidant here, because there is no existence of impurity metal ions. To obtain better surface topography and roughness, acetic acid and citric acid as complexing agents were compared. The concentration effects of complexing agent on surface roughness was also discussed.

EXPERIMENT

DC magnetron sputtering 1000nm Co film used for CMP process was deposited on 4-inch silicon wafers at room temperature with metal targets. The slurry was composed of abrasives (5 wt% colloidal silica with average particle size of ~60 nm), oxidant (1 wt% H_2O_2), complexing agent (acetic acid and citric acid) and deionized (DI) water. All the reagents used in the experiment are pure analytical reagents. The slurry flow rate was maintained at 150 ml/min, the down force was 1 psi, the table/chuck rotate speed was 60 rpm, and polishing time was fixed at 30s. After polishing, the surface of film was rinsed with DI water and blown dry with nitrogen. The thickness and surface topography of Co films pre-CMP and post-CMP were measured by step-profiler and AFM respectively.

Table 1. Parameter ranges of CMP for Cobalt thin film

Polishing Parameter	value
Down force (psi)	1
Table/chuck rotate speed (rpm)	60/60
slurry flow (ml/min)	150
Slurry pH	4~10
concentration of citric acid (mol/L)	0.01~0.06
concentration of acetic acid (wt%)	0.2~1.0

RESULTS AND DISCUSSION

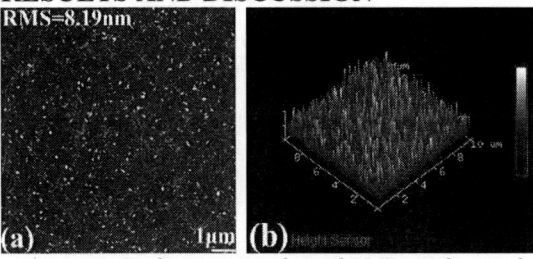

Figure 1: Surface topography and RMS roughness of pre-polishing Co films (a) 2D image and (b) 3D image

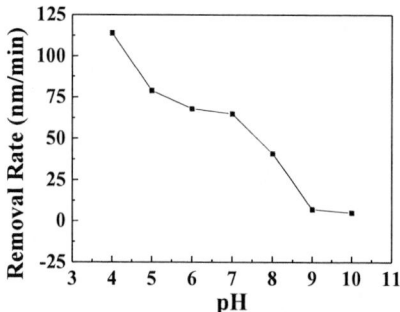

Figure 2: Relationship of removal rate and pH

As shown in Figure 1, the surface of samples fluctuated with a large RMS roughness (8.19 nm). Before the experiment, the pH value of slurry was adjusted by KOH and HNO₃, which consists of 5 wt% colloidal silica and 1 wt% H_2O_2. RR was measured among the pH range of 4 ~ 10 and the results are shown in Figure 2. With the pH increasing, the RR reduced gradually approaching to zero. The result speculates that as the pH increasing there is a passivation film attached on the surface of the Co film which is difficult to polishing under the condition of a certain concentration H_2O_2.

Effect of citric acid concentration on cobalt removal rate and RMS roughness.

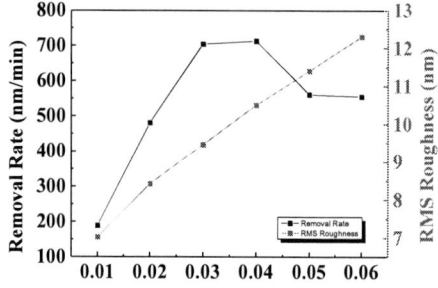

Figure 3: Relationship of removal rate and RMS roughness with citric acid concentration

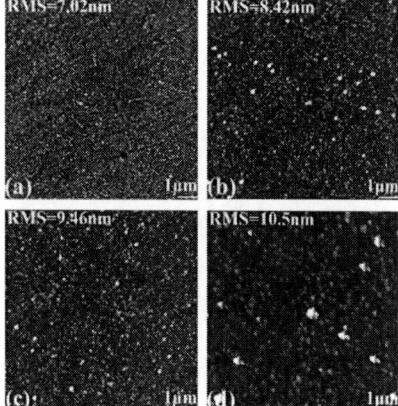

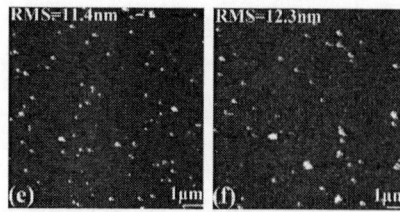

Figure 4: The surface topography of post-polishing Co film under the citric acid concentration of (a) 0.01 mol/L, (b) 0.02 mol/L, (c) 0.03 mol/L, (d) 0.04 mol/L, (e) 0.05 mol/L, (f) 0.06 mol/L

The effect of citric acid concentration on RR and RMS roughness of Co films was investigated under the circumstance of citric acid chosen as complexing agent. 5 wt% of colloidal silica and 1 wt% H_2O_2 was added in the slurry which selected is based on recently report [9] , the results are shown in Figures 3, 4.

As shown in Figure 3, RR peaked at 721.4 nm/min when the concentration of citric acid is 0.04 mol/L. RR is positive linear to concentration which the range is 0.01 ~ 0.03 mol/L, RR is gradually stable when the concentration is higher than 0.04 mol/L.

Its inferred that when the citric acid concentration is lower than the peak value, Co^{2+} can be fully complexed with citric acid further removed by mechanical action [10]. Correspondingly when the citric acid concentration is higher than the peak value, the formed complex can't be completely removed by mechanical action.

Relationship of RMS roughness and citric acid concentration was shown in Figures 4. The RMS roughness was about 7.02 nm when the concentration of citric acid was 0.01 mol/L consequently enhanced along with the concentration of citric acid increased, which proved RMS roughness wasn't improved with the addition of citric acid.

Effect of acetic acid concentration on cobalt removal rate and RMS roughness.

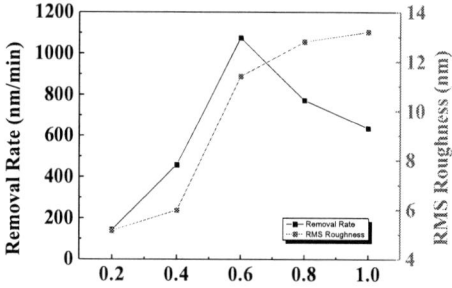

Figure 5: Relationship of removal rate and RMS roughness with acetic acid concentration

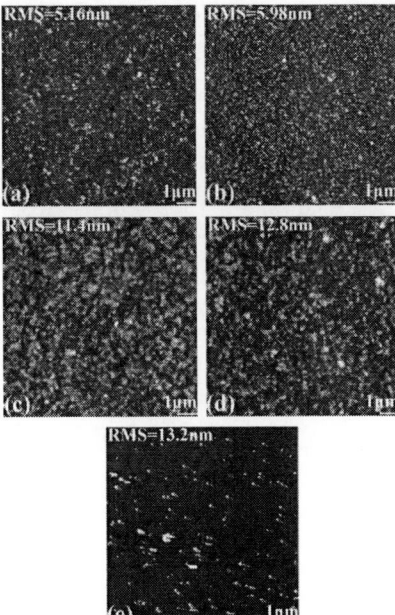

Figure 6: The surface topography of post-polishing Co film under the acetic acid concentration of (a) 0.2 wt%, (b) 0.4 wt%, (c) 0.6 wt%, (d) 0.8 wt%, (e) 1.0 wt%

Series of experiments were performed under the condition that type of complexing agent was changed in order to achieve a higher removal rate and obtain better surface roughness.

It was found that when acetic acid was utilized as complexing agent, the surface topography was improved. When the acetic acid concentration was 0.4 wt%, the surface roughness of the cobalt film was reduced from 8.19 to 5.98, and the removal rate reached 457.8 nm/min (Figures 5, 6). When the concentration reached to 0.6wt%, the removal rate of maximum value can be obtained. As the concentration reached to 1.0 wt%, the surface roughness of 13.2 nm was acquired. The result demonstrates that low concentration of acetic acid can reduce RMS effectively.

CONCLUSION

Exploration on slurry including two different concentration of complexing agents on the Co-CMP has been carried out in this manuscript. It was found that low concentration of acetic acid is more adaptable for Co-CMP. Polishing experiments can be concluded by adjusting the concentrations of two complexing agents under adaptable process parameters. The acetic acid is more appropriate for Co-CMP than citric acid, the removal rate stabilizes at 457.8 nm/min and the RMS roughness is reduced to 5.98 nm when the concentration of acetic acid is 0.4wt%. The mechanism of Co-CMP needs further research in order to get better application in copper interconnect technology.

ACKNOWLEDGEMENTS

This work was supported by Natural Science Foundation of Tianjin City (Grant Nos.18JCZDJC30500, 17JCYBJC16100 and 17JCZDJC31700)，National Key Research and Development Program of China (Grant No.2017YFB0405600) and National Natural Science Foundation of China (Grant Nos.61404091, 61274113, 61505144, 51502203 and 51502204).

REFERENCE

[1] P. C. Andricacos, C. Uzoh, J. O. Dukovic, J. Horkans, and H. Deligianni, "Damascene copper electroplating for chip interconnections," *IBM J. Res. Dev.*, 1998.

[2] R. Huo, F. Wang, Y. Feng, Y. Han, Y. Yuan, and K. Zhang, "Optimization of slurry and process parameter on chemical mechanical polishing of CR-doped Sb2Te3 thin film," in *China Semiconductor Technology International Conference 2017, CSTIC 2017*, 2017.

[3] K. Zhang *et al.*, "Optimization and mechanism on chemical mechanical planarization of hafnium oxide for RRAM devices," *ECS J. Solid State Sci. Technol.*, 2014.

[4] K. Zhang, T. Zhang, F. Wang, Y. Yuan, and Y. Li, "Exploration on chemical mechanical planarization of ZnO functional thin films for novel devices," *Microelectron. Eng.*, 2013.

[5] L. Wang, F. Wang, Y. Li, J. Huang, W. Li, and K. Zhang, "Optimization on chemical mechanical planarization of chromium doped antimony telluride (Cr-SbTe) for PCM Devices," in *China Semiconductor Technology International Conference 2019, CSTIC 2019*, 2019.

[6] M. Lane, R. H. Dauskardt, N. Krishna, and I. Hashim, "Adhesion and reliability of copper interconnects with Ta and TaN barrier layers," *J. Mater. Res.*, 2000.

[7] N. A. Lanzillo *et al.*, "Exploring the Limits of Cobalt Liner Thickness in Advanced Copper Interconnects," *IEEE Electron Device Lett.*, 2019.

[8] J. Zhou, X. Niu, Z. Wang, Y. Cui, J. Wang, and R. Wang, "Study on effective methods and mechanism of inhibiting cobalt removal rate in chemical mechanical polishing of GLSI low-tech node copper film," *ECS J. Solid State Sci. Technol.*, 2019.

[9] H. S. Lu *et al.*, "The Effect of H 2O 2 and 2-MT on the chemical mechanical polishing of cobalt adhesion layer in acid slurry," *Electrochem. Solid-State Lett.*, vol. 15, no. 4, pp. 97–100, 2012.

[10] R. Popuri *et al.*, "Citric acid as a complexing agent in chemical mechanical polishing slurries for cobalt films for interconnect applications," *ECS J. Solid State Sci. Technol.*, 2017.

EFFECT OF POTASSIUM OLEATE AS INHIBITOR ON COPPER CHEMICAL MECHANICAL POLISHING

Chenghui Yang[1,2], Xinhuan Niu[1,2*], Jiakai Zhou[1,2], Zhaoqing Huo[1,2] and Yanan Lu[1,2]*

[1] School of Electronics and Information Engineering, Hebei University of Technology Tianjin 300130, People's Republic of China

[2] Tianjin Key Laboratory of Electronic Materials and Devices, Tianjin 300130, People's Republic of China

1058108928@qq.com, xhniu@hebut.edu.com

ABSTRACT

The effect of potassium oleate (PO, $C_{18}H_{33}KO_2$) in a glycine-based weakly alkaline slurry on copper chemical mechanical polishing (CMP) process was discussed. The corrosion inhibitor in the slurry could balance the over etching to realize the global planarization of the copper layers. The experimental results verified PO was indeed effective in inhibiting copper removal rate. The corrosion and passivation mechanism were also discussed. SEM and XPS test results confirmed that PO can adsorb on the copper surface to form a passivation film.

Keywords—Potassium oleate; corrosion inhibitor; global planarization; removal rate; chemical mechanical planarization (CMP)

INTRODUCTION

With the development of giant-large scale integrated circuits (GLSI), the feature size drops down to 14 nm and even smaller, and the manufacturing process becomes more and more complex[1]. Copper (Cu) with its highly favorable electrical and electro migration resistance characteristics has been chosen to be a substitute for aluminum in interconnections. In the conventional Cu CMP, 1-benzotriazole (BTA) is the normally used corrosion inhibitor, which can be strongly absorbed on the copper surface and promote realization of planarization[2]. Many literatures have claimed that BTA could be used as an effective inhibitor to reduce the etching rate and thus avoid isotropic etching on the Cu surface and ensure good surface planarization in aqueous acidic, neutral, and alkaline solutions[2,3,4]. However, BTA has great damage to the environment and will cause trouble for following other processes. Hence, it is necessary to develop novel environment friendly inhibitors used in semiconductor and microelectronics industries.

Previous studies reported strong binding between cobalt (Co) and potassium oleate (PO) which has a longer 18-carbon hydrophobic chain and a hydrophilic carboxylate end group[5,6,7]. Because of that, PO is chosen as a dissolution and corrosion inhibitor to inhibit the Cu corrosion, which is proposed to substitute BTA in copper CMP slurry. X-ray photoelectron spectroscopy (XPS) and electrochemistry methods were presented to investigate the inhibitor property of PO for the copper layer[8]. Figure 1 shows the formula of PO.

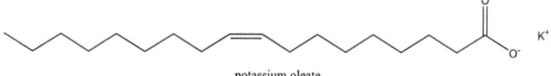

potassium oleate

Figure 1: The formula of PO

EXPERIMENTAL

The slurry contains colloidal silica (mean particle size about 100 nm), complexing agent glycine, oxidizing agent H_2O_2 (the mass fraction of 30 wt %), inhibitor agent PO, pH was adjusted by KOH.

3 inch diameter Cu blanket wafer and patterned wafer were polished on Alpsitecin-E460E polisher made in France. IC 1000 polishing pad was used in the all polishing experiments. The polishing parameters were shown in Table Ⅰ. All experiments were performed in a super clean laboratory controlled at 25℃. After polishing, every wafer was taken out, washed in purified de-ionized (DI) water, and dried in N_2 air steam.

Table I. Process parameters of CMP

Parameters	Conditions
Down force(psi)	1.5
Back pressure(psi)	0
Polishing head speed(r/min)	87
Platen rotation speeds(r/min)	93
Slurry flow rate(ml/min)	300

The removal rates of Cu were determined by measuring the difference of film thickness before and after polishing by an aver 333A four-point probe. Each specimen was weighed several times to get the average. A CHI660E electrochemical workstation with a three-electrode glass cell was used to acquire dynamic potential curves (Tafel plots). A high-resolution stylus profiler Ambios XP-2 was used to measure the step height before and after polishing. Scanning electron microscopy (SEM, Zeiss Sigma 500/VP) was performed to identify the effect of PO on Cu wafer surface. XPS was applied to analyze the surface characteristics of Cu films.

RESULTS AND DISCUSSION

The effects of PO concentration on Cu electrochemical properties were shown in Figure 2. It can be seen that with the PO concentration increasing from 0 g/L to 0.5 g/L the *Icorr* (corrosion current density) reduced and the *Ecorr* (corrosion potential) rose, which indicated the reactions occurring on Cu surface was inhibiting. It can be inferred that the introduction of PO may inhibit the removal rate of Cu. Under the guidance of electrochemical experiment results, the effects of PO concentration on Cu RR were studied and the results were shown in Figure 3. It indicated that with the PO concentration increasing from 0 g/L to 0.5 g/L, Cu RR dramatically descended from 15626 Å/min to 6406 Å/min. This result is consistent with the electrochemical conclusions, confirming that PO can indeed act as a corrosion inhibitor for Cu.

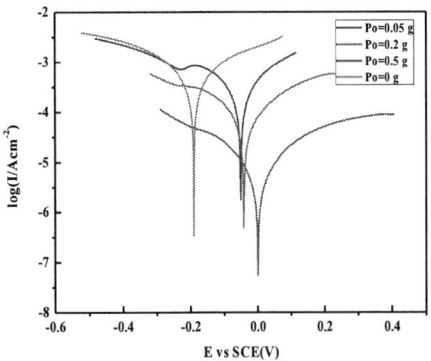

Fig 2. Polarization curves of slurry with different concentration of PO.

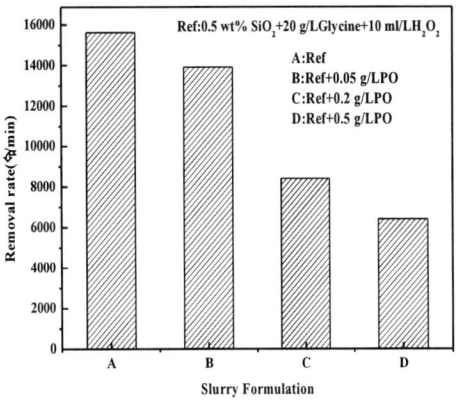

Fig 3. Effect of PO concentration on Cu removal rate.

In order to figure out the reaction mechanism of Cu

during CMP, Cu surface topographies scanned by SEM after polishing by the slurry without and with PO were shown in Figure 4. Compared with Figure 4a, with the introduction of PO, the number of corrosion pits significantly reduced and the surface became smoother (shown in Figure 4b). It indicated that with the introduction of PO, the reactions occurring on Cu surface was inhibited. Hence, PO surfactant molecular preferentially adsorbed on the copper surface and a protective layer was formed, which can prevent the chemical agent contacting with copper wafer, and inhibit the chelation of glycine in a certain extent. At the same time, surfactant as a surfactant has a good encapsulation effect on particles and products, which reduces the roughness of the surface.

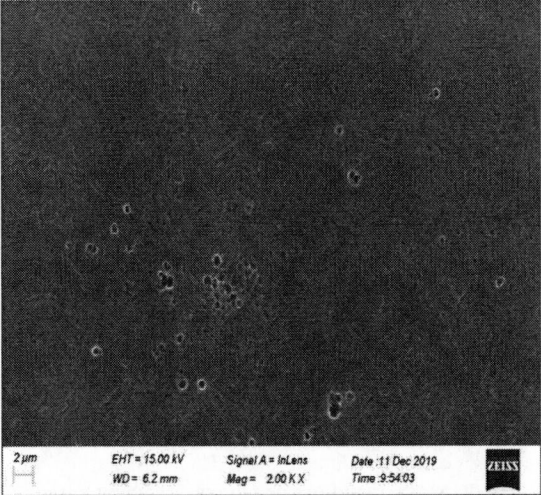

(a)

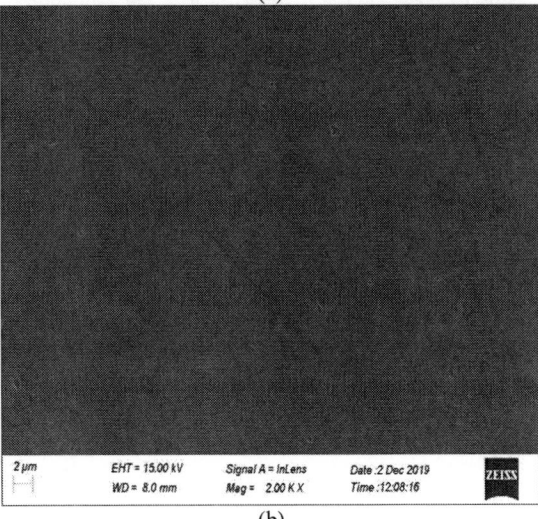

(b)

Figure 4: The SEM measurements results. (a) pH=9.5 H_2O_2+glycine (b) pH=9.5 H_2O_2+glycine+PO

To further confirm the previous conjecture, Figure 5

shows the XPS Cu 2p spectra after immersing in solution containing PO. It can be seen that compared to the Ref, the quantity of Cu 2p decreased dramatically with the introduction of PO. This result indicated that PO surfactant molecular preferentially adsorbed on the copper surface and a protective layer was formed, which leads to the drop in quantity of Cu 2p. It was consistent with the results of the above discussion.

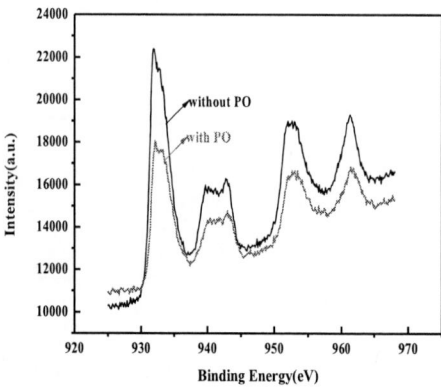

Figure 5. Cu 2p spectra of Cu surface after immersing in solution containing PO.

CONCLUSIONS

Polishing behavior of Cu using PO as corrosion inhibitor was investigated. It can be concluded that in a glycine-based weakly alkaline slurry, PO has a strong inhibiton effect on Cu corrosion. The corrosion inhibitor PO in the slurry could balance the over etching to realize the global planarization of the Cu layers. The reduction of removal rate was due to PO's adsorption on the surface of Cu film to restrict the reactions, which was verified by SEM and XPS measurements. It has important reference value for the research of weakly alkaline Cu CMP in the future.

ACKNOWLEDGMENTS

This work was supported by the Major National Science and Technology Special Projects (No. 2016ZX02301003-004-007), Natural Science Foundation of Tianjin China (16JCYBJC16100, 18JCTPJC57000), and the Key Laboratory of Electronic Materials and Devices of Tianjin, China. The authors also thank the teachers and classmates for their helpful suggestions.

REFERENCES

[1] J. K. Zhou, X. H. Niu, J. C. Wang, K. Zhang, Y. Q. Cui, Z. Wang. *2019 China Semiconductor Technology International Conference (CSTIC), Shanghai, China*, 2019, pp. 1-3.
[2] C. Yan, Y. Liu, J., Zhang, C. Wang, W. Zhang, P. He, G. Pan. *ECS Journal of Solid State Science and Technology,* vol.6, pp.1–6, 2016.
[3] Y. Hong, V. K. Devarapalli, D. Roy, S. V. Babu. *Journal of The Electrochemical Society,* vol.154, H444, 2007.
[4] B.-J. Cho, S. Shima, S. Hamada, J.-G. Park. *Applied Surface Science,* vol.384, pp.505–510, 2016.
[5] R. Popuri, H. Amanapu, C. K. Ranaweera, S. V. Babu. *ECS Journal of Solid State Science and Technology,* vol.6 pp.845–852, 2017.
[6] C. K. Ranaweera, N. K. Baradanahalli, R. Popuri, J. Seo, S. V. Babu. *ECS Journal of Solid State Science and Technology,* vol.8 pp.3001–3008, 2018.
[7] V. S. Molchanov, Y. A. Shashkina, O. E. Philippova, A. R. Khokhlov. *Colloid Journal,* vol. 67, pp.606–609, 2005.
[8] S. Kondo, Y. Ichige, Y. Otsuka. *Japanese Journal of Applied Physics,* vol.56, 07KA01. 2017.

EFFECT OF CHELATORS ON THE REMOVAL OF BTA IN POST-CMP CLEANING

Mengrui Liu[1,2], Baimei Tan[1,2*], Xiaoqin Sun[1,2], Pengcheng Gao[1,2], Qi Wang[1,2], Siyu Tian[1,2]

[1]School of Electronic Information Engineering, Hebei University of Technology,
Tianjin 300130, China
[2]Tianjin Key Laboratory of Electronic Materials and Devices, Tianjin, China
*Email: bmtan@hebut.edu.cn

ABSTRACT

During Cu CMP process, BTA is widely used as a corrosion inhibitor. After Cu CMP process, the residual BTA leads to hydrophobic copper surface, which seriously affects the device stability. Therefore, BTA, as the main organic residue on Cu surface, needs to be effectively removed in the post-CMP cleaning process. In this paper, the removal of organic residual contaminants BTA on polished copper surface by different concentrations of chelating agent FA/OII and ethylene diamine tetraacetic acid (EDTA) was studied. The removal of organic residues was characterized by contact angle measurement and electrochemical techniques. The results show that the cleaning effect of 2mM EDTA is better than 200ppm FA/OII.

Keywords—FA/OII chelating agent; EDTA; Cu-BTA; Post-CMP cleaning

INTRODUCTION

Chemical mechanical Planarization (CMP) is one of the key steps in IC manufacturing. Copper is widely used as an interconnect material due to its excellent high conductivity, and high resistance to electromigration[1]. In the Cu CMP process, it is necessary to add inhibitor benzotriazole (BTA) into the polishing slurry to form Cu-BTA passivation film on the copper surface to protect the concave copper surface from chemical corrosion, so as to achieve global or local planarization[2]. BTA can easily react with copper ions on the copper surface to form a thin Cu-BTA polymer layer [3], which is adsorbed on the surface of the wafer and extremely difficult to be desorbed[4]. As the residual BTA makes the surface highly hydrophobic in nature, which causes severe drying issues and also poor adhesion of the stacking layers[5,6]. Therefore, BTA residue must be removed as a serious defect in the post-Cu CMP cleaning process. The cleaning and removal of BTA has become the focus of research.

In this paper, the effects of two kind of chelating agent on removal BTA contamination were compared and analyzed. The contact angle and electrochemical experiment were carried out after BTA-contaminated Cu samples cleaning with different chelating agents, and the removal effect was characterized by the change of contact angle and the magnitude of corrosion current.

EXPERIMENT

Commercial 8inch copper blanket wafer was used for the cleaning experiments. Because that the copper wafer is exposed in the air, it is easy to react with the water and oxygen in the air to form a dense oxidation layer. The copper surface was treated with 25mM citric acid to remove the surface oxides and the fresh copper wafer were immersed in saturated BTA solution for 10 min to be contaminated with BTA. For the cleaning experiment, BTA-contaminated Cu samples were immersed with different concentrations of chelating agent for 10 minutes, then washed with PVA brush for 3 minutes and dried with N_2. KOH was used as the pH regulator to adjust the pH to 10.5. Cu disks (99.99% purity) were sealed in epoxy with an exposure area of 6 mm in diameter for electrochemical experiments. The effect of chelating agent with different concentration in cleaning solution on the removal of organic residues was studied by contact angle measurement and electrochemical technology.

RESULTS AND DISCUSSIONS

The effect of FA/OII concentration on BTA removal

The alkaline chelating agent (FA/OII, shortened as R $(NH_2)_2$) is a multi-hydroxyl and multi-amine organic molecule, which contains more than 13 chelating rings with strong chelating ability for metal ions.

The contact angle of pure copper is 25°, indicating that copper was hydrophilic. The contact angle is changed above 70° after the copper sample was immersed in BTA. The contact angle results of copper surface cleaned by chelating agent with concentration of 100, 150, 200, 250 and 300ppm are shown in Figure 1. It can be seen that when the BTA-contaminated Cu samples were cleaned with different concentrations of FA/OII, the contact angle decreases significantly. When the FA/OII concentration is 200 ppm, the contact angle is decreased from 77.5°to 41.9°, and it is considered that the removal effect of BTA is the optimal at this concentration. It indicates that FA/OII can effectively complex copper ions, break the chemical bond between Cu-BTA complexes, and achieve the stripping removal of BTA.

In this experiment, 0.05mol/L potassium sulfate solution was used as the electrolyte. The Cu-BTA passivation film generated by the reaction between BTA and copper surface could effectively inhibit the corrosion of copper surface. From the potentiodynamic curves in Figure 2, it is found that the I_{corr} value of fresh copper is $3.472 \times 10^{-7} A/cm^2$, while the I_{corr} of copper sample contaminated by BTA is $8.481 \times 10^{-8} A/cm^2$, which indicates that BTA has formed an effective passivation film on the copper surface to provide sufficient chemical protection. Therefore, by comparing the corrosion currents of copper samples cleaned with FA/OII chelating agent of different concentrations, the optimal concentration for BTA removal can be determined. Potentiodynamic polarization of copper after cleaning with different concentrations of FA/OII was shown in Figure 3. The corresponding I_{corr} values are listed in Table 1. The copper samples cleaned with FA/OII all showed higher corrosion current density, especially when the concentration of FA/OII was 200 ppm, the I_{corr} value was up to $8.794 \times 10^{-7} A/cm^2$, indicating that BTA was effectively removed. It is consistent with the results of contact angle measurement.

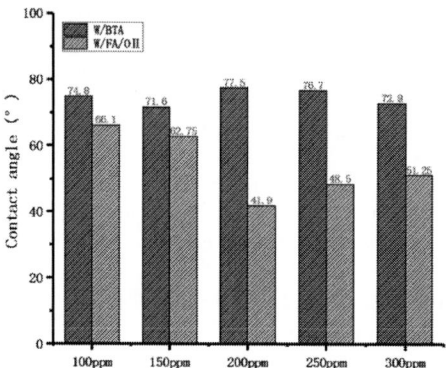

Figure 1: Changes of contact angle of Cu surface before and after cleaning with different concentrations of FA/OII.

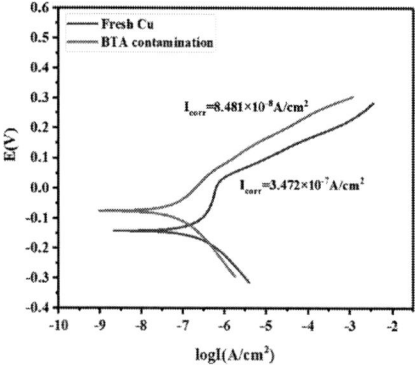

Figure 2: Curve of corrosion current density of Cu surface after BTA contamination.

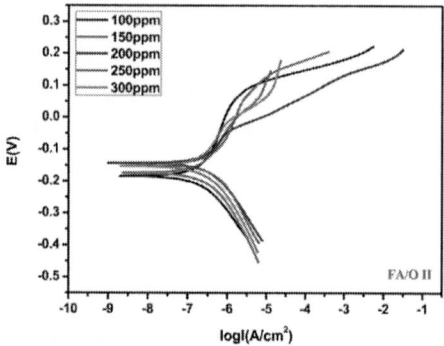

Figure 3: Potentiodynamic polarization curves of Cu surface after cleaning with different concentrations of FA/OII.

Table 1: Corresponding corrosion potential E_{corr} and corrosion density I_{corr} of the potentiodynamic polarization curves of the copper electrode in Figure 3.

FA/OII/ppm	$I_{corr}(10^{-7}A/cm^2)$	$E_{corr}(V)$
100	3.609	-0.198
150	5.588	-0.183
200	8.794	-0.149
250	8.504	-0.151
300	7.570	-0.159

The effect of EDTA concentration on BTA removal

EDTA is a chelating agent containing carboxyl and amino functional groups. It has a wide range of coordination abilities and can form stable water-soluble chelates with almost all transition heavy metal ions. However, EDTA is extremely difficult to dissolve in water in the general environment.

The contact angle of Cu surface before and after cleaning with different concentrations of EDTA was measured and the results are presented in Figure 4. When the concentration of EDTA was further increased to 2.0mM, the contact angle decreased significantly from 84.2° to 45.9°. The contact angle measurement results indicate that EDTA can effectively remove BTA adhered on Cu surface, leading to a change of Cu surface from hydrophobic to hydrophilic. Nevertheless, when the chelating agent concentration was higher than 2.0mM, the contact angle increased, implying the optimum BTA removal performance obtained in 2.0mM chelating agent involved cleaning solution.

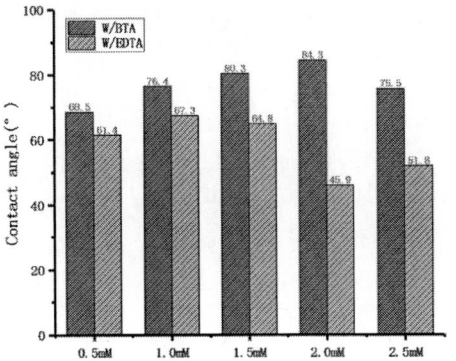

Figure 4: Changes of contact angle of Cu surface before and after cleaning with different concentrations of EDTA.

The electrochemical test results at the concentrations of 0.5, 1.0, 1.5, 2.0 and 2.5mM of EDTA are shown in Figure 5 and the corresponding I_{corr} values are listed in Table 2. The optimal test results show that when the concentration of EDTA is 2.0mM, the electrochemical results are $8.920 \times 10^{-7} A/cm^2$, BTA can be considered to be almost eliminated. The conclusion is the same as that of contact angle measurement.

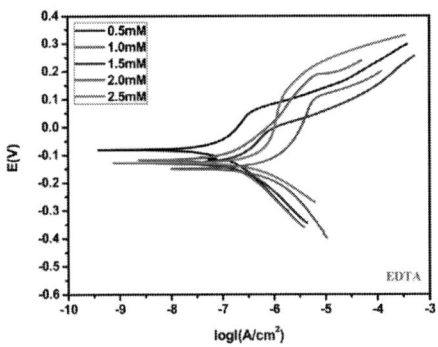

Figure 5: Potentiodynamic polarization curves of Cu surface after cleaning with different concentrations of EDTA.

Table 2: Corresponding corrosion potential E_{corr} and corrosion density I_{corr} of the potentiodynamic polarization curves of the copper electrode in Figure 5.

EDTA/mM	$I_{corr}(10^{-7}A/cm^2)$	$E_{corr}(V)$
0.5	2.438	-0.081
1.0	2.686	-0.128
1.5	5.217	-0.139
2.0	8.920	-0.152
2.5	6.148	-0.135

CONCLUSION

It can be seen from the above results that both 200ppm FA/OII and 2mM EDTA can effectively remove BTA from the copper surface. In comparison, the contact angle change value and corrosion current value of the EDTA chelating agent with the optimal concentration are slightly better than FA/OII, but the corrosion and surface roughness of the copper surface after cleaning need to be further studied.

ACKNOWLEDGEMENTS

This work was supported by the Major National Science and Technology Special Projects (No. 2016ZX02301003-004-007), the Natural Science Foundation, China (No. 61704046), and the Hebei Natural Science Foundation Project (No. F2018202174). The authors thank the teachers and classmates for helpful discussions.

REFERENCES

[1]H-Y Ryu, B-J Cho, N. P. Yerriboina, C-H Lee, J-K Hwang, S. Hamada, et al. *J. ECS Journal of Solid State Science and Technology*, 2019, 8(5):3058-3062.

[2]B. H. Gao, B. M. Tan, Y. L. Liu, C.W. Wang, Y. G. He, Y. Y. Huang. *J. Surface Interface Analysis*, vol.51,2014, pp.566-575.

[3]Y. X. Miao, S. L. Wang, C. W. Wang, Y. L. Liu, M. B. Sun, Y. Chen. *J. Microelectronic Engineering*, vol. 130, 2014, pp. 18–23.

[4]J. Hong, X. H. Niu, Y. L. Liu, Y. G. He, B. G. Zhang, J. Wang, et al. *J. Applied Surface Science*, vol.378, 2016, pp. 239–244.

[5]R. Prasanna Venkatesh, T-Y Kwon, Y. N. Prasad, S. Ramanathan, J-G Park. *J. Microelectronic Engineering*, vol.102,2013, PP.74-80.

[6]L. Yang, B. M. Tan, Y. L. Liu, B. H. Gao, Y. L. Liu, et al. *J. Journal of Semiconductors*,vol.39, 2018,pp.1-5.

HIGH-K METAL GATE AL-CMP WITHIN DIE UNIFORMITY AND SELECTIVITY STUDY

Ziheng_Li [1*], Baicen_Wan[1], Hongdi_Wang [1], Andy Wang[1], Pujia_Shan [1], Zhiyang_Liang[1], Jian_Li[1], Zhijie_Zhang [1]

[1]Semiconductor Manufacturing North China (Beijing) Corp, Beijing, China,
*Corresponding Author's Email: liziheng1025@163.com

ABSTRACT

High-K Metal Gate (HKMG) is one of the most significant steps in CMOS manufacturing for 28nm node process and beyond. For Metal Gate step to be accurately controlled the Chemical-mechanical planarization (CMP) method is required for surface planarization. In Al-CMP, the control of metal residue defect and thickness uniformity were crucial to influence the device and yield performance.

In this work, different slurry was investigated to control different pattern selectivity and within die uniformity. We found that, different selectivity slurry combined with different polish pad and disk have different effect in within die uniformity. With the same pad disk, for Al/Poly selectivity, slurry A was the twice of slurry B, and Al/Oxide selectivity didn't change at the same time. As a result, poly thickness was improved by 7% when gate height meet target, and over polish risk can be reduced. With another Pad/disk, poly thickness can be improved by 10% when Al residue was all removed clear. Besides, chemical rinse treatment were also investigated to remove Al residue..

Keywords—High-K Metal Gate, Al-CMP, Selectivity, Within Die Uniformity

INTRODUCTION

As the MOSFET dimensions have shrunk over the past years for increase speed at constant power density, the gate oxide thickness has historically reduced. The reduce gate oxide thickness caused a serious gate leakage current due to tunneling and gate oxide breakdown [1]. Therefore, replacing the conventional gate oxide with high-k dielectric materials was introduced to reduce gate leakage for further reduce dimension at gate. The first HKMG processor using a RMG approach was manufactured by Intel in 2007[2–4]. For RMG approach, two CMP process, which were poly opening polishing (POP) CMP and Al-CMP, were introduced to RMG process. For Al CMP, the control of metal gate height and defect were crucial to influence the device and yield performance of the HKMG with RMG structures. It is a challenge to have a good balance between metal gate height and metal residue and over polish defect free.

Polishing pad, as a critical role in CMP process, contacts with wafer directly and affects the remove rate, wafer uniformity and defect[5]. The different rigidity, thickness and compression ratio of pad has the different polish performance, especially for WEE performance.

In this research, in order to enlarge the process window of Al-CMP and get a better balance between gate height and defect, different selectivity slurry and different type polish pad were developed.

EXPERIMENT

A standard 5 Zones tool was used for 300mm Al-CMP process by using a three-step polishing approach.

2.1 Slurry selectivity improve test

Three types of control wafers have been used for selectivity test. For Oxide remove rate rest, 300mm blank wafers were constructed with annealed High Aspect Ratio Process (HARP) oxide layer. For Al remove rate, also 300mm blank wafers were constructed with standard Al film which was constructed by PVD process. For poly remove rate test, a standard blank poly film was prepared by diffusion process.

Two different slurry(shown in table I) with different formula was developed for selectivity improve. In Al polish slurry, the oxidizer and inhibitor was two key components to decide slurry remove rate and selectivity.

TABLE I.   PARAMETERS OF TWO DIFFERENT SLURRY

slurry	Component		
	Abrasive	*Oxidizer*	*Inhibitor*
Slurry A	Alumina	Oxidizer A	Inhibitor A
Slurry B	Alumina	Oxidizer B	Inhibitor B

2.2 Pattern wafer within die uniformity improve

In order to monitor within die uniformity, several pieces of standard 300mm pattern wafer were also prepared.

2.3 Pad test for selectivity and within die uniformity improve

Two different types of pads (shown in table II) with different rigidity, thickness and compression ratio were tested to check the selectivity and within die uniformity.

TABLE II.   PARAMETERS OF TWO DIFFERENT TYPE PADS

Pad type	Characteristic		
	Rigidity	*Thickness*	*Compression ratio*
Pad A	80	0.8	2
Pad B	70	1.27	5

RESULT AND DISCUSSION

3.1 Different slurry selectivity for within die uniformity improvement

Firstly, we use the different slurry to test the three different blanket wafer remove rate and the slurry selectivity between different materials.

The result was shown in tableⅢ. All data was collected in same tool, close time and same polish head to make sure the reasonable result.

TABLE III.   SELECTIVITY TEST RESULT IN BLANK WAFER

Slurry	Selectivity		
	Al/poly	*Al/HARP*	*Poly/HARP*
slurry A	<5	>10	<10
slurry B	>6	>10	<10

The most crucial factor was Al to Poly selectivity. For gate height and defect concern, Al CMP need remove Al clear and insure no metal residue in both iso and dense area. At the same time, Al-CMP need keep more poly remain to protect the underneath device.

With this two different slurry, we find that Al remove rate slurry A was about twice of slurry B, Poly remove rate slurry A was higher than slurry B about 3 times. Compare with slurry A, slurry B can gained Al to Ploy selectivity about 90.0%.

Pattern wafer were also tested with different slurry. Different area thickness measurement was shown in figure1.When we keep the same gate height, with slurry B can gained poly thickness, and helpful for prevent over polish risk.

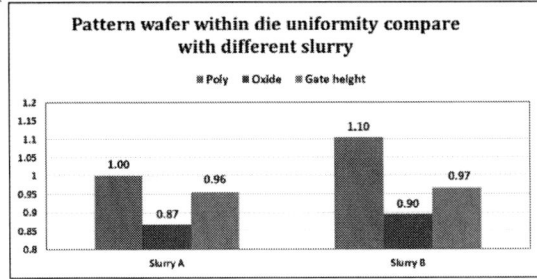

Figure 1: Pattern wafer within die uniformity compare with different slurry

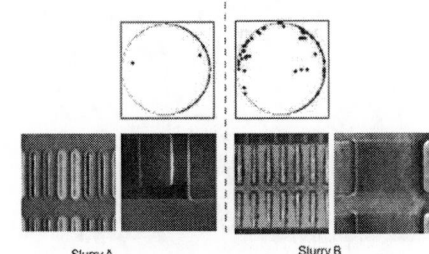

Figure 2: Pattern wafer defect compare with different slurry

When we compared with the defect with different slurry (shown in figure2), more metal residue was find in slurry B both in iso or dense area.

3.2 Different types of pad for within die uniformity improvement

To further study the selectivity in different system and try to have a better balance in thickness and metal residue or over polish defect. Two different polish pad was studied bounded with slurry B. With the different pad rigidity, thickness and compression ratio, we have both blank and pattern wafer result.

TABLE IV.   SELECTIVITY TEST RESULT IN BLANK WAFER

Pad	Selectivity		
	Al/poly	*Al/HARP*	*Poly/HARP*
Pad A	>6	>10	<10
Pad B	>20	>10	<10

The blank wafer selectivity was shown in table IV. It indicate that with the slurry B and combined Pad B, the Al to poly selectivity can gained about 4times.

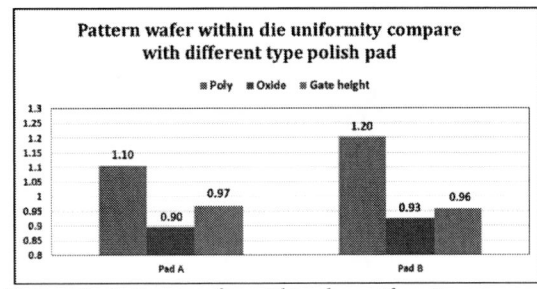

Figure 3: Pattern wafer within die uniformity compare with different type polish pad

Combined with the new polish pad B, metal residue defect have been improved, iso area metal was rare while dense area still can't clear.(Shown in figure 4).

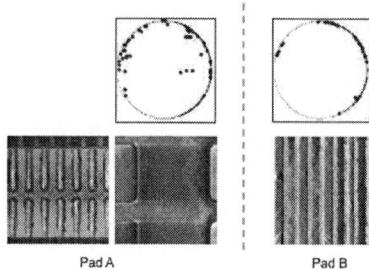

Figure 4: Pattern wafer defect compare with different type polish pad

3.3 Using chemical rinse to reduce Al residue

Different slurry and polish pad were tested to balance gate height and defect. The thickness result was ideal when we combined with slurry B and Pad B. While it meets new trouble that the Al residue was not easy to removed clear in dense area.

We suspect that on the one hand, Al slurry inhibitor was too strong and could protect Al very well, on the other hand, Al surface was more easier absorb inhibitor in dense area. One solution is that before the thickness meet target, a chemical was used to rinse Al surface, the inhibitor was removed as a result. Then keep polish to remove Al residue in dense area.

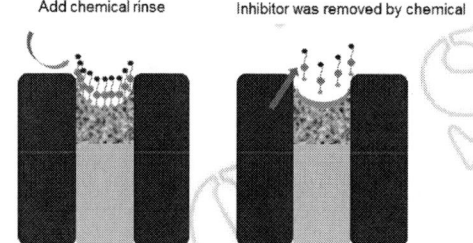

Figure 5: A illustration of chemical surface treatment

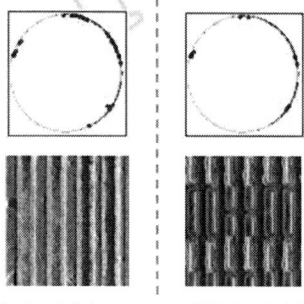

Figure 6: Pattern wafer defect compare with chemical treatment

With the chemical surface treatment, part of the metal residue could be removed in iso area, but still have some metal residue remain. Chemical rinse can give a direction in metal residue remove and need further study to remove all metal residue clear.

CONCLUSION

In this study, we test different slurry and different type polish pad selectivity with blank wafer, and achieved pattern wafer within die uniformity fine tune to have a better balance between metal gate height and defect.

By slurry formula tuning and combine different polish pad can improve Al to poly selectivity about total 5 times.

During the study, an innovation chemical rinse was applied to solve the Al residue issue in dense area.

REFERENCES

[1] S. Guha et aI., Annu. Rev. Mater. Res., Vol. 39, 2009, p I 81-202.
[2] K. Mistry et aI., IEDM Tech. Dig., 2007, pp.247-25.
[3] C. Auth et a\., Symp. on VLSI Tech., 2008, pp. 128-129.
[4] P. Packan et aI., IEDM Tech. Dig., 2009, pp.659-662.
[5] Q. Bainian, J George J and L Zhan, CMP polishing pad solutions for advanced technology nodes, ICPT, 2016:17-20.

IMPACT OF PAD MICRO CONTACT SIZE AND DISTRIBUTION ON THE PLANARIZATION IN CMP

Lin Wang[1], Haipeng Li[1], Ping Zhou[1], Ying Yan[1]*

Key Laboratory for Precision and Non-traditional Machining Technology of Ministry of Education, Dalian University of Technology, Dalian 116024, China
*Corresponding Author's Email: pzhou@dlut.edu.cn

ABSTRACT

Optimal planarization of the overburdened copper layers is a critical objective in the integrated circuit (IC) field and is mainly realized via chemical mechanical planarization (CMP) technique. As this kind of mechanical related performance is strongly related to the action of pad asperities, understanding the effect of the micro-contact conditions of pad asperities on the planarization process will be meaningful. In this study, a millimeter-sized polishing pad sample with a known contact condition is used and conducted on a copper patterned surface. Based on the real contact area image, the relationships between planarization efficiency and real contact condition are established. The results indicate that high material removal in up areas is necessary but is not a sufficient condition for material removal in down areas. Both contact spot size and distribution have an obvious effect on the planarization process. This study is expected to provide some insights into the planarization process from the microscale.

Keywords—Chemical mechanical planarization; Copper; Pad asperity; planarization efficiency; Microscale

INTRODUCTION

Copper has been widely used as interconnects material due to its low resistivity and high conductivity in IC fabrication [1]. To achieve both local and global planarization, the CMP technique is adopted to remove the excess burden of the copper film. As the name implies, CMP involves the removal of materials by a unique combination of chemical and mechanical actions to achieve highly planar and very smooth surfaces. For a typical CMP process, the polishing pad, which is commonly made of polyurethanes, is rougher compared to the wafer surface. Only pad asperities that are tall enough can reach the wafer surface and form massive micro contact spots in the wafer-pad contact interface. It is generally agreed that most material removal related to the mechanical action occurs under the micro-contact spots. Thus, clarify the relationships between planarization efficiency and these micro contact spots will be meaningful.

However, due to the complicated polishing surroundings, it is still impossible to observe the real contact interface directly during the real CMP process. More importantly, the trajectories of pad asperities have a high overlap with each other. Effect of micro contact spots size and distribution on material removal and planarization process is very hard to obtain. Nguyen [2] found that different polishing pads have different planarization efficiency. The planarization rates of small and large copper features were also different. Xie [3] and Wei [4] further established physical based die-level and wafer-level models for the planarization process. However, the effects of contact spots size and distribution on planarization was still lacking. Vasilev [5] characterized the pad surface

texture and modeled its impact on the planarization process. In their study, the contact between the pad-asperities and the up and down areas of the patterned wafer was depicted in a specific height distribution function. However, a direct relation between contact spots and planarization behavior was also not involved.

In this study, a specially designed experiment was carried out on a copper patterned surface. The real contact area between the pad and the patterned surface was measured using an *in-situ* contact area measurement method. Incorporating the real contact area image, the effects of contact spot size and distribution on the copper planarization process were discussed. The evolution of step height in different removal depths was also compared.

EXPERIMENTAL PROCEDURES

As shown in Fig.1, a millimeter-sized IC1000 polishing pad sample (1185×1300 μm^2) was fixed to a micro force sensor. During polishing, the pad sample was pressed against a copper patterned surface (with three different structural features) under $F=10$ g and then slid in a reciprocating motion (1200 cycles). A microscope is settled underneath to get the image of the real contact area *in-situ*. Detail descriptions about the contact area measurement method can be found in our previous studies [6]. The slurry used in this study contained 5wt% hydrogen peroxide (H_2O_2), 1wt% glycine ($C_2H_5NO_2$) and 0.1wt% 1,2,4-Triazole. Commercial colloidal silica with average particle size of 72 nm was added to the slurry at a concentration of 3wt%. The pH was adjusted to 4 by adding nitric acid.

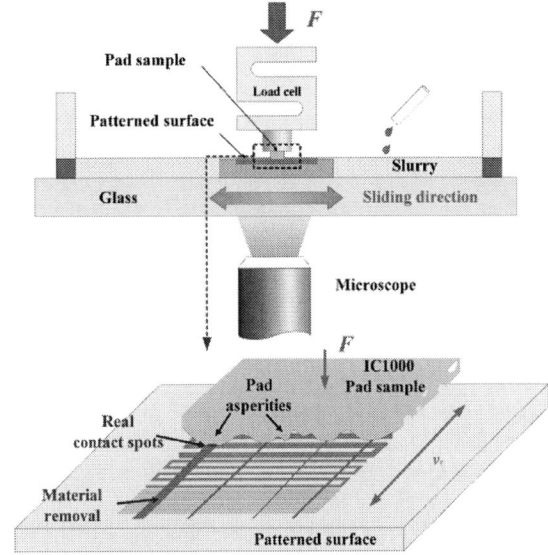

Figure 1: Schematic diagram of the experimental device and the polishing process.

Fig.2 shows the three structural features using a laser confocal microscope (KEYENCE, VK-X250). Tab.1 shows a summary of the feature parameters. Noting that the three regions are polished together in each cycle, and the size of the polishing pad sample is smaller than the size of each feature region.

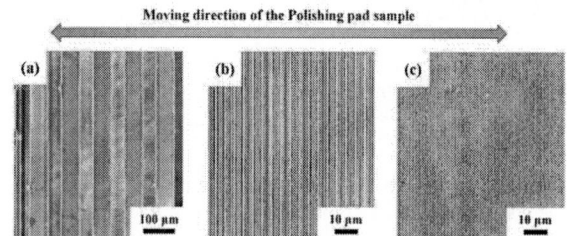

Figure 2: Image of the three different features on a patterned wafer after copper deposition: (a) region1; (b) region2; (c) region3.

Table1: Feature parameters

	line (µm)	space (µm)	h (µm)
region1	50	50	0.3
region2	1	1	0.1
	1	1	0.07
region3	0.5	0.5	0.03

RESULTS AND DISCUSSION

The surface topography of region1 after CMP is shown in Fig.3. The grooves are produced by pad asperities, where massive of abrasive particles are embedded in during the polishing process. Obviously, grooves topography in the up (protruding) areas are quite different from grooves topography in the down (recessed) areas.

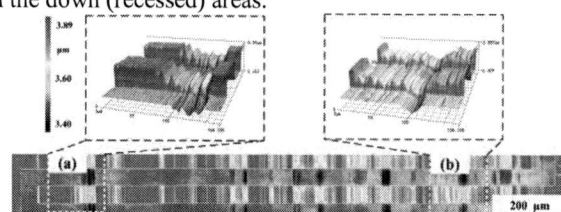

Figure 3: Surface topography of region1 after CMP.

Using the *in-situ* contact area measurement method, the real contact area between the IC1000 polishing pad sample and the copper patterned surface is obtained. Fig.4 shows a comprehensive image incorporating the real contact area image, the surface topography of Region-1 as well as the cross-sectional profiles of both up areas and down areas. The real contact area (RCA) is calculated as 0.96%. It should be noted that all components in Fig.4 are aligned in the direction of motion, which means that the grooves on the patterned surface are produced by the corresponding contact spots in the real contact area image. In this study, seven different lines, *i.e.* Baseline, B, C, D, b, c, d, which parallel to the direction of motion, are selected to investigate the relationships between the real contact area and the planarization process.

The step height evolution at the selected lines is shown in Fig.5 and Fig.6, where *Baseline* is a reference area that has no polished grooves. It can be found that there have no corresponding contact spots for *Baseline* in the contact area

image. Therefore, the direct contact of pad asperities is considered as one of the necessary factors for the planarization. As pure chemical corrosion have no contribution to the planarization process, it is not discussed in this study. The pure chemical corrosion is about 40 nm after the experiment in this study.

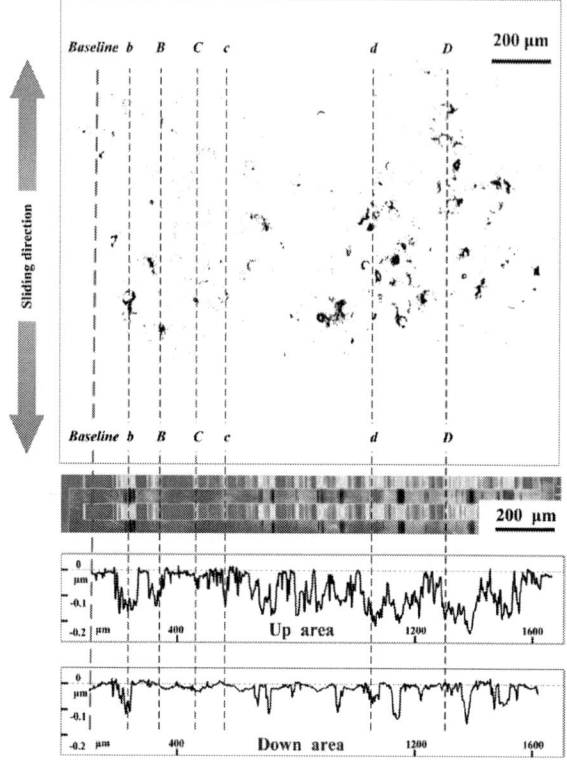

Figure 4: The real contact area image under 10 g (top); the surface topography of Region-1 after CMP under 10 g (middle); the cross-sectional profiles of both up areas and down areas (bottom).

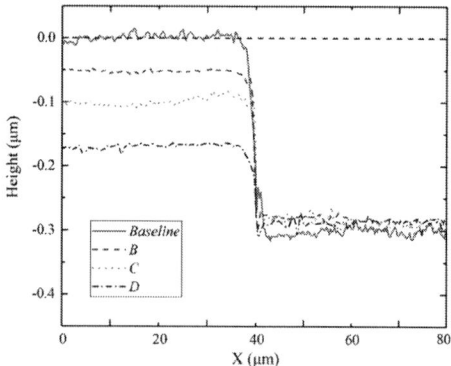

Figure 5: Step height evolution for B, C, D, and Baseline.

For *B*, *C*, and *D*, the groove depth in up areas is quite different, *i.e.* 50 nm, 100 nm, and 170 nm, respectively. However, there was almost no material removal in down areas. The results indicate that despite there are many pad asperities in contact with up areas, only a few of them are tall enough to reach the down areas and induce material removal.

While for *b*, *c*, and *d*, despite almost the same groove depth in up areas, the groove depths in down areas are quite different, *i.e.* 70 nm, 40 nm, and 100 nm, respectively. The results indicate that despite the small and dispersed distributed pad asperities have the ability in creating a deep groove in up areas, they are far from in creating a deep groove in down areas. In other words, a deep groove in up areas is necessary but is not a sufficient condition for material removal in down areas. On the other hand, a deep groove in up areas does not always correspond to a big contact spot (tall asperity), while a groove in down areas necessarily corresponds to a big contact spot (tall asperity). However, the precondition of this conclusion is that the size of the contact spot is smaller than the space size. More importantly, the micro-contact spots with different size and distribution can still yield the same performance or material removal in up areas, but the planarization efficiency will differ significantly. Both size and distribution of micro contact spots play key roles in planarization.

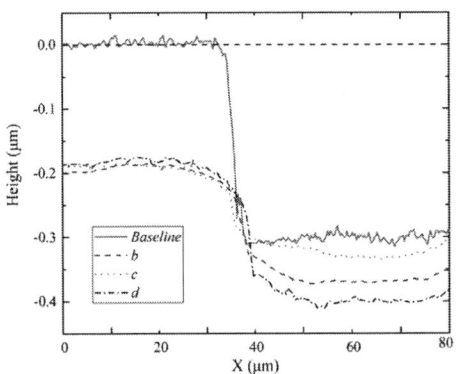

Figure 6: Step height evolution for b, c, d, and Baseline.

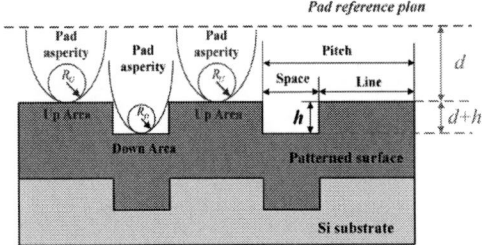

Figure 7: Contact of a randomly rough pad surface with a patterned surface.

To model the planarization process, Vasilev [7] proposed a parabolic shape approximation, where effective curvatures are adopted to describe the contact surfaces of up (κ_U) and down (κ_D) areas, as shown in Fig.7. Merged with a Greenwood-Williamson model and Preston's law, the final material removal rates in up RR_U and down areas RR_D are as follows:

$$RR_U = \frac{e^{\frac{h}{\lambda}}\kappa_U\sqrt{\kappa_D}}{\kappa_{asp}\left(\sqrt{\kappa_U}(1-\rho)+e^{\frac{h}{\lambda}}\sqrt{\kappa_D}\rho\right)}\frac{F_T}{A}K_pV \quad (1)$$

$$RR_D = \frac{\kappa_D\sqrt{\kappa_U}}{\kappa_{asp}\left(\sqrt{\kappa_U}(1-\rho)+e^{\frac{h}{\lambda}}\sqrt{\kappa_D}\rho\right)}\frac{F_T}{A}K_pV \quad (2)$$

where κ_{asp} is the curvature of pad asperities, κ_U and κ_D are effective curvatures of up areas and down areas, respectively.

$$\kappa_U = \kappa_{asp} + \frac{4\alpha h}{line^2}; \quad \kappa_D = \kappa_{asp} - \frac{4\alpha h}{space^2} \quad (3)$$

α is a geometric fit parameter, h is step height, ρ is pattern density, λ is asperity height distribution, F_T is total force, A is total contact area, K_p is the Preston coefficient, V is relative velocity. Therefore, the ratio of material removal in up areas and down areas is:

$$\frac{RR_U}{RR_D} = e^{\frac{h}{\lambda}}\sqrt{\frac{\kappa_U}{\kappa_D}} \quad (4)$$

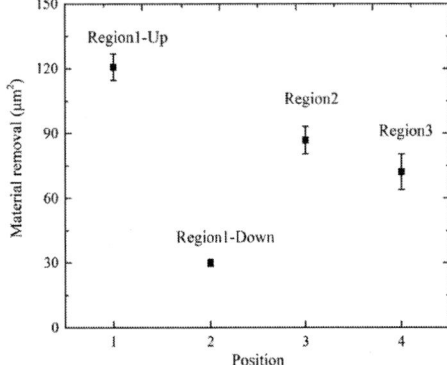

Figure 8: The final material removal of different regions, where chemical corrosion is not included.

Fig.8 shows the final material removal of different regions. The material removal in this study refers to the areas of the cross-sectional profiles after CMP (as shown in Fig.4). For region2 and region3, the features (step height) have been completely flattened after the experiment. This result agrees well with previous studies: small feature planarization is faster than large feature because no down area removal occurs until a later polishing stage [4]. Therefore, the material removal of region2 and region3 in Fig.8 contains two stages, *i.e.* planarization stage and polishing stage. The difference comes from the early planarization stage.

For region1, the material removal is about 120 µm² in up areas and about 30 µm² in down areas. Thus, the ratio of material removal is 4. On the other hand, by substituting parameters in Tab.1 and Ref [7] into Eqs.(3) and (4), the model predicted ratio is calculated as 11.6.

Therefore, there are great differences between the experimental results and the model predicted results. This may because the planarization is very sensitive to the size and distribution of micro contact spots, which is not considered in this model. Therefore, an advanced model that can involve both contact spots size and distribution is under demand for chemical mechanical planarization predicting.

CONCLUSIONS

In this study, by incorporating the real contract area image with the material removal profilers and step height evolutions, the effects of the contact spot size and distribution on the planarization behavior were investigated. The relationships between micro contact condition and planarization can be summarized as follows. First, a deep groove in up areas is necessary but is not a sufficient condition for material removal in down areas. Only a big contact spot (tall asperity) can reach the down areas and induce material removal. Second, both the size and distribution of real contact spots have a significant impact on planarization efficiency. A physical model that can involve both contents is under demand.

ACKNOWLEDGE

The authors would like to acknowledge the support of National Natural Science Foundation of China (51991373, 51875078, 51475076) and the Science Fund for Creative Research Groups of NSFC of China (51621064).

REFERENCES

[1] Steigerwald J M, Murarka S P, Gutmann R J, et al. Chemical processes in the chemical mechanical polishing of copper [J]. Materials Chemistry and Physics, 1995, 41(3): 217-228.

[2] Nguyen V H, Daamen R, Hoofman R. Impact of different slurry and polishing pad choices on the planarization efficiency of a copper CMP process [J]. Microelectronic Engineering, 2004, 76(1-4): 95-99.

[3] Xie X. Physical understanding and modeling of chemical mechanical planarization in dielectric materials [D]. Massachusetts Institute of Technology, 2007.

[4] Fan W. Advanced modeling of planarization processes for integrated circuit fabrication [D]. Massachusetts Institute of Technology, 2012.

[5] Vasilev B, Bott S, Rzehak R, et al. A method for characterizing the pad surface texture and modeling its impact on the planarization in CMP [J]. Microelectronic Engineering, 2013, 104: 48-57.

[6] Wang L, Zhou P, Yan Y, et al. Physically-based modeling of pad-asperity scale chemical-mechanical synergy in chemical mechanical polishing [J]. Tribology International, 2019.

[7] Vasilev B, Rzehak R, Bott S, et al. Greenwood–Williamson model combining pattern-density and pattern-size effects in CMP [J]. IEEE Transactions on Semiconductor Manufacturing, 2011, 24(2): 338-347.

Role of AEO-9 as nonionic surfactant in Barrier Slurry for Copper Interconnection CMP

Guoqiang Song[1,2] ; Baimei Tan[1,2*]; Yuling Liu[1,2*]; Chenwei Wang[1,2];

1School of Electronic Information Engineering, Hebei University of Technology, Tianjin 300130, China

2Tianjin Key Laboratory of Electronic Materials and Devices, Tianjin 300130, China Tianjin 300130, China

E-mail：bmtan@hebut.edu.cn liuyl@jingling.com.cn;

Biography

Guoqiang Song(1995-), a native of Handan, Hebei, China. Now studying in Hebei University of Technology, Tianjin, the second year of master's degree. Research work was carried out in Tianjin Key Laboratory of Electronic Materials and Devices. Research tutor is Professor Baimei Tan of School of Electronics and Information Engineering, Hebei University of Technology.

ABSTRACT

In this paper, the effect of Alcohol Ethoxylate (AEO-9) in barrier slurry for chemical mechanical planarization(CMP) of copper interconnection was studied. The results showed that the surface roughness of copper can be reduced by AEO-9. The TEOS(tetraethylorthosilicate) removal rate(RR) decreased with the increase of AEO-9 concentration, and the removal rate of copper was essentially unchanged. The influence mechanism of AEO-9 on surface roughness was tested and analyzed with atomic force microscopy (AFM), surface tensiometer tester, large particle counts (LPC) and contact angle tester.

Keywords—Copper Interconnection CMP; surface roughness; surface tension; contact angle; AEO-9;

INTRODUCTION

Recent advances of integrated circuit (IC) technology have led to a significant increase of the number of the active components and a significant decrease of feature size. Because the critical features key characteristics of semiconductor devices have decreased to nanoscale and the implementation of additional levels led to multilevel interconnection, the required flatness has become more challenging. The results of chemical mechanical planarization(CMP) of barrier layer determine the performance of the device, such as the electrical characteristics of the following layer line depends on the dishing and corrosion after the barrier layer CMP, and the surface roughness of copper lines is one of the important parameters to evaluate CMP performance. High roughness will lead to electromigration and wire holes of the device line, which then cause device failure and reduce product yield [1-2]. Copper interconnection CMP has developed into the key technology to achieve local and global planarization of integrated circuits in semiconductor manufacturing industry [3-4].

Higher surface roughness will lead to increased noise and leakage current, inconsistent electrical characteristics, and influence frequency characteristics of devices, such as RC delay [5-6]. Therefore, reducing the surface roughness and the surface defects of copper wire is of great significance to improve the performance and reliability of the circuit.

Polishing slurry is an important part of CMP process, and the surfactant is an important additive in slurry, AEO-9 was studied as an additive to copper slurry which could decrease surface tension and contact angle. With the increase of AEO-9 concentration, the surface roughness decreased and the influence mechanism of AEO-9 on surface roughness and removal rate of TEOS were analyzed with atomic force microscopy (AFM), surface tensiometer tester, large particle counts (LPC) and contact angle tester.

EXPERIMENT

Polishing machine— All polishing experiments were carried out on E460E polishers manufactured by Alpsitec Inc, France, and IC politex conventional polishing pads by Rohm & Haas. Each wafer used in the polishing experiment was made by PVD method with an initial thickness of ~1μm. The Cu blanket wafer and TEOS was polished for 60 s, at a pressure of 1.5 psi, a head/pad speed of 57/63 rpm and a slurry flow rate of 300 ml/min. All experiments were performed at least three times to ensure accuracy. In order to eliminate the influence of the specialty of polishing pad characteristics on the experimental results, after replacement of the new pad and turning on the CMP machine, a diamond conditioner (SEASOL AM02BSL8031E7-PM) was used to perform break in operation.

Thickness determination— Polishing was performed on different 300mm blanket wafers (including Cu and TEOS. The material removal rate (MRR) of Cu and oxide was measured through calculating the difference thickness before and after polish 60s on the blanket wafer. The sheet resistance of the Cu film was measured by the four-point probe (Resistivity Test System, Kokusai Electric, VR-1201085) to determine the thickness. The MRR was taken as the average of 81 measurement points, and the was calculated according to the following equation (Eq. (1))

$$WIWNU = X/M_{avg} \times 100(\%) \qquad (1)$$

Where M_{avg} and X are average MRR and standard deviation of MRRs.

Slurry preparation— Slurry components were colloidal silica with a mean particle diameter of 84.2nm, FA/O chelating agent (produced by Institute of Microelec tronics, Hebei University of technology), nonionic surfactant but free of oxidizers and inhibitors. In the experiments, the pH value of the slurry was kept constant at 8.9 by adding phosphoric acid as a pH adjustment in small amounts.

Surface Analysis— An Agilent 5600LS taping mode of atomic force microscopy (AFM) was used for surface roughness examination of the copper blanket wafer after CMP, and the detection range was 10 μm × 10 μm. The root mean square roughness (Sq) was achieved by the Pico Image software with an accuracy of 0.01 nm.

RESULTS AND DISCUSSIONS

Polishing results —Figure 1 shows the difference of the material removal rate of copper polished by four different groups of slurry. The material removal rate of Cu in S1#−S4# Group can be regarded as basically unchanged. As can be seen from the polishing rate of S1#-S4# , with the addition of AEO-9 surfactants , the TEOS MRR gradually decreased.

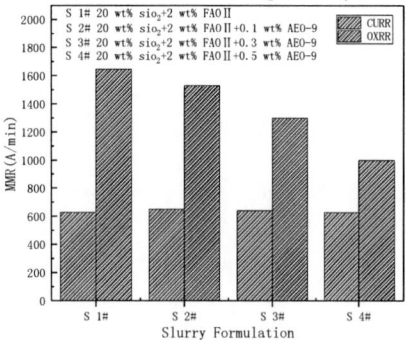

Fig.1. Effect of different concentration of AEO-9 on the RRs of Cu/TEOS.

Figure 2 shows the results of AFM measurement to evaluate the surface quality after 60s polishing. (the marginal scanned area is 10μm ×10μm)with four different colloidal silica-based slurries S1#-S4# under the same CMP parameters. As the surfactant AEO-9 is added to the slurry, the surface roughness Sq decrease gradually. After adding 0.1wt% of AEO-9, the surface roughness decreased by 49% compare with previous. While adding 0.3wt% of the AEO-9, the surface roughness decreased by 80% compare with previous. As a result, AEO-9 can effectively decrease the surface roughness.

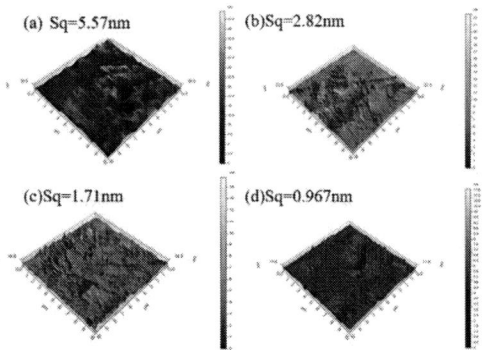

Figure 2. AFM images of copper coupon (a) pre-polishing, (b) polished by S1 # slurry, (c) polished by S2 # slurry, (d) polished by S3 # slurry.

The Mechanism of the Effect of Nonionic Surfactants on Cu CMP- Surfactant has the strong permeability and wettability. Nonionic surfactant is not ionized and is unlikely to introduce additional ionic contamination in the wafer cleaning process. Figure 3 shows the effect of different concentration of surfactant in the slurry on surface tension and contact angle on the copper surface. It can be seen from Figure 3(a) and 3(b), with the increasing of concentration of AEO-9, surface tension and contact angle gradually decreased. The slurry with AEO-9 is fully spread on the copper surface, which can help to improve the wetability of slurry on the

wafer surface in the polishing process, thus improving the flow performance and decreasing the polishing friction. Therefore, the addition of AEO-9 to the CMP slurry is helpful to improve surface flatness in the CMP process.

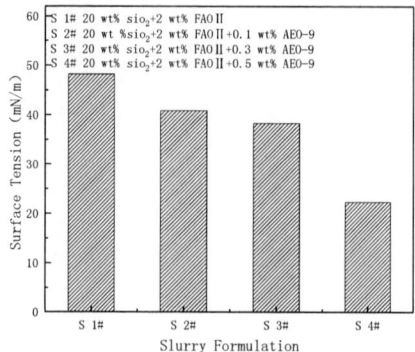

(a). The surface tension of the slurry on copper surface.

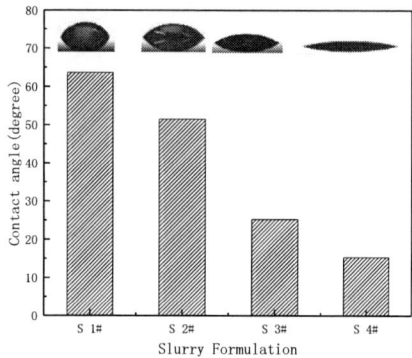

(b). The contact angle of the slurry on copper surface.
Fig. 3. Surface tension and contact angle of the four slurries.

In order to further analyze the mechanism of surfactant in CMP, large particle counts (LPC)was used to measure the number of large particles in polishing solution. Figure 4 showed that with the addition of surfactant AEO-9, the number of large particles ($\geq$ 0.5 μ m) in the slurry decreases effectively. It can be seen from the Figure 4, with the increase of AEO-9 concentration, the number of LPCs gradually decreased. From previous studies and experimental data, large particles are the main cause of scratch defects.

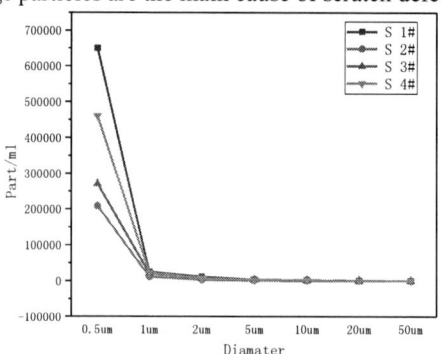

Fig. 4. The relationship between the number and size of large particles of four different slurries.

Based on the above results, the hydrophilic head (polar) is adsorbed onto the dispersed particles surface, while the hydrophobic tail (non-polar) is extend into aqueous solution. Therefore, the addition of AEO-9 nonionic surfactants in

978-1-7281-6559-2/20 $31.00 © 2020 IEEE

silica slurry can provide repulsive electrostatic interactions between particles to prevent coagulation, and higher repulsive barrier forces can stabilize the suspension of colloidal particles (as shown in Fig.5). AEO has a long polyoxyethylene chain, which curls into the solution to provide entropy repulsion. The AEO-9 plays a steric hindrance to the collision between particles. When the AEO-9 surfactant is added into the slurry, the surfactant molecular is adsorbed on the surface of silica abrasives and formed repulsion force between the particles to prevent agglomeration, so as to improve the dispersity of the silica abrasives,avoid agglomeration of particles, and reduce the large particles in the slurry.

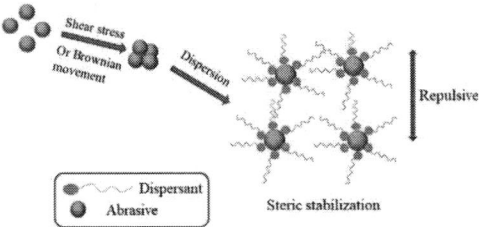

Fig. 5. Schematic of steric stabilization of AEO-9 nonionic surfactnat to stabilize colloidal particles.

CONCLUSION

In this paper, the role of AEO-9 surfactant in the roughness reduction of copper CMP was studied synthetically. The results showed that AEO-9 were added to the slurry, which succeeded in improving the surface quality. And then explained that the surfactant affected LPC by increasing the surface wettability and preventing particle agglomeration synergy, thereby reducing the mechanism of scratching. These conclusions are of great significance to industrial production and research in the future.

References

[1] Chen Su, Zhang Kailiang, Song Zhitang. Advances inchemical mechanical polishing of copper in mutilevel interconnect. Semicond Technol, 2005, 30(8): 21

[2] Liu Yuling, Li Weiwei, Zhou Jianwei, et al. Microelectronics chemical technology foundation.

[3] J. Cheng et al. Material removal K. Shan, P. Zhou, J. Cai, R. Kang, K. Shi, and D. Guo, Electrogenerated chemical polishing of copper, Precision Engineering, 39, 161 (2015).

[4] Mechanism of copper chemical mechanical polishing in a periodate-based slurry,Appl. Surf. Sci., 337, 130 (2015).

[5] J. Bao, H. Shi. J. Liu, H. Huang, P. S. Ho, M. D. Goodner, M. Moinpour, andG. M. Kloster. J. Vac. Sci. Technol. B. 26. 219 (2008).

[6] O. T. Le. J-F. de Marneffe. T. Conard, I. Vaesen, H. Struyf, and G. Vereecke. J.Electrochem. Soc., 159, H208 (2012).

Effect of TT-LYK on Copper CMP with Ru/Ta as barrier/liner

Xue Zhang[1,2] ; Jianwei Zhou[1,2*] ; Chenwei Wang[1,2]; Chao Wang[1,2]

[1]School of Electronic Information Engineering, Hebei University of Technology, Tianjin 300130, China
[2]Tianjin Key Laboratory of Electronic Materials and Devices, Tianjin 300130, China Tianjin 300130, China
Email: *jwzhou@hebut.edu.cn

ABSTRACT

This paper investigates the effect of TT-LYK as an inhibitor on the copper chemical mechanical planarization with Ru/Ta as barrier layer. The results show that the TT-LYK can obviously reduce the dissolution rate of Cu. It can effectively passivate the copper surface. The results also show that the passivation capability was decreased as pH increase from 8 to 10. The effect of TT-LYK on dishing and erosion with different pattern density was also studied. However, the addition of TT-LYK is almost unchanged the removal rate of Ru. The rate of Ru is almost zero compared to the rate of copper. The passivation mechanism of TT-LYK on Cu CMP was revealed by electrochemistry.

Keywords—chemical mechanical planarization, ruthenium, removal rate, inhibitor, dishing and erosion, removal rate

INTRODUCTION

As technology node scales down, lowering the thickness of barrier/liner without sacrificing its integrity and achieving voids-free gap-filling of Cu become the two major challenges for the back end of line (BEOL) Cu/low-k dielectric interconnects[1]. Hence, new barrier is need to support direct electroplating of Cu onto it, which is known as "seed enhancement layers.

Ru is a good barrier to displace Ta/TaN because of its low resistance, good Cu wettability and the ability to deposit copper films directly.[1~3] Since Ru will fail after high temperature annealing and its adhesion for dielectric materials is very terrible, it alone is insufficient to be used as a barrier layer. Hence Ru/Ta are used as liner to afford good property. At present, most of investigations were carried out on blanked Ru wafer, instead of the patterned wafers with Cu interconnect. But it is reported that blanket and patterned are different during CMP. The pattern geography has significant effect on CMP results, which cannot reflect on blanket wafers.

The slurry composed of 0.5%wt SiO_2, 20ml/L H_2O_2,10g/L glycine and different concentration 2,2' -[[(methyl-1H -benzotriazol-1-yl)methyl]imino]diethanol (TT-LYK). The effect of TT-LYK on copper/ruthenium removal rate (RR) and copper dissolution rate (DR) was investigated. The effect of TT-LYK on copper dishing and oxide erosion of typical features during CMP was also studied. The dishing of the 100/100 μm and the erosion of the 9/1μm array were measured. Finally, the inhibition mechanism of TT-LYK on removal rates of Ru was analyzed by electrochemistry .

EXPERIMENTAL

Materials—200 nm Ru film was deposited on TaN film, again by plasma-enhanced atomic layer deposition. Subsequently, 76.2 mm diameter sample was cut from the wafer with Ru/Ta layer and used in the polishing experiments.

Polishing experiments—All polishing experiments were performed on Alpsitecin-E460E polisher made in France and the polishing parameters were set as follows: the polishing pressure was 1.5 psi; the carrier and platen rotational speeds was 87/93 rpm; the fade rate of slurry was 300 mL/min and all the experiments were conducted at laboratory temperature (25±1℃).The basical component of slurry was colloidal silica (mean particle size was about 85 nm,the pH nearly 9.6), hydrogen peroxide (H_2O_2,the mass fraction of 30 wt%, semiconductor grade), ammonium salt. The removal rates of Ru were determined by measuring the difference of film thickness before and after polishing by an aver333A four-point probe from Four Dimensions of company and by taking the averaged result.

Electrochemical experiment—Potentiodynamic polarization curves tests were carried out to describe the electrochemical property of the Cu and Ru samples immersed into different solutions (without silica particles) by using a Shanghai Chenhua-CHI660E electrochemical workstation with a three-electrode glass cell (100 ml volume, with a Pt counter electrode and saturated calomel electrode). Ru and Cu film electrodes (1 × 2 cm^2) were used as a working electrode. Prior to measurement, the working electrode was wiped with 1wt% citric acid solution, followed by rinsing with deionized water for 2 minutes, then dried with high purity nitrogen, and finally sealed with a tape (exposed area of 1 cm^2).

Static etch rate experiment— The static etch rate experiment were carried out in a 500ml glass beaker containing test solutions without abrasive particles. The Cu samples were immersed in citric acid solution for five minutes, and then dried with high purity nitrogen gas. Afterwards, the Cu were dipped into test solutions for 3 minutes respectively, and the static etch rates were obtained by measuring the difference of film thickness before and after.

RESULTS AND DISCUSSION

Effect of pH on removal rate of Cu and Ru

Fig.1 showed the effect of PH on removal rate of Cu and Ru in the slurries containing 0.5%wt SiO_2, 20ml/L H_2O_2,10g/L glycine. It can be seen that the polishing rates of Cu are 7732 Å/min,7760 Å/min and 7887 Å/min severally at pH 8~10.The static corrosion rates of Cu are 2317 Å/min,2346 Å/min and

2089 Å/min respectively. It can be seen from the above data that pH has little effect on the removal rate of copper. The polishing rates of Ru are 17 Å/min,31 Å/min and 155 Å/min respectively. Compared with Cu removal rate, Ru removal rate is almost zero.

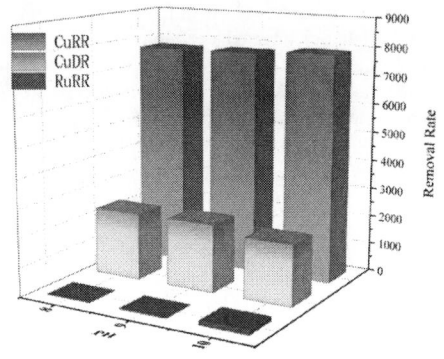

Fig.1. The removal rate and dissolution rate of Cu/Ru as a function of pH.

Effect of TT-LYK on the removal rate of Cu/Ru as a function of pH

In the process of Chemical Mechanical Polishing of inlaid Copper Structures with Ru/Ta as Barrier/Liner Layer, the high rate of copper and the almost zero rate of ruthenium are very favorable for flattening. However, the static corrosion rate of copper is so high that it maybe led to form erosion which is not conducive to planarization. Therefore, the TT-LYK is selected as inhibitor to add to the slurry for reducing static corrosion.

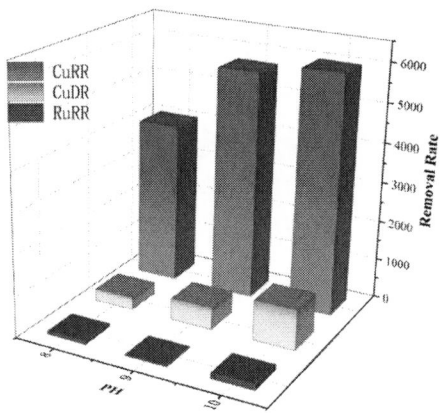

Fig. 2 Effect of 200ppm TT-LYK on removal rate and dissolution rate of Cu/Ru as a function of pH.

The effect of TT-LYK on Cu removal rate and dissolution rate are showed in fig.4.The slurry is composed of 0.5%wt SiO_2,20ml/LH_2O_2,10g/L glycine and 200ppm TT-LYK.As can be seen from fig.4,the removal rate and dissolution rate of Cu reduced obviously compared to Fig.1 after adding TT-LYK to slurry. As the PH increase, the Cu removal rate increase from 4019 Å/min to 5973 Å/min and the dissolution rate increased

from 299 Å/min to1000 Å/min.e[-4]

Surface chemistry analysis.

The surface film formation state and chemical reaction rate of Cu and Ru in the solution can be reflected by potentiodynamic polarization curves. Fig. 3 shows the potential polarization curves of Cu in different solutions containing (1)20ml/LH_2O_2, (2)20ml/L H_2O_2 and 10g/L glycine, (3)20ml/L H_2O_2,10g/L glycine and 200ppm TT-LYK, (4) 20ml/LH_2O_2, 10g/L glycine and 1000ppm TT-LYK. The corrosion potential (Ecorr) and corrosion current density (Icorr) of Cu was obtained by linear fitting of the Tafel potential dynamic polarization curve, and the results are shown in Table I.

Table I. The calculated I$_{corr}$ (corrosion current density), E$_{corr}$ (corrosion potential).

Solution system	Ecorr(V)	Icorr(A/cm^2)
H2O2	0.178	3.348*e^{-5}
H2O2+glycine	0.077	1.964*e^{-4}
H2O2+glycine+200ppmTT-LYK	0.17	1.034*e^{-4}
H2O2+glycine+1000ppmTT-LYK	0.179	5.43e^{-6}

It can be seen from fig.3 that the Ecorr of Cu increase from 0.077V to 0.179V, and the Icorr decreased from 1.964*e-5 to 5.43*e-6 after adding TT-LYK. The Ecorr of Cu increased as the TT-LYK concentration increasing. It is probably due to the formation of insoluble compact passivation layer Cu-TT on the Cu surface. There was a stable current between a rage of potentials, proving that the effect of the corrosion inhibition of Cu was related to the formation of a protective thin film. The chemical reactions generated on the Cu surface can be depicted as follows. The main cathodic reactions were the reactions of oxygen reduction as follows:

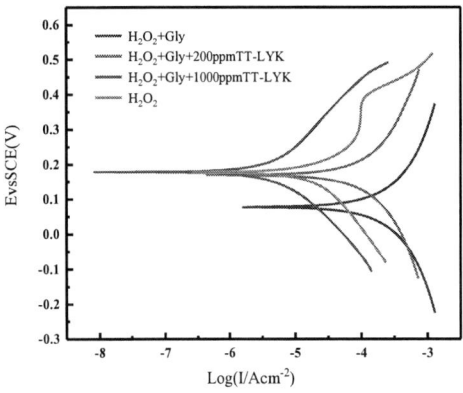

Figure 3. Effect of TT on the dynamic polarization curves of Cu in different solutions containing (1)20ml/LH2O2, (2) 20ml/L H2O2 and 10g/L glycine, (3) 20ml/L H2O2, 10g/L glycine and 200ppm TT-LYK, (4) 20ml/L H2O2, 10g/L glycine and 1000ppm TT-LYK.

$$2Cu+2OH- \leftrightarrow Cu2O+H2O+2e- \qquad [1]$$
$$Cu2O+2OH- \leftrightarrow 2CuO+H2O+2e- \qquad [2]$$
$$Cu+2OH- \leftrightarrow Cu(OH)2+2e- \qquad [3]$$

$$Cu2{+}{+}TT\text{-}LYK \leftrightarrow Cu\text{-}TT\text{-}LYK \qquad [4]$$

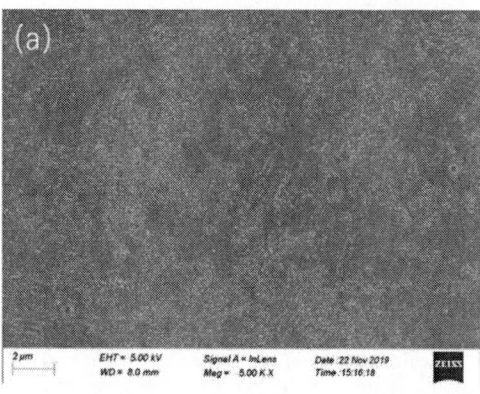

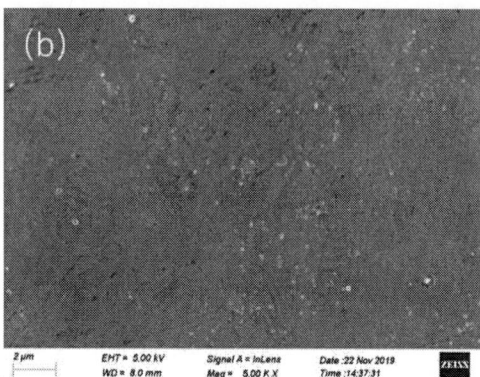

Figure 4. SEM image of Cu in different solutions (a) 20ml/L H2O2,10g/L glycine and 300ppm TT-LYK. (b) 20ml/L H2O2 and 10g/L glycine

Figs. 2a and 2b show the SEM images of the Cu wafers in the different solutions. It can be seen that many deep holes are formed by corrosion on the copper surface when the slurry is composed of H_2O_2 and glycine. However, it disappeared after adding 300ppm TT-LYK which proved that inhibitor can reacted with Cu to form effective protection film on Cu surface.

CONCLUSIONS

The effect of TT-LYK on copper CMP performance with Ru/Ta as barrier/liner was investigated. It is found that glycine is a good complexing agent for copper CMP which gives copper a high polishing rate while the removal rate of Ru tends to zero. The all above result showed that corrosion can be effective suppressed by adding TT-LYK to slurry as inhibitor and at this situation, both of the removal rate of Cu and dissolution rate of Cu are increased as PH increasing. However, it is not remarkable in the slurry without TT-LYK.

References

[1] B. J. Cho, S. Shima, S. Hamada, and J. G. Park, Applied Surface Science, 384, 505(2016).

[2] L. Peedikakkandy, L. Kalita, P. Kavle et al., Applied Surface Science, 357, 1306(2015).

[3] Chenwei Wang Jiaojiao Gao et al., Microelectronic Engineering, 108, 71 (2013).

[4] Cheng J . Galvanic Corrosion Inhibitors for Cu/Ru Couple During Chemical Mechanical Polishing of Ru[J]. 2018..

[5] Liang, J., He, Y., Li, Y., and Luo, J. "Effect of ionic strength on ruthenium CMP in H2O2-based slurries ," Applied Surface Science,vol.317,pp.332–337,2014.

[6] Amanapu, H. P., Sagi, K. V., Teugels, L. G., and Babu, S. V. "Role of Guanidine Carbonate and Crystal Orientation on Chemical Mechanical Polishing of Ruthenium Films,"Journal of Solid S tate Science and Technology, vol,2,pp.445–451,2013.

[7] Zeng X, Wang J X, Lu H S. Improved removal selectivity of Ruthenium and copper by glycine in potassium periodate (KIO_4) based slurry [J]. Journal of the Electrochemical Society, 2012,159(11):525-529.

[8] Du, Y., Wang, C., Zhou, J., Zhang, W., Ji, J., and Han, L., et al. "Effect of Guanidinium Ions on Ruthenium CMP in H2O2-Based Slurry," Journal of Solid State Science and Technology, vol.6,pp.521–525,2017.

[9] PEETHALA B C, BABU S V. "Ruthenium Polishing Using Potassium Periodate as the Oxidizer and Silica Abrasives," Journal of the Electrochemical Society, vol. 158, pp. 271–276, 2011.

Effect of Various Complexing Agents for Cobalt "bulk step" Chemical Mechanical Planarization

Yundian Yang[1,2], Shengli Wang[1,2] *, Chenwei Wang[1,2] , Yuanshen Cheng [1,2]

[1]School of Electronic Information Engineering, Hebei University of Technology, Tianjin 300130, China

[2]Tianjin Key Laboratory of Electronic Materials and Devices, Tianjin 300130, China

*Corresponding Author's Email: shlwang@hebut.edu.cn

ABSTRCT

In this paper, various complexing agents were evaluated for cobalt chemical mechanical planarization(CMP) in H_2O_2 based slurry, during "bulk step" process. Firstly, effect of different complexing agents on polishing removal rate(RR) and dissolution rate (DR) of Co were studied. The results showed that glycine has the highest Co RR compared with other complexing agents, such as citric acid(CA), tartaric acid(TA), potassium citrate(PC), potassium tartrate(PT). Comparison of corrosion current of different complexing agents was observed by electrochemical measurement. However, the glycine has a has higher Co DR without inhibitor. Thus，Carboxyl benzotriazole (CBT) was used as inhibitor in glycine-H_2O_2 based slurry. The results showed that CBT can effectively reduce the Co DR and passivate the Co surface, achieve a good surface quality and maintain a desirable Co RR with a few amount of CBT added into the slurry .

INTRODUCTION

In the development of integrated circuit manufacturing, cobalt is one of the new materials, was originally proposed as a barrier to copper interconnects.[1] As the feature size is reduced to 7 nm, Co is used as a new interconnect material due to its excellent properties such as low resistance and high aspect ratio trench[2].

Chemical mechanical planarization (CMP) is a critical step in achieving global and local planarization. The slurry is a major consumable material in the CMP process. The slurry was consist of abrasives, oxidants, inhibitors and complexing agents. All of slurry components direct impact on cobalt removal rate (RR)

and dissolution rate (DR).[3] Nevertheless, the complexing agent plays an important role in the high removal rate of cobalt. There are many reports about complexing agents. T R. Popuri et al.[4] the effect of citric acid as a complexing agent on the cobalt film was investigated. C.K.Ranaweera et al.[1] The effect of ammonium persulfate as a complexing agent on the RR of Co was investigated. Liang Jiang et al.[5] The synergistic effect of H_2O_2 and glycine was studied to improve the RR and DR of cobalt, but the change of dissolution rate and the surface morphology were not studied after increasing the inhibitor.

In this paper, the effects of various complexing agents on Co RR and DR were studied. The corrosion performance was characterized by electrochemical experiments. The surface morphology was studied by scanning electron microscopy (SEM) measurements and atomic force microscopy (AFM).

EXPERIMENTAL

Polishing experiments.—All the polishing experiments were performed on a France-E460E polisher from Alpsitec Inc and the polishing pad was IC1000 (purchased from Dow Electronic Materials,USA) using 3 inches Co disk(99.99% pure). The polishing experiments were carried out at a polishing pressure of 1.5 psi with a slurry flow rate of 300ml/min and platen/carrier speeds of 97/83 rpm for 1 min. After polishing, Co disk was rinsed with DI water and then dried with N_2 gas.

Electrochemical experiments——A CHI660E electrochemical workstation (CH Instrument Inc.) with a Pt counter electrode and a SCE reference electrode were used to acquire potentiodynamic polarization cures .New

Co (99.99% pure,10*20*1.5mm^3) coupons were used in all the electrochemical experiments as the working electrode.

Surface quality measurements.—In order to observe the surface topography of Co wafers, ZEISS SIGMA 500/VP Scanning electron microscopy (SEM) and atomic force microscopy (AFM) were used.

RESULTS AND DISCUSSION

Effect of Complexing Agents on Co RR and DR.

All the slurries are at pH 8. From Fig 2, at concentrations, the complexing agent glycine can make Co have a high removel rate and a high dissolution rate. Different complexing agents have different mechanisms of action on cobalt, which can be interpreted as follows. In the glycine and hydrogen peroxide solution, the removal mechanism of cobalt is considered to be (1)-(2). [5] It can be shown that glycine has a synergistic effect with hydrogen peroxide, and cobalt is first oxidized and then complexed. For alkaline solution citric acid-hydrogen peroxide system, the reaction equations (3)-(4), citric acid is synthesized before the Co^{2+} ion complex $[Co\,(C_6H_5O_7)_2^{4-}]$, and then oxidized by hydrogen peroxide $[Co\,(C_6H_5O_7)_2^{3-}]$.[4]

$$Co^{2+} + H_2O_2 \rightarrow Co^{3+} + 2OH^- \qquad (1)$$
$$Co^{3+} + {}^{3+}H3NCH2COO^- \rightarrow Co(H_2NCH_2CO)_3 + H_2O \qquad (2)$$
$$Co^{2+} + C_6H_5O_7^{3-} \rightarrow [Co\,(C_6H_5O_7)_2^{4-}] \qquad (3)$$
$$[Co\,(C_6H_5O_7)_2^{4-}] + H_2O_2 \rightarrow [Co\,(C_6H_5O_7)_2^{3-}] + 2OH^- \qquad (4)$$

Electrochemical Experiments

In order to further study the effect of different complexing agents on Co, electrochemical experiments were performed. Silica particles are removed from the slurry to prevent the surface of the cobalt from contaminating the particles. All the slurries are at pH 8. Figure 4 shows the obtained Tafel plots of Co in different solutions. The corrosion current density (Icorr) and the corrosion potential (Ecorr) are calculated from the Tafel plot using the CHINSTR software as shown in Table I. It can be seen from the Potentidynamic plots that in addition

to glycine, other complexing agents a passivation region appears at the rear end of the anodic polarization current., thereby reducing the removal rate of cobalt.

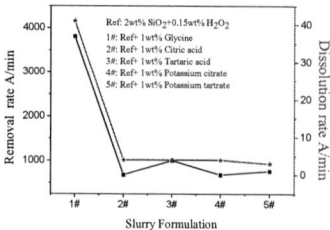

Fig 2. Effect of various complexing agents on Co removal rate and dissolution rate at different concentrations

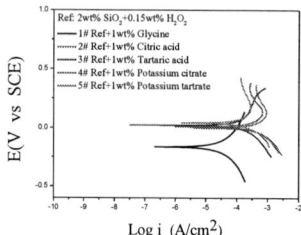

Fig 3. Potentidynamic plots for Co in 1wt% complexing agents at pH=8

Table I. Corrision potentials and corrision current densities for Co in different complexing agents at pH=8

pH=8	E_{corr}(V)	I_{corr}(mA/cm^2)
1#	-0.166	0.03
2#	-0.002	0.63
3#	0.022	0.16
4#	0.009	0.48
5#	0.043	0.26

Inhibition effect of CBT on SER and RR of Co

Fig 4, in slurry containing 0.15wt% H_2O_2, 1wt% glycine at pH 8, after addition of 5mM CBT the RR of Co decreased sharply from to 1182 Å/min and 7Å/min. SEM and AFM measurements of Co wafers after different treatments were used with the purpose of further investigating the inhibition effect of CBT on Co, shown in Fig 5. Fig.5c indicated that after adding additional 5mM CBT to the slurry , the corrosion on the surface of Co wafer was significantly inhibited, which may be due to the adsorption of CBT on Co surface. Fig.5d, the RMS and P/V of Co wafer was reduced to 35.5 nm and 20 nm, respectively. The results indicated that the CBT was an inhibitor of great performance on Co.

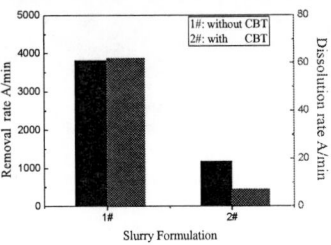

Fig 4. DR and RR of Co in slurry containing 0.15wt% H_2O_2, 1wt% glycine at pH 8 without and with of CBT.

CONCLUSIONS

The effects of different complexing agents on cobalt removal rate and static corrosion rate were compared. At the same concentration, glycine can give cobalt a high removal rate of 3815 Å/min and a high dissolution rate of 62 Å/min. After the inhibitor CBT was added, the removal rate and dissolution rate decreased to 1182 Å/min and 7 Å/min, respectively. And through SEM and AFM tests, the surface corrosion was significantly reduced, and the RMS and P/V were reduced after the inhibitor was added.

ACKNOWLEDGEMENTS

Financial supports for this research work is provided by Natural Science Foundation of Hebei Province (E201920367), Major National Science and Technology Special Projects (No.2016ZX02301003-004-007), and the Key Laboratory of Electronic Materials and Devices of Tianjin, China. Natural Science Foundation of Tianjin, China (14JCYBJC18500)

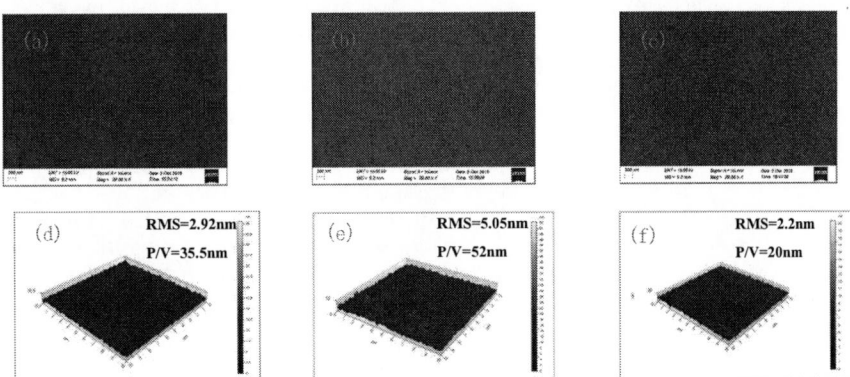

Fig 5. SEM and AFM images of Co wafers: (a) and (d) without any treatment; (b) and (e) immersed in 0.15wt% H_2O_2, 1wt% glycine solution; (c) and (f) immersed in 0.15wt% H_2O_2, 1wt% glycine and 5mM CBT solution

REFERENCES

[1] C. K. Ranaweera, et.al, Ammonium Persulfate and Potassium Oleate Containing Silica Dispersions for Chemical Mechanical Polishing for Cobalt Interconnect Applications [J]. ECS Journal of Solid State Science and Technology, 2019,8 (5) P3001-P3008 .

[2] Qiyuan Tian, et al. Effect of Amine Based Chelating Agent and H_2O_2 on Cobalt Contact Chemical Mechanical Polishing [J]. ECS Journal of Solid State Science and Technology, 2018, 7(8): 1-7.

[[3] Wenhu Xu , et.al, Mechano-oxidation during cobalt polishing [J]. Wear 416–417 ,2018, 36–43.

[[4] R. Popuri, et.al, Citric Acid as a Complexing Agent in Chemical Mechanical Polishing Slurries for Cobalt Films for Interconnect Applications [J]. ECS Journal of Solid State Science and Technology, 2017, 6(9) P594-P602.

[[5] Liang Jiang , Yongyong He , et.al, Synergetic effect of H_2O_2 and glycine on cobalt CMP in weakly alkaline slurry [J]. Microelectronic Engineering ,2014, 82–86

Lead the Future
Semiconductor Evolution as Seen by CMP Manufacturers

Manabu Tsujimura

Ebara Corporation, 11-1 Haneda Asahi-cho, Ohta-ku, Tokyo 144-8510, Japan,
tsujimura.manabu@ebara.com

Abstract

The global situation has become increasingly uncertain and it has been a long time since the so-called VUCA era began. The semiconductor market, however, is now booming and semiconductor technology seems to have overcome its stagnation. The market has now entered the so-called Multi-driver era, driven by the IoT, Cloud, AI, Car, and 5G applications, collectively referred to as ICAC5, in addition to the former single set applications, such as PCs and mobile phones. Technologies adapted to the ICAC5 market have also been developed, and device development has been active in three directions: MM (More Moore), MtM (More than Moore), and BC (Beyond CMOS). With the development of semiconductor devices, new technologies are also required for manufacturing equipment such as CMP. Based on the uncertain global outlook, this paper provides an overview of device technology up to 2030 by showing the uniqueness of the semiconductor market, and presents as an example how the CMP process should evolve. Semiconductors are immortal as long as there is human desire. Semiconductors will continue to lead the future of technologies. And semiconductor manufacturing equipment will continue to support the evolution of semiconductor devices.

I. Introduction

Figure 1 provides a fairly broad overview of the future of the semiconductor market and technology based on today's uncertain global situation.

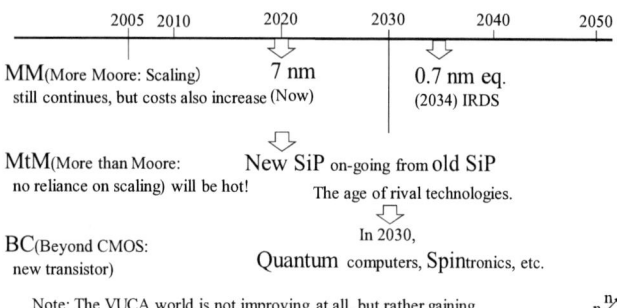

Figure 1 Broad Overview of Global Situation, Semiconductor Market and Technology

Although I am an outsider as far as my world view is concerned, and although my view may be irrelevant, I nevertheless agree with the public view that today is the VUCA era, marked by Volatility, Uncertainty, Complexity, and Ambiguity, and feel that VUCA is increasingly gaining momentum. Even confronted with such a gloomy global outlook, I still feel that the future of semiconductors has not lost its luster, but is shining even brighter.

As will be described in detail later, the semiconductor market has now entered the so-called Multi-driver era, driven by the IoT, Cloud, AI, Car, and 5G applications, collectively referred to as ICAC5, from the Single Driver era, which featured the former single set applications, such as PCs and mobile phones. Not only does this mean that the semiconductor market no longer sees large ups and downs, but also that the level itself is growing significantly.

Driven by the ICAC5 market, semiconductor technology has also emerged from its stagnation and has been revitalized. In fact, at most academic conferences in 2019, similar views were announced with confidence.

The MM (More Moore) roadmap indicates successful scaling to 7 nm at present with prospects to achieve 1 nm or less by 2030.

In the MtM (More than Moore) roadmap, SiP (System in Package), which was discussed at the time of Paradigm Shift 45, as described below, is now blossoming and thriving in competition in different forms.

In the BC (Beyond CMOS) roadmap, new transistor principles and structures have already been developed. Along with these new transistors, the development of quantum computers, which are expected to grow with new applications, has also become active.

The semiconductor market, which has been expanding according to Moore's Law supported by Dennard's theory on scaling (and modified versions thereof), has appeared to be immortal, as Manabu's Principles[1] suggested.

The development roadmap has already been presented.

All that remains is to push forward for the development of each element technology.

It is the frontier spirit of developers that paves the way for the future and leads the future. Thanks to the efforts of engineers and scientists, semiconductors, which were born in the 20th century, will lead the future as the most advanced technology in the 21st and 22nd centuries.

II. Semiconductor Market Driven by ICAC5

In discussing the market, let's first understand the market characteristics of the semiconductor supply chain (Set, Device, and Tool) and then look at how it will change in the future.

II-1 Market Characteristics of Semiconductor Supply Chain

It is important to understand the characteristics of the Set, Device, and Tool markets from the three perspectives of "size", "ratio : golden ratio", and "market instability: bullwhip".

1. Market size: Currently, the sizes of the Set, Device, and Tool markets are about US$2,000B, US$500B, and US$70B, respectively.

2. Ratio : Golden ratio: There is a golden ratio in the supply chain market. If the Set market is 100%, the Device market is 20% based on the device mounting ratio in the set. If the device manufacturer's investment ratio is 10%, the Tool market will be 2%. In other words, the golden ratio is (100: 20: 2) and this trend has not changed much in

the past 20 years.

3. Market instability: Bullwhip: It is said that the Set market for semiconductors is almost linked to GDP, the Device market expands in conjunction with the fluctuation of the Set market, and the Tool market changes greatly depending on the management decisions of the Device manufacturers. The margin of fluctuation in the Tool market is usually ±20% based on past results, and even ± 50% can be reached as we have experienced twice in the past. The market becomes less stable gradually from the upstream Set market toward the downstream Device and Tool markets. If the margin of fluctuation is ± 2% in the Set market, it will be ± 20% in the Device market and ± 50% in the Tool market, which is why the supply chain market is called the bullwhip market.

II-2 Future Market Led by ICAC5

We have understood the characteristics of each market from the past experiences. Next is the future. Figure 2 shows a part of the future led by ICAC5.

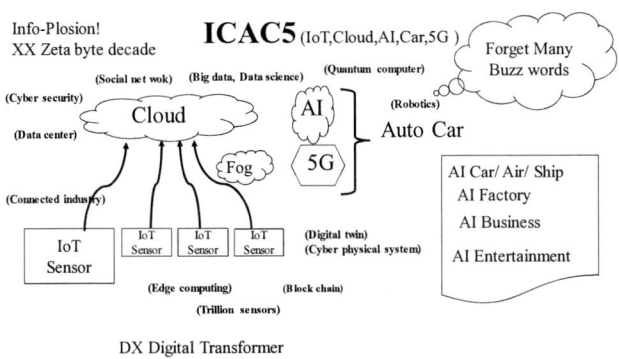

DX Digital Transformer
DX S Digital Transformer Society

Figure 2 Market Led by ICAC5

Other than IoT, Cloud, AI, Car, and 5G, as represented by ICAC5, various technologies have emerged in our time, making the market very difficult to understand. It may be rude to call them the Buzz Words, but it would be easier to understand the market if the discussion focused only on ICAC5 without using other words. For example,

"More and more IoT sensors are now used in the world, and information from the sensors can be put into the Cloud. The amount of information is expected to reach a zettabyte in the so-called information explosion era (Information Explosion: Info-Plosion). The information is managed by AI at 5G speed, making possible, for example, auto drive Cars."

Moving forward, this concept will create new markets and technologies such as AI Factories, AI Business, AI Entertainment, etc. These are sometimes called DX (Digital Transformers), and the societies featuring them are referred to as DXS (Digital Transformer Societies). Semiconductors will still play a major role in these societies, and it is now a consensus that they will continue to grow with ICAC5 in the DXS.

III. Semiconductor Technology; Very Sure, No Surprises

We predicted the future of the market by learning from experience. We will do the same for the technology. First of all, let's examine the turning points of past technologies and predict the future.

III-1 Turning Points of Technology

Figure 3 shows the Paradigm Shift 45 experienced in 2005, the major technological challenges expected in 2020 (called the Paradigm Cliff),

and technological revolutions expected in 2030. Various attempts have been made to overcome Paradigm Shift 45 as described later, and device development has largely diverged in multiple directions. They are:

· MM (More MM): Scaling of interconnects will further develop.

· More than Moore (MtM): Devices will develop without depending on scaling of interconnects.

· Beyond CMOS (BC): Invention of a new transistor

In the MM roadmap, various three-dimensional technologies have emerged. They are 3D transistor construction (FiNFET, etc.), 3D transistor layout (3DNAND), and 3D chips (3D packages).

In the MtM roadmap, SiP (System in Package), SoC (System on Chip), and other technologies have been developed, and the development efforts are still ongoing.

In 2030, new transistors, such as quantum computers and spin transistors, are anticipated.

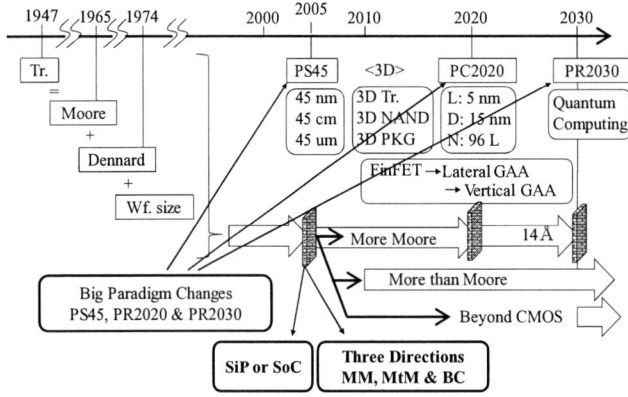

Figure 3 Turning Points of Technology

III-2 Paradigm Shift 45 Again [2]

Although the technology overview up to 2030 is shown, in fact, the development challenges experienced in Paradigm Shift 45 in 2005 will emerge again in the current scaling to 7 nm and the scaling to 1 nm or less expected in 2030."

Figure 4 looks back on the three paradigm shifts and shows what challenges lie ahead.

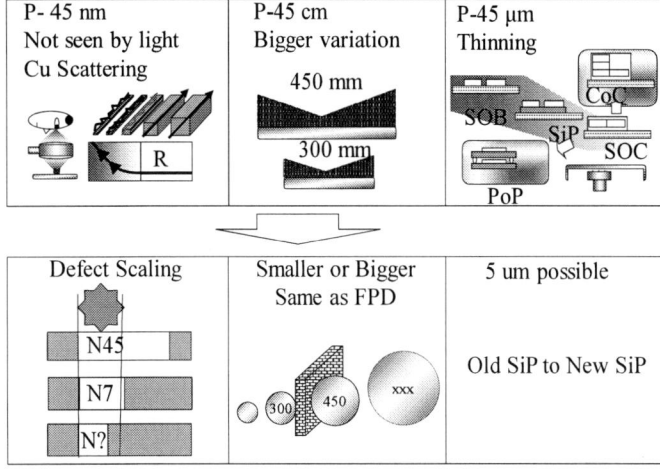

Figure 4 PS 45 Persists Forever

PS 45 nm (Scaling problems): Defects are not scaled down even if the device is scaled down. Therefore, the problem that appears to have been solved will inevitably reappear in the new generation of scaling. This is easier to understand if an FM (Foreign Material) is used as an example, as shown in Figure 4. The size and number of FMs that could have been finalized with 45 nm are already problems at 7 nm.

PS 45 cm (Larger wafer problems): It is true that the shift to a 45 cm diameter is stagnant, but 45 cm wafers are now possible to handle, and systems are already used to handle FPDs as large as 3 m. Recently, systems for interposers of 50 cm or more have appeared. Even if the wafer size does not jump to 45 cm, the substrate size is increasing.

PS 45 μm (Film thinning problems): In 2005, film thinning to 45 μm was difficult, but now thinning down to about 5 μm is possible. The only issue is how to implement the technology in packaging. In packaging, the new SiP, which has evolved from the former SiP, is now blossoming and thriving in competition.

III-3 Technology; Very Sure, No Surprises

Looking back on the 2019 academic year, one trend has come to light.
MM: 7 nm Now, 5 nm On-going, 3 nm Promising, and 1 nm Beyond
MtM: As mentioned earlier, many new SiP technologies have been announced.
BC: IBM, Qualcomm, Google and other companies have made announcements about various quantum computers, and more are coming about new transistors.

Until last year, there was a sense of stagnation in the technology; it seemed that the BC roadmap was running away to MtM without showing the future of scaling, and that BC was still a long way off. This year, however, many groups are talking about the year 2030 with confidence. I dare to describe this trend as "Very Sure, No Surprises (VSNS)".

IV. International Roadmap Now [3]

The ITRS, which ceased activity after 2015, has resumed activity in a different form. Here I discuss challenges in the past and future only in the multi-layer interconnects.

IV-1 Multi-layer Interconnect Problems in 2005

Figure 5 shows potential problems with multi-layer interconnects raised in the year 2005.
I do not go into detail, but it should be noted that these problems may recur in the future.

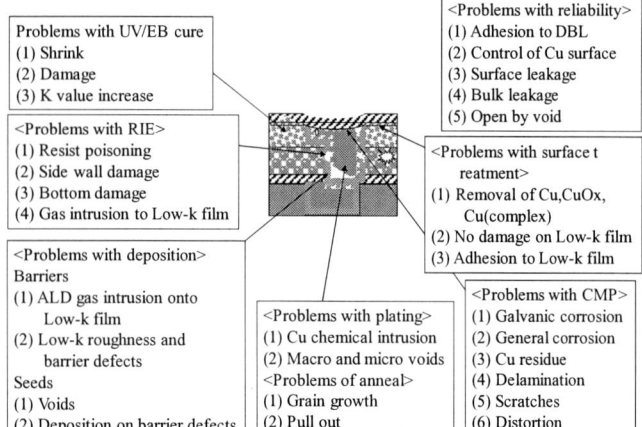

Figure 5 Potential Problems with Multi-layer Interconnects

IV-2 Roadmap Now : From ITRS to IRDS

Roadmaps are important for equipment manufacturers as a means to understand development trends of customers. They obtained new roadmaps from IRDS (International Roadmap for Devices and

Systems), which started activity after ITRS (International Technology Roadmap for Semiconductors) was suspended. The focus of the roadmap has shifted from devices to sets and applications, but it is still an important source of information for where the traditional MM, MtM, and BC roadmaps are going.

Figure 6 shows the relationship between device development trends and IRDS roadmaps. The figure features the new roadmaps, such as AB (Application Benchmarking), SA (System Architecture), OSC (Outside System Connectivity), etc.

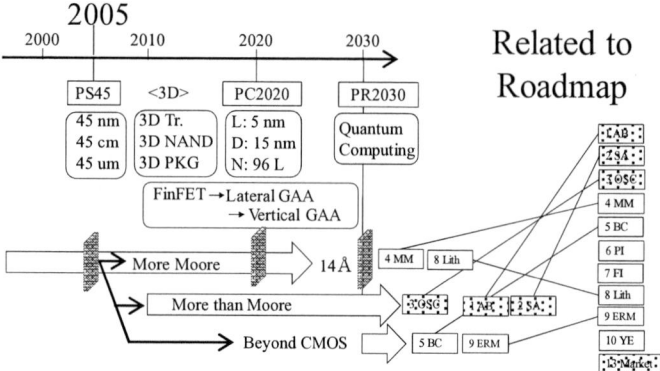

Figure 6 Roadmap and Development Directions

V. CMP Leading the Future [4]

I have argued that the semiconductor market has grown and the technology has advanced with ICAC5 in an unstable global background, and now the trend is "VSNS". Let's prove "VSNS," taking CMP as a specific example.

V-1 CMP for MM, MtM, and BC

Figure 7 shows CMP evolution keeping up with the ICAC5 market. That is the CMP for MM, MtM, and BC.

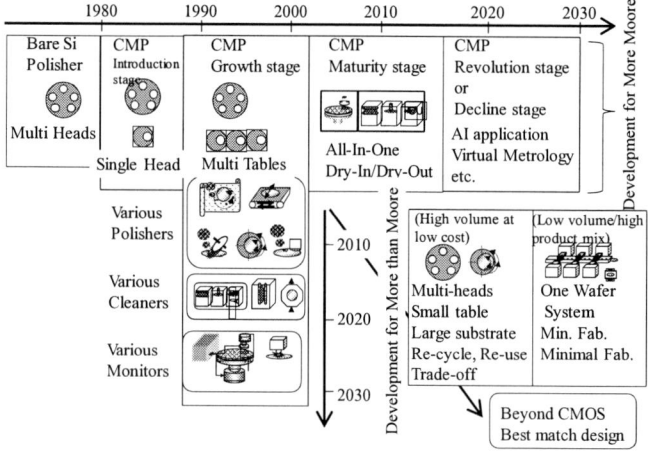

Figure 7 CMP Development for MM, MtM & BC

As mentioned earlier, devices have evolved in three directions: MM, MtM, and BC. So have the systems.

CMP for MM: This has developed up to now, and will continue to develop in response to further scaling of interconnects.

CMP for MtM: This is a new idea. Various ideas of high throughput with cost priorities are considered.

CMP for BC: We should always pay attention to the roadmap and foresee what kind of BC technology will appear and what processes will be required.

V-2 MM: Shift of CMP Processes

As an example, Figure 8 shows how CMP processes shift as scaling of interconnects advances.

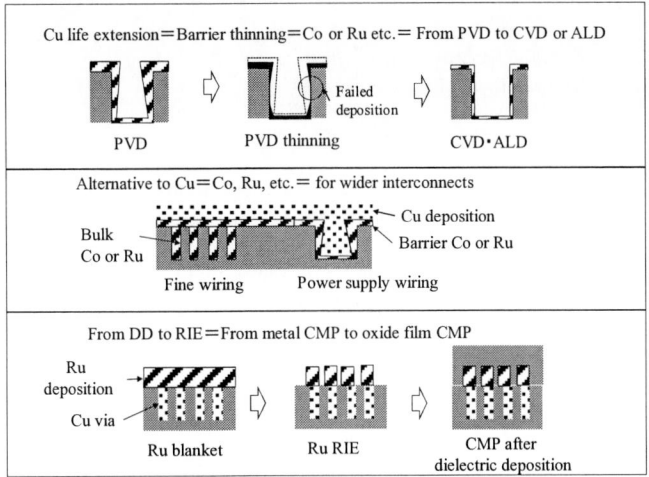

Figure 8 Examples of CMP's Latest Challenges

Cu life extension: It is natural to thin barriers in order to increase the effective area of Cu, but this should be decided in consideration of Cu embedability and subsequent CMP adaptability. This is under development with Co and Ru. While Ru is better at wettability for Cu embedding, Co is better at CMP adaptability. Currently, Co is ahead of Ru in development.

Alternatives to Cu: Co, Ru, Mn, etc. are promising based on the evaluation using the formula, (resistance value) x (mean free path).

RIE (Reactive Ion Etching) again: Interconnects using Ru or others is also considered using RIE without CMP, like the former Al interconnect. In this case, the inter-layer dielectric CMP will come back instead of the metal CMP.

Inter-layer dielectric: Development to reduce the dielectric constant as a countermeasure against RC delay has continued, but now the dielectric constant has been slightly in a reverse trend. The reason should be that a dielectric layer having a (strong) structure with a high elastic modulus had to be adopted because the structure itself became weaker due to scaling of interconnects.

In this way, taking CMP alone, there are many things to be done, such as addressing again the problems experienced in the PS 45 era as well as scaling, and going back to the RIE technology from the damascene technology. We should just "Revisit Old Wisdom to find New Knowledge."

VI. Whose Law Next?

Finally, let's take a look at how the business model and device technology have evolved along with the advancement of transistors, according to "someone's laws." In Figure 9, I personally picked major shifts in the laws that have significantly impacted the business model. After the invention of the transistor in 1947, Intel was founded in 1968 following Moore's Law. In 1987, Dr. Morris of TSMC introduced the Fabless-Foundry (FF) model. In 2007, I announced "Manabu's principles" to encourage the stagnant semiconductor industry. These laws appeared about every 20 years, but whose law will govern the semiconductor industry in the coming 20 years and in the year 2027? Let's look forward to the answer.

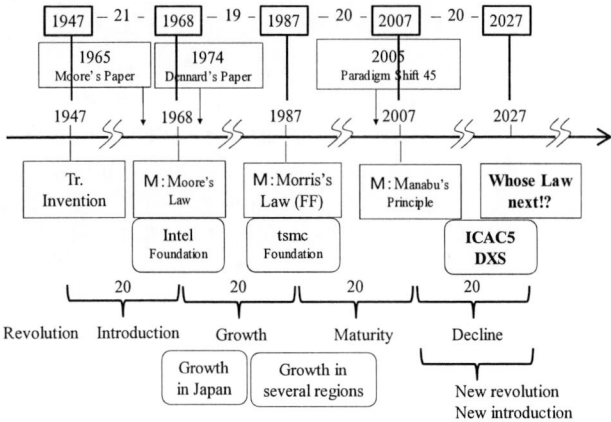

Figure 9 Laws and History

VII. Conclusion:

1. In the VUCA era, the global situation is becoming more and more uncertain.
2. Looking forward, the semiconductor market, driven by multiple applications, such as IoT, Cloud, AI, Car, and 5G as represented by ICAC5, as well as conventional PCs and mobile phones, will be less volatile and grow explosively.
3. Semiconductor technology has also been developed to meet the ICAC5 market, and has overcome its stagnation. Presentations in academic conferences also stress "VSNS (Very Sure, No Surprises)."
4. The device roadmap, which ceased activity after 2015, has resumed activity in a different form. The new roadmap has provided system manufacturers with an overview of the future of sets and devices.
5. Chances are high that challenges experienced in the past will recur as scaling of interconnects advances. Let's note the challenges experienced in Paradigm Shift 45.
6. Semiconductor devices have developed in three directions: MM, MtM, and BC roadmaps. So have the systems. CMP systems for MM, MtM and BC are required.
7. This paper introduced the process shifting with the advancement of scaling of interconnects, and some examples of the most advanced challenges that CMP is facing.
8. I put together the laws that appeared about every 20 years since the invention of the transistor in 1947, and expressed a hope for what laws will govern in the next 20 years.

Everyone can open up and lead the future. I will continue to look forward to the frontier spirits of semiconductor engineers.

REFERENCES

1) M.Tsujimura. "Innovations to Support Semiconductor Technology "2016 International Conference on Planarization/CMP Technology ICPT2016 Proceedings.Beijing China, 2016-10-17/19, 2016, p58-63.
2) M. Tsujimura: Paradigm Cliff 2020- The future of semi-conductor device and CMP technologies -, ICPT2014, Kobe Japan, (2014) 8.
3) ITRS（International Technology Roadmap for Semiconductors）,SIA,
4) M.Tsujimura. "Ahead CMP, Beyond CMP- More CMP, More than CMP, Beyond CMP -", 2019 International Conference on Planarization/CMP Technology ICPT2019 Proceedings

MULTI-GRANULARITY RECONFIGURATION BASED PHYSICAL UNCLONABLE FUNCTION DESIGN

Jianan Mu[1,2], Jing Ye[1,2], Xiaowei Li[1,2], Huawei Li[1,2,3], Yu Hu[1,2]

[1] State Key Laboratory of Computer Architecture, Institute of Computing Technology, Chinese Academy of Sciences

[2] University of Chinese Academy of Sciences, Beijing, China

[3] Peng Cheng Laboratory, Shenzhen, China

{mujianan19s, yejing, lxw, lihuawei, huyu}@ict.ac.cn

ABSTRACT

This paper proposes a multi-granularity reconfiguration based physical unclonable function, which reduces the hardware cost of the adjustable PUF in FPGA. In comparison with existing works, the uniformity and the reliability remain same, while the average hardware cost is reduced 24%.

Keywords—arbiter physical unclonable function, FPGA, self-adjustment, hardware resource

INTRODUCTION

The Physical Unclonable Function (PUF) is an emerging hardware security primitive [1]. It exploits the random process variations to produce particular responses for input challenges, which are called the Challenge-Response Pairs (CRPs). Even with the same design, different manufactured PUFs will have different CRPs. In general, PUFs can be classified into two categories: the weak PUF such as the SRAM PUF [2] and the strong PUF.

The feature of a weak PUF circuit is similar to that of a unique fingerprint. The circuit can only generate one or a small number of corresponding bits. It is often loaded in a chip and used as a key [3-5].

A strong PUF is similar to a cipher table. Compared with a weak PUF, a strong PUF circuit with the same hardware overhead can generate a larger number of challenge-response pairs, which makes it applicable to the authentication field [3-5]. A typical strong PUF is the arbiter PUF [6]. Fig.1 shows its architecture. The arbiter PUF compares the delays of two paths to produce a response bit. Each path is consisted of path segments selected by a challenge.

One of the most important and critical metrics to evaluate a PUF, including the arbiter PUF, is the uniformity [1]. Uniformity refers to the distribution characteristic of response bits of a PUF circuit. For a PUF circuit with an output bit of 0 or 1, when the PUF output is evenly distributed, the probability of output 0 and output 1 is equal, and the uncertainty of its output is the highest, so is the corresponding security. The uniformity is calculated as the percentage of response bit 1s among all the response

bits For the arbiter PUF, an adjustment strategy was proposed to adjust the uniformity to around 0.5 [7].

In practice, due to the uncontrollable process deviation, the uniformity of PUF cannot be totally controlled. In order to adjust the uniformity of the output bits, an adjustable strategy of the arbiter PUF is proposed by [7]. It adds an adjustment module at the back end of the two transmission paths. The adjustment module balances the delay difference between the two transmission paths of the arbiter PUF by inserting different delay buffers to the transmission paths, so to adjust the uniformity of the response bits.

The architecture of the arbiter PUF with an adjustment module is shown in Fig.2. However, when the adjustable arbiter PUF is implemented on the FPGA, in order to allow uniformity adjustment of the arbiter PUF on different chips, the adjustment module needs to prepare more redundant adjustment particles. Previous work adopts small granularity. This eventually causes a large waste of hardware resources.

To solve this issue, this paper proposes a multi-granularity reconfiguration based physical unclonable function design. Our major contributions include:

1) The average hardware cost is reduced 24%.

2) The uniformity and the reliability of PUF remain the same.

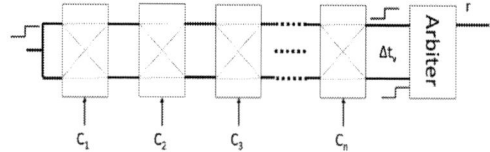

Fig.1 architecture of Arbiter PUF

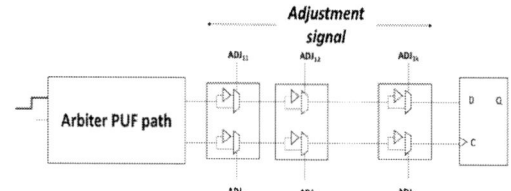

Fig.2 architecture of Arbiter PUF with an adjustment module

PROPOSED DESIGN

The overview of the proposed method is shown in Fig.3. It contains two key steps:

1) To solve the problem of wasting hardware resources due to the usage of redundant adjustment modules, a "tailing" technology based on reconfiguration is proposed.

2) To solve the problem that different delay buffers are required by the adjustment module for different FPGA chips, a technique of adding multi-granularity delay particles is proposed.

"Tailing" technology based on reconfiguration:

In traditional adjustable PUF design, the adjustment signal determines whether to insert the delay buffer adjustment particles to achieve good uniformity. By observing the adjustment values, it is found that the distribution of 0 and 1 in the adjustment values is sparse. In other words, there are many redundant particles.

We propose to use the greedy algorithm and the simulated annealing algorithm to obtain the effective bits of the adjustment signal which are as continuous as possible. Then the particles controlled by invalid signal bits can be deleted through reconfiguration. This reduces the amount of redundant particles, thereby reducing the hardware cost.

Adjustment technique using multi-granularity particles:

Previous work uses small delay buffer particles, and such adjustment values have highest performance for adjusting the uniformity. However, when the delay difference between the two paths is large, it has to use a lot of delay buffer particles to adjust. This paper proposes to use multi-granularity particles. These particles have different size of delays. In fact, the hardware overhead required to implement these particles with different granularity are the same. If several particles with small delays are used, then one or more particles with longer delays can replace them to reduce the hardware cost.

For example, ideally, if the delay difference of the two paths to be adjusted is 10 ns, and if there are only particles in the same size-1ns, it takes ten particles to achieve the purpose of adjustment. But if there are two more particles of 4ns, they can be used to replace 8 particles of 1ns, which ultimately reduces the hardware overhead.

However, it is not possible to arrange all particles into 4ns or larger. If only large-size adjustment particles are used, because the step size is too large, it is possible to skip the adjustment of the expected value during the adjustment process.

The specific implementation is illustrated in Fig.4. There are a lot of connection resources on the FPGA to achieve the same function, and different connections have different lengths and transmission delays. In this paper, when implementing the adjustable arbiter PUF, different routing resources are used to implement particles with different delays. The Fig.5 shows the corresponding routing. One LUT on the FPGA has several pins, and each pin can use different routing resources to achieve the same function.

However, the actual situation is not as ideal. If we need to adjust 150ns delay for a PUF, it is often not feasible to use a delay particle of 120ns and a delay of 30ns. This is because 120ns and 30ns are the design values. The process variations make the actual values different from the design values. However, this does not affect the adjustment process.

In addition, it can be seen that the hardware overhead required to achieve delay particles in different granularity is actually the same, because only routing resources are different.

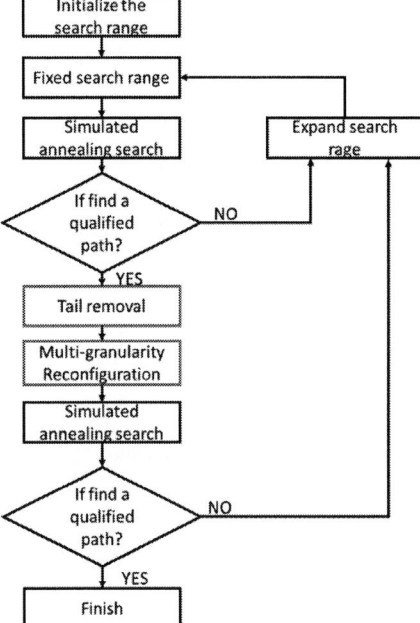

Fig.3 Design Flow

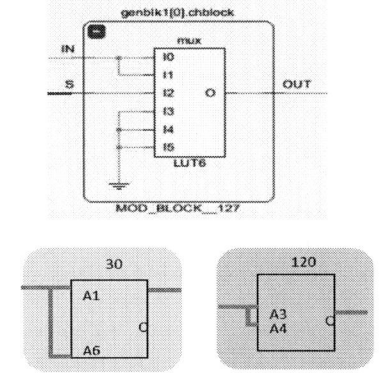

Fig.4 delay particles with different granularities

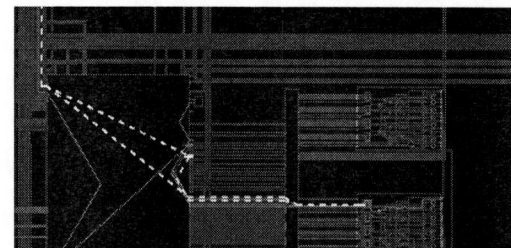

Fig.5 implementation figure of Arbiter PUF on FPGA (local)

EXPERIMENTAL RESULTS

The proposed method is evaluated on Xilinx Zynq7000 chips. Twenty different adjustable 32-bit arbiter PUF circuits are generated at 20 different locations in a chip. 1000 CRPs are tried each time and each challenge is repeated for 1000 times for evaluation. The experiment environment is shown in TABLE I.

The results prove that the hardware cost of the adjustable arbiter PUF implemented in this paper is 23.96% lower than [8] as shown in Fig.6.

At the same time, as shown in Fig.7, the uniformity of PUF implemented in this paper is 49.71%, and the reliability is 93.4%, which keep similar as [8].

ACKNOWLEDGEMENTS

This paper is supported in part by National Natural Science Foundation of China (NSFC) under grant No. (61704174, 61532017, 61432017, 61521092). The corresponding authors are Jing Ye and Xiaowei Li.

REFERENCES

[1] U. Ruhrmair, D. E. Holcomb, "PUFs at a Glance," DATE, 2014.

[2] J. Guajardo, S. S. Kumar, G.-J. Schrijen, P. Tuyls, "FPGA Intrinsic PUFs and Their Use for IP Protection," CHES, 2007.

[3] Gassend B, Clarke D, Dijk M V, et al. Controlled Physical Random Functions[C] //Annual Computer Security Applications Conference. Los Alamitos: IEEE Computer Society Press, 2002: 149-160.

[4] Aysu A, Gulcan E, Moriyama D, et al. Compact and Low-Power ASIP Design for Lightweight PUF-based Authentication Protocols[J]. IET Information Security, 2016, 10(5): 232-241.

[5] Katashita T, Sasaki A, Hori Y. A Novel Smart Card Development Platform for Evaluating Physical Attacks and PUFs[C] //Global Conference on Consumer Electronics. New Jersey: IEEE Transactions on consumer electronics, 2013: 37-39.

[6] D. Lim, J. W. Lee; B. Gassend, G. E. Suh, M. van Dijk, S. Devadas, "Extracting Secret Keys from Integrated Circuits," TVLSI, 2005.

[7] Jing Ye, Xiaowei Li, Huawei Li, Yu Hu, "Adjustable Arbiter Physical Unclonable Function with Flexible Response Distribution," SEMICON China Semiconductor Technology International Conference (CSTIC), paper 6.1, 2019

[8] J. Ye, Y. Hu, X. Li, "VPUF: Voter based Physical Unclonable Function with High Reliability and Modeling Attack Resistance," IOLTS, 2017.

TABLE I
EXPERIMENTAL ENVIRONMENT

Host PC	Hardware	Intel(R) Core(TM) i7-6700K
		8G RAM
	Software	Vivado 2016.4
		Vivado Tcl shell
FPGA	Xilinx ZC702 Board	#LUT = 53200
		#Register = 106400
		#IO = 200

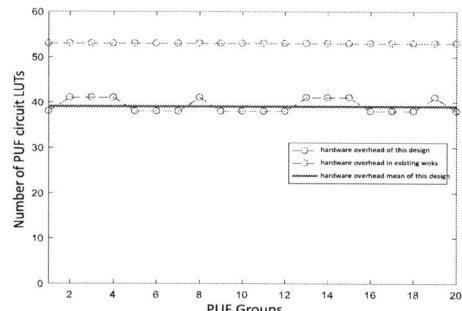

Fig.6 Hardware Cost Comparison

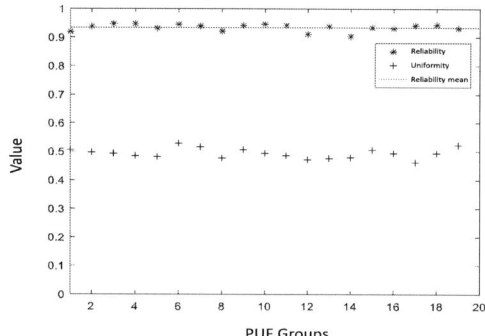

Fig.7 Uniformity and Reliability

978-1-7281-6559-2/20 $31.00 © 2020 IEEE

A PROGRAMMING FRAMEWORK OF CONCURRENT TEST ON SMT7 FOR IPS WHICH SHARE SAME ACCESS PORT

Tianyu Zhang[1], Fang Yanfen[2] and Kai Zhou[3]*

[2]Advantest, Shanghai 201210, China

[3] Advantest, Shanghai 201210, China

*Corresponding Author's Email: tianyu.zhang@ advantest.com.cn

ABSTRACT

With the development of the microelectronic technique, SOC integrates dozens or even hundreds of IP, which brings long test time and also significant test costs. This paper proposes a new method of concurrent testing using wait time of BIST of each IPs which share same access port. Meanwhile, this paper designs a software framework to implement the concurrent test rapidly and keep maintenance of test methods convenient. In this framework, test method need be divided into a set of functions marked by related type. Actions which could be run parallel is packed into these functions, then these functions are run serially at first, test time of each one will be recorded by the framework. Finally, the framework will call these functions according to their type and test time to implement concurrent test. The actual result show that more than 90% concurrent test efficiency was achieved on a 2 IP device which share same access port based on this framework.

INTRODUCTION

Current Situation

The trend toward higher System-on-Chip (SOC) integration of many independent functional system blocks into a single packaged device [1], SOC becomes more and more integrated and complex. In Terms of chip packaging technology, multi-die sealing of SOC has become a mainstream trend, the way of multi-die sealing is to encapsulate multiple dies on the wafer onto a single chip. This technology makes chip design and manufacture easier because it divides the various complex functions in the SOC into their respective dies, and the application industry including telephone\Server\Cloud\AI. Nevertheless, it takes more time to test a SOC with the development of this technology due to more IP is integrated into the SOC. Therefore, finding a solution to solve the problem of the high test time and high test cost caused by this complex SOC helps to win the market as ATE venders. In most cases, these IPs share one access port (e.g. JTAG, SPI, etc.). Moreover, BIST of these IPs include a lot of wait time (>20% test time) to get BIST result. During the wait time, access port is released, another IP could be tested.

Existing CCT Solutions

The existing concurrent test solutions in 93K including CCF and ISR as figure1 shows [2]. In CCF mode, the original execution mode which single-port patterns executed serially changed into new execution mode which multiport burst pattern executed one time, the way of CCF is merge several single-port patterns together into one multiport burst. CCF mode can be selected for DFT concurrent test due to only one sequencer is supported in this scenario. This mode can't be implemented in scenario which sequencer keeps starting and stopping.

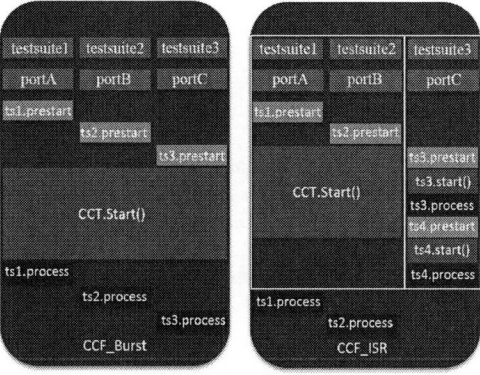

Figure 1: 93K CCT mode

In ISR mode, an independent sequencer is added based on the original CCF mode, which can realize the constant start and stop of pattern. However, neither CCF nor ISR modes in existing solutions can meet the demand to implement concurrent testing when sharing same test interface.

NEW FRAMEWORK SOLUTION

Feasibility Analysis

A SOC model with IP1 and IP2 is established, both two IPs have test suite A\test suite B\test suite C. The execution sequence of the entire test flow is shown in figure2.

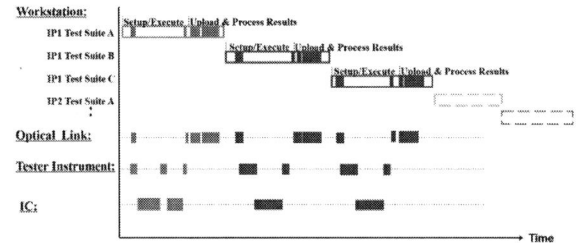

Figure 2: Test Time Analysis

The order of execution is IP1 suite A\suite B\suite C, and then IP2 suite A\suite B\suite C. Break down the test time of one suite, the test time consumed in workstation/optical link/tester instrument/DUT. When the test suite is executed, the workstation start to run and gives instructions, the instructions transmitted through optical link, then the electronic board start to configure the chip through the protocol interface(e.g. JTAG) according to the received instructions. Once the configuration is complete, the chip enters a self-test state. After the self-test is completed, the tester obtain the test result and return the result to the workstation for processing and judging. Through analysis, it can be found that the overall thread is occupied when workstation, optical link, electronic cards are in operation. The thread and test interface is released only when the IP enters self-test mode, the tester resources and interfaces can be utilized to test another IP at the same time, thus realizing concurrent testing when two IP sharing same test interface.

Framework Feature

The new framework has the following five features. Feature I, supports splitting the test programs into blocking and non-blocking modules. Blocking means execute the next module until the completion of current module, the common action of blocking module including pattern execution, DC test and so on. Non-blocking means the other module can be executed when the current module is being executed, the common action of non-blocking module including build in self-test, wait time and so on.

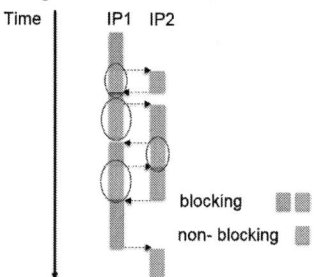

Figure 3: Blocking & Non-Blocking

As figure 3 shows, IP1 start to execute a non-blocking module and IP1 enters self-test state, then the thread assigned to the execution of IP2. Feature II, supports automatic recording test time of each blocking and non-blocking modules when the entire test flow is executed serially at the first time. The execution time of each module is the input of the sort algorithm. Feature III, supports automatic sorting to find optimal test execution sequence during concurrent testing. The final mass production program execution needs to achieve the thread jump between different modules in different IP. The performance of the sort algorithm determines the final test time and concurrent test efficiency. Feature IV, supports one button switch between serial execution and parallel execution. The whole process of concurrent test can be divided into two steps, the first step is execute the whole test flow serially and recording the execution time of each module, the second step is switch to concurrent execution mode and do the test. One button switch facilitates the seamless connection between program debugging and mass production. Feature V, high usability and strong scalability of the new framework. The usability of the framework is an important consideration during design the framework, it determines the development cost.

Sort Algorithm

The best concurrent test efficiency can be derived from optimal execution order, the sort algorithm is designed based on dynamic programming. Establishing one SOC model contains n IPs and each IP contains a different number of blocking modules and non-blocking modules. The total test time of model can be expressed as equation 1, T represents the total test time, i represents the number of modules per IP, t represents the test time of each module, k represents weighting coefficient, k equals zero when the module belongs to non-blocking. Otherwise, k equals one.

$$Min\ T = \sum_{i=1}^{i=x} K_{1i}\, t_{1i} + \sum_{i=1}^{i=y} K_{2i}\, t_{2i} + \cdots + \sum_{i=1}^{i=z} K_{ni}\, t_{ni} \quad (1)$$

Based on the idea of dynamic programming, regarding the all module ordering in total IP as a complete set of problems, then break the whole set down into small problem subsets. The subset problem consists of two types, one is the running order of IP in the initial state the other is the running order of IP when the non-block is filled. The optimal solution of the complete set can be obtained by the optimal solution of the subset. The output of subset is based on the current state and the decision mechanism of the subset, the output of the current subset is the input of next subset. The process keeps going around and obtain the optimal solution finally.

The current subset is the execution order of each IP in the initial state, as two IPs example shown in figure4, serial number I/II/III represents three different situations.

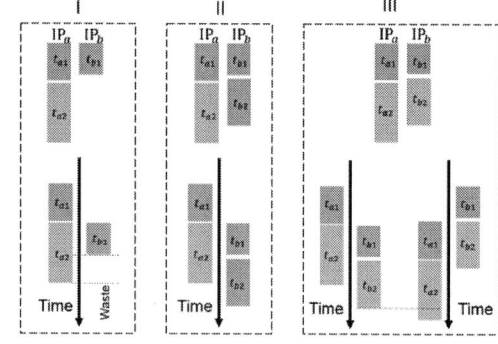

Figure 4: Two IP Sorting Diagrams

I indicates that IP_a contains non-blocking module and IP_b

978-1-7281-6559-2/20 $31.00 © 2020 IEEE

contains blocking module, and the total test time is shortest when IP_a run first, and then IP_b is started when the non-blocking module in IP_a start. But time is wasted due to the non-blocking module is not filled, then run module in IP_b to fill the time of non-blocking in IP_a as shown in II, this results in higher time utilization. III indicates that both IP_a and IP_b contain non-blocking module, and the total test time is shortest when IP_a run first which with longer non-blocking time. Upgrade the problem complexity, analyzing the complex case with n IP, the following conclusion can be obtained: the time length of the first non-blocking module of each IP is the ordering basis, and the longer the time, the higher the priority. Based on the above analysis, the sort algorithm applied in this programming framework is simplified in real engineering. By calculating the mean value of all IP's respective non-blocking modules, the higher the mean value, the higher the priority. Figure 5 shows the algorithm flow chart with 2 IP case.

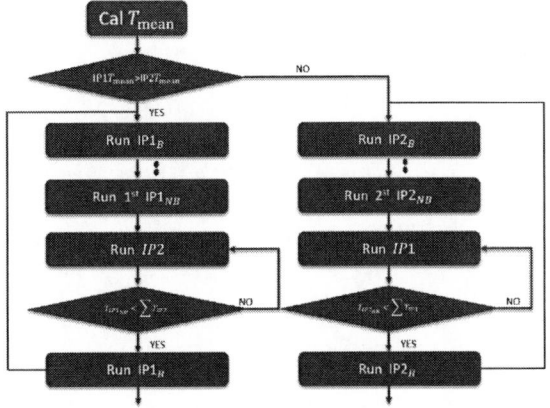

Figure 5: Algorithm Flow Chart

Programming Framework

The program framework consists of four layers as shown in figure6. The top-level class "User_Layer" provides the API interface for registering the module, user only need to splits the original test program into blocking or non-blocking module. The second-level class "YaCCT_Common_Layer" provides common functions. The third-level class "Executor_Layer" is the core layer of the entire framework, including algorithm implementation, automatic time recording, serial execution and concurrent execution. The bottom-class "TestMethod_Layer" provides the 93K underlying API, drive the tester hardware execution.

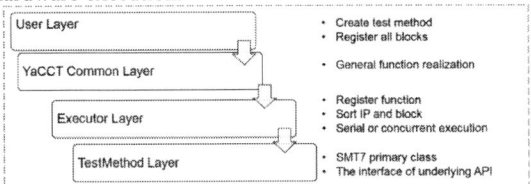

Figure 6: Program Software Framework

Framework Application Results

A real project was developed for concurrent testing based on this program framework, and the result is shown as figure7.More than 90% concurrent test efficiency was achieved, the suite name was removed for confidential reasons.

Framework Application Results Demo				
Test Suite	Serial	Concurrent	Saved Time	Concurrent test efficiency(CTE)
IP1_Suite1	308ms	310ms	70ms	97.22%
IP2_Suite1	72ms			
IP1_Suite2	467ms	467ms	232ms	100%
IP2_Suite2	232ms			
IP1_Suite3	1769ms	1778ms	550ms	98.38%
IP2_Suite3	559ms			
IP1_Suite4	185ms	482ms	169ms	90.37%
IP2_Suite4	466ms			

Figure 7: Application Results

The concurrent test efficiency can be calculated as eqution2.

$$CTE = 1 - \frac{\text{Actuall Parallel Test Time} - \text{Ideal Parallel Test Time}}{\text{Actuall Serial Test Time} - \text{Ideal Serial Test Time}} \quad (2)$$

CONCLUSION

This framework gives a solution to implement concurrent test between IPs which share same access port. The real application project results shows that high concurrent test efficiency is achieved according to this framework, and the total test time significantly reduced compared with normal test time reduction.

ACKNOWLEDGEMENTS

I would like to extend my sincere gratitude to Pan Yuanyuan and Fang Yanfen, for their instructive advice and useful suggestions on my thesis. I am deeply grateful of their help in the completion of this thesis.

REFERENCES

[1] SmarTest Documentation Center, Topic 105652.
[2] SmarTest Documentation Center, Topic 122230.

LOW VOLTAGE TIME-RESOLVED EMISSION (TRE) MEASUREMENTS OF VLSI CIRCUIT

Shang Chih Lin and Frank Yong
Gallant Precision Machining Co., Ltd
No. 5-1, Innovation 1st Rd., Hsinchu 30076, Taiwan Tel: +886-3-563-9999 Ext. 3776
E-mail : nicholaslinn@gpmcorp.com.tw.,

ABSTRACT

As a process node is getting smaller, the types of failure mechanisms are increasing.

New EFA technologies and methods are constantly development. One of the main changes EFA analyses is an enhancement of dynamic EFA in circuit failed in functional test.

We propose a technique for advanced Electrical Failure Analysis (EFA) tool with a Picosecond Imaging Circuit Analysis (PICA) detector with enhanced sensitivity for discussing Time Resolved Emission (TRE).

The key applications where the time-resolved imaging capability is very effective in reducing the debug time and improving the understanding the failure behaviors of VLSI chip for fault characteristics

INTRODUCTION

Time-Resolved Emission (TRE) measurement, is known as Picosecond imaging circuit analysis (PICA), based on A collection of near-infrared light (NIR) emitted by a hot source Carriers in transistor channels are an invaluable method Widely adopted for testing and failure analysis field. This technology can be detected non-invasive Probe switching activity inside VLSI circuits for measurements Skew, propagation delay, duty cycle, etc. [1-9]. In recent the years, its capabilities have been continuously expanded.

Use of light due to off-state leakage current (LEOSLC) [10-15] Apply to brand new applications, such as Logic state mapping [10], mode debugging [12], latch Ignition study [16,17], power supply noise [14,18,19] and Slew rate measurement [20], self-heating estimation [21], System and random device variability characterization [15],, etc.

Continuing trends in the semiconductor industry for smaller devices and lower supply voltages [22] causes major changes in intensity and spectrum light from IC [23]: detectable light decreases is exponentially related to the electric field in the transistor, and it has a linear relationship with the lateral dimension of the CMOS. In addition, A decrease in circuit voltage also results in longer wavelengths of the spectral distribution of the emitted light.

For these reasons, the detectors have higher sensitivity at these longer wavelengths, as well as lower intrinsic noise (dark counts), are needed.

In this paper, we will discuss the performance of a Single-Photon Detector that system developed by GPM(Gallant Precision Machining).

EXPERIMENTS

Custom-designed test chip manufactured using 32 nm low power technology was used to characterize PICA detector. The chip contains many designed experiments such as delay lines, PLLs, scan chains, and other circuits. In terms of size and pitch, the chip can perform general evaluations of detector sensitivity, jitter, and lateral spatial resolution, respectively.

For the purposes of this paper, most of the data has been collected from dedicated inverter-based delay line devices.

Low voltage circuit measurement

To simplify the analysis, a simple inverter with a 5 x 120 nm CMOS, switching at a fixed 15.625 ns (64 MHz) period was used in Figure. 1. The timing electronics was setup for a 125ns trigger loop. Figure. 2 shows the first TRE waveform obtained from an CMOS when the power supply voltage is set to 1.5 V, it is 1 minute in this case collect, then 8X periodic folding, and low-pass filtering for clear identification switching emission peak corresponding to falling edge by N-MOS and P-MOS, as well as the low and high LEOSLC regions of the waveform.

Figure.3 shows an example of a 10 min TRE waveform acquired from the same inverter gate at 1 V, the reduction 33% voltage as presented for a TRE waveform.

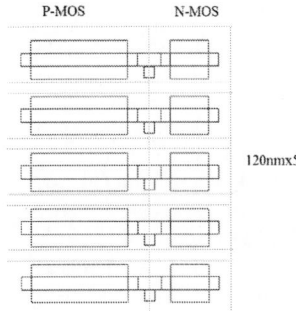

Figure. 1: Inverter gate layout used for the TRE measurements.

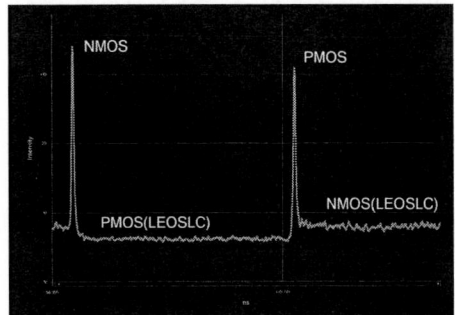

Figure. 2: 1.5 V TRE waveform acquired in 1 min.

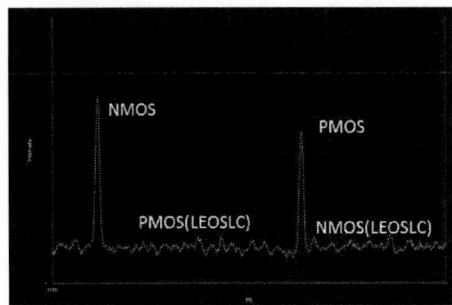

Figure. 3: 1 V TRE waveform acquired in 10 min.

Propagation delay measurements at low voltage

To demonstrate the basic function of the PICA system in actual PICA measurements, the signal propagation delay along the inverter chain described in Figure. 4 was measured at nominal 1.5 V and reduced 1.25 V.

Figure. 4 (a) shows an zoom in view of the switching transition events along the odd-numbered gates of the inverter chain at 1.5 V. The result was collected in 3 minutes using a 15.625 ns (64 MHz) clock and a 125ns trigger loop. The switched emission peaks are fitted to a

low-pass filter and Gaussian, respectively, and the calculated centroids are reported in Figure. 4 (b). A gate propagation delay of ~5.7 ps and ~8.4 ps are estimated for 1.5V and 1.25 V conditions, respectively.

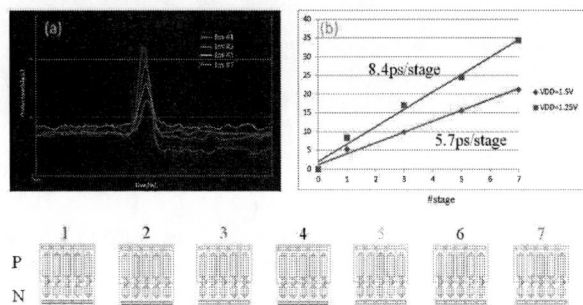

Figure. 4: (a) TRE waveforms at different position along a delay line and (b) calculated propagation delay.

Conclusions

In this paper, we propose a novel time-resolved emissions (TRE) system and its application in low-voltage measurement to scale VLSI circuits. We will demonstrate TRE measurements down to 1V in commercial tools.

We hope to show better results as soon as possible so that non-invasive TRE and PICA technologies can continue to be applied to future expansion nodes with smaller gates and lower supply voltages.

ACKNOWLEDGEMENTS

The authors would like to acknowledge and thank many other people from both IBM T.J. Watson Research Center for useful discussions, suggestions, and support, Including: Franco Stellari, Peilin Song.

REFERENCES

[1] J.A. Kash and J.C. Tsang, "Dynamic Internal Testing of CMOS Circuits Using Hot Luminescence", IEEE Electron Dev. Let., vol. 18, no. 7, 1997, pp. 330-332.

[2] P. Song et al., "Timing Analysis of a Microprocessor PLL using High Quantum Efficiency Superconducting Single Photon Detector (SSPD)", ISTFA, 2004, pp. 197-202.

[3] E.B. Varner et al., "Single Element Time Resolved Emission Probing for Practical Microprocessor Diagnostic Applications", ISTFA, 2002, pp. 741-746.

[4] D. Bodoh et al., "Defect Localization Using Time-Resolved Photon Emission on SOI Devices That Fail Scan Tests", ISTFA, 2002, pp. 655-661.

[5] J.S. Vickers et al., "Time-Resolved Photon Counting System Based on a Geiger-Mode InGaAs/InP APD and a Solid Immersion Lens", LEOS, 2003, pp. 600-601.

[6] G.L. Woods and S. Kasapi, "Spectrally- and temporally-resolved dynamic emission from CMOS ICs", LEOS, 2003, pp. 598-599.

[7] R.R. Goruganthu et al., "Spray Cooling for Time Resolved Emission Measurements of ICs", ISTFA, 2004, pp. 18-23.

[8] P. Ouimet et al., "Analysis of 0.13 m CMOS Technology Using Time Resolved Light Emission", ISTFA, 2004, pp. 203-209.

[9] H.L. Marks et al., "PC Card Based Optical Probing of Advanced Graphics Processor Using Time Resolved Emission", ISTFA, 2004, pp. 36-39. resolved dynamic emission from CMOS ICs", LEOS, 2003, pp. 598-599.

[7] R.R. Goruganthu et al., "Spray Cooling for Time Resolved Emission Measurements of ICs", ISTFA, 2004, pp. 18-23.

[8] P. Ouimet et al., "Analysis of 0.13 m CMOS Technology Using Time Resolved Light Emission", ISTFA, 2004, pp. 203-209.

[9] H.L. Marks et al., "PC Card Based Optical Probing of Advanced Graphics Processor Using Time Resolved Emission", ISTFA, 2004, pp. 36-39.

[10] F. Stellari et al., "Testing and Diagnostics of CMOS Circuits Using Light Emission from Off-State Leakage Current", IEEE Trans. on Electron Dev., vol. 51, no. 9, 2004, pp. 1455-1462.

[11] S. Polonsky et al., "Picosecond Imaging Circuit Analysis of Leakage Currents in CMOS Circuits", ISTFA, 2002, pp. 387-390.

[12] P. Song et al., "A Novel Scan Chain Diagnostics Technique Based on Light Emission from Leakage Current", ITC, 2004, pp. 140-147.

[13] S. Polonsky and A.J. Weger, "Off-state luminescence in metal-oxide-semiconductor field-effect transistors and its use as on-chip voltage probe", Appl. Phys. Lett., vol. 85, no. 12, 2004, pp. 2390-2392.

[14] F. Stellari et al., "Local Probing of Switching Noise in VLSI Chips using Time Resolved Emission (TRE)", NATW, 2005, pp. 130-137.

[15] S. Polonsky et al., "Photon emission microscopy of inter/intra chip device performance variations", ESREF, 2005, pp. 1471-1475..

[16] A. Weger et al., "Transmission line pulse picosecond imaging circuit analysis methodology for evaluation of ESD and latchup", IRPS, 2003, pp. 99-104.

[17] F. Stellari et al., "Study of critical factors determining latchup sensitivity of ICs using emission microscopy", ISTFA, 2003, pp. 19-24.

[18] S. Kasapi and G.L. Woods, "Voltage Noise and Jitter Measurement using Time-Resolved Emission", ISTFA, 2006, pp. 438-443.

[19] F. Stellari et al., "On-chip power supply noise measurement using Time Resolved Emission (TRE) waveforms of Light Emission from Off-State Leakage Current (LEOSLC)", ITC, 2009, paper 8.1, pp. 1-10.

[20] F. Stellari et al., "Switching time extraction of CMOS gates using time-resolved emission (TRE)", IRPS, 2006, pp. 566-573.

[21] S. Polonsky and K.A. Jenkins, "Time-resolved measurements of self-heating in SOI and strained-silicon MOSFETs using photon emission microscopy", IEEE Electron Dev. Lett., vol. 25, no. 4, 2004, pp. 208-210.

[22] S. Thompson et al., "A 90 nm Logic Technology Featuring 50nm Strained Silicon Channel Transistors, 7 Layers of Cu Interconnects, Low k ILD, and 1um² SRAM Cell", IEDM, 2002, pp. 61-64.

[23] A. Tosi et al., "Characterization of backside hot-carrier luminescence in scaled CMOS technologies", IRPS, 2006, pp 595-601.

UNIFYING YIELD ENHANCEMENT AND MANUFACTURING INTELLIGENCE WITH SMART SAMPLING

Yan-Qiu Zhang, Haw-Jyue Luo

Fujian Jinhua Integrated Circuit Co., Ltd., Jinjiang, Quanzhou 362200, China

Email: tc.zhang@jhicc.com

ABSTRACT

This paper presents Optimized Dynamic Sampling application of smart method based on Global Sampling Indicator (GSI). The optimized sampling method that detects the abnormal products and tool proposed in this paper can reduce risk and improve yield in semiconductor manufacturing process. It proposes several optimize algorithms to accurately filter out candidate lots, maximize sampling capacity and select samples with minimized risk and detect abnormal lots from fault detection and classification (FDC) or statistical process control (SPC) with real-time adjusting mechanism. The optimized sampling application has been validated on simulation data that transformed from production data.

Keywords—Smart sampling; GSI; Minimized risk; Abnormal; Sampling capacity

INTRODUCTION

Intelligent manufacturing is complex and important for semiconductor fab. In this domain, early identification and correction of the defect wafer or abnormal tool can reduce risk and improve yield. To overcome limited measurement capacity, it's critical to develop a smart sampling application. The concept of adaptive and dynamic sampling method is proposed [1]. This paper presents a dynamic sampling method based on GSI which aims at the risk of tool in each process step with high risk assessed by wafer quantity [2].

In this paper, we propose a smart sampling application that aims to minimize the tool risks and detect abnormal wafers. The Optimized Dynamic Sampling application includes 5 steps in Figure 1:

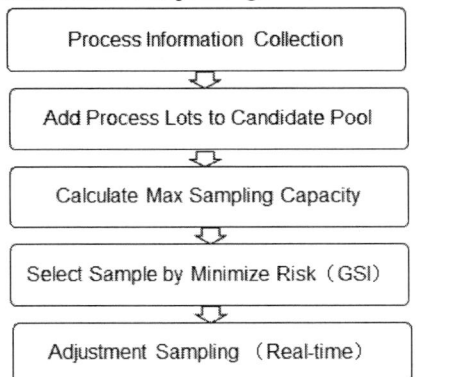

Figure 1: Process flow of Optimized Dynamic Sampling application

This paper provides a dynamic method to classify lots. And the loop time is estimated according to minimum process time of candidate lots in related process steps. The dynamic smart sampling includes real time sampling sequence adjustment based on risk of tool, FDC and SPC data.

THE OPTIMIZED DYNAMIC SAMPLING APPLICATION

Process Information Collection

It's important to collect general information, such as inspection tool list and related process steps, estimate process lots and information (Lot ID, Wafer counts of each lot, Product ID, Recipe Name, Process Time of each lot in each process step), process tool list of related process steps and information (Tool ID, the related inspection tools), and inspection time in related inspection step, the lot of inspection queue and information, the process step list of each process tool. Figure 2 shows the relationship of process tool and process steps. The process step P includes several process tools. And the process tool 1 can produce products of different steps. Then we can arrange lots to inspect according to general process information.

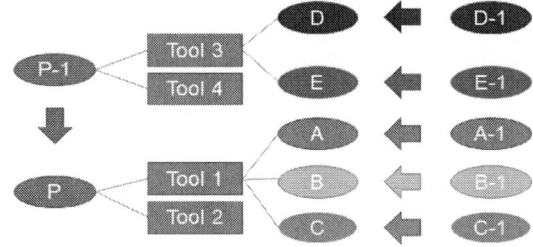

Figure 2: The relationship of process tool and process steps (Process Step: P-1, P, A, B, C, D, E, A-1, B-1, C-1, D-1, E-1)

Add related lots into candidate pool

This paper defines c number of considered process steps according to minimum estimate process time and estimate process time from $P - m$ to P process step. The estimate process time is calculated based on cycle time [3].

$$T_{C,min} = \max\left(T_{p,min} - T_{(P-i)\sim P,Estimate}\right)$$
$$and\ T_{c,min} \leq 0$$

$T_{p,min}$: The minimum estimate process time of related process steps.

$T_{(P-i)\sim P,Estimate}$: The process time of lots (lots of $P - i$

process step) from $P - i$ to P step.

$$T_{(P-i) \sim P, Estimate} = \max(T_{(P-(i-1)) \sim P, Estimate},$$
$$T_{(P-i) \sim P-1, Estimate} + T_{(P-1 \sim P), P-i})$$
$$1 \leq i \leq m$$

$T_{(P-1 \sim P), P-i}$: The delta time of lots (lots of process $P - i$) between department time at $P - i$ step and department time at P step.

Then add process lots of $P, P - 1 \ldots P - c$ process steps into candidate pool. Figure 3 shows the lots that will be considered as candidate lots.

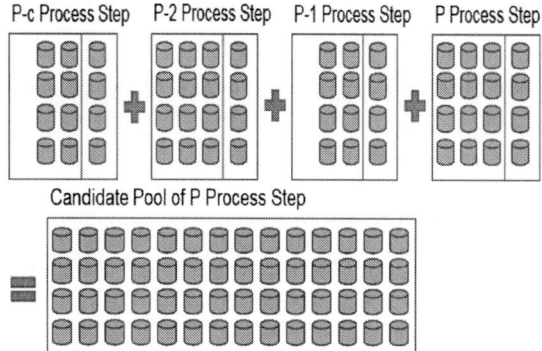

Figure 3: The process of selecting lot into Candidate pool of P process step

Calculate maximum sampling capacity

It's critical to effectively evaluate the inspection tact time in a loop for calculating sampling capacity. This paper proposes a method that assigns capacity to each process step with different weightage and then all process steps are calculated independently.

The maximum sampling capacity:

$$n_p = \frac{\beta_p N_{L,P} * \sum_{m=1}^{M} d * T_{v,m}}{\left(\sum_{p=1}^{a} \beta_p N_{L,P}\right) * T_{L,P}}$$

β_p: The weightage of the p process step

$T_{v,m}$: The valid inspection time of tool m

$T_{L,P}$: The inspection time of lots processed in P step

$T_{L,P}$: The Amplification factor

$N_{L,P}$: The number of candidate lots of P Process Step

Select samples by minimized risk

This paper uses iterative addition algorithm to obtain largest GSI reduction and then selects samples by minimized risk base on GSI introduced by Mr. Good and Mr. Purdy [4]. The iterative addition algorithm is to choose lot into the sampling pool one by one. The GSI is defined as:

$$GSI(S) = \sum_{r_{tool,PS}=1}^{R} \left(\frac{NRV_r(S)}{IL_r}\right)^{\alpha}$$
$$GR(S, I) = GSI(I) - GSI(S \cup I)$$
$$GR(\{l_1\}, I) = max_{(l)} \left(GR(\{l\}, I) \right)$$

$$GR(S_{Best}, I) = max_{(l)}(GR(\{l_1\} \cup \{l_2\} \cdots \cup \{l_k\}, I))$$

$NRV_{r,l}$: New risk value of $r_{tool,PS}$ if lot is inspected.

$$NRV_{r,l} = RV_r - GV_r \quad NRV_r(S) = min_{l \in I} NRV_{r,l}$$

$r_{tool,PS}$: The process tool in Process Step

IL_r: Inhibit Limit of Process tool

S_{Best}: The best sampling Lots by Smart Sampling

Figure 4 below shows the lot classifier takes based sampling mechanism into consideration. First, select the lots from the candidate pool according to the max GSI Reduction algorithm. Then adjustment of the based sampling mechanism lots and other candidates is made before process end.

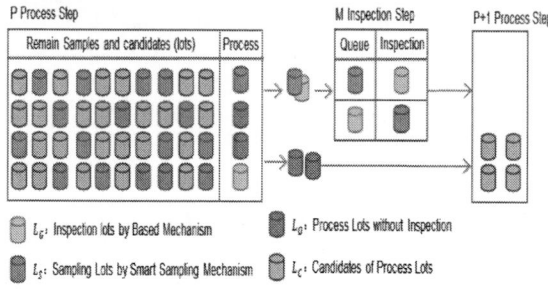

Figure 4: The lots are classified by considering sampling sequence

Real-Time adjustment mechanism

Real time adjustment mechanism is developed and implemented into dynamic smart sampling by considering the max GSI Reduction Mechanism, Real Time Risk Adjustment Mechanism, FDC and SPC data influence mechanism and Maximum Waiting Inspection Time Mechanism. The sampling sequence will be adjusted in real time [5].

The FDC&SPC Data Influence Mechanism is to define the abnormal process tool or abnormal lots according to the information of FDC & SPC, add the abnormal lot and the next lot processed by the same tool to S_{Best} or add the process lots of abnormal tool to S_{Best}.

The Real Time Risk Adjustment Mechanism is to define risk tool in related process steps and add process lot of risk tool in process step to S_{Best}.

If sampling pool is full, remove the lot of S_{Best} according to the remove max GSI reduction mechanism. Figure 5 illustrates the adjustment of selecting lots to sample.

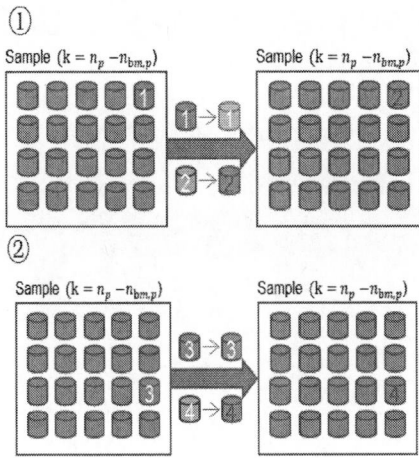

Figure 5: The adjustment sampling sequence

SIMULATION AND RESULTS

The simulation experiment presented in this section is performed on data that transformed from actual data of a production line in semiconductor factory. This experiment applied the max GSI Reduction Mechanism, Real Time Risk Adjustment Mechanism and FDC & SPC data influence mechanism to select samples.

Table 1 below shows that the smart sampling mechanism can detect more abnormal wafers with more GSI reduction compared to the base sampling approach. Properly increasing inspection utilization can effectively monitor product. It's necessary to strike a balance between ensuring effective monitoring and avoiding overload of measuring tool. The smart sampling mechanism can alleviate the problem of overburden on the measuring tool while ensuring effective monitoring. The optimized dynamic sampling mechanism can detect more abnormal rate and reduce more GSI based on same inspection utilization.

The anomalies are divided into two groups: random and non-random. The random anomalies can interfere with simulation results. The algorithm proposed by this paper could reduce interference from random events by repeated trials and large amounts of lots.

ACKNOWLEDGEMENTS

Under the guidance of the leader, Haw-Jyue Luo, this paper was successfully completed, and I would like to express my deep gratitude to him.

Thanks to my colleagues in the department for their encouragement and support, and for assisting in collating and obtaining manufacturing data. Thanks to other members of the JHICC for their help and assistance with this project.

Thanks to my family and my friend for preparing delicious food and spending good time with me.

REFERENCES

[1] J. Nduhura Munga, G. Rodriguez-Verjan, S. Dauz`ere-P´er`es, C. Yugma, P. Vialletelle, and J. Pinaton, "A Literature Review on Sampling Techniques in Semiconductor Manufacturing," in IEEE Transactions on Semiconductor Manufacturing, vol. 26, 2013, pp. 188–195.

[2] S. Dauz`ere-P´er`es, J.-L. Rouveyrol, C. Yugma, and P. Vialletelle, "A Smart Sampling Algorithm to Minimize Risk Dynamically," In Proceedings of the IEEE/SEMI Advanced Semiconductor Manufacturing Conference, 2010, pp. 307-310.

[3] C. Yu and H. Huang, "On-line learning delivery decision support system for highly product mixed semiconductor foundry," IEEE Trans. Semicond. Manuf., vol. 15, no. 2, pp. 274–278, 2002.

[4] R. P. Good and M. A. Purdy, "An MILP Approach to Wafer Sampling and Selection," IEEE Transactions on Semiconductor Manufacturing, vol. 20, no. 4, pp. 400–407, 2007.

[5] S. Housseman, S. Dauz`ere-P´er`es, Gloria Rodriguez-Verjan and Jacques Pinaton, "Smart Dynamic Sampling for Wafer at Risk Reduction in Semiconductor Manufacturing," IEEE International Conference on Automation Science and Engineering (CASE) Taipei, Taiwan, August 18-22, 2014

TABLE I
The results of simulation when $\alpha = 2$

Base Sampling Mechanism		Smart Sampling Mechanism		
Inspect utilization	GSI Reduction (Base/Base)	Inspect utilization	Detect abnormal Rate (SS-Base)	GSI reduction (SS/Base)
60%	1	100%	16.84%	2.35
80%	1	100%	22.33%	2.14
100%	1	100%	18.27%	1.41
120%	1	120%	8.42%	1.43
120%	1	100%	0.86%	1.02

978-1-7281-6559-2/20 $31.00 © 2020 IEEE

AN ADAPTIVE DENOISING SYSTEM FOR SUB-NM SCALE FAILURE ANALYSIS BASED ON TEM IMAGE

Chang Xu, Yi-Fu Zhang, and Chi-Ren Luo

Fujian Jinhua Integrated Circuit Co., Ltd., Jinjiang, Quanzhou 362200, China

Email: xander.xu@jhicc.com

ABSTRACT

TEM (Transmission Electron Microscopy) image-based failure analysis is one of the key Physical Failure Analysis (PFA) methods in DRAM (Dynamic Random Access Memory) manufacturing. Noise caused by surface damage or front material will corrupted the small features of defect which increases the difficulty for analyzing it. The system described in this paper is designed to enhance small features and reduce background noise for TEM images. The system is formed by 2-Dimensional Fourier Transformation (2D FT) with orientation and other features detection, Gabor filter and Wiener filter.

Keywords—Denoise; TEM; 2-Dimensional Fourier Transform; Wiener Filter; Gabor Filter; Nanomaterial; Sub-nm Scale; Failure Analysis; DRAM Manufacturing

INTRODUCTION

Over the past decades the semiconductor manufacturing industry has been developed rapidly. The device size has followed Moore's Law [1] undergoes tremendous shrinkage.

Failure analysis of sub-nm scale TEM image has become much more frequently used. However, for sub-nm scale failure analysis for polycrystalline silicon or single-crystalline silicon area, the fine silicon crystal structure will form a rigid grid mask which covered all details below. This disturbance will cause significant risk during failure analysis. The judgment of defect size and severity can vary significantly from engineer to engineer. Hence, one can see in Fig. 1-1 an illustration of defect covered by crystal structure while in Fig. 1-2 is locally magnified detail image of the defect.

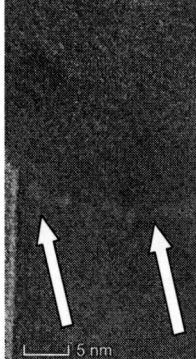

Fig. 1-1: As arrowed pointed, the defect area is covered by crystal structure

Fig. 1-2: Locally magnified defect area. The defect cluster has mixed with foreground silicon texture.

Previous research on TEM denoise is mainly focus on real image noise. [2][3][4] By far, little research about removing surface mask of crystal structure during sub-nm scale failure analysis is conducted.

PROPOSED SYSTEM OUTLINE

This paper now focuses on denoise for sub-nm scale TEM failure analysis. The system proposed by this paper is shown as Fig 2-0:

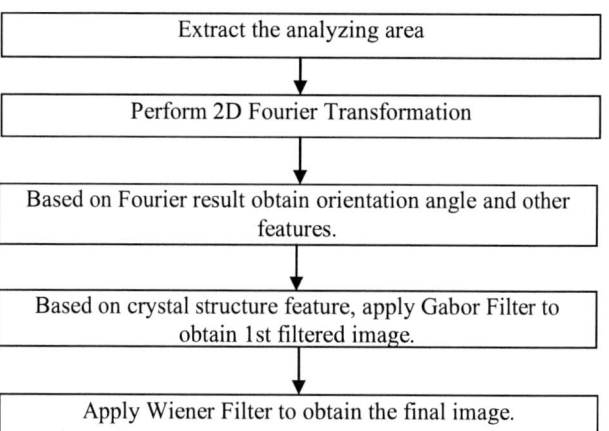

Fig. 2-0: Workflow of the adaptive denoising system

2D-FT based orientation angle analysis

The efficiency of Gabor filter is largely depends on the correct features of the texture. During PFA process, in order to make better failure analysis, the rotating of TEM specimen is inevitable which will result in orientation angle varies from specimen to specimen. Meanwhile different deposition methods of silicon will result in structure texture variation. Hence to optimize the Gabor filter, the correct orientation angle and other features must be obtained. The image Fig. 2-1 shows is the original image and its 2D FT. The 2D FT is defined as:

$$F(u,v) = \sum_{x=1}^{N} \sum_{y=1}^{M} F(x,y) e^{-\left(j\frac{2\pi u x}{N} + j\frac{2\pi v y}{M}\right)}$$

where
- F(u,v) is after 2D FT image
- N,M is image size
- F(x,y) is original image

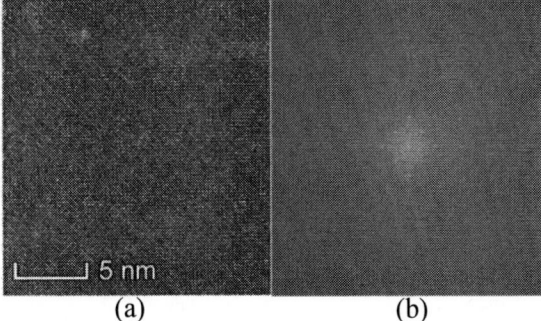

(a)   (b)

Fig. 2-1: Original image (a) and 2D FT (b).

From Fig 2-1 (b) 2D FT of the silicon crystal structure, we can obtain the structure feature based on bright points location on the 2D FT image. There are various ways to obtain the orientation angle and other features. [5][6] All these features will be applied later when using the Gabor filter.

Gabor filter application

Gabor filter is widely used for image denoise and contrast enhancement. [7][8] However, research about sub-nm scale TEM image processing with Gabor filter is hardly found. By performing Gabor filter, one can observe in Fig. 2-2. The covering crystal structure has reduced to only one direction.

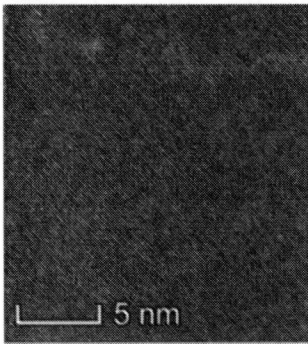

Fig. 2-2: After Gabor filter out other direction of texture.

Apply Wiener filter

Wiener filter is widely used in image processing. Previous reach on TEM image denoise had drawn conclusion that Wiener filter has better performance compare with some commonly used filters.(4) The wiener filter is defined as:

$$f(u,v) = \left| \frac{1}{H(u,v)} \frac{|H(u,v)|^2}{|H(u,v)|^2 + \frac{F_n(u,v)}{F_s(u,v)}} \right| F(u,v)$$

Where
F(u,v) is original image
f(u,v) is image after filtered.
H(u,v) is degradation function
Fn(u,v) is power spectral density of noise
Fs(u,v) is power spectral density of un-degraded image.

After applying Wiener filter and contrast adjustment. The final image shows as in Fig. 2-3 below.

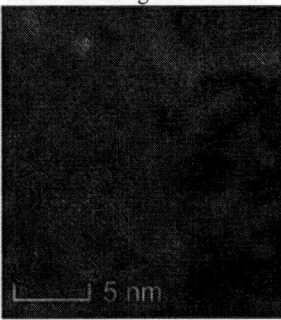

Fig. 2-3: After Wiener filter the defect area shows stronger contrast.

RESULT AND COMPARISON

In the section, the paper will perform comparison between proposed denoise system and other frequently used simple filters. The defect area from original image, after system, after median, after bilateral filter is shown in Fig. 3 below

(a)

(b)

(c)

(d)

Fig. 3: Defect area magnifying: (a) original image, (b) after proposed system, (c) after median filter, (d) after bilateral filter

The result shows image (b) denoised after proposed system the contrast between defect and surrounding area is much more significant. While for image (c) denoised after median filter, the defect area is accompanied with grid mask and sandy noise. This disturbance will bias the human expert judgment on defect size and affected area. And for image (d) denoised by bilateral filter the defect area become blur and mixed with surrounding silicon area. This will result in miss-judgment on defect severity level.

ACKNOWLEDGEMENTS

We would like to show our gratitude to the (Haw-Jyue Luo, Manager，JHICC) for sharing his wisdom and expertise with us during the course of this research.

We also thank our colleagues from JHICC who provided insight and expertise that strongly supported the research.

REFERENCES

[1] G. E. Moore, "Cramming more components onto integrated circuits," Electronics, vol. 38, No. 8, pp. 114-117, Apr. 1965

[2] T. Buchholz, M. Jordan, G. Pigino and F. Jug, "Cryo-CARE: Content-Aware Image Restoration for Cryo-Transmission Electron Microscopy Data," 2019 IEEE 16th International Symposium on Biomedical Imaging (ISBI 2019), Venice, Italy, 2019, pp. 502-506.

[3] B. Bajić et al., "Denoising of short exposure transmission electron microscopy images for ultrastructural enhancement," 2018 IEEE 15th International Symposium on Biomedical Imaging (ISBI 2018), Washington, DC, 2018, pp. 921-925.

[4] H. S. Kushwaha, S. Tanwar, K. S. Rathore and S. Srivastava, "De-noising Filters for TEM (Transmission Electron Microscopy) Image of Nanomaterials," 2012 Second International Conference on Advanced Computing & Communication Technologies, Rohtak, Haryana, 2012, pp. 276-281.

[5] B. Verma, V. Muthukkumarasamy and Changming He, "Unsupervised clustering of texture features using SOM and Fourier transform," Proceedings of the International Joint Conference on Neural Networks, 2003., Portland, OR, 2003, pp. 1237-1242 vol.2.

[6] Feng Zhou, Ju Fu Feng and Qing Yun Shi, "Texture feature based on local Fourier transform," Proceedings 2001 International Conference on Image Processing (Cat. No.01CH37205), Thessaloniki, Greece, 2001, pp. 610-613 vol.2.

[7] N. Nezamoddini-Kachouie and P. Fieguth, "A Gabor based technique for image denoising," Canadian Conference on Electrical and Computer Engineering, 2005. Saskatoon, Sask., 2005, pp. 980-983.

[8] G. Hemalatha and C. P. Sumathi, "Preprocessing techniques of facial image with Median and Gabor filters," 2016 International Conference on Information Communication and Embedded Systems (ICICES), Chennai, 2016, pp. 1-6.

Optical Scatterometry Modeling of 5 nm Logic Metal Gate Structures

Qi Wang, Aihua Yang, Yanli Li, Yushu Yang, Qiang Wu, Shoumian Chen

Shanghai IC R&D Center
497 Gaosi Road, Zhangjiang Hi-Tech Park, Shanghai 201210, PR China
Tel: +86-15201926624 E-mail: wangqi@icrd.com.cn

ABSTRACT

Optical scatterometry, owning to its noncontact and nondestructive nature, has become one of the most important metrology techniques in semiconductor manufacturing. At 5 nm technology node, as the high-k metal gate structure consists of nanometer scale layers, the detection of such small structure is very challenging. In this paper, we built a scatterometry model based on Rigorous Coupled Wave Analysis (RCWA) method, and explored the parameter space to find an optimized combination for grid size, illumination angle, illumination wavelength, number of orders, etc. From our study, we recommend a set of parameter settings which will provide a balance between sensitivity and efficiency for 5 nm FinFET structures.

Keywords—Metrology, Optical scatterometry, Optical critical dimension (OCD), Hi-k metal gate, Fin-FET

1. INTRODUCTION

Over the past decades, rapid development in semiconductor industry has brought continuous and aggressive shrinkage in dimensions. At 5 nm technology node, the metal gate length can be less than 20 nm. Within such narrow structure, there is a combination of several different high-k work function metal compounds, such as Titanium Nitride (TiN), Tantalum Nitride (TaN), Titanium Aluminum Oxide (TiAlO), etc. Although the penetration depth of visible light in the metals are around 20 nm to 60 nm, which are large enough compared to the size of the metal gate structures, the contrast between different metals can be very weak, especially for those metal structures that are "buried" under other metal layers. So we need an effective method to collect optical information from the metal gate structure.

The Optical Critical Dimension (OCD) metrology has proven to be one of the most rapidly evolving techniques for collecting high-resolution optical signals from surfaces non-destructively [1, 2]. Different from the image-based Scanning Electron Microscopy (SEM) and Atomic Force Microscopy (AFM), OCD metrology is a model-based metrology. By collecting the optical response from light-nanostructure interaction, OCD could collect information related to Critical Dimension (CD), overlay, and Line Edge Roughness (LER) or microelectronic devices at nanometer scale. As one key of OCD is in the modeling, we built a scattering model here based on the popular Rigorous Coupled Wave Analysis (RCWA) method [3, 4], and studied the metal gate structure with the model under various settings.

2. SIMULATIONS AND SETUPS

In our simulations, one-dimensional binary gratings structures are used as a reference for our models in Matlab. We built a typical FinFET structure depicted in Figure 1 [5]. The metal gate structure has an 33.2 nm length, 40 nm width, and 55 nm height. From core to edge, each layer has the thickness of 2.6, 2.6, 3.6, 0.8, 0.8, 1.8, 1.4 nm for Tungsten (W), TiN, TiAlO, TaN, TiN, Hafnium diOxide (HfO$_2$), SiOCN in Figure 1 (b).

First, for the implementation of RCWA, we build the device's dielectric constant map on high resolution grid and stratify the FinFET geometry profile along the z direction. The constraint of lateral electric and magnetic field continuity is imposed to the interfaces bewteen the vertical layers. Then, the dielectric functions in each layer are expanded into Fourier series, and the S-matrix calculation process for multilayers is applied to compute layer scattering matrix [6-8]. By iterating scattering matrix through all layers, the global reflection and transmission scattering fields can be obtained. During the implementation of RCWA, we can choose incident wavelength with associated refractive index n and extinction coefficient k for all materials. Illumination angle can be altered by tuning the incidence polar angle θ and azimuthal angle φ.

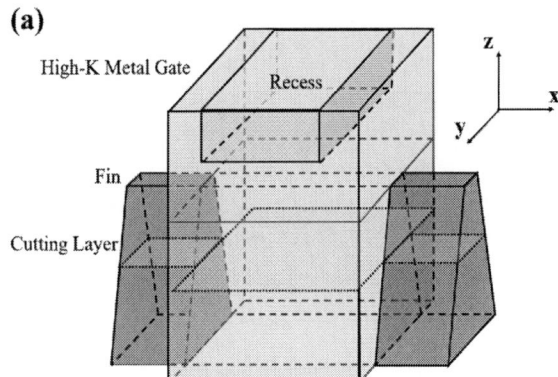

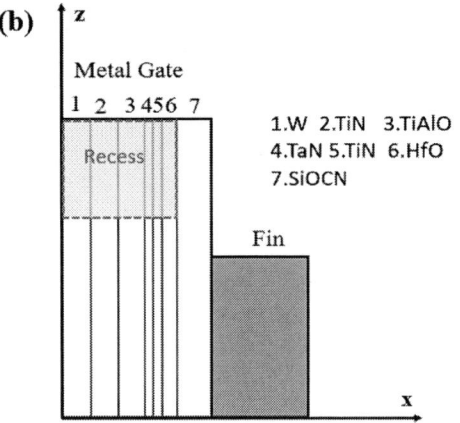

Figure 1. Diagram of 5 nm FinFET model structure (a) 3D (b) X-Z cross-section.

Figure 2 shows a schematic diagram of a typical OCD angular scatterometry. We choose a comparatively larger pitch (for example 300 nm) for FinFET structures to enable the

978-1-7281-6559-2/20 $31.00 © 2020 IEEE

detection of non-zeroth order diffracted light: oblique incidence with high incident angle (above 60°) and 300 nm period can provide the first order diffraction below 529 nm wavelength with a numerical aperture of 0.9.

As we have found that, a grid size of 2.5 nm or 1.25 nm can not only maintain the accuracy to reflect the tiny signal difference between different function metals layers with 2~4 nm thickness, but also complete the simulation within a short period of time (2.5 nm segmentation only needs several minutes while 1.25 nm segmentation consumes near an hour).

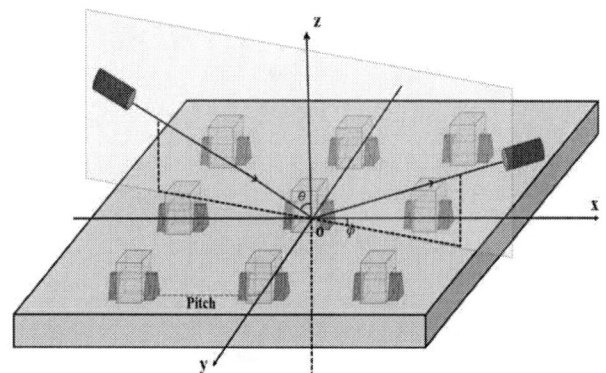

Figure 2. Schematic of optical scatterometery for periodic FinFET array.

3. RESULTS AND ANALYSIS

Figure 3 indicates that the shrinkage of FinFET dimensions will significantly reduce the signal sensitivity for OCD metrology. For example, at the condition of normal incident light in Figure 3, with recess depth varying from 0.5 nm to 2.5 nm, from the zeroth diffraction order signal, the sensitivity of OCD signals to the metal gate recess depth behaves almost linearly, in which when the structure expands to its 1.3 times in volume, the signal is 3 times larger.

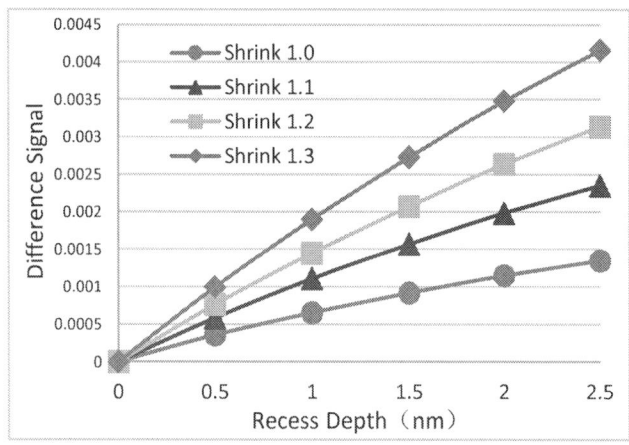

Figure 3. Calculated difference signal sensitivity when recess depth varies under different FinFET sizes (in volume).

Another quality index for OCD metrology is intermixing and correlation between different input parameters [9]. It refers to the ability in the discrimination between different input

variables in the output signal, or the ability to recognize the variable changes from different individual dimensions. In Figure 4, we set the recess depth, layer thickness 1, 2, 3, and 7 for W, TiN, TiAlO and SiOCN as five input variables and they are all varied at steps of 2.5 nm. The signal are collected by changing incident angle and diffraction orders.

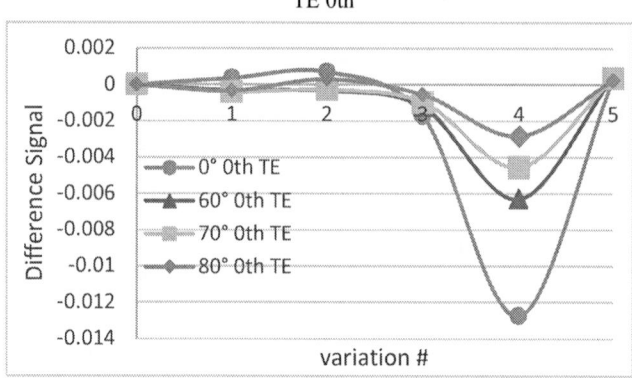

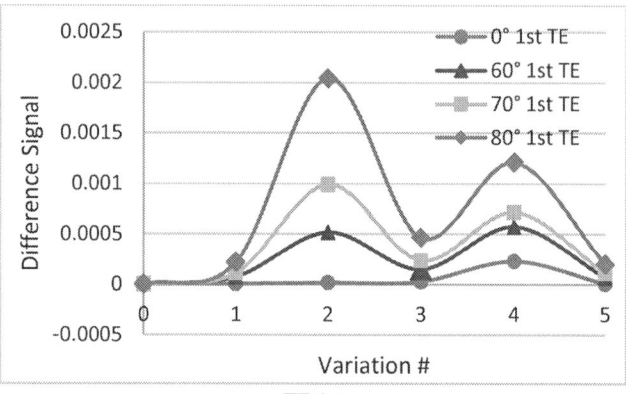

Figure 4. Difference signals (variations of recess depth, layer 1,2,3,7 for W, TiN, TiAlO and SiOCN) under the channels of incident polar angle θ (0°,60°,70°, and 80° at +X direction) and diffraction orders.

Based on the results, we can decide whether it is necessary to collect high order diffraction signals. The basic OCD metrology usually focuses on the zeroth-order diffracted signal because zeroth-diffraction has a much larger intensity which normally leads to a much higher signal-to-noise ratio, and it always exists for any incident angle, wavelength and pitch. However, the high order diffraction can provide extra signal channels to bring more versatility and alternativity. In some circumstances, these signal channels still remain substantial diffraction efficiency thus detectability. In our settings in Figure 4, it is shown that the 1st TE signals can be significant with large incident polar angle θ. Some input variables only have significant reponse to 1st order diffractions. As evidence, there already have been many other studies on OCD with near-ultraviolet light or even EUV light source and large numerical aperture lens to capture the high-order diffraction [10].

Also it is noteworthy that, by scanning the incident wavelength for 5 nm FinFET structure, the difference signals in zeroth and first order diffractions were collected respectively. In Figure 5, the X axis represents 5 input variables as above (X=1 means no variation) and the Y axis represents incident wavelength from 200 nm to 250 nm with 10 nm interval. Color gradient represents the signal intensities' fluctuation. From Figure 5, it is shown that both of polarizations and diffraction orders can provide the signatures of input variable in each pixel. By comparing these signatures with OCD library, the structural parameters can be obtained by solving the inverse diffraction problem [11].

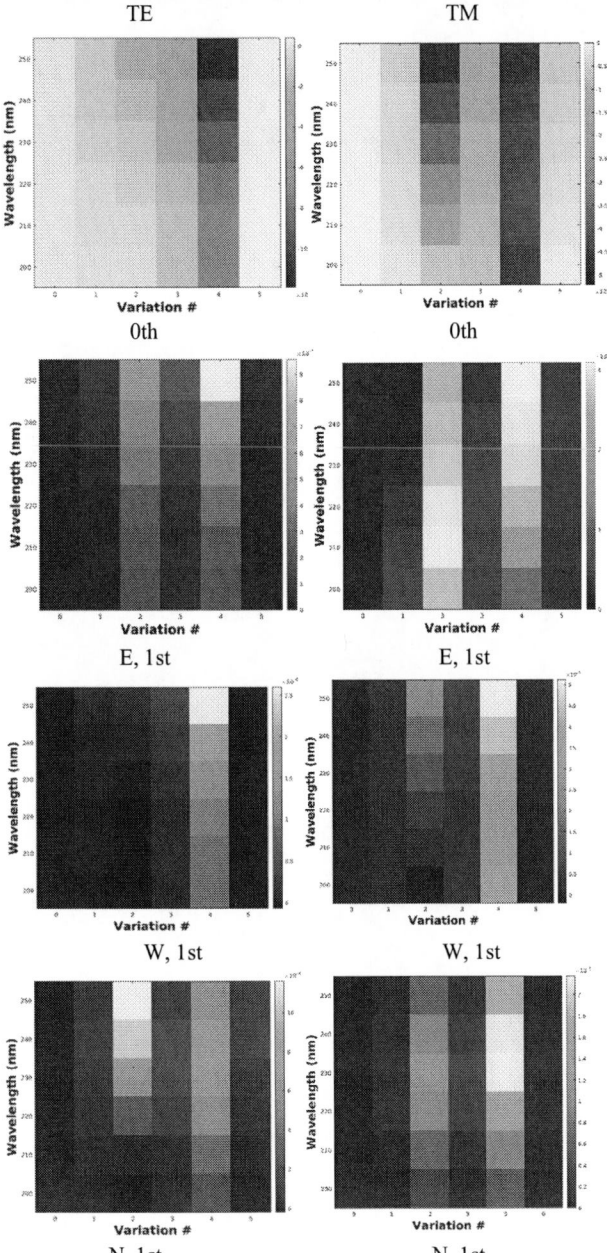

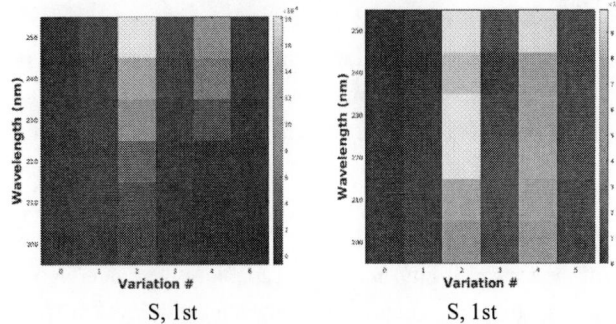

Figure 5. Comparison of difference signals for different incident wavelength, polarizations and diffraction orders at incident polar angle θ of 60°.

The direction of oblique incident beam also seems crucial to the FinFET optical scatterometry. The east (+X) and north (+Y) titled beam brings totally different signals because FinFET geometry is not symmetric along X and Y directions. As the metal gate layer boundary is oriented along the Y direction in our model, it is possible that the east (+X) titled

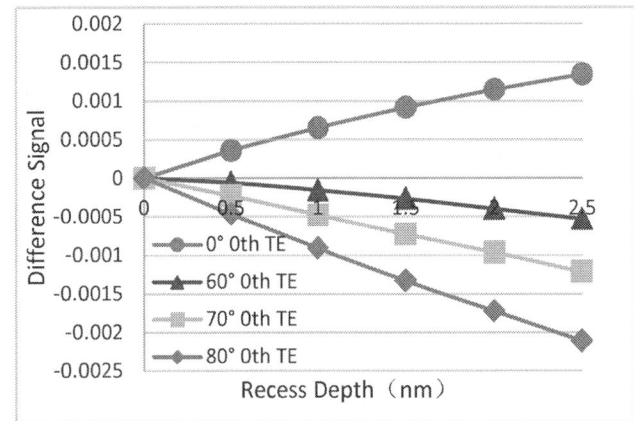

beam can feel more of scattering along X direction and has a

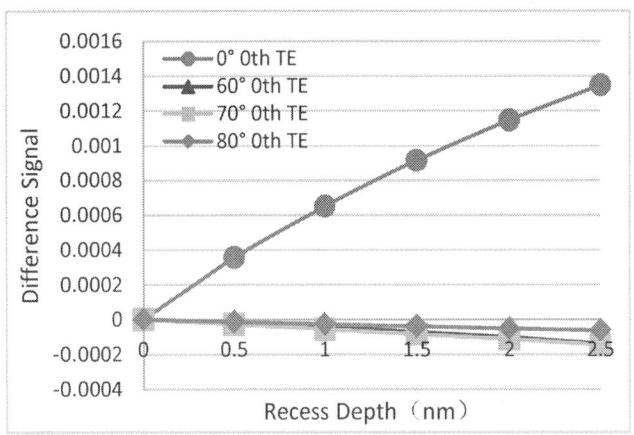

(a) East incidence

(b) North incidence

Figure 6. Difference signals for (a) east (b) north incident directions.

much larger signal intensity shown in Figure 6.

4. CONCLUSION

In this paper, we have done an OCD simulation for 5 nm FinFET structure. By making full use of angular and spectroscopic scattering measurements for metal gate structure, we have found that the sensitivity of optical scatterometry can be significantly improved. Our next step is to solve the diffraction inverse problem and improve OCD simulations with better sensitivity, modeling capability. The overlay and line edge roughness metrology is also on the way [12, 13].

5. ACKNOWLEDGEMENTS

This work was supported by Yi Huang, Qingyun Zuo and Shanghai Integrated Circuit Research & Development Center. The authors would like to acknowledge the help from them.

6. REFERENCES

[1] Madsen MH, Hansen PE, "Scatterometry-fast and robust measurements of nano-textured surfaces". Surf Topogr Metrol Prop 4:023003, 26pp. (2016).

[2] Raymond CJ,"Overview of scatterometry applications in high volume silicon manufacturing". AIP Conf Proc 788:394–402, (2005).

[3] M. G. Moharam and T. K. Gaylord. "Rigorous coupled-wave analysis of planar-grating diffraction". J. Opt. Soc. Am. A. Vol.71 (7), (1981).

[4] Moharam MG, Grann EB, Pommet DA, "Gaylord TK, Stable implementation of the rigorous coupled-wave analysis for surface-relief gratings: enhanced transmittance matrix approach". J Opt Soc Am A 12:1077–1086, (1995).

[5] Techsights Inc.Summary of Critical Device Metrics, TEM-EDS Results. Report code: ACE-1809-801.

[6] Li L, "Use of Fourier series in the analysis of discontinuous periodic structures". J Opt Soc Am A 13:1870–1876, (1996).

[7] Boher P, Petit J, Leroux T, Foucher J, Desières Y, Hazart J, Chaton P,"Optical Fourier transform scatterometry for LER and LWR metrology". Proc SPIE 5752:192–203, (2005).

[8] Li L, "Formulation and comparison of two recursive matrix algorithms for modeling layered diffraction gratings". J Opt Soc Am A 13:1024–1035, (1996).

[9] C. J. Raymond, M. R. Murnane, S. L. Prins, S. S. H. Naqvi, J. W. Hosch, and J. R. McNeil, "Multiparameter grating metrology using optical scatterometry," J. Vac. Sci. Technol. B 15, 361–368, (1997).

[10] Gross H, Rathsfeld A, Bär M, "Profile reconstruction in extreme ultraviolet (EUV) scatterometry: modeling and uncertainty estimates". Meas Sci Technol 20:105102, 11pp, (2009).

[11] Raymond CJ, Littau M, Chuprin A, Ward S, "Comparison of solutions to the scatterometry inverse problem". Proc SPIE 5375:564–575, (2004).

[12] Huang H-T, Kong W, Terry FL Jr, "Normal-incidence spectroscopic ellipsometry for critical dimension monitoring". Appl Phys Lett 78:3983–3985, (2001).

[13] Ko C-H, Ku Y-S, "Overlay measurement using angular scatterometer for the capability of integrated metrology". Opt Express 14:6001–6010, (2006).

QUALITY CONTROL IN SAPPHIRE GROWING: FROM AUTOMATED DEFECT DETECTION TO BIG DATA APPROACH

Dr. Ivan Orlov[1]* and Frédéric Falise[1]

[1] Scientific Visual, Lausanne, Switzerland

*Corresponding Author's Email: Ivan.Orlov@ScientificVisual.ch

ABSTRACT

We illustrate how automated scanners visualise internal defects in raw sapphire *prior* to its processing, and present some defect statistics that Scientific Visual has collected over five years of serving key sapphire suppliers in Europe and Asia. The article illustrates use of defect location and morphology data to reveal trends in sapphire quality, compare production modes, and to find out the optimal parameters for sapphire growth.

INTRODUCTION

In order to improve the crystal quality, it is important to know specific reason of defects formation, the causes of their propagation and their distribution in the crystal. The most abundant defects in sapphire crystals are gaseous and solid inclusions and block boundaries. Their typical origin is loss of the crystallisation front stability [1]. It leads to capture of impurities, inclusions and gas bubbles from melt, which is accompanied by crystal structure deterioration. Another defect origin is the seed orientation, which defines direction of dislocation lines and orientation of small-angle boundaries and, therefore, probability of their inheritance.

It is equally important to obtain the data on defectiveness as early as possible in the production cycle. Today many manufacturers evaluate crystal defectiveness after costly coring or wafering because rough surface of raw crystals does not allow to precisely locate and quantify defects in the crystal volume. At that moment up to 20% of the processed material is rejected. Such approach actually means that a typical sapphire factory is working a day per week machining initially defective material.

Modern quality control tools allow to see complete defect distribution in raw crystals *prior* to its processing. Unlike human control, such data are not limited by the acuity of the operator's eye and subjective bias. For the first time manufacturers can accumulate large amount of objective quality data that are suitable for scientific processing, notably for Big Data and artificial intelligence algorithms.

STEP 1. AUTOMATED DEFECT DETECTION

Among three most popular sapphire growing techniques – Edge-Defined Film-Fed Growth (EFG), Verneuil and Kyropolous – below we give examples for the former two; the latter is the subject of next article.

Edge-Defined Film-Fed Growth (EFG)

EFG is a method of film replenishment with edge limitation of the growth. The melt is introduced into the crystallisation zone by capillary forces that arise while using capillaries wetted by the melt. It is ideal for producing crystals with small cross-section and defined shape, such as ribbons for substrates for integrated circuits or tubes for lasers and high-pressure Na lamps. Currently, a clear tendency exists to expand the applicability of EFG sapphire in the watch and armour industries.

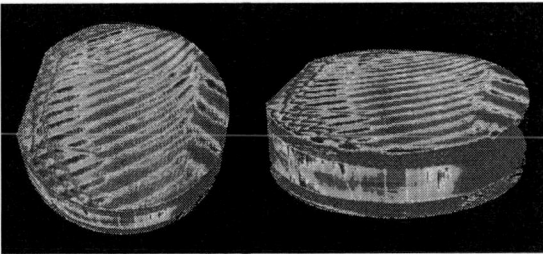

Figure 1: Defect pattern in non-processed (raw) disc cored from an EFG-grown plate. Disk ∅44.5mm, thickness 10.6mm. Colour code marks sapphire defect density: from deep blue (non-defective material) to deep red (highest defectiveness). © SapphiroScan™ image.

Figure 1 shows defects recognised in non-polished item cored from EFG ribbon. Well visible are:

1. Typical wavy pattern of surface bubbles (bubble walls) – large clouds of um-size bubbles directly under the plate surface, protruding up to 500um. Waves are oriented along pull direction and have oscillating thickness.
2. Sandwich structure in the volume: high defect density layer equidistant from ribbon surfaces - see intermediate layer on the right of Figure 1.

Verneuil

In Verneuil furnace a fine dry alumina powder of microns size is shaken through the wire mesh and allowed to fall through the oxygen-hydrogen flame. The powder melts and ~50um-thick film of liquid is formed on the top of the seed crystal. It then freezes progressively as the seed crystal is slowly lowered. By this method sapphire crystals, so called "carrots", are grown up to ∅50 mm that makes them widely used in watch industry, where carrots are sliced to 1-5mm thick disks and then polished into transparent watch covers.

Figure 2 shows a typical defect pattern in Verneuil-grown carrot. Large fraction of our data denotes three zones of quality (listed by order of defectiveness):

1. Top of the carrot (end opposite to its seed – the bottom side on Figure 2) – due to flame turned off at the end of the growth cycle, causing abrupt stop of the growth.
2. Seed end of the carrot – due to seed impact on the growth process
3. Gaz bubbles spread along the carrot volume, that often decorate stress lines or crystal domain borders.

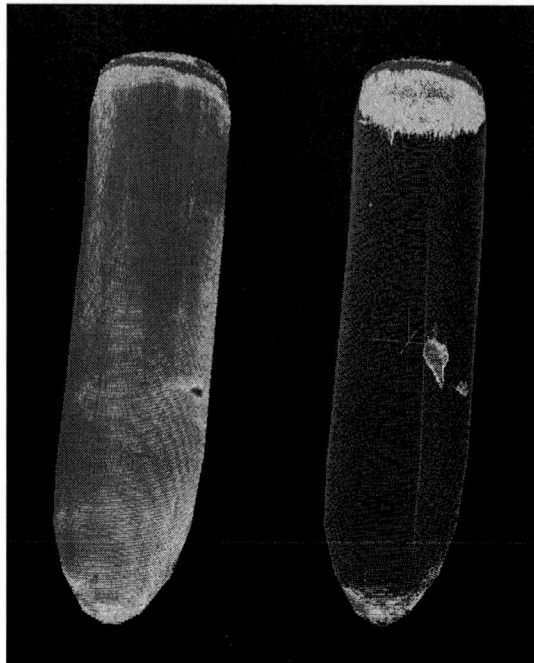

Figure 2: Defect pattern in non-processed (raw) Verneuil carrot ⌀34mm, length 220mm. Skin visualisation (left) and inside view (right) show the same carrot in the same orientation. For colour code see Figure 1 caption. © SapphiroScope™ image.

Note the big internal 3D defect that translate very differently at the surface. Without the early-stage 3D tomography such defect can't be properly assessed until post-slicing stage. Given the upward direction of growth, a local non-uniform condition appeared at the surface gave rise to aggregation of gas bubbles which 'migrated' into the crystal and got healed over time.

Irrelevant to growth method, the precise location, geometry and morphology of defects are automatically stored for a statistical analysis.

STEP 2. FROM DEFECTS TO PATTERNS

Defect morphology and defect distribution in the crystal is a manifest of the crystal growth process issues. Therefore, improving crystallisation process as a whole requires complete and objective data about defect patterns.

Analysis of defect geometry

Each defect geometry is automatically captured by the Quality Control system, which allows to compare occurrence and morphology of defects in different furnaces or manufacturers. Below we take an example of Verneuil crystals to show how statistics of defect size and morphology can help to trace a crystallisation issue.

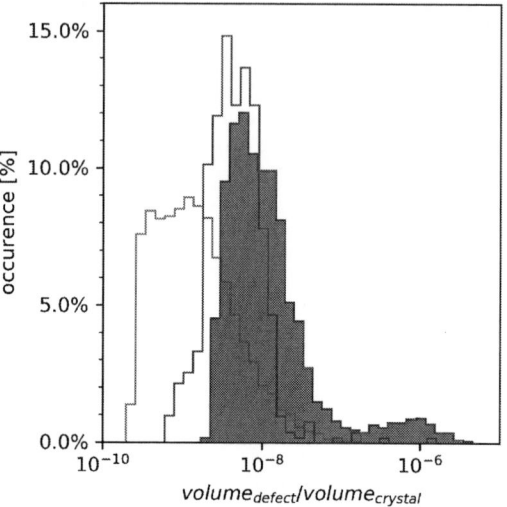

Figure 3: Occurrence of defect of different size in Verneuil carrots from three manufacturers (red, green and blue). Horizontal axis shows volume occupied by defects of a given size over full crystal volume.

Figure 3 illustrates size distribution of defects in Verneuil crystals produced by three companies. Note the tail corresponding to bigger defect size in Manufacturer 3 (blue filled). Generally speaking it has two groups of defects characterised by different sizes: < 15um and >25um. Morphology statistics (not shown here) confirms that defects of larger size originate from the dissociation of molten alumina, and their diameter increase as the pulling rate decrease, as mentioned by Hui [2].

Regarding EFG sapphire, an excellent example of correlating defect patterns with relevant process parameters - although with human analysis only - is recently presented by Stoddard et al. [3].

Quality Signature

Clear advantage of an instrumental quality control is the ability to visualise typical defect patterns in a batch of items. 'Batch' is a set of identical items produced and/or processed in similar conditions, for example originating from a specific furnace or a specific supplier.

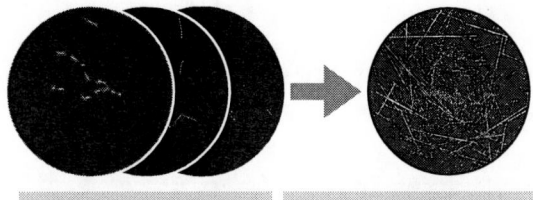

Individual defect images Batch Quality Signature™

Figure 4: Batch Quality Signature: digitally superimposing defects in individual items. Imagine that you look through a thousand of non-polished sapphire pieces together and see all defects in them at once.

Software option *Quality Signature* allows to superimpose defect patterns of individual items into a single digital "quality fingerprint". Generally, such batch fingerprint contains all defects from all controlled items.

STEP 3. FROM PATTERNS TO BIG DATA

Modern quality control tools allow to store defect patterns and to validate production improvement over time, or compare various production modes objectively.

However, growth of industrial crystals is accompanied by hundred thousand data points for temperature, pressure, chemical atmosphere, speed etc. Human brain can only capture relationships, establish correlations and find the root cause analysis between a very limited amount of inputs and associated outputs. The ultimate step in finding optimal growth parameters is the application of Big Data and deep learning algorithms to multitude of different set of growth parameters to find out dependencies that human brain cannot discover.

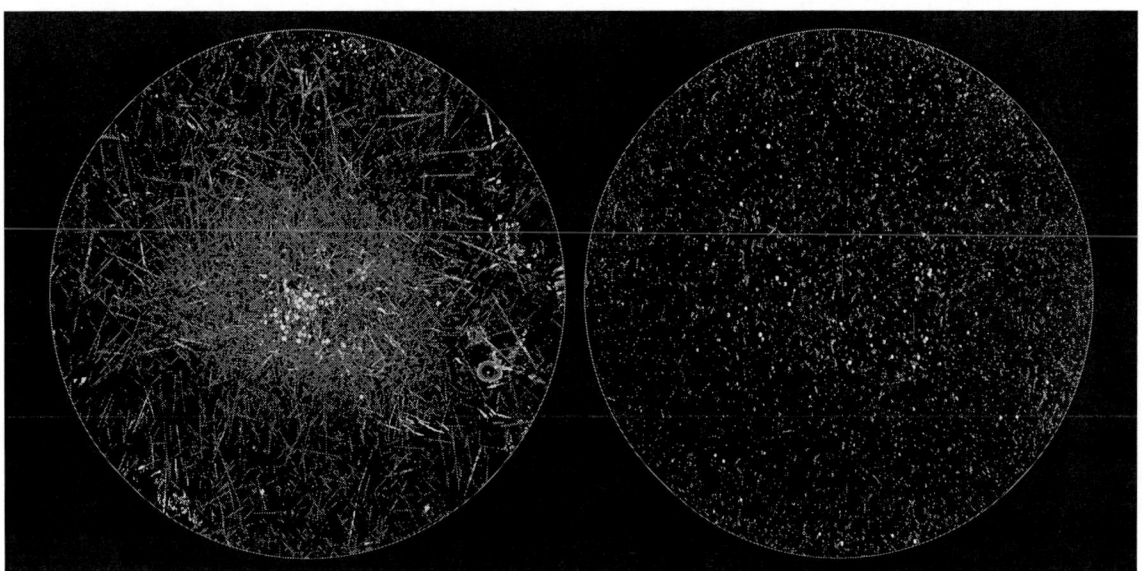

Figure 5: Two Quality Signatures comprised of 1000 non-polished Verneuil carrot slices of ⌀32 mm each. Colour encode both defect morphology and density: blue for structures, red for bubbles. The denser the defect - the whiter its colour.

Quality Signatures (QS) are essential for optimisation of crystal growth. By comparing QS taken with time interval or at different production lines, the production team can effectively measure whether its growth process tuning efforts are bearing fruits.

QS on Figure 5 represent two sapphire production lines located in Europe. The left one reveals high concentration of large bubbles in the centre (red to white colour) and substantial number of elongated defects (blue colour). It is a sign of too high radial temperature gradient, due to insufficient thermal insulation of furnace active area. Signature on the right exhibits a well tuned production - while defects are unavoidable, they are uniformly distributed and their size is under control. There are only few dense (white coloured) defects.

Accumulation of defect patterns allows to measure impact of process parameters and, with computer power, find their combination that leads to minimum defectiveness.

We believe that the combination of immersive confocal tomography and the deep learning technologies will create a significant disruption in improving industrial crystal growth yield.

Scientific Visual mission is to empower every crystal manufacturer to achieve this important milestone.

REFERENCES

1. Kurlov, Vladimir. *Sapphire: Properties, Growth, and Applications. In: Reference Module in Materials Science and Materials Engineering.* Elsevier (2016)
2. Hui Li. *Bubbles propagation in undoped and Titanium (Ti3+)-doped sapphire crystals grown by Czochralski (Cz) technique.* Crystallography. Université Claude Bernard - Lyon I (2014).
3. Stoddard, N., Seitz, M., Seitz, M., & Mushock, W. *Surface Defects in EFG Sapphire Single Crystals.* Journal of Crystal Growth, 125306 (2019).

TOWARDS UNDERSTANDING INTERACTION BETWEEN HOT CARRIER AGEING AND PBTI

M. Duan, J. F. Zhang, Z. Ji, W. Zhang

Department of Electronics and Electrical Engineering, Liverpool John Moores University,
Byrom Street, Liverpool L3 3AF, UK
E-mail: j.f.zhang@ljmu.ac.uk

Abstract

Early works on device ageing often focus on one source, while devices in a circuit suffer degradation from different sources. There are only limited information on the impact of ageing from one source on ageing from a different source. This work researches into the interaction of ageing induced by Hot Carriers with that by Positive Bias Temperature Instability (PBTI). It will be shown that one can slow down the other and the ageing can be substantially overestimated without considering their interaction. Although a PBTI after Hot Carrier Ageing (HCA) will increase the degradation, a HCA following a PBTI can result in a reduction in ageing for long channel devices. The defect responsible for their interaction will be explored.

Introduction

There are a number of sources causing the ageing of MOSFETs, including Hot Carrier Ageing (HCA) [1-4], Positive/Negative Bias Temperature Instabilities (PBTI/NBTI) [4-12], and Time Dependent Dielectric Breakdown (TDDB) [13]. Early works typically investigate them separately and develop models without considering their interaction. For example, As-grown-Generation (AG) model has been proposed for NBTI [7-12] and PBTI [5,8] and power law based lifetime prediction has been used for HCA [1,2].

In real circuit operation, device ageing can be dominated by different sources under different operating conditions and ageing from one source can affect ageing from another source in the subsequent operation. For example, the access transistor of a Static Random Access Memory (SRAM) cell in Fig. 1a suffers from HCA during 'Read 0', as shown in Fig. 1b. The same transistor is subjected to PBTI stress in the subsequent 'Write 0' operation. At present, knowledge on the potential interaction between HCA and PBTI is limited [14] and the objective of this work is to research into it. It will be shown that one can slow down the other and ageing will be overestimated without considering this interaction.

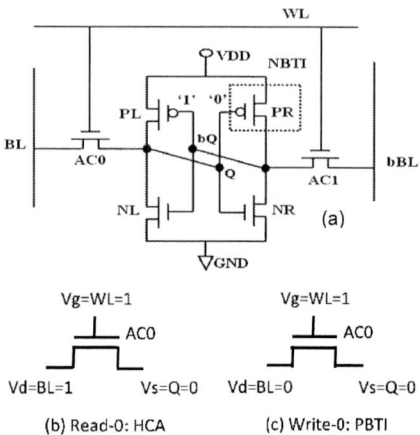

Fig. 1. (a) The six-transistor SRAM cell. (b) When 'Read 0', the access nMOSFET AC0 is under hot carrier ageing. (c) When 'Write 0', AC0 is under PBTI stress.

Devices and Experiments

nMOSFETs used in this work were fabricated by a commercial 28 nm CMOS process. They have metal gate and the high-k/SiON dielectric stack has an equivalent oxide thickness of 1.2 nm. To vary the relative strength of HCA versus PBTI, different channel lengths have been used, ranging from 27 to 225 nm. A relatively wide channel width of 900 nm is used to minimize the device-to-device variation [15].

PBTI was carried out with source and drain grounded, while HCA was under Vg=Vd. The ageing was monitored from the threshold voltage shift, measured under a given drain current of 100 nA × W/L [16]. All stresses and measurements were carried out at 125 °C.

Over-estimation of ageing

If one assumes that HCA and PBTI are independent processes and there is no interaction between them, the total ageing will be the sum of HCA and PBTI (The symbol 'Δ' in Fig. 2a), which can be obtained by performing HCA on one device and PBTI on another device.

To investigate the potential interaction, we used the waveform in Fig. 2b to stress one device alternately by HCA and PBTI. Fig. 2a shows that the ageing under the alternating HCA/PBTI stress (line) is considerably lower than the sum of HCA and PBTI. This conforms that HCA and PBTI affect each other and their interaction will be further investigated next.

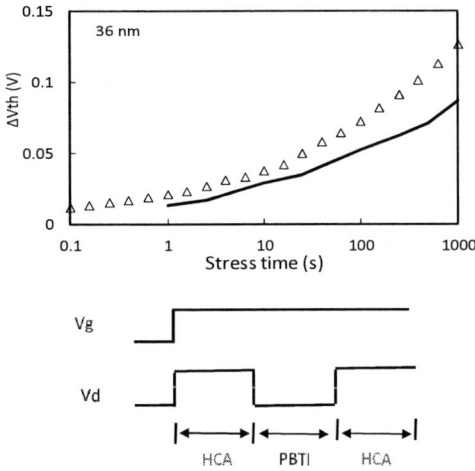

Fig. 2. (a) A comparison of HCA+PBTI (Symbol 'Δ'), when carried out independently on two different devices, with that when HCA and PBTI were carried out alternatively on the same device (the line). (b) Voltage waveforms for the line.

Interaction between PBTI and HCA

Fig. 3a shows the results when devices of different channel lengths were stressed by PBTI, HCA, and PBTI again in sequence. During the first PBTI, the ageing is independent of channel length, confirming that PBTI is a uniform process. As expected, the follow-on HCA is more severe for shorter channels, because of the higher lateral field over a shorter channel length for the same drain voltage. After HCA, the ageing during the 2nd PBTI becomes channel length dependent. To show this clearly, the HCA phase was removed in Fig. 3b and the two PBTIs were joined together. It can be seen that the shorter the channel, the less the 2nd PBTI ageing is. We conclude that HCA slows down the subsequent PBTI.

Fig. 4 shows the ageing following a sequence of HCA-PBTI-HCA. After HCA, PBTI ageing is more in longer channel device, in agreement with Figs. 3a&b. In the 2nd HCA post PBTI, ageing increases for the short channel device, but decreases for the long channel device. To explain this behavior, we study the defects responsible for the ageing next.

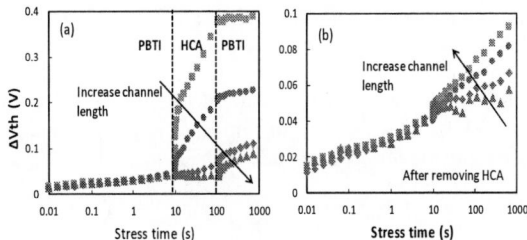

Fig. 3. (a) Ageing under a stress sequence of PBTI, HCA, and PBTI for different channel lengths. (b) A replot of (a) by removing the HCA phase.

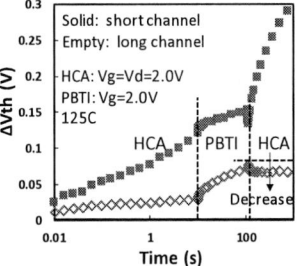

Fig. 4. Ageing under a stress sequence of HCA, PBTI, and HCA for different channel lengths.

Defects

Fig. 5 shows that the ΔVth is cyclic-able by alternating the gate bias polarity, when the PBTI Vg is relatively low at +1.5 V. After a trap captures an electron, the trapping is not permanent and it can be detrapped. This type of traps are referred to as 'Cyclic Electron traps (CET)'.

When stress voltage increases to Vg=+2.0 V, ΔVth is higher, as expected. The amount of CET is also higher. Moreover, some traps can remain charged at the end of the discharge phase and they are called as 'anti-neutralization electron traps (ANET)'. Based on this understanding of defects, we explain the interaction between PBTI and HCA observed in Figs. 3 and 4 next.

Physical processes

In Fig. 3, HCA can charge some of the ANET. As these ANET is already charged, they are not available in the subsequent PBTI. The shorter the channel, the more ANET is charged by HCA, so that the less the ageing in the subsequent PBTI, as shown in Fig. 3b. This indicates that the same ANET can be charged by either HCA or PBTI.

The reduction for the longer channel device during the 2nd HCA in Fig. 4 is caused by detrapping of CET. As illustrated by Fig. 6, the vertical electrical field is progressively reduced when moving from the source towards the pinch-off point during HCA. Some CET charged under higher vertical field during the PBTI can be detrapped during HCA. For shorter channel, HCA is strong and over-compensates the detrapping. For longer channel, HCA is too weak to compensates the detrapping, leading to the observed reduction in ageing.

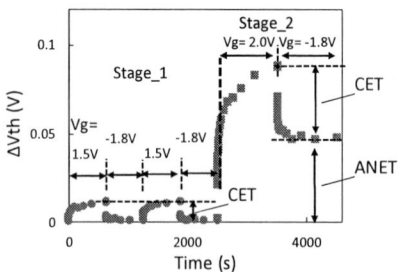

Fig. 5. In the stage 1, the cyclic electron traps (CET) can be charged by PBTI at Vg=+1.5 V and discharged under Vg=-1.8 V. In the stage 2, PBTI was under Vg=+2 V. CET increases and the anti-neutralization electron traps (ANET) remain charged at the end of discharge phase.

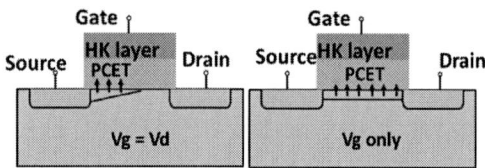

Fig. 6. An illustration of the reduction of vertical oxide field between source and pinch-off point during HCA, when compared with PBTI.

Conclusions

The interaction between PBTI and HCA is investigated in this work. It is shown that one can slow down the other substantially and the overall ageing will be overestimated without considering this interaction. Although PBTI is uniform without HCA, it becomes channel-length dependent after HCA. HCA can pre-charge the traps, making them unavailable to the subsequent PBTI.

For long channel devices, ageing during HCA after PBTI can even reduce. This is because the reduction of vertical field between source and pinch-off point during HCA results in a partial detrapping of the cyclic electron traps filled by the preceding PBTI.

Acknowledgements

The authors thank D. Vigar for supply of test samples used in this work. This work was supported by the EPSRC of UK under the grant no. EP/L010607/1.

References

[1] C. Hu, S. C. Tam, F. C. Hsu, P. K. Ko, T. Y. Chan, and K. W. Terrill, IEEE Trans. Elec. Dev., Vol. ED-32, pp. 375–385, 1985.

[2] M. Duan, J. F. Zhang, A. Manut, Z. Ji, W. Zhang, A. Asenov, L. Gerrer, D. Reid, H. Razaidi, D. Vigar, V. Chandra, R. Aitken, B. Kaczer, and G. Groeseneken, Proc. IEDM, pp. 547-550, 2015.

[3] M. Duan, J. F. Zhang, Z. Ji, W. Zhang, B. Kaczer, and A. Asenov IEEE Trans. Elec. Dev., Vol. 64, pp. 2478–2484, 2017.

[4] M. Duan, J. F. Zhang, Z. Ji, W. Zhang, D. Vigar, A. Asenov, L. Gerrer, V. Chandra, R. Aitken, and B. Kaczer IEEE Trans. Elec. Dev., Vol. 63, pp. 3642-3648, 2016.

[5] R. Gao, Z. Ji, J. F. Zhang, J. Marsland, and W. D. Zhang, IEEE Trans. Elec. Dev., Vol. 65, pp. 3662–3668, 2018.

[6] B. Kaczer, T. Grasser, P. J. Roussel, J. Franco, R. Degraeve, L. A. Ragnarsson, E. Simoen, G. Groeseneken, and H. Reisinger, Proc. IRPS, pp. 26-32, 2010.

[7] J. F. Zhang, Z. Ji, and W. Zhang, Microelectronics Reliability, Vol. 80, pp. 109–123, 2018.

[8] R. Gao, Z. Ji, S. M. Hatta, J. F. Zhang, J. Franco, B. Kaczer, W. Zhang, M. Duan, S. De Gendt, D. Linten, G. Groeseneken, J. Bi and M. Liu, Proc. IEDM, pp. 778-781, 2016.

[9] Z. Ji, S. F. W. M. Hatta, J. F. Zhang, J. G. Ma, W. Zhang, N. Soin, B. Kaczer, S. De Gendt, and G. Groeseneken, Proc. IEDM, pp. 413-416, 2013.

[10] M. Duan, J. F. Zhang, Z. Ji, W. Zhang, B. Kaczer, T. Schram, R. Ritzenthaler, A. Thean, G. Groeseneken, and A. Asenov, Proc of Symp. VLSI Technol., pp.74-75, 2014.

[11] R. Gao, A. B. Manut, Z. Ji, J. Ma, M. Duan, J. F. Zhang, J. Franco, S. W. M. Hatta, W. Zhang, B. Kaczer, D. Vigar, D. Linten, and G. Groeseneken, IEEE Trans. Elec. Dev., Vol. 64, pp. 1467–1473, 2017.

[12] R. Gao, Z. Ji, A. B. Manut, J. F. Zhang, J. Franco, S. W. M. Hatta, W. D. Zhang, B. Kaczer, D. Linten, and G. Groeseneken, IEEE Trans. Elec. Dev., Vol. 64, pp. 4011–4017, 2017.

[13] W. D. Zhang, J. F. Zhang, C. Z. Zhao, M. H. Chang, G. Groeseneken, and R. Degraeve, IEEE Electron Dev.Lett., vol.27, no.5, pp.393-395, 2006.

[14] M. Duan, J. F. Zhang, J. C. Zhang, W. Zhang, Z. Ji, B. Benbakhti, X.F. Zheng, Y. Hao, D. Vigar, V. Chandra, R. Aitken, B. Kaczer, G. Groeseneken and A. Asenov, Proc. IRPS, pp. XT-5.1-XT.5.7, 2017.

[15] M. Duan, J. F. Zhang, Z. Ji, W. D. Zhang, B. Kaczer, T. Schram, R. Ritzenthaler, G. Groeseneken, and A. Asenov, IEEE Trans. Elec. Dev., Vol. 61, pp. 3081–3089, 2014.

[16] M. Duan, J. F. Zhang, Z. Ji, J. G. Ma, W. Zhang, B. Kaczer, T. Schram, R. Ritzenthaler, G. Groeseneken, and A. Asenov, Proc. IEDM, pp. 774-777, 2013.

COMPREHENSIVE COMPARISON OF THE WIRE BOND RELIABILITY PERFORMANCE OF CU, PDCU AND AG WIRES

Liao Jinzhi Lois[1], Yu Minglang[2], Tee Weikok[3], Wang Bisheng[4], Jia Wenping[5], Yee Boonhwa[6], Zheng Haipeng[7], Zhang Xi[8], Fu Chao[9], Li Xiaomin[10], Hua Younan[11]*

[1,7,8,9,10,11] WinTech Nano-Technology Services Pte. Ltd.,10 Science Park Road, #03-26, The Alpha Science Park II, Singapore 117684

[2,4,5] Huawei Technologies Co Ltd, Bantian Huawei Base, Longgang District, Shenzhen, China 518129.

[3,6] Sumitomo Bakelite Singapore Pte Ltd, 1 Senoko S Rd, Singapore 758069

*Corresponding Author's Email: lois@wintech-nano.com; yuminglang@huawei.com; jiawenping@huawei.com; wangbisheng@huawei.com

ABSTRACT

Wire bond is the most common inter-connection method used to connect microchips to the terminals of a chip package. Wire bond reliability is vital to the packaging device performance. The work here investigated and compared the wire bond reliability performances of bare copper (Cu), palladium coated copper (PdCu) and silver (Ag) wires on aluminum pad. For PdCu, different Pd distributions on bonded ball and their effects on PdCu corrosion resistance were studied in detail. Chlorine (Cl) and sulfur (S) contaminations were purposely introduced to epoxy molding compound (EMC) to accelerate the corrosion process. bHAST (biased highly accelerated stress test) was conducted to see the reliability performance of Cu, PdCu, Ag wires under Cl and S environment. The results showed that Cl played a significant role on wire bond reliability. Whilst, S did not impact much on wire bond reliability. The failure mechanisms were studied. It was found that poor Pd coverage on bonded ball led to poor reliability performance. On the other hand, excessive Pd accumulation on the bonded ball bottom will deteriorate the reliability performance.

INTRODUCTION

Wire bond is the most common inter-connection method used to connect microchips to the terminals of a chip package. Wire bond reliability is vital to the packaging device performance. In current years, the wide usage of PdCu wires already exceeded the usage conventional bare Cu in electronic industry due to its good performance, like better bondability on lead surface and longer floor lift etc. Besides PdCu wire, Ag wires have become a novel bonding material in recent years. The usage of Ag wires is greatly increasing due to the high demand of LED and memory devices. However, the wire bond reliability failures of PdCu and Ag wires occasionally occur in the fields, especially with the presence of moisture and halogen element Cl.

It is important to study the reliability of Ag and PdCu,

especially on how the bonded ball Pd coverage effect on wire bond reliability. This work compared the wire bond reliability of bare 4N Cu, PdCu and 1N Ag. The Pd coverage on PdCu reliability was studied in details. The purpose of this work is to provide a reference to semiconductor industry on wire selection.

EXPERIMENT

In this study, 0.8mil 4N Cu (99.99% Cu) wire, 1N Ag (95% Ag) wire, and PdCu wires were used. The wires were bonded to SOP leadframe with 0.6um thick Al-0.5Cu bond pad using K&S ProCu bonder. Different EFO currents were adjusted to obtain different FAB Pd coverage. The FAB was cold mounted using epoxy. The samples were mechanically ground and polished. The Pd coverage of the FABs was checked by SEM/EDX. The Pd coverage calculation method is shown in Fig. 1. The targeted FAB Pd coverage were 50%, 75%, 85%, 100%.

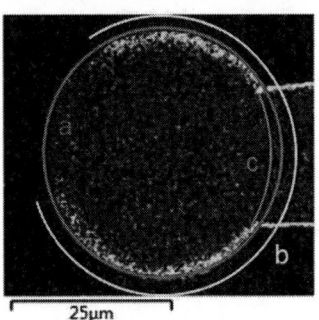

a = FAB perimeter
b = perimeter covered by Pd
c = perimeter covered by wire

Calculation method Pd%= (b-c)/(a-c) x100%

Figure 1: FAB Pd coverage calculation method.

The bonded devices were epoxy molded by Sumitomo. During the molding process, contamination of Cl and S ions (50ppm) were purposely introduced into the

EMC. Control samples, that is Cl and S ions <20ppm, were prepared. The molded device ran through reliability tests bHAST (130^0C/85%RH/20V). The materials and key factors were list in Table 1. There are total 18 legs in the experiment. 4N Cu and 1N Ag were added to compared with PdCu reliability performance. During bHAST test, sample electricity resistance was tested at time 0h, 96h. 192h, 288h, 584h, respectively. Two samples were taken out at each time point for analysis purpose. The device was considered failed if its resistance is larger than 20% of the time 0 samples. The experiment legs were shown in Table 2.

TABLE I.   EXPERIMENTAL LEGS

Leg	Bonding wire	EMC contamination
1	Bare Cu	< 20ppm Cl&S
2	Bare Cu	50ppm Cl
3	Bare Cu	50ppm S
4	PdCu-50% Pd coverage	< 20ppm Cl&S
5	PdCu-50% Pd coverage	50ppm Cl
6	PdCu-50% Pd coverage	50ppm S
7	PdCu-75% Pd coverage	< 20ppm Cl&S
8	PdCu-75% Pd coverage	50ppm Cl
9	PdCu-75% Pd coverage	50ppm S
10	PdCu-85% Pd coverage	< 20ppm Cl&S
11	PdCu-85% Pd coverage	50ppm Cl
12	PdCu-85% Pd coverage	50ppm S
13	PdCu-100% Pd coverage	< 20ppm Cl&S
14	PdCu-100% Pd coverage	50ppm Cl
15	PdCu-100% Pd coverage	50ppm S
16	1N Ag	< 20ppm Cl&S
17	1N Ag	50ppm Cl
18	1N Ag	50ppm S

TABLE II.   KEY MATERIALS AND FACTORS

#	Tests	Description
1	Wire type	4N Cu, 1N Ag, PdCu A, PdCu B
2	Wire diameter	0.8mil
3	Device type	SOP (lead frame), Daisy chain
4	Bond pad	0.6um Al-0.5Cu
5	Contamination elements in EMC	50ppm Cl⁻, S²⁻ respectively
6	Reliability tests	bHAST (130℃, 85%RH, 20V)

RESULTS AND DISCUSSION

Table 3 shows the bHAST results of the bare Cu, PdCu, 1N Ag wire bonds. There was no failure of all the wire bonds with <20ppm Cl & S in EMC and with 50ppm S in EMC up to bHAST 384h. It is obvious that there were failures of the wire bonds with 50ppm Cl in EMC, except for PdCu-75%Pd coverage. For 50ppm Cl EMC, bare Cu had most failures, followed by 1N Ag, PdCu-50%Pd

coverage, PdCu-100%Pd coverage, PdCu-85%Pd coverage. Based on the bHAST performance, the 18 legs can be divided into 3 groups: 1) bare Cu, PdCu-50%Pd coverage, PdCu-100%Pd coverage, 1N Ag, 2) PdCu-85%Pd coverage, and 3) PdCu-75%Pd coverage.

It has been reported by researchers that Pd coverage on the PdCu ball will act as protection layer to halogen and moisture attack and results in better reliability [1-3]. But in this study, it is not simple that good Pd coverage leads to good reliability performance. The best reliability performance is PdCu-75%Pd coverage. Poor Pd coverage (i.e. 50%) and excessive Pd coverage (i.e. 100%) both showed deteriorative reliability performance.

TABLE III.   BHAST RESULTS

Leg	Bonding wire	EMC contamination	bHAST (130C, 85%RH, 20V) Results					
			0 hr	96 hr	192 hr	288 hr	384 hr	Remark
1	Bare Cu	< 20ppm Cl&S	0/13	0/13	0/11	0/9	0/7	Pass
2	Bare Cu	50ppm Cl	0/17	6/17	15/15	13/13	11/11	Fail
3	Bare Cu	50ppm S	0/17	0/17	0/15	0/13	0/11	Pass
4	PdCu-50% Pd coverage	< 20ppm Cl&S	0/17	0/17	0/15	0/13	0/11	Pass
5	PdCu-50% Pd coverage	50ppm Cl	0/17	4/17	7/15	9/13	11/11	Fail
6	PdCu-50% Pd coverage	50ppm S	0/17	0/17	0/15	0/13	0/11	Pass
7	PdCu-75% Pd coverage	< 20ppm Cl&S	0/17	0/17	0/15	0/13	0/11	Pass
8	PdCu-75% Pd coverage	50ppm Cl	0/17	0/17	0/15	0/13	0/11	Pass
9	PdCu-75% Pd coverage	50ppm S	0/17	0/17	0/15	0/13	0/11	Pass
10	PdCu-85% Pd coverage	< 20ppm Cl&S	0/17	0/17	0/15	0/13	0/11	Pass
11	PdCu-85% Pd coverage	50ppm Cl	0/17	0/17	1/15	1/13	3/11	Fail
12	PdCu-85% Pd coverage	50ppm S	0/17	0/17	0/15	0/13	0/11	Pass
13	PdCu-100% Pd coverage	< 20ppm Cl&S	0/17	0/17	0/15	0/13	0/11	Pass
14	PdCu-100% Pd coverage	50ppm Cl	0/17	0/17	9/15	7/13	11/11	Fail
15	PdCu-100% Pd coverage	50ppm S	0/17	0/17	0/15	0/13	0/11	Pass
16	1N Ag	< 20ppm Cl&S	0/17	0/17	0/15	0/13	0/11	Pass
17	1N Ag	50ppm Cl	0/17	4/17	14/15	13/13	11/11	Fail
18	1N Ag	50ppm S	0/17	0/17	0/15	0/13	0/11	Pass

Fig. 2 shows the SEM image of bare Cu-EMC 50ppmCl (leg2) after bHAST 384h, and Fig. 3 shows the SEM image of 1N Ag-EMC 50ppmCl (leg3) after bHAST 384h. There was clear interfacial crack in the Cu ball-pad interface (Fig. 2). The crack occurred more on the Cu rich side IMCs (intermetallic compounds). The failure rates of Cu-Al and Ag-Al are 100% at bHAST 384h. But compared to Cu wire bond, the interfacial crack at the Ag ball-pad is more severe. It may be due to the thicker IMC of Ag-Al than that of Cu-Al.

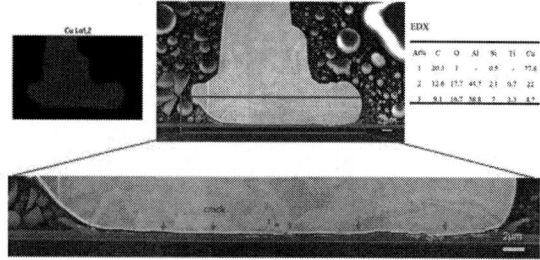

Figure 2: SEM image of Bare Cu-EMC 50ppmCl (leg2) after bHAST 384h.

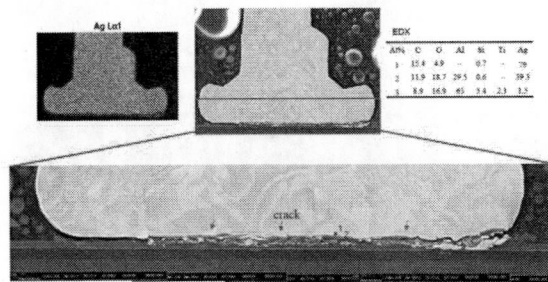

Figure 3: SEM image of 1N Ag-EMC 50ppmCl (leg17) after bHAST 384h.

Fig. 4 shows the SEM image of PdCu-50%Pd-EMC 50ppmCl (leg5) after bHAST 384h. There was obvious interfacial crack. EDX mapping on the top left side shows poor Pd coverage on the EDX point data shows there was Cl at the interface. Thin IMC formation was observed. The crack occurred more on the Cu rich side IMCs.

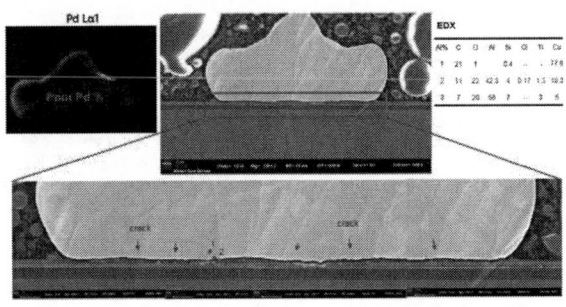

Figure 4: SEM image of PdCu-50%Pd-EMC 50ppmCl (leg5) after bHAST 384h.

Fig. 5 shows the SEM image of PdCu-75%Pd-EMC 50ppmCl (leg8) after bHAST 384h. There was no abnormality observed. Thin and continuous IMC formation was observed at the interface.

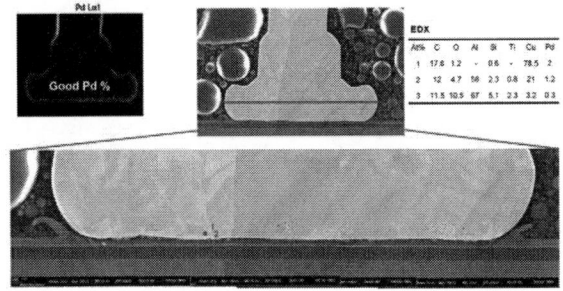

Figure 5: SEM images of PdCu-75%Pd-EMC 50ppmCl

(leg8) after bHAST 384h

Fig. 6 shows the SEM image of PdCu-85%Pd-EMC 50ppmCl (leg11) after bHAST 384h. There was thin interfacial crack observed under SEM. It is noted that there were areas with no/little IMC formation at the interface, as the yellow frame highlighted. Fig. 7 shows the SEM image of PdCu-100%Pd-EMC 50ppmCl (leg14) after bHAST 384h. There was obvious interfacial crack observed under SEM. It is noted that there were areas with no/little IMC formation at the interface, as the yellow frame highlighted. These areas without IMC formation were more than that of PdCu-85%Pd-EMC 50ppmCl (leg11), by comparing Fig. 6 with Fig. 7.

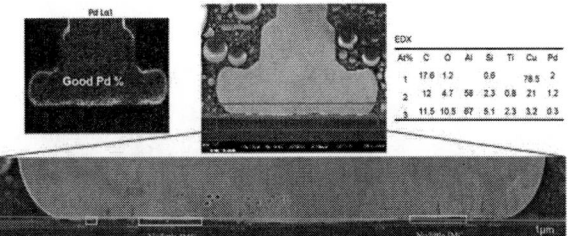

Figure 6: SEM images of PdCu-85%Pd-EMC 50ppmCl (leg11) after bHAST 384h.

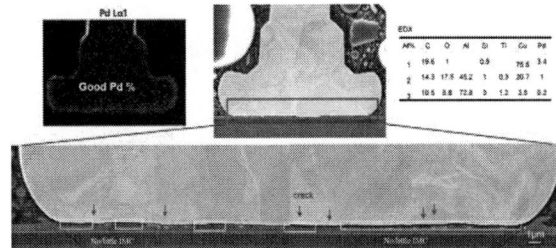

Figure 7: SEM images of PdCu-100%Pd-EMC 50ppmCl (leg14) after bHAST 384h.

Fig. 8 shows the TEM image of PdCu-50%Pd-EMC 50ppmCl (leg5) after bHAST 384h. There was obvious interfacial delamination. The crack occurred more on the Cu rich side IMCs. EDX mapping (Fig. 9) show the corrosion products were Cu and Al_2O_3. It is noted that Cl were detected at the interface.

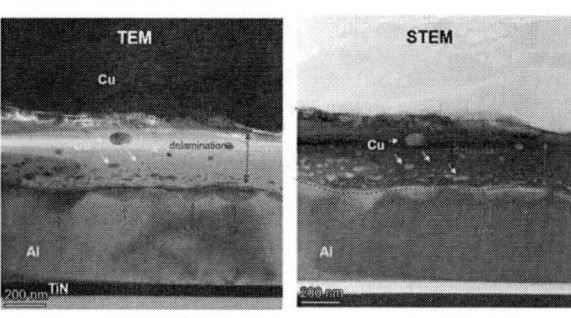

Figure 8: TEM images of PdCu-50%Pd-EMC 50ppmCl (leg5) after bHAST 384h.

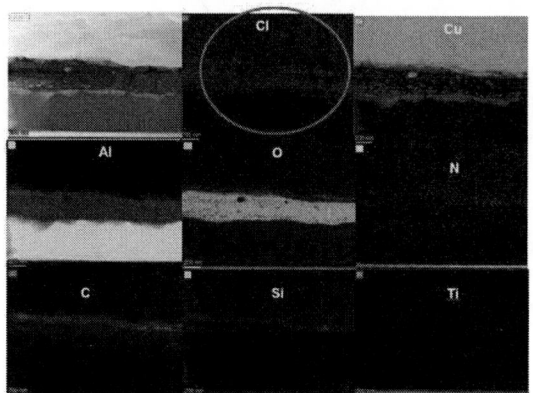

Figure 9: EDX mapping of PdCu-50%Pd-EMC 50ppmCl (leg5) after bHAST 192h.

It has been reported by researchers that Pd coverage on the PdCu ball acts as protection layer to halogen and moisture attack [1-3]. In this study, it was found that Pd coverage on PdCu wire bond is not simple linear relationship (see Fig. 10). More Pd coverage on the bonded ball may not necessarily lead to good reliability performance. On the contrary, excessive Pd coverage on the bonded ball bottom will hinder the IMC formation, resulting in weak adhesion strength and deteriorate reliability. Moderate Pd coverage on bonded ball (in this study 75%) not only generates protection layer Pd on the ball shell, but also forms continuous IMC. The cross-section comparison of PdCu with Pd coverage (50%, 75%, 85%, 100%) after bHAST 384h were showed in Fig. 11. It can be seen that the combination of moderate Pd coverage and the good IMC coverage is the key to good reliability performance. Either poor Pd coverage or poor IMC coverage will lead to poor reliability performance.

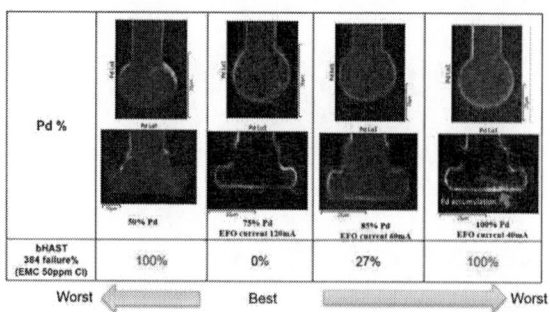

Figure 10: bHAST results of different Pd coverage samples.

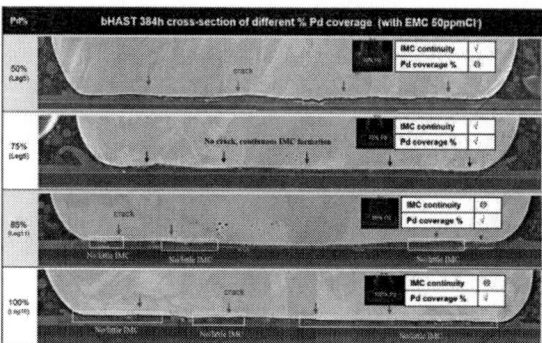

Figure 11: bHAST results of different Pd coverage samples.

CONCLUSIONS

The work compared the wire bond bHAST reliability performances of bare 4N Cu, PdCu and 1N Ag wire. The results showed that:

1) There was no failure of all the wire bonds with <20ppm Cl & S in EMC and with 50ppm S in EMC up to bHAST 384h.

2) For 50ppm Cl EMC, reliability performance (from best to worst): PdCu-75%Pd coverage > PdCu-85%Pd coverage > PdCu-100%Pd coverage > PdCu-50%Pd coverage > 1N Ag > Bare 4N Cu.

3) For PdCu, Pd coverage on PdCu wire bond is not simple linear relationship. It was found that poor Pd coverage on bonded ball led to poor reliability performance. On the other hand, excessive Pd accumulation on the bonded ball bottom will deteriorate the reliability performance. The combination of moderate Pd coverage and the good IMC coverage is the key to good reliability performance.

978-1-7281-6559-2/20 $31.00 © 2020 IEEE

ACKNOWLEDGEMENTS

The authors would like to thank Huawei, Sumitomo, Tanaka and Wintech-nano for the support of this study.

REFERENCES

[1] L. Jinzhi, and Z, Xi, et. al. 2016 IEEE 18th Electronics Packaging Technology Conference (EPTC), 30 Nov.-3 Dec. 2016, Singapore.

[2] N. Seokho, and J. Sungho. 2018 IEEE 19th International Conference on Electronic Packaging Technology (ICEPT), 8-11 Aug. 2018, Shanghai, China. [3] R. P. Feynman, *Lectures on Physics*, Addison Wesley, 1989.

[3] S.C, Preeti, C. Anopam, Z. Zhaowei, P. Michael. Springer, ISBN978-1-4614-5760-2, 2013, pp. 13.

[4] G. Chong Leong, F. Classe, C. Baklee, H. Uda. Gold *Bulletin*, vol. 46, 2013, pp.103-115.

A STUDY OF LOW TEMPERATURE AL SPUTTER PROCESS ELECTROMIGRATION LIFETIME

Jun Liu, Lei Zhang, Jianmin Wang, Qinghua Liu, Di Lou

Shanghai Huahong Grace Semiconductor Manufacturing Corporation
Shanghai, P.R.C.
Tel:86-21-20828888-52345 Email:john.liu@hhgrace.com

Abstract

As 8 inch fab moves to 0.13μm process and beyond, backend Aluminum line width also shrinks to 0.14μm or below with much tightened overlay spec. Low temperature Al sputter process shows very smooth metal surface, which significantly improves overly mark recognition. Thus cold Al is preferred. Meanwhile Cold Al electromigration lifetime is worse than that of hot Al due to smaller aluminum grain size. Cold Al EM must be well controlled for production. In this article, we study the backend HDP oxide deposition temperature and its influence on Metal-1 EM lifetime based on fab 0.13μm process, and a strong correlation has been found. From which, a suitable HDP temp control can be set for cold Al mass production.

Keywords—Cold Al ; EM; HDP temperature control;

INTRODUCTION

As 8 inch fab moves to 0.13μm process and beyond, backend Aluminum line width also shrinks to 0.14μm or below with much tightened overlay spec.

Hot Al uses 420℃, and Cold Al uses 270℃ aluminum sputter process in Applied Materials Endura system. Both metals are same Ti/TiN/Al(with 0.5% Cu)/Ti/TiN sandwich structure. Figure 1 is Box in line metal overlay mark top view.

Figure 2 is the magnified image of the line edge in figure 1 overlay mark. Cold Al sputter process shows much more smooth line edge than that of hot Al. The reason is that cold Al process has significantly smaller aluminum grain size.

From the metal overlay performance in figure 3, cold Al shows much more stable lot level 3 sigma performances in Y direction. This data indicates more accurate overlay measurement is achieved with cold Al process. Therefore, cold Al is preferred especially for small metal line inline overlay control.

On the other hand, due to small DOF, thin photoresist is used for small metal line process. Thus thin aluminum has to be used due to less etch resistance from the thin photoresist. As a result, metal electromigration life time becomes worse. And it

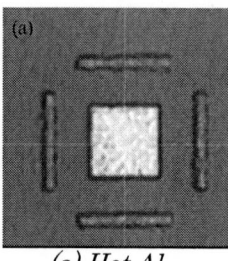

(a) Hot Al   (b) Cold Al

Figure 1: BOX in line Metal overlay mark top view

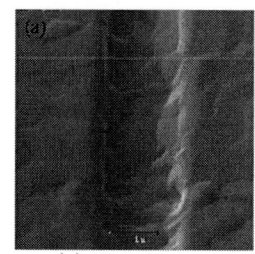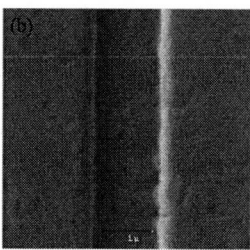

(a) Hot Al   (b) Cold Al

Figure 2: magnified line edge from figure 1

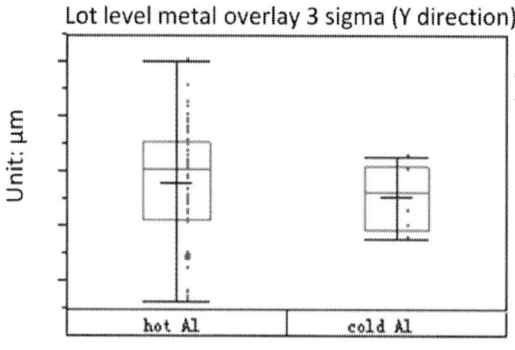

Figure 3: metal overlay performance comparison

gets even worse, when we use cold Al sputter process, as we can see from figure 2 that cold Al has smaller aluminum grain size. It is critical for us to find out the main factor of metal EM lifetime and get it in well control.

Aluminum EM failure is well understood that both current density and high temperature induce failure. In fact, high temperature itself can also make aluminum move and both hillocks and voids can be formed. As a result, EM performance will be downgraded. In this work, we study the backend HDP oxide deposition temperature and its influence on Metal-1 EM lifetime based on fab 0.13μm process.

EXPERIMENTAL RESULTS

Full loop fab 0.13μm process wafers are fabricated for metal-1 EM evaluation. For each IMD HDP oxide deposition step, same HDP dep. temperature split condition is executed on the same split wafer. The HDP dep. temperature splits are achieved by changing of HDP chamber IHC (Independent Helium Cooling) pressure in Applied Materials Centura system. During HDP oxide dep. process, both the maximum and the average wafer temperature are collected. Figure 4 shows that the maximum wafer temperature has very strong correlation with the average wafer temperature. Either temperature can be used for EM lifetime study. What we choose is the average HDP dep. temperate. Since there are several IMD HDP oxide dep. steps for each wafer, and for each step there is an average wafer temperature. So there are several average wafer temperatures for each wafer. And we select the maximum value from these average IMD HDP oxide dep. temperatures for each wafer to do the correlation.

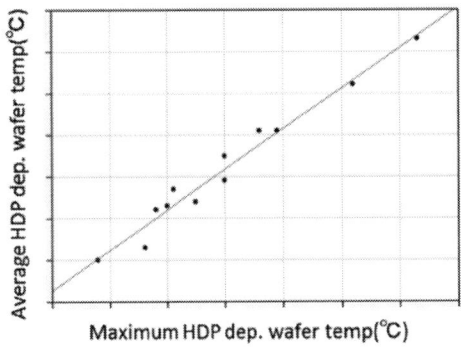

Figure 4: HDP oxide dep. Ave. Temp vs., Max temp

There are 2 metal EM test structures: M1N(narrow metal structure with 0.14μm metal line width) and M1W(wide metal structure with 8μm metal line width). Due to bamboo effect, M1N has obvious longer EM life time than M1W structure. It is verified in figure 5 by our experimental data. So M1W EM life time data is selected for EM correlation study.

Figure 6 clearly shows strong correlation between maximum average HDP dep. temperate and wide metal-1 EM life time. The maximum average HDP dep. temperate should be kept below grey area in figure 6 to keep wide metal-1 EM life time meet 10 years' spec. (the grey area in the right means IMD

HDP Oxide dep. temperature too high for Metal-1 EM.) Since there are 7 full loop lots data in figure 4, lot to lot variation is also included.

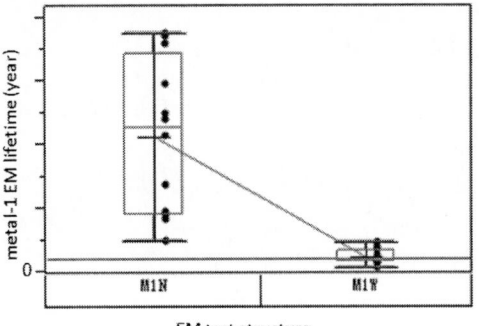

Figure 5: metal -1 EM life time vs. EM test structure (horizontal line represents 10 year's lower spec limit)

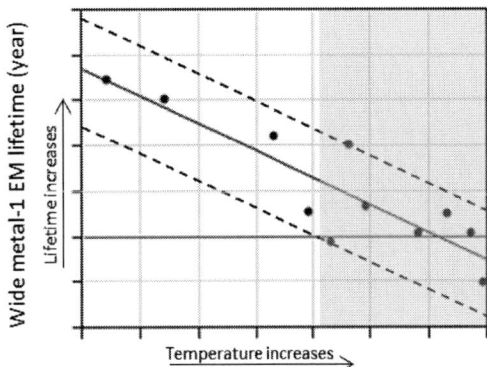

Figure 6: wide metal -1 EM life time vs. Max. IMD HDP oxide dep. average wafer temperature. (horizontal line represents 10 year's lower spec limit)

This strong correlation also means that one of the main variation factors of aluminum metal EM life time is HDP oxide dep. process.

EM failure is a reliability issue that cannot be find out by PCM or sort yield result. In order to do mass production, we need set up offline SPC control for this average HDP oxide dep. temperate, and the average HDP oxide dep. temperature upper spec limit should be set with enough margin.

CONCLUSION

One main factor of fab 0.13μm process Metal-1 EM lifetime has been found to be IMD HDP oxide dep temperature. And upper spec limit and SPC control are set for mass production release of cold Al sputter process. It should be noted that for each baseline process, due to different metal and IMD stacks/processes, the maximum IMD HDP oxide dep.

temperature to meet EM life time control requirement should be verified one by one.

REFERENCES

[1] David Burgess, "Electromigration history and failure analysis," Electronic Device Failure Analysis, volume 16 No.3 2014, pp14-19

RELIABILITY IMPROVEMENT BY 0.153UM CMOS USING HDP-CVD AT STI EDGE SIN LINER

WeiYang Zhang[1], RenGang Qin[1]*, XiaoFeng Sun[1], DeJin Wang[1]*, HaoYu Wen[1], YaoHui Zhou[1]

[1]Technology Development, Central Semiconductor Manufacturing Corporation (CSMC), Wuxi, China

*Corresponding Author's Email: ganghua2001@126.com, wangdj@csmc.crmicro.com

Abstract

We investigated MOSFET shallow trench isolation (STI) process with SIN liner and found no "double hump" effect in the Id–Vg curves and solution leakage. Furthermore, the yield was good in the memory. In the STI corners, a sharp angle was jacked up by SIN. Although the yield result was very good, it was still a hidden danger and could affect GOX. We utilized the high-density plasma chemical vapor deposition (HDP-CVD) process to clip a certain number of SiN liners. From the experiment results, we found that SiN liner and sidewall oxidation significantly improved the time-dependent dielectric breakdown (TDDB).as a result of test, new technology does not generate hot carrier injection (HCI) effects.

Keywords—HDP-CVD; SiN liner; TDDB; HCI

INTRODUCTION

The 0.153um MOSFET technology process has a leakage problem, so we need to add a SiN liner solution, which is applied to LED displays and technology [1]. Thus far, the shallow trench isolation (STI) technique has been used in the MOSFET technology because of its superior isolation capability and process topology. Furthermore, the sharp top corner of the STI profile is acquired with the conventional STI fabrication process; such sharp corners result in a thin gate oxide and a high electric field at the corners, causing leakage currents and device reliability problems [2]. In this study, we investigated the role of sidewall oxidation, including the removal of plasma damage during trench etching, active edge rounding for the suppression of a parasitic channel, and SiN liner reduction of the interface traps for decreasing the junction leakage [3]. Figure 1(b) shows the improved electrical properties caused by the additional liner oxidation and SiN liner.

We used the HDP-CVD technology to clip a certain number of SiN liners. However, the main advantages of the CVD methods stimulated by plasma, in comparison with the PVD methods are as follows: conformal relief reproduction, absence or considerable reduction of radiation damage, high film adhesion to the underlying layers, and a wide range of chemical reactions and sources for reactions with favorable thermodynamics and kinetics [4-5]. In the CVD process, the chemisorption, diffusion, hydrogen abstraction, and chemical bonding of radicals occur. The easiest method to enhance the deposition rate in a CVD system is to increase the RF power. High-density plasma (HDP) is a type of advanced CVD technology in which the plasma density is higher than that used in conventional PECVD. This high ion density enables the HDP-CVD technology to use electron cyclotron resonance to obtain high plasma densities [6-7]. An inherent drawback of this technique is the limited uniformity of the magnetic field, and consequently, it is not possible to deposit very uniform films over large areas. In HDP-CVD, the gap fill capability for a given aspect ratio depends on the ratio between the two processes: deposition and sputtering.

In this study, the different deposition/sputter rate behavior of the HDP-CVD was attributed to the different facility. In the light of the above, the objective of the present study was to clip a certain number of SiN liner applied ACHD12 devices. The experimental results showed an effective improvement in the GOI in the 0.153um MOSFET process.

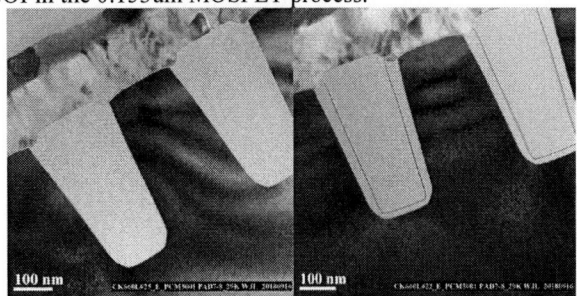

Figure 1: Cross-sectional SEM micrographs of sample: (a) STI with only sidewall oxidation and (b) STI with sidewall oxidation and SIN liner.

EXPERIMENT

In this study, the shallow trench isolation was performed by using a high-density plasma chemical vapor deposition (HDP-CVD) technique. An HDP-CVD machine of side RF and top RF was used to tune the plasma density uniformity for optimized puttering uniformity, and the sputter rate was directly proportional to the bias RF. For throughput considerations, we maximized the sputtering rate by running as high a bias RF power as possible (limited by the hardware reliability and device damage) and minimizing the pressure to keep the mean free path sufficiently high to maximize the sputtering rate and anisotropy. Argon enhanced the sputtering rate, but the flow was limited by the pumping speed and pressure constraints.

The deposition/sputter rate is defined as the ratio between the blanket deposition rate and the blanket sputtering rate. The latter is determined by sputtering a blanket oxide film under the actual process conditions, including the chamber pressure, but without silane. A subsequent measurement of the net gap fill deposition rate allows the determination of the blanket deposition rate as follows:

Dep Rate (net) = Dep Rate (blanket) − Sput Rate (blanket) (1)

D/S ratio follows from the following relationship:

D/S = [Dep Rate (net) + Sput Rate (blanket)]/ [Sput Rate (blanket)] (2)

The HD04 machine used a 9090T7000 recipe. Firstly, there were two types of HDP deposition in the later process. One was a two-step deposition process: The D/S ratio of the plasma was 8.3 in the pre-deposition step, and in the main deposition step, the D/S ratio was reduced to 5.4, which implied that the plasma had strong sputter ability and had benefits with respect to the gap filling. The ACHD12 machine used the 51T7000 recipe in which the deposition process was directly completed with a D/S ratio of 5.4. It had the following characteristics: sputter with strong ability in the process of D/S = 5.4; therefore, it resulted in active corner cutting. However, because in the entire process, D/S=5.4, it had good gap filling ability. In the ACHD12 machine clip application on the SiN finer, the sputter with D/S = 5.4 had strong ability and clip the SiN liner in the corner. After this clipping, the gate oxide active corner was not affected.

So, we know ACHD12 machine had only one deposition process, and the D/S ratio was 5.4. The main deposition process of the first type had no effect on the SiN liner, because there was already an HDP-OX layer deposited in the pre-deposition process on the SiN liner. Moreover, the second type clipped the SiN liner with strong sputter ability. A comparison of the HD04 and ACHD12 deposition/sputter rates is presented in Table I.

TABLE I. HD04 AND ACHD12 D/S RATE COMPARSION

Machine	Recipe	Deposition/sputter ratio
HD04	9090T7000	Step 1: D/S = 8.3 Step 2: D/S = 5.4
ACHD12	51T7000	Step: D/S = 5.4

Therefore, the following spilt was designed in this study to clarify which active has not SIN Liner; active has SiN Liner; active SiN Liner uses ACHD12 machine. Because before the reliability of 1.8 V, all pass whether HD04 or ACHD12 machine; Therefore, we only discuss the 5V reliability.

RESULTS AND DISCUSSION

Time-dependent dielectric breakdown (TDDB) is a known mechanism responsible for an increase in the defect concentration and the subsequent premature failure of the dielectrics used in transistors. The time to failure (TTF) was found to be Weibull distributed (similar to what happens in dielectrics), even if in these cases, no dielectric was directly exposed to a high electric field and the applied voltage was relatively low [8]. In this study, TDDB experiments were performed on 1.8/5V MOSFET test structures. At room temperature, a constant voltage stress (CVS) was observed at various stress voltages. Model to be used the lifetime of the test structure at use conditions is then given by

$$tBD(oper) = tBD(test)*exp \{-\gamma [EOX (oper)-EOX (test)] \quad (3)$$

After the TDDB test, we inferred the Weibull distribution of the spilt scheme from Figs. 2 and 3. It was obvious that the active did not have a SIN liner, which was the baseline, and the active SIN liner using the ACHD12 machine was relatively straight. In contrast, the Weibull distribution obtained by the active with the SIN liner scheme was curved. Therefore, in the case of a strong sputter power of the ACHD12 machine, the reliability was relatively good in the 5V MOSFET.

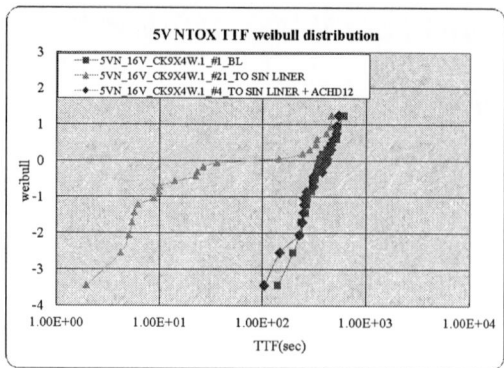

Figure 2: Cross-sectional Weibull distribution of 5V NTOX TTF Weibull distribution.

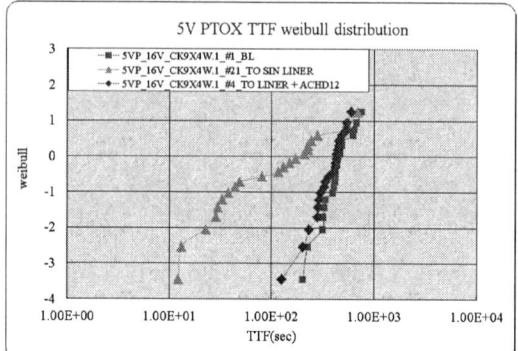

Figure 3: Cross-sectional Weibull distribution of 5V PTOX TTF Weibull distribution.

We verified whether the SiN liner thickness value was about 60 Å for different machines by using the SEM corresponding to the reliability passing. The HD04 machine clipped about 60 Å off the SiN liner. Figure 4 shows that that the SiN liner already gets to the gate oxide above of device with poor TDDB of about 12.34 s.

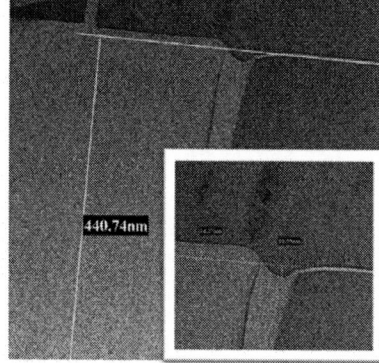

Figure 4: Cross-sectional SEM micrographs obtained at STI depth of 440.74 nm.

According to the current SEM results, the ACHD12 machine could clip about 60 Å SIN in the corner of most of the source area; The ACHD12 machine clipped about 60 Å SiN liner. Figure 5(a) shows that at the STI depth of 465.71 nm, it corresponded to TDDB map good results for 295.14 s. Figure 5(b) shows that at the STI depth of 444.94 nm, it corresponded to TDDB map good results for 338.85 s.

Figure 5: Cross-sectional SEM micrographs obtained at (a) STI depth of 465.71 nm and (b) STI depth of 444.94 nm.

The above results showed that the ACHD12 machine clipped about 60 Å of the SiN liner on the 5V MOSFET, making the TDDB pass. In order to ensure that the new process did not cause the gate oxide other problems, we performed the hot carrier injection (HCI) test，which test methods and their results are described below [9-10]. As shown in Figs. 6 and 7, the HCI screening provided a correlation to the conventional constant voltage stress (CVS) methodology. Whether 5V MOSFET, which baseline and adopt ACHD12 machine clip about 60Å SiN Liner with increase of stress time along, both of basically remain in same Idsat degradation. So, we can determine HCI does not because new technology has the effects.

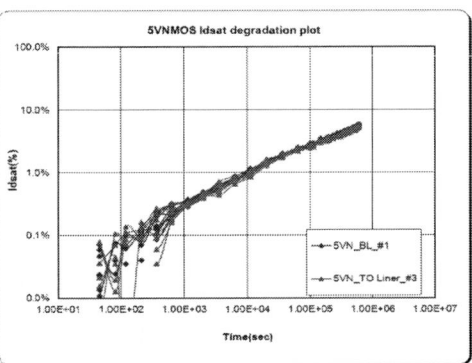

Figure 6:5V NMOSFET Idsat degradation.

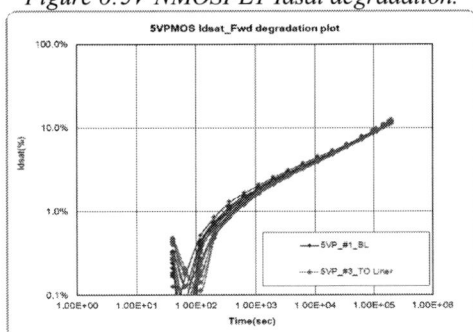

Figure 7:5V NMOSFET Idsat degradation.

CONCLUSION

Because of the device leakage problem, we added to the original device a SiN liner, but the SiN liner was too high, possible approaching the gate oxide, which caused the TDDB failure of the 5V MOSFET device. Because of the difference in the HD04 and ACHD12 machines, the clip had differences. So, we used the ACHD12 machine clipping SiN liner of about 60 Å. The new process for the 5V MOSFET had good reliability. Further, we considered that the new process might affect the gate oxide and conducted additional HCI tests to ensure device stability. We thus proved that the results of the tests were close to our original baseline value.

REFERENCES

[1] M. S. Kim et al., "Reduction of oxide leakage currents of EEPROM at STI corners using sacrificial oxide/liner SiN/LPCVD MTO," Surface and Coatings Technology, vol. 202, no. 22, pp. 5697–5700, Aug. 2008.

[2] C. H. Li et al., "A robust shallow trench isolation (STI) with SiN pull-back process for advanced DRAM technology," in 13th Annual IEEE/SEMI Advanced Semiconductor Manufacturing Conference. Advancing the Science and Technology of Semiconductor Manufacturing. ASMC 2002 (Cat. No.02CH37259), Boston, MA, USA, 2002, pp. 21–26.

[3] A. S. Teng et al., "Gate bias temperature stress-induced off-state leakage in nMOSFETs: Mechanism, lifetime model and circuit design consideration," in 2014 IEEE International Reliability Physics Symposium, Waikoloa, HI, USA, 2014, p. XT.4.1-XT.4.6.

[4] A. P. Mousinho, R. D. Mansano, and P. Verdonck, "High-density plasma chemical vapor deposition of amorphous carbon films," Diamond and Related Materials, vol. 13, no. 2, pp. 311–315, Feb. 2004.

[5] M. A. Abdelgadir, "HDP-CVD STI oxide process with in situ post deposition laterally enhanced sputter etchback for the reduction of pattern-dependent film topography in deep submicron technologies," IEEE Transactions on Semiconductor Manufacturing, vol. 19, no. 1, pp. 130–137, Feb. 2006.

[6] W.-C. Hsiao, C.-P. Liu, and Y.-L. Wang, "Thermal properties of hydrogenated amorphous silicon prepared by high-density plasma chemical vapor deposition," Journal of Physics and Chemistry of Solids, vol. 69, no. 2–3, pp. 648–652, Feb. 2008.

[7] C. C. Lai, L. Y. Li, T. B. Huang, H. J. Chien, and T. H. Ying, "Influence of plasma power and sputtering agent on gap-fill and MOSFET performances in HDP-CVD STI oxide process," in 2014 International Symposium on Next-Generation Electronics (ISNE), 2014, pp. 1–2.

[8] Seung-Ho Pyi, In-Seok Yeo, Dae-Hee Weon, Young-Bog Kim, and Sahng-Kyoo Lee, "Roles of sidewall oxidation in the devices with shallow trench isolation," IEEE Electron Device Lett., vol. 20, no. 8, pp. 384–386, Aug. 1999.

[9] A. Kerber, W. McMahon, and E. Cartier, "Voltage Ramp Stress for Hot-Carrier Screening of Scaled CMOS Devices," IEEE Electron Device Lett., vol. 33, no. 6, pp. 749–751, Jun. 2012.

[10] Yuhao Luo, D. Nayak, D. Gitlin, Ming-Yin Hao, Chia-Hung Kao, and Chien-Hsun Wang, "Oxide reliability of drain engineered I/O NMOS from hot carrier injection," IEEE Electron Device Lett., vol. 24, no. 11, pp. 686–688, Nov. 2003.

NON LINEAR GROWTH OF VARIANCE IN THE PROCESS GAPS. A CAUSE OF ADVERSE CYCLE TIMES

George W Horn*, William Podgorski Ph.D, Middlesex Industries SA; *Corresponding Author's Email. gwhorn@midsx.ch

ABSTRACT

Variability in the manufacturing process (the coefficient of the OC curve), together with its counter measures, is the primary cause of long cycle times and reduced fab throughput. In an IC fab model, where substrate lots are processed through steps, and where each step is followed by a gap in processing, the wafer lots will spend more time in the accumulated transitions through the gaps than in-process. In these gaps non-linear growth of manufacturing variability occurs, through multiplication. Then, as this growth of variability is countered by stabilizing measures, such as wafer lot buffering and storage, cycle times are increased (and capacity reduced). An analysis of the above growth of variability is presented. Alternate product handling methods in the gap are studied to minimize this growth of variance. and so reduce fab cycle times (increase throughput).

VARIANCE AND THE OC

In 1997, the IBM Consulting Group introduced the concept of an Operating Characteristic Curve for semiconductor manufacturing, which relates Fab cycle time to Fab throughput. Today, the generalized Operating Characteristics curves (OC) of a factory are approximated with the following formula:

$$FF = \alpha \frac{U}{1-U} + 1 \qquad (1)$$

Where FF is the normalized cycle time, plotted on the vertical axis: Cycle time/raw process time; U is the normalized fab throughput: throughput/capacity; α is the coefficient of overall variability, containing the factors V for arrivals, and v for process related. V is determined by what happens between two value-add process steps. At the same time Cycle time :

$$CT = \frac{Q}{C} \qquad (2)$$

Where CT is cycle time; Q is mean wafer content of the fab; and C is capacity (ideal tool processing capacity). But, Q, the mean wafer content of the fab is the sum of : (tool capacity)(total ideal tool cycle time) + (AMHS capacity)(mean delivery time) + (mean tool buffer capacity)

+ (mean live stocker capacity). Clearly, the nominator in equation (2) requires live buffer and stocker content, which increase fab cycle times. It is important to recognize, however, that the buffer and stocker requirements are a consequence of process and AMHS variability.

The tool to tool move variability, is the subject of this study. Extensive data exists on specific fab simulations with various AMHS. And these indicate a major influence of V, arrival variance, on overall manufacturing variability α. When managing Fab operations, however, the multitude of difficulties arising from the physical processes' stability and upsets in scheduling, seems overwhelming in comparison to problems of AMHS logistics. As a result, little attention has been given to AMHS issues. This is a mistake, since any variance that occurs at the processing nodes is amplified in the wafer lot moves (not an addition of variance) between the nodes in the gap. Indeed, the ideal AMHS would be one, that does not amplify the variability of the process nodes, but instead, reduces them. In today's Fab, the task of reducing the variance occurring at process nodes is dealt with the temporary storage of the wafer lots in between the process steps. And that increases cycle times. And more, such measures shifts the OC curve towards reduced fab throughput (Ref. 1).

VARIABILITY IN THE FAB PROCESS

Wafer lots emerge from processing tools with a variance. (Such variance may partly be created by the process tool, and may partly be the burden of variance with which the substrate entered that tool). Then, with that exit variance, the substrate lot is handed over to the AMHS for managing. The AMHS, however, will manage that wafer lot with its own variance. The evolution of wafer lots from a process, and their service by the AMHS, each proceed with their own independent variables. And this independence causes a nonlinear growth of variance for the combined process, through multiplication. Then, further down the line, the next process step will inherit that combined variance. A chain of process steps, followed by gaps of AMHS handling of the lot, would normally create exponential growth of manufacturing variability, were it not for pausing a wafer lot in the gap via buffering or storage. This buffer or storage in the gap is today the only tool afforded in fab operations for the elimination of potentially catastrophic growth of manufacturing variability. Yet, with buffering and storage the fab cycle time will increase and fab throughput decrease. Countering the intent to increase throughput and reduce cycle time.

Ideally, extended cycle times of a fab (in steady state) come about due to the inability to synchronize the tool process steps with zero transport variance in between. In steady state, throughput = wafer start rate. In which case U of equation (1) becomes an indicator of fab loading showing wafer lot density in the fab (i.e. how many lots are expected to idle on live buffers at tool input ports and stockers). Managing U means to manage lean manufacturing. Then, how to design wafer handling in the gap (AMHS) if we know that in that gap the variance will be amplified or dampened? Variance of the AMHS, in the process gap occurs 1) due to lot pickup variance, 2) due to transit time variance because of lot to lot interference or storage, and 3) due to variance in placing the lot onto tool ports. The successful design of an AMHS will minimize all the above, and consequently, the growth of variability of lot arrivals to the next tool. As a consequence buffering and storage of wafer lots between process nodes can be reduced, shortening cycle times.

TRANSFORMING WAFER RATES

The physical process nodes of a fab discharge wafer lots into a queue. This queue is subsequently serviced by the AMHS, delivering wafer lots to the next process step. Arrivals into the next process will be the result of the independent discharge rates of the tool into the discharge queue and the independent service rates of the AMHS. The AMHS server acts as a transfer function of the wafer lot discharge rate of a process into the arrival rate to the next process step (Fig 1). Thus, the variance of a physical process node is transformed by the AMHS.

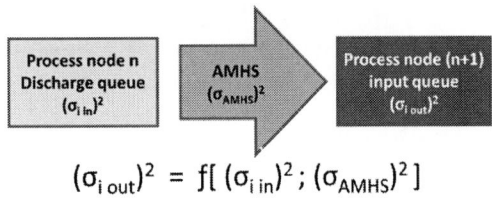

$$(\sigma_{i\ out})^2 = f[\ (\sigma_{i\ in})^2\ ;\ (\sigma_{AMHS})^2\ ]$$

The variability of lot arrivals to the (n+1)th tool is a function of the discharge rate variability of the nth tool and the AMHS variability

Fig 1: the AMHS server transforms discharge rates of a process node into arrival rates to the next process node.

What is, then, the probability distribution of arrival intervals to the next process? And what is its variance?

$$P[\Delta t \cap \Delta T] = P[\Delta t] \cdot P[\Delta T \mid \Delta t]$$

Where Δt and ΔT are independent variables, thus:

$$P[\Delta T \mid \Delta t] = P[\Delta T]$$

$$P[\Delta t \cap \Delta T] = P[\Delta t] \cdot P[\Delta T] \qquad (3)$$

Alternately, if the probability of servicing the discharge queue is conditioned on the queue's prior probability:

$$P[\Delta T \mid \Delta t\ ] = P[\Delta T \cap \Delta t] / P[\Delta t] \qquad (4)$$

This probability algebra in independent variables, and its prediction of variance transformation by the AMHS (Fig 2), is overlooked in managing today's fabs. Instead, this natural transform of variance is substituted by heuristics, thus masking the underlying physics, and its need for understanding.

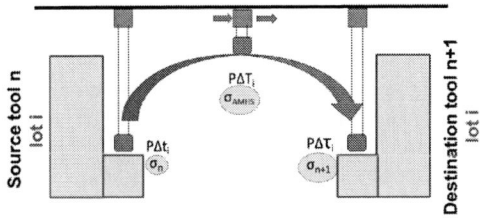

$$P(\Delta\tau_i) = P(\Delta t_i) \times P(\Delta T_i)$$

Fig 2: The nonlinear growth of variance in the gap due to multiplication

SIMULATING ARRIVAL VARIANCE

The distributions of Δt (discharge) and ΔT (transport) are combined via simulation (using heuristics instead of the probability algebra above by: adding arrival times of the wafer lot into a process' output queue to the transit time of the same lot through the gap, Fig 3). The simulation output is the variance and probability distribution of the resultant arrival to the next process step.

This illustrates how these independent rates, with their interactions generate the growth of variance. Each process is modeled with a Poisson process, with a likely (expected) distribution of its variable, followed by a simulation of their interactions. Arrivals into the discharge queue are forced into exponential probability of the variable (Δt). While the service of the AMHS, with its independent variable (ΔT) forced into a normal distribution Fig 4, 5, 6).

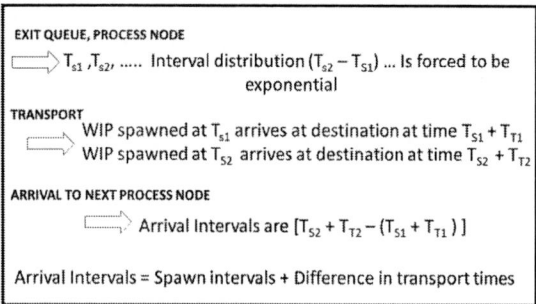

FIG 3: Combining variance of the process-node-exit queue with the variance of the AMHS. The result is the variance of arrivals to the next process node

MATERIAL EFFECT OF VARIANCE GROWTH

This study analyzes the growth of variance in a typical gap between two process nodes. The magnitude of transformed variance to be compared between a discrete move type of AMHS (OHT-vehicle type, Fig 5), and a hybrid type, (third generation AMHS, Fig 6), where major inter and intra bay moves are of network conveyor type combined with the OHT vehicle type in the bays.

The study offers fab designers the means to increase throughput through reduced variability in the IC manufacturing process. (For overall effects on fab throughput and cycle time see reference article 1.)

APPLICATION

In new fabs Conveyor network is placed to connect tools, bay to bay, ending in in-bay I/O buffers. OHT

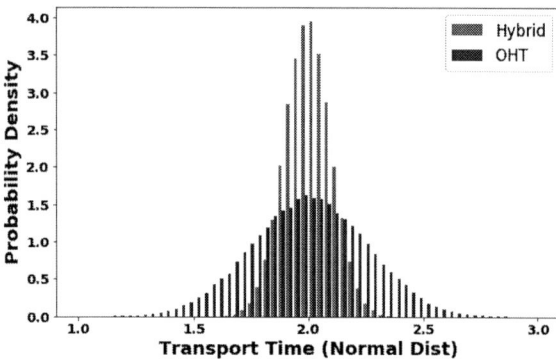

Fig 4: Distribution of AMHS transport times between process nodes, shows smaller variance for Hybrid;

vehicle lines are then inserted into each bay to manage moves of wafer lots from AMHS conveyor output ports to tool ports. (Confer reference article 1.)

On the other hand, to improve fab throughput of existing fabs, built with pure OHT-vehicle type AMHS, the process "to-from" table of wafer lot moves is examined and high density line segments (process node to process node) are identified. Onto these lines conveyor lines are added, as an overlay, incrementally, removing variance of the AMHS in degrees desired.

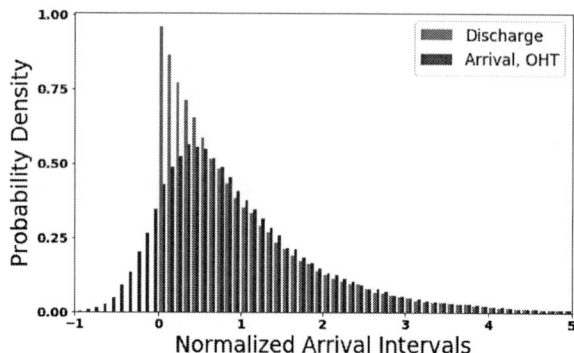

Fig 5 Probability density of arrivals shows large variance increase due to Pure OHT vehicle AMHS handling

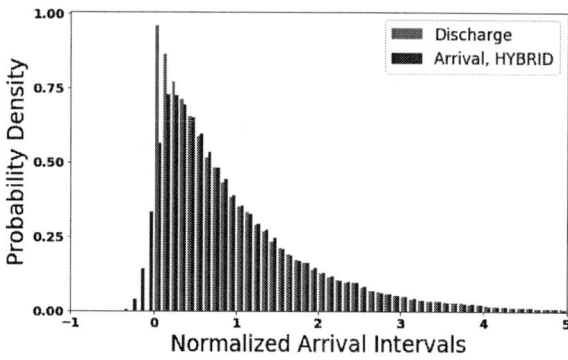

Fig 6: Probability density of arrivals show minor increase of variance due to hybrid AMHS handling

REFERENCES

[1] Cycle Time of Front End IC Manufacturing and AMHS Variability. *George W Horn, William Podgorski PHD,* IEEE proceedings, Cystic 2019.

[2] Distribution of Products of Independent Random Variables. *Springer, M.D; Thompson, W.E.* SIAM Journal of Applied Mathematics, 1966 14 (3).

Research on improvement of reference voltage shift of wire-bound products

Yang Chen[1]*, Na Mei[1], and Tuobei Sun[2]
[1]Zhangjiang Hi-tech Park ,pudong district,shanghai,P.R.China
[2]The department of packing and Testing of ZTE corporation.
chen.yang184@zte.com.cn

ABSTRACT

The paper should start with a brief abstract of approximately 100 words summarizing the main goals, developments, and achievements of the work. Consider that the abstract may be included in abstract search databases. Think of what requirements the abstract should fulfill in view of this perspective, taking into account the fact that the main text part will not be accessible to the searching person. For wafer-level package and flip chip package, bump connection reliability may be caused by overall package stress. Therefore, it is well known in the industry that PI will be used as the buffer layer, and low-stress assembly materials include substrate materials will be used to improve the yield and reliability of subsequent package and application due to high stress. For traditional WB products, considering that the chip size is smaller and most of them are wire bond products without soft bump, they are less sensitive to stress, also due the PI cost is relative high, so few researches will focus on the stress improvement of WB products. However, our research shows that some stress-sensitive WB products may cause high yield loss of reference Voltage(Vr) shift due to high stress. The Vr yield loss exceeds 10% or even 20%. In this paper, DOE of the stress effect of PI materials, different EMC materials and package structures on WB products is studied to help select the best production process and material parameters to obtain the highest yield for our proudcts.

INTRODUCTION

In traditional WB products, the electrical yield of general products has littlerelation with chip stress, but in some chip designs that are particularly sensitive to stress, the chip reference voltage and other indicators will greatly fluctuate with the overall stress of WB products. On some of our developed products, we found that the factors that affect chip stress mainly include chip packaging structure, EMC materials, wafer materials, etc. The packaging structure is mainly reflected in the asymmetric effect of the upper and lower plastic seals. The packaging material is manifested in the thermal stress and shrinkage stress of the plastic packaging material, which leads to the uneven stress in the chip. The use of PI materials in wafers will help to relieve the stress on the chip itself. All of the above thre factors will lead to the deviation of the electrical test

results. We analyzed the yield performance through different experiments.

EXPERIMENT

1.change to low stress EMC material

EMC is a kind of thermosetting chemical material. The thermal expansion coefficient of the EMC material is more than ten times of the silicon wafer. When the EMC material melts from high temperature to low temperature, the shrink pressure of the EMC material is very large for the silicon wafer. For the sensitive silicon wafer, only the low stress EMC material can be used to reduce the expansion coefficient of the EMC material to reduce the pressure on the silicon wafer. We have used normal EMC materials for engineering verification, and the test yield is basically kept at about 85%, most of the failures are voltage reference offset. After a variety of DOE verification, it is found that the yield of low stress EMC materials has increased by 10%, and the voltage reference failure has been greatly improved. the reusult is as below table showed.

DOE	Wafer type	EMC	Yield
Leg 1	Normal	Normal	--
Leg 2	Normal	Modified	10% ⬆

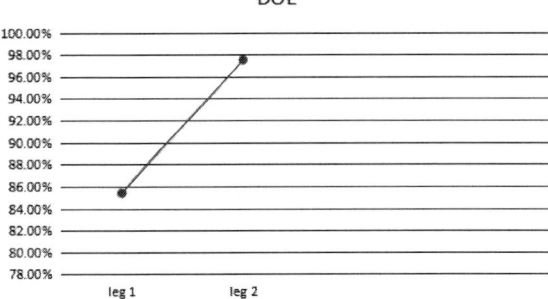

2.Add polyimide on the wafer

As mentioned above, for sensitive chips, the pressure of direct contact between EMC material and silicon wafer is very large, and because of the existence of metal circuit on the surface of silicon wafer, the area of contact between EMC and wafer fluctuates, so it is difficult to ensure the uniformity of stress on silicon wafer, so we considered adding a layer of insulating material between EMC material and silicon wafer, polyimide to buffer The direct impact of the EMC material on the silicon wafer, the stress of the EMC material is transferred on the PI, so that the direct stress on the silicon wafer is effectively reduced. Polyimide is one of the best organic polymer buffer materials. Its high temperature resistance is more than 400 ℃, long-term use temperature range is - 200 ~ 300 ℃, some parts have no obvious melting point, high insulation performance. We carried out engineering verification for the conventional wafer and the wafer added PI material, and the test yield is shown in the table below. Compare to normal wafer，the wafer with PI yield improved 10%，but the difference between chips with low stress EMC and with PI waferwas not significant..

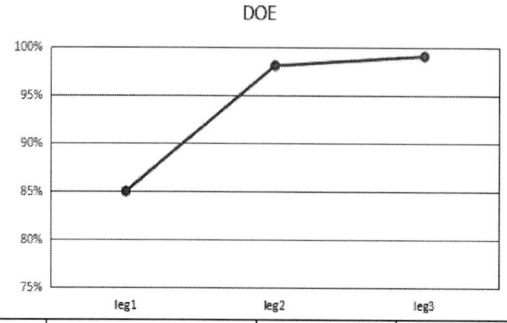

DOE

DOE	Wafer type	Molding compound	Yield
Leg 1	Normal	normal	--
Leg 2	Add polyimide wafer	normal	10% ⬆
Leg 3	Add polyimide wafer	Low stress	10% ⬆⬆

3.Optimize the structure of package

Based on the research and analysis of a SSOT product with reference voltage shift, we found that the design of upper and lower die is asymmetric, as shown in the figure below .

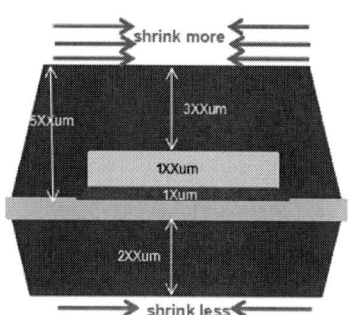

The thickness difference between the upper mold and the lower mold is large. When the EMC material is cooled and formed, the shrinkage difference between the upper mold and the lower mold is large, which leads to the stress imbalance and the deviation of the reference voltage shift.

We improve the stress by increasing the thickness of silicon wafer, balancing the thickness of upper and lower die, and the volume of EMC material. As shown in the following table, we have verified the different silicon thickness.

Compound	DIE thickness	YIELD
Normal EMC	1xx	86.00%
Normal EMC	2xx	87.16%
Low stress EMC	1xx	99.00%

Through verification, it can be concluded that increasing the thickness of silicon wafer, slightly improving the yield but not significant, maybe it was related to the size of silicon wafer. Through the verification of low stress EMC materials , it can be concluded that the performance of EMC materials is the best.

CONCLUSION

There are manyfactors impact the chip stress, and the factors that can be related with electrical performance is more divergent. We only verify the yield through three experiments, include changing the low stress EMC material and adding the PI material on wafer, changing the die thickness. Although we can achieve expected yield through DOE, We also need overall consider the balance between the technical solution and the total cost.

A NOVEL VERTICAL CLOSED-LOOP CONTROL METHOD FOR HIGH GENERATION TFT LITHOGRAPHY MACHINE

Dan Chen[*]

Shanghai Micro Electronics Equipment (Group) CO., LTD., Shanghai 201203, China

* Email: chend@smee.com.cn

ABSTRACT

A prominent problem of TFT lithography machine is that its plate stage's size and weight increase rapidly as it develops to higher generation. Because of plate stage's large size and heavy weight, it could be extremely difficult to achieve high bandwidth. Hence, it could be technical complicatedly and economic expensively to compensate vertical control error by the movement of plate stage. In this paper, a novel vertical closed-loop control method using the idea of coordinated movement of reticle stage and plate stage has been proposed. Experiment results show that, the total vertical control error based on our method has better performance, and CDU results based on these two methods are similar. However, our method has much higher technical feasibility and lower cost.

Keywords—focal plane; object plane; coordinated movement; vertical control error; bandwidth

INTRODUCTION

CDU is one of three top specifications of lithography machine, which reflects the uniformity of exposed line widths directly [1] [2]. Vertical control error (**VCE**) has significant influence on CDU [3]. The aim of vertical control is that, overall exposure plate area is controlled to coincide with the best focal plane, so that VCE can meet its specification. Glass plate's surface shape contains global surface shape and local surface shape. Global surface shape reflects the overall trend of plate surface, which corresponds to the spatial DC and low frequency components of plate surface shape; and local surface shape describes irregular ups and downs in the range of exposure field view, which corresponds to the spatial medium & high frequency components of plate surface shape.

Moving-stage subsystem, which contains reticle stage (**RS**) and plate stage (**PS**), is one of core subsystems of lithography machine [4] [5]. As for traditional vertical control method, the vertical position of mask, which is carried by reticle stage, is almost fixed. Meanwhile, the glass plate is carried by plate stage and moved vertically [6], so that the surface of plate coincides with the best focal plane, see fig. 1. When the size of plate is small, and the focal depth of objective lens is large, VCE can satisfy its specification by vertical movement of plate stage easily. With the development of TFT lithography machine to higher generation, it is more and more difficult to design

plate stage with high bandwidth. To control VCE using this traditional method, it could be extremely difficult technically and exceedingly expensive economically.

Our proposed vertical closed-loop control method has been described in section 2; in section 3, experiment results are given and discussed. Finally, conclusions to the whole paper have been given.

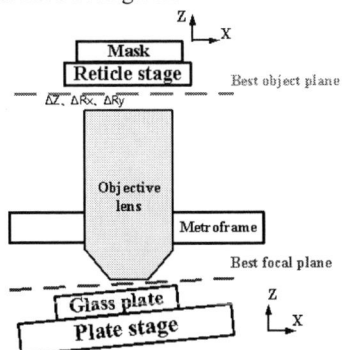

Figure 1: Vertical relationship sketch

Proposed Method

Basic Idea

As for TFT lithography machine, plate stage is much heavier than reticle stage, and the latter has higher bandwidth than the former. The basic idea of our method is that, during the exposure time, plate stage moves vertically along the fixed global leveling plane, which is a bevel plane and reflects the overall trend of plate shape, so that plate shape's DC and low frequency components can be compensated. Meanwhile, reticle stage moves vertically based on our proposed method to compensate plate shape's medium & high frequency components.

Our Proposed Method

Our proposed vertical closed-loop control model can be described in fig.2 below.

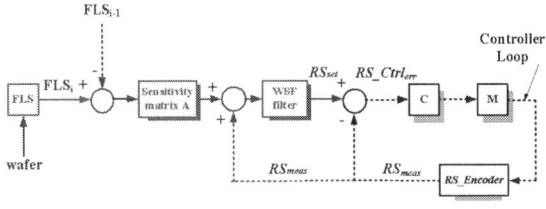

Figure 2: Proposed vertical closed-loop control diagram

In this figure, *FLS* is the vertical measurement sensor of plate, and *Encoder* is the vertical measurement sensor of reticle stage. Sensitivity matrix A is used to transfer image-side vertical increments to object-side vertical increments. Wafer shape filter (***WSF***) is a composite of filters, which is used to filter out high and specific frequencies. C is a controller, and M is a motor.

Our proposed method can be described in model below, shown in formulae (1) to (3):

$$BF_{Die} = BF_{mc} + UserOffset + AO_{cor} \qquad (1)$$

therein, *BFmc* is a machine constant of the best focal plane, *UserOffset* is a user defined offset, *AOcor* is the compensation of thermal effect, *BFDie* is the actual best focal plane of exposure field.

$$RS_{set\_1} = (RS_{meas\_1} + A * (FLS_1 - BF_{Die})) \ominus WSF \qquad (2)$$

$$RS_{set\_i} = (RS_{meas\_i-1} + A * (FLS_i - FLS_{i-1})) \ominus WSF \qquad (3)$$

therein, $A = \begin{pmatrix} a_{11} & a_{12} & a_{13} \\ a_{21} & a_{22} & a_{23} \\ a_{31} & a_{32} & a_{33} \end{pmatrix}$ is the sensitivity

coefficient matrix, $RS_{set\_i} = \begin{pmatrix} RS.Z_{set\_i} \\ RS.Rx_{set\_i} \\ RS.Ry_{set\_i} \end{pmatrix}$ is the vertical

set value of reticle stage during exposure time except for the first scan servo period. FLS_1 is the vertical value in the first scan servo period during exposure time, RS_{meas_1} is the vertical measurement value of encoder in the first scan servo period, and RS_{set_1} is the vertical set value of reticle stage in the first scan servo period. FLS_{i-1} is the vertical measurement value of *FLS* in the last servo period, while FLS_i is the vertical value in the current servo period. All reticle stage's vertical set values should been filtered by WSF filter.

Experiments
Experiment Design
Tests have been carried out on TFT lithography machine, using traditional and our proposed vertical control methods respectively. Supposed the VCE specification of test scanner is E0. In our experiments, total VCEs are divided by benchmark E0, and are expressed in percentage (%).

Experiment Results Discuss
Fig. 3 to 5 show the comparisons of FLS height values, plate stage's height values, reticle stage's height values using two methods. Because of the similarity of vertical tilt comparisons and space constraints, only Ry comparisons are given, shown in fig. 6 to 8, and comparisons of Rx are omitted. Finally, the comparisons of total VCE during scanning exposure time is given in fig. 9. In these figures, *PS-loop-mode* represents the traditional method, while *RS-loop-mode* is our proposed method.

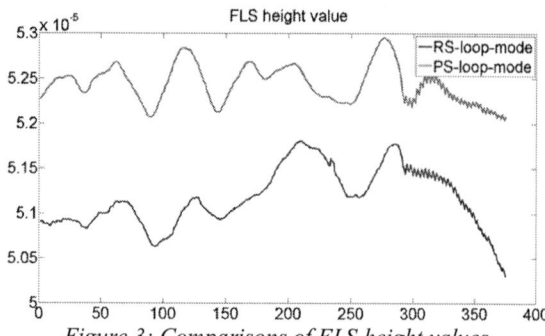

Figure 3: Comparisons of FLS height values

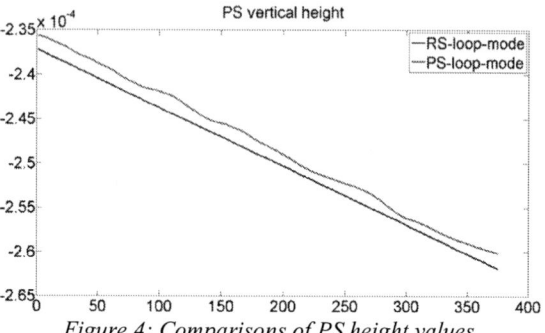

Figure 4: Comparisons of PS height values

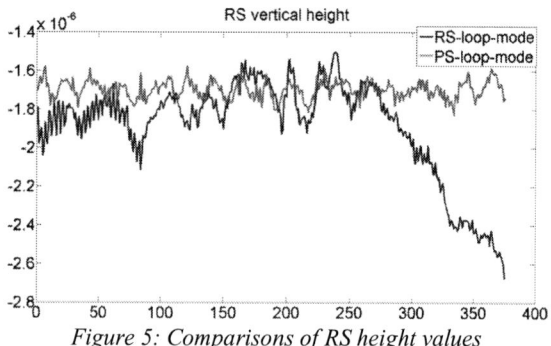

Figure 5: Comparisons of RS height values

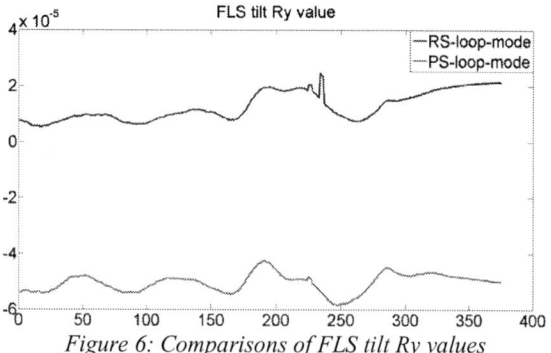

Figure 6: Comparisons of FLS tilt Ry values

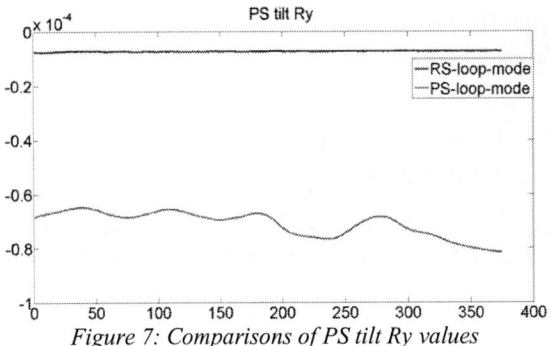

Figure 7: Comparisons of PS tilt Ry values

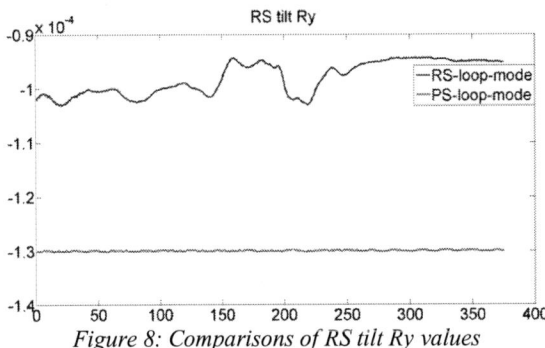

Figure 8: Comparisons of RS tilt Ry values

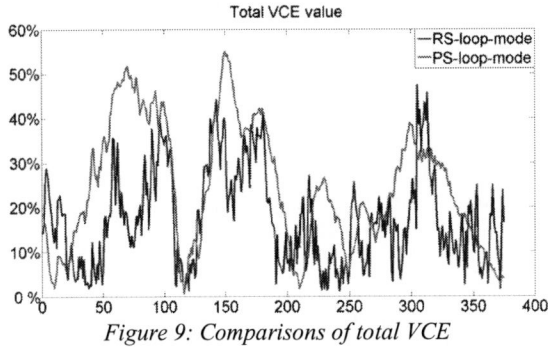

Figure 9: Comparisons of total VCE

From table 1 below, we can find that total VCE (47.4%) by using our method is superior to total VCE (55.1%) by using traditional method. However, by using our method, mask could not coincide with the best object plane, which brings the problem of optical distortion. This has bad effect on CDU. The final CDUs by using these two methods are very close to each other.

TABLE I COMPARISONS OF TWO METHODS

Item	PS-loop-mode	RS-loop-mode
FLS_Z(um)	PV: **0.92**	PV: **1.5**
PS_Z(um)	PV: **24.5**	PV: **24.7**
RS_Z(um)	PV: **0.25**	PV: **1.2**
FLS_Ry(urad)	PV: **16.0**	PV: **19.4**
PS_Ry(urad)	PV: **16.8**	PV: **0.48**
RS_Ry(urad)	PV: **0.5**	PV: **8.79**
Total VCE	Max: **55.1%**	Max: **47.4%**
CDU	9.7%	9.5%

Conclusions

In this paper, a novel vertical control method based on coordinated movement of reticle stage and plate stage has been proposed. As for CDU, the results based on two methods are quite similar. But based on our method, vertical control of high generation TFT scanner can be realized with much fewer technical difficulties and lower economic cost. To solve the problem of optical distortion, the study of coordinated movement model of reticle stage, plate stage and objective lens is our future work.

REFERENCES

[1] K. He, and Q. Wu. "Calculation method of intra-field CDU and inter-field CDU revisited for advanced immersion lithography," *2015 China Semiconductor Technology International Conference*, Shanghai, March 15-16, 2015.

[2] W.H. Arnold. "Toward 3nm overlay and critical dimension uniformity: an integrated error budget for double patterning lithography," *Proceedings of SPIE – The International Society for Optical Engineering*, 2008, Vol. 6924.

[3] S.D. Hector, S.V. Postnikov, and J. Cobb. "Evaluation of the critical dimension control requirements in the ITRS using statistical simulation and error budgets," *Optical Microlithography XVII. International Society for Optics and Photonics*, 2004.

[4] D. Liu. "Process of Wafer Stage and Reticle Stage for Step-and-Scan-Lithography System," *Laser & Optronics Progress*, 40(5), 2003, pp. 14-20.

[5] C. Wang, W. Yin, G. Duan. "Cross-coupling control for synchronized scan of experimental wafer and reticle stage", *2006 International Technology and Innovation Conference*, Hangzhou, Nov. 6-7, 2006.

[6] M. Quirk, J. Serda. *Semiconductor Manufacturing Technology*, Pearson Schweiz Ag, 2000.

PROBE CARD LIFETIME CONTROL AND ABRASION COEFFICIENT STUDY

Lei Wang[1]*, Song Ma[1]

[1]Shanghai Huahong Grace Semiconductor Manufacturing Corporation, Shanghai, China

*Corresponding Author's Email: ray.wang@hhgrace.com

ABSTRACT

With continued scaling of deep-submicron CMOS technology, more and more transistors are integrated within one die. Both the function verification and reliability performance are taken into account. In order to decrease unnecessary cost on backend package and assembling, variety of the test items and flows are transferred from FT (Final Test) level to CP (Chip Probing) level. However, the probe card is the key while wafer sort is in processing. Clean sheet, abrasion coefficient, clean frequency and overdrive impact the test stability and the cost of test directly. The balance on the above critical factors is necessary to be analyzed. The paper focuses on the abrasion coefficient model establishment and the probe card lifetime control for the specific probe card. The consumption algorithm model could be applied to improve the efficiency and control the cost of the test.

INTRODUCTION

With the evolution on Moore's and more than Moore's law, wafer level test is becoming critical for IC (Integrated Circuit) design and verification. During CP test, higher C_{RES} (Contact Resistance) is inevitable and undesirable. The efficient way to control the C_{RES} is doing probe cleaning. However, the excessive probe cleaning reduces the probe card lifetime. Meanwhile, the abrasion residue left would also have great influence on the chip quality and test stability. The way on clean sheet selection and scientific probe cleaning setting is crucial. On right way, not only loose debris could be collected efficiently in polymer, but also the weld nuggets could be removed from contact area to prevent the high C_{RES} issue. What's more, the original probe shapes are validly maintained and the probe card life time could be extended. Following paper analyzes the theory of contaminations effect and types of probe cleaning model. At last, the practical probe card consumption algorithm model is proposed to forecast and control the probe card lifetime.

CONTAMINATIONS EFFECT

The probe tips contacts chip bonding pads continuously while doing wafer sort. The metallic contact and film resistance act two leading roles on the impacts of intrinsic C_{RES} [1]. With the repetitive of touchdowns, the higher contact force required for reduced scrub length has the effect of shaving the bond pads. It leads to aluminum flaking and the contaminations will be increased by the friction and electromagnetic field. The contaminations are accelerated by the field emission effect. Main factors affecting the C_{RES} are the overdrive, probe tip planarity, contact pressure, tip surface cleanness, plastically deformation regions and so on. The oxidized non-conducting Al_2O_3 makes the C_{RES} increased rapidly. The parametric measurement accuracy and wafer yield are seriously impacted. The basic model on probe contact interface is demonstrated as figure 1 shown.

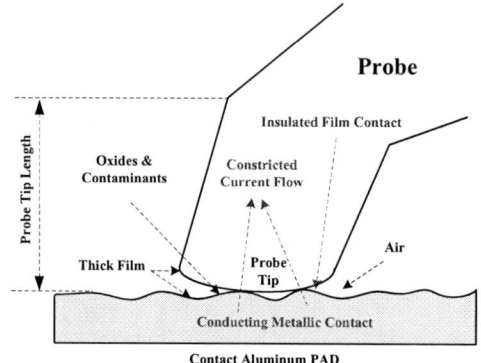

Figure 1: Probe Contact Interface Model

PROBE CLEANING

In order to solve the contaminations effect, probe card cleaning is essential to high volume wafer production test. It could be divided into offline cleaning and online cleaning [2].

Offline cleaning is suitable for probe card maintenance or special issue. If online cleaning is not sufficient to remove debris from the probe tips, it should be necessary to perform offline cleaning. Offline cleaning could use nozzle holder to blow easy debris off the probe tips using compressed dry air or nitrogen (25-30 PSI). Some strong adhesion contamination should be removed by camel hair dry brushing or IPA wet brushing. If brushing is in vain, using various solvents in ultrasonic tank is the smart solution.

Online cleaning is applied in the routine mass production. The wafer yield and test performance depend on correct cleaning materials, cleaning frequency, cleaning operation and so on. According to the degree of the abrasion and wear, probe card cleaning could be divided to clean, polish, scrub and lap. Clean unit and wafer are the traditional means to do the online probe card clean. The clean sheet laminated on the unit and wafer is

978-1-7281-6559-2/20 $31.00 © 2020 IEEE

critical point. Table 1 shows the most common types and applications.

TABLE I.   CLEAN SHEET CLASSIFICATION

Sheet Type	Base Material	Adhesive	Recommended
PET Type	Polyethylene Terephthalate	Room temperature	Cantilever with flat tips
PF3 Type	Foamed Polyurethane	Heat-Resistant Cover 130C	Cantilever Vertical MEMS
SWE Type	Foamed Polyurethane	Room Temperature	Cantilever Vertical POGO pin
Silicone Type	Silicone	Heat-Resistant Cover 160C	Vertical MEMS POGO pin
BC3 Type	Polyolefin	Room temperature	Cantilever MEMS

Probe card has many different types. The paper only focuses on cantilever epoxy probe card, because it is extensively used. Meanwhile, the common clean sheets (PET and PF3 type) are selected as experimental carries. Figure 2 shows the different structure on two types' cross section.

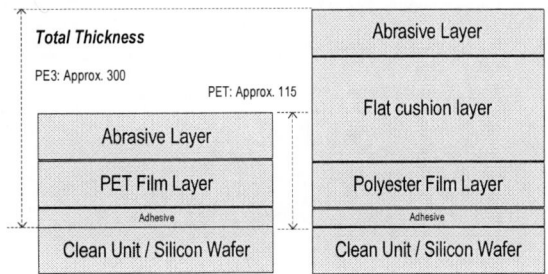

Figure 2: PET Type and PE3 Type Cross-section

PET type with lapping film and the adhesive layer is recommended for cantilever with flat tips. It performs greatly especially for tough debris. PF3 type's abrasives are coated on the flat cushion layer directly. It could perform effective cleaning with less damage on the tips. What's more, its heat-resistant is up to 130C and the special cushion structure works for any probe card type or any probe tips' shape.

ABRASION AND LIFETIME CONTROL

In this section, one of the cantilever probe cards which have reached the scraped spec is selected to do the abrasion and lifetime experiments. The standard scraped target is 100K TDS (Touchdowns). The probe card cleaning study is on the same prober with the same BCF (Balance Contact Force). The cleaning OD (Over Drive) is set as 50/60/70 um separately. The temperature control covers both room and high temperature.

The experimental product requires 72 TDS per wafer. Every 35 TDS do one time probe cleaning. Also every time wafer changes, one more probe cleaning will be done. One time probe cleaning would be executed 30 contacts.

TABLE II.   ABRASION COEFFICIENT RESULTS ON SPLIT TABLE

Type	Temp	OD	Abrasion Size(um)	TDS per Wafer	Clean Count per Wafer	Clean Contact	30K Clean Contact Max Abrasion(um)	30K Clean Contact Average Abrasion(um)	100K TDS Max Abrasion(mil)	100K TDS Average Abrasion(mil)
PF3	RT	50	1	72	1	30	5	2.1	0.27	0.11
PF3	HT	50	1	72	3	30	3	1.47	0.49	0.24
PF3	HT	60	1	72	3	30	5	2.1	0.82	0.34
PF3	HT	70	1	72	3	30	6	2.8	0.98	0.46
PET	HT	60	3	72	3	30	26	22.33	4.27	3.66

TABLE III.   ABRASION RESULTS WITHOUT PROBE CLEANING

	Unit	P1	P2	P3	P4	P5	P6	P7	P8	P9	P10	P11	P12	P13	P14

Tip Length before CP	um	104	113	115	126	127	125	127	121	139	144	130	132	115	113
Tip Length after CP	um	102	111	113	124	125	123	124	118	137	140	126	129	112	112
Unit Abrasion	um	2	2	2	2	2	2	3	3	2	4	4	3	3	1
Max Abrasion	um	4													
Average Abrasion	um	2.5													
Touchdown	count	20700													
Equivalent 100K TDS Max Abrasion	mil	0.76													
Equivalent 100K TDS Average Abrasion	mil	0.48													

From the experiment results shown as table 2, it is found that the max average abrasion wear is 0.82 mil/100K TDS. The abrasion results without probe clean are picked in random 14 probes. Table3 shows the maximum equivalent 100K TDS abrasion is 0.76 mil/100K. It is recommended to use PF3 type clean sheet for lifetime extension. What's more, the lifetime could be concluded as follow equation (1):

$$N_{LF} = \frac{TL_{FC} - S_{PCS} - AWPM_{max}}{ACPC_{max} + ACNC_{max}} \times 100 \quad (1)$$

N_{LF}: Number of the Probe Card Life Time
TL_{FC}: Tip Length on Factory Configure
S_{PCS}: Spec on Probe Card Scape
$AWPM_{max}$: Maximum of Abrasion Wear during probe Plant Maintenance
$ACPC_{max}$: Maximum of Abrasion Coefficient with Probe Cleaning per 1K TDS
$ACNC_{max}$: Maximum of Abrasion Coefficient without Probe Cleaning per 1K TDS

$$N_{LF} = \frac{12 - 4 - 0.7}{0.82 + 0.76} \times 100 = 462K\ TDS \quad (2)$$

Base on above algorithmic model and the evaluation data, the experimental probe card life time could be calculated as 462K TDS as equation (2) shown. Therefore, the predicted goal 462K TDS may be set to the new baseline for the experimental probe card and clean condition.

CONCLUSION

Considering that probe tip length couldn't be monitored in real time, phased sampling and algorithm forecast is rationalizing and practical. Moreover, the methodology could be applied in different clean sheet, probe clean type and temperature control easily. Through the study of abrasion coefficient model and the improvement of the probe card lifetime, the cost of CP test could be reduced significantly. On later phase study, it still needs further exploration on reducing the probing OD and BCF spec to improve the test stability and setup the alarm rules for card PM to monitor needle status to reduce the PM frequency.

ACKNOWLEDGEMENTS

I would like to express my gratitude to my corporation HHGrace that provides the experimental platform. Also I acknowledge the contribution of my team members who support me to develop the fundamental and essential academic competence. Last but not least, I want to thank my family especially for their encouragement and support.

REFERENCES

[1] Penghui ZHANG, Kaihong ZHANG. *Influence of Probe Contact Resistance and Improvement Method* [J]. Electronics & Packaging, 2018,18(01):8-11
[2] Otto Weeden. *Probe Card Tutorial,* Keithley Instruments, Inc:30-31

THE INSPECTION SOLUTIONS AND REDUCTION OF EXTREME TINY POLY RESIDUE

Jianye Song[1]*, Qiliang Ni [1], Xiaofang Gu[1], Chao Han[1] and Lijing Huang [1]

[1] Shanghai Huali Microelectronics Corporation, Shanghai 201203, China

Corresponding Author's Email: songjianye@hlmc. cn

ABSTRACT

As design rule shrinks, it is essential that the capability to detect smaller defects should be improved. In this paper, extreme tiny poly-residue defects are found after poly-etch process. In order to improve the signal to noise rate, poly-residue is detected and monitored by using KLA 29XX BBP inspection system post poly-etch process. Finally, the recipe is optimized by using NanoPoint to remove the noise of poly grain. The purpose of these studies is to capture defects more efficiently and also provides a new way to reduce noise. After defect detecting failure models of these defects were established and the effective defect reduction actions were carried out.

INTRODUCTION

Wafer inspection for defects in semiconductor industry has been more difficult because critical dimension shrinks during semiconductor process development. However, yield loss induced by defects becomes more seriously significant impact on the semiconductor products as device shrunk. There for, yield enhancement engineers are facing with an important topic and challenge to improve the sensitivity of capturing the defect of interests (DOI) in the critical layers.[1] There are many new advanced inspection tools included bright-field (BF), dark-field (DF), and e-beam inspection (EBI) are developed to meet the requirement for challenging defect inspection work. Optical inspection tools (BF&DF) are widely used in inline defect monitor because of their high throughput.

BFI is more suitable for pattern layer because of its incident light perpendicular to the wafer, while the DFI's incident light oblique incident to the wafer.

When traditional scan recipe was established, the care area was simply divided into Full Die, Logic and SRAM according to the density and characteristics of the patterns. As defects become smaller and smaller with design nodes shrink, this traditional method can no longer cover all defect detection. Especially in the pattern etch-related layer, we usually care more about the defect in the etching open area than the defect in the etching remain area where defect detection that needs to be avoided. The BBP NanoPoint function can generate a care area specific to the Design Rule Check (DRC) or GDS.[2][3] In this paper, we found a very small poly residue defect that the traditional care area could not catch because of the noise of poly grain defect. According to GDS, NanoPoint is used to generate

care area, and Etch and Non-etch areas are detected separately. On this basis, scan recipe can remove the influence of poly grain noise. Finally, the formation mechanism and influencing factors of poly residue were found through a series of experiments.

CHARACTERIZATION

In this paper, the poly residue defect is first detected during post spacer deposition inspection. Detailed characterization including top down SEM, Side view SEM, high resolution cross sectional TEM and EELS analysis was performed on the defects at their point of detection, to understand morphology and chemistry. Figure1 (c)and(d) show the Top down SEM image and Side view SEM image of poly residue defect post spacer deposition(SPA DEP). From the top down SEM image we find that the defect size is about 0.16μm and has a round outline. The side SEM image shows the defect is cylindrical and standing vertically on the substrate. The bottom of the defect is tightly connected with the substrate.

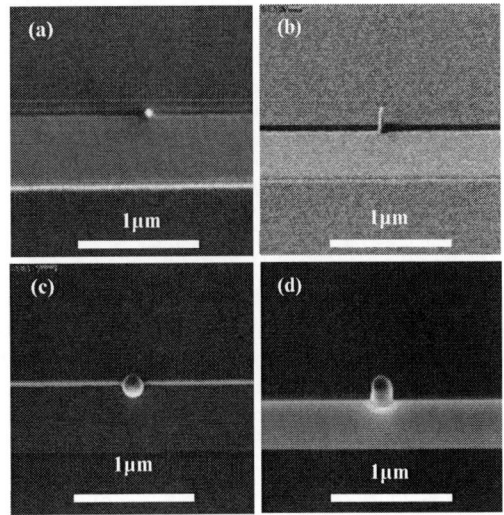

Figure 1: (a)Top down SEM image and(b) Side view SEM image of poly residue, (c) Top down SEM image and(d) Side view SEM image of poly residue defect post spacer deposition

A wafer post poly-silicon etching was inspected by KLA 29XX BBP inspection system with a hot scan recipe. The poly residue，which has similar size about 30nm (figure1(a),(b)) with TEM result, was found through the SEM review with the scan map. Form the Defect Pareto for the hot scan wafer we indicated that the capture rate of poly residue defect only ~3.15%. The hot scan recipe is not suitable for monitor and split test because of the high defect count and low capture rate. So the scan recipe needs to be optimized to catch more cared defect.

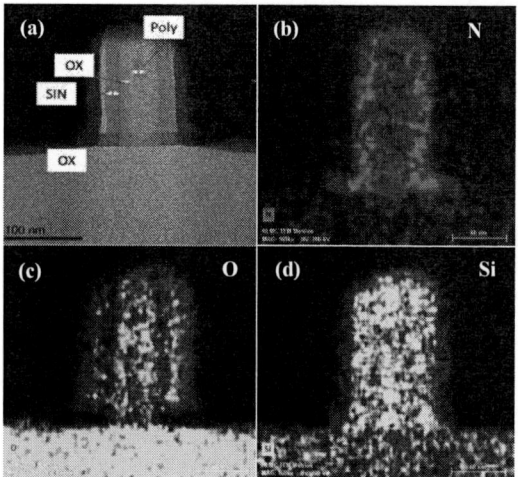

Figure 2: (a) cross sectional TEM parallel to the defect axis, and EDS analysis(b,c,d) of poly residue defect after SPA DEP

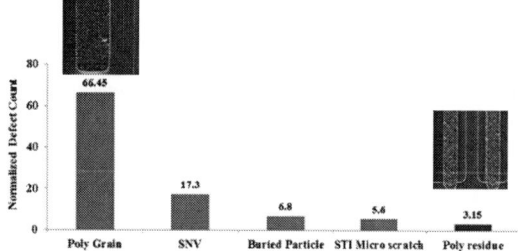

Figure 3: Defect Pareto for hot scan wafer

In this case, it is found that poly grain has the biggest influence on defect detection. After poly etch, poly grain is mainly distributed in poly remained area. Through NanoPoint function, care area was established according to GDS to filter the poly grain noise, which could improve the detection rate of poly residue.

Figure 4 shows the steps to generate the care area using the GDS and NanoPoint function. Inspection of GDS generated hotspots using NanoPoint function on a BBP tool is a three-step process. During the first step, GDS files are changed to Reticle design files. In the second step, the reticle design files are used to generate NanoPoint care areas. In the third step, these NanoPoint care areas are used by the BBP tool to perform a full wafer scan. Finally, care areas were obtained as shown in figure 5. The green areas is the scan area and black is the mask area.

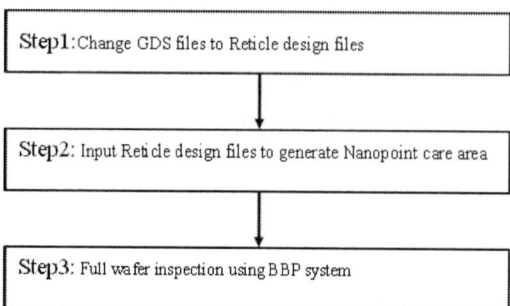

Figure 4: Metrology of poly etch area using BBP Nanopoint care area

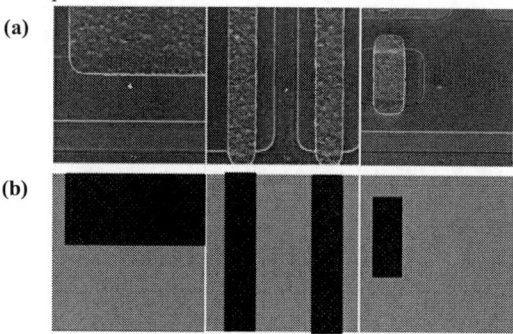

Figure 5: (a)The SEM image and (b)corresponding care area (green is the scan area and black is the mask area)

RESULTS AND DISCUSSION

Scan recipes are established using the optimal conditions with NanoPoint care areas. Review data shows that the poly grain noise completely disappears and the nuisance rate drops to about 10%.

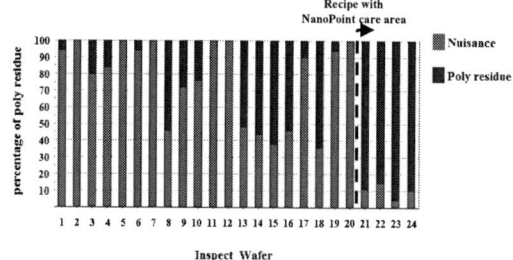

Figure 6: The trend chart of poly residue catch ratio

Polysilicon gate is widely used in semiconductor fabrication and usually formed by plasma etching. During the etching process, if the etching is not complete, the polysilicon residue will cause a short circuit between the electrodes and cause device failure. Therefore, it is very important to avoid the formation of polysilicon residue in the etching of polysilicon. There are generally two types of polysilicon residues, one is the incomplete etching caused by the front layer particles or residual shielding, the other is the etching ability is not enough to form etching residue. The NanoPoint care area is used to set up scan recipe and tiny residue defect which may cause poly residue is detected post poly photo. It was suspected that the block of residue during poly etch caused the formation of the poly residue defect.

TABLE I THE POLY RESIDUE DEFECT RESULT OF DIFFERENT SPLIT CONDITION

Split Condition	Poly Residue Defect Ratio
Baseline	1.00
New Develop1	0.90
New Develop2	1.10
HMDS	0.98
SPM&SC1	0.15

TABLE II THE POLY RESIDUE DEFECT RESULT OF DIFFERENT SPLIT CONDITION

Split Condition	Poly Residue Defect Ratio
Old PR Batch	1.00
New PR Batch	0.10

The residue defect post photo process may come from the residue of photoresist, which may be not totally removed after lithography develop process. Since it is suspected to be residue defect after photoresist development, the effect of development on the defect is considered first. The spin speed and develop time are carried out to optimized develop recipe. From the experimental results in table I, it can be seen that the poly residue defect has not been significantly improved.

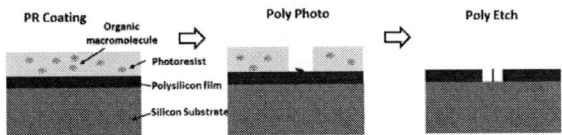

Figure 7: The mechanism of poly residue formation

Based on the above defect results, the tiny residue defect reduced when the wafer on the condition of reworking. The main components of PR wet striping solution are SPM (H_2SO_4 and H_2O_2) and SC1(NH_3 and H_2O_2). During the development process, the characteristics of the wafer surface will also affect the formation of the residue defect. The hydrophobic surface is more prone to make the photoresist adhesion on the wafer surface than the hydrophilic surface. The surface become hydrophobic after treatment with hexamethyl disilazane(HMDS) and hydrophilic after PR rework. [4] In the split condition of lithography, the wafers are pretreated with HMDS and PR wet striping solution before PR coating. As the table I suggests, the poly residue defect is obviously reduced in the condition of pre-clean with SPM&SC1.

From a series of experiments, it is also found that the defect level of different photoresist batches is quite different, as shown in table II. According to the results of previous experiments，the poly residue defect formation mechanism was stated in figure 7. The organic macromolecules in the photoresist cannot be completely removed during the development process to form residue defects, which block the subsequent ion etching process to form poly residue defect. The hydrophobic surface of the wafer can inhibit the aggregation of large molecules to form residue defects and improve the poly residue defect. It can be seen from the defect formation mechanism that controlling the quality of the photoresist and pretreating the wafer can effectively reduce the defect.

CONCLUSION

In this paper, we successfully set up a recipe based on NanoPoint function to filter out noise defects. Based on the recipe established by NanoPoint care area, we carried out experiments and found the formation mechanism of poly residue. It is found that the defects can be improved by photoresist quality control and wet pretreatment. As the critical dimension shrinks, the defect size also gets smaller and smaller. From this case, it can be seen that the method of detecting etched region and non-etched region separately can reduce noise, thus obtaining higher defect detection rate.

REFERENCES

[1]A. Srivastava, H. Nguyen, T. Herrmann, R. Kirsch and R. M. Kini, *In-Line Inspection of Hotspots and Monitoring Strategies*, in IEEE Transactions on Semiconductor Manufacturing, vol. 29, no. 4, Nov. 2016,pp. 299-305.

[2] Julie L. Lee, Brian M. Trapp, and John A. Rudy, *Quality Metric for Defect Inspection Recipes*, IEEE Transactions on Semiconductor Manufacturing, Volume 26 , Issue 1 , Feb. 2013, pp. 3-10.

[3] M Daino, G Jensen, A Jain, S Kini, *Line End Voids defectivity improvement on 64 pitch Cu wire interconnects of 14 nm technology*, IEEE Transactions on Semiconductor Manufacturing Volume 29 , Issue 4 , Nov. 2016, PP. 299 – 305.

[4] D Li, ZF Gan, YY Wang, ZK Yang, ZB Mao, Y Zhang, *The problems and solutions in 40 nm node dual gate lithography process development*, China Semiconductor Technology International Conference(CSCIT), 2015.

AN APPLICATION OF ADAPTIVE GENETIC ALGORITHM COMBINING MONTE CARLO METHOD

Wei Yu[1]*, Xu Chen[2],Jingjing Lu[3], Zhengying Wei[4]

[1, 2, 3, 4] E1, HLMC, Zhangjiang Hi-Tech Park, Shanghai 200120, People's Republic of China
*Corresponding Author's Email: yuwei_e1@hlmc.cn

ABSTRACT

Over the past decades, semiconductor manufacturing has been drawing more and more attention. The procedures it involves could be considered as one of the most complicated processes in manufacturing. Owing to this, cases caused by abnormal machines happened from time to time. It requires a large amount of time and experience to solve this problem manually. Meanwhile, state of machines changes as time goes by. As a result, engineers are expected to find out root causes as soon as possible. In this paper, adaptive genetic algorithm is introduced to identify common bad tools and provide suggestions for case study.

INTRODUCTION

For semiconductor manufacturing with mature procedures, the quality of products (CP/FT) mainly depends on the status of process machines they are delivered to. Hundreds of process steps through various processing tools are involved. Any tool excursion could lead to costly yield problems on product. SPC and in line process control is applied to avoid serious yield loss. However, some low-level intermittent problem is hard to detect and can only be found after yield or electrical parametric data is collected. Tool commonality is thus a proven technique to figure out the root cause [1].

George Kong has provided a methodology of systematic tool commonality analysis [2]. Classification for several key elements like sample size and metrology data are introduced. Statistical analyses are widely discussed as well. Engineers look for commonality among the "BAD" lots which does not exist on "GOOD' lots [2]. By configuring data in different ways, they may finally rise up some meaningful ideas. However, because of the complexity of process and huge number of process steps, it is challenging to pin point the source tool and at which step it occurs. Moreover, it relies on the experience of engineers and the final results may not be objective.

Genetic algorithm (GA) was firstly raised by J. Holland in 1970s. It was aimed to study the self-adaptation behavior in natural system [3]. It is an algorithm that includes random search and global optimization. It achieves high fault-tolerance ability and strong robustness. Moreover, it has a strong flexibility which means it could be easily combined to other methods and is suitable for complex combinatorial optimization problem system.

The adaptive genetic algorithm (AGA) is an efficient approach for multimodal function optimization which evolved from GA [4]. It uses adaptive probabilities of crossover and mutation so that the diversity in the population could be maintained. The convergence capacity of the GA could be sustained at the same time.

Monte Carlo methods are a wide class of computational algorithms that rely on a vast number of repeated random sampling to compute numerical results. Optimization is one of the problem classes they are mainly used in [5]. The underlying concept is to use randomness to figure out problems which might be deterministic in principle. In application to engineering problems, Monte Carlo–based predictions are usually better than human intuition or alternative "soft" methods [6].

In this paper, we propose a hybrid algorithm that combines adaptive genetic algorithm (AGA) and Monte Carlo method to overcome the weaknesses of manual calculation. Instead of analyzing data by human, we simply choose index for algorithm to evolve. Via mimicking the evolution of natural system, the final output would be the common bad tools that resulted in the yield loss.

METHODS

Data Collection

Semiconductor manufacturing process is complex, usually can be split into four processes: Wafer Fabrication, electrical performance test, assembly and final test [7]. For tool commonality analysis, we only took processing steps into consideration and used fail rate related to a specific bin as index.

Process machines of each step are compiled in binary code and then connected into a path following the order of process flow. The path of an individual lot is coded as a chromosome, representing a solution to the problem. We chose the mean fail rate of a lot to represent the performance of this chromosome.

Adaptive Genetic Algorithm

The GA is an outstanding stochastic search technique raised to mimic some of the phenomena observed in natural evolution. Chromosomes evolve through successive iterations which are made up of crossover, mutation and selection. After several iterations, the algorithm converges to the best chromosome, which stands for the optimum solution [8], [9].

Adaptive genetic algorithm is a variant of genetic

algorithms, which was raised owing to increasingly complex computing expectations. Instead of fixed parameters, AGA evaluates with adaptive parameters, which have a great influence on the degree of solution accuracy and the converge speed. To be specific, the population information in each generation is utilized to adjust the probabilities of crossover (p_c) and mutation (p_m) in eq.1. Concretely, f_{max} indicates the largest individual fitness in population; f_{avg} indicates the average individual fitness in population; f' indicates the higher fitness value of the two chromosomes which are selected to crossover; f indicates the individual fitness value which is selected to mutation. Parameters are set as follows: $P_{c1} =0.9$, $P_{c2} =0.6$, $P_{m1} =0.1$, $P_{m2} =0.001$ as recommended [4]. In fact, the exact values of these parameters affect the progression of evolution to a small degree.

$$P_c = \begin{cases} P_{c1} - \frac{(P_{c1}-P_{c2})(f'-f_{avg})}{f_{max}-f_{avg}}, f' \geq f_{avg} \\ P_{c1}, f' < f_{avg} \end{cases} \quad (1a)$$

$$P_m = \begin{cases} P_{m1} - \frac{(P_{m1}-P_{m2})(f_{max}-f)}{f_{max}-f_{avg}}, f \geq f_{avg} \\ P_{m1}, f < f_{avg} \end{cases} \quad (1b)$$

Monte Carlo Method

Monte Carlo methods are a wild class of computational algorithms which are quite good at modeling phenomena with significant uncertainty. In manufacturing, the final performance of a process machine is not only decided by its intrinsic status, but some accidental elements (e.g. defect) as well. As a result, we applied inner-built random number generator in Python for each chromosome. The distribution of random value is set to be Gaussian. It is commonly applied to natural and social sciences on behalf of real-valued random variables whose distributions are unknown [10]. In this paper, we adopt the mean value and standard deviation (SD) of each machine to simulate random state.

Modeling

The process of our model follows the rule of usual GA. It contains selection, crossover, and mutation in a single iteration. All the data were collected as illustrated above. Mean and SD values of each machine were then summed up for $g(x)$ (eq.2) where x represents the chromosome x; i represents the i-th machine in x; $mean_i$ represents the mean fail rate of i-th machine; $random_i$ represents the random value generated from the Gaussian distribution of $mean_i$ and SD_i. β denotes the significance of random values to the final result.

$$g(x) = \sum_{i=1}^{n}(mean_i + \beta * random_i) \quad (2)$$

The fitness function is evaluated for each individual (eq. 3), providing fitness values. Usually, these values need to be normalized. However, the differences between values were small (usually at the order of 0.0001). To amplify it, we add a modulating factor α (eq.3). It has been proved to increase the performance. The value of α

depends on the specific value of g. In this paper, we make α equaling to 100.

$$f(x) = \begin{cases} \alpha * g(x), maximum\ problem \\ -\alpha * g(x), minimum\ problem \end{cases} \quad (3)$$

We chose roulette wheel for selection operator. The population is sorted by descending fitness values and the probability is calculated following rank-based fitness assignment [11]. Linear ranking (eq. 4) was applied where i stands for the serial number; N stands for the total number; η^+ stands for selective pressure, usually in the range of 1.0 and 2.0 [12]. In this paper, we chose $\eta^+=1.5$. While there are many possible choices, final results are not particularly sensitive to exact values of most design parameters. The experiment begins with two-point cross over and single point mutation. The model will not stop unless it meets one of these criterions: reaching the iterations of 200 or the residual between maximum fitness and average fitness less than 10^{-7}.

$$p_i = \frac{1}{N}\left[\eta^+ - (\eta^+ - \eta^-)\frac{i-1}{N-1}\right] \quad (4a)$$
$$\eta^- = 2 - \eta^+ \quad (4b)$$

Each time we run the model, it would output a population. The chromosome with highest fitness value pops up to be the final result. Considering the random factor added, we run the model N times and find out the common tools among N outputs. N can be any number larger than 10 (see Appendix for detail).

To demonstrate the feasibility of this hybrid algorithm, we conducted experiments on the fail rate of white pixel collected from 150 lots with same manufacturing process.

RESULTS
Evolution

We started the experiment with the simple evolution task: finding out the maximum value it can achieve. Values and procedures follow what have mentioned above and summarized in Appendix.

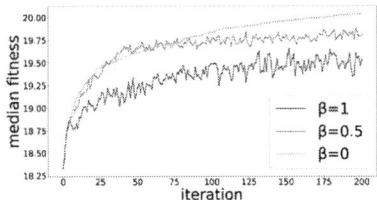

Figure 1: Results of evolutions.

Fig.1 manifests the evolution of fail rate. It is obvious that the median value of the population ascends as iteration increases. This is assistant with the phenomenon in basic simple genetic algorithm [13]. Although the value of β does not affect the trend of evolution, it do bring out differences. Specifically, when $\beta>0$, the median value evolves with fluctuations. The larger the β is, the greater variations it will experience. Moreover the value of β has an influence on the final value achieved. Usually, the final value decreases with the growth of β.

Common Bad Tools

In this section, we demonstrate the application in finding common tools. Instead of giving sole path, we intend to find the root causes for serious yield loss. To achieve this goal, stochastic factors are introduced when calculated $g(x)$. It will not only better represent the actual status of machines, but lead to different final chromosomes each time as well.

TABLE I.   RESULTS FOR DIFFERENT INDEX VALUES

β	Accuracy (%)	F1 Score
1	99.1	0.75
0.5	98.2	0.60
0	91.0	0.23

We use N (number of running times) = 10 for illustration. Our results show (Table I) that the accuracy keeps high (over 90%) regardless of the exact number of β. To be specific, $\beta=1$ obtains the highest accuracy (99.1%) while $\beta=0$ receives lowest value (91.0%). The high accuracy may result from the fact that good tools are in the majority. While it is time-consuming to check the authenticity, we do not want too many false alarms. F1 score is thus presented in case the high accuracy is caused by class imbalance. It can be seen from the table that the evolution with $\beta=1$ achieves the highest score and the performance gets worse while β gets smaller. This phenomenon is within expectation. When the proportion of randomness ascends, it will preclude disturbing terms. Consequently, non-key machines will less likely be highlighted.

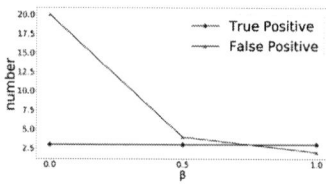

Figure 2: Results of various β values.

The line chart in Fig. 2 gives specific numbers of true positive alarms and false positive alarms. It is demonstrated that the number of false positive alarms decreases dramatically when the value of β ascends from 0 to 0.5. The trend keeps when β continues to increase. However, the trend becomes mild in the second half. Different from false positive alarms, the number of true positive alarms keeps steady when β changes. In spite of this, we do not recommend to make β larger than 1.0. Excessively large value of β is harmful to the alarm of key

factors. Some root causes might be missed if too much disturbance is involved. Similar experiment has been conducted on another real case happened in a different process flow B. The results denote the applicability of this algorithm as well (see Appendix for detail).

CONCLUSION

In general, we proposed a hybrid stochastic algorithm that combines adaptive genetic algorithm and Monte Carlo method. AGA can find the worst path in manufacturing by evolution while Monte Carlo method excludes the influence of insignificant factors. The algorithm is then utilized in cases with known root causes. It has been proved that this approach is simple and effective both in finding the maximum value and analyzing tool commonality.

REFERENCES

[1] Dan Malinaric, Richard R.H, Cindy Sun, *Advanced Semiconductor Manufacturing Conference*, 2000, P8-13.

[2] Kong, G. Y. *advanced semiconductor manufacturing conference* (2002): 202-205.

[3] Sampson, Jeffrey R. *Siam Review* 18.3 (1976): 529-530.

[4] Srinivas, M., and L. M. Patnaik. *systems man and cybernetics* 24.4 (1994): 656-667.

[5] Kroese, Dirk P., et al. *Wiley Interdisciplinary Reviews: Computational Statistics* 6.6 (2014): 386-392.

[6] Hubbard, Douglas; Samuelson, et al. (October 2009). *Modeling Without Measurements* OR/MS Today: 28–33.

[7] R. Uzsoy, C. Lee, L. A. Martin-Vega, *IEEE Transactions*, vol. 24, no. 4, pp. 47-60, 1992.

[8] Guo, Pengfei, Xuezhi Wang, et al. *biomedical engineering and informatics* (2010): 2990-2994.

[9] T. G. Campbell, R. A. Nicolaides, et al. *Computational Electromagnetics and Its Applications*. Boston, MA, USA: Kluwer,1997.

[10] Lyon, Aidan. *The British Journal for the Philosophy of Science* 65.3 (2014): 621-649.

[11] Goldberg, David E., and Kalyanmoy Deb. *foundations of genetic algorithms* (1991): 69-93.

[12] Whitley, L. Darrell.*international conference on genetic algorithms* (1989): 116-123.

[13] Barker, J. S F. *Australian Journal of Biological Sciences* 11.4 (1958): 603-612.

APPENDIX

Values for parameters

TABLE I.   SYMBOLS AND VALUES OF PARAMETERS

Symbol	Interpretation	Values	Comments
P_{c1}	Probability of cross over for chromosome with large fitness	0.9	Usually $0.5 < P_{c1} < 1.0$[4]
P_{c2}	Probability of cross over for chromosome with small fitness	0.6	Less than P_{c1}
P_{m1}	Probability of mutation for chromosome with large fitness	0.1	Larger than P_{m2}
P_{m2}	Probability of mutation for chromosome with small fitness	0.001	Usually $0.001 < Pc1 < 0.05$[4]
f_{avg}	Average fitness in population	variable	/
f_{max}	Largest fitness in population	variable	/
f'	Larger value of two individual fitness	variable	Used in cross over
f	Individual fitness	variable	Used in mutation
x	Representing an individual chromosome	/	A path composing of process machines
α	Modulating factor	100	Modifying the scale of fitness
η^+	Selective pressure	1.5	Linear ranking allows values in [1.0, 2.0][12]
η^-	$2 - \eta^+$	0.5	/
β	significance of random value	/	Using $\beta=0$, $\beta=0.5$, $\beta=1$ for comparison
N	Times of running models	10	/

Choice of N

TABLE II.   VARIOUS F1 SCORES FOR DIFFERENT N

N	Accuracy (%)	F1 Score
5	98.7	0.67
10	99.1	0.75
15	99.1	0.75
20	99.6	0.86

We show the feasibility of this algorithm in finding tool commonality with $N=10$. Actually, we have experimented on different values of N. The value of each parameter is shown in Table I. Specially, we fixed β to be 1.0 in this section. As shown in Table II and Fig. 3, the performance actually gets better when we repeated the model for more times. However, the improvement is not significant. With the increase of N, the number of true positive alarms is steady while the number of false positive alarms decreased a little bit. To be specific, when N inclines from 5 to 10, the number of false positive alarms declines from 3 to 2, which, in turn, results in a slight increase of F1 score. There is no variation when N changes from 10 to 15. The trend when N grows from 15 to 20 is similar to the trend from 5 to 10. Taking the cost of computation into consideration, we finally chose N to be 10.

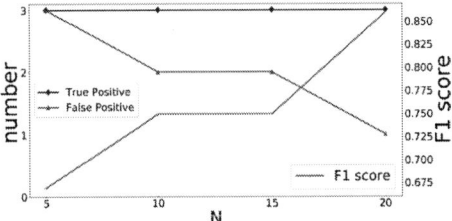

Figure 3: Results of various N values. Green line represents the number of false positive alarms; blue line represents the number of true positive alarms; red line represents the value of F1 score.

Results of process flow B

TABLE III.   VARIOUS F1 SCORES FOR DIFFERENT INDEXES

N	β	Accuracy (%)	F1 Score
10	0.5	97.7	0.68
10	1.0	99.2	0.83
15	0.5	98.5	0.76
15	1.0	99.2	0.83

This hybrid algorithm is applied to another case of different process flow to investigate whether it can be generalize to all situations. Based on previous results shown in RESULTS, we only compared the results of $\beta=1.0$, $\beta=0.5$, $N=10$ and $N=15$. From Table III, it can be seen that the increase of N can make up for the shortcoming of small β. Additionally, when β reaches 1.0, a larger N will not improve the performance.

INVESTIGATION AND REDUCTION OF SYSTEMATIC DEFECTS BY WAFER BACKSIDE PROCESS IN NANOMETER SEMICONDUCTOR MANUFACTURING

JianGang Zhou[1*], *Hungling Chen, Yin Long, Kai Wang, Hao Guo*
Shanghai Huali Integrated Circuit Manufacturing Co., Ltd, Shanghai, China.
*Corresponding Author's Email: zhoujiangang@hlmc.cn

ABSTRACT

The research aims at wafer front side defects causing by faulty back sides during the semiconductor manufacturing. The rapid semiconductor development of semiconductor technology comes with the extremely tight control of SPC and small tolerance of defects even the defects in wafer backside could cause the fail of front side chips while technologies entered nanometer node. The wafer backside be polluted and damaged repeatedly by mechanical chucks or robots during manufacturing processes, and which combined with gas or liquid chemical processes could make uneven produce marks. The formation mechanism of the abnormal backside condition was investigated, and the corresponding front side's defect condition was drawn out. In Cu backend, backside clean is required to remove metal ions, however, the difference of acid etching rate on poly silicon and silicon nitride would lead to the abnormal leveling of lithography. Optimizing chemical formulation can improve backside leveling obviously.

Key words: backside clean, leveling, defect, backend, chemical formulation

INTRODUCTION

As the shrinkage of the geometry size of the semiconductor, higher quality flatness is needed by photo process, as well as leveling of backside. It is well known that the backside without clean can add defectively on the front side [1]. Research on backside condition seems particularly important.

More specifically, deeply impact will formed by backside worse condition;

Phenomenon	Result
Wafer's backside vacuum abnormal	Tool alarm, production is hampered
Litho leveling abnormal	Wafer's front side defocus
Backside scratch by parts which can touch during process	Particle source and impact wafer's front side

This paper describes wafer's front side defocus caused by dirty backside. As shows in the fig.1, there is obvious special yield loss map which can match with backside image. We design some experiments to verify the fail mode and reduce this kind of defect.

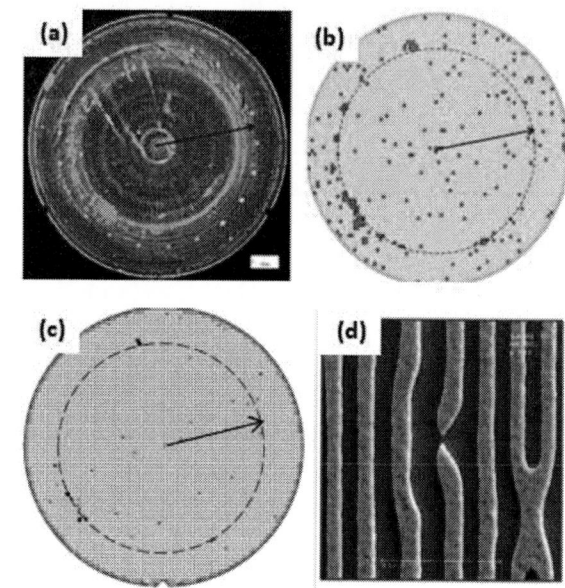

Fig.1 (a) Backside defect image; (b) CP map with low yield ring special map; (c) Front side ring special map; (d) Review image of front side ring map, all defocus.

EXPERIMENT

In this paper, the Si-Si structure will be discussed. Wafer run to BELO, metal ion should be removed by hydrogen nitrate for contamination concern. Acid can remove partial of poly silicon. Some particles on wafer's backside will turn into the mask during wet process, which will induce block-etch.

Fig 2 shows the formation mechanism during backside clean process.

Fig 3(a) shows the Particles which are produced by rubbing the backside with tool's chuck or robot.

Fig 3(b) shows the wafer profile after acid clean.

978-1-7281-6559-2/20 $31.00 © 2020 IEEE

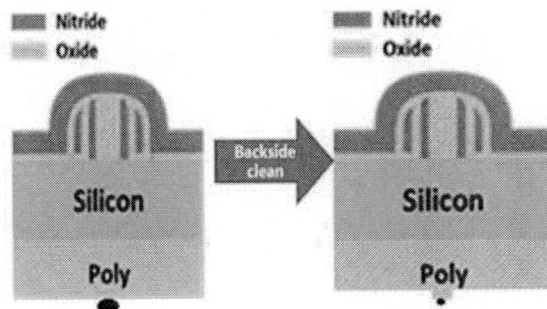

Fig. 2 Acid clean etch is blocked by the particle on the backside

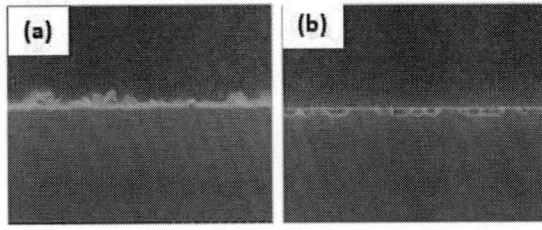

Fig. 3 (a) Before acid clean, particles are objected on backside; (b) Block-etch after acid clean

As we known, acid can expend silicon but SiN. Backside clean experiment on different material of backside is designed. The leveling value measured by litho tool can represent the uniformity. The result shows that backside with silicon nitride will be protected in case of additional etched.

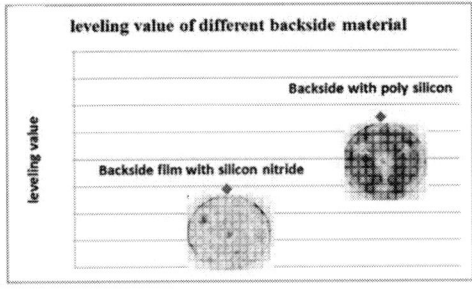

Fig.4 Leveling value after acid clean under different backside materials

Obviously, as Fig.4 shows, nitride on wafer's backside as a protect film can keep the acid from etching effectively.

As this paper introduce above, acid can remove backside's poly silicon. The moving trajectory of wet process tool's nozzle is designed always from one location to another (center location to edge is used generally in the industry), as Fig.5(a) and (b) shows, this kind of process mode tend to cause the very center's silicon loss be heavy than edge's.

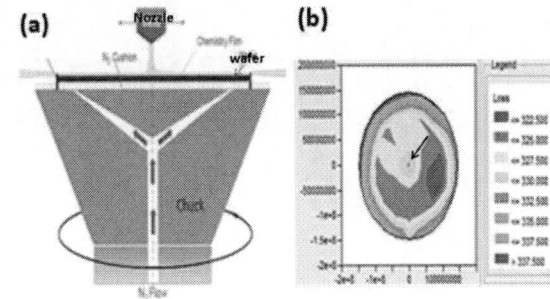

Fig.5 (a) Clean tool's structural representation ; (b)Contour map of backside after backside clean with acid chemical

An experiment is designed to solve this problem. Different chemical formulation is used to run on backside clean process tool. The amount of metallic ion residue is measured by inductively coupled plasma mass spectrometry (ICPMS) and the leveling range value is collected by litho scanner unit.

As the Fig.6 shows, group I ~IV represent different acid concentration gradients. Group I is the lowest and Group IV is the highest. The result shows that the lower acid concentration used, the leveling performance is better, conversely, the metallic ion residue is worse.

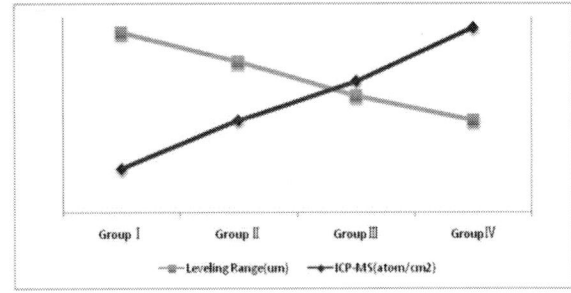

Fig.6 Acid concentration gradients' correlation with leveling and metallic ion

From fig.7 (a) and (b), wafer center 3D topographic shows that the concentration of acid is higher, the silicon loss is heavier.

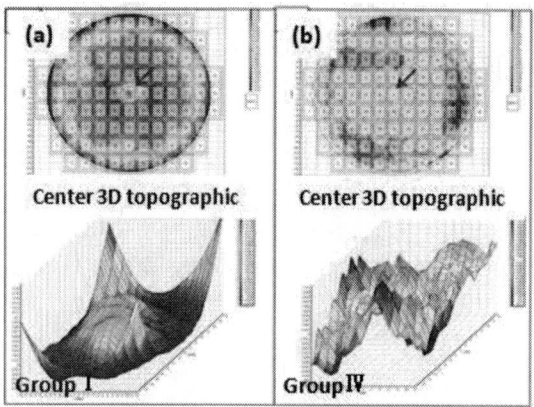

Fig.7 (a) Group I :backside leveling map and wafer center 3D topographic; (c) Group IV backside leveling map and wafer center 3D topographic

CONCLUSION

Backside of wafer will be damaged by backside clean process. The impacted map is always similar with some parts of process tool which can touch wafer's backside. Particles generated during rub with backside, some of the particles may turn into the mask during wet clean process which can lead to backside block-etch. SiN film as the wafer's backside material can protect backside from acid's etching. Wafer center low yield map is often found for wet clean tool's design flaw. Appropriate diluted acid can improve the leveling performance under lower amount of metallic ion residue

REFERENCES

[1] Neng-Cheng Wang, Hui-An Chang, Chung-I Chang, Ting *Wang, /ProMOS Technologies Inc."Backside Wafer Damage Include Wafer Front Side Defect and Yield Impact"* 2007 IEEE/SEMI Advanced Semiconductor Manufacturing Conference

STUDY OF HIGH-PRECISION INTERFEROMETER DYNAMIC SWITCHING FOR TFT LONG-TRAVEL-RANGE MOVING STAGE

Dan Chen [*], Zhiyong Yang, and Yuebin Zhu

Shanghai Micro Electronics Equipment (Group) CO., LTD., Shanghai 201203, China

* Email: chend@smee.com.cn

ABSTRACT

Interferometer dynamic switching (**IDS**) is one of core technologies of long-travel-range moving stage of high-generation TFT scanner. The accuracy and repeat-ability of IDS have a major influence on the servo performance of moving stage, further, on the hierarchical overlay error of lithography machine. Zeroing model and dynamic switching strategy are of great significance to IDS. To solve IDS problem efficiently, in this paper, a novel zeroing model as well as a dynamic switching strategy have been proposed. Experiment results show that, the IDS error specification has been satisfied in both single and multiple IDS test scenarios by using our proposed method.

Keywords—zeroing model; switching strategy; weighted curve; overlapped zone; dynamic switching

INTRODUCTION

Nowadays, display screen has been widely used in our daily life. To produce a display screen, patterns on mask have been exposed to glass plate in the production [1]. Glass plate is carried by plate stage for six-degree-freedom movement, while mask is carried by reticle stage mainly for scan direction movement [2] [3]. With the size of display screen gets larger, the size of glass plate and mask has also become larger. Accordingly, the size of moving stage, including plate stage and reticle stage, and their travel ranges have also been increased. As for scan lithography machine, plane mirror interferometer has been widely used for the measurement of hierarchical position of moving stage [4], and the measurement results have been used for the feedback of moving stage control system [5].

Under the premise of high standard smoothness, according to known information, the length of plane mirror can reach up to c.a. 1.8 meters at most. The full-travel-range of moving stage of high-generation TFT scanner has already beyond the maximum measurement range of single interferometer. Thus, arranging multiple interferometers along the scan direction (**Y**) to satisfy full range hierarchical position measurement has been considered. An interferometer layout is shown in fig.1, interferometer switches from one to another during moving stage's full-range-movement, which brings the problem of IDS.

Our proposed zeroing model and dynamic switching strategy have been described in section 2; in section 3, experiment results are given and discussed. Finally, conclusions to the whole paper have been given.

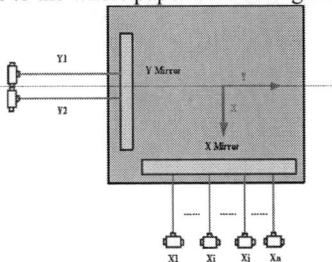

Figure 1: An Interferometer layout

PROPOSED METHOD

Zeroing Model

Because of local environmental differences of multiple interferometers, such as temperature, pressure differences, the measurement baseline for interferometer X_i and X_j is different, which means the measurement values of X_i and X_j are different for the actual same X position of moving stage. To solve this problem, the zeroing process has to be done before interferometer X_i switches to interferometer X_j and the raw fringe number of interferometer X_j has to be modified based on formula (1).

$$\begin{cases} \Delta h_j = g(\lambda_j) * \Delta X_{ij} \\ \Delta X_{ij} = X_i - X_j \end{cases} \quad (1)$$

therein, $\triangle h_j$ is the raw fringe number increment of interferometer X_j; λ_j is the laser wave length of X_j; $\triangle X_{ij}$ is the difference of measurement values between X_i and X_j.

Dynamic Switching Strategy

After the zeroing process has been finished, the switching process has been done in the overlapped measurement zone (**OMZ**) of both interferometers X_i and X_j, see fig. 2. During the process, the control power has been switched from one interferometer to another smoothly. Meanwhile, the X position of moving stage is determined by the combination of weighted interferometer model outputs of both X_i and X_j, see formula (2). In this paper, we have adopted polynomial weighted curve, see fig.3. In this figure, interferometer X_i has two different weighted curves, depends on scan direction of moving stage, see formula (3).

$$X = \omega_i * X_i + \omega_j * X_j \quad (2)$$

978-1-7281-6559-2/20 $31.00 © 2020 IEEE

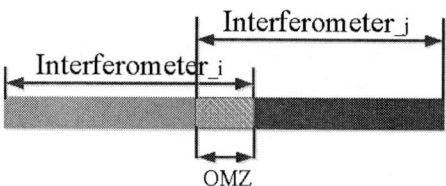

Figure 2: Overlapped measurement zone (OMZ)

$$
\omega_j = \begin{cases} 1 - \displaystyle\sum_{k=0}^{n} p_i * Y^k, & \vec{d} > 0 \\[2ex] \displaystyle\sum_{k=0}^{n} p_i * Y^k, & \vec{d} < 0 \end{cases} \quad (3)
$$

therein, ω_i is the weight of interferometer X_i; ω_j is the weight of interferometer X_j; X is the x position of moving stage; Y is the y position of moving stage; $\vec{d}$ is the scan direction of moving stage; p_i is the polynomial coefficient of weighted curve. In formula (3), Y position has been normalized to zone [-Zd/2, Zd/2], Zd is the range of OMZ.

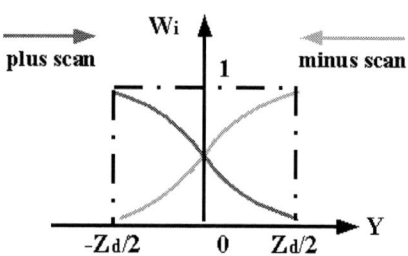

Figure 3: Weighted curve in plus/minus scan direction

EXPERIMENTS

Experiment Design

Tests based on single and multiple IDS test scenarios have been done using our proposed method. All tests have been carried out on TFT lithography machine. In our experiments, the stage moves in low, constant speed in OMZ. Because of IDS, according to our known information, the moving stage couldn't move steadily in a random OMZ in its full-travel-range before our method was adopted. Hence, there is no comparison of IDS errors before and after our method is adopted. Supposed the IDS error specification of test TFT scanner is E_0. In our experiments, IDS errors are divided by benchmark E_0, and are expressed in percentage (%).

Experiment Results Discuss

In single IDS test scenario, the control power of hierarchical X position has been switched from interferometer X_i to X_j in OMZ smoothly with the length of 60 servo periods. IDS error during a single switching process is shown in fig.4.

In multiple IDS test scenario, the stage moves back

TABLE I
IDS ERROR (%) IN DIFFERENT TEST SCENARIOS

IDS error	Max	3Sigma
Single IDS error	4.53%	2.86%
After 20 IDS times error	39.0%	6.11%
After 40 IDS times error	52.05%	6.59%

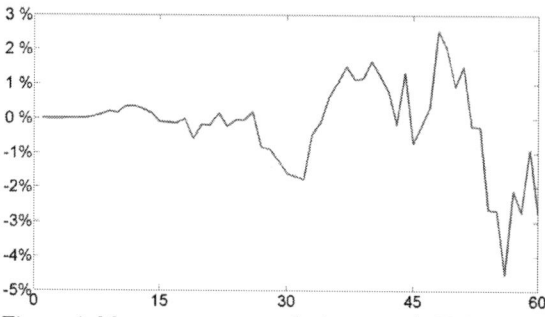

Figure 4: Measurement error during a single IDS process

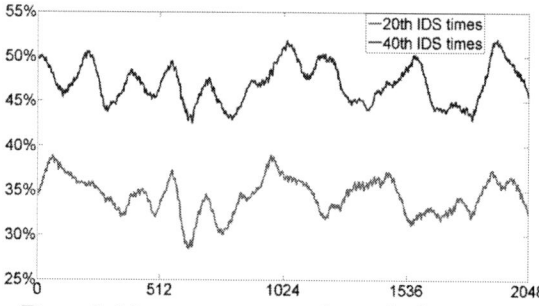

Figure 5: Measurement error after multiple IDS times

and forth continuously, while IDS has been taken place in a OMZ for multiple times. IDS errors after 20 IDS times and 40 IDS times are compared in fig.5 with 2048 sample points. From the figure, we can find that the cumulative IDS error is drifted, but the drift rate gets slower as the number of IDS times increases. Statistical analysis is given in table 1 above. It is shown in the table that, maximum IDS errors of single, 20, 40 IDS times, in our experiments, are 4.53%, 39.0%, and 52.05% respectively. These maximum IDS errors are less than IDS error specification, and the drift velocity of cumulative IDS error is convergent.

CONCLUSIONS

Based on our proposed zeroing model and switching strategy, IDS errors in both single and multiple IDS test scenarios have met the requirement of IDS error specification of test TFT scanner. More importantly, IDS error growth rate is convergent with the increase of IDS times. By using our method, moving stage can move in the full-travel-range with good servo performance.

It is a trend of TFT lithography machine to develop

to higher generation. To achieve overlay specification, high-precision and robust IDS is a significant technical problem, which needs to be studied deeply, for instance, more smart switching strategy. Meanwhile, with the increase of optical length of interferometer, the influence of environment on IDS needs to be analyzed carefully.

REFERENCES

[1] G. Ning, et al. "Reticle and wafer CD variation for different dummy pattern," *Photomask Technology International Society for Optics and Photonics*, 2012.

[2] L. He, et al. "Method for measuring the granite surface topography of wafer stage with laser interferometer," *Optical Test & Measurement Technology & Equipment*, 2005.

[3] H. Zhang, B. Kou, and Y. Jin. "Modeling and Analysis of a new Cylindrical Magnetic Levitation Gravity Compensator With Low Stiffness for the 6-DOF Fine Stage," *IEEE Transactions on Industrial Electronics*, 62(6), 2015, pp. 3629 – 3639.

[4] Z. Bo, W. Lei, and T. Jiu-Bin. "Design and Realization of a Three Degrees of Freedom Displacement Measurement System composed of Hall Sensors based on Magnetic Field Fitting by an Elliptic Function," *Sensors* 15.9, 2015, pp. 22530 – 22546.

[5] Y. Zhou, and B. Yu. "Mirror Mapping of Laser Interferometer Measurement System on Ultra-precise Movement Stage," *Journal of Southwest Jiaotong University*, Vol. 04, 2006, pp.57 – 64.

FAULT DETECTION OF SENSOR DATA IN SEMICONDUCTOR PROCESSING WITH VARIATIONAL AUTOENCODER NEURAL NETWORK

Wang Yong, Chen Xu, Wei Zhengying

Shanghai Huali Microelectronics Corporation, Shanghai, China

(86)21-61871212, wangyong@hlmc.cn

*Corresponding Author's Email: chenxu@hlmc.cn

ABSTRACT

Tremendous amounts of equipment parameters and sensor values were generated during modern semiconductor wafer manufacturing. These process data were utilized for early diagnosis of anomalies to prevent subsequent yield loss. However, process data from different steps and machines were highly customized that it is of great difficulties for traditional fault detection and classification (FDC) analysis to find a universal model to identify process excursions. In this paper, we present a neural network method with deep convolutional variational autoencoder structures which used reconstruction error as abnormal scores. This method exhibited a more reliable detection precision than FDC method and traditional anomaly detection algorithms. Furthermore, the reconstruction error of different sensors could be used to identify the root cause of abnormal processes. It was found that similarly structured networks could be applied to different processing steps, which enable this method to be accepted as a standard method to process data.

INTRODUCTION

Semiconductor wafer manufacturing includes complicated and lengthy processes with hundreds of process steps and thousands of different machines. Sensors were used commonly in foundry both in machines and environments to collect real-time machine and wafer states, such as temperature, pressure, and gas flow. Sensor data is generated and recorded continuously, resulting in high dimensions voluminous trace data. For analysis purposes, sensor data is usually transformed into much more condensed form like informative statistics such as max, min, length, and slope, which could be examined against user-defined limits to identify deviations. However, this transformation process would lead to information loss in a large amount, so more and more corporations try to use the sensor's raw data directly.

There were several studies using machine learning techniques to perform fault detection and classification analysis from raw sensor data. A rich body of literature existed using supervised methods to classify semiconductor sensor data, using decision tree and neural network algorithms [1–3]. It is still of considerable difficulty in detecting multi-dimensional sensor abnormal data, which could not be categorized into known types. Shroff et al. [4] used LSTM based autoencoder neural network to detect faults in electrocardiogram signals, and Xu et al. [5] used convolutional neural network (CNN) based autoencoder to analysis KPIs of internet web servers. These studies only use anomaly detection algorithms on single dimension sensor data. It had not been ordinarily used to review semiconductor wafer fabricating sensor data with unsupervised methods.

In this paper, we introduced a deep neural network model with convolutional variational autoencoder structures, which is trained using unsupervised learning methods. It is observed that the reconstruction error could be used to identify deviations from normal processing sensor data. Furthermore, Z-score criterions of reconstruction errors in different processes had similar values that could be used as a standard abnormal score to distinguish anomalies.

METHODOLOGY

Data Preprocessing

There were hundreds of sensors in each machine in modern semiconductor wafer manufacturing. Typically, sensor data were gathered at a specific sampling rate of about 0.1-2 sites/second. The processing procedure of a single wafer could be divided into different stages, and characteristics of sensor data varied between the stages. It was discerned that the data of different Wafers in the same Lot could have differences like optional stages omitting or different sampling rates.

The data used in this paper was obtained from the etching processing step in a 300mm advanced semiconductor foundry. Known abnormal data was removed to train models which only generalize characteristics of normal data. Secondly, non-critical processing stages data was removed to avoid additional noise. Also, the sensors with only null value or with little variations were excluded, and about 60 sensor values are retained. At last, the data is resampled or interpolated to the same data length to simply CNN based network architectures.

Model description

As depicted in Figure 1, the model was a combination of CNN and a variational autoencoder network (VAE) [6].

The VAE network was applied to reconstruct the sensor data, and reconstruction error was utilized to calculate the anomaly score. Specifically, the encoder part of the VAE network composed of 14 one-dimensional convolutional layers, followed by a flatten layer and three layers of fully connected layers. The structure of the decoder part was exactly mirrored from the encoder network. Data from different sensors fit into the channels of the convolutional operation, and the convolution was calculated through the time axis.

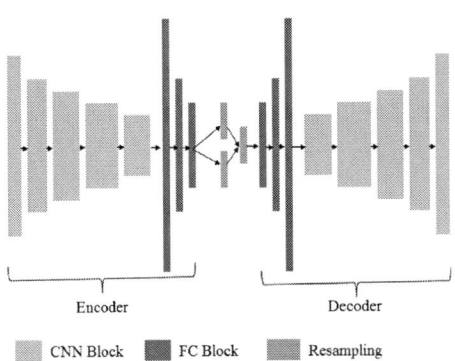

Figure 1: The architecture of proposed model

Experimental Settings

The data used in this paper contained 500 wafers processing data and cross-trained using a k-fold method. Two kinds of anomaly data were artificially constructed. Single-site anomalies were constructed by single-site manual modification. Multi-site anomalies were data obtained from a different machine in the training set.

$$Loss = L_{REC} + \lambda L_{KL} \qquad (1)$$

As in Eqn (1), the loss of the VAE model was a sum of reconstructed loss and KL-divergence. The reconstructed loss was calculated by the mean square error between the original data and their reconstruction. λ was used to balance a better reconstruction result or a better generalization capability. In this paper, λ is 0.1 and the learning rate is 0.001. As in Eqn (2), Learning rate scheduled with exponential decay with k = 0.95 in every 30 epochs.

$$LR = LR * k \qquad (2)$$

RESULTS AND DISCUSSTION

Figure 2 was a comparison of the original sensor data and its reconstruction of multi-site and single-site anomalies. It can be clearly seen in the figure that there are certain differences between the reconstructed data and the original data in specific positions, and these differences reflect the difference of the data compared to

the normal data. The reconstruction of VAE network on normal data is quite ideal that the curve of normal data and its reconstruction are almost completely overlapped. In Figure 2 (B) and (D), the non-overlapping part is the difference between abnormal data and normal data. The comparison between sensor data and its reconstruction could be utilized to identify the deviated location from normal patterns.

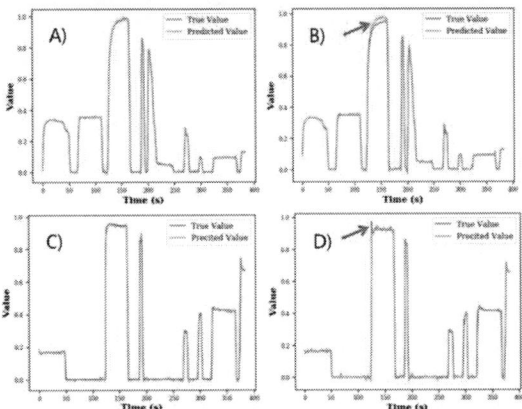

Figure 2: The original sensor data and its reconstruction (A. normal data, B. multi-site anomalies, C. normal data, D. single-site anomalies)

Figure 3(A) and (B) were the losses and ROC AUC scores of the validation set using reconstruction losses as anomaly scores during the training process. It was observed that the ROC AUC value of the model in anomaly detection continues to increase as the error decreases except for the initial instability phase. Therefore, the neural network model trained through unsupervised learning to reduce VAE errors also improved the model's anomaly detection capabilities.

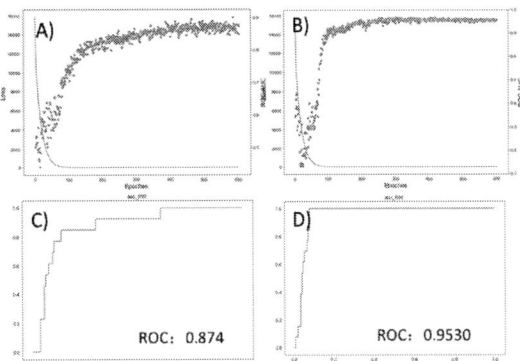

Figure 3: The Loss and ROC AUC Score during training(A. Multi-site anomalies loss and ROC AUC score; B. Single-site anomalies loss and ROC AUC score; C.

Multi-site anomalies ROC curve; D. Single-site anomalies ROC curve)

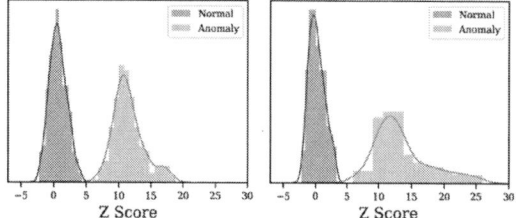

Figure 4: Distribution of reconstruction loss of normal and anomalies of two different processes.

It is of great adversity in utilizing the reconstruction error value as anomaly detection criterions because these criterions varied between processes and machines. The reconstruction error was converted to Z-Score to overcome this problem. In detail, mean and variation value was calculated from the training set and applied the standardization of the validation set. It can be discerned from Figure 4 that the distribution of reconstruction error of the normal data and the anomalies data separated widely. The anomaly detection criterions of two different processes are almost the same. It is believed that converting the reconstruction error into Z-Score can be used as a standard method for anomaly detection to overcome the differences between models.

Table 1: ROC AUC score of VAE and other anomaly detection algorithms

Model	Multi Site	Single Site
ABOD	0.659	0.909
COF	0.582	0.844
CBLOF	0.676	0.818
HBOS	0.792	0.923
IForest	0.782	0.925
KNN	0.732	0.848
OCSVM	0.760	0.629
PCA	0.659	0.908
SOS	0.734	0.694
VAE	0.874	0.953

The anomaly detection performance of VAE model was compared with nine different anomaly detection algorithms in Table 1. All anomaly detection algorithms detect single-point anomalies better than multi-point anomalies. Most algorithms have poor multi-point anomaly detection results, and the VAE algorithm is superior to other commonly used algorithms in both single-point and multi-point anomalies.

CONCLUTION

Efficient use of sensor data is crucial in semiconductor wafer fabricating, which helps to identify the deviation of wafer processing states, detecting a faulty process, thus improve wafer yield. In this paper, we have presented a deep neural network model with convolutional variational autoencoder structures. The VAE model is trained with only normal sensor data and could be used to identify which deviated from common patterns. This model achieved ROC AUC score at 0.953 for single-site anomalies and 0.874 for multi-site anomalies on plasma etching sensor data. The Z-score of reconstruction loss of sensor data could be used in statistical process control methods as anomaly indicator.

REFERENCES

[1] S. J. Hong, W. Y. Lim, T. Cheong, and G. S. May, *IEEE Trans. Semicond. Manuf.*, vol. 25, no. 1, pp. 83–93, Feb. 2012.

[2] C.-F. Chien, W.-C. Wang, and J.-C. Cheng, *Expert Syst. Appl.*, vol. 33, no. 1, pp. 192–198, Jul. 2007.

[3] C.-F. Chien, C.-Y. Hsu, and P.-N. Chen, *Flex. Serv. Manuf. J.*, vol. 25, no. 3, pp. 367–388, Sep. 2013.

[4] P. Malhotra, A. Ramakrishnan, G. Anand, L. Vig, P. Agarwal, and G. Shroff, *ArXiv160700148 Cs Stat*, Jul. 2016.

[5] H. Xu *et al.*, *Proc. 2018 World Wide Web Conf. World Wide Web - WWW 18*, pp. 187–196, 2018.

[6] D. P. Kingma and M. Welling, *ArXiv13126114 Cs Stat*, May 2014.

NEW PRECISION JITTER MEASUREMENT SOLUTION ON TMU -- CHALLENGE ON PRBS RECONSTRUCTION

Kai Zhou[1]*, Tianyu Zhang[1], Xurong Cao[1], and Yanyan Chang[1]

[1]SA, Advantest (China) Co., Ltd, Shanghai 201203, China

*Corresponding Author's Email: kai.zhou@advantest.com

ABSTRACT

Jitter Measurement is an important part of High Speed test. With customer's test requirement rapidly growing, result only include RJ and DJ is not acceptable. The solution of test specific kinds of jitter (such as DDJ, DCD, ISI...) to verify the IC transform performance is strongly demanded by customers.

So far, the industry of existing jitter measurement method include strobe method and Time-stamp method on TMU. However, the limitations are either complex or just measure several types of jitter.

An industry leading solution of high accuracy jitter measurement by TMU directly sampling with new PRBS pattern reconstruction method is proposed in this paper. All types of jitter can be skillfully tested with the condition of no cost increase.

Keywords—TMU (Time Measurement Unit); Jitter; PRBS reconstruction

INTRODUCTION

Background of Semiconductors and ATE

In recent years, integrated circuits has a rapid development toward complexity, which is immensely promoting the research and innovation of test method. More integrated SOC device, especially SerDes module, more test parameters are concerned. In the semiconductor industry, ATE (Automatic Test Equipment) plays an important role in meeting the complicated parameters test requirements.

ATE system performs various tests on UUT (unit under test). In general, an ATE test system is the result of this merging of test instrumentation (test card, such as digital card, high-speed card) with a computer. And the computer controls the test hardware by executing a set of instructions called the test program [1]. Moreover, ATE calibration reference sources are used by the tester as the "golden" standard for the volt, ohm, ampere, second [2], which guarantee the highest test reliability.

High-Speed Test on ATE

The high-speed test of the industry covers SerDes, high-speed ADC/DAC, protocol test (PCIE/SAS/SATA). In these different test applications, jitter measurement is indispensable. Figure 1 illustrates the ATE test solution with high-speed test card (PSSL). TMU is an excellent resource integrated in the PSSL. It records timestamps the instant that a signal crosses over a specified threshold.

Figure 1: ATE Tester and High-Speed Test Card (PSSL)

TEST CHALLENGE AND SOLUTION

Traditional jitter measurement in industry include strobe test method based on over sampling, time-stamp interpolation test method on TMU and Frequency test method on TMU. However, the limitations of strobe is complex and a long time data process. And for the existing TMU test method, the former is limited by accuracy because of interpolation technology, the latter could not measure ISI/DCD directly. Therefore, all of them could not meet the new test requirement described in abstract very well.

Time-stamp based on TMU

Figure 2 demonstrates the TMU's internal frame from the programming point of view. "RISE_FALL" mode captures rising and falling edges of input signal under test while sweeping threshold, which is applied in this paper to recognize all types of jitter. The Prescaler parameter defines how much the frequency of the signal is to be reduced before it is passed on to the event speed block. Possible Prescaler values (PSSL) are up to 49 under "RISE_FALL" mode. A value of 1 means no prescaling. InitialDiscard parameter is the event to be discarded at the beginning of sampling. Inter SampleDiscard parameter represents the number of events to discard between two samples. After these parameters of TMU are configured, the input signal is captured accurately to acquire time-stamp values of each edge.

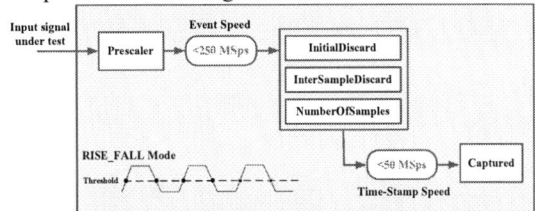

Figure 2: TMU internal frame

978-1-7281-6559-2/20 $31.00 © 2020 IEEE

Figure 3 illustrates a TMU setting case recording the time-stamp value (T[0], T[1], T[2]). The Pink-marked is rising edge. Green-marked is falling edge. When Prescaler is programmed to 2, that means four edges are omitted after one time-stamp is recorded. It is noteworthy that only rising edges or falling edges are captured when InterSampleDiscard is configured to 1. For this reason, the parameter should be set to an even number.

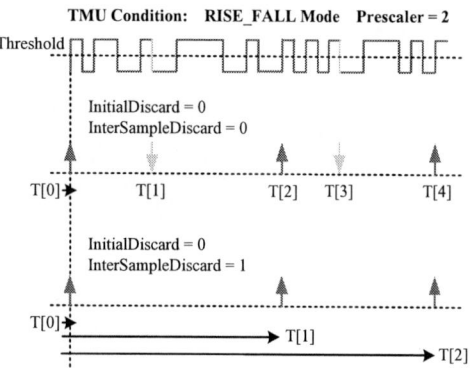

Figure 3: Time-Stamps

PRBS Reconstruction Solution

In order to acquire much more signal characteristic, rising edge position and falling edge position relative to input signal under test should be recognized from the captured time-stamp data. Accordingly, a kind of edge reconstruction solution based on PRBS is provided.

For simplicity, prescaling is ignored and PRBS4 signal is regarded as input signal under test. Two cases are explained here. For case1, InterSampleDiscard is set to 2 and two sampling action are skipped between two capture position of TMU. PRBS4 is consist of 15 bit streams, which is named as segment in this paper. Segment also describes the unique edge position. For PRBS4, there are 8 unique edge position from e1 to e8, including rising and falling edge. As demonstrated in Figure 4, when three PRBS4 segments are continuously captured by TMU, the captured time-stamps should be e1, e4, e7, e2, e5, e8, e3, e6, e1. Obviously, the final edge number is the same as the beginning edge number. Moreover, there are 8 unique patterns (UP) corresponding to 8 unique edge number in this case.

However, time-stamps captured by TMU is just a set of number in real case. And edge number in PRBS segment is not directly known. If the edge number is n and time-stamp value is $T(n)$ as assumed in Figure 4, it can be truly reconstructed by integral multiple of PRBS segment and do not need interpolation technology, as expressed in Eq. (1).

$$n = round\left(\frac{T(n) - (N_{segment} - 1) * UI_{ps} * L_{PRBSLength}}{UI_{ps}}\right) + 1 \quad (1)$$

Where $N_{segment}$, $L_{PRBSLength}$ and UI_{ps} represent the PRBS segment number of n, bit stream length of PRBS and the unit interval widely used in jitter measurement respectively.

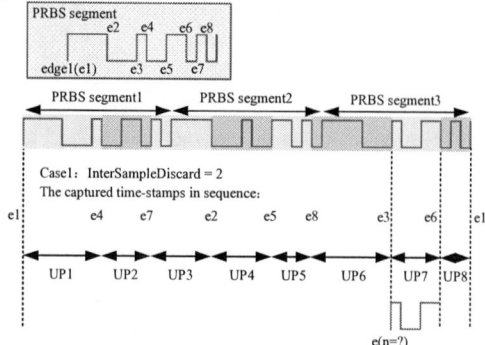

Figure 4: TMU case1 (InterSampleDiscard=2)

So far, the rising and falling position edge in PRBS segment is extracted accurately, which is meaningful to research signal performance with specific bit combination.

Case2 is similar to case1. In Figure 5, it is certain that the unique pattern number is also 8. Generally, for PRBS N, the bit length of PRBS, the number of edge in PRBS segment and the number of unique pattern should be 2^N-1, 2^{N-1} and 2^{N-1} respectively.

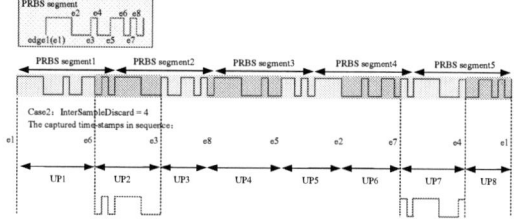

Figure 5: TMU case2 (InterSampleDiscard=4)

Jitter Separation and Computation Solution

In this paper, the measurement is based on TIE (Time Interval Error), which defines the difference between measured arrival time of an edge and expected arrival time for the edge. The expected arrival time of each edge position can be estimated from the above PRBS reconstruction solution. Then TIE of each edge can be obtained as described in Figure 6.

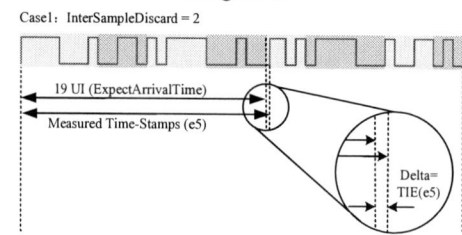

Figure 6: TIE computation

If each edge number is sampled by 128 times, the TIE mean of each edge number can be computed as explained in Figure 7.

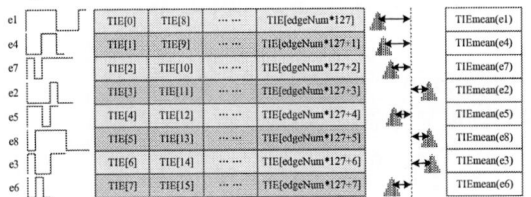

Figure 7: TIE mean of each edge number

Then In time domain, DCD and ISI can be calculated as in Eq. (2) and (3), which are generally represented with peak to peak (pp).

$$DCD = \left| \left| \frac{\sum_{n=1}^{N_{odd}} TIE[2n-1]}{N_{odd}} \right| - \left| \frac{\sum_{n=0}^{N_{even}} TIE[2n]}{N_{even}} \right| \right| \quad (2)$$

$$ISI = \max\left(\max_{1 \le n \le N_{odd}} TIE[2n-1] - \min_{1 \le n \le N_{odd}} TIE[2n-1], \max_{0 \le n \le N_{even}} TIE[2n] - \min_{0 \le n \le N_{even}} TIE[2n] \right) \quad (3)$$

Where N_{odd} and N_{even} represent the number of TIE corresponding to falling and rising edge respectively.

In case1, 8 edges are captured during three PRBS4 segments, which is equal to be sampled 8 times at the equivalent sample rate, i.e. 5.625 UI. Then in frequency domain, jitter spectrum of TIE by applying FFT to TIE can be analyzed to get other common types of jitter, i.e. Total Jitter (TJ), Random Jitter (RJ), Periodic Jitter (PJ) and Data Dependent Jitter (DDJ). The square-root of total jitter spectrum power is used to calculate TJ, which usually convert to unit of peak to peak so that the computational result is multiplied by 2.

In Figure 7, TIE means in each row is based on TIE data of the same edge number, which actually is caused by DDJ. In order to recognize the RJ and PJ in jitter spectrum, DDJ need to be removed in TIE data firstly by Eq. (4).

$$TIE_{RJ+PJ}[i] = TIE[i] - TIE_{mean} \quad (4)$$

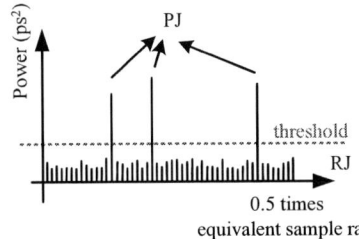

Figure 8: New Jitter Spectrum (DDJ is removed)

TABLE I.   MEASUREMENT RESULT

Test Method	Jitter Parameter (ps)					
	ISI	DCD	DDJ	RJ	PJ	TJ
Strobe	27.1	0.97	27.32	5.9	99.42	104.8
Proposed	26.0	1.02	26.1	6.2	98.9	103.4
Error (%)	4	5.1	4.5	5	0.5	1.3

Then new jitter spectrum based on TIE data of Eq. (4) should only include RJ and PJ, which can be separated by designated threshold as shown in Figure 8. The computation of PJ is similar to TJ. While the RJ is usually represented with rms so that the power of RJ is square-rooted and divided by square-root of 2. Finally, the DDJ power can be acquired from the difference between two jitter spectrums.

The proposed technique is realized on Advantest SOC test platform for hardware validation. Result is based on PSSL external loopback mode and a 2 MHz 100 ps-pp sinusoidal jitter is injected. 4Gbps PRBS7 is applied to input signal under test and 64 unique patterns including 89 PRBS7 segment are sampled 128 times by TMU. Then equivalent sample rate is 375.046875 UI, which means approximately 10.7 Msps for 4Gbps. Table I shows the measurement result. The first row lists strobe result for reference, which is based on over sampling. In last row, the relative measurement error is within 6%, which is acceptable in most project development. Furthermore, less sample is required in proposed method and no need a long time data process.

LIMITATION OF THE TECHNIQUE

Proposed technique in this paper is based on TMU hardware. The signal frequency under test is thus limited to Prescaler parameter and event speed of 250 Msps.

CONCLUSION

This paper introduces a new precision jitter measure solution using TMU intended to get all types of jitter and easy to use in lab and production line. Moreover, due to acceptable measurement error, less samples and no cost increase, this solution can meet the challenge of the industry such as SerDes.

ACKNOWLEDGEMENTS

Authors acknowledge the contribution of Advantest on the project.

REFERENCES

[1] Mark Burns and Gordon W. Roberts. *An Introduction to Mixed-Signal IC Test and Measurement2001*, New York Oxford, 2001.

[2] Guy Perry, *The Fundamentals of Digital Semiconductor Testing*, Soft Test, 1996.

INVESTIGATION AND DISCOVERY OF THE INTEGRATION OF FEOL PROCESS BY ELECTRON BEAM INSPECTIONS

Fengjia Pan*, Hunglin Chen, Yin Long, Kai Wang and Hao Guo
Shanghai Huali Integrated Circuit Corporation, 201314, China
*Corresponding Author's Email: panfengjia@hlmc.cn

ABSTRACT

A novel inspection method is proposed for checking the integration of FEOL (front-end-of-line) device fabrication. As the designed electron-beam (as e-beam in the following text) inspection methodology applies to the last step of FEOL device and prior to MEOL interconnection fabrication, the capability of both voltage contrast and physical feature detection discovered the surface and underneath defects in the very narrow space of Nickel Silicide formation.

Experiments showed the variation of multiplex parameters involving poly critical dimension, spacer and SMT film thickness with dry, wet, furnace and plasma ashing processes would lead to invisible change of Nickel Silicide formation and can be detected by the designed inspection. Defect count would be high while those majority pre-steps process windows being marginal. After all, the cumulative effect would lead to electrical failures of the device.

Keywords — e-beam inspection methodology; defect detection; EBI (electron beam inspection); poly CD (Critical Dimension), process window control; integration of FEOL (front-end-of-line).

INTRODUCTION

With the aggressive scaling down progress of critical dimension size in semiconductor device manufacturing, we are facing more and more challenges of process window control. Along with inline measurement metrology, inline defect detection is playing a further important role as the index of process stability and integrity.

Nickel Silicide (As NiSi in the following text) is widely used to reduce the metal-semiconductor contact resistance of the device source/drain beyond 65nm generation. [1] The quality of NiSi formation defines the contact resistance. According to failure analysis data, bad NiSi formations can lead to electrical failures in the final performance of the devices. With a narrower NiSi forming space for 28nm process or beyond generation, NiSi poor formation is becoming a major type of defects during the FEOL process of the device formation.

DEFECT DETECTION BY EBI

As this kind of NiSi poor formation defect is physically located down below in the gap of dense poly lines, and no obvious height difference in topography, this type of defect can hardly be detected with optical defect inspection tools, even the most advanced bright-field inspection tool. E-beam inspection tool has better resolution without the barrier of optical wavelength limitation. [2] E-beam inspection collects both SE and BSE signals to detect voltage contrast, or topographical, or material contrast. Because of high yield, in conventional application of e-beam, SE dominates the signal and shows strong topographical contrast for the top poly patterns.

According to SE yield theory, the higher energy of the primary e-beam, the less the SE yield becomes. The relationship of Landing Energy and electron yield (SE, BSE and their combinations) is shown in Figure 1. [3]

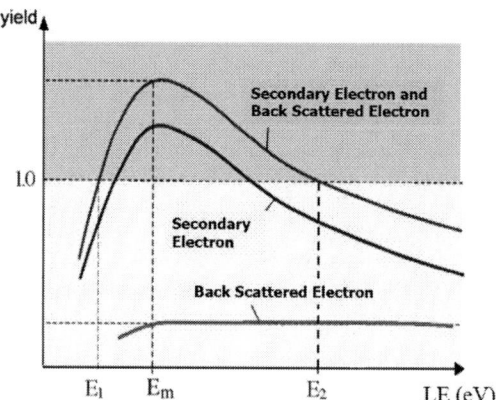

Figure 1. Relationship of Landing Energy of primary beam and SE/BSE yields.

In this case, we tried different Landing Energy condition combinations, and then fixed on a BSE mode with less SE collected. Test results showed that we managed to detect the NiSi poor formation signal on NMOS Active Area. Figure 2 is EDS image of the plane view TEM sample that shows the discontinuity of the silicide formation.

978-1-7281-6559-2/20 $31.00 © 2020 IEEE

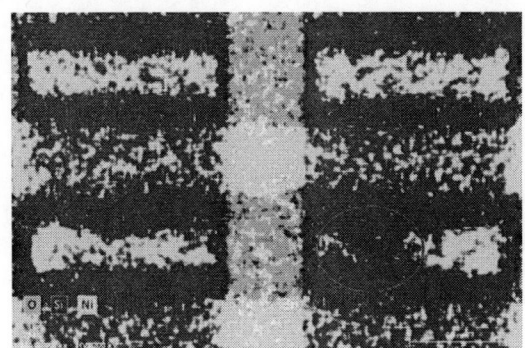

Figure 2. EDS image of defect shows that Nickel Silicide not formed on NMOS active area (as marked with red circle).

MECHANISM AND EXPERIMENTS

With the steady and repeatable detection ability of this kind of defect, we did some studies about the relationship of different process steps and NiSi poor formation. Spacer residue or spacer merge by narrower NiSi forming space is suspected to cause NiSi formation block.

Experiments showed the variation of multiplex parameters can lead to invisible change of Nickel Silicide formation.

Poly CD

Poly line CD (critical dimension) is a key numerical value which defines the narrow spacing of device formation. Experiments of photo exposure CD splits showed that the larger poly CD (with narrower spaceing between poly lines) causes more NiSi poor formation defects, and it is very marginal as the CD difference is quite small. The relationship of poly CD and defect count is shown in Figure3.

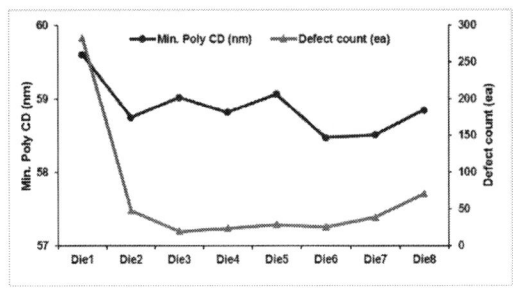

Figure 3. Defect count related with Poly CD within wafer.

Spacer thickness

Spacer process is essential as the separation and protection of device gates. The thickness of the thin spacers also has a marginal window, as our thickness split experiments showed that tiny variation of the thickness can lead to high counts of defects as Figure4 shows.

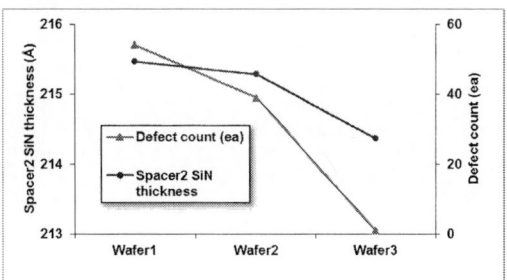

Figure 4. Defect count related with Spacer Silicon Nitride thickness.

These two key process values show an interactive effect of a combined process window, which requests a much more precise inline process control to keep it steady and healthy.

SMT (Stress Memorization Technique) process

In addition, SMT Silicon Oxide deposition thickness and Nitride removal wet process, and asher processes would lead to invisible change of Nickel Silicide formation and can be detected by the designed inspection. These process window variations lead to high defect counts as shown in Figure 4, 5, 6and 7, respectively.

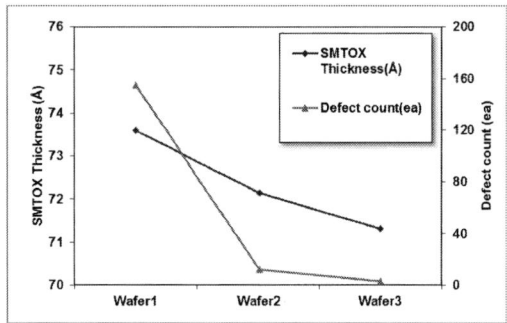

Figure 5. Defect count related with SMT silicon oxide thickness.

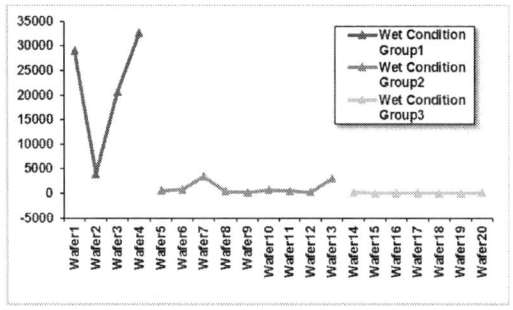

978-1-7281-6559-2/20 $31.00 © 2020 IEEE

Figure 6. Defect count related with silicon nitride removal wet process condition.

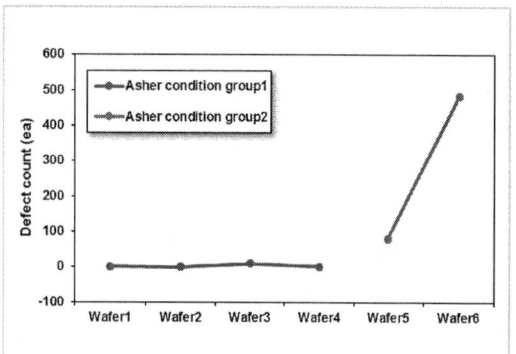

Figure 7. Defect count related with Asher condition groups.

Defect Improvement

After process window improvement tuning of all these previous steps, this kind of NiSi poor formation defect trended down and kept stable in a low level, which means a healthy and steady FEOL device formation process integration. Defect trend chart is as shown in Figure 8.

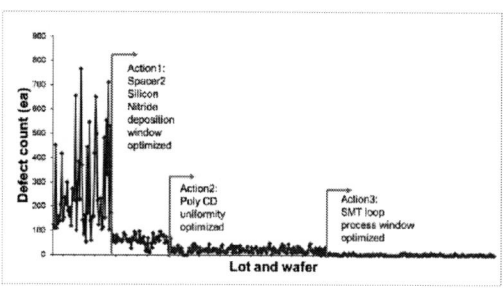

Figure 8. Defect count trended down when process window optimized.

CONCLUSIONS

The study presents the optimized inspection method for checking and confirming the integration status of FEOL device fabrication. This kind of defect as NiSi poor formation acts as an index of process window health. After a series of improvement actions including the optimizations of different process windows, the defect count trended down and kept stable.

ACKNOWLEDGEMENTS

The authors would like to thank the co-work of all the members of E1YE team and co-work members from E1PIE/E2/E3 department of Huali.

REFERENCES

[1] T.Morimoto et al., *Self-aligned nickel-mono-silicide technology for high-speed deep submicrometer logic CMOS ULSI*, IEEE Trans. E.D., 42, pp.915 (1995)

[2] Rob Cappel and Jay Rathert, *The Advantages of In-Line Electron-Beam Wafer Inspection*, Yield Management Solutions, Summer 2000.

[3] Lreimer, *Scanning Electron Microscopy, Physics of Image Formation and Microanalysis*, Second Edition, Springer, New York, 1998.

One Comprehensive Method to Analyze Semiconductor Manufacturing Data by "Piecewise" Regression

Lin Gu, Wei Yu

Shanghai Huali Microelectronics Corporation
Shanghai, China
Gulin@hlmc.cn, Yuwei@hlmc.cn

Abstract

Big data analytics is a powerful tool in smart manufacturing (SM). Semiconductor manufacturing data is complex, and also need big data methods to realize correlation analysis. Traditional regression, such as fitting data with certain formula, can't reveal the potential correlation between some semiconductor data. Here, "piecewise" regression (PWR) is proposed, in which raw data is divided into groups, to enhance the potential correlation. This method can be applied to different kinds of semiconductor, such as inline, WAT and CP, to highlight potential correlation and realize predictive analysis.

Keywords—big data; "piecewise" regression; semiconductor manufacturing; prediction analysis; automatic analysis system

Introduction

Big data analytics is applied to handle manufacturing data with big volume, velocity, and variety features. Semiconductor manufacturing data is also complex and use big data methods to realize fault detection (FD) and predictive maintenance (PdM) [1-3] are studied. The current big data analytics in semiconductor manufacturing is normally deal with in-line and off-line cases. As is known to all, in FAB, the final wafer will do electrical test (WAT), and its pass or fail directly determine if the wafer can ship to custom. What's more, CP test and product yield will determine if the quality of the wafer meets custom's request. All the above WAT and CP data is related to in-line and off-line data during manufacturing and CP fail bin may also be related to certain WAT parameter. Because of semiconductor data's dynamic and complex nature, the latent relationship between different data may be ignored.

The traditional degree of correlation between two sets of data is characterized by correlation coefficient (R^2), obtained by regression with certain equation. As show in Fig.1, (a) and (b) are scatter plots of certain CP fail Bin-1 with two WAT parameter (WAT-1 and 2). It's obvious that, it shows some correlation between Bin-1 and WAT-1 (bin fail more with WAT parameter increase), while there's no obvious correlation between Bin-1 and WAT-2 (the scatter shows normal distribution). However, regression with proper function, the R^2 is 0.24 and 0.03 respectively for Bin-1 with WAT-1 and 2. Normally, $R^2 < 0.5$ means no obvious correlation. It can be seen that traditional regression method can't reveal the potential correlation, which can only be judged by engineer's experience,

inefficiently. Here a new method named as "piecewise" regression (PWR) is proposed to explore the potential correlation between different kinds of semiconductor data, such as Inline to WAT, WAT to CP, etc. And PWR can also be optimized to do prediction analysis.

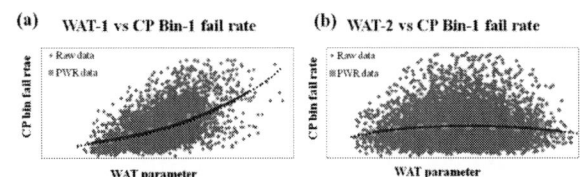

Fig.1. Correlation scatter plot between WAT-1 (a) /WAT-2 (b) and CP Bin-1.

Methods and discussion

A. " Piecewise " Regression (PWR) method

" Piecewise " Regression (PWR) is realized by split the raw data into certain groups. For example, as show in Fig.2, 90 pairs of data are firstly sorted from small to large of X. Then the data is spitted into several pieces, such as 9 pieces here. Each group will calculate the average of X and Y data:

$$Xi=Average (Xi1+Xi2+...+Xi8+Xi9)$$
$$Yi=Average (Yi1+Yi2+...+Yi8+Yi9)$$

Then regression is carried on with piecewise data, rather than raw data. Therefore the PWR is doing based on less and averaged data. With PWR, the R^2 of Bin-1 with WAT-1 and 2 is 0.969 and 0.580, respectively, as shown in Fig.1(blue scatter). The regression curve (Fig.1 navy blue) of PWR data is similar to that of raw data (Fig.1 black dotted), but the R^2 is enlarged. The advantage of PWR is reduction of raw data's noise and enhancement of the positional correlation.

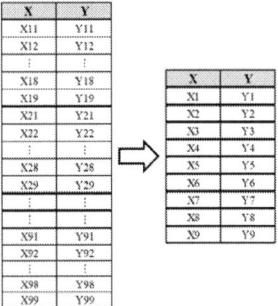

Fig.2 Diagram of PWR.

978-1-7281-6559-2/20 $31.00 © 2020 IEEE

B. Optimization of PWR

By set a proper spec. of PWR R^2, such as 0.9 for the above sample, correlation between WAT-1 and Bin-1 can be highlighted, while that between WAT-2 and Bin-1 is filtered. As is known to all, correlation coefficient not means degree of dependent variable changing with independent variable. Fig.3 shows one sample of this case. WAT-1 and 3 all show correlation with Bin-1 ($R^2>0.9$), but Bin-1 ramps up more rapidly with the increase of WAT-1, while Bin-1 changes only a little with WAT-3. To distinguish these two situations, one method of Std_PWR/Std_Raw is introduced. Though the standard deviation of certain CP Bin-1 data is fixed, when do PWR correlation with different WAT parameters, standard deviation of RWR Bin-1 data is different. As shown in Fig.2, Bin-1 changes more rapidly with WAT-1 and the standard deviation of RWR Bin-1 data is also larger. With a proper spec. of Std_PWR/Std_Raw, such as 0.5 for this case, a more meaningful correlation can be highlighted.

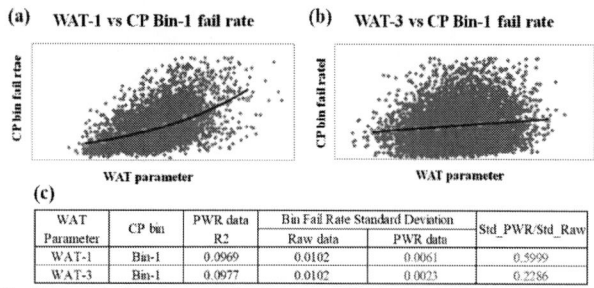

Fig.3 Correlation scatter plot between WAT-1 (a) (WAT-3 (b)) and CP bin-1; (c) is Std_PWR/Std_Raw table.

C. Automatic PWR analysis

Automatic PWR analysis is realized based on Hadoop and Python. Different regression functions and piecewise methods (split into groups with different count from 10 to 50). The system can find the best solution automatically. Fig.4 show that the best PWR solution for this sample is logarithm function with 10 PWR groups.

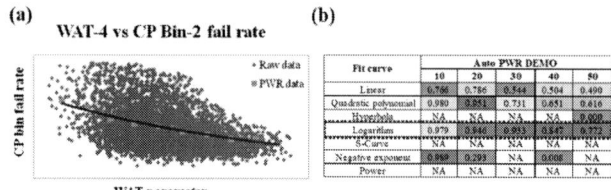

Fig.4 Automatic PWR analysis sample.

D. Prediction analysis with PWR

The best solution equation obtained by PWR can be used to do prediction. For example, given a WAT data, a prediction averaged Bin fail rate can be obtained. That means PWR can be used to do yield prediction and so on. It must be noted that, the equation obtain by PWR can't predict dependent variable accurately, if independent variable deviates from training set. One important reason is that the equation may show inappropriate trend outside training set, for example in Fig.5.

Beyond training set, a tangent fitting is proposed. Though we can't predict accurately, a trend prediction can be achieved.

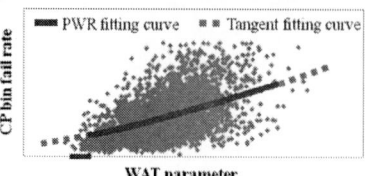

Fig.5 Prediction with RWP and tangent fitting curve.

Conclusions

A novel "piecewise" regression method (PWR) is proposed to analyze the potential correlation between different semiconductor data. With Hadoop and Python, an automatic regression system is established, which can do prediction analysis effectively, such as inline to WAT prediction, WAT to CP prediction, and so on.

References

[1] James Moyne and Jimmy Iskandar, Big Data Analytics for Smart Manufacturing: Case Studies in Semiconductor Manufacturing, Processes 2017, 5(3), 39

[2] Moyne, J., Iskandar, J. and Armacost, M., Big Data Analytics Applied to Semiconductor Manufacturing. In Proceedings of the AIChE 3rd Annual Big Data Analytics Conference, San Antonio, TX, USA, 28 March 2017

[3] Iskandar, J., Moyne, J., Subrahmanyam, K., Hawkins, P. and Armacost, M., Predictive Maintenance in Semiconductor Manufacturing: Moving to Fab-Wide Solutions. In Proceedings of the 26th Annual Advanced Semiconductor Manufacturing Conference (ASMC 2015), Saratoga Springs, NY, USA, 3–6 May 2015

THE INSPECTION SOLUTIONS OF 3BAR STRUCTURE CU VOID IN BEOL ADVANCED SEMICONDUCTOR PROCESS

Xingdi Zhang[1], Hunglin Chen[1]*, Yin Long[1], and Kai Wang[1]

[1]Shanghai Huali Microelectronics Corporation, Shanghai 201210, China

*Corresponding Author's Email: chenhunglin@hlmc.cn

ABSTRACT

In this paper, 3bar structure Cu void defects are detected and monitored by using brightfiled inspection system in 28nm back-end-of-the-line (BEOL) processes. 3bar structure Cu void defects were studied by a novel combination of dedicated scan settings which can help capture defects more efficiently. The spectrum mode, directional electrical field (DEF) and focus offset were chosen as critical parameters. After defect detecting, failure models were studied and the effective defect reduction actions were carried out. Combining effective defect monitor and defect reduction actions can make BEOL yield improve rapidly.

INTRODUCTION

The tolerance of tiny physical defects in the integrated circuit (IC) manufacturing becomes smaller and smaller, since advanced semiconductor process continually drive the line width shrinking. In 55nm technology, tiny Cu void defects are gentle, but they would become fatal killers in 28nm node. The study proposed effective methods allowed inline detection of tiny Cu void defects in back-end-of-the-line (BEOL) processes. Because of the transparent inter metal dielectric (IMD) and rough copper grains, the tiny defects in the BEOL contenting multi stacks of repeated metal lines can be detected difficultly. A novel combination of dedicated scan settings on the equipment and designed layout patterns on the wafer is proposed to achieve the inline monitor [1]-[6].

Cu plating process (ECP) is very important in BEOL processes. As the CD of integrated circuit scaling down, the Cu gap filling became a big challenge. The Cu void can lead to metal open which can cause end-of-line (EOL) yield failures. Cu void can be caused by many factors such as Cu plating process, Cu barrier/seed process, pre layer CD and so on [7]-[9].

In this paper, 3bar structure Cu void were effectively detected by series of studies. After defect detecting, failure models were established and the effective defect reduction actions were carried out. Combining effective defect monitor and defect reduction actions make average count of 3bar Cu void defects improve from 2000 to 10.

3BAR STRUCTURE CU VOID DEFECT DETECTION AND IMPROVEMENT

3bar Structure Cu Void Defects

3bar structure Cu void defects were found post Cu Chemical Mechanical Polishing (CMP). As shown in Fig.1a, Cu void defects were very tiny and less than 50 nm, and they all locate in special 3bar structure. Fig.1b shows that the 3bar structure which aren't suffered by Cu void is normal. According to image post Cu CMP, it is difficult to capture the Cu void because of small size and rough copper grains.

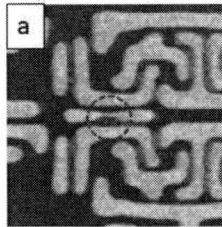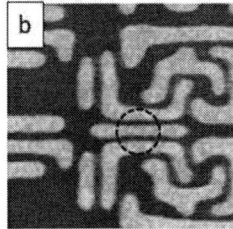

Fig. 1 (a) SEM image of bridge defect at AEI, (b) SEM image of bridge transferred to CMP layer

Bridge defect detection and monitoring methodology

The study of defect performed for effective detection. The spectrum mode, directional electrical field (DEF) and focus offset were chosen as critical parameters. As shown in Fig 2a, shortwave has good performance in detection of bridge defect and blueband shows the highest signal to noise ratio(SNR). Fig 2b shows that positive focus offset value has good performance and 0.1 um shows the highest signal to noise ratio. Fig 3c shows that horizontal DEF is better than vertical DEF. After determining the relevant parameters, the 3bar structure Cu void defects can be effectively detected post Cu CMP.

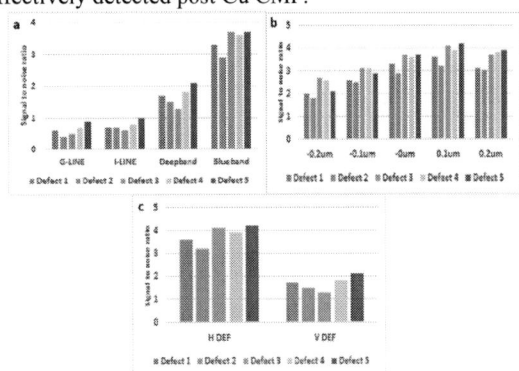

Fig.2 (a) SNR of void defects in different spectrum mode, (b)SNR of void defects in different offset,(c) SNR of void defects in different DEF

3bar Cu void defects failure model

Based on past experience, Cu plating process and Cu barrier/seed process became the main suspect which caused the Cu void defects. A series of experiments were conducted around these two processes. However, the results showed that Cu void defects can't be improved (Table I).

Table I Results of experiments conducted around Cu plating process and Cu barrier/seed process

	Cu plating			Cu barrier/seed		
	Split1	Split2	Split3	Split1	Split2	Split3
Count of Cu void defects on one wafer	2000	1800	1900	2100	1700	2200

The results showed that these two processes Maybe not really murderous. The real reason needs to be found. Because these Cu void defects all locate in special 3bar structure, the topography of 3bar structure in pre-layer were suspected. The SEM images of 3bar structure were collected step by step to verify this conjecture. Fig3a and b shows that these 3bar structures are normal post all-in-one (AIO) etch porcess.

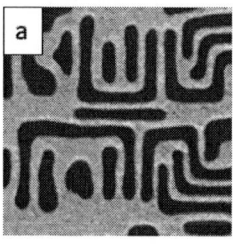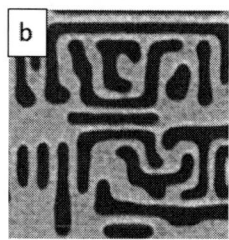

Fig.3 (a) and (b) SEM images of 3bar structures post all-in-one etch process

However, Fig4a and b shows that two lines are severely merged together in 3bar structure post Cu barrier/seed process. Because of merging of lines, Cu is difficultly filled in these locations which lead to Cu void.

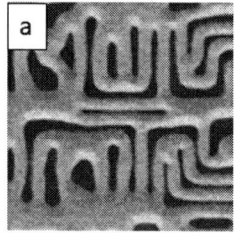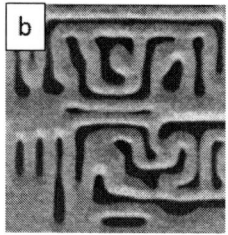

Fig.4 (a) and (b) SEM images of 3bar structures post Cu barrier/seed process

Stress of TiN is considered the main reason which

cause the merging of lines. Eliminating stress to improve 3bar structure Cu void is considered a good solution. Using wet cleaning post all-in-one etch process were chosen to eliminate stress. Experiments of wet cleaning (Table II) shows that the count of 3bar Cu void defects can be reduced to less 10 on one wafer.

Table II Results of experiments of wet clean

	Wet cleaning post all-in-one etch process				
	Baseline	Split1	Split2	Split3	Split4
Count of Cu void defects on one wafer	2300	900	500	200	<10

The 3bar structure Cu void defects trend chart of Fig.5 shows that the 3bar structure Cu void defects were effectively detected with the developed defect scan recipe, and the defect was trend to less10 post wet new condition.

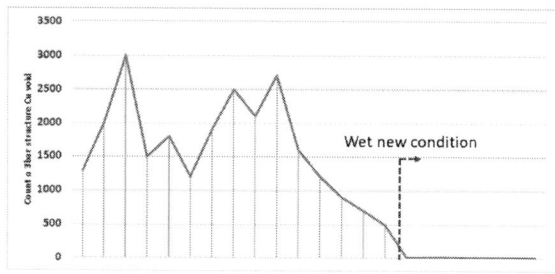

Fig.5 The trend chart of 3bar structure Cu void defects

CONCLUSION

In this paper, 3bar structure Cu void defects were studied by a novel combination of dedicated scan settings which can help capture defects more efficiently. The results show that blueband, 0.1um offset and H DEF are the best condition of detection recipe for 3bar structure Cu void. According to collecting of SEM images, it is verified that the stress of TiN is the main reason which can cause the merging of line to lead to Cu void. Finally, experiments of wet cleaning shows that the count of 3bar Cu void defects can be reduced to less 10 on one wafer.

ACKNOWLEDGEMENTS

Thank you very much for the leadership and colleagues for their support of my work

REFERENCES

[1] H.Chen, R.Fan, H.Lou, M.Kuo, Y.Huang, Mechanism and application of NMOS leakage with intra-well isolation breakdown by voltage contrast detection, J. Semicond. Technol.Sci. 13(2013)402–409.

[2] H.Chen, R.Fan, H.Lou, Y.Huang, Alternative voltage contrast inspection for pMOS leakage due to adjacent

nMOS contact-to-poly misalignment, Mater.Sci. Semicond. Process.16(2013)1873–1878.

[3] M.Kuo, H.Chen, R.Fan, D.Zhang，Mechanism and detection of poly gate leakage with nonvisual defects by voltage contrast inspection, Mater.Sci. Semicond. Process. 56(2016) 362-367.

[4] H.Chen, R.Fan, H.Lou, Y.Huang, Electrochemical mechanism of layout-dependent corrosion of tungsten in contact plugs, Mater.Sci. Semicond. Process. 20(2014)17-22.

[5] I. Malik and B. Pinto, "Immersion Changes Litho Cluster Qualification," Semiconductor International, September 2006.

[6] L. Peters, "Defectivity Issues Drive Immersion Lithography," Semiconductor International, April 2006.

[7] Weiye He, Beichao Zhang, Jian Kang et al., "The contributions of barrier resputter for BEOL integration", *ECS Transactions*, vol. 44, no. 1, pp. 487-492, 2012.

[8] Yu Bao, Xuezhen Jing, Jingjing Tan et al., "Optimization of Metallization Processes for 28-nm-node Low-k /Cu Multilevel Interconnects", ECS Transactions, vol. 44, no. 1, pp. 477-480, 2012.

[9] Xuezhen Jing, Jingjing Tan, Jiquan Liu, 32/28NM BEOL CD GAP-FILL CHALLENGES FOR METAL FILM, CSTIC, 2015.

THE INSPECTION AND SOLUTION OF INLINE CT DEFECT FOR 28NM PROCESS IMPROVEMENT

Min Wang[1], Hunglin Chen[1], Yin Long[1], and Hao Guo[1]

[1] Shanghai Huali Microelectronics Corporation

Shanghai, China

Author's Email: Wangmin_E1@hlmc.cn

ABSTRACT

The systematic defect in the CT holes of the wafer edge are always observed in the advanced semiconductor process, which will directly result in chip yield loss or reliability issue. In this study, the novel bright field inspection (BFI) and electron-beam inspection (EBI) were applied to enhance the monitoring of the inline CT defect, including CT open, over polish and W_pits, so that the process window and stability can be verified and examined instantly. Furthermore, a series of process evaluation were carried out, and the results showed that the failure mode contained the poor uniformity of CT hole CD and film thickness. Interestingly, the processes were related to each other during ILD~CTW loop, but meanwhile they exhibited weak stability in the wafer edge and narrow window in the advanced process, as shown in Figure 1. On the basis of this reason, the root cause of these defects was very intricately. Therefore, the corresponding improvement actions for removing these CT defects were executed through a comprehensive and deep discussion of the defects formation mechanism. In detail, the W_pits was fixed by optimizing the uniformity of CT hole CD and controlling the uniformity of film thickness, which were impacted by photo and etch process, chemical and mechanical polish (CMP) process, respectively. Meanwhile, the CT open defect in the wafer edge were improved significantly based on plenty of CT etch split experiments.

Keywords: 28 nm technology node, defect inspection, ILD~CT Loop, CT defects, yield improvement

INTRODUCTION

In the advanced integrated circuit (IC) manufacturing, Tungsten (W) has been widely applied for filling memory and logic contact holes due to its superior step coverage capability and low resistivity [1-2]. However, with the development of semiconductor process, the critical dimension (CD) of local interconnect is shrinking continually [3]. The CD has even scaled down to 40 nm in the 28 nm technology-node, meanwhile the aspect ratio (AR) of contact hole has increased to 4:1. Although, W deposition process itself has excellent step coverage property, it is still a big challenge for W gap-filling with such smaller CD and higher AR. Therefore, the W defects in CT holes often occurred, especially in wafer edge dice.

In order to avoid or remove the W defects in CT holes, we need adjust the process windows and its uniformity, which contain the CT holes CD and film thickness. These process windows are controlled by the process of lithology (Litho), Etch, and chemical mechanical polishing (CMP) during ILD~CT Loop. More interestingly, each process step is critical and can effect or limit the window of the next step, thus causing a chain reaction during this loop and forming serious defect problems after process integration. What's more, these process all exhibited weak stability in the wafer edge due to the marginal process window for the 28 nm logical IC. As we all known, the CT photo and etching directly determine the characteristic dimensions of contact holes (CD). Furthermore, the CMP of inter-layer dielectric (ILD) and Tungsten determines the uniformity of film thickness. Especially, the CMP process of ILD, an oxide material removal process of chemical reaction with slurry chemicals and mechanical abrasion with abrasives, would have indirect effect to CT photo and etching [5-6].

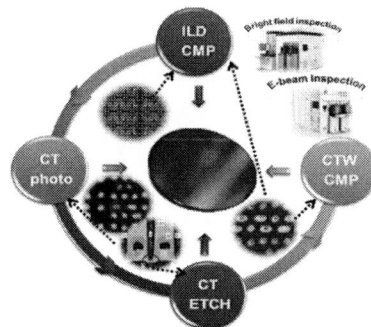

Figure1. Illustration for the CT defects inspected by the BFI and EBI in the ILD~CT Loop.

In this work, we use the novel bright field inspection (BFI) and electron-beam inspection (EBI) to monitor the inline CT defect, including CT open, over polish and W_pits. A serious of process experiments was adopted to tune the process window and improve the weak process stability in the wafer edge, in order to removal the CT defects. Moreover, the CT defects' failure mode and formation mechanism were investigated comprehensively and deeply to optimize the process conditions and process window. The results showed that these correlative process, such as ILD CMP and CTW CMP would impact the film thickness and its uniformity, CT photo and etch would impact the CT holes CD and its uniformity. What's more important is that these four processes are not independent of each other, but closely interrelated to each other, so we need make a comprehensive and systematic adjustment for the ILD~CT loop.

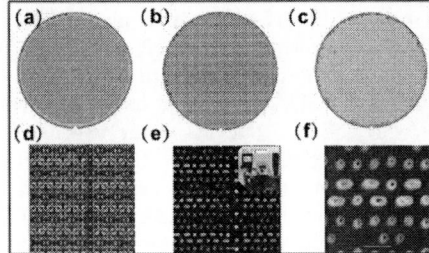

Figure2. The wafer map and corresponding image of the inline CT defects: (a), (d) over polish; (b), (e) CT open; (c), (f) W_pits.

EXPERIMENTS AND RESULTS

ILD CMP process window

The ILD CMP process is a crucial step in this loop, and the process window is very narrow by the limitation of the CT etching and photo. As shown in Table 1 and Figure3, the experiments of different ILD CMP target film thickness were designed to estimate the safe process window. The corresponding defect results indicated that if the THK>1811 A, it have high risk to suffer W open, this is because that the thick ILD film thickness would cause the etching cannot etch the bottom of the CT hole completely, especially in the wafer edge (as displayed in figure 2 b and 2 e). In contrast, if the THK<1156 A, the ILD film thickness is too thin, it have high risk to suffer over polish (as shown in figure 2a and 2c). The ILD THK range from 1732 A to 1811 A, and from 1156 A to 1280 have low risk suffer the same type defect, respectively. Thus, the safe ILD THK window is 1280~1732 A。

Table Ⅰ: The split condition of ILD THK and the results for corresponding defect count by radius

Split Condition		#15	#16	#18	#25
		BSL (target 1450)	BSL (target 1350)	BSL (target 1650)	BSL (target 1550)
ILD CMP THK Mean		1479	1329	1633	1535
ILD CMP THK Range		485	460	269	412
Over polish	radius 142-146 mm	0	0	0	0
	radius 146-147 mm	4	0	0	0
W open	radius 142-144 mm	0	0	0	0
	radius 144-146 mm	0	1	2	0
	radius 146-147 mm	4	0	0	0

ILD CMP process condition

Moreover, the film thickness of ILD in the wafer edge is very difficult to control. As displayed in Figure3, all wafers film thickness drop sharply in wafer edge and the range were very large (>400 A), indicating that the CMP process performance was worse. The results showed that the condition is easy to suffer ATPG fail in the wafer edge, due to the M1 to poly short. Thus, many of conditions were studied to tune the wafer edge thickness by controlling the CMP pressure of zoom. As shown in Figure 4, through tuning the wafer edge polish pressure by zoom, the thickness of wafer edge was obviously drive up. Meanwhile, figure 4e demonstrated that the full map range was decreased gradually (from 450 A to 250 A) for the 4 split conditions, and the split 3 N02 condition have the smallest range 250A, which indicated that ILD_CMP N02 condition have the best performance.

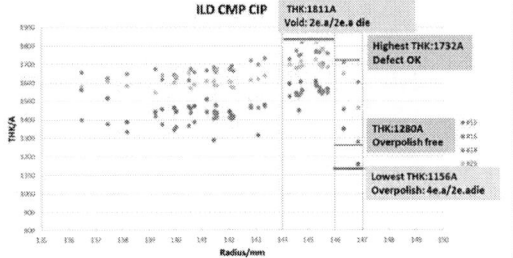

Figure3: The ILD THK by radius of split condition

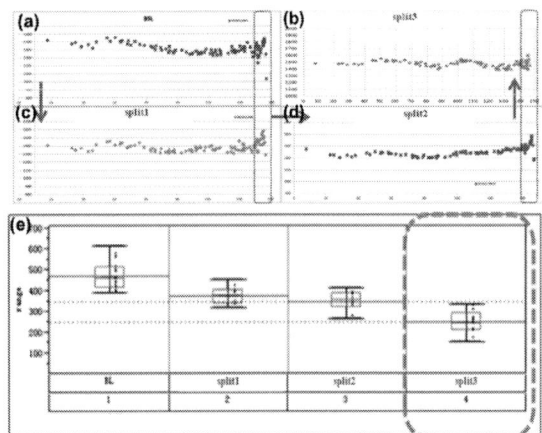

Figure 4: the full map film thickness of ILD by wafer radius of BSL (a), split 1 (b), split 2 (c), split 3 (d); the range of four split conditions (e).

However, the inline defect showed that (figure 5a) when the ILD film thickness was higher than 1450 A, the probability of suffering the W_pits defect in the wafer edge (figure 2 c and 2f) was increased. Meanwhile, the CT etching and photo all have weak working point in the wafer edge. The raised ILD THK in the wafer edge would leading to the smaller far edge CT CD, and thus increased the risk of suffering CT open defects. Moreover, the ILD THK window split showed that the lower ILD THK target would result in the worse THK range. Thus, ILD CMP process condition is N02 and film thickness target is 1450 A. Based on this result, the CT photo and etching, CTW CMP process condition might need to be further adjusted for avoiding the CT open or W_pits defects.

978-1-7281-6559-2/20 $31.00 © 2020 IEEE

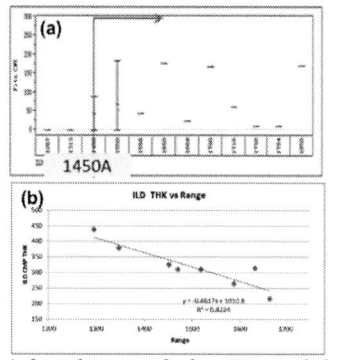

Figure 5: (a) the relativity of inline W_pits defect with ILD THK; (b) the linear correlation of ILD THK and Range.

CT photo and etching, CTW CMP process condition

As we all known, the CT CD was controlled by CT photo and CT etching. When the ILD THK rise to 1450 A in the wafer edge, the inline measurement data displayed that the far edge (~147 mm) CT CD decrease ~ 3 nm. Thus, the CT photo of BSE condition would be implemented to increasing the wafer edge CT CD, which focus on these partial shot and bring minimum impact on inner location. In order to further increase the CT CD of far edge, the CT etching carried out the new condition of edge +0.5 O_2. As shown in Figure 6, the CT open defects were free with the BSE & edge +0.5 O_2 condition.

For W_pits issue, we propose to execute a new CTW CMP condition (E1), which would reduce 40A oxide loss for wafer edge through tuning pressure, and the range also decreased from 87 A to 40 A, as shown in Figure 7a. Based on the new CTW CMP condition, the W_pits defect was removed completely (Figure 7b).

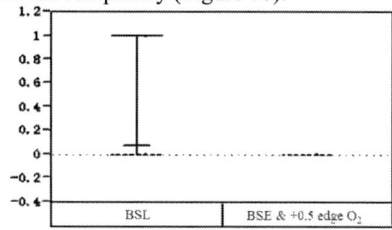

Figure 6: CT open defect performance with different photo and etching split condition.

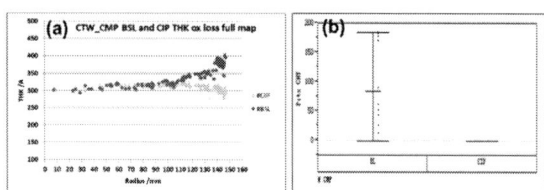

Figure 7: (a) the full map ox loss THK after CTW CMP by radius; (b) W_pis defect performance with different CTW_CMP split condition.

CONCLUSION

In this work, the systematic CT defects of CT open and over polish, W_pits, which could be monitored by the novel BFI and EBI inspection tool, were comprehensively investigated. Through the deep analysis of scan map, defect image, TEM results, we proposed the potential defect formation mechanism. Then, a series of process experiment were designed to verify the defect failure mode, and the results showed that the poor uniformity of CT hole CD and film thickness were two main factors contribute to the CT defects. Moreover, during the ILD~CT loop, each process step is critical and can effect or limit the window of the next step, thus causing a chain reaction during this loop and forming serious defect problems after process integration, but meanwhile they all have weak stability in the wafer edge. Thus, the corresponding improvement actions of optimizing the uniformity of film thickness and enlarging CT CD were executed to avoid or removal CT defects. Briefly, the film thickness was controlled by the CMP process, and the CT CD was main impacted by the CT photo and etching. In which, the ILD CMP is a particularly crucial step. Finally, based on the new process conditions of N02 (ILD CMP, target 1450 A), E1 (CTW CMP, edge ox loss -40 A), BSE (CT photo), edge +0.5 O_2 (CT etching), the yield loss in the wafer edge were improved signifinally.

REFERENCES

[1] J.K. Huang, C.L. Huang, S. C. Chang, Y.L. Cheng, Y.L Wang, *Thin Solid Films,* vol 519, 2011 , pp: 4948-4951.

[2] M.S Kuo, H. Chen, R.W. Fan, *Materials Science in Semiconductor Processing,* vol. 20, 2014, pp. 17-22.

[3] Y. Park, H.Jeong, S. Choi, H. Jeong, *Int. J. Precis. Eng. Manuf.*, vol. 14, 2013, pp. 11-15.

[4] S.H. Whang, J.K. Kim, J.W. Park, D.H. Kim, D.L. Cho, W.J. Lee, Jpn. J. *Appl. Phys.* Vol. 40, 2001, pp: 265.

[5] B. Egan, H.J. Kim, *ECS Journal of Solid State Science and Technology,* vol. 18, 2019, pp: 3206-3211.

[6] H. Lee. D. Lee. H. Jeong, International journal of precision engineering and manufacturing, vol. 17, 2016, pp: 525`-536.

THE STUDY AND INVESTIGATION OF INLINE E-BEAM INSPECTION FOR 28NM PROCESS WINDOW MONITOR

Yin Long[1]*, Zengyi Yuan[1], Fengjia Pan[1], Kai Wang[1] and Hunglin Chen[1]
[1] Shanghai Huali Integrated Circuit Corporation, 201314, China
*Corresponding Author's Email: longyin@hlmc.cn

ABSTRACT

The research aims at the VC (voltage contrast) of contact-loop defects in 28nm processes. A new type defect, marginal via open, was found on 28nm logic product wafer and the designed inline E-beam inspection was used to detect those defects by their VC features. The mechanism of defects and the method of the enhancement of the detection were described. A golden image E-beam scan recipe is required in corresponding to the unique defect wanted and, in doing so, marginal Via open defects can be found. Since the defect can be detected instantly by E-beam inspection, and therefore an inline monitoring index can be set up. Compared to the end-of-line electrical test, this inline monitor is very much closer to the trouble process and shrinks the response time. The following process experiments and evaluation can be instantly verified and examined. Finally, the Via open defect was fixed by an optimization of VIA Etch process. Instead to the debug method of failure analysis, E-beam inspection can speed up the trouble shooting cycle time.
Key word: 28nm technology node, E-beam scan, voltage contrast, metal via open.

INTRODUCTION

The tolerance of the defects becomes very small as the continuous shrinkage of the geometric size of the device for various applications involving new processes, new materials and new equipments. Defects from those different process sources could result in end-of-line yield failure.

Traditionally, optical inspection system including bright-field and dark-field inspection is the prior choice for defect detection and monitoring, while the electron-beam inspection (EBI) system plays as the assistant scheme. However, the scenario goes in opposite way while semiconductor technology entered 28 nanometer scale. E-beam inspection changed to be the indispensable approach during the development and research of 28nm product, because of optical inspection's limitation especial on non-physical defects.

In this paper, the e-beam scan methodology was studied to detect marginal metal via open and identify etch process window on 28nm technology node wafers.

MARGINAL VIA OPEN DETECTION

During the period of 28nm low power logic production yield ramping up, metal via open failure was found on yield lot. By end-offline analysis, marginal metal via open defects were found, shown in Fig. 1 (a). All of these structures of metal via open induced more than 5% yield loss. TEM results by Physical failure analysis (PFA) also showed that open failure was induced by metal via open defect, shown in Fig. 1(b).

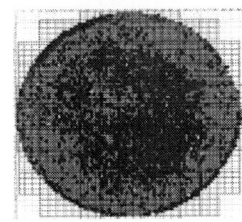 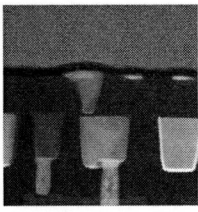

(a)   *(b)*

Figure 1: Wafer map for (a) wafer CP test ATPG bin fail and (b) TEM image of defect

Traditionally, e-beam scan method is implemented to detect the metal via open defects on SRAM or logic area in chips. The first scanning was the traditional electron beam scanning, selecting the SRAM or logic region for scanning detection. Due to the limited scanning speed of e-beam, only about 5% area of the wafer could be selected. Even if the e-beam inspection was completed in this way, the scanning time was more than 3 hours. However, no via open defect was detected. Therefore, it is suspected that the via open defect was caused by insufficient process window. Since the failure rate is low to the wafer area, it is difficult to catch this defect by sampling scan mode. But it is impossible to scan the whole wafer because it will take more than 30 days.

Therefore, MA (Misalignment) area, the special region of wafer test key is selected for detection. MA area is designed to make the overlap area distribute regularly by adjusting metal line and via hole CD, shown in Fig. 2.

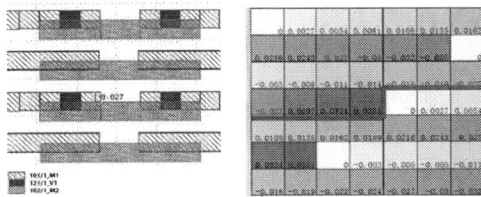

Fig. 2: MA area of test key

So it is more sensitive to reflect via open defects. In this way, the defects were detected and consistent with the map of yield failure. As shown in the Fig3.

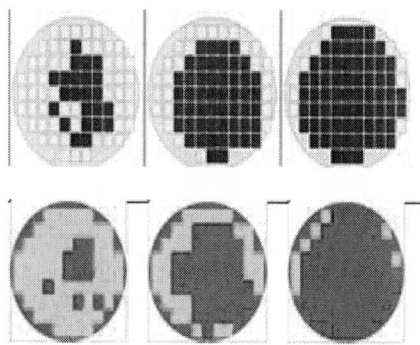

Fig. 3: Yield map vs. e-beam scan map

EXPERIMENTS AND RESULTS

After the detection method is determined, a series of process experiments were then carried out to fix via open defects.

From some of the experiment results, high correlation between via open count and AIO（all in one）etch process was found, shown in Fig.4. It means that, when via partial etch increased 10%, the process window was enlarged.

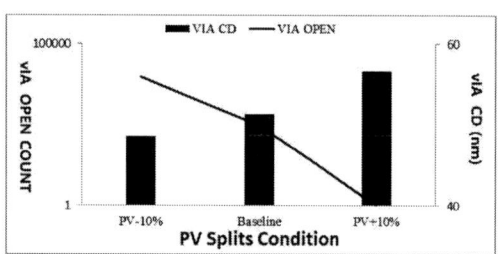

Fig. 4: PV split vs. via open defects

The TEM Image also showed no via open was found, shown in Fig.5.

Finally, with the optimized AIO etch processes; MA process window was acceptable and metal via open defects was solved.

CONCLUSIONS

The study builds up an inline index for metal via open defects and speed up the ramp up of 28nm technology

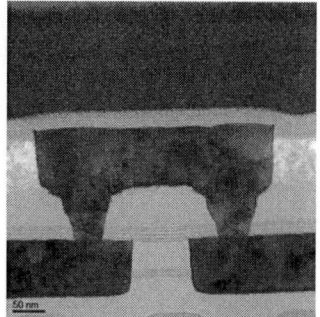

Fig. 5: TEM Image of optimized etch process

node. A series of experiments were carried out and showed the defect was obviously reduced by AIO etch process optimized. The failure model is process window is not enough which induced marginal via open. Furthermore, the corresponding improvement actions of partial via etch increasing were executed. The via open defects were solved accordingly.

ACKNOWLEDGEMENTS

The author would like to thank Fiona Pan and Zengyi Yuan of HLIC YE for the support of performing the e-beam inspection. Kqiqu Aang of HLIC Etch for coordinating the split experiments and process direction.

REFERENCES

[1] X.P. Wang, et al, "Dry Etching Solutions to Contact Hole Profile Optimization for Advanced Logic Technologies", *CSTIC*, 2012.

[2] Jing-Yong Huang, "Contact Etch Schemes at Advanced Logic Technology Nodes", *CSTIC*, 2014.

[3] T.R. Cass, D. Hendricks, J. Jau, H.J. Dohse, A.D. Brodie, W.D. Meisburger, Application of the SEMSpec electron-beam inspection system to in-process defect detection on semiconductor wafers, *Microelectronic Engineering*, Volume 30, Issues 1–4, January 1996, Pages 567-570.

[4] L. Lin, J.Y. Chen, S.D. Luo, W. Y. Wong, S. Oestreich, A. Tsai, I. Yao, and L. Grella, "Residual oxide detection with automated E-beam inspection", *IEEE International Symposium on Semiconductor Manufacturing*, Sep. 2005, pp. 241-244.

[5] O.D. Patterson, K. Wu, D. Mocuta, and K. Nafisi, "Voltage Contrast Inspection Methodology for Inline Detection of Missing Spacer and Other Nonvisual Defects ", *IEEE Transactions on Semiconductor Manufacturing*, Vol. 21, No. 3, Aug. 2008 pp. 322-328.

A UNIFIED 4H-SIC MOSFETS TDDB LIFETIME MODEL BASED ON LEAKAGE CURRENT MECHANISM

Hua Chen[1]*, Pan Zhao[1], Jiahao Liu[1], Yusen Su[1], Tuo Zheng[1], Hao Ni[1] and Liang He[1]
[1]School of Advanced Materials and Nanotechnology, Xidian University, Xi'an , China
*Corresponding Author's Email: hchen@xidian.edu.cn

ABSTRACT

The leakage currents of 4H-SiC MOSFET were measured at different gate voltages and temperatures, which revealed the critical condition of differentiating FN tunneling current from Ohmic current and FP emission. By assuming that the critical conditions indicated the applicable conditions of E model and 1/E model, a unified time-dependent-dielectric-breakdown (TDDB) model was proposed, which predicted a TDDB lifetime longer than that of E model, and lower than that of 1/E model.

Keywords—TDDB lifetime model; FN tunneling; Ohmic current; FP emission; 4H-SiC MOSFETs ; E model; 1/E model

INTRODUCTION

4H-SiC MOSFETs, as a kind of power device, usually work at high temperature and high voltage pressure. Time-dependent-dielectric-breakdown (TDDB) is one of the most important failure mechanisms of 4H-SiC MOSFETs[1-4], which means the TDDB lifetime should be fully considered in the reliability prediction and design. Bond breaks and traps on the gate oxide layer and its interface with 4H-SiC, always lead to the degradation of the oxide layer. Generally speaking, hole capturing occurs first, followed by electron capturing, and continues. After a relatively long time of degradation, electron capturing continues until localized Joule heat forms a conductive fuse in the dielectric material, resulting in short circuiting between the anode and cathode of the MOS structure, and eventually breakdown of the dielectric layer[2-4].

For 4H-SiC MOSFET with thick oxide, two models are used to describe TDDB mechanism. One is the E model, based on the electric field driving theory. The electric field weakens the polar molecular bonds and accelerates the thermal fracture of the molecular bonds in the standard Boltzmann heat process[2,4]. The other is the 1/E model, based on the current driven theory, assuming that the FN tunneling current wears oxide layer to initiate final breakdown[3,4]. Electrons, which are F-N injected from the cathode into the conduction band of SiO_2, are accelerated toward the anode. When these accelerated electrons finally reach the anode, hot holes can be produced which may tunnel back into the dielectric causing damage (hot hole anode-injection model). E model is more conservative than 1/E model, that is, for the same set of accelerated TDDB test data, E model will give a shorter failure time. E model is often used for insurance.

But, too conservative estimates often lead to increased manufacturing or maintenance costs. Thus, how to effectively select or develop a new TDDB model is a very important issue.

In 1/E model, TDDB originates from FN tunneling, and therefore, studying of leakage current mechanism may provide a new way to explore TDDB. Ref. [5] suggests that FN tunneling is the main mechanism of leakage current in 4H-SiC MOSCAP. The current caused by FN tunneling hardly varies with temperature, and is mainly affected by electric field. Ref. [6] and [7] suggests that FP emission contributes much more than FN tunneling in leakage current. FP emission is affected not only by electric field, but also by temperature.

EXPERIMENTAL DETAILS

The type of SiC MOSFETs under study is C2M0160120D manufactured by Cree Company, n-type channel with planar architecture. The gate dielectric is SiO_2. The variance of leakage current with voltage was examined by Keithley 4200-SCS, at temperature of 25°C-75°C.

RESULTS AND DISCUSSION

The leakage currents of 4H-SiC MOSFET were measured at different gate voltages and temperatures, as shown in Fig.1. In order to analyze the components of the leakage currents, the figures from 2 to 4 were drawn.

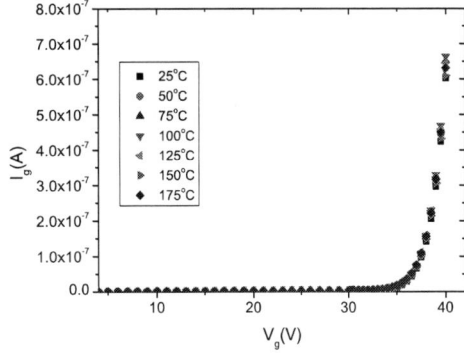

Figure 1: Gate leakage current I_g with gate voltage V_g at different temperatures

Fig.2 and Fig.3 show when gate voltage is higher than

35V, $\ln(I_g/V_g^2) \propto 1/V_g$ and currents nearly do not change with temperature, indicating FN tunneling as dominant leakage current mechanism in 4H-SiC MOSFET in high electric field.

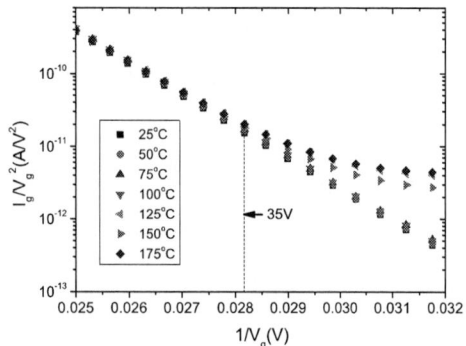

Figure 2: I_g/V_g^2 with $1/V_g$ at different temperatures

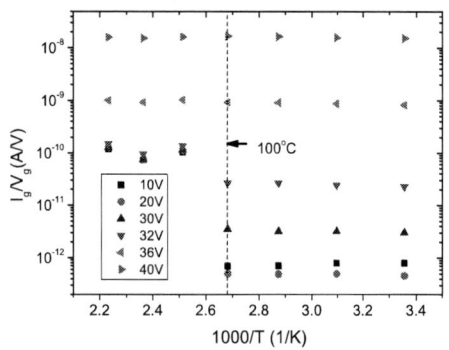

Figure 3: I_g/V_g with $1000/T$ at different voltages

Fig.4 shows when gate voltage is lower than 31.5V and temperature is higher than 100℃, I_g/V_g change very slowly with $\sqrt{V_g}$ but vary obviously with temperature, indicating Ohmic current as the dominant leakage current mechanism. Fig.4 also shows when gate voltage is lower than 31.5V and temperature is lower than 100 ℃, $\ln(I_g/V_g) \propto \sqrt{V_g}$ and I_g/V_g increases slightly with temperature, the same trend with Fig.3, indicating that FP emission possibly is present. Both Ohmic current and FP emission increase with temperature, different from FN tunneling.

In the transition from high electric field (corresponding to 35V voltage) to low electric field (corresponding to 31.5V voltage), FN tunneling, FP emission, Ohmic current may all be present, and none of them is dominant.

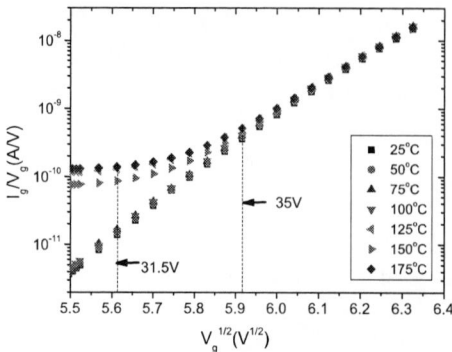

Figure 4: I_g/V_g^2 with $V_g^{1/2}$ at different temperatures

Compared with FP emission and Ohmic current, FN tunneling is hardly affected by temperature, and is recognized as the origin of TDDB in 1/E model. Therefore, the condition distinguishing FN tunneling form FP emission and Ohmic current, indicates the critical condition of E model and 1/E model. For more insurance, the high bound of the transition is set as the critical condition of two models. In electric field higher than critical condition, FN tunneling is dominant leakage current, which is so large that the current-based 1/E-Model physics (hole-catalyzed bond breakage mechanism) could be dominating. So, 1/E model is employed. Whereas, in electric field lower than critical condition, anode hole-injection is relatively small, heat effect become more important. So, the field-based E-Model physics (thermal breakage of field-stretched bonds) could be dominating, and E model is employed. 1/E model and E model are unified in one model, which is called unified model.

The accelerated TDDB test results for a Silicon Metal-Oxide-Semiconductor Capacity (MOSCAP) with the SiO$_2$ gate oxide, is recorded in Tab. 1[4]. Origin software is used for data fitting. As shown in Fig.5, data are fitted by E Model, 1/E Model and the unified model, respectively.

The formula of E Model is

$$T = A_0 \exp(Q/k_B T) \exp(-\gamma E_{ox}) = b \exp(-\gamma E_{ox}), \quad (1)$$

where, γ is the field-acceleration parameter, E_{ox} is the electric field in the oxide, Q is the activation energy, and A_0 is a process/material dependent coefficient that varies from device to device. Among them, $b = A_0 \exp(Q/k_B T)$ and γ are fitting parameters, which are fitted by "ExpDec1" of exponential fit in Origin. The offset is fixed to zero, and the fitting results are $A_0 = 1.27 \times 10^{20}$, and $\gamma = 4.10$.

TABLE I
ACCELERATED TDDB TEST RESULTS FOR A SILICON
MOSCAP[4]

No.	E	Lifetime
	MV/cm	s
1	10	203
2	9.8	438
3	9.6	973
4	9.4	2240

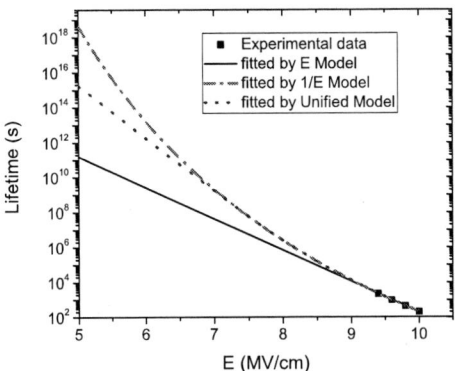

Fig.5 TDDB lifetime of a Silicon MOSCAP

The TDDB lifetime of 1/E Model is given by

$$T = \tau_0(T)\exp\left[G(T)/E_{ox}\right], \quad (2)$$

where, $\tau_0(T)$ is a temperature dependent prefactor, and $G(T)$ is a temperature dependent field acceleration parameter. We take logarithm on both sides of eq. (2), make $y = \ln T$ and $x = 1/E_{ox}$, and convert it into linear equation

$$y = \ln \tau_0 + Gx, \quad (3)$$

where, $\ln \tau_0$ and G are fitting parameters, which are fitted by linear fit in Origin. The fitting results are $\ln \tau_0 = -32.30$, and $G=376.09$.

In unified model, if the critical condition is assumed as 7.5 MV/cm, different models are used in different electric field ranges. When electric field is higher than 7.5 MV/cm, 1/E model is adopted, and thus, unified model has the same result with 1/E model. When electric field is lower than 7.5 MV/cm, E model is employed. The data of 1/E Model around the 7.5 MV/cm are used for fitting. During fitting, the offset is no longer set to zero and the fitting result is

$$y = 1.60\times10^6 + 1.76\times10^{30}\exp(-x/0.14) \quad (4).$$

Each model could fit the data quite well. While in lower electrical field stress, E Model and 1/E Model give quite different predictions, and unified model predicted a TDDB lifetime longer than that of E model, and lower than that of 1/E model.

Of course, the correctness of the unified model is needed to be further confirmed by the 4H-SiC MOSCAP TDDB data.

CONCLUSION

The leakage currents of 4H-SiC MOSFET were measured at different gate voltages and temperatures. According to the relationship between current, voltage and temperature, the critical condition of differentiating FN tunneling current from Ohmic current and FP emission was revealed. When the electric field was larger than the critical value, FN tunneling was the dominant leakage mechanism; when the electric field was smaller than the critical value, FN tunneling was no longer dominant.

In 1/E model, FN tunneling current was recognized to bring out the degradation of oxide layer. Thus, when FN tunneling was the dominant leakage current mechanism, 1/E-Model physics could be the dominant TDDB mechanism. By assuming that the critical conditions indicated the applicable conditions of E model and 1/E model, a unified TDDB model was proposed, which predicted a TDDB lifetime longer than that of E model, and lower than that of 1/E model.

In the next step, it is necessary to carry out 4H-SiC MOSCAP TDDB accelerated life test on different types of samples, and complete the verification of the model.

ACKNOWLEDGEMENTS

This work was supported by the National Natural Science Foundation of China (Grant No. 61504099), and the Fundamental Research Funds for the Central Universities of Ministry of Education of China (Grant Nos.JB151403, JB181409).

REFERENCES

[1] H. A. Moghadam, S. Dimitrijev, J. S. Han, and D. Haasmann. *Microelectron. Reliab.*, vol. 60, 2016, pp.1-9.

[2] Z. Chbili, A. Matsuda, J. Chbili, J. T. Ryan, J. P. Campbell, M. Lahbabi, et al. *IEEE Trans. Electron devices*, vol. 63, 2016, pp.3605-3613.

[3] K. P. Cheung, *IEEE International Reliability Physics Symposium (IRPS) 2018*, Burlingame, March 11-15, 2018, pp. 2B.3.

[4] J. W. Mcpherson, *Reliability Physics and Engineering: Time-To-Failure Modeling*, Springer Science Business Media, 2009.

[5] S. R. Kodigala, S. Chattopadhyay, C. Overton, and I. Ardoin. *Solid State Electronics*, Vol. 114, 2015, pp.104-110.

[6] M. Sometani, D. Okamoto, S. Harada, H. Ishimori, S. Takasu, T. Hatakeyama, et al. *J. Appl. Phys.*, vol. 117, 2015, pp.024505.

[7] A. Xiang, X. Xu, L. Zhang, Z. Li, J. Li, and G. Dai. *Appl. Phys. Lett.*, vol. 112, 2018, pp.062101.

STUDY ON WAFER EDGE TEST WITH OPTIMIZED TEST SOLUTION

Yuxiang Zhang[*], Yuanyuan Zhu, Zhimin Zeng

Shanghai Huahong Grace Semiconductor Manufacturing Corporation, Shanghai 201206, China

*Corresponding Author's Email: yuxiang.zhang@hhgrace.com

ABSTRACT

With the increase of the process complexity, the layered problem of stack film on wafer edge, especially on ugly dice (incomplete dice), is becoming more and more serious, and ultimately affect the test yield. Therefore, improving the wafer edge process becomes more and more important to enhance the yield and test stability. On the other hand, we need to optimize the test scheme to bypass the ugly dice or to improve the tolerance of the ugly dice. This paper will focus on the test optimization. Several approaches will be discussed, and some corresponding suggestions for improvement are put forward.

Keywords — wafer edge, ugly dice, test optimization

INTRODUCTION

As the diameter of the wafer gets larger and the process complexity increases, there are more and more process issue around the wafer edge of silicon wafer, which leads to the unexpected abnormal and yield loss in wafer testing. Absolutely, the improvement of the frontend manufacture process is the necessary countermeasure to solve the root problem. However, it will be of great positive significance if we can think some ways to improve the tolerance of the ugly dice, and mitigate the influence of the frontend wafer edge process issue by some test approach.

EXPERIMENTS AND DISCUSSIONS

Case 1

By touch Open/Short failure was found, more serious at wafer edge. The needle tip was extraordinary dirty by inspection. We found the punctured bonding pad on ugly dice by wafer edge. By FIB, we confirmed that no surface metal layer covered on the pad area of ugly dice, so the Tungsten plugs were exposed, spaded up and adsorbed on the tip of the needle while testing. Test failure started from these pads, and Tungsten plugs were constantly brought to the pads of next touch, affecting the results of subsequent dice under test (as shown in Figure 1).

Figure 1: Bin fail due to ugly dice at wafer edge

If the needle tip is dirty, the most direct way to improve the test is to clean the needle. Furthermore, increasing the frequency of cleaning needles is effective in improving this test problem.

As shown in Figure 2, zone-01 is the region in which all dice are complete dice in any test touch, while zone-02 is the region in which contain both complete dice and incomplete dice in any test touch. According to the statistics result of a certain amount of wafers, it was found that the failure rate of zone-02 was significantly higher than that of zone-01.

After we increase the needle clean frequency, the failure rate of zone-02 reduced obviously (as shown in Figure 2).

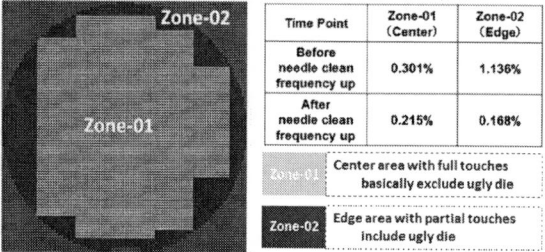

Figure 2: Open/Short fail rate comparison by Zone-01(Center) and Zone-02(Edge)

In addition, the selection of the optimal test movement path on wafer will help reduce the impact of wafer edge issue on yield results. For example, first test zone-01, then test zone-02, and then clean the needles, so as to avoid bring the dirty things from zone-02 to zone-01.

However, increasing the needle clean frequency will shorten the lifetime of the probe card and increase the test cost. And it may not apply in all cases for the yield rescue.

Case 2

Parallel testing is generally used to improve the test efficiency. Sharing power supply for multi device under test (DUT) is generally adopted because of the limitation of the number of the power supply channels by parallel testing. Therefore, target dice (complete dice, effective dice) and ugly dice (incomplete dice) may share power supply on wafer edge. Due to the special formation of the layer stack on the edge of the wafer, ugly dice often produce unexpected large current. That will induce the failure of normal target dice, which share power supply with ugly dice, finally affect the wafer yield (as shown in Figure 3). In addition, unexpected large current from ugly dice will cause abnormal loss of the needle, which will shorten the life time of the probe card. Serious current even cause the test machine fault down.

Figure 3: Ugly dice induce unexpected large current

We can skip testing of target dice which share power with ugly dice around the wafer edge, so that the tester can work smoothly without the impact of unexpected large current from ugly dice. However, the good dice among the target dice will be overkilled. And this will result in unnecessary yield loss.

In fact, not all ugly dice will have bad impact on the test. Therefore, adding special leakage test item for power pin before high-voltage test can screen out the ugly dice which may generate abnormal large current. Once the current value exceeds a certain setting spec, the following test item will be skipped and the dice (ugly dice and target dice that share same power) under test will be judged fail finally. If the dice under test pass the special leakage test item, test procedure will go on. This will protect the tester meanwhile minimize the impact to the yield.

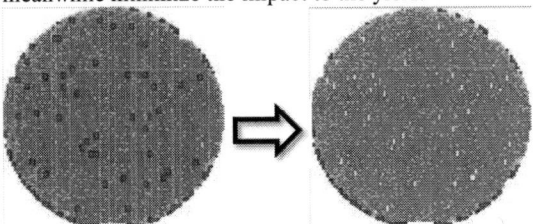

(a) apply on full wafer   (b) apply only on wafer edge
Figure 4: Yield overkill reduced after scope of special leakage test item application optimized

However, it is difficult to precisely set the spec for screening leakage current by which will result in the test issue. Around the wafer edge, when we detect the pin leakage of the ugly die, it will inevitably kill the target dice that share power with ugly dice; And also on the center of the wafer, it will over kill the target dice when same level leakage current (by another cause not from ugly dice) is detected. In order to reduce the overkill, we change the power pin leakage current monitor scope from full wafer to the certain region only at wafer edge which include of ugly dice. You can see the effect of this change by Figure 4.

Another approach by test solution is to add relay on probe card. If unexpected large current was detected, the current path can be cut off independently by relay only on ugly dice. This will reduce the over kill to the target good dice and save the yield.

CONCLUSION

Many test issue and low yield are caused by layered problem of stack film on wafer edge with the increase of the process complexity. The improvement of the frontend manufacture process is the necessary countermeasure to solve the root problem. Some test approaches are presented in this paper to improve the tolerance of the ugly dice, minimize the influence of wafer edge process issue, prevent from the yield overkill, and ultimately ensure the test accuracy.

ACKNOWLEDGEMENTS

The authors would like to thank all members of testing group of HHGrace for their great support on this work.

REFERENCES

[1] Hong Xiao, "Introduction to Semiconductor Manufacturing Technology", Prentice Hall (2000)
[2] Ankun Mao, "Study on the fluctuation of photolithography on the edge of silicon wafer in IC manufacturing", Fu Dan University (2012)
[3] Xiaofeng Yuan, Qiang Zhang, Jing'an Hao, "Wafer edge treatment in lithographic process for peeling defect reduction", IEEE/CSTIC (2017)
[4] Yuxiang Zhang, Qin Huang, Yun Xu, "Effect of wafer edge cut on testing and yield", IEEE/CSTIC (2018)

Full Metrology Solutions for Advanced RF with Picosecond Ultrasonic Metrology

Johnny Dai[1], Priya Mukundhan[1], Johnny Mu[2], Frank Zheng[2], Cheolkyu Kim[3]
[1]Onto Innovation, 550 Clark Drive, Budd Lake, NJ 07828, USA
[2]Onto Innovation, Room 103-2, Building 1, No. 690, Bibo Road, Pudong, Shanghai 201203, China
[3]Onto Innovation, 16-6, Sunae-dong, Bundang-gu, Sungnam-si,Gyunggi-do, 3965 Korea
*Corresponding Author's Email: johnny.dai@ontoinnovation.com

ABSTRACT

Picosecond Ultrasonics (PULSE[TM] Technology) [1] has been widely used in thin metal film metrology because of its unique advantages, such as being a rapid, non-contact, non-destructive technology and its capabilities for simultaneous multiple layer measurement. Measuring velocity and thickness simultaneously for transparent and semi-transparent films offers a lot of potential for not only monitoring process but offers insight into the device performance. In this paper, we show Picosecond Ultrasonics provides a complete metrology solution in advanced radio frequency (RF) applications. This includes measurement of various thin metal films for wide thickness ranges with extremely excellent repeatability which could meet stringent process control requirements, simultaneous multilayer measurement capability, and simultaneous measurement of sound velocity and thickness for piezoelectric films which play a key role in the performance of RF devices.

INTRODUCTION

All signal processing requires filters that evaluate signals and remove undesirable frequencies while preserving desirable frequencies. Modern smartphones are required to filter, transmit, and receive paths for 2G, 3G, and 4G in up to 15 bands, as well as support Bluetooth, Wi-Fi, and other wireless communications. Phones such as these could require up to 40+ filters. The revolution of communication technology is driving a dramatic increase in the number of RF bands that smartphones and other mobile devices must support which significantly increases the number of RF filters. Not surprisingly, with the move to higher frequencies and 5G, the complexity of the devices is expected to increase as well as the performance requirements. At these higher frequencies, surface acoustic wave (SAW) filters require smaller width and pitch of the interdigital transducers, which limits their performance. Bulk acoustic wave (BAW) filters is the primary technology employed above 2.5GHz [2]

METROLOGY REQUIREMENTS AND CHALLENGES

Such advances in filter technologies will place stringent demands on manufacturing which in turn will require very accurate metrology techniques.

The thickness for the full stack can shift the center frequency and affect the device performance, and piezoelectric layer thickness control is key for SAW and BAW devices. The frequency accuracy (3σ) of 0.1% requires film thickness control within the same accuracy or better. Thin film deposition systems with wafer uniformity of (3σ <2%) cannot meet these standards. To overcome this limitation, semiconductor equipment manufacturers have developed a trimming process. Monitor wafer thickness measurements are helpful for characterizing deposition chambers and process qualification but do not help with device-level process control. RF filter manufacturers require the ability to not only measure the thickness but would like to be able to adjust the thickness via a trimming process as thickness is directly correlated to the filter characteristics. It becomes important to accurately measure thickness on multiple sites on production wafers. Typically, the measured thickness is forwarded to a trimming tool to adjust the thickness profile across the wafer and improve the thickness uniformity to enhance device performance and yield. Another parameter that is also helpful for process control is the ability to monitor the acoustic velocity. Velocity variations can also shift center frequency and affect RF device performance.

We have previously demonstrated the application of Picosecond Ultrasonic Technology for characterizing the inter-digital transducer (IDT) thickness and sound velocity of SiO_2 [3]. IDT thickness needs to be controlled at Angstrom level to achieve the desired filter frequency in the final product. The excellent repeatability (3 sigma < 0.15Å) and accuracy of the technique enabled measurements on device wafers and met the tight process control requirements.

In this paper, we discuss other use cases of PULSE Technology for RF filter process monitoring and control.

PICOSECOND ULTRASONIC TECHNOLOGY: BASICS ABOUT PICOSECOND ULTRASONIC AND APPLICATIONS

Picosecond Ultrasonic Technology is a non-contact, non-destructive pump-probe laser acoustic technique for film thickness, sound velocity, Young's Modulus, density, and roughness measurement. It has been widely adopted as the tool-of-record for metal film thickness metrology in semiconductor fabs around the world. An acoustic wave is launched in a film by a 100fs laser pulse (pump) focused onto the film surface. The acoustic wave travels away from the surface through the film at the speed of sound in the film. At the interface with another material, a portion of the acoustic wave is reflected and comes back to the surface while the rest is transmitted. The probe pulse detects this reflected acoustic wave as it reaches the wafer surface. One can detect the change of optical reflectivity that is caused by the strain of acoustic wave. Using standard sound velocity in the material, thickness can be readily extracted using first principles technique. In addition to thickness, depending on applications and the stack up of films, film density, sound velocity, Young's Modulus, and surface roughness can also be measured.

Picosecond Ultrasonic Technology provides excellent repeatability and stability for single layer and full stack thickness measurements. The small beam spot and rapid measurement time can enable direct measurement on actual device structures and allows measurement of multiple die. The capability of measuring both thickness and sound velocity at the same time gives the Picosecond Ultrasonic technique unique technological advantages.

ADVANTAGES OF PICOSECOND ULTRASONICS FOR ADVANCED RF APPLICATIONS

The following sections are some of the recent advanced applications in RF filter process control.

1.1. Typical filmstack and tight control for film thickness

Figure 1 and table 1 show typical measurement signal (left) and performance (right) of Mo, Mo/AlN, Mo/AlN/Mo, and Pt of Picosecond Ultrasonics used for a BAW RF device. As we can see the piezoelectric layer has the repeatability of 0.03% for 3 sigma while the repeatability for both top electrode and bottom electrode metal layers are better than 0.04%.

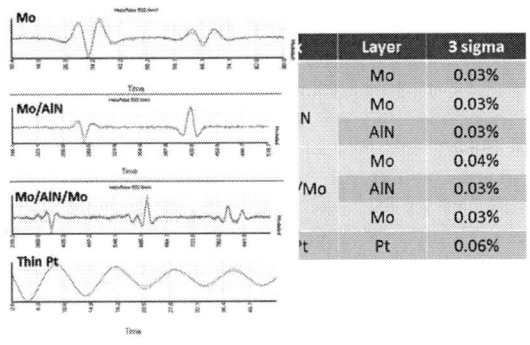

Figure 1. Typical measurement raw signal from RF stacks, and Table 1. Typical repeatability performance for different stacks of common RF applications

1.2. Device level process control for RF devices

More and higher frequency bands require tighter process control to improve device performance. 3σ frequency accuracy must be < 0.1%. To meet this target, the uniformity of the deposited thin films must be roughly < 0.1% as well. But the best thin film deposition system achieves 3σ uniformity across the wafers of no better than 2%. To overcome this limitation, equipment vendors have developed a trimming process. Picosecond Ultrasonic Technology has been adopted to monitor the pre- and post- trimming process because of its accuracy and robustness.

The following two figures show measurement of SiO_2 thickness at pre- and post- trimming. The thickness is scaled to protect customer confidentiality. The piezoelectric layer film thickness was measured by Picosecond Ultrasonics before trimming and after trimming. For both wafers, as shown in left and right, you can see that thickness varies in a wide range and is well controlled as designed using trimming process for all measured sites.

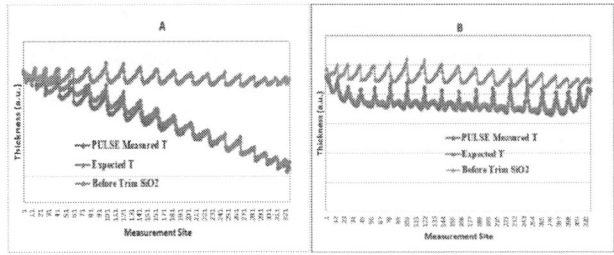

Figure 2. Pre- and post-trimming process monitored by Picosecond Ultrasonics

1.3. Wide thickness range

With Picosecond Ultrasonic Technology, depending on tool configuration, we can measure a film thickness range from 50Å to ~20μm depending on material parameters. Figure 3 shows typical repeatability performance for Mo, W, AlCu, and TiW films ranging from 100Å to 15000Å. We can see excellent repeatability for all thickness range. Because of its first principle and standard-less nature of Picosecond Ultrasonics, we can use one single recipe to cover the whole thickness range from tens of Angstroms to tens of microns with excellent repeatability. We can see that 3σ of dynamic repeatability is below 0.20% for these films.

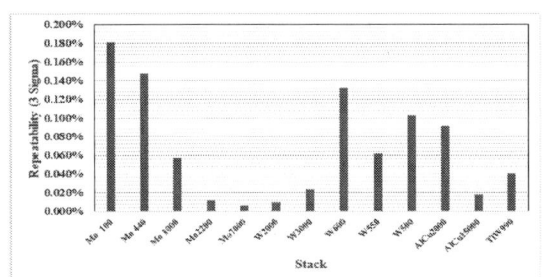

Figure 3. 3σ repeatability for typical filmstack used in RF

1.4. Multilayer measurement capability and accuracy

Tight control of frequency for RF filters relies on even tighter thickness control for every layer in the stack although the piezoelectric layer plays the most critical role. Figure 4 shows a typical measurement signal from a BAW device using Picosecond Ultrasonics. From the measurement, we report the thickness of six layers in the stack. The thickness and repeatability are shown in table 2. We rescaled the thickness to protect customer confidentiality. The accuracy has been confirmed with TEM with R^2 linear correlation higher than 0.97 for all layers. For the piezoelectric layer, the linear correlation R^2 between Picosecond Ultrasonics and TEM cross section is about 0.998% with the slope about 1.0.

978-1-7281-6559-2/20 $31.00 © 2020 IEEE

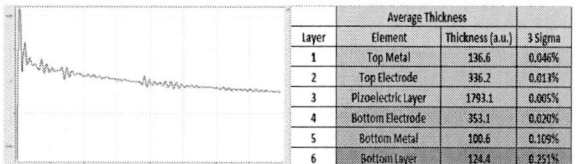

	Average Thickness		
Layer	Element	Thickness (a.u.)	3 Sigma
1	Top Metal	136.6	0.046%
2	Top Electrode	336.2	0.013%
3	Piezoelectric Layer	1793.1	0.005%
4	Bottom Electrode	353.1	0.020%
5	Bottom Metal	100.6	0.109%
6	Bottom Layer	124.4	0.251%

Figure 4. Picosecond Ultrasonic measurement signal of a multilayer stack and Table 2. Thickness (arbitrary unit) and dynamic repeatability (3σ) for each layer

1.5. Simultaneous measurement of thickness and sound velocity

Piezoelectric layer thickness and sound velocity variation and uniformity is critical for center frequency control in RF devices. For piezoelectric films, such as oxide or AlN films on silicon or other metal transducers, the opaque substrate absorbs energy from the pump pulse, launching a sound wave that travels up through the transparent film at the speed of sound. The strain causes a local change in the index of refraction of the film. The partial reflection of the probe beam from the moving sound wave, combined with the partial reflection from the film surface, leads to destructive and constructive interference at the detector [4]. As a result of this time dependent interference, the measured signal oscillates with a period, τ, from which the sound velocity (V) in the material can be determined by

$$ V = \frac{\lambda}{2n\tau cos\varphi} $$

where n is the index of refraction, λ is the wavelength, and φ is the angle of refraction. We can also report Young's Modulus calculated from measured sound velocity. Figure 5a and 5b show the typical performance for thickness and sound velocity measurement on oxide films. We can see that 3σ of repeatability at site level for both thickness and sound velocity is below 0.06%. For wafer average of nine site measurements, 3σ of the repeatability is below 0.02%. The excellent repeatability makes it possible for tighter control of the piezoelectric layer and then adjust the center frequency of filters.

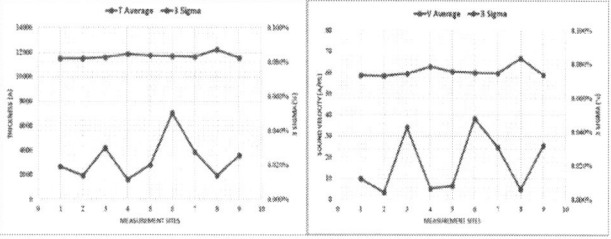

Figure 5a). Example of SiO_2 thickness and repeatability by Picosecond Ultrasonic. 5b). Example of SiO_2 sound velocity and repeatability by Picosecond Ultrasonic.

1.6. Long term stability and tool matching

Long term stability of the tool and tool-to-tool matching are extremely critical for process control in a high volume manufacturing environment. First principle nature of Picosecond Ultrasonics and its intrinsic standard enables its long-term stability without standard calibration as most metrology tools require. Its long-term stability is better than

1.0% for 3σ and tool-to-tool matching is better than 0.5% at site-level. Typical performance for within wafer average is ~ 3σ < 0.3% for both long-term stability and tool-to-tool matching for wafer average thickness. Figure 6 shows dynamic repeatability and tool-to-tool matching for three layers, top layer Mo, piezoelectric layer AlN, and bottom Mo of six wafers. We can see that dynamic repeatability is well below 0.05% for all three layers of the six wafers, and tool-to-tool matching is well below 0.25%.

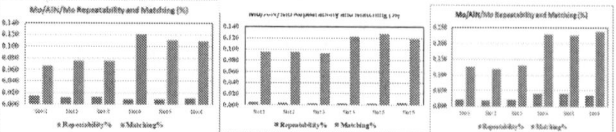

Figure 6. Mo, AlN and Mo dynamic repeatability and tool to tool matching for the tri-layer stack Mo/AlN/Mo.

1.7. Concentration measurement

Doping into piezoelectric layer can increase piezoelectric coefficients, soften the material, increase permittivity, and boost electromechanical coupling K^2 significantly. The PULSE technique has also shown sensitivity to detect concentration changes by monitoring the changes in velocity. Figure 7 shows sound velocity of AlN has a clear correlation with element X doping level with $R^2 = 0.97$.

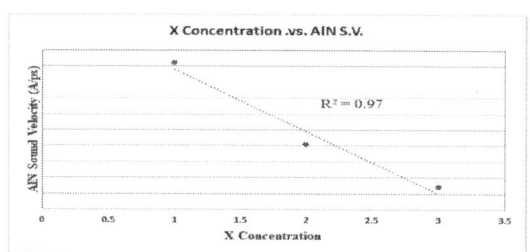

Figure 7. Correlation of element X concentration and AlN sound velocity

RF filter process control requires stringent metrology due to tight process tolerances. PULSE technology can measure thickness, sound velocity, Young's Modulus and these are critical to the RF filter process that has very tight process control limits. The PULSE technique can also simultaneously measure full stacks for multilayer metal stack measurements with excellent repeatability and long-term stability. In RF filter applications, temperature compensation top SiO_2 for SAW and sacrificial top metal layer for BAW are monitored to adjust center frequency. PULSE technology is widely used to monitor the device level trimming process to enhance yield and reduce cost. With the implementation of Discover® data analytics software, information turnaround time for process control and monitoring is enhanced.

REFERENCES

[1] C. Thomsen, H. T. Grahn, H. J. Maris, J. Tauc, Phys. Rev. B, vol. 34, 1986, pp. 4129-4138

[2] R. Aigner, 2008 IEEE Ultrasonics Symposium, 2008.

[3] J. Dai, R. Mair, K. Park, X. Zeng, P.Mukundhan, C. Kim, and T. Kryman, March 18-19, 2019 CSTIC, Shanghai, China

[4] J. L. Arlein, S. E. M. Palaich, B. C. Daly, P. Subramonium, and G. A. Antonelli, J. of Appl. Physics, vol 104, 2008, pp 033508 1-6

Monitoring Critical Process Steps in 3D NAND using Picosecond Ultrasonic Metrology with both Thickness and Sound Velocity Capabilities

Johnny Dai[1], Priya Mukundhan[1], Robin Mair[1], Manjusha Mehendale[1], Calvin Wang[2], Ewen Wang[2], Cheolkyu Kim[3]

[1]*Onto Innovation, 550 Clark Drive, Budd Lake, NJ 07828, USA*
[2]*Onto Innovation, Room 103-2, Building 1, No. 690, Bibo Road, Pudong, Shanghai 201203, China*
[3]*Onto Innovation, 16-6, Sunae-dong, Bundang-gu, Sungnam-si,Gyunggi-do, 3965 Korea*
Corresponding Author's Email: johnny.dai@ontoinnovation.com

ABSTRACT

Amorphous carbon (a-C) based hard masks provide superior etch selectivity, chemical inertness, are mechanically strong, and have been used for etching deep, high aspect ratio features that conventional photoresists cannot withstand. Picosecond Ultrasonic Technology (PULSE™ Technology) has been widely used in thin metal film metrology because of its unique advantages, such as being a rapid, non-contact, non-destructive technology and its capabilities for simultaneous multiple layer measurement [1]. Simultaneous measurement of velocity and thickness for transparent and semi-transparent films offers a lot of potential for not only monitoring the process but offers insight into the device performance. In this paper, we show successful applications of Picosecond Ultrasonics in 3D NAND. This includes measurement of various thin metal films and simultaneous measurement of sound velocity and thickness for amorphous carbon films which has been widely used as hard mask materials.

INTRODUCTION

3D NAND, driven by data intensive applications, changes the paradigm for manufacturing by providing an opportunity for vertical scaling using a highly repetitive and precise deposition and etch process. Currently, commercially available 3D NAND products have 64-layer and 96-layer tier stacks in high volume manufacturing and a 128-layer stack is in pilot production. To make 3D NAND, the most difficult and critical process is the high-aspect ratio (HAR) etching step. In this process, an etch tool drills tiny circular holes from the device top to the bottom substrate. A device may have up to 2.5 million tiny channels in one chip. Hard masks are used for etching deep, high aspect ratio features that conventional photoresists cannot withstand. Amorphous carbon (a-C)-based hard masks provide superior etch selectivity, chemical inertness and are mechanically strong [2]. Monitoring a-C thickness is critical to the 3D NAND process as it goes through an iterative etch process. Film thickness and sound velocity (mechanical property) and repeatability affect the active area of a cell and consistency of the deposition/etch performance. Picosecond Ultrasonic Technology (PULSE™ Technology), implemented in the MetaPULSE® G system, is a non-contact, non-destructive pump-probe laser acoustic technique for the measurement of metal film thickness. It is a proven workhorse in semiconductor fabs around the world. A 0.1ps laser pulse (pump) is focused to a small ($\sim 10 \times 15 \mu m^2$) spot onto a wafer surface to create a sharp acoustic wave. The acoustic wave travels away from the surface through the film at the speed of sound. At the interface with another material, a portion of the acoustic wave is reflected and comes back to the surface while the rest is transmitted. The probe pulse detects this reflected acoustic wave as it reaches the wafer surface. One can detect the change of optical reflectivity that is caused by the strain of acoustic wave or alternatively detect the deflection of reflected probe beam that is caused by the deformation of the surface due to the acoustic wave using a position sensitive detector (PSD). Both of these modes, reflectivity and PSD, are used in characterizing metal films. Knowing the speed of sound in the material, and the arrival time of the echoes, thickness is readily extracted using the first principles technique. Information on film density and surface roughness, depending on the application, can also be obtained by fitting the damping rate of the echoes and the width of the echoes, respectively. The latest system improvement included some additional modifications to the experimental setup to enhance the signal to noise (SNR). A detailed description of the modification and the resulting benefits will be discussed in the paper. Semi-transparent films such as a-C films can absorb energy from the pump pulse, launching a sound wave that travels down through the semi-transparent film at the speed of sound. The strain causes a local change in the index of refraction of the film. The partial reflection of the probe beam from the moving sound wave, combined with the partial reflection from the film surface, leads to destructive and constructive interference at the detector [3]. As a result of this time-dependent interference, the measured signal oscillates with a period, τ, from which the sound velocity (V) in the material can be determined by

$$V = \frac{\lambda}{2n\tau cos\varphi}$$

where n is the index of refraction, λ is the wavelength, and φ is the angle of refraction.

Using the PULSE technique, we have performed high resolution line scans (0.5mm EE) on different types of amorphous carbon films and demonstrated 3σ repeatability performance for thickness and velocity of $< 0.5\%$. Accuracy of the technique has been correlated to cross-section scanning electron microscopy (SEM) with $R^2 > 0.95$. Velocity values provided by the technique served a two-fold purpose. First, when used in the calculation of the thickness, it provided a more accurate representation of wafer level variation and second, the velocity values were found to have a correlation to the etch process enabling a direct monitor of this process.

APPLICATION IN 3D NAND

1. Metal film thickness measurement using PULSE Technology

PULSE Technology has been widely adopted as the process of record tool for metal thickness measurement. In 3D NAND, it has been used for the measurement of AlCu, W, Cu, Cu Seed, Ta, TaN, Ti, TiN, WSi, and etc. Figure 1a shows the typical measurement of a 350Å Ta films and its modeling example. Figure 1b shows the thickness profiles of 13 points across the wafer. PULSE Technology can be used to measure films down to 50Å and ~20μm depending on the material's property. It can also be used to measure up to eight layers of the stack simultaneously with excellent repeatability.

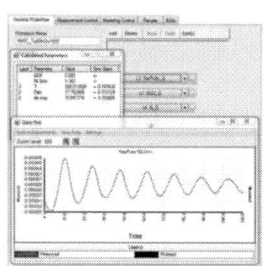 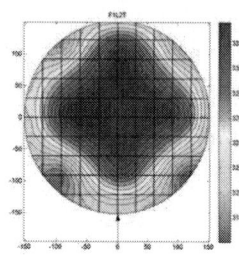

Figure 1a. Typical measurement of a 350Å Ta films and its modeling example. Figure 1b. Thickness profiles of 13 points across the wafer.

2. Repeatability of thickness measurement by PULSE Technology

PULSE Technology has been widely used for the measurement of AlCu, W, Cu, Cu Seed, Ta, TaN, Ti, TiN, WSi, and etc. Table 1 shows the typical dynamic thickness repeatability (3σ) performance for a 13 point measurement.

Film	Thickness	Repeatability (3 sigma)
ACL	10k~17k	<0.3%
AlCu	4k~8k	<0.3%
Thin W	250	<0.6%
Thick W	400~3000	<0.3%
Cu Seed	200~500	<0.3%
Cu	500~6000	<0.3%
Ta, TaN,TiN & Ti	50~200	<0.6%
	200~1000	<0.3%
WSi	600~1000	<0.3%

Table 1. Typical dynamic repeatability performance for PULSE measurement for commonly used metal films in 3D NAND

3. Simultaneous sound velocity and thickness measurement by PULSE Technology

The following figure shows a standard a-C measurement and modeling example using PULSE Technology. The top red curve shows the raw signal, and the following bottom two graphs show sound velocity and thickness reported from the measurement.

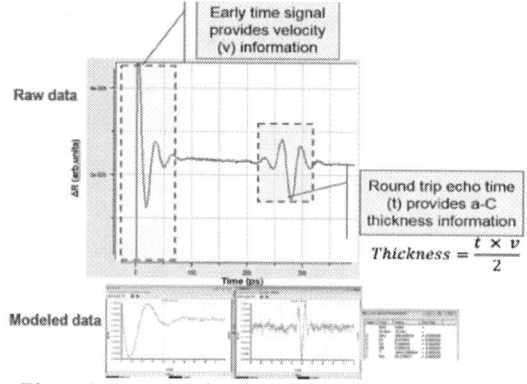

Figure 2. Raw signal and modeling example of measurement on a-C films using PULSE Technology

4. 49-point measurement for thickness and sound velocity

Through measurement and modeling algorithm improvement, we have significantly improved the repeatability for both thickness and velocity while maintaining high throughput. Table 2 shows the typical dynamic repeatability (3σ) for sound velocity and thickness based on the average for 49 points per wafer.

49 points		
	Sound Velocity (3σ)	Thickness (3σ)
Wafer 1	0.40%	0.18%
Wafer 2	0.37%	0.17%

Table 2. Dynamic repeatability for sound velocity and thickness measurement based on 49 points

5. 49-point profile across the wafer for thickness and sound velocity

Figure 3 shows a typical 49-point profile for thickness and velocity. We can see that thickness and sound velocity have different profiles. Because both thickness and sound velocity play important roles in HAR etching. Monitoring thickness and sound velocity are very critical for the HAR etching process.

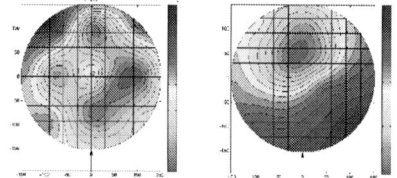

Figure 3. 49 point profiles for thickness (left) and sound velocity (right)

6. PULSE thickness measurement correlation with SEM

Figure 4 shows the SEM correlation of PULSE thickness measurement with input from simultaneous sound velocity measurement. We can see a correlation with R^2=0.95 that indicates very strong correlation of the PULSE measurement with the SEM measurement. In order to protect the confidentiality of the data, we have not shown the actual thickness values but suffice it to say that the excellent correlation was validated across the process window.

The technique has also been proven on next generation a-C materials, including metal-doped hard masks.

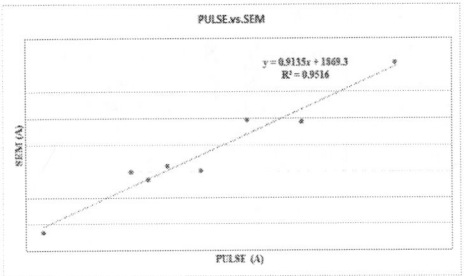

Figure 4. Correlation of PULSE thickness measurement with SEM thickness measurement

CONCLUSIONS

In summary, PULSE Technology has been successfully used for thickness measurement of both metal and amorphous carbon. Excellent repeatability of this technique can meet the stringent demands for process control. Besides film thickness, it can also measure velocity of semi-transparent and transparent films simultaneously, and a-C films for etching process control. We have demonstrated both thickness and velocity measurement capability for amorphous carbon based hard masks, and sound velocity has been found to be very well correlated to etch characteristics.

REFERENCES

[1] J. Dai, R. Mair, K. Park, X. Zeng, P.Mukundhan, C. Kim, and T. Kryman, March 18-19, 2019 CSTIC, Shanghai, China
[2] H. Singh, Solid State Technology, July 2017, pp 18-21.
[3] J. L. Arlein, S. E. M. Palaich, B. C. Daly, P. Subramonium, and G. A. Antonelli, J. of Appl. Physics, vol 104, 2008, pp 033508 1-6

DESIGN AND DEVELOPMENT OF 3D WLCSP FOR CMOS IMAGE SENSOR USING VERTICAL VIA TECHNOLOGY

Tianshen Zhou[1,2], Shuying Ma[2], Fengxia Zheng[2] and Tao Hang[1]*

[1] School of Materials Science and Engineering, Shanghai Jiao Tong University, Shanghai 200240, China

[2] Huatian Technology (Kunshan) Electronics Co., Ltd., Kunshan 215300, China

*Corresponding Author's Email: shuying.ma_ks@ht-tech.com

ABSTRACT

Wafer-level chip scale packaging (WLCSP) using tapered through silicon via (TSV) has been widely applied in CMOS image sensor (CIS) in mass production. The mass production technology now is using laser drill and surface landing, which has been shown in Fig. 1. However, as the increase of IO numbers and reduction in chip size, there is no enough space for ball placement when using these two structures. Meanwhile, side wall delamination and function layer crack are often found in strict reliability conditions. To satisfy small chip size packaging requirement and pass higher reliability test, we propose a new design of WLCSP with vertical TSV which can effectively reduce the stress of pad in temperature cycle (TC) test. The finite element simulation evaluation and key processes including via etch, redistribution layer (RDL) lithography and solder mask formation are also presented.

INTRODUCTION

3D integration using through silicon via (TSV) is considered to be a promising solution to extend Moore's law[1] for the growth of IC industry. Today TSV technology has widely applied in different type of products such as memory stacking and MEMS, which takes advantages in wide bandwidth of I/O, short interconnecting length, and reduction of power consumption[2,3]. Nevertheless, one of the most advanced TSV application is the package of WLCSP for CMOS image sensor (CIS) [4,5].

For CIS-WLCSP packaging, the via last TSV technology using the tapered and low aspect ratio TSV has been introduced in mass production for several years. However due to the fragile structure, the products sometimes fail in OS in reliability test as Fig. 1 shows. What's more, with the increase of the resolution and I/O number, the low aspect ratio (AR) TSV design can't meet the requirement for high-end products, not to speak of passing the reliability test.

Thus, in this paper we presented a design of WLCSP for CIS products using vertical via (the schematic view is shown in Fig. 2) and especially made a bubble inside the via. We could see the finite element simulation results indicate that the bubble can reduce the stress of pad in TC test for more than 60%. At present, we have realized the design with dummy wafer by optimizing several processes like via etch, RDL lithography and solder mask formation to ensure a controllable bubble. The SEM cross-section images of the package were also presented. The cross-section of TC 1000 results showed that there were no any deformation and crack in the via bottom in daisy-chain pad.

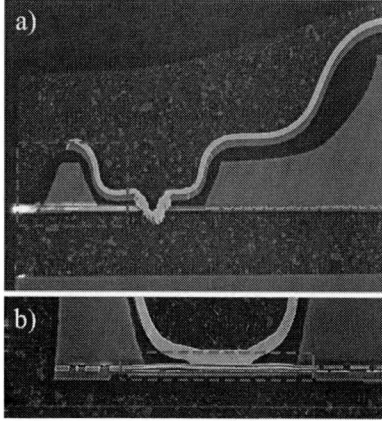

Figure 1: failure after reliability test: a) silicon crack; b) pad delamination

DESIGN AND SIMULATION

Design of the structure

The Figure 1 showed the schematic view of the CIS package. The design of the TSV had to be compatible with the Pad size in the front side of wafer. The via should open within the pad area that its diameter needs to be smaller than pad size. Generally, a higher aspect ratio could bring better mechanical performance, while increase the process difficulty like PECVD and PVD step coverage. To balance the two contradictory factors, we fixed the via diameter of 70um and the silicon thickness of 120um. And to reduce the pad stress in the whole process and reliability test, the partial fill of metal and solder mask film were proposed.

Finite element simulation

Generally speaking, the CTE mismatch between different materials of each layer will incite a high stress. In both high or low AR TSV structure used in CIS products, when there is much glue upon the pad, the local stress may drag the pad up and bring about the pad delamination. To solve this problem, making a bubble inside the via for reducing stress is proposed. To evaluate the function of this design, we used finite element simulation tool to simulate the pad stress during TC test between two types of design that inside the via with or without a bubble.

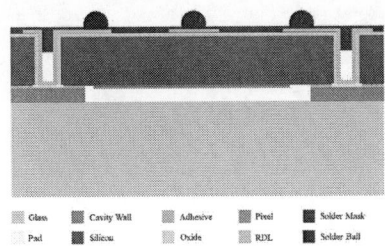

Figure 2: schematic view of the CIS package design

The strip model developed in ABAQUS was shown in Fig. 3. From the top of via, there were solder mask layer, RDL layer, pad, cavity wall and glass. There dimensions were set the same as the designed structure. To simplify the model, we applied the symmetrical constraints on the relevant facets, and the element type was set in C3D8R and C3D4. The TC-B test condition from -55°C to 125°C was loaded.

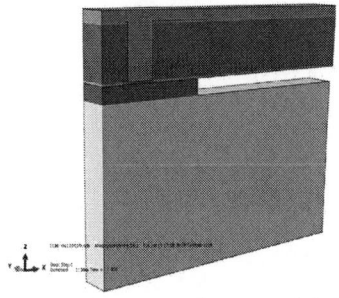

Figure 3: finite element simulation model of CIS chip

Fig. 3b) and d) presented the max S33 stress distribution inside the TSV sidewall and on the metal Pad. According to the simulation results shown in table I, the bubble inside the via could effectively reduce the max S33 stress upon the Pad for more than 60%. Since the CTE of solder mask material and silicon were 58 and 2.8, it could be predicted that the thermal stress caused by CTE mismatch between solder mask glue and RDL would be a high risk for the fragile pad. For a higher reliability concern, it's necessary to ensure a controllable bubble inside the TSV.

PROCESS DEVELOPMENT

The whole process flow for this package structure was shown in Fig. 4. At first, we formed the cavity wall (CV) on the glass which provides the cavity for the sensor area. Then, the CIS wafer in a diameter of 300mm was bonded

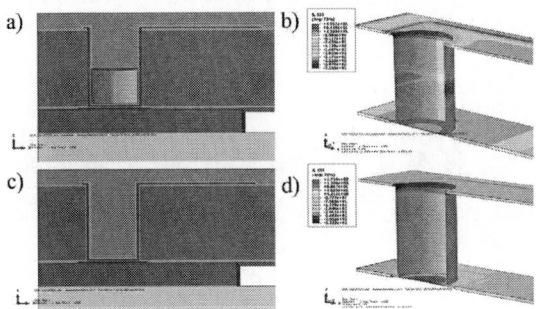

Figure 4: simulation results of the Pad S33 stress in TC-B reliability test condition: a) structure of SMF with bubble; b) Simulation stress diagram of Structure with TSV bubble; c) structure of SMF without bubble; d) Simulation stress diagram of Structure without TSV bubble;

TABLE I. Result of the Pad S33 stress of Simulation

SMF bubble	Largest S33 stress in Pad	Largest stress location
Yes	64.38Mpa	Edge
No	172.6Mpa	Center

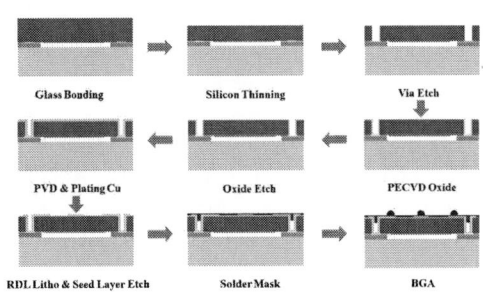

Figure 5: process flow of WLCSP with TSV assembly

together with the glass. After bonding, the silicon wafer was thinning to 120um by grinding and dry etched.

After a step of positive photo lithography, the vertical TSV were formed using deep reactive ion etching (DRIE) by Bosch process. To remove the polymer inside the via wall and bottom induced in via etching, the single wafer ultrasonic cleaning was introduced before the oxide deposition process. Since the restrict temperature control within 200°C for image sensor, in this study we chose plasma-enhanced chemical vapor deposition (PECVD) process in 150°C to deposit silicon oxide as an isolation layer. The thickness of oxide was 3um on the surface and the step coverage in the bottom corner of the via was about 20%. To remove the silicon oxide formed in BEOL and last PECVD process upon the Al/Cu pad, a directly oxide etch using dry reactive ion etching was applied.

After that the backside redistribution layer (RDL) was formed in a special way. At first, a seed layer contained 0.15um Ti and 1um Cu was deposited by sputtering and the Cu layer was thicken to 3um by electroplating. Then we developed a lithography method to protect the metal inside the via and form RDL pattern on the die surface. By etching the Ti/Cu at exposed area and removing the photoresist, the RDL was formed. And electroless-plating nickel (Ni) and gold (Au) processes were followed showed in Fig. 6(a). Instead of bottom to up copper electroplating, a partial filled plating inside the via took advantages in lower cost and smaller residual stress.

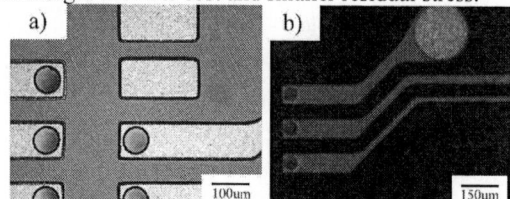

Figure 6: a) RDL after electroless Ni & Au; b) Die surface view after solder mask process

To protect the RDL trace, the negative solder mask film was achieved by spin-coating on wafer with opening of under-bump metallization (UBM) presented in Fig. 6(b). By adjusting the process parameters, we successfully realized the controllable bubble design inside the via. The solder balls (BGA) were formed by stencil printing of solder paste and 260°C reflow processes. Finally, the wafer was dicing to each die using 70um diamond blade for silicon grooving and 40um for glass dicing.

RESULT AND DISCUSSION

The cross-section of single via was showed in Fig. 7. The electrical signal was transmitted from pad in the front of chip to the solder ball in the backside by RDL. There was a bubble inside the TSV and the solder mask thickness upon the via was controlled more than 30um to ensure enough mechanical performance and acted as vaper-diffusion barrier.

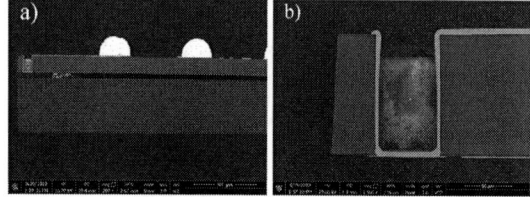

Figure 7: a) SEM cross-section image of the package; b) SEM cross-section image of the TSV;

To evaluate the TSV-bubble design, the 45ea samples were put into pre-condition and TC-B (55°C to 125°C) test. There were no abnormal occasion after 1000 cycles by visual check and 10ea were checked by cross-section. Fig.8(a) showed the cross-section of via bottom after TC 1000, it was clear that there is no any crack and deformation. By comparison, a pad deformation was found after the solder mask process using via filling method presented in Fig.(b), not to mention the strict reliability test. All in all, a bubble inside the TSV design could improve the reliability performance for pad and coincided with the simulation result.

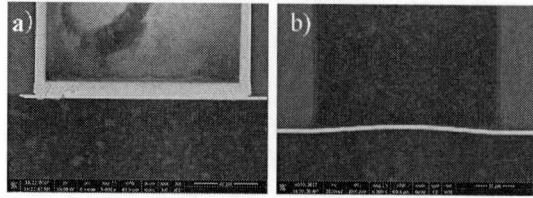

Figure 8: SEM cross-section of the via a) After TCT 1000 cycle; b) After via filling process

CONCLUSION

In this paper, the design and development of 3D WLCSP for image sensor using vertical via technology was presented. The main process like via etch, RDL lithography and solder mask formation were also introduced. To meet the higher reliability requirement, a TSV-bubble design was evaluated by finite element simulation. Both simulation and experience results verified the conclusion that a bubble inside the via can effectively reduce the pad stress and avoid crack and deformation in strict TC reliability test.

ACKNOWLEDGEMENTS

The authors would like to express their gratitude to the support from the RD and PE team within Huatian Group.

REFERENCES

[1] Moore, Gordon E. "Cramming more components onto integrated circuits." (1965): 114-117.

[2] Charbonnier, J., Henry, D., Jacquet, F., Aventurier, B., Brunet-Manquat, C., Enyedi, G., ... & Sillon, N. (2008, September). Wafer level packaging technology development for CMOS image sensors using Through Silicon Vias. In 2008 2nd Electronics System-Integration Technology Conference (pp. 141-148). IEEE.

[3] Hsien Chung, Ching-Yu Ni, Che-Min Tu, Yu-Yao Chang, Yao-Te Haung et al., 2010 Proceedings 60th Electronic Components and Technology Conference (ECTC), Las Vegas, 2010, pp. 297-302.

[4] Liu Chen, Chen Jie, Mairui Huang, Mark Huang, 2016 China Semiconductor Technology International Conference (CSTIC), Shanghai, 2016, pp. 1-4.

[5] Ping, W., Bangxu, W., Jun, L., Huang, M., & Lai, C. (2015, March). TSV fabrication for image sensor packaging. In 2015 China Semiconductor Technology International Conference (pp. 1-4). IEEE.

Laser-Based Full Cut Dicing Evaluations for Thin Si wafers

Peter Dijkstra, Jeroen van Borkulo, Richard van der Stam

ASM Laser Separation International B.V.

Beuningen, The Netherlands

E-mail: pdijkstra@alsi.asmpt.com

Abstract—Over the last years, singulation of thin semiconductor wafers with (ultra) low-K top layer has become a challenge in the production process of integrated circuits. The traditional blade dicing process is encountering serious yield issues. These issues can be addressed by applying a laser grooving process prior to the blade dicing, which is the process of reference nowadays. However, as wafers are becoming thinner, this process flow is not providing the yield and cost required. This paper will discuss the results of a study done on several multi beam laser ablation technologies on thin Si wafers and describe the pro's and con's for each of them.

Index Terms—3D packaging, dicing, die strength, laser, thin wafer, multiple beams, diffractive optics

I. INTRODUCTION

Full cut separation of semiconductor substrates has been the domain of a saw blade process for several decades [1]. Advances in the technology node, which require the use of low-K materials, have introduced laser technology in the IC wafer separation process [1]. The recent trend [2] to also reduce the substrate thickness drives the industry towards different separation technologies. We can distinguish two main areas which drive the substrate reduction [3]. The first one is the transition from 2-D to 3-D IC-packages that require thinner wafers (<50 um), which is already commonplace for memory and mobile applications. The second one is substrate thickness reduction to manage the heat dissipation and reduce internal resistance, which is typical for power devices such as IGBTs and MOSFETs.

The mechanical forces of the blade dicing process pose a serious challenge for thin wafer dicing [4, 5]. Wafers with low-K layers, that first require a laser grooving step [1], are even more challenging as the groove weakens the wafers resulting in a serious risk of wafer breakage. The semiconductor industry is still looking for a separation solution which provides high yield and productivity at low cost.

Section II starts by describing the method how die strength is measured. This is followed by Section III which describes the standard multi beam dicing process, benchmarked against the typical PoR's for thin wafer separation such as blade saw and Stealth dicing. An introduction and description of the processes developed for thin wafer dicing can be found in sections IV to VI. These processes are V-DOE, multi beam short pulse and nanosecond multi beam in combination with plasma etching. Each technology is reviewed for the die strength achieved and the overall dicing quality. Section VII compares the productivity for the different technologies. Finally, the overall comparison between the separation technologies and the conclusions are presented in section VIII.

II. DIE STRENGTH

All results mentioned in this paper have been measured on a 3 or 4-point die strength tool, both active structure up (SU) and structure down (SD) (see figure 1). The part of the sample that is facing down is subject to tensile stresses, which are generally responsible for the failure of the dies. To obtain good position accuracy of the support and loading edges while preventing friction effects, a measurement jig with an air bearing was employed.

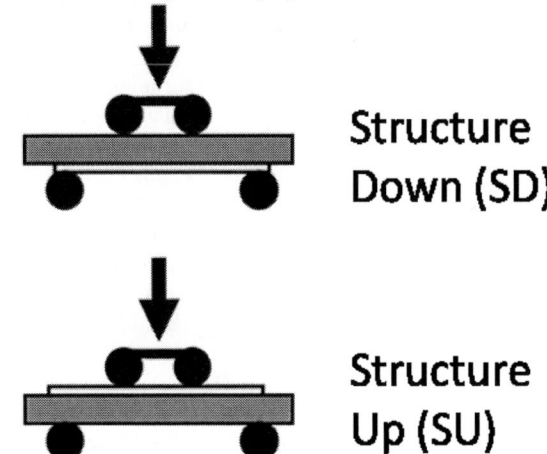

Figure 1. 4-point bending test for die strength measurements

III. MULTI BEAM LASER DICING

ASMPT's multi beam laser ablation dicing technology utilizes the patented multibeam technology. This technology has proven to achieve high yield and reliability, high productivity, a narrow kerf width, and a cost effective dicing solution at various assembly and packaging companies [6]. All of these criteria are advantages over the traditional separation technique; the blade dicing technique that is currently used in the industry. However, there is one criterion at which the standard multi beam dicing technology is not fulfilling the required specification which is the die strength.

978-1-7281-6559-2/20 $31.00 © 2020 IEEE

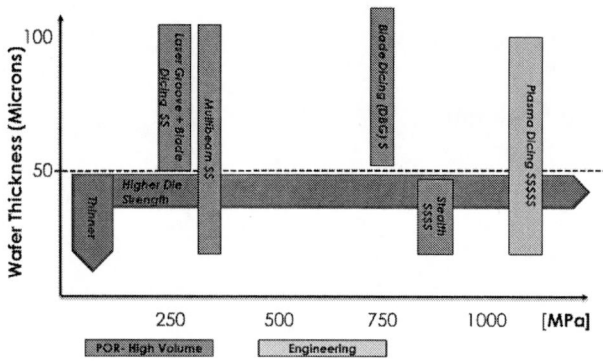

Figure 2. Die strength for various thin Si dicing technologies.

As shown in figure 2, the die strength of the multi beam dicing technology is on par with low-K grooving followed by blade dicing. But it is significantly lower compared to blade dicing or stealth dicing. In general, the target die strength value which a wafer dicing technology should achieve for thin Si wafers is 800 MPa or higher.

In the paper presented at ECTC in 2017 [7] we analyzed the root cause for the reduced die strength. It was found that during the rapid cooldown of the Si melt once the laser beams have passed by, various defects are formed. These defects range from voids and cracks in the micrometer range, to lattice defects in crystalline Si with otherwise the same orientation as the nearly defect-free monocrystalline substrate. When these defects are put under tension during die strength testing, we believe that they can cause early cracking of the die and therefore a reduction of the die strength.

To be able to address the issue of the reduced die strength, we investigated three different multi beam laser dicing technologies, each with their own characteristics.

IV. V-DOE LASER DICING

During studies performed last year [7] we found that by using low power laser beams along the edge of the dicing kerf, a laser annealing process can be established which partially removes the Heat Affected Zone (HAZ). Performing this type of laser annealing process with a single beam laser process requires a significant amount of time.

Figure 3. V-Shape spot pattern showing the central forward positioned spots which are used to dice through the wafer and the outer spots which clean the edges of the die.

Through using the multi beam technology a multi beam configuration, known as V-DOE, is established which allows both the dicing to take place as well as the cleaning/removal of the HAZ which the full cut process creates (see figure 3). When comparing the die strength results obtained from the multi beam and V-DOE processes on a 75-um thick polished Si wafer which was taped on a 20 um DAF tape a factor two improvement in the front side (structure down) die strength is found in favor of the V-DOE process. The die strength for the SU situation is already high due to the dicing parameters chosen for the dice through process. Therefore, the laser cleaning is only required on the top side of the die sidewall as is shown in figure 4.

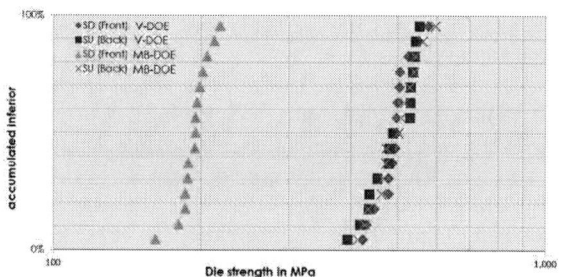

Figure 4. Die strength results SU & SD for standard multi beam (MB) laser dicing process and V-DOE process for a 75 um Si wafer on 20 um DAF measured on a 4-point bending tool.

Although this V-DOE process has shown a significant improvement in the die strength (factor 2x) and has been successfully qualified at several customers, it could not comply to the die strength established by PoR's such as saw blade or stealth dicing. This means that if customers maintain the adagio that a new wafer separation process should be able to meet the current die break strength level of a PoR, the process is not meeting this requirement.

V. SHORT PULSE DICING

As reported in several papers [8-11] short pulse lasers (femtosecond regime) are capable to dice wafer material with a minimized HAZ and therefore maintaining a good die strength. For many years, the laser power and pulse energy available from these lasers were insufficient to achieve the required throughput. In recent years, the power and pulse energy of short pulse lasers has significantly increased which motivated us to combine the multi beam dicing concept with these lasers. We performed a large range of dicing experiments in which we compared the die strength and dicing quality of a short pulse laser against more conventional nanosecond UV lasers.

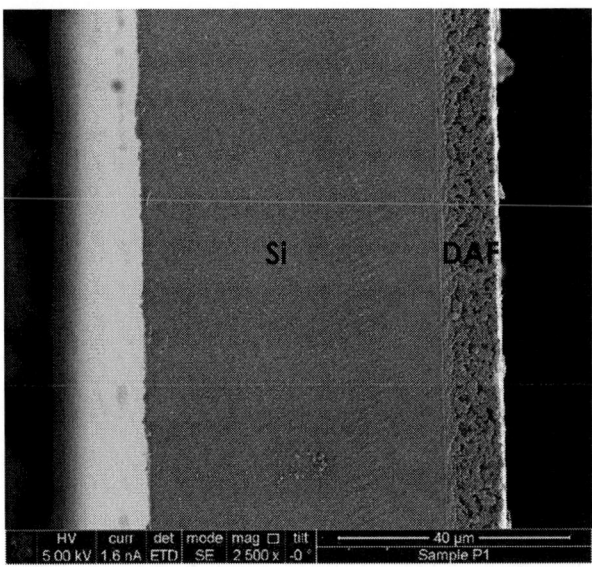

Figure 5. SEM micrograph of the sidewall dicing quality of a multi beam femtosecond process.

As shown in figure 5, the sidewall quality of the multi beam femtosecond laser shows a very smooth and well-controlled structure of both the Si and the Die Attach Foil (DAF).

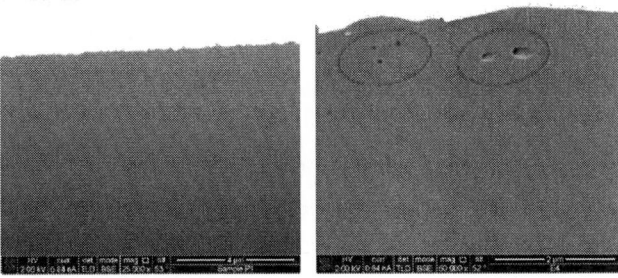

Figure 6. SEM micrographs of cross-sections of the sidewalls from femtosecond (left) and nanosecond (right) dicing processes.

Cross-section of sidewalls of laser-diced dies can be found in figure 6. No damage is visible by scanning electron microscopy for the femtosecond process while the nanosecond process shows voids. A reduction of the damage below the sidewall of the dies can result in an improved mechanical strength.

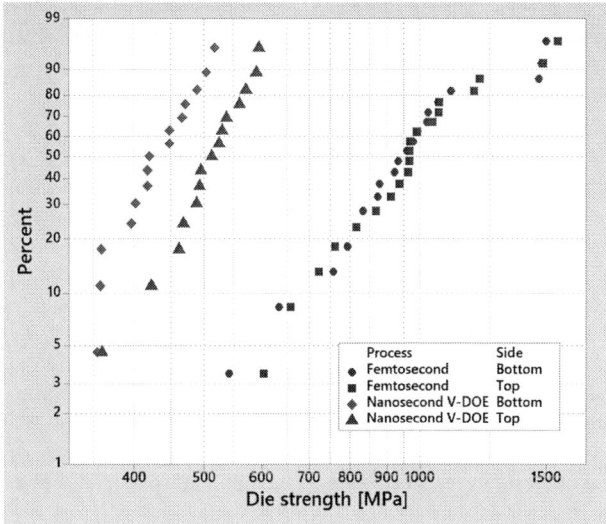

Figure 7. Comparison between the die strength (front side and backside) of nanosecond and femtosecond multi beam dicing.

Figure 7 demonstrates the die strength achieved on a 50 um polished Si wafer but the trend is similar for wafers with lithography active structures. Both the front side and backside die strength of the femtosecond dicing process are significantly improved compared to the nanosecond process.

VI. LASER AND PLASMA ETCHING

While the short pulse process is using cold ablation to prevent the formation of a HAZ and therefore maintains a good die strength, an alternative approach is to accept that it is created and remove it later on. ASMPT holds patents for this process [6] and for compound materials this process is applied in high volume production since 2006. For thin Si wafers, a similar process flow can be used in which the removal of the HAZ is done using a dry plasma etch. The main benefit of this process flow is that the strong point of the laser is used, the ability to dice through a stack of multiple layers (oxides, low-K, polymers, metal and Si), while the plasma etch process is used to etch away the 2-3 um wide HAZ which is created during the dicing process.

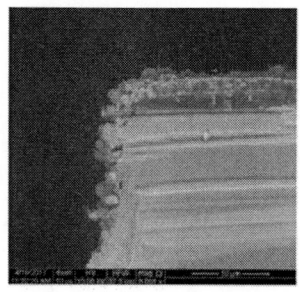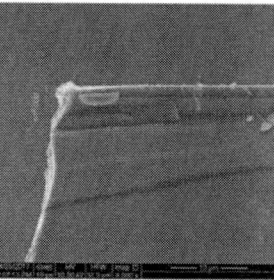

Figure 8. SEM micrographs of cross-sections of sidewalls before (left) and after plasma etching (right).

In figure 8, cross-sections are shown of the sidewall and top side of the die prior to and after the plasma etch process. As clearly shown, the plasma etch process removes the 2-3 um HAZ effectively and results in a HAZ-free sidewall. In the process flow, no photoresist is applied on the wafer prior to the plasma process. The standard polymer-based coating material used during the laser dicing step will also be used to protect the wafer surface during the etching process. The etching time will be shorter than the laser dicing allowing both process steps to take place in parallel without impacting the overall productivity of the laser dicing process.
The die strength recovery of this process outperforms any of the other dicing technologies reviewed in this paper as shown in figure 9.

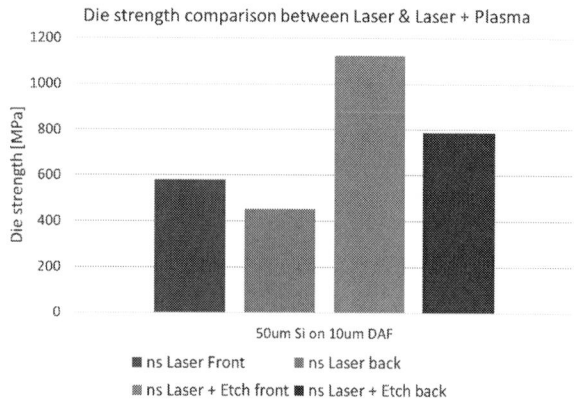

Figure 9. Die strength comparison between nanosecond laser dicing and nanosecond + plasma etch.

VII. PRODUCTIVITY

Apart from the fact that the wafer dicing technology has to meet the die strength and quality requirements, it also needs to be cost effective and therefore maintain a certain productivity.
Figure 10 provides an overview of the approximate productivity for the various dicing technologies reviewed.
A significant reduction in productivity can be observed for the femtosecond dicing process for larger wafer thicknesses (> 40 um). Further upscaling is required to obtain a competitive throughput for these wafer thicknesses. As the laser power development for short pulse lasers is increasing the plan is to combine ASMPT's patent for multi lane dicing

with the multi beam dicing process. This allows to utilize the surplus of power and convert it into productivity.

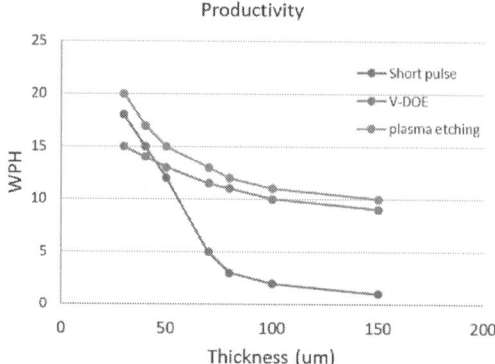

Figure 10. Productivity vs thickness for the various wafer separation technologies.

The benefit of the nanosecond process is that a significant part of the material removal process is through melt ejection, which is a more efficient way of removing material. Therefore, also for thicker substrates, a relatively high productivity can be achieved especially when comparing it to the short pulse femtosecond process.

VIII. CONCLUSIONS

All three different dicing technologies that were developed fill different market segments. As shown in figure 11, the V-DOE process achieves a die strength twice above the standard multi beam laser full cut dicing or laser grooving process. The V-DOE as reported [9] fulfills the qualification and reliability criteria and provides a high productivity with the minimum capex required. However, if the die strength achieved with this process does not meet the requirement of the product, two alternative dicing technologies can be used. First, the short pulse process which achieves a good die strength without the requirement of a post-process step. However the wafer thickness regime in which it can currently be applied is limited to approx. 40 um. For memory wafers, the volume is already below 40 um wafer thickness. Having a dicing technology which can cover this and be able to dice the increased active layer thickness (specifically for 3D NAND) while maintaining sufficient die strength, gives it a strong potential. However, if the application requires thicker wafers than 40 um which also have a multi-layer stack of materials, the nanosecond laser dicing process in combination with plasma etching provides a good productivity and die strength. This process flow does require additional capex however this will be significantly lower (factor 5 to 6) compared to the plasma dicing systems available in the market which can only dice Si.

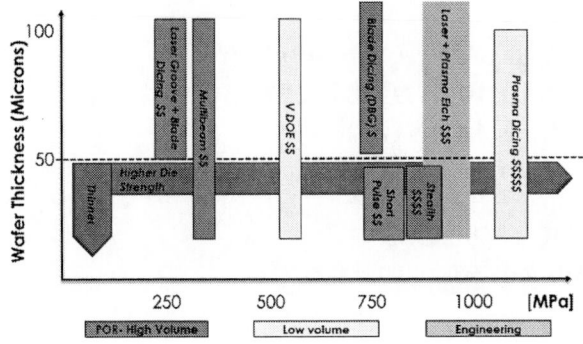

Figure 11. Die strength for various Thin Si dicing technologies including V-DOE, short pulse and laser plus plasma etch.

We therefore believe that with the three wafer separation technologies, a broad spectrum of the thin wafer dicing market can be served.

REFERENCES

[1] W.-S. Lei, A. Kumar, and R. Yalamanchili, "Die singulation technologies for advanced packaging: A critical review," *Journal of Vacuum Science and Technology B*, vol. 30, no. 4, p. 040801, 2012.

[2] "Thin wafer processing and dicing equipment market 2016," Yole Development, Tech. Rep., 2016.

[3] J. Burghartz, Ed., *Ultra-thin Chip Technology and Applications*. Springer, 2011.

[4] M. Hong, Q. Xie, K. Tiaw, and T. Chong, "Laser singulation of thin wafers & difficult processed substrates:A niche area over saw dicing," *Journal of Laser Micro/Nanoengineering*, vol. 1, no. 1, pp. 84–88, 2006.

[5] L. K. Shiuann, "Laser as a future direction for wafer dicing: Parametric study and quality assessment," in *Electronics Manufacturing and Technology, 31st International Conference on*, Nov 2006, pp. 506–509.

[6] R. Hendriks, J. van Borkulo, and M. Mueller, "Paradigm shift in compound semiconductor production since the introduction of laser dicing," in *Proceedings of Compound Semiconductor Manufacturing Technology Conference, Tampa, FL., 18-21 May 2009*.

[7] J. van Borkulo, Paul Verburg and Richard van der Stam, "Multi Beam Full Cut Dicing of Thin Si IC wafers," in *Electronic Components and Technology Conference,2017 ECTC 2017. 67th*, May 2017

[8] M. Domke, B. Egle, S. Stroj, M. Bodea, E. Schwarz, and G. Fasching, "Ultrafast-laser dicing of thin silicon wafers: strategies to improve front- and backside breaking strength," *Applied Physics A*, vol. 123, no. 12, p. 746, Nov 2017.

[9] O. Haupt, F. Siegel, A. Schoonderbeek, L. Richter, R. Kling, and A. Ostendorf, "Laser dicing of silicon: Comparison of ablation mechanisms with a novel technology of thermally induced stress," *Journal of Laser Micro/Nanoengineering*, vol. 3, pp. 135–140, 2008.

[10] D. S. Finn, Z. Lin, J. Kleinert, M. J. Darwin, and H. Zhang, "Study of die break strength and heat-affected zone for laser processing of thin silicon wafers," *Journal of Laser Applications*, vol. 27, no. 3, pp. –, 2015.

[11] N. Sudani, K. Venkatakrishnan, and B. Tan, "Laser singulation of thin wafer: Die strength and surface roughness analysis of 80 μm silicon dice," *Optics and Lasers in Engineering*, vol. 47, no. 7–8, pp. 850 – 854, 2009.

A SINGLE-LAYER SOLUTION WITH LASER DEBONDING TECHNOLOGY FOR TEMPORARY BOND/DEBONDING APPLICATIONS IN WAFER-LEVEL PACKAGING

Xiao Liu*, Lisa Kirchner, Luke Prenger, Wenkai Cheng, Rama Puligadda

Brewer Science, Inc.

2401 Brewer Drive, Rolla, MO 65401, USA

*Corresponding Author's Email: xliu@brewerscience.com

ABSTRACT

Performance and cost are the two main driving forces for high demand of new advanced packaging technologies. It's not only OSATs and packaging houses that investigate advanced packaging systems anymore. Foundries and IDMs now also recognize that performance of devices can be significantly improved by advanced packaging technologies, which facilitate the evolution to next generation of process nodes, especially in the post–Moore's Law era. In this paper, several different innovative materials are presented as cost-efficient solutions to enable advanced wafer-level packaging applications. All of these materials are single-component, but with multiple functionalities, to facilitate wafer processes. The laser-sensitive thermoplastics perform as both bonding and laser debonding materials, and have been demonstrated in applications like wafer thinning, chip mounting, wafer flip, and epoxy compound wafer handling. The materials show good thermal stability (>300ºC), high adhesion to various substrates, good response to different UV laser sources, and ease of cleaning with commonly used industry organic solvents. On other hand, a laser-sensitive thermosetting material is presented as well. In addition to the features mentioned above, the material can be wet-cleaned after debonding, which is a very unique property compared to other thermosetting materials, where a dry-etching process is required for cured material removal. A typical application for this material is to handle highly stressed substrates, like epoxy compound wafers (EMCs) such as in fan-out wafer-level packaging (FOWLP). With the unique combination of properties in these materials, a versatile toolbox is available to enable wafer processes in different advanced packaging applications.

INTRODUCTION

For decades, the semiconductor industry has been guided by Moore's Law to move forward from one node to the next through the ideology of decreasing gate pitch in an orderly way. This step to achieve smaller, higher-density, better-performing and more cost-effective devices has never come to an end even though Moore's Law is being challenged to reach its limit in recent years [1]. To compensate for the difficulty of continuing Moore's Law in the traditional fashion, various wafer-level packaging techniques have been widely used since 2001 due to their capability of size reduction [2,3,4]. In the area of wafer-level packaging, ultra-thin wafers need to be handled, which makes having high adhesion and thermally stable materials critical to bond thicker substrates for thin wafer handling. The market currently utilizes temporary bonding and debonding materials [5]. This method requires coating and baking for both bonding and release materials. Due to two materials typically being used it will become more cost effective if there is a single material that can fulfill both bonding and debonding functions [6]. In this paper, several innovative materials are presented to meet these market needs, and most importantly, all of them are single-layer solutions.

LASER-SENSITIVE THERMOPLASTIC

Laser debonding technology can provide several advantages over other debonding methods with regard to high throughput, low-stress release, and force-free separation from a carrier at room temperature [7]. It has demonstrated potential for high-volume production and been actively adopted in various advanced packaging platforms. Currently, most of laser-debonding materials in temporary bonding/debonding applications consist of two-layer structure, one as bonding adhesive, and the other as laser release layer. In order to further reduce cost and increase throughput in high volume production, it makes sense to combine the two functions in a single layer. Based on this concept, Material A was invented as both bonding and laser release material.

Table 1. Property Summary of Material A

	$T_{d\,2\%,N2}$ (°C)	$T_{d\,2\%,air}$ (°C)	T_g (°C)	Complex Viscosity at 210°C (Pa·s)
Material A	334	339	101.6	337

The properties of Material A were summarized in Table 1. Material A exhibits very good thermal stability with 334ºC and 339ºC for 2% weight loss from TGA in nitrogen and air, respectively. In addition, the moderate T_g (101.6ºC) makes it possible to process this material at commonly used temperatures for wafer bonding processes. The material shows a complex viscosity of 337 Pa·s at 210°C (Figure 1), which indicates that a good bond line can be obtained at this temperature during the wafer bonding process.

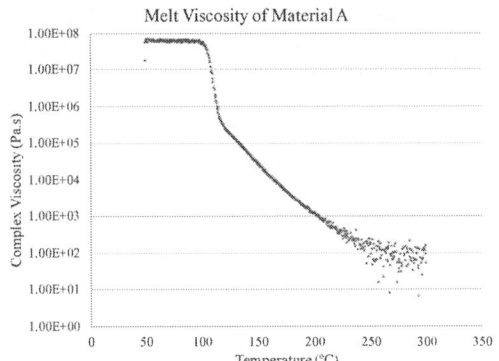

Figure 1: Melt viscosity of Material A.

In addition, Material A exhibits high absorbance in the UV range, especially at the wavelength of commonly used laser sources. Table 2 provides the UV percent transmittance at different wavelength with 20-μm film thickness. The great UV absorbance can not only ensure good laser debonding performance, but also protects the device wafer surface by completely blocking laser energy.

Table 2. UV Percent Transmittance of Material A

Wavelength (nm)	248	308	343	355
Percent Transmittance (%)	0	0	0	0.28

Furthermore, good adhesion is another critical factor in order to secure the wafer bond line, especially for substrates with high stress. Table 3 shows the adhesion of Material A on different substrates. The data indicate that Material A has high adhesion to various surfaces.

Table 3. Adhesion of Material A

	Si	Glass	Si_3N_4	SiOx
Material A	> 39 psi	> 31 psi	> 27 psi	37 psi

A bonded wafer pair on 8" blank Si and glass wafers was also prepared by using Material A with a commonly used thermal compression bonding method utilizing the following conditions: 210°C at 2000 N for 3 minutes. No defects or voids were detected after bonding. The wafer pair was then subjected to grind down to 50 μm, followed by thermal simulation at 250°C for 30 minutes. Again, no defects waere observed. Figure 2 shows the MicroPROF® FRT images of the thinned wafer after thermal simulation. Afterwards, the wafer was successfully debonded with a

308-nm excimer laser with 230 mJ/cm_2 debonding dose. After laser debond, the thinned device wafer can be cleaned using a spray clean method with cyclopentanone within 2 minutes to remove the 22-μm thick layer of Material A.

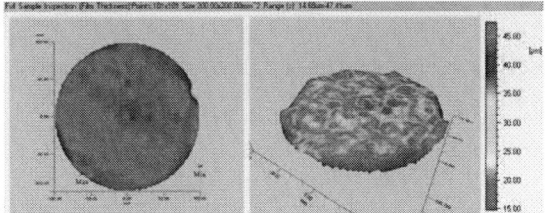

Figure 2: MicroPROF® FRT images of thinned Si wafer with Material A.

LASER-SENSITIVE THERMOSET

Based on the same concept as in laser-sensitive thermoplastic, one extra property was added to create a laser-sensitive thermoset material. Material B was invented as a new multifunctional bonding material. The properties of Material B are summarized in Table 4. In addition to good thermal stability and moderate bonding temperature, Material B can be thermally cured at 250°C (Figure 3), which can further secure the bond line by reducing the risk of material bleeding and squeeze out at high temperature because of high flowability of a thermoplastic-based bonding material. The cured Material B will also have high resistance to deformation caused by externally applied forces, such as in processes like die-to-wafer bonding.

Table 4. Property Summary of Material B

	$T_{d\,2\%,N2}$ (°C)	Weight loss at 250°C for 1 hr	T_g (°C)	Complex Viscosity at 145°C (Pa·s)
Material B	370	0.78%	25	883

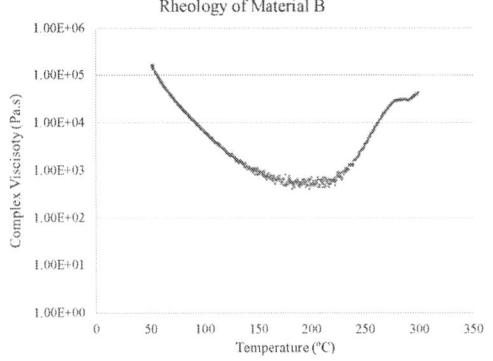

Figure 3. Melt Viscosity of Material B

Another unique property of Material B is that the material can be wet-cleaned even after thermal cure. Unlike most thermosetting materials, the crosslinking reaction occurred in Material B is designed to be reversible and it is possible to be removed with a milder wet-cleaning process instead of dry etching or by mechanically grinding the material off. Therefore, the thermosetting Material B can be used as temporary bonding/debonding material with more benefits compared to the standard thermoplastic materials.

Similar to Material A, Material B can also provide good adhesion to various substrates as shown in Table 5, especially for EMC wafers, one of the most stressed wafer types to handle due to the high warpage. Hence, Material B is promising to handle different wafer types as needed.

Table 5. Adhesion of Material B

	Si	Glass	EMC	SiOx
Material B	28 psi	23 psi	15 psi	31 psi

Wafer bonding/debonding performance for Material B was also demonstrated on 8" blank Si and glass wafers, with commonly used thermal compression bond conditions, 145°C at 2200 N for 3 minutes, followed by curing at 250°C for 10 minutes. No defects or voids were detected after bonding. The wafer pair was then subjected to thermal simulation at 250°C for 1 hour, and no defects were observed as shown in Figure 4. Afterwards, the wafer was successfully debonded in 71 seconds with a 308-nm excimer laser or 355-nm solid state laser with 275 mJ/cm2 and 4 W debonding energy, respectively. In the end, the cured Material B on the device wafer can be cleaned by a specially developed remover within 7 minutes using a soaking process.

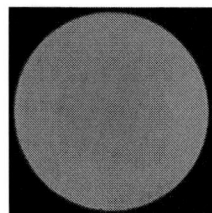 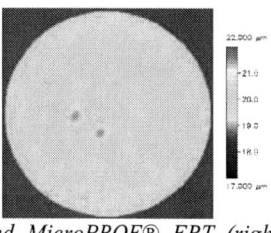

Figure 4. Photo (left) and MicroPROF® FRT (right) images of bonded wafer pair with Material B after 250°C for 1 hour.

DRY FILM OF MATERIAL A

A dry film made from Material A was demonstrated with dimensions of 250 mm width and 21 μm, 45 μm, and 75 μm thicknesses. Due to the great mechanical properties, a transparent, flexible, and tough dry film can be fabricated from Material A as shown in Figure 5 (left). The major advantage for the application of dry film

focuses on temporary bonding/debonding handling on panel level with lamination technique, in addition to wafer processes. The bonding performance was demonstrated on 8" Si and glass wafers. The dry film was laminated on top of a glass wafer at 120°C and bonded to blank Si wafer using the same bonding conditions of Material A mentioned before. The photo image in Figure 5 (right) shows no defect or void after bonding and heat treatment, exhibiting the similar performance as in a spin coating process.

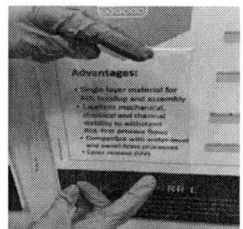 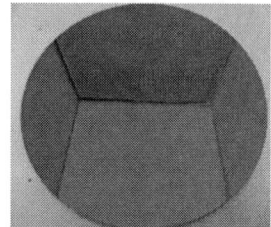

Figure 5. Photo image of Material A freestanding dry film (left) and bonded wafer pair (right).

SUMMARY

Two new multifunctional materials were introduced in this work as a way to enable future advanced wafer-level packaging applications. Both materials can be used as a bonding adhesive and laser debonding layer as single-layer solutions, and demonstrate good thermal stability and bonding/laser debonding performance. In addition, Material B can also be thermally cured to further secure the bond line to enable the use for more challenging processing conditions. Due to the unique chemistry and molecular design, the cured Material B can be wet-cleaned after being through wafer processing.

ACKNOWLEDGEMENTS

We would like to thank SUSS and Kingyoup for their support for this work.

REFERENCES

[1] R. Trichur, T. Flaim, *Chip Scale Review*, November-December 2015, pp38-41.

[2] Hedler, H.,T. Meyer, and B. Vasquez. 2004. US Patent 6,727,576, Filed on October 31, 2001, Patented on April 27, 2004.

[3] Tseng, CF et al., *Electronic Components and Technology Conference, 2016 IEEE 66th*, Las Vegas, May 31-June 3, 2016.

[4] Yu, C.H. et al., *Electronic Components and Technology Conference, 2018 IEEE 68th*, San Diego, May 29-June 1, 2018.

[5] Dongshun Bai, et al., *China Semiconductor Technology International Conference*, Shanghhai, March 11-12, 2018.

[6] Xiao et al., *China Semiconductor Technology International Conference*, Shanghai, March 18-19, 2019.

[7] R. Trichur, T. Flaim, *Chip Scale Review*, September-October 2016, pp12-18.

ETCHING POLYMER TECHNOLOGY FOR LARGE NUMBER OF SMALL VIA

Kyu Jin Lee, Moon Sang You, Gun Woo Kim, Hyun Chul Han, Sang Ki Ahn, Ken Lee
Woo Jae Jeong, Jeong Hyuk Ahn, Jeong Wook Moon, Jee Hyeon Hwang, Kwang Joo Lee
Simmtech Co., Ltd, 73, Sandan-ro, Heungdeok-gu, Cheongju, Chungcheongbuk-do, Korea
LG Chem Co., Ltd, 188 Munji-ro, Yuseong-gu, Daejeon, Korea
gwkim@simmtech.co.kr

ABSTRACT

In order to increase data capacity and improve signal processing, the IC package is changing to a module package in which an application processor, high bandwidth memory (HBM), and a controller are mounted together. Substrate of module package requires more and smaller vias than single package's substrate to handle high I/O of several chips, and also requires mechanical properties that can withstand structural deformation. Since the formation of a large amount of small vias excessively increases the laser drill cost. Photosensitive materials for low cost are bad in properties. In this study, we will introduce a new photosensitive material of high mechanical strength at low cost.

Keywords — *small via; etching type polymer; laser free substrate; photo via*

INTRODUCTION

Mobile substrates are required to reduce PCB areas due to increased battery capacity and expanded camera modules, as well as technical challenges to realize high performance and low-power together [1].

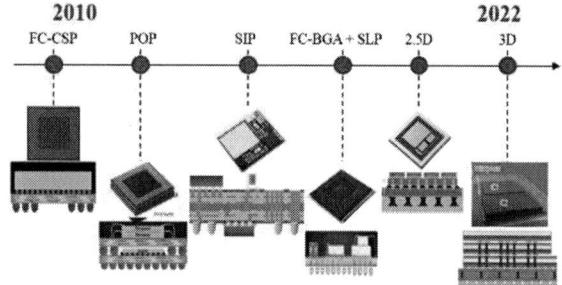

Figure 1: Package Trend

This led to an increase in chip field density within the same area and is changing from the existing package FC-CSP format to the Module PKG format such as SiP, Interposer (2.5D, 3D package), and FC-BGA + SLP products[2]. It is expected that these package types will be adopted and applied as better technologies for next generation products such as AI, Autonomous Driving (Automotive business), and IoT [3]. Meanwhile, module packages, unlike conventional single packages, require more I/O terminals on a substrate, with various chips loaded together on one substrate. This will require a large number of small via and fine pattern line compared to the same area of the existing Substrate. However, in general, many small via and fine patterns cause increased manufacturing costs. In particular, small via cannot collectively process insulation at a time due to the laser technique/facility characteristics, and each hole must be machined through several laser shot. This causes problems in laser processing time and facility investment. In contrast, the via open method using exposure / developing (polymer etching type) enables via open in a package, and it has the advantage of fixing manufacturing costs even when the number of via increases.

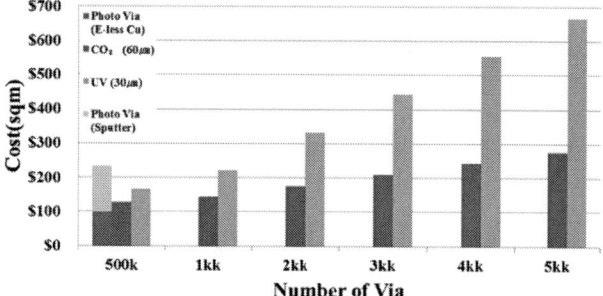

Figure 2: Cost Effect (Photo Via & Laser Drill)

In addition, the cost increase due to via size reduction is cheaper than laser processing, and the lithography method compared to laser machining method has a high alignment of pad to via, which makes it advantageous for multi-layer products such as SiP and FC-BGA. Therefore, the process suitability was assessed using the developer resin of the polymer etching type with the above advantages, and the possibility of replacing the laser technique was to be verified. Placing a variety of chips in a limited substrate space can cause various problems. Moreover, most polymer materials used in substrates differ in mechanical properties from silicon die, which causes the following problems when the substrate is mounted on the semiconductor [4, 5]

1) Bumping miss-align by dimension deformation of substrates
2) Damage: Substrates crack according to silicon load
3) Warpage behavior problem after room temperature and reflow
4) Reliability test fail

Thus, the mechanical properties of the developer resin were reinforced to evaluate the warpage behavior, dimensional deformation, and reliability items between layers of material from room temperature to high temperature.

The figure 3 below is a schematic diagram of photo via processing using etching resin.

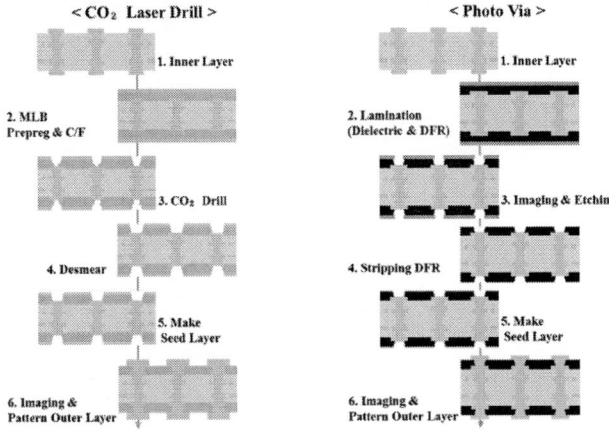

Figure 3: Schematic diagram(Photo Via & Laser Drill)

Unlike the conventional laser treatment method, the photo via processing method coats the photosensitive film as a mask on the etching resin and uncures the mask and the etching resin only in the area to be opened using the exposure process. The polymer etching solution dissolves the uncured mask film, and then penetrates to the lower etching resin to form vias by isotropically etching the etching resin [6]. Through mask stripping, polymer curing, seed layer forming, pattern electroplating, and seed etching processes on the etched resin area, copper wiring circuit for interlayer connection between insulation can be made.

The formed photo vias showed the same process capability level for the via size deviation and via vertical shape (Taper) compared to the laser drill, and it was confirmed that they were superior to the laser drill in the open capability of small vias.

Table[I] is shown that laser drill (CO_2 with PPG), problems such as hole separation and bottom residue occurred during 40 micro via open.

TABLE I
SIZE OF PHOTO VIA/LASER DRILL

Item	Via Size(μm)			
	Avg.	*Min*	*Max*	*CpK*
Photo Via	39.96	38.07	43.22	1.58
Laser Drill	41.01	37.42	44.92	0.77

On the other hand, the via open method using polymer etching resin showed that there was no unetchable polymer residue in the via bottom area that could cause reliability failure through EDX component analysis and X-Section analysis after plating. Figure 4 and Table [II] shows that.

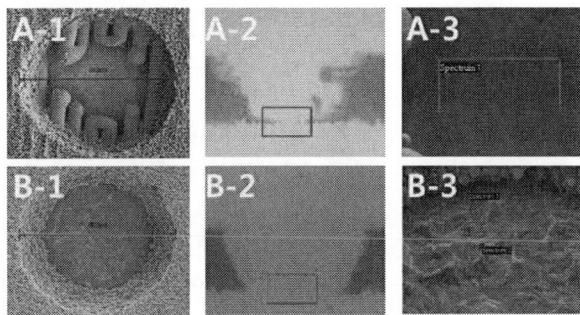

Figure 4: Photo Via & Laser Drill X-Section & EDX
A-1,2,3: Laser Drill Via / B-1,2,3 Photo Via

TABLE II
EDX FOR RESIDUE ON CU PAD

Item	C	O	Al	Mg	Si	Cu
On Cu Pad	8.72	2.52	-	-	-	85.76
Laser Drill	32.14	41.15	3.28	1.45	17.52	4.46
Photo Via	9.90	3.55	-	-	-	86.55

In addition, the under-cut risk at the corner edge of the via bottom was common. It was able to achieve minimum via size, shape and size deviation by using base resin and etching resin composition ratio, mask open and etching process capability. After forming the photo via, the adhesion characteristics of the seed layer were evaluated to form a pattern on the insulation.

The seed layer was sputter deposited, and the deposited metal was composed of a titanium layer for improving adhesion between the insulation and the metal and a copper layer for electrical wiring. The deposited seed layer was subjected to electroplating 20 micrometers and the adhesive properties were measured using a vertical UTM instrument. As a result of the measurement, the average peel strength of the seed layer and the insulation was 0.45kgf/cm².

And Table [III]&Figure 5 was confirmed that the line / space that can be implemented by this adhesive force can be equal to a fine pattern (SAP method) of up to 20pitch at 11/9.

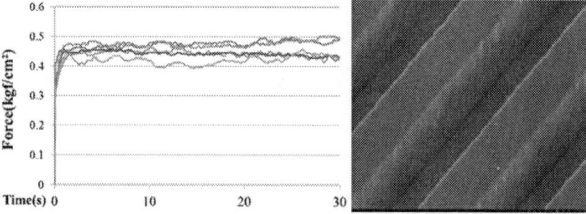

Figure 5: Graph Of Peel Strength & Trace Image(FE-SEM)

TABLE III. PATTERN SCALE ON PHOTO VIA DIELECTRIC

	Spec	Avg.	Min	Max	CpK
Line(μm)	10±5	11.56	10.16	13.01	1.67
Space(μm)	10±5	8.63	7.30	9.84	2.04

Table [IV] is a table comparing the physical properties of each material. Compared to prepreg impregnated with glass fiber, CTE and modulus were somewhat lacking, but the elongation characteristics were excellent. And when compared with ABF (Ajinomoto Build-up Film) and RCC (Resin Coated Copper) materials, which are composed only of filler, the mechanical properties (CTE, Modulus, Elongation) were excellent [7].

TABLE IV
PROPERTY OF DIELECTRIC MATERIAL

Item	Unit	Photo Via	ABF	RCC
CTE	α1	13	39	16
Tg	℃(TMA)	197	153	230
Y's Modulus	Gpa	11.7	5.0	15
Elongation	%	1.6	5.6	0.5
Df	10 GHz	0.011	0.018	0.006

It is believed that the crack, interlayer delamination, and bumping characteristics are good during PKG because of excellent physical properties such as toughness, handling stiffness, and dimensional deformation even with packaging a large number of silicon dice on a single substrate. Based on the evaluation result, reliability, warpage and structural evaluation were applied to the actual product. The applied product was 8L multi-layered product (Body Size 260 μm, Figure 7) and etching polymer was applied to outermost layer (1L, 8L) respectively.

Figure 6: X-Section image Of Photo Via Dielectric

As a result the structural analysis is shown Table [V], the overall thickness met the specification, but the outer layer thickness with etching polymer had a rather large deviation range, so it seemed necessary to improve polymer flowability and vacuum lamination conditions in the future. There were no specific factors, such as dimensional deformation and breakage between processes, and the black type material properties did not produce such defects as copper splashes, despite its low insulation thickness.

TABLE V
TOTAL THICKNESS OF ACTUAL PRODUCT APPLIED PHOTO VIA DIELECTRIC

Thickness	Target	Avg.	Max.	Min	Cpk
Total	240±30	249.28	251.52	246.12	4.37
Photo Via Dielectric #1	10±4	8.78	11.85	7.02	0.99
Photo Via Dielectric #2		11.43	13.17	9.66	1.15

In order to verify the interlayer adhesion reliability and electrical resistance characteristics of internal hole crack risk in high temperature and high humidity environment, the

Table [VI] reliability evaluation was performed and all the reliability specifications were satisfied. (Also, warpage is max. 2.3micron, so there will be no problem during package process. It is shown that Table [VII], Figure 9)

TABLE VI
RELIABILITY RESULTS

Item	Condition		Judgement (Pass/Fail)
HAST	▶ 125℃, 24hrs ▶ 130℃, 85%RH	288hrs	Pass
B-HAST	▶ 125℃, 6hrs ▶ 130℃, 85%RH, DC 3.6V	288hrs	Pass
MSL2a	▶ 60℃, 60%RH, 120hrs ▶ Peak Temp. 260℃, x3	144hrs	Pass
HTST	▶ 150℃, 1000hrs	1000hrs	Pass
TC	▶ 125℃, 24hrs ▶ -65℃ ↔ 150℃, 15min.,	1000 Cycles	Pass

TABLE VII
WARPAGE DATA OF ACTUAL PRODUCT

	Shape	Avg	Min	Max
Data	Cry	1.15	0.1	2.32

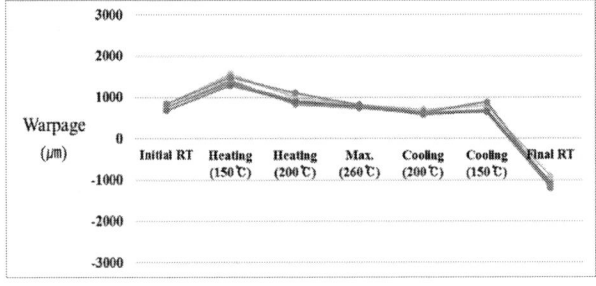

Figure 7: Warpage by Temperature

CONCLUSION

In this study, epoxy-resin-base resin was synthesized with etching resin to evaluate processability and reliability. Small vias can be fabricated up to 40 microns with a 15 micron isolation distance and 20 microns with circuitry. In addition, the filler content was able to achieve low CTE, high modulus. It has been confirmed that there is no problem with reliability and warpage characteristics.

Lastly, we are currently working on improving insulation distance deviation by improving the resin flow rate and the minimal viscosity. Also, it is necessary to develop process and resin for the improved adhesion characteristics of the seed layer on the insulation material and the small vias of 30microns or less.

REFERENCE

[1] Min Sung Kim ; Jong Tae Lee ; Dong Ju Jeon ; Eun Ju Jang ; Kyu Jin Lee ; Ken Lee, "Cavity Substrate Technology for SiP Application with Passive Components", CSTIC 2019, 2019.

[2] Anysilicon, "Semiconductor Packaging History and Trends", Feb 27, 2016

[3] ASM, "ASM Pacific Technology ADR 2017 – Result – Earnings Call Slides", 2017 Annual Result Anoouncemend, Mar 1, 2018

[4] Huili Fu, "Embedded Packaging Technology – The Opportunities and Challenges", SEMICON Taiwan, Sep, 2019

[5] Gil Sharon, " Understanding and Mitigating Chip-PackageBoard Interactions", Design for Reliability Conference, Mar 26, 2019

[6] LG CHEM, Ltd, "METHOD FOR MANUFACTURING INSULATING FILM AND MULTILAYERED PRINTED CIRCUIT BOARD",10-2040224, filed July 31, 2017, and issued October 29, 2019

[7] LG CHEM, "LG Line-Up Property", Sep,2018; Doosan, "Doosan Substrate Package Material Propery", Nov,2017; Ajinomoto,"ABF Update", June, 2018

A PROMISING EMBEDDED SILICON FAN-OUT WAFER LEVEL PACKAGE WITH LASER RELEASABLE TEMPORARY BONDING TECHNOLOGY

Chengqian Wang*,[1,2] Aibing Zhang,[1] Zhengfeng Li,[1] Shouwei Li,[1] Yang Li,[1] Yong Ji,[1] Xuefei Ming,[1] Daquan Yu*[2]

[1]The 58th Research Institute of China Electronics Technology Group Corporation
Wuxi, Jiangsu Province, China
[2]Xiamen University
Xiamen, Fujian Province, China
*Corresponding Author's Email: yudaquan@ime.ac.cn, chengqiankk@yeah.net

ABSTRACT

Fan-out wafer level package with silicon substrate is attracting more and more attention because of the compatible production process. In this paper, a new embedded silicon fan-out package scheme was proposed. SiO_2 stop layer deposited by PECVD and 355 nm laser releasable temporary bonding technology were applied to solve silicon etching uniformity, footing and grass issues. With 12% over-etching, there is no footing and grass found after silicon cavity etching through optical microscopy and SEM inspections. The new scheme can achieve thinner package and better heat dissipation performance owing to introducing of temporary bonding.

INTRODUCTION

As Moore's law is gradually closing physical limits, More-than-Moore attracts more attention. It aims to improve system performance not only by decreasing transistor size but also depending on system optimization design and advanced packaging technologies. In advanced packaging technologies, Fan-Out Wafer Level Package (FOWLP) has emerged as a successful technology to satisfy this goal. FOWLP can miniaturize package size, reduce power consumption, increase integration density and decrease manufacturing cost, which correspondingly meets the trends and requirements of 5G, internet of things (IoT), artificial intelligence (AI), smart wearables and automotive applications [1, 2].

There are mainly two kinds of FOWLP schemes according to the reconstruct methods, molding compound substrate and silicon substrate. Molding compound substrate is widely applied in mass production such as embedded wafer level ball grid array (eWLB), integrated fan out (InFO) and fan-out panel level package (FoPLP) [3-5]. Recent years, silicon substrate has been successfully developed for single die and multiple dies integration to realize fan-out package of application processors (AP), capacitive fingerprint sensors, mmWave, and 3D applications [2, 6, 7]. It exhibits some advantages like low cost, simple process and low warpage which is conducive

to large-scale commercial applications. However, further research of embedded silicon fan-out package found that some issues such as etching uniformity, grass, and footing resulted in unacceptable yield loss. To solve these problems, a new silicon substrate fan-out package scheme with laser releasable temporary bonding technology was introduced in this paper.

PROCESS DEVELOPMENT

Fig. 1 shows the silicon etching uniformity results. There are 18 points collected by depth measuring microscope from the left edge to the right edge of silicon wafer as indicated in the lower left. From the data we can see that the silicon etching depth is deeper in the wafer edge. The average depth is 70.22 μm for 70 μm target etching depth while it is 91.83 μm for 90 μm target etching depth. Meanwhile, the etching depth gaps between deepest and shallowest silicon etching depth are 12 μm for 70 μm target etching depth and 15 μm for 90 μm target etching depth respectively. More results indicate that the etching depth gap will become bigger as increasing of etching depth.

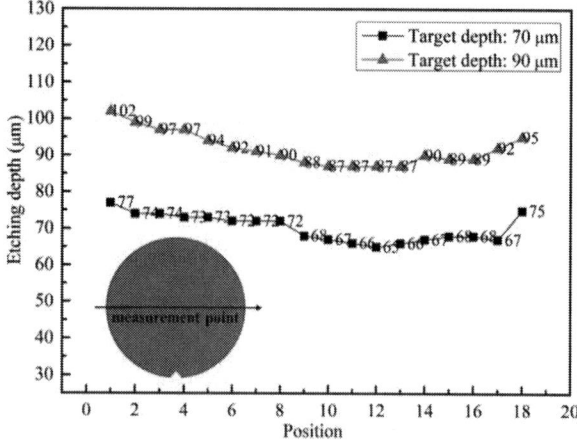

Fig. 1 Silicon etching uniformity data and schematic of measurement point in the lower left.

978-1-7281-6559-2/20 $31.00 © 2020 IEEE

The silicon etching uniformity is closely related to reconstruction coplanarity after die attaching and dry film filling processes. Poor etching uniformity will result in serious yield loss. The etching uniformity issue is mainly caused by array of gas outlet during dry etching process. Generally, etching uniformity can be controlled in a reasonable value by about 10%~30% over-etching. As for embedded silicon fan-out package, etching uniformity can't be solved by this method because it is blind cavity etching (no stop layer) during silicon substrate etching process.

Fig. 2 and Fig. 3 show grass issue after silicon substrate etching process. Fig. 2 presents a grass point in the bottom of silicon cavity. The grass point diameter is 44.16 μm and its height is 28.65 μm. Fig. 3 exhibits more serious grass issue in the bottom of silicon cavity characterized by 3D scanning microscope. The grass usually causes two kinds of abnormalities. The first is die crack, and the second is die tilt during die attaching process. From DOE studies we found that the grass is resulted from particles and photoresist residuals in silicon

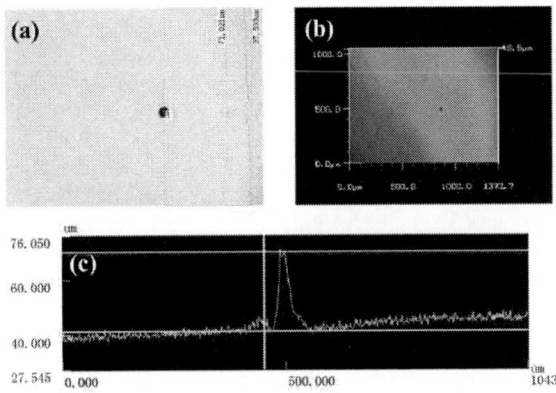

Fig. 2 (a) optical microscope image of a grass point in the bottom of silicon cavity, and (b)(c) 3D scanning results.

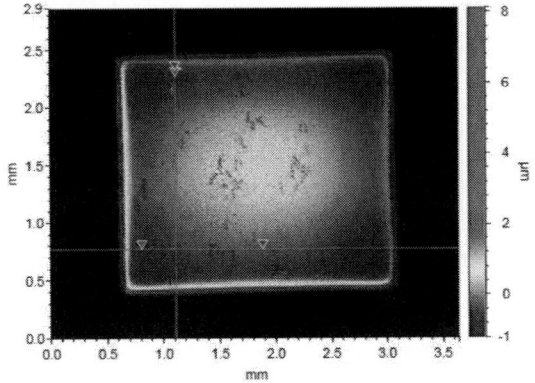

Fig. 3 Grass in the bottom of silicon cavity characterized by 3D scanning microscope.

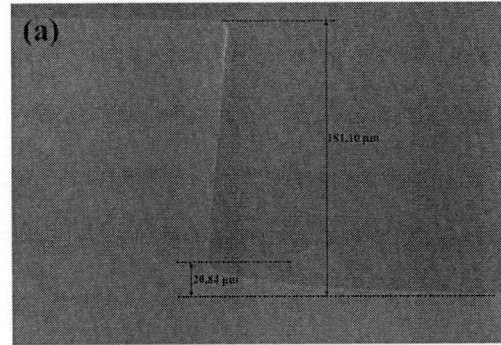

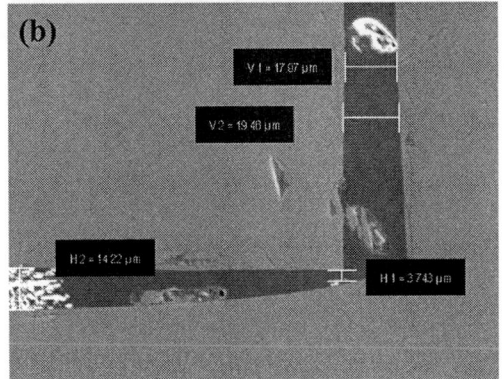

Fig. 4 (a) footing after silicon cavity etching and (b) high risk after die attaching.

wafer surface. Therefore, the key factors to decrease grass ratio are improving cleaning level and developing ability.

Fig. 4 shows footing issue after silicon cavity etching process. There is 20.84 μm high footing for 181.10 μm silicon etching depth as presented in Fig. 4(a). Fig. 4(b) shows a high risk crash between embedded die and silicon substrate. The gap is only 3.743 μm. And the further studies indicate that the footing size is larger as increasing of the silicon etching depth and cavity area. So as for the large fan-out package size, it is difficult to satisfy high yield production using blind cavity etching for embedded silicon substrate.

Obviously, no matter silicon etching uniformity, grass or footing are all dry etching issue. And they can be completely solved by over-etching. Based on this, a new scheme was proposed as presented in Fig. 5. SiO_2 stop layer deposited by PECVD was introduced. Meanwhile, laser releasable temporary bonding technology was also implemented to achieve thinner packaging [8]. As shown in Fig. 5, 1~5 μm thick SiO_2 was firstly deposited in the frontside of silicon substrate and did the temporary bonding process. Then silicon cavity etching process was finished with 10%~30% over-etching to improve silicon etching uniformity, grass and footing problems. The following is dry film filling, litho and plating processes.

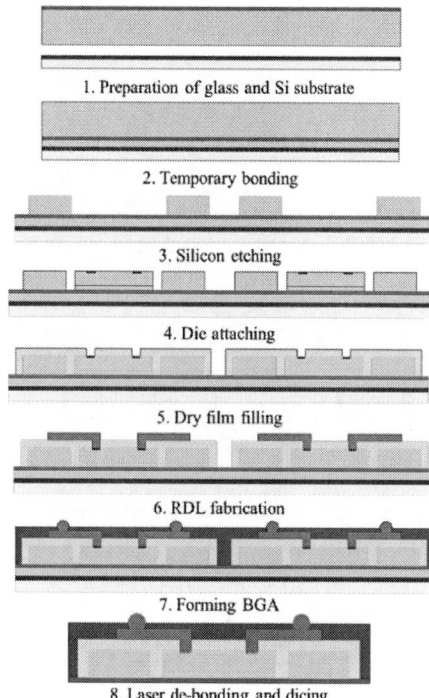

1. Preparation of glass and Si substrate

2. Temporary bonding

3. Silicon etching

4. Die attaching

5. Dry film filling

6. RDL fabrication

7. Forming BGA

8. Laser de-bonding and dicing

Fig. 5 Process follow of embedded silicon fan-out wafer level packaging with laser releasable temporary bonding technology.

Finally, the glass carrier was removed by laser de-bonding method, and segmentation was completed.

In order to verify the new scheme, an 8 inch silicon wafer deposited with 2 μm thick SiO_2 as stop layer was bonded with a piece of glass carrier (400 μm) through temporary bonding technology as shown in Fig. 6(a). Then, the silicon wafer was grinded to 100 μm thick, and silicon cavity etching process was completed as exhibited in Fig.

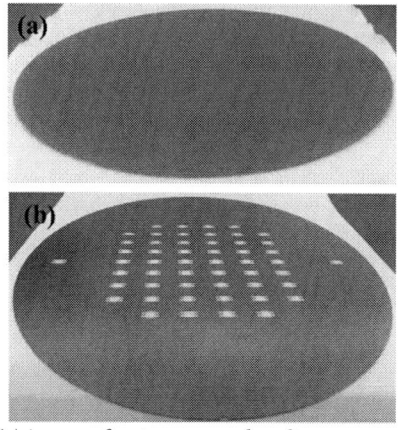

Fig. 6 (a) image after temporary bonding process and (b) image after silicon cavity etching process.

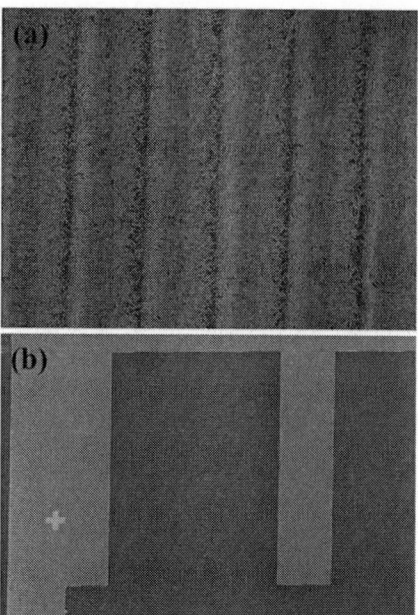

Fig. 7 (a) line patterns of laser de-bonding and (b) SEM picture of silicon cavity.

6(b). To achieve an accurate result, a special-shaped cavity with package size of 8.96mm * 5.76mm was used for verification sample. Meanwhile, 12% over-etching was introduced to ensure the etching uniformity as well as remove grass and footing. As shown in Fig. 6(b), it becomes completely transparent. And further SEM analysis found that no footing and grass were found as shown in Fig. 7(b). In addition, with the help of 355 nm UV laser, the glass carrier was easily removed from silicon wafer and Fig. 7(a) presents the laser de-bonding line after laser de-bonding process.

CONCLUSIONS

A new embedded silicon fan-out package scheme with assistance of SiO_2 stop layer deposited by PECVD and laser releasable temporary bonding technology were proposed to solve silicon etching uniformity, footing and grass issues. Through the experiment results, no footing and grass were found when the silicon wafer deposited with 2 μm thick SiO_2 as stop layer. Meanwhile, the glass carrier can be easily removed using 355 nm UV laser. The new scheme can achieve thinner package and better heat dissipation performance owing to introducing of temporary bonding technology. Next, complete development process will be carried out to verify this new scheme.

ACKNOWLEDGEMENTS

The authors appreciate the support from Samcien, NAURA Technology and HANS Laser.

REFERENCES

[1] Chen Cheng, Yu Daquan, Wang Teng et al., *IEEE Transactions on Components, Packaging and Manufacturing Technology*, vol. 9, 2019, pp. 845-853.

[2] Ma Shuying, Wang Chengqian, Zheng Fengxia et al., *2019 IEEE 69th Electronic Components and Technology Conference (ECTC)*, Las Vegas, 2019, pp. 28-34.

[3] Brunnbauer, M., Furgut E., Beer G., et al., *56th Electronic Components and Technology Conference*, San Diego, 2006, pp. 547-551.

[4] Braun T., Becker K-F, Voges S., et al., *2013 IEEE 63rd Electronic Components and Technology Conference*, Las Vegas, 2013, pp. 1235-1242.

[5] Tseng Chien-Fu, Liu Chung-Shi, Wu Chi-Hsi et al., *2016 IEEE 66th Electronic Components and Technology Conference (ECTC)*, Las Vegas, 2016, pp. 1-6.

[6] Yu Daquan, Huang Zhenrui, Xiao Zhiyi et al., *2017 IEEE 67th Electronic Components and Technology Conference (ECTC)*, Orlando, 2017, pp.28-34.

[7] Ma Shuying, Wang Jiao, Zheng Fengxia et al., *2018 IEEE 68th Electronic Components and Technology Conference (ECTC)*, San Diego, 2018, pp. 1493-1498.

[8] Wang Chengqian, "A kind of embedded silicon fan-out package structure and method", *Chinese Patent 201910701985.3*, filed on Jul. 31, 2019.

IMPROVEMENT OF HEAT DISSIPATION IN IPM PACKAGING STRUCTURE

Wenjie Xia[1], Jie Bao[1,2], Yuan Xu[2], Renxia Ning[2], Li Hou[2], Zhenhai Chen[2]

[1]College of Mechanical and Electrical Engineering, Huangshan University, Huangshan, China

[2]Engineering Technology Research Center of Intelligent Microsystem of Anhui Province, Huangshan, China

E-mail: hsbonnie@163.com

ABSTRACT

In recent years, with the development of electronic integration technology, high-speed railway transportation, new energy development, smart grid, smart furniture and other emerging industries, higher requirements have been put forward for intelligent power modules, and improvement measures have been proposed. This article will delve into the thermal management optimization of a small out-line package (SOP) typed integrated power module (IPM) packaging structure through simulation. Four optimization schemes are used to analyze the feasible conditions for improving the heat dissipation of the IPM packaging structure. They are separately from the point of view as the peripheral packaging structure, the internal layout, the internal three-dimensional stack structure and double copper substrates. It can be found that in the last scheme, the maximum temperature of the IPM packaging structure can be effectively reduced .

Keywords—IPM, heat dissipation, packaging structure, simulation

INTRODUCTION

IPM packaging structure includes drive unit and power unit composed of insulated gate bipolar transistor (IGBT) and fast recovery diode (FRD) chips. It has the unparalleled advantages of modular construction, high efficiency, high frequency, large capacity and intelligence. However, at high temperature, the heat transfer speed of the power unit is obviously lower than the case at room temperature. If the heat can not be dissipated in time, it will result in thermal failure of the device[1-3]. Therefore, thermal management of IPM becomes an important research field. Common ways are:

First, reasonable external heat dissipation conditions of the module should be designed; Second, optimal module's package structure which can improve the heat dissipation path inside the module is needed[4-6].

OPTIMIZED STRUCTURE
Building the original model

It is necessary to understand the packaging structure and working principle of the commercial IPM. Figure 1 is

a schematic diagram of the IPM packaging structure.

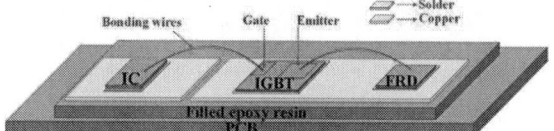

Fig.1. Schematic diagram of the main components of the IPM.

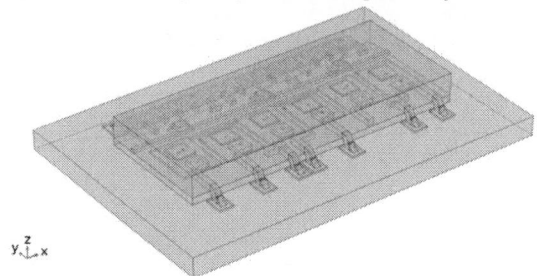

Fig.2. 3D modeling of IPM package

3D modeling is used to simulate the original IPM and analyze the impact of internal structure on the thermal effects. Based on the model in Figure 2, finite element analysis can be used to investigate the IPM temperature with different structures. It was referred to Infineon IRSM506-0507 for modeling. There are three drive chips, six IGBT chips and six FRD chips connected by gold bonding wires.

Peripheral structure optimization

In IPM, most of the heat generated from the chip must be released to the environment through an effective heat flow path. The basic law of heat conduction is often described by one-dimensional Fourier equation:

$$Q = -kA\frac{\Delta T}{\Delta x} \qquad (1)$$

In formula (1), Q is the heat flux (W); k is the thermal conductivity of the material (W/(m·K)); A is the cross-sectional area (m^2) perpendicular to the direction of heat flow; $\frac{\Delta T}{\Delta x}$ is the temperature gradient in the heat flow direction.

In order to improve the thermal management of IPM,

978-1-7281-6559-2/20 $31.00 © 2020 IEEE

besides material performance optimization, packaging structure adjustment, such as peripheral structure optimization is currently a very mainstream solution to enhance heat transfer.. A small piece of copper is placed between the PCB board and the IPM device for enhancing heat conduction, which is a very cheap and excellent solution.

In the simulation experiment, the heat flux coefficient h at room temperature (293.15K) is set as 500 W/(m^{2}K). It can be seen that the temperature of the IPM has an obvious decline by using the peripheral structure optimization, which is about 60 °C when working at 84W, as shown in Fig.3.

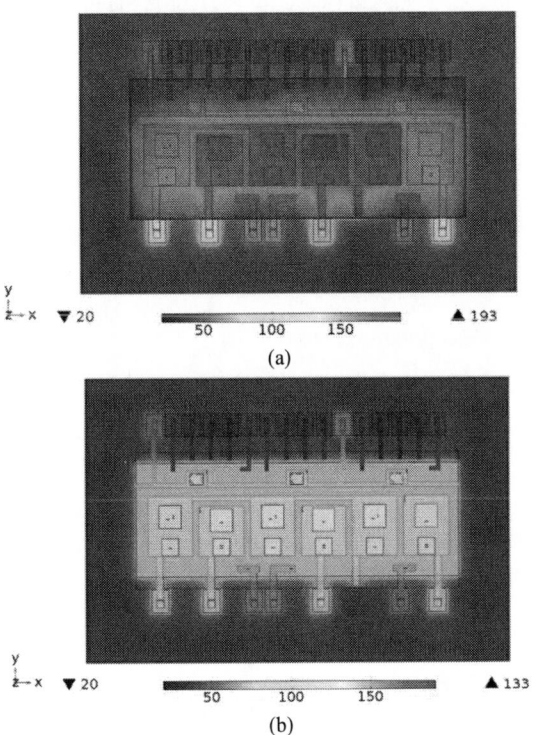

(a)

(b)

Fig.3. Temperature distribution of IPM packaging structure. (a) traditional structure (b) peripheral optimized structure.

Internal structure optimization

Internal layout optimization, internal 3D stack and double copper substrates are simulated in turn to have a further improvement in the cooling effect. The simulated results are shown in Fig.3. The internal layout optimization is a scheme of plane position exchange between the high-heat IGBT chips and the low-heat FRD chips. The purpose is to reduce the local temperature, and distribute the heat evenly in all parts of the IPM. The Internal 3D stack scheme is optimized for the connection mode between different phase power chips to improve performance of IPM device by reducing the use of

bonding wires. Double copper substrates scheme is a solution based on another kind of 3D stack. It uses double DBC substrates, still hoping to reduce the use of bonding wires, while also improving heat dissipation.

Among these different IPM structures, it is desirable to study the performance differences of these structures. Therefore, a power of 84W was applied to these three structures. As shown in Fig.4, the temperature changes and differences between each structure can be seen. Internal layout optimization improves heat distribution, but has little impact on the overall IPM. Although the internal 3D stack increases the heat transfer area, heat cannot be effectively transferred through the outer casing, resulting in an increase in temperature. The double copper structure optimizes heat distribution and enhances heat transfer, so the effect is the best.

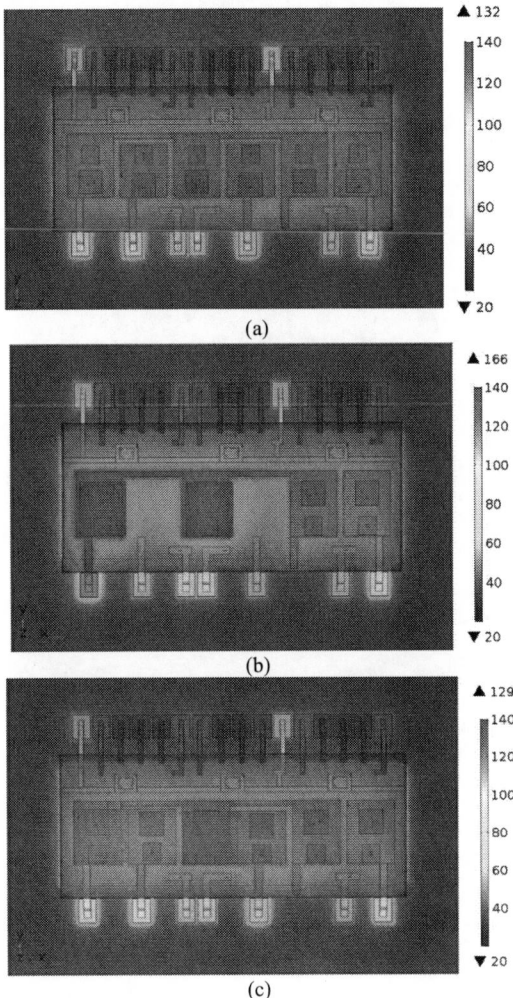

(a)

(b)

(c)

Fig.4. Temperature distribution of IPM packaging structure. (a) internal layout optimization (b) internal 3D stack (c) double copper substrates

It can also be found in the isotherm diagram (Fig.5) that the double copper substrates optimization scheme has a more uniform and reasonable heat transfer than the peripheral structure optimization. The thermal conductivity of the main internal power devices is enhanced, making the heat dissipation in the high heat parts more rapid and uniform. In terms of improving heat dissipation efficiency, this is an advantage of the double copper substrates optimization versus the Peripheral optimization structure.

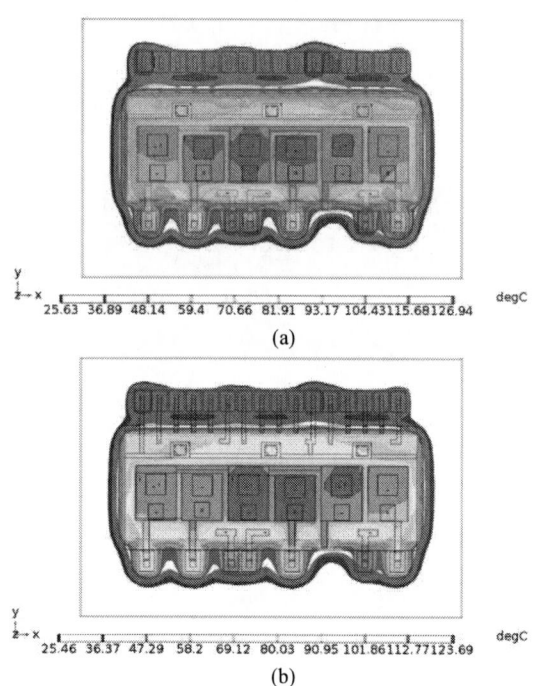

(a)

(b)

Fig.5. Isotherm diagram of the IPM packaging structure. (a) Peripheral optimization (b) double copper substrates

TABLE 1

Power Improved Method	60W	72W	84W
Traditional structure	144°C	169°C	193°C
Peripheral optimization	101°C	117°C	133°C
Internal layout	100°C	117°C	132°C
Internal 3D stack	101°C	145°C	166°C
Double copper	98.1°C	114°C	129°C

As concluded in Table1, the internal layout optimization scheme has no substantial impact, the internal 3D stack structure scheme may even affect heat dissipation, and the double-layer copper substrate scheme can play an effective role. With a power of 84W, the maximum temperature of the IPM package structure can be further reduced by about 4 °C based on peripheral optimization.

SUMMARY AND CONCLUSION

Based on a large number of finite element model simulation experiments, it can be understood that the peripheral structure optimization is a very efficient and economical method to improve heat dissipation performance of the IPM packaging. Furthermore, using of double copper substrates can further reduce the maximum temperature of the module and improve the reliability of the device.

ACKNOWLEDGEMENTS

This project has been funded by Major Science and Technology Projects of Anhui Province (18030901006), Anhui Excellent Young Talents Project (gxgwfx2019054, gxyqZD2019069). We also acknowledge the program from National Natural Science Foundation of China (61704161), Anhui Key Research and Development Plan Project (201904b11020007), Anhui Natural Science Foundation (1908085MF178).

REFERENCES

[1] S. Xu. "Study on condition monitoring and reliability assessment of IGBT module," Chongqing: Chongqing University, 2013, pp. 1-20.

[2] Y. Xu, J. Bao, R. Ning, Z. Chen and W. Xu. "Heat dissipation study of graphene-based film in single tube IGBT devices," AIP Advances, vol.9, 2019, pp.035103.

[3] J. Bao, Y. Xu, N. Jing, H. Zhao, Y. Fu, Y. Zhang, et al. " Thermal Management Technology of IGBT Modules Based on Two-Dimensional Materials," International Conference on Electronic Packaging Technology, Shanghai, China, August 2018, pp. 585-588.

[4] H. Zhao, J. Bao, Y. Xu and W. Xu, "Optimal design of heat dissipation structure of IGBT Modules based on graphene," China Semiconductor Technology International Conference, Shanghai, China, March 2019.

[5] J. Wang. "The thermal design for intelligent power module package," Nanjing: Southeast University, June 2016, pp. 11-30.

[6] Y. Xu, J. Bao, R. Ning, Z. Chen, H. Li, W. Xu, B Zhou. "Heat Dissipation Simulation of Double-sided Liquid-cooled IGBT Module Package," International Conference on Electronic Packaging Technology, Hong Kong, China, August 2019

CHARACTERTERIZATION OF MULTI-SCALE NANOSILVER PASTE REINFORCED WITH SIC PARTICLES

Ziwei Jiang [1], Yongqian Sun [1], Qiaoran Zhang [1], Zhen Lv [1], Yanpei Wu [2], Weijuan Xi [2], Cheng Zhou [2], Shujing Chen [1], Maomao Zhang [1], Johan Liu [1,3], Xiuzhen Lu* [1]*

[1] SMIT Center, School of Mechatronics Engineering and Automation Shanghai University, Shanghai, China
[2] Space Research Institute of Electronics and Information Technology, Aerospace Science and Technology Corporation, Xi'an, China
[3] Department of Microtechnology and Nanoscience Chalmers University of Technology, Gothenburg, Sweden
* Corresponding Author's Email: xzlu@shu.edu.cn; jliu@chalmers.se

ABSTRACT

Nanosilver paste with high operation temperature and low sintering temperature has attracted more and more attention for its promising application in high power devices. In this paper, the thermal properties of multi-scale nanosilver paste composed of nanometer and micrometer silver particles, and Ag-coated SiC particles were investigated. The thermal conductivity of multi-scale nanosilver paste increases with the increasing amount of SiC particles with Ag coating. The maximum value of Vickers hardness for multi-scale nanosilver paste with 0.5 wt.% Ag-coated SiC particles were 24.

INTRODUCTION

With the rapid development of electronic technology, the electronic products are developing towards integration, miniaturization and multifunction. Consequently, the operation temperature of electronic products is becoming higher and higher [1-2]. The traditional interconnection materials are failed to meet the requirements of high operation temperature for high density and powerful devices [3]. For example, the operation temperature of some electronic device has exceeded the glass transition temperature of conductive adhesive and the melting point of solder [4]. A new-type interconnect material is urgently needed to ensure high performance and reliability of devices with higher operation temperature. Micrometer sliver paste based on low temperature jointing technique (LTJT) is used in electronic devices packaging to solve this problem at first. However, the micrometer silver paste is difficult to be implemented for industrialization due to the high pressure up to 40 MPa applied to electronic devices during the sintering process [5]. It has long been known that the particle-size-dependent melting point depression occurs when the particle size is on the order of nanometers since it was first reported by Pawlow [6]. Moreover, researchers have found that the melting point of 2.4 nm silver particles is less than 400°C estimated by molecular dynamic simulation [7]. Adequate driving forces were supplied in the sintering process by reducing the size of Ag particles to nanometer [8]. Therefore, nanosilver paste has become the most promising interconnection materials applied in electronic devices owing to its excellent performance of its low sintering temperature (<300°C) and high operation temperature (961°C) [9]. Furthermore, it is reported that using fillers with different sizes particles can reduce porosity and increase heat dissipation path of sintered silver structure, which will lead to excellent mechanical and thermal properties of sintered silver structure [10-11]. Cracks will occur between chip and sintered silver joints with the change of working temperature due to the difference of thermal expansion coefficient (CTE) between chips and silver paste. SiC particles with high thermal conductivity and low CTE are supposed to be used as filers to improve the mechanical and thermal properties of silver paste [12].

In this paper, the properties of multi-scale nanosilver paste with nano-Ag particles, micro-Ag particles, and with reinforced Ag-coated SiC particles were studied.

EXPERIMENT

The multi-scale nanosilver paste was composed of nano-Ag particles, micro-Ag r particles, Ag-coated SiC particles, dispersant, binder and thinner. The nano-Ag and micro-Ag particles accounted for 82 wt.% of multi-scale nanosilver paste. All the samples were divided into two groups. As shown in TABLE I, the proportions of nano-Ag and micro-Ag particles were varying with the total filler proportion of 82 wt.% in group one. The Ag-coated SiC particles were used to replace a small amount of micro-Ag particles by keeping the total proportion of micro–Ag and Ag-coated SiC particles with 18 wt.% in group two. The proportions of nano-Ag particles, micro-Ag particles, and Ag-coated SiC particles are shown in TABLE II. The proportions of dispersant, binder and thinner were 1 wt.%, 10 wt.% and 7 wt.%, respectively.

TABLE I   PROPORTIONS OF NANO-SILVER AND MICRO-SILVER PARTICLES

	Component Proportions (wt.%)						
Micro-Ag particles	0	5	10	15	18	20	30
Nano-Ag particles	82	77	72	67	64	62	52

978-1-7281-6559-2/20 $31.00 © 2020 IEEE

TABLE II   PROPORTIONS OF NANO-AG AND MICRO-AG PARTICLES, Ag-COATED SiC PARTICLES

	Component Proportions (wt.%)				
Ag-coated SiC particles	0	0.5	1	1.5	2
Micro-Ag particles	18	17.5	17	16.5	16
Micro-Ag particles	64	64	64	64	64

Micro-Ag particles, nano-Ag particles and SiC particles, with diameter of 1-2 μm, 20 nm and 10-200 nm, respectively, were selected as fillers. SiC particles were difficult to melt and form sintering neck with silver particles at low sintering temperature due to its high melting temperature of 2700 °C. The nanometer silver was plated on the surface of SiC by electroless plating to solve this problem. A preparation process of multi-scale silver paste included the following steps. Firstly, the alcohol was heated up to 60°C. Then the Polyvinyl butyral (PVB) as organic vehicle was mixed with the alcohol slowly by stirring. After cooling to room temperature, triethylene glycol and terpineol were added into this solution as dispersant and thinner, respectively. Nano-Ag particles, micro-Ag particles, and Ag-coated SiC particles were added into the mixture subsequently. Finally, the mixture was stirred for 30 minutes and subjected to ultrasonic treatment for 30 minutes to make the particles distributed uniform [13].

A mould with diameter of 16 mm and height of 3.5 mm was used to prepare the samples for thermal conductive testing. Firstly, the mould was filled with the multi-scale nanosilver paste. Then the mould with paste was dried at 110 °C for 15 minutes in the blast drying box to remove ethyl alcohol from the silver paste. Then the mould was filled with silver paste and dried again. The above steps were repeated until the mould was full of the silver paste. Finally, the multi-scale silver paste in the mould was sintered at 250 °C for 120 minutes. The sintered samples were ground to standard size for thermal conductive testing. The preparation process of samples for thermal expansion coefficient testing with the diameter and height of 6 mm and 25 mm is the same as that of thermal conductivity samples.

RESULTS AND DISCUSSION

As shown in Fig. 1, the thermal conductivity of the sintered silver increases at first and then decreases with the increase of micro-Ag particles proportions. It reaches the maximum value of 18.84 W/ (m · K) for the multi-scale nanosilver paste with micrometer silver particles of 5 wt.%. As the proportion of micro-Ag particles increased from 20 wt.% to 30 wt.%, the thermal conductivity drops sharply from 15.61 to 5.26 W/ (m · K). Fig. 2 shows the SEM images of the sintered multi-scale nanosilver structure with different proportions of micro-Ag particles.

The porosity of silver paste decreases firstly with the increase of micrometer silver particles proportion. However, the porosity of silver paste increases with proportion of micrometer silver particles increase from 20 wt.% to 30 wt.%. It is due to the high melting temperature of micro-Ag particles. Consequently, the sintering quality of multi-scale nanosilver paste reduces with micro-Ag particles of 30 wt.% (shown in Fig. 2(h)). It can be concluded that the increase of thermal conductive of multi-scale nano-silver paste owe to the increase of heat dissipation path.

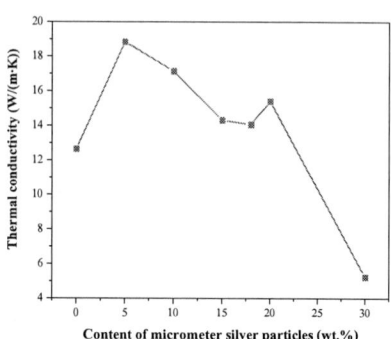

Fig. 1. The thermal conductivity of the samples with different proportions of micro-Ag particles

As shown in the Fig. 3, the thermal conductivity of silver paste increases from 14.07 to 23.77 W/ (m · K) with the proportion of Ag-coated SiC particles increasing from 0 wt.% to 2 wt.%. Earlier research has shown that the shear strength increased firstly and then decreased with the increasing of Ag-coated SiC particles proportions. And the largest shear strength of 9.22 MPa was gained in sintered joints with 1.5 wt.% Ag-coated SiC particles [14]. Fig. 4 show the SEM images of sintered multi-scale nanosilver structure with different proportions of Ag-coated SiC particles. Those SEM images show the porosity of silver paste decreases with the increase of Ag-coated SiC particles proportion. It can be concluded that the thermal conductivity of sintered silver structure increased by adding Ag-coated SiC particles, while the porosity reduced.

The hardness of multi-scale nanosilver paste without Ag-coated SiC particles is 16-17 as shown in Fig. 5. The hardness of silver paste is obviously improved by adding 0.5wt% Ag-coated SiC. However, the hardness of multi-scale nanosilver paste does not change obviously by adding 1 wt.%, 1.5 wt.% and 2 wt.% Ag-coated SiC particles.

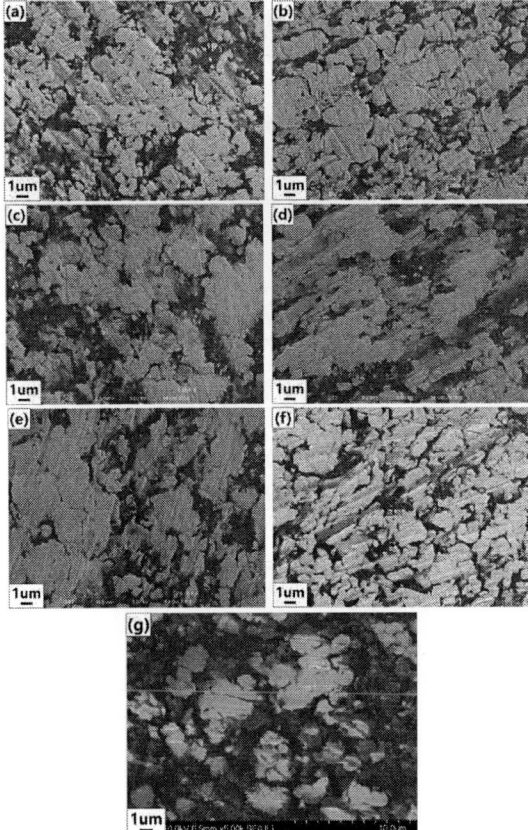

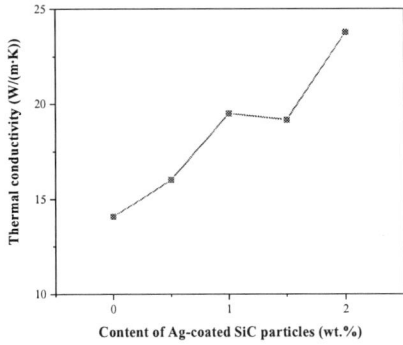

Fig. 2. SEM image of sintered multi-scale nanosilver structure with different proportions of micro-Ag particles, (a) 0 wt.%, (b) 5 wt.%, (c) 10 wt.%, (d) 15 wt.%, (e) 18 wt.%, (f) 20 wt.%, (g) 30 wt.%.

Fig. 3. The thermal conductivity of the samples with different proportions of Ag-coated SiC particles

The failure of the chips could be caused by the stress concentration at the interface, which is caused by the mismatch of the CTE between interconnection materials and chips. The CTE of silver paste is different from that of chips. Cracks will occur between chip and sintered silver joints with the change of working temperature. And the cracks gradually expand and eventually lead to interconnection failure. The CTE of SiC is $4.63 \times 10^{-6}/°C$, much lower than that of silver [15]. Ag-coated SiC particles can reduce the thermal expansion coefficient of silver paste while the reliability of interconnection was improved. Fig. 6 is the CTE curves of multi-scale nanosilver paste with different proportion of Ag-coated SiC particles.

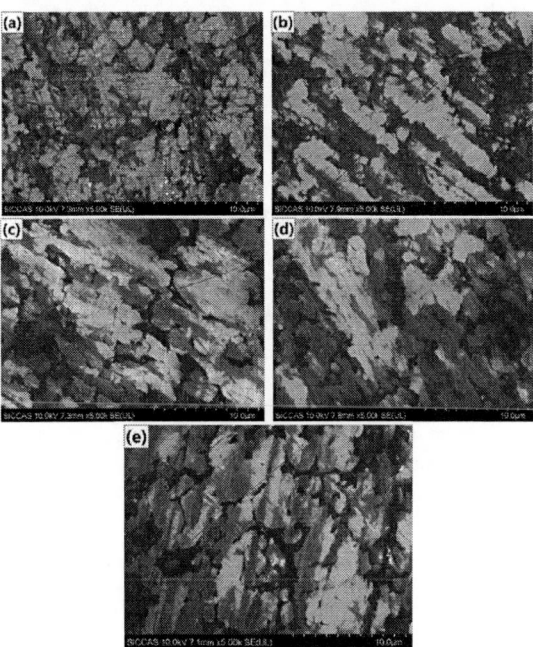

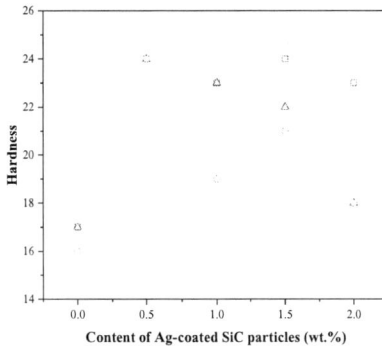

Fig. 4. SEM image of the multi-scale nanosilver paste with different proportions of Ag-coated SiC particles, (a) 0 wt.%, (b) 0.5 wt.%, (c) 1.0 wt.%, (d) 1.5 wt.%, (e) 2.0 wt.%

Fig. 5. Vickers hardness of multi-scale nanosilver paste with different proportions of Ag-coated SiC particles

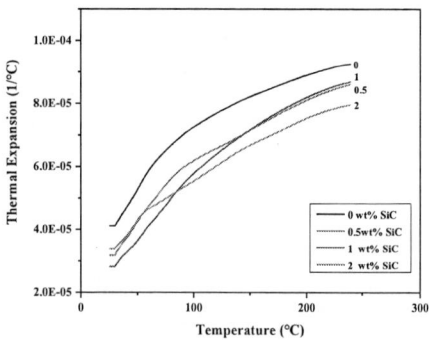

Fig. 6. CTE of multi-scale nanosilver paste with different proportions of Ag-coated SiC particles

CONCLUSION

The thermal conductivity of the multi-scale nanosilver paste increased at first and then decreased with the increasing of micrometer silver particles proportions. The thermal conductivity of silver paste increases from 14.07 to 23.77 W/ (m · K) with the increasing of Ag-coated SiC particles from 0 wt.% to 2 wt.%. The Vickers hardness of multi-scale nanosilver paste with 0.5 wt.% Ag-coated SiC reaches the maximum value of 24. Silver paste with Ag-coated SiC particles can reduce the thermal expansion coefficient of silver paste.

ACKNOWLEDGMENTS

The authors acknowledge the financial support by the Key R & D Development Program from the Ministry of Science and Technology of China with the contract No. 2017YFB0406000, the National Natural Science Foundation of China (No. 51872182, and No. U1537104). J.L. also acknowledges the financial support from the Swedish Board for Strategic Research (SSF) with the contract No GMT14-0045, from Formas with the contract No. FR-2017/0009, and from the Swedish National Science Foundation with the contract No. 621-2007-4660, as well as from the Production Area of Advance at Chalmers University of Technology, Sweden.

REFERENCE

[1] K. Shenai, R. S. Scott, et al. IEEE Transactions on Electron Devices, vol. 36, no. 9, pp. 1811-1823, 1989.

[2] R. Khazaka, et al. Journal of Electronic Materials, vol. 43, no. 7, pp. 2459-2466, 2014.

[3] Peng, W. Q. and Marques, M.E., Journal of Electronic Materials, vol. 36, No. 12, pp. 1679-1690, 2007.

[4] J. R. Groza, Nanostructured Materials, vol. 18, No. 26, pp. 173-233, 2007.

[5] Schwarzbauer, H., et al. IEEE Transactions on Industry Applications, vol. 27, No. 1, pp. 93-95, 1991.

[6] Pawlow, P. Z. Phys. Chem. 1909, 65, 1.

[7] S. J. Zhao, S. Q. Wang, D. Y. Cheng. Journal of Physical Chemistry B, vol. 105, no. 51, pp. 12857-12860, 2001.

[8] Z. Zhang and G. Q. Lu. IEEE Transactions on Electronics Packaging Manufacturing, vol. 25, No. 4, pp. 279-283, 2002.

[9] Bai, J. G., Zhang, Z. Z., et al. IEEE Transactions on Components and Packaging Technologies, vol. 29, No. 3, pp. 589-593, 2006.

[10] Maruyama, M., et al. Applied Physics A, vol. 93, pp. 467-470, 2008.

[11] Liu, H.J., et al. Industrial & Engineering Chemistry Research, vol. 50, pp. 198-206, 2011.

[12] Saheb, N. and Hayat, U., Ceramics International, vol. 43, No. 7, pp. 5715-5722, 2017.

[13] Bai, J G., et al. IEEE Transactions on Electronics Packaging Manufacturing, vol. 30, No. 4, pp. 241-245, 2007.

[14] Zhang Q.R., Abdelhafid Zehri, et al. Soldering & Surface Mount Technology, vol. 31, No. 4, pp. 193-202, 2018.

[15] Wu, Q. R. and Wen, B. X., Journal of South China University of Technology, vol. 24, No. 3, pp. 11-15, 1996.

INVESTIGATION OF BOND PAD CRYSTAL DEFECT FOR DIFFERENT COVER TRANSMISSION RATE

Chengyang Sun
Shanghai Huali Microelectronics Corporation (HLMC)
Shanghai, China
sunchengyang654@163.com

ABSTRACT

With the advancement of VLSI technology and the continuous development of metal oxide semiconductor field effect transistors (MOSFET), process nodes are constantly improving, integrated circuit package precision requirements are also increasing, and the difficulty of controlling bonding quality and reliability is increasing, the crystal of the pad on the surface of the aluminum (Al) pad has become a real problem in the semiconductor industry. When doing a shear test, this kind of defect will cause the package to fail[Fig.1]. The essay proposes that different products have different cover transmission rate, which will affect the degree of pad crystal a degree of influence of different cover T/R on the crystal of the pad and the solution to provide a strong evidence for the subsequent solution of the pad crystal problem. The experimental results show that the product cover T/R is the smaller, the greater the chances crystal of the pad.

Keywords—MOSFET;Pad crystal; Cover transmission rate

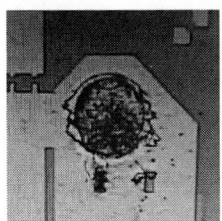

Figure 1:Bonding failure model-shear test

INTRODUCTION

In modern integrated circuit technology, the manufacture of aluminum pad must go through process including Cover-PH, Cover-ET, WET Clean and so on. The dry-etching (Cover-ET) includes 4 steps. Firstly, it uses plasma to etch and remove silicon nitride and silicon oxide film covered on pad sequentially and the reaction gases include CHF3, CF4, and Ar+. Next CF4 with a higher fluorocarbon ratio is utilized to increase the fluorine concentration in the plasma for subsequent over-etching. Finally, the photoresist is ashed by high-active oxygen, and the surface of Al pad is bombarded by argon ion at the same time. After the etching is completed, fluorinated residue will remain on the surface of the Al pad.

When the residue combines with water vapor in the atmosphere, it will gradually react with aluminum and finally form pad crystal defects, commonly known as pad crystal (PDCY) defects [Fig.2]. This procedure is called electrochemical reaction model. In this model theory, the formation of PDCY defect is strongly related to F ions and moisture around the aluminum plate. Therefore, controlling the F element and water concentration is a common method to eliminate such pad defects. However, a lot of improvements have been made in FAB, such as dry etching and wet cleaning, but the improvement is not very obvious. With the fluctuation of the external environment, this PDCY defect is still easy to happen.

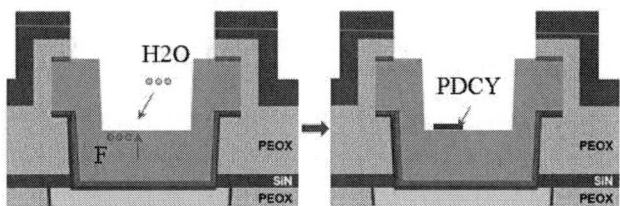

Figure 2: Produce model of PDCY

Based on the results of a large number of FAB experiments, this paper proposes a new idea and improvement for PDCY defect. From the perspective of product design, this paper studies the influence of the passivation transmission rate between different products on PDCY and the subsequent solutions are of high practical value. The research shows that the generation of PDCY is strongly correlated with the transmission rate of passivation layer. The lower the transmission rate of the passivation layer, the more easily PDCY is produced.

EXPERIMENT AND CHARACTERIZATION

In order to verify the influence of the light transmission rate of passivation layer between different products on the crystal defects of pad, a group of controlled experiments were designed. Three products (product1 ~ 3) with different light transmission rate of passivation layer on three different platforms were selected to go through Cover-ph, Cover-et, Wet clean, O2 treatment, Final Annual and so on process at the same time, so as to ensure that the products pass through the same machine at the same time and reduce the influence of other factors. In the end, 3pcs wafer stay at the dummy step and in the same crystal box nearby to ensure that the external environment is almost same. The wafers were stored for different days correspondingly (5 days, 10 days, 15 days, 20 days, etc.). To confirm the condition of bond pad crystal defects, inline detection technology is used for scanning the wafer surface and, scanning electron microscopy (SEM) is used for defecting morphology observation.

RESULTS AND DISCUSSION

a) It can be seen clearly that the product 1 defect map has obvious intensive defects after 10 days of static state. From the enlarged SEM micrograph (Fig. 3), it can be seen that the defect is an irregular leaf characteristic pattern whose size is about 3 um. EDX spectrum shows that F element is detected in the defect area(Fig. 4), but product 2 and product 3 PDCY are clear. We can get by comparing Cover T/R that Product 1 T / R is the smallest, only 1.33, and firstly suffers PDCY, which indicates that there is certain correlation between cover T / R and PDCY. What's more, this phenomenon confirms the electrochemical model principle of PDCY defect generation.

Condition	Cover T/R	wait 5D	wait 10D
Product1	1.33	●	●
Product2	3.42	●	●
Product3	6.49	●	●

Record: ● PDCY Clear ● Suffer PDCY

Table 1: wait 10 days PDCY condition

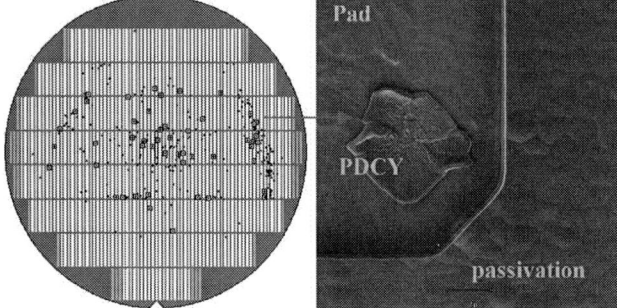

Figure 3: Product 1 PDCY defect

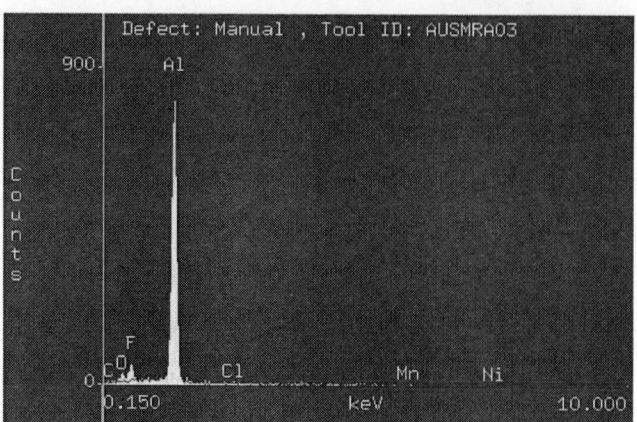

Figure 4: Product 1 PDCY EDX

b) PDCY was not observed on the other two products in the following static experiment process till the 30th day.

Condition	Cover T/R	5D	10D	15D	20D	25D	30D
Product1	1.33	●	●				
Product2	3.42	●	●	●	●	●	●
Product3	6.49	●	●	●	●	●	●

Table 2: wait 30 days PDCY condition

Figure 5: Product 2/3 PDCY clear

c) In order to get more accurate data, Auger analysis was conducted on the samples of the three different products, and the concentration of F element in the surface of the product pad and the deep layer was detected to compare the difficulty for product 2 and product 3 to suffer PDCY, product 1 As a reference (already suffer PDCY), the results show that the concentration and depth of F element on the surface of product 1 is significantly higher than that of product 2 and product 3. This further illustrates that element F is the main factor that induces PDCY (under the same external environment)); Product 2 and product 3 have the same F element on the pad surface, but the depth of the F element of product 2 is significantly higher than that of product 3. The experimental derivation and induction summarizes that product 3 is the least likely to suffer PDCY, followed by product 2 and the worst. It is product 1, which confirms that the product cover T/R is the smaller, the greater the chances crystal of the pad.

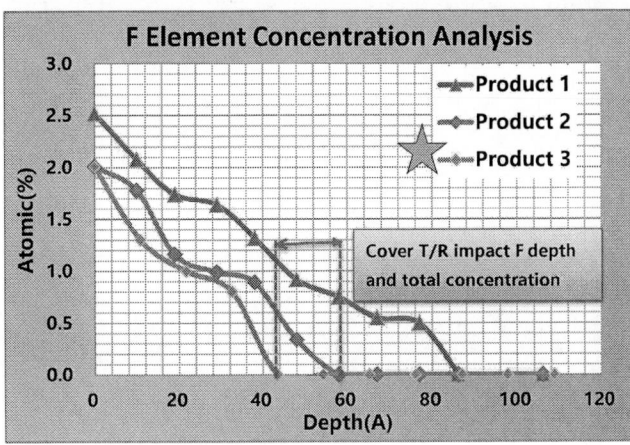

Figure 6: Fluorine element analysis

d）By summarizing and analyzing the PDCY situation of a large number of mass-produced products in the HLMC FAB, we found that products with a small Cover T / R (≤3%) are indeed more prone to suffer PDCY, and found a root cause of PDCY which provides a clear direction for the following research to find the improving methods.

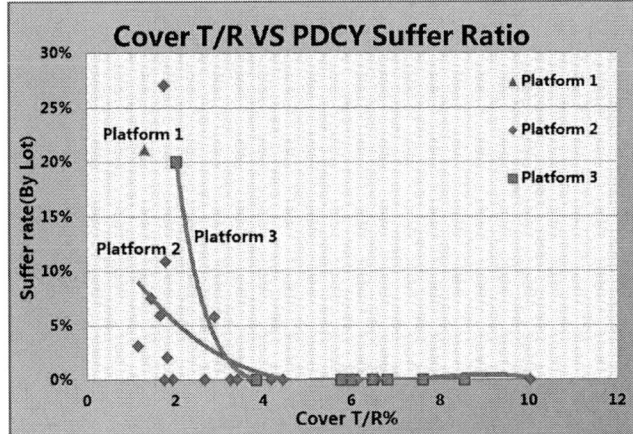

Figure 7: FAB product cover T/R summary

IMPROVEMENT AND SOLUTION

Based on the conclusions of this experiment, an innovative idea to improve PDCY arises. Since Cover T / R affects the difficulty for product to suffer PDCY, it is possible to imprve PDCY by increasing the Cover T / R size. One idea is to increase the Cover T / R through the customer's optimized design, and the other is to increase the transmission rate of the Cover layer by increasing the number of Al Pad openings in the Scribe line, thereby effectively solving the PDCY defect. By comparing PDCY condition of products before and after the increase of Cover T/R area in FAB, it was found that the Cover T / R of some products of Platform2 can meet> 3% after the increase, and the T / R of other products after the increase is still <3%, but it can also improve PDCY. If the product is improved, it will still suffer PDCY. This kind of defect can also be removed by wet cleaning . After several times of

cleaning (900DC1 * 12 cycles) with NE111 solution, the pad crystal can be removed and the defect will not continue to grow in the next 30 days, which means completely resolving PDCY defect completely. The formula of NE111 solution is not described in detail. It obtain customer satisfaction, and creating unlimited value for FAB.

Platform	Product	TR%-now	TR%-Add Pad	Comment
1	1	1.33	1.92	PDCY improve
2	1	3.42	4.01	PDCY solve
	2	2.89	3.51	
	3	2.67	3.42	
3	1	2.03	2.85	PDCY improve

Table 3: T/R compare

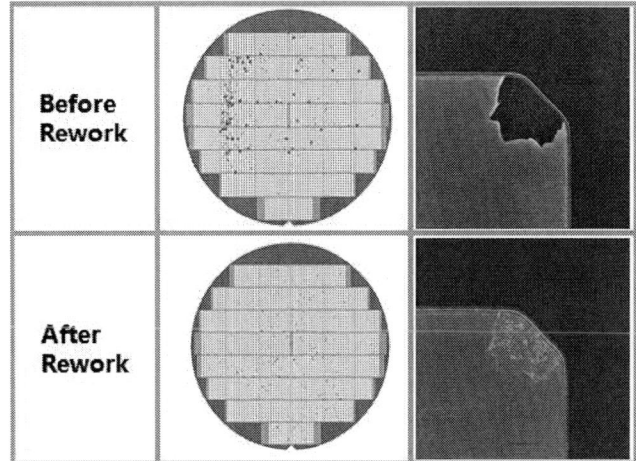

Figure 8 : Pad Crystal rework summary

CONCLUSIONS

This paper mainly describes the electrochemical reaction model of PDCY, showing that the F content is strongly related to PDCY, and water vapor will aggravate PDCY generation. It is proposed that different products have different Cover T / R with subsequent effect on the generation of PDCY. The results show that Cover T / R affects the generation of PDCY by affecting the F content on the pad surface and the deep layers. The smaller the Cover T / R, the easier the product is to produce PDCY. When T / R <3%, the product is prone to produce PDCY. This can effectively restrict the PDCY defect by increasing the transmission rate of the Cover layer by increasing the number of Al Pads in the Scribe line.

REFERENCES

[1] Y. N. Hua, Z. X. Xing, X. M. Li, "Characterization studies of fluorine-induced corrosion crystal defects on microchip Al bondpads using X-ray photoelectron spectroscopy", *Proceedings of the 21th International Symposium on the Physical and Failure Analysis of Integrated Circuits*, pp. 90-93, 2014.

[2] G. C. Fu, L. Ni, X. M. Zhang, "A new method to prevent aluminum pad corrosion", *International Workshop on*

Junction Technology, pp. 1-4, 2014.

[3] R. J. Qi, S. Q. Duan, M. Li et al., "Study on a leaf-like bonding pad defect", *Proceedings of the 20th IEEE International Symposium on the Physical and Failure Analysis of Integrated Circuits*, pp. 481-484, 2013.

[4] M. Li, W. T. K. Chien, Q. Q. Yu et al., "A study on the relationship between pad surface fluorine concentration and the formation of pad corrosion defect", *Proceedings of the 20th IEEE International Symposium on the Physical and Failure Analysis of Integrated Circuits*, pp. 311-315, 2013.

[5] J. S. Luo, H. M. Lo and J. D. Russell, The microstructure evolution of corrosion phenomenon on Aluminum bond pads, ISTFA 2005, 2005, pp. 266-273.

[6] J. S. Chen, L. K. Wei, Y. P. Chang, C. C. Huang, Bond pad F-crystal defect control and monitoring, ISMS 2001,2001, pp. 297-299.

[7] T. H. Nguyen, R. T. Foley, "The Chemical Nature of Aluminum Corrosion", *Electrochemical Soc.*, pp. 2563-2567.

[8] Y. N. Hua, "A study on Al bondpad grain boundaries and galvanic corrosion in wafer fabrication", *International Conference on Semiconductor Electronics*, pp. 83-85, 2004.

[9] S. H. Ahn, T. J. Cho, Y. S. Kim, S. Y. Oh, "Prevention of Aluminum Pad Corrosion by UV/Ozone Cleaning", *Proceedings of Electronic Components and Technology Conference 46 th* , pp. 107-112, 1996.

EFFECT OF BONDED BALL SHAPE ON GOLD WIRE BONDING QUALITY BASED ON ANSYS/LS-DYNA SIMULATION

Weidong Huang*, Wei Wu, Jacky Wu, Grass Dong, and CF Oo

Department of Research and Development, Diodes Shanghai Co., LTD., Shanghai, China

*E-mail: vic_huang@cn.diodes.com

ABSTRACT

Gold wire bonding processes on Cu/low-K dies are simulated with two sequential bonding processes: capillary lowering down and USG power bonding. The major purpose of this modeling is to verify one engineering fact in wire bonding: bonded ball shape has significant impact on the IMC performance, i.e., the bond quality. In this study BBR is defined as the ratio of the bond ball height (BBH) to bond ball diameter (BBD) and regarded as the basic feature of bonded ball shape. The simulation results indicate that BBR at 25% has the highest appearance frequency of tensile strain in bond interface, and the appearance frequency of tensile strain decreases with BBR increasing from 25% to 45%. Since high appearance frequency of tensile strain in bond interface may cause good IMC, the BBR at 25% could be the best for IMC performance and bond quality. This conclusion from simulation is almost coincident with all gold wire bonding practices where the BBR needs to be kept around 25% for the best bonding performance.

INTRODUCTION

Wire bonding has been an established packaging technology for decades as a reliable electrical connecting method. Compared with other interconnect technology, wire bonding is much easier to be accomplished. Engineering experiences on wire bonding practice conclude that wire bonding quality is quite related with the final bonded ball shape. The ball deformation during wire bonding may play a main role on bonding energy distribution on bonding interface, thus impact the IMC formation and wire bonding quality. Up to now, all conclusions about relationship between bonding ball deformation and wire bonding quality are come from the engineering experience. Several papers have reported the deformation during wire bonding process from Finite Element Simulation and data comparison with the engineering results of wire bonding quality at various bonded ball deformation [1-3]. But there are still no theoretical studies reported to explain how the bonded ball shape affects the wire bonding quality. This paper will study the mechanism of the effect of bonded ball shape on wire bonding quality based on ANSYS/LS-DYNA simulation. However the inspection on bonded ball shape has been introduced in new generation machine of K&S as the process monitor item for wire bonding quality control [4]. The study in this paper will provide the theoretical support for good bonding quality using state-of-the-art machine with bonded shape control.

PROCESS DESCRIPTION AND FINITE ELEMENT MODEL

For the first bonding on the die pad surfaces, the experience from wire bond engineers indicates that the bond ball shape has a direct relationship with the IMC performance, i.e., the bond quality. In general, the BBR (Bond ball height to Bond ball diameter Ratio) needs to be kept around 25% to get good IMC "coverage". So far, no study has been reported to explain why there is a suchlike link. In this modeling, gold wire bond process was simulated with including two sequential bonding stages: capillary push down and USG power bonding. Since Cu/low-K wafer technology has been extensively adopted in industry, this modeling work will simulate the gold wire bond process on Cu/low-K dies to get the real responses with low-K material. The primary purpose of this modeling is to verify the BBR's impact on bonding quality with comparison to wire bond engineers' experiences.

The 3D finite element model is built with commercial software ANSYS LS-DYNA. The ½ FE model is illustrated in Figure 1 (a), where FAB (Free air ball), HAZ (heat affected Zone), capillary and Cu/low-k structure are included. The FAB diameter is 40um, the copper wire diameter is 0.9mil. The low K structure consists of various layers including Aluminum pad, passivation, TEOS dielectric, low K material, copper interconnections and the contact material bonded to silicon bulk as shown in Figure 1 (b). The silicon bulk hasn't been considered into the model because of assuming it as the fixed boundary condition during whole bonding process.

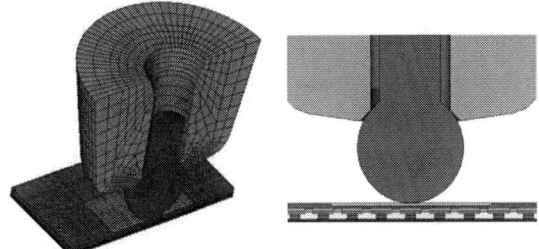

(a) 1/2 FE model of wire bonding   (b) Capillary, FAB and Cu/low-k structure

Figure 1: 3d FE model and Cu/low-k structure

Elastic properties (Young Modulus **E**, Poisson's ratio

v and density ρ) of the constituents in the model are listed in Table I. In this study, the FAB is assumed bilinear plastic. The yield stress and the tangent modulus for the FAB are 117MPa and 450MPa respectively.

TABLE I.   MATERIAL PROPERTIES FOR SIMULATION

Material	E(GPa)	v	ρ(g/cm3)
Copper interconnections	121	0.38	8.91
Passivation	32	0.24	1.31
TEOS dielectric	80	0.23	2.00
Low K	11	0.30	2.00
Contact material	80	0.23	2.00
Al Pad	69	0.33	2.71
FAB	39.5	0.42	19.3

The capillary is defined as rigid, the bond force and the USG power are loaded on it. The capillary dimension is listed in Table II:

TABLE II.   CAPILLARY DIMENSION

Capillary	Dimension
Hole size	30um
Cap CD (Chamfer Diameter)	38um
Cap T(Tip)	99um
Cap OR (Outer Radius)	12.7um
Face angle	8 deg
Inner Chamfer Angle	90 deg

The force and the sine wave vibration are applied on the capillary to simulate the whole bonding process. The force loading is divided to two stages: (1) first the force ramped up to the maximum value (called as initial force) in 2 micro-seconds; (2) then the force was kept at the constant value 10g (10g is the bond force adopted in this wire bonding process). In the first stage, no USG power was loaded on the capillary. In the second stage, a sine wave vibration is synchronously applied on the capillary to simulate USG power loading. The amplitude of the vibration is defined to 1um. The above loading conditions are illustrated in Figure 2.

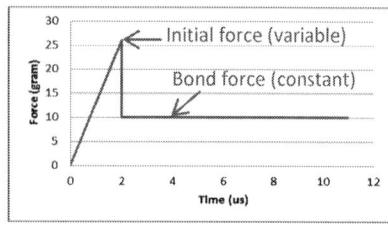

(a) Force loading curve

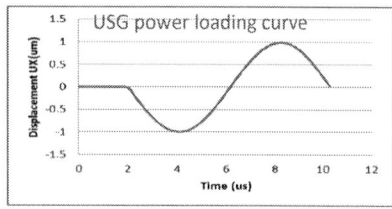

(b) USG power loading curve (displacement vs time)

Figure 2: Wire bonding loading curves for simulation

BBR is defined as the ratio of the bond ball height (BBH) to bond ball diameter (BBD) as illustrated in Figure 3:

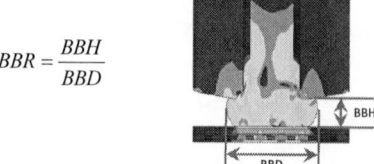

$$BBR = \frac{BBH}{BBD}$$

Figure 3: Illustration for BBH and BBD

BBR has been regarded as the main criterion to judge the wire bond performance in WB engineers' long time experiences. It is curious that BBR has a relationship with the IMC performance. The stress/strain response on bond pad will be sought by this modeling to try to explain the immanent link between BBR and IMC performance.

ANALYSIS ON SIMULATION RESULTS

As shown in Figure 4, the modeling results show that maximum stress is always inside the bond ball. In Cu/low-K structure, the stress response comes up to the maximum value when the maximum initial force is achieved at the end of the push down stage.

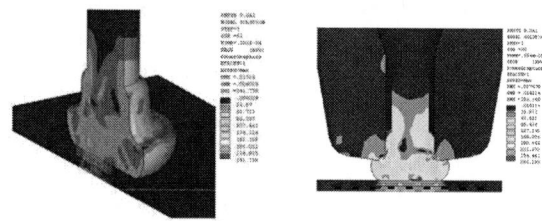

(a) Maximum stress in ball for all stages (b) Maximum stress at push down

Figure 4: Maximum Von Mises stresses in wire bonding

There are about 1,000 vibration cycles during bonding process. In this modeling, only one vibration cycle is calculated (as shown in Figure 2) because it's impossible to calculate 1,000 vibration cycles with consuming very huge computer time. To study the BBR impact on IMC, the stress/strain status on bond pad surface is investigated to try to find the response parameter related to IMC performance. Various resulted responses for different BBRs are investigated. No significant differences are found for most stress/strain responses. The only reasonable finding that could be available to explain the BBR impact on IMC is the normal strain in vibration direction.

Figure 5 show normal strains in vibration direction on bond pad surface for different BBRs. All pictures illustrate the strain distributions at 6.54us. The areas bounded by the dashed lines represent the bonding zones on bond pads. The strain patterns for BBR 25%

978-1-7281-6559-2/20 $31.00 © 2020 IEEE

and 30% are quite different from those for BBR 35%, 40% and 45%. As the positive value means the tensile strain, it's found that BBR at 25% has the maximum area occupancy of tensile strain. Assuming high tensile strain area causes good IMC coverage, we deduce that BBR at 25% could be the best for IMC performance.

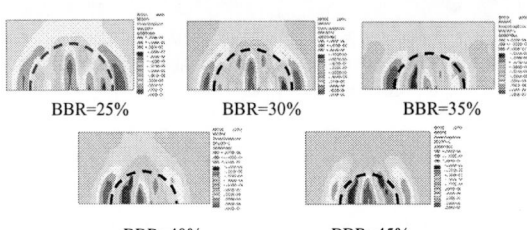

BBR=25% BBR=30% BBR=35%

BBR=40% BBR=45%

Figure 5: Normal strains on bond pad with various BBRs

For history responses of normal strain at different locations on bond pad surface with various BBRs, the 5 typical positions are marked as (see Figure 6): 1-Bond Edge, 2-Bond Mid, 3-Bond Center, 4-Bond Edge, 5-Bond Edge.

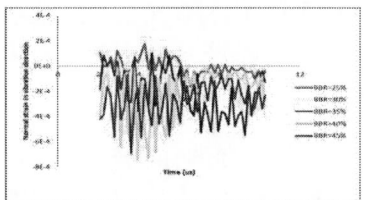

1-Bond Edge 2-Bond Mid 3-Bond Center 4-Bond Edge 5-Bond Edge

Figure 6: 5 typical positions are marked on bond area

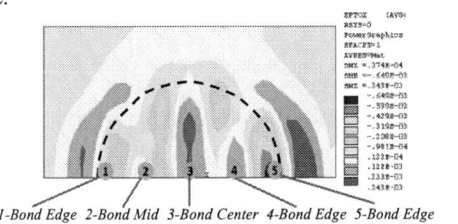

(a) History of Normal Strain at Position 1-Bond Edge

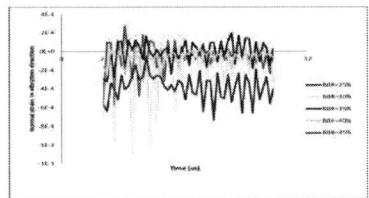

(b) History of Normal Strain at Position 2-Bond Mid

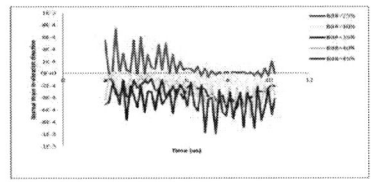

(c) History of Normal Strain at Position 3-Bond Center

(d) History of Normal Strain at Position 4-Bond Mid

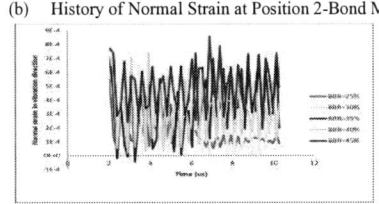

(e) History of Normal Strain at Position 5-Bond Edge

Figure 7: Normal strain history with various BBRs at 5 marked positions (a)-(e)

Figure 7 (a-e) illustrates the normal strain history with various BBRs at 5 marked positions. In each picture from (a) to (e), BBR at 25% always has the highest appearance frequency of tensile strain in bond interface. Generally, appearance frequency of tensile strain decreases with BBR increasing from 25% to 45%. Assuming high appearance frequency of tensile strain in bond interface will cause good IMC, one can deduce that BBR at 25% should be the best for IMC and bond quality.

CONCLUSIONS

The modeling results prove that BBR has significant impact on bonding quality. BBR at 25% has the highest appearance frequency of tensile strain in bond interface, which causing good IMC and bond quality. Modeling results are coincident with the engineering practice of gold wire bonding which indicates BBR needs to be kept around 25% for the best bond performance.

REFERENCES

[1] W. Huang, *2009 International Conference on Electronic Packaging Technology & High Density Packaging (ICEPT-HDP),* Beijing, China, August 10-13, 2009, pp. 344-352.

[2] W. Huang, *2010 International Conference on Electronic Packaging Technology & High Density Packaging (ICEPT-HDP),* XiAn, China, August 16-19, 2010, pp. 679-682.

[3] W. Huang, D. Bai, *2011 International Symposium on Advanced Packaging Materials (APM),* Xiamen, China, October 25-28, 2011, pp. 134-140.

[4] I. Qin, A. Shah, B. Milton, *2018 IEEE Electronics Packaging Technology Conference (EPTC),* Singapore, December 4-7, 2018, pp. 396-399.

RELIABILITY SIMULATION AND LIFE PREDICTION OF SN$_{63}$PB$_{37}$ BGA SOLDER JOINT UNDER THERMAL CYCLING LOAD

Jiahao Liu [1]*, Liang He [1], hua Chen [1], pan Zhao [1], yahui Su [2], Li chao [2], Qin Pan[2]

[1] School of Advanced Materials and Nanotechnology, Xidian University, Xi'an 710126, China
[2] Hua Dong Institute of Optoelectronics Devices, Bengbu 233030, China
*liujiahao951213@163.com

ABSTRACT

The reliability of solder joint determines whether the electronic system operates steadily and lastingly. As a common type of load, thermal stress has a non-ignorable effect on the reliability of electronic products. During the thermal cycles, the solder joints are subjected to periodic tensile and compressive stresses since the difference in thermal expansion coefficient of adjacent materials, whereby causing creep and thermal fatigue damage. As fatigue damage builds up, cracks generate and propagate continuously, which will lead to solder joint failure. In this paper, the finite element simulation method is used to simulate the stress-strain distribution in the Sn$_{63}$Pb$_{37}$ BGA solder joint array under temperature cyclic loading, the position of dangerous solder joint is located, and the lifetime of solder joints is predicted through the Manson-coffin model. Furthermore, the influence of PCB thickness and solder joint height on lifetime are analyzed.

INTRODUCTION

With the development of electronic industry, modern electronic products tend to miniaturized and multi-functional, which require a high-integration packaging, and it will lead to a greatly increased probability of reliability problems of solder joints. As a connection between the chip and the substrate, solder joints not only provide mechanical support, but also play an important role in electronic connection. The lifetime of solder joint will directly affect the service life of electronic product [1].

Among all the factors that affect the reliability of solder joints, temperature plays a dominate role. During the operation, electronic products are often disturbed by periodic switching, as well as the periodic variation of ambient temperature. Ordinarily, these conditions can be equivalent to corresponding temperature cyclic loads [2,3]. Due to the different thermal expansion coefficient of each material, the solder joints are affected by thermal stress under temperature cyclic loading. The alternating thermal stress causes deformation, crack generation, propagation and fracture, so that solder joints cannot transmit signals, causing electronic products failure [4,5]. The aim of this work is to investigate the reliability of SnPb solder in airborne oxygen monitor circuit under alternating temperature. Through finite element modeling, the hazardous solder joints in BGA array were positioned, lifetime prediction was performed and the influencing factors were analyzed.

ESTABLISHMENT OF SOLDER JOINT ARRAY FINITE ELEMENT MODEL

For the typical operating environment and application requirements of airborne electronic equipment, the finite element simulation process in this paper is based on the following assumptions:

(1) In the model, the residual stress and strain values in the initial state are zero;

(2) All materials behave isotropically when pressed or pulled them;

(3) Do not consider the hygroscopicity of the materials;

(4) Interface contacts are completely sealed for all materials;

(5) Under the temperature cycling condition, the temperature of the entire package rises and falls evenly during the heating or cooling process;

(6) The materials of the chip and the substrate are linear elastic, and the material of the solder joint is viscoplastic;

(7) Do not consider defects caused by any process of the product.

BGA solder joint material is Sn$_{63}$Pb$_{37}$ eutectic alloy. The melting point of this material is relatively lower (about 183℃, 456K), and the temperature of electronic components can reach 0.48~0.87 times of this melting point under service or typical thermal cycle test conditions (-55~125℃). Under these conditions, the deformation behavior of SnPb solder is related to time and temperature, and has obvious creep and superplastic characteristics[6,7]. Fig 1 shows a three-dimensional view of the BGA packaging model used in the simulation.

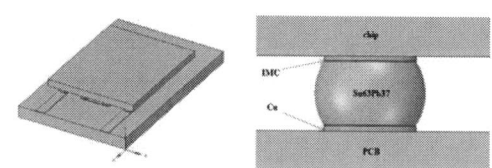

Fig.1 Three-dimensional view of the BGA package and solder joint model

FINITE ELEMENT ANALYSIS OF SOLDER JOINT ARRAY'S RELIABILITY

During operation, electronic products may undergo periodic power on/off and ambient temperature fluctuation, which will induce alternating thermal stress. In our simulation, the thermal cycling loading are selected according to China national military standard GJB150.5A-2009: the temperature ranges are from -55℃ to + 125℃, the soaking time at the highest and lowest temperature are 10 minutes. The heating and cooling rates are 20℃/ min. One cycle is subjected to 38min. The reference temperature with zero stress-strain is 22℃.

Figure 2 shows the equivalent plastic deformation nephogram of the solder joint array after five thermal cycles, It can be seen that the maximum amount of plastic deformation appears at the four solder joints located at the corner of the array, they can be considered as the most dangerous solder joints in this BGA array structure.

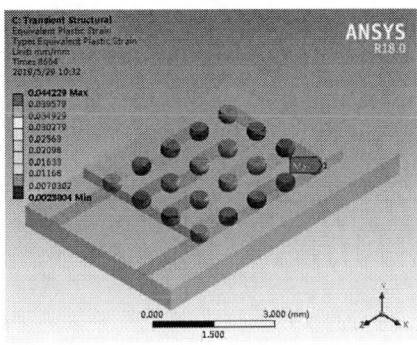

Fig.2 Plastic deformation of solder joint array after five thermal cycles

Since the thermal expansion coefficients of the selected PCB material (FR4), copper pad and IMC layer are less than that of the solder joint, the plastic strain will be induced by alternating thermal stress which generated from the thermal mismatch of the solder joint, PCB and copper pad materials during the thermal cycle process. This stress will cause the deformation of the structure, thereby applying a periodic force on the solder joint.

The change of the stress on the most dangerous solder joint during the process of temperature alternation is shown in Figure 3. It can be found that because of the thermal mismatch among different materials, during the calefactive phase, the PCB, copper pad and IMC layer will prevent the solder joint expanding outward, so that axial compressive stress is applied on the interface of the solder joint, while due to no constraint ,the middle part of the solder joint is subjected to the tensile stress caused by the expansion. During the cooling phase, the PCB, copper pad and IMC layer prevent the solder joint shrinking inward, so that the axial tensile stress is applied on the solder joint interface, and the middle part of the solder ball is subjected to compressive stress from outside to inside due to the tensile stress at the upper and lower interfaces. During thermal cycles process, because of the alternation of these two kinds of stress mentioned above, plastic deformation will accumulate at the interface where stress changes periodically.

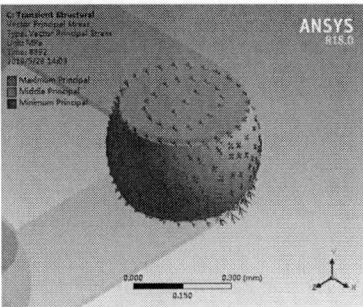

(a) Heating stage

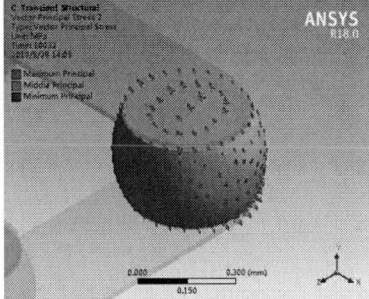

(b) Cooling stage

Fig.3 Stress patterns of the most dangerous solder joints during the calefactive phase (8892s) and the cooling phase (10032s)

The time-stress relationship of the most dangerous solder joints is extracted, and the result is shown in Fig 4. It can be seen that the equivalent stress exhibit periodic variation with the time. Comparing it with the variation of temperature, it is found that this periodicity originates from the load. During the high temperature and low temperature soaking stages, the stress are gradually reduced, resulting in a significant stress relaxation, and the stress relaxation extent in the high temperature soaking stage is less than the low temperature soaking stage. During the calefactive phase, the stress decreases, which indicates that part of the stress is released by the thermal expansion of the solder ball. During the cooling phase, the stress increases, indicating that the drop of temperature induces the solder joint to shrink, causing the stress to rise again [8]. Furthermore, during the five cycles, the stress curves are parallel to each other in the soaking stage, which reflects the steady-state creep characteristics of the solder joint. From the fourth cycle, the maximum stress inside the solder joint has stabilized.

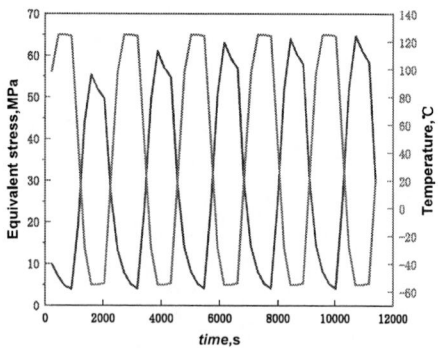

Fig.4 Time-stress diagram

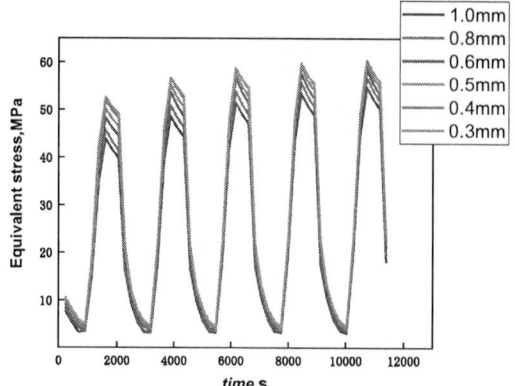

Fig.5 The equivalent stress curves of dangerous solder joints for different PCB thickness

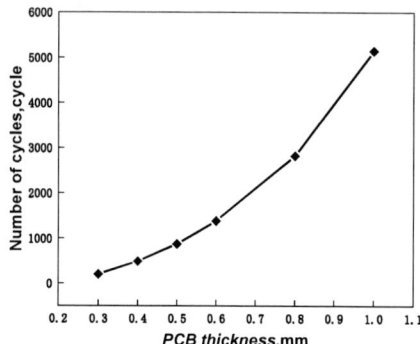

Fig.6 The fatigue life of dangerous solder joints corresponding to different PCB thicknesses

EFFECT OF PCB THICKNESS ON SOLDER JOINT LIFETIME

As an indispensable component of electronic packaging, PCB has a significant impact on the lifetime of solder joints. If the PCB was relatively thicker, then it has a larger thermal expansion coefficient, and this will introduce a greater thermal stress on the solder joints. Therefore, the impact of PCB thickness on solder joint life cannot be ignored.

In this simulation, the diameter of the solder joint is selected to be 0.48mm and the thickness of the IMC layer is set to 5μm. A finite element simulation was performed on PCBs with five different thicknesses, which are 0.3mm, 0.4mm, 0.5mm, 0.6mm, and 0.8mm respectively. According to the results, for five different PCB thickness, the most dangerous solder joints still locate at four corners of the arrays, the equivalent stress curve of these dangerous solder joints are extracted, as shown in Fig 5. It can be seen that with the increasement of PCB thickness, the equivalent stress is reduced. This is because the thinner the PCB is, the easier it will warp, and more likely to induce stress concentration at the dangerous solder joints, making them more susceptible to deform.

Based on the analysis above, the Manson-coffin model is adopted to calculate the fatigue life of dangerous solder joints, the result is shown in Fig 6. It can be seen that solder joints lifetime increases with the PCB thickness, in other words, under alternating temperature condition, increasing the thickness of PCB can reduce the warpage and the stress level on board. Therefore, during the design and preparation processes of a circuit, the device failure rate can be reduced effectively by selecting an appropriate PCB thickness.

EFFECT OF SOLDER JOINT HEIGHT ON SOLDER JOINT LIFETIME

For a traditional spherical solder joint, if its diameter was kept as a constant, the lifetime will be affected by the height of the solder joint, it is because the contact area above and below the solder joint must be decreased when the height increases[9].

In this simulation, the diameter of the solder joint is selected as 0.48mm and the thickness of the IMC layer is 5 μm. The finite element simulations are performed on the solder joints with height of 0.27mm, 0.29mm, 0.31mm, and 0.33mm respctively. The simulation results show that the most dangerous solder joints all appeared at the edge of the array. The equivalent stress curves of the most dangerous solder joint are shown in Fig 7. It can be seen that the equivalent stress still changes periodically with the load, and the stress values increase with the decline of solder joint height.

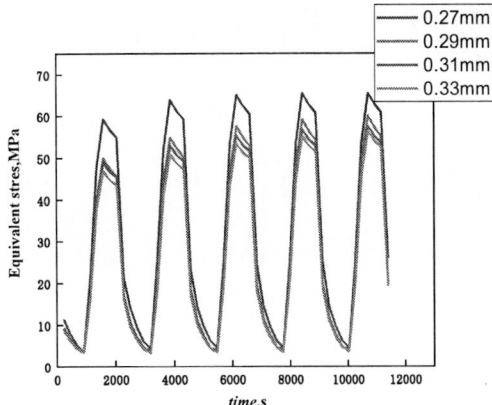

Fig.7 The equivalent stress curves of dangerous solder joint for different solder joint heights

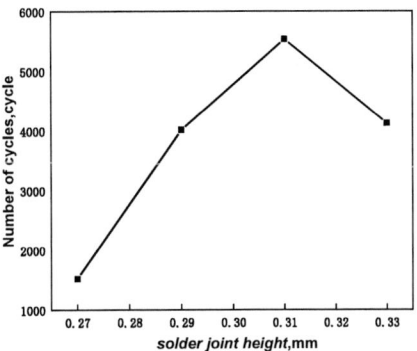

Fig.8 The fatigue life of dangerous solder joints corresponding to different solder joint heights

Similarly, use the Manson-coffin model to calculate the fatigue life of dangerous solder joints, the result is shown in Fig 8. It can be concluded that under the premise of a certain diameter of the solder ball, when the height of solder joint increases, the fatigue life does not rise or fall monotonously, but has an optimum value, the solder joint lifetime will be the longest at this point.

CONCLUSION

This paper conducts a finite element simulation analysis of SnPb BGA solder joints in an airborne oxygen monitor circuit which undergoes thermal alternating stress. The main conclusions obtained from this article are as follows:

(1) Under thermal cycling conditions, the solder joints exhibit a good steady-state creep characteristics. Fatigue failure is the main failure mechanism of solder joints. The simulation results show that the failure manifests as cracks in the solder near the area of the connection between the IMC layer and the solder joints. The propagation of these cracks will lead to fatigue

fracture eventually .

(2) Both PCB thickness and solder joint height will affect the fatigue life of solder joints. The calculation results show that optimizing the thickness of the PCB board and the height of the solder joint can improve the lifetime of the solder joint.

REFERENCES

[1] Dalton E, Ren G, Punch J, et al. Accelerated temperature cycling induced strain and failure behaviour for BGA assemblies of third generation high Ag content Pb-free solder alloys[J], 2018: S0264127518304155.

[2] Eckert T, Kruger M, Muller W H, et al. Investigation of the solder joint fatigue life in combined vibration and thermal cycling tests[C]. Electronic Components and Technology Conference, 2010: 1209-1216.

[3] Li H, An T, Bie X, et al. Thermal fatigue reliability analysis of PBGA with Sn63Pb37 solder joints[C]. International Conference on Electronic Packaging Technology, 2016: 1104-1107.

[4] Le V N, Benabou L, Etgens V, et al. Finite element analysis of the effect of process-induced voids on the fatigue lifetime of a lead-free solder joint under thermal cycling[J], 2016, 65: 243-254.

[5] Wang W, Ma X H, Qin L J, et al. Analysis of the influence of temperature cycling on SMT solder joints[J]. Electronic Devices, 2018 ,41,(1):1.

[6] Surendar A , Akhmetov L G , Ilyashenko L K , et al. Effect of thermal cycle loadings on mechanical properties and thermal conductivity of a porous lead-free solder joint[J]. IEEE Transactions on Components, Packaging and Manufacturing Technology, 2018:1.

[7] Ma L , Zuo Y , Guo F , et al. Effects of current densities on creep behaviors of Sn–3.0Ag–0.5Cu solder joint[J]. Journal of Materials Research, 2014, 29(22):2738.

[8] Zheng X, Wang J, Wei W, et al. Elastic-plastic-creep response of multilayered systems under cyclic thermo-mechanical loadings[J], 2018, 32(3): 1227-1234.

[9] Yang J S. Effect of Solder Joint Shape and Height on Thermal Fatigue Life[J]. Special Equipment for Electronics Industry, 2017(3): 6.

YIELD IMPROVEMENT AND COST OF TEST REDUCTION VIA AUTOMATED SOCKET CLEANING

Jerry J. Broz* and Bret A. Humphrey

International Test Solutions, Inc.

Reno, Nevada 89502 USA

*Corresponding Author's Email: jerryb@inttest.net

ABSTRACT

Accurate testing of advanced devices using sockets is the primary method of assuring that the final assembled devices meet performance and reliability specifications. During any device test operation, contact is made with the device in a socket and, consequently, contamination from the package accumulates into the socket and onto the contactor surfaces. To maintain high yields during test operations, the sockets and contactors must be regularly cleaned. Modern production handlers are equipped for auto-contactor cleaning (ACC) functions to reduce downtime and maintain high throughput. In this paper, implementation of cleaning units used for in-situ cleaning execution are presented; and production test results are presented with an emphasis on the overall performance for the long-term cleaning effects and reduced total test time. A successful customer implementation shows the benefits of this approach within a high-volume testing environment.

Keywords— first pass yield improvement; cost of test; test socket cleaning; auto-contactor cleaning

INTRODUCTION

With the continuous and rapid growth of the device types and stacked packages needed for 5G, mobile, IoT, and IoV devices (such as Infotainment, ADAS, drivetrain sensors, etc.), there is a high demand for accurate and stable package testing to attain high first pass yields and reduced total test times. Contactor cleanliness adversely affects socket performance and can impact the device first pass yields of high value devices such that recovery testing is required, especially for automotive devices [1, 2].

During high volume package testing, yield metrics are typically monitored and when performance specification limits are exceeded, the testing operations must be paused to facilitate some level of cleaning execution [2, 3]. Reactive manual cleaning (i.e., cleaning after yield drops below the specification limit) is time intensive and results in decreased or variable yields as well as reduced throughput. Proactive manual cleaning (i.e., cleaning at some predetermined interval before the yield drops below the specification limit) results in increased test time reduced overall throughput, but increased composite yields; however, different devices can have varying sensitivities to contamination [4].

Some socket cleaning operations can be performed manually; however; optimized automated contactor cleaning (ACC) using handler functions will provide fast proactive and effective cleaning execution. Auto-contactor cleaning (ACC) is available on modern production handlers that are tray fed or have a pick-and-place functionality [5]. Other handler types, such as bowl-feed, do not have an automated cleaning functionality and these features need to be developed.

In order to attain desired process improvements, International Test Solutions engaged with a key end-customer that was working to develop a cleaning solution which could be implemented via software and hardware enhancements into an SRM XD248 Bowl-Fed handler. The primary project objectives were to develop and incorporate process improvements that would maintain high first pass and composite yields without impacting the overall total test times.

METHODS AND MATERIALS

Device Handler

The SRM XD248 handler was used for testing small packages sizes ranging from 1.0mm x 0.6mm to 12mm x 12mm [7]. Various hardware enhancements and software improvements were developed by the end-customer and implemented by SRM to facilitate auto-cleaning functionalities.

Leadless IC Package Devices

Three high volume leadless packages were selected by the end-customer for this project.

TABLE I – Device Overview
Leadless Package Types and Dimension

Device	Dimensions
A	4.5mm x 3.5mm x 0.89mm
B	4.5mm x 5.0mm x 0.91mm
C	4.5mm x 5.0mm x 0.88mm

All three devices were being regularly tested in very high volumes with approximately 600 to 1M devices tested per handler per month. Due to the very high production volumes, the impact of downtime associated with manual cleaning operations will significantly impact production volumes.

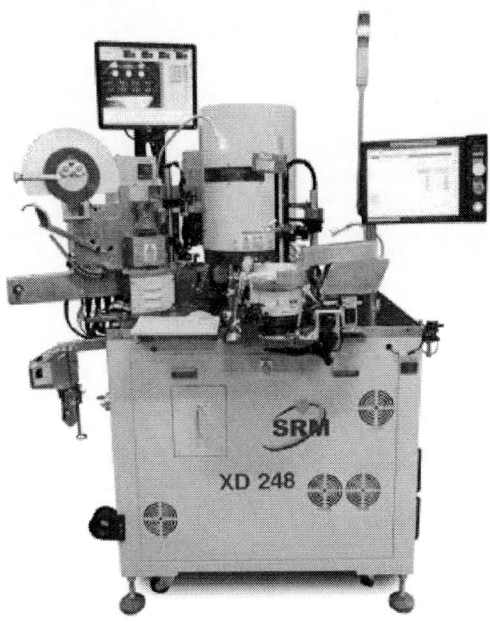

FIG. 1. SRM XD248 high-volume bowl feed handler for small package testing operations (Image from SRM website).

Test Contactor Cleaning Device

Test Contactor Cleaning (TCC) Devices are "turn-key" surrogate packages fabricated to match device package geometries. As shown in Fig. 2, these cleaning units are built to emulate the device package type and geometry. Depending on the customer application requirements, the cleaning efficiency of these materials can be optimized in order to maximize the performance across a wide temperature range [6].

Two basic types of cleaning materials are available for on-line cleaning socket cleaning applications. The cleaning materials are (1) abrasive-polymer-based materials; or (2) tacky-abrasive-materials [4,6]. Both material types are engineered to collect debris from the pin contact area, remove debris accumulated within the bed of the socket, and polish the contactor surface to recover electrical performance. The cleaning devices can be regularly cycled through a handler to remove contaminants with minimal downtime.

Depending on the customer application requirements, the cleaning efficiency of these materials can be optimized in order to maximize the performance across a wide temperature range.

Manual and Auto-Cleaning Process Assessment

Three production systems for testing Device-A were selected such that 1-system was enabled with the ACC function; and the other 2-systems used manual cleaning. Production testing was monitored for a one-month period during which approximately 700K devices were tested per system [7].

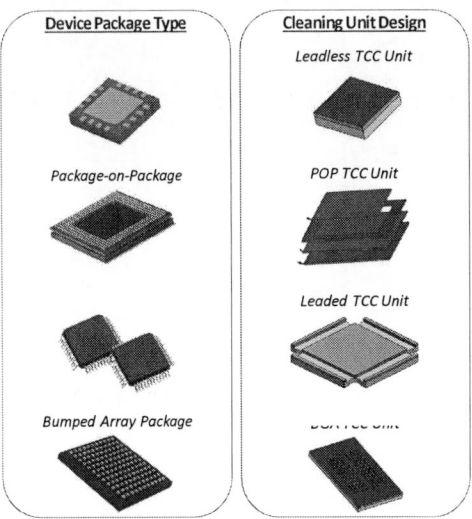

FIG. 2. Examples of Device Packages and associated Cleaning Unit Designs to facilitate Auto Contactor Cleaning execution [4].

Six production systems for testing Device B and Device C were selected such that 4-systems were enabled with the ACC function; and the other 2-systems used manual cleaning. Production testing was monitored for a one-month period during which approximately 1-million devices were tested per system.

For all three devices (A, B, and C), the first pass yield and retest yield metrics were monitored as well as the first pass test-time, retest time, and total test test-time, respectively.

RESULTS

In Table II, the average monthly yield and retest yield metrics for all three devices are summarized [7]. Implementation of the in-situ cleaning using the ACC function resulted in average first pass yield gains of 1.11%, 0.77%, and 3.23%, respectively, and for the three devices the re-test or recovery yields with ACC enabled were consistently lower. This is indicative of the overall effectiveness during the first pass test.

In Table III, the average test times per system for the three devices are summarized. Implementation of in-situ cleaning using ACC function resulted in consistent reductions in first pass test and retest times for all three devices. The total test time improvements ranged from 20-minutes for Device B to as much as 73-minutes for Device C [7].

*TABLE II – **Average Monthly Yield***
First Pass, Retest, and Yield Gain

Device A	First Pass Yield	Retest Yield
Manual Clean	97.25 %	1.40 %
ACC Clean	98.36 %	0.57 %
Yield Gain	*1.11 %*	*0.83 %*

Device B	First Pass Yield	Retest Yield
Manual Clean	92.49 %	3.04 %
ACC Clean	93.26 %	2.48 %
Yield Gain	*0.77 %*	*0.56 %*

Device C	First Pass Yield	Retest Yield
Manual Clean	88.77 %	3.86 %
ACC Clean	91.99 %	2.61 %
Yield Gain	*3.23 %*	*1.25 %*

*TABLE III – **Test Time Hours per System***
First Pass, Retest, and Test Time Reduction

Device A	First Pass Test Time	Retest Test Time	Total Test Time
Manual Clean	10.57	0.41	10.98
ACC Clean	10.16	0.22	10.38
Time Reduction	*0.40*	*0.19*	*0.60*

Device B	First Pass Test Time	Retest Test Time	Total Test Time
Manual Clean	10.15	0.89	11.26
ACC Clean	10.37	0.77	10.92
Time Reduction	*0.22*	*0.12*	*0.34*

Device C	First Pass Test Time	Retest Test Time	Total Test Time
Manual Clean	13.36	1.56	14.93
ACC Clean	12.64	1.06	13.71
Time Reduction	*0.72*	*0.50*	*1.22*

A combination of higher total yields and reduced overall total test times were obtained with all three devices upon implementation of the ACC in-situ cleaning.

COST BENEFITS

Device retest increases the overall Cost of Test. Major causes behind device retest are the electrical contact reliability and the socket cleanliness. With consistent ACC implementation, contact resistance is controlled to increase first pass yields and test system uptime for improved UPH. As a result of this project and cleaning process implementation, cost benefits of approximately US $500K per device per system per month were attained.

SUMMARY AND CONCLUSIONS

Historically, sockets have been maintained with off-line cleaning and pin replacement [1]. Ideally, socket cleaning should be performed in-situ to reduce system downtime. In some handlers, reactive or proactive manual cleaning methods require the system to pause and remain idle while the cleaning is executed. Manual cleaning methods increase unscheduled downtime will negatively affect UPH, throughput and consequently increase the overall Cost of Test (COT). Programmed in-situ preventative cleaning using the handler for "on-demand" cleaning execution will improve test cell efficiency [3,4]

In this project, surrogate cleaning devices were designed and built to match high volume leadless devices. All major handler suppliers have developed programmable auto contactor clean (ACC) functionalities capable of regular socket cleaning with minimal downtime [5]. In order to implement the benefits of on-line cleaning for high volume production with a bowl-feed handler configuration, the end-customer worked with the handler company to develop hardware and control software. Upon deployment of the cleaning function, significant improvements in first pass yield and total test time were attained.

Significant reductions in test time enable higher UPH values and reduce the need for additional test hardware as well as lowering the overall cost of test (COT). Depending upon the commercial value of the IC packages and the ASPs, small improvements in first pass yield and test time reductions will have immediate benefits for net revenue.

ACKNOWLEDGEMENTS

Authors would like thank Brent Edington, formerly of TriQuint Semiconductor, for his hard work on this project.

REFERENCES

[1] J. Broz, and G. Humphrey, "Controlling test cell contact resistance with non-destructive conditioning practices". *IEEE Burn-In Test Socket BiTS Workshop*, March, 2004.

[2] B. Gibbs and K. McNamamera, "Auto contact cleaning engineering study applied to package test". *Burn-In Test Socket BiTS Workshop*, March, 2007.

[3] B. Gibbs and K. McNamamera, "Auto contact cleaning engineering study applied to package test". *Semicon TechXPOT*, July 2007

[4] J. Broz and B. Humphrey, "Consistent online test socket cleaning for first pass yield stability and reduced retest.", *Burn-In Test Socket BiTS Workshop*, March 2012. Attendee Choice Award.

[5] J. Broz, "Package test is a dirty business - Socket cleaning strategies to reduce cost of test and improve overall equipment effectiveness (OEE)", *Burn-In Test Socket BiTS Workshop*, March 2013.

[6] J. Broz, S. Khavandi, and B. Humphrey, "Unique methodologies for investigating on-line cleaning process parameters and recipe optimization", *Burn-In Test Socket BiTS Workshop*, March 2014.

[7] B. Edington. "Yield and test time improvement via automated online cleaning". *Burn-In Test Socket BiTS Workshop*, March 2014.

A New CIS Packaging Process and Structure to Improve Die Chipping

Yuan Hu
Semiconductor Manufacturing International Corporation.
Shenzhen, Guangdong, China.
Tel: 0755-28615003, Email: Rainy_Hu@smics.com

ABSTRACT

The CIS product performs easily to receive customer complaints for a variety of screen issue. As a result, great effort was made by failure analysis team and found that the most of NG samples suffered die chipping during the module assessment. This paper presents a new CIS packaging process and structure which can effectively improve the film stress on the backside silicon of the CIS product, according to reduce the failure risk of the chip. The packaging process with the new structure has the character of simple craft and low cost as well as large-scale and roboticized production.

Keywords—CIS, packaging process, chipping.

INTRODUCTION

CIS chips are widely used in modern electronic products and related industrial systems. Increasingly demand on the high quality and low cost of the chips from customer and user makes more important to ensure CIS technology process including fabrication and packaging. Lately the CIS product performs easily to receive customer complaints for a variety of screen issue and it makes all the end customers and users confused that these fail products all passed the test by fabrication including wafer acceptance test, chip prober test, final test and module level functional test.

However, it's worth us to appreciate all the engineering team members to find the reason for the high failure level of CIS product in customer site is the chip suffered die chipping during the module assessment by a large number of FA analysis. This problem has a great impact on the quality and reliability of the final product, and it also influence customer satisfaction. To fundamentally overcome the aforementioned problems, a new CIS packaging process and structure which can effectively improve the film stress on the backside silicon of the CIS product, according to reduce the failure risk of the chip.

BACKGROUND

At present, chip sale package (CSP) technology is commonly used for CIS product. The process flow is as follows: Firstly, backside grinding and die saw, then form a certain pattern layout with bumping ball which can connect the die pad

in the backside of the chip. And then the chip can achieve the image sensing function of the product by using the CF (color filter) as the sensitive window. But the wafer grinding thickness is very thin as the product requirement from the market. What's more, the original wafer stress of backside is little higher and the punch force is much larger, it leads to the wafer easily suffer die chipping during the key process of setting bumping ball. The failure analysis team found there is abnormality from the NG samples appearance and then they confirmed it resulted in die silicon chipping at the wafer backside, as show in Fig. 1 and 2.

Fig. 1. NG Sample Appearance

Fig. 2. NG Sample FA Result

NEW STRUCTURE

To fundamentally overcome the aforementioned problems, a new CIS chip packaging process and structure which can effectively improve the film stress on the backside silicon of the CIS product, according to reduce the failure risk of the chip.

Described new CIS process methods include at least five steps. Step 1: Provided CIS chip which have not yet been encapsulated and the chip include middle device area and peripheral pad area, as show in Fig. 3.

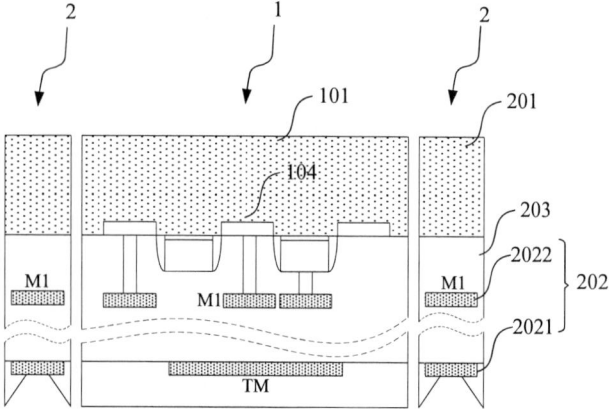

Fig. 3. Step 1 – Provided CIS Chip

Step 2: Form several trenches at wafer backside through etching process and fill insulating materials in the trenches, as show in Fig. 4.

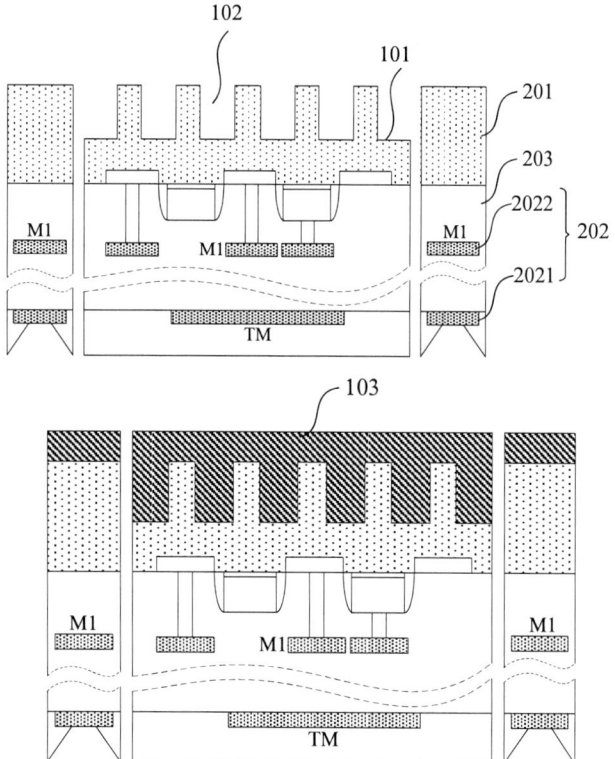

Fig. 4. Step 2 – Form and Fill Trench

Step 3: Etch the backside die silicon to form the open area in the peripheral pad area, so that the bottom metal layer which connected pad through the layout pattern design will be exposed, as show in Fig. 5.

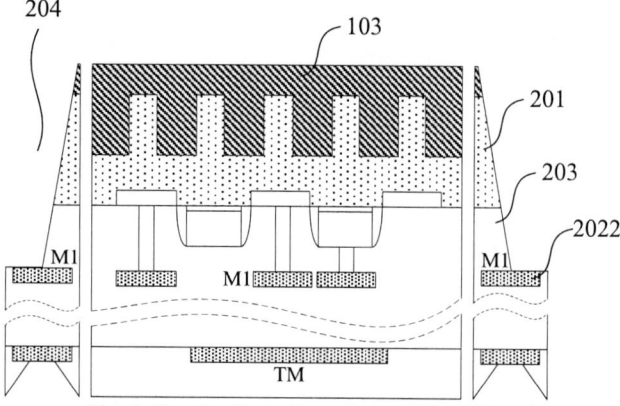

Fig. 5. Step 3 – Form Open Bottom Metal Trench

Step 4: Form inslulating layer at open bottom metal trench, and make the metal connective line which need cover the open bottom metal layer and partial backside, as show in Fig. 6.

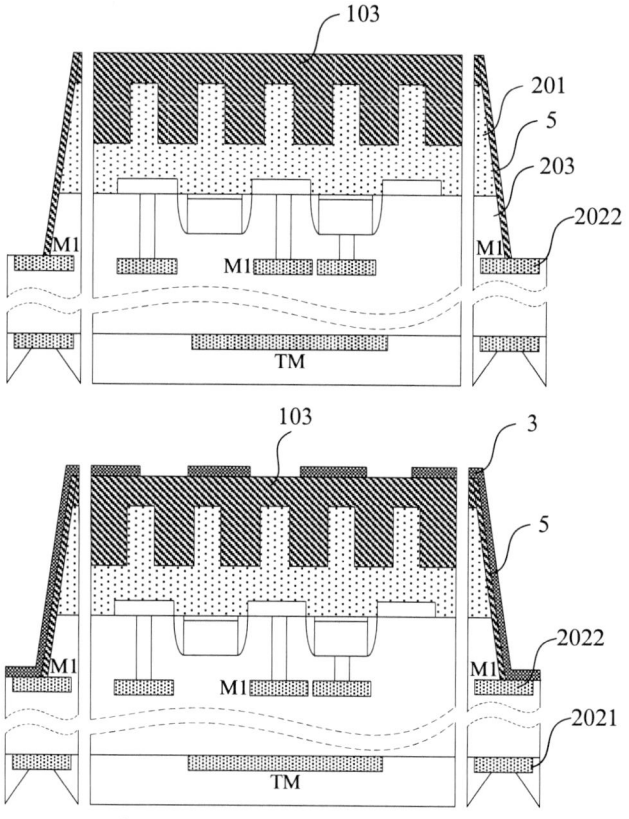

Fig. 6. Step 4 – Form Metal Connective Line

Step 5: Form protective layer at chip backside and set bumping ball, as show in Fig. 7.

978-1-7281-6559-2/20 $31.00 © 2020 IEEE   610

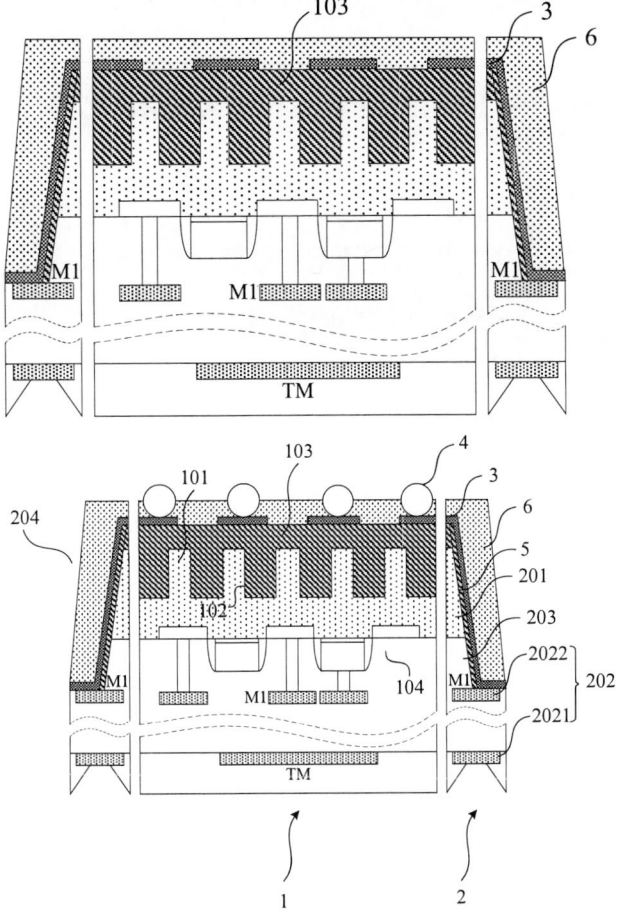

Fig. 7. Step 5 – Form Metal Connective Line

CONCLUSION

The new CIS process is used to fill the insulating material layer in the backside of the middle part of device area and form substrate insulation material structure. It can effectively relieve the internal stress of the substrate itself and the pressure of bumping ball at the back of the substrate to reduct the die chipping failure and improve product encapsulation yield. Besides, the process is simple to operate, low cost and suitable for commercialization.

References

[1] Erhu Zheng, Yi Huang, and Haiyang Zhang,"Applying optimal experiment design in reversed self-aligned contact etch of nor flash for profile performance improvement", China Semiconductor Technology International Conference (CSTIC), Shanghai, pp.1-3, March 2012.

[2] Atkinson, A.C. and Fedorov, V.V., "Optimal design: Experiments for discriminating between several models." Biometrika, 62, pp.289-303, 1975.

[3] F.Z. Zeng, "Robust Design Principle Technology", Method and Practice, Beijing, Weapon Industry Press, 2004.

A NOVEL DIE SORTER BASED ON MICRO TWEEZER FOR TERAHERTZ SCHOTTKY BARRIER DIODES

Li He[1,2], Yang Kai[1,2], Zhang Jie[1,2], Zhang Hao[1,2], Zeng Jianping[1,2], An Ning[1,2], Jiang Jun[1,2], Wang Xi[1,2]*

[1]Microsystem & Terahertz Research Center, China Academy of Engineering Physics (CAEP), Chengdu, 610200, China
[2] Institute of Electronic Engineering, CAEP, Mianyang, 621900, China
*Corresponding Author's Email: wangxi@mtrc.ac.cn

ABSTRACT

A novel die sorter based on mechanical gripper was developed for picking and placing diced terahertz Schottky barrier diodes (SBD). The designing and working principle of die sorter are based on flexible hinge and linear-actuator. Dimensions of dies gripped by the sorter can be as small as 100 μm * 50 μm* 15 μm without damage to its air-bridge. For flipping upside down dies, the rotation motion was obtained by rotating linear actuator about its shaft. It is demonstrated that the placement accuracy of the sorter can achieve less than 9.5 μm and 3.0 μm in x-y direction respectively.

INTRODUCTION

To reduce parasitic parameters and improve cut-off frequency, air bridge is used as connector between junction region and pad in terahertz SBD [1-4]. However, during assembly process these fragile air bridges result great challenges to conventional vacuum nozzle based die-sorting technique. By vacuum nozzle picking dies from dicing tape would result in damage to air bridge due to direct contact between nozzle tip and dies top surface. An effective approach to avoid such issue is to hold die by its side edges. However, due to thin thickness of SBD die (down to 15 μm), the vacuum nozzle type picking by its side edge is even more difficult. Hence a micro tweezer instead of a vacuum nozzle would be more suitable for picking dies from side edges. Motivated by such requirement, a micro tweezers based mechanical gripper was invented to sort terahertz SBD dies [5, 6].

DESIGNING AND WORKING PRINCIPLE OF GRIPPER

The mechanical gripper mainly consists of linear-actuator, coupling, connecting rod, collet sleeve, collet and micro tweezer, as illustrated by Figure 1.

The collet is fixed to outer shell of actuator by coupling which is a steel tube with two symmetry and parallel rectangular holes. Non-rotating rectilinear motion of actuator shaft can be transmitted to collet sleeve, another steel tube outside coaxial to coupling, by a connecting rod through the rectangular holes of coupling and square holes of collet sleeve. The connecting rod is fixed to actuator shaft by screw and screw hole.

Collet is a steel tube with increasing outer diameter with distance far end from coupling, and split longitudinally along its most length into two equal sections. Two microgrooves were cut at the collet's tip for installing of the micro tweezer shown in Figure 1C, a laser-cut sheet steel with intentionally designed shape for fixing in microgrooves.

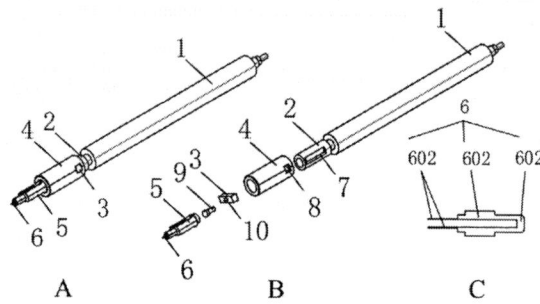

Figure 1: A, assembled gripper. B, exploded view of gripper. C, details of micro tweezer. 1, actuator. 2, coupling. 3, connecting rod. 4, collet sleeve. 5, collet. 6, micro tweezer. 7, rectangular holes of coupling. 8, square holes of collet sleeve. 9, screw. 10, screw hole.

As computer-controlled actuator shaft pushes the connecting rod and collet sleeve forward, internal diameter of collet sleeve would squeeze the collet. Consequently, two spaced split sections of collet pinch inward the micro tweezer and thus provide a clamping mechanism for terahertz SBD positioned inside the micro tweezer. Contrarily, actuator shaft's backward motion would release the clamped SBD.

SBD dies are as thin as 15 μm which is extremely fragile. An advantage of using such mechanism is the gripping force can be precisely controlled by adjusting forward distance of actuator shaft in step size of 0.05 μm. In this approach, such gripping mechanism enables gripping force control by neither cracking die caused by exceed an amount of large force nor dropping off caused by insufficient force in pick-and-place procedure.

As shown in Figure 2, the gripper was installed on a 4 degrees of freedom (DOF) (x, z, rotation about z and rotation about actuator shaft) mechanical arm. Diced dies for picking were placed on a precision 3 DOF (x, y and rotation about z) stage. Two long working distance microscope cameras for vertical positioning and side inspection were equipped to this die sorter.

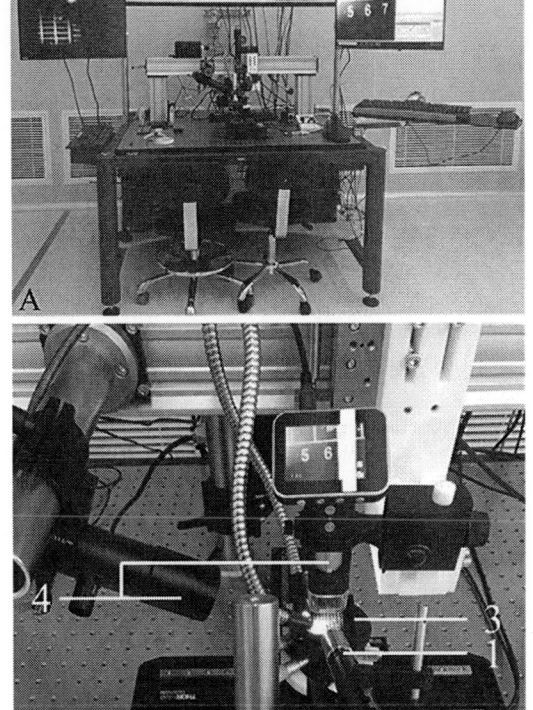

Figure 2: A, overall view of die sorter. B: 1, the gripper; 2, the mechanical arm; 3. three DOF stage; 4, two microscope cameras.

RESULT AND DISCUSSION
Capacity for Sorting Terahertz SBD Dies

Three types of terahertz SBD dies with different dimensions, 240 μm * 70 μm, 200 μm * 50 μm, and 100 μm * 50 μm, were all sorted un-damaged by this die sorter.

Figure 3A and B show diode dies holding by the gripper. Post-sorting inspection results in Figure 3C show the die sorter can effectively avoid air bridge cracking as the gripping of die by its side edges without contacting top surface.

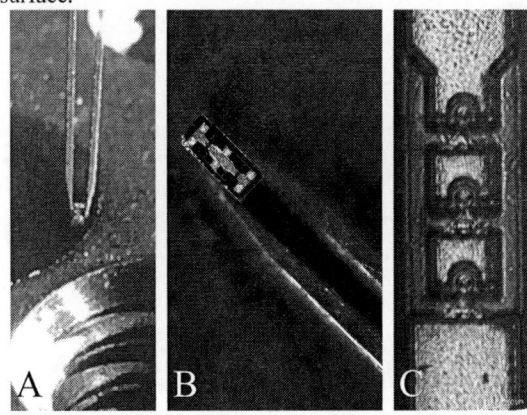

Figure 3: A and B, diode dies gripped by the sorter. C, typical post-sorting inspection results: intact air bridges on the smallest (100 μm * 50μm * 15 μm) SBD dies picked by the sorter.

Flipping of SBD Dies for Assembly

The fabrication process of sub-100 μm scale SBD dies involves die splitting and filtering selection process. All SBD dies were dispersed in acetone after split by chemical etch from a 4-inch wafer. Then, during the filtering selection process, numbers of these scattered dies randomly were left up-side down. To flip SBD dies over for following positioning and assembly process, a bevel gear box was installed on the mechanical arm to achieve the fourth DOF by rotating linear-actuator about its shaft. In addition, flipping motion allows the feasibility of inspecting SBD die from multi-perspective angles under microscope camera, as shown by Figure 4.

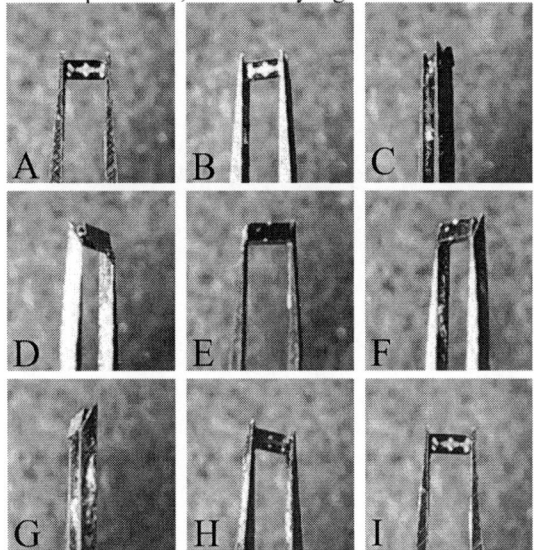

Figure 4: microscope camera was inspecting the die flipped over by the sorter from multi-perspective angles.

Placement Accuracy

As shown in Figure 5, placement accuracy was examined by aligning and stacking a cross-marked square silicon slice onto reference mark. Cross-marked square silicon slice was gripped by the sorter by its sides. With the controlled movement of 3 DOF stage, the cross-marked square silicon slice was placed carefully onto the reference mark, finally inspected by microscope. It is demonstrated that x and y placement accuracy can achieve 9.5 μm and 3.0 μm respectively. This specification could be further optimized by equipping higher magnification and resolution microscope camera to monitor aligning operation clearer.

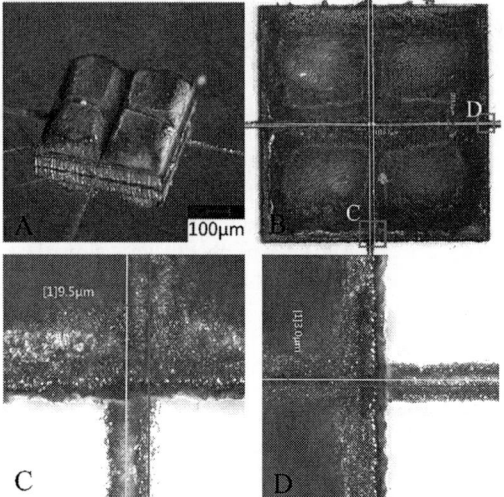

Figure 5: A and B, two aligned crosses. C and D, enlarged red boxes in B for measurement of deviation between two crosses. Placement accuracy can achieve 9.5 μm and 3.0 μm in x-y direction respectively.

CONCLUSION

In this paper, we demonstrated a novel mechanical gripper-based sorter could provide a damage-free picking and placing approach for micron scale (100 μm * 50 μm* 15 μm) terahertz SBD dies with flipping motion and sub-10 μm placement accuracy. Future works would focus on integration of secondary mechanical arm with dispensing module to this sorter and further application on die bonding such as surface mounting techniques and flip chip.

ACKNOWLEDGEMENTS

This work is supported by Science Challenge Project, No. TZ2018006-0205-05.

REFERENCES

[1] Shixiong Liang, Yulong Fang, Dong Xing, Junlong Wang, Lisen Zhang, Dabao Yang, Xiongwen Zhang, Zhihong Feng. 2014 12th IEEE International Conference on Solid-State and Integrated Circuit Technology (ICSICT), Guilin, 2014, pp. 1-3.

[2] Chang Xiong, David Massoubre, Erdan Gu, Martin D. Dawson, Ian M. Watso. Applied Physics A, 2009. 96: pp. 495-501.

[3] Tianhao Ren, Yong Zhang, Shuang Liu, Fangzhou Guo, Zhi Jin, Jingtao Zhou, Chengyue Yang. Journal of Infrared, Millimeter, and Terahertz Waves, 2017. 38(2): pp. 143-154.

[4] E. Giovine, R. Casini, D. Dominijanni, A. Notargiacomo, M. Ortolani, V. Foglietti. Microelectronic Engineering, 2011. 88(8): pp. 2544-2546.

[5] John T. Feddema, Ron Simon, Marc Polosky, Todd Christenson. Sandia Report, SAND99-0746, Printed April 1999.

[6] John Feddema, Marc Polosky, Todd Christenson, Barry Spletzer, Ron Simo. World Automation Congress '98, Anchorage, Alaska, May 10-14, 1998

A NOVEL DISPENSING HEAD FABRICATION METHOD FOR PRECISE EPOXY DISPENSING

Zhang Jie[1,2], Li Zheng[1,2], Li ruoxue[1,2], Li He[1,2], Wang Xi[1,2]*

[1]Microsystem & Terahertz Research Center, China Academy of Engineering Physics (CAEP),
Chengdu, 610200, China

[2]Institute of Electronic Engineering, CAEP, Mianyang, 621900, China

*Corresponding Author's Email: wangxi@mtrc.ac.cn

ABSTRACT

Dimensions of high frequency dies are typically less than 50 μm * 200 μm. It requires a micron scale adhesive dots dispensing method to the bonding area for micro assembly. In this paper, electrochemical-polishing and picosecond laser micro-machining were used to prepare such novel dispensing head. These needle-like dispensing heads with tip diameter of 5 μm can achieve 20 μm adhesive dot far smaller than 100 μm produced by standard industry equipment. The dispensing module was integrated into a custom-made motion stage with microscopy system to achieve higher dispensing accuracy.

INTRODUCTION

Micro assembly for high frequency die requires high precision bonding process.[1] With the reducing dimension of high frequency die, the dispensing error and dispensing dot size becomes extremely important for functioning of fabricated devices [2]. There are multiple methods for bonding points formation. Conventionally, jet dispensing method is often used for the dot size around 100 μm. To achieve dot size less than 100 μm, smaller needle tube will be used, however, the clogging issue is difficult to resolve [3]. Another dispensing method is daubing dispensing [4], which is developed for dots diameter around 80 μm. It is to dip epoxy with dispensing head and then dip onto the bonding area to form the adhesive point. To further achieve smaller dispensing dots for micron scale high frequency dies, it is purposed to use electro-chemical polishing and picosecond laser machining for preparing a novel dispensing head with tip diameter less than 5 μm for daubing dispensing method. The dispensing module was integrated into a custom-made motion stage with microscopy system to achieve higher accuracy than the standard industry equipment which can only achieve 5 μm [6]. Different epoxy dispensing dots diameters and dispensing accuracy were investigated in this paper.

DISPENSING HEAD FABRICATION

The fabrication of needle-like dispensing heads is using electrochemical polishing method. The working principle for such method is based on the current density and electro-field potential differences in the polishing electrolyte, the anodization reaction and oxide etching speed on the stainless-steel rod tip and body significantly different. Faster anodization reaction and oxide etching speed at the tips due to higher current density generated the needle-like dispensing heads. The schematic diagram of electrochemical polishing setup was shown in Figure 1. The setup includes external voltage supply, stainless steel rod, graphite plate, magnetic stirrer and electrochemical polishing electrolyte.

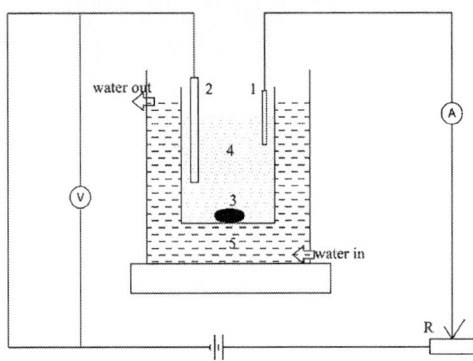

Figure 1: Schematic diagram of electrochemical polishing process. (1) stainless steel rod, (2) graphite electrode, (3) magnetic stirrer, (4) electrochemical polishing electrolyte, (5) water circulating bath

In the experiment, stainless steel rod with diameter of 1mm was used as anode, graphite plate was used as cathode. The anodization voltage used was 25 V, at current of 0.5 A. The electrochemical polishing reaction was carried out in polishing electrolyte of 11 vt.% DI water, 47 vt.% glycerol and 42 vt.% phosphoric acid. In the reaction process, magnetic stirrer was used to stir polishing solution, stirring speed was 600 rpm. In order to keep the reaction process at a relatively stable temperature, the process was carried out in a water circulating bath. After 250 minutes of electrochemical polishing, the stainless-steel rod is polished at the distance of 1-2 mm from the electrolyte surface until it breaks. The needle tip shown in Figure 2 is the final needle-like dispensing heads. Then the tip is further processed by picosecond laser to optimize the tip surface morphology. The tip diameter is less than 5 μm.

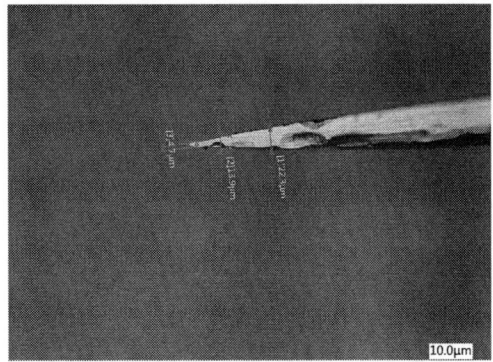

Figure 2: Optical image of needle-like dispensing head optimized by pico-second laser. The tip diameter is 4.7 μm

EXPERIMENTAL AND RESULT

As shown in Figure 3, the prepared dispensing head was installed on a 3D printed chuck to become a dispensing module, this dispensing module was installed on a motion stage with 3 degrees of freedom(DOF)(x,y and z) and two long working distance microscope cameras were used for process monitoring and visual positioning. Digital images captured by these two cameras (vertical and 45 degree) were shown in Figure 4.

Figure 3: Dispensing head installed on a 3D printed chuck

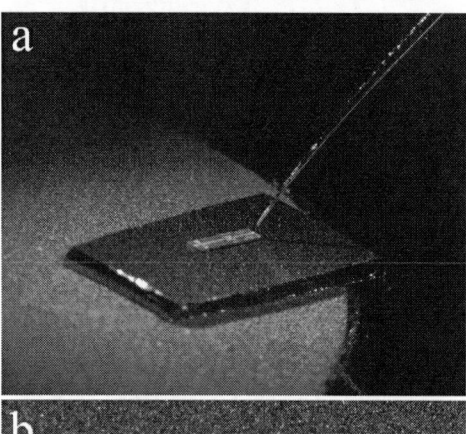

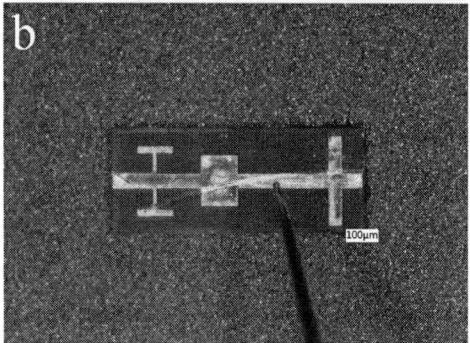

Figure 4: a, Optical image of dispensing head and substrate captured by process monitoring camera (45 degree); b, overlook image of dispensing head and substrate captured by vertical camera

A commercially available two-component 100% solids silver-filled epoxy system (EPO-TEK H20E) was used to verify the performance of this dispensing head. Different sizes of dots from 20 μm to 100 μm can be stably obtained by adjusting the process parameters such as contact time and distance between the needle tip and the substrate. It is demonstrated in Figure 5 that this head could dispense dots with diameter less than 20 μm.

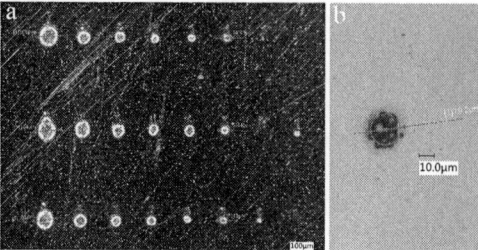

Figure 5: a, Different sizes of epoxy dots from 20 μm to 100 μm; b, Zoomed-in view of dot A, this dot is 19.2 μm in diameter

In order to investigate the accuracy of this dispensing module, the offset error between the center of dot and the center of reference cross was measured. As shown in Figure 6, these round and uniform dots were about 400 μm in diameter, and the measured offset error between the achieved result and the target is less than 3 μm.

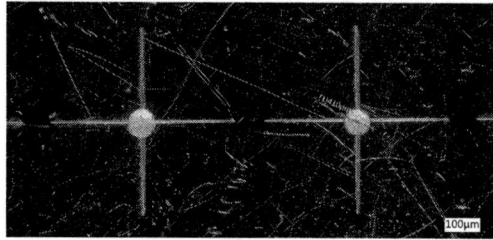

Figure 6: Optical image of epoxy dots on cross mark, these dots were 400 μm in diameter and offset error is less than 3 μm

To evaluate the mechanical properties of these epoxy dots, die shear tests were performed after die attach process using a shear-testing machine (Royce, UK). These dies were 600 μm square and 200 μm thick. The shear speed was 500 μm/s, and the shear height was 30 μm. Seven attached dies were tested to ensure accuracy. As shown in Figure 7, the shear strengths of most samples were between 50 and 60 MPa, it achieved the industry standards of electronic packaging. However, the shear strength of sample 4 is only 27.37 MPa .

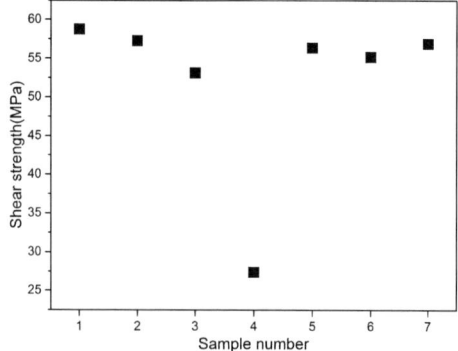

Figure 7: The shear strength of die attached samples. The shear strengths were between 50 and 60 MPa except the sample 4

The morphology of the interface after shear test was further observed under the microscope, as shown in Figure 8. It shows that the effective attached area is only half of the die area. The reason for the abnormal shear strength of the sample is due to insufficient applying downward pressure result in less coverage of adhesive dot during the die attach. In the future, integrating the dispensing module with the pressure sensor can effectively avoid this issue, and the shear strength can be further improved.

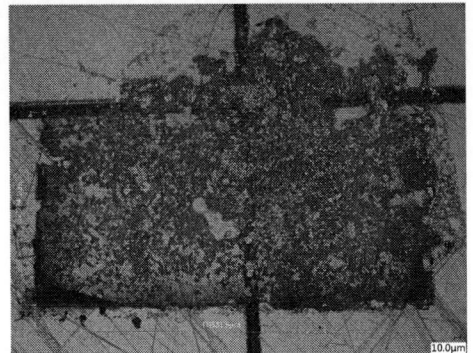

Figure 8: The morphology after die shear test of sample 4, its effective attached area is only half of the die area

CONCLUSION

In summary, we demonstrated the fabrication and performance of a novel dispensing module can achieve relatively higher precision than standard industry equipment.

The needle-like dispensing head with a diameter of less than 5 μm can be produced by a simple electrochemical polishing method. The minimum epoxy dot of 20 μm diameter was obtained by daubing with this dispensing head. The dispensing module integrated into a custom-made motion stage with microscopy system is feasible for typical micro assembly applications include solder paste, silver epoxy, and dispensing micro bumps. In future, micro scale pressure sensor will be developed and introduced into the system for further closed loop assembly force controlling to achieve more reliable die attachment.

ACKNOWLEDGEMENTS

This work is supported by Science Challenge Project, No. TZ2018006-0205-05.

REFERENCES

[1] Khanal, Subash, et al. Global Symposium on Millimeter Waves (GSMM) & ESA Workshop on Millimetre-Wave Technology and Applications , Espoo, June 6-8,2016.

[2] J. Vivari. Electronic Packaging Technology Conference, Singapore, December, 2005.

[3] S. J. Adamson, Asymtek, Carlsbad, CA, USA, MEPTEC Rep., 2003, pp. 1–6.

[4] Kolbe J, Arp A, Calderone F, et al. Microelectronics Reliability, 2007, vol. 47, Issues 2-3, pp. 331-334.

[5] H. L. Chen. International Conference on Digital Printing Technologies, 2003, pp. 333-337(5)

[6] Bursik M., et al. Microelectronics Packaging Conference IEEE, 2014.

VOLUME RESISTANCE OF EPOXY MOLDING COMPOUND

Hongjie Liu, Wei Tan, Xingming Cheng, Lanxia Li, Yangyang Duan, Dandan Fan, Xiaojuan Jiang, Liang Cui, Jianglong Han

JiangSu HuaHaiChengKe Advanced Materials Co., Ltd, Lianyungang, Jiangsu, 222047, China
*Corresponding Author's Email: Hongjie.liu@hhck-em.com

ABSTRACT

Epoxy molding compound, with excellent electrical insulation and mechanical properties as well as low water absorption, has been widely used in insulation application of generators, motors, transformers, LED drivers and other electrical apparatus. With the power increase of the semiconductor device, the rigid requirement of electrical behavior of packaging materials becomes necessary. Here, the electrical performance of epoxy molding compound have been investigated from aspect of volume resistance. Types of epoxy resin, harder, and ion catchers on the influence of electrical performance of epoxy molding compound were carried out. And the results disclosed that epoxy molding compound with epoxy resin D and phenol hardener H possess the highest volume resistance, epoxy molding compound combined with epoxy resin C and phenol hardener G showed the lowest data, Higher Tg contribute to higher volume resistance. Ion catchers showed positive effect on the volume resistance of epoxy molding compound with small amount, and there is a download trend of EMC volume resistance with the ion catcher content increasing from 1% to 2%. And ion catcher C made greater contribution to the volume resistance of epoxy molding compound compared to ion catcher A and ion catcher B when adding at the same level.

INTRODUCTION

Epoxy molding compound, with excellent electrical insulation and mechanical properties as well as low water absorption, has been widely used in insulation application of generators, motors, transformers, LED drivers and other electrical apparatus. With the power increase of the semiconductor device, the rigid requirement of electrical behavior of packaging materials becomes necessary, and the electrical performance of epoxy resin has become one of the major topic researchers pay close attention to. As epoxy resin will be exposed to various environmental conditions, B. Lutz, etc. al.[1] have investigated the Influence of absorbed water on volume resistivity of epoxy resin insulators, and found that the volume resistivity of the epoxy resin samples decreases over 7 orders of magnitude with increasing water content. Z. L. Xing, etc. al.[2]. have Studied on DC breakdown characteristics of epoxy insulating materials, they disclosed that with the increase of the dielectric constant, the DC breakdown strength of epoxy insulating material decreases; With the increase of the volume resistivity, the DC breakdown strength of epoxy insulating material

increases. Y. Xia. Etc. al.[3] have developed the newly toughening modified Epoxy-Anhydride VPI resin which is applied in the evaluation of electrical insulation of coils for HV motors and generators. And it expressed good insulation properties, including dielectric loss, breakdown strength, partial discharge, and electrical endurance compared with Unsaturated Polyester imide-Vinyl toluene and commercial epoxy-anhydride VPI resin system. B. S. Kong, etc. al.[4] have tried to improve the electrostatic characteristics of EMC applied package and reduced the electrostatic damage of IC device by changing curing accelerator and modifying the volume resistivity of epoxy molding compound (EMC), and found that EMC with phosphonium salt accelerator results in much lower ESD failure than EMC with phosphine salt accelerator. M. Funabashi, etc. al.[5] have investigated the volume resistivity of carbon-fibre-filled epoxy-resin samples under shear flow by a rheometer with double cylindrical sample cells, and they discovered that by replacing silicone oil with epoxy resin, reliable and reproducible data of the volume resistivity of carbon-fibre-filled suspensions under shear flow could be derived over a wide range of shear rate and a wide range of fibre content. Here, we tried to explore the electrical performance of epoxy molding compound from the volume resistance point of view. Different types of epoxy resin, hardener, as well as ion catchers and catalyst have been chosen as factors influencing the volume resistance of the epoxy molding compound.

EXPERIMENTAL

Materials

Cresol novolac type epoxy resin EOCN-1020, Xylylene type type epoxy resin NC2000, Biphenyl contained epoxy resin NC3000, Triphenylmethane Type epoxy resin EPPN-501 were got from Nippon Kayaku Co., Ltd. Phenol hardeners Crseol Novolac Type HF-1M, Xylylene Type MEHC-7800SS, Biphenylene type MEHC-7851SS and Triphenylmethane Type MEH-75003S were all obtained from MeiWa Plastic Industries, Ltd. Triphenylphosphine as accelerator was got from HOKKO's Fine Chemicals. Filler was chosen as the spherical fused silica from Novoray Corporation.with the mean diameter as 23μm and maximum filler size as 75μm. Silanes were form shin-Etsu Chemical Co. Ltd., mold releasing agent was carnauba wax obtained from Foncepi Comercial Exportadora Ltd. Ion catcher A, Ion catcher B and Ion catcher C were gained from Kyowa Chemical Industrial Co. Ltd..

TABLE I.   STRUCTURE OF EPOXY AND HARDENER

Code	Epoxy and hardener	
	Name	*Structure*
A	EOCN-1020	
B	NC2000	
C	NC3000	
D	EPPN-501	
E	HF-1M	
F	MECH-7800SS	
G	MECH-7851SS	
H	MEH-75003S	

EMC preparation

All the ingredients were weighted up according to the formulation and mixed in the high speed mixing machine, after this, the mixture was feeding to double screw extruder under the heating condition for extrusion, and then the discharged material go through sheet formation, cooling, pre-braker, granulation, post blend, and storage at 5 °C.

Measurements

Volume resistance test

The specimens with the diameter of 100mm and thickness of 2mms were prepared through transfer molding method and post mold cure (PMC) at 175°C for 6 hours. And then samples were tested at 30°C, 100 °C, 120°C and at 150°C separately by Ultra High Resistance Meter 5450 from ADCMT with the high temperature oven under the testing voltage of 500V.

Glass Transition Temperature (Tg)

The specimens of 50mm×13mm×3mm were prepared through transfer molding method and post mold

cure (PMC) at 175°C for 6 hours. The Dynamic Mechanical Analysis (DMA) of the cured specimens was performed in the three point bending mode under 1Hz sinusoidal strain loading on the Dynamic Mechanical Analyzer (DMA, TA Instruments, Q800) with the temperature ramping from room temperature to 265°C at the rate of 5°C/min. Peak of tanδ were recorded as Tg.

RESULTS AND DISCUSSION

Influence of different resin on volume resistance

Epoxy resin and hardeners are the two most important ingredients in the epoxy molding compound. Thus the influence of structures of epoxy resin and hardeners on the volume resistance of EMC have been discussed here. EMC samples with different resin combination from Table I have been prepared with the filler content as 88% and catalyst as TPP. The volume resistance, Tg were tested. The Detailed results were disclosed in Table II and Fig. 1.

TABLE II.   RESIN COMBINAITON AND TG OF EMC

Num.	Resin combination and Tg		
	Epoxy	*hardener*	*Tg/℃*
1	A	E	187
2	A	F	131
3	B	F	125
4	C	F	122
5	C	G	120
6	D	H	210

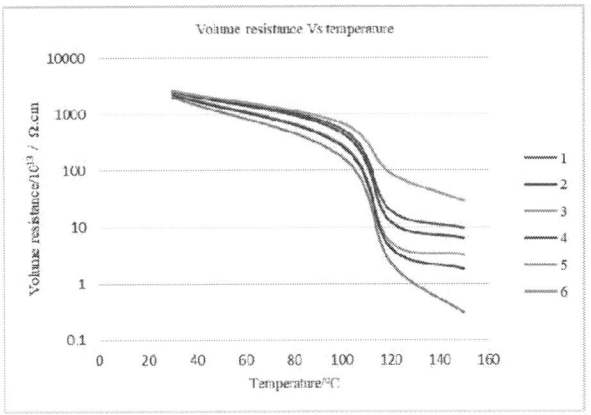

Figure 1: Volume resistance of EMC with different resin combination vs temperature

From the above results, we can see that, the volume resistance of epoxy molding compound decreased with the increasing of the temperature. EMC samples possess higher Tg displayed higher volume resistance under the equal condition compared with that with low Tg. Epoxy molding compound combined with epoxy resin D and hardener H disclosed highest Tg and volume resistance, and epoxy molding compound with epoxy C and hardener G displayed the lowest data both at low and high temperature.

Influence of different ion catcher on volume resistance

Ion content is one of the most significant factors influencing the electrical performance of epoxy molding compound, so here we tried three different types of ion catcher: ion catcher A (ICA), ion catcher B (ICB) and ion catcher C (ICC) in epoxy molding compound, and the influence of the ion catcher content on the volume resistance of epoxy molding compound have been tested, the results were revealed in Fig. 2.

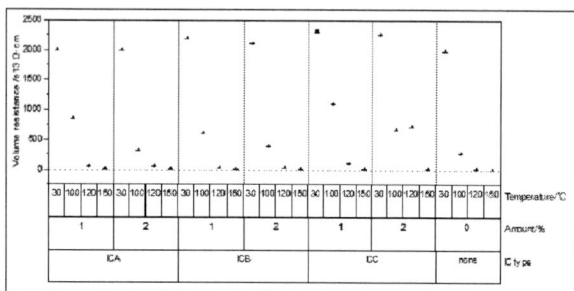

Figure 2: Volume resistance of EMC with different types and content of ion catcher at different temperature

From the above figure we can see that: whatever type ion catcher in, the volume resistance of epoxy molding compound has been improved in the testing temperature range. And there is a download trend of EMC volume resistance with the ion catcher content increasing from 1% to 2%. And the ion catcher C made greater contribution to the volume resistance of epoxy molding compound compared to ion catcher A and ion catcher B when adding at the same level.

CONCLUSIONS

The influence of different types of epoxy resin, hardener and ion catcher on the volume resistance of epoxy molding compound have been investigated here. And the results disclosed the both epoxy resin and hardener, which possess higher Tg make greater contribution to volume resistance. Perhaps higher Tg with high cross-linking density can prevent the movement of the ions effectively in the testing temperature range. Meanwhile, the ion catcher show positive effect on the volume resistance of epoxy molding compound with small amount, and there is a download trend of EMC volume resistance with the ion catcher content increasing from 1% to 2%. And the ion catcher C made greater contribution to

the volume resistance of epoxy molding compound compared to ion catcher A and ion catcher B when adding at the same level. Here, only several typical types of epoxy resin, hardener, and ion catchers were investigated, there are more new types of epoxy resin, hardener, ion catcher and other additives influencing the electrical performance of epoxy molding compound, more detailed work will be done in the later days.

REFERENCES

[1] B. Lutz, J. Kindersberger. "Influence of Absorbed Water on Volume Resistivity of Epoxy Resin Insulators," *2010 International Conference on Solid Dielectrics,* Potsdam, Germany, July 4-9, 2010, pp. 1-4.

[2] Z. L. Xing, W. J. Lu, C. P. Li, C. Zhang, X. N. Wan, Z. Zhang, X. N. Shi, W. K. Li, Y. ZH. Zhou, J. SH. Ru, and X. Chen. "Study on DC Breakdown Characteristics of Epoxy," *12th IEEE International Conference on the Properties and Applications of Dielectric Materials,* Xi'an, China.

[3] Y. Xia, W. Wang, C. C. Tao, C. C. Li, S.B. He and W. Chen. "Application of Newly Developed Epoxy-Anhydride VPI Resin for High Voltage Motors and Generators," *2015 Electrical Insulation Conference (EIC),* pp. 511-514.

[4] B. S. Kong, S. S. Lee, D. F. Lee, H. O. Choi and H. W. Kim. "Electrostatic Discharge Failure control of IC Package by Epoxy Molding compound modification", *China Semiconductor Technology International Conference (CSTIC) 2017.*

[5] M. Funabashi. "Volume Resistivity of Carbon-fibre-filled Epoxy-resin under Shear Flow," *Polymer International, vol. 45, 1998,* pp. 303-307.

WARPAGE CONTROL METHOD IN EPOXY MOLDING COMPOUND

Wei Tan[1], Cheng Cheng[2], Hongjie Liu[1], Yangyang Duan[1], Linlin Liu[1], Xingming Cheng[1], Lanxia Li[1]*
[1]Jiangsu HHCK Advanced Materials Co., Ltd
[2] School of Language and Business, Jiangsu Normal University,
Lianyungang, Jiangsu,2220007,China
wei.tan@hhck-em.com

ABSTRACT

For asymmetric packaging, especially for one-side molded packages, the mismatch between EMC's CTE and substrate (PCV or leadframe) can lead to different degrees of package warpage. The extent of warpage also depends on package designs like package size, package thickness, chip size and etc. This paper discusses the factors that affecting the package warpage, the warpage control method from epoxy molding compound, and the packaging process parameters. The results show the package mold cap, transfer pressure and the weight of the post-curing blocks can affect the final warpage. Epoxy molding compound can control the package warpage by using stress modifier, amount of filler loading and Tg value.

Keywords—Epoxy molding compound; Warpage; Stress; Glass transition; CTE

INTRODUCTION

Warpage is one of the major concerns on the asymmetric packages such as BGA and QFN. It is very important to maintain a reasonable flat package for the singulation and board level soldering process. Warpage can be defined as a crying face or a smiling face when the epoxy molding compound is molded upward.

The mismatch of EMC's CTE and leadframe is the root cause of warpage. The package warpage with the same EMC aslo depends on different package design like package size, package thickness, chip size and etc.[1-2].

Commonly, the warpage shape depends on both geometries and properties of package's components as well as process conditions. In package geometries, the thickness of molding compound and leadframe are the main factors. Molding compound as the largest component in package usually controls the package warpage performance. Molding compound is a composite material consists of resins, hardeners, fillers, catalysts, coupling/release agents, and small percentage of other ingredients. The formulation percentage can be adjusted and material properties will change accordingly. Molding compound's properties, such as Tg, CTE and stress can control the warpage. CTE of molding compound is mainly controlled by filler contents. Tg usually depends on the stiffness of polymer chains, chain lengths, and free volume between the chains. The stress is controlled by the resin structure, the type and amount of stress releasing agent. In term of transfer molding process, injection pressure is the most significant factor control package warpage, Sometimes transfer pressure increased dramatically to eliminate the voids in the package can change the extent of warpage.

In this paper, the effects of Tg, mold cap, transfer pressure and weight of post-cure-block on the warpage were studied.

EXPERIMENTS

The compound was prepared through dry blending, followed by melt-mixing and extrusion into sheet form, It was then fine-ground into powder, and then pelletized into performs.

Standard transfer molding techniques were used to fabricate the test specimens. All test specimens were prepared with in mold cure time of 120sec at175°C followed by a six-hour post mold cure at the same temperature.

Tg and modulus were measured by DMA Q800, TA instrument model, from 25°C to 280°C at 5°C per minute. The peak of tanδ were considered as the Tg of epoxy molding compound.

Coefficient of thermal expansion (CTE) was measured by using TA Instrument TMA Q400 from 25°C to 280°C at 10°C per minute.

Molded array strips were prepared by ASM machine using a 2-cavities auto-mold on HFBP-06L without dies with standard molding process at 175°C. The cure time is 120s, the clamp pressure was 30ton, the transfer time is 12s. figure 1 showed one strip after post cured. The warpage of strips were measured after post cured at 175 °C for 4 hours under a 2 kg or 4 kg block.

Figure1 the HFBP-06L strip after post cured

Warpage measurements were made by high sensitivity Shadow Moire' method. The heights of four corners of each array strip were measured with averaging results and subtracting the average from the center height of the strip. The record values were all in positive and warpage directions ("crying" or "smiling") were noted.

RESULTS AND DISCUSSION.

Figure2 shows the effect of different transfer pressure and different weight of post-cure-block on the warpage. The Tg of epoxy molding compound used in these two experiments is 150℃, the filler loading is 86.5%. Table I is the data's analysis of variance.

In table I, both the Probability of transfer pressure and post-cure-block are larger than 0.05, it shows that the transfer pressure and post-cure-block are the significant factors that affect the warpage in the process parameters. Figure1 also shows when transfer pressure and the weight of post-cure-block increase, the warpage changes to cry face.

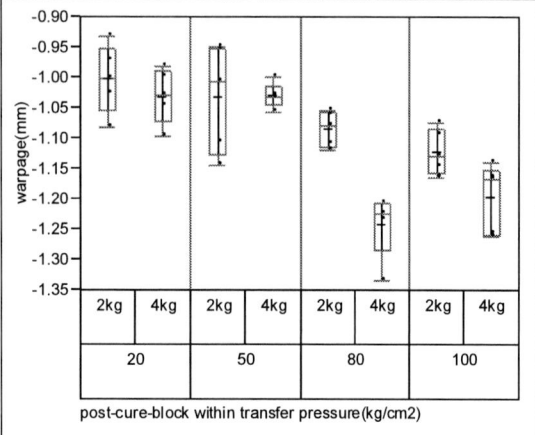

Figure 2 Strip warpage with different process parameters

Table I Analysis of Variance of transfer pressure and weight of post-cure-block

Source	DF	SS	Mean Square	F Ratio	Prob > F
transfer pressure(kg/cm2)	3	0.189	0.063	5.208	0.1043
post-cure-block	1	0.041	0.041	3.421	0.1615
transfer pressure(kg/cm2)*post-cure-block	3	0.036	0.012	4.420	0.0104
Within	32	0.087	0.0027		
Total	39	0.354	0.0091		

Figure 3 shows the effect of different thickness of mold cap on warpage. The thickness of mold cap is 0.5 mm and 0.75 mm respectively. It also shows mold cap is a significant factor controlling warpage with the same epoxy molding compound and process conditions. With the thickness of mold cap increases, warpage become more crying. It is because when the thickness of mold cap increases, the ratio of mold cap to lead frame thicknesses also increase, and the effect of EMC on warpage will gradually increase which eventually control the warpage.

Table II is a 2-level fractional DOE design with three factors and one center point experiment. The factors are filler loading(86.5% and 88%), amount of stress modifier (0.8% and 0.2%) and the glass transition temperature (125℃ and 160℃).

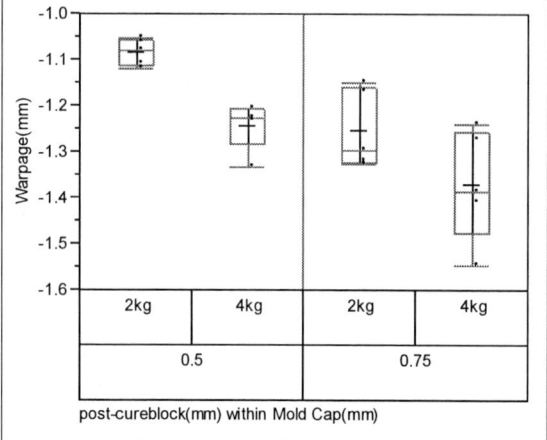

Figure 3 Strip warpage with different mold cap

Table II The effect of different EMC design on warpage

2-level Factors design	A	B	C	D	E
Filler loading(%)	86.5	86.5	88	88	87.25
Amount of stress modifier (%)	0.8	0.2	0.2	0.8	0.5
Tg (℃)	125	160	125	160	142.5
Warpage(mm)	1.33	-1.05	0.65	-1.62	-0.58

Figure 4 is the Pareto chart of the DOE factors effects. It shows that Tg(B), filler loading(A), interaction of 3 factors - Tg, filler loading and interaction of Tg, filler loading and quantity of stress modifier(ABC), are the significant ($\alpha = 0.05$) factors.

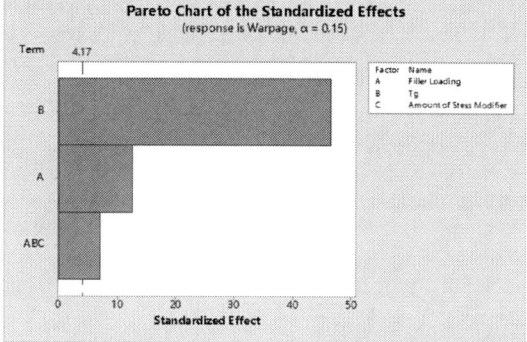

Figure 4 Pareto chart of the DOE factors effects

In addition, from the figure 4, filler loading (A) is the most significant factor, follow by Tg of molding compound. The interaction among the three factors- ABC filler loading, Tg and amount of stress modifier) is the smallest. From the study the amount of stress modifier has no significant effect on the warpage individually.

Figure 5 is the main effects plot of Tg and filler loading for the warpage.it shows that the increase of both Tg and filler loading lead to the development of warpage to cry face.

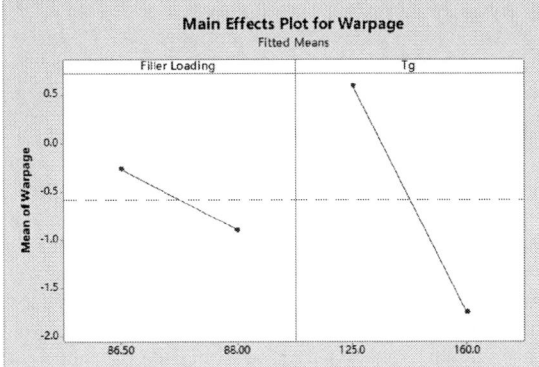

Figure 5 Main effects plot for warpage

Considering the problem of wiresweep, the filler loading of epoxy molding compound should be below 87% to ensure low enough viscosity. On the other hand, considering the slight crying warpage needed in the subsequent cutting process, the author uses the response optimizer function of MINITAB by setting the target warpage value to -0.3mm and the filler loading to less than 87% to predict the optimal value of factors which is shown in Figure 6.

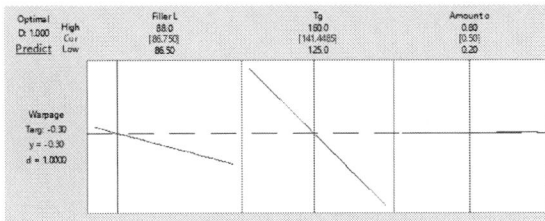

Figure 6 the warpage prediction optimization result

Figure 6 shows epoxy molding compound with filler loading of 86.75, Tg of 141.4 degrees and amount of stress modifer is 0.5 can obtain a package with warpage of -0.3 mm.

According to the requirements of Fig. 6, the author develops a formulation of epoxy molding compound. The basic information and performance of the formulation are shown in Table III. Table III shows that the formulation basically meets all the requirements especially the warpage requirement.

Table III shows the basic information and performance of Epoxy molding compound

Formulation details		
Resin	MAR+MFN	
Filler loading	86.7	
Filler type	Spherical silica	
Stress modifier	Silicon resin	
Catalyst	P type	
Basic properties		
Items	Unit	Value
Gel Time	S	35
Spiral flow	Cm	160
Tg	°C	145
CTE1	ppm	7.2
CTE2	ppm	35
Performance		
Mold cap	0.5mm	0.75mm
Transfer pressure	$80kg/cm^2$	$80kg/cm^2$
Weight of post-cure-block	4kg	4kg
Warpage(mm)	-0.53mm	-0.35mm
Wiresweep (%)	<13%	<10%

CONCLUSION

Epoxy molding compound, and the process parameters could affect the warpage of the package.

From epoxy molding compound, Tg, filler loading and the interaction of Tg with filler loading, and the amount of stress modifier statistically ($\alpha = 0.05$) shows that they are the factors which contribute to package warpage.

From process parameters, mold cap, transfer pressure and the weight of post-cure-block are the significant factors affecting package warpage.

REFERENCES

[1] Wei Tan, Fang Zhou, Xingming Cheng, etc.2009 International Conference on Electronic Packaging Technology and High Density Packaging, ICEPT-HDP 2009,p722-724

[2] K Irving Y. Chien, Jack Zhang, Lou Rector, Michael Todd. Low Warpage Molding Compound Development for Array Packages, 2006 Electronics System integration Technology Conference, p 1001-1006

SURFACE ANALYSIS AND POST THERMAL TREATMENT PROCESS OPTIMIZATION OF GRAPHENE OXIDE THIN FILM FOR HUMIDITY SENSOR APPLICATION

Xiaoxu Kang[*], Ruoxi Shen, Xiaolan Zhong

Process Technology Department, Shanghai IC R&D Center, Shanghai, 201210, China

*Corresponding Author's Email: kangxiaoxv@icrd.com.cn

ABSTRACT

Graphene Oxide (GO) has the two-dimensional (2D) layered structure with lots of oxygen containing groups. In this work, GO thin film was coated on substrate with carefully prepared GO dispersion. The GO film was thermally treated by different process condition, and characterized by surface analysis methods to get the optimized process condition. After that, GO based capacitive humidity sensor structure was designed and fabricated, and the capacitance of the sensor structure was increased about seven times from ~22.5% RH% to ~85% RH%, which shows excellent sensitivity performance.

INTRODUCTION

As a 2D layered inorganic nanomaterial, Graphene is one of the most promising materials developed in this century with its exceptional mechanical, thermal and electrical properties [1–2]. Graphene-based 2D material is also attracting great interest for their potential application in various sensors. Because of the 2D layered structure and variously functional groups attached to the layered structure, more and more research work is beginning to focus on these materials, such as doped Graphene, GO, reduced GO (rGO), etc. [3-4]

GO is made up of single or several heavily oxidized graphene sheets, and has sp^2 and sp^3 hybridized carbon atoms which is typically insulating. GO has shown promising application value for gas sensor because of its excellent properties, such as the 2D layered structure, higher surface to volume ratio, soluble in water and ease for coating processing, high solubility in various solvents, and having lots of oxygen containing functional groups or defects [5]. These functional groups or defects will greatly influence the sensing properties of GO based sensors, because these defects or functional groups are normally acting as reactive or absorption sites for gas or humidity adsorption [6-7].

In this study, GO dispersion was prepared by dispersing high-purity GO nanosheets into water. Isopropyl alcohol (IPA) and related solvent were used to adjust its viscosity and surface tension. Spin-coating process was developed to get a uniform GO thin film. Fourier transform infrared spectroscopy (FTIR) were used to study chemical bonds of GO material. After film formation, the samples were thermally treated at different process condition, and XRD was done to investigate the influence of different process treatment on GO film. Finally a CMOS capacitive humidity sensor structure was fabricated and evaluated with optimized GO film as sensing material.

EXPERIMENTAL DETAILS

High-purity GO nanosheets (>99%) with thickness of 0.8–1.2 nm and diameter of 0.5–5 um purchased from Nanjing XFNANO Materials Tech. Co. Ltd (Nanjing, China) were dispersed in a mixed solvent which contains deionized water and isopropyl alcohol (IPA) with fixed volume ratio to form GO dispersion with optimized surface tension and viscosity. GO dispersion with desired concentration was prepared by sonicated treatment at fixed temperature for 60 min, which was shown in Figure 1. Then GO dispersion was filtered by filter membrane to control GO nanosheet size distribution for certain application requirements. The surface tension and viscosity of the GO dispersion was adjusted to optimized value by changing volume ratio of IPA and water. In order to get a uniform GO film and avoid thickness variation influence on material analysis of different process treatment, dedicated coating process was developed for GO film deposition. And after GO film was coated on substrate, thermal treatment was done to remove residue solvent and get a more stable film. X-SEM of GO film was shown in Figure 2.

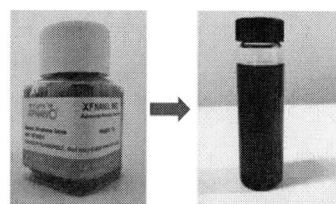

Figure 1: Prepared GO Dispersion

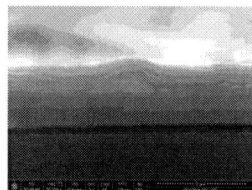

Figure 2: X-SEM GO film on Si substrate

GO film was characterized by different surface analysis methods to investigate its humidity sensing mechanism and find the optimized process condition. After that, GO-based capacitive humidity sensor structure was designed, fabricated and evaluated.

RESULTS AND DISCUSSIONS

GO has plenty of defects and oxygen-containing functional groups, which can act as reactive or absorption center for gas adsorption and can influence the selective and response properties of GO-based sensors. Characteristic FTIR analysis was done for GO film to check these defects and functional groups, shown in Figure 3. It can be seen that the absorption band at 3300 cm^{-1} and 1436 cm^{-1} are assigned to the –OH group stretching vibrations. The absorption peak at 1722 cm^{-1} and 1606 cm^{-1} can be assigned to C=O stretching of carboxylic and carbonyl moiety functional groups. The two absorption peaks at about 1258 cm^{-1} and 1097 cm^{-1} are assigned to the C–O stretching vibrations. From the FTIR spectrum analysis, it can be seen that plenty of oxygen-containing functional group can be found in GO film, such as epoxy, hydroxyl (–OH), and carboxyl (–COOH) groups.

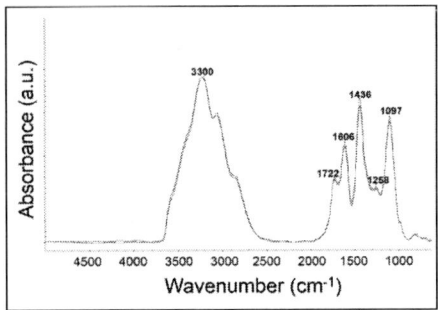

Figure 3: FTIR spectrum of GO film

To remove residue solvent and get a stable film, the GO film was thermally treated by different annealing temperature (60/80/100/120/140 ℃), different time (30/60/90 min), and different atmosphere (Air/N_2/Vacuum). XRD was done to check the thermal treatment influence on GO film. As shown in Figure 4 and Figure 5, 2theta of XRD diffraction peak is increasing with increasing annealing temperature and time.

According to Bragg's law, when the incident X-ray is at certain specific wavelength, the parallel atom plane distance is inversely proportional to the diffraction peak angle. With the temperature and time increasing, the increasing of 2theta peak angle means decreasing of the interlayer distance of GO film. As most of oxygen containing groups were located between the GO nanosheet layers, decreasing of the interlayer distance of GO film may indicate the decomposition and loss of these

oxygen-based functional groups.

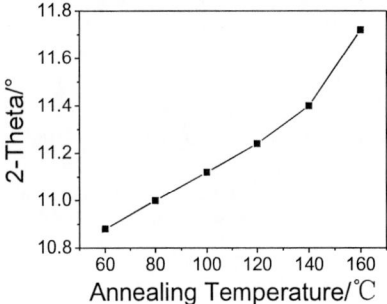

Figure 4: XRD diffraction peak (2theta) of GO film under different annealing temperature and fixed time

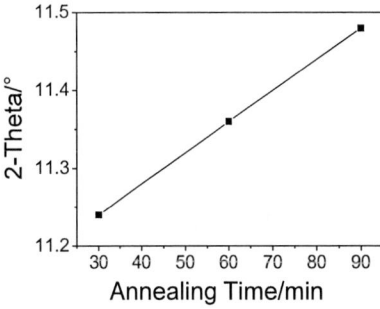

Figure 5: XRD diffraction peak (2theta) of GO film under different annealing time and fixed temperature

The influence of different annealing ambience on GO film was checked by XRD. As shown in Figure 6, the interlayer distance of GO was decreasing with less ambient oxygen concentration, which means oxygen ambience can reduce decomposition and loss of oxygen-based functional group during thermal treatment.

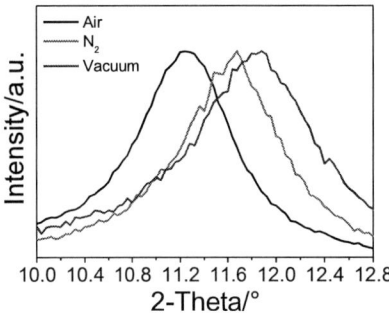

Figure 6: XRD of GO film under different annealing ambience

Finally, 60°C/30 min with air-ambience was chosen as optimized thermal treatment condition, which is the tradeoff between film stability and the higher concentration of oxygen-containing groups.

Next, GO-based capacitive sensor structure was designed and fabricated with a comb/serpent capacitor structure by standard 0.35μm CMOS compatible process on 200mm CMOS FAB to check its sensing performance. CMOS BEOL Al metal layer was used as capacitor electrode and the structure was designed with consideration of least fringing field and parasitic capacitance. As shown in Figure 7, the sensor was fabricated by filling GO material in the capacitor structure. GO film of humidity sensor was formed by spin-coating at a spinning speed of 800 rpm and annealed in air condition at 60℃ for 30 min, and the coating/annealing process was done for three cycles to form GO film.

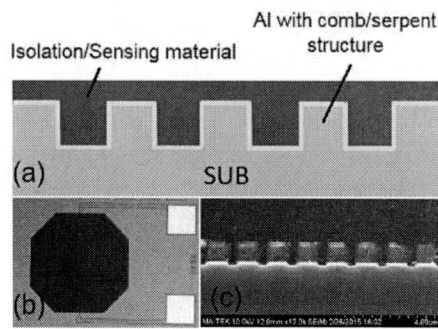

Figure 7: Schematic (a), top view (b) and X-SEM (c) of capacitor structure

Leakage is very critical to any capacitive sensor application, because leakage current will induce error in capacitance measurement. In this work, leakage of the sensor device was measured in voltage sweeping mode from 0 V to 5 V. As shown in Figure 8, the sensor leakage can be controlled below $1E^{-11}$ A at about 50% RH level, which can well meet the sensor application requirements.

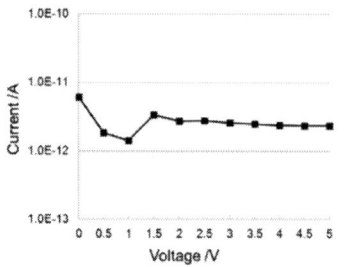

Figure 8: I/V curve of the capacitive sensor in this work

The saturated salt solution was used to create mini-environment with fixed Relative Humidity (RH%) value. As shown in Figure 9, the capacitance of the sensor device was increased about seven times (C/C0, where C0 is the capacitance at minimal RH%) from ~22.5% RH% to ~85% RH%.

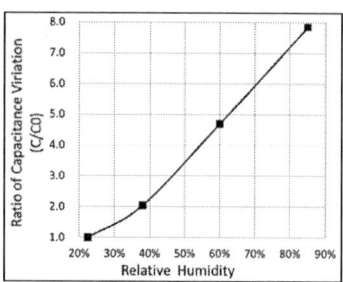

Figure 9: capacitance variation ratio of C/C0 with different RH% of the sensor in this work

The recovery time was evaluated by quickly taking the sensor out from high RH% mini-environment to the outside atmosphere. And the recovery time of this work is less than 3s, which is much faster than standard sensor.

CONCLUSION

GO film was formed and characterized by different surface analysis methods, and optimized thermal treatment condition was obtained based on the analysis results. It was found that the GO interlayer distance was decreasing with increasing annealing temperature and time, which may be induced by the decomposition and loss of oxygen containing functional group. GO-based capacitive humidity sensor structure was designed, fabricated and evaluated. Higher humidity sensitivity and faster recovery time can be achieved for the sensor device, which indicated potential value of GO material in humidity and related gas sensor application. .

ACKNOWLEDGEMENTS

This work was supported in part by the National Key Research Plan of China under Grant 2018YFB0407500. The author would like to thank R&D department of Shanghai IC R&D Center for the support in this work. Special thanks to Weijun Wang, Qingyun Zuo and Bin Jiang for their strong support in this work.

REFERENCES

[1] K. S. Novoselov, et al. *Nature*, vol. 490, no. 7419, pp. 192–200, 2012.
[2] X. M. Li, et al, *Applied Physics Reviews*, vol. 4, no. 2, pp. 021306, 2017
[3] D. C. Marcano, et al. *ACS Nano*, vol. 4, no. 8, pp. 4806-4814, 2010.
[4] D. Lei, et al. *Chemical Society Reviews*, vol. 46, no. 23, pp. 7306-7316, 2017
[5] D. Chen, et al. *Chemical Reviews*, vol. 112, no. 11, pp. 6027-6053, 2012.
[6] J. F. Feng, et al. *Sensors*, vol. 16, no. 3, pp. 314, 2016.
[7] X. X. Kang, et al. *IEEE International Conference on Asic*, 2015.

ARBITRARILY POLARIZED CMOS TERAHERTZ DETECTOR WITH SILICON-BASED PLASMONIC ANTENNA

Yiming Liao[1], Ke Wang[2], Jingyu Peng[2], Yaozu Guo[2], Feng Yan[2], and Xiaoli Ji[2]*

[1]School of Electronic and Optical Engineering, Nanjing University of science and technology, China

[2] D School of Electronic Science and Engineering, Nanjing University, China

*Corresponding Author's Email: xji@nju.edu.cn

ABSTRACT

In this paper, we demonstrate the idea to combine silicon-based plasmonic antennas with MOSFETs to form an arbitrarily polarized terahertz (THz) detector based on standard CMOS technology. For an optimized detector with a size of 162 μm × 161 μm, the simulation results show that the field response to the vertical polarized wave has a radio of 0.91 with 90 degree delay compared to the response to the horizontal polarized wave at 0.25 THz. The proposed arbitrarily polarized on chip detector occupies a small area size, which offers a novel approach to THz array imaging and sensing.

INTRODUCTION

CMOS THz detectors with on chip antennas have shown great application potentials in THz imaging and sensing systems due to their excellent advantages such as fast on-state response time, low noise equivalent power (NEP), low cost and high integration level [1-2]. The key element is the arbitrarily polarized THz detector unaffected by the alignment between the transmitter and receiver in order to achieve high efficient and low disturbance THz imaging. It is reported that the linearly polarized wave can be converted completely into its cross-polarized wave by a chiral spiral slot structure [3], which provides a way for the linear polarization detector to realize arbitrary polarization. In addition to the polarization conversion method, another more commonly used methods are designed circularly polarized antennas, such as double-fan-shaped slot antenna [4], patch antenna array [5], annular slot antenna [6] and fractal antenna [7]. Nevertheless, some of the antennas' requirements on the dielectric layer make it difficult to realize in the standard CMOS process while others may need differential quadrature couplers and transmission lines to ensure phase delay, which occupies much more chip area and inevitably increases cost and integration difficulty. Fortunately, silicon can be used to manufacture the optical plasma antenna [8], which is more compact and compatible with standard CMOS technology.

From the discussion above, a circularly polarized plasmonic antenna for CMOS THz detector is presented in this paper. Meanwhile, the influence of antenna structure on antenna performance is studied deeply. The presented modeling results are achieved through High-Frequency Structure Simulator (HFSS).

DESIGN

According to Maxwell's equations, the propagation constant K_{spp} at the infinite interface between semiconductor and dielectric can be calculated by the following formula:

$$K_{SPP} = \frac{2\pi}{\lambda}\left(\frac{\varepsilon_{semi}\varepsilon_{dielec}}{\varepsilon_{semi} + \varepsilon_{dielec}}\right)^{1/2} \qquad (1)$$

In the THz detector, ε_{semi} and ε_{dielec} are the frequency dependent dielectric constant of poly-silicon and dielectric layer, respectively. λ is the THz wavelength in air. The complex dielectric constant of ε_{semi} can be characterized by the Drude model [7]. The estimated plasmonic wavelength $\lambda_{spp} = 2\pi/K_{SPP}$ is much shorter than THz wavelength λ, which allows the plasmonic antenna to have a smaller size based on this condition.

Two sets of the silicon-based plasmonic antennas placed in an orthogonal way and a MOSFET are used to achieve the detection of arbitrary polarization of THz waves as shown in Fig. 1. Two 0.25 THz silicon antennas with fan-rod structure are realized with poly-silicon doping with 1×10^{20} cm^{-3} and thickness of 0.2 μm, where L_1 is the half length of the antenna, d is the length of the bar, θ is the fan angle and the width of the rod is w_1 in 0.18 μm standard CMOS technology. The antenna is fabricated above the shallow trench isolation structure, below which is a Si substrate. The antenna is embedded in silicon dioxide with a thickness of 10.7 μm. The insets show the 45-degree inclined MOSFET mixer placed in the center of the antennas and the phase-shifting structure combined with rod. The unique phase-shifting part is a rectangular ring of equal width w_1, whose outer length is L_2 and outer width is w_2.

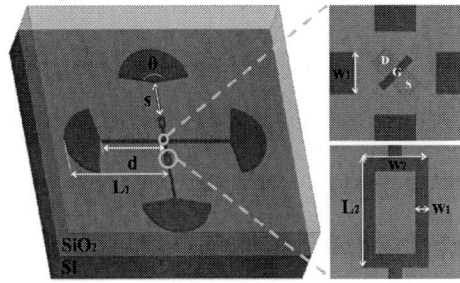

Figure 1: The schematic illustration of arbitrarily polarized CMOS THz detector

RESULTS AND DISCUSSION

The THz antenna acting as the key component of the THz detector, plays an important role in receiving and transmitting THz signals. The better performance of the antenna, the stronger feed signal received by the MOSFET mixer, which leads to the higher responsivity of the detector. In order to obtain the optimal performance and minimize the size of the antenna, the antenna structures are designed and simulated by means of HFSS.

Fig 2 (a) shows the simulated result of local electric field enhancement of the silicon antenna, when the electric field directions are respectively parallel to the x-axis (horizontal polarized) and y-axis (vertical polarized). The field response of the antenna to the vertical polarized wave has a 90 degree delay compared with the response to the horizontal polarized wave, and the ratio of two field peak is about 0.91 When the geometric parameters of antenna are optimized as $L_1 = 80$ μm, $L_2 = 8$ μm, $d = 52$ μm, $\theta = 160°$, $g = 2$ μm, $s = 32$ μm, $w_1 = 1$ μm, $w_2 = 5$ μm. The typical electric field distributions on the antenna surface shown in Fig 2 (b) also illustrates that the antenna structure parallel to the polarization direction of the incident wave could cause strong local field enhancement, but it is contrary in the case of vertical. In addition, it is worth noting that the phase-shifting structure can cause not only the phase change of the electric field, but also the slight enhancement of the electric field.

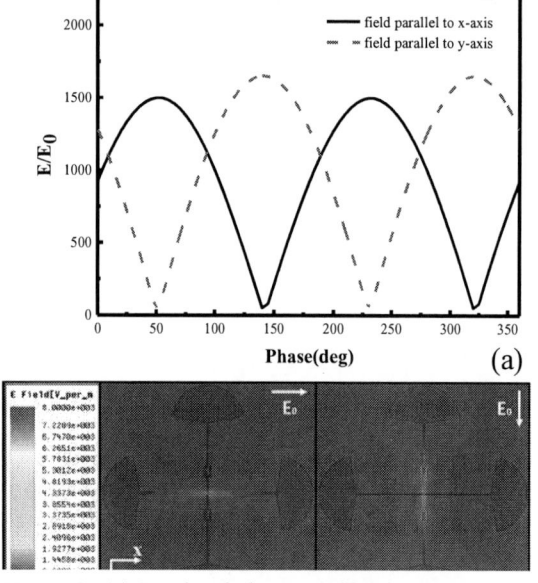

Figure 2: (a) Simulated electric field enhancement of the silicon antenna and (b) the typical electric field distributions on the antenna surface with horizontal and vertical polarized THz wave at 250 GHz frequency

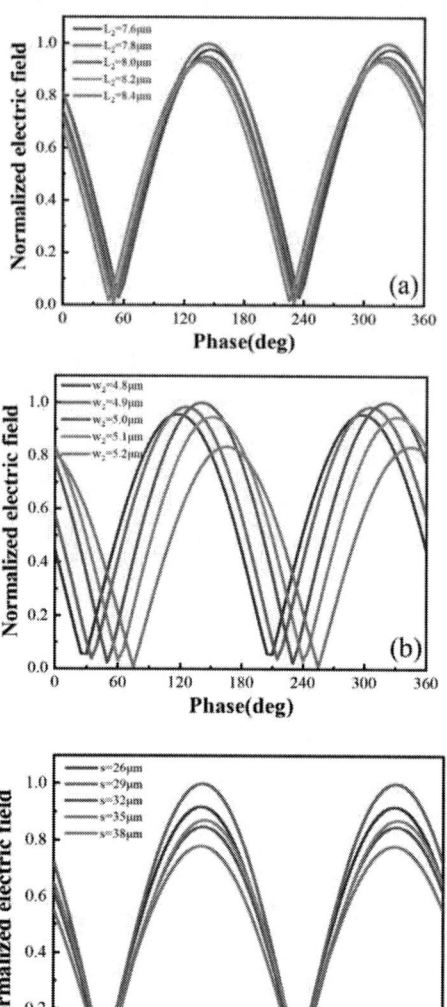

Figure 3: Simulated phase dependence of the normalized electric field with vertical polarized THz wave at different (a) L_2; (b) w_2; (c) s

The main antenna geometric parameters have been studied to further explore the influence of antenna structure on antenna performance. First of all, the value of L_2 only changed based on the parameters of the optimal structure. As shown in Fig 3 (a), it can be seen that the peak field response of the antenna to the vertical polarized wave and the corresponding phase are slowly decreasing with L_2 increase from 7.6 μm to 8.4 μm gradually. Then, the same method is used on w_2 for analysis and the result is shown in Fig 3 (b). The phase increases about 50 degree when w_2 increases from 4.8μm to 5.2μm, which indicating that the increase of w_2 will adds the phase delay. However, the peak field response increases first and then decreases, and there seems to be no obvious linear change rule. By contrast, w_2 is the main parameter to adjust the phase delay. This can be well explained by the fact that the increase of w_2 rather than L_2 causes the extension of the current path on the antenna surface, which acts like a transmission line leading to increase of the effective length. Thus, a 90 degree delay of transmission can be introduced through optimizing w_2. Furthermore the effect of parameter s is discussed and the simulated result is shown in Fig 3 (c). The phase dependence of electric field varies when the value of s is modified from 26 to 38 μm. The almost unchanged phase is far less than the variation of the peak value. Therefore, it can be considered that s has no regulating effect on the phase and the position of the phase-shifting structure can be optimized to gain a suitable filed response.

According to the analysis results, the arbitrarily polarized CMOS THz detector can be well adjusted by optimizing the geometric parameters of the circularly polarized plasmonic antenna.

CONCLUSION

In this work, we proposed a novel silicon-based plasmonic antenna for arbitrarily polarized CMOS THz detector. As compared to conventional metal antenna, our design can meet the requirements of arbitrary polarization, and also promising for a higher gain with smaller area.

ACKNOWLEDGEMENTS

This work was supported by the National Key Research and Development Program of China (No. 2016YFB0402403 and No. 2016YFB0400402).

REFERENCES

[1] D. Coquillat, J. Marczewski, P. Kopyt, N. Dyakonova, B. Giffard, W. Knap, "Improvement of terahertz field effect transistor detectors by substrate thinning and radiation losses reduction," *Opt. Express.,* vol. 24, 2016, pp. 272-281.

[2] Y. Shang, H. Yu, C. Yang, Y. Liang and W. M. Lim, "A 239–281GHz Sub-THz imager with 100MHz resolution by CMOS direct-conversion receiver with on-chip circular-polarized SIW antenna," *Proceedings of the IEEE 2014 Custom Integrated Circuits Conference,* San Jose, Sept. 15-17, 2014, pp. 1-4.

[3] J. Tang, Z. Xiao, K. Xu, X. Ma, Z. Wang, "Cross polarization conversion based on a new chiral spiral slot structure in THz region," *Optical and Quantum Electronics,* vol. 48, 2016, pp. 111.

[4] P. Zhao, Y. Liu, H. Lu, Y. Wu and X. Lv, "Experimental Realization of Terahertz Waveguide-Fed Circularly Polarized Double-Fan-Shaped Slot Antenna," *IEEE Antennas and Wireless Propagation Letters,* vol. 16, 2017, pp. 2066-2069.

[5] X. Bai, S. Qu and C. Chan, "Circularly polarized series-fed patch array for THz applications," *2016 IEEE International Symposium on Antennas and Propagation (APSURSI),* Fajardo, 26 June-1 July, 2016, pp. 595-596.

[6] J. Grzyb, K. Statnikov, N. Sarmah and U. R. Pfeiffer, "A wideband 240 GHz lens-integrated circularly polarized on-chip annular slot antenna for a FMCW radar transceiver module in SiGe technology," *2015 SBMO/IEEE MTT-S International Microwave and Optoelectronics Conference (IMOC),* Porto de Galinhas, Nov. 3-6, 2015, pp. 1-4.

[7] J. Pourahmadazar, R. Karimian and T. A. Denidni, "A High Data-Rate Kiosk Application circularly polarized fractal antenna for mm-wave band radio with 0.18μm CMOS technology," *2016 10th European Conference on Antennas and Propagation (EuCAP),* Davos, April 10-15, 2016, pp. 1-4.

[8] A. Berrier, R. Ulbricht, M. Bonn, J. Rivas, "Ultrafast active control of localized surface plasmon resonances in silicon bowtie antennas," *Opt. Express.,* vol. 18, 2010, pp. 23226-23235.

GATE TUNABLE MEMTRANSISTOR BASED ON MONOLAYER MOLYBDENUM DISULFIDE

Meng Yan, Fang Wang*, Jiaqiang Shen, Xichao Di, Xin Lin, Huanhuan Di, Wei Mi, Kailiang Zhang*

Tianjin Key Laboratory of Film Electronic & Communication Devices
School of Electrical & Electronic Engineering, Tianjin University of Technology, Tianjin 300384, China.
*Corresponding Author's Email: fwang75@163.com; kailiang_zhang@163.com

ABSTRACT

As a typical representative of two-dimensional (2D) materials, recently MoS_2 was considered as the candidate to the development of electronic synaptic devices due to its ultrathin thickness and special properties. However, dual-terminal artificial synapse devices still exit the challenges about the simulation of biological synapses, which is hard for two-terminal devices to update and read the synaptic weight at the same time. In this work, MoS_2 films were grown by chemical vapor deposition (the sample's largest single triangular size is about 83μm), and three-terminal synaptic devices based on back-gate FETs on Si/SiO_2 substrate were fabricated. MoS_2 sample's morphology and device's structure were characterized by Raman spectroscopy and optical microscope (OM). The memtransistor has excellent resistive switching (RS) behavior. By optimizing the pulse, the memtransistor showed a better conductivity linearity, and typical synaptic characteristics were mimicked, such as short-term/long-term plasticity (STP/LTP), excitatory post-synaptic current (EPSC)/inhibitory post-synaptic current (IPSC) and paired-pulse facilitation (PPF).

INTRODUCTION

Due to the separation of the computing unit from the storage unit in conventional von-Neumann computing architecture, the data access speed of the in-memory cannot be significantly improved. A series of complicated tasks are simultaneously managed in the human nervous system, which include learning, memory, recognition and control, with extremely low power consumption. Due to the advantages of synaptic plasticity, nanoscale sizes and low power consumption, the synaptic devices based on 2D materials have gotten great attentions.

Here, we reported a memtransistor based on monolayer MoS_2 for neuromorphic computing. The high quality MoS_2 films were grown by chemical vapor deposition (CVD) method and characterized using Raman spectroscopy and Raman mapping. Then, the artificial synapse based on back gate transistor was structured by electron beam lithography (EBL), electron beam evaporation and lift-off process. By optimizing pulse interval, pulse amplitude and pulse width, it can efficiently emulate the key synaptic activities including short-term plasticity and long-term plasticity (STP/LTP), paired-pulse facilitation (PPF) and so on. Furthermore, compared with traditional two-terminal devices, the memtransistors offered another feasible way for synaptic weight modulation, exhibiting prominent potential for achieving brain-like artificial intelligence targets.

EXPERIMENTAL

The samples of monolayer MoS_2 were grown as following. MoS_2 samples were grown on the Si/ SiO_2 substrate by atmospheric pressure CVD with two-zone tube furnace. The substrate was ultrasonically cleaned in acetone, alcohol and DI water, then was dried by nitrogen gas. The crucible boat with 30mg sulfur powder were placed in upstream temperature-zone I (90℃) ,and another boat with 0.3mg MoO_3 powder in downstream temperature-zone II（800℃）. During the experimental process, the growth time was 20 minutes and the gas flow rate of high-purity argon was 30sccm.

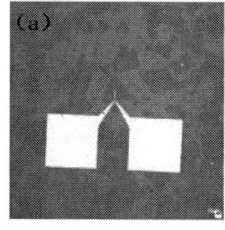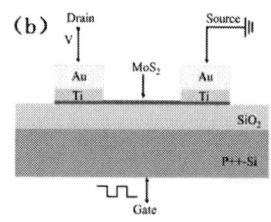

Figure 1: (a) Optical micrograph of a MoS_2 memtransistor built on 270-nm-thick thermal SiO_2 on doped Si (gate), the length of MoS_2 channel are 2μm.(b) Schematic of a MoS_2 electric-gate memtransistor

The memtransistors were constructed as following. The electrode was patterned by EBL. The Ti/Au (10/80 nm) electrode was attached by the e-beam evaporation and lift-off process. Actual structure of MoS_2 memtransistor with 2μm channel length was observed in Optical micrograph (in Figure 1a).Electrical measurements were carried out by Agilent Semiconductor Analyzer B1500 and wave function generator through a probe station. For this

978-1-7281-6559-2/20 $31.00 © 2020 IEEE

test, a series of pulses were applied to the gate P++-Si electrode, while a constant electric bias V_{ds} was applied on the left Au electrode as drain , and the right Au electrode as drain was ground (in Figure 1b).

RESULTS AND DISCUSSION

The great advantage of few MoS_2 layers is high mobility and tunable bandgap that modulate the channel conductance in a large scale by the gate voltage. The sample of uniform monolayer MoS_2 films is characterized by Raman spectra and Raman mapping (in Figure 2). E^1_{2g} at 383.39 cm^{-1} and A_{1g} at 402.23 cm^{-1} is two characteristic peaks of MoS_2 generated at the laser wavelength of 532nm, which represent respectively the in-plane reverse vibration mode of S and Mo atoms and the out-of-plane vertical vibration mode of S atoms. It indicates that the MoS_2 sample is monolayer, attributing to wavenumber difference of two peaks is approximately 18.84 cm^{-1} (in Figure 2a). The Raman mapping shows the peak intensity of A_{1g} peak at multi-point scanning (in Figure 2b). The consistent color indicates that the MoS_2 sample has good thickness uniformity.

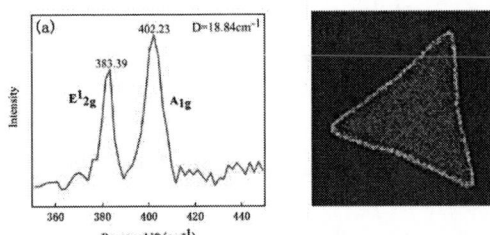

Figure 2: (a) Raman spectra of monolayer MoS_2 (b) Raman mapping of monolayer MoS_2

The memtransistor based on MoS_2 enables a tunable memristive behavior, indicating potential in neuromorphic computing. Figure 3 shows I_{ds}–V_{ds} curves under different bias V_{gs} (-30V,-10V,-20V, 0V, 10V, 20V, 50V) for the same device, by sweeping V_{ds} in range from -6V to 6V. Generally, the tendency of I_{ds}–V_{ds} curves rises with gate voltage increasing. Owing to the n-type MoS_2 channel, a negative gate voltage would bring about a lower carrier concentration and a higher resistance at HRS. For V_g=0, the hysteretic curve presents that the memtransistor hardly has a switching window. In the reverse-biased (V_g<0), the memtransistor is gradually switched from a high resistance state (HRS) to a low resistance state (LRS) during sweeping V_{ds} from 0V to 6V and retains its LRS during sweeping V_{ds} back to 0V. A similar switching from HRS to LRS occurs for sweeping V_{ds} as 0V→ -6V → 0V. This phenomenon indicates a volatile resistive switching and proves potentiation feasibility. The on/off ratio and switching window continues to increase obviously with improving negative V_{gs} from -10V to -30V. The

phenomenon is just opposite in the forward-biased (V_{gs}>0), the memtransistor gradually switches from LRS to HRS (Along the arrow) for sweeping V_{ds} of 0V → 6V → 0V. The switching window almost be increasing constantly under positive V_g from 10V to 50V. This phenomenon indicates a volatile resistive switching and proves depression feasibility. Typical transfer characteristics for different sweep range of V_{gs} (-50V~50V, -30V~30V, -10V~10V) are shown in Figure 4, the hysteresis transfer curves are measured by direct current double sweep. As the V_{gs} sweep range increases, the switching window get bigger visibly, which consistent with the behavior of potentiation or depression.

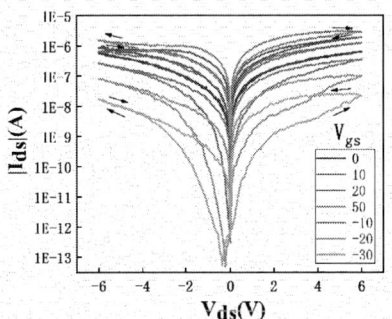

Figure 3: I_{ds}–V_{ds} curves for different bias V_{gs}

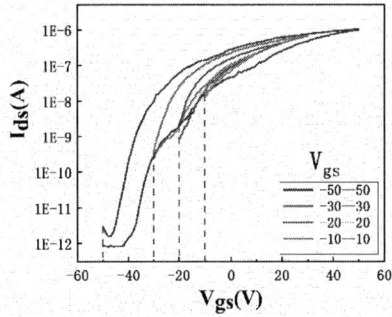

Figure 4: Transfer characteristics of a memtransistor at V_{ds}= 1 V and different sweep range of V_{gs}

MoS_2 memtransistors demonstrate STP/STD, which imitate excitatory and inhibitory behavior of biological synapse. The post-synaptic current gradually increase when apply 25 consecutive negative V_g pulses that pulse amplitude (in Figure 5), pulse interval and pulse width interval are −10V, 100ms and 30ms, respectively, which can be regarded as mimicking excitatory behavior. Figure 5b shows that the current gradually decreases when applying 25 consecutive positive V_g pulses (10V, 100ms, 30ms), which can be deemed to mimicking inhibitory behavior. It shows a good conductivity linearity in Figure 5a and Figure 5b. Paired-pulse facilitation (PPF) displays

978-1-7281-6559-2/20 $31.00 © 2020 IEEE

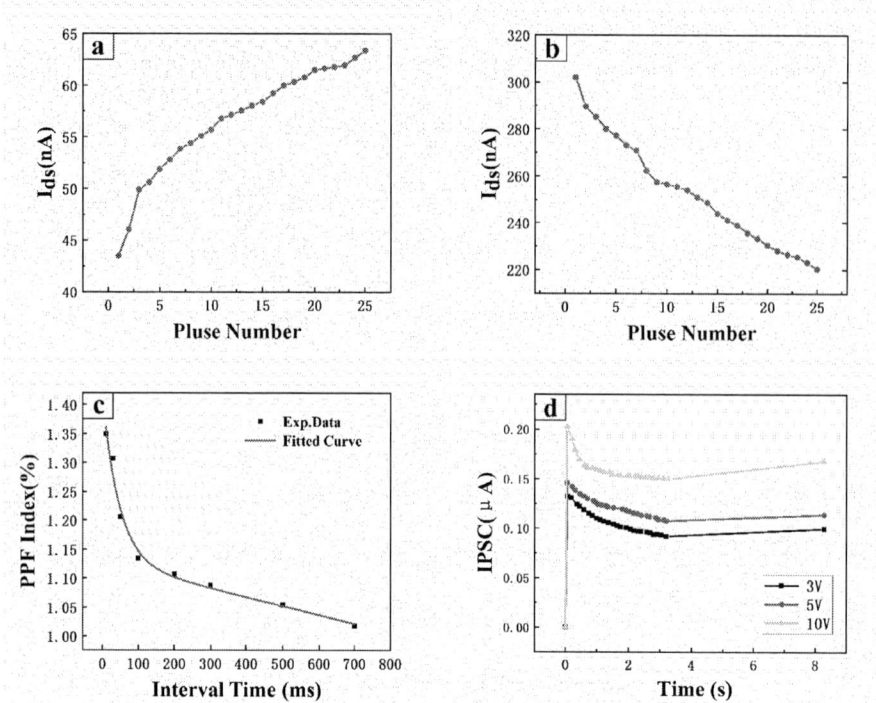

Figure 5: (a) Post-synaptic current versus pulse number, showing potentiation under 25 successive gate positive pulses (10 V) at V_{ds} = 1 V. (b)Post-synaptic current versus pulse number, showing depression under 25 successive gate negative pulses (-10 V) at V_{ds} = 1 V. (c) PPF index (defined as 100% × A_2/A_1) as a function of interval time of paired presynaptic spikes. (d) IPSC responses on presynaptic spikes with different spike amplitudes ranged from 3 to 10V.

the short-term plasticity (STP) in artificial synapses. The trend is displayed in Figure 5c under a constant V_{ds} of 1 V, when paired presynaptic spikes (-10V, 600ms) were applied on the bottom gate. PPF index decreases with the increased interval time, that PPF index is ~135% for 10ms interval time and the PPF index decreases ~102% for 700ms interval time.

IPSC are recorded by presynaptic spikes with different spike amplitudes and multiple pulses as shown in Figure 5d. When spike amplitude is 3V, 5V and 10V, the IPSC is respectively about 92nA, 107nA and 150nA for 25 spikes, and the last IPSC current do not decays back to the initial current after 5s. It is demonstrated that MoS$_2$ memtransistor forms a decreased channel conductivity, which results in a nonvolatile long-term plasticity.

CONCLUSION

In conclusion, based on a scalable monolayer MoS$_2$, we have established successfully a three-terminal memtransistor with a wide tunable gate voltage and analog RS behavior. The PPF index was measured, in order to observe the STP operation of the device, and the behavior of LTP was imitated through the number or intensity of applied pulses on the memtransistor. Furthermore, a large area of high-quality MoS$_2$ growth and synaptic device fabrication process that are compatible with existing technologies are developed, demonstrating foundation for efficient brain-like chips.

ACKNOWLEDGEMENT

This work was supported by Natural Science Foundation of Tianjin City (Grant Nos.18JCZDJC30500, 17JCYBJ C16 100 and 17JCZDJC31700), National Key Research and Development Program of China (Grant No.2017YFB0405600) and National Natural Science Foundation of China (Grant Nos.61404091, 61274113, 61505144, 51502203 and 51502204).

REFERENCES

[1] D, X, Tang, F. Wang, B.J. Zhang and K. L. Zhang. J, et al. *Journal of Materials Science* 53.20, 2018, 14447-14455.

[2] Li. Y, Wang. F, Tang. D, Wei. J, Li. Y, Xing. Y, & Zhang, K. *Materials Letters* 2018, 216, 261-264.

[3] Sangwan, Vinod K, et al. *Nature* 554.7693 (2018): 500.

[4] Yin, Siqi, et al. *ACS applied materials & interfaces* 11.46 (2019): 43344-43350.

[5] Arnold, Andrew J, et al. *ACS nano* 11.3 (2017): 3110-3118.

AMBIENT-STABLE AND HIGH ON/OFF RATIO NEAR-INFRARED PHOTODETECTOR BASED ON PEROVSKITE-TREATED PBS COLLOIDAL QUANTUM DOTS

*Qingqing Wu[*1], Yajie Yan[2], Ziqi Liang[2], Shaojian Hu, Jianjun Zhu[1] and Shoumian Chen[1]*

[1] Shanghai IC R&D Center, 497 Gaosi Road, Zhangjiang Hi-Tech Park, Shanghai, China.

[2] Department of Materials Science, Fudan University, Shanghai, China.

*Corresponding Author's Email: wuqingqing@icrd.com.cn

ABSTRACT

Due to their low-cost in manufacturing, size-tunable spectral sensitivity, and flexible substrate compatibility, lead sulphide colloidal quantum dots (PbS CQDs) are increasingly regarded as promising active material candidates for next-generation NIR photodetectors. In this study, the effective passivation on the surface of PbS CQDs with perovskites is demonstrated, and the perovskite-treated PbS CQDs show improved ambient stability and reduced agglomeration. The PbS CQDs photodiode detectors are self-powered and exhibit a high on/off ratio up to 3×10^4 along with a large photocurrent density of 2 mA cm^{-2}. Meanwhile, the photoconductor structured photodetectors based on the same materials, displays a responsivity of 0.48 A W^{-1} and excellent long-term stability in ambient atmosphere.

INTRODUCTION

The growing interest in photodetectors in the near infrared (NIR) is related to the demanding application in the spectral region, including imaging [1], process monitor [2], night vision [3] and remote sensing [4]. The photosensitivity of silicon deteriorate rapidly beyond 800 nm. InGaAs is the current choice for NIR detection while challenges still exist in epitaxial growth process. In the past decade, among all candidate materials studied, lead sulfide Colloid Quantum Dots (PbS CQDs) offer several advantages including low-cost in manufacturing, size-tunable spectral sensitivity, and compatibility with flexible substrates. Besides, whose achievable figures of merit outperforms conventional photodetectors [5-6]. However, there remain challenges waiting to be resolved to further improve the photodetector performance. Since the property of PbS CQDs such as the carrier mobility and stability can be modified by changing surface ligands, various surface modifications have been investigated [7-8]. For example, replacing long aliphatic chain ligands with short chain ligands enhances electronic coupling between CQDs and thus improve the carrier mobility of PbS CQDs [9]. Besides, the introduction of strongly bound ligands could protect the surfaces of CQDs against incursion of oxygen for better air stability [10].

Herein, we introduce an approach to effectively passivate the surface of PbS CQDs with perovskites, forming perovskite-treated PbS CQDs for the application in high performance NIR photodetector. The resulting perovskite-treated PbS CQDs show improved ambient stability and reduced agglomeration compared to other short ligands. The photodiodes incorporating the perovskite-treated PbS CQDs film as a light absorber, are self-powered and exhibit a high on/off ratio up to 3×10^4 along with a large photocurrent density of 2 mA cm^{-2}, under 41 mA cm^{-2} polychromatic light with $\lambda >800$ nm. Meanwhile, we have fabricated a photoconductor structured photodetector based on the same materials, which displayed a responsivity of 0.48 A W^{-1} and on/off ratio of 190 under the intensity of 0.18 mW cm^{-2} NIR light with excellent long-term stability in ambient atmosphere.

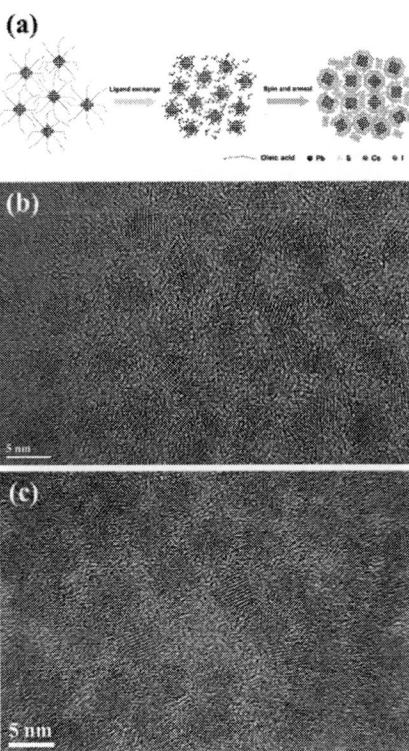

Figure 1: Solution-phased ligand exchange. (a) Schematic illustration of the ligand exchange process of PbS CQDs. HRTEM images of PbS CQDs before (b) and after (c) ligand exchange, respectively.

The schematic illustration of the solution-phased ligand exchange process of PbS CQDs is depicted in Figure 1(a). The exchange process was carried out by mixing the oleic acid capped PbS CQDs in octane with the perovskite precursors in DiMethylFormamide (DMF). During the mixing and stirring, the oleic acid ligands were gradually replaced by perovskite precursors, such as CsI and PbI_2, until the PbS CQDs were completely transferred to the DMF phase. Then, the PbS CQDs were washed, precipitated, dried, and finally dissolved in mixed solvent of Dimethyl sulfoxide (DMSO)/Butyl-amine (4/1, vol. %). The exchanged CQDs were spin-coated and annealed to form the perovskite-treated PbS CQDs film. We used High Resolution Transmission Electron Microscopy (HRTEM) to confirm the nanocrystal structure and spacing before and after ligand exchange, respectively (Figure 1(b)-1(c)). The HRTEM images show that replacing the oleic acid ligands with perovskite clearly reduces the spacing between adjacent nanocrystals with no obvious aggregation. Since most of the perovskites were washed away in the final colloidal solution, only PbS lattice fringes were observed in Figure 1(c).

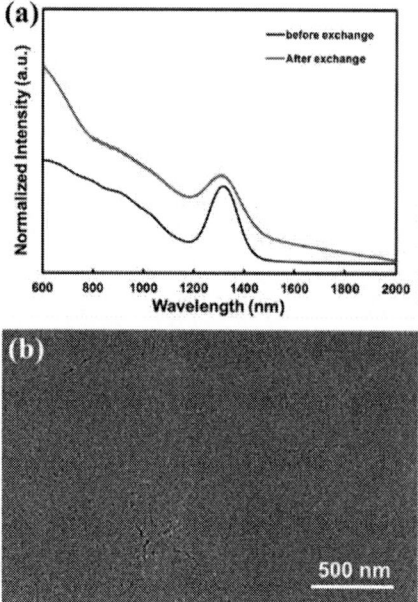

Figure 2: (a) Absorbance spectra of PbS CQDs films before ligand exchange (oleate-capped, black), after ligand exchange (perovskite-treated, red), respectively. (b) SEM image of perovskite-treated PbS CQDs film.

Then we carried out absorption spectra of the PbS CQDs films before and after the ligand exchange as revealed in Figure 2(a), with a peak at 1360 nm. After the exchange, the absorption peak has a slight blue shift. The exchanged film has a stronger absorption in visible region

than a pristine one, which is ascribed to the enhanced response of residual perovskite ligands. Field-Effect Scanning Electron Microscope (FESEM) was conducted to characterize the morphology of the perovskite-treated PbS CQDs film on device substrate as shown in Figure 2(b), the CQDs film showed highly uniform over large area. Compared to solid-phase ligand exchange, the solution-phase perovskite exchange remarkably suppresses the formation of pinholes and film cracking caused by the free volume loss.

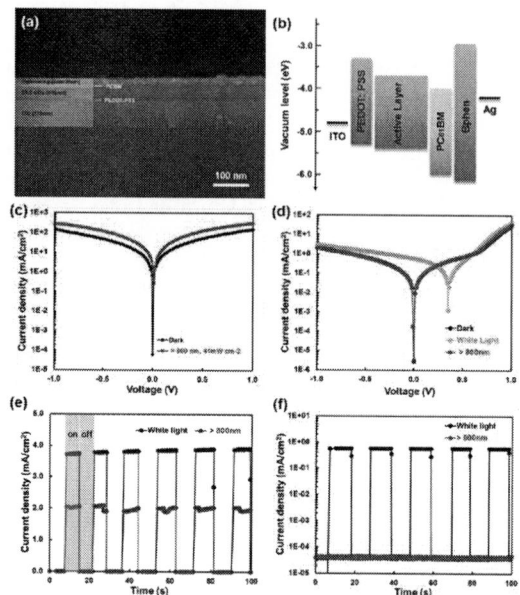

Figure 3: Photodiode architecture and performance. (a) Cross-sectional SEM image of PbS CQDs photodiode with device architecture of ITO/PEDOT: PSS/PbS CQDs/ PC$_{61}$BM/Bphen/Ag. (b) Band energy diagrams of PbS CQDs photodiode. I-V characteristics of perovskite-treated PbS CQDs (c) and pure perovskite (d) photodiodes in dark and under 41 mW cm^{-2} polychromatic light with λ > 800 nm. Temporal photocurrent response of perovskite-treated PbS CQDs (e) and pure perovskite (f) photodiodes under 100 mW cm^{-2} white light (black line) and 41 mW cm^{-2} polychromatic light with λ > 800 nm (red line), at bias of 0 V.

To probe into the application of such perovskite-treated PbS CQDs, photodiode configured photodetectors were fabricated with the device architecture, which consists of a PbS CQDs layer deposited onto Poly(3,4-EthyleneDiOxyThiophene):Poly(StyreneSulfonate) (PEDOT:PSS) Hole-Transporting Layer (HTL) -coated Indium Tin Oxide (ITO) electrodes, a thin layer [6,6]-Phenyl C$_{61}$ Butyric acid Methyl (PC$_{61}$BM) as an

Electron-Transporting Layer (ETL) and Bphen/Ag deposition as the top metal electrode (Figure 3(a)). Figure 3(b) shows the energy level landscape of the device. Neat perovskite based photodiodes were also fabricated for comparison.

Figure 3(c) presents the I-V characteristics of photodiodes based on perovskite-treated PbS CQDs measured in dark and under 41 mW cm^{-2} polychromatic light with λ > 800 nm, respectively. At a bias of 0V, the photocurrent density reaches the value of 1.9 mA cm^{-2}, and dark current density maintains at 6.2×10^{-5} mA cm^{-2}, delivering an on/off ratio up to 3×10^{4}. Then the response of device based on pure perovskites (Figure 3(d)) is investigated showing a photocurrent density of 0.57 mA cm^{-2} under white light with 100 mW cm^{-2}, and no response under 41 mW cm^{-2} polychromatic light with λ > 800 nm, which indicates that the PbS CQDs is playing a key role under the light with λ > 800 nm rather than perovskite.

We then explored the cycling stability of perovskite treated PbS CQDs photodiodes, the temporal photocurrent responses have been carried out under λ > 800 nm polychromatic light with an intensity of 41 mW cm^{-2} and under a white light with an intensity of 100 mW cm^{-2}, at 0V bias (Figure 3(e)-3(f)). The photocurrent density of CQDs remains at 3.7 mA cm^{-2} after several on/off switches under white light, and still keeps 2 mA cm^{-2} under white light with 800 nm filter with the intensity down to 41 mW cm^{-2}. In contrast, the photocurrent density of pure perovskites shows 0.58 mA cm^{-2} under white light, and down to 3.7×10^{-5} mA cm^{-2} under light with λ > 800 nm, which means there is no response for perovskites in light with λ > 800 nm.

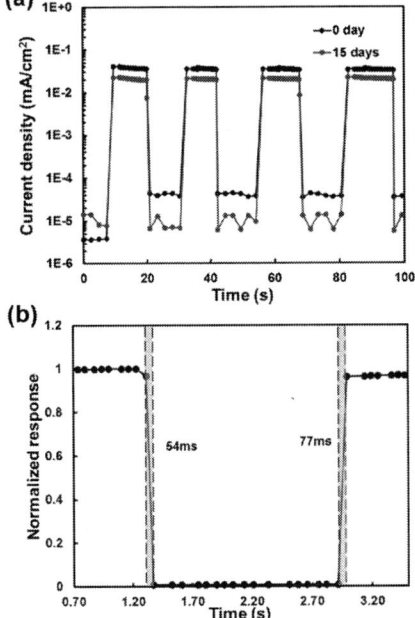

The ambient stability of perovskite capped PbS CQDs photodiode is also evaluated shown in Figure 4(a). After storing in glove box for 15 days, the photocurrent still maintains at 60 % of the initial value under 4.1 mW cm^{-2} polychromatic light with λ > 800 nm, and the on/off ratio was up to 1.5×10^{3} due to the decrease of dark current. The magnification of photo response in Figure 4(b) gives the decay and rise time of the photodiodes 54 ms and 77 ms, respectively, limited by the test equipment, under 13.5 mW cm^{-2} polychromatic light with λ > 800 nm.

Figure 4: (a) Device stability of perovskite-treated PbS CQDs photodiode storing in glove box. (b) The rise and decay time of perovskite-treated PbS CQDs photodiodes under 13.5 mW cm^{-2} polychromatic light with λ > 800 nm.

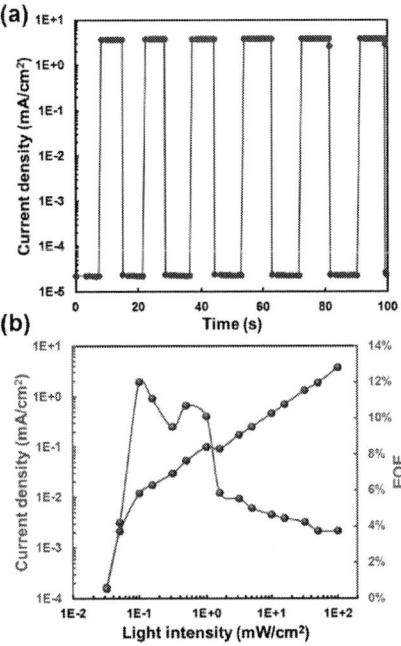

Figure 5: (a) Temporal photocurrent response of perovskite-treated PbS CQDs photodiodes under 100 mW cm^{-2} white light. (b) Current density and external quantum efficiency (EQE) of perovskite-treated PbS CQDs photodiode versus light intensity under white light.

Figure 5(a) shows the on/off switches of PbS CQDs photodiodes under white light with 100 mW cm^{-2}. The photocurrent density of CQDs maintains at 3.7 mA cm^{-2} under white light, and down to 2.2×10^{-5} mA cm^{-2} in darkness, so the on/off ratio reaches the value of 1.6×10^{5}. Current density and External Quantum Efficiency (EQE) of the PbS CQDs photodiode versus different light intensity was shown in Figure 5(b). The EQE value obtains a maximum value of 12 % as the light intensity approaches 0.1 mW cm^{-2}. The photocurrent density increases linearly with the light intensity, which shows the PbS CQDs photodiode has a high saturation light

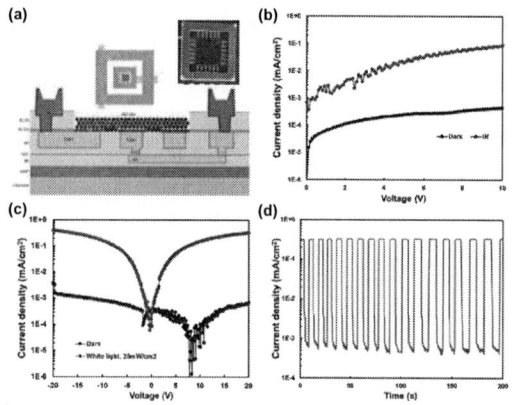

intensity.

Figure 6: Photoconductor detector and performance. (a) Schematic illustration of perovskite-treated PbS QDs photoconductor, the inset shows a photograph of device. (b) I-V curves of device under dark and an IR light with illumination of 0.18 mW cm^{-2}. (c) I-V curves of device under darkness and a white light illumination of 25 mW cm^{-2}. (d) I-t curves of device upon on-off switching with a white light of 25 mW cm^{-2} at 20 V, after storing in air for a week.

The device architecture of PbS QDs photoconductor is shown schematically in Figure 6a, in which the electrode structure and interconnection are compatible with 55 nm CMOS process and hold great potential for practical deployments. The spacing between the inner and outer ring electrodes was 4 μm, and there were thousands of ring electrodes connected in series in the photoconductor device. Figure 6(b) shows the I-V curves of photoconductor in darkness and under NIR illumination with 0.18 mW cm^{-2}. The photocurrent density reaches 0.087 mA cm^{-2}, and on/off ratio of 190 at 10 V bias was achieved, the responsivity (R) reaches a value of 0.48 A W^{-1}. The fluctuation of photocurrent was caused by environmental disturbances. Meanwhile, we have investigated the responsivity of the photoconductor in the visible spectrum. Figure 6(c) shows I-V curves of the device in darkness and under a white light illumination of 25 mW cm^{-2}. The photocurrent density is up to 0.33 mA cm^{-2} and on/off ratio reaches 475 at 20 V. Then we studied the ambient stability of the device, after the device being stored in ambient condition for a week, in Figure 6(d), shows the I-t curves of device upon on-off switching with a white light of 25 mW cm^{-2} at 20 V. Therefore, the photocurrent and the dark current remains stable after being stored in air for a week.

In conclusion, we have successfully fabricated perovskite-treated PbS CQDs based photodetector in photodiode and photoconductor structures, respectively. The perovskite-treated PbS CQDs film shows good flatness and obvious an absorption peak in 1360 nm. As a result, the as-prepared self-driven near-infrared photodiode detectors exhibit a high on/off ratio of 3×10^{4} and photocurrent of 2 mA cm^{-2} under $\lambda > 800$ nm polychromatic light with 41 mW cm^{-2}. Meanwhile, the PbS CQDs photoconductors show a responsivity of 0.48 A W^{-1} at NIR intensity of 0.18 mW cm^{-2}. Furthermore, both of detectors show the excellent ambient and photoelectrical cycling stabilities.

EXPERIMENTAL SECTION

PbS CQD synthesis and solution perovskite treatment: A mixture of PbO (0.45 g, 2 mmol), oleic acid (1.11 g, 4 mmol) and ODE (14.8 g, 58.6 mmol) was heated to 100 ℃ under vacuum for 15 min and placed under Ar. The flask temperature was increase to 110 ℃~180 ℃, and the Hexamethyldisilathiane/1-Octadecene (210 uL/5 mL) mixture was injected. When the mixture was cooled gradually to 100 ℃, the 20 mL n-hexane was injected for quenching. After the injection, the flask was allowed to cool gradually to 30 ℃. The PbS quantum dots were purified twice with toluene/acetone, and redispersed in octane at a concentration of 40 mg/ml. 3 mL of DMF solvent containing 138 mg PbI$_2$, 77.6 mg CsI and 11.6 mg Ammonium Acetate, was added to 3 mL 5~7 mg/mL PbS QDs solution. The mixture was stirred for 10 min at room temperature for liquid ligand exchange. Toluene was added to precipitate the QDs, and the mixture was centrifuged to separate the QDs powder from solution. The QDs were finally dispersed in DMSO/N-butylamine (4:1, vol. %) with a concentration of 20mg/ml.

Preparation of photodetector: The patterned ITO coated glass substrates were cleaned sequentially in an ultrasonic bath with water, acetone and isopropanol for 20 min, and dried with nitrogen flow. The substrates were treated with UV-ozone for 20 min before PEDOT: PSS layer deposition. The PEDOT:PSS solution was spin-coated on the top of ITO substrate at 3000 rpm for 60 s, and annealing at 130 ℃ for 30 min. The quantum dots solution was filtered through 0.22 μm Poly Tetra FluoroEthylene (PTFE) filters and spin-coated on the top of PEDOT:PSS layer at 2000 rpm for 60 s, and then the film was annealed at 80 ℃ in an nitrogen box for 2 min to remove solvent, repeating the steps to obtain different thickness. The final thickness was achieved by depositing 9~11 layers of quantum dots. PC$_{61}$BM dissolved in chlorobenzene (20 mg/mL) was spin-coated on the top of QDs layer at 3000 rpm for 60 s. Finally, Bphene and Ag were evaporated onto the as-prepared samples in vacuum in sequence with thickness of 5 and 100 nm, respectively.

Characterization: The active area as defined by the shadow mask was ≈ 0.04 cm^2. I-V curves were measured using a Keithley 4200 semiconductor characterization system connected to a probe station from -1 V to 1 V at a scan rate of 10 mV s^{-1}. Illumination test was conducted under 1 sun equivalent illumination with $\lambda > 800$ nm filter or other bandpass filters.

ACKNOWLEDGEMENTS

This work was the collaborative project with Fudan University.

REFERENCES

[1] T. Rauch, M. Böberl, S. F. Tedde, J. Fürst, M. V. Kovalenko, G. Hesser, U. Lemmer, W. Heiss, O. Hayden. *Nature Photonics*, vol. 3, 2009, pp. 332-336.

[2] J. B. Barton, R. F. Cannata, S. M. Petronio, *Proc.SPIE*, 2002, pp 37-42.

[3] J. Källhammer. *Nature Photonics*, 2006, pp 12-13.

[4] F. Pelayo Garcia de Arquer, A. Armin, P. Meredith, E. H. Sargent. *Nat. Rev. Mater*, 2017, 2, 16100.

[5] S. A. Mcdonald, G. Konstantatos, S. Zhang, P. W. Cyr, E. Klem, L. Levina, E. H. Sargent. *Nature,* 2005, vol. 4, pp 138-142.

[6] G. Konstantatos, J. Clifford, L. Levina, E. H. Sargent. *Nature Photonics*, 2007, vol. 1, pp 531-534.

[7] Y. Kim, F. Che, J. W. Jo, J. Choi, F. Pelayo Garcia de Arquer, O. Voznyy, B. Sun, J. Kim, M. Choi, R. Quintero-Bermudez, F. Fan, C. Tan, E.Bladt, G. Walters, A. H.Proppe, C. Zou, H. Yuan, S. Bals, J. Hofkens, M. Roeffaers, S. Hoogland, E.H.Sargent. *Advanced Materials*, 2019, 1805580.

[8] X. Zhang, J. Zhang, D. Phuyal, J. Du, L. Tian, V. Öberg, M. B. Johansson, U. B. Capper, O. Karis, J. Liu, H. Rensmo, G. Boschloo, E. M. Johansson. *Advanced Energy Materials*, 2017, 1702049.

[9] G. Sarasqueta, K. Choudhury, F. So. *Chem. Mater,* 2010, vol. 22, pp 3496-3501.

[10] Z. Ning, O. Voznyy, J. Pan, S. Hoggland, V. Adinolfi, J. Xu, M. Li, A. Kirmani, J. Sun, J.Minor, K. Kemp, H. Dong, L. Rollny, A. Labelle, G. Carey, B. Surtherland, I. Hill, A. Amassian, H. Liu, J. Tang, O. M. Bakr, E. H. Sargent. *Nature Materials,* 2014, vol. 13, pp 822-828.

SYNTHESIS OF MoS$_2$/WS$_2$ VERTICAL HETEROSTRUCTURE AND ITS PHOTOELECTRIC PROPERTIES

Xin Lin, Fang Wang*, Jiaqiang Shen, Xichao Di, Huanhuan Di, Meng Yan, Kailiang Zhang*

Tianjin Key Laboratory of Film Electronic & Communication Devices, School of Electrical & Electronic Engineering, Tianjin University of Technology, Tianjin 300384, China

Corresponding Author's Email: fwang75@163.com(Fang Wang);
kailiang_zhang@163.com(Kailiang Zhang) and 86-22-60214229

ABSTRACT

Two-dimensional (2D) heterostructures based on transition metal dichalcogenides have sparked significant attention due to their excellent electrical and optical properties. However, synthesis of heterojunction is still a challenge. In this work, MoS$_2$/WS$_2$ vertical heterostructure was achieved on SiO$_2$/Si substrate via transferring CVD-grown MoS$_2$ onto WS$_2$. The morphology and structure properties of the MoS$_2$/WS$_2$ heterostructure were characterized by optical microscope (OM), atomic force microscope (AFM), scanning electron microscope (SEM), Raman spectroscopy. Compared with individual MoS$_2$ or WS$_2$, the slight offset of the Raman peaks in the MoS$_2$/WS$_2$ heterostructure was observed, which implies the charge transfer of the heterojunction. A photodetector based on MoS$_2$/WS$_2$ heterostructure was fabricated with a channel length of 2μm. High on/off ratio (>10^7) and electron mobility of 10 cm^{2}V^{-1}S^{-1} of the photodetector were achieved. This work plays an active role in the development of photoelectronic devices based on 2D heterostructures.

INTRODUCTION

Two-dimensional transition metal dichalcogenides (TMDCs), such as MoS$_2$, WS$_2$, MoSe$_2$, WSe$_2$, have attracted much attention because of their interesting physical and chemical properties in the past years. And TMDCs have a wide range of application, for example: field effect transistors, optoelectronic devices, photocatalysts and so on. Moreover, 2D vertical van der Waals heterostructures stacked by different TMDCs materials have been built. Ultrafast charge transfer in different stacking layers and modulation of bandgap have been achieved in 2D heterostructures, which are contributed to optoelectronic devices.

In this work, MoS$_2$/WS$_2$ vertical heterostructure was achieved on SiO$_2$/Si substrate via transferring CVD-grown MoS$_2$ onto WS$_2$. Then the morphology of the MoS$_2$/WS$_2$ heterostructure was characterized by OM, SEM, Raman, AFM. Compared with individual MoS$_2$ or WS$_2$, the slight offset of the Raman peaks in the MoS$_2$/WS$_2$ heterostructure was observed, which implies the charge transfer of the heterojunction. A photodetector based on MoS$_2$/WS$_2$ heterostructure was fabricated with a channel length of 2μm. High on/off ratio and electron mobility was

achieved. Our work plays an active role in next-generation optoelectronic devices.

EXPERIMENTAL SECTION

Synthesis of WS$_2$: WS$_2$ films were first grown via CVD method on SiO$_2$/Si substrate using sulfur (S) and Tungsten trioxide(WO$_3$) powder as the precursors, as shown in Figure 1. S powder was positioned in the upstream of quartz tube with temperature of 120 ℃. Argon (Ar) gas with a flow rate of 50 sccm was used as the carrier gas. The downstream WO$_3$ powder was heated to 900 ℃ with a ramping rate of 20 ℃ min^{-1} and maintained for 30 min for the growth of WS$_2$ films. After the growth was completed, the tubular furnace was naturally cooled to room temperature.

Figure 1: Schematic illustration for the growth of MoS$_2$ and WS$_2$

Synthesis of MoS$_2$: MoS$_2$ films were also grown via CVD method using sulfur (S) and molybdenum oxide (MoO$_3$) powder as the precursors, which is similar with the growth process of WS$_2$. The difference is that the MoO$_3$ powder is heated to 750 ℃ for the growth of MoS$_2$.

Synthesis of MoS$_2$/WS$_2$ vertical heterostructure: the MoS$_2$/SiO$_2$/Si substrate was spin-coated with polymethacrylate (PMMA) and immersed in a hydrofluoric acid (HF) solution. Then PMMA/MoS$_2$ film can float on the surface of the solution. The HF solution was diluted ten times and PMMA/MoS$_2$ film was picked up with the WS$_2$/SiO$_2$/Si substrate. Finally acetone and deionized water were used to remove PMMA and clean. Then MoS$_2$ was transferred to the WS$_2$/SiO$_2$/Si substrate to form the MoS$_2$/WS$_2$ vertical heterostructure.

Fabrication of the photodetector: Metal electrodes (80 nm Ti) were deposited via e-beam evaporation after the electrode regions were defined by e-beam lithography, utilizing PMMA as the resist.

RESULTS AND DISCUSSIONS

The morphology and structure properties of the MoS_2/WS_2 heterostructure were characterized by OM, AFM, SEM, Raman spectroscopy. Figure 2(a-c) show optical images of WS_2, MoS_2 and MoS_2/WS_2 heterostructure respectively. From figure 2c, overlap region (yellow circle part) can be clearly seen, demonstrating the formation of vertical heterostructure. Figure 2e shows the SEM image of the as-transferred MoS_2/WS_2 sample, a distinct stacking-layered structure can be obviously seen from the SEM image, and the color of the framed region is darker than that of the outside region. Figure 2f shows the AFM image of the MoS_2/WS_2 heterostructure, which reveals the morphology of the MoS_2/WS_2 clearly.

In order to further explore the properties of MoS_2/WS_2 heterostructure, Raman spectroscopy was shown with 532 nm incident laser in the figure 2d. Figure 2d shows the Raman peak for monolayer MoS_2, monolayer WS_2 and MoS_2/WS_2 vertical heterostructure respectively. The monolayer MoS_2 shows two characteristic peaks around 384 cm^{-1} and 403 cm^{-1}, corresponding to the in-plane E_{2g}^1 vibration mode and out-of-plane A_{1g} vibration mode. For the monolayer WS_2, E_{2g}^1 peak and A_{1g} peak have been observed at 351 cm^{-1} and 417 cm^{-1}, which is consistent with previous reports. The Raman spectrum obtained from the MoS_2/WS_2 heterostructure was mostly composed of the vibrational modes of both MoS_2 and WS_2 domains. It can be clearly seen that there are four peaks which are originated from the vibration peaks of MoS_2 and WS_2. Compared with individual MoS_2 or WS_2, the slight offset of the Raman peaks in the MoS_2/WS_2 heterostructure was observed, which implies the charge transfer of the heterojunction.

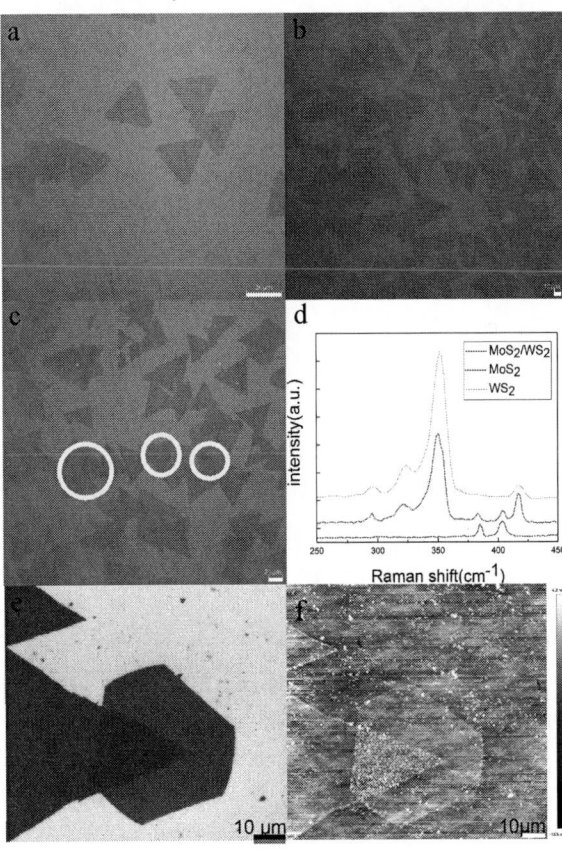

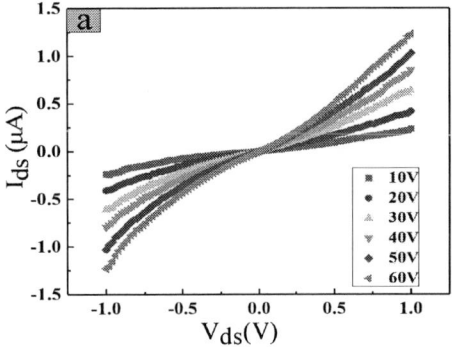

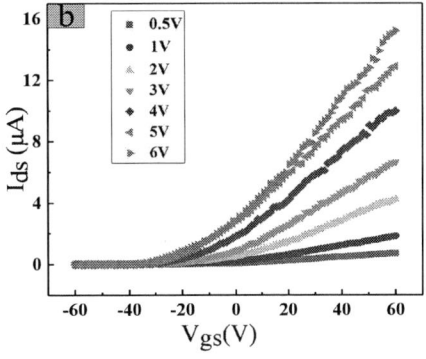

Figure 2: the morphology and structure properties of the MoS_2/WS_2 heterojunction. a, b, c) the optical images of the monolayer WS_2, monolayer MoS_2 and MoS_2/WS_2 heterojunction with the size of 50 μm, 58 μm and 45 μm, respectively. d) Raman spectrum of the WS_2, MoS_2 and MoS_2/WS_2 heterostructure with slight offset of the heterostructure. e) the SEM image of the MoS_2/WS_2 sample. f) AFM image of the MoS_2/WS_2.

Figure 3: electric characteristics of the MoS_2/WS_2 heterojunction. a) experimental output (I_{ds}-V_{ds}) curve of the MoS_2/WS_2 heterojunction photodetector under different V_{gs}. b) the I_{ds}-V_{gs} curve of the device with varied source-drain voltage in steps of 1 V.

A photodetector based on the MoS_2/WS_2 vertical heterojunction was successfully achieved via e-beam lithography and e-beam evaporation with the channel width of 2 μm. The output and transfer characteristics of the heterojunction device were plotted in figure 3a, b respectively. In the I_{ds}-V_{ds} curve (figure 3a), the source-drain current (I_{ds}) increases with the rising of source-drain voltage （V_{ds}）during 0 V to 1 V. And for different positive gate voltage，the I_{ds} also raises with it because the increase of V_{gs} could enhance the carrier density and I_{ds}. Figure 3b illustrates that I_{ds}-V_{gs} curve of the vertical MoS_2/WS_2 heterojunction under different V_{ds}. Obviously, from the curve, on/off ratio (>10^7) and electron mobility of 10 $cm^2V^{-1}S^{-1}$ was achieved, which is higher than pure MoS_2 or WS_2.

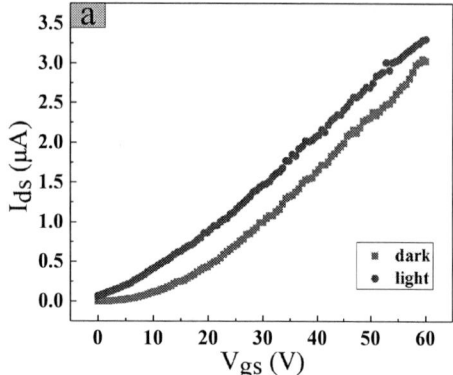

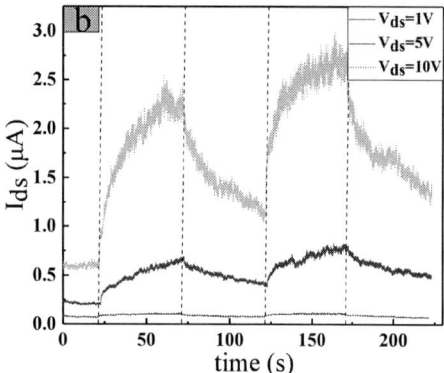

Figure 4: optical characteristics of the MoS₂/WS₂ heterojunction. a) gating response (I_ds-V_gs) of the photodetector in the dark (red line) and under light illumination (blue line) with V_ds=2V. b) time-resolved photoresponse of the device, recorded for different values of V_ds.

The photoresponse of MoS_2/WS_2 heterostructure device was shown in figure 4a,b. Then the I_{ds}-V_{gs} curve of the photodetector was depicted in figure 4a with and without light illumination when V_{ds} = 2V. Compared with

I_{ds} of the device in the dark state, the I_{ds} with light illumination is higher, which reveals that 2D heterostructures have great application prospect in optoelectronic devices. Light can influence the value of current, then current can be studied using time-resolved measurements. In figure 4b, the light is first turned on for a period of 50 s, then turned off for 50 s. For three different bias voltages, there is one thing in common that when light is turned on, the current will rise and the current will drop slowly when the light is turned off. With the emergence of light, the amount of photogenerated carriers increased, including the electrons in the conduction band and holes in the valence band, which led to an increase in the photocurrent.

CONCLUSION

In summary, pure high-quality MoS_2 and WS_2 were successfully achieved on SiO_2/Si substrates via CVD method. And MoS_2/WS_2 heterojunction was also synthesized by transferring MoS_2 to WS_2. Then The morphology and structure properties of the MoS_2/WS_2 heterostructure were characterized by OM, SEM, AFM and Raman spectroscopy, which confirm high quality of the MoS_2/WS_2 heterostructure. A photodetector based on the heterojunction was fabricated. The transfer and output characteristics of the photodetector are achieved. At the same time, when there is light irradiation, the output current will increase, which is important for the photoelectronic devices. This work may be extendable to the various heterojunctions and provides an opportunity to develop optoelectronic devices in the future.

ACKNOWLEDGEMENTS

This work was supported by Natural Science Foundation of Tianjin City (Grant Nos.18JCZDJC30500, 17JCYBJC16100 and 17JCZDJC31700)，National Key Research and Development Program of China (Grant No.2017YFB0405600) and National Natural Science Foundation of China (Grant Nos.61404091, 61274113, 61505144, 51502203 and 51502204).

REFERENCES

[1] D. X. Tang, F. Wang, B. J. Zhang and K. L. Zhang. *J. JOURNAL OF MATERIALS SCIENCE* 53.20, 2018, 14447-14455.

[2] J. Zhang, L. Du, S. Feng and R. Zhang. *J. Nature Communications*, 10.1, 2019, pp. 1-8.

[3] K. Chen, X. Wan, J. Wen and W. Xie. *J. Acs Nano* 9.10, 2015, pp. 9868-9876.

[4] F. Gong, W. Luo, J. Wang and P. Wang. *J. Advanced Functional Materials.* 26.33, 2016, 6084-6090.

[5] J. Lee. S. Pak, Y. W. Lee, Y. Park and A. R. Jang. *J. ACS nano,* 13.11, 2019, 13047-13055.

EFFECTIVE SPARSITY-PRIOR IMAGE DENOISING ALGORITHM FOR CMOS IMAGE SENSOR IN ULTRA-LOW LIGHT IMAGING APPLICATIONS

Tao Zhou[1], Chen Li[1,]*, Jiebin Duan[1], Xuan Zeng[2] and Yuhang Zhao[1]

[1]Shanghai Integrated Circuits R&D Center Co., Ltd., Shanghai, China
[2]State Key Lab of ASIC & System, School of Microelectronics, Fudan University, Shanghai, China
*Corresponding Author's Email: lichen@icrd.com.cn

ABSTRACT

An effective algorithm is designed for incorporating in a 3D stacked CMOS image sensor for image denoising in ultra-low light conditions. The algorithm originates from sparsity-prior of image and non-locally clustered sparse representation. The simulation results of the CIS image signal processing (ISP) demonstrate high-performance at intense noise level of ultra-low light images, which reveal a great potential of the CIS design in various applications such as night shot, security monitoring, machine-state inspection, medical imaging, biophysics detection, etc.

INTRODUCTION

Noise exists in any imaging devices and systems. When encounters low light, imaging becomes a particular challenging task in consideration of heavy noise. Therefore, the noise suppression is of great importance for CIS applications in such conditions and for demonstration, Figure 1 shows some typical low light scenes which are collected by some popular datasets for verifying the effectiveness of denoising algorithms. Obviously, these conditions and alike are very common in daily life.

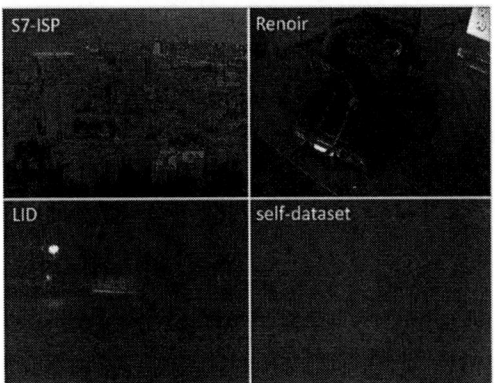

Figure 1: Low light scenes taken from public self-made datasets. S7-ISP is from a samsung smart phone[1], Renoir from Canon ES90 camera (ISO 1000, auto exposure time)[2], LID from a Sony α 7S camera (luma < 0.1 lux, exposure time=1/30 s, ISO 8000)[3], self-dataset from our reported CIS(exposure time=1/25 s, luma ~0.003 lux).

Traditionally, there are two aspects for imaging to alleviate the noise problem: one is on pixel improvement, which is the fundamentally solution but also the most difficult. Many works have been reported [4-6], but trade-offs between key parameters are ineluctable; the other is on image/signal processing, combining a denoising function to improve image quality, but to find an adaptive and effective method is also not an easy task [7-9]. Besides the above mentioned, normal operation is to increase the ISO or prolong the exposure time for CIS, however, they are either noise-sensitive or time-comsuming. In this work, a compositive scheme for the ultra-low light imaging is realized by a stacked design, and the effectiveness of the proposed solution is verified by the test and simulation results.

ARCHITECTURE OF IMAGE SENSOR

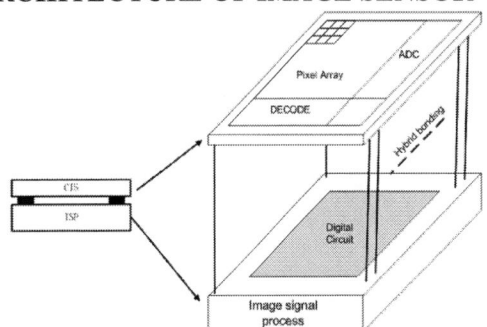

Figure 2: Structure of the designed CMOS image sensor.

The designed structure, shown in Figure 2, is made up of an image sensor chip stacked with an ISP chip using 3D hybrid bonding technology. Each row of ADC output of the sensor chip on the upper floor is connected to the input of the lower IPS chip via a contact. The advantages are, on the one hand, the output transmission rate could be greatly improved through 3D interconnection, and on the other hand, higher chip integration can be obtained to enable more powerful image signal processing functions.

SPARSITY-PRIOR IMAGE DENOISING ALGORITHMS

When the luminance of imaging is ultra-low and in a

978-1-7281-6559-2/20 $31.00 © 2020 IEEE

real-time mode, the photons received by a single pixel is too low to produce a typical electric signal for detection. The quantum noise and thermal noise overwhelms the image signal, and as a result the SNR is extremely low and the major features of image are corrupted, normal denoising techniques are less effective. In this work, an effective image algorithm based on sparsity-prior for noise suppression is presented, which contains two parts: image pre-processing and image non-locally clustered sparse representation. The algorithm flow is shown in Figure 3.

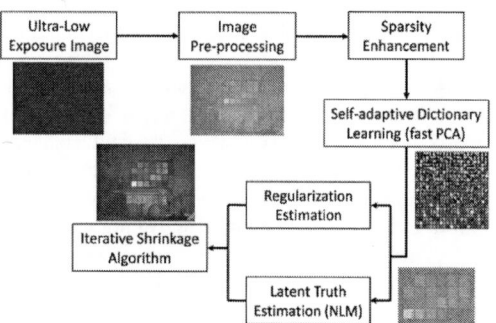

Figure 3: The flow of the sparsity-prior image denoising algorithms incorporated with CIS.

Image pre-processing

There are several types of noise in imaging, the pre-processing mainly includes two steps: the defect pixel correction (DPC) and the histogram matching (HM). DPC is an essential step for CIS and realized by a simple and effective median filter. The HM algorithm is based on histogram equalization by shifting the related gray level to a pre-set target.

Image non-locally clustered sparse representation

After the pre-processing, the dominated noise left is the thermal noise and the quantum noise, which are subject to Gaussian and Poisson distribution, respectively. Simple denoising algorithms, such as Gaussian filter, Wiener filter, wavelet filter, etc., are not competent for the job. A more sophisticated method based on the successful non-local means (NLM) algorithm is widely employed as non-local clustered sparse representation (NCSR). The NCSR model is first reported in Ref [10]:

$$\alpha = \arg\min_{\alpha}\left\{\left\|Y - HD\alpha\right\|_2^2 + \lambda\sum_i\left\|\alpha_i - \beta_i\right\|_p\right\}$$

where Y is noisy image, H is degradation factor, D is an adaptive dictionary, λ is regularization parameter, α is the latent clean image and β is the estimation α with a proper method based on Y. The key idea of NCSR is to use the sparsity prior of image to recover the latent truth due to compressed sensing (CS) theory which lays a foundation of reconstruction based on sparse signals. The NCSR algorithm is mainly composed of three parts: sparse

representation (sparsity enhancement & self-adaptive dictionary learning), model parameters estimation (regularization parameter and latent truth estimations) and model solution (iterative shrinkage algorithm). Different from Ref. 10, two improvements are made in this work: the sparsity enhancement and latent truth estimation.

The sparse representation is of the most important because it directly influences the quality of reconstructed image according to the CS theory, the sparser the better [11]. The common-used Euclidean Distance is replaced by cosine distance, the patches searched could make a better sparsity enhancement because of higher similarity. Then the dictionary is computed with a fast principle component analysis (PCA) method. As for the latent truth estimation, the latent clean image is computed by importing external similar clean images into NLM algorithm to guide a better estimation. With the optimization of key parameters D and β in NCSR model, better denoising effect is expected.

RESULTS AND DISCUSSIONS

Table I Performance Characteristics of CIS

Parameter	Value
Technology	55 nm 1P4M CIS Process
No. of effective pixels	800 (H) * 600 (V)
Image Size	12.67 mm (H) * 9.50 mm (V)
Pixel Size	15.84 um * 15.84 um
Fill Factor	81%
Power Supply	1.2 V (Digital), 3.3 V (Analog)
Max. Sensitivity	131 V/lux-sec
Max. Conversion Gain	129 µV/e
Dynamic Range	70.1 dB for Pixel side
ADC Resolution	12 bit

In the 3D stacked design, the upper floor is a 6-Transistor and 1-Capacitor active pixel sensor (APS) with pinned photodiode (PPD), described in our previous work [12]. The customized performance characteristics are list in Table I. High-sensitive performance is achieved: maximum sensitivity of 131 V/lux-sec, maximum conversion gain of 129 µV/e, large of dynamic range of 70.1 dB for pixel side.

For the lower ISP chip, the processing pipeline is constructed and simulated, as shown in Figure 3. The images at ultra-low light level ($<= 3\times10^{-3}$ lux @ 25 frames/s) have been adopted for pipeline simulate. The subjective effect of denoising are compared in Figure 4, the first row is the raw image, the middle is the image after pre-processing and the bottom is the denoised image. As we can find in the enlarged areas (green square), the noise intensity has been suppressed effectively while the details are retained successfully (the curves are clearly revealed in the denoised results).

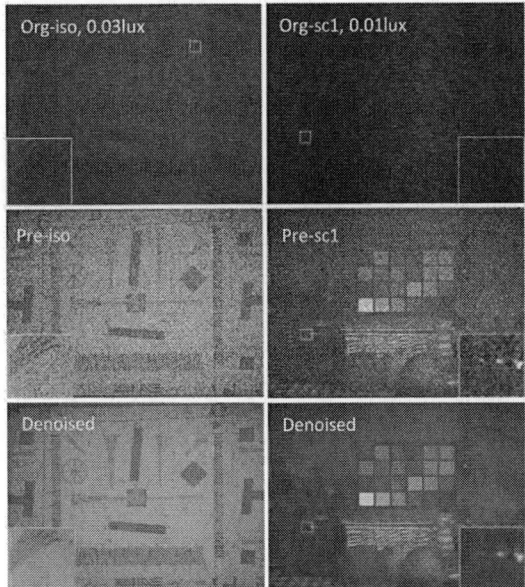

Figure 4: The comparison of the original image and the corresponding denoised results.

To quantify the effect of denoising, two popular parameters are adopted, Standard Noise (SN) and Peak Signal Noise Ratio (PSNR), and for comparison, several popular algorithms are chosen (the bold number indicates the superior). The results are shown in the Table II (4 scenes are adopted: cc, gray, iso, sc1). SN represents noise intensity level and is equal to the standard deviation of a flat area. In this work, the average value of SN is suppressed < 3@12bit ADC, which provides a pleasant visual effect for the smooth area. As for PSNR, it represents the similarity between the original image and the reference image (clean image is captured in the same scene) at pixel level. The value of PSNR is increased to an average value > 25 dB, which is a remarkable improvement and could be found out obviously from the overall and details of images in Figure 4.

Table II Comparison of the denoising algorithms.

PSNR (dB)	Pre	K-SVD [13]	EPLL [14]	BM3D [15]	NCSR [10]	Self
cc	22.67	26.25	26.34	26.49	26.54	**26.60**
gray	27.62	32.41	32.73	32.97	32.94	**33.01**
iso	16.66	18.80	18.97	18.90	18.94	**19.07**
sc1	19.97	23.54	23.76	23.78	23.79	**23.88**
Ava.	21.73	25.25	25.45	25.53	25.55	**25.64**
SN	Pre	KSVD	EPLL	BM3D	NCSR	Self
cc	17.38	6.54	6.85	3.50	6.32	**3.28**
gray	13.52	3.94	3.43	3.42	**3.22**	3.30
iso	22.21	3.59	3.21	2.47	**1.80**	2.26
sc1	24.56	5.34	4.50	3.19	3.48	**2.97**
Ava.	19.41	4.85	4.49	3.16	3.70	**2.95**

CONCLUSION

This work presents a design of stacked CMOS image sensor for ultra-low light condition, which consists of an image sensor chip stacked with an ISP chip using 3D hybrid bonding technology. Taking advantage of 3D stacked structure, an effective and sophisticated image denoising algorithm is designed and could be incorporated into the CIS chip to achieve advanced functionality in a compact size. The simulated results demonstrate the effectiveness of the sparsity-prior based algorithm for image denoising, and reveal the potentials for various applications range from scientific to industry fields. The next step of our work is to optimize the stacked design as well as hardware implementation of image processing modules for chip tape out.

REFERENCES

[1] E. Schwartz and R. Giryes, *Trans. Image Process.*, vol. 28, 2019, pp. 912-923.

[2] J. Anaya, A. Barbu, *J. Vis. Commun. Image R.*, vol. 18, 2018, pp.144-154.

[3] C. Chen and Q. F. Chen, *arXiv*, 2018, 1805.01934.

[4] Y. Chen, et al., *IEEE Int. Solid-State Circuits Conf. Dig. Tech. Papers (ISSCC)*, 2012, pp. 384–386.

[5] C. Lotto, et al., *IEEE Int. Solid-State Circuits Conf. Dig. Tech. Papers (ISSCC)*, 2011, pp. 402–404.

[6] Y. Lim et al., *IEEE Int. Solid-State Circuits Conf. Dig. Tech. Papers (ISSCC)*, 2010, pp. 396–397.

[7] C. Tomasi and R. Manduchi, *ICCV*, 1998, pp. 839-846.

[8] A. Buades and B. Coll, *CVPR*, vol. 2, 2005.

[9] K. Dabov and A. Foi, *IEEE Trans. Image Process.*, vol. 16, 2007, pp. 2080–2095.

[10] W. Dong, et al., *IEEE Trans. Image Process.*, vol. 22, 2013, pp. 700–711.

[11] L. Zhang, et al., *IEEE Sig. Proc. Maga.*, vol.34, pp.172-179.

[12] C. Li, et al., *IEEE Xplore*, CSTIC. 2019. 8755712.

[13] M. Elad, et al., *IEEE Tran. Image Process.*, vol. 15, 2006, pp. 4311-4322.

[14] D. Zoran, et al., *ICCV*, 2011:479-486.

[15] K. Dabov, et al., *IEEE Tran. Image Process.*, vol.16, 2007, pp. 2080-2095.

MONOLITHIC 3D ENABLED PROCESSING-IN- SRAM MEMORY

Vijaykrishnan Narayanan[*], Nagadastagiri Challapalle, Ikenna Okafor, Srivatsa Srinivasa, and Nicholas Jao

School of Electrical Engineering and Computer Science, The Pennsylvania State University,
University Park, PA, USA, 16801.
*Corresponding Author's Email: vxn9@psu.edu

ABSTRACT

This work will provide an overview of recent advances in enabling SRAM-based compute fabrics leveraging monolithic 3D (M3D). It will highlight that the fine grain connectivity enabled by M3D, enables to embed computations close to the memory cells significantly reducing the data transfer costs. The application level benefits to emerging workloads will also be presented.

M3D-BASED PROCESSING-IN-MEMORY

In the current era of big data and artificial intelligence (AI), today's computing systems which are heavily optimized for bringing the data closer to the compute units suffer from data movement bottlenecks. The computer vision and deep learning applications deployed in autonomous driving systems, smart cameras, and mobile phones etc. generate large amounts of data. These applications also need these massive data to be processed quickly to make real time decisions and predictions. Moving these large volumes of data back and forth between various levels in memory hierarchy and processors requires lot of energy and time [1]. Also, as the applications become more data-intensive, data access latency and the memory bandwidth become key bottlenecks in improving performance. These energy, latency, and bandwidth constraints limit the efficiency of energy constrained systems such as mobile processors and edge processors in leveraging the benefits of AI algorithms for user experience. To alleviate these data movement bottlenecks, researchers are exploring ways to process the data where it resides. Several works have demonstrated significant performance improvements by leveraging the 3D stacked logic and memory layers [2], multi-row activations coupled with sense amplifiers augmented with additional functionality [3], emerging architectures such as crosspoint memories to process the data in memory [1].

Monolithic 3D integration enables vertically stacked layers of logic and memories with nano-scale inter-layer vias providing large degree of connectivity between the layers. We leverage the M3D integration capabilities to augment the conventional SRAM designs with multidimensional access capabilities [4], support for in-situ Boolean operations such as AND/NAND, OR/NOR, XNOR/XOR [5], arithmetic operations such as addition and hybrid content addressable-random access

memory capabilities [6].

We enhance the single layer (layer 1) 6T SRAM design with two additional transistors in second layer grown monolithically over the first layer to enable multi dimensional access capabilities [4]. We place two nMOS transistors exactly above the 6T SRAM layer using the M3D to enable the column wise single ended read. We use access transistors in layer 1 to perform the write and row wise read operations. The access transistors in second layer are used for doing the column wise read operation. The multidimensional access capability achieves 2.15x savings in execution time and 7.81% savings in access energy in integral image algorithm which is a representative algorithm requiring access to both row and column data of an image.

We further enhance the capabilities of 6T SRAM design by adding six transistors in the second layer using four M3D vias [5]. These additional transistors provide the flexibility to enhance the stability of read and write operations depending on the application and also to provide faster data access mechanism in turbo mode. The strengths of pull down and pull up transistors are controlled based on the given mode using the additional transistors in layer 2. The transistors in top layer are further leveraged to perform several in-situ Boolean bitwise operations on the stored data by selectively enabling sensing of multiple rows. We achieve ≈17% improvement in write stability and write latency. We achieve 1.5x improvement in read noise margins and 23% improvement in energy-delay product for the in-memory Boolean operations. For incorporating the in-memory add operation [6], we leverage the transistors in the layer 2 to perform bitwise operations to compute sum and carry.

We also enable in-memory associative and approximate bit search operations [6] by placing the additional transistors required for enabling content addressability (CAM) in layer 2. Both search operations use CAM functionality as the fundamental operation. In associative search operation, the encoder output from the CAM operation is routed back to the decoder in layer 2 to read the search results. In the approximate search operation, pulsed data is provided as the input to search lines which enables selective discharge for approximate search.

ACKNOWLEDGEMENTS

This work was supported in part by Semiconductor Research Corporation (SRC) Center for Brain-inspired Computing Enabling Autonomous Intelligence (C-BRIC) and Center for Research in Intelligent Storage and Processing in Memory (CRSP).

REFERENCES

[1] P. Chi, S. Li, C. Xu, T. Zhang, J. Zhao, Y. Liu, Y. Wang, and Y. Xie, *Proceedings of ISCA*, 2016, pp. 27-39.

[2] X. Xin, Y. Zhang, and Z. Yang, *Proceedings of DAC*, 2019, pp. 29:1-29:6.

[3] S. Aga, S. Jeloka, A. Subramaniyan, S. Narayanasamy, D. Blaauw and R. Das, *Proceedings of HPCA*, 2017, pp. 481-492.

[4] S. Srinivasa, X. Li, M. Chang, J. Sampson, S. K. Gupta and V. Narayanan, *IEEE Trans. VLSI*, 2018, pp. 671-683.

[5] S. Srinivasa, A. Ramanathan, X. Li, W. Chen, F. Hsueh, C. Yang, C. Shen, J. Shieh, S. Gupta, M. Chang, S. Ghosh, J. Sampson, and V. Narayanan, *Proceedings of ISPLED*, 2018, pp. 34:1-34:6.

[6] N. Jao, A. K. Ramanathan, S. Srinivasa, S. George, J. Sampson and V. Narayanan, *Proceedings of ISVLSI*, 2018, pp. 447-452.

POWER ORIENTED CMOL DEFECT-TOLERANT MAPPING WITH AVAILABLE NANODEVICES

Shangluan Xie, Yinshui Xia*, Xiaojing Zha, and Xiangui Gu

EECS, Ningbo University, Ningbo 315211, China

* Email: xiayinshui@nbu.edu.cn

ABSTRACT

Due to the high defect rate of nano-technologies, the defect-tolerant mapping of CMOS/nanowire/molecular hybrid (CMOL) circuit becomes critical. In this paper, a simplified power consumption model is established first to guide the exploration of power-optimized mapping methods. Considering the additional energy loss caused by the activation of extra nanodevices during the process of tolerating stuck-at-close defects, a trim method is proposed to extract available nanodevices by bypassing defects on the CMOL array. Evolutionary algorithm (EA) is adopted to complete mapping. The experiment shows the proposed method has a better performance in power consumption with higher mapping speed.

INTRODUCTION

CMOS/nanowire/molecular hybrid (CMOL) circuit is a Field Programmable Gate Array (FPGA)-like structure. As shown in Figure 1 (a),(b), a CMOL cell consists of two perpendicular nanowires with nanodevices at the cross-points, a CMOS inverter and interface pins [1].

Nanodevices can be programmed into state "on" (low resistance, R_{ON}) or "off" (high resistance, R_{OFF}), "on" nanodevices transmit signals from the top layer nanowires to the bottom layer nanowires, while "off" nanodevices block signal transmission.

The periodic fracture structure of nanowires causes each cell can only connect with $M = 2r(r-1)-1$ other cells, which forms so-called "cell connectivity domain" and r represents the connectivity domain radius [2]. Since connections between cells are completed through a nanowire-nanodevice-nanowire link, when a cell's available nanodevices are searched, its connectivity domain is determined.

The defect rate of nano-technologies can reach 10^{-3} to 10^{-1} as indicated in [3]. The defects mainly consist of nanowires irregularly broken, nanodevices stuck-at-open or stuck-at-close. The first two defects which classified as open-based defects will reduce cells' connectivity domain, and the last will lead to incorrect pre-implementation logic functions on the CMOL array. Figure 1 (c) shows the mapping result of the function $F = \overline{A+B}$ after bypassing defects.

Accurate synthesis [4] and heuristic search [5, 6] are used in the existing defect-tolerant mapping methods. In [6], blocking method for stuck-at-close defect propagation is implemented by inserting a pair of inverse signals to defective cells. Since only method [6] explores the mechanism of stuck-at-close defect in detail, it is chosen as the comparative sample.

In this paper, a simplified power consumption model is established and the fact is pointed out that additional activated nanodevices during the blocking process in [6] will increase the static power. Targeting to reduce extra "on" nanodevices, a trim method is proposed to extract available nanodevices by bypassing defects on the CMOL array.

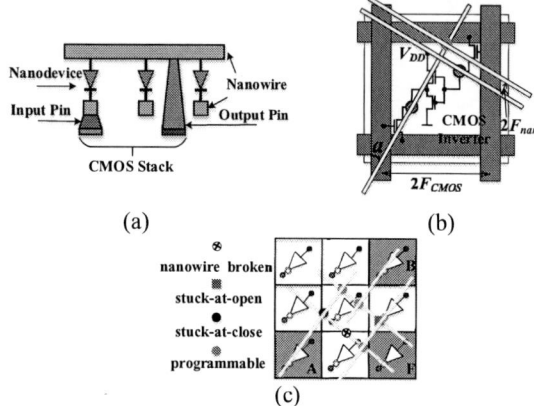

Figure 1: (a) Cross-sectional view of a CMOL cell; (b) Top view of a CMOL cell; (c) $F = \overline{A+B}$ mapping diagram

POWER CONSUMPTION MODEL

The power consumption of CMOL can be estimated as a sum of the static power P_{ON} due to the current through "on" nanodevices; leakage power P_{Leak} due to the leakage current through "off" nanodevices in parallel with mapped cells; and the dynamic power P_{dyn} due to recharging of nanowire capacitances [1]. The equivalent circuit of a CMOL logic stage is shown in Figure 2.

Strukov [2] gave a set of classic circuit parameters and pointed out that the static power is the main power consumption of CMOL arrays. In this paper, P_{ON} is calculated as formulae (1), where N_D is the number of "on" nanodevices; R_{Pin} is the pin-to-nanowire contact resistance; R_{Wi} is the nanowire resistance of each connection. By analyzing the optimization parameters

978-1-7281-6559-2/20 $31.00 © 2020 IEEE

given in [2], the value of R_{Pin} and the change of R_{Wi} can be ignored compared with R_{ON}. A simple form of P_{ON} is proposed as formulae (2), constant $R_{W1/2}$ equals to half of the resistance of a complete nanowire.

$$P_{ON} = \sum_{i=1}^{N_D} \frac{V_{DD}^2}{2(R_{ON} + 2R_{Pin} + R_{Wi})} \qquad (1)$$

$$P_{ON} = P_0 \cdot N_D, \ P_0 = \frac{V_{DD}^2}{2(R_{ON} + 2R_{W1/2})} \qquad (2)$$

Since the power consumption of CMOL mainly consists of static power P_{ON} and N_D determines P_{ON}, the reduction of nanodevices utilization leads to the reduction of power consumption.

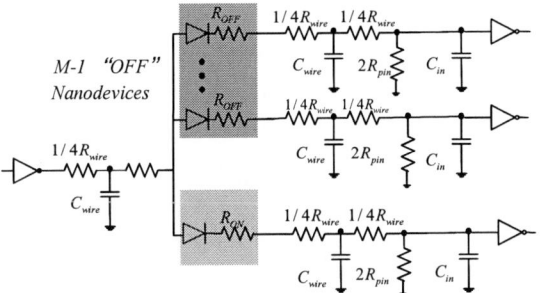

Figure 2: The equivalent circuit of a CMOL logic stage.

CONNECTIVITY DOMAIN TRIM METHOD

First of all, the definition of "trim" is given: Trimming a nanodevice means the use of this nanodevice is prohibited during the mapping process. In order to implement the correct logic operation on the CMOL array, three constraints must be observed:

$$f(i,j) = 1, \ g(a,b) \neq 0 \qquad (3)$$
$$f(i,j) = 0, \ g(a,b) \neq 1 \qquad (4)$$
$$g(a,b) = 1, \ i \neq null \qquad (5)$$

Where i, j are logic gates or null elements which will be mapped to cells a, b separately. $f(i,j)$ is a binary value(1/0) represents whether i is a fan-in of j. $g(a,b) = 0$ represents cell a can't connect with cell b due to the connectivity domain constraint or open-based defects, $g(a,b) = 1$ represents there is a stuck-at-close nanodevice between a and b.

After analysis, it is found that the existing methods judge violations of constraints separately, which will increase the solution time of the algorithm, and the blocking approach for constraint (5) will increase the static power.

To optimize these, a trim method which increases the solution speed at the cost of reduced mapping resources and avoid extra power consumption is proposed: first, trim the nanodevices on the path where the open-based defects occurred; then find out the cells with stuck-at-close defects in the output connectivity domain and trim related nanodevices. The pseudo code of the method is as follows:

pseudo code of connectivity domain trim method

1. if stuck-at-open then
     delete the nanodevice;
 2. if nanowire irregularly broken then
     delete nanodevices behind the breakpoint;
 3. if stuck-at-close then
     find the related outgoing signal cell
     delete all nanodevices around it.

After above operation, possible error connections have been pruned before the mapping, and the algorithm only needs to make one judgment according to formulae (6) during the mapping process:

$$f(i,j) = 1, p(a,b) = 1 \qquad (6)$$

$p(a,b)$ is a binary value(1/0) represents whether there is an available nanodevice between cells a, b.

OPTIMIZATION ALGORITHM

Evolutionary algorithm (EA) is adopted to complete mapping. EA simulates the process of "natural selection, survival of the fittest" in nature and can be generalized as five steps as below which includes reproduction, evaluation, selection, mutation and reallocation.

Step1: Complete the trim method and establish the constraint (6); generate initial population of *psize* individuals randomly. A trait of an individual represents a mapping location of a logic gate.

Step2: $fitness = violations + (N_D - N_E)$ is defined, *violations* represents the number of traits that violate the constraint (6) in the individual, N_D is the number of "on" nanodevices which reflects the size of static power consumption. And N_E is the number of connecting edges in the gate-level netlists to be mapped. The larger the *fitness*, the worse the quality of the solution.

TABLE I. COMPARISON WITH METHOD [6]

Benchmark	Cells	Area	Method[6]		Proposed method	
			CPU (sec)	Power (μW)	CPU (sec)	Power (μW)
s27	19	7*7	0.02	3.29	0.01	3.00
s208	136	15*15	4.08	34.59	1.17	28.80
s298	122	15*15	4.91	40.68	0.95	34.50
s344	179	18*18	4.75	42.30	3.47	39.60
s349	184	18*18	6.06	43.50	4.65	40.80
s382	175	17*17	3.95	47.55	2.96	44.85
s386	164	17*17	2.59	53.55	3.72	50.85
s400	188	18*18	5.58	51.00	4.31	48.30
s420	299	23*23	43.15	69.30	32.86	64.50
s444	187	18*18	6.73	52.50	4.18	49.80
s510	304	27*27	153.03	78.60	128.04	73.80
s526	273	23*23	34.10	79.35	31.32	75.15
s641	301	25*25	119.63	65.55	63.91	61.05
s713	321	26*26	134.74	70.95	73.05	66.15
Average					**31.07%**	**-7.60%**

Step3: Sort individuals according to *fitness* , eliminate the individuals with poor quality; use roulette algorithm to select several parents with high quality to reproduce by cross-interchanging traits.

Step4: After step3, if individuals' *fitness* don't decrease, randomly pick two traits for information exchange.

Step5: Repeat step 3 & 4 until the optimal solution appears.

EXPERIMENTAL RESULT

The algorithm is implemented using C programming language and tested on a Windows machine with a 3.40 GHz Intel Core processor and 2048MB memory. ISCAS benchmarks are used and each benchmark is tested under different seeds for 20 times due to the randomness of EA, and the average result value is taken. For a fair comparison with method [6], radius r =18 is chosen, and the defect rates of stuck-at-open and stuck-at-close are set as 10% and 0.1%, respectively. The uniform distribution of defects is assumed. Circuit parameters in [2] are adopted, and $P_0 = 0.15 \mu W$.

In Table 1, Column "cells" shows the number of components in circuits. Column "CPU" is the runtime of the algorithm. Column "Area" stands for the scale of CMOL arrays. Column "power" is the value of P_{ON} , $P_{ON} = P_0 \cdot N_D$. In the same environment, the experiment shows the method proposed in this paper can increase 31.07% mapping speed and reduce 7.60% static power consumption on average compared with method [6].

SUMMARY

In this paper, a trim method is proposed to extract available nanodevices by bypassing defects on the CMOL

array. At the cost of reduced mapping resources, additional static power consumption caused by traditional mapping methods is avoided and the mapping speed is improved. The experiment shows 7.60% static power consumption is reduced and 31.07% mapping speed is increased on average compared with method [6].

However, because the mapping resources are greatly reduced, the algorithm is liable to fall into a local optimal solution and can't to solve large-scale circuits. Therefore, the method which can solve large-scale circuits should be studied. Since this paper only analyzes the impact of traditional mapping methods on power consumption, the impact of various defects on power consumption and its elimination methods are also waiting to be discovered.

ACKNOWLEDGEMENTS

This research was supported by the National Natural Science Foundation of China under Grants 61571248, Zhejiang Provincial Xinmiao Talent Project under Grants 2019R405078, and the Scientific Research Foundation of Graduate School of Ningbo University.

REFERENCES

[1] D. B. Strukov and K. K. Likharev. *Nanotechnology*, vol.16, no.6, 2005, pp. 888-900.

[2] D. B. Strukov. Diss. Stony Brook University, 2006.

[3] H. Yan, H.S. Choe, S.W. Nam, et al. *Nature 470*, 2011, pp.240-244.

[4] W. Hung, C. Gao, X. Song and D. Hammerstrom. *IEEE Sensors Journal*, vol.8, no.6, 2008, pp. 823-830.

[5] J. Wang, Y. Xia and Z. Chu. *Wireless Communication Technology*, 2016(3). (in Chinese)

[6] D. Chen, Y. Xia, and Z. Wang. *Microelectronics Journal 72*, 2018, pp.100-108.

Signoff-level full-chip ESD/reliability design verification using logic-driven layout static approach

Li Li, Yi-Ting Lee, and Sridhar Srinivasan

Mentor, a Siemens Business

8005 SW Boeckman Rd, Wilsonville, Oregon 97070 USA

Email: li_li@mentor.com

ABSTRACT

An automated signoff-level full-chip design verification methodology for electrostatic discharge (ESD) protection is established in foundries, integrated device manufacturers (IDMs), and semiconductor design houses. This methodology employs a static rule check approach implemented by electronic design automation (EDA) tools using logic-driven layout (LDL) check functionality [1]. It can not only examine ESD protection circuits on logical circuitry, but also correlate check criteria to corresponding layout geometrical data and electrical resistor networks. This methodology aligns well with foundry/IDM/design house ESD design guidelines, enabling ESD design kits to provide practical ESD signoff verification for tapeout. In this paper, we describe topological IO-ESD, cross-power-domain ESD, and physical ESD path robustness design verification using the LDL check methodology.

INTRODUCTION

Designing a chip in advanced technology nodes and/or containing complex functionality for modern communication, computing, gaming, and many other emerging applications is often a multi-million dollar project requiring high integrity in design methodology and extensive cross-team collaboration.

To ensure good chip yield, design rule checking (DRC) and design for manufacturing (DFM) electronic design automation (EDA) tools are required to thoroughly verify the geometrical property/interaction of each polygon and clustered polygons across the entire chip. To ensure circuit functionality operates as designed on implemented layout data, layout vs. schematic (LVS) EDA tools are required to extract the layout netlist and compare it against the design schematic. To ensure reliability issues are minimized or eliminated in a chip, reliability EDA tools must carry out reliability verification on relevant circuitry and interconnect across the entire chip [2].

Electrostatic discharge (ESD) is one of oldest recognized reliability issues in CMOS technology. ESD failure is often only detected post-silicon, with the resulting impacts to both cost and time-to-market negatively impacting companies in a highly competitive environment. Until recently, there was no proven, qualified automated chip-level ESD verification [3,4]

available when implementing a layout. Manual examination was typically used during tapeout, with resulting signoff quality uncertain with respect to ESD design verification. To meet rigorous signoff criteria for silicon chip design tapeout, full-chip ESD protection verification can no longer be performed using either a logical circuitry approach or manual inspection.

We introduce a signoff-quality full-chip ESD design verification methodology that is qualified and used in reference flows by major foundries [5,6], and has also been proven in use by many design houses [4,7], integrated device manufacturers (IDMs), and intellectual property (IP) vendors over multiple technology generations of chip tapeout.

LDL CHECK METHODOLOGY FOR ESD

Semiconductor foundries, IDMs, and design houses all have defined ESD design guidelines that must be followed through chip design to tapeout. However, not all of these guidelines clearly reflect ESD protection physics, or are correlated to appropriate electrical/geometrical properties that can be consistently measured. One significant reason these ESD design guidelines have not evolved into more realistic design rules that match ESD protection physics and can be consistently used across different chip projects is the lack of EDA tools that can accurately and consistently verify chip-level ESD design rules. ESD design guidelines are often based on design logic (circuitry) and design margins deemed acceptable for a particular use condition [8], often referred to as a "mission profile." A few can be checked using just the schematic or extracted layout netlist. Others must be checked against ESD devices, power clamping devices, and interconnects along the ESD paths on layout data.

Automated ESD check methodology using logic-driven layout (LDL) functionality can verify ESD design rules on either schematics or layout netlist, and further drive verification through identification of ESD paths on logical circuitry (including IO/power/ground nets, nets bridging IO to internal gates, cross-power-domain nets, and stage devices such as ESD diode/MOS, power clamping MOS, back-to-back (B2B) diode devices, etc.) to their physical representatives on layout data for geometrical/electrical properties checking.

In general, LDL verification methodology can verify the following characteristics:

- ESD protection circuits are in place
  - device parameters/geometrical properties meet the required parameters
  - ESD path interconnect robustness is satisfied
    - effective resistance along defined ESD paths is less than the defined effective resistance constraint value
    - current density on each wire segment/via area of defined ESD paths is less than defined current density constraint values

VERIFY ESD PROTECTION CIRCUITS

A required step for ESD verification at the block or chip level is to check that appropriate ESD/power clamping protection circuits are connected between each IO port and internal gates, between paired power and ground ports, between local and common ground ports, and on cross-power-domain paths between driver and receiver gates [9]. These protection circuits consist of different devices, such as pull-up/pull-down diodes, NMOS, PMOS, ESD resistors, and B2B diodes, as shown in Figure 1. Depending on the voltage values applied to IO/power/ground ports, or when used to protect high voltage applications, they could be connected in a cascaded structure or tied to RC circuits [10].

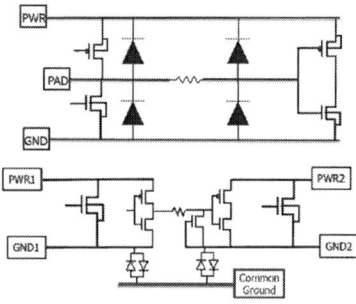

Figure 1: Typical IO ESD protection circuits (top), and cross-domain ESD protection circuits (bottom).

Understanding the connectivity structure is very important to ensure the ESD protection circuit functions as designed. For example, tracing from each IO port, the negative pin of a pull-down diode or source/drain pin of ESD-purposed NMOS must be tied to the IO port, and the positive pin of a pull-down diode or source/drain pin of ESD purposed NMOS must be tied to the ground port. The gate pin of an ESD-purposed NMOS must directly connect to a ground port, or indirectly connect to a ground port through resistors. In another connectivity layout, a pull-up diode or ESD-purposed PMOS is used between IO and power ports.

Driven by the connectivity structure, the conjoined geometrical/electrical constraints of matched ESD devices on layout are equally important to ensure the ESD protection circuit functions as designed. The individual and total strength of all ESD devices in a protection circuit must both be verified. That means the gate channel width/length of placed MOS devices or the area/perimeter of placed pull-up and pull-down diode devices used for ESD protection purposes must be examined.

For ESD protection between IO and internal gates, some designs use secondary charge device model (CDM) ESD protection, requiring resistors and another stage of pull-up/pull-down diode or PMOS/NMOS devices. With regard to these ESD resistors, they may be clustered, with one end tied to the primary ESD protection circuit on the IO net, and the other end tied to the secondary ESD protection circuit on the net connecting to the gate pin of the internal gate. The effective resistance of the ESD resistors cluster between the IO and internal gate nets must be analyzed. The effective resistance is simulated by solving a matrix of connected resistors [11], which may require the resistance value of each resistor device to be calculated based on layout geometrical properties.

For ESD protection between power and ground ports, tracing from each power port, power-clamping-purposed NMOS, or cascaded NMOS devices tied to an RC trigger circuit should exist and be connected to the ground port. For designs with multiple power domains, a B2B diode circuit structure is generally required between common ground and local ground ports.

In the interface of different functional circuitry operating in different power domains, the cross-power-domain paths are likely to require ESD protection similar to an IO ESD protection circuit. The cross-power-domains path is identified by its connectivity structure, as follows: starting from the source/drain pin of a PMOS receiver gate connected to one power domain net/path, search across the net (or ESD resistors) through the PMOS gate pin to the source/drain pin of a second PMOS connected to a different power domain net/path. When this connectivity structure is satisfied, the net/path across the gate pin of one PMOS to the source/drain pin of another PMOS is a candidate for a cross-power-domain net/path.

Due to the physical area required for a cross-domain ESD protection circuit, additional criteria are used to examine the status of ground domains and power clamping protection between each power/ground pair to narrow down the number of cross-power-domain paths that are at high risk and require ESD protection. These high risk cross-power-domain net/path ESD protection circuits require checks similar to those used for IO ESD protection verification. Driven by the required connectivity structure, the geometrical/electrical constraints of matched ESD devices on layout must also be checked to ensure the ESD protection circuit functions as designed.

The previous descriptions are general guidelines for verifying ESD protection on block/chip-level designs. To implement these guidelines as rule checks that can be consistently applied for ESD design verification during the chip design process, engineers require a tool capable of both traversing the transistor-level netlist of a design based on logical requirements (as defined by rule checks written in tool syntax), and accessing the device's physical layout properties. Rule checks can be executed by starting from a group of user-defined nets and/or devices/instances that are filtered out from the complete design circuitry based on user-defined net type, path type, device type, or device subtype. The tool's functionality should be programmable to enable engineers to handle the complexity of both the logical (connectivity structure) and layout (geometrical/electrical constraints) requirements.

Using the IO ESD check as an example, the tool first loops through each IO net. Tracing along each IO net, the tool collects the connected devices under certain connectivity constraints (such as some specific device "type" and "subtype"). On each device of interest, the tool can further examine associated pins, or corresponding nets. This collection of specific devices, nets, or pins continues until all defined logical requirements have been assessed, and then the check results with their associated circuit elements can be output for debugging of any check failures.

If possible, the relevant physical or electrical properties are stored inside the extracted layout netlist, enabling the LDL check to be fully implemented based on layout netlist. If the physical properties can't be feasibly collected inside the layout netlist, the devices and/or nets collected by the defined "connectivity scheme" should be exported, and then converted to polygon shapes for further geometrical operations to complete check requirements. In Figure 2, the flow diagram (on the left) illustrates the LDL geometrical check methodology.

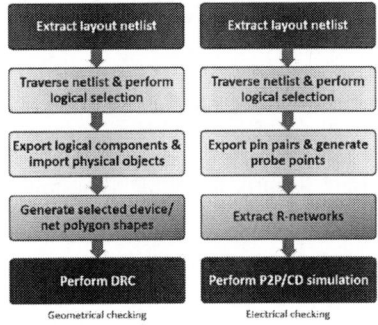

Figure 2: LDL geometrical check flow (left), and LDL electrical check flow (right).

VERIFY INTERCONNECT ROBUSTNESS

The previous section described the methodology to check that required IO ESD, power clamping, and cross-power-domain ESD protection circuits are in place. How about the electrical properties of the interconnect wires connecting these circuits? In advanced technology, the influence of interconnect parasitics is becoming significant. One concern is their impact on as-designed ESD paths along IO and power/ground nets across the entire chip, even when ESD protection circuits are in place. ESD paths usually extend from pad to pad, with a pad being the physical representative of a port. An IO port usually has one corresponding physical pad, while a power or ground port could have one to many physical power or ground pads [12].

A typical ESD path might run from an IO pad to a cluster of ESD devices to a cluster of power clamping devices to one power or ground pad. Along this path, the power clamping devices (connected to the same power or ground net) can be clustered in many physical locations, with each cluster relatively close to one of many power or ground pads. Eventually, such an ESD path can devolve into many physical ESD paths, each of them passing through one cluster of power clamping devices and ending on one power or ground pad (or grouped power or ground pads, depending on the designer's preference). To ensure the ESD path functions as designed, it is critical to examine not only the placement of the ESD circuits, but also the effective resistance values along each ESD path. If the interconnect along an ESD path has a high effective resistance, then the ESD path becomes ineffective. Instead of diverting any ESD surge through the as-designed ESD path, the ESD path will fail to protect internal function circuits, leaving them vulnerable to damage from an ESD surge.

Measuring the effective resistance of an ESD path is challenging, requiring the ability to derive each path from layout database, and extract huge power/ground resistors networks for simulation. In the LDL methodology, every ESD path is identified through the layout netlist, then integrated with parasitic extraction and static simulation flows, enabling the ESD path effective resistance measurement to be realized as a rule check. The electrical check is executed in the following steps (as shown on the right in Figure 2):

1) Extract layout netlist from GDS/OASIS database.
2) Based on connectivity scheme of ESD and power clamping circuit, traverse the connectivity and locate ESD and power clamping devices/pins to form logical pin pairs (point-to-point (P2P) resistance segments) to combine into an ESD path.
3) Convert logical pin pairs into physical probe points.
4) For the nets on which the logical pin pairs reside, extract corresponding resistors networks.
5) Simulate resistor network to calculate whether or not the effective resistance of ESD paths meets criteria.

Using the IO pad to ground pad example, a ESD path of "IO pad to a cluster of ESD devices to a cluster of power clamping devices to one ground pad" typically contains multiple P2P segments (e.g., P2P segment from IO pad to ESD devices on IO net, P2P segment of ESD devices to power clamping devices on power net, P2P segment of power clamping devices to ground pad on ground net). The "P" in P2P can be either a device pin or IO/power/ground pad. Each P2P segment is formed by pin pairs on the same net, and these pin pairs can be exported for simulation. The pin pairs are transformed into a network, and used when running static simulation to solve a V=IR matrix equation to calculate the effective resistance of the P2P segments that form the ESD path.

Even when the interconnect effective resistance of an ESD path complies with the design constraints, a huge ESD surge current (with different dynamic natures for different ESD modes) can still burn a weak interconnect spot. To locate these weak interconnect spots, engineers should also examine the current density (CD) through all wire segments and via arrays along ESD paths to ensure CD is less than some specified value. Measuring current density along an ESD path is similar to calculating the effective resistance measurement of ESD paths. However, engineers should define an injected current value corresponding to a human body model (HBM) or CDM ESD mode, Figure 3 illustrates what can be expected from current density measurement. The top left picture shows a list of CD violations, which can be histogrammed to identify CD error frequency (bottom left). A color mapping assigned to the current density of each segment and via area when the electrical current of ESD path can also be shown (bottom right). The top right picture shows the current density violations (over some defined current density criteria per layer), highlighted to help engineers debug the interconnect spots at greatest risk of being burned by an extreme ESD surge.

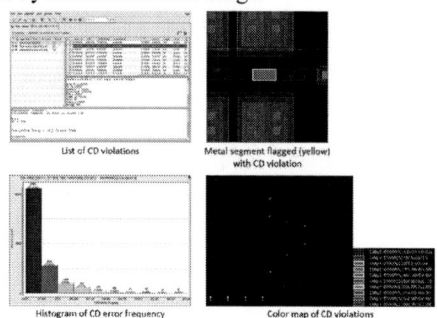

Figure 3: Current density analysis enables debugging of weak interconnect spots to prevent ESD failures [13].

SUMMARY

Since the introduction of LDL functionality as part of design verification methodology in EDA, it has enabled the implementation of robust and repeatable ESD design checks that can be directly correlated to both logical ESD prevention circuitry and physical ESD path interconnect resistance. Major foundries, IDMs, and design houses have tuned their ESD design verification guidelines/rules to align and benefit from the LDL methodology, supporting the creation of comprehensive ESD design

rules to verify the reliability of chip design (IP/block/full chip) for each layout design iteration stage, complete through final tape-out and signoff. Many of these ESD rules are assigned as signoff rules for tape-out. A critical factor for the practicality of such design rules is that of performance. Runtime, and memory consumption of a full chip ESD run must be considered. As IC technology process nodes have progressed, so too has the EDA technology and best practices to verify these full chip designs with reasonable runtime and memory for today's most demanding production designs.

REFERENCES

[1] P. Gibson, et al., "A framework for logic-aware layout analysis," 2010 11th International Symposium on Quality Electronic Design, pp.171-175, 2010.

[2] H. Wagieh, "Solving Electrostatic Discharge Design Issues with Calibre PERC," Mentor, a Siemens Business. June 2012.

[3] MD Ker, "Whole-chip ESD protection design with efficient VDD-to-VSS ESD clamp circuits for submicron CMOS VLSI," IEEE Transitions on Electron Devices 46(1), 173-183, 1999.

[4] Mentor, a Siemens Business, "MediaTek Adopts Mentor Graphics Calibre PERC as its ESD and Circuit Reliability Verification Solution," 2010.

[5] Mentor, a Siemens Business, "Mentor Graphics and TSMC Collaborate to Improve and Expand 20nm IC Physical Verification Offering," 2013.

[6] Mentor, a Siemens Business, "GLOBALFOUNDRIES Improves IC Reliability with Customized Circuit Checks Using Mentor Graphics Calibre PERC," 2012.

[7] Mentor, a Siemens Business, "VIA Technologies Adopts Mentor Graphics Calibre PERC for Critical ESD Checking," 2011.

[8] M. Hogan, "Migrating Consumer Electronics to the Automotive Market with Calibre PERC," Mentor, a Siemens Business, Nov. 2014.

[9] EDA Tool Working Group, ESD Association Technical Report for ESD Electronic Design Automation Checks, Technical Report 18, 2014 (ESD TR18.0-01-14).

[10] CC Yen, MD Ker, "Failure of On-Chip Power-Rail ESD Clamp Circuits During System-Level ESD Test," 2007 IEEE International Reliability Physics Symposium Proceedings. 45th Annual (2007), pp. 598–599.

[11] CW Ho, et al., "The modified nodal approach to network analysis", IEEE Transactions on Circuits and Systems, vol. 22, issue 6, pp. 504–509.

[12] D. Yan, "How Robust Is Your ESD Protection? Are You Sure?," SemiEngineering, Feb. 8, 2018.

[13] D. Yan, "Ensuring Robust ESD Protection in IC Designs," Mentor, a Siemens Business, Oct. 2017.

Analysis of ESD Effect and Ionizing Radiation Particles in Gate Oxide

C.-Z. Chen[a*], David Y. Hu[b] and Hanming Wu[a]

[a]EtownIP Microelectronics, Beijing, China 100176
[*]Mobile: (+86) 13910199301, Email: czchen126@126.com
[b]MetroSilicon Microsystems, Kunshan, Jiangsu, China 215300
Email: David.yu.hu@gmail.com

ABSTRACT

Non catastrophic electro-static discharge (ESD) events can cause latent damage to the gate silicon oxide (SiO_2) of the transistor in an integrated circuit by entrapped charges, and create an interface state that bears the similarity to the oxide damage by ionizing radiation particles, such as from the space. Shift of threshold voltage of the gate and saturation current of the drain in a CMOS transistor have been reported to examine such effect in the literatures. In current work for the ionizing radiation particles of interest, the energy loss of protons and electrons and the derived mean free path (MFP) are calculated using stopping power and linear energy transfer (LET) to analyze the charges trapping inside SiO_2, which caused the interface state. The charged device model (CDM) based devices in ESD protection can effectively provide a few kV high voltage breakdown, thus it can also serve radiation hardening (RadHard) purpose in the I/O pads for system-on-chip (SoC) design. The RadHard study includes the total ionizing dose (TID) effect and the single event effects (SEE), such as single event transient (SET), thus the linkage of electro-charges between CDM and SET is discussed and concluded.

Keywords: electro-static discharge (ESD), charged device model (CDM), ionizing radiation, stopping power, linear energy transfer (LET), mean free path (MFP), radiation hardening (RadHard), total ionizing dose (TID), single event effects (SEE).

INTRODUCTION

Due to CMOS technology scaling, the integrated circuit (IC), typically system-on-chip (SoC) designs are facing new reliability challenges due to the complex geometries. The challenges include protection of electro-static discharge (ESD); prevention of single event effects (SEE) induced by ionizing radiation particles coming from solar or galactic cosmic rays; and the mixed reliability concern between ESD and SEE [1,2]. In ESD protection, especially the most common type modeled by the charged device model (CDM), and in SEE prevention such as single event upset (SEU), it has been qualitatively described that ESD and SEE display similar damage phenomenon by causing voltage pulses or glitches that propagate the circuit [3,4].

It is known that in CDM, a short pulse containing a few pF electric charges and up to kV voltage can occur [1-3]. Similarly, SEE caused by a high energy ionizing radiation charged particle, e.g. a proton having energy of 60 MeV, can lose or transfer similar magnitude of a few keV energy in a micrometer (μm) distance in a transistor gate of silicon dioxide (SiO_2) media. However, to relate a specified ionizing radiation particle to SEE, the stopping power or linear energy transfer (LET) exhibits a non-monotonic behavior as a function of its energy, and shows little relationship of the transistor sizes (thickness, width, height etc.). While the mean free path (MFP) of radiation particles has a unit of length, it provides a track length of ionizing process, which can be an alternative way to interpret SEE.

In this study, we intend to look into the behavior of the discharge or energy transfer (dissipation) of ionizing radiation particles in a transversal path of gate oxide, the critical thin layer of the physical formation of a CMOS transistor. Some physical parameters, stopping power, LET, Range and MFP, of ionizing particles are first calculated and then used to visualize at a CMOS gate dimension, to illustrate and establish the linkage between the CDM and the SET effects.

The particles are first to be identified for the work. The primary types of ionizing radiation particles are protons and electrons in the Van-Allen belts; and protons, alphas and heavy (HZE) ions from the Solar and Galactic cosmic rays (Table 1). In the literatures, therefore, the protons are widely studied for IC designs subject to the measured SEE.

Table 1: Ionizing radiation particles in/from the Space

Environment	Composition	Energy	Flux, $1/(cm^2 \cdot s)$
Inner Van Allen belt Alt. (1,000-6,000) km	protons (99%) electrons (1%)	10-50 MeV ≥100 keV	$(1-2) \times 10^5$ (≥50MeV) 3×10^6 (≥1 MeV)
Outer Van Allen belt Alt. (13,000-60,000) km	protons (1%) electrons (99%)	1 MeV (0.1-10) MeV	2×10^6
Solar/Galactic cosmics At Earth's surface	protons (90%) alphas (9%) HZE ions (1%) photons (x-, γ-)	1×10^9 (1 GeV) 1×10^{12} (1 TeV) 1×10^{16} (10 PeV) 1×10^{20} (100 EeV)	1×10^4 1×10^0 1×10^{-7} [a] 1×10^{-9} [b]

[a] a few times a year; [b] once a century

The concern of the impact of ionizing radiation on IC or SoC includes SEE and total ionizing dose (TID, its SI unit is Gy, 1 Gy=1 J/kg), which is an accumulated (energy absorption) effect over a period of time. The SEE includes the following types of potential damages to a SoC target:

SEU, single event upset; SET, single event transient;
SEL, single event latchup; SEB, single event burnout;
SEGR, single event gate rupture.

In radiation hardening (RadHard) technology to increase the reliability of a SoC design, most of reported SEE studies [5-7] have either focused on non-destructive SEU or on destructive SEL. While SEL can be prevented with epitaxial substrates, such as silicon on insulator (SOI) or silicon in sapphire (SOS), both SEB and SEGR types can cause permanent damages by large dose of ionizing radiation, thus we limit the discussion below on SEU and SET.

METHODOLOGY
Stopping Power and Linear Energy Transfer

The ionizing radiation of charged particles (protons, alphas, heavy ions), when passing through matter, can be described with the _linear_ stopping power $S(E)$, or L, the linear energy transfer (LET, keV/μm). The unit of L is the energy loss per unit length [8], conventionally expressed as $S(E)= -dE/dx$, where the minus sign means energy loss.

When solid media are studied, the _mass_ stopping power S_m or simply S ($MeV\text{-}cm^2/g$) can be obtained from $S=S(E)/\rho$, where ρ is the density of the stopping material. The classic Bethe formula [8,9] with corrections is used to calculate the S,

$$S = k_1 \frac{Z z^2}{M_a \beta^2}[ln\frac{2\mu\beta^2\Delta}{I^2(1-\beta^2)} - \frac{(1-\beta^2)\Delta}{2\mu} - \beta^2 - 2\frac{c}{z} - \delta] \quad (1)$$

Note: To be simple, unrestricted LET, or simply L is used for current work, to avoid separately the calculation on the secondary ionization particles (delta rays), i.e. the restricted LET, L_Δ. Similarly, in the following, for the _mass_ stopping power S, we use the total stopping power [10] S_{tot} (the sum of the electronic or collision stopping power S_e, and the nuclear stopping power S_n), for the calculations of the particle Range and the mean free path. It is noted that the contribution of S_n to S_{tot} is minimal for protons at higher energy, e.g. it is about 0.05% at 100 MeV, 0.08% at 1 MeV and 2% at 10 keV (see Fig. 1).

We can define the unrestricted LET, or simply L the product of S_{col} and the density ρ,

$$L = S_{col} \cdot \rho \quad (2)$$

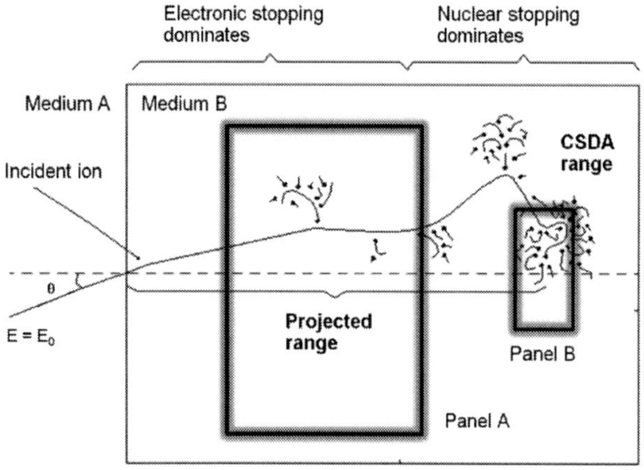

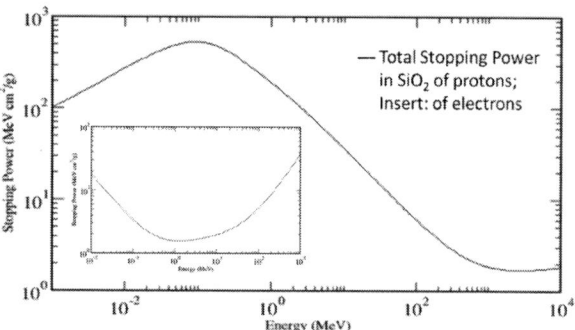

Figure 1: _Illustration of interactions and stopping of ionizing particle entering from medium A to medium B on a presumed CMOS gate oxide geometry. The total stopping power is used to include especially energy deposition at low energy such as electrons. CSDA Range is preferred to include the effect (see text). Different thicknesses of a CMOS transistor gate is presumed to subject to be irradiated. Panel A: test process A, Panel B: test process B._

Range and Mean Free Path

Charged particles having initial energy E_0 lose energy through many ionizing interactions in passing through the matter, due to the straggling [11], in a continuous slowing down approximation (CSDA), until their energy is about zero, the total distance travelled which may not be in a straight line path, the CSDA Range, R_{CSDA} (g/cm^2), or Δx is

$$\Delta x = \int_0^{E_0}[1/S(E)]dE \quad (3)$$

The projected Range (R_p) of the particle stands for the distance between the point where the particle enters the stopping medium and the point where the particle is absorbed (or comes to rest), projected onto the original direction of travel. The R_{CSDA} is always higher than R_p. The ratio R_d of R_p/R_{CSDA} is called detour factor [10], at low energy, e.g. R_d =0.5 for 10 keV proton.

The number of ion (or electron-hole) pairs produced per unit track length, or the specific ionization, ($I_{spec} = L/w_{exp}$) ($1/\mu m$) is determined from L and w_{exp}, while MFP or δx, is defined the reciprocal of I_{spec} [12], hence

$$\delta x = 1/I_{spec} = w_{exp}/L = w_{exp}/(S_{tot} \cdot \rho) \quad (4)$$

In the above, w_{exp} is the average energy needed to create an ion pair (w_{exp} =17eV in SiO_2). At much lower energy of electrons, the inelastic MFP is presented as a reading reference [13]. At or below 10 keV, the electrons lose energy much faster, and contribute much more electric charges in CSDA (Fig. 3).

RESULTS AND DISCUSSION

From the computing programs available at NIST [9], we have performed several calculation works, the results are presented below [Figs. 2 & 3].

Results

The total stopping power (electronic and nuclear in SiO_2 using Eq. (1) is shown in Fig. 2, for protons and electrons (insert).

Figure 2: _Total Stopping Power in SiO_2 of Protons at 1keV-10GeV; Insert: Electrons at10keV-1GeV._

Similarly, the CSDA Range (see Fig. 2, for maximum Range versus projected Range) in SiO_2 using Eq. (3) is calculated for protons and electrons (insert) in Fig. 3. The LET values L using Eq. (2), R_{CSDA}/ρ (in unit of μm) and MFP values δx using Eq. (4) are derived in SiO_2 (w_{exp}=17eV), for protons of 0.01-1,000 MeV (Table 2) and electrons. Only a few values are selected and listed in these tables for the brief discussion and analysis purpose.

Discussion

The stopping power and/or LET data for protons and electrons in SiO_2 are presented as references for RadHard study such that energy deposition can be evaluated for either SEE or TID. The Range and derived MFP data are presented as a comparison to visualize the physical dimensions of protons and electrons may travel in a CMOS device (see Fig.

2 and Fig. 3) at known energies. For example, for proton of 60 MeV, its total stopping power S_{tot} is 8,874 MeV-cm2/g, *LET* is 2.04 keV/µm, the CSDA Range is 1.65 cm, and MFP of 2.94 cm. As both Range and/or MFP are monotonic as a function of the particle energy, it is more meaningful to use them to relate to physical dimensions of CMOS gate.

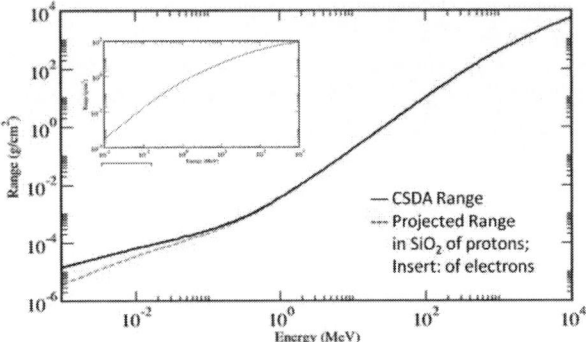

Figure 3: CSDA Range (Δx) in SiO_2 of Protons at 1keV-10GeV; Insert: Electrons at10keV-1GeV.

Observing the LET values Fig. 2, it shows that proton lose more energy at the lower energy region. At higher energies, most of the primary particles will generate very a few ionizations within the trajectory inside the silicon. Overall, the damage can also be caused by secondary (delta rays) or tertiary particles of protons (Fig. 2). This indicates that RadHard study can explore more with low energy proton, such as protons of 0.1 MeV that has a maximum of LET entering the SiO_2 medium, as a comparison with 60 MeV protons for a comparative study of SEE. The secondary (and tertiary) particles are electrons whose LET and MFP are also calculated and shown as references.

Though methods of TID and SEE are *not* established in our work, existed RadHard studies on radiation damage on SoC or memory such as SRAM design shows that with reduced gate thickness (using 1.6µm, 1.2µm, 0.8µm, 0.5µm), TID tolerance that measured with gate threshold voltage (V_{th}) is increased [5]. Recent studies [6,7] also show 60 MeV protons can induce mobility degradation in advanced CMOS processes, such as FinFET, bulk CMOS as well as in the fully depleted SOI (FD-SOI). As illustrated in Fig. 2, when a CMOS encompasses larger volume (assuming Panel A is a 0.18 µm process), then the probability of accumulated electric charges on the average can be several times of smaller CMOS volume (assuming Panel B a 45 nm process). On the other hand, if the energy depositions with a CMOS happen to be at the end of the CSDA Range of the particle, then the advanced device may still show increased sensitivity towards TID and SEE as reported. SEU may further generate a single event function interrupt (known as SEFI), it can be geometrically dependent as shown in a comparative study between planar CMOS and 3D structural FinFET. In these studies, as the electric charges arisen from ionization radiation such as high energy protons in the gate oxide (SiO_2) and isolation oxide are critical as they have direct impact on threshold voltage, thus transconductance (g_m), and saturation drain current ($I_{d,sat}$), as measures of TID impact, are used and analyzed in various devices [6-7].

Thus far, SEL is reported immune in FD-SOI due to its inherent structure, as a comparison with bulk CMOS. More recent works on the mobility has been carried out with 62 MeV protons on FinFET and FD-SOI [6], and bulk CMOS with respect to FinFET [7] adopted in advanced SoC designs. For an SET which may propagate a pulse signal as in a coupling effect in a pair of parallel routing paths, to result in a signal integrity issue, if to result in an incorrect value being latched in a sequential logic unit, it is then considered an SEU. Therefore understanding the differences between SET and SEU impose significant meaning in preventing them. Both SET and CDM may exhibit sudden short pulse rise in electric charges or current to cause a CMOS device failure. It is anticipated that the SET type of SEE has a similar behavior as CDM of ESD. These glitches can cause a failure in a SoC design which can lead to life threatening accent if it is applied in an automotive electronic system, or aviation disaster such as in a civil air flight or advanced satellite [2,3].

Conclusion

The monotonic values of Range and/or MFP of protons and electrons can be used to relate to CMOS gate geometries (not the stopping power or LET). In these TID studies using 60 MeV protons, the dependency of the mobility, threshold voltage (V_{th}) and other parameters on both CMOS gate thickness and FinFET structure offer two folds of meanings, i.e. the deposition of radiation charges in SiO_2 interface state has established a linkage for the study between CDM of ESD and SET of SEE; and secondly the analysis of electric charges using CDM in relation to Range and/or MFP of ionizing radiation particles can be further proceeded. (*Note: Paper of extended discussion is available upon request.*)

REFERENCES

[1] D. Y. Hu, SMIC Tech. Forum, 23-25 Jun., Shanghai, China (2012).

[2] D.Y. Hu, and C. -Z. Chen, *ECS Trans.*, **60**(1), 1185-1190 (2014).

[3] N. Wakai *et. al*, RAMS 2009. Annual Conference 26-29 Jan. 2009, pp.509-514.

[4] A. Griffoni, Universita di Padova, PhD Thesis (2009).

[5] J.V. Osborn, R.C. Lacoe, D.C. Mayer and G. Yabiku, *IEEE Trans. Nucl. Sci.*, **45**(3), 1458-1463 (1998)

[6] D. Kobayashi, E. Simoen and S. Put et al., *IEEE Trans. Nucl. Sci.*, **58**(3), 800-807 (2011).

[7] M. Bertoldo, A. V. de Oliveira, P. G. Der Agopian, E. Simoen, C. Claeys and J. A. Martino, *ECS Trans.,* **66**(5), 295-301 (2015).

[8] ICRU Report 16, *Linear Energy Transfer*. Washington D.C. (1970).

[9] H. Bethe und J. Ashkin in "Experimental Nuclear Physics, ed. E. Segré, J. Wiley, New York, p. 253 (1953).

[10] NIST, National Institute of Standards and Technology, http://www.physics.nist.gov/PhysRefData/Star/Text/programs.html

[11] N.J. Carron, An Introduction to the Passage of Energetic Particles through Matter. *CRC Press*, Tayler & Francis Group, (2007).

[12] C.-Z. Chen and D.E. Watt, *Int. J. Radiat. Biol.* **49**, 131-142 (1986).

[13] J.C. Ashley and V.E. Anderson, *J. Elect. Spectrosc.* **24**, 127 (1981).

978-1-7281-6559-2/20 $31.00 © 2020 IEEE

STATISTICAL WEAR-LEVELING FOR PHASE CHANGE MEMORY

Chien Wang, Chengyu Xu

Jiangsu Advanced Memory Semiconductor Corporation, Beijing, China

Ste. 802, Unit 4, Bldg. 2, Zhongguancun IC Park, Haidian District, Beijing, China

{chienwang,peterxu}@amtpcm.com

BIOGRAPHY

Chien Wang graduated MSEE from UC Santa Barbara in 1995. Currently with Jiangsu Advanced Memory Semiconductor Corporation.

Chengyu Xu graduated with BSEE from Harbin University of Science and Technology in 2014. Currently with Jiangsu Advanced Memory Semiconductor Corporation.

ABSTRACT

Wear leveling techniques have been successfully used in increasing the useful life of NAND flash devices. Although Phase-change Memory's endurance is much higher than NAND, and can reach up to a range from 10^6 to 10^9, it still lacks the endurance needed for use as system main memory such as DRAM, which has nearly $\sim 10^{14}$ re-write endurance. In this paper, we propose and have developed a novel statistical and hierarchical wear-leveling technique to be used in a 4Gb DRAM-like PCM chip. The technique uses real-time memory address statistics to compute the physical-to-device address mapping using an embedded CPU. The CPU automatically adjust for different system workloads based on current address statistics. Results from system simulations shows the techniques to be effective in our tile-based memory architecture while requiring relatively low computational overhead.

Keywords—Phase Change Memory; PCM; Wear-leveling

INTRODUCTION

Phase-change Memory has been recognized as one of the most promising next-generation memories for use as main system memory. However, due to its mechanical wear out from crystallization cycles, the PCM exhibits much lower endurance than DRAM. The endurance can be increased through proper wear leveling, and bit failures due to mechanical wear-out can be effectively mitigated.

This work presents a novel hierarchical and statistically adaptive wear leveling algorithm used in our 4Gb PCM chip design. The DRAM-like chip is based on 4096 individual memory tiles, simulation using real application trace data has shown very uneven tile wear.

By using this method, the hot areas are shown to be effectively re-distributed throughout the entire memory array, and the simulated media lifetime was increased from about 10^7 to 10^9 - 10^{10} average Set-Reset cycles.

PCM CHIP ORGANIZATION

The design of the 4Gb PCM chip is based on a fine-grained architecture, where the entire memory array is built by using 4096 individual memory tiles. Each tile is 1Mb (1024x1024) in density and self-contained, i.e. all peripherals and drive/sense circuits are built into each tile. The 4096 tiles are arranged in a 64x64 array, with the data, address and control buses forming a grid structure around each tile. Each PCM tile also contains extra non-volatile memory that keeps record of the number of PCM cell writes for each 1024-bit row.

A central processing unit (CPU) is used in the memory chip to perform chip initialization and tile management tasks including statistic collection and mapping table updates. Additionally, each tile is interconnected to form a ring network, called the Tile-Interconnect Bus (TIB), with the CPU as the bus master.

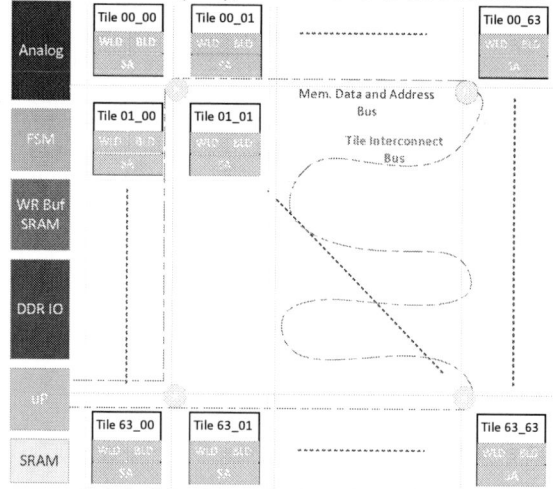

Figure 1: 4Gb PCM chip block diagram.

STATISTICAL WEAR-LEVELING

Generally speaking, wear-leveling algorithms can be divided into two categories: Dynamic and Fixed. Dynamic methods are done using real-time address calculations, the mapping algorithm can be based on Algebraic equation. Fixed methods, as the name implies, uses a look-up table generated ahead of time to perform the mapping. Either the algebraic or the fixed mapping methods are sub-optimal due to their strong dependency on a single workload model.

In our design, a statistical approach was used to generate the mapping function. This dynamic mapping function is based on real-time statistical address distribution, which we model using a probability density function, that is, we model the input addresses as a random variable characterized by some arbitrary probability density function.

It can be proved that for any random variable x, with probability density function (pdf) $f(x)$ within a region of support R, we can map x to another random variable y with uniform probability density function $f(y)=1/R$. The mapping function T[.] always exists and is obtained as the cumulative distribution function (cdf) of the random variable x:

$$y = T[x] = (R - 1) \int_0^x f_X(u)\,du \qquad (1)$$

Since in our case, the address can only take on discrete values, we obtain the mapping function by using summation instead of the integral:

$$\hat{y} = \hat{T}[x] = (R - 1) \sum_0^x P_X \qquad (2)$$

Where P_X is the probability mass function (pmf) of the discrete random variable x. Note that since the output $\hat{y}$ is also discrete valued, we will need incorporate some extra processing in the mapping function $\hat{T}[.]$ to make the mapping unique.

By using the above relationship, we generate our mapping function using a piecewise linear approximation of T, with points A, B, C, … N, and their associated slopes ma, mb, … mn of the new mapping function.

The input address pmf is kept in the CPU local memory and as more and more address data are gathered, the pmf converges to the "true" statistical pmf of the input memory address distribution. After the mapping function has been computed, address mapping will begin. The actual address mapping is implemented using an SRAM look-up table to allow for fast operation. Initially, The SRAM will contain a straight-line input-to-output mapping. As new memory accesses are counted and updated, and when a pre-determined write-count threshold has been reached, the look-up table will begin an update process using the most current statistical data beginning with the next incoming memory write transaction. Each new write will be mapped to the newly updated address location, and the associated SRAM entry will be updated. The updating process keeps going using the current statistical data until the next pre-determined

threshold value has been reached, then the above cycle is repeated again. A second and completely different updating mechanism is also needed in order to handle changing system workloads, and will be described in the next section.

In order to guarantee uniqueness of the input-to-output address mapping, a sub-LUT exists in each tile to further check and remap a range of input addresses with tags to perform address range compression.

Pseudocode for computing the mapping function T[.]:

```
1: for each (Rn)
2: if (mn == 1) then {
3:   addrOut = addrIn + bn; }
4: else if (mn > 1) then {
5:   addrOut = floor(addrIn * mn) +
mod(i/Rn);
6:   i++; }
7: else {
8:   addrOut = floor(addrIn * mn) +
tagaddr; }
```

Where bn is the offset of the linear equation of the nth linear segment $y_n=m_n x+b_n$
Rn is the range of the nth mapped linear-segment
And i is an integer value that is persistent.

Figure 2: Pseudo code of the mapping function T[.]

CHANGING WORKLOADS

Since the memory address statistics likely changes by a large margin when the system workload changes, we need to monitor such condition. Each tile stores the write count data, and these data are used by the CPU to compute the cumulative mass function (cmf) of the mapped device address. When the cmf diverges from the desired straight-line form, giving indication that the input statistics no longer matches the mapping function, the CPU resets the input address pmf counter to re-start collection of new statistical data.

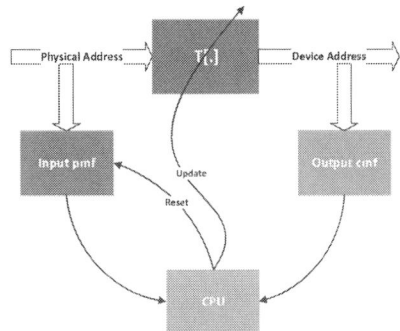

Figure 3: Adaptive mapping table update via CPU.

HIERARCHICAL WEAR-LEVELING

While the statistic-based adaptive wear-leveling keeps the tiles at even wear, intra-tile wear leveling must also be done. At the tile level, intra-tile wear-leveling aims to make sure that within each tile, each row is evenly written to. To achieve this, a static wear-leveling method is used. The static wear-leveling algorithm consists of relocating data from extremely low-wear row groups to rows with high-wear. This operation is relatively simple as it's done locally at the tile level as a background operation. Correct data moving is done by commands from CPU via the Tile Interconnect Bus (TIB). Each tile only needs to report the wear data of each row and perform row-remapping as instructed by the CPU.

RESULTS

We used several different application-based memory trace data to test the effectiveness of our wear leveling algorithm. Fig. 3 shows a comparison between the input and output memory tile addresses. It can be seen that from a highly skewed input address distribution, using the presented method, the output address gets mapped to have a near-uniform distribution. Also note the maximum single tile write count has been reduced from 5000 to below 200, when simulated using 100-thousand address traces.

The mapping could be further optimized using additional static wear leveling that target only the peak write-counts. These peaks were formed at the beginning of workload change, during the adaptation period of the mapping algorithm. Simulation has shown that, as time progresses, the mapping algorithm eventually adapts to the new input statistics. Fig. 4 shows the convergence of the output address cumulative mass function to a uniform cmf as time progresses from t1, t2, … to t6, with around 10,000 memory accesses between each tn.

With our current statistical wear-leveling mechanism, the increased endurance allows our PCM chip to be used in DRAM-like applications, such as the main memory of a computational system.

REFERENCES

[1] B. Peleato, H. Tabrizi, R. Agarwal, J. Ferreira, *"BER-based Wear Leveling and Bad Block Management for NAND Flash"*, IEEE Intl. Conference on Communications (ICC), 2015.

[2] M. Qureshi, V. Srinivasan, J. Rivers, *"Scalable High-Performance Main Memory System Using Phase-Change Memory Technology"*, ACM/IEEE Intl. Symposium on Computer Architecture (ISCA), 2009.

[3] P. Gao, D. Wang, H. Wang, *"Increasing PCM Lifetime by Using Pipelined Pseudo-Random*

Encoding Algorithm", http://crad.ict.ac.cn/fileup/HTML/2017-6-1357.shtml

[4] B. Lee, P. Zhou, J. Yang et al, *"Phase-change technology and the future of main memory"*, ACM/IEEE Intl. Symposium on Microarchitecture (Micro), 2010.

[5] Chung H. Lam, *"Phase Change Memory and its intended Applications"*, IEEE Intl. Electron Devices Meeting (IEDM), 2014.

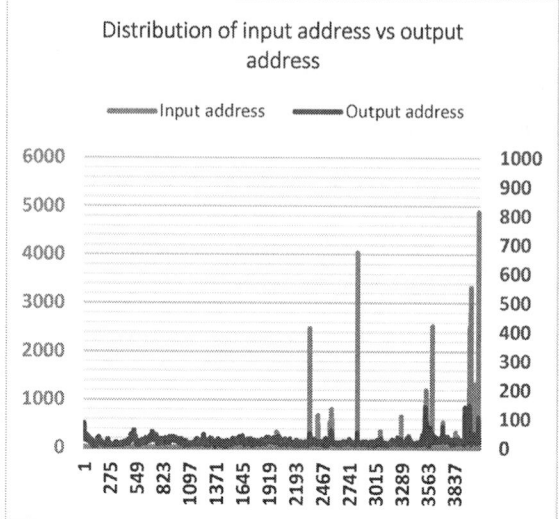

Figure 4: Statistical distribution of the input vs. output Tile address

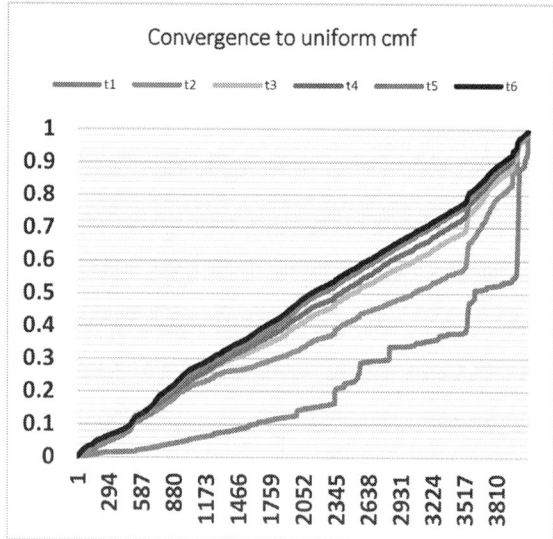

Figure 5: Convergence trend of the physical tile address cmf

THE STUDY OF DEFECTS AUTO-CLASSIFICATION SYSTEM IN SEMICONDUCTOR MANUFACTURING

Pengfei Wang[1], Chen Li[1*], Hao Fu[1], Zhengying Wei[2], Xu Chen[2], Zhounan Wang[2], Shoumian Chen[1], and Yuhang Zhao[1]

[1]Shanghai Integrated IC R&D Center, No.497 Gaosi Road, Shanghai 201210, China

[2] Shanghai Huali Microelectronics Corporation, No.568 Gaosi Road, Shanghai 201210, China

*Email: lichen@icrd.com.cn

ABSTRACT

In semiconductor manufacturing, defects detecting and analyzing are one of the key contributors for process yield improvement. A system is proposed to classify the defects automatically, which is more efficient than traditional manual work. It becomes possible to control the process promptly. The proposed system has three main parts, the defect image pre-process module, the defect classification module and the network results post-process module. The convolutional neural network (CNN) is used as the base algorithm for defect classification. And the post-process part is developed to refine the confusing images results. The dataset in this work are all collected from the 12-inch process line and the results show that the system proposed is a good candidate for saving massive human work.

Keywords—defect classification; semiconductor manufacturing; convolutional neural network; image process algorithm

I. INTRODUCTION

Image processing has a long history and it has proven to be powerful in many applications [1][2][3][4]. But due to the complexity of images in semiconductor manufacturing, such as various defects features and background graphics and textures, it is very hard to classify the defects with rule based image process approaches. As a result this job is done manually in foundry up till now. It is very resource consuming. And this job is not experience accumulated, so it is hard to have enough full time skilled workers on the manufacturing lines.

In the past decade, computer visions make big success because of the breakthrough of deep learning technologies, especially the developing of effective training algorithms for convolutional neural networks (CNN) [5][6][7][8]. CNN based deep learning becomes the prevailing method in image recognition and object detection, which has already succeeded in a lot of application areas.

Unfortunately, in semiconductor industry the situation is more complicated. The technology advances rapidly as indicated by Moore's Law. Hundreds of new products are invented every year. It is impossible to collect dataset that can cover the whole sampling space. This scenario does not meet the mathematical requirements of deep learning, which may result in poor generalization property. But up

till today, the design of neural network is still a work of state-of-art. Take into account the classification rules and circuit process information that workers use to classify the defect images to combine neural network and traditional image process methods together, it may help to make the auto-classification software deployable.

II. SYSTEM ARCHITECTURE

The defect images are very complicated compared to the dataset commonly used in researches, like ImageNet [9]. The defect size varies a lot, from several pixels to almost 80% of the image. Even for the images in the same defect code, the defect sizes and features differ vastly. And the relationship between the defects themselves and the image backgrounds are also divided into two categories. The classification criteria of most codes aren't related with the image backgrounds, while the other defects correlate with the background features deeply, such as the damage on metal or oxide layer. In the latter case, the process step may provide additional information to classify the defects correctly as well as the background textures analyses.

In order to overcome the above difficulties, besides the neural network for defects classification, rule based image process approaches are developed. It can detect the defects area and extract the features including both defects and image background, such as circuit graphics. It can provide auxiliary information for classification when necessary.

The proposed system architecture is shown in Fig.1.

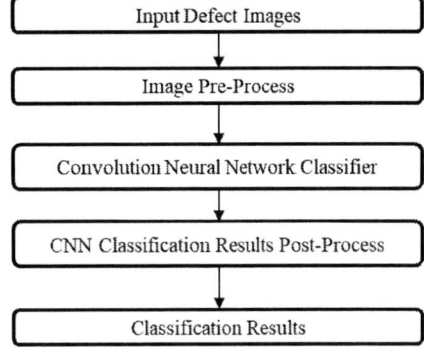

Fig.1 *The defects auto-classification system architecture*

The image pre-process step is optional. Neural network

trained for classification work is highly sensitive to local structure in the image and is not just using broad scene context [10]. So it is helpful to extract image details for better understanding the features. This pre-process module uses the image process approaches described above. The circuit graphics in the image can be removed for cleaner input dataset, since the graphics are not related with the code identification. And it can help CNN to concentrate on the defect features.

Then the dataset is fed into CNN. Since this system is to be deployed in foundry, the purity has the highest priority, which is set to be 90% for each defect code, otherwise the system can't be qualified. In order to ensure purity, only top1 result is used and a belief confidence threshold is specified for each defect code separately by users. If belief confidence of certain code from CNN output is larger than its threshold, the image will be marked as the defect code indicated by network output, otherwise the image will be put into the category named others for analyzing manually later. The most popular CNN architectures are tested [11] [12], and so do the novel architectures [13][14]. No big overall accuracy or purity difference is discovered for the tested networks. The result shows that each architecture has advantage in certain defect codes, but inferior in some other defect codes. To take advantage of this, an ensemble of the tested CNNs is also tested and it shows the potential for system performance improvement.

The post-process module is to analyze slight differences between the confusing defect images. For example, think about the injury to the metal layer, if it is severe, the defect code is A, otherwise the code is B. The defect code names are hidden for confidential considerations. The boundary between this two kinds of defect codes are hard to tell even for experienced employee. Therefore the rule based image process approaches described above are needed to analyze the images in detail based on network outputs.

III. DATASET AND TRAINING METHOD

All defect images are collected from 12 inch process lines. The dataset has 54000 images in total with 24 defect codes. The image number for each code ranges from 1130 to 4050. Over 90% defect images generated by the process lines daily belong to these 24 codes.

From the dataset, 80% are selected randomly as training set and the remains to be test set. Regular augmentation is applied to enlarge the training set, including flip, rotation, scale, crop and translation.

Considering the deploy requirements, only top1 result is taken into account. The belief confidence thresholds are set to be 0.9 for each code during training and testing in this work for simplicity. When the system is deployed, the thresholds should be adjusted through enough online test to meet the purity requirements.

As discussed in part II, for single CNN, the most popular or novel architecture doesn't make big difference

in the overall accuracy and purity result. Therefore in this work, ResNet is used for illustration purpose. For network training, ReLU is adopted as the activation function. As for the other training aspects, such as optimizer, learning rate, this work uses the common settings. The purpose of this work is to build up a robust system, not the highest accuracy and purity result with certain magic settings.

IV. RESULTS

The top1 purity and accuracy results are illustrated in Fig.2 and Fig.3 respectively. The defect code names are hidden for confidential considerations. There is only about 14 codes meet the purity requirement. And the accuracy of these codes range from 75% to 85%, except three codes with distinctive features whose accuracy are above 95%.

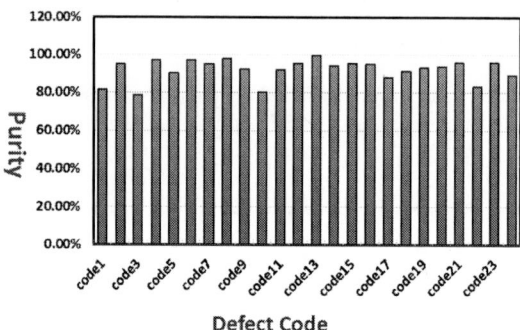

Fig.2 the purity of the 24 defect codes classification

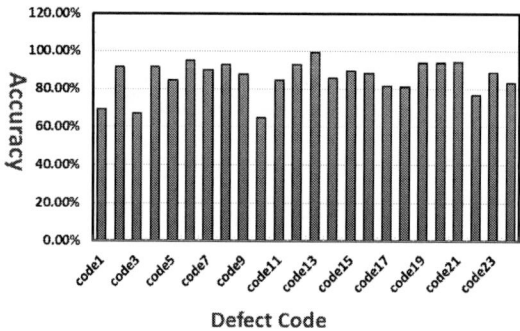

Fig.3 the purity of the 24 defect codes classification

The classification model with ensemble CNNs is also tested. The result is depicted in Fig.4. It shows that with compound models, the overall purity is improved about 5%. Most defect codes purity now meet the requirements. But the accuracy drops about 10%, which needs to be improved when keeping the purity high.

Through analyzing the misclassified images, it shows that most failures are due to image background graphics and textures, which mainly has two impacts. Firstly, image features of some codes are hard to be distinguished. Workers use additional information to determine which defect code it is, such as the process step. It is impossible for software to tell the right category without this auxiliary information. Secondly, for most codes, the classification criteria has no relation with the background. But in many images, the defect itself is very small and looks like noises. Thus the background takes the spot of neural network. In order to improve the classification performance, the background interferences should be limited to minimum, which is under investigation now.

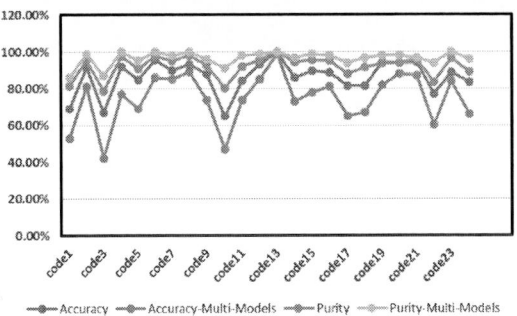

Fig.4 the model with ensemble models performance

V. CONCLUSION

This work indicates that deep learning has potential advantage in semiconductor manufacturing to save cost and improve efficiency. But up till now only half of the defect codes can meet release requirements. Many areas related to computer vision have similar challenges, such as medical diagnosis. Image recognition and object detection in industry applications are very complicated. In our work, we are trying to combine neural network with rule based image process algorithms, and all auxiliary information in reach together to make the auto-classification system practical as soon as possible.

REFERENCES

[1] R. C. Gonzalez and R. E. Woods, *Digital Image Processing 4ᵗʰ Edition*, Pearson, 2018.

[2] H. Bay, A. Ess, T. Tuytelaars, and L. Van Gool, "Speeded-Up Robust Features (SURF)," *Comput. Vis. Image Underst.*, vol.110, no.3, 2008, pp.346-359.

[3] N. Dalal and W. Triggs, "Histograms of Oriented Gradients for Human Detection," *2005 IEEE Comput. Soc. Conf. Comput. Vis. Pattern Recognit. CVPR05*, vol.1, no.3, 2004, pp. 886-893.

[4] T. Ojala, M. Pietikäinen, and D. Harwood, "A comparative study of texture measures with classification based on feature distributions," *Pattern Recognit.*, vol. 29, no.1, 1996, pp. 51-59.

[5] Y. LeCun, K. Kavukcuoglu, and C. Farabet, "Convolutional networks and applications in vision," *Circuits and Systems (ISCAS), Proceedings of 2010 IEEE International Symposium*, 2010, pp. 253-256.

[6] A. Krizhevsky, I. Sutskever and G. E. Hinton, "Imagenet classification with deep convolutional neural networks," *Adv. Neural Inf. Process. Syst.*, 2012, pp. 1-9.

[7] Y. Bengio, "Learning Deep Archiectures for AI," *Found. Trends® Mach. Learn.*, vol. 2, no.1, 2009, pp. 1-127.

[8] Y. LeCun, "Effcient BackPrp," *J. Exp. Psychol. Gen.*, vol. 136, no. 1, 2007, pp.23-42.

[9] J. Deng, W. Dong, R. Socher, L -J. Li, K. Li and F. F. Li, "Imagenet: A large-scale hierarchical image database," *Computer Vision and Pattern Recognition 2009. CVPR 2009. IEEE Conference on,* 2009, pp. 248-255.

[10] M. D. Zeiler and R. Fergus, "Visualizing and Understanding Convolutional Networks," *arXiv: 1311.2901v3*, 2013

[11] K. M. He, X. Y. Zhang, S. Q. Ren and J. Sun, "Deep Residual Learning for Image Recognition," *IEEE Conference on Computer Vision and Pattern Recognition*, 2016.

[12] C. Szegedy, W. Liu, Y. Q. Jia, P. Sermanet, S. Reed, D. Anguelov, D. Erhan, V. Vanhoucke and A. Rabinovich, "Going Deeper with convolution," *IEEE Conference on Computer Vision and Pattern Recognition*, 2015.

[13] A. Khan, A. Sohail, and A. Ali, "A New Channel Boosted Convolutional Neural Network using Transfer Learning," *arXiv: 1804.08528v4*, 2019.

[14] J. Hu, L. Shen, S. Albanie, G. Sun and E. Wu, "Squeeze-and-Excitation Networks," *arXiv: 1709.01597v4*, 2019.

A NEURAL-NETWORK APPROACH TO BETTER DIAGNOSIS OF DEFECT PATTERN IN WAFER BIN MAP

Junjun Zhuang[1], Guiyun Mao[1], Yong Wang[1], Xu Chen[1]* and Zhengying Wei[1]
[1] Shanghai Huali Microelectronics Corporation, Shanghai, China
*Corresponding Author's Email: chenxu@hlmc.cn

ABSTRACT

Wafer bin map (WBM) represents specific defect patterns that provide information for diagnosing root causes of low yield in semiconductor manufacturing. In practice, most semiconductor engineers use subjective and time-consuming eyeball analysis to assess defect patterns. Given shrinking feature sizes, various types of WBMs with different defect patterns occur; therefore, relying on human vision to judge defect patterns become more complicated, inconsistent, and unreliable. To bridge the gap, a system is proposed to facilitating WBM patterns extraction and assisting engineer to recognize defect patterns efficiently. We propose an individual classification model which is trained by Deep Belief Network (DBN) to diagnose the patterns on wafer bin maps. By setting up six single classifiers with different thresholds, the individual classifiers were combined for both single and mixed-type patterns recognition. The numerical results showed that the single classifiers outperform MLP method both on single and mix-type patterns. Simultaneously, the results have shown the validity and practical viability of the proposed combined classifiers.

INTRODUCTION

In semiconductor manufacturing, the wafer fabrication process typically involves hundreds of complicated steps, which becomes more lengthy and complex in recent years owing to the technologically advances[1], in which the wafer test [2]is conducted to analyze and evaluate the electrical properties of the dies in a wafer after, in which the defective dies are filtered out based on the results of this step, and the resulting spatial data of a wafer is called a wafer bin map (WBM).

There are two types of defects in WBMs: global random defects and local systematic defects [3]. In general, local systematic defects are generated by assignable causes, and their patterns are related to different root causes of failure. Figure 1(a) illustrates four typical patterns of systematic defects in WBMs. For example, a zone pattern is generated by non-uniformity or uneven cleaning [4]. Therefore, detecting systematic defects in WBMs and classifying them by different patterns is important to identifying root causes of failure and providing appropriate remedies.

Recently, the probability of mixed-type defects has increased because of the more complicated fabrication process. Figure 1(b) illustrates mixed-type defect patterns.

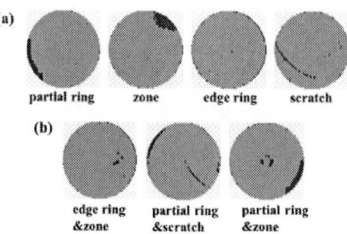

Figure 1 The typical WBMs with single type pattern (a) and mixed-type patterns (b)

The classification of mixed-type defect patterns can be more challenging than that of single-type defect patterns, particularly if the number of defect patterns that are mixed over a wafer is unknown. Various methods have been proposed for the classification of mixed-type defect patterns. Wang et al. [5] identified different mixed-type patterns by applying a hybrid method combining hierarchical clustering with k-means, then the patterns were classified into linear, elliptic, and ring patterns. Jeong et al. [6] detected the presence of spatial autocorrelation by spatial correlogram, and then classified the patterns into one of known patterns with dynamic time warping. More recently, Kim et al. [7] proposed connected-path filtering to remove random defects, following with the infinite warped mixture model for clustering mixed-type defect patterns.

In this article, we propose a new approach to classify mixed-type defect patterns in WBMs using Deep Belief Network (DBN). Individual classifiers for six single type patterns are established. Then, each classifier estimates if the corresponding pattern exists when several defect patterns are mixed over a wafer. Our numerical study is discussed in Section EXPERIMENT.

METHODOLOGY

The research framework of our study is shown as Figure 2, and will be illustrated in this section.

Spatial Randomness Test (SRT)

For better extraction of wafer bin map features, we divided WBM into three categories roughly by the spatial randomness test following Taam and Hammda[8]. For given WBM, die i will denote as $Y_i=1$(Bad); $Y_i=0$(Otherwise), and for this method, a binary bin map will be generated. Four statistics defined as follows:

$$Ngg = \sum\sum \sigma_{ij}(1 - Y_i)(1 - Y_j)$$
$$Ngb = \sum\sum \sigma_{ij}(1 - Y_i)Y_j$$
$$Nbg = \sum\sum \sigma_{ij}(1 - Y_j) \qquad (1)$$
$$Nbb = \sum\sum \sigma_{ij}Y_iY_j$$

where σ_{ij} is a neighboring index. σ_{ij} equals 1 if die i and die j are neighbors; otherwise, σ_{ij} equals 0.

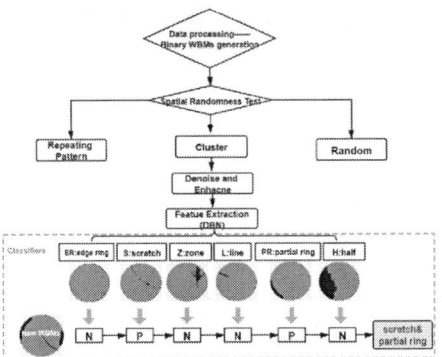

Figure2. The research framework of this study

The spatial correlation of two groups of data can be tested by the odd ratio hypothesis test:

$$\varepsilon = \frac{N_{gg} * N_{bb}}{N_{gb} * N_{bg}} \qquad (2)$$

when the number of dies on a wafer is large, $\log\varepsilon$ is almost equals 0, indicating the random defects distributed on the wafer. A positive value of $\log\varepsilon$ indicates the cluster defects on the wafer. Otherwise, a negative value of $\log\varepsilon$ represents repeating pattern with no bad-bad neighboring dies.

Denoise and Enhance (DE)

To obtain good clustering result, we used a data preparation procedure to enhance the signal and remove the noise as described in [9].

Deep Belief Network (DBN)

A DBN contains a sequence of Restricted Boltzmann Machine (RBMs). The hidden layer of each RBM is connected to the visual input layer of the next RBM.

The RBM is an undirected graphical model that has visible units $v \in \{0, 1\}$ and hidden units $h \in \{0, 1\}$, with each visible unit connected to each hidden unit (Figure 3a).

For a given set of values of (v, h), the model can be defined in an energy function as follows:

$$E(v, h) = -\sum_{i \in visible} a_i v_i - \sum_{j \in hidden} b_j h_j - \sum_{i,j} v_i h_j w_{ij} \qquad (3)$$

where h_j and v_i are the binary states of hidden unit j and visible unit i, respectively; b_j and a_j are the biases of h_j and v_j, respectively; and w_{ij} is the weight between h_j and

v_i.

Feature extraction based on maximum-likelihood learning is intractable for all RBM forms. Therefore, efficient learning can be performed by following an approximation to the gradient of the contrastive divergence objective.

We adopted five hidden layers to capture the features and a classifier on top of the hidden layer to carry out classification (Figure 3b). The output layer had two nodes, indicating the presence and absence of the tested pattern, respectively. We employed the Adam optimizer, with a learning rate of 0.01 and decay parameters for the moments of the gradients, 0.9, with the batch size of 64 for 300 epochs. The hyper-parameters were optimized in a heuristic manner by evaluating reconstruction performance.

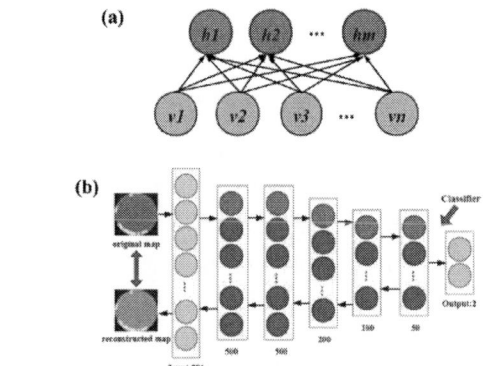

Figure 3 (a) The basic structure of RBM, (b) DBN structure with five hidden layers used in our study

Proposed Classification Model

For our proposal, we suggest an individual classification model for each pattern. For a single WBM, a total of six classification models are separately applied. Then, the final classification results can be determined by combining the results of each classification model, just as the part in the dotted box of Figure 2.

EXPERIMENT

Data and Setup

We used real WBM data to train and evaluate the proposed classification models. The map size of each wafer was resized to 28×28. A total of real 9000 WBMs contains six single types of patterns as shown in Figure 4, among which 1800 were randomly selected for testing and other 7200 WBMs for training. Also, another 110 mixed-type WBMs were used for test.

The result of spatial randomness test was shown in Table 1. WBM1 failed to reject the null hypothesis at the significance level. WBM2, WBM3 and WBM4 rejected the null hypothesis and showed a specific spatial pattern. WBM5 also rejected the null hypothesis and represented a repulsive pattern.

978-1-7281-6559-2/20 $31.00 © 2020 IEEE

Table 1 The result of the spatial randomness test

N_{gg}	16784	23649	24561	27473	28094
N_{bb}	1818	1493	2175	322	5
N_{gb}	5408	2231	1360	831	650
N_{bg}	5406	2190	1360	830	625
$\log\varepsilon$	0.040	1.973	3.362	2.551	-0.968

Accuracy (A), precision (P), recall (R) were considered to evaluate the performance of the proposed method. Multi-layer perceptron (MLP), a widely used classification method, was taken as control. For the MLP, we assumed five hidden layers with 500, 500, 200,100 and 50 hidden units, respectively.

Results for single-type patterns classification

Table 2 summarized the classification results for the single classifiers trained by MLP and DBN, respectively. For each case, the results from the MLP, and DBN were compared, in terms of the average of accuracy, recall, and precision. The values in parentheses are the standard errors. For all models for six types, DBN achieves the higher accuracy, recall, and precision compared with MLP. MLP showed a tendency where performance decreased if a scratch pattern was included in a WBM. In contrast, DBN shows good performance consistently for various types defect patterns.

Table 2 The performance of the single classifiers

	MLP			DBN		
	A	R	P	A	R	P
PR	0.77(0.020)	0.81(0.012)	0.80(0.030)	0.97(0.011)	0.92(0.029)	0.97(0.001)
S	0.75(0.034)	0.72(0.051)	0.74(0.027)	0.94(0.021)	0.85(0.007)	0.92(0.035)
Z	0.82(0.021)	0.70(0.025)	0.83(0.054)	0.95(0.052)	0.86(0.047)	0.90(0.014)
L	0.83(0.016)	0.78(0.022)	0.86(0.014)	0.91(0.041)	0.96(0.015)	0.92(0.023)
H	0.83(0.009)	0.75(0.031)	0.82(0.041)	0.85(0.011)	0.91(0.023)	0.89(0.038)
ER	0.84(0.037)	0.86(0.007)	0.82(0.016)	0.94(0.016)	0.96(0.033)	0.89(0.042)

Results for mixed-type patterns classification

We validated the proposed method for diagnosing mixed-type WBMs as illustrated in Figure 1, the mixed-type WBMs were judged by six single classification models and obtained the six results. The performance of the combined classifiers was shown in Table 4. Our proposed individual DBN classification model achieves better accuracy than that of MLP.

Table 4 The results of classifiers on mixed-type patterns

	Accuracy	Recall	Precision
MLP	0.756(0.025)	0.718(0.016)	0.807(0.105)
DBN	0.912(0.011)	0.924(0.020)	0.903(0.004)

The detailed results of typical mixed-type WBMs were summarized in Table 5. For WBM1, MLP and DBN both predicted the label correctly. For WBM2, only DBN performed correctly; MLP predicted the "partial ring" correctly but missed the "zone" pattern. Also for WBM3 and WBM4, MLP missed "partial ring" and "edge ring"

instead. For the last WBM5, MLP wrongly identified a zone pattern, instead of the line pattern. Overall, only DBN performed correctly for all five WBMs.

Table 5 Detailed results of mixed-type WBMs diagnosis

		PR	S	Z	L	H	ER
	True		√				
	MLP		√				
	DBN		√				
	True	√		√			
	MLP	√					
	DBN	√		√			
	True	√			√		
	MLP				√		
	DBN	√			√		
	True	√					√
	MLP	√					
	DBN	√					√
	True		√	√			
	MLP		√	√			
	DBN		√	√			

CONCLUSIONS

Due to WBMs are the key evidence to find the root causes which make low CP yield, a systematical WBMs analysis method is necessary. In this study, we presented a hybrid algorithm that integrates spatial statistics and DBN to automatically extract patterns from WBMs. An empirical study showed that the proposed framework can effectively improve the efficiency and accuracy of WBM pattern diagnosis. Specifically, building individual classification models for each pattern separately is more effective for classifying mixed-type defect patterns. Using real data examples, DBN showed significantly better performance, compared to competing method.

REFERENCES

[1] N. Kumar. *International Journal of Production Research*, vol. 44, 2006, pp. 5019–5036

[2] R. Uzsoy, C. Y. Lee, and L. A. Martin-Vega, *IIE Transactions*, vol. 24, 1992, pp. 47–60.

[3] W. Taam and M. Hamada, *Technometrics*, vol. 35, 1993, pp. 149–160.

[4] J. Kim, Y. Lee, and H. Kim, *IISE Transactions*, vol. 50, 2018, pp. 99–111.

[5] Y. S. Jeong, S. J. Kim, and M. K. Jeong, *IEEE Transactions on Semicondutor Manufacturing*, vol. 21, 2008, pp. 625–637.

[6] M. S. Yang and K. L. Wu, *IEEE Transactions on Pattern Analysis and Machine Intelligence*, vol. 26, 2004, pp. 434–448.

[7] I. Goodfellow, Y. Bengio, and A. Courville, MIT Press, 2016.

[8] W. Taam, M. Hamada, *Technometrics*, vol. 35, 1993, pp.149–160.

[9] S. C. Hsu and C. F. Chien, *International Journal of Production Economics*, vol. 107, 2007, pp.88–103.

978-1-7281-6559-2/20 $31.00 © 2020 IEEE

Perceptron Algorithm and Its Verilog Design

Kainan Wang [1,2,], Yingxuan Zhu [1,2], C.-Z. Chen [2,3]*

[1]State Key Laboratory of Information Security, Institute of Information Engineering, CAS, Beijing, China
*Mobile: (+86) 17801000887, Email: wangkainan@iie.ac.cn
[2]University of Chinese Academy of Sciences, Huairou, Beijing, China 101408
[3]EtownIP Microelectronics, Beijing, China 100176

ABSTRACT

In artificial neural network (ANN), the basic perceptron algorithm plays a significant role in supervised machine learning due to its simple structure. Though it cannot solve some non-linear problems like XOR, however, this feature offers a possibility to build perceptron on a hardware design. Due to high efficiency and defect tolerant, researchers have proposed some ANN accelerators with complicated memory units and specific registers. In this work, we focus on a simplest perceptron and accomplish its hardware design using Verilog HDL. The design module includes one core for learning and four memory units for storing the training data. The study shows that the proximate floating-point simulation of the simple perceptron design can replace the defect-tolerant registers and the simple memory units, thus to make the accelerator a tiny scale, it also demonstrates that the accuracy rate on test set achieved at 98% and the total area cost is only $0.0078 \ mm^2$.

Keywords—Perceptron, Verilog, hardware design, machine learning, ANN, CNN

INTRODUCTION

With the big progress of CNN (convolution neural network), artificial intelligence application has stepped into a new climax, especially in the supervised machine learning. Not only can the software languages be used to run the algorithm on CPU, but also the hardware be designed to co-work with CPU to accelerate the calculation. In this case, the design of the hardware is highly decisive on performance.

Although the complexity of machine learning algorithm is increasing, many AI-specific accelerator designs have been accomplished. For example, some designs accelerate the calculation on GPU [5] or FPGA [6], others propose the ASIC implementation of neural network [10]. However, the performance of the accelerators is always compromised by the communication with memory via DMA (Direct Memory Access) [6, 10]. Considering all the drawbacks, Tianshi Chen et al. proposed the 1st version of state-of-the-art AI-chip in 2014 [4]. Following that, many high-level AI-specific chips with better accuracy and more functions have been reported [7, 8, 9].

All of the fore-going new AI-specific chips or accelerators are complicated to implement as they have to satisfy most of the algorithms such as CNN, RNN (recurrent neural network) and so on. The complex calculation, such as in the convolution and in the differential, imposes one key factor in the hardware design. Nevertheless, most of the neural network can be seen as improved version of ANN (artificial neural network) and the most fundamental ANN algorithm is perceptron. Therefore, if the perceptron accelerator with high scalability and compatibility is available, it will be easier to implement the high-level hardware. The ANN hardware design was commonplace in the past [2, 3]. On account of the deficiency of spatially storage on neurons, Olivier Temam proposes the better design [1]. In this work, we report a kind of perceptron design with simpler memory and less dimensionality, which in [1] is the complicated DMA and 90 inputs. On the other hand, the activation function is a sign activation function as it is the simplest one. The 2-input port will be adopted.

METHODOLOGY

Implement the floating computing in hardware

There are many floating-point numbers in perceptron calculation. Unfortunately, the variables in Verilog are combined with 1 and 0, in such case, the floating numbers cannot be implemented. There are two methods can be used to nearly simulate the floating-point numbers. The first one is enlarging all the numbers simultaneously. For instance, multiply every number 512 times and store the results in memory. The calculating function calls the resulted numbers in memory divided by 512 as their approximated numbers. In fact, if the sign function is used, all the formulas can use the enlarged numbers with no impact on results which will be discussed later. The second method is using the IEEE 754 standard. Every floating-point number can be expressed as a 32-bit binary number or 64-bit. The second method is more complex than the first one because the extra summing unit has to be designed. On the other hand, all the training data for the perceptron have to be transferred to IEEE format before calculating. Given the original design intention, the first one is the best method and chosen for this work.

The details of the multiplication will be discussed in memory design and the enlarged numbers have no effect on the result is proved below.

Firstly, the activation function is the sign meaning the decision formula is (1), the F means the sign function, and the updating formula is (2) and (3). [11]

$$y = F (w_1 {}^* x_1 + w_2 {}^* x_2 + b) \quad (1)$$

$$w_i = w_i + \eta * (label - y) * x_i \quad (2)$$

$$b = b + \eta * (label - y) \quad (3)$$

The data which needs to be stored is w_i, x_i, and label. If all the stored data are multiplied by 512, formula (1) is the same as the original. But the formula (2) will change because the (label-y) will multiply with other enlarged data. Noticing the structure of formulae (2) and (3), the label should be stored as original and other data should be stored as multiples of 512. In this case, all have to change is enlarging the difference between label and y when calculating formula (3). Then all the formula will be enlarged 512 times at the same time. When the computing is over, divide the new weights by 512, the floating-point results become the final results.

The simple memory unit design

The DMA is widely used in hardware storage. It has accurate control and independent operation to efficiently co-work with other units like CPU. Nevertheless, the complicated structure will make the design burdensome. As a result, the simple memory unit which can be written and read must be used. The details are listed in *Table I*.

TABLE I. Details of memory unit

Ports	Bit width	I/O	Description
ena	1	I	Enable
w	1	I	Write(1) or Read(0)
Data-in	16	I	Input data
Data-out	16	O	Output data
clk	1	I	Clock
rst	1	I	Reset
addr	8	I	Address

As *Table I* shows, the stored data is 16-bit. As the number is multiplied by 512, the actual value cannot be larger than 128. Most of the perceptron 2-dimension input data is smaller than it. On the other hand, since the purpose of perceptron is to find a classified interface, the larger data can be zoomed out simultaneously. The "w" port controls the memory unit to write or read. If "w" is 1, the memory will write the data-in to memory [address]. If "w" is 0, the memory will output the data-out from memory [address] to request. Considering the total number of data, the 8-bit address is big enough to store 256 numbers.

In the memory unit, all the data will be read from four text files at the initialization. The training data and the weight data will be stored separately in the text files.

The top-level structure of accelerator

The perceptron schematic is shown in *Fig. 1*.

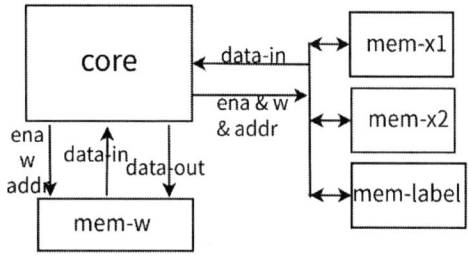

Figure 1: The top structure of the accelerator

As the figure shows, the perceptron has only 5 modules, one core module and four memory modules. The core module read the weight data and one of the training data. The core will calculate the real-time output of this point and update the weights depend on the difference between output and label. The new weight will be output to mem-w unit. The first updating is over now. After that, the core will read the new weights and training data for updating. Now there is a cycle. The computation will continue until there is no more training data input.

RESULT AND DISCUSSION

Accuracy Result

To test the accuracy of this accelerator, one specific dataset is used for training and testing. The dataset contains 700 lines of data. Every line represents the attributes of one data point: x1 data and x2 data and label data. All the data is divided into two parts, training set and testing set. The training set contains 500 data and testing set contains 200. When the training is over, the updated weight data will be output through mem-w-data-out port in serial. The final result can also be viewed in *-data-out ports. The output wave graph is depicted below. To be clear, the graph only shows some critical ports. The output data must be shown and other enable signals or middle variables can be left out.

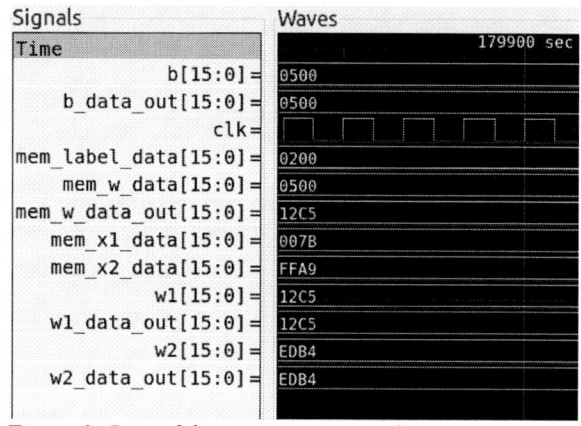

Figure 2: Part of the output wave graph

At the end of simulation time, the final b is 0500, the final w2 is EDB4, the final w1 is 12C5. To be more accurate, the signed variables are used in Verilog design. The transformation is shown in *Table II*.

TABLE II. Data transformation

	W1	W2	B
Signed value	12C5	EDB4	0500
Truly value	12C5	-124C	0500
Decimal value	4805	-4684	1280
Float value	9.38	-9.15	2.5

By using the weight data and perceptron formulae, the classify interface can be used to test the other data. The confusion matrix is shown in *Table III*.

TABLE III. Confusion matrix

Confusion Matrix		Label	
		0	1
Predicted	0	92	0
	1	4	104

In the graph, the accuracy rate is 196/200, 98%; the precision is 92/92, 100%; the sensitivity is 92/96, 96%; the specificity is 104/104, 100%.

Synthesis Result

To test the area and power cost, the hardware design was synthesized using Synopsys Design Compiler using a 25nm TSMC ASIC library. The synthesis started from 1.25 GHz and the highest frequency can be up to 1.9 GHz. The summary of consumption result can be seen in *Table IV*. As the table shows, the higher frequency means larger area and higher power. Despite this, the total area is only 0.0078 mm^2, which is only 5.5% of a 4-way 32KB Cache. The small scale and low power give the hardware design much more portability.

TABLE IV. Preliminary synthesis results

Frequency[GHz]	Area[mm^2]	Average power[mW]
1.25	0.0049	0.974
1.91	0.0078	1.798

Discussion

Since there is less two-category data with two dimensioning input, we generate our dataset using Python script. The ideal classification interface is x1-x2=0. In the transformation table, we can see the interface is pretty close to this. That is the reason why the performance is great (96%-100%). The bias, which is b, should close to zero, but the result is 2.5. In this case, the interface is moving upward a little bit. As a result, the points labelled 0 is predicted to 1, which can be seen in *Table III*.

Another interesting thing in the simulation is the loss of overflowed data. The registers hold the product of two 16-bit data should be at least 32-bit. Otherwise, the overflowed data is abandoned and the weight data cannot be convergent.

All the test result is based on one specific dataset and the accuracy may be lower in other data. Nevertheless, because of the robustness of perceptron algorithm, the performance could keep in a good level.

CONCLUSION

In this paper, we report a simple accelerator design to simulate a perceptron algorithm. Since the machine learning accelerators have been widely studied, many comprehensive chips have been invented, some fundamental machine learning algorithms like perceptron also needs more attention. Towards the goals two key problems are resolved. The first is to have floating-point calculation accomplished in hardware. The second is using simple memory units the design module. By achieving these, a fast and simple accelerator has been developed.

FUTURE WORK

The hardware design is done, there are additional works to proceed. Firstly, the input port is not big enough. To support the multidimensional data, the core module needs to be extended itself. In addition, the memory unit can have more control ports and functions so that all the training data can be in a unity. Finally, when this accelerator communicates with other standard chips, some ports need to reconfigured. In other words, the compatibility is also the key factor in the future design.

ACKNOWLEDGEMENTS

We would like to thank Li Sen for his help in paper writing. We also wish to thank the anonymous reviewers for their valuable comments and suggestions.

REFERENCE

[1] Temam O. A defect-tolerant accelerator for emerging high-performance applications. international symposium on computer architecture, 2012, 40(3): 356-367.

[2] M. Holler, S. Tam, H. Castro, and R. Benson. An electrically trainable artificial neural network (ETANN) with 10240 "Floating Gate" synapses. IEEE Press, Piscataway, NJ, USA, 1990.

[3] N. Mauduit, M. Duranton, J. Gobert, and J.-A. Sirat. Lneuro 1.0: a piece of hardware lego for building neural network systems. Neural Networks, IEEE Transactions on, 3(3):414–422, May 1992.

[4] Chen T, Du Z, Sun N, et al. DianNao: a small-footprint high-throughput accelerator for ubiquitous machine-learning. architectural support for programming languages and operating systems, 2014, 49(4): 269-284.

[5] A. Coates, B. Huval, T. Wang, D. J. Wu, and A. Y. Ng. Deep learning with cots hpc systems. In International Conference on Machine Learning, 2013.

[6] S.Chakradhar, M.Sankaradas, V.Jakkula, and S.Cadambi. A dynamically configurable coprocessor for convolutional neural networks. In International symposiumon Computer Architecture, page247, SaintMalo, France, June 2010. ACM Press.

[7] Liu D, Chen T, Liu S, et al. PuDianNao: A Polyvalent Machine Learning Accelerator. architectural support for programming languages and operating systems, 2015, 43(1): 369-381.

[8] Du Z, Fasthuber R, Chen T, et al. ShiDianNao: shifting vision processing closer to the sensor. international symposium on computer architecture, 2015, 43(3): 92-104.

[9] Jouppi N P, Young C S, Patil N, et al. In-Datacenter Performance Analysis of a Tensor Processing Unit. international symposium on computer architecture, 2017, 45(2): 1-12.

[10] C. Farabet, B. Martini, B. Corda, P. Akselrod, E. Culurciello, and Y. LeCun. NeuFlow: A runtime reconfigurable dataflow processor for vision. In CVPR Workshop, pages 109-116. IEEE, June 2011.

[11] Rosenblatt F. The perceptron: a probabilistic model for information storage and organization in the brain. Psychological Review, 1958, 65(6): 386-408.

A CLOCK JITTER TOLERANT ΣΔ MODULATOR EMPLOYING A HYBRID LOOP FILTER IN CMOS 40NM TECHNOLOGY

Negar Rashidi, Sungjun Yoon, and Jose Silva-Martinez*
Department of ECE, Texas A&M University, College Station, Texas, USA
*Corresponding Author's Email: jose-silva-martinez@tamu.edu

ABSTRACT

A Sigma-Delta Modulator (ΣΔ) that employs a hybrid loop filter to reduce its sensitivity to clock jitter is proposed. Part of the loop filter is implemented in the digital domain. The clock jitter effects are significantly reduced without changing the loop gain properties. The ΣΔ Modulator is implemented in TSMC 40 nm CMOS technology and occupies a 0.06 mm². The solution consumes 6.9mW and operates at a clock rate of 500 MS/s. In the presence of 20psrms clock jitter (peak to peak jitter of almost 215 ps) and bandwidth of 10 MHz, the measured peak signal-to-noise-ratio (SNR) and peak signal-to-noise+distortion ratio (SNDR) are 65 dB, and 64 dB, respectively.

INTRODUCTION

The performance of the ADCs is ultimately limited by the quality of the clock and sampling circuits. It is well known that all Nyquist ADCs require a sample and hold (S/H) circuit to complete the signal conversion, mainly in high performance solutions. The clock jitter performance is a major stopper for high resolution broadband circuits; Fig. 1 shows the ADCs reported in both ISSCC and VLSI symposiums during the last few years as well as the theoretical limits to Nyquist ADCs due to the use of jittered clocks (aperture noise) [1]. Notice that high resolution broadband ADCs demand high performance clocks, that can only be obtained using very clean reference clocks and high performance phase-locked loops with low-noise VCOs, increasing the cost of the final solution. According to Fig. 1, if the quality of the sampling clock is relaxed to 1psrms jitter, the Nyquist ADCs are bounded by aperture noise to SNDR=70 dB for the case of signal bandwidth of 80 MHz; in practice, the inherent limitations of the ADCs such as additional thermal and quantization noise components lower the achievable SNDR. When the clock jitter is limited to around 4psrms, the signal bandwidth is bounded to around 10 MHz for the Nyquist ADCs with SNDR around 70 dB.

Unfortunately, the cost of the clock generator is not included in the traditional ADC figure of merit (FoM), reported in scientific publications. Moreover, without considering the power consumption of the S/H and its limitations due to clock jitter, the conventional figures of merit (FoM) could be misleading as well. The SAR-based architectures which shows extremely low power for the ADC itself, require significant calibration efforts and

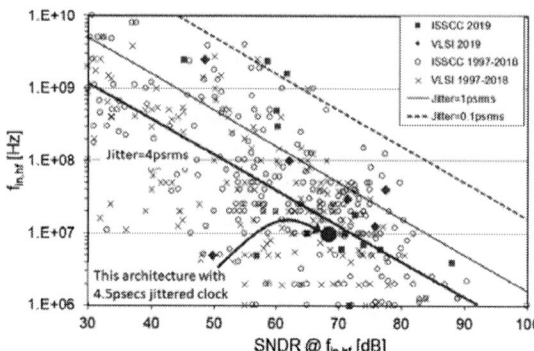

Fig. 1. ADC's performance and limitations due to clock jitter.

digital resources which increase silicon area and power consumption. The conventional pipeline ADC employs power hungry operational amplifiers but it is inherently more linear and accurate than SAR based ADCs. When considering the cost of the S/H, it is the major part of the power consumption of SAR-based ADCs, but it is just a fraction of the total power for the case of the conventional pipeline ADCs. When accounted the S/H, the FoM of SAR based and pipeline ADCs are not quite different.

On the other hand, the continuous-time sigma delta modulator (CT-ΣΔM), employed for high resolution ADCs, does not require the S/H circuitry at the input. This architecture relies on the operation of a low-resolution ADC, properties of linear digital-to-analog converters (DAC), as well as signal oversampling. Inherently, CT-ΣΔMs present low-sensitive to clock jitter effects generated in the quantizer, but it is sensitive to errors, including noise-induced by the clock jitter, present in the feedback DACs. Linear and low-noise feedback DACs are needed for the design of high performance CT-ΣΔMs.

In this paper, the proposed CT-ΣΔM architecture targets the reduction of clock induced errors due to the use of a low-cost jittered clock. The proposed architecture implements part of the loop filter in the digital domain after the quantizer. The digital filter reduces the power of the high frequency quantization noise before convolving with the jitter present in the clock that drives the feedback DAC. The resulting topology shows more tolerance to clock jitter. Results for a prototype fabricated in TSMC 40 nm achieves SNDR of 67dB with the bandwidth of 10MHz even if the clock jitter is as high as 5psrms.

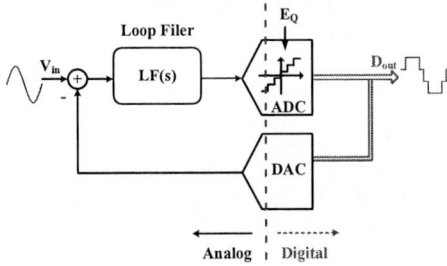

Fig. 2. Conventional Sigma-Delta Modulator.

CONVENTIONAL ΣΔ ARCHITECTURE

The schematic of the conventional ΣΔ modulator is displayed in Fig. 2. It consists of an analog loop filter, a signal quantizer (ADC) and the feedback DAC. It is well known that the continuous time sigma-delta modulator is tolerant to quantizer errors and its noise since the baseband information arises after the filter and inband signal that arises at modulator output is high-pass shaped by the high-gain low-pass loop filter. However, this is not the case of the baseband DAC noise and distortion since these signals appear at the input of the modulator and they are processed by the loop in a similar fashion as the incoming (desired) signal. Thus, very linear low-noise DACs are required for the realization of high performance ΣΔ modulators. The DAC linearity problem can be solved through analog or digital calibration or adopting dynamic element matching (DEM) techniques [2-6]. In any case, the DACs and its calibration engine should be very agile, since the extra feedback delay may result in stability issues.

The CT-ΣΔM is sensitive to clock jitter, and the problem becomes critical when the sampling rate increases. In high resolution data converters, the problem of clock-jitter can significantly limit the achievable SNR. The jitter-induced error CT-ΣΔM is approximately given by $e(t) = (D_{out}[nT] - D_{out}[nT - T]\frac{\beta_n}{T}$. Thermal noise present in the clock is irrelevant in this case since DAC is a switched like device. In the frequency domain it can be computed as [6]:

$$J_{error}(\omega) = (1 - Z^{-1}).\{(V_{in}(\omega) + V_B(\omega)).STF(\omega) + Q_N(\omega).NTF(\omega)\}\otimes J_n(\omega) \quad (1)$$

where the error signal includes the in-band desired signal $V_{in}(\omega)$, the high-pass shaped quantization noise coming from the quantizer $Q_N(\omega)$, and the remaining out-of-band thermal noise and blocker signal represented by $V_B(\omega)$. Also, $STF(\omega)$ and $NTF(\omega)$ stand for the signal and noise transfer function, respectively. The in-band signal and blockers are shaped by $1 - Z^{-1}$ before convolving with clock jitter; only the low-frequency

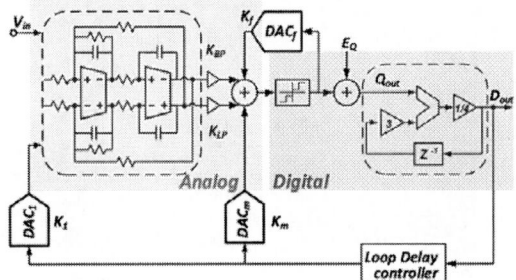

Fig. 3. Proposed low jitter sensitive Sigma-Delta Modulator with Hybrid loop filter.

clock jitter convolves with them and the result falls in-band, so its effect usually is not critical since clock's phase noise at low frequency is usually very small. The quantization noise near to the clock frequency is also shaped by $1 - Z^{-1}$. The noise $Q_N(\omega)$ around the Nyquist frequency is even amplified by 6dB. Indeed, the main issue in continuous-time ΣΔ modulators arises because the out-of-band quantization noise and high frequency blockers are convolved with the high frequency clock phase noise in the feedback DAC, and part of these components fall over the desired band, rising the in-band noise level. This issue is critical for high-resolution modulators using high clock frequencies because the out-of-band noise level is well above the in-band noise level.

PROPOSED ΣΔ MODULATOR

In order to decrease the jitter effects, the quantizer output is processed through a digital low-pass filter which is a part of the original loop filter (replaces part of the analog filter), but the loop gain remains the same. In the proposed system the output is taken after the digital filter; the STF remains the same but the NTF is subject to an additional low-pass filtering. The high frequency components are attenuated before convolving with the jittered clock, making the proposed ΣΔ modulator less sensitive to jittered clocks. Fig. 3 shows the schematic of the proposed architecture.

The third order loop filter is realized with a 2nd order state variable active RC filter and the third pole is realized employing a digital topology. The system has 3 loops: the first one composed by DAC1 with the signal fed back to modulator's input; this loop determines the inband quantization noise. The second loop is realized through DAC_m, which determines most of the compensating zeros and determines loop gain during the transition from passband to stop band. The loop is stabilized through the fast feedback path realized through DAC_f. This loop does not include any filter pole, which makes it very fast;

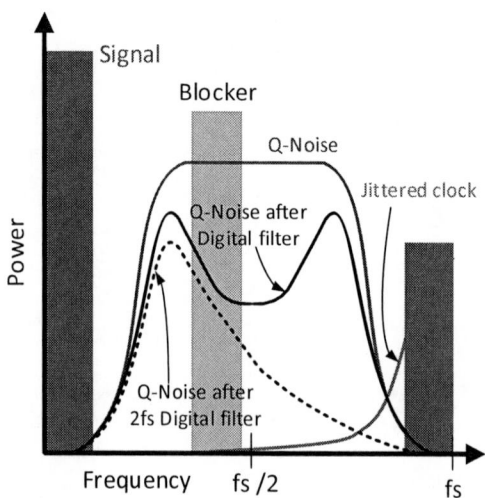

Fig. 4. Effect of the digital filter on output quantization noise and out-of-band blockers. Significant benefits are obtained if filter's clock frequency is doubled.

loop's stability is determined by this DAC_f. The jitter error can be described as

$$J_{error}(\omega) = (1 - Z^{-1}).\{(v_{in}(\omega) + B(\omega)).STF(\omega) + Q_N(\omega).NTF(\omega).\boldsymbol{H_{LP}(z)}\} \otimes J_n(\omega) \quad (2)$$

According to Eq. 2, the input signal's effect on jitter error is the same in comparison with the conventional architecture shown in Eqn. 1, except for the shaped quantization noise present in the digital low-pass filter transfer function H_{LP}. The effect of the first order digital filter on quantization noise, $Q_N(\omega)$, is depicted in Fig. 4. The out-of-band quantization noise is quasi flat after loop filter transition; the first order digital filter provides out-of-band attenuation that suppress both quantization noise and blockers power. As a result, the convolution of clock jitter with the out-of-band noise reduces $H_{LP}(z)$ in equation 2 and their effects are then minimized.

In case the digital filter runs at the clock rate, the low-pass filter reaches the maximum attenuation at the Nyquist frequency (50% of the clock frequency) and the spectrum repeats after that frequency. The benefits of the digital filter increase if the filter is handled with twice the clock frequency; e.g. 1GHz in this case. The digital filter drastically attenuates all out-of-band noise until the clock frequency. Since the quantization noise around and after the Nyquist frequency is severely attenuated, clock jitter effects on modulator's SNDR will be significantly reduced. The architecture discussed in this paper, however, employs the digital filter operating at clock rate. As a final remark, it should be mentioned that the digital filter does not provide additional attenuation to the blocker's power; it only helps reducing the effects of the quantization noise. Blockers arise at the input of the modulator and then they are treat in a similar fashion as

the input signals. Pre-filtering is needed to mitigate the effect of the blockers [7].

A 3rd order $\Sigma\Delta$ modulator that employs 4-bit quantizer was simulated and used as a comparison tool. The proposed $\Sigma\Delta$ modulator employs a 4-bit quantizer and 6-bit DAC (4bit quantizer/6bit DAC); the extra bits in the DAC are necessary because the digital low-pass filter receives 4-bits and delivers 8bits, from which the 6 most significant bits are fed back to DAC input. Therefore, the proposed architecture is compared with the performance obtained from a conventional 4-bit quantizer and 4bits DAC, and with the architecture employing 6-bit quantizer and 6bits DAC. The in-band input signal is set at 9.6MHz with a power that corresponds to -6dBFS, altogether with a blocker tone placed at 55MHz with a power of -20dBFS.

For the simulations, the loop gain remains similar for all architectures. The simulated (Cadence) results are displayed in Fig. 5. As expected, the ADC with 6-bit quantizer/DAC outperform the modulators with 4-bit quantizer. However, for the case of a jittered clock, the best results are obtained with the proposed architecture if a reference clock with over 1% clock jitter is used. In that case, the SNDR of the proposed ADC surpass by 6 dB to the conventional 4-bits quantizer/DAC. More notably, the proposed architecture outperforms the SNDR by 2 dB when compared to the 6-bits quantizer/DAC. Notice that Fig. 5 does not correspond to a peak SQNR since the input signal was set to -6 dBFS for the comparison. However, the proposed architecture operates properly until the input of -2 dBFS.

At clock jitter of 1%, the SNDR of the proposed architecture outperforms the $\Sigma\Delta$ modulator with 6-bit quantizer/DAC by 2dB, and by more than 6dB when compared with the conventional case using 4-bit

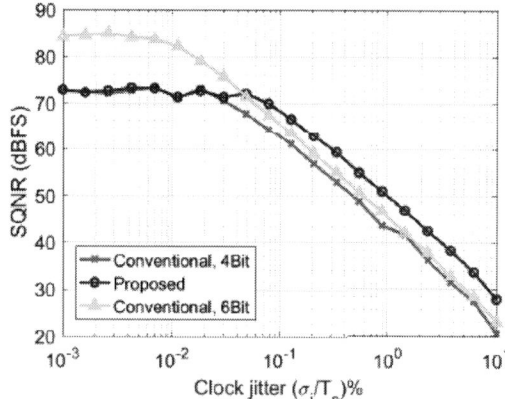

Fig. 5. Simulation results: SQNR for the conventional 4 and 6 bits $\Sigma\Delta$ modulators and the proposed 4-bit quantizer-bit DAC $\Sigma\Delta$ modulator. A 9.6MHz -6dBFS in-band tone altogether with a -20dBFS blocker at 55MHz were employed.

978-1-7281-6559-2/20 $31.00 © 2020 IEEE

quantizer/DAC. Notice in Fig. 5 that the SQNR does not correspond to peak SQNR since the input signal was set at -6dBFS, but the proposed architecture operates properly until -2dBFS.

EXPERIMENTAL RESULTS

The proposed low-jitter sensitive architecture was fabricated in 40nm CMOS through TSMC; the chip was biased with 1.1V. The power consumption is 6.9mW from which the analog power is 5.7mW and digital power is 1.2mW. The digital filter saves over 1mW when compared with the case of full analog loop filter.

The chip was measured with jittered clock for the extreme case of 215ps peak to peak jitter (which corresponds to 20psrms) operating at the clock rate of 500MHz. The linearity of the proposed CT-$\Sigma\Delta$ modulator was tested employing two tones at 3.8MHz and 4.5MHz with a power of -7dBFS each, leading to a total power of -4dBFS. In addition, several clocks with different clock jitter was used. Fig. 6 shows the case of 20psecs RMS jittered clock. The DAC was carefully designed and layout, leading to an outstanding linearity; each one of the

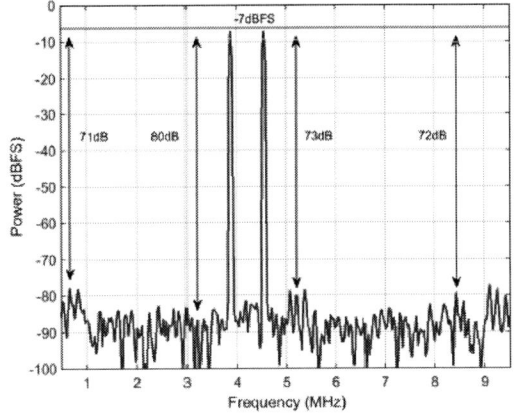

Fig. 6. Measured two-tone test with RMS power of -4dB with RMS clock jitter of 20psrms (1% of clock period).

spurious tones and intermodulation products are under 78 dB with respect to full-scale.

Fig. 7 shows the measured SNR and SNDR in the presence of 1% Ts RMS jitter (20 psrms). Peak SNR and peak SNDR are 65 dB and 64 dB, respectively. The $\Sigma\Delta$ modulator shows superior linearity even if the input signal is as large as -2dBFS.

CONCLUSIONS

A CT-$\Sigma\Delta$M that splits the loop filter to an analog and a digital sections is presented which reduces the quantization noise in the feedback path. One of the 3rd

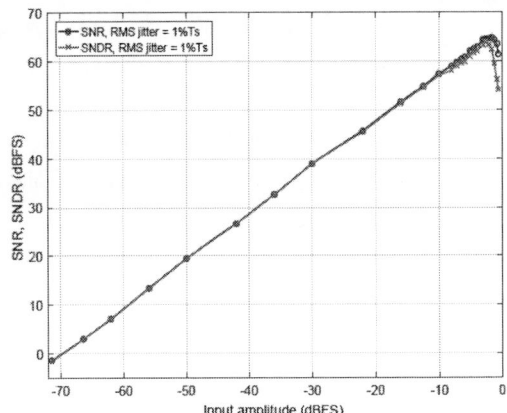

Fig. 7. Measured SNR and SNDR as function of input power; clock jitter is 20psrms.

order sections is realized in the digital domain; this thermal noise free filter section does not require any calibration since it is highly precise. Since the first order filter is placed after the quantizer to filter the quantization noise, the high frequency quantization noise is then attenuated, and clock jitter effects are effectively diminished; jitter noise skirts are mainly concentrated

A chip prototype was fabricated in TSMC 40nm CMOS technology. The prototype achieves a spurious free dynamic range of 75 dB and a peak SNDR of 67 dB in presence of 5psrms jittered clock. The chip consumes a total of 6.9 mW, and achieves a Schreier FoM of 166.5 dB with jittered clock of 0.2% Ts.

ACKNOWLEDGEMENTS

Authors would to recognize TSMC for fruitful discussions and chip fabrication.

REFERENCES

[1] B. Murmann, ADC Performance Survey 97-2019, http://web.stanford.edu/~murmann/adcsurvey.html.

[2] James A. Cherry and W. Martin Snelgrove, Kluwer Academic Publishers, Norwell, MA, USA, 2000.

[3] R. Schreier and G. C. Temes, John Wiley and Sons, 2005.

[4] Y. Chang, C. Lin,W.Wang, C. Lee, and C. Shih, IEEE Transactions on Circuits and Systems I: Regular Papers, pp. 1861–1868, Sep. 2006.

[5] Z Li and T.S. Fiez, IEEE Journal of Solid-State Circuits, pp. 1873 - 1883, Sept. 2007.

[6] C. Briseno-Vidrios, et.al., IEEE Journal of Solid-State Circuits, pp. 3280 –3292, Nov. 2018.

[7] H. Geddada, et.al., IEEE Trans on Very Large Scale Integration Systems, pp. 54–67, Jan 2015.

TOWARDS OPTIMAL LOGIC REPRESENTATIONS FOR IMPLICATION-BASED MEMRISTIVE CIRCUITS

Lin Chen, and Zhufei Chu*

EECS, Ningbo University, Ningbo 315211, China

* Email: chuzhufei@nbu.edu.cn

ABSTRACT

Memristive circuits natively perform material implication (IMPLY) operation, IMPLY together with FALSE (by setting a signal of the IMPLY to '0') is a complete set of operators. As one promising approach for in-memory computing, memristive circuits allow for both data-storing and logic-operation. Logic synthesis is essential for the design of emerging technologies. Instead of using well-known logic synthesis data structures to derive an implication logic network, the paper presents an exact synthesis method to obtain an optimal IMPLY logic network, which is a dedicated homogeneous network by using IMPLY as its only logic primitives. By synthesizing all the 256 three-input Boolean functions, the experimental results show 74 of these have better size compared with one-to-one mapping from optimal And-Inverter Graph (AIG) representations.

INTRODUCTION

The memristor was first predicted by Chua to describe the relationship between charge and magnetic flux. However, it was not until 2008 that Hewlett Packard (HP) laboratories first discovered the physical objects of memristors based on TiO_2. The memristor has a unique high and low resistance value state, which can perform the digital logic '0' and '1'. This nonvolatile two-terminal device can natively execute material implication (IMPLY) operation, which is a universal Boolean logic operation [1]. By using IMPLY operation to represent the Boolean function, the memristive circuits can perform in-memory computations based on these IMPLY operation sequences.

The well-known logic synthesis data structures are graph-based logic representations, such as Binary Decision Diagram (BDD), And-Inverter Graph (AIG), and Majority-Inverter Graph (MIG). The previous work [2], [3] adopts AIG in ABC tool[1] to obtain an IMPLY logic network. By reading into a circuit and storing it as the AIG network, the IMPLY logic network is obtained by using a technology mapping process. However, this method cannot guide to obtain the best cost trade-offs, since one needs an independent node to realize an inverter while the inversion in AIG is not accounted for the network size. In this paper, we aim to obtain the

[1] https://github.com/berkeley-abc/abc

IMPLY logic network from Boolean functions without intermediate steps. By exact synthesis, we can obtain the optimal IMPLY logic representations for the Boolean functions. By synthesizing all 256 three-input functions, 74 of these are optimized using our method compared with the exact synthesis approach using AIG as the underlying logic representations.

BACKGROUND

The typical well-known fundamental logic operations are AND, OR, XOR, and NOT, etc. Implication logic is known as implies or is typically explained as "only if" or "if ... then". The symbol of the IMPLY gate is denoted as '$\rightarrow$'. For a two-input IMPLY gate $p \rightarrow q$, it can be realized by two memristors and a resistor as shown in Figure 1(a). And its truth table is shown in Figure 1(b). When p is true '1' but q is false '0', $p \rightarrow q$ is false '0'. For the other cases, $p \rightarrow q$ is always true. Hence, $p \rightarrow q$ is logically equivalent to $\bar{p} + q$,

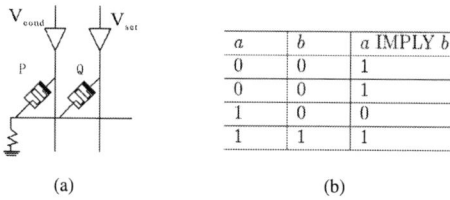

a	b	a IMPLY b
0	0	1
0	0	1
1	0	0
1	1	1

(a)　　　　　　　　　　　　(b)

Figure 1: (a) IMPLY gate, (b) Truth table of IMPLY gate.

In Figure 1(a), to perform the $p \rightarrow q$ operation, signals p and q need to be present on the memristors P and Q. Then, V_{cond} and V_{set} voltages are imposed on the memristors P and Q, respectively.

We consider the homogeneous network by using IMPLY operation as its only logic primitives for the synthesis of Boolean functions. In our case, we make use of IMPLY Graphs (IMG) as the underlying data structure for exact synthesis. The IMPLY logic function together with FALSE comprises a computationally complete logic structure. Some basic logic identities are shown as follows $\bar{a} = a \rightarrow 0$; $a \vee b = (a \rightarrow 0) \rightarrow b$; $a \wedge b = ((b \rightarrow (a \rightarrow 0)) \rightarrow 0)$.

SAT-BASED EXACT SYNTHESIS

Exact synthesis aims to find the optimal logical network to represent a Boolean function by giving a set of primitives. The objectives can be the size or logic level of the network, which corresponds to the area and delay of the circuit, respectively. Take the Boolean Satisfiability (SAT) technique as an example, exact synthesis is executed by solving sequences of SAT formulations [4]. Given a set of primitives and the required number of logic nodes (r) or levels (l), the exact synthesis will answer the question: "Is there exist a logic network of r gates or l levels to realize the given functions?".

Variables in SAT Formulation

A network of Boolean function which has n inputs $(x_1, ..., x_n)$ is a series of gates $(x_{n+1}, ..., x_{n+r})$ [5].

$$x_i = x_j \circ_i x_k, n+1 \le i \le n+r, j < k < i. \quad (1)$$

It means that each gate has two inputs which are previous gates or primary inputs. To express exact synthesis problem, variables in the SAT formula are defined. In our design, we investigate a dedicated homogeneous network structure that uses IMPLY as the only logic primitive. For $1 \le h \le m$, where m is the number of the outputs, $n < i \le n+r$, and $0 < t < 2^n$, the variables used in the SAT formulation are defined in the following:

$x_{it}:$ t^{th} bit of x_i's truth table

$g_{hi}:$ $[g_h = x_i]$

$s_{ijk}:$ $[x_i = x_j \to x_k]$ for $1 \le j < i, 0 \le k < i, j \ne k$ (2)

If g_{hi} is true, it means function g_h is represented by gate x_i. The variable s_{ijk} is a selection variable, which evaluates to true if the two inputs of gate x_i are x_j and x_k. Since $a \to b \ne b \to a$ and $a \to 0 = \bar{a}$, we allow the selection variables s_{i12} and s_{i21} as well as s_{i10}, which represent $x_i = x_1 \to x_2$, $x_i = x_2 \to x_1$, and $x_i = x_1 \to 0 = \bar{x}_1$, respectively.

Encoding of the Clauses

The SAT-based exact synthesis algorithm can convert the logic gates and the topological relationships of the logic gates to conjunction normal formal (CNF) forms. To get the CNF clauses of the IMPLY gate, we should first introduce the Boolean variables, then the next step is to add clauses based on these variables as constraints to make the logic network work correctly.

Since every operator performs an implication operation, if step x_i has fan-in x_j and x_k, then x_{it} must be equivalent to $x_{jt} \to x_{kt}$. Hence, the main operation constraints can be written as:

$$(s_{ijk} \to (x_{it} \Leftrightarrow (x_{jt} \to x_{kt}))) \quad (3)$$

Then it's necessary to convert IMPLY gate ($x_{it} \Leftrightarrow (x_{jt} \to x_{kt})$) in CNF forms. Take an AND gate ($c \Leftrightarrow a \wedge b$) in a combination circuit as an example, when the

AND gate works, then

$$c = a \wedge b$$
$$c \Leftrightarrow a \wedge b$$
$$c \to (a \wedge b)$$
$$(a \wedge b) \to c$$
$$(c \to (a \wedge b)) \wedge ((a \wedge b) \to c)$$

Similarly, the equation (4) can be obtained for $c = a \to b$

$$(c \to (a \to b)) \wedge ((a \to b) \to c)$$
$$(x_{it} \to (x_{jt} \to x_{kt})) \wedge ((x_{jt} \to x_{kt}) \to x_{it}) \quad (4)$$

In summary, the equation (3) can be further transposed to

$$s_{ijk} \to ((x_{it} \to (x_{jt} \to x_{kt})) \wedge ((x_{jt} \to x_{kt}) \to x_{it})) \quad (5)$$

In IMPLY, $((x_{jt} \to x_{kt}) \to x_{it}) == (\bar{x}_{it} \to (\overline{x_{jt} \to x_{kt}}))$. It can be further simplified to

$$(s_{ijk} \to (\bar{x}_{it} \to (\overline{x_{jt} \to x_{kt}}))) \wedge (s_{ijk} \to (x_{it} \to (x_{jt} \to x_{kt}))) \quad (6)$$

Next, we can convert above constraints into CNF by converting the implication $(a \to b) = (\bar{a} + b)$.

$$(\bar{s}_{ijk} \vee x_{it} \vee \overline{\bar{x}_{jt} \vee x_{kt}}) \wedge (\bar{s}_{ijk} \vee \bar{x}_{it} \vee \bar{x}_{jt} \vee x_{kt}) \quad (7)$$

Based on the Tseitin's transformation, for the NOR gate

$$c = \overline{a \vee b} \Leftrightarrow (a \vee b \vee c) \wedge (\bar{a} \vee \bar{c}) \wedge (\bar{b} \vee \bar{c})$$

Hence, we introduce a new Boolean variable $c_{jkt} = \overline{\bar{x}_{jt} \vee x_{kt}}$ to represent NOR gate in (7).

$$c_{jkt} = \overline{\bar{x}_{jt} \vee x_{kt}} \Leftrightarrow (\bar{x}_{jt} \vee x_{kt} \vee c_{jkt}) \wedge (x_{jt} \vee \bar{c}_{jkt}) \wedge (\bar{x}_{kt} \vee \bar{c}_{jkt}) \quad (8)$$

In summary, we can encode the main clauses of operation as follows.

$$\bar{s}_{ijk} \vee x_{it} \vee c_{jkt} \quad (9)$$
$$\bar{s}_{ijk} \vee \bar{x}_{it} \vee \bar{x}_{jt} \vee x_{kt} \quad (10)$$
$$\bar{x}_{jt} \vee x_{kt} \vee c_{jkt} \quad (11)$$
$$x_{jt} \vee \bar{c}_{jkt} \quad (12)$$
$$\bar{x}_{kt} \vee \bar{c}_{jkt} \quad (13)$$

Besides, let $(t_1, ..., t_n)_2$ be the binary encoding of t, such that t_i refers to the ith bit of t. In order to fix the proper output values, we add the clauses $(\bar{g}_{hi} \vee \bar{x}_{it})$ or $(\bar{g}_{hi} \vee x_{it})$ depending on the value $g_h(t_1, \cdots, t_n)$. then the clauses

$$(\bar{g}_{hi} \vee (\bar{x}_{it} \oplus g_h(t_1, ..., t_n))) \quad (14)$$

constrain the output values to the gates they point to. Moreover, the constraints $\bigvee_i g_{hi}$ ensure that each output is realized by the network and the constraints $\bigvee_{j,k} s_{ijk}, j \ne k, j \ne 0$ ensure that each gate has exactly two inputs. The above-mentioned clauses are essential to make the algorithm work.

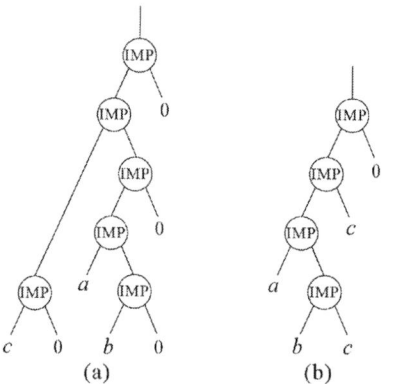

Figure 2: IMG of the function $(00000111)_2$, (a) AIG-based synthesis, (b) our method.

ALGORITHMS

Based on the encoding methods described above, we can call an SAT solver to answer the question: "Is there exist an IMPLY logic network to realize Boolean function f with r gates?". The initial value of r is set as 0 since some functions are trivial as constants or projection. If we get a satisfiable solution, we finished the exact synthesis; otherwise, we increase r and resolve the problem until the solution is returned. In this way, we can guarantee the solution is optimal in size.

EXPERIMENTAL RESULTS

The method is implemented in an open-source logic synthesis tool ALSO[2], by the command "`exact_imply`". As a comparison, we also implemented the exact synthesis using AIG as the underlying data structures by adding an augment as "`exact_imply -a`". The main difference is that the inversion in AIG is represented as a complemented edge, while it accounts for one node in the IMPLY logic network. The computation complexity is sensitive to the number of nodes used in the logic network due to the nonlinear increase of Boolean variables. Therefore, the exact synthesis using AIG is faster than our method.

Once the optimal-size AIG representations are enumerated, by one-to-one mapping into the IMPLY logic networks, we can find the IMPLY representation of function f with minimum size, denoted $N_{aig}(f)$. For example, the function with the truth table value of $(00011010)_2$ has four different IMPLY networks by the mapping method based on AIG-based synthesis, which results in the number of IMPLY nodes 6, 6, 8, and 9. Hence, we select the best one in size and $N_{aig}((00011010)_2) = 6$. In contrast, the proposed method can obtain the optimal

[2]https://github.com/nbulsi/also

IMPLY logic representations directly. The optimal size by our method is recorded as $N_{img}(f)$.

By enumerating all 256 three-input Boolean functions, we compare the results of both AIG-based exact synthesis and our method. The result are shown in Table I. It indicates 74 out of 256 three-input Boolean functions can have fewer IMPLY nodes than the AIG-based synthesize method. This is because the way of AIG-based synthesis does not guarantee the optimal. The function represented by the truth table $(00000111)_2$ (its expression is $((\bar{a} \vee \bar{b}) \wedge \bar{c})$) achieves the most significant improvement. By AIG-based synthesis, the resulted IMPLY graph is shown in Figure 2(a). Therefore, $N_{aig}((00000111)_2) = 6$ and the logic level is 5. Its IMPLY expression is :

$$(((c \rightarrow 0) \rightarrow ((a \rightarrow (b \rightarrow 0)) \rightarrow 0)) \rightarrow 0)$$

Correspondingly, Figure 2(b) is the IMPLY graph by applying our method. Hence, $N_{img}((00000111)_2) = 4$ and the logic level is 4. Our method can reduces the number of nodes and levels by 2 and 1, respectively. Its corresponding IMPLY expression can be written as:

$$(((a \rightarrow (b \rightarrow c)) \rightarrow c) \rightarrow 0)$$

TABLE I. COMPARISONS WITH AIG-BASED EXACT SYNTHESIS

	$N_{aig}(f) = N_{img}(f)$	$N_{img}(f) < N_{img}(f)$
# functions	182	74

SUMMARY

In this paper, we proposed an SAT-based exact synthesis method to directly obtain an optimal IMPLY logic network without intermediate steps. By synthesizing all 256 three-input Boolean functions, the experimental results show that compared with the way of AIG-based synthesis, 74 of the 256 three-input Boolean functions have a size optimization.

ACKNOWLEDGMENTS

The work was supported by NSFC under Grant 61871242.

REFERENCE

[1] J. Borghetti, G. S. Snider, P. J. Kuekes, J. J. Yang, D. R. Stewart and R. S Williams. *Nature*, vol. 464, NO. 7290, 2010, pp. 873-876.

[2] H. Wang, C. Lin, C. Wu, Y. Chen and C. Wang. *IEEE Transactions on Very Large Scale Integration (VLSI) Systems*, Vol. 26, 2018, pp. 2842–2852.

[3] F Lalchhandama, B. G. Sapui and K. Datta Pittsburgh, *IEEE Computer Society Annual Symposium on VLSI (ISVLSI)*, Pittsburgh, July 11-13, 2016, pp. 319-324.

[4] Z. Chu, M. Soeken, Y. Xia, L. Wang and G. D. Micheli. *IEEE Transactions on Computer-Aided Design of Integrated Circuits and Systems*, 2019, pp. 1-1.

[5] D. E. Knuth, *The Art of Computer Programming, Volume 4, Fascicle 6: Satisfiability.* Addison Wesley, 2015.

TIMING VIOLATION AS DOMINANT REASON FOR FAILURE OF CLOCKED DIGITAL CIRCUIT DUE TO RF INTERFERENCE IN SUPPLY

Shanshan Nong and Tao Su*

School of Electronics and Information Technology, Sun Yat-sen University
Guangzhou, Guangdong, China
*Corresponding Author's Email: sutao@mail.sysu.edu.cn

ABSTRACT

This paper covers our observations of the failure behavior of a clocked digital circuit with sinusoidal interference acting on its supply. Conventionally, it has been thought that interference causes mainly logic-level errors in digital circuits, with the average value of the interference determining the circuit delay. As the interference cycle time is much shorter than both the data path delay and clock cycle time, the average value of the interference is almost zero. However, it still causes a timing violation, rather than a logic-level error, in the circuit. This observation was at odds with conventional thinking. This behavior was confirmed with both transistor-level simulations and board-based measurements. The findings of the present study are important for determining the frequency response of the maximum tolerable interference amplitude of a digital circuit in the design phase

Keywords—Timing Violation; RFI; Supply; Digital circuit

INTRODUCTION

The large-scale use of radio, the large-scale increase and intensive placement of electronic equipment and components, harsh environment, and intentional electromagnetic attacks, etc., all of these will lead to electromagnetic interference waves at the power, signal or other locations of integrated circuits. These electromagnetic waves will have a certain negative impact on the stable working state of the digital circuit. So the electromagnetic immunity of integrated circuits (ICs) has been a focus of concern for a considerable time.

In the processes of modelling, simulation, and testing, the main point is determining whether the circuit has failed. It has been believed that the interference mainly causes logic-level errors in digital circuits, with the average value of the interference determining the circuit delay [1]. When the average value of the interference is almost zero, the path jitter is also almost zero [2]. Another opinion is that path jitter will arise because the supply voltage levels may differ in successive clock cycles. Discretized power supply is used to compute the cell delay and analyze the timing.

An important form of interference that has attracted widespread attention in industry and academia is sinusoidal interference in the MHz-GHz radio frequency band [3]. This paper is aimed at the situation that the power supply is subject to sinusoidal interference in the MHz-GHz radio frequency band. It is found that for low-frequency and high-frequency interference, the circuit is more prone to timing violations than logic level errors.

TIMING VIOLATION

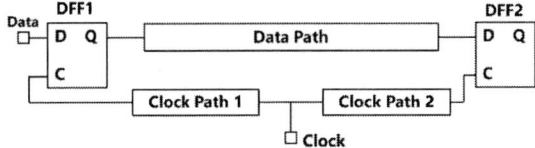

Fig. 1 *Example of synchronous digital circuit*

In the case of the circuit shown in Fig. 1, the timing of the synchronous digital circuit can be divided into two parts: the delay (t_{cq}) from C to Q of D-flip-flop 1 (DFF1) and the delay incurred in the data path (t_{dp}) between the two DFFs. Here, t_{clk} is the clock latency, t_{setup} represents the setup time of the DFF, and t_{hold} represents the hold time of DFF. The timing requirements for a synchronous digital circuit can be described by Eqs. (1) and (2).

$$t_{cq} + t_{dp} < t_{clk} - t_{setup} \qquad (1)$$

$$t_{cq} + t_{dp} > t_{hold} \qquad (2)$$

The setup time is more sensitive than the hold time when the supply [2] is subject to RF interference (RFI), so we focused on the verification of the setup time. When the changes in the data path latency (Δt_{dp}) and clock path latency (Δt_{cp}) are greater than the margin of the setup time (mt_{setup}), a timing violation will occur. The condition corresponding to a timing violation can be expressed as:

$$\Delta t_{dp} + \Delta t_{cp} \geq mt_{setup} \qquad (3)$$

Using Eq. (3), we can predict whether RFI acting on the power supply is likely to cause a timing violation.

SIMULATION

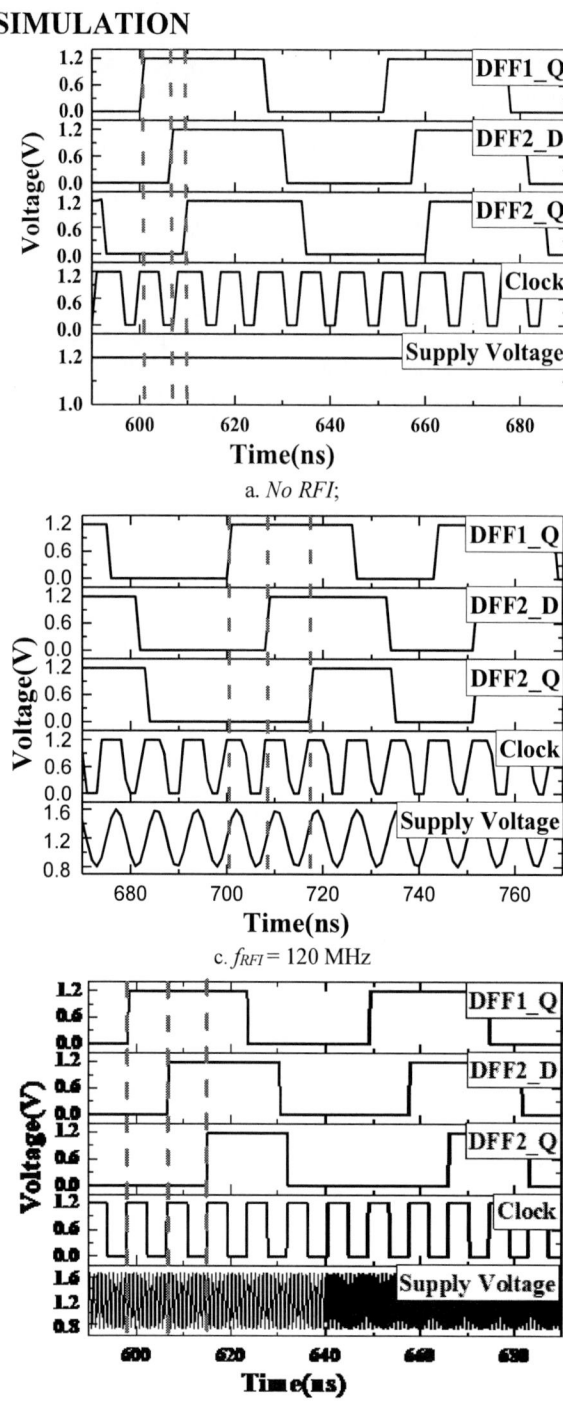

a. *No RFI*;

c. f_{RFI} = 120 MHz

d. f_{RFI} = 1200 MHz

Fig. 2: *Timing diagrams of HSpice simulation results for different values of f_{RFI}*

Using the SMIC 130-nm technology library of HSpice,

a random path extracted from the Loongson GS232 CPU was taken as the data path to be examined in the simulation. This data path contained 90 logic cells: AND, OR, NOR, XOR, INV, etc. Clock paths 1 and 2 each contained 100 inverters. The standard supply voltage was 1.2 V. The maximum RFI amplitude was 0.4 V. With no RFI acting on the supply, the data path delay was 8 ns, the clock period was 8.5 ns, and the setup time for DFF2 was 0.05 ns [4]. Therefore, the margin was 0.45 ns. This corresponds to about 5% of the clock cycle, which is much larger than the timing margin of the design.

Fig. 2 shows the simulation results obtained for different values of f_{RFI} (the frequency of RFI), namely, f_{RFI} =120MHz, and f_{RFI} = 1200 MHz . When we consider t_{dp}, the delay between the rising edge of DFF1_Q and that of DFF2_D, is 8.6 and 8.5 ns for f_{RFI} values of 120MHz and 1200 MHz respectively(see Fig. 2 b, c). The corresponding values of Δt_{dp} (compare with Fig.2a) were 0.6 and 0.5ns All the values are larger than the 0.45 ns which satisfies Eq. (3). This still causes a timing violation when the interference cycle time is much smaller than either the data path delay or the clock cycle time (Fig. 2 d).

MEASUREMENT

Fig. 3: *Test board*

A test board (shown in Fig. 3) was built with a 90-stage data path and a 100-stage clock path. We used 74-series 1-gate logic chips to build the data path, and 74HC04D inverter chips for the clock path. The supply voltage was 5 V. A filter network was inserted between the path and the supply to ensure that the RFI was applied to the supply pin of the design being tested. The maximum amplitude of the RFI was 1.1 V. With no RFI, the delay incurred in the data path was 191 ns, the clock period was 204 ns, and the setup time for DFF2 was 3 ns [5]. Therefore, the timing margin was 10 ns, that is, about 6% of the clock cycle.

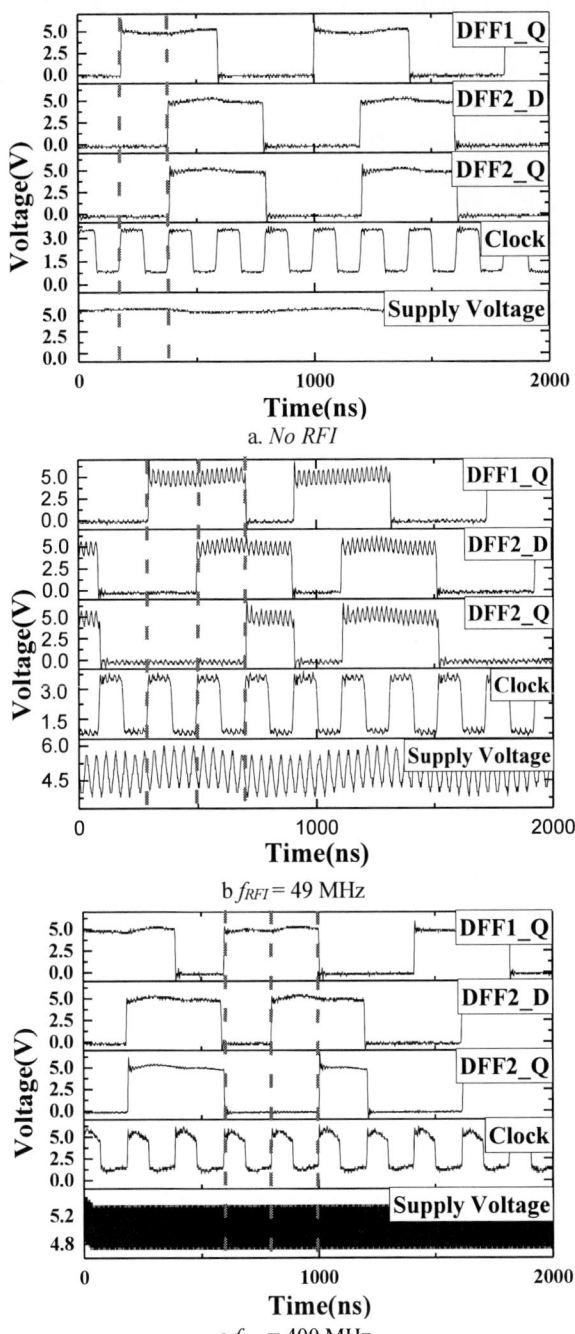

a. *No RFI*

b f_{RFI} = 49 MHz

c f_{RFI} = 400 MHz

Fig. 4: *Timing diagrams of test results obtained with different values of f_{RFI}*

Fig. 4 shows the results obtained for No RFI, f_{RFI} =49MHz, and f_{RFI} = 400 MHz . The measured results

were very similar to the results of the simulation. The corresponding values of t_{dp} were 203, and 204 ns, respectively (see Fig. 4b, c). The values of Δt_{dp} (compare with Fig. 4a) were 12 and 13 ns, respectively, which indicates that there is a timing violation. This also proves that, even when the average value of the interference is almost zero, a timing violation will nevertheless occur.

CONCLUSION

It has been proven, through both simulation and measurement, that when the amplitude of the RFI acting on a power supply does not exceed the conventionally maximum allowable interference amplitude, the output of the circuit will be affected. In the present study, it was revealed that this was caused by a timing violation. The circuit is more likely to exhibit a timing violation than a logic-level error, not only when low-frequency interference is applied, but also in the case of high-frequency interference. This timing violation cannot be explained by the linear model of jitter. Our future work will focus on building an accurate model.

ACKNOWLEDGEMENTS

Project 61471402 supported by NSFC. This work is also supported in part by the Science and Technology Program of Guangdong Province under Grant 2017B090909005 and 2019B010140002

REFERENCES

[1] Synopsys. PrimeTime® User Guide: Version P-2019.03[EB/OL]. http://www.synopsys.com, 2019.03

[2] Masanori Hashimoto R N. Power Integrity For Nanoscale Integrated Systems[M]. United Stated of America: Cenveo Publisher Services, 2014.

[3] IEC62433-2:2017 EMC IC modelling – Part 2: Models of integrated circuits for EMI behavioural simulation – Conducted emissions modelling (ICEM-CE)[S]

[4] Artisan Components. SMIC 130-nm Logic013G Process 1.2-Volt SAGE-XTM v2.0 Standard Cell Library Databook[DB/OL].http://www.smics.com/site/multi_project,2015.

[5] TI: SNx4AHC374 Octal Edge-Triggered D-Type Flip-Flops With 3-State Outputs[DB/OL].http://www.ti.com/sc/docs/stdterms.htm, accessed March, 2019

DREAMPlace 2.0: Open-Source GPU-Accelerated Global and Detailed Placement for Large-Scale VLSI Designs

Yibo Lin[1*], David Z. Pan[2], Haoxing Ren[3], and Brucek Khailany[3]

[1]CS Department, Peking University
[2]ECE Department, The University of Texas at Austin
[3]NVIDIA, Inc.
[*]yibolin@pku.edu.cn

Abstract—Modern backend design flow for very-large-scale-integrated (VLSI) circuits consists of many complicated stages and requires long turn-around time. Among these stages, VLSI placement plays a fundamental role in determining the physical locations of standard cells. Due to increasingly large design sizes, placement algorithms usually require long execution time to achieve high-quality solutions. Meanwhile, developing a placer often needs huge coding effort and tedius tuning, raising the bar of further researches. In this work, we present an open-source placement framework, *DREAMPlace 2.0*[1], with deep learning toolkit-enabled GPU acceleration for both global and detailed placement optimization to tackle the issues of efficiency and development overhead.

I. INTRODUCTION

Placement plays critical role to design closure of the VLSI backend flow. It decides the physical locations of standard cells in the layout, which will significantly affect the solution space of the succeeding routing stages.

The placement problem takes a circuit netlist and a standard cell library as input. It needs to place the cells in a fixed-outline layout region with no overlaps between cells and all the design rules satisfied. In modern designs, cells have to be placed in discrete placement sites. The objective of placement includes wirelength, routability, timing, and so on. As the problem is $\mathcal{NP}$-hard [1], [2], placement is often divided into three optimization steps: global placement (GP), legalization (LG), and detailed placement (DP). GP relaxes the discrete cell locations to continuous ones and roughly distributes cells in the layout. LG removes all the overlaps between cells and clean all design rule violations. DP is a refinement step to improve the objective with incremental movement of cells.

As a classic problem, there are many existing efforts. The algorithms for GP can be categorized into quadratic placement and nonlinear placement. Quadratic placement iterates between two phases: an unconstrained quadratic programming phase to minimize wirelength, and a heuristic spreading phase to remove overlaps. Typical quadratic placers include FastPlace [3], Polar [4], [5], Ripple [6], SimPL/ComPlx [7], [8], etc. Nonlinear placement formulates a nonlinear optimization problem and tries to directly solve it with gradient descent methods. There are many nonlinear placers such as mPL6 [9], APlace [10], NTUplace families [11]–[13], ePlace/RePlAce families [14],

[15]. Generally speaking, nonlinear placement can achieve better solution quality, while quadratic placement is more efficient.

As GP takes a majority of the runtime in placement, there are also works to accelerate GP algorithms such as POLAR 3.0 [5] and UTPlaceF 3.0 [16] with multi-threading. The acceleration ratio is around $5\times$ with $2 - 6\%$ quality degradation. Cong et al also explored GPU acceleration for multi-level GP algorithms [17], where $15\times$ speedup was achieved with less than 1% quality degradation.

Besides GP, algorithms for DP are mostly based on heuristic approaches with local searching. Widely used DP algorithms include independent set matching, local reordering, global swap/move, row-based techniques [11], [12], [18]–[20], etc. Dhar et al [21] explored GPU acceleration for a row-based interleaving algorithm.

Despite the previous works, current placement engines suffer from following issues: 1) poor quality-efficiency tradeoff; 3) high development effort. As mentioned, multi-threading on CPU can only achieve limited speedup with high quality gradation [5], [16]. GPU acceleration requires huge development effort and lacks systematic frameworks [17], [21].

To tackle these challenges and stimulate researches on placement, we present an open-source *DREAMPlace 2.0* framework with GPU acceleration enabled by deep learning toolkit `PyTorch` for both global and detailed placement. The major contributions can be summarized as follows.

- We develop the placement algorithms with deep learning toolkit to decouple the algorithmic design and the kernel operator development with reduced coding overhead.
- We present a full placement flow including GP, LG, and DP with both multi-threading and GPU acceleration.
- Experimental results demonstrate that the framework can accelerate the entire placement flow by $14\times$, while matching the state-of-the-art solution quality [11], [15].

Meanwhile, the framework offers an easy way to explore new solvers developed in the deep learning toolkit, e.g., Adam [22], and stochastic gradient descent (SGD) with momentum. The rest of the paper is organized as follows. Section II explains the formulation of placement and the DREAMPlace framework in details. Section III validates the framework with experimental results. Section IV concludes the paper.

---

[1]https://github.com/limbo018/DREAMPlace

978-1-7281-6559-2/20 $31.00 © 2020 IEEE

II. DREAMPLACE FRAMEWORK

A. Placement Problem

In this work, we consider wirelength as the objective.

Problem 1. *Given a netlist, a standard cell library, and a fixed-outline layout, determine the physical locations of movable cells with minimum wirelength.*

In the GP step, we solve a nonlinear placement problem,

$$\min_{\mathbf{x},\mathbf{y}} \quad WL(\mathbf{x},\mathbf{y}),$$
$$\text{s.t.} \quad D_i(\mathbf{x},\mathbf{y}) \le t_d, \quad \forall i \in B, \tag{1}$$

where $\mathbf{x}, \mathbf{y}$ denote the coordinates of cells in the layout, which is divided into a set of bins B uniformly. $WL(\cdot)$ denotes the wirelength cost, $D_i(\cdot)$ denotes the density at bin i, and t_d is a given target density. Intuitively, if the density constraints are satisfied at all bins, it indicates that cells have already spread out with very small overlaps. The LG step then legalizes the solution and removes all overlaps between cells. The DP step further refines the solution while maintaining the legality.

B. Software Architecture

As the placement framework is developed with deep learning toolkit, we adopt the same software architecture that separates high-level optimization algorithms with low-level operators, as shown in Figure 1. The algorithms are developed and assembled in `Python`, while low-level operators are highly optimized in `C++/CUDA`.

For GP, we rely on two important operators, wirelength and density, with forward and backward functions to compute the cost and the gradients, respectively. We propose both multi-threading and GPU accelerations for these operators. The gradient descent solvers can be implemented in `Python` with the automatic gradient derivation package in `PyTorch`. Solvers in `PyTorch` like Adam [22] and SGD can be used for optimization. We further develop a custom solver based on Nesterov's method with line search [15], which is among the state-of-the-art optimization techniques for placement.

For LG and DP, each technique is developed as an operator in `C++/CUDA` and assembled in `Python`. For LG, we implement a greedy legalization technique in NTUplace3 [11] and a row-based Abacus algorithm [23] on CPU. For DP, three techniques have been developed, independent set matching, local reordering, and global swap [11], [19]. To accelerate the techniques, we propose parallel variations based on batch execution for multi-threading and GPU.

C. Main Features

As an open-source project, we are actively incorporating new features. With the release of DREAMPlace 2.0, the main features so far are summarized as follows.

- Multi-threading and GPU accelerated GP and DP for weighted wirelength minimization.
- Movable macros supported in GP and LG.
- Bookshelf and LEF/DEF formats supported.

Users can specify whether running on CPU only or GPU only according their machines. In other words, GPU is not mandatory.

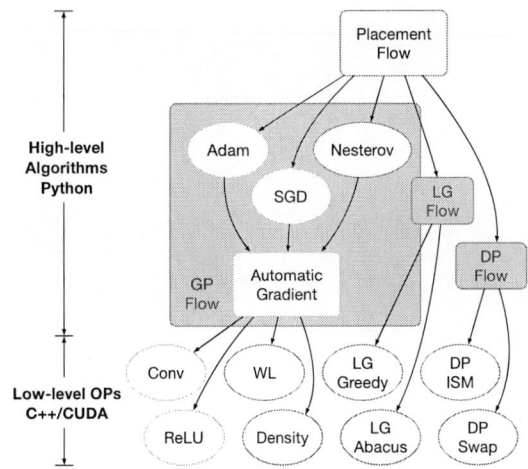

Fig. 1 Software architecture of DREAMPlace. The circles and rectangles with blue boundaries denote the portions we need to implement, while those with green boundaries are offered by `PyTorch`.

D. Global Placement Algorithm

The GP algorithm adopts the state-of-the-art family of placers ePlace/RePlAce, which model the layout as an electrostatic system [14], [15]. It uses the weighted average wirelength for the $WL(\cdot)$ term to approximate the nonsmooth half-perimeter wirelength (HPWL) [24], [25].

$$\mathrm{WA}_e = \frac{\sum_{i \in e} x_i e^{\frac{x_i}{\gamma}}}{\sum_{i \in e} e^{\frac{x_i}{\gamma}}} - \frac{\sum_{i \in e} x_i e^{-\frac{x_i}{\gamma}}}{\sum_{i \in e} e^{-\frac{x_i}{\gamma}}}, \tag{2}$$

where γ is a parameter to control the smoothness and accuracy of the approximation.

Its density term $D(\cdot)$ comes from the analogy to an electrostatic system, where cells are modeled as charges, density penalty is modeled as potential energy, and density gradient is modeled as the electric field. By solving Poisson's equation, the electric potential and field distribution can be computed from the charge density distribution.

$$\nabla \cdot \nabla \psi(x,y) = -\rho(x,y), \tag{3a}$$
$$\hat{\mathbf{n}} \cdot \nabla \psi(x,y) = \mathbf{0}, \quad (x,y) \in \partial R, \tag{3b}$$
$$\iint_R \rho(x,y) = \iint_R \psi(x,y) = 0, \tag{3c}$$

where R denotes the placement region, ∂R denotes the boundary to the region, $\hat{\mathbf{n}}$ denotes the outer normal vector of the placement region, ρ denotes the charge density, and ψ denotes the electric potential. To solve the constrained optimization in Equation (1), we relax the constraints into the objective with λ as the Lagrangian multiplier,

$$\min_{\mathbf{x},\mathbf{y}} \quad WL(\mathbf{x},\mathbf{y}) + \lambda D(\mathbf{x},\mathbf{y}), \tag{4}$$

where by gradually increasing the λ, the overlaps between cells can be eliminated.

E. Legalization Algorithm

In the LG step, we first perform macro legalization by ignoring the standard cells. Two macro legalization techniques

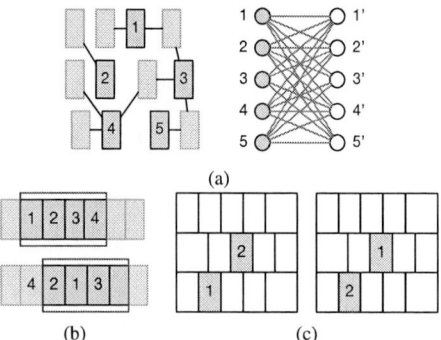

Fig. 2 Three DP techniques: (a) independent set matching; (b) local reordering; (c) global swap.

are developed. The first one is based on greedy spiral search of legal locations for movable macros. Once the overlaps between macros are resolved, we perform min-cost flow based refinement to minimize the displacement following the relative orders between macros [26]. After macro legalization, we fix the movable macros and legalize standard cells based on the Tetris-like approach in NTUplace3 [11], where cells are sorted from left to right and we find spaces to host cells one by one. Once a legal solution is found, we perform row-based Abacus refinement to shift cells for minimum displacement subjecting to the relative orders [23].

F. Detailed Placement Algorithm

In the DP step, three batch-based concurrent detailed placement techniques are proposed: independent set matching, local reordering, and global swap. Figure 2 illustrates the sequential version of the techniques [11], [19]. Independent set matching extracts a small set of cells that are independent (do not share common nets) to each other and perform permutation to the locations with bipartite matching. Local reordering slides a small window within each row and enumerates all the permutations for the best cost. Global swap performs pair-wise swapping of cells to improve the objective. As cells are connected, these techniques can only work on a small set of cells each time, assuming all other cells fixed. Thus, it is difficult to parallelize them.

In DREAMPlace, we modify the algorithms by incorporating batched execution to enable massive parallelization. For independent set matching, instead of extracting independent cells within a small window, we extract a maximal independent set from the entire netlist and partitions the set into many small subsets based on locality. Then, the bipartite matching problems of all the subsets can be solved independently. Be aware that the partitioning is not constrained by any physical window. For local reordering, instead of sliding a window within a row sequentially, we first construct a dependency graph for all rows, where two rows are dependent if any cell in a row is connected with another cell in the other row. With the dependency graph, we can extract independent sets of rows and perform the algorithm on these rows in parallel. For global swap, instead of finding a swap pair of cells one by one, we simultaneously search for swap candidates for a batch of cells and compute the best swapping

candidates without considering any conflict. When realizing the swap, we adopt sequential execution with a predetermined order and omit the swap candidates that have conflicts with previous swaps. In this way, we avoid possible data race in parallelization.

III. EXPERIMENTAL RESULTS

The framework was developed in `Python` with `PyTorch` for optimizers and API, and `C++`/`CUDA` for low-level operators. `OpenMP` was adopted to support multi-threading on CPU. All experimental results are collected from a Linux machine with two Intel 20-core Gold 6230 CPUs @ 2.10GHz (40 cores in total) and one NVIDIA RTX 2080Ti GPU. Due to page limit, we only show the experiments on ISPD 2005 contest benchmarks [27]. More results can be found in [28].

We compare with the state-of-the-art placer RePlAce [15] (use NTUplace3 [11] for legalization and detailed placement) in Table I. Results of 40 threads and GPU with different solvers are reported. Without any quality degradation, DREAMPlace can achieve $1.3\times$ speedup over RePlAce with 40 threads on CPU, and $14\times$ speedup on GPU. The major speedup comes from the GP step. Solvers like Nesterov, Adam and SGD with momentum are also compared, where Nesterov can provide similar solution quality to Adam, but around $2\times$ faster in GP. SGD momentum leads to 1.2% worse wirelength than the other two.

Table II shows the lines of code summarized from `cloc` [29] under Linux. As RePlAce only implemented GP and NTUplace3 is not open-source, we can only compare the lines of code for GP. We can see that DREAMPlace-GP only requires two-third of lines of code compared with RePlAce, while we support both CPU and GPU. It needs to mention that among the 20K lines in DREAMPlace-GP, more than 8K lines are for IO and database construction. Although we count these lines into DREAMPlace-GP for fair comparison with RePlAce, it indicates that the lines for core algorithms are actually even fewer.

IV. CONCLUSION

In this work, we present *DREAMPlace 2.0*, an open-source placement framework with multi-threading and GPU acceleration enabled by the deep learning toolkit `PyTorch`. Experimental results demonstrate that with GPU acceleration, more than $14\times$ speedup over the state-of-the-art RePlAce using 40 CPU threads can be achieved on the entire placement flow. With the decoupled algorithmic design and kernel operator development, we not only show reduced coding effort, but also easy adoption of new solvers from the deep learning community. Future work includes routability and timing consideration.

ACKNOWLEDGEMENT

This is support in part by NVIDIA, Inc.

REFERENCES

[1] S. Chowdhury, "Analytical approaches to the combinatorial optimization in linear placement problems," *IEEE TCAD*, vol. 8, no. 6, pp. 630–639, 1989.

[2] R. G. Michael and S. J. David, *Computers and Intractability: A Guide to the Theory of NP-Completeness.* W. H. Freeman & Co., 1979.

[3] N. Viswanathan, M. Pan, and C. Chu, "FastPlace 3.0: A fast multilevel quadratic placement algorithm with placement congestion control," in *Proc. ASPDAC*, 2007, pp. 135–140.

TABLE I Comparison on ISPD 2005 Contest Benchmarks with `float32`.

Design	#cells	#nets	RePlAce (40 threads)						DREAMPlace (40 threads)					
			HPWL	Runtime (s)					HPWL	Runtime (s)				
				GP	LG	DP	IO	Total		GP	LG	DP	IO	Total
adaptec1	211K	221K	73.23	89	4	20	5	118	73.19	67	0.6	9	4	84
adaptec2	254K	266K	81.86	175	6	25	7	214	82.11	167	0.5	14	5	190
adaptec3	451K	467K	193.32	342	19	44	13	419	193.22	338	1	18	9	372
adaptec4	495K	516K	175.17	411	18	51	14	494	174.08	274	1	23	9	313
bigblue1	278K	284K	89.86	156	3	25	8	192	89.39	92	0.4	11	5	113
bigblue2	535K	577K	138.14	354	21	72	14	461	136.86	350	8	15	10	390
bigblue3	1093K	1123K	304.94	971	39	110	28	1147	304.05	984	3	46	20	1067
bigblue4	2169K	2230K	743.60	2171	50	274	60	2555	744.11	1361	8	91	44	1536
ratio			1.002	79.478	9.485	9.965	1.018	14.099	1.000	66.023	1.041	3.832	0.725	10.743

Design	DREAMPlace (GPU-Nesterov)						DREAMPlace (GPU-Adam)			DREAMPlace (GPU-SGD)		
	HPWL	Runtime (s)					HPWL	Runtime (s)		HPWL	Runtime (s)	
		GP	LG	DP	IO	Total		GP	Total		GP	Total
adaptec1	73.18	2	0.5	3	8	15	73.04	4	17	73.75	4	17
adaptec2	82.11	3	0.5	4	8	18	82.51	5	19	83.65	5	20
adaptec3	193.33	4	1	6	12	30	190.76	8	32	197.55	8	33
adaptec4	174.08	5	2	6	13	31	172.61	9	35	175.98	10	36
bigblue1	89.36	3	0.4	3	9	18	90.04	6	21	89.57	5	20
bigblue2	136.97	4	8	6	13	38	136.96	9	42	137.84	10	44
bigblue3	304.40	10	3	10	23	59	302.99	25	75	313.28	19	69
bigblue4	744.01	21	8	15	48	122	742.54	57	157	745.03	51	151
ratio	1.000	1.000	1.000	1.000	1.000	1.000	0.998	2.153	1.156	1.012	2.114	1.151

TABLE II Comparison of lines of code using `cloc` [29].

	C/C++ Header	C++	CUDA	Python	Total
RePlAce	2174	27930	-	-	30104
DREAMPlace-GP	7320	7485	3555	3499	21859
DREAMPlace-LG	1814	863	-	336	3013
DREAMPlace-DP	2404	2390	7150	632	12576

[4] T. Lin, C. Chu, J. R. Shinnerl, I. Bustany, and I. Nedelchev, "POLAR: A high performance mixed-size wirelengh-driven placer with density constraints," *IEEE TCAD*, vol. 34, no. 3, pp. 447–459, 2015.

[5] T. Lin, C. Chu, and G. Wu, "POLAR 3.0: An ultrafast global placement engine," in *Proceedings of the IEEE/ACM International Conference on Computer-Aided Design*. IEEE Press, 2015, pp. 520–527.

[6] X. He, T. Huang, W.-K. Chow, J. Kuang, K.-C. Lam, W. Cai, and E. F. Y. Young, "Ripple 2.0: High quality routability-driven placement via global router integration," in *Proc. DAC*, 2013, pp. 152:1–152:6.

[7] M.-C. Kim, D.-J. Lee, and I. L. Markov, "SimPL: An effective placement algorithm," *IEEE TCAD*, vol. 31, no. 1, pp. 50–60, 2012.

[8] M.-C. Kim and I. L. Markov, "Complx: A competitive primal-dual lagrange optimization for global placement," in *DAC Design Automation Conference 2012*. IEEE, 2012, pp. 747–755.

[9] T. F. Chan, K. Sze, J. R. Shinnerl, and M. Xie, "mpl6: Enhanced multilevel mixed-size placement with congestion control," in *Modern Circuit Placement*. Springer, 2007, pp. 247–288.

[10] A. B. Kahng, S. Reda, and Q. Wang, "Aplace: A high quality, large-scale analytical placer," in *Modern Circuit Placement*. Springer, 2007, pp. 167–192.

[11] T.-C. Chen, Z.-W. Jiang, T.-C. Hsu, H.-C. Chen, and Y.-W. Chang, "Ntuplace3: An analytical placer for large-scale mixed-size designs with preplaced blocks and density constraints," *IEEE Transactions on Computer-Aided Design of Integrated Circuits and Systems*, vol. 27, no. 7, pp. 1228–1240, 2008.

[12] C.-C. Huang, H.-Y. Lee, B.-Q. Lin, S.-W. Yang, C.-H. Chang, S.-T. Chen, Y.-W. Chang, T.-C. Chen, and I. Bustany, "NTUplace4dr: a detailed-routing-driven placer for mixed-size circuit designs with technology and region constraints," *IEEE Transactions on Computer-Aided Design of Integrated Circuits and Systems*, vol. 37, no. 3, pp. 669–681, 2017.

[13] W. Zhu, Z. Huang, J. Chen, and Y.-W. Chang, "Analytical solution of poisson's equation and its application to vlsi global placement," in *2018 IEEE/ACM International Conference on Computer-Aided Design (ICCAD)*. IEEE, 2018, pp. 1–8.

[14] J. Lu, P. Chen, C.-C. Chang, L. Sha, D. J.-H. Huang, C.-C. Teng, and C.-K. Cheng, "eplace: Electrostatics-based placement using fast fourier transform and nesterov's method," *ACM Transactions on Design Automation of Electronic Systems (TODAES)*, vol. 20, no. 2, p. 17, 2015.

[15] C.-K. Cheng, A. B. Kahng, I. Kang, and L. Wang, "RePlAce: Advancing solution quality and routability validation in global placement," *IEEE Transactions on Computer-Aided Design of Integrated Circuits and Systems*, 2018.

[16] W. Li, M. Li, J. Wang, and D. Z. Pan, "UTPlaceF 3.0: A parallelization framework for modern fpga global placement," in *Proceedings of the 36th International Conference on Computer-Aided Design*. IEEE Press, 2017, pp. 922–928.

[17] J. Cong and Y. Zou, "Parallel multi-level analytical global placement on graphics processing units," in *2009 IEEE/ACM International Conference on Computer-Aided Design-Digest of Technical Papers*. IEEE, 2009, pp. 681–688.

[18] S. W. Hur and J. Lillis, "Mongrel: hybrid techniques for standard cell placement," in *Proc. ICCAD*, 2000, pp. 165–170.

[19] M. Pan, N. Viswanathan, and C. Chu, "An efficient and effective detailed placement algorithm," in *Proc. ICCAD*, 2005, pp. 48–55.

[20] W.-K. Chow, J. Kuang, X. He, W. Cai, and E. F. Y. Young, "Cell density-driven detailed placement with displacement constraint," in *Proc. ISPD*, 2014, pp. 3–10.

[21] S. Dhar and D. Z. Pan, "GDP: Gpu accelerated detailed placement," in *2018 IEEE High Performance extreme Computing Conference (HPEC)*. IEEE, 2018, pp. 1–7.

[22] D. P. Kingma and J. Ba, "Adam: A method for stochastic optimization," *CoRR*, 2014.

[23] P. Spindler, U. Schlichtmann, and F. M. Johannes, "Abacus: fast legalization of standard cell circuits with minimal movement," in *Proc. ISPD*, 2008, pp. 47–53.

[24] M.-K. Hsu, Y.-W. Chang, and V. Balabanov, "Tsv-aware analytical placement for 3d ic designs," in *Proceedings of the 48th Design Automation Conference*. ACM, 2011, pp. 664–669.

[25] M.-K. Hsu, V. Balabanov, and Y.-W. Chang, "Tsv-aware analytical placement for 3-d ic designs based on a novel weighted-average wirelength model," *IEEE Transactions on Computer-Aided Design of Integrated Circuits and Systems*, vol. 32, no. 4, pp. 497–509, 2013.

[26] X. Tang, R. Tian, and M. D. F. Wong, "Optimal redistribution of white space for wire length minimization," in *Proc. ASPDAC*, 2005, pp. 412–417.

[27] G.-J. Nam, C. J. Alpert, P. Villarrubia, B. Winter, and M. Yildiz, "The ispd2005 placement contest and benchmark suite," in *Proc. ISPD*. ACM, 2005, pp. 216–220.

[28] Y. Lin, S. Dhar, W. Li, H. Ren, B. Khailany, and D. Z. Pan, "Dreamplace: Deep learning toolkit-enabled gpu acceleration for modern vlsi placement," in *Proc. DAC*. Las Vegas, NV: ACM, June 2019.

[29] "Aldanial/cloc," https://github.com/AlDanial/cloc.

A REAL-TIME VISUAL TRACKING FOR UNMANNED AERIAL VEHICLES WITH DYNAMIC WINDOW

Jia Zhang[1*], Tianrun Chen[1], and Zhiguo Shi[1]

[1]College of Information Science & Electronic Engineering,
Zhejiang University, Hangzhou 310027, China
*Corresponding Author's Email: jiazhang@zju.edu.cn

ABSTRACT

Visual object tracking is one of the basic topic in computer vision that has been studied for decades. As deep neural network has been applied in trackers, their tracking accuracy has been greatly improved and achieved state-of-the-art performance in recent years. Due to the special viewing-angle of UAV, the video sequences captured are high resolution and large in size. So, when applying these trackers to UAVs for tracking a target, the processing speed is hard to reach real-time requirement. In this paper, we propose a visual object tracking framework, DyWinSiam, which can speed up the tracking process significantly compared with state-of-the-art deep neural network trackers, yet it still achieve compatible accuracy.

Keywords — visual object tracking; UAV; real-time

INTRODUCTION
Visual Object Tracking

Visual object tracking is one of the basic topic in computer vision that has been studied for decades. In this research, we were focusing on single object tracking task, which is aiming at the prediction of the target location in the successive frames of video sequence, given its location in the first frame of the video sequence. [1]

As deep neural network has been applied in trackers, tracking accuracy has been greatly improved and these trackers have achieved state-of-the-art performance in recent years. Among these trackers, there are mainly two types of tracking strategies: tracking-by-detection method [2], and template-matching method [3]. Tracking-by-detection strategies distinguish target from the background by training a classifier online, which is first learned from the bounding box in first frame and fine-tuned from the newly predicted bounding box in the subsequent frames. These methods are usually quite slow, since the online fine-tune process is computational expensive. The template-matching methods extract feature of target from the first frame, and match it with frame image to locate the target. Since these models are not updating the parameters of the neural network online, they run faster comparing with tracking-by-detection models. However, the performance of template-matching methods is not as good as tracking-by-detection methods, due to the lack of template update process, especially in the situation where the appearance of the target varies significantly through the video sequences.

Tracking for UAV

One of the major applications for visual object tracking is to track target using an UAV. Since the UAV is viewing the target and its surroundings from high above, the video captured by an UAV have particular characteristics, which usually have high-resolution frame images with a large size. When applying visual object tracking models to UAV captured video sequences for tracking target, the requirement of real-time is not easy to be satisfied, due to the high demand of processing time for high-resolution images.

Acceleration and Efficiency

Although the application of deep neural networks has improved the accuracy of visual object tracking greatly, these trackers also pay the price for getting slow and less efficient. Recent years, researchers in computer vision area and circuit design area have devoted themselves to the research of accelerating deep neural network, either by redesigning convolution calculation [4], or by accelerating computation process from hardware level [5-6].

The progress in deep neural network acceleration and efficiency improvement is especially beneficial in the situation when computation capacity is limited, for example, computing using internet of things (IoT) devices e.g. UAVs [7-8]. When apply visual object tracker on UAVs, real-time processing with high tracking speed is crucial for tracking algorithm design.

In this paper, we propose a visual object tracking framework, DyWinSiam, which can track a target from UAV captured video sequences in real time, according to our experimental results. Meanwhile, it still reaches a compatible accuracy compared with state-of-the-art trackers.

METHODOLOGY
Framework Overview

Our proposed tracker, DyWinSiam, consists of two main stages: feature extraction stage and region proposal stage, as illustrated in Fig. 1.

As introduced earlier, in the task of single object tracking, target location is given in the first frame of the video sequence, which is the only information about the target that we could trust. So, in feature extraction stage, feature map of the target patch is extracted using a convolutional neural network (CNN) from the template frame, which is the first frame in the video sequence. Later on, in the successive frames, called detection frame in the figure, the region of interest (ROI) is selected through the dynamic window selection strategy, which will be introduced in detail in the next subsection. From ROI, feature map of detection frame is extracted using the same CNN with the same parameters.

Then, in the second stage, the region proposal subnetwork is adopted from the method in SiamRPN [9] to predict the target bounding box by comparing the two feature maps from template frame and detection frame.

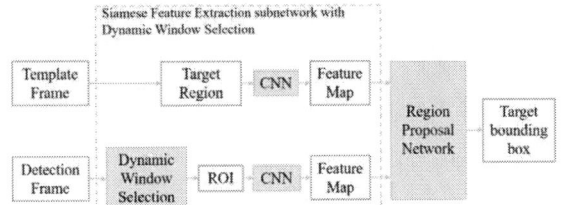

Figure 1: DyWinSiam architecture. The left part shows the pipeline for extracting feature maps, which is the feature extraction stage. The right part indicates the region proposal stage for predicting the target bounding box.

Dynamic Window Selection Strategy

In this work, we aim to build an efficient and fast tracker for UAVs. When walking through the video sequences captured by UAVs, we noticed that, due to the particular viewing angle of the video capture devices carried by an UAV, the frame pictures from these video sequences has the following special characteristics in common.
• images captured are high resolution in large size;
• viewing area is much larger in size compared to the size of the tracking target;
• target variation in size and appearance is limited in two successive frames.

Based on these characteristics as our assumption, we design a dynamic window selection strategy, which is the crucial part for speeding up our tracker.

Figure 2 illustrates the processing pipeline for dynamic window selection strategy in detail. We adopted bounding box center from previous frame as the selection window center. Let p represents the size of ROI, which is a preset number that needs to be carefully selected. For a large p, the speed increase would be limited, while for a small p, the accuracy might drop in a large amount. When p is set, the selection window size is determined dynamically by comparing p with bounding box size from previous frame. After that, we crop the detection frame with the selection window center and size, and resize the cropped image to size p * p. Then the image of size p * p is the ROI, which is the output of dynamic window selection process.

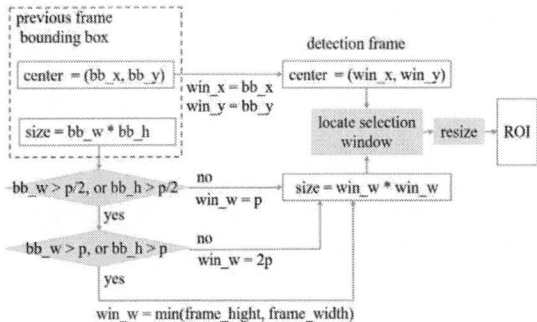

Figure 2: processing pipeline illustrating dynamic window selection strategy. Size and location of selection window is determined by those of the bounding box from previous frame dynamically. ROI is obtained by cropping and resizing the selection window from detection frame.

PERFORMANCE EVALUATION
Datasets

In our experiment, by excluding the low-resolution video sequences and simulated video sequences, we selected 61 high resolution video sequences from UAV123 dataset for testing the performance of our tracker. Figure 2 shows a collection of example frames from these sequences, and red boxes indicate the location of tracking target from ground-truth data.

These video sequences are captured at 30/60/96 fps, and the size of these frame images is 1280 pixels by 720 pixels. The lengths of sequences range from 109 frames to 3085 frames, with a mean of 839.52 frames. The tracking targets include person, boat, bike, car, and etc.

Figure 3: A collection of example frames in UAV captured video sequences for evaluating tracking performance in our experiments. Red boxes indicate the target location according to ground-truth data.

Experiment Setup

In our experiment, we test our tracker using a NVIDIA GeForce GTX 1080Ti. The ROI size p is set to be 200 in our tracker. Performance is evaluated against two state-of-the-art trackers, SiamRPN and DaSiamRPN [10] in the same environment using the same datasets.

Experimental Results

We use success rate to evaluate tracking accuracy, which measures the percentage of successfully tracked frames where the overlap exceeds a certain threshold. Here the overlap is calculated by the intersection over union (IoU) of ground-truth bounding box and predicted box. By setting the overlap threshold from low to high, the success rate is dropping from high to low, as shown in Fig.4.

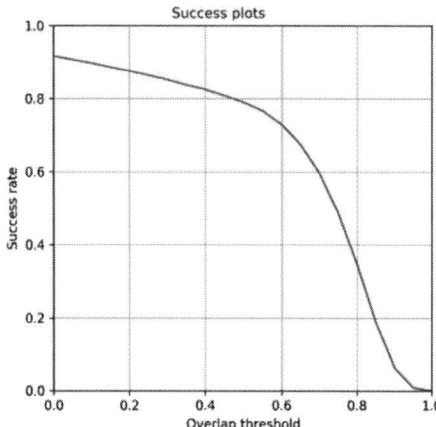

Figure 4: Success plot of DyWinSiam. Horizontal axis is the threshold of overlap, while vertical axis is success rate under certain overlap threshold.

Performance comparison for DyWinSiam, SiamRPN and DaSiamRPN is listed in Table 1. Accuracy is the success rate at the threshold of 0.5. Our proposed tracker, speeds up by 3.92/1.78 times compared with SiamRPN and DaSiamRPN, respectively. The accuracy is dropped by 0.49% compared with DaSiamRPN, and increased by 5.30% compared with SiamRPN.

TABLE I.   PERFORMANCE COMPARISON FOR DYWINSIAM WITH CURRENT STATE-OF-THE-ART TRACKERS

Tracker	performance	
	Speed (fps)	*Accuracy*
SiamRPN	25.23	75.05%
DaSiamRPN	55.44	79.42%
DyWinSiam	**98.78**	**79.03%**

CONCLUSIONS

To design a real-time visual object tracking for UAV captured video sequences, we proposed a high speed tracker, DyWinSiam. In its framework, a dynamic window selection strategy is designed, which significantly speeds up the tracking process. Experimental results show that our proposed tracker can run at the speed of 98.78 fps, which is 3.92/1.78 times compared with SiamRPN and DaSiamRPN, respectively, yet it still reaches compatible accuracy.

REFERENCES

[1] M. Kristan, et. al., "The Visual Object Tracking VOT2018 Challenge Results," in *Proc. of European Conference on Computer Vision Workshops*, pp.3-53, 2018.

[2] H. Nam, B. Han, "Learning Multi-Domain Convolutional Neural Networks for Visual Tracking," in *Proc. of IEEE Conference on Computer Vision and Pattern Recognition*, pp.4293-4302, 2016.

[3] L. Bertinetto, J. Valmadre, J. Henriques, et. al., "Fully Convolutional Siamese Networks for Object Tracking," In *Proc. of European Conference on Computer Vision Workshops*, pp.850-865, 2016.

[4] A. Howard, M. Zhu, B. Chen, et. al., "MobileNets: Efficient Convolutional Neural Networks for Mobile Vision Applications," In *arXiv preprint arXiv: 1704.04861*, 2017.

[5] C. Zhuo, K. Unda, Y. Shi, and W.-K. Shih, "From Layout to System: Early Stage Power Delivery and Architecture Co-Exploration," IEEE Transactions on Computer-Aided Design of Integrated Circuits and Systems, vol. 38, issue 7, pp. 1291-1304, 2019.

[6] Z. Liu, S. Luo, X. Xu, Y. Shi, and C. Zhuo, "A Multi-Level Optimization Framework for FPGA-Based Cellular Neural Network Implementation," ACM Journal on Emerging Technologies in Computing Systems, Vol. 14, Issue 4, pp. 47:1-47:17, Dec. 2018.

[7] C. Zhuo, S. Luo, H. Gan, J. Hu, and Z. Shi, "Noise-Aware DVFS for Efficient Transitions on Battery-Powered IoT Devices," IEEE Transactions on Computer-Aided Design of Integrated Circuits and Systems, DoI:10.1109/TCAD.2019.2917844, 2019.

[8] Z. Liu, C. Zhuo, X. Xu, "Efficient Segmentation Method Using Quantised and Non-Linear CeNN for Breast Tumour Classification," Electronics Letters, volume 54, Issue 12, p.737-738, June 2018.

[9] B. Li, J. Yan, W. Wu, Z. Zhu, X. Hu, "High Performance Visual Tracking with Siamese Region Proposal Network," In *Proc. of IEEE Conference on Computer Vision and Pattern Recognition*, pp.8971-8980, 2018.

[10] Z. Zhu, Q. Wang, L. Bo, W. Wu, J. Yan, and W. Hu, "Distractor-aware Siamese Networks for Visual Object Tracking," in *Proc. of European Conference on Computer Vision*, pp.103-119, 2018

SCALABLE MULTI-SESSION TCP OFFLOAD ENGINE FOR LATENCY-SENSITIVE APPLICATIONS

Jingbo Gao, Wenbo Yin, Wai-Shing Luk, and Lingli Wang*

School of Microelectronics, Fudan University, Shanghai 201203, China

*Corresponding Author's Email: luk@fudan.edu.cn

ABSTRACT

Latency-sensitive applications, such as Network File System (NFS) and High-Frequency Trading (HFT), demand ultra-low latency in network communications. These applications usually need more than one TCP session to guarantee Quality of Service (QoS) in case of communication interruption. This paper introduces a scalable multi-session TCP Offload Engine (TOE) for latency-sensitive applications which reduces the delay using the kernel bypass approach. The input-output receiving latency of a 48-byte-payload packet is 262.4 *ns*, and the sending latency of the same size packet is 179.3 *ns*. The latencies grow linearly with the amount of data at the rate of 12.8 *ns* per 8 bytes. The latencies are irrelevant to the TCP session number, which shows the scalability of our implementation.

INTRODUCTION

As the most commonly used transport layer protocol in Ethernet, TCP sessions consume around one-third time in the network communication of latency-sensitive applications [1]. These delays can be attributed to the features of TCP, such as connection establishment handshake, connection termination handshake, and checksum calculation. The software implementation of these features requires huge CPU usage and results in unsatisfying network latency. Offloading the management of TCP sessions to dedicated hardware, which is called TCP Offload Engine (TOE), solves the problem of excessive CPU overhead and makes progress in optimizing network latency. This paper introduces a TOE implementation that balances both scalability and performance.

RELATED WORK

TOEs have been implemented on embedded systems using the hardware-software co-design approach [2][3]. These implementations can only support several Ethernet protocols without complex control logic, such as Address Resolution Protocol (ARP), Internet Control Message Protocol (ICMP) and User Defined Protocol (UDP). A fully implemented TOE for Gigabit Ethernet is proposed [4], but the performance is limited by the bottleneck of low-speed Ethernet. An ultra-low latency TOE designed for 10-Gigabit Ethernet is presented [5]. However, it can only support a single TCP session. High-Level Synthesis (HLS) is used to implement a scalable multi-session TOE [6]. Although HLS brings high flexibility in hardware

design, the performance is still inferior to the conventional Hardware Description Language (HDL) method. The sending latency and the receiving latency of our HDL implementation are merely 13.8% and 18.2% of the TOE proposed in [6] respectively.

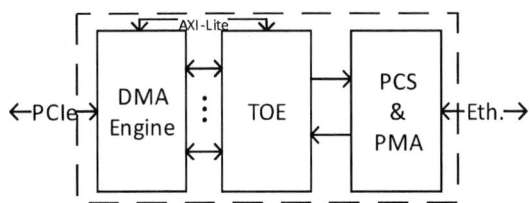

Figure 1: System architecture

SYSTEM ARCHITECTURE

Figure 1 illustrates the system architecture of our design. The Direct Memory Access (DMA) engine handles data transfer and control signals between the system and the server via a PCIe interface. DMA channel pairs for data transfer between the DMA engine and the TOE are configured as the streaming mode. A separate DMA channel is responsible for configuring the TOE via AXI-Lite. The Physical Coding Sublayer (PCS) and Physical Media Attachment (PMA) interface encodes and decodes the packets transmitted, which includes both the Media Access Control (MAC) and the physical layer in Ethernet.

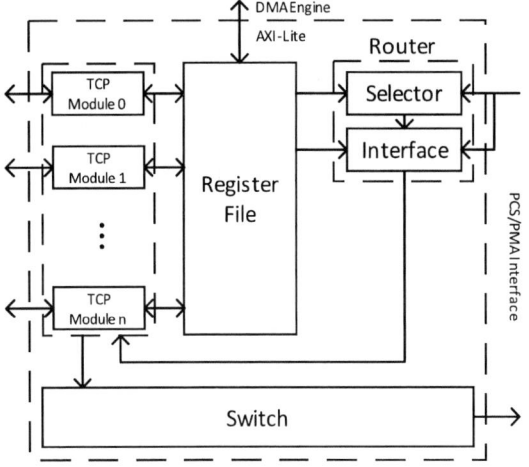

Figure 2: The proposed TOE architecture

The proposed TOE architecture is shown in Figure 2. Each TCP module manages a TCP session. All control and session state registers are in the register file. They can be accessed or modified by the DMA engine and the TCP modules. These registers can also be read by the router. The switch gathers the outputs of the TCP modules and sends the data to the PCS/PMA interface. The router connects all the TCP modules. It routes the received packets to the corresponding TCP module or drops them. The following sections provide the detail design of the proposed TOE.

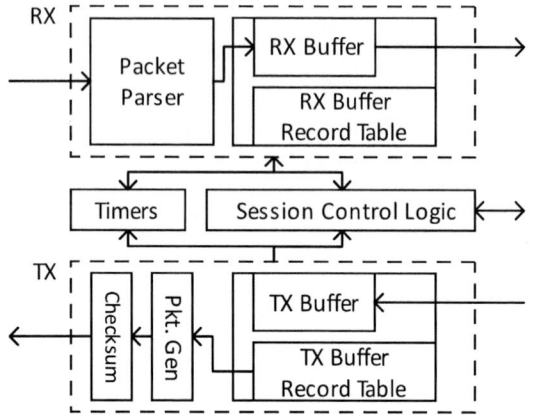

Figure 3: TCP module architecture

TCP MODULE ARCHITECTURE

The TCP module architecture is similar to the architecture proposed in [5] as shown in Figure 3. The RX engine and the TX engine are similar in architecture. The Buffer Record Table (BRT) is designed to record the information of the data entries stored in the data buffer. BRT provides *search*, *insert* and *delete* operations, and can be applied in both RX and TX engines for different purposes.

RX Engine

Received packets are parsed by the packet parser, and the extracted data is stored in the RX buffer after duplication check, sequence number validation and write address search in the RX BRT. Since the sequence number validation is limited by the receive window in the RX BRT, packets outside the receive window are discarded. Entries are removed by the RX BRT after they are read by the DMA engine.

TX Engine

New data from the DMA engine is stored in the TX buffer. A local counter in the TX BRT provides the data length to be transmitted after the data is written into the TX buffer, which brings an unavoidable latency proportional to the data length. Acknowledged entries are

removed from the TX BRT.

Both TCP timeout retransmission and TCP fast retransmission are supported. When packet retransmission happens, the congestion window is reduced to Maximum Segment Size (MSS), which can be set in the register file, until all the packets in the congestion window are acknowledged.

Session Control Logic and Timers

Session control logic manages and migrates the session state of the TCP modules. All TCP session state migrations are supported in the control logic. It always starts with an active open setup or a passive open setup, which corresponds to the client mode and the server mode respectively. Standard TCP timers are implemented.

REGISTER FILE

Registers of control signals, session information, MAC address and TCP session quad-tuples (source IP address, source port number, destination IP address and destination port number) are stored in the register file. The register file provides an AXI-Lite interface for the DMA engine to read and write the registers. Users can open, close and configure the TOE by editing some of the registers via the TOE driver.

ROUTER

Packets received from the PCS/PMA interface enter the selector and the TOE interface simultaneously as shown in Figure 2.

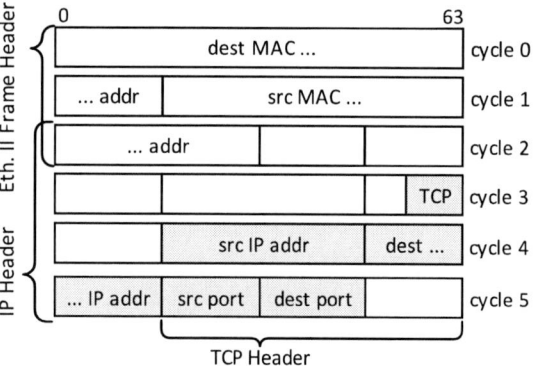

Figure 4: Ethernet stream format

The selector matches the session quad-tuples and the packet header in six clock cycles as shown in Figure 4. The transport layer protocol type is checked in the meantime. The interface routes the packet following the instruction given by the selector if the packet quad-tuple matches one of them in the register file. The new packet is dropped if there is no match.

EVALUATION

The implementation is evaluated in regard to throughput, latency and resource utilization. We implement the design on Xilinx Kintex UltraScale FPGA KCU105 evaluation board using the HDL approach. The implementation utilizes a 10-Gbps network interface, which corresponds to two TCP modules. The clock frequency is 156.25 MHz due to the 10 Gbps bandwidth and the 64-bit data width.

Throughput

The theoretical throughput of the implementation can be calculated by the following equation:

$$Throughput = 10 \ Gbps \times \frac{MSS}{MSS + Header \ Length} \qquad (1)$$

As shown in (1), a large MSS can result in high throughput. The typical number of MSS in Ethernet is 1460 bytes and the header length is 54 bytes. Therefore, the theoretical maximum throughput is 9.64 Gbps. To overcome the bottleneck of throughput, MSS can be configured larger than 1460 bytes thus jumbo frame is supported.

TABLE I. COMPARISON WITH OTHER MULTI-SESSION TOES

Design	Data Length/byte(s)	TX Latency/ns	RX Latency/ns
Sider [6]	1	838.4	1088
Easics [7]	N/A	496	480
Ours	1	115.3	198.4
	48	179.3	262.4

Latency

Ultra-low latency is the primary target in our TOE implementation. The measured input-output receiving latency with a 48-byte payload and a 54-byte header is 262.4 *ns*. And the sending latency of the same size packet is 179.3 *ns* because of the data length counter and checksum calculation. The latency grows linearly with the amount of data at the rate of 12.8 *ns* per 8 bytes, which is two clock cycles of the 156.25 MHz clock rate. TABLE I lists the latency comparison with other TOEs.

Utilization

TABLE II. RESOURCE UTILIZATION ON KCU105 (2 UNITS)

Resource	DMA	TOE		PCS/PMA	Total
		TCP Module	Others		
LUT	42684	17466	58	10987	71195
FF	38089	11464	8547	9980	68080
BRAM	73	75	69	0	217

TABLE II lists the resource utilization of our two-TCP-module prototype. The utilization rate of Flip-Flops and LUTs is 14% and 29.4% of the overall available resource. A TCP module utilizes 8733 LUTs and 5732 Flip-Flops. Since each TCP module manages a single TCP session, we can further increase the number of the TCP module and support more TCP sessions.

CONCLUSION

In this work, we have presented a scalable TOE for latency-sensitive applications based on the single-session TCP module in [5]. The implementation achieves 179.3 *ns* input-output sending latency and 262.4 *ns* input-output receiving latency for 48-byte payload and a 54-byte header. These latency numbers are independent of TCP session numbers and grow linearly with the amount of data at the rate of 12.8 *ns* per 8 bytes.

ACKNOWLEDGEMENT

This work is supported by the National Natural Science Foundation of China under Grant 61971143.

REFERENCES

[1] J. W. Lockwood et al., "A low-latency library in FPGA hardware for high-frequency trading (HFT)," in *IEEE Symposium on High-Performance Interconnects*, Santa Clara, CA, 2012, pp. 9-16.

[2] Z. Wu and H. Chen, "Design and implementation of TCP/IP offload engine system over gigabit ethernet," in *International Conference on Computer Communications and Networks (ICCCN)*, Arlington, VA, 2006, pp. 245-250.

[3] S. Chung et al, "Design and implementation of the high speed TCP/IP offload engine," in *International Symposium on Communications and Information Technologies (ISCIT)*, Sydney, NSW, 2007, pp. 574-579.

[4] T. Uchida, "Hardware-based TCP processor for gigabit ethernet," in *IEEE Transactions on Nuclear Science*, vol. 55, no. 3, pp. 1631-1637, June 2008.

[5] L. Ding, P. Kang, W. Yin and L. Wang, "Hardware TCP offload engine based on 10-Gbps ethernet for low-latency network communication," in *International Conference on Field-Programmable Technology (FPT)*, Xi'an, 2016, pp. 269-272.

[6] D. Sidler et al., "Scalable 10Gbps TCP/IP stack architecture for reconfigurable hardware," in *IEEE International Symposium on Field-Programmable Custom Computing Machines (FCCM)*, Vancouver, BC, 2015, pp. 36-43.

[7] Easics, "TCP Offload Engine". [Online]. Available: https://www.easics.com/products/tcp-offload-engine.

Advanced MOSFET Model Based on Artificial Neural Network

JH. Wei[1], W. Mao[1], H. Fang[2], Z. Zhang[2], JX. Zhang[2], BJ. Lan[2] and J. Wan[1]*

[1]State key lab of ASIC and System, School of Information Science and Engineering, Fudan University, Shanghai, China

[2]Suzhou Foohu Technology Co., Ltd.

*Corresponding Author's Email: jingwan@fudan.edu.cn

ABSTRACT

In this work, we develop a novel MOSFET model for circuit simulation purpose. Instead of using traditional physics-driven model, such as BSIM model, our work uses ANN to model the electrical behavior of the transistor. With unique pre and post-processing procedures, the ANN is trained to model the drain current precisely under various applied voltage, device size and temperature. The model is further successfully implemented in SPICE through Verilog-A language. Both n-type and p-type MOSFETs show good fitting between the BSIM and ANN models. Eventually, an inverter and ring oscillator based on ANN model are demonstrated with static and transient simulations, showing good agreement with the results from BSIM model.

INTRODUCTION

The integrated circuit (IC) has been developing rapidly in last decades following the Moore's law [1]. It has been playing important role in modern information technology fields. As the Moore's law proceeds, the transistor in the IC continues scaling down. In order to boost the transistor performance in small size, many new technologies have been used, such as high-k/metal gate, FinFET, FD-SOI, nanowire and nanosheet. As the structure of the MOSFET evolves, modeling of the MOSFET with new elements are constantly needed. Initial level 1 model of the MOSFET based on an ideal device physics has long been outdated. More precise BSIM models, such as BISM 3 and BSIM 4, have been developed for advanced technology nodes [2]. There are also special models developed for FinFET and FD-SOI technologies [3]. Besides, many novel semiconductor devices emerge recent years as complement or replacement to conventional MOSFET, such as tunneling field effect transistor (TFET) and negative capacitance FET (NC-FET) [4]. In order to be used in SPICE simulation, compact models are needed for these devices. (such as NC-FET, TFET) without studying their complex physical characteristics [5].

Conventional model is mainly based on device physics. However, as the MOSFET structure becomes more and more complicated, developing models for advanced MOSFET is difficult and time consuming. Besides, novel devices with totally different device physics need enormous endeavor and good understanding on its physics. Thus, models based purely on data and involves no device physics have been attracting lots of interests[6], mainly due to their precision and ease of development[7]. Artificial neural network, which is used in many different fields[8][9], has also been used in MOSFET modeling, which shows good precision [10][11]. In this work, we develop a data-driven MOSFET model based on ANN. With pre and post processing on the training data, the model precisely models the relation between the output drain current and six variants, including applied voltage on four terminals, gate length and width and the operating temperature. The model is further plugged in Cadence through the Verilog-A code. The simulation results using conventional BSIM model and ANN model are compared on MOSFET, inverter and ring oscillator, all shows very good agreement.

DEVELOPMENT OF ANN MODEL

Figure 1 shows the procedure of developing the ANN model and implementation of it in Cadence for circuit simulation. A complete six parameters are used as input, including gate voltage (V_G), drain voltage (V_D), bulk voltage (V_B), gate length (L_G), width (W_G) and temperature. The input parameters are pre-processed before they are fed into ANN whose output is post-processed in order to obtain the drain current (I_D). The training data, which includes I_D values under various six parameters, is generated from SPICE simulation. The ANN has three layers, each of which has 32 neurons. The training of the model takes place in python.

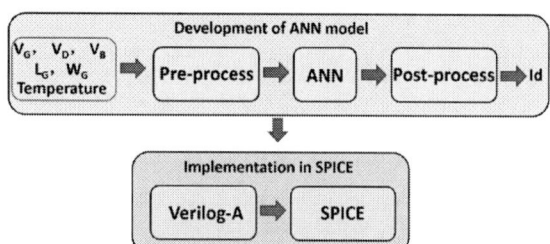

Figure1. A schematic view showing the development of ANN model and its implementation in Cadence for circuit simulation

After trained, the ANN model is translated into Verilog-A language which can be executed in Cadence, as shown schematically in Fig. 1. For simplicity, the BSIM charge model is preserved in AC simulation, and only I_D is replaced by the ANN model. Both n-type and p-type transistor models are obtained in such similar way.

Figure 2 shows the simulation results of a n-type MOSFET with conventional BSIM model and ANN model under a temperature of 27°C. The simulated MOSFET has $L_G = 1\mu m$ and $W_G = 1\mu m$. Fig. 2(a) compares the transfer (I_D-V_G) characteristics of the MOSFET with three different models under a fixed V_D=1V, BSIM 3, ANN without pre/post-process and ANN with pre/post process. It is clearly observed that the ANN model with the pre/post process fits the BSIM 3 model very well in both logarithm and linear scales. However, without pre/post process, the ANN model fails to fit the current under low V_G when the transistor is in subthreshold region. Thus, the pre/post process is mandatory here to have good fitting in large dynamic range of I_D.

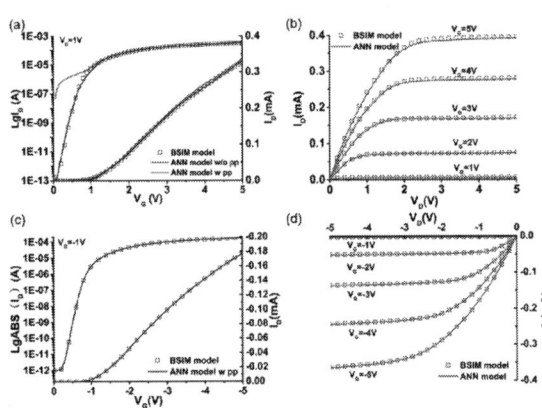

Figure 2. (a) Comparison of I_D-V_G characteristics on nFET. (b) I_D-V_D characteristics between ANN model and BSIM model on a nFET. (c) Comparison of I_D-V_G characteristics on pFET. (d) I_D-V_D characteristics between ANN model and BSIM model on a pFET.

Figure 2(b) shows the output characteristics (I_D-V_D) of the n-type transistors under various V_G values. The ANN model captures the trend very well in both linear and saturation regions with neglected error. For example, when $V_{GS} = 3V$, the saturated I_D value of ANN model is 0.168mA, which is very close to that obtained by BSIM 3

model (0.17mA). Figure 2 (c) and (d) compare simulation results on the pFET between BSIM and ANN models, which also show good fitting. This fully demonstrates the precisely fitting capability of our developed ANN model in the transistor level.

CIRCUIT SIMULATION BASED ON ANN MODEL

The ANN models developed for n-type and p-type MOSFETs are further used for circuit simulation in SPICE. Both inverter and ring oscillator are simulated to demonstrate the application of ANN model in circuit simulation. Figure 3(a) shows the simulated inverter circuit with a supply voltage of 5V. The pull-up transistor (pFET) has $L_G = 1\mu m$ and $W_G = 2.5\mu m$, and the pull-down transistor (nFET) has $L_G = 1\mu m$ and $W_G = 1\mu m$. Figure 3(b) shows the simulated voltage transfer characteristics of this inverter. The simulation results with BSIM model and ANN model show exactly the same voltage transfer characteristics, as the input voltage (Vin) sweeps from 0V to 5V. The switching point is also precisely predicted by the ANN model.

The same nFET and pFET are used to form a ring oscillator which consists of three inverters. Loading capacitors with capacitance of 1pF are placed at the output port of each inverter. The transient simulation is performed to obtain the output waveform of the ring

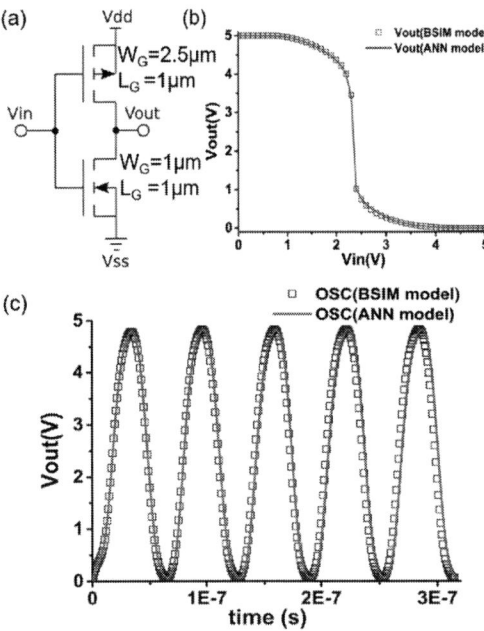

Figure 3. (a) Schematic view of the simulated inverter

structure. (b) Comparison of voltage transfer characteristics of the inverter between the ANN model and the BSIM model. (c)Comparison of ring oscillator output characteristics between ANN model and transistor model.

oscillator. Figure 3(c) compares the output waveform of the ring oscillator with BSIM and ANN models. The oscillating curve with ANN model fits the BSIM model very well. The oscillating frequency of ANN model is 15.810MHZ, while that of transistor model is 15.873MHZ. The relative error is only 0.39%.

CONCLUSION

We have successfully developed an ANN-based MOSFET model. With unique pre and post process procedures , the transfer and output characteristics of both n-type and p-type MOSFETs show good fitting between BSIM and ANN models. The ANN model is further implemented in Verilog-A and used in circuit simulation. The simulated inverter and ring oscillator with ANN model show the same voltage transfer characteristics and output waveform as those with the BSIM model. Since the developed ANN model is purely based on data without involving any device physics, it has not only high fitting precision but also great potentiality in the modeling of advanced MOSFET and novel semiconductor devices.

ACKNOWLEDGEMENTS

The work at Fudan University is sponsored by National Natural Science Foundation of China (61904032).

REFERENCES

[1] D. Root, "Future device modeling trends, " *Microwave Magazine, IEEE*, vol. 13, no. 7, pp. 45-59, nov.-dec. 2012.

[2] Y. Cheng, M.-C. Jeng, Z. Liu, J. Huang, M. Chan, K. Chen, P. K. Ko, C. Hu, " A physical and scalable I-V model in BSIM3v3 for analog/ digital circuit simulation ", *IEEE Trans. Electron Devices*, vol. 44, pp. 277-287, Feb. 1997.

[3] J. Song, Y. Yuan, B. Yu, W. Xiong , Y. Taur, "Compact modeling of experimental n- and p-channel FinFets ", *IEEE Transactions on Electron Devices*, vol. 57, no. 6, pp. 1369-1374, Jun 2010.

[4] Yunpeng Dong, Lining Zhang, Xiangbin Li, Xinnan Lin, Mansun Chan, "A Compact Model for Double-Gate Heterojunction Tunnel FETs", *Electron Devices IEEE Transactions on*, vol. 63, no. 11, pp. 4506-4513, 2016.

[5] C. Enz , Y. Cheng, "MOS transistor modeling for RF IC design", *Solid-State Circuits IEEE Journal of*, vol. 35, no. 2, pp. 186-201, 2000.

[6] Lining Zhang, Mansun Chan, "Artificial neural network design for compact modeling of generic transistors", *Journal of Computational Electronics*, 2017.

[7] Yazi Cao, Xi Chen, and Gaofeng Wang, " Dynamic Behavioral Modeling of Nonlinear Microwave Devices Using real-Time Recurrent neural Netwrok," *IEEE Transactions on Electron Devices*, Vol. 56, No. 5, pp. 1020-1026, May 2009.

[8] M. H. Weatherspoon, H. A. Martinez, D. Langoni, S. Y. Foo, "Small-signal modeling of microwave MESFETs using RBF-ANNs", *IEEE Trans. Instrum. Meas.*, vol. 56, no. 5, pp. 2067-2072, Oct. 2007.

[9] J. Xu, D. Gunyan, M. Iwamoto, A. Cognata, D. Root, "Measurement-based non-quasi-static large-signal FET model using artificial neural networks", *IEEE MTT-S Int. Microw. Symp. Dig.*, pp. 469-472, 2006.

[10] Litovski, J.I. Radjenovic, Z.M. Mrcarica, S.L. Milenkovic, "MOS transistor modelling using neural network", *Electron. Lett.*, vol. 28, no. 18, pp. 1766-1768, 1992.

[11] Youngseo Ko, Patrick Roblin, Andrés Zarate-De Landa, J. Apolinar Reynoso-Hernandez, Dan Nobbe, Chris Olson, Francisco Javier Martinez, "Artificial Neural Network Model of SOS-MOSFETs Based on Dynamic Large-Signal Measurements", *IEEE T Transactions On Microwave Theory and Techniques*, vol. 62, no. 3, marchal 2014.

A HYBRID DOMAIN FRAMEWORK FOR PREDISTORTER MODELING AND ADAPTIVE DIGITAL PREDISTORTION REALIZATION

Hairui Wang, Junyao Wang, and Bo Wang[*]

The Key Lab of IMS, School of ECE, Peking University Shenzhen Graduate School, Shenzhen, China
*Corresponding Author's Email: wangbo@pkusz.edu.cn

ABSTRACT

In this paper, we propose a hybrid domain framework to model the predistorter (PD) of a power amplifier (PA). It is available for obtaining the PD not only through a PA model but also a PA circuit on ADS. Simulation results show that with predistortion, the NMSE of the linearity can reach -50 dB and the ACPR can reduce around 17dB. What's more, a processing unit is designed to work with the PD, so that the PD's coefficients can adjust to the changes of the PA adaptively, making the PA perform great linear in the transmitter system. The proposed framework is a helpful guidance for the design of circuits.

INTRODUCTION

With the rapid development of modern wireless communication technology, it leaves great challenges to the communication system. As the indispensable part of a transceiver system, radio frequency (RF) power amplifier (PA) suffers from the tradeoff among linearity, efficiency and so on long before, whose working status has a great impact on the communication quality and energy consumption of the whole system. Except that improving the design of the PA itself, there are some other ways through system assistance to get good performance, like digital predistortion (DPD), which is a high cost performance method to linearize a PA and has been researched a lot [1].

As we know, the research on PA linearization is usually based on the PA model obtained through physical PA test data, and the simulations of DPD are usually carried out on MATLAB or ADS separately [2]. In this paper, we establish a hybrid domain framework so that it's easy to obtain the predistorter (PD) of a PA circuit designed on ADS. What's more, by means of ADS Ptolemy MATLAB Cosimulation, we combine the two

tools with each other, thus the constructed system can cooperate with the presented PD and processing unit to realize an adaptive DPD.

THE PROPOSED QAM SYSTEM

At the beginning, we construct a hybrid domain transmitter simulation system on ADS platform applying the quadrature amplitude modulation (QAM) signal, as shown in Fig. 1. A bit sequence is generated randomly at first and then mapped to the complex plane. Then the complex signal is transformed into two parts, in-phase (I) and quadrature (Q). Filtered by the raised cosine filter respectively, these two signals are up converted by the QAM module. Finally, the obtained signal is applied to the PA. To observe the results, we place the Timedsink and SpectrumAnalyzer at the input and output of the PA.

It's noting that the PA module in the simulation system is commonly a model identified from the data of a physical PA. However, the PA cell used in this system is a subcircuit that contains a PA circuit set in envelop simulation mode. The PA circuit is designed with Cree's power transistor, CGH40010F, whose bias circuit and match network are well designed. Simulation results show that at the operation frequency of 2.5 GHz, the gain is about 20 dB, and the 1dB compression point is 35 dBm. The PA circuit is given in Fig. 2.

DIGITAL PREDISTORTER MODELING

From the simulation of the QAM system, we can acquire the input and output data of the PA cell in time domain. With these data, we can build the PA model. More importantly, it makes us model the PD of the PA circuit conveniently. In this part, we concentrate on the PD modeling itself and more related details like time delay and normalized gain will be discussed in the next section. Here,

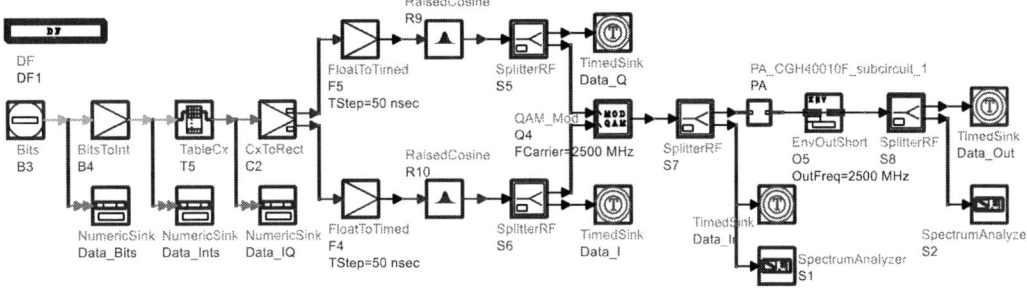

Figure 1: The schematic of the QAM system

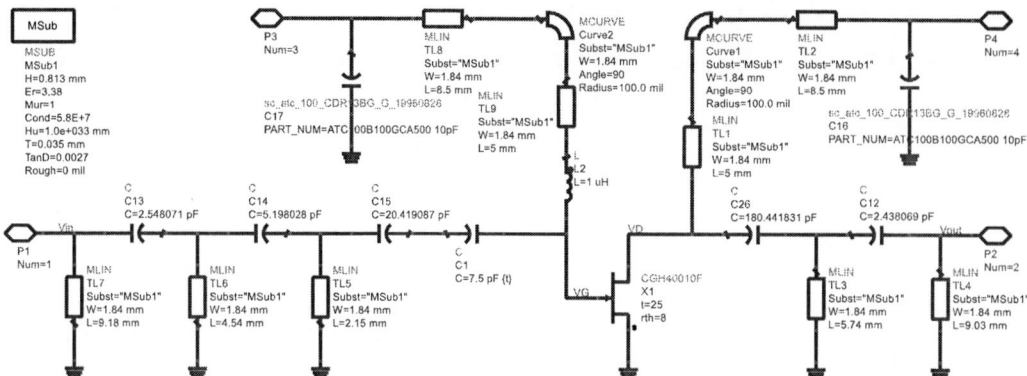

Figure 2: PA circuit

we utilize the memory polynomial (MP) model to describe the PD and its formula is as follows:

$$y_{MP} = \sum_{m=0}^{M} \sum_{k=1}^{K} W_{mk} \cdot x(n-m) \cdot |x(n-m)|^{k-1} \quad (1)$$

where M, K, W_{mk} are respectively the memory depth, nonlinear order and model coefficients.

Thus, our aim is to determine the unknown parameters of the MP model. Through simulation, we assign 4 and 6 to M and K. In addition, the model coefficients are identified using the recursive least square (RLS). By means of ADS simulation, it is proved that the PD compensates the nonlinearity of the PA effectively. The nonlinearity compensation process can be explained with Fig. 3. The upper curve shows the PA's input output characteristic while the lower expresses the PD's. The original input and the final output behave great linearity, as the middle curve presented. The NMSE between it and the ideal one is -51.37 dB, that means the linearity is well.

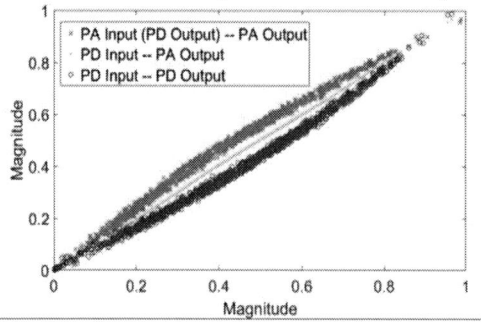

Figure 3: Nonlinearity compensation process

Fig. 4 demonstrates the results of the power spectrum. Without DPD, the spectrum spreads severely, while with DPD, the spectrums of the input and output are almost consistent. Measured from the ADS simulation results, the ACPR reduces about 17 dB at the frequency offset of -30 to -20 and 20 to 30 MHz. The performance comparison results between the proposed PD and others are listed in table I, and all of them are simulation results with the linearization method of DPD.

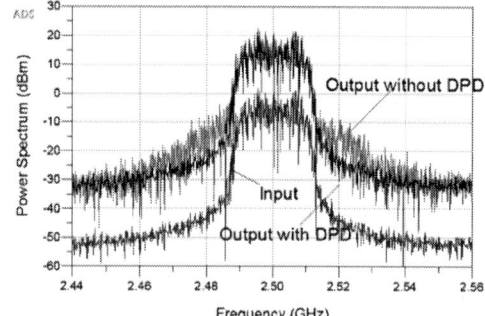

Figure 4: Power spectrum

ADAPTIVE DIGITAL PREDISTORTION

Based on the discussion above, we propose an adaptive processing unit realized by MATLAB, which works as a cell in ADS, together with the DPD system to form a structure similar to the indirect learning architecture (ILA), which can keep the PA work in a relative linear way adaptively. Before that, the time delay of the transmitter and the linearization gain of the PA should be considered first [7].

TABLE I. PERFORMANCE COMPARISON

Ref.	2012 [3]	2013 [4]	2016 [5]	2017 [2]	2018 [6]	This work
Frequency (GHz)	2.5	2.1	2.5	2.1	2.1	2.5
Bandwidth (MHz)	20	5	5	15	10	20
Improvement in ACPR (dB)	8	15	35	15	16	17

Time Delay Estimation

As we are going to deal with the input and output data of the PA and obtain the post-inverse model coefficients, the data must be aligned. Here, we compute the cross-correlation of differential magnitudes, and a digital differentiator is served as a peak detector to find its peak values [8]. Then by comparing these local maximum values, we can acquire the maximum value in the whole range and its index, which is exactly the delay of sample points that we want to know.

Normalized Linear Gain

From [9], it's known that different normalized gains we choose, the impact on system performance is different. In other words, we need a tradeoff between valid working range and the power gain. In this paper, to achieve full DPD range, we linear the PA to its saturation region gain.

Adaptive DPD Process

The procedure of the proposed adaptive DPD is shown in Fig. 5. It contains three parts, preparation, extraction and working mode. Preparation mode comes first and in this part, we correct the time delay, then get the normalized linear gain, both of which are significant for correctly extracting the PD coefficients. After that, the coefficients are identified in extraction mode until the modeling accuracy is acceptable, so the working mode begins. The coefficients are copied to the PD module, so signals will be pre-distorted by the PD before given to the PA. As a consequence, the final output can be amplified linearly relative to the initial input. When the cumulative error between the normalized output and input exceeds a setting threshold, it will come back to the extraction mode to update the PD coefficients.

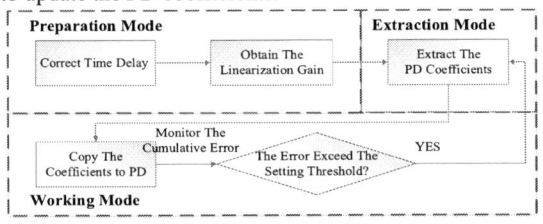

Figure 5: Procedure of adaptive DPD

To imitate the real condition that a PA's characteristic may gradually change after long time use, a PA model is used here, whose input-output characteristic is set to change during the process, and we'll observe the error variation between the normalized output and the initial input, which is presented in Fig. 6. The system is on extraction mode at first and get into working mode after N1. At N2, the PA's characteristic is changed and the error increases but still in the controllable range. At N3, the error exceeds our setting threshold and the system comes to extraction mode again. Finally, coefficients are refreshed at N4. From this process, it is evident that the proposed adaptive DPD is valid.

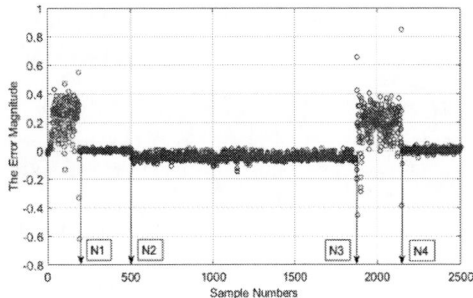

Figure 6: Error between normalized output and input

CONCLUSION

A hybrid domain framework is proposed for modeling PD and realizing adaptive DPD in this paper. On ADS, we design a PA circuit and construct a QAM system. Then, we achieve a fix coefficients PD, which can compensate the PA's nonlinearity well. By designing the processing unit in the form of MATLAB cell, cooperating with the QAM system, an adaptive DPD process is proved to be valid. In fact, this design and simulation process of the prototype is helpful to the design of physical circuits.

ACKNOWLEDGEMENTS

This work is supported by R&D projects of Shenzhen Government (Project JCYJ20170412151226061 and Project JCYJ20170818085827131).

REFERENCES

[1] M. A. Hussein, O. Venard, B. Feuvrie and Y. Wang. *NEWCAS*, Paris, 2013, pp. 1-4.

[2] A. R. Belabad, S. A. Motamedi and S. Sharifian. *Integration, the VLSI Journal*, vol. 57, 2017, pp. 184-191.

[3] M. G. Hernandez, A. P. Guerrero, G. L. Sanchez, J. M. Valencia, J. S. Garcia. *Procedia Engineering*, vol. 35, 2012, pp. 118-125.

[4] G. Karimi, A. Lotfi. *AEU - International Journal of Electronics and Communications*, vol. 67, 2013, pp. 723-728.

[5] A. Rahati Belabad, S. Ahmad Motamedi and S. Sharifian. *ICSPIS*, Tehran, 2016, pp. 1-5.

[6] A. Rahati Belabad, S. Sharifian and S. Ahmad Motamedi. *Analog Integrated Circuits and Signal Processing*, vol. 95, 2018, pp. 231-247.

[7] H. Enzinger, K. Freiberger, G. Kubin and C. Vogel. *ICECS*, Monte Carlo, 2016, pp. 285-288.

[8] S. Tang, K. Gong, J. Wang, K. Peng, C. Pan and Z. Yang. *WCNC*, Kowloon, 2007, pp. 1987-1990.

[9] S. Wang, M. A. Hussein, O. Venard and G. Baudoin. *EuMC*, Nuremberg, 2017, pp. 1050-1053.

A COMPILER DESIGN FOR A PROGRAMMABLE CNN ACCELERATOR

Jiadong Qian, Zhongcheng Huang, and Lingli Wang[*]
School of Microelectronics, Fudan University, Shanghai 201203, China
*Corresponding Author's Email: llwang@fudan.edu.cn

ABSTRACT

Convolutional Neural Networks (CNNs) are widely used in many AI applications, such as image classification, target detection, and target tracking. Due to the increase of CNN computational complexity, hardware acceleration is necessary for inference. Programmable accelerators are promising because of their support for different CNN models. To program an existing programmable accelerator, dedicated instructions need to be generated. In this paper, a compiler is designed to generate the instructions. The compiler explores the best partition of CNN models, schedules the sequence of computing, and generates the instructions automatically. With the proposed compiler, the instruction-driven CNN accelerator achieves the throughput varied from 114 FPS (ResNet152) to 1130 FPS (AlexNet).

INTRODUCTION

High performance has been achieved in multiple AI applications based on CNNs, such as computer vision, robotics, and natural language processing. However, the accuracy and capability of CNNs come at the expense of computational complexity. Therefore, some researches focus on the deployment of CNNs on hardware platforms such as GPUs, FPGAs or CGRAs[1] for acceleration. Corresponding configuration for programmable accelerators on these platforms must be provided according to different CNN models. In [1], configuration for an accelerator based on CGRA is generated manually, which costs much time and effort. Automatic compilers, such as TVM[1], DLA[3], fpgaConvNet[4], are also compared in [5]. TVM is a powerful compiler which can deploy CNNs to different hardware platforms, such as CPUs, GPUs and VTA[6], a programmable accelerator on FPGAs. However, VTA on the FPGA Ultra-96 only outperforms the Cortex-A53 by 3.8x on ResNet50. DLA presents a compiler and FPGA overlay for Neural Networks acceleration. It proposes a very long instruction word (VLIW) but introduces some overhead.

A compiler which can generate different hardware configuration sequences according to different network structures is proposed in this paper with the following contributions:

- The general accelerator architecture is abstracted in the analysis of the existing CNN accelerators [1] [6].
- The method of partitioning and scheduling is proposed according to the methodology of CNN calculation.

- The functional of the compiler is verified and the performance of the accelerator is evaluated on Xilinx VCU118.

THE HARDWARE ARCTHITECTURE

The high-level overview of the programmable accelerator architecture is proposed as shown in *Figure 1*. The accelerator is composed of four modules:

- The *instruction schedule* module fetches the instructions from the DRAM, decodes and passes them to the other three modules according to the type of the instructions.
- The *load* module loads input feature maps, weights, and biases from the DRAM.
- The *compute* module can perform various calculations such as *Convolution, Pooling, Activation,* etc.
- The *store* module stores results produced by the compute module back to the intermediate buffer or the DRAM.

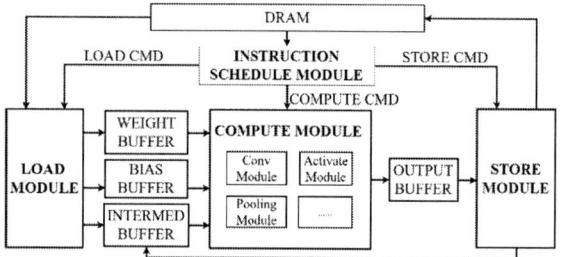

Figure 1: The hardware architecture of a programmable CNN accelerator

THE SOFTWARE FLOW

According to the function, the compiler can be divided into the following two parts:

- Partitioning: The partitioner divides layers of CNN into multiple computing blocks.
- Scheduling: The scheduler schedules the block sequence and generates specific instructions.

The compiling flow is shown in *Figure 2*.

Partitioning

Because of the limited hardware computing and storage resources, the hardware modules have to be reused during the inference of CNN. Data needs to be loaded, computed and stored multiple times. The input feature map of each layer is divided into small Data Blocks(DBs) of the

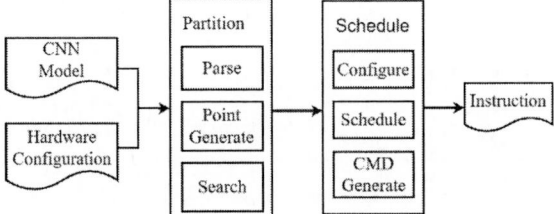

Figure 2: The flow of the CNN compiler

same size, which is called partitioning. As shown in *Figure 3*, the notations are defined as the height of input blocks(*bih*), the channel of input blocks(*bic*), the height of output blocks(*boh*), the channel of output blocks(*boc*).

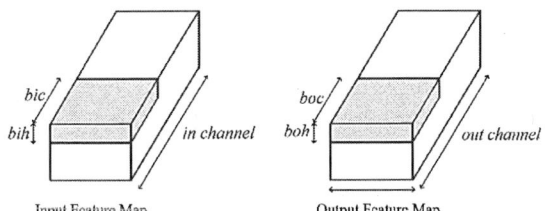

Figure 3: Data partition of the input/output feature map

Partitioning can be described as an optimization problem: with a set of variables, find the lowest cost solution under a series of constraints. The variables, which are represented by a quad-tuple *<mode, bic, boc, boh>*, include the degree of parallelism and the size of the DBs. The constraint in the partitioning is that the size of a DB should be smaller than the size of the on-chip buffers. The cost to be minimized is the total time to calculate a convolution layer, while the cost of computing a single DB is defined as:

$$Cost_{block} = Cost_{load} + Cost_{compute} + Cost_{store} \quad (1)$$

the cost of load and store could be defined as:

$$Cost_{io} = \frac{DB\ Size}{BW} \quad (2)$$

where *DB Size* is the total amount of data and *BW* is the IO bandwidth. The cost of compute is defined as:

$$Cost_{compute} = \frac{T_c}{f} \quad (3)$$

where T_c is the clock cycle number of computing a DB and *f* is the clock rate of the computing core in the accelerator.

Finally, the partitioning flow is summarized in *Figure 4*. A quad-tuple set, which represents the search space, is generated by an iterator in the compiler. Then the tuple set, the constraint function and the cost function are passed to the searcher. The searcher traverses the tuple set to find the lowest cost solution under the constraints.

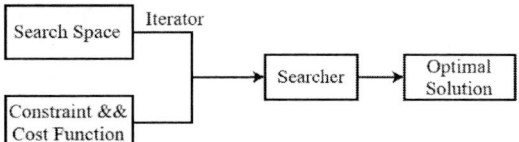

Figure 4: Partitioning flow

Scheduling

The goal of scheduling is to increase throughput. The purposed runtime distribution of the instructions is shown in *Figure 5*. Three types of instructions of the accelerator are described as follows:

- *Load*: fetches the data (feature maps, biases, weights) from DRAM and writes it to on-chip buffers.
- *Store*: writes the data in the output buffer to DDR.
- *Compute*: performs computing functions.

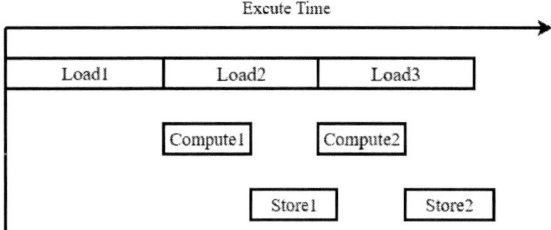

Figure 5: Purposed runtime distribution of the instructions

The scheduling is implemented with two steps. The first step is the Computing Blocks (CBs) generation: according to the partition results, the scheduler generates a series of CBs which includes some information needed for computing. The second step is to take the CBs as input and perform the scheduling. This step mainly considers the logic of the runtime, and launches the instructions to the accelerator as soon as possible. Hence, the hardware performance can be maximized.

EXPERIMENTAL RESULT

This paper implements a compiler with an existing programmable CNN accelerator which matches all the descriptions of the previous hardware architecture. For example, the partial partitioning result of ResNet50 is shown in *Figure 6*.

In order to verify the compiler, this paper uses the instructions generated by the compiler to emulate on the CNN accelerator. All of the simulation results are the same as the results inferred by Caffe. To evaluate the performance of scheduling, we test the idle ratio of Resnet50,

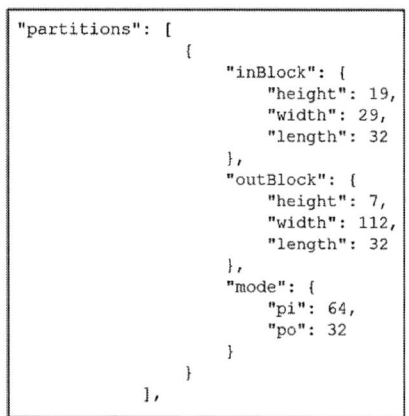

```
"partitions": [
    {
        "inBlock": {
            "height": 19,
            "width": 29,
            "length": 32
        },
        "outBlock": {
            "height": 7,
            "width": 112,
            "length": 32
        },
        "mode": {
            "pi": 64,
            "po": 32
        }
    }
],
```

Figure 6: Partitioning result of one layer of ResNet50

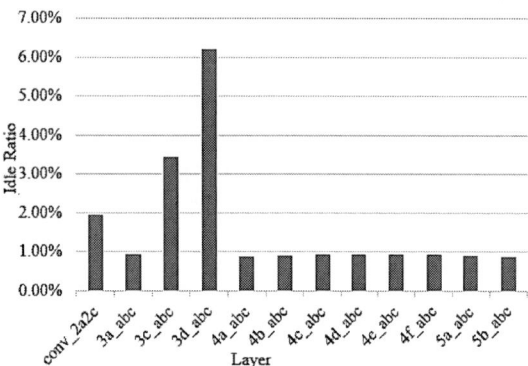

Figure 7: Idle Cycle Ratio of ResNet50

the partial result is shown in *Figure 7*, where the idle ratio is defined as follow:

$$Idle\ ratio = \frac{total\ cycles - effective\ cycles}{total\ cycles} \quad (4)$$

Layer conv_2a2c in *Figure 7* includes conv1, pool1, res2a_abc, res2b_abc, res2c_abc.

The programmable accelerator is implemented on Xilinx VCU118 at the clock rate of 400 MHz. The data is quantized to 8-bit fixed-point representation. The performance of the accelerator and the evaluation results on the ImageNet dataset are shown in Table I. We also compare our compiler with other compilers for FPGA-based CNN accelerator. The result is shown in Table II, with DNNVM for Deephi DPU on ZCU102 at 330 MHz, xfDNN of Xilinx on VU9P at 450 MHz, and fpgaConvNet on ZC706 at 125 MHz.

CONCLUSION

In this paper, the partitioning and scheduling methods of CNN calculation are proposed, and the corresponding compiler, which can automatically deploy CNNs on the specific programmable accelerator, is realized. Using this

TABLE I. PERFORMANCE OF DIFFERENT CNN MODELS

Model	Top-1 (%)	Top-5 (%)	FPS
ResNet50	72.6	90.9	222
ResNet101	73.7	91.6	151
ResNet152	74.7	92.3	114
AlexNet	58.7	81.2	1130

TABLE II. COMPARISON OF DIFFERENT COMPILERS

Model	DNNVM [5]	xfDNN [5]	fpgaConvNet [5]	This paper
ResNet50	74	80.5	N/A	222
ResNet152	27.5	28.7	6.5	114

(Unit: FPS)

compiler, the implementation process of CNN hardware acceleration can be automatic, and the accelerator can compute efficiently after the scheduling. A potential direction for further research is to develop the compiler and accelerator to support more different models.

ACKONWLEDGE

This work is supported by the National Natural Science Foundation of China under grant 61971143.

REFERENCES

[1] X. Fan et al., "Stream Processing Dual-Track CGRA for Object Inference," in *IEEE Transactions on Very Large Scale Integration (VLSI) Systems*, 2018, pp. 1098-1111.

[2] T. Chen et al., "TVM: An Automated End-to-End Optimizing Compiler for Deep Learning," in *Proceedings of the 12th USENIX conference on Operating Systems Design and Implementation*, Berkeley, 2018, pp. 579-594.

[3] M. S. Abdelfattah et al., "DLA: Compiler and FPGA Overlay for Neural Network Inference Acceleration," in *2018 28th International Conference on Field Programmable Logic and Applications (FPL)*, Dublin, 2018, pp. 411-4117.

[4] S. I. Venieris and C. Bouganis, "fpgaConvNet: Mapping Regular and Irregular Convolutional Neural Networks on FPGAs," in *IEEE Transactions on Neural Networks and Learning Systems*, 2019, pp. 326-342,

[5] Y. Xing et al., "An In-depth Comparison of Compilers for Deep Neural Networks on Hardware," in *2019 IEEE International Conference on Embedded Software and Systems*, Las Vegas, 2019, pp. 1-8.

[6] Moreau T. et al., "A Hardware-Software Blueprint for Flexible Deep Learning Specialization," in *arXiv*, 2018, 1807.04188

Crystal Oscillator Frequency compensation technology of High Precision Clock Synchronization for Time-triggered Ethernet

Haiying Yuan[*], Kai Zhang, Tong Zheng, Yichen Wang

Faculty of Information Technology, Beijing University of Technology, Beijing 100124, PR China

yhyingcn@gmail.com

ABSTRACT

Temperature change, aging and voltage fluctuation also cause clock deviation of the crystal oscillator frequency, which makes a great impact on the real-time communication and network stability in the distributed control system. While local clock value is periodically corrected based on SAE AS6802 clock synchronization protocol in Time-Triggered Ethernet, but a large cumulative error occurs to the calibration cycle, which reduces the clock synchronization accuracy. According to the severe influence of temperature on crystal oscillator frequency, a digital frequency calibration circuit is designed in Local_clock module of TTEthernet nodes, and it is modeled based on the frequency error Look-Up table generated by Temperature-Frequency characteristics. The maximum clock deviation between network nodes was analyzed to verify the synchronization performance of the TTEthernet. Numerous experimental results show that proposed digital frequency calibration scheme achieves high clock synchronization accuracy.

Introduction

Measurement and Control Bus plays a control center role in aerospace electronic system, mobile system and vehicle system, it is responsible for task scheduling, timing control and data transmission during system operation, so the real-time and stability is crucial to the communication system [1-3]. Time Triggered Ethernet (TTEthernet) bus [4-5] combines real-time performance, fault-tolerant capability with Ethernet flexibility and large communication bandwidth, it is widely applied to aerospace electronics system and autopilot technology. A globally unified synchronous clock is established with SAE AS6802[6] protocol in TTEthernet, it ensures critical data messages correctly transmitted and received by each node at the right time [7]. Hence, High-precision clock synchronization techniques fundamentally decides time determinacy and network stability of TTEthernet.

Time determinacy and network stability fundamentally depends on clock synchronization accuracy in distribution real-time control system[8]. Strictly speaking, global clocks in TTEthernet are unified only in synchronous operation, and phase deviation is caused by frequency deviation at other times.

Considering that clock deviation must be compensated as much as possible in Improved-TTEthernet, a high precision clock synchronization technology is applied to calibrate crystal oscillation frequency of local clock for all nodes. In order to Simultaneously eliminating clock phase deviation and frequency deviation, Improved-TTEthernet high precision clock synchronization scheme based on crystal oscillation frequency digital calibration is derived from Default-TTEthernet clock synchronization model, it is verified by hardware modelling and simulation well.

Time Synchronization Technology

TTEthernet Model is shown in Figure 1, a network topology shown in Figure 1(a) consists of 13 end systems and 3 switches, these terminal nodes are abstracted as Synchronization Master (SM), Synchronization Client (SC) and Compression Master (CM), they are composed of a sub-network model shown in Figure 1(b). Since local clocks of all nodes calling the same simulation time, it indicates these is a same crystal oscillation periods.

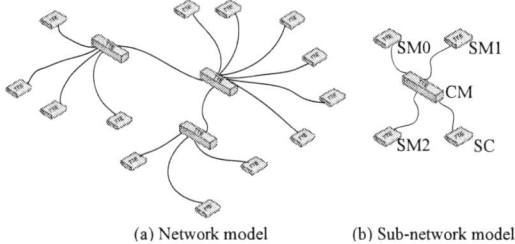

(a) Network model   (b) Sub-network model

Figure 1 : Default-TTEthernet Model

Clock Synchronization Model

During TTEthernet transmits time-triggered message, local clocks of all terminal nodes need to synchronize with each other. TTEthernet synchronization approach is demonstratsed in Figure 2, during clock synchronization cold start (CS) frames and cold start acknowledge (CA) frames are exchanged between SM node and CM node to achieve fault-tolerant handshake. When SM node enters the synchronous process and then sends an Interference (IN) frame, once CM node receives a matching IN frame, TTEthernet enters synchronous steady state.

Protocol control frame (pcf) consists of IN/CS/CA frames during clock synchronization process. SM node sends pcf0~pcf2 frames to CM node. Once CM node receives these frames, and then it performs permanence function and starts the compression algorithm. CM node integrates time information to pcf3 frame and then sends to SM and SC node, these two nodes performs permanence function to restore the sending order of pcf3 fame from different links. The deviation between local clock and average clock is calculated to calibrate local clock. CM node also calibrates its local clock synchronously.

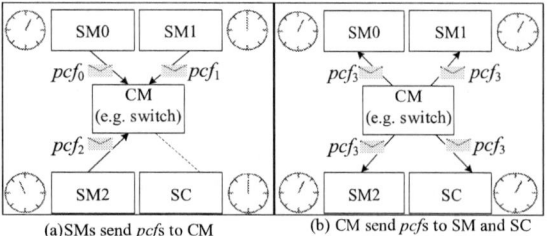

(a)SMs send *pcf*s to CM   (b) CM send *pcf*s to SM and SC

Figure 2: TTEthernet synchronization approach

High Precision Crystal Oscillation Frequency Digital Calibration

Although the frequency deviations on oscillators caused by temperature, voltage and aging, etc. Temperature factor with serious influence on crystal oscillator frequency is considered in Improved-TTEthernet. Temperature differences from all end systems lead to the deviation of local clocks. A crystal oscillator temperature |ppm| absolute value curve is shown in Figure 3. Supposing ideal clock is at ambient temperature 20°C, the corresponding clock accuracy reach to +70ppm at 60°C and -70ppm at -20°C respectively. 70ns phase deviation occurs in each integration cycle (1ms) during Default-TTEthernet clock synchronization. Owe to a look-up table is adopted to periodically compensate local clock counts, clock synchronization accuracy is significantly increased in Improved-TTEthernet.

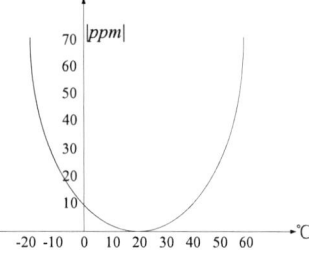

Figure 3: Crystal oscillator temperature |ppm| curve

SM/SC/CM nodes are established as hardware models in VerilogHDL. Take SM node as an example in Figure 4(a), it includes four sub-modules: Pcf receiver, Clock synchronizer, Pcf sender and Improved-local_clock module. The latter module is used to correct the frequency

deviation in Local_clock module. Significantly, Freq_corre module marked in Figure 4(b) is used to eliminate the influence of temperature on frequency deviation, which consists of a look-up table (LUT) and a correction logic, the correction frequency value is calculated from crystal oscillator temperature |ppm| curve and then stored in LUT, the circuit periodically compensate the local clock counts and effectively improves the clock frequency accuracy.

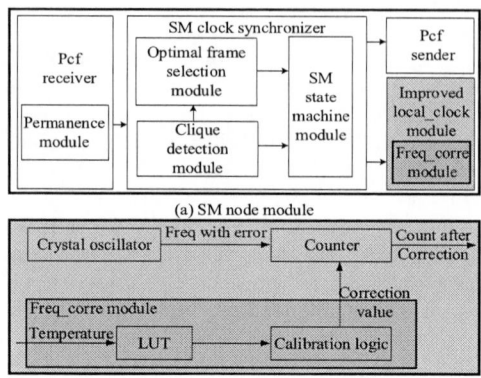

(a) SM node module

(b) Improved local_clock module

Figure 4 : SM node Hardware block diagram

Network Simulation and Hardware Validation

Opnet and VCS simulation scenes are set to a sub-network model shown in Figure 1(b). The experimental parameters are set as: (1) SM0 and SM2 are respectively simulated as fast and slow clock. The frequency deviations are respectively set to +70ppm and -70ppm under 60°C and 20°C. The initial phase deviations are set to +100ns and -100ns. (2) SM1, CM and SC are simulated as standard clocks, the clock frequency reaches 1Ghz through frequency multiplication, integration period is set to 1ms.

Default-TTEthernet performs network calculation on time parameter by Opnet simulator, the node deviation of local clock is shown in Figure 5(a). Since the clock synchronization is not established between nodes, clock deviation caused by initial phase deviation and frequency deviation is occurred. The clock deviation gradually increases to about 700ns after 9 integration cycles. As a result, the time trigger traffic in TTEthernet conflicts during the transmission process, which seriously affects the time deterministic and communication real-time.

To achieve the clock synchronization of Default-TTEthernet, local clock is calibrated by SAE AS6802 protocol, the node deviation is shown in Figure 5(b). The first two integration cycles for the cold start process realizes fault tolerance among network nodes. Clock deviations of SM0 and SM2 are both accumulated to 240ns. The clock calibration is periodically performed in the third integration cycle. Due to the crystal oscillator in different nodes has the frequency deviation, clock deviation is accumulated to 70ns at the end of each integration cycle.

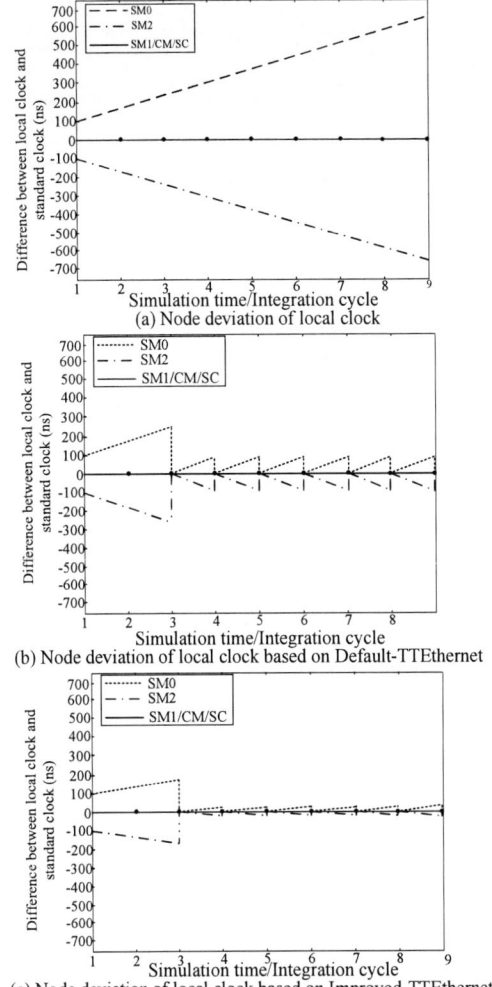

(a) Node deviation of local clock

(b) Node deviation of local clock based on Default-TTEthernet

(c) Node deviation of local clock based on Improved-TTEthernet

Figure 5 : Hardware block diagram

To improve the clock synchronization accuracy of Default-TTEthernet, Freq_corre module is added to Local_clock module for crystal oscillator frequency calibration, RTL-level code of node model for clock synchronization is verified by VCS simulator, and local clock data from each node are collected. The frequency is corrected by the Freq_corre module within 1ms, so the maximum clock deviation is not increased during the calibration cycle. Compared to Figure 5(b), the phase deviation caused by the frequency deviation is eliminated in Figure 5(c), local clock achieves high precision synchronization.

Conclusion

High-precision clock synchronization techniques is critical to time determinacy and network stability of TTEthernet. Clock synchronization parameters of Default-TTEthernet are calculated through network modeling and calculation performed on Opnet simulator, local clock is periodically calibrated using SAE AS6802 clock synchronization protocol, and then clock deviation is gradually diverged to a larger value during each integration cycle. Hence, an accurate clock model is constructed to achieve crystal oscillator frequency digital calibration for Improved-TTEthernet, and a high-precision Local_clock module is design to improve the real-time communication between network nodes. Node models were constructed in Verilog HDL and verified by VCS simulator. It is obviously noted that Improve-TTEthernet has high clock synchronization accuracy. More factors results in crystal oscillator frequency drift deviation will be further considered, the corresponding crystal oscillator frequency compensation scheme is designed to fit for more application scenarios in distributed real-time systems.

Acknowledgments

This research work was supported by Joint Fund for Advanced Equipment Research and Aerospace Science and Technology (6141B060914), Beijing Natural Science Foundation (4172010), Research Fund from Beijing Innovation Center for Future Chips (KYJJ2018009).

REFERENCES

[1] R. Solis, V. S. Borkar, P. R. Kumar, et al. *Proceedings of IEEE Conference on Decision and Control*, New York, Dec13-15, 2006, pp. 2737.

[2] Charara. H, Scharbarg. J, et al. *Proceedings of 18th IEEE Conference on Real-time System,* Dresden, July 05-07, 2006, pp. 192-202.

[3] Li. Jifeng, M. Chai, et al. *Proceedings of IEEE International Conference on Electronics, Communications and Control*, Ningbo, Sep 09-11, 2011, pp. 3642-3645.

[4] Kopetz. H, Ademaj. A, et al. Grillinger. P, *Proceedings of IEEE International Symposium on Object-Oriented Real-Time Distributed Comput*ing, Seattle, May18-20, 2005, pp. 22-23.

[5] M. Abuteir, R. Obermaisser, et al. *Proceedings of IEEE International Conference on Industrial Informatics*, New York, Jul 29-31, 2013, pp. 642-648.

[6] SAE. AS6802, Time Triggered ethernet, *Society of Automotive Engineers*, 2011.

[7] C. Fetzer, and F. Cristian, et al. *J. Real-Time Systems*, vol. 12, no. 2, Mar 1997, pp. 123-171.

[8] A. M. Toufik, J. G. Yao, Y. Jin, et al. *J. IEEE Access*, vol. 6, 2018, pp. 8412-8425.

A 5.5nW Voltage Reference Circuit

Kaixuan Du[1], Ziyuan Xu[3], Xiulong Wu[1], Libo Yang[2], Hao Zhang[2], Zhixuan Wang[2], Le Ye[2]*

[1]School of Electronics and Information Engineering, Anhui University, Hefei 230601, China
[2]Laboratory of Microelectronic Devices and Circuits (MOE)
Institute of Microelectronic, Peking University, Beijing 100871, China
[3]Jiangsu Union Technical Institute, China
*E-mail: yele@pku.edu.cn

ABSTRACT

This paper proposed a nano-watt voltage reference circuit was implemented in a 0.18um CMOS process with trim techniques. In order to reduce power consumption, a MOS-Only Voltage Reference is presented, which is based on the threshold voltage, However, the deviation of Vref because of process variation is large. We use the difference of Vth instead of Vth to improve the stability of output voltage at different process corner. The simulation results show that under 27°C and 0.5V supply voltage, the output reference voltage is 236mV, the temperature coefficient is 30.8 ppm/°C over temperature range of 125°C (-40°C to 85°C) and only consume 5.5nW at 0.5V supply voltage.

Keywords—subthreshold voltage reference; Native NMOS; low power; process variation; trimming.

INTRODUCTION

With the development of the Internet of things and medical electronics, low power consumption becomes the most important factor. The traditional voltage reference circuit uses the negative temperature coefficient of V_{BE} and the positive temperature coefficient of ΔV_{BE}, by adjusting the magnitude of one of the temperature coefficient terms, and combining the two terms in an adder circuit, the output of the adder will be temperature insensitive to a first order [1]. This kind of reference circuit has low temperature coefficient and good process stability, but the power is very high. If you want to reduce the power, you need a large resistance, which will occupy a large area and increase the cost of the chip.

In recent years, a MOS-only voltage reference has been designed by using MOSFET threshold voltage to implement the CTAT function [2] [3]. In [4], the reference voltage is generated by the native MOS working in the subthreshold region. The reference voltage in this circuit is mainly determined by the threshold voltage. but most of them present variability issues due to high process spread of the threshold voltage, and the accuracy of the reference voltage is low, the difference Vth is used as the basis of the voltage reference to reduce the process variation [5]. they used the reverse short channel effect, and narrow-width effect to obtain different Vth while using the same type of device. However, the output voltage of this voltage reference circuit is only 26mV, difficult to use for most application.

This paper presents a voltage reference circuit based on the different Vth by using different types of NMOS, compared to the same type of device, this circuit increases the output voltage. To further reduce the impact of the process, we designed trimming circuits.

CIRCUIT DESCRIPTIONS

Figure 1 is a conventional voltage reference circuit [6]. Uses the negative temperature coefficient of V_{BE} and the positive temperature coefficient of ΔV_{BE}, by adjusting the magnitude of one of the temperature coefficient terms, and combining the two terms in an adder circuit, the output of the adder will be temperature insensitive to a first order. Power consumption is mainly determined by resistance, we must use resistors with a high resistance of several hundred megaohms to achieve low-current, which will occupy a large area.

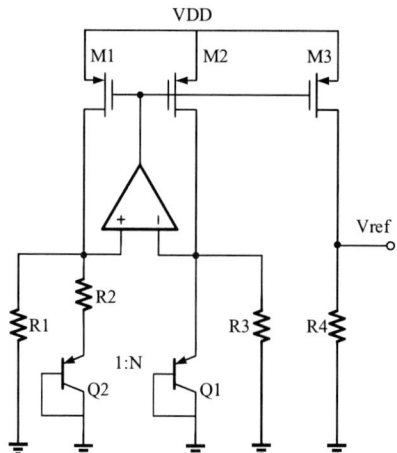

Figure 1: Schematic of conventional voltage reference

The proposed voltage reference circuit are shown in Figure 2. The MN1 is a native NMOS, operating in the subthreshold region. The MN2 is thin-oxide transistors and the MN3 is thick-oxide NMOS. If the drain-source voltage (V_{DS}) of a sub-threshold NMOS is much larger than V_T, the drain current can be expressed as

978-1-7281-6559-2/20 $31.00 © 2020 IEEE

$$I_D = KI_0 \exp\left(\frac{V_{GS} - V_{TH}}{\eta V_T}\right) \qquad (1)$$

$$I_0 = \mu C_{ox}(\eta - 1)V_T^2 \qquad (2)$$

$$V_{GS} = V_{TH} + \eta V_T \ln\left(\frac{I_D}{KI_0}\right) \qquad (3)$$

$$V_{ref} = V_{GS3} - V_{GS2} \qquad (4)$$

$$V_{ref} = \Delta V_{TH} + \eta V_T \ln\left(\frac{K_2}{K_3}\right) \qquad (5)$$

where K is the aspect ratio (= W/L) of the MOS, V_{GS} is the gate-source voltage, V_{TH} is the threshold voltage, n is the subthreshold slope factor, and V_T is the thermal voltage. Form (1) - (4) we obtain an analytical solution for Vref in (5). we can adjust the aspect ratio to get the temperature-independent voltage value.

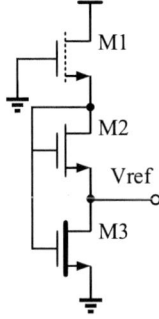

Figure 2: Schematic of proposed voltage reference circuit

Although the proposed voltage reference circuit uses the difference value of Vth to offset the influence of process variation, the deviation of Vref caused by process cannot be completely offset due to the different types of M2 and M3. Therefore, we have designed a trimming circuit. Figure 3 is a reference source circuit with the trimming circuit. The MOS size can be adjusted by S7-S0.

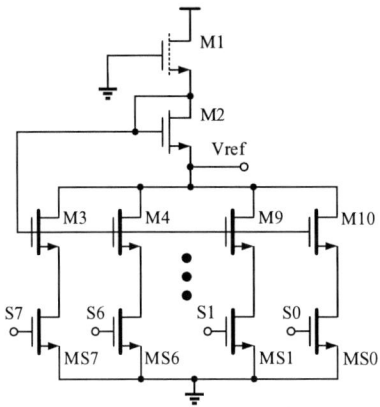

Figure 3: Complete schematic

SIMULATION RESULTS

Figure 4 shows the curve of output voltage Vref changing with temperature. the temperature coefficient is 30.8 ppm/°C within the range of -40°C to 85°C. Figure 5 shows the curve of output voltage Vref changing with power supply voltage, and Figure 6 is the layout of this circuit.

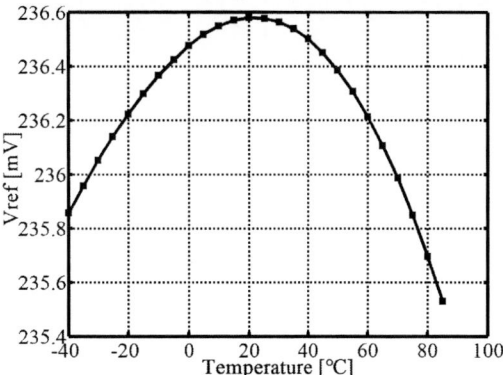

Figure 4: Temperature dependence of the circuit

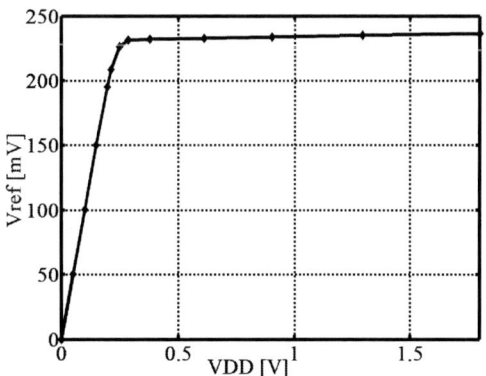

Figure 5: Variation of Vref versus VDD at T=27°C

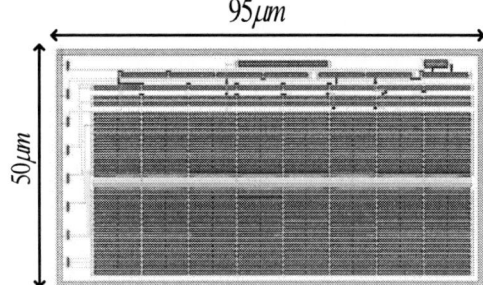

Figure 6: Layout of the proposed voltage reference

TABLE I. COMPARISON RESULTS WITH PREVIOUS WORK

	This work	[5]	[7]	[8]
Process(um)	0.18	0.13	0.18	0.18
VDD(V)	0.5-1.8	0.3-1.2	0.55	0.7-1.8
Vref(mV)	236	26	460	548
Temp.Range(°C)	-40-85	-25-125	-45-120	-40-120
TC(ppm/°C)	30.8	208	28	114
PSR@100Hz(dB)	-50	-67	-62	-56
Power(nW)	5.5	0.040	83	52.5

CONCLUSION AND FUTURE WORK

This paper presents a low power voltage reference circuit, according to the simulation results, the power consumption is only 5.5nw when VDD is 0.5V, and the temperature coefficient is 30.8ppm in the temperature range of -40-85 °C. Due to the addition of the trimming circuit, this circuit need fixed through external control signals. In the future work, we can design an automatic trimming circuit to realize the automatic adjustment of the circuit.

ACKNOWLEDGMENT

This work is supported by National Key R&D Program of China (No. 2019YFB2204900), National Science Foundation of China (No. 61722401), and Beijing New-star Plan of Science and Technology (No. Z181100006218047)

REFERENCE

[1] K. E. Kuijk, "A precision reference voltage source," in *IEEE J. Solid-State Circuits*, vol. 8, no. 3, pp. 222-226, June 1973.

[2] H. Zhang *et al.*, "A Nano-Watt MOS-Only Voltage Reference With High-Slope PTAT Voltage Generators," in *IEEE Trans. Circuits Syst. II: Express Briefs*, vol. 65, no. 1, pp. 1-5, Jan. 2018.

[3] E. M. Camacho-Galeano, C. Galup-Montoro and M. Cherem Schneider, "Design of an ultra-low-power current source," *2004 IEEE Int. Symp. Circuits Syst. (ISCAS)*, Vancouver, BC, 2004, pp. I-I.

[4] M. Seok, G. Kim, D. Blaauw and D. Sylvester, "A Portable 2-Transistor Picowatt Temperature Compensated Voltage Reference Operating at 0.5 V," in *IEEE J. Solid-State Circuits*, vol. 47, no. 10, pp. 2534-2545, Oct. 2012.

[5] A. C. de Oliveira, D. Cordova, H. Klimach and S. Bampi, "A 0.12–0.4 V, Versatile 3-Transistor CMOS Voltage Reference for Ultra-Low Power Systems," in *IEEE Trans. Circuits Sys. I: Reg. Papers*, vol. 65, no. 11, pp. 3790-3799, Nov. 2018.

[6] H. Banba *et al.*, "A CMOS bandgap reference circuit with sub-1-V operation," in *IEEE J. Solid-State Circuits*, vol. 34, no. 5, pp. 670-674, May 1999.

[7] L. Liu, J. Mu and Z. Zhu, "A 0.55-V, 28-ppm/°C, 83-nW CMOS Sub-BGR With Ultra Low Power Curvature Compensation," in *IEEE Trans. Circuits Syst. I: Reg. Papers*, vol. 65, no. 1, pp. 95-106, Jan. 2018.

[8] Y. Osaki, T. Hirose, N. Kuroki and M. Numa, "1.2-V Supply, 100-nW, 1.09-V Bandgap and 0.7-V Supply, 52.5-nW, 0.55-V Sub bandgap Reference Circuits for Nanowatt CMOS LSIs," in *IEEE J. Solid-State Circuits*, vol. 48, no. 6, pp. 1530-1538, June 2013.

A HIGH LINEARITY READOUT INTEGRATED CIRCUIT FOR UNCOOLED IR DETECTOR

Chang Liu, Kai Wang, Mingcheng Luo, Yaozu Guo, Feng Yan and Xiaoli Ji[*]
Institute of the electronic Science and Engineering, Nanjing University, China
*Corresponding Author's Email: xji@nju.edu.cn

ABSTRACT

In this paper, we presented a CMOS microbolometer design integrated with a linearity readout circuit using 0.18μm CMOS process. The air-bridge microbolometer was designed with polycrystalline silicon film while a capacitive trans-impedance amplifier (CTIA) was used for detecting the weak current from IR sensors. Simulation results shows that the noise equivalent power of detector is about 1nW and the static power consumption is 163μW for CTIA with a small area of 30×40 μm^2.

INTRODUCTION

Uncooled infrared detectors have been widely used in military, medical, and meteorological fields due to their advantages such as low cost and easy design. Amorphous silicon and vanadium oxide materials [1-4] are now widely used because of the benefit of the high resistance and large temperature coefficient of resistance (TCR). However, there exists the barrier to be fully integrated with standard CMOS technology. Previously, metal Aluminum films had been adapted for the resistance-type CMOS microbolometer and an amplifier was designed to amplify the weak current of the microbolometer. CTIA which consists of an operational amplifier and a feedback integral capacitor [5, 6] is one of the most widely used one. However, a high-performance operational amplifier needs to be designed carefully in order to gain lower noise and higher linearity.

In this paper, a ROIC for the microbolemeter with polycrystalline silicon multilayer sandwich structures is presented. The performance of the circuit is simulated, which shows that it is a low noise and high linearity readout circuit with reasonably power consumption, indicating the designed ROIC is compatible with the microbolometer.

DESIGN AND SIMULATION

Fig 1 (a) shows 3D view of the polycrystalline silicon resistance-type sensors. The absorber is made up of polycrystalline silicon embedded in SiO$_2$ and supported by the suspended microbridge that linked with the anchors and an air gap between the absorber and SiO$_2$ on the silicon substrate. When IR light illuminates on the surface of the absorber, IR energies are absorbed and converted to the distribution of heat, thus, the temperature of the structure is changed and resulting in resistance shift.

The photocurrent signal of the sensors can be obtained under a certain bias voltage. Fig 1 (b) shows the designed pixel cell. The designed microbolometer is equivalent to a current source and a resistor in parallel. Its resistance changes inducing a photo-current shifts when IR light illuminates on the poly-Si sensors. CTIA circuit is then used to collect and amplify the current data. The output voltage of the CTIA is compared and the condition of the counter is determined, finally, the response to the infrared radiation is saved as digital signal in register.

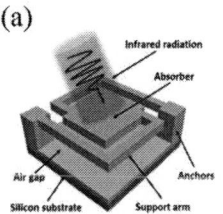

(a)

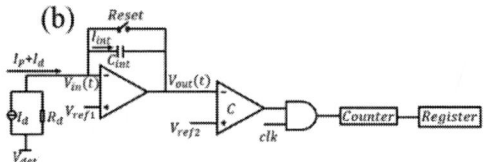

(b)

Fig 1 (a) and (b) show the structure of microbolometer and the designed pixel cell.

CTIA circuit as shown in Fig 1 (b), the output voltage of the CTIA is expressed as the following formula at the integration time:

$$V_{out}(t) = V_{in}(t) - \frac{I_{int}t}{C_{int}} \qquad (1)$$

Where $V_{int}(t)$ is the voltage at the inverting input node, I_{int} is the integration current and C_{int} is the integration capacitance. $V_{int}(t)$ should keep constant in order to get a linearity output $V_{out}(t)$. The equation (1) can be rewritten as following formula when the amplifier has a high gain:

$$V_{out}(t) = V_{ref1} - \frac{I_{int}t}{C_{int}} \qquad (2)$$

Where V_{ref1} is the reference voltage, a two-stage amplifier [7] is designed to obtain high gain in this study as shown in Fig 2. The first stage is differential amplifier

978-1-7281-6559-2/20 $31.00 © 2020 IEEE

with active load in which the differential input unit can suppress the influence of common-mode noise and active load unit can increase the output impedance while increasing circuit gain. The second stage is a common source amplifier which can improve the output swing while increasing the gain. Miller compensation capacitor and resistor adjust the zero and pole of the circuit to ensure the stability of the circuit. The current bias is provided by the current mirror.

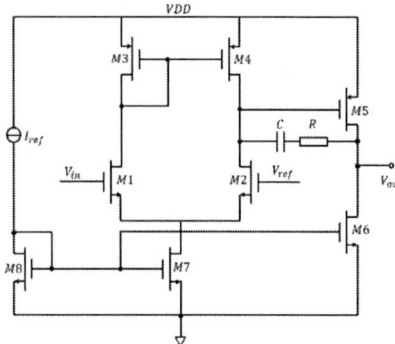

Fig 2: The two-stage amplifier architecture.

Fig 3 shows the schematic diagram of comparator. Here a high speed and precision comparator that can distinguish the difference of 1 mV is used. It is composed of three stages: the input preamplifier stage, a latch stage and an output buffer stage, respectively. The preamplifier stage amplifies the input signal to improve the comparator sensitivity while isolating the input of the comparator from kickback noise coming from the latch stage. The latch stage is used to determine which of the input signals is larger and extremely amplifies their difference. The output buffer stage converts the output of the latch stage into a full scale digital level output (0 V or 1.8 V). After the comparator, it follows the counter which counts one if $V_{out}(t)$ is greater than V_{ref2} otherwise it counts zero, then the obtained digital signal are saved in register followed the counter as shown in Fig 1 (b).

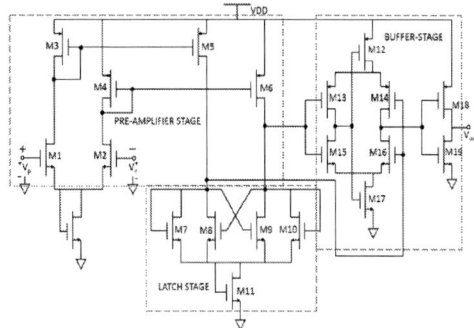

Fig 3: The schematic diagram of comparator.

SIMULATION RESULTS

Cadence is used to simulate the performance of CTIA.

Fig 4 (a) shows the CTIA output voltage with integration time under different integration current. The reference voltage, integration time and the integration capacitor are 2.2 V, 700 ns, 500 fF respectively. It is found that the output voltage has a very good linearity with integration time. Fig 4 (b) shows current dependence of $V_{in}(t)$, it is found that $V_{in}(t)$ is in good agreement with V_{ref1} with integration time, but $V_{in}(t)$ would shift thereby leads the non-linearity of $V_{out}(t)$ when the photo-current of the sensor is larger. Therefore, the bias of the microbolometer should be set reasonably.

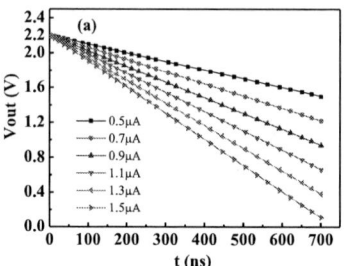

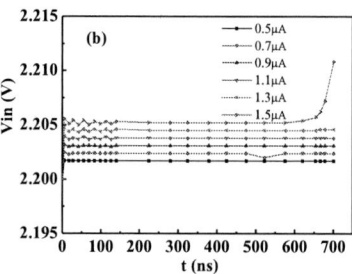

Fig 4: (a) and (b) show the output voltage and inverting input node voltage with integration time.

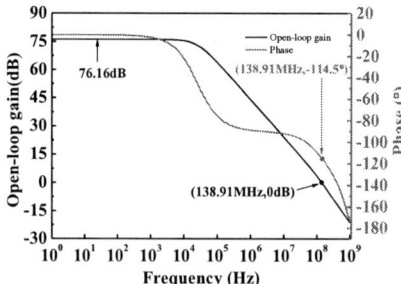

Fig 5: The amplitude-frequency response of the two-stage amplifier.

Fig 5 shows the amplitude-frequency response of the amplifier. It can be seen that the two-stage amplifier has a high open-loop gain of 76.16 dB and the phase margin is

65.5 degrees which ensures the stability of the amplifier. The main parameters of the two-stage amplifier have been summarized, as shown in Table I. It is found that the performance of the operational amplifier is good, such as high open-loop gain and low noise, which ensures the good performance of CTIA.

Table I. Simulation results of the two-stage amplifier.

Parameter	Value
Open-loop gain (dB)	76.16
Phase margin (°)	65.5
Unity gain bandwidth (MHz)	138.91
Area (μm×μm)	30×40
Static power consumption (mW)	0.163
Output swing (V)	1-3.2
CMRR (dB)	80
Equivalent input noise	54nv@0.1MHZ

CONCLUSION

A readout circuit for microbolemeter with polycrystalline silicon multilayer sandwich structures using standard 0.18μm CMOS process is proposed in this paper. The linearity of the circuit is very high and the power consumption is low. Simulation results show that the amplifier has the features of small area and low noise which is suitable for microbolometer array.

ACKNOWLEDGEMENT

This work was supported by the National Key R&D Program of China (No. 2016YFA0202102 and No. 2016YFB0400402).

REFERENCE

[1] M. Mario, J. Ricardo, T. Alfonso and A. Roberto. "Microbolometers based on amorphous Silicon-Germanium films with embedded nanocrystals," IEEE Transaction on Electron Devices, vol. 62, 2015, pp. 2120-2127.

[2] C. Ozer, D. Memed. "High temperature coefficient of resistance and low noise tungsten oxide doped amorphous vanadium oxide thin films for microbolometer applications," Thin Solid Films, vol. 691, 2019.

[3] S. Eminoglu, M. Y. Tanrikulu and T. Akin. "A low-cost 128×128 uncooled infrared detector array in CMOS process," Journal of Microelectromechanical Systems, vol. 17, 2008, pp. 20-30.

[4] C. Lamsal, N. M. Ravindra. "Simulation of spectral emissivity of vanadium oxides (VO_x)-based microbolometer structures," Emerging Materials Research, vol. 3, 2014, pp. 194-202.

[5] S. L. Liu, Y. C. Zhang, X. Y. Meng, W. G. Lu and Z. J. Chen. "A design of readout circuit for 384× 288 uncooled microbolometer infrared focal plane array." IEEE 11th International Conference on Solid-State and Integrated Circuit Technology, 2012 pp. 1-3.

[6] L. Jian. Y. D. Jiang, D. L. Zhang, Y. Zhou. "An ultra low noise readout integrated circuit for uncooled microbolometers," Analog Integrated Circuits and Signal Processing, vol. 63, 2010, pp. 489-494.

[7] B. Razavi, Design of Analog CMOS Integrated Circuits, McGraw Hill, 2003.

THERMAL MODELING OF MONOLITHIC 3D ICS

Baoli Peng[1], Vasilis F. Pavlidis[2], and Yuanqing Cheng[1]*

[1]School of Microelectronics, Beihang University, Beijing 100191, China

[2] School of Computer Science, University of Manchester, Manchester, United Kingdom

*Corresponding Author's Email: yuanqing@ieee.org

ABSTRACT

Monolithic 3D integration can approach ultra-high device density compared to TSV-based integration owing to the sequential process. So it can effectively sustain Moore's law without resorting to costly technology shrinking. Nevertheless, heat dissipation problem in M3D ICs poses a big challenge, and is different from TSV-based counterparts due to close thermal coupling between neighboring tiers, which requires further investigations. In this work, we compare the thermal characteristics of M3D ICs to those of 2D ICs in 45nm technology node with a thermal model based on the finite element method. Experimental results show that the average and maximum temperature of M3D ICs is higher. We expect this work can invoke more research interests in thermal modeling and thermal aware physical design of M3D ICs.

Keywords—Monolithic 3D ICs; thermal modeling; 45nm technology node;

INTRODUCTION

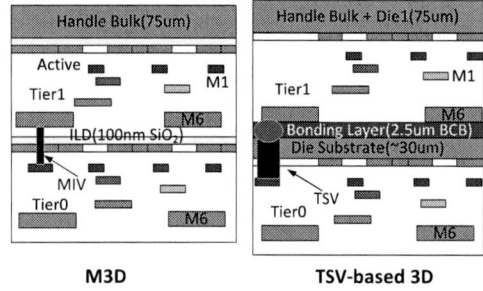

Figure 1: *2-tier 3D IC layer structure of Monolithic 3D IC and TSV-based 3D IC*

Recently monolithic 3D integration technology[1] is proposed to enable sequential integration of device layers instead of wafer-to-wafer tacking by TSVs. Monolithic 3D ICs uses nano-scale monolithic inter-tier vias (MIVs) to connect the different device layers. MIVs have similar feature size as regular metal-layer vias, whose capacitance and area are negligible compared to those of TSVs owing to sequential fabrication process, as shown in Fig. 1. This allows a large number of MIVs for vertical connections, so its integration density is significantly higher than that of TSV-based 3D ICs.

Monolithic 3D ICs can overcome the shortcomings of TSV-based 3D ICs. However, one major problem with 3D ICs in general is the increase in power density, leading to high temperature, because of the shrinking footprint. Although 3D technology can reduce interconnect power effectively, the increased power density aggravates thermal dissipation problem, especially in the layers far away from the heat sink or other equivalent cooling structures. Therefore, thermal-aware design techniques become more and more critical in 3D ICs. The major bottleneck of carrying out thermal analysis within the physical design process is the long runtime required for accurate thermal analysis. The inclusion of such timing-consuming analysis within the iterative physical design process is not feasible especially for the giga-scale integrated circuits [2].

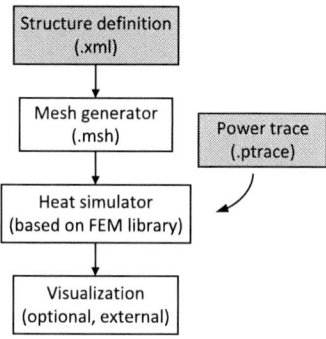

Figure 2: *The MTA flow with user-supplied input files highlighted in grey*

In this paper, we study the thermal characteristics of monolithic 3D ICs by extending an open source thermal modeling tool[3], i.e., the Manchester Thermal Analyzer (MTA), whose flow is shown in Fig. 2. MTA is an efficient academic thermal analysis tool for 2D and TSV-based 3D circuits based on finite element method. Taking the circuit layout and power trace of each functional block as inputs, MTA can generate adaptive mesh grids efficiently. Then, MTA employs a unified methodology that involves advanced spatiotemporal refinement techniques and fast preconditioned iterative solvers that can be applied to solve the discretized thermal equations efficiently for highly accurate thermal analysis.

EXPERIMENT METHODOLOGY

978-1-7281-6559-2/20 $31.00 © 2020 IEEE

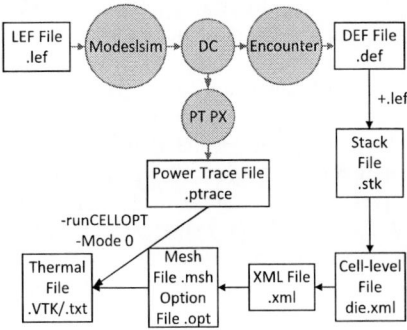

Figure 3: cell-level thermal evaluation steps

Open source FreePDK45[4] and Mono3D[5] cell libraries are used to in our experiments. FreePDK45 is a free and widely used 45nm PDK developed by NCSU and Mono3D PDK is released for transistor-level M3D IC design based on FreePDK45. We use these two cell libraries to synthesize benchmark circuits for cell-level and block-level thermal analysis and comparisons. The cell-level experimental steps are shown in Fig. 3. With the Library exchange format (LEF) file provided by cell library, we synthesized HDL and then perform P&R to generate design exchange format (DEF) files with commercial EDA tools. On the other hand, power trace file is generated by Primetime PX[6]. Then, the DEF file and power trace were input to MTA. Besides, due to the nanoscale feature size of the technology library, a cell-level mesh can contain tens (even hundreds) of millions of nodes, which can make thermal simulations computationally infeasible. Spatially weighted grid lines and power density weighted grid lines algorithms[7] are used to reduce the size of the computational mesh while still maintaining the accuracy of the circuits temperature profile.

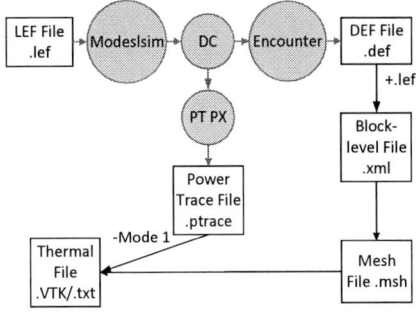

Figure 4: block-level experimental steps

As shown in Fig. 4, in the coarse grain block-level

thermal simulations, instead of calculating the power of each standard cell in the circuit, the overall power consumption of the circuit block was used. Thus, the number of the mesh grid cells can be greatly reduced. This method is suitable for large scale integrated circuits with a millions or tens of millions of cells.

Steady-state and cell-level thermal analysis were performed for light-weight circuits, and transient-state and block-level thermal analysis were performed for larger circuits, which will be presented in the next section.

RESULTS AND DISCUSSION

The open source FreePDK45 and Mono3D cell libraries were used to investigate the footprint, average power (AP), and thermal characteristics of several benchmarks with various number of gates (No. gates). The selected benchmarks (BM) are SIMON (light-weight encryption core), S series circuits (academic benchmark) and larger FFT cores (64-, 128-, and 256-point[9]). They are synthesized using Synopsys Design Compiler at 1 GHz clock frequencies. The results are shown in Table 1.

TABLE 1.   BENCHMARK PARAMETERS

BM	No. Gates	Footprint/ 2D (mm²)	Footprint/ M3D (mm²)	AP/2D (W)	AP/M3D (W)
Steady-state & Cell-level					
SIMON	1278	0.0142	0.0089	2.31e-4	3.60e-4
S13207	2611	0.0247	0.0152	3.374e-4	7.732e-4
S15850	3178	0.0263	0.0157	3.430e-4	9.72e-4
S35932	6892	0.0619	0.0382	1.520e-3	3.563e-3
S38417	9880	0.0648	0.0354	9.216e-4	2.509e-3
Transient-state & Block-level					
FFT64	136398	0.631	0.310	3.43e-2	5.82e-2
FFT128	265287	1.20	0.574	6.47e-2	0.103
FFT256	338923	1.54	0.717	7.07e-2	0.101

Then, MTA was used for thermal analysis. Since the power value of each benchmark is so small that the temperature variation is not obvious. To make our experimental settings match with power consumptions of typical ICs, the order of magnitude of power value were scaled accordingly. That is to say, for steady-state & cell-level thermal analysis, the power values were multiplied by 10^4 and for transient-state & block-level thermal analysis, the power values were multiplied by 10^2.

The average and maximum temperature of each benchmark is shown in Fig. 5 and Fig. 6. It shows that the average and maximum temperature of M3D is higher. The difference between the maximum temperature and the initial temperature (318.15K) of M3D is 3.0-11.0 times higher than that of 2D counterparts, and that of average temperature is 1.0-5.4 times higher.

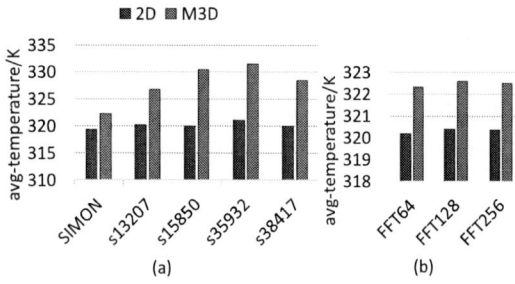

Figure 5: The average temperature of some benchmarks under (a) steady-state and cell-level (b) transient-state and block-level experiments respectively

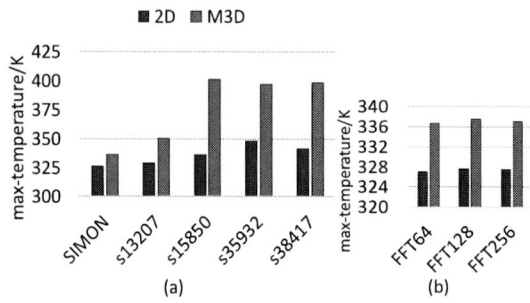

Figure 6: The maximum temperature of some benchmarks under (a) steady-state and cell-level (b) transient-state and block-level experiments respectively

Furthermore, the temperature variations were also compared. Fig. 7 shows that the M3D implementation has a wider temperature distribution than its 2D counterpart. It is 1.0-3.6 times more than that of 2D.

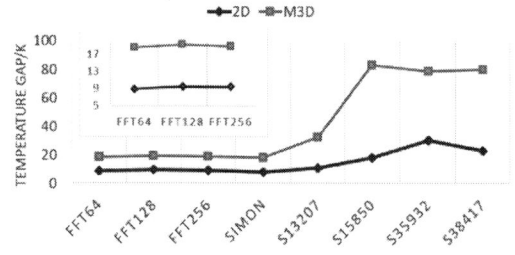

Figure 7: Temperature variations of different benchmarks.

CONCLUSIONS

With diminishing return of the planar scaling of semiconductor devices, monolithic 3D IC technology was proposed to sustain Moore's law more effectively. However, due to higher integration density, M3D ICs face severe thermal dissipation problem, which necessitates the accurate thermal modeling and thermal aware physical design. In this experiment, we extended the MTA tool to build the thermal model of monolithic 3D ICs, and compared the temperature with 2D implementations. The experimental results shown that the average and maximum temperature of M3D is higher. In particular, the difference between the maximum temperature and the surrounding temperature (318.15 K) of M3D is 3.0-11.0 times higher than that of the 2D design, and the average temperature is 1.0-5.4 times higher than that of the 2D design, which highlights the thermal dissipation problem of M3D IC design.

ACKNOWLEDGEMENTS

Thanks for the experimental setup help provided by professor Yi-Chung Chen from Tennessee State University.

REFERENCES

[1] Batude P, Ernst T, Arcamone J, et al. 3-D sequential integration: A key enabling technology for heterogeneous co-integration of new function with CMOS[J]. IEEE Journal on Emerging and Selected Topics in Circuits and Systems, 2012, 2(4): 714-722.

[2] Samal S K, Panth S, Samadi K, et al. Fast and accurate thermal modeling and optimization for monolithic 3D ICs[C]//2014 51st ACM/EDAC/IEEE Design Automation Conference (DAC). IEEE, 2014: 1-6.

[3] Ladenheim S, Chen Y C, Mihajlović M, et al. The MTA: An Advanced and Versatile Thermal Simulator for Integrated Systems[J]. IEEE Transactions on Computer-Aided Design of Integrated Circuits and Systems, 2018, 37(12): 3123-3136.

[4] NCSU E D A. FreePDK45[J]. 2011.

[5] Yan C, Salman E. Mono3D: Open source cell library for monolithic 3-D integrated circuits[J]. IEEE Transactions on Circuits and Systems I: Regular Papers, 2017, 65(3): 1075-1085.

[6] Haider S. Virginia Tech,"Power estimation with osu standard cell library and synopsys tools (primetime-px),"[J]. 2008.

[7] Chen Y C, Ladenheim S, Kalargaris H, et al. Computationally efficient standard-cell FEM-based thermal analysis[C]//Proceedings of the 36th International Conference on Computer-Aided Design. IEEE Press, 2017: 490-495.

[8] Milder P, Franchetti F, Hoe J C, et al. Computer generation of hardware for linear digital signal processing transforms[J]. ACM Transactions on Design Automation of Electronic Systems (TODAES), 2012, 17(2): 1-33.

A 2-D CAPACITANCE SOLVER WITH FINITE DIFFERENCE METHOD

Wenjie Liang*, and Wenjian Yu

BNRist, Dept. Computer Science & Tech., Tsinghua University, Beijing 100084, China

*Corresponding Author's Email: liang-wj18@mails.tsinghua.edu.cn

ABSTRACT

In this paper, we present a capacitance solver based on finite difference method (FDM). It simulates the cross section of interconnect structures and computes the capacitances per unit length. The techniques of forming symmetric coefficient matrix and nonuniform FDM grids are developed. And, with a sparse direct solver based on Cholesky factorization the presented solver exhibits high runtime efficiency with good accuracy. Experiments on pattern structures show that the presented solver is 3X faster than Raphael rc2, and is capable of accurately extracting structures with trapezoidal cross-section conductors and conformal dielectrics.

INTRODUCTION

Nowadays, with the development of nanometer technology, the interconnect wires are very densely routed in integrated circuits (ICs). As a consequence, the parasitic capacitance among interconnects becomes increasingly important for IC design, where signal delay and other performance metrics need to be verified [1]. Accurate layout parasitic extraction (LPE) tool and capacitance field solver are much demanded [2].

Domain discretization method [3, 4] and boundary element method (BEM) [5, 6] are two major methods for the field solver, especially useful for building capacitance library for LPE tools. The former includes finite difference method (FDM) [3] and finite element method (FEM) [4]. Based on the latter, a fast solver was proposed in [5] for 2-D capacitance extraction. However, its accuracy is sensitive to the boundary discretization. Compared with BEM, the domain discretization in FDM is much easier so that the FDM based capacitance solver is robust on accuracy and widely used in industry.

In this paper, we present a 2-D capacitance solver based on FDM, which simulates the cross section of interconnect structures and obtains the capacitances per unit length. Several techniques have been developed, which make the solver more efficient than the golden tool Raphael rc2 without accuracy loss. A key contribution is that we produce symmetric coefficient matrix so that the sparse direct solver based on Cholesky factorization [7] can be employed to solve the FDM equations. We have also developed the nonuniform gridding approach and the measures for handling trapezoidal cross-section conductors and conformal dielectrics. Numerical experiments with thousands of test cases have validated the accuracy and efficiency of the presented solver.

PROBLEM FORMULATION

Consider a structure with m conductors embedded in M dielectric layers. The relation between the potentials of m conductors, denoted by $V \in \mathbb{R}^m$, and their charges, denoted by $Q \in \mathbb{R}^m$, is given by $Q = CV$, where $C \in \mathbb{R}^{m \times m}$ is the capacitance matrix [5]. To calculate the capacitances, we set the potential of a conductor, called master conductor, to be 1 Volt, and the potential of rest conductors to be 0 Volt. Then, the capacitances between the master conductor and other conductors can be obtained by solving the charge vector Q with FDM, etc.

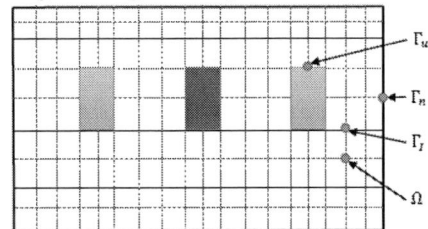

Figure1: A simple 2-D model for capacitance extraction.

The typical structure considered in 2-D capacitance solver is illustrated in Fig. 1. The red block denotes the cross section of the master conductor. The gray blocks are two nearby parallel wires. As shown in Fig. 1, the 2-D grids are imposed, and there are four kinds of grid points where different equations apply (u denotes potential):

$$\begin{cases} \nabla^2 u = \dfrac{\partial^2 u}{\partial x^2} + \dfrac{\partial^2 u}{\partial y^2} = 0, & \text{in } \Omega \\[2mm] u = u_0, & \text{on } \Gamma_u \\[2mm] \dfrac{\partial u}{\partial \vec{n}} = 0, & \text{on } \Gamma_n \\[2mm] \varepsilon_a \dfrac{\partial u_a}{\partial \vec{n}_a} = -\varepsilon_b \dfrac{\partial u_b}{\partial \vec{n}_b}, & \text{on } \Gamma_l \end{cases} \quad (1)$$

For the points on Γ_u (surface of conductors), u satisfies the Dirichlet boundary condition with predefined voltage 0 or 1. For the points on Γ_n (outer boundary of the whole domain Ω), the Neumann boundary condition is set, i.e. the derivative of potential along normal direction is equal to 0. For the points on Γ_l (interface of two adjacent dielectrics) the last equation in (1) holds, where ε_a and ε_b represent the permittivities of adjacent dielectric a and b respectively. The 2-D Laplace equation holds in Ω.

With FDM, the partial differential equations in (1) can be transformed to linear equations (2)~(4).

$$\frac{1}{d_l(d_l + d_r)} u_l + \frac{1}{d_r(d_l + d_r)} u_r + \frac{1}{d_d(d_d + d_u)} u_d$$
$$+ \frac{1}{d_u(d_d + d_u)} u_u - \left(\frac{1}{d_l d_r} + \frac{1}{d_u d_d} \right) u = 0, \quad \text{P in } \Omega \quad (2)$$

978-1-7281-6559-2/20 $31.00 © 2020 IEEE

$$\frac{1}{d}u - \frac{1}{d}u_n = 0, \qquad\qquad \text{P on } \Gamma_n \qquad (3)$$

$$\left(\frac{\varepsilon_a}{d_a} + \frac{\varepsilon_b}{d_b}\right)u - \frac{\varepsilon_a}{d_a}u_a - \frac{\varepsilon_b}{d_b}u_b = 0, \qquad \text{P on } \Gamma_l \qquad (4)$$

Here, u is the potential of target point (hereinafter called P). Eq. (2) describes the relationship between u and the potential of its neighbor grid points. The subscripts l, r, u and p differ the distances from P to its neighbors along left, right, up and down directions. If P is on Γ_n, then u_n is the potential of point nearest to P in the normal direction and d is the distance between the two points. If P is on Γ_l, u_a and u_b are potentials of the points nearest to P on the both sides of dielectric interface. d_a and d_b are the corresponding distances from them to P. ε_a and ε_b are permittivities of the two adjacent dielectrics.

TECHNIQUES BASED ON FDM

Symmetric Matrix Formation

According to (2)~(4), there is a linear equation for each grid points. Overall, the equations form a linear equation system for the unknown potentials on grid points. After solving the equation system to get the potentials, the conductor charges or capacitances can be calculated with the difference of potentials around conductor surfaces. In practice, the size of linear equation system can be very large. So, it costs much computational time and memory storage for solving the linear equation system.

From (2)~(4), we see that every point is only related to its surrounding points. So, we can modify them to guarantee the symmetry of coefficient matrix in the linear equation system. The modified equations are (5)~(8):

$$\varepsilon\left(\frac{1}{d_l d_r} + \frac{1}{d_u d_d}\right)(d_l + d_r)(d_u + d_d)u - \frac{\varepsilon}{d_l}(d_u + d_d)u_l -$$
$$\frac{\varepsilon}{d_r}(d_u + d_d)u_r - \frac{\varepsilon}{d_u}(d_l + d_r)u_u - \frac{\varepsilon}{d_d}(d_l + d_r)u_d = 0, \text{P in } \Omega \qquad (5)$$

$$\frac{\varepsilon}{d}(d_u + d_d)u - \frac{\varepsilon}{d}(d_u + d_d)u_n = 0, \qquad \text{P on } \Gamma_n \qquad (6)$$

$$\frac{\varepsilon}{d}(d_l + d_r)u - \frac{\varepsilon}{d}(d_l + d_r)x_n = 0, \qquad \text{P on } \Gamma_n \qquad (7)$$

$$\left(\frac{\varepsilon_u}{d_u} + \frac{\varepsilon_d}{d_d}\right)(d_l + d_r)u - \frac{\varepsilon_u}{d_u}(d_l + d_r)u_u$$
$$- \frac{\varepsilon_d}{d_d}(d_l + d_r)u_d = 0, \qquad \text{P on } \Gamma_l \qquad (8)$$

There are two equations for P on Γ_n, reflecting the situations of top/bottom boundary and left/right boundary, respectively. With these equations, the linear equation system obviously has a symmetric sparse coefficient matrix. So, the advanced sparse direct solver CHOLMOD [7] can be employed for reliably solving the FDM equation.

Nonuniform FDM Grids for Complex Structures

Nonuniform FDM gridding is necessary to reduce the number of unknowns without scarifying accuracy. Considering the model structure in Fig. 1, we can impose several strategies to design the desirable nonuniform grids. The key point is to decide which parts of the domain should be discretized into dense or coarse grids. Based on the knowledge of electrostatic field, we should consider

the shielding effect, the radiation of electric field from the master conductor. Basically, denser grids should be set in the region with large variation of electric field. On the other hand, the aspect ratio of the grid cell should also be within a reasonable range to ensure the accuracy.

For the model structure, i.e. that in Fig. 2(a), the basic FDM grid is depicted in Fig. 2(b). Two rules for choosing the region for which denser grids are applied are illustrated in Fig. 2(c) and 2(d), respectively. In Fig. 2(c), the effect of master conductor is considered. In Fig. 2(d), the regions with large variation of electric field are discretized with denser grids. More sophisticated gridding strategy has been developed for handling structure with conformal dielectric. They ensure accurate computational results with a relatively small number of unknowns.

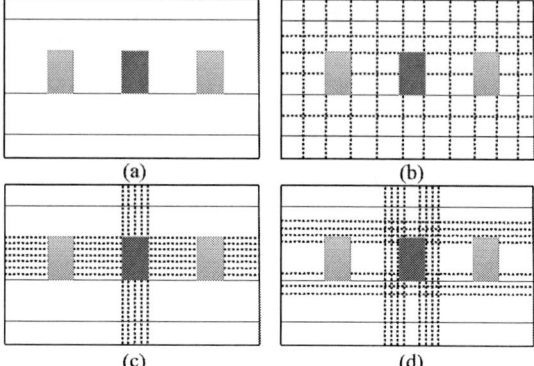

Figure 2: Nonuniform FDM grids for capacitance extraction

Implementation details

From Fig. 2 we can see that the whole domain is divided into neat grids. For each grid point, we store the attributes including boundary type, position coordinates and neighboring points. The attributes of all grid points form a 2-D array where the points are numbered from bottom to top and from left to right in the model. The algorithm of forming the coefficient matrix for FDM is implemented by scanning the 2-D array. Once a grid point is processed, with its attributes we choose an equation from (5)~(8) for it and fill the coefficient matrix with correct entries. The overall flow of this FDM based capacitance solver is described as Algorithm 1.

Algorithm 1 2-D capacitance solver based on FDM

Step 1. Extract all the boundary of conductors and dielectrics and form the original gridlines.

Step 2. Generate all the gridlines according to nonuniform FDM gridding approach.

Step 3. Traverse all grid points from bottom to top, left to right, and store their attributes (boundary type, coordinates, neighbors, etc.) into a 2-D array.

Step 4. Scan the 2-D array and choose suitable linear equation in (5)~(8) for each point to construct the coefficient matrix for the linear equation system.

Step 5. Solve the symmetric sparse linear equation system with CHOLMOD.

Step 6. Calculate conductor charges and capacitances.

NUMERICAL RESULTS

With the presented techniques, we have developed a capacitance extraction program called FDCap2d. We have tested the program with structures of interconnect patterns, and compared the results with Raphael rc2 [8]. Raphael rc2 is a widely used commercial finite difference solver using the techniques of advanced nonuniform meshing and the preconditioned conjugate gradient (PCG) iterative solver. The CPU time for the programs is recorded on a Linux server with Intel Xeon E5-2650 2.0 GHz CPU.

Firstly, we show the results of several pattern structure without conformal dielectrics in Table I. From it we see that while the number of grids is similar to Raphael, FDCap2d obtains accurate results with less than 3% error.

Table I. Results of total capacitances for pattern structures

Case	# of grid points		Capacitance (F/um)		Error
	Raphael	FDCap2d	Raphael	FDCap2d	
1	199892	193638	1.69e-16	1.67e-16	-1.2%
2	199980	195839	9.37e-17	9.22e-17	-1.6%
3	199992	194448	1.49e-16	1.48e-16	-0.7%
4	199938	194406	8.07e-16	7.86e-16	-2.6%
5	199752	196128	1.53e-16	1.52e-16	-0.7%

We have tested three batches of cases named iscas89, 45nm and ITF with 500, 1200 and 3024 structures respectively. Under the condition of similar number of grids, the results show that the average relative errors of FDCap2d are within 2% with the max relative errors no more than 3.69%, as shown in Table II. With the computational results for all the 4724 cases, we find out that for around 95% of them that FDCap2d's error is within 3%. In Table II, the total computational time of the both programs are also shown. We see that our FDCap2d is about **3X** faster than Raphael rc2. This suggests that the direct sparse solver is advantageous over the iterative solver for 2-D capacitance extraction problems.

Table II. Results of three batches of test structures

Tech.	Error_avg	Error_max	Time (FDCap2d)	Time (Raphael)
Iscas89	0.93%	2.59%	9m46s	16m44s
45nm	0.91%	2.72%	23m24s	33m6s
ITF	1.31%	3.69%	61m	202m55s

Furthermore, FDCap2d is able to handle conductor wires with trapezoidal cross section. Under nanometer technology, the side tangent, i.e. $\tan(\alpha)$ shown in Fig. 3(a), is a small value. So, we can adjust the length of its top and bottom edges to generate a new rectangle with area unchanged (see figure 3(a)). The conformal dielectrics are special-shape dielectrics around conductors, as shown in Fig. 3(b). For structure with conformal dielectrics, FDCap2d can also be used to calculate the capacitances. The experimental results of several such structures are given in Table III. The first three include conductors with

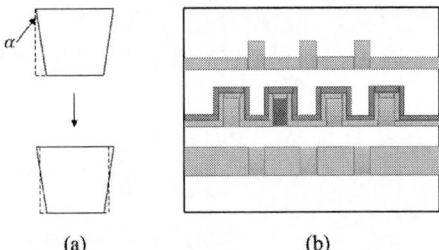

(a)   (b)

Figure 3: (a) Trapezoidal conductor, (b) conformal dielectrics.

trapezoidal cross section (with $\tan(\alpha)$=0.05), while the last three are structures with conformal dielectrics. From the results the good accuracy of FDCap2d is validated again.

Table III. Results of total capacitances for cases in Fig. 3

Case	# of grid points		Capacitance (F/um)		Error
	Raphael	FDCap2d	Raphael	FDCap2d	
1	199980	193368	1.71e-16	1.70e-16	-0.6%
2	199836	195076	1.36e-16	1.35e-16	-0.7%
3	199992	192878	3.29e-16	3.22e-16	-2.1%
4	249956	253600	5.43e-16	5.35e-16	-1.5%
5	249956	256692	6.19e-17	5.92e-17	-4.4%
6	249956	254446	5.43e-16	5.35e-16	-1.5%

CONCLUSIONS

In this work, a 2-D capacitance solver based on FDM is developed. It utilizes a sparse direct solver and nonuniform gridding strategies, exhibiting better efficiency than Raphael rc2 based on iterative solver. The experiments validate the benefit of the presented solver, and show that it is able to handle structures with trapezoidal cross-section conductors and conformal dielectrics. In the future, the solver will be employed to build the capacitance library for LPE tools.

REFERENCES

[1] K. Nabors K and J. White, "FastCap: A multipole accelerated 3-D capacitance extraction program," *IEEE Trans. Computer-Aided Design.*, 10(11): 1447-1459, 1991.

[2] W. Yu and X. Wang, *Advanced Field-Solver Techniques for RC Extraction of Integrated Circuits*, Springer Inc., Apr. 2014.

[3] P. Sumant and A. Cangellaris, "Algebraic multigrid Laplace solver for the extraction of capacitances of conductors in multi-layer dielectrics," *Int. J. Numer. Model.*, 20:253-269, 2007.

[4] C. D. Taylor, G. N. Elkhouri and T. E. Wade, "On the parasitic capacitances of multilevel parallel metallization lines," *IEEE Trans. Electron Devices*, 32(11): 2408-2414, 1985.

[5] K. Zhai and W. Yu, "The 2-D boundary element techniques for capacitance extraction of nanometer VLSI interconnects," *Int. J. Numer. Model.*, 27: 656-668, 2014.

[6] X. Wang, D. Liu, W. Yu, and Z. Wang, "Improved boundary element method for fast 3-D interconnect resistance extraction," *IEICE Trans. Electronics*, E88-C(2): 232-240, 2005.

[7] Y. Chen, T. A. Davis, et al., "Algorithm 887: CHOLMOD, supernodal sparse cholesky factorization and update/downdate," *ACM Trans. Math. Soft.*, 35(3): 22, 2008.

[8] Synopsys Inc., Raphael: 2D, 3D resistance, capacitance and inductance extraction tool. https://www.synopsys.com/

METAL TRENCH CRITICAL DIMENSION AND OVERLAY MINOR VARIATION MONITORING METHOD WITH VOLTAGE CONTRAST INSPECTION

Lijing Huang, Qiliang Ni, Xiaofang Gu, Chao Han, Jiansi Yuan

Shanghai Huali Microelectronics Corporation

Shanghai, China

*Corresponding Author's Email: huanglijing@hlmc.cn

ABSTRACT

The investigation aims at the metal trench critical dimension (CD) and overlay variation monitoring methodology with voltage-contrast (VC) inspection. The VC inspection with negative charging mode was performed to detect metal trench CD and overlay variation issue, at NDC film deposition layer post second metal layer chemical and mechanical polish. Dark VC (DVC) defects were found at extreme wafer edge, which would cause end of line (EOL) yield loss by data retention soft bin failure. DVC defects were identified by inline SEM review results with high voltage and PFA analysis results. It was demonstrated that the defects were induced by metal trench CD and overlay variation. Furthermore, the defects impacted factors, including uniformity of thin film deposition and chemical and mechanical polish (CMP), the etching rate performance of all in one etch process, and even the e-chuck accumulated contamination of lithography, were also investigated. By increasing the relative process tools' offline monitor frequency and optimizing the prevent maintenance method of lithography tool, defects were fixed and trend low. The study here extended the usage of VC inspection to detect CD and overlay minor variation.

INTRODUCTION

An e-beam scan with VC images comparison is an effective inspection method and a good alternative to bright and dark field ones as tolerance of defects in the semiconductor process decreases [1]. With the SEM (scanning electron microscope) image observation, VC inspection is able to detect tiny defects nonvisual to optical inspection [2], furthermore, the under layer defect is possible to be found by the surface charging [3]. Some applications even use dedicated test structures [4] for VC inspection as a routine monitoring vehicle to sweep the line defects. The VC inspection also be used to verify NMOS leakage caused by broken intra-well isolation of the SRAM. [5]. Usual cases of those applications are mostly aimed at detecting physical defects or short open defect post CTW CMP at MEOL of the SRAM. The study herein further extends VC inspection to detect defect at the BEOL steps.

In the process step of post contact METAL CMP, each independent copper conductive plug is isolated by the surrounding oxide insulating film(Fig.1), connected with tungsten through copper wire, and then falls in different areas of the device, such as source/drain and poly gates in either N or PMOS, which constructs a pure metric condition for distinguishing VC.

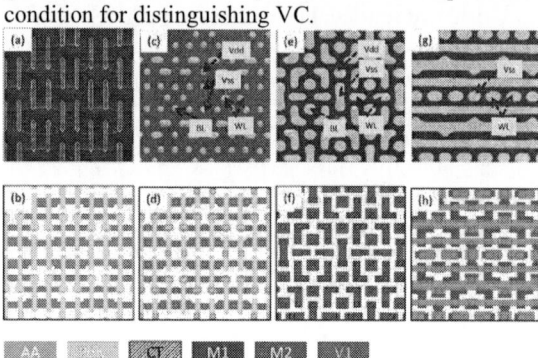

Fig.1. (a) image after P1_ASI;(b) AA and Poly layer out;(c) image after CTW_CMP;(d) AA, Poly and CT layer out; (e) image after M1_CMP;(f) CT and M1 layer out; (g) image after M2_CMP;(h) CT ,M1,M2and V1 layer out

In this paper, a negative mode detection method is mainly proposed; it is used to detect the leakage of BEOL metal from defocus which is not easily detected by optical defect inspection. Metal leakage defect's mechanism and the impacted factors were analyzed and inline verified by VC inspection. By optimizing the cleaning method of e-chuck on photolithography machine, the metal leakage defect was solved.

The M2 CMP layer was selected to do VC inspection, which can detect the short between metal and via open defect at the same time. As shown in Fig.4, under normal status without metal short, VC difference can be seen using negative mode: WL is bright while Vss is dark (Fig.2(a)). However, when metal short occurs, the VC change can be observed in negative mode: WL appears dark (Fig.2(b)). Copper properties were exposed to air for too long and crystal was easy to grow, which will have an impact on chip performance. Therefore, this paper for safety reasons selected to test at NDC film deposition layer post M2 CMP.

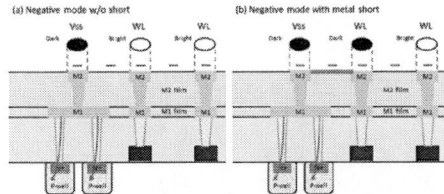

Fig.2. VC images on top of the copper plugs for Vss and WL. (a) The Vss and WL without short in negative modes; (b) The Vss and WL with metal short in negative modes

Theoretically, if metal trench CD or overlay is variable, it will lead to metal exceptional contact on the surface, which will lead to metal short. As mentioned above, metal short can be detected by VC detection. In this study, it was found that this variation was manifested as DVC by VC inspection.

DEFECT ANALYSIS

Defect Map Distribution Feature

Through inspection, the defects are mainly distributed in the direction of 3 o 'clock and 9 o 'clock on the wafer, and are also regular in the die, shown in Fig.3. This phenomenon indicates that the closer to the position of the wafer edge, the more likely defects would occur. The map distribution was consistent with the large variation of wafer edge overlay by lithography.

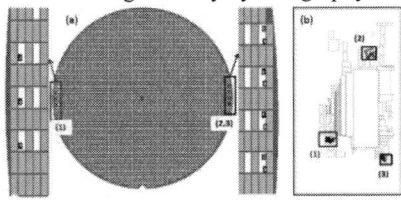

Fig.3. (a) defect distribution in wafer; (b) defect distribution in die;(1,2,3)E-Beam scan care area

Defect Review Result and PFA Analysis

Worse WL metal dark VC (DVC) defect was found from inline VC inspection result with negative charging mode, post M3LINER DEP layer, as shown Fig.4 (a). This defect was invisible using normal voltage (~1000V) condition with review tool of SEMvision, shown in Fig.4 (b). Then high voltage (>3000V) was selected to do defect review. And this condition could penetrate the film surface; therefore pre layer metal pattern could be seen. From the review result with high voltage condition, slight metal spot enlarge was found, as shown in Fig.4(c). Then PFA analysis was carried out to identify defect cross section performance. The metal spot of WL DVC's CD was larger than the normal one, shown in Fig.4 (d). Therefore, second metal trench CD was exceptional.

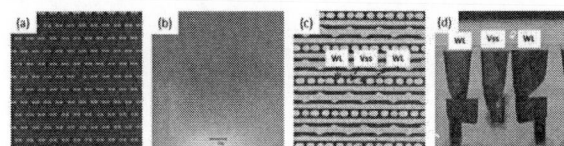

Fig.4. WL leakage defect.(a) defect image with E-beam review; (b) defect image with SEMvision review(1000V); (c) defect image with SEMvision review(3000V); (d) defect TEM

Impacted Factors of the metal short

The metal short defect was occurred on the wafer edge, which distribution feature was influenced by metal trench CD and overlay variation. Furthermore, the defects impacted factors, including uniformity of thin film deposition and chemical and mechanical polish (CMP), the etching rate performance of all in one etch process, and even the e-chuck accumulated contamination of lithography, were also causes.

The impact of lithography

The ability of the photolithography process itself is poor on the wafer edge. Shown as Fig. 5(a), from the overlay map, the more on the wafer edge, the lager variation of overlay would be. Meanwhile, there are problems on the wafer edge on the hardware of the machine. Flatness of the test machine would show poor performance on the wafer edge, shown in Fig. 5(b).

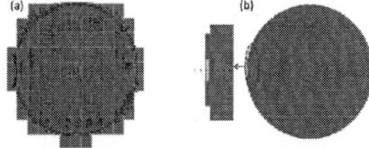

Fig.5. (a) photolithography tool overlay(OVL) map;(b) suffer WL leakage defect tool flatness

The impact of thin film deposition, etching rate and CMP

In the thin film process in front of the photolithography process, its characteristics of deposition chamber cause different deposition thickness of the wafer edge and the wafer center, as shown in Fig.6(a). Thickness differences would enhance the differences between wafer edge of lithography technology and center overlay.

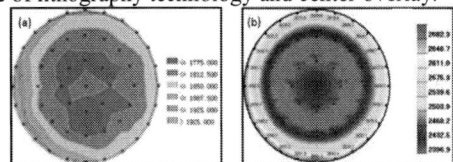

Fig.6.(a)film process deposition rate map;(b)Etch process etch rate map

In addition, there is a difference in the etching speed between wafer edge and center in the dry etching process on the wafer edge after the lithography pattern is displayed.

If edge has a slight overlay shift, the etching rate will be amplified due to the fast etching rate, and then there is leakage after copper filling, also exacerbates CD anomalies.

The wafer profile was seen from the thickness monitor trend after copper CMP, the edge thickness is thinner than the wafer center, which is caused by faster grinding speed of the wafer edge ,shown in Fig.7. This would also affect the leveling of the wafer in lithography.

From what has been discussed above, the reasons for the metal short defect come from many aspects. As uniformity of thin film deposition and chemical and mechanical polish (CMP), the etching rate performance of all in one etch process all lead to different uniformity between wafer edge and center, edge metal trench CD and overlay variation in lithography is intensified

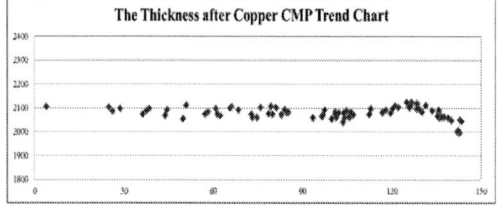

Fig.7 the thickness after copper CMP trend in one wafer

Defect Reduction Solutions

The difference between front and back layer processes on the wafer edge and center is the process characteristics, so improvement cannot be made temporarily. This paper mainly aims to improve defects by optimizing e-chuck cleaning mode of lithography machine. The old method mainly uses chemical solution cleaning and low-frequency physical contact grinding method, but this grindstone is made of SiC material as shown in Fig.8 (a). It is hard and will damage e-chuck if it is grinded for a long time. After that, a soft Si grindstone is selected as shown in Fig.8 (b) to reduce the risk of hardware damage. The surface roughness is large and the cleaning effect is obvious as shown in Fig.8(c)(d)

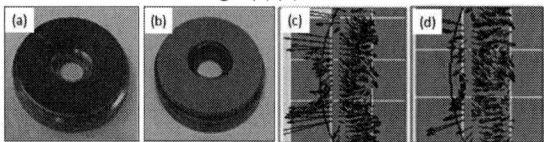

Fig.8. (a) old grindstones: Sic; (b) old grindstones: Si; (c) edge overlay before clean; (d) edge overlay after clean;

Finally, the DVC defect of metal short was reduced shown in Fig.9

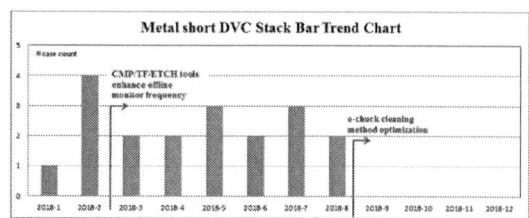

Fig.9 The metal short DVC defects stack bar trend chart

SUMMARY

The investigation aimed at the metal trench CD and overlay variation monitoring methodology with VC inspection. Inline index of metal short defect detection by negative charging mode of VC inspection was established. And the failure model that the soft bin failure was induced by uniformity of thin film deposition and CMP, the etching rate performance of all in one etch process, and even the e-chuck accumulated contamination of lithography. By increasing the relative process tools' offline monitor frequency and optimizing the prevent maintenance method of lithography tool, defects were fixed and trend low.

REFERENCES

[1] T.R. Cass, D. Hendricks, J. Jau, H.J. Dohse, A.D. Brodie, W.D. Meisburger, Application of the SEMSpecelectron-beam inspection system to in-process defect detection on semiconductor wafers, Microelectronic Engineering, Volume 30, Issues 1–4, January 1996, Pages 567-570, ISSN 0167-9317, 10.1016/0167-9317(95)00311-8.

[2] C. Boye, "Industry survey on nonvisual defect detection," Proc. SPIE: Process Mater. Characterization Diagnostics in IC Manufacturing, Vol. 5041, Feb. 2003.

[3] L. Lin, J.Y. Chen, S.D. Luo, W. Y. Wong, S. Oestreich, A. Tsai, I. Yao, and L. Grella, "Residual oxide detection with automated E-beam inspection", IEEE International Symposium on Semiconductor Manufacturing, Sep. 2005, pp. 241-244.

[4] O.D. Patterson, K. Wu, D. Mocuta, and K. Nafisi, "Voltage Contrast Inspection Methodology for Inline Detection of Missing Spacer and Other Nonvisual Defects", IEEE Transactions on Semiconductor Manufacturing, vol. 21, no. 3,, Aug. 2008 pp. 322-328.

[5] Hunglin Chen, Rongwei Fan, Hsiaochi Lou, Mingsheng Kuo, and Yiping Huang. Mechanism and 281 Application of NMOS Leakage with Intra-Well Isolation Breakdown by Voltage Contrast Detection: 282 Journal of semiconductor technology and science, Vol.13, NO.4, August, 2013

A

Ahn Sang Ki	7-10
Alian AliReza	6-7
An Xia	1-81, 3-40

B

Bai Wenqi	1-17, 1-22
Bao Jie	7-5
Bao Shengyu	1-77, 1-78
Bao W Z	1-29
Bian Yuyang	2-28
Brown James	1-96
Broz Jerry J	7-3

C

Cai Yimao	1-77, 1-78
Cao Steam	1-45
Cao Yanpeng	1-40, 4-35
Cao Yurong	5-2, 6-24
Caowenjie zhouchun	1-38
Challapalle Nagadastagiri	9-4
Chan Mansun	4-10
Chang Hsueh-Er	4-39
Chang Richard	1-23
Chang Yanyan	6-24
Chao Li	7-14
Chasin A	1-33
Che Dongchen	3-52
Chen Cheng	1-84, 1-88
Chen C-Z	9-1, 9-26
Chen Dan	6-19, 6-20
Chen Daqin	1-76, 1-78
Chen Gim	1-20
Chen Haoyu	1-18, 1-26, 1-27, 1-28
Chen Hua	6-39, 7-14
Chen Hualun	1-8, 1-9, 1-10, 1-11, 1-13, 1-24
Chen Hunglin	1-48,6-28, 6-32, 6-33, 6-37
Chen Hungling	6-18
Chen Jin	5-2
Chen Kecheng	4-25, 4-27, 4-33
Chen Li	8-2
Chen Liang	1-82
Chen Liming	1-43
Chen Lin	9-23
Chen Lu	3-52
Chen Rui	2-7

Chen Shanshan	1-48
Chen Shoumian	1-44, 1-50, 1-54, 2-20, 2-34, 3-11, 3-21, 3-24, 4-45, 6-25, 8-14, 9-29
Chen Shujing	7-13
Chen Tianrun	9-24
Cheng Xinhua	4-33
Chen Xu	6-15, 6-23, 9-12, -29
Chen Yang	6-38
Chen Yanpeng	2-29
Chen Yaoyu	1-19
Chen Ying	2-7
Chen Yongyue	4-29, 4-31
Chen Yu	1-8, 1-9, 1-10, 1-11
Chen Zhenzai	7-5
Cheng Cadie	1-83
Cheng Cheng	7-22
Cheng Wenkai	7-6
Cheng Xingming	7-18, 7-22
Cheng Xinhua	4-25, 4-27
Cheng Yuanqing	9-16
Cheng Yuanshen	5-34, 5-41
Chi Min-Hwa	1-23
Chi Yu Shan	3-17
Chong J	3-8
Chu Zhufei	9-23
Collaert Nadine	6-7
Cristoloveanu S	1-29
Cui Liang	7-18
Cui Yaqi	5-38

D

Dai Johnny	6-34, 6-36
Dai Shugang	1-28
Dang Bingjie	1-71
De Chen Yu	1-68
De Keersgieter A	1-33
Deng Guogui	1-40
Deng Hai	2-44
Deng JN	1-29
Deng Meng	4-43
Di Hunhuan	4-43, 8-12, 8-13
Di Xichao	8-12, 8-13
Dijkstra Peter	7-2
Ding Alice	1-23
Ding Yu	1-44, 1-50, 1-54
Dong Grass	7-11
Dong Liqun	1-26, 1-28

Dong Lisong	2-7
Dong Qingwei	4-25, 4-27, 4-33
Dong Zhigang	5-24
Dong Zihan	3-20
Du Kaixuan	9-27
Duan Jiebin	8-2
Duan M	6-16
Duan Qingxi	1-71, 1-83
Duan Wenting	1-25, 1-41, 1-42, 1-51
Duan Yangyang	7-18, 7-22
Dube Belinda Langelihle	1-34

E

Eneman G	1-33, 6-7

F

Falise Frédéric	6-41
Fan Dandan	7-18
Fang H	9-11
Fang Jingxun	1-52, 4-25, 4-27, 4-29, 4-30, 4-31, 4-33, 5-14, 5-15
Fang Ziquan	1-25, 1-41, 1-42, 1-51, 1-76
Feng Zhe	5-2
Fu Chang	4-17
Fu Chao	4-36, 6-11
Fu Hao	9-29
Fu Zhiyuan	1-84, 1-88
Fuh Yiin-Kuen	4-39
Fujimori Toru	2-30

G

Gao Baohong	
Gao Bin	1-79
Gao Jingbo	9-14
Gao Jun	1-71
Gao Pengcheng	5-28, 5-37
Gao Song	2-27
Gao Xing	1-6, 2-3
Gaudestad J	2-1
Gong X	1-29
Gong Yiqu	4-30
Green Michael	2-38
Gu Jinglun	1-7
Gu Lin	6-29
Gu Xiao Fang	1-36, 6-13
Gu Xiangui	9-30
Gu Zhiqiang	3-52

Guan Xijun	2-28
Gui Yan	4-25, 4-27, 4-33
Guo Dongming	5-24, 5-25
Guo Hao	1-36, 6-18, 6-28, 6-33
Guo Xiaobo	2-26, 2-27, 2-28
Guo Xiaoguang	5-25
Guo Yaozu	8-10, 9-28

H

Ham Young	2-38
Hamada Satomi	5-47
Han Chao	6-13
Han Han	6-7
Han Hyun Chul	7-10
Han Jianglong	7-18
Han Leng	4-45
Han Penggang	3-33
Han Xiaojing	1-30
Han Yemei	1-53, 1-74
Hang Mingguang	3-25
Hang Tao	7-8
Hao Yanxia	1-52
Hao Zhang	7-16
He Guang	1-36
He Keqiang	3-18
He Li	7-16,
He Liang	6-39, 7-14
Heyns Marc	6-7
Hiyama Hirokuni	5-47
Hong Jaiaqi	4-29, 4-31
Horiguchi N	1-33
Horn George W	6-6
Hou Li	7-5
Hsu Chien-Pin Sherman	3-49
Hsu Po-Chun (Brent)	6-7
Hu David Y	9-26
Hu Dongdong	3-52
Hu Haoru	2-7
Hu Jun	1-24, 1-25, 1-42
Hu Kailiang	5-26
Hu Shaojian	1-44, 1-50, 1-54, 8-14
Hu Shijie	1-80
Yu Yidan	1-40
Hu Yongjie	3-52
Hu Yu	6-30
Hu Yuan	7-1

Hu Zhanyuan	1-17, 1-22, 2-31
Hua Younan	4-36, 6-11
Huang Chong	1-45
Huang Guanqun	1-26, 1-27, 1-28
Huang Jacky	1-68
Huang Jun	2-26, 2-27, 2-28, 3-15, 3-20,3-26, 3-27, 3-28, 3-29, 3-31
Huang Kang	3-18, 4-17
Huang Lijing	6-13
Huang Qianqian	1-81, 1-82, 1-84, 1-88, 1-89
Huang Qiumin	4-29, 4-31
Huang Ru	1-71, 1-73, 1-77, 1-78, 1-80, 1-81, 1-82, 1-83, 1-84, 1-86, 1-88, 1-89, 3-4
Huang Weidong	7-11
Huang Zhisen	1-17, 1-22
Huang Zhoncheng	9-17
Huo Zhaoqing	5-31, 5-36, 5-38
Humphrey Bret A	7-3
Huynh-Bao T	1-33

I

J

Jang D	1-33
Jao Nicholas	9-4
Ji Feng	5-2
Ji Xiaoli	8-10, 9-28
Ji Yong	7-21
Ji Z	6-16
Ji Zhigang	1-96
Ji Zujun	5-2
Jia Changzhen	5-16, 5-20
Jia Lili	3-25, 3-27
Jia Wenpeng	4-36, 6-11
Jiang Lan	4-25, 4-27, 4-33
Jiang Lingpeng	3-29
Jiang Xiaojuan	7-18
Jiang Y L	1-29
Liang Yu	3-18
Jiang Ziwei	7-13
Jiaping Zeng	7-16
Jie Zhang	7-16, 7-17
Jin Zhuji	5-25
Jing Zhaokun	1-83
Jones Christopher P	4-22
Joseph Ervin	1-68
Ju Xiaohua	1-19
Jun Jiang	7-16

K

Kai Jang	7-16
Kai Wang	1-48
Kang Junlong	1-52
Kang Renke	5-24, 5-25
Kang Xiaoxu	8-8
Kashkoush Ismael	1-20
Khailany Brucek	9-10
Kim Cheolkyu	6-34, 6-36
Kim Gun Woo	7-10
Kim Tae-Gon	5-47
Kirchner Lisa	7-6

L

Lai Lulu	2-26
Lai Walter	4-39
Lan B J	9-11
Lau W S	4-18, 4-19
Lee Boon-Chye	4-36
Lee Boon-Seong	4-36
Lee H J	2-38
Lee Jong-Ho	1-93
Lee Ken	7-10
Lee Kunghong	1-17, 1-22
Lee Kyu Jin	7-10
Lee Sung-Tae	1-93
Lee Yi-Ting	9-7
Lei Haibo	1-56
Li Andrew	3-17
Li Baoxuan	2-29
Li Chen	2-7, 2-34, 9-29
Li Chuang	1-74
Li Ding	3-8
Li Fang	3-25, 3-26, 3-27
Li Haipeng	5-22
Li Haixa	1-73, 4-30
Li Hong	3-26
Li Hu	5-2
Li Hui	1-53
Li Huawei	6-30
Li Jian	5-12
Li Jing	4-45
Li Juanjuan	1-18, 1-19
Li Kaixuan	4-43
Li Lanxia	7-18, 7-22

Li Li	9-7
Li Ming	1-73, 1-80, 3-10, 3-40
Li Quanbo	3-28, 3-29, 3-30, 3-31
Li Shoutian	5-16, 5-20
Li Shouwei	7-21
Li Ting	1-89
Li Tomi T	4-39
Li Wei	4-43
Li Weiwei	5-3
Li Xiaomin	4-36, 6-11
Li Xiaowei	6-30
Li Xixiang	1-73
Li Xuemiao	2-44
Li Yan	1-27
Li Yang	7-21
Li Yanli	2-20, 2-25, 3-11, 3-21, 3-24, 6-25
Li Yibin	5-15
Li Yiqing	1-89
Li Yu	1-100
Li Xiaoting	2-7
Li Yimei	1-81, 1-89
Li Yue	5-31
Li Ziheng	5-12
Li Zhengfeng	7-21
Liang Wenjie	9-36
Liang Zhen	1-82, 5-3
Liang Zhiyang	5-12
Liang Zhongxin	1-89
Liang Ziqi	8-14
Liao Jinzi Lois	4-36, 6-11
Liao Yiming	8-10
Lim Kong Tjien	1-23
Lin Nan	1-72
Lin Shang Chih	6-17
Lin Weihao	3-21, 3-24, 4-45
Lin Xin	8-12, 8-13
Lin Yibo	9-10
Lin Yuanwei	3-16, 3-20
Lin Zhiting	1-71
Ling Zhiting	1-86
Liu Biqiu	2-26, 2-27, 2-28
Liu Chang	9-28
Liu Chunling	1-43
Liu Chunwen	4-30
Liu Donghua	1-24, 1-25, 1-42, 1-51
Liu Hongjie	7-18, 7-22
Liu J	1-16, 1-29

Liu Jia	5-31, 5-31
Liu Jihao	6-39, 7-14
Liu Johan	7-13
Liu Jun	1-72, 6-35
Liu Junwen	1-11
Liu Keqin	1-71, 1-83
Liu Lingling	7-18
Liu Linlin	7-22
Liu Mengrui	5-28, 37
Liu Qinhua	6-35
Liu Shaoxiong	3-31
Liu Shuhan	1-84
Liu Shuo	2-27
Liu Tao	1-19
Liu Tianyi	1-76, 1-84
Liu Weili	5-43
Liu Wenyan	3-25, 3-26, 3-27
Liu Wei	1-56
Liu Wuping	1-68
Liu Xiao	7-6
Liu Yuling	5-32, 5-33
Liu Zhenghong	1-26, 1-27,
Liu Zhengwu	1-79
Liu Zhunhua	3-10, 3-21
Long Yin	1-48, 6-18, 6-28, 6-32, 6-33, 6-37
Lou Di	6-35
Lu Colbert	2-38
Lu David	1-85
Lu Guangyuan	1-10
Lu Jingjing	6-15
Lu Lian	3-29, 3-31
Lu Menggang	5-25
Lu Peiming	1-73
Lu Shenzhou	4-17
Lu Te-Yun	4-39
Lu Xiuzhen	7-13
Lu Yanan	5-36, 5-38
Lu Yetao	2-3
Lu Yingming	1-83
Luk Wai-Shing	9-14
Luo Chi-Ren	6-22
Luo Chong	5-32
Luo Jin	1-81, 1-88
Luo Jyue	6-21
Luo Mingcheng	9-28
Luo Xin	1-44, 1-50, 1-54
Luo Yongjian	3-10, 3-21, 3-24

Luo Zhihong	2-31
Lv Jian	2-3
Lv Zhen	7-13

M

Ma Shuying	7-8
Ma Song	6-9
Ma Weiwei	3-46
Ma Yanfei	4-41, 4-42
Mair Robin	6-36
Mao Guiyun	9-12
Mao W	9-11
Matagne P	1-33
Mehendale Manjusha	6-36
Mei Na	6-38
Meng Renyang	2-22
Meng Yuanyuan	5-14, 5-15
Merckling Clement	6-7
Mi Wei	8-13
Ming Xuefei	7-21
Mols Yves	6-7
Morinaga Hitoshi	5-18
Mu Jianan	6-30
Mu John	6-34
Mukundhan Priya	6-34, 6-36
Muller Chris	1-85

N

Narayanan Vijaykrishnan	9-4
Neoh Din-Ghee	4-36
Ni Guangyu	1-52
Ni Hao	6-39
Ni Qi Liang	1-36, 6-13
Ning An	7-16
Ning Renxia	7-5
Niu Feng	4-41
Niu Xinhuan	5-36
Niua Xinhuan	5-38
Nong Shanshan	9-25

O

Okafor Ikenna	9-4
Oo CF	7-11
Orlov Ivan	6-41
Ou Liwei	5-24

P

Pan David	9-10
Pan Fengjia	1-48, 6-28, 6-37
Pan Honggang	1-74
Pan Qin	7-14
Park Jin-Goo	5-47
Pavlidis Vasilis F	9-16
Peng Baoli	9-16
Peng Jingyu	8-10
Peng Li	1-6
Peng Xiang	5-2
Podgorski William	6-8
Prawoto Clarissa	4-10
Prenger Luke	7-6
Progler Chris	2-38
Puligadda Rama	7-6

Q

Qi Jiacheng	
Qi Ruisheng	1-26, 1-27, 1-28
Qi Xuxin	1-46
Qian He	1-79
Qian Jiadong	9-17
Qian Kai	3-31
Qian Rui	2-26
Qian Wensheng	1-24, 1-25, 1-41, 1-42, 1-51, 1-76
Qiao Fulong	3-18, 3-33
Qiao Yanhui	2-29
Qiao Zhenjie	4-17
Qin RenGang	6-42
Qin Youhua	1-18
Qu Lei	2-7
Qu Wanyuan	

R

Ramadan Mohamed	2-38
Ramos-Rodriguez JM	2-1
Rashidi Negar	9-32
Ren Haoxing	9-10
Ren Xianming	1-53, 5-20
Ren Xiaobing	1-13
Ren Xiaoming	5-16
Ruoxue Li	7-17
Ryu Heon-Yul	5-47

S

Shan Duo	4-17
Shan Pujia	5-12
Shang Enming	1-44, 1-50, 1-54
Shao Chris	1-26, 1-27, 1-28
Shen Jiaqiang	8-12, 8-13
Shen Luhang	4-30
Shen Ruoxi	8-8
Shen Yaoting	4-25, 4-27, 4-33
Shen Yiijiang	1-35
Shi Kang	5-24
Shi Xuelong	2-34
Shin Zhiguo	9-24
Silva-Martinez Jose	9-32
Simoen E	1-33, 6-7
Song Guoqiang	5-31, 5-33
Song Jianye	6-13
Song Yang	4-41, 4-42
Song Zhitang	5-43
Srinivasa Srivatsa	9-4
Srinivasan Sridhar	9-7
Strauch Mario	4-36
Su Chang	1-82
Su Tao	9-25
Su Xiaojing	2-7
Su Yahui	7-14
Su Yusen	6-39
Sun Chao	3-46
Sun Chengyang	7-9
Sun Fangce	1-45
Sun Hongxu	2-28
Sun Lei	3-17, 3-28, 3-30
Sun Li Fei	3-17
Sun Qin	4-17
Sun Qing	2-3
Sun Qing Qing	1-6
Sun Shuang	3-40
Sun Tao	5-29
Sun Tiantuo	1-6
Sun Tuobei	6-38
Sun Wenyan	3-33
Sun XiaoFeng	6-42
Sun Xiaoqin	5-28, 5-37
Sun Yiling	3-18
Sun Yongqian	7-13
Sun Yuming	2-27
Sun Yunqian	5-3

T

Tan Andy	4-17
Tan Baimei	5-28, 5-33, 5-37
Tan Jun	4-29, 4-31
Tan Li	1-100
Tan Wei	7-18, 7-22
Tan Yiqun	2-22
Tang Jiale	3-52
Tang Jianshi	1-79
Tang Xia	4-41
Tao Zhi	1-53
Tee Weikok	4-36, 6-11
Tian Siyu	5-28, 5-37
Trujillo-Sevilla JM	2-1
Tsuji Naoki	1-27
Tsujimura Manabu	5-51

U

V

van Borkulo Jeroen	7-2
van der Stam Richard	7-2
Veloso A	1-33

W

Wada Yutaka	5-47
Wan Baicen	5-12
Wan Chuangyun	5-20
Wan Dan	2-29
Wan J	1-16, 1-29, 9-11
Wan Ke	8-10
Wang Andy	5-12
Wang Bisheng	4-36, 6-11
Wang Bowen	9-15
Wang Bowen	3-21, 3-24
Wang C	3-8
Wang Calvin	6-36
Wang Chanfeng	4-41, 4-42
Wang Chao	5-39, 5-40
Wang Chenqian	7-21
Wang Chenwei	5-28, 5-31,5-33, 5-34, 5-39, 5-40, 5-41
Wang Chien	9-13
Wang Dan	1-40
Wang David H	3-26
Wang DeJin	6-42
Wang Ewen	6-36

Wang Fang	1-53, 1-74, 4-43, 5-26, 8-12, 8-13
Wang Hairui	9-15
Wang Hongdi	5-12
Wang Jianmin	6-35
Wang Jintao	2-27
Wang Juyao	9-15
Wang Kai	4-17, 6-18, 6-28, 6-32, 6-37, 9-28
Wang Kainan	9-1
Wang Lei	6-9
Wang Li	1-8

Wang Lipeng	2-28
Wang Luguang	5-16
Wang Min	6-33
Wang Mudan	2-22
Wang Ning	1-24
Wang Pengfei	9-29
Wang Peter J	4-39
Wang Ping	4-17
Wang Qi	5-28, 5-37, 6-25
Wang Qimeng	4-17
Wang Qing	2-3
Wang Qing Peng	3-17
Wang Qingpeng	1-68
Wang Qiwei	1-18
Wang Runsheng	1-86
Wang Shenli	5-34, 5-41
Wang Shiming	1-17, 1-22
Wang Song-Hi	4-39
Wang Weijun	3-10
Wang Wenjin	3-26
Wang Xiaopeng	1-35
Wang Xingjie	1-43
Wang Xuejiao	1-56
Wang Yi	3-33
Wang Yibo	3-21, 3-24
Wang Yichen	9-19
Wang Yingshuai	1-52
Wang Yong	6-23, 9-12
Wang Zhixuan	9-27
Wang Zhongwei	1-76, 1-78
Wang Zhouman	9-29
Waugh Darian	1-20
Wei J H	9-11
Wei Juan	2-22
Wei Yayi	2-7
Wei Zhengying	6-15, 6-23, 9-12, 9-29
Wen Hao Yu	6-42
Wu Huaqiang	1-79
Wu Hanming	9-26
Wu Jacky	7-11
Wu Qiang	2-20, 2-25, 3-11, 3-21, 3-24, 4-45, 6-25
Wu Qingqing	3-21, 3-24, 4-45, 8-14
Wu Sen	3-33
Wu Wei	1-16 , 7-11
Wu Xiulong	9-27
Wu Yanpei	7-13

X

Xi Wang	7-16, 7-17
Xi Weijuan	7-13
Xia Wenjie	7-5
Xia Yinshui	9-30
Xian Yi	5-14
Xiao K	1-16, 1-29
Xiao Ying	4-10
Xie Q	3-8
Xie Shangluan	9-30
Xiong Wei	1-9, 1-13
Xu Aoxue	5-43
Xu Chang	6-22
Xu Chen	6-23
Xu Chengyu	9-13
Xu Dongyu	2-31
Xu Fan	5-43
Xu Jiangrong	1-21
Xu Jianhua	4-35
Xu Kaidong	3-52
Xu Lingzhi	4-17
Xu Liying	1-71, 1-83
Xu Pengkai	3-33
Xu Renhui	3-30
Xu Wen	3-10
Xu Xiaolin	3-46
Xu Xiaoyan	3-40
Xu Youfeng	5-2
Xu Yuan	7-5
Xu Zhaozhao	1-25, 1-42, 1-76
Xu Zheng	1-16
Xu Ziyuan	9-27
Xue Xingkun	4-35
Xun Fangjing	1-38

Y

Yan Feng	8-10, 9-28
Yan Liang	2-7
Yan Meng	8-13
Yan Qiang	4-29, 4-31
Yan Shuo	4-43
Yan Ying	5-22
Yanfen Fang	6-14
Yang Aihua	6-25
Yang Chenghui	5-36, 5-38
Yang Guanghua	1-45

Yang Haewan	1-19
Yang Huishan	1-17, 1-22
Yang Jianguo	3-28, 3-30
Yang Jingjing	1-71
Yang Ke	1-83
Yang Libo	9-27
Yan Meng	8-12
Yang Mengxuan	1-82, 1-89
Yan Yajie	8-14
Yang Yu-Pu	4-39
Yang Yuancheng	1-73, 3-40
Yang Yuchao	1-71, 1-83
Yang Yundian	5-34, 5-41
Yang Yushu	2-20, 3-11, 3-21, 3-24, 4-45, 6-25
Yang Zhanhai	4-41
Yang Zhenyu	2-44
Yang Zhiyong	6-19
Yao Shaokang	1-19
Yao Ting	3-26
Ye Jing	1-72, 6-30
Ye Le	1-89, 9-27
Yee Boonhwa	4-36, 6-11
Yerriboina Nagendra Prasad	5-47
Yi Chunyan	3-10
Yin Jun	1-52
Yin Wenbo	9-14
Yong Frank	6-17
You Moon Sang	7-10
Yoon Sungjun	9-32
Yu Bao	4-30, 4-35
Yu Daquan	7-21
Yu Henry	1-85
Yu Jun	2-27
Yu Minglang	4-36, 6-11
Yu Shirui	1-40, 2-22, 2-29
Yu Wei	6-15, 6-29
Yu Wenjian	9-36
Yuan Haiying	9-19
Yuan Renzhi	3-20
Yuan Rui	1-83
Yuan Song	5-25
Yuan Yujie	5-26
Yuan Zengyi	6-37
Yushan	3-18

Z

Zaslavsky A	1-29
Zeng Xuan	8-2
Zeng Zhaqin	4-35
Zeng Zhimin	6-40
Zha Xiaojing	9-30
Zhang Aibing	7-21
Zhang Baichun	3-28
Zhang Baotong	1-73, 3-40
Zhang Cong	2-26, 2-27, 2-28
Zhang Fan	4-17
Zhang Hao	9-27
Zhang J X	9-11
Zhang Ji Hong	3-17, 3-18
Zhang Jia	9-24
Zhang Jian	1-9
Zhang Jianhua	4-45
Zhang Jianfu	1-96
Zhang Yiayang	1-86
Zhang Jie	3-30
Zhang Kai	9-19
Zhang Kailiang	1-53, 1-74, 4-43, 5-26, 8-12, 8-13
Zhang Kegang	1-24
Zhang Lei	5-14, 5-15, 6-35
Zhang Libin	5-20
Zhang Maomao	7-13
Zhang Ming	1-45
Zhang Qiaoran	7-13
Zhang Teng	1-71, 1-83
Zhang Tianyu	6-14, 6-24
Zhang W	6-16
Zhang Wei	5-14, 5-15,
Zhang Wei Yang	6-42
Zhang Xi	4-36, 6-11
Wang Xiaoyan	3-26
Zhang Xingdi	6-32
Zhang Xue	5-39, 5-40
Zhang Y	3-8
Zhang Yan-Qiu	6-21
Zhang Yanyan	4-35
Zhang Yi-Fu	6-21
Zhang Yijun	1-100
Zhang Ying	3-10, 3-21, 3-24
Zhang Yu	2-26, 2-27, 2-28, 3-26, 3-28, 3-29, 3-30, 3-31, 4-35
Zhang Yudong	3-52
Zhang Yuwei	1-86
Zhang Yuxiang	6-40
Zhang Z	9-11

Zhang Zhe	1-86
Zhang Zhenzhong	1-74
Zhang Zhigang	3-18, 4-17
Zhang Zhijie	5-12
Zhang Zuodong	1-86
Zhao Bin	3-33
Zhao Gaoyung	5-43
Zhao Pan	7-14
Zhao Xin	5-38
Zhao Yuhang	1-44, 1-50, 1-54, 2-34, 2-25, 3-11, 8-12, 9-29
Zhao Zhaoyuan	4-17
Zhao Zhengyuan	5-2
Zhao Zhilin	5-3
Zhen Gu	1-18
Zheng Fengxia	7-8
Zheng Frank	6-34
Zheng Haipeng	4-36, 6-11
Zhen Li	7-17
Zheng Susanna	1-45
Zheng Tong	9-19
Zheng Tuo	6-39
Zhi Tian	1-18, 1-19
Zhong Lichao	1-43
Zhong Xiaolan	8-8
Zhong Yuan	1-89
Zhou Cheng	7-13
Zhou Haifeng	4-29, 4-31, 5-14, 5-15
Zhou Jiakai	5-36
Zhou Jianwei	5-39, 5-40
Zhou Jipeng	4-30
Zhou Kai	6-14, 6-24
Zhou KianGang	6-18
Zhou Piao	5-29
Zhou Ping	5-22
Zhou Tao	8-2

Zhou Wenzhan	2-31
Zhou Xiaoqiang	3-10, 4-45
Zhou YaoHui	6-42
Zhu Jiadi	1-71
Zhu Jianjun	3-24, 4-45, 8-14
Zhu Yingxuan	9-1
Zhu Yizheng	3-29
Zhu Yongwei	5-29
Zhu Yuanyuan	6-40
Zhu Yuebin	6-23
Zhu Yuning	3-21, 3-24
Zhuang Junjun	9-12
Zhuang Shiwei	3-52
Zuo Jinsong	5-26

IEEE
445 Hoes Lane
Piscataway, NJ 08854-4141

ISBN 978-1-7281-6559-2